MATHEMATICAL FORMULAS*

QUADRATIC FORMULA

If $ax^2 + bx + c = 0$, then $x = \dfrac{-b \pm \sqrt{b^2 - 4ac}}{2a}$.

BINOMIAL THEOREM

$$(1 + x)^n = 1 + \frac{nx}{1!} + \frac{n(n-1)x^2}{2!} + \ldots \qquad (x^2 < 1)$$

PRODUCTS OF VECTORS

Let θ be the smaller of the two angles between $\mathbf{a}$ and $\mathbf{b}$. Then

$$\mathbf{a} \cdot \mathbf{b} = \mathbf{b} \cdot \mathbf{a} = a_x b_x + a_y b_y + a_z b_z = ab \cos \theta$$

$$\mathbf{a} \times \mathbf{b} = -\mathbf{b} \times \mathbf{a} = \begin{vmatrix} \mathbf{i} & \mathbf{j} & \mathbf{k} \\ a_x & a_y & a_z \\ b_x & b_y & b_z \end{vmatrix}$$

$$= \mathbf{i} \begin{vmatrix} a_y & a_z \\ b_y & b_z \end{vmatrix} - \mathbf{j} \begin{vmatrix} a_x & a_z \\ b_x & b_z \end{vmatrix} + \mathbf{k} \begin{vmatrix} a_x & a_y \\ b_x & b_y \end{vmatrix}$$

$$= (a_y b_z - b_y a_z)\mathbf{i}$$

$$+ (a_z b_x - b_z a_x)\mathbf{j} + (a_x b_y - b_x a_y)\mathbf{k}$$

$$|\mathbf{a} \times \mathbf{b}| = ab \sin \theta$$

TRIGONOMETRIC IDENTITIES

$$\sin \alpha \pm \sin \beta = 2 \sin \tfrac{1}{2}(\alpha \pm \beta) \cos \tfrac{1}{2}(\alpha \mp \beta)$$
$$\cos \alpha + \cos \beta = 2 \cos \tfrac{1}{2}(\alpha + \beta) \cos \tfrac{1}{2}(\alpha - \beta)$$

* See Appendix E for a more complete list.

DERIVATIVES AND INTEGRALS

$$\frac{d}{dx} \sin x = \cos x \qquad \int \sin x \, dx = -\cos x$$

$$\frac{d}{dx} \cos x = -\sin x \qquad \int \cos x \, dx = \sin x$$

$$\frac{d}{dx} e^x = e^x \qquad \int e^x \, dx = e^x$$

$$\int \frac{dx}{\sqrt{x^2 + a^2}} = \ln(x + \sqrt{x^2 + a^2})$$

$$\int \frac{x \, dx}{(x^2 + a^2)^{3/2}} = -\frac{1}{(x^2 + a^2)^{1/2}}$$

$$\int \frac{dx}{(x^2 + a^2)^{3/2}} = \frac{x}{a^2(x^2 + a^2)^{1/2}}$$

CRAMER'S RULE

Two simultaneous equations in unknowns x and y,

$$a_1 x + b_1 y = c_1 \qquad \text{and} \qquad a_2 x + b_2 y = c_2,$$

have the solutions

$$x = \frac{\begin{vmatrix} c_1 & b_1 \\ c_2 & b_2 \end{vmatrix}}{\begin{vmatrix} a_1 & b_1 \\ a_2 & b_2 \end{vmatrix}} = \frac{c_1 b_2 - c_2 b_1}{a_1 b_2 - a_2 b_1}$$

and

$$y = \frac{\begin{vmatrix} a_1 & c_1 \\ a_2 & c_2 \end{vmatrix}}{\begin{vmatrix} a_1 & b_1 \\ a_2 & b_2 \end{vmatrix}} = \frac{a_1 c_2 - a_2 c_1}{a_1 b_2 - a_2 b_1}.$$

THE GREEK ALPHABET

Alpha	A	α	Iota	I	ι	Rho	P	ρ
Beta	B	β	Kappa	K	κ	Sigma	Σ	σ
Gamma	Γ	γ	Lambda	Λ	λ	Tau	T	τ
Delta	Δ	δ	Mu	M	μ	Upsilon	Y	υ
Epsilon	E	ϵ	Nu	N	ν	Phi	Φ	ϕ, φ
Zeta	Z	ζ	Xi	Ξ	ξ	Chi	X	χ
Eta	H	η	Omicron	O	o	Psi	Ψ	ψ
Theta	Θ	θ	Pi	Π	π	Omega	Ω	ω

FUNDAMENTALS OF
PHYSICS

FIFTH EDITION

FUNDAMENTALS OF
PHYSICS

VOLUME 1

DAVID HALLIDAY
University of Pittsburgh

ROBERT RESNICK
Rensselaer Polytechnic Institute

JEARL WALKER
Cleveland State University

JOHN WILEY & SONS, INC.

New York • Chichester • Brisbane • Toronto • Singapore

ACQUISITIONS EDITOR Stuart Johnson
DEVELOPMENTAL EDITOR Rachel Nelson
SENIOR PRODUCTION SUPERVISOR Cathy Ronda
PRODUCTION ASSISTANT Raymond Alvarez
MARKETING MANAGER Catherine Faduska
ASSISTANT MARKETING MANAGER Ethan Goodman
DESIGNER Dawn L. Stanley
MANUFACTURING MANAGER Mark Cirillo
PHOTO EDITOR Hilary Newman
COVER PHOTO William Warren/Westlight
ILLUSTRATION EDITOR Edward Starr
ILLUSTRATION Radiant/Precision Graphics

This book was set in Times Roman by Progressive Information Technologies, and printed and bound by Von Hofmann Press. The cover was printed by Phoenix Color Corp.

Hello There!

You are about to begin your first college level physics course. You may have heard from friends and fellow students that physics is a difficult course, especially if you don't plan to go on to a career in the hard sciences. But that doesn't mean it has to be difficult for you. The key to success in this course is to have a good understanding of each chapter before moving on to the next. When learned a little bit at a time, physics is straightforward and simple. Here are some ideas that can make this text and this class work for you:

- Read through the **Sample Problems** and solutions carefully. These problems are similar to many of the end-of-chapter exercises, so reading them will help you solve homework problems. In addition, they offer a look at how an experienced physicist would approach solving the problem.

- Try to answer the **Checkpoint** questions as you read through the chapter. Most of these can be answered by thinking through the problem, but it helps if you have some scratch paper and a pencil nearby to work out some of the harder questions. Hold the answer page at the back of the book with your thumb (or a tab or bookmark) so you can refer to it easily. The end-of-chapter **Questions** are very similar—you can use them to quiz yourself after you've read the whole chapter.

- Use the **Review & Summary** sections at the end of each chapter as the first place to look for formulas you might need to solve homework problems. These sections are also helpful in making study sheets for exams.

- The biggest tip I can give you is pretty obvious. Do your homework! Understanding the homework problems is the best way to master the material and do well in the course. Doing a lot of homework problems is also the best way to review for exams. And by all means, consult your classmates whenever you are stuck. Working in groups will make your studying more effective.

The study of basic physics is required for degrees in Engineering, Physics, Biology, Chemistry, Medicine, and many other sciences because the fundamentals of physics are the framework on which every other science is built. Therefore, a solid grasp of basic physical principles will help you understand upper-level science courses and make your study of these courses easier.

I took introductory physics because it was a prerequisite for my B.S. in Mechanical Engineering. Even though engineering and pure physics are worlds apart, I find myself using this book as a reference almost every day. I urge you to keep it after you have finished your course. The knowledge you will gain from this book and your introductory physics course is the foundation for all other sciences. This is the primary reason to take this class seriously and be successful in it.

Best of luck!

Josh Kane

Josh Kane

Preface

For four editions, *Fundamentals of Physics* has been successful in preparing physics students for careers in science and engineering. The first three editions were coauthored by the highly regarded team of David Halliday and Robert Resnick, who developed a groundbreaking text replete with conceptual structure and applications. In the fourth edition, the insights provided by new coauthor, Jearl Walker, took the text into the 1990s and met the challenge of guiding students through a time of tremendous advances and a ferment of activity in the science of physics. Now, in the fifth edition, we have expanded on the conventional strengths of the earlier editions and enhanced the applications that help students forge a bridge between concepts and reasoning. We not only *tell* students how physics works, we *show* them, and we give them the opportunity to show us what they have learned by testing their understanding of the concepts and applying them to real-world scenarios. Concept checkpoints, problem solving tactics, sample problems, electronic computations, exercises and problems—all of these skill-building signposts have been developed to help students establish a connection between conceptual theories and application. The students reading this text today are the scientists and engineers of tomorrow. It is our hope that the fifth edition of *Fundamentals of Physics* will help prepare these students for future endeavors by contributing to the enhancement of physics education.

CHANGES IN THE FIFTH EDITION

Although we have retained the basic framework of the fourth edition of *Fundamentals of Physics,* we have made extensive changes in portions of the book. Each chapter and element has been scrutinized to ensure clarity, currency, and accuracy, reflecting the needs of today's science and engineering students.

Content Changes

Mindful that textbooks have grown large and that they tend to increase in length from edition to edition, we have reduced the length of the fifth edition by combining several chapters and pruning their contents. In doing so, six chapters have been rewritten completely, while the remaining chapters have been carefully edited and revised, often extensively, to enhance their clarity, incorporating ideas and suggestions from dozens of reviewers.

- *Chapters 7 and 8 on energy* (and sections of later chapters dealing with energy) have been rewritten to provide a more careful treatment of energy, work, and the work–kinetic energy theorem. As the same time, the text material and problems at the end of each chapter still allow the instructor to present the more traditional treatment of these subjects.

- *Temperature, heat, and the first law of thermodynamics* have been condensed from two chapters to one chapter (*Chapter 19*).

- *Chapter 21 on entropy* now includes a statistical mechanical presentation of entropy that is tied to the traditional thermodynamical presentation.

- *Chapters on Faraday's law and inductance* have been combined into one new chapter (*Chapter 31*).

- *Treatment of Maxwell's equations* has been streamlined and moved up earlier into the chapter on magnetism and matter (*Chapter 32*).

- *Coverage of electromagnetic oscillations and alternating currents* has been combined into one chapter (*Chapter 33*).

- *Chapters 39, 40, and 41 on quantum mechanics* have been rewritten to modernize the subject. They now include experimental and theoretical results of the last few years. In addition, quantum physics and special relativity are introduced in some of the early chapters in short sections that can be covered quickly. These early sections lay some of the groundwork for the ''modern physics'' topics that appear later in the extended version of the text and add an element of suspense about the subject.

New Pedagogy

In the interest of addressing the needs of science and engineering students, we have added a number of new pedagogical features intended to help students forge a bridge between concepts and reasoning and to marry theory with practice. These new features are designed to help students test their understanding of the material. They were also developed to help students prepare to apply the information to exam questions and real-world scenarios.

- To provide opportunities for students to check their understanding of the physics concepts they have just read, we have placed **Checkpoint** questions within the chapter presentations. Nearly 300 Checkpoints have been added to help guide the student away from common errors and misconceptions. All of the Checkpoints require decision making and reasoning on the part of the student (rather than computations requiring calculators) and focus on the key points of the physics that students need to understand in order to tackle the exercises and problems at the end of each chapter. Answers to all of the Checkpoints are found in the back of the book, sometimes with extra guidance to the student.

- Continuing our focus on the key points of the physics, we have included additional **Checkpoint-type questions** in the Questions section at the end of each chapter. These new questions require decision making and reasoning on the part of the student; they ask the student to organize the physics concepts rather than just plug numbers into equations. Answers to the odd-numbered questions are now provided in the back of the book.

- To encourage the use of computer math packages and graphing calculators, we have added an **Electronic Computation problem section** to the Exercises and Problems sections of many of the chapters.

These new features are just a few of the pedagogical elements available to enhance the student's study of physics. A number of tried-and-true features of the previous edition have been retained and refined in the fifth edition, as described below.

CHAPTER FEATURES

The pedagogical elements that have been retained from previous editions have been carefully planned and crafted to motivate students and guide their reasoning process.

- *Puzzlers* Each chapter opens with an intriguing photograph and a ''puzzler'' that is designed to motivate the student to read the chapter. The answer to each puzzler is provided within the chapter, but it is not identified as such to ensure that the student reads the entire chapter.

- *Sample Problems* Throughout each chapter, sample problems provide a bridge from the concepts of the chapter to the exercises and problems at the end of the chapter. Many of the nearly 400 sample problems featured in the text have been replaced with new ones that more sharply focus on the common difficulties students experience in solving the exercises and problems. We have been especially mindful of the mathematical difficulties students face. The sample problems also provide

an opportunity for the student to see how a physicist thinks through a problem.

- *Problem Solving Tactics* To help further bridge concepts and applications and to add focus to the key physics concepts, we have refined and expanded the number of problem solving tactics that are placed within the chapters, particularly in the earlier chapters. These tactics provide guidance to the students about how to organize the physics concepts, how to tackle mathematical requirements in the exercises and problems, and how to prepare for exams.

- *Illustrations* Because the illustrations in a physics textbook are so important to an understanding of the concepts, we have altered nearly 30 percent of the illustrations to improve their clarity. We have also removed some of the less effective illustrations and added many new ones.

- *Review & Summary* A review and summary section is found at the end of each chapter, providing a quick review of the key definitions and physics concepts *without* being a replacement for reading the chapter.

- *Questions* Approximately 700 thought-provoking questions emphasizing the conceptual aspects of physics appear at the ends of the chapters. Many of these questions relate back to the checkpoints found throughout the chapters, requiring decision making and reasoning on the part of the student. Answers to the odd-numbered questions are provided in the back of the book.

- *Exercises & Problems* There are approximately 3400 end-of-chapter exercises and problems in the text, arranged in order of difficulty, starting with the exercises (labeled ''E''), followed by the problems (labeled ''P''). Particularly challenging problems are identified with an asterisk (*). Those exercises and problems that have been retained from previous editions have been edited for greater clarity; many have been replaced. Answers to the odd-numbered exercises and problems are provided in the back of the book.

VERSIONS OF THE TEXT

The fifth edition of *Fundamentals of Physics* is available in a number of different versions, to accommodate the individual needs of instructors and students alike. The Regular Edition consists of Chapters 1 through 38 (ISBN 0-471-10558-9). The Extended Edition contains seven additional chapters on quantum physics and cosmology (Chapters 1–45) (ISBN 0-471-10559-7). Both editions are available as single, hardcover books, or in the alternative versions listed on page ix:

- Volume 1—Chapters 1–21 (Mechanics/Thermodynamics), cloth, 0-471-15662-0
- Volume 2—Chapters 22–45 (E&M and Modern Physics), cloth, 0-471-15663-9
- Part 1—Chapters 1–12, paperback, 0-471-14561-0
- Part 2—Chapters 13–21, paperback, 0-471-14854-7
- Part 3—Chapters 22–33, paperback, 0-471-14855-5
- Part 4—Chapters 34–38, paperback, 0-471-14856-3
- Part 5—Chapters 39–45, paperback, 0-471-15719-8

The Extended edition of the text is also available on CD ROM.

SUPPLEMENTS

The fifth edition of *Fundamentals of Physics* is supplemented by a comprehensive ancillary package carefully developed to help teachers teach and students learn.

Instructor's Supplements

- *Instructor's Manual* by J. RICHARD CHRISTMAN, U.S. Coast Guard Academy. This manual contains lecture notes outlining the most important topics of each chapter, as well as demonstration experiments, and laboratory and computer exercises; film and video sources are also included. Separate sections contain articles that have appeared recently in the *American Journal of Physics* and *The Physics Teacher*.

- *Instructor's Solutions Manual* by JERRY J. SHI, Pasadena City College. This manual provides worked-out solutions for all the exercises and problems found at the end of each chapter within the text. *This supplement is available only to instructors.*

- *Solutions Disk.* An electronic version of the Instructor's Solutions Manual, for instructors only, available in TeX for Macintosh and Windows™.

- *Test Bank* by J. RICHARD CHRISTMAN, U.S. Coast Guard Academy. More than 2200 multiple-choice questions are included in the Test Bank for *Fundamentals of Physics*.

- *Computerized Test Bank.* IBM and Macintosh versions of the entire Test Bank are available with full editing features to help you customize tests.

- *Animated Illustrations.* Approximately 85 text illustrations are animated for enhanced lecture demonstrations.

- *Transparencies.* More than 200 four-color illustrations from the text are provided in a form suitable for projection in the classroom.

Student's Supplements

- *A Student's Companion* by J. RICHARD CHRISTMAN, U.S. Coast Guard Academy. Much more than a traditional study guide, this student manual is designed to be used in close conjunction with the text. The Student's Companion is divided into four parts, each of which corresponds to a major section of the text, beginning with an overview "chapter." These overviews are designed to help students understand how the important topics are integrated and how the text is organized. For each chapter of the text, the corresponding Companion chapter offers: Basic Concepts, Problem Solving, Notes, Mathematical Skills, and Computer Projects and Notes.

- *Solutions Manual* by J. RICHARD CHRISTMAN, U.S. Coast Guard Academy and EDWARD DERRINGH, Wentworth Institute. This manual provides students with complete worked-out solutions to 30 percent of the exercises and problems found at the end of each chapter within the text.

- **Interactive Learningware** by JAMES TANNER, Georgia Institute of Technology, with the assistance of GARY LEWIS, Kennesaw State College. This software contains 200 problems from the end-of-chapter exercises and problems, presented in an interactive format, providing detailed feedback for the student. Problems from Chapter 1 to 21 are included in Part 1, from Chapters 22 to 38 in Part 2. The accompanying workbooks allow the student to keep a record of the worked-out problems. The Learningware is available in IBM 3.5" and Macintosh formats.

- **CD Physics.** The entire Extended Version of the text (Chapters 1–45) is available on CD ROM, along with the student solutions manual, study guide, animated illustrations, and Interactive Learningware.

Acknowledgments

A textbook contains far more contributions to the elucidation of a subject than those made by the authors alone. J. Richard Christman, of the U.S. Coast Guard Academy, has once again created many fine supplements for us; his knowledge of our book and his recommendations to students and faculty are invaluable. James Tanner, of Georgia Institute of Technology, and Gary Lewis, of Kennesaw State College, have provided us with innovative software, closely tied to the text's exercises and problems. J. Richard Christman, of the U.S. Coast Guard Academy, and Glen Terrell, of the University of Texas at Arlington, contributed problems to the Electronic Computation sections of the text. Jerry Shi, of Pasadena City College, performed the Herculean task of working out solutions for every one of the Exercises and Problems in the text. We thank John Merrill, of Brigham Young University, and Edward Derringh, of the Wentworth Institute of Technology for their many contributions in the past. We also thank George W. Hukle of Oxnard, California, for his check of the answers at the back of the book.

At John Wiley, publishers, we have been fortunate to receive strong coordination and support from our former editor, Cliff Mills. Cliff guided our efforts and encouraged us along the way. When Cliff moved on to other responsibilities at Wiley, we were ably guided to completion by his successor, Stuart Johnson. Rachel Nelson has coordinated the developmental editing and multilayered preproduction process. Catherine Faduska, our senior marketing manager, and Ethan Goodman, assistant marketing manager, have been tireless in their efforts on behalf of this edition. Jennifer Bruer has built a fine supporting package of ancillary materials. Monica Stipanov and Julia Salsbury managed the review and administrative duties admirably.

We thank Lucille Buonocore, our able production manager, and Cathy Ronda, our production editor, for pulling all the pieces together and guiding us through the complex production process. We also thank Dawn Stanley, for her design; Brenda Griffing, for her copy editing; Edward Starr, for managing the line art program; Lilian Brady, for her proofreading; and all other members of the production team.

Stella Kupferburg and her team of photo researchers, particularly Hilary Newman and Pat Cadley, were inspired in their search for unusual and interesting photographs that communicate physics principles beautifully. We thank Boris Starosta and Irene Nunes for their careful development of a full-color line art program, for which they scrutinized and suggested revisions of every piece. We also owe a debt of gratitude for the line art to the late John Balbalis, whose careful hand and understanding of physics can still be seen in every diagram.

We especially thank Edward Millman for his developmental work on the manuscript. With us, he has read every word, asking many questions from the point of view of a student. Many of his questions and suggested changes have added to the clarity of this volume. Irene Nunes added a final, valuable developmental check in the last stages of the book.

We owe a particular debt of gratitude to the numerous students who used the fourth edition of *Fundamentals of Physics* and took the time to fill out the response cards and return them to us. As the ultimate consumers of this text, students are extremely important to us. By sharing their opinions with us, your students help us ensure that we are providing the best possible product and the most value for their textbook dollars. We encourage the users of this book to contact us with their thoughts and concerns so that we can continue to improve this text in the years to come. In particular, we owe a special debt of gratitude to the students who participated in a final focus group at Union College in Schenecdaty, New York: Matthew Glogowski, Josh Kane, Lauren Papa, Phil Tavernier, Suzanne Weldon, and Rebecca Willis.

Finally, our external reviewers have been outstanding and we acknowledge here our debt to each member of that team:

MARIS A. ABOLINS
Michigan State University

BARBARA ANDERECK
Ohio Wesleyan University

ALBERT BARTLETT
University of Colorado

MICHAEL E. BROWNE
University of Idaho

TIMOTHY J. BURNS
Leeward Community College

JOSEPH BUSCHI
Manhattan College

PHILIP A. CASABELLA
Rensselaer Polytechnic Institute

RANDALL CATON
Christopher Newport College

J. RICHARD CHRISTMAN
U.S. Coast Guard Academy

xi

ROGER CLAPP
University of South Florida

W. R. CONKIE
Queen's University

PETER CROOKER
University of Hawaii at Manoa

WILLIAM P. CRUMMETT
Montana College of Mineral Science and Technology

EUGENE DUNNAM
University of Florida

ROBERT ENDORF
University of Cincinnati

F. PAUL ESPOSITO
University of Cincinnati

JERRY FINKELSTEIN
San Jose State University

ALEXANDER FIRESTONE
Iowa State University

ALEXANDER GARDNER
Howard University

ANDREW L. GARDNER
Brigham Young University

JOHN GIENIEC
Central Missouri State University

JOHN B. GRUBER
San Jose State University

ANN HANKS
American River College

SAMUEL HARRIS
Purdue University

EMILY HAUGHT
Georgia Institute of Technology

LAURENT HODGES
Iowa State University

JOHN HUBISZ
North Carolina State University

JOEY HUSTON
Michigan State University

DARRELL HUWE
Ohio University

CLAUDE KACSER
University of Maryland

LEONARD KLEINMAN
University of Texas at Austin

EARL KOLLER
Stevens Institute of Technology

ARTHUR Z. KOVACS
Rochester Institute of Technology

KENNETH KRANE
Oregon State University

SOL KRASNER
University of Illinois at Chicago

PETER LOLY
University of Manitoba

ROBERT R. MARCHINI
Memphis State University

DAVID MARKOWITZ
University of Connecticut

HOWARD C. MCALLISTER
University of Hawaii at Manoa

W. SCOTT MCCULLOUGH
Oklahoma State University

JAMES H. MCGUIRE
Tulane University

DAVID M. MCKINSTRY
Eastern Washington University

JOE P. MEYER
Georgia Institute of Technology

ROY MIDDLETON
University of Pennsylvania

IRVIN A. MILLER
Drexel University

EUGENE MOSCA
United States Naval Academy

MICHAEL O'SHEA
Kansas State University

PATRICK PAPIN
San Diego State University

GEORGE PARKER
North Carolina State University

ROBERT PELCOVITS
Brown University

OREN P. QUIST
South Dakota State University

JONATHAN REICHART
SUNY—Buffalo

MANUEL SCHWARTZ
University of Louisville

DARRELL SEELEY
Milwaukee School of Engineering

BRUCE ARNE SHERWOOD
Carnegie Mellon University

JOHN SPANGLER
St. Norbert College

ROSS L. SPENCER
Brigham Young University

HAROLD STOKES
Brigham Young University

JAY D. STRIEB
Villanova University

DAVID TOOT
Alfred University

J. S. TURNER
University of Texas at Austin

T. S. VENKATARAMAN
Drexel University

GIANFRANCO VIDALI
Syracuse University

FRED WANG
Prairie View A&M

ROBERT C. WEBB
Texas A&M University

GEORGE WILLIAMS
University of Utah

DAVID WOLFE
University of New Mexico

We hope that our words here reveal at least some of the wonder of physics, the fundamental clockwork of the universe. And, hopefully, those words might also reveal some of our awe of that clockwork.

DAVID HALLIDAY
6563 NE Windermere Road
Seattle, WA 98105

ROBERT RESNICK
Rensselaer Polytechnic Institute
Troy, NY 12181

JEARL WALKER
Cleveland State University
Cleveland, OH 44115

HOW TO USE THIS BOOK:

You are about to begin what could be one the most exciting course that you will undertake in college. It offers you the opportunity to learn what makes our world "tick" and to gain insight into the role physics plays in our everyday lives. This knowledge will not come without some effort, however, and this book has been carefully designed and written with an awareness of the kinds of difficulties and challenges you may face. Therefore, before you begin, we have provided a visual overview of some of the key features of the book that will aid in your studies.

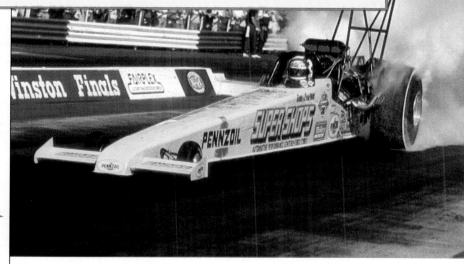

Chapter Opening Puzzlers

Each chapter opens with an intriguing example of physics in action. By presenting high-interest applications of each chapters concepts, the puzzlers are intended to peak your interest and motivate you to read the chapter.

In 1977, Kitty O'Neil set a dragster record by reaching 392.54 mi/h in a sizzling time of 3.72 s. In 1958, Eli Beeding Jr. rode a rocket sled from a standstill to a speed of 72.5 mi/h in an elapsed time of 0.04 s (less than an eye blink). How can we compare these two rides to see which was more exciting (or more frightening)—by final speeds, by elapsed times, or by some other quantity?

SAMPLE PROBLEM 2-6

(a) When Kitty O'Neil set the dragster records for the greatest speed and least elapsed time, she reached 392.54 mi/h in 3.72 s. What was her average acceleration?

SOLUTION: From Eq. 2-7, O'Neil's average acceleration was

$$\bar{a} = \frac{\Delta v}{\Delta t} = \frac{392.54 \text{ mi/h} - 0}{3.72 \text{ s} - 0}$$

$$= +106 \frac{\text{mi}}{\text{h} \cdot \text{s}},$$

(Answer)

where the motion is taken to be in the positive *x* direction. In

Answers to Puzzlers

All chapter-opening puzzlers are answered later in the chapter, either in text discussion or in a sample problem.

If the car
r, the bob
oninertial

CHECKPOINT **1:** In the figure, two perpendicular forces F_1 and F_2 are combined in six different ways. Which ways may be used to correctly determine the net force ΣF?

w wish to
accelera-
the stan-
e use) the
been as-

ictionless
at by trial
accelera-
efinition,
ody has a

body by

F_1 F_1 F_1
F_2 F_2 F_2
(a)   (b)   (c)

F_2
F_1 F_1 F_1
          F_2 F_2
(d)   (e)

Checkpoints

Checkpoints appear throughout the text, focusing on the key points of physics you will need to tackle the exercises and problems found at the end of each chapter. These checkpoints help guide you away from common errors and misconceptions.

Checkpoint Questions

Checkpoint-type questions at the end of each chapter ask you to organize the physics concepts rather than plug numbers into equations. Answers to the odd-numbered questions are provided in the back of the book.

ey puck in

$v = -2t\mathbf{i}$

ponents of
ion vector
nd and t is
-2 and 3?

al projec-
t identical
n the same
nal speeds

(c)

peed (a) a

$-4.9\mathbf{j}$ (x is
). Has the

9. Figure 4-25 shows three paths for a kicked football. Ignoring the effects of air on the flight, rank the paths according to (a) time of flight, (b) initial vertical velocity component, (c) initial horizontal velocity component, and (d) initial speed. Place the greatest first in each part.

FIGURE 4-25 Question 9.

10. Figure 4-26 shows the velocity and acceleration of a particle at a particular instant in three situations. In which situation, and at that instant, is (a) the speed increasing, (b) the speed decreasing, (c) the speed not changing, (d) $\mathbf{v} \cdot \mathbf{a}$ positive, (e) $\mathbf{v} \cdot \mathbf{a}$ negative, and (f) $\mathbf{v} \cdot \mathbf{a} = 0$?

FIGURE 4-26 Question 10.

Sample Problems

The sample problems offer you the opportunity to work through the physics concepts just presented. Often built around real-world applications, they are closely coordinated with the end-of-chapter Questions, Exercises, and Problems.

SAMPLE PROBLEM 4-1

The position vector for a particle is initially

$$\mathbf{r}_1 = -3\mathbf{i} + 2\mathbf{j} + 5\mathbf{k}$$

and then later is

$$\mathbf{r}_2 = 9\mathbf{i} + 2\mathbf{j} + 8\mathbf{k}$$

(see Fig. 4-2). What is the displacement from $\mathbf{r}_1$ to $\mathbf{r}_2$?

SOLUTION: Recall from Chapter 3 that we add (or subtract) two vectors in unit-vector notation by combining the components, axis by axis. So Eq. 4-2 becomes

$$\Delta\mathbf{r} = (9\mathbf{i} + 2\mathbf{j} + 8\mathbf{k}) - (-3\mathbf{i} + 2\mathbf{j} + 5\mathbf{k})$$
$$= 12\mathbf{i} + 3\mathbf{k}. \qquad \text{(Answer)}$$

The displacement vector is parallel to the xz plane, because it lacks any y component, a fact that is easier to pick out in the numerical result than in Fig. 4-2.

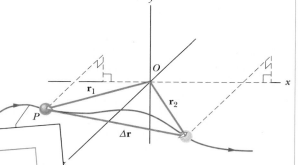

Sample Problem 4-1. The displacement $\Delta\mathbf{r} = $
d of $\mathbf{r}_1$ to the head of $\mathbf{r}_2$.

PROBLEM SOLVING TACTICS

TACTIC 1: *Reading Force Problems*
Read the problem statement several times until you have a good mental picture of what the situation is, what data are given, and what is requested. In Sample Problem 5-1, for example, you should tell yourself: "Someone is pushing a sled. Its speed changes, so acceleration is involved. The motion is along a straight line. A force is given in one part and asked for in the other, and so the situation looks like Newton's second law applied to one-dimensional motion."

If you know what the problem is about but don't know what to do next, put the problem aside and reread the text. If you are hazy about Newton's second law, reread that section. Study the sample problems. The one-dimensional-motion parts of Sample Problem 5-1 and the constant acceleration should send you back to Chapter 2 and especially to Table 2-1, which displays all the equations you are likely to need.

TACTIC 2: *Draw Two Types of Figures*
You may need two figures. One is a rough sketch of the actual real-world situation. When you draw the forces on it, place the tail of each force vector either on the boundary of or within the body feeling that force. The other figure is a free-body diagram in which the forces on a *single* body are drawn, with the body represented with a dot or a sketch. Place the tail of each force vector on the dot or sketch.

TACTIC 3: *What I   em?*

Problem Solving Tactics

Careful attention has been paid to helping you develop your problem-solving skills. Problem-solving tactics are closely related to the sample problems and can be found throughout the text, though most fall within the first half. The tactics are designed to help you work through assigned homework problems and prepare for exams. Collectively, they represent the stock in trade of experienced problem solvers and practicing scientists and engineers.

REVIEW & SUMMARY

Conservative Forces

A force is a **conservative force** if the net work it does on a particle moving along a closed path from an initial point and then back to that point is zero. Or, equivalently, it is conservative if its work on a particle moving between two points does not depend on the path taken by the particle. The gravitational force (weight) and the spring force are conservative forces; the kinetic frictional force is a **nonconservative force**.

Potential Energy

A **potential energy** is energy that is associated with the configuration of a system in which a conservative force acts. When the conservative force does work W on a particle within the system, the change ΔU in the potential energy of the system is

$$\Delta U = -W. \qquad (8\text{-}1)$$

If the particle moves from point x_i to point x_f, the change in the potential energy of the system is

$$\Delta U = -\int_{x_i}^{x_f} F(x)\, dx. \qquad (8\text{-}6)$$

Gravitational Potential Energy

The potential energy associated with a system consisting of the Earth and a nearby particle is the **gravitational potential energy**. If the particle moves from height y_i to height y_f, the change in the gravitational potential energy of the particle–Earth system is

$$\Delta U = mg(y_f - y_i) = mg\,\Delta y. \qquad (8\text{-}7)$$

If the **reference position** of the particle is set as $y_i = 0$ and the corresponding gravitational potential energy of the system is set as $U_i = 0$, then the gravitational potential energy U when the particle is at any position y is

$$U = mgy. \qquad (8\text{-}9)$$

in which the subscripts refer to different instants during an transfer process. This conservation can also be written as

$$\Delta E = \Delta K + \Delta U = 0.$$

Potential Energy Curves

If we know the **potential energy function** $U(x)$ for a sy which a force F acts on a particle, we can find the force

$$F(x) = -\frac{dU(x)}{dx}.$$

If $U(x)$ is given on a graph, then at any value of x, the fo the negative of the slope of the curve there and the kinetic of the particle is given by

$$K(x) = E - U(x),$$

where E is the mechanical energy of the system. A **turnin** is a point x where the particle reverses its motion (there, The particle is in **equilibrium** at points where the slope $U(x)$ curve is zero (there, $F(x) = 0$).

Work by Nonconservative Forces

If a nonconservative applied force F does work on particl part of a system having a potential energy, then the wo done on the system by F is equal to the change ΔE in the m ical energy of the system:

$$W_{\text{app}} = \Delta K + \Delta U = \Delta E. \qquad (8\text{-}24$$

If a kinetic frictional force f_k does work on an obj change ΔE in the total mechanical energy of the object system containing it is given by

$$\Delta E = -f_k d,$$

in which d is the displacement of the object during the wo

Review and Summary
Review & Summary sections at the end of each chapter review the most important concepts and equations.

FIGURE 10-44   Problem 56.

57P.   Two 22.7 kg ice sleds are placed a short distance apart, one directly behind the other, as shown in Fig. 10-45. A 3.63 kg cat, standing on one sled, jumps across to the other and immediately back to the first. Both jumps are made at a speed of 3.05 m/s relative to the ice. Find the final speeds of the two sleds.

FIGURE 10-45   Problem 57.

58P.   The bumper of a 1200 kg car is designed so that it can just absorb all the energy when the car runs head-on into a solid wall at 5.00 km/h. The car is involved in a collision in which it runs at 70.0 km/h into the rear of a 900 kg car moving at 60.0 km/h in the same direction. The 900 kg car is accelerated to 70.0 km/h as a result of the collision. (a) What is the speed of the 1200 kg car immediately after impact? (b) What is the ratio of the kinetic energy absorbed in the collision to that which can be absorbed by the bumper of the 1200 kg car?

59P.   A railroad freight car weighing 32 tons and traveling at 5.0

FIGURE 10-46   Exercise 62.

63E.   In a game of pool, the cue ball strikes anothe at rest. After the collision, the cue ball moves at 3.5 line making an angle of 22.0° with its original dir tion, and the second ball has a speed of 2.00 m/s angle between the direction of motion of the secon original direction of motion of the cue ball and (b speed of the cue ball. (c) Is kinetic energy conserv

64E.   Two vehicles A and B are traveling west and tively, toward the same intersection, where they co together. Before the collision, A (total weight 2700 with a speed of 40 mi/h and B (total weight 3600 lt of 60 mi/h. Find the magnitude and direction of the v (interlocked) vehicles immediately after the collisi

65E.   In a game of billiards, the cue ball is given a V and strikes the pack of 15 stationary balls. All engage in nume   and ball–cushion col time lat   same accident) all

Exercises and Problems
A hallmark of this text, nearly 3400 end-of-chapter exercises and problems are arranged in order of difficulty, starting with the exercises (labeled "E"), followed by the problems (labeled "P"). Particularly difficult problems are identified with an asterisk (*). Answers to all the odd-numbered exercises and problems are provided in the back of the book. New electronic computation problems, which require the use of math packages and graphing calculators, have been added to many of the chapters.

Brief Contents

Contents

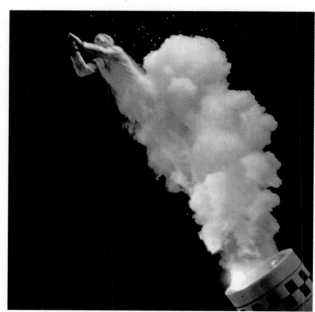

CHAPTER 5

FORCE AND MOTION—I   *81*

Can a man pull two railroad passenger cars with his teeth?

CHAPTER 6

FORCE AND MOTION—II   *108*

Why do cats sometimes survive long falls better than shorter ones?

CHAPTER 7

KINETIC ENERGY AND WORK   *130*

How much work is required in lifting great weights?

CHAPTER 8

POTENTIAL ENERGY AND CONSERVATION OF ENERGY   *155*

How far will a bungee-cord jumper fall?

CHAPTER 9

SYSTEMS OF PARTICLES   *186*

How does a ballerina seemingly "turn off" gravity?

CHAPTER 10

COLLISIONS *214*

*Is a board or a concrete block
easy to break in karate?*

CHAPTER 11

ROTATION *238*

*What advantages does physics offer
in judo throws?*

CHAPTER 12

ROLLING, TORQUE, AND ANGULAR MOMENTUM *268*

*Why is a quadruple somersault
so difficult in trapeze acts?*

CHAPTER 13

EQUILIBRIUM AND ELASTICITY *297*

*Can you safely rest in a fissure
during a chimney climb?*

CHAPTER 14

GRAVITATION *322*

How can a black hole be detected?

CHAPTER 19

TEMPERATURE, HEAT, AND THE FIRST LAW OF THERMODYNAMICS *453*

Why are black robes worn in extremely hot climates?

CHAPTER 20

THE KINETIC THEORY OF GASES *484*

When a room's air is heated, does the internal energy of the air increase?

CHAPTER 21

ENTROPY AND THE SECOND LAW OF THERMODYNAMICS *509*

What in the world gives direction to time?

1
Measurement

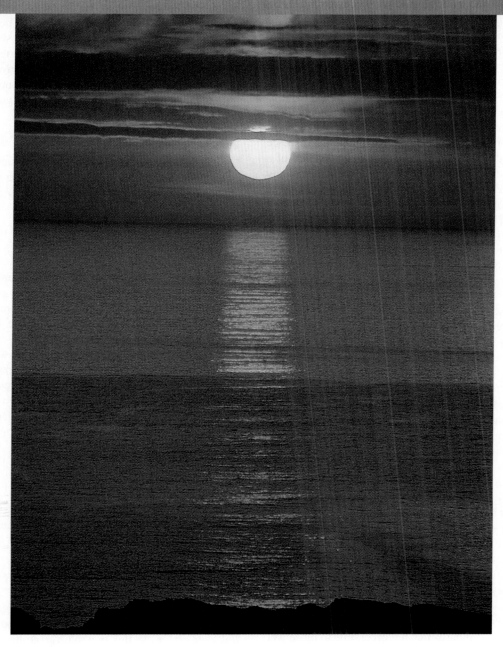

You can watch the Sun set and disappear over a calm ocean once while you lie on a beach, and once again if you stand up. Surprisingly, by measuring the time between the two sunsets, you can approximate the radius of the Earth. How can such a simple observation be used to measure the Earth?

1-1 MEASURING THINGS

Physics is based on measurement. We discover physics by learning how to measure the quantities that are involved in physics. Among these quantities are length, time, mass, temperature, pressure, and electrical resistance.

To describe a physical quantity we first define a **unit,** that is, a measure of the quantity that is defined to be exactly 1.0. Then we define a **standard,** that is, a reference to which all other examples of the quantity are compared. For example, the unit of length is the meter, and, as you will see, the standard for the meter is defined to be the distance traveled by light in vacuum during a certain fraction of a second. We are free to define a unit and its standard in any way we care to. The important thing is to do so in such a way that scientists around the world will agree that our definitions are both sensible and practical.

Once we have set up a standard, say, for length, we must work out procedures by which any length whatever, be it the radius of a hydrogen atom, the wheelbase of a skateboard, or the distance to a star, can be expressed in terms of the standard. Rulers, which approximate our length standard, give us one such procedure for measuring length. But many of our comparisons must be indirect. You cannot use a ruler, for example, to measure either the radius of an atom or the distance to a star.

There are so many physical quantities that it is a problem to organize them. Fortunately, they are not all independent. Speed, for example, is the ratio of a length to a time. So what we do is pick out—by international agreement—a small number of physical quantities, such as length and time, and assign standards to them alone. We then define all other physical quantities in terms of these *base quantities* and their standards. Speed, for example, is defined in terms of the base quantities length and time and the associated base standards.

The base standards must be both accessible and invariable. If we define the length standard as the distance between one's nose and the index finger on an outstretched arm, we certainly have an accessible standard—but it will, of course, vary from person to person. The demand for precision in science and engineering pushes us to aim first for invariability. We then exert great effort to make duplicates of the base standards that are accessible to those who need them.

1-2 THE INTERNATIONAL SYSTEM OF UNITS

In 1971, the 14th General Conference on Weights and Measures picked seven quantities as base quantities, thereby forming the basis of the International System of Units, abbreviated SI from its French name and popularly known as the *metric system.* Table 1-1 shows the units for the three base quantities—length, mass, and time—that we use in the early chapters of this book. The units for the quantities were chosen to be on a "human scale."

Many SI *derived units* are defined in terms of these base units. For example, the SI unit for power, called the **watt** (abbreviated as W), is defined in terms of the base units for mass, length, and time. Thus, as you will see in Chapter 7,

$$1 \text{ watt} = 1 \text{ W} = 1 \text{ kg} \cdot \text{m}^2/\text{s}^3. \qquad (1\text{-}1)$$

To express the very large and very small quantities that we often run into in physics, we use the so-called *scientific notation,* which employs powers of 10. In this notation,

$$3{,}560{,}000{,}000 \text{ m} = 3.56 \times 10^9 \text{ m} \qquad (1\text{-}2)$$

and $\qquad 0.000\ 000\ 492 \text{ s} = 4.92 \times 10^{-7} \text{ s}. \qquad (1\text{-}3)$

Since the advent of computers, scientific notation sometimes takes on an even briefer look, as in 3.56 E9 m and 4.92 E −7 s, where E stands for "exponent of ten." It is briefer still on some calculators, where E is replaced with an empty space.

As a further convenience when dealing with very large or very small measurements, we use the prefixes listed in Table 1-2. As you can see, each prefix represents a certain power of 10, as a factor. Attaching a prefix to an SI unit has the effect of multiplying by the associated factor. Thus we can express a particular electric power as

$$1.27 \times 10^9 \text{ watts} = 1.27 \text{ gigawatts} = 1.27 \text{ GW} \quad (1\text{-}4)$$

or a particular time interval as

$$2.35 \times 10^{-9} \text{ s} = 2.35 \text{ nanoseconds} = 2.35 \text{ ns}. \quad (1\text{-}5)$$

Some prefixes, as used in milliliter, centimeter, kilogram, and megabyte, are already familiar to you.

1-3 CHANGING UNITS

We often need to change the units in which a physical quantity is expressed. We do so by a method called *chain-link conversion.* In this method, we multiply the original measurement by a **conversion factor** (a ratio of units that is equal to unity). For example, because 1 min and 60 s are identical time intervals, we can write

$$\frac{1 \text{ min}}{60 \text{ s}} = 1 \quad \text{and} \quad \frac{60 \text{ s}}{1 \text{ min}} = 1.$$

This is *not* the same as writing $\frac{1}{60} = 1$ or $60 = 1$; the *number* and its *unit* must be treated together.

TABLE 1-1 SOME SI BASE UNITS

QUANTITY	UNIT NAME	UNIT SYMBOL
Length	meter	m
Time	second	s
Mass	kilogram	kg

Because multiplying any quantity by unity leaves it unchanged, we can introduce such conversion factors wherever we find them useful. In chain-link conversion, we use the factors to cancel the unwanted units. For example, to convert 2 min to seconds, we have

$$2 \text{ min} = (2 \text{ min})(1) = (2 \text{ min})\left(\frac{60 \text{ s}}{1 \text{ min}}\right)$$

$$= 120 \text{ s}. \tag{1-6}$$

If you introduce the conversion factor in a way that the units do *not* cancel, invert the factor and try again. Units obey the same algebraic rules as variables and numbers.

Appendix D and the inside back cover give conversion factors between SI and other systems of units. The United States is the only major country (in fact, almost the only country) that has not officially adopted the International System of Units.

SAMPLE PROBLEM 1-1

The research submersible ALVIN is diving at a speed of 36.5 fathoms per minute.

(a) Express this speed in meters per second. A *fathom* (fath) is precisely 6 ft.

SOLUTION: To find the speed in meters per second, we write

$$36.5 \frac{\text{fath}}{\text{min}} = \left(36.5 \frac{\text{fath}}{\text{min}}\right)\left(\frac{1 \text{ min}}{60 \text{ s}}\right)\left(\frac{6 \text{ ft}}{1 \text{ fath}}\right)\left(\frac{1 \text{ m}}{3.28 \text{ ft}}\right)$$

$$= 1.11 \text{ m/s}. \qquad \text{(Answer)}$$

(b) What is this speed in miles per hour?

SOLUTION: As above, we have

$$36.5 \frac{\text{fath}}{\text{min}} = \left(36.5 \frac{\text{fath}}{\text{min}}\right)\left(\frac{60 \text{ min}}{1 \text{ h}}\right)\left(\frac{6 \text{ ft}}{1 \text{ fath}}\right)\left(\frac{1 \text{ mi}}{5280 \text{ ft}}\right)$$

$$= 2.49 \text{ mi/h}. \qquad \text{(Answer)}$$

(c) What is this speed in light-years per year?

SOLUTION: A light-year (ly) is the distance that light travels in 1 year, 9.46×10^{12} km.

We start from the result of (a) above:

$$1.11 \frac{\text{m}}{\text{s}} = \left(1.11 \frac{\text{m}}{\text{s}}\right)\left(\frac{1 \text{ ly}}{9.46 \times 10^{12} \text{ km}}\right)$$

$$\times \left(\frac{1 \text{ km}}{1000 \text{ m}}\right)\left(\frac{3.16 \times 10^7 \text{ s}}{1 \text{ y}}\right)$$

$$= 3.71 \times 10^{-9} \text{ ly/y}. \qquad \text{(Answer)}$$

We can write this in the even more unlikely form of 3.71 nly/y, where "nly" is an abbreviation for nanolight-year.

SAMPLE PROBLEM 1-2

How many square meters are in an area of 6.0 km²?

SOLUTION: Each kilometer in the original measure must be converted. The surest way is to separate them:

TABLE 1-2 PREFIXES FOR SI UNITS[a]

FACTOR	PREFIX	SYMBOL	FACTOR	PREFIX	SYMBOL
10^{24}	yotta-	Y	10^{-24}	yocto-	y
10^{21}	zetta-	Z	10^{-21}	zepto-	z
10^{18}	exa-	E	10^{-18}	atto-	a
10^{15}	peta-	P	10^{-15}	femto-	f
10^{12}	tera-	T	10^{-12}	**pico-**	**p**
10^9	**giga-**	**G**	10^{-9}	**nano-**	**n**
10^6	**mega-**	**M**	10^{-6}	**micro-**	**μ**
10^3	**kilo-**	**k**	10^{-3}	**milli-**	**m**
10^2	hecto-	h	10^{-2}	**centi-**	**c**
10^1	deka-	da	10^{-1}	deci-	d

[a]The most commonly used prefixes are shown in bold type.

$$6.0 \text{ km}^2 = 6.0 \text{ (km)(km)}$$

$$= 6.0 \text{ (km)(km)} \left(\frac{1000 \text{ m}}{1 \text{ km}} \right) \left(\frac{1000 \text{ m}}{1 \text{ km}} \right)$$

$$= 6.0 \times 10^6 \text{ m}^2.$$

PROBLEM SOLVING TACTICS

TACTIC 1: *Significant Figures and Decimal Places*

If you calculated the answer to (a) in Sample Problem 1-1 without your calculator automatically rounding it off, the number 1.112804878 might have appeared in the display. The precision implied by this number is meaningless. We have properly rounded the number to 1.11 m/s, which is the precision that is justified by the precision of the original data: The given speed of 36.5 fath/min consists of three digits, called **significant figures.** Any fourth digit that might appear to the right of the 5 is not known, and so the result of the conversion is not reliable beyond three digits, or three significant figures.

> In general, no final result should have more significant figures than the original data from which it was derived.

If multiple steps of calculation are involved, you should retain more significant figures than the original data have. However, when you come to the final result, you should round off according to the original data with the least significant figures. (The answers to sample problems in this book are usually presented with the symbol = instead of ≈ even if rounding off is involved.)

When a number such as 3.15 or 3.15×10^3 is provided in a problem, the number of significant figures is apparent. But how about the number 3000? Is it known to only one significant figure (it could be written as 3×10^3)? Or is it known to as many as four significant figures (it could be written as 3.000×10^3)? In this book, we assume that all the zeros in such provided numbers as 3000 are significant, but you had better not make that assumption elsewhere.

Don't confuse *significant figures* with *decimal places*. Consider the lengths 35.6 mm, 3.56 m, and 0.00356 m. They all have three significant figures but, in sequence, they have one, two, and five decimal places.

1-4 LENGTH

In 1792, the newborn Republic of France established a new system of weights and measures. Its cornerstone was the meter, defined to be one ten-millionth of the distance from the north pole to the equator. Eventually, for practical reasons, this Earth standard was abandoned and the meter came to be defined as the distance between two fine lines engraved near the ends of a platinum–iridium bar, the **standard meter bar,** which was kept at the International Bureau of Weights and Measures near Paris. Accurate copies of the bar were sent to standardizing laboratories throughout the world. These **secondary standards** were used to produce other, still more accessible standards so that ultimately every measuring device derived its authority from the standard meter bar through a complicated chain of comparisons.

In 1959, the yard was legally defined to be

$$1 \text{ yard} = 0.9144 \text{ meter (exactly)}, \tag{1-7}$$

which is equivalent to

$$1 \text{ inch} = 2.54 \text{ centimeters (exactly)}. \tag{1-8}$$

Table 1-3 lists some interesting lengths. One corresponds to a virus; an example is shown in Fig. 1-1.

Eventually, modern science and technology required a more precise standard than the distance between two fine scratches on a metal bar. In 1960, a new standard for the meter, based on the wavelength of light, was adopted. Specifically, the meter was redefined to be 1,650,763.73 wavelengths of a particular orange-red light emitted by atoms of krypton-86 in a gas discharge tube.* This awkward number of wavelengths was chosen so that the new standard would be as consistent as possible with the old meter-bar standard.

TABLE 1-3 SOME APPROXIMATE LENGTHS

LENGTH	METERS
Distance to the farthest observed quasar (1996)	2×10^{26}
Distance to the Andromeda galaxy	2×10^{22}
Distance to the nearest star (Proxima Centauri)	4×10^{16}
Distance to the farthest planet (Pluto)	6×10^{12}
Radius of the Earth	6×10^6
Height of Mt. Everest	9×10^3
Thickness of this page	1×10^{-4}
Wavelength of light	5×10^{-7}
Length of a typical virus	1×10^{-8}
Radius of a hydrogen atom	5×10^{-11}
Radius of a proton	1×10^{-15}

---

*The number 86 in the notation krypton-86 identifies a particular one of the five stable isotopes (or types) of this element. An equivalent notation is ^{86}Kr. The number is called the *mass number* of the isotope in question.

FIGURE 1-1 A ''false-color'' electron micrograph of influenza virus particles. Lipoproteins (colored yellow) taken from the host surround the core (colored green) of the virus. Together they are less than 50 nm in diameter.

The krypton-86 atoms of the atomic length standard are available everywhere, are identical, and emit light of precisely the same wavelength. As Philip Morrison of MIT has pointed out, every atom is a storehouse of natural standards, more secure than the International Bureau of Weights and Measures.

By 1983, the demand for higher precision had reached such a point that even the krypton-86 standard could not meet it, and in that year a bold step was taken. The meter was redefined as the distance traveled by a light wave in a specified time interval. In the words of the 17th General Conference on Weights and Measures:

> The meter is the length of the path traveled by light in vacuum during a time interval of 1/299,792,458 of a second.

This number was chosen so that the speed of light c could be exactly

$$c = 299,792,458 \text{ m/s}.$$

Measurements of the speed of light had become extremely precise, so it made sense to adopt the speed of light as a defined quantity and to use it to redefine the meter.

SAMPLE PROBLEM 1-3

Both 100 yd and 100 m are used as distances for dashes in track meets.

(a) Which is longer?

SOLUTION: From Eq. 1-7, 100 yd is equal to 91.44 m, so that 100 m is longer than 100 yd.

(b) By how many meters is it longer?

SOLUTION: We represent the difference as ΔL, where Δ is the capital Greek letter delta. Thus

$$\Delta L = 100 \text{ m} - 100 \text{ yd}$$
$$= 100 \text{ m} - 91.44 \text{ m} = 8.56 \text{ m}. \quad \text{(Answer)}$$

(c) By how many feet is it longer?

SOLUTION: We can express this difference by the method used in Sample Problem 1-1,

$$\Delta L = (8.56 \text{ m})\left(\frac{3.28 \text{ ft}}{1 \text{ m}}\right) = 28.1 \text{ ft}. \quad \text{(Answer)}$$

1-5 TIME

Time has two aspects. For civil and some scientific purposes, we want to know the time of day (see Fig. 1-2) so that we can order events in sequence. And in much scientific work, we want to know how long an event lasts. Thus any time standard must be able to answer two questions: ''*When* did it happen?'' and ''What was its *duration*?'' Table 1-4 shows some time intervals.

Any phenomenon that repeats itself is a possible time standard. The rotation of the Earth, which determines the length of the day, has been used in this way for centuries. A quartz clock, in which a quartz ring continuously vibrates, can be calibrated against the rotating Earth via astronomical observations and used to measure time intervals in the laboratory. However, the calibration cannot be carried out

FIGURE 1-2 When the metric system was proposed in 1792, the hour was redefined to provide a 10-hour day. The idea did not catch on. The maker of this 10-hour watch wisely provided a small dial that kept conventional 12-hour time. Do the two dials indicate the same time?

TABLE 1-4 **SOME APPROXIMATE TIME INTERVALS**

TIME INTERVAL	SECONDS
Lifetime of the proton (predicted)	1×10^{39}
Age of the universe	5×10^{17}
Age of the pyramid of Cheops	1×10^{11}
Human life expectancy (U.S.)	2×10^{9}
Length of a day	9×10^{4}
Time between human heartbeats	8×10^{-1}
Lifetime of the muon	2×10^{-6}
Shortest laboratory light pulse (1989)	6×10^{-15}
Lifetime of most unstable particle	1×10^{-23}
The Planck time[a]	1×10^{-43}

[a]This is the earliest time after the ''Big Bang'' at which the laws of physics as we know them can be applied.

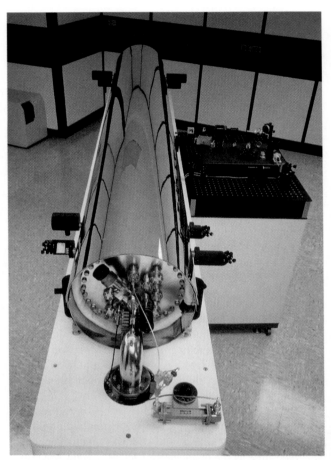

FIGURE 1-3 The cesium atomic frequency standard at the National Institute of Standards and Technology in Boulder, Colorado. It is the primary standard for the unit of time in the United States. Dial (303) 499-7111 and set your watch by it. Call (900) 410-8463 or http://tycho.usno.navy.mil/time.html for Naval Observatory time signals.

with the accuracy called for by modern scientific and engineering technology.

To meet the need for a better time standard, atomic clocks have been developed. Figure 1-3 shows such a clock, based on a characteristic frequency of the isotope cesium-133, at the National Institute of Standards and Technology (NIST). It forms the basis in this country for Coordinated Universal Time (UTC), for which time signals are available by shortwave radio (stations WWV and WWVH) and by telephone. (To set a clock extremely accurately at your particular location, you would have to account for the travel time that is required for these signals to reach you.)

Figure 1-4 shows variations in the length of one day on Earth over a 4-year period, as determined by comparison with a cesium clock. Because of the seasonal variation displayed by Fig. 1-4, we suspect the rotating Earth when there is a difference between Earth and atom as timekeepers. The variation is probably due to tidal effects caused by the Moon and to large-scale atmospheric winds.

The 13th General Conference on Weights and Measures in 1967 adopted a standard second based on the cesium clock:

> One second is the time taken by 9,192,631,770 vibrations of the light (of a specified wavelength) emitted by a cesium-133 atom.

In principle, two cesium clocks would have to run for 6000 years before their readings would differ by more than 1 s. Even such accuracy pales in comparison to that of clocks currently being developed; their precision may be as fine as 1 part in 10^{18}, that is, 1 s in 1×10^{18} s (about 3×10^{10} y).

FIGURE 1-4 Variations in the length of the day over a 4-year period. Note that the entire vertical scale amounts to only 3 ms (= 0.003 s).

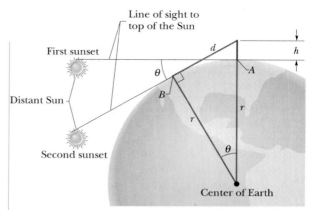

SAMPLE PROBLEM 1-4*

Suppose that you watch the Sun set over a calm ocean while lying on the beach, starting a stopwatch just as the top of the Sun disappears. You then stand, elevating your eyes by a height $h = 1.70$ m, and stop the watch when the top of the Sun again disappears. If the elapsed time on the watch is $t = 11.1$ s, what is the radius r of the Earth?

SOLUTION: As shown in Fig. 1-5, your line of sight to the top of the Sun, as the top first disappears, is tangent to the Earth's surface at your location, at point A. The figure also shows that your line of sight to the top of the Sun as the top again disappears is tangent to the Earth's surface at point B. Let d represent the distance between point B and the location of your eyes when you are standing, and draw radii r as shown in Fig. 1-5. From the Pythagorean theorem, we then have

$$d^2 + r^2 = (r + h)^2 = r^2 + 2rh + h^2,$$

or
$$d^2 = 2rh + h^2. \qquad (1\text{-}9)$$

Because the height h is so much smaller than the Earth's radius r, the term h^2 is negligible compared to the term $2rh$, and we can rewrite Eq. 1-9 as

$$d^2 = 2rh. \qquad (1\text{-}10)$$

In Fig. 1-5, the angle between the two tangent points A and B is θ, which is also the angle through which the Sun moves about the Earth during the measured time $t = 11.1$ s. During a full day, which is approximately 24 h, the Sun moves through an angle of 360° about the Earth. This allows us to write

$$\frac{\theta}{360°} = \frac{t}{24 \text{ h}},$$

which, with $t = 11.1$ s, gives us

$$\theta = \frac{(360°)(11.1 \text{ s})}{(24 \text{ h})(60 \text{ min/h})(60 \text{ s/min})} = 0.04625°.$$

From Fig. 1-5 we see that $d = r \tan \theta$. Substituting this for d in Eq. 1-10 gives us

$$r^2 \tan^2\theta = 2rh$$

or
$$r = \frac{2h}{\tan^2\theta}.$$

Substituting $\theta = 0.04625°$ and $h = 1.70$ m, we find

$$r = \frac{(2)(1.70 \text{ m})}{\tan^2 0.04625°} = 5.22 \times 10^6 \text{ m}, \quad \text{(Answer)}$$

which is within 20% of the accepted value (6.37×10^6 m) for the (mean) radius of the Earth.

*Adapted from "Doubling Your Sunsets, or How Anyone Can Measure the Earth's Size with a Wristwatch and Meter Stick," by Dennis Rawlins, *American Journal of Physics*, Feb. 1979, Vol. 47, pp. 126–128. This technique works best at the equator.

FIGURE 1-5 Sample Problem 1-4. Your line of sight to the top of the setting Sun rotates through the angle θ when you move from a prone position at point A, elevating your eyes by a distance h. (Angle θ and distance h are exaggerated here for clarity.)

1-6 MASS

The Standard Kilogram

The SI standard of mass is a platinum–iridium cylinder (Fig. 1-6) kept at the International Bureau of Weights and Measures near Paris and assigned, by international agreement, a mass of 1 kilogram. Accurate copies have been sent to standardizing laboratories in other countries, and the masses of other bodies can be determined by balancing them against a copy. Table 1-5 shows some masses expressed in kilograms.

The U.S. copy of the standard kilogram is housed in a vault at NIST. It is removed, no more than once a year, for the purpose of checking duplicate copies that are used elsewhere. Since 1889, it has been taken to France twice for recomparison with the primary standard. The procedure of comparison with the standard kilogram will someday be

FIGURE 1-6 The international 1 kg standard of mass, a cylinder 39 cm in height and in diameter.

TABLE 1-5 **SOME APPROXIMATE MASSES**

OBJECT	KILOGRAMS
Known universe	1×10^{53}
Our galaxy	2×10^{41}
Sun	2×10^{30}
Moon	7×10^{22}
Asteroid Eros	5×10^{15}
Small mountain	1×10^{12}
Ocean liner	7×10^{7}
Elephant	5×10^{3}
Grape	3×10^{-3}
Speck of dust	7×10^{-10}
Penicillin molecule	5×10^{-17}
Uranium atom	4×10^{-25}
Proton	2×10^{-27}
Electron	9×10^{-31}

replaced with a more reliable and accessible procedure involving a base unit of mass that is the mass of an atom.

A Second Mass Standard

The masses of atoms can be compared with each other more precisely than they can be compared with the standard kilogram. For this reason, we have a second mass standard. It is the carbon-12 atom, which, by international agreement, has been assigned a mass of 12 **atomic mass units** (u). The relation between the two units is

$$1 \text{ u} = 1.6605402 \times 10^{-27} \text{ kg}, \qquad (1\text{-}11)$$

with an uncertainty of ± 10 in the last two decimal places. Scientists can, with reasonable precision, experimentally determine the masses of other atoms relative to the mass of carbon-12. What we presently lack is a reliable means of extending that precision to more common units of mass, such as a kilogram.

REVIEW & SUMMARY

Measurement in Physics

Physics is based on measurement of physical quantities. Certain physical quantities have been chosen as **base quantities** (such as length, time, and mass); each has been defined in terms of a **standard** and given a **unit** measure (such as meter, second, and kilogram). Other physical quantities (such as speed) are defined in terms of the base quantities and their standards and units.

SI Units

The unit system emphasized in this book is the International System of Units (SI). The three physical quantities displayed in Table 1-1 are used in the early chapters of this book. Standards, which must be both accessible and invariable, have been established for these base quantities by international agreement. These standards are used in all physical measurement, for both the base quantities and the quantities derived from them. Scientific notation and the prefixes of Table 1-2 can be used to simplify measurement notation in many cases.

Changing Units

Conversion of units from one system to another (for example, from miles per hour to kilometers per second) may be performed by using *chain-link conversions* in which the units are treated as

algebraic quantities and the original data are multiplied successively, by *conversion factors* written as unity, until the desired units are obtained.

Length

The unit of length—the meter—is defined as the distance traveled by light during a precisely specified time interval. The yard, together with its multiples and submultiples, is legally defined in this country in terms of the meter.

Time

The unit of time—the second—was formerly defined in terms of the rotation of the Earth. It is now defined in terms of the vibrations of the light emitted by an atomic (cesium-133) source. Accurate time signals are sent worldwide by radio signals keyed to atomic clocks in standardizing laboratories.

Mass

The unit of mass—the kilogram—is defined in terms of a particular platinum–iridium prototype kept near Paris, France. For measurements on an atomic scale, the atomic mass unit, defined in terms of the atom carbon-12, is usually used.

EXERCISES & PROBLEMS

SECTION 1-2   The International System of Units

1E.   Use the prefixes in Table 1-2 to express (a) 10^{6} phones; (b) 10^{-6} phone; (c) 10^{1} cards; (d) 10^{9} lows; (e) 10^{12} bulls; (f) 10^{-1}

mate; (g) 10^{-2} pede; (h) 10^{-9} Nannette; (i) 10^{-12} boo; (j) 10^{-18} boy; (k) 2×10^{2} withits; (l) 2×10^{3} mockingbirds. Now that you have the idea, invent a few similar expressions. (See, in this connection, p. 61 of *A Random Walk in Science,* compiled by R.

L. Weber; Crane, Russak & Co., New York, 1974.)

2E. Some of the prefixes of the SI units have crept into everyday language. (a) What is the weekly equivalent of an annual salary of $36 K (= 36 kilobucks)? (b) A lottery awards 10 megabucks as the top prize, payable over 20 years. How much is in each monthly check? (c) The hard disk of a computer has a capacity of 30 MB (= 30 megabytes). At 8 bytes per word, how many words can it store? [In computerese, *mega* means 1,048,576 (= 2^{20}), not 1,000,000.]

SECTION 1-4 Length

3E. A space shuttle orbits the Earth at an altitude of 300 km. What is this altitude in (a) miles and (b) millimeters?

4E. What is your height in meters?

5E. The micrometer (1 μm) is often called the *micron*. (a) How many microns make up 1.0 km? (b) What fraction of a centimeter equals 1.0 μm? (c) How many microns are in 1.0 yd?

6E. The Earth is approximately a sphere of radius 6.37 $\times$ 10^6 m. (a) What is its circumference in kilometers? (b) What is its surface area in square kilometers? (c) What is its volume in cubic kilometers?

7E. Calculate the number of kilometers in 20.0 mi using only the following conversion factors: 1 mi = 5280 ft, 1 ft = 12 in., 1 in. = 2.54 cm, 1 m = 100 cm, and 1 km = 1000 m.

8E. Give the relation between (a) a square yard and a square foot; (b) a square inch and a square centimeter; (c) a square mile and a square kilometer; and (d) a cubic meter and a cubic centimeter.

9P. A unit of area, often used in measuring land areas, is the *hectare,* defined as 10^4 m². An open-pit coal mine consumes 75 hectares of land, down to a depth of 26 m, each year. What volume of earth, in cubic kilometers, is removed in this time?

10P. The *cord* is a volume of cut wood equal to a stack 8 ft long, 4 ft wide, and 4 ft high. How many cords of wood are in 1.0 m³?

11P. A room is 20 ft, 2 in. long and 12 ft, 5 in. wide. What is the floor area in (a) square feet and (b) square meters? If the ceiling is 12 ft, 2$\frac{1}{2}$ in. above the floor, what is the volume of the room in (c) cubic feet and (d) cubic meters?

12P. Antarctica is roughly semicircular, with a radius of 2000 km (Fig. 1-7). The average thickness of its ice cover is 3000 m. How many cubic centimeters of ice does Antarctica contain? (Ignore the curvature of the Earth.)

FIGURE 1-7 Problem 12.

13P. A typical sugar cube has an edge length of 1 cm. If you had a cubical box that contained a mole of sugar cubes, what would its edge length be? (One mole = 6.02 $\times$ 10^{23} units.)

14P. Hydraulic engineers often use, as a unit of volume of water, the *acre-foot,* defined as the volume of water that will cover 1

acre of land to a depth of 1 ft. A severe thunderstorm dumps 2.0 in. of rain in 30 min on a town of area 26 km². What volume of water, in acre-feet, fell on the town?

15P. A certain brand of house paint claims a coverage of 460 ft²/gal. (a) Express this quantity in square meters per liter. (b) Express this quantity in SI base units (see Appendixes A and D). (c) What is the inverse of the original quantity, and what is its physical significance?

16P. Astronomical distances are so large compared to terrestrial ones that much larger units of length are needed. An *astronomical unit* (AU) is equal to the average distance from the Earth to the Sun, about 92.9 $\times$ 10^6 mi. A *parsec* (pc) is the distance at which 1 AU would subtend an angle of exactly 1 second of arc (Fig. 1-8). A *light-year* (ly) is the distance that light, traveling through a vacuum with a speed of 186,000 mi/s, would cover in 1.0 year. (a) Express the distance from the Earth to the Sun in parsecs and in light-years. (b) Express 1 ly and 1 pc in miles. Although "light-year" appears frequently in popular writing, the parsec is preferred by astronomers.

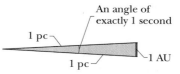

FIGURE 1-8 Problem 16.

17P. During a total solar eclipse, your view of the Sun is almost exactly replaced by your view of the moon. Assuming that the distance from you to the Sun is about 400 times the distance from you to the moon, (a) find the ratio of the Sun's diameter to the moon's diameter. (b) What is the ratio of their volumes? (c) Hold up a dime (or another small coin) so that it would just eclipse the full moon, and measure the angle it subtends at the eye. From this experimental result and the given distance between the moon and the Earth (= 3.8 $\times$ 10^5 km), estimate the diameter of the moon.

18P*. The standard kilogram (see Fig. 1-6) is in the shape of a circular cylinder with its height equal to its diameter. Show that, for a circular cylinder of fixed volume, this equality gives the smallest surface area, thus minimizing the effects of surface contamination and wear.

SECTION 1-5 Time

19E. Express the speed of light, 3.0 $\times$ 10^8 m/s, in (a) feet per nanosecond and (b) millimeters per picosecond.

20E. Enrico Fermi once pointed out that a standard lecture period (50 min) is close to 1 microcentury. How long is a microcentury in minutes, and what is the percent difference from Fermi's approximation?

21E. How many seconds are in 1 year (= 365.25 days)?

22E. A certain pendulum clock (with a 12-hour dial) happens to gain 1.0 min/day. After setting the clock to the correct time, how long must one wait until it again indicates the correct time?

23E. What is the age of the universe (see Table 1-4) in days?

24E. (a) A unit of time sometimes used in microscopic physics is the *shake*. One shake equals 10^{-8} s. Are there more shakes in a second than there are seconds in a year? (b) Humans have existed for about 10^6 years, whereas the universe is about 10^{10} years old. If the age of the universe is taken to be 1 "day," for how many "seconds" have humans existed?

25E. The maximum speeds of various animals are roughly as follows, in miles per hour: (a) snail, 3.0×10^{-2}; (b) spider, 1.2; (c) human, 23; and (d) cheetah, 70. Convert these data to meters per second. (All four calculations involve the identical conversion factor. You might want to determine that factor first and store it in the memory of your calculator, to use when needed.)

26P. An astronomical unit (AU) is the average distance of the Earth from the Sun, approximately 1.50×10^8 km. The speed of light is about 3.0×10^8 m/s. Express the speed of light in terms of astronomical units per minute.

27P. Until 1883, every city and town in the United States kept its own local time. Today, travelers reset their watches only when the time change equals 1 h. How far, on the average, must you travel in degrees of longitude until your watch must be reset by 1.0 h? *Hint:* The Earth rotates 360° in about 24 h.

28P. Assuming the length of the day uniformly increases by 0.001 s per century, calculate the cumulative effect on the measure of time over 20 centuries. Such slowing of the Earth's rotation is indicated by observations of the occurrences of solar eclipses during this period.

29P. On two *different* tracks, the winners of the mile race ran their races in 3 min, 58.05 s and 3 min, 58.20 s. In order to conclude that the runner with the shorter time was indeed faster, what is the maximum error, in feet, that can be permitted in laying out the mile distances?

30P. Five clocks are being tested in a laboratory. Exactly at noon, as determined by the WWV time signal, on successive days of a week the clocks read as in the following table. Rank the five clocks according to their relative value as good timekeepers, best to worst. Justify your choice.

CLOCK	SUN.	MON.	TUES.	WED.	THURS.	FRI.	SAT.
A	12:36:40	12:36:56	12:37:12	12:37:27	12:37:44	12:37:59	12:38:14
B	11:59:59	12:00:02	11:59:57	12:00:07	12:00:02	11:59:56	12:00:03
C	15:50:45	15:51:43	15:52:41	15:53:39	15:54:37	15:55:35	15:56:33
D	12:03:59	12:02:52	12:01:45	12:00:38	11:59:31	11:58:24	11:57:17
E	12:03:59	12:02:49	12:01:54	12:01:52	12:01:32	12:01:22	12:01:12

31P*. The time it takes the Moon to return to a given position as seen against the background of the fixed stars is called a *sidereal* month. The time interval between identical phases of the Moon is called a *lunar* month. The lunar month is longer than the sidereal month. Why, and by how much?

SECTION 1-6 Mass

32E. Using conversions and data in the chapter, determine the number of hydrogen atoms required to obtain 1.0 kg of hydrogen. A hydrogen atom has a mass of 1.0 u.

33E. One molecule of water (H_2O) contains two atoms of hydrogen and one atom of oxygen. A hydrogen atom has a mass of 1.0 u and an atom of oxygen has a mass of 16 u, approximately. (a) What is the mass in kilograms of one molecule of water? (b) How many molecules of water are in the world's oceans, which have an estimated total mass of 1.4×10^{21} kg?

34E. The Earth has a mass of 5.98×10^{24} kg. The average mass of the atoms that make up the Earth is 40 u. How many atoms are in the Earth?

35P. What mass of water fell on the town in Problem 14 during the thunderstorm? One cubic meter of water has a mass of 10^3 kg.

36P. A person on a diet might lose 2.3 kg (about 5 lb) per week. Express the mass loss rate in milligrams per second.

37P. (a) Assuming that the density (mass/volume) of water is exactly 1 g/cm³, express the density of water in kilograms per cubic meter (kg/m³). (b) Suppose that it takes 10 h to drain a container of 5700 m³ of water. What is the "mass flow rate," in kilograms per second, of water from the container?

38P. Grains of fine California beach sand are approximately spheres with an average radius of 50 μm; they are made of silicon dioxide, 1 m³ of which has a mass of 2600 kg. What mass of sand grains would have a total surface area equal to the surface area of a cube 1 m on an edge?

39P. The density of iron is 7.87 g/cm³, and the mass of an iron atom is 9.27×10^{-26} kg. If the atoms are spherical and tightly packed, (a) what is the volume of an iron atom and (b) what is the distance between the centers of adjacent atoms?

2
Motion Along a Straight Line

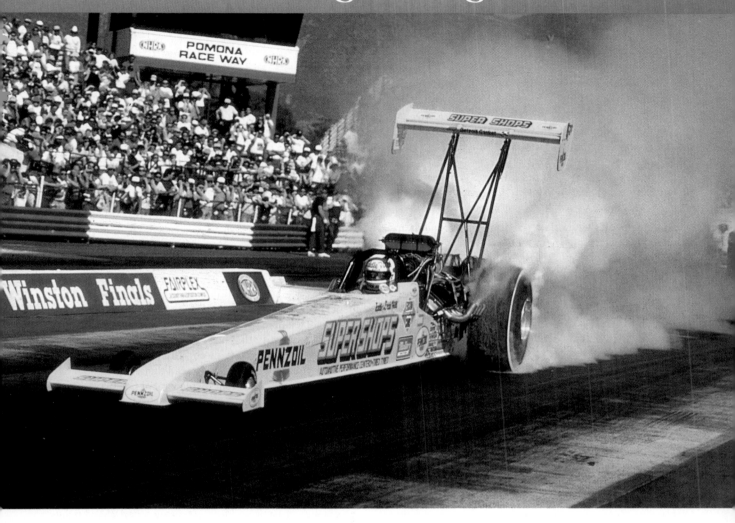

In 1977, Kitty O'Neil set a dragster record by reaching 392.54 mi/h in a sizzling time of 3.72 s. In 1958, Eli Beeding Jr. rode a rocket sled from a standstill to a speed of 72.5 mi/h in an elapsed time of 0.04 s (less than an eye blink). How can we compare these two rides to see which was more exciting (or more frightening)—by final speeds, by elapsed times, or by some other quantity?

2-1 MOTION

The world, and everything in it, moves. Even seemingly stationary things, such as a roadway, move with the Earth's rotation, the Earth's orbit around the Sun, the Sun's orbit around the center of the Milky Way galaxy, and that galaxy's migration relative to other galaxies. The classification and comparison of motions (called **kinematics**) is often challenging. What exactly do you measure, and how do you compare?

Before we attempt an answer, we shall examine some general properties of motion that is restricted in three ways.

1. The motion is along a straight line only. The line may be vertical (that of a falling stone), horizontal (that of a car on a level highway), or slanted, but it must be straight.

2. The cause of the motion will not be specified until Chapter 5. In this chapter you study only the motion itself. Does the object speed up, slow down, stop, or reverse direction; and, if the motion does change, how is time involved in the change?

3. The moving object is either a **particle** (by which we mean a pointlike object such as an electron) or an object that moves like a particle (such that every portion moves in the same direction and at the same rate). A pig slipping down a straight playground slide might be considered to be moving like a particle; however, a rotating merry-go-round would not, because different points around its rim move in different directions.

2-2 POSITION AND DISPLACEMENT

To locate an object means to find its position relative to some reference point, often the **origin** (or zero point) of an axis such as the *x* axis in Fig. 2-1. The **positive direction** of the axis is in the direction of increasing numbers (coordinates), which is toward the right as Fig. 2-1 is drawn. The opposite direction is the **negative direction.**

For example, a particle might be located at $x = 5$ m, which means that it is 5 m in the positive direction from the origin. If it were at $x = -5$ m, it would be just as far from the origin but in the opposite direction. On the axis, a coor-

dinate of -5 m is less than one of -1 m, and both coordinates are less than a coordinate of $+5$ m.

A change from one position x_1 to another position x_2 is called a **displacement** Δx, where

$$\Delta x = x_2 - x_1. \tag{2-1}$$

(The symbol Δ, which represents a change in a quantity, means the final value of that quantity less the initial value.) When numbers are inserted for the position values x_1 and x_2, a displacement in the positive direction (toward the right in Fig. 2-1) always comes out positive, and one in the opposite direction (left in the figure) negative. For example, if the particle moves from $x_1 = 5$ m to $x_2 = 12$ m, then $\Delta x = (12\text{ m}) - (5\text{ m}) = +7$ m. The positive result indicates that the motion is in the positive direction. If the particle then returns to $x = 5$ m, the displacement for the full trip is zero. The actual number of meters covered for the full trip is immaterial; displacement involves only the original and final positions.

If we ignore the sign of a displacement (and thus the direction), we are left with the **magnitude** of the displacement Δx, which is always positive.

Displacement is an example of a **vector quantity,** which is a quantity that has both a direction and a magnitude. We explore vectors more fully in Chapter 3 (in fact, some of you may have already read that chapter), but here all we need is the idea that displacement has two features: (1) its magnitude is the distance (such as the number of meters) between the original and final positions, and (2) its direction on an axis, from an original position to a final position, is represented by a plus sign or a minus sign.

What follows is the first of many checkpoints that you will see in this book. Each consists of one or more questions whose answers require some reasoning or a mental calculation, and each gives you a quick check of your understanding. The answers are listed in the back of the book.

CHECKPOINT **1:** Here are three pairs of initial and final positions, respectively, along an *x* axis. Which pairs give a negative displacement: (a) -3 m, $+5$ m; (b) -3 m, -7 m; (c) 7 m, -3 m?

2-3 AVERAGE VELOCITY AND AVERAGE SPEED

A compact way to describe position is with a graph of position *x* plotted as a function of time *t*—a graph of $x(t)$. As a simple example, Fig. 2-2 shows $x(t)$ for a stationary jack rabbit (which we treat as a particle) at $x = -2$ m.

FIGURE 2-1 Position is determined on an axis that is marked in units of length and that extends indefinitely in opposite directions.

Figure 2-3*a*, also for a rabbit, is more interesting, because it involves motion. The rabbit is apparently first noticed at $t = 0$ when it is at the position $x = -5$ m. It moves toward $x = 0$, passes through that point at $t = 3$ s, and then moves on to increasingly larger positive values of x.

Figure 2-3*b* depicts the actual straight-line motion of the rabbit and is something like what you would see. The graph in Fig. 2-3*a* is more abstract and quite unlike what you would see, but it is richer in information. It also reveals how fast the rabbit moves. Actually, several quantities are associated with the phrase "how fast." One of them is the **average velocity** $\bar{v}$, which is the ratio of the displacement Δx that occurs during a particular time interval Δt to that interval:*

$$\bar{v} = \frac{\Delta x}{\Delta t} = \frac{x_2 - x_1}{t_2 - t_1}. \tag{2-2}$$

On a graph of x versus t, $\bar{v}$ is the **slope** of the straight line that connects two particular points on the $x(t)$ curve: one is the point that corresponds to x_2 and t_2, and the other is the point that corresponds to x_1 and t_1. Like displacement, $\bar{v}$ has both magnitude and direction. (Average velocity is another vector quantity.) Its magnitude is the magnitude of the line's slope. A positive $\bar{v}$ (and slope) tells us that the line slants upward toward the right; a negative $\bar{v}$ (and slope), that the line slants upward to the left. The average velocity $\bar{v}$ always has the same sign as the displacement Δx because Δt in Eq. 2-2 is always positive.

Figure 2-4 shows how to find $\bar{v}$ for the rabbit of Fig. 2-3 for the time interval $t = 1$ s to $t = 4$ s. The average velocity during that time interval is $\bar{v} = 6$ m/3 s $= +2$ m/s; it is computed as the slope of the straight line that connects the point on the curve at the beginning of the interval and the point on the curve at the end of the interval.

SAMPLE PROBLEM 2-1

You drive a beat-up pickup truck down a straight road for 5.2 mi at 43 mi/h, at which point you run out of fuel. You walk 1.2 mi farther, to the nearest gas station, in 27 min ($= 0.450$ h). What is your average velocity from the time you started your truck to the time you arrived at the station? Find the answer both numerically and graphically.

SOLUTION: To calculate $\bar{v}$ we need your displacement Δx, from start to finish, and the elapsed time Δt. Assume, for convenience, that your starting point is at the origin of an x axis (so $x_1 = 0$) and that you move in the positive direction.

---

*In this book, a bar over a symbol means an average value of the quantity that the symbol represents.

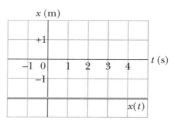

FIGURE 2-2 The graph of $x(t)$ for a jack rabbit that is stationary at $x = -2$ m. The value of x is -2 m for all times t.

(*a*)

(*b*)

FIGURE 2-3 (*a*) The graph of $x(t)$ for a moving jack rabbit. (*b*) The path associated with the graph. The scale below the x axis shows the times at which the rabbit reaches various x values.

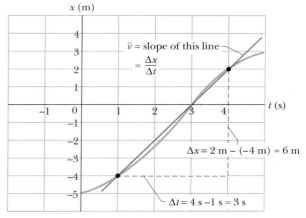

FIGURE 2-4 Calculation of the average velocity between $t = 1$ s and $t = 4$ s as the slope of the line that connects the points (on the curve) representing those times.

You end up at x_2 = 5.2 mi + 1.2 mi = +6.4 mi, and so, $\Delta x = x_2 - x_1$ = +6.4 mi. To get the driving time, we rearrange Eq. 2-2 and insert the data about the driving:

$$\Delta t = \frac{\Delta x}{\bar{v}} = \frac{5.2 \text{ mi}}{43 \text{ mi/h}} = 0.121 \text{ h},$$

or about 7.3 min. So the total time, start to finish, is

$$\Delta t = 0.121 \text{ h} + 0.450 \text{ h} = 0.571 \text{ h}.$$

Finally, we insert Δx and Δt into Eq. 2-2:

$$\bar{v} = \frac{\Delta x}{\Delta t} = \frac{6.4 \text{ mi}}{0.571 \text{ h}} \approx +11 \text{ mi/h}. \quad \text{(Answer)}$$

To find $\bar{v}$ graphically, we must first plot $x(t)$, as in Fig. 2-5 where the start and finish points on the curve are the origin and P, respectively. Your average velocity is the slope of the straight line connecting those points. The dashed lines show that the slope also gives $\bar{v}$ = 6.4 mi/0.57 h = +11 mi/h.

FIGURE 2-5 Sample Problem 2-1. The lines marked "Driving" and "Walking" are the position–time plots for the driver–walker in Sample Problem 2-1. The slope of the straight line joining the origin and point P is the average velocity for the trip.

SAMPLE PROBLEM 2-2

Suppose that you next carry the fuel back to the truck, making the return trip in 35 min. What is your average velocity for the full journey, from the start of your driving to your arrival back at the truck with the fuel?

SOLUTION: As previously, we must find your displacement Δx from start to finish and then divide it by the time interval Δt between start and finish. In this problem, however, the finish is back at the truck. You started at x_1 = 0. Back at the truck you are at position x_2 = 5.2 mi. And so Δx is 5.2 − 0 = 5.2 mi. The total time Δt you take in going from start to finish is

$$\Delta t = \frac{5.2 \text{ mi}}{43 \text{ mi/h}} + 27 \text{ min} + 35 \text{ min}$$

$$= 0.121 \text{ h} + 0.450 \text{ h} + 0.583 \text{ h} = 1.15 \text{ h}.$$

So $$\bar{v} = \frac{\Delta x}{\Delta t} = \frac{5.2 \text{ mi}}{1.15 \text{ h}} \approx +4.5 \text{ mi/h}. \quad \text{(Answer)}$$

This is slower than the average velocity computed in Sample Problem 2-1 because here the displacement is smaller and the time interval longer.

CHECKPOINT **2:** In Sample Problems 2-1 and 2-2, suppose that after refueling the truck, you drive back to x_1 at 52 mi/h. What is your average velocity for the entire trip?

Average speed $\bar{s}$ is a different way of describing "how fast" a particle moves. Whereas the average velocity involves the particle's displacement Δx, the average speed involves the total distance covered (for example, the number of meters run), independent of direction. That is,

$$\bar{s} = \frac{\text{total distance}}{\Delta t}. \quad (2\text{-}3)$$

Average speed also differs from average velocity in that average speed does *not* include direction and thus lacks any algebraic sign. Sometimes $\bar{s}$ is the same (except for the absence of a sign) as $\bar{v}$. But, as demonstrated in Sample Problem 2-3, when an object doubles back on its path, the results can be quite different.

SAMPLE PROBLEM 2-3

In Sample Problem 2-2, what is your average speed?

SOLUTION: From the beginning of your drive to your return to the truck with the fuel, you covered a total of 5.2 mi + 1.2 mi + 1.2 mi = 7.6 mi, taking 1.15 h; so

$$\bar{s} = \frac{7.6 \text{ mi}}{1.15 \text{ h}} \approx 6.6 \text{ mi/h}. \quad \text{(Answer)}$$

PROBLEM SOLVING TACTICS

TACTIC 1: *Do You Understand the Problem?*

For beginning problem solvers, no difficulty is more common than simply not understanding the problem. The best test of understanding is this: Can you explain the problem, in your own words, to a friend? Give it a try.

Write down the given data, with units, using the symbols of the chapter. (In Sample Problems 2-1 and 2-2, the given data allow you to find your net displacement Δx and the corresponding time interval Δt.) Identify the unknown and its symbol. (In these sample problems, the unknown is your average velocity, symbol $\bar{v}$.) Then find the connection between the unknown and the data. (The connection is Eq. 2-2, the definition of average velocity.)

TACTIC 2: *Are the Units OK?*

Be sure to use a consistent set of units when putting numbers into the equations. In Sample Problems 2-1 and 2-2, which involve a truck, the logical units in terms of the given data are miles for distances, hours for time intervals, and miles per hour for velocities. You may need to make conversions.

TACTIC 3: *Is Your Answer Reasonable?*

Look at your answer and ask yourself whether it makes sense. Is it far too large or far too small? Is the sign correct? Are the units appropriate? In Sample Problem 2-1, for example, the correct answer is 11 mi/h. If you find 0.00011 mi/h, -11 mi/h, 11 mi/s, or 11,000 mi/h, you should realize at once that you have done something wrong. The error may lie in your method, in your algebra, or in your arithmetic.

TACTIC 4: *Reading a Graph*

Figures 2-2, 2-3a, 2-4, and 2-5 are graphs that you should be able to read easily. In each graph, the variable on the horizontal axis is the time t, the direction of increasing time being to the right. In each, the variable on the vertical axis is the position x of the moving particle with respect to the origin, the direction of increasing x being upward.

Always note the units (seconds or minutes; meters, kilometers, or miles) in which the variables are expressed, and note whether the variables are positive or negative.

2-4 INSTANTANEOUS VELOCITY AND SPEED

You have now seen two ways to describe how fast something moves: average velocity and average speed, both of which are measured over a time interval Δt. But the phrase "how fast" more commonly refers to how fast a particle is moving at a given instant—and that is its **instantaneous velocity** (or simply **velocity**) v.

The velocity at any instant is obtained from the average velocity by shrinking the time interval Δt closer and closer to 0. As Δt dwindles, the average velocity approaches a limiting value, which is the velocity at that instant:

$$v = \lim_{\Delta t \to 0} \frac{\Delta x}{\Delta t} = \frac{dx}{dt}. \qquad (2\text{-}4)$$

Velocity is another vector quantity and thus has an associated direction.

In the language of calculus, the instantaneous velocity is the rate at which a particle's position x is changing with time at a given instant. According to Eq. 2-4, the velocity of a particle at any instant is the slope of its position-time curve at the point representing that instant.

Speed is the magnitude of velocity; that is, speed is velocity that has been stripped of any indication of direction, either in words or via an algebraic sign.* A velocity of $+5$ m/s and one of -5 m/s both have an associated speed of 5 m/s. The speedometer in a car measures the speed, not the velocity, because it cannot ascertain anything about the direction of motion.

SAMPLE PROBLEM 2-4

Figure 2-6a is an $x(t)$ plot for an elevator cab that is initially stationary, then moves upward (which we take to be the positive direction), and then stops. Plot v as a function of time.

SOLUTION: The slope, and so also the velocity, is zero in the intervals containing points a and d, when the cab is stationary. During the interval bc the slope is constant and nonzero; so the cab moves with a constant velocity. We calculate the slope of $x(t)$ as

$$\frac{\Delta x}{\Delta t} = v = \frac{24 \text{ m} - 4.0 \text{ m}}{8.0 \text{ s} - 3.0 \text{ s}} = +4.0 \text{ m/s}.$$

The plus sign indicates that the cab is moving in the positive x direction. These values ($v = 0$ and $v = 4$ m/s) are plotted in Fig. 2-6b. In addition, as the cab initially begins to move and then later slows to a stop, v might vary as indicated in the intervals 1 s to 3 s and 8 s to 9 s. (Figure 2-6c is considered in Section 2-5.)

Given a $v(t)$ graph such as Fig. 2-6b, we could "work backward" to produce the shape of the associated $x(t)$ graph (Fig. 2-6a). However, we would not know the actual values for x at various times, because the $v(t)$ graph indicates only *changes* in x. To find such a change in x during any interval, we must, in the language of calculus, calculate the area "under the curve" on the $v(t)$ graph for the same interval. For example, during the interval 3 s to 8 s in which the cab has a velocity of 4.0 m/s, the change in x is given by the "area" under the $v(t)$ curve:

$$\text{area} = (4.0 \text{ m/s})(8.0 \text{ s} - 3.0 \text{ s}) = +20 \text{ m}.$$

(This area is positive because the $v(t)$ curve is above the t axis.) Figure 2-6a shows that x does indeed increase by 20 m in the interval. But Fig. 2-6b does not tell us the *values* of x at the beginning and end of the interval. For that we need additional information.

---

*Speed and average speed can be quite different, so you must be careful solving problems that involve either quantity.

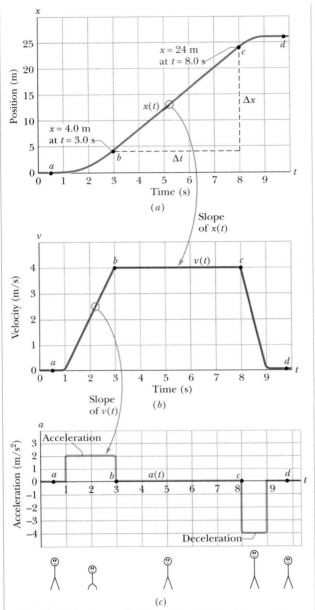

FIGURE 2-6 Sample Problem 2-4. (*a*) The *x*(*t*) curve for an elevator cab that moves upward along an *x* axis. (*b*) The *v*(*t*) curve for the cab. Note that it is the derivative of the *x*(*t*) curve (*v* = *dx/dt*). (*c*) The *a*(*t*) curve for the cab. It is the derivative of the *v*(*t*) curve (*a* = *dv/dt*). The stick figures along the bottom suggest how a passenger's body might feel during the accelerations.

SAMPLE PROBLEM 2-5

The position of a particle moving on the *x* axis is given by

$$x = 7.8 + 9.2t - 2.1t^3. \qquad (2\text{-}5)$$

What is its velocity at *t* = 3.5 s? Is the velocity constant, or is it continuously changing?

SOLUTION: For simplicity, the units have been omitted but you can insert them if you like by changing the coefficients to 7.8 m, 9.2 m/s, and −2.1 m/s³. To find the velocity, we use Eq. 2-4 with the right side of Eq. 2-5 substituted for *x*:

$$v = \frac{dx}{dt} = \frac{d}{dt}(7.8 + 9.2t - 2.1t^3),$$

which becomes

$$v = 0 + 9.2 - (3)(2.1)t^2 = 9.2 - 6.3t^2. \qquad (2\text{-}6)$$

At *t* = 3.5 s,

$$v = 9.2 - (6.3)(3.5)^2 = -68 \text{ m/s}. \quad \text{(Answer)}$$

At *t* = 3.5 s, the particle is moving toward decreasing *x* (note the minus sign) with a speed of 68 m/s. Since the quantity *t* appears in Eq. 2-6, the velocity *v* depends on *t* and so is continuously changing.

CHECKPOINT **3:** The following equations give the position *x*(*t*) of a particle in four situations (in each equation, *x* is in meters, *t* is in seconds, and *t* > 0): (1) *x* = 3*t* − 2; (2) *x* = −4*t*² − 2; (3) *x* = 2/*t*²; and (4) *x* = −2. (a) In which situation is the velocity *v* of the particle constant? (b) In which is *v* in the negative *x* direction? (c) In which is the particle slowing?

PROBLEM SOLVING TACTICS

TACTIC 5: *Derivatives and Slopes*

Every derivative can be interpreted as the slope of a curve at a point. In Sample Problem 2-4, for example, the velocity of the cab at any instant (a derivative; see Eq. 2-4) is the slope of the *x*(*t*) curve of Fig. 2-6*a* at that instant. Here's how you can find a slope at a point (and thus a derivative) graphically.

Figure 2-7 shows an *x*(*t*) plot for a moving particle. To find the velocity of the particle at *t* = 1 s, put a dot on the curve at the point that represents *t* = 1 s. Then draw a line tangent to the curve through the dot (*tangent* means *touching*; the tangent line touches the curve at a single point, the dot), judging carefully by eye. Then construct a right triangle *ABC* with sides parallel to the axes. (Although the slope is the same no matter what the size of this triangle, the larger the triangle, the more precise will be your measurement of the slope.) Find Δ*x* and Δ*t*, using the vertical and horizontal scales. The slope (derivative) is the quotient Δ*x*/Δ*t*. In Fig. 2-7,

$$\text{slope} = \frac{\Delta x}{\Delta t} = \frac{5.5 \text{ m} - 2.3 \text{ m}}{1.8 \text{ s} - 0.3 \text{ s}} = \frac{3.2 \text{ m}}{1.5 \text{ s}} = +2.1 \text{ m/s}.$$

As Eq. 2-4 tells you, this slope is the velocity of the particle at $t = 1$ s. If you change the scale on either axis of Fig. 2-7, the appearance of the curve and the angle θ will change, but the value you find for the velocity at $t = 1$ s will not.

If you have a mathematical expression for the function $x(t)$, as in Sample Problem 2-5, you can find the derivative dx/dt by the methods of calculus and avoid this graphical method.

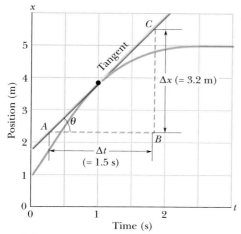

FIGURE 2-7 The derivative of a curve at any point is the slope of its tangent line at that point. At $t = 1.0$ s, the slope of the tangent line (and thus dx/dt, the instantaneous velocity) is $\Delta x/\Delta t = +2.1$ m/s.

2-5 ACCELERATION

When a particle's velocity changes, the particle is said to undergo **acceleration** (or to accelerate). The **average acceleration** $\bar{a}$ over an interval Δt is computed as

$$\bar{a} = \frac{v_2 - v_1}{t_2 - t_1} = \frac{\Delta v}{\Delta t}. \quad (2\text{-}7)$$

The **instantaneous acceleration** (or simply **acceleration**) is the derivative of the velocity:

$$a = \frac{dv}{dt}. \quad (2\text{-}8)$$

In words, the acceleration of a particle at any instant is the rate at which its velocity is changing at that instant. According to Eq. 2-8, the acceleration at any point is the slope of the curve of $v(t)$ at that point.

We can combine Eq. 2-8 with Eq. 2-4 to write

$$a = \frac{dv}{dt} = \frac{d}{dt}\left(\frac{dx}{dt}\right) = \frac{d^2x}{dt^2}. \quad (2\text{-}9)$$

In words, the acceleration of a particle at any instant is the second derivative of its position $x(t)$ with respect to time.

A common unit of acceleration is the meter per second per second: m/(s · s) or m/s². You will see other units in the problems, but they will each be in the form of distance/(time · time) or distance/time². Acceleration has both magnitude and direction (it is yet another vector quantity). The algebraic sign represents the direction on an axis just as it does for displacement and velocity.

Figure 2-6c is a plot of the acceleration of the elevator cab discussed in Sample Problem 2-4. Compare this $a(t)$ curve with the $v(t)$ curve—each point on the $a(t)$ curve is the derivative (slope) of the corresponding point on the $v(t)$ curve. When v is constant (at either 0 or 4 m/s), the derivative is zero and so also is the acceleration. When the cab first begins to move, the $v(t)$ curve has a positive derivative (the slope is positive), which means that $a(t)$ is positive. When the cab slows to a stop, the derivative and slope of the $v(t)$ curve are negative; that is, $a(t)$ is negative.

Next compare the slopes of the $v(t)$ curve during the two acceleration periods. The slope associated with the cab's stopping (commonly called ''deceleration'') is steeper, because the cab stops in half the time it took to get up to speed. The steeper slope means that the magnitude of the deceleration is larger than that of the acceleration, as indicated in Fig. 2-6c.

The sensations you would feel while riding in the cab of Fig. 2-6 are indicated by the sketched figures. When the car first accelerates, you feel as though you are pressed downward; when later the cab is braked to a stop, you seem to be stretched upward. In between, you feel nothing special. Your body reacts to accelerations (it is an accelerometer) but not to velocities (it is not a speedometer). When you are in a car traveling at 60 mi/h or an airplane traveling at 600 mi/h, you have no bodily awareness of the motion. But if the car or plane quickly changes velocity, you may become keenly aware of the change, perhaps even frightened by it. Part of the thrill of an amusement park ride is due to the quick changes of velocity that you undergo. A more extreme example is shown in the photographs of Fig. 2-8, which were taken while a rocket sled was rapidly accelerated and then rapidly braked to a stop.

Large accelerations are sometimes expressed in terms of ''g'' units, with

$$1g = 9.8 \text{ m/s}^2 \quad (g \text{ unit}). \quad (2\text{-}10)$$

(As we shall discuss in Section 2-8, g is the magnitude of the acceleration of a falling object near the Earth's surface.) On a roller coaster, you have brief accelerations up to 3g, which is (3)(9.8 m/s²) or about 29 m/s².

FIGURE 2-8 Colonel J. P. Stapp in a rocket sled as it is brought up to high speed (acceleration out of the page) and then very rapidly braked (acceleration into the page).

SAMPLE PROBLEM 2-6

(a) When Kitty O'Neil set the dragster records for the greatest speed and least elapsed time, she reached 392.54 mi/h in 3.72 s. What was her average acceleration?

SOLUTION: From Eq. 2-7, O'Neil's average acceleration was

$$\bar{a} = \frac{\Delta v}{\Delta t} = \frac{392.54 \text{ mi/h} - 0}{3.72 \text{ s} - 0}$$

$$= +106 \frac{\text{mi}}{\text{h} \cdot \text{s}}, \qquad \text{(Answer)}$$

where the motion is taken to be in the positive x direction. In more conventional units, her acceleration was 47.1 m/s², which is 4.8g.

(b) What was the average acceleration when Eli Beeding Jr. reached 72.5 mi/h in 0.04 s on a rocket sled?

SOLUTION: Again from Eq. 2-7,

$$\bar{a} = \frac{\Delta v}{\Delta t} = \frac{72.5 \text{ mi/h} - 0}{0.04 \text{ s} - 0}$$

$$= +1.8 \times 10^3 \frac{\text{mi}}{\text{h} \cdot \text{s}} \approx +800 \text{ m/s}^2, \quad \text{(Answer)}$$

or about 80g.

Recall our question in the chapter opener, where O'Neil and Beeding were introduced. How can we tell who had the more exciting ride? Should we compare final speeds, elapsed times, or some other quantity? You now can answer that question. Because the human body senses acceleration rather than speed, you should compare accelerations, and so Beeding wins out, even though his final speed was considerably slower than O'Neil's. In fact, Beeding's acceleration could have been lethal, had it continued for much longer.

PROBLEM SOLVING TACTICS

TACTIC 6: *An Acceleration's Sign*

Look again at the algebraic sign for the accelerations that are calculated in Sample Problem 2-6. In many common examples of acceleration, the sign has its nonscientific meaning: positive acceleration means that the speed of an object is increasing, and negative acceleration means that the speed is decreasing (the object is decelerating).

Such meanings cannot be interpreted without some thought, however. For example, if a car with an initial velocity $v = -27$ m/s ($= -60$ mi/h) is braked to a stop in 5.0 s, $\bar{a} = +5.4$ m/s². The acceleration is *positive,* but the car has slowed. The reason is the difference in signs: the direction of the acceleration is opposite that of the velocity.

Here then is the proper way to interpret the signs:

> If the signs of the velocity and acceleration of a particle are the same, the speed of the particle increases. If the signs are opposite, the speed decreases.

The interpretation will have more meaning in Chapter 4, where we explore the vector nature of velocity and acceleration.

CHECKPOINT 4: A wombat moves along an x axis. What is the sign of its acceleration if it is moving (a) in the positive direction with increasing speed, (b) in the positive direction with decreasing speed, (c) in the negative direction with increasing speed, and (d) in the negative direction with decreasing speed?

FIGURE 2-8 *Continued*

SAMPLE PROBLEM 2-7

A particle's position is given by

$$x = 4 - 27t + t^3,$$

where the units of the coefficients are meters, meters per second, and meters per second cubed, respectively, and the x axis is shown in Fig. 2-1.

(a) Find $v(t)$ and $a(t)$.

SOLUTION: To get $v(t)$, we differentiate $x(t)$ with respect to t:

$$v = -27 + 3t^2. \qquad \text{(Answer)}$$

To get $a(t)$, we differentiate $v(t)$ with respect to t:

$$a = +6t. \qquad \text{(Answer)}$$

(b) Is there ever a time when $v = 0$?

SOLUTION: Setting $v(t) = 0$ yields

$$0 = -27 + 3t^2,$$

which has the solution $t = \pm 3$ s. (Answer)

(c) Describe the particle's motion for $t \geq 0$.

SOLUTION: To answer, we examine the expressions for $x(t)$, $v(t)$, and $a(t)$.

At $t = 0$ the particle is at $x = +4$ m, is moving leftward with a velocity of -27 m/s, and is not accelerating.

For $0 < t < 3$ s, the particle continues to move to the left, but at decreasing speed, because it is now accelerating to the right. (Check $v(t)$ and $a(t)$ for, say, $t = 2$ s.) The rate of the acceleration is increasing.

At $t = 3$ s, the particle stops momentarily ($v = 0$) and is as far to the left as it will ever get ($x = -50$ m). It continues to accelerate to the right at an increasing rate.

For $t > 3$ s, its acceleration to the right continues to increase, and its velocity, which is now also to the right, increases rapidly. (Note that now the signs of v and a match.) The particle moves continuously to the right.

2-6 CONSTANT ACCELERATION: A SPECIAL CASE

In many common types of motion, the acceleration is either constant or approximately so. For example, you might accelerate a car at an approximately constant rate when a traffic light turns from red to green. (Graphs of your position, velocity, and acceleration would resemble those in Fig. 2-9.) If you later had to brake the car to a stop, the deceleration during the braking might also be approximately constant.

Such cases are so ubiquitous that a special set of equations has been derived for dealing with them. One approach to the derivation of the equations is given in this section. A second approach is given in the next section.

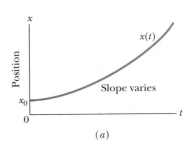

FIGURE 2-9 (*a*) The position $x(t)$ of a particle moving with constant acceleration. (*b*) Its velocity $v(t)$, given at each point by the slope of the curve in (*a*). (*c*) Its (constant) acceleration, equal to the (constant) slope of $v(t)$.

Throughout both sections and later when you work on the homework problems, keep in mind that *the equations are valid only for constant acceleration (or situations in which you can approximate the acceleration as being constant).*

When the acceleration is constant, the average acceleration and instantaneous acceleration are equal and we can write Eq. 2-7, with some changes in notation, as

$$a = \frac{v - v_0}{t - 0}.$$

Here v_0 is the velocity at time $t = 0$, and v is the velocity at any later time t. We can recast this equation as

$$v = v_0 + at. \tag{2-11}$$

As a check, note that this equation reduces to $v = v_0$ for $t = 0$, as it must. As a further check, take the derivative of Eq. 2-11. Doing so yields $dv/dt = a$, which is the definition of a. Figure 2-9*b* shows a plot of Eq. 2-11, the $v(t)$ function.

In similar manner we can rewrite Eq. 2-2 (with a few changes in notation) as

$$\bar{v} = \frac{x - x_0}{t - 0}$$

and then as

$$x = x_0 + \bar{v}t, \tag{2-12}$$

in which x_0 is the position of the particle at $t = 0$, and $\bar{v}$ is the average velocity between $t = 0$ and a later time t.

If you plot v against t using Eq. 2-11, a straight line results. Under these conditions, the *average* velocity over any time interval (say, $t = 0$ to a later time t) is the average of the velocity at the beginning of the interval ($= v_0$) and the velocity at the end of the interval ($= v$). For the interval $t = 0$ to the later time t then, the average velocity is

$$\bar{v} = \tfrac{1}{2}(v_0 + v). \tag{2-13}$$

Substituting the right side of Eq. 2-11 for v yields, after a little rearrangement,

$$\bar{v} = v_0 + \tfrac{1}{2}at. \tag{2-14}$$

Finally, substituting Eq. 2-14 into Eq. 2-12 yields

$$x - x_0 = v_0 t + \tfrac{1}{2}at^2. \tag{2-15}$$

As a check, note that putting $t = 0$ yields $x = x_0$, as it must. As a further check, taking the derivative of Eq. 2-15 yields Eq. 2-11, again as it must. Figure 2-9*a* is a plot of Eq. 2-15.

Five quantities can possibly be involved in any given problem regarding constant acceleration, namely, $x - x_0$,

v, t, a, and v_0. Usually, one of these quantities is *not* involved in the problem, *either as a given or as an unknown*. We are then presented with three of the remaining quantities and asked to find the fourth.

Equations 2-11 and 2-15 each contain four of these quantities, but not the same four. In Eq. 2-11, the "missing ingredient" is the displacement, $x - x_0$. In Eq. 2-15, it is the velocity v. These two equations can also be combined in three ways to yield three additional equations, each of which involves a different "missing ingredient." First, we can eliminate t to obtain

$$v^2 = v_0^2 + 2a(x - x_0). \tag{2-16}$$

This equation is useful if we do not know t and are not required to find it. Second, we can eliminate the acceleration a between Eqs. 2-11 and 2-15 to produce an equation in which a does not appear:

$$x - x_0 = \tfrac{1}{2}(v_0 + v)t. \tag{2-17}$$

Finally, we can eliminate v_0, obtaining

$$x - x_0 = vt - \tfrac{1}{2}at^2. \tag{2-18}$$

Note the subtle difference between this equation and Eq. 2-15. One involves the initial velocity v_0; the other involves the velocity v at time t.

Table 2-1 lists Eqs. 2-11, 2-15, 2-16, 2-17, and 2-18 and shows which one of the five possible quantities is missing from each. To solve a constant acceleration problem, you must decide which of the five quantities is *not* involved in the problem, either as a given or as an unknown. Select the correct equation from Table 2-1 and substitute for the three given quantities to find the unknown. Or, instead of using the table, you might find it easier to remember only Eqs. 2-11 and 2-15, and solve them as simultaneous equations when needed.

TABLE 2-1 EQUATIONS FOR MOTION WITH CONSTANT ACCELERATION[a]

EQUATION NUMBER	EQUATION	MISSING QUANTITY
2-11	$v = v_0 + at$	$x - x_0$
2-15	$x - x_0 = v_0 t + \tfrac{1}{2}at^2$	v
2-16	$v^2 = v_0^2 + 2a(x - x_0)$	t
2-17	$x - x_0 = \tfrac{1}{2}(v_0 + v)t$	a
2-18	$x - x_0 = vt - \tfrac{1}{2}at^2$	v_0

[a]Make sure that the acceleration is indeed constant before using the equations in this table. Note that if you differentiate Eq. 2-15 you get Eq. 2-11. The other three equations are found by eliminating one or another of the variables between Eqs. 2-11 and 2-15.

CHECKPOINT **5:** The following equations give the position $x(t)$ of a particle in four situations: (1) $x = 3t - 4$; (2) $x = -5t^3 + 4t^2 + 6$; (3) $x = 2/t^2 - 4/t$; (4) $x = 5t^2 - 3$. To which of these situations do the equations of Table 2-1 apply?

SAMPLE PROBLEM 2-8

Spotting a police car, you brake a Porsche from 75 km/h to 45 km/h over a displacement of 88 m.

(a) What is the acceleration, assumed to be constant?

SOLUTION: In this problem we are given v_0, v, and $x - x_0$ and need to find a. The time is not involved, being neither given nor required. Table 2-1 then leads us to Eq. 2-16. Solving this equation for a yields

$$a = \frac{v^2 - v_0^2}{2(x - x_0)} = \frac{(45 \text{ km/h})^2 - (75 \text{ km/h})^2}{(2)(0.088 \text{ km})}$$
$$= -2.05 \times 10^4 \text{ km/h}^2 \approx -1.6 \text{ m/s}^2. \quad \text{(Answer)}$$

(In converting hours squared to seconds squared in the last step, we must convert *both* the hour units.) Note that the velocities are positive and the acceleration is negative, which is consistent with a slowing of the car.

(b) What is the elapsed time?

SOLUTION: Now time is involved in the problem, but the acceleration is not. Table 2-1 suggests Eq. 2-17. Solving that equation for t, we obtain

$$t = \frac{2(x - x_0)}{v_0 + v} = \frac{(2)(0.088 \text{ km})}{(75 + 45) \text{ km/h}}$$
$$= 1.5 \times 10^{-3} \text{ h} = 5.4 \text{ s}. \quad \text{(Answer)}$$

(c) If you continue to slow down with the acceleration calculated in (a), how much time will elapse in bringing the car to rest from 75 km/h?

SOLUTION: The quantity not given or asked for here is the displacement, $x - x_0$. Table 2-1 then suggests that we use Eq. 2-11. Solving for t gives

$$t = \frac{v - v_0}{a} = \frac{0 - (75 \text{ km/h})}{(-2.05 \times 10^4 \text{ km/h}^2)}$$
$$= 3.7 \times 10^{-3} \text{ h} = 13 \text{ s}. \quad \text{(Answer)}$$

(d) In (c), what distance will be covered?

SOLUTION: From Eq. 2-15, we have, for the displacement of the car,

$$x - x_0 = v_0 t + \tfrac{1}{2}at^2$$
$$= (75 \text{ km/h})(3.7 \times 10^{-3} \text{ h})$$
$$+ \tfrac{1}{2}(-2.05 \times 10^4 \text{ km/h}^2)(3.7 \times 10^{-3} \text{ h})^2$$
$$= 0.137 \text{ km} \approx 140 \text{ m}. \quad \text{(Answer)}$$

(Neglecting the sign on the acceleration would give a wrong result—you should always be alert to signs.)

(e) Suppose that later, using the acceleration calculated in (a) but a different initial velocity, you bring your car to rest after traversing 200 m. What is the total braking time?

SOLUTION: The missing quantity here is the initial velocity, so we use Eq. 2-18. Setting v (the final velocity) equal to zero (''rest'') and solving this equation for t, we obtain

$$t = \left(-\frac{(2)(x - x_0)}{a}\right)^{1/2} = \left(-\frac{(2)(200 \text{ m})}{-1.6 \text{ m/s}^2}\right)^{1/2}$$
$$= 16 \text{ s}. \quad \text{(Answer)}$$

PROBLEM SOLVING TACTICS

TACTIC 7: *Check the Dimensions*

The dimension of a velocity is L/T, that is, length L divided by time T. The dimension of an acceleration is L/T^2; and so on. In any equation, the dimensions of all terms must be the same. If you are in doubt about an equation, check its dimensions.

To check the dimensions of Eq. 2-15 ($x - x_0 = v_0 t + \tfrac{1}{2}at^2$), we note that every term must be a length, because that is the dimension of x and of x_0. The dimension of the term $v_0 t$ is $(L/T)(T)$, which is L. The dimension of $\tfrac{1}{2}at^2$ is $(L/T^2)(T^2)$, which is also L. This equation checks out. A pure number such as $\tfrac{1}{2}$ or π has no dimension.

2-7 ANOTHER LOOK AT CONSTANT ACCELERATION*

The first two equations in Table 2-1 are the basic equations from which the others are derived. Those two can be obtained by integration of the acceleration with the condition that a is constant. The definition of a (in Eq. 2-8) is

$$a = \frac{dv}{dt},$$

which can be rewritten as

$$dv = a \, dt.$$

To take the *indefinite integral* (or *antiderivative*) of both sides, we write

$$\int dv = \int a \, dt.$$

---

*This section is intended for students who have had integral calculus.

Since acceleration a is a constant, it can be taken outside the integration. Then we obtain

$$\int dv = a \int dt$$

or

$$v = at + C. \tag{2-19}$$

To evaluate the constant of integration C, we let $t = 0$, at which time $v = v_0$. Substituting these values into Eq. 2-19 (which must hold for all values of t, including $t = 0$) yields

$$v_0 = (a)(0) + C = C.$$

Substituting this into Eq. 2-19 gives us Eq. 2-11.

To derive the other basic equation in Table 2-1, we rewrite the definition of velocity (Eq. 2-4) as

$$dx = v \, dt$$

and then take the indefinite integral of both sides to obtain

$$\int dx = \int v \, dt.$$

There is no reason to believe that v is constant, so we cannot move it outside the integration. But we can substitute for v with Eq. 2-11:

$$\int dx = \int (v_0 + at) dt.$$

Since v_0 is a constant, this can be rewritten as

$$\int dx = v_0 \int dt + a \int t \, dt.$$

Integration now yields

$$x = v_0 t + \tfrac{1}{2} at^2 + C', \tag{2-20}$$

where C' is another constant of integration. At time $t = 0$, we have $x = x_0$. Substituting these values in Eq. 2-20 yields $x_0 = C'$. Replacing C' with x_0 in Eq. 2-20 gives us Eq. 2-15.

2-8   FREE-FALL ACCELERATION

If you tossed an object either up or down and could somehow eliminate the effects of air on its flight, you would find that the object accelerates downward at a certain rate. That rate is called the **free-fall acceleration** g. The acceleration is independent of the object's characteristics, such as mass, density, or shape; it is the same for all objects.

Two examples of free-fall acceleration are shown in Fig. 2-10, which is a series of stroboscopic photos of a feather and an apple. As these objects descend, they accelerate downward—both at the same rate g. Their speeds

FIGURE 2-10   A feather and an apple, undergoing free fall in a vacuum, move downward at the same acceleration g. The acceleration causes the increase in distance between images during the fall.

increase together. The value of g varies slightly with latitude and with elevation. At sea level in the midlatitudes the value is 9.8 m/s^2 (or 32 ft/s^2), which is what you should use for the problems in this chapter.

The equations of motion in Table 2-1 for constant acceleration apply to free fall near the Earth's surface. That is, they apply to an object in vertical flight, either up or down, when the effects of the air can be neglected. However, we can make them simpler to use with two minor changes. (1) The directions of motion are now along the vertical y axis instead of the x axis, with the positive direction of y upward. (This change will reduce confusion in later chapters when combined horizontal and vertical motions are examined.) (2) The free-fall acceleration is now negative, that is, downward on the y axis, and so we replace a with $-g$ in the equations.

With these small changes, Eqs. 2-11 and 2-15 to 2-18 of Table 2-1 become, for free fall, Eqs. 2-21 to 2-25 in Table 2-2. Note carefully that:

The free-fall acceleration near the Earth's surface is $a = -g = -9.8$ m/s^2, and the *magnitude* of the acceleration is $g = 9.8$ m/s^2. Do not substitute -9.8 m/s^2 for g.

TABLE 2-2 EQUATIONS FOR FREE FALL

EQUATION NUMBER	EQUATION	MISSING QUANTITY
2-21	$v = v_0 - gt$	$y - y_0$
2-22	$y - y_0 = v_0 t - \frac{1}{2}gt^2$	v
2-23	$v^2 = v_0^2 - 2g(y - y_0)$	t
2-24	$y - y_0 = \frac{1}{2}(v_0 + v)t$	g
2-25	$y - y_0 = vt + \frac{1}{2}gt^2$	v_0

Suppose that you toss a tomato directly upward with an initial velocity v_0 and then catch it when it returns to the level of release. During its *free-fall flight* (just after its release and just before it is caught), the equations of Table 2-2 apply to its motion. The acceleration is always $a = -g = -9.8$ m/s² and thus downward. The velocity, however, changes, as indicated by Eqs. 2-21 and 2-23: during the ascent, the magnitude of the (positive) velocity decreases, until it is momentarily zero. Because the tomato has then stopped, it is at its maximum height. During the descent, the magnitude of the (negative) velocity increases.

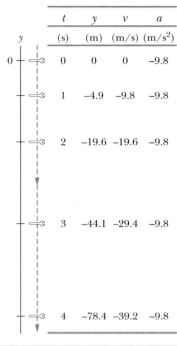

	t	y	v	a
	(s)	(m)	(m/s)	(m/s²)
	0	0	0	−9.8
	1	−4.9	−9.8	−9.8
	2	−19.6	−19.6	−9.8
	3	−44.1	−29.4	−9.8
	4	−78.4	−39.2	−9.8

FIGURE 2-11 Sample Problem 2-9. The position, velocity, and acceleration of a freely falling object.

SAMPLE PROBLEM 2-9

A worker drops a wrench down the elevator shaft of a tall building.

(a) Where is the wrench 1.5 s later?

SOLUTION: Here we know the time t and the acceleration magnitude g, and we can assume the wrench is initially at rest ($v_0 = 0$). We want the displacement. So the missing quantity is the velocity v, which is neither given nor required. This suggests Eq. 2-22 of Table 2-2. Choose the release point of the wrench to be at the origin of the y axis. Setting $y_0 = 0$, $v_0 = 0$, and $t = 1.5$ s in Eq. 2-22 gives

$$y = (0)(1.5 \text{ s}) - \frac{1}{2}(9.8 \text{ m/s}^2)(1.5 \text{ s})^2$$
$$= -11 \text{ m}. \qquad \text{(Answer)}$$

The minus sign means that after falling 1.5 s the wrench is below its release point, which we certainly expect.

(b) How fast is the wrench falling just then?

SOLUTION: The wrench's velocity is, by Eq. 2-21,

$$v = v_0 - gt = 0 - (9.8 \text{ m/s}^2)(1.5 \text{ s})$$
$$= -15 \text{ m/s}. \qquad \text{(Answer)}$$

Here the minus sign means that the wrench is falling downward. Again, no great surprise. Figure 2-11 displays the important features of the motion up to $t = 4$ s.

SAMPLE PROBLEM 2-10

In 1939, Joe Sprinz of the San Francisco Baseball Club attempted to break the record for catching a baseball dropped from the greatest height. Members of the Cleveland Indians had set the record the preceding year when they caught baseballs dropped about 700 ft from atop a building. Sprinz used a blimp at 800 ft. Ignore the effects of air on the ball and assume that the ball falls 800 ft.

(a) Find its time of fall.

SOLUTION: Mentally erect a vertical y axis with its origin at the point of the ball's release in the blimp, which means that $y_0 = 0$. The initial velocity v_0 is zero. The missing ingredient is v, and so Eq. 2-22 is required:

$$y - y_0 = v_0 t - \frac{1}{2}gt^2$$
$$-800 \text{ ft} = 0t - \frac{1}{2}(32 \text{ ft/s}^2)t^2$$
$$16t^2 = 800$$
$$t = 7.1 \text{ s}. \qquad \text{(Answer)}$$

When taking a square root, we have the option of attaching a plus or minus sign to the square root. Here we choose the plus sign, since the ball reaches the ground *after* it is released.

(b) What is the velocity of the ball just before it is caught?

SOLUTION: To get the velocity from the original data, rather than from the result of (a), we use Eq. 2-23:

$$v^2 = v_0^2 - 2g(y - y_0)$$
$$= 0 - (2)(32 \text{ ft/s}^2)(-800 \text{ ft})$$
$$= 5.12 \times 10^4 \text{ ft}^2/\text{s}^2$$
$$v = -226 \text{ ft/s} (\approx -154 \text{ mi/h}). \quad \text{(Answer)}$$

Since the ball is moving downward, we choose the minus sign in our option of signs in taking the square root.

Neglecting the effects of air is actually unwarranted in such a fall. If you included them, you would find that the fall time was longer and the final speed was smaller than the values calculated above. Still, the speed must have been considerable, because when Sprinz finally managed to get a ball in his glove (on his fifth attempt) the impact slammed the glove and hand into his face, fracturing the upper jaw in 12 places, breaking five teeth, and knocking him unconscious. And he dropped the ball.

SAMPLE PROBLEM 2-11

A pitcher tosses a baseball straight up, with an initial speed of 12 m/s. See Fig. 2-12.

(a) How long does the ball take to reach its highest point?

SOLUTION: The ball is at its highest point when its velocity v becomes zero. From Eq. 2-21, we have

$$t = \frac{v_0 - v}{g} = \frac{12 \text{ m/s} - 0}{9.8 \text{ m/s}^2} = 1.2 \text{ s}. \quad \text{(Answer)}$$

(b) How high does the ball rise above its release point?

SOLUTION: We take the release point of the ball to be the origin of the y axis. Putting $y_0 = 0$ in Eq. 2-23 and solving for y, we obtain

$$y = \frac{v_0^2 - v^2}{2g} = \frac{(12 \text{ m/s})^2 - (0)^2}{(2)(9.8 \text{ m/s}^2)}$$
$$= 7.3 \text{ m}. \quad \text{(Answer)}$$

If we wanted to take advantage of the fact that we also know the time of flight, having found it in (a), we could also calculate the height of rise from Eq. 2-25. Check it out.

(c) How long will it take for the ball to reach a point 5.0 m above its release point?

SOLUTION: Inspection of Eqs. 2-21 to 2-25 suggests that we try Eq. 2-22. With $y_0 = 0$, we have

$$y = v_0 t - \tfrac{1}{2}gt^2,$$

so $5.0 \text{ m} = (12 \text{ m/s})t - (\tfrac{1}{2})(9.8 \text{ m/s}^2)t^2.$

If we temporarily omit the units (having noted that they are consistent), we can rewrite this as

$$4.9t^2 - 12t + 5.0 = 0.$$

Solving this quadratic equation for t yields*

$$t = 0.53 \text{ s} \quad \text{and} \quad t = 1.9 \text{ s}. \quad \text{(Answer)}$$

There are two such times! This is not really surprising because the ball passes twice through $y = 5.0$ m, once on the way up and once on the way down.

We can check our findings because the time at which the ball reaches its maximum height should lie halfway between these two times, or at

$$t = \tfrac{1}{2}(0.53 \text{ s} + 1.9 \text{ s}) = 1.2 \text{ s}.$$

This is exactly what we found in (a) for the time to reach maximum height.

FIGURE 2-12 Sample Problem 2-11. A pitcher tosses a baseball straight up into the air. The equations of free fall apply for rising as well as for falling objects, provided any effects from the air can be neglected.

CHECKPOINT **6:** In Sample Problem 2-11, what is the sign of the ball's displacement (a) during its ascent and (b) during its descent? (c) What is the ball's acceleration at its highest point?

PROBLEM SOLVING TACTICS

TACTIC 8: *Meanings of Minus Signs*
In Sample Problems 2-9, 2-10, and 2-11, many answers emerged automatically with minus signs. It is important to know what these signs mean. For falling-body problems, we established a vertical axis (the y axis) and we chose—quite arbitrarily—its upward direction to be positive.

---

*See Appendix E for a general solution to quadratic equations.

We then chose the origin of the y axis (that is, the $y = 0$ position) to suit the problem. In Sample Problem 2-9, the origin was the worker's hand; in Sample Problem 2-10 it was at the blimp; in Sample Problem 2-11 it was the pitcher's hand. A negative value of y means that the body is below the chosen origin.

A negative velocity means that the body is moving in the direction of decreasing y, that is, downward. This is true no matter where the body is located.

We have taken the acceleration to be negative ($= -9.8$ m/s^2) in all problems dealing with falling bodies. A negative acceleration means that, as time goes on, the velocity of the body becomes either less positive or more negative. This is true no matter where the body is located and no matter how fast or in what direction it is moving. In Sample Problem 2-11, the acceleration of the ball is negative throughout its flight, whether the ball is rising or falling.

TACTIC 9: *Unexpected Answers*

Mathematics often generates answers that you might not have thought of as possibilities, as in Sample Problem 2-11c. If you get more answers than you expect, do not discard out of hand the ones that do not seem to fit. Examine them carefully for physical meaning; it is often there.

If time is your variable, even a negative value can mean something; negative time simply refers to time before $t = 0$, the (arbitrary) time at which you decided to start your stopwatch.

2-9 THE PARTICLES OF PHYSICS

As we progress through the book, we plan to step aside occasionally from the familiar world of large, tangible objects and look at nature on a much finer scale. The "particles" that we have dealt with in this chapter, for example, have included pigs, baseballs, and dragsters. In the spirit of our plan, we ask: How small can a particle be? What are the *ultimate* particles of nature? **Particle physics**—for so the field that relates to our inquiry is called—attracts the attention of many of the best of today's physicists.

The realization that matter, on its finest scale, is not continuous but is made up of lumps of material called atoms was the beginning of understanding for physics and chemistry. With modern microscopes, we can "photograph" these atoms, as Fig. 2-13 makes clear. We describe the "lumpiness" of matter by saying that matter is *quantized,* using a word that comes from the Latin word *quantus,* meaning "how much." Quantization is a central feature of nature, and as we go along you will see other physical quantities that are quantized when looked at on a fine enough scale. This pervasiveness of quantization is reflected in the name we give to physics at the atomic level—**quantum physics.**

There is no sharp discontinuity between the quantum world and the world of large-scale objects. The quantum world and the laws that govern it are universal but, as we move from atoms to baseballs and automobiles, the fact of quantization becomes less noticeable and finally totally undetectable. The "graininess" effectively disappears, and the laws of **classical physics** that govern the motions of large objects emerge as special limiting forms of the more general laws of quantum physics.

The Structure of Atoms

An **atom** consists of a central, almost unimaginably compact and dense **nucleus** that is surrounded by one or more lightweight **electrons.** An atom is usually considered to be spherical; so is the nucleus. The radius of a typical atom is on the order of 10^{-10} m; the radius of a nucleus is 100,000 times smaller, about 10^{-15} m. An atom is held together by electrical attraction between the electrons, which are electrically negative, and **protons,** which are electrically positive and reside within the nucleus. The nature of that attraction is explored later in this book, but for now you might realize that were it not present, atoms could not exist, and so neither could you.

The Structure of Nuclei

The simplest nucleus, that of common hydrogen, has a single proton. There are two other, rare, versions of hydrogen: they differ from the common version by the presence

FIGURE 2-13 A hexagonal array of uranium atoms is revealed in this image from a scanning transmission electron microscope. The color has been added by a computer.

of one or two **neutrons** (electrically neutral particles) in-side the nucleus. Hydrogen, in any of its versions, is an example of an **element;** each element is distinguished from all the others by the number of protons in the nucleus. When there is only one proton, the element is hydrogen. When, instead, there are six, the element is carbon. The various versions of each element are called **isotopes;** they are distinguished by the number of neutrons.

Roughly speaking, the purpose of the neutrons is to glue together the protons, which, being all electrically pos-itive and closely packed, strongly repel one another. If the neutrons did not provide the glue, the only type of atom that could exist would be common hydrogen; all others would blow apart.

Such instability can be found in many isotopes of common elements, but thankfully not the elements on which your existence depends. For example, 19 of the 21 isotopes† of copper are unstable and undergo transforma-tions to become other elements. The two stable isotopes are the ones used in electronics and other technology.

The Structure of the Particles Within Atoms

The electron is simple but perplexingly so. When detected, it appears to be infinitesimal in size; that is, it has no size and no structure. It is a member of a family of pointlike particles called **leptons;** there are six basic types, each with an **antiparticle** version.

FIGURE 2-14 A representa-tion of the nucleus of an atom, showing the neutrons and pro-tons that make it up. These par-ticles, in turn, are composed of "up" (*u*) and "down" (*d*) quarks.

Protons and neutrons are believed to be different from electrons and the other leptons, because each of the former appears to be a bundle of three simpler particles called **quarks.*** A proton consists of two "up" quarks and one "down" quark, and a neutron two "down" quarks and one "up" quark (Fig. 2-14). Other particles that were first thought to be fundamental appear to be similar bundles.

Provocatively, quarks come in six basic types† (each with its antiparticle) just as do leptons. Here physicists wonder: Is there a basic reason for the match in number of types? Or is the match simply coincidence? We do not know.

---

*On a quirk, the word "quark" was lifted from James Joyce's *Finnegans Wake*:

> Three quarks for Muster Mark.
> Sure he hasn't got much of a bark
> And sure any he has it's all beside the mark.

†The other types are called charm, strange, top, and bottom (even phys-icists have their moments).

REVIEW & SUMMARY

Position

The *position* x of a particle on an axis locates the particle with respect to the **origin,** or zero point, of the axis. The position is either positive or negative, according to which side of the origin the particle is on, or zero if the particle is at the origin. The **positive direction** on an axis is the direction of increasing posi-tive numbers; the opposite direction is the **negative direction.**

Displacement

The *displacement* Δx of a particle is the change in its position:

$$\Delta x = x_2 - x_1, \qquad (2\text{-}1)$$

Displacement is a vector quantity. It is positive if the particle has moved in the positive direction of the x axis, and negative if it has moved in the negative direction.

Average Velocity

When a particle has moved from position x_1 to x_2 during a time interval $\Delta t = t_2 - t_1$, its *average velocity* is

$$\bar{v} = \frac{\Delta x}{\Delta t} = \frac{x_2 - x_1}{t_2 - t_1}. \qquad (2\text{-}2)$$

The algebraic sign of $\bar{v}$ indicates the direction of motion ($\bar{v}$ is a vector quantity). Average velocity does not depend on the actual distance a particle covers, but instead depends on its original and final positions.

On a graph of x versus t, the average velocity for a time interval Δt is the slope of the straight line connecting the points on the curve that represent the ends of the interval.

Average Speed

The *average speed* $\bar{s}$ of a particle depends on the full distance it covers in a time interval Δt:

$$\bar{s} = \frac{\text{total distance}}{\Delta t}. \qquad (2\text{-}3)$$

Instantaneous Velocity

If we allow Δt to approach zero in Eq. 2-2, then $\bar{v}$ will approach a limiting value v, the *instantaneous velocity* (or simply **velocity**) of the particle at the time in question; that is,

$$v = \lim_{\Delta t \to 0} \frac{\Delta x}{\Delta t} = \frac{dx}{dt}. \qquad (2\text{-}4)$$

The instantaneous velocity (at a particular time) may be found as the slope (at that particular time) of the graph of x versus t. **Speed** is the magnitude of instantaneous velocity.

Average Acceleration

Average acceleration is the ratio of a change in velocity Δv to the time interval Δt in which the change occurs:

$$\bar{a} = \frac{\Delta v}{\Delta t}. \qquad (2\text{-}7)$$

The algebraic sign indicates the direction of $\bar{a}$.

Instantaneous Acceleration

Instantaneous acceleration (or simply **acceleration**) is the rate of change of velocity with time:

$$a = \frac{dv}{dt}. \qquad (2\text{-}8)$$

The acceleration can also be written as the second derivative of the position $x(t)$ with respect to time:

$$a = \frac{d^2x}{dt^2}. \qquad (2\text{-}9)$$

On a graph of v versus t, the acceleration a at any time t is the slope of the curve at the point that represents t.

Constant Acceleration

Figure 2-9 shows $x(t)$, $v(t)$, and $a(t)$ for the important case in which a is constant. In this circumstance, the five equations in

Table 2-1 describe the motion:

$$v = v_0 + at, \qquad (2\text{-}11)$$

$$x - x_0 = v_0 t + \tfrac{1}{2}at^2, \qquad (2\text{-}15)$$

$$v^2 = v_0^2 + 2a(x - x_0), \qquad (2\text{-}16)$$

$$x - x_0 = \tfrac{1}{2}(v_0 + v)t, \qquad (2\text{-}17)$$

$$x - x_0 = vt - \tfrac{1}{2}at^2. \qquad (2\text{-}18)$$

These are *not* valid when the acceleration is not constant.

Free-Fall Acceleration

An important example of straight-line motion with constant acceleration is that of an object rising or falling freely near the Earth's surface. The constant acceleration equations describe this motion, but we make two changes in notation: (1) we refer the motion to the vertical y axis with $+y$ vertically *up*; (2) we replace a with $-g$, where g is the magnitude of the free-fall acceleration. Near the Earth's surface, $g = 9.8 \text{ m/s}^2 (= 32 \text{ ft/s}^2)$. The free-fall equations are those shown as Eqs. 2-21 to 2-25.

The Structure of Atoms

All ordinary matter is composed of **atoms,** which, in a simple model, consist of **electrons** that surround a highly compact central core, the **nucleus. Neutrons** and **protons** reside inside nuclei. Each **element** is distinguished by the number of protons in its nucleus. Variations of an element, differing in the number of neutrons, are **isotopes** of the element.

Quarks and Leptons

Electrons appear to be pointlike particles with no size or internal structure, but protons and neutrons appear to have size and to contain more elementary particles, called **quarks.** There are six basic types of quarks, each with an antiparticle version. Electrons are members of a family of particles, **leptons,** that also come in six basic types, each with an antiparticle version.

QUESTIONS

1. (a) Can an object have zero velocity and still be accelerating? (b) Can an object have constant velocity and still have a varying speed? (c) Can the velocity of an object reverse direction when the object's acceleration is constant? (d) Can an object be increasing in speed as its acceleration decreases?

2. Figure 2-15 is a graph of a particle's position along an x axis versus time. (a) At time $t = 0$, what is the sign of the particle's position? Is the particle's velocity positive, negative, or zero at (b) $t = 1$ s, (c) $t = 2$ s, and (d) $t = 3$ s? (e) How many times does the particle go through the point $x = 0$?

3. At $t = 0$, a particle moving along an x axis is at position $x_0 = -20$ m. The signs of the particle's initial velocity v_0 (at time t_0) and constant acceleration a are, respectively, for four situations: (1) +, +; (2) +, −; (3) −, +; (4) −, −. In which situation will

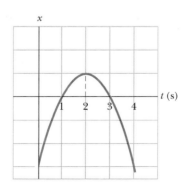

FIGURE 2-15 Question 2.

the particle (a) undergo a momentary stop, (b) definitely pass through the origin (given enough time), and (c) definitely not pass through the origin?

4. Figure 2-16 gives the velocity of a particle moving on an x axis. What are (a) the initial and (b) the final directions of travel? (c) Does the particle stop momentarily? (d) Is the acceleration positive or negative? (e) Is it constant or varying?

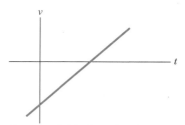

FIGURE 2-16   Question 4.

5. The initial and final velocities, respectively, of a particle in four situations are: (a) 2 m/s, 3 m/s; (b) −2 m/s, 3 m/s; (c) −2 m/s, −3 m/s; (d) 2 m/s, −3 m/s. The magnitude of the particle's constant acceleration is the same in all four situations. Rank the situations according to the magnitude of the particle's displacement, greatest first, during the change from initial to final velocity.

6. The following equations give the velocity $v(t)$ of a particle in four situations: (a) $v = 3$; (b) $v = 4t^2 + 2t − 6$; (c) $v = 3t − 4$; (d) $v = 5t^2 − 3$. To which of these situations do the equations of Table 2-1 apply?

7. Suppose that a hot-air balloonist drops an apple over the side while the balloon is accelerating upward at 4.0 m/s² during liftoff. (a) What is the apple's acceleration once it has been released? (b) If the velocity of the balloon is 2 m/s upward at the instant of release, what is the apple's velocity just then?

8. (a) Graph y, v, and a versus t for a cream tangerine that is thrown straight up from a cliff but which, upon falling back down, barely passes by the cliff edge. (b) On the same figures, graph the same quantities for a cream tangerine that is dropped (released with no initial speed) from the cliff edge.

9. You throw a ball straight up from the edge of a cliff, and it lands on the ground below the cliff. If you had, instead, thrown the ball down from the cliff edge with the same speed, would the ball's speed just before landing be larger than, smaller than, or the same as previously? (*Hint:* Consider Eq. 2-23.)

10. The driver of a blue car, moving at a speed of 60 mi/h, suddenly realizes that she is about to rear-end a red car, moving at a speed of 40 mi/h. To avoid a collision, what is the maximum speed the blue car can have just as it reaches the red car? (Warmup for Problem 57)

11. At $t = 0$ and $x = 0$, an initially stationary blue car begins to accelerate at the constant rate of 2.0 m/s² in the positive direction of the x axis. At $t = 2$ s, a red car, traveling in an adjacent lane and in the same direction, passes $x = 0$ with a speed of 8.0 m/s and a constant acceleration of 3.0 m/s². What pair of simultaneous equations should be solved to find when the red car passes the blue car? (Warmup for Problem 56)

12. Figure 2-17 shows that a particle moving along an x axis undergoes three periods of acceleration. Without written computation, rank the acceleration periods according to the increases they produce in the particle's velocity, greatest first.

FIGURE 2-17   Question 12.

13. A second ball is dropped from a cliff 1 s after a first ball was dropped. As both fall, does the distance between them increase, decrease, or stay the same?

EXERCISES & PROBLEMS

In several of the problems that follow you are asked to graph position, velocity, and acceleration versus time. Usually a sketch will suffice, appropriately labeled and with straight and curved portions apparent. If you have a computer or programmable calculator, you might use it to produce the graph.

SECTION 2-3   Average Velocity and Average Speed

1E. Carl Lewis ran the 100 m dash in about 10 s, and Bill Rodgers ran the marathon (26 mi, 385 yd) in about 2 h 10 min. (a) What are their average speeds? (b) If Lewis could have maintained his sprint speed during a marathon, how long would he have taken to finish?

2E. During a hard sneeze, your eyes might shut for 0.50 s. If you are driving a car at 90 km/h, how far does it move during that time?

3E. On average, a blink lasts about 100 ms. How far does a MiG-25 "Foxbat" fighter travel during a pilot's blink if the plane's average velocity is 2110 mi/h?

4E. Boston Red Sox pitcher Roger Clemens could routinely throw a fastball at a horizontal speed of 160 km/h. How long did the ball take to reach home plate 18.4 m away?

5E. Figure 2-18 is a plot of the age of ancient seafloor material, in millions of years, against the distance from a particular ocean ridge. Seafloor material extruded from this ridge moves away from it at approximately uniform speed. Find that speed in centimeters per year.

FIGURE 2-18 Exercise 5.

6E. When the legal speed limit for the New York Thruway was increased from 55 mi/h (= 88.5 km/h) to 65 mi/h (= 105 km/h), how much time was saved by a motorist who drove the 435 mi (= 700 km) between the Buffalo entrance and the New York City exit at the legal speed limit?

7E. Using the tables in Appendix D, find the speed of light (= 3×10^8 m/s) in miles per hour, feet per second, and light-years per year.

8E. An automobile travels on a straight road for 40 km at 30 km/h. It then continues in the same direction for another 40 km at 60 km/h. (a) What is the average velocity of the car during this 80 km trip? (Assume that it moves in the positive x direction.) (b) What is its average speed? (c) Graph x versus t and indicate how the average velocity is found on the graph.

9P. Compute your average velocity in the following two cases. (a) You walk 240 ft at a speed of 4.0 ft/s and then run 240 ft at a speed of 10 ft/s along a straight track. (b) You walk for 1.0 min at a speed of 4.0 ft/s and then run for 1.0 min at 10 ft/s along a straight track. (c) Graph x versus t for both cases and indicate how the average velocity is found on the graph.

10P. A car travels up a hill at a constant speed of 40 km/h and returns down the hill at a constant speed of 60 km/h. Calculate the average speed for the round trip.

11P. You drive on Interstate 10 from San Antonio to Houston, half the *time* at 35 mi/h (= 56 km/h) and the other half at 55 mi/h (= 89 km/h). On the way back you travel half the *distance* at 35 mi/h and the other half at 55 mi/h. What is your average speed (a) from San Antonio to Houston, (b) from Houston back to San Antonio, and (c) for the entire trip? (d) What is your average velocity for the entire trip? (e) Graph x versus t for (a), assuming the motion is all in the positive x direction. Indicate how the average velocity can be found on the graph.

12P. The position of an object moving in a straight line is given by $x = 3t - 4t^2 + t^3$, where x is in meters and t in seconds. (a) What is the position of the object at $t = 1, 2, 3,$ and 4 s? (b) What is the object's displacement between $t = 0$ and $t = 4$ s? (c) What is the average velocity for the time interval from $t = 2$ s to $t = 4$ s? (d) Graph x versus t for $0 \le t \le 4$ s and indicate how the answer for (c) can be found on the graph.

13P. The position of a particle moving along the x axis is given in centimeters by $x = 9.75 + 1.50t^3$, where t is in seconds. Consider the time interval $t = 2.00$ s to $t = 3.00$ s and calculate (a) the average velocity; (b) the instantaneous velocity at $t = 2.00$ s; (c) the instantaneous velocity at $t = 3.00$ s; (d) the instantaneous velocity at $t = 2.50$ s; and (e) the instantaneous velocity when the particle is midway between its positions at $t = 2.00$ s and $t = 3.00$ s. (f) Graph x versus t and indicate your answers graphically.

14P. A high-performance jet plane, practicing radar avoidance maneuvers, is in horizontal flight 35 m above the level ground. Suddenly, the plane encounters terrain that slopes gently upward at 4.3°, an amount difficult to detect (see Fig. 2-19). How much time does the pilot have to make a correction to avoid flying into the ground? The speed of the plane is 1300 km/h.

FIGURE 2-19 Problem 14.

15P. Two trains, each having a speed of 30 km/h, are headed at each other on the same straight track. A bird that can fly 60 km/h flies off the front of one train when they are 60 km apart and heads directly for the other train. On reaching the other train it flies directly back to the first train, and so forth. (We have no idea *why* a bird would behave in this way.) (a) How many trips can the bird make from one train to the other before they crash? (b) What is the total distance the bird travels?

SECTION 2-4 Instantaneous Velocity and Speed

16E. (a) If a particle's position is given by $x = 4 - 12t + 3t^2$ (where t is in seconds and x is in meters), what is its velocity at $t = 1$ s? (b) Is it moving toward increasing or decreasing x just then? (c) What is its speed just then? (d) Is the speed larger or smaller at later times? (Try answering the next two questions without further calculation.) (e) Is there ever an instant when the velocity is zero? (f) Is there a time after $t = 3$ s when the particle is moving toward decreasing x?

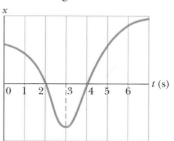

FIGURE 2-20 Exercise 17.

17E. The graph in Fig. 2-20 pertains to an armadillo that scampers left (direction of decreasing *x*) and right along an *x* axis. (a) When, if ever, is the animal to the left of the origin on the axis? When, if ever, is its velocity (b) negative, (c) positive, or (d) zero?

18E. Sketch a graph of *x* versus *t* for a mouse that is constrained in a narrow corridor (the *x* axis) and that scurries in the following sequence: (1) runs leftward (the direction of decreasing *x*) with a constant speed of 1.2 m/s, (2) gradually slows to 0.6 m/s toward the left, (3) gradually speeds up to 2.0 m/s toward the left, (4) gradually slows to a stop and then speeds up to 1.2 m/s toward the right. Where is the curve steepest? The least steep?

19P. How far does the runner whose velocity–time graph is shown in Fig. 2-21 travel in 16 s?

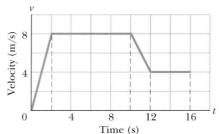

FIGURE 2-21
Problem 19.

SECTION 2-5 Acceleration

20E. A particle had a velocity of 18 m/s and 2.4 s later its velocity was 30 m/s in the opposite direction. (a) What was the magnitude of the average acceleration of the particle during this 2.4 s interval? (b) Graph *v* versus *t* and indicate how the average acceleration can be found on the graph.

21E. An object moves in a straight line as described by the velocity–time graph of Fig. 2-22. Sketch a graph that represents the acceleration of the object as a function of time.

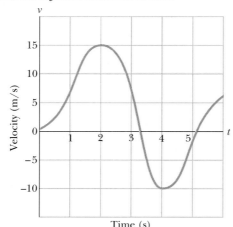

FIGURE 2-22
Exercise 21.

22E. The graph of *x* versus *t* in Fig. 2-23*a* is for a particle in straight-line motion. (a) State, for each of the intervals *AB*, *BC*, *CD*, and *DE*, whether the velocity *v* is positive, negative, or 0 and whether the acceleration *a* is positive, negative, or 0. (Ignore the end points of the intervals.) (b) From the curve, is there any inter-

val over which the acceleration is obviously not constant? (c) If the axes are shifted upward together such that the time axis ends up running along the dashed line, do any of your answers change?

23E. Repeat Exercise 22 for the motion described by the graph of Fig. 2-23*b*.

FIGURE 2-23
Exercises 22 and 23.

24E. A particle moves along the *x* axis with *x*(*t*) as shown in Fig. 2-24. Make rough sketches of velocity versus time and acceleration versus time for this motion.

FIGURE 2-24
Exercise 24.

25E. Sketch a graph that is a possible description of position as a function of time for a particle that moves along the *x* axis and, at *t* = 1 s, has (a) zero velocity and positive acceleration; (b) zero velocity and negative acceleration; (c) negative velocity and positive acceleration; (d) negative velocity and negative acceleration. (e) For which of these situations is the speed of the particle increasing at *t* = 1 s?

26E. Consider the two quantities $(dx/dt)^2$ and d^2x/dt^2. (a) Are these two equivalent expressions for the same thing? (b) What are the SI units of these two quantities?

27E. A particle moves along the *x* axis according to the equation $x = 50t + 10t^2$, where *x* is in meters and *t* is in seconds. Calculate (a) the average velocity of the particle during the first 3.0 s of its motion, (b) the instantaneous velocity of the particle at *t* = 3.0 s, and (c) the instantaneous acceleration of the particle at *t* = 3.0 s. (d) Graph *x* versus *t* and indicate how the answer to (a) can be obtained from the plot. (e) Indicate the answer to (b) on the graph. (f) Plot *v* versus *t* and indicate on it the answer to (c).

28E. (a) If the position of a particle is given by $x = 20t - 5t^3$, where *x* is in meters and *t* is in seconds, when, if ever, is the particle's velocity zero? (b) When is its acceleration *a* zero? (c) When is *a* negative? Positive? (d) Graph *x*(*t*), *v*(*t*), and *a*(*t*).

29P. A man stands still from $t = 0$ to $t = 5.00$ min; from $t = 5.00$ min to $t = 10.0$ min he walks briskly in a straight line at a constant speed of 2.20 m/s. What are his average velocity and average acceleration during the time intervals (a) 2.00 min to 8.00 min and (b) 3.00 min to 9.00 min? (c) Sketch x versus t and v versus t, and indicate how the answers to (a) and (b) can be obtained from the graphs.

30P. If the position of an object is given by $x = 2.0t^3$, where x is measured in meters and t in seconds, find (a) the average velocity and (b) the average acceleration between $t = 1.0$ s and $t = 2.0$ s. Then find (c) the instantaneous velocities and (d) the instantaneous accelerations at $t = 1.0$ s and $t = 2.0$ s. (e) Compare the average and instantaneous quantities and in each case explain why the larger one is larger. (f) Graph x versus t and v versus t, and indicate on the graphs your answers to (a) through (d).

31P. In an arcade video game, a spot is programmed to move across the screen according to $x = 9.00t - 0.750t^3$, where x is distance in centimeters measured from the left edge of the screen and t is time in seconds. When the spot reaches a screen edge, at either $x = 0$ or $x = 15.0$ cm, t is reset to 0 and the spot starts moving again according to $x(t)$. (a) At what time after starting is the spot instantaneously at rest? (b) Where does this occur? (c) What is its acceleration when this occurs? (d) In what direction is it moving just prior to coming to rest? (e) Just after? (f) When does it first reach an edge of the screen after $t = 0$?

32P. The position of a particle moving along the x axis depends on the time according to the equation $x = et^2 - bt^3$, where x is in feet and t in seconds. (a) What dimensions and units must e and b have? For the following, let their numerical values be 3.0 and 1.0, respectively. (b) At what time does the particle reach its maximum positive x position? (c) What distance does the particle cover in the first 4.0 s? (d) What is its displacement from $t = 0$ to $t = 4.0$ s? (e) What is its velocity at $t = 1.0, 2.0, 3.0,$ and 4.0 s? (f) What is its acceleration at these times?

SECTION 2-6 Constant Acceleration: A Special Case

33E. The head of a rattlesnake can accelerate 50 m/s² in striking a victim. If a car could do as well, how long would it take to reach a speed of 100 km/h from rest?

34E. An object has a constant acceleration of +3.2 m/s². At a certain instant its velocity is +9.6 m/s. What is its velocity (a) 2.5 s earlier and (b) 2.5 s later?

35E. An automobile increases its speed uniformly from 25 km/h to 55 km/h in 0.50 min. A bicycle rider uniformly speeds up to 30 km/h from rest in 0.50 min. Calculate their accelerations.

36E. Suppose a rocketship in deep space moves with constant acceleration equal to 9.8 m/s², which gives the illusion of normal gravity during the flight. (a) If it starts from rest, how long will it take to acquire a speed one-tenth that of light, which travels at 3.0×10^8 m/s? (b) How far will it travel in so doing?

37E. A jumbo jet must reach a speed of 360 km/h ($= 225$ mi/h) on the runway for takeoff. What is the least constant acceleration needed for takeoff from a 1.80 km runway?

38E. A muon (an elementary particle) enters an electric field with a speed of 5.00×10^6 m/s, whereupon the field slows it at the rate of 1.25×10^{14} m/s². (a) How far does the muon take to stop? (b) Graph x versus t and v versus t for the muon.

39E. An electron with initial velocity $v_0 = 1.50 \times 10^5$ m/s enters a region 1.0 cm long where it is electrically accelerated (Fig. 2-25). It emerges with velocity $v = 5.70 \times 10^6$ m/s. What was its acceleration, assumed constant? (Such a process occurs in the electron gun in a cathode-ray tube, used in television receivers and oscilloscopes.)

FIGURE 2-25
Exercise 39.

40E. A car can be braked to a stop from 60 mi/h in 43 m. (a) What is the magnitude of the acceleration in SI units and g units? (Assume that a is constant.) (b) What is the stopping time? If your reaction time T for braking is 400 ms, to how many "reaction times" does the stopping time correspond?

41E. A world's land speed record was set by Colonel John P. Stapp when, on March 19, 1954, he rode a rocket-propelled sled that moved down a track at 1020 km/h. He and the sled were brought to a stop in 1.4 s. (See Fig. 2-8.) What acceleration did he experience? Express your answer in g units.

42E. On a dry road a car with good tires may be able to brake with a deceleration of 4.92 m/s² (assume that it is constant). (a) How long does such a car, initially traveling at 24.6 m/s, take to come to rest? (b) How far does it travel in this time? (c) Graph x versus t and v versus t for the deceleration.

43E. A rocket-driven sled running on a straight, level track is used to investigate the physiological effects of large accelerations on humans. One such sled can attain a speed of 1600 km/h in 1.8 s starting from rest. Find (a) the acceleration (assumed constant) in g units and (b) the distance traveled.

44E. The brakes on your automobile are capable of creating a deceleration of 17 ft/s². (a) If you are going 85 mi/h and suddenly see a state trooper, what is the minimum time in which you can get your car under the 55 mi/h speed limit? (The answer reveals the futility of braking to keep your high speed from being detected with a radar gun.) (b) Graph x versus t and v versus t for such a deceleration.

45E. A motorcycle is moving at 30 m/s when the rider applies the brakes, giving the motorcycle a constant deceleration. During the 3.0 s interval immediately after braking begins, the speed decreases to 15 m/s. What distance does the motorcycle travel from the instant braking begins until it comes to rest?

46P. A hot rod can accelerate from 0 to 60 km/h in 5.4 s. (a) What is its average acceleration, in m/s², during this time? (b) How far will it travel during the 5.4 s, assuming its acceleration is constant? (c) How much time would it require to go a distance of 0.25 km if its acceleration could be maintained at the value in (a)?

47P. A train started from rest and moved with constant acceleration. At one time it was traveling 30 m/s, and 160 m farther on it was traveling 50 m/s. Calculate (a) the acceleration, (b) the time required to travel the 160 m mentioned, (c) the time required to attain the speed of 30 m/s, and (d) the distance moved from rest to the time the train had a speed of 30 m/s. (e) Graph x versus t and v versus t for the train, from rest.

48P. A car traveling 56.0 km/h is 24.0 m from a barrier when the driver slams on the brakes. The car hits the barrier 2.00 s later. (a) What was the car's constant deceleration before impact? (b) How fast was the car traveling at impact?

49P. A car moving with constant acceleration covers the distance between two points 60.0 m apart in 6.00 s. Its speed as it passes the second point is 15.0 m/s. (a) What is the speed at the first point? (b) What is the acceleration? (c) At what prior distance from the first point was the car at rest? (d) Graph x versus t and v versus t for the car, from rest.

50P. Two subway stops are separated by 1100 m. If a subway train accelerates at $+1.2$ m/s² from rest through the first half of the distance and decelerates at -1.2 m/s² through the second half, what are (a) its travel time and (b) its maximum speed? (c) Graph x, v, and a versus t for the trip.

51P. To stop a car, you require first a certain reaction time to begin braking; then the car slows under the constant braking deceleration. Suppose that the total distance moved by your car during these two phases is 186 ft when its initial speed is 50 mi/h, and 80 ft when the initial speed is 30 mi/h. What are (a) your reaction time and (b) the magnitude of the deceleration?

52P. You are driving toward a traffic signal when it turns yellow. Your speed is the legal speed limit of $v_0 = 35$ mi/h; your best deceleration rate is $a = 17$ ft/s². Your best reaction time to begin braking is $T = 0.75$ s. To avoid having the front of your car enter the intersection after the light turns red, should you brake to a stop or continue to move at 35 mi/h if the distance to the intersection and the duration of the yellow light are (a) 40 m and 2.8 s, and (b) 32 m and 1.8 s?

53P. When a driver brings a car to a stop by braking as hard as possible, the stopping distance can be regarded as the sum of a "reaction distance," which is initial speed times the driver's reaction time, and a "braking distance," which is the distance covered during braking. The following table gives typical values:

INITIAL SPEED (m/s)	REACTION DISTANCE (m)	BRAKING DISTANCE (m)	STOPPING DISTANCE (m)
10	7.5	5.0	12.5
20	15	20	35
30	22.5	45	67.5

(a) What reaction time is the driver assumed to have? (b) What is the car's stopping distance if the initial speed is 25 m/s?

54P. (a) If the maximum acceleration that is tolerable for passengers in a subway train is 1.34 m/s², and subway stations are located 806 m apart, what is the maximum speed a subway train can attain between stations? (b) What is the travel time between stations? (c) If the subway train stops for 20 s at each station, what is the maximum average speed of a subway train, from one start-up to the next? (d) Graph x, v, and a versus t.

55P. An elevator cab in the New York Marquis Marriott has a total run of 624 ft. Its maximum speed is 1000 ft/min. Its acceleration and deceleration both have a magnitude of 4.0 ft/s². (a) How far does the cab move while accelerating to full speed from rest? (b) How long does it take to make the nonstop 624 ft run, starting and ending at rest?

56P. At the instant the traffic light turns green, an automobile starts with a constant acceleration a of 2.2 m/s². At the same instant a truck, traveling with a constant speed of 9.5 m/s, overtakes and passes the automobile. (a) How far beyond the traffic signal will the automobile overtake the truck? (b) How fast will the car be traveling at that instant?

57P. When a high-speed passenger train traveling at 100 mi/h rounds a bend, the engineer is shocked to see that a locomotive has improperly entered onto the track from a siding 0.42 mi ahead; see Fig. 2-26. The locomotive is moving at 18 mi/h. The engineer of the passenger train immediately applies the brakes. (a) What must be the magnitude of the resulting constant acceleration if a collision is to be just avoided? (b) Assume that the

FIGURE 2-26 Problem 57.

engineer is at $x = 0$ when, at $t = 0$, he first spots the locomotive. Sketch the $x(t)$ curves representing the locomotive and passenger train for the situations in which a collision is just avoided and not quite avoided.

58P. Two trains, one traveling at 72 km/h and the other at 144 km/h, are headed toward one another along a straight, level track. When they are 950 m apart, each engineer sees the other's train and applies the brakes. The brakes decelerate each train at the rate of 1.0 m/s². Is there a collision?

59P. Sketch a $v(t)$ graph that would be associated with the $a(t)$ graph shown in Fig. 2-27.

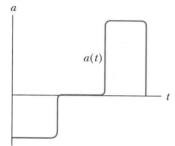

FIGURE 2-27
Problem 59.

SECTION 2-8 Free-Fall Acceleration

60E. At a construction site a pipe wrench strikes the ground with a speed of 24 m/s. (a) From what height was it inadvertently dropped? (b) How long was it falling? (c) Sketch graphs of y, v, and a versus t for the wrench.

61E. (a) With what speed must a ball be thrown vertically from ground level to rise to a maximum height of 50 m? (b) How long will it be in the air? (c) Sketch graphs of y, v, and a versus t for the ball. On the first two graphs, indicate the time at which 50 m is reached.

62E. Raindrops fall to Earth from a cloud 1700 m above the Earth's surface. If they were not slowed by air resistance, how fast would the drops be moving when they struck the ground? Would it be safe to walk outside during a rainstorm?

63E. The single cable supporting an unoccupied construction elevator breaks when the elevator is at rest at the top of a 120-m-high building. (a) With what speed does the elevator strike the ground? (b) How long was it falling? (c) What was its speed when it passed the halfway point on the way down? (d) How long had it been falling when it passed the halfway point?

64E. A hoodlum throws a stone vertically downward with an initial speed of 12.0 m/s from the roof of a building, 30.0 m above the ground. (a) How long does it take the stone to reach the ground? (b) What is the speed of the stone at impact?

65E. The Zero Gravity Research Facility at the NASA Lewis Research Center includes a 145 m drop tower. This is an evacuated vertical tower through which, among other possibilities, a 1 m diameter sphere containing an experimental package can be dropped. (a) How long is the sphere in free fall? (b) What is its speed just as it reaches a catching device at the bottom of the tower? (c) When caught, the sphere experiences an average de-

celeration of 25g as its speed is reduced to zero. Through what distance does it travel during the deceleration?

66E. A model rocket, propelled by burning fuel, takes off vertically. Plot qualitatively (numbers not required) graphs of y, v, and a versus t for the rocket's flight. Indicate when the fuel is exhausted, when the rocket reaches maximum height, and when it returns to the ground.

67E. A rock is dropped from a 100 m high cliff. How long does it take to fall (a) the first 50 m and (b) the second 50 m?

68P. A startled armadillo leaps upward (Fig. 2-28), rising 0.544 m in 0.200 s. (a) What was its initial speed? (b) What is its speed at this height? (c) How much higher does it go?

FIGURE 2-28 Problem 68.

69P. An object falls from a bridge that is 45 m above the water. It falls directly into a model boat, moving with constant velocity, that was 12 m from the point of impact when the object was released. What was the speed of the boat?

70P. A model rocket is fired vertically and ascends with a constant vertical acceleration of 4.00 m/s² for 6.00 s. Its fuel is then exhausted and it continues as a free-fall particle. (a) What is the maximum altitude reached? (b) What is the total time elapsed from takeoff until the rocket strikes the Earth?

71P. A basketball player, standing near the basket to grab a rebound, jumps 76.0 cm vertically. How much (total) time does the player spend (a) in the top 15.0 cm of this jump and (b) in the bottom 15.0 cm? Does this help explain why such players seem to hang in the air at the tops of their jumps?

72P. At the National Physical Laboratory in England, a measurement of the free-fall acceleration g was made by throwing a glass ball straight up in an evacuated tube and letting it return. Let ΔT_L in Fig. 2-29 be the time interval between the two passages of the ball across a certain lower level, ΔT_U the time interval between the two passages across an upper level, and H the distance between the two levels. Show that

$$g = \frac{8H}{\Delta T_L^2 - \Delta T_U^2}.$$

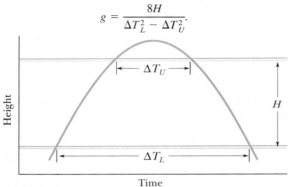

FIGURE 2-29   Problem 72.

73P. A ball of moist clay falls to the ground from a height of 15.0 m. It is in contact with the ground for 20.0 ms before coming to rest. What is the average acceleration of the clay during the time it is in contact with the ground? (Treat the ball as a particle.)

74P. A ball is thrown *down* vertically with an initial *speed* of v_0 from a height of h. (a) What is its speed just before it strikes the ground? (b) How long does the ball take to reach the ground? (c) What would be the answers to (a) and (b) if the ball were thrown *upward* from the same height and with the same initial speed? Before solving any equations, decide if the answers here should be greater than, less than, or the same as in (a) and (b).

75P. Figure 2-30 shows a simple device for measuring your reaction time. It consists of a cardboard strip marked with a scale and two large dots. A friend holds the strip *vertically,* with thumb and forefinger at the dot on the right in Fig. 2-30. You then position your thumb and forefinger at the other dot (on the left in Fig. 2-30), being careful not to touch the strip. Your friend releases the strip, and you try to pinch it as soon as possible after you see it begin to fall. The mark at the place where you pinch the strip gives your reaction time. (a) How far from the lower dot should you place the 50.0 ms mark? (b) How much higher should the marks for 100, 150, 200, and 250 ms be? (For example, should the 100 ms marker be two times as far from the dot as the 50 ms marker? Can you find any pattern in the answers?)

Reaction time (ms)

FIGURE 2-30   Problem 75.

76P. A juggler tosses balls vertically a certain distance into the air. How much higher must they be tossed if they are to spend twice as much time in the air?

77P. A stone is thrown vertically upward. On its way up it passes point A with speed v, and point B, 3.00 m higher than A, with speed $\frac{1}{2}v$. Calculate (a) the speed v and (b) the maximum height reached by the stone above point B.

78P. To test the quality of a tennis ball, you drop it onto the floor from a height of 4.00 m. It rebounds to a height of 3.00 m. If the ball was in contact with the floor for 10.0 ms, what was its average acceleration during contact

79P. Water drips from the nozzle of a shower onto the floor 200 cm below. The drops fall at regular (equal) intervals of time, the first drop striking the floor at the instant the fourth drop begins to fall. Find the locations of the second and third drops when the first strikes the floor.

80P. A lead ball is dropped into a lake from a diving board 5.20 m above the water. It hits the water with a certain velocity and then sinks to the bottom with this same constant velocity. It reaches the bottom 4.80 s after it is dropped. (a) How deep is the lake? (b) What is the average velocity of the ball? (c) Suppose that all the water is drained from the lake. The ball is now thrown from the diving board so that it again reaches the bottom in 4.80 s. What is the initial velocity of the ball?

81P. An object falls from height h from rest. If it travels $0.50h$ in the last 1.00 s, find (a) the time and (b) the height of its fall. Explain the physically unacceptable solution of the quadratic equation in t that you obtain.

82P. A woman fell 144 ft from the top of a building, landing on the top of a metal ventilator box, which she crushed to a depth of 18 in. She survived without serious injury. What acceleration (assumed uniform) did she experience during the collision? Express your answer in terms of g units.

83P. A stone is dropped into a river from a bridge 144 ft above the water. Another stone is thrown vertically down 1.00 s after the first is dropped. Both stones strike the water at the same time. (a) What was the initial speed of the second stone? (b) Plot velocity versus time on a graph for each stone, taking zero time as the instant the first stone is released.

84P. A parachutist bails out and freely falls 50 m. Then the parachute opens, and thereafter she decelerates at 2.0 m/s². She reaches the ground with a speed of 3.0 m/s. (a) How long was the parachutist in the air? (b) At what height did the fall begin?

85P. Two objects begin a free fall from rest from the same height 1.0 s apart. How long after the first object begins to fall will the two objects be 10 m apart?

86P. As Fig. 2-31 shows, Clara jumps from a bridge, followed closely by Jim. How long did Jim wait after Clara jumped? Assume that Jim is 170 cm tall and that the jumping-off level is at the top of the figure. Make scale measurements directly on the figure.

87P. A hot-air balloon is ascending at the rate of 12 m/s and is 80 m above the ground when a package is dropped over the side. (a) How long does the package take to reach the ground? (b) With what speed does it hit the ground?

88P. An elevator without a ceiling is ascending with a constant speed of 10 m/s. A boy on the elevator throws a ball directly

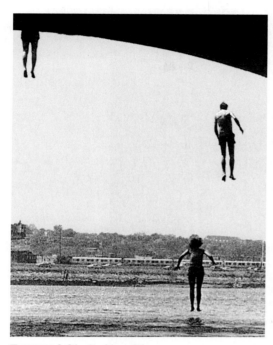

FIGURE 2-31 Problem 86.

upward, from a height of 2.0 m above the elevator floor, just as the elevator floor is 28 m above the ground. The initial speed of the ball with respect to the elevator is 20 m/s. (a) What is the maximum height attained by the ball? (b) How long does it take for the ball to return to the elevator floor?

89P. A steel ball is dropped from a building's roof and passes a window, taking 0.125 s to fall from the top to the bottom of the window, a distance of 1.20 m. It then falls to a sidewalk and bounces ''perfectly'' back past the window, moving from bottom to top in 0.125 s. (The upward flight is a reverse of the fall.) The time spent below the bottom of the window is 2.00 s. How tall is the building?

90P. A drowsy cat spots a flowerpot that sails first up and then down past an open window. The pot was in view for a total of 0.50 s, and the top-to-bottom height of the window is 2.00 m. How high above the window top did the flowerpot go?

ELECTRONIC COMPUTATION

Problems in this section are to be solved with a graphing calculator or with a mathematics package on a computer.

91. A driver is to test the braking of a car equipped with antilock brakes, which can decelerate the car at the maximum possible rate without the tires skidding. The test is made on a dry straight section of a test track. In response to a visual signal, the driver stops the car in the shortest possible distance. The stopping distances for six initial speeds are given in the following table. (a) Find an expression for the stopping distance d in terms of the initial speed v_i, the response time T_R of the driver to the visual signal, and the magnitude a' (assumed constant) of the car's deceleration. Using a least squares fit for the data, find (b) a' and (c) T_R.

v_i (m/s)	5.00	10.0	15.0	20.0	25.0	30.0
d (m)	4.6	12.1	22.3	35.2	51.0	69.5

92. A motorcyclist starts from rest and accelerates along a horizontal straight track. Photogates are attached to six posts evenly spaced every 10.0 m along the track; the first post is at the starting point. The photogates on each post measure the time required by the motorcyclist to reach that post. The following table gives the results for one test. (a) Find an expression for the distance d to each post in terms of the time t to reach that post and the acceleration a' of the motorcyclist (assumed constant). (b) Using the data of the table, graph d versus t^2. (c) Using a linear regression fit of the data, find the acceleration of the motorcyclist.

Post number	1	2	3	4	5	6
Distance traveled (m)	0	10.0	20.0	30.0	40.0	50.0
Time required (s)	0	1.63	2.33	2.83	3.31	3.79

93. The motorcyclist of Problem 92 next starts from rest and accelerates uniformly along another straight test track. A series of six posts equally spaced by a distance $d_0 = 10.0$ m line the track, beginning an unknown distance d from the motorcyclist's starting point. The following table gives the speed v at each post. (a) Find an expression for the square of the speed v_j at post j in terms of the speed v_1 at post 1, the distance d_0, the constant acceleration a', and the post number j. (b) Graph v_j^2 versus $(j - 1)$. Using a linear regression fit of the data, find (c) acceleration a' and (d) distance d.

Post number j	1	2	3	4	5	6
Speed v (m/s)	14.0	18.3	21.7	24.6	27.5	30.0

94. Suppose the coordinate of a particle moving along the x axis is given as a function of time t by

$$x(t) = -32.0 + 24.0t^2 e^{-0.0300t},$$

where x is in meters and t is in seconds. (a) Write expressions for the velocity v and acceleration a as functions of the time. (b) Plot graphs of the coordinate, velocity, and acceleration as functions of time from $t = 0$ to $t = 100$ s. (c) Find the time when the coordinate of the particle is zero; then find the velocity and acceleration at that time. (d) Find the time when the velocity of the particle is zero; then find the coordinate and acceleration at that time.

3
Vectors

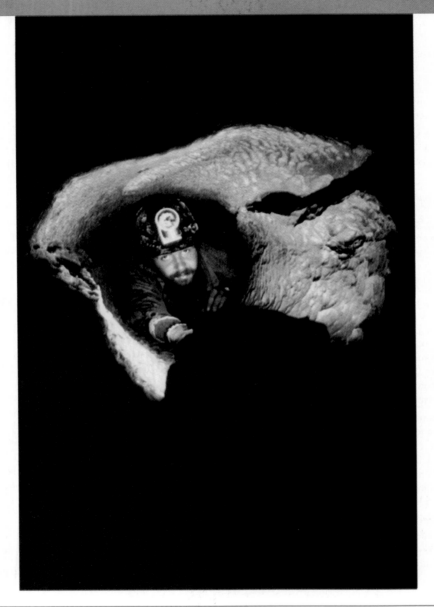

For two decades spelunking teams crawled, climbed, and squirmed through 200 km of Mammoth Cave and the Flint Ridge cave system, seeking a connection. The photograph shows Richard Zopf pushing his pack through the Tight Tube, far inside the Flint Ridge system. After 12 hours of "caving" along a labyrinthine route, Zopf and six others waded through a stretch of chilling water and found themselves in Mammoth Cave. Their breakthrough established the Mammoth–Flint cave system as being the longest cave in the world. How can their final point be related to their initial point other than in terms of the actual route they covered?

3-1 VECTORS AND SCALARS

A particle confined to a straight line can move in only two directions. We can take its motion to be positive in one of these directions and negative in the other. For a particle moving in three dimensions, however, a plus sign or minus sign is no longer enough to indicate the direction of the motion. Instead, we must use a *vector*.

A **vector** has magnitude as well as direction, and vectors follow certain rules of combination, which we examine below; a **vector quantity** is a quantity that has both a magnitude and a direction. Some physical quantities that can be represented by vectors are displacement, velocity, and acceleration.

Not all physical quantities involve a direction. Temperature, pressure, energy, mass, and time, for example, do not "point" in the spatial sense. We call such quantities **scalars,** and we deal with them by the rules of ordinary algebra. A single value, with a sign (as in $-40°F$), specifies a scalar.

The simplest vector quantity is displacement, or change of position. A vector that represents a displacement is called, reasonably, a **displacement vector**. (Similarly, we have velocity vectors and acceleration vectors.) If a particle changes its position by moving from A to B in Fig. 3-1a, we say that it undergoes a displacement from A to B, which we represent with an arrow pointing from A to B. The arrow specifies the vector graphically. To distinguish vector symbols from other kinds of arrows, we use the outline of a triangle as the arrowhead.

The arrows from A to B, from A' to B', and from A'' to B'' in Fig. 3-1a represent the same *change of position* for the particle and we make no distinction among them. All three arrows have the same magnitude and direction and thus specify identical displacement vectors.

The displacement vector tells us nothing about the actual path that the particle takes. In Fig. 3-1b, for example, all three paths connecting points A and B correspond to the

same displacement vector, that of Fig. 3-1a. Displacement vectors represent only the overall effect of the motion, not the motion itself.

Another way to specify a vector is to give its magnitude and the angles it makes with two reference lines. We do that in Sample Problem 3-1.

SAMPLE PROBLEM 3-1

The 1972 team that connected the Mammoth–Flint cave system went from the Austin Entrance in the Flint Ridge system to Echo River in Mammoth Cave (see Fig. 3-2a), traveling a net 2.6 km westward, 3.9 km southward, and 25 m upward. What was their displacement vector?

SOLUTION: We first take an overhead view (Fig. 3-2b) to find their horizontal displacement d_h. The magnitude of d_h is given by the Pythagorean theorem:

$$d_h = \sqrt{(2.6 \text{ km})^2 + (3.9 \text{ km})^2} = 4.69 \text{ km}.$$

The angle θ of their displacement vector relative to the west is given by

$$\tan \theta = \frac{3.9 \text{ km}}{2.6 \text{ km}} = 1.5,$$

or $\qquad \theta = \tan^{-1} 1.5 = 56°.$

We next take a side view (Fig. 3-2c), to find the magnitude of the overall displacement d,

$$d = \sqrt{(4.69 \text{ km})^2 + (0.025 \text{ km})^2} = 4.69 \text{ km} \approx 4.7 \text{ km},$$

and the angle ϕ,

$$\phi = \tan^{-1} \frac{0.025 \text{ km}}{4.69 \text{ km}} = 0.3°.$$

So the team's displacement vector had a magnitude of 4.7 km and was at an angle of 56° south of west and at an angle of 0.3° upward. The net vertical motion was, of course, insignificant compared to the horizontal motion, but that fact would have been no comfort to the team as they climbed up and down countless times. The route they actually covered was quite different from the displacement vector, which merely points from start to finish.

3-2 ADDING VECTORS: GRAPHICAL METHOD

Suppose that, as in the vector diagram of Fig. 3-3a, a particle moves from A to B and then later from B to C. We can represent its overall displacement (no matter what its actual path) with two successive displacement vectors, AB and BC. The net effect of these two displacements is a single displacement from A to C. We call AC the **vector sum** (or **resultant**) of the vectors AB and BC. This sum is

(a)   (b)

FIGURE 3-1 (a) All three arrows represent the same displacement. (b) All three paths connecting the two points correspond to the same displacement vector.

(a)

(b)

(c)

FIGURE 3-2   Sample Problem 3-1. (a) Part of the Mammoth–Flint cave system, with the spelunk-ing team's route from the Austin Entrance to Echo River indicated. (b) An overhead view of the team's displacement. (c) A side view. (Adapted from map by Cave Research Foundation.)

not the usual algebraic sum, and we need more than simple numbers to specify it.

In Fig. 3-3b, we redraw the vectors of Fig. 3-3a and relabel them in the way that we shall use from now on, namely, with boldface symbols such as **a**, **b**, and **s**. In handwriting, you can place an arrow over the symbol, as in $\vec{a}$. If we want to indicate only the magnitude of the vector (a quantity that is always positive), we shall use the italic symbol, such as *a*, *b*, and *s*. (You can use just a handwritten symbol.) The boldface symbol always implies both prop-erties of the vector, magnitude and direction.

We can represent the relation among the three vectors in Fig. 3-3b with the vector equation

$$\mathbf{s} = \mathbf{a} + \mathbf{b}, \tag{3-1}$$

which says that the vector **s** is the vector sum of vectors **a** and **b**. The symbol + in Eq. 3-1 and the words "sum" and "add" have different meanings for vectors than they do in the usual algebra.

Figure 3-3 suggests a procedure for adding two-di-mensional vectors **a** and **b** graphically. (1) On a sheet of paper, lay out vector **a** to some convenient scale and at the proper angle. (2) Lay out vector **b** to the same scale, with its tail at the head of vector **a**, again at the proper angle. (3) Construct the vector sum **s** by drawing an arrow from the tail of **a** to the head of **b**. Note that this procedure takes into account both the magnitudes and the directions of the vec-tors; you can easily generalize it to add more than two vectors.

Vector addition, defined in this way, has two impor-tant properties. First, the order of addition does not matter. That is,

$$\mathbf{a} + \mathbf{b} = \mathbf{b} + \mathbf{a} \quad \text{(commutative law).} \tag{3-2}$$

Figure 3-4 should convince you that this is the case.

Second, if there are more than two vectors, it does not matter how we group them as we add them. Thus if we want to add vectors **a**, **b**, and **c**, we can add **a** and **b** first and then add their vector sum to **c**. On the other hand, we can add **b** and **c** first, and then add *that* sum to **a**. We get the same result either way. In equation form,

$$(\mathbf{a} + \mathbf{b}) + \mathbf{c} = \mathbf{a} + (\mathbf{b} + \mathbf{c}) \quad \text{(associative law).} \tag{3-3}$$

Figure 3-5 should convince you that Eq. 3-3 is correct.

(a)

(b)

FIGURE 3-3   (a) AC is the vector sum of the vectors AB and BC. (b) The same vectors relabeled.

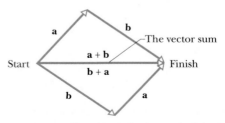

FIGURE 3-4   The two vectors **a** and **b** can be added in either order; see Eq. 3-2.

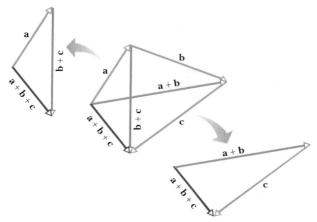

FIGURE 3-5 The three vectors **a**, **b**, and **c** can be grouped in any way as they are added; see Eq. 3-3.

The vector −**b** is a vector with the same magnitude as **b** but the opposite direction (see Fig. 3-6). If you try to add the two vectors in Fig. 3-6, you will see that

$$\mathbf{b} + (-\mathbf{b}) = 0.$$

Adding −**b** has the effect of subtracting **b**. We use this property to define the difference between two vectors: let **d** = **a** − **b**. Then

$$\mathbf{d} = \mathbf{a} - \mathbf{b} = \mathbf{a} + (-\mathbf{b}) \qquad \text{(subtraction).} \qquad (3\text{-}4)$$

That is, we find the difference vector **d** by adding the vector −**b** to the vector **a**. Figure 3-7 shows how this is done graphically.

Remember, although we have used displacement vectors as a prototype, the rules for addition and subtraction hold for vectors of all kinds, whether they represent forces, velocities, or anything else. However, as in ordinary arithmetic, it is still true that we can add only vectors of the same kind. We can add two displacements, for example, or

FIGURE 3-6 The vectors **b** and −**b**.

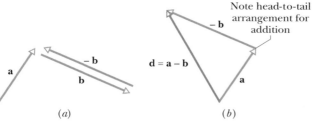

(a)   (b)

FIGURE 3-7 (a) Vectors **a**, **b**, and −**b**. (b) To subtract vector **b** from vector **a**, add vector −**b** to vector **a**.

two velocities, but it makes no sense to add a displacement and a velocity. In the world of scalars, that would be like trying to add 21 s and 12 m.

CHECKPOINT **1:** The magnitudes of displacements **a** and **b** are 3 m and 4 m, respectively, and **c** = **a** + **b**. Considering various orientations of **a** and **b**, what is (a) the maximum possible magnitude for **c** and (b) the minimum possible magnitude?

SAMPLE PROBLEM 3-2

In an orienteering class, you have the goal of moving as far (straight-line distance) from base camp as possible by making three straight-line moves. You may use the following displacements in any order: (a) **a**, 2.0 km due east; (b) **b**, 2.0 km 30° north of east; (c) **c**, 1.0 km due west. Or you may substitute either −**b** for **b** or −**c** for **c**. What is the greatest distance you can be from base camp at the end of the third displacement?

SOLUTION: Using a convenient scale, we draw vectors **a**, **b**, **c**, −**b**, and −**c** as in Fig. 3-8a. We then mentally slide the vectors over the page, connecting three of them in head-to-tail arrangements to find their vector sum **d**. The tail of the first vector represents base camp. The head of the third vector represents the point at which you stop. The vector sum **d** extends from the tail of the first vector to the head of the third vector. Its magnitude d is your distance from base camp.

We find that distance d is greatest for a head-to-tail arrangement of vectors **a**, **b**, and −**c**. They can be in any order, because their vector sum is the same for any order. The order shown in Fig. 3-8b is for the vector sum

$$\mathbf{d} = \mathbf{b} + \mathbf{a} + (-\mathbf{c}) = \mathbf{b} + \mathbf{a} - \mathbf{c}.$$

Using the scale given in Fig. 3-8a, we measure the length d of this vector sum, finding

$$d = 4.8 \text{ m.} \qquad \text{(Answer)}$$

(a)   (b)

FIGURE 3-8 Sample Problem 3-2. (a) Displacement vectors; three are to be used. (b) Your distance from base camp is greatest if you undergo displacements **a**, **b**, and −**c**, in any order. One choice of order is shown; it gives vector sum **d** = **b** + **a** − **c**.

3-3 VECTORS AND THEIR COMPONENTS

Adding vectors graphically can be tedious. A neater and easier technique involves algebra but requires that the vectors be placed on a rectangular coordinate system. The x and y axes are usually drawn in the plane of the page, as in Fig. 3-9a. The z axis comes directly out of the page at the origin; we ignore it for now and deal only with two-dimensional vectors.

Vector **a** in Fig. 3-9 is in the xy plane. If we drop perpendicular lines from the ends of **a** to the coordinate axes, the quantities a_x and a_y so formed are called the **components** of the vector **a** in the x and y directions. The process of forming them is called **resolving the vector**. In general, a vector will have three components, although for the case of Fig. 3-9a the component along the z axis is zero. As Fig. 3-9b shows, if you move a vector in such a way that it remains parallel to its original direction at all times, the values of its components remain unchanged.

We can easily find the components of **a** in Fig. 3-9a from the right triangle there:

$$a_x = a \cos \theta \quad \text{and} \quad a_y = a \sin \theta, \quad (3\text{-}5)$$

where θ is the angle that the vector **a** makes with the direction of increasing x, and a is its magnitude. Figure 3-9c shows that the vector and its x and y components form a right triangle. It also shows how we can reconstruct a vector from its components: we arrange the components head to tail. Then we complete a right triangle with the vector being along the hypotenuse, from the tail of one component to the head of the other component.

Depending on the value of θ, the components of a vector may be positive, negative, or zero. In a figure, we use small, solid triangles as arrowheads to indicate the signs of the components, according to the usual conven-

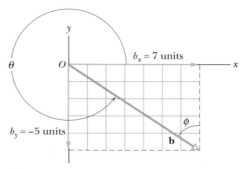

FIGURE 3-10 The components of **b** are positive on the x axis and negative on the y axis.

tion: positive in the direction of increasing coordinate values, and negative in the opposite direction. Figure 3-10 shows a vector **b** for which b_y is negative and b_x is positive.

Once a vector has been resolved into its components, the components themselves can be used in place of the vector. For example, the two numbers a and θ specify the magnitude and direction of the two-dimensional vector **a** in Fig. 3-9a. But instead of a and θ, we can specify the vector with the two other numbers a_x and a_y. Both sets of numbers contain exactly the same information, and we can pass back and forth readily between the two descriptions. To obtain a and θ if we are given a_x and a_y, we note (see Fig. 3-9c) that

$$a = \sqrt{a_x^2 + a_y^2} \quad \text{and} \quad \tan \theta = \frac{a_y}{a_x}. \quad (3\text{-}6)$$

In solving problems, you may use *either* the a_x, a_y notation or the a, θ notation.

CHECKPOINT 2: In the figure, which of the ways indicated for combining the x and y components of vector **a** are proper to determine that vector?

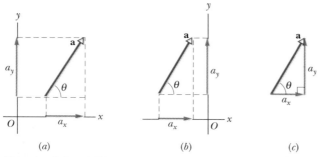

FIGURE 3-9 (a) The components of vector **a**. (b) The components are unchanged if the vector is shifted, as long as the magnitude and orientation are maintained. (c) The components form the legs of a right triangle whose hypotenuse is the magnitude of the vector.

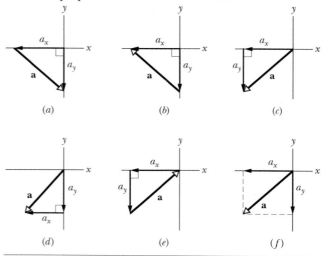

SAMPLE PROBLEM 3-3

A small airplane leaves an airport on an overcast day and is later sighted 215 km away, in a direction making an angle of 22° east of north. How far east and north is the airplane from the airport when sighted?

SOLUTION: On an xy coordinate system the situation is as shown in Fig. 3-11, where for convenience the origin of the system has been placed at the airport. The airplane's displacement vector **d** points from the origin to where the airplane is sighted.

To answer the question, we find the components of **d**. With Eq. 3-5 and the angle $\theta = 68°$ ($= 90° - 22°$), we have

$$d_x = d \cos \theta = (215 \text{ km})(\cos 68°)$$
$$= 81 \text{ km} \qquad \text{(Answer)}$$

and $\quad d_y = d \sin \theta = (215 \text{ km})(\sin 68°)$
$$= 199 \text{ km}. \qquad \text{(Answer)}$$

So the airplane was spotted 199 km north and 81 km east of the airport.

FIGURE 3-11 Sample Problem 3-3. A plane takes off from an airport at the origin and is later sighted at P.

PROBLEM SOLVING TACTICS

TACTIC 1: *Angles—Degrees and Radians*

Angles that are measured relative to the positive direction of the x axis are positive if they are measured in the counterclockwise direction, and negative if clockwise. For example, 210° and −150° are the same angle. Most calculators (try yours) will accept angles in either form when a trig function is taken.

Angles may be measured in degrees or radians (rad). You can relate the two measures by remembering that one full circle is equivalent to 360° and to 2π rad. So if you needed to convert, say, 40° to radians, you would write

$$40° \, \frac{2\pi \text{ rad}}{360°} = 0.70 \text{ rad}.$$

Is the answer reasonable? Quickly check by realizing that 40° is one-ninth of a full circle, and with a full circle equivalent to 2π rad, which is about 6.3 rad, the angle should be $\frac{1}{9}(6.3)$, which it is. Or, check by remembering that 1 rad $\approx 57°$.

Most calculators are in the degree mode when they are turned on, so that angles must be entered in degrees. However, you may be able to change yours to the radian mode. Check it before your first exam.

TACTIC 2: *Trig Functions*

You need to know the definitions of the common trig functions—sine, cosine, and tangent—because they are part of the language of science and engineering. They are given in Fig. 3-12 in a form that does not depend on how the triangle is labeled.

FIGURE 3-12 A triangle used to define the trigonometric functions. See also Appendix E.

You should also be able to sketch how the trig functions vary with angle, as in Fig. 3-13, in order to be able to judge whether a calculator result is reasonable. Even knowing the signs of the functions in the various quadrants can be of help.

TACTIC 3: *Inverse Trig Functions*

The most important of the inverse trig functions are $\sin^{-1}$, $\cos^{-1}$, and $\tan^{-1}$. When they are taken on a calculator, you must consider the reasonableness of the answer you get, because there is usually another possible answer that the calculator does not give. The range of operation for a calculator in taking each inverse trig function is indicated in Fig. 3-13. As an example, $\sin^{-1} 0.5$ has associated angles of 30° (which is displayed by the calculator, since 30° falls within its range of operation) and 150°. To see both angles, draw a horizontal line through 0.5 in Fig. 3-13a and note where it cuts the sine curve.

How do you distinguish a correct answer? As an example, reconsider the calculation of θ in Sample Problem 3-1, where $\tan \theta = 1.5$. Taking $\tan^{-1} 1.5$ on your calculator tells you that $\theta = 56°$, but $\theta = 236°$ ($= 180° + 56°$) also has a tangent of 1.5. Which is correct? From the physical situation (Fig. 3-2b), 56° is reasonable and 236° is clearly not.

TACTIC 4: *Measuring Vector Angles*

The equations for $\cos \theta$ and $\sin \theta$ in Eq. 3-5 and the equation for $\tan \theta$ in Eq. 3-6 are valid only if the angle is measured relative to the positive direction of the x axis. If it is measured relative to some other direction, then the trig functions in Eq. 3-5 may have to be interchanged, and the ratio in Eq. 3-6 may have to be inverted. A safer method is to convert the given angle into one that is measured from the positive direction of the x axis, as shown in Sample Problem 3-3.

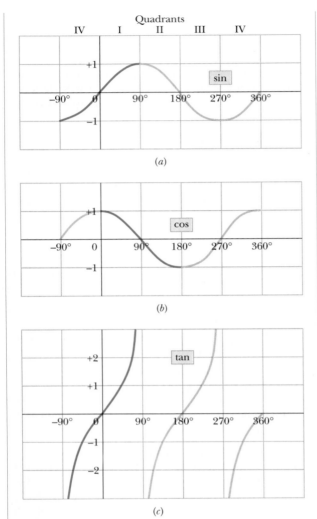

FIGURE 3-13   Three useful curves to remember. A calculator's range of operation for taking *inverse* trig functions is indicated by the darker portions of the colored curves.

3-4 UNIT VECTORS

A **unit vector** is a vector that has a magnitude of exactly 1 and points in a particular direction. It lacks both dimension and unit. Its sole purpose is to point, that is, to specify a direction. The unit vectors in the positive directions of the x, y, and z axes are labeled **i**, **j**, and **k**, as shown in Fig. 3-14.* The arrangement of axes in Fig. 3-14 is said to be a **right-handed coordinate system.** The system remains right-handed if it is rotated rigidly to a new orientation. We use such coordinate systems exclusively in this text.

*To distinguish handwritten unit vectors from other symbols, including other vectors, you might top them with a "hat": $\hat{i}, \hat{j}, \hat{k}$.

FIGURE 3-14   Unit vectors **i**, **j**, and **k** define the directions of a right-handed rectangular coordinate system.

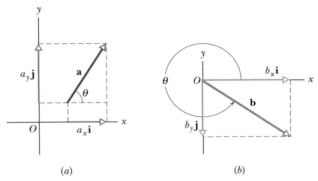

FIGURE 3-15   (*a*) The vector components of vector **a**. (*b*) The vector components of vector **b**.

Unit vectors are very useful for expressing other vectors; for example, we can express **a** and **b** of Figs. 3-9 and 3-10 as

$$\mathbf{a} = a_x\mathbf{i} + a_y\mathbf{j} \tag{3-7}$$

and

$$\mathbf{b} = b_x\mathbf{i} + b_y\mathbf{j}. \tag{3-8}$$

These two equations are illustrated in Fig. 3-15. The quantities $a_x\mathbf{i}$ and $a_y\mathbf{j}$ are the **vector components** of **a**, in contrast to a_x and a_y, which are its **scalar components** (or, as before, simply its **components**).

Look back for a moment at Sample Problem 3-1. If you superimpose the coordinate system of Fig. 3-14 at the Austin Entrance of Fig. 3-2a, with **i** eastward, **j** northward, and **k** upward, then the displacement **d** to Echo River is neatly expressed as

$$\mathbf{d} = -(2.6 \text{ km})\mathbf{i} - (3.9 \text{ km})\mathbf{j} + (0.025 \text{ km})\mathbf{k}.$$

3-5 ADDING VECTORS BY COMPONENTS

As we have noted, adding vectors graphically is tedious; it also has limited accuracy and is challenging in three dimensions. Here we find a more direct technique—adding vectors by combining their components, axis by axis.

To start, consider the statement

$$\mathbf{r} = \mathbf{a} + \mathbf{b}, \tag{3-9}$$

which says that the vector **r** is the same as the vector (**a** + **b**). If that is so, then each component of **r** must be the same as the corresponding component of (**a** + **b**):

$$r_x = a_x + b_x, \tag{3-10}$$

$$r_y = a_y + b_y, \tag{3-11}$$

$$r_z = a_z + b_z. \tag{3-12}$$

In other words, two vectors are equal if their corresponding components are equal. Equations 3-10 to 3-12 tell us that to add vectors **a** and **b**, we must (1) resolve the vectors into their scalar components; (2) axis by axis, combine these scalar components to get the components of the sum **r**; and (3) if necessary, combine the components of **r** to get **r** itself. We have a choice in step 3. We can express **r** in unit-vector notation, or we can give the magnitude of **r** and its direction (by stating one angle when we are working in two dimensions or two angles when we are working in three dimensions).

$\mathcal{C}$HECKPOINT **3:** (a) In the figure, what are the signs of the x components of d_1 and d_2? (b) What are the signs of the y components of d_1 and d_2? (c) What are the signs of the x and y components of $d_1 + d_2$?

Next, we use Eq. 3-6 to find the magnitude and direction of **d**:

$$d = \sqrt{d_x^2 + d_y^2} = \sqrt{(18.3 \text{ km})^2 + (59.7 \text{ km})^2}$$

and $= 62 \text{ km}$ (Answer)

$$\theta = \tan^{-1} \frac{d_y}{d_x} = \tan^{-1} \frac{59.7 \text{ km}}{18.3 \text{ km}} = 73°, \quad \text{(Answer)}$$

where θ is the angle shown in Fig. 3-16.

(b) Write **d** in unit-vector notation.

SOLUTION: We simply write **d** as

$$\mathbf{d} = (x \text{ component})\mathbf{i} + (y \text{ component})\mathbf{j}$$

$$= (18.3 \text{ km})\mathbf{i} + (59.7 \text{ km})\mathbf{j}. \quad \text{(Answer)}$$

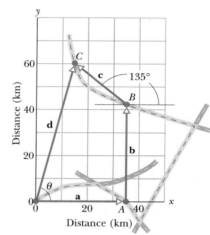

FIGURE 3-16 Sample Problem 3-4. A rally route, showing the origin, checkpoints Able (*A*), Baker (*B*), and Charlie (*C*), and the road network.

SAMPLE PROBLEM 3-4

In a road rally, you are given the following instructions: from the starting point use available roads to drive 36 km due east to checkpoint "Able," then 42 km due north to "Baker," and then 25 km northwest to "Charlie." (The roadway and checkpoints are shown in Fig. 3-16.)

(a) At "Charlie," what are the magnitude and orientation of your displacement **d** from the starting point?

SOLUTION: Figure 3-16 also shows a convenient orientation for an *xy* coordinate system, as well as vectors representing the three displacements you have undergone. The scalar components of **d** are

$$d_x = a_x + b_x + c_x = 36 \text{ km} + 0 + (25 \text{ km})(\cos 135°)$$

$$= (36 + 0 - 17.7) \text{ km} = 18.3 \text{ km}$$

and

$$d_y = a_y + b_y + c_y = 0 + 42 \text{ km} + (25 \text{ km})(\sin 135°)$$

$$= (0 + 42 + 17.7) \text{ km} = 59.7 \text{ km}.$$

SAMPLE PROBLEM 3-5

Here are three vectors, each expressed in unit-vector notation:

$$\mathbf{a} = 4.2\mathbf{i} - 1.6\mathbf{j},$$

$$\mathbf{b} = -1.6\mathbf{i} + 2.9\mathbf{j},$$

$$\mathbf{c} = -3.7\mathbf{j}.$$

All three lie in the *xy* plane, and none of them has a *z* component. Find the vector **r** that is the sum of these three vectors. For convenience, the units have been omitted from these vector expressions; you may take them to be meters.

SOLUTION: From Eqs. 3-10 and 3-11 we have

$$r_x = a_x + b_x + c_x = 4.2 - 1.6 + 0 = 2.6$$

and

$$r_y = a_y + b_y + c_y = -1.6 + 2.9 - 3.7 = -2.4.$$

Thus
$$\mathbf{r} = 2.6\mathbf{i} - 2.4\mathbf{j}. \qquad \text{(Answer)}$$
Figure 3-17a shows the three vectors and their sum. Figure 3-17b shows $\mathbf{r}$ and its vector components.

(a)

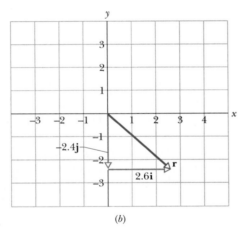

(b)

FIGURE 3-17   Sample Problem 3-5. Vector $\mathbf{r}$ is the vector sum of the other three vectors.

3-6 VECTORS AND THE LAWS OF PHYSICS

So far, in every figure that includes a coordinate system, the x and y axes are parallel to the edges of the book page. And so, when a vector $\mathbf{a}$ is included, its components a_x and a_y are also parallel to the edges (as in Fig. 3-18a). The only reason for that orientation of the axes is that it looks proper: there is no deeper reason. We could, instead, rotate the axes (but not the vector $\mathbf{a}$) through an angle ϕ as in Fig. 3-18b, in which case the components would have new values, call them a_x' and a_y'. Since there are an infinite number of choices of ϕ, there are an infinite number of different pairs of components for $\mathbf{a}$.

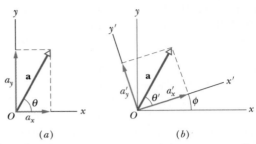

FIGURE 3-18   (a) The vector $\mathbf{a}$ and its components. (b) The same vector, with the axes of the coordinate system rotated through an angle ϕ.

Which then is the "right" pair of components? The answer is that they are all equally valid because each pair (with its axes) just gives us a different way of describing the same vector $\mathbf{a}$; all produce the same magnitude and direction for the same vector. In Fig. 3-18 we have
$$a = \sqrt{a_x^2 + a_y^2} = \sqrt{a_x'^2 + a_y'^2} \qquad (3\text{-}13)$$
and
$$\theta = \theta' + \phi. \qquad (3\text{-}14)$$

The point is that we have great freedom in choosing a coordinate system, because the relations among vectors (including, for example, the vector addition of Eq. 3-1) do not depend on the location of the origin of the coordinate system or on the orientation of the axes. This is also true of the relations of physics; they are all independent of the choice of coordinate system. Add to that the simplicity and richness of the language of vectors and you can see why the laws of physics are almost always presented in that language: one equation, like Eq. 3-9, can represent three (or even more) relations, like Eqs. 3-10, 3-11, and 3-12.

3-7 MULTIPLYING VECTORS*

There are three ways in which vectors can be multiplied. None is exactly like the usual algebraic multiplication.

Multiplying a Vector by a Scalar

If we multiply a vector $\mathbf{a}$ by a scalar s, we get a new vector. Its magnitude is the product of the magnitude of $\mathbf{a}$ and the absolute value of s. Its direction is the direction of $\mathbf{a}$ if s is positive, but the opposite direction if s is negative. To divide $\mathbf{a}$ by s, we multiply $\mathbf{a}$ by $1/s$.

In either multiplication or division, the scalar may be a pure number or a physical quantity; in the latter case, the

---

*This material will not be employed until later (Chapter 7 for scalar products and Chapter 12 for vector products), and so your instructor may wish to postpone assignment of the section.

physical nature of the product differs from that of the original vector **a**.

The Scalar Product

There are two ways to multiply a vector by a vector: one way produces a scalar and the other produces a new vector. Students commonly confuse the two ways, and so starting now, you should carefully distinguish between them.

The **scalar product** of the vectors **a** and **b** in Fig. 3-19a is written as **a · b** and defined to be

$$\mathbf{a} \cdot \mathbf{b} = ab \cos \phi, \qquad (3\text{-}15)$$

where a is the magnitude of **a**, b is the magnitude of **b**, and ϕ is the angle* between **a** and **b**. Note that there are only scalars on the right side of Eq. 3-15 (including the value of $\cos \phi$). Thus **a · b** on the left side represents a *scalar* quantity; because of the notation, it is also known as the **dot product** and is spoken as ''a dot b.''

A dot product can be regarded as the product of two quantities: (1) the magnitude of one of the vectors and (2) the scalar component of the second vector along the direction of the first vector. For example, in Fig. 3-19b, **a** has a scalar component $a \cos \phi$ along the direction of **b**; note that a perpendicular dropped from the head of **a** to **b** determines that component. Similarly, **b** has a scalar component $b \cos \phi$ along the direction of **a**. If ϕ is 0°, the component of one vector along the other is maximum, and so also is

(a)

Component of **b** along direction of **a** is $b \cos \phi$

Component of **a** along direction of **b** is $a \cos \phi$

(b)

FIGURE 3-19 (a) Two vectors **a** and **b**, with an angle ϕ between them. (b) Each vector has a component along the direction of the other vector.

*In Fig. 3-19a, there are actually two angles between the vectors: ϕ and $360° - \phi$. Either can be used in Eq. 3-15, because their cosines are the same.

the dot product. If, instead, ϕ is 90°, the component of one vector along the other is zero, and so is the dot product.

Equation 3-15 can be rewritten as follows to emphasize the components:

$$\mathbf{a} \cdot \mathbf{b} = (a \cos \phi)(b) = (a)(b \cos \phi). \qquad (3\text{-}16)$$

The commutative law applies to a scalar product, so we can write

$$\mathbf{a} \cdot \mathbf{b} = \mathbf{b} \cdot \mathbf{a}.$$

When two vectors are in unit-vector notation, we write their dot product as

$$\mathbf{a} \cdot \mathbf{b} = (a_x\mathbf{i} + a_y\mathbf{j} + a_z\mathbf{k}) \cdot (b_x\mathbf{i} + b_y\mathbf{j} + b_z\mathbf{k}), \qquad (3\text{-}17)$$

for which the distributive law applies: each component of the first vector is to be dotted into each component of the second vector.

Cʜᴇᴄᴋᴘᴏɪɴᴛ **4:** Vectors **C** and **D** have magnitudes of 3 units and 4 units, respectively. What is the angle between **C** and **D** if **C · D** equals (a) zero, (b) 12 units, and (c) −12 units?

SAMPLE PROBLEM 3-6

What is the angle ϕ between $\mathbf{a} = 3.0\mathbf{i} - 4.0\mathbf{j}$ and $\mathbf{b} = -2.0\mathbf{i} + 3.0\mathbf{k}$?

SOLUTION: By Eq. 3-15, the dot product is

$$\mathbf{a} \cdot \mathbf{b} = ab \cos \phi = \sqrt{3.0^2 + 4.0^2} \sqrt{2.0^2 + 3.0^2} \cos \phi$$
$$= 18.0 \cos \phi. \qquad (3\text{-}18)$$

We next find the dot product with Eq. 3-17:

$$\mathbf{a} \cdot \mathbf{b} = (3.0\mathbf{i} - 4.0\mathbf{j}) \cdot (-2.0\mathbf{i} + 3.0\mathbf{k}).$$

With the distributive law, this yields

$$\mathbf{a} \cdot \mathbf{b} = (3.0\mathbf{i}) \cdot (-2.0\mathbf{i}) + (3.0\mathbf{i}) \cdot (3.0\mathbf{k})$$
$$+ (-4.0\mathbf{j}) \cdot (-2.0\mathbf{i}) + (-4.0\mathbf{j}) \cdot (3.0\mathbf{k}).$$

We next apply Eq. 3-15 to each term. The angle between vectors for the first term is 0°, and for the other three terms it is 90°. We then have

$$\mathbf{a} \cdot \mathbf{b} = -(6.0)(1) + (9.0)(0) + (8.0)(0) - (12)(0)$$
$$= -6.0. \qquad (3\text{-}19)$$

Setting the results of Eqs. 3-18 and 3-19 equal to each other, we find

$$18.0 \cos \phi = -6.0$$

or

$$\phi = \cos^{-1} \frac{-6.0}{18.0} = 109° \approx 110°. \qquad \text{(Answer)}$$

The Vector Product

The **vector product** of **a** and **b**, written **a** × **b**, produces a third vector **c** whose magnitude is

$$c = ab \sin \phi, \tag{3-20}$$

where ϕ is the *smaller* of the two angles between **a** and **b**.* Because of the notation, **a** × **b** is also known as the **cross product** and in speech it is "a cross b." If **a** and **b** are parallel or antiparallel, **a** × **b** = 0. The magnitude of **a** × **b**, which can be written as |**a** × **b**|, is maximum when **a** and **b** are perpendicular to each other.

The direction of **c** is perpendicular to the plane that contains **a** and **b**. Figure 3-20a shows how to determine the direction of **c** = **a** × **b** with what is known as the **right-hand rule.** Place the vectors **a** and **b** tail to tail without altering their orientations and imagine a line that is perpendicular to their plane where they meet. Pretend to grasp the line with your *right* hand in such a way that your fingers would sweep **a** into **b** through the smaller angle between them. Your outstretched thumb points in the direction of **c**.

The order of the vector multiplication is important. In Fig. 3-20b, we are determining the direction of **c**′ = **b** × **a**, so the fingers are placed to sweep **b** into **a** through the smaller angle. The thumb ends up in the opposite direction from previously, and so it must be that **c**′ = −**c**; that is,

$$\mathbf{b} \times \mathbf{a} = -(\mathbf{a} \times \mathbf{b}). \tag{3-21}$$

In other words, the commutative law does not apply to a vector product.

In unit-vector notation, we write

$$\mathbf{a} \times \mathbf{b} = (a_x\mathbf{i} + a_y\mathbf{j} + a_z\mathbf{k})$$
$$\times (b_x\mathbf{i} + b_y\mathbf{j} + b_z\mathbf{k}), \tag{3-22}$$

and the distributive law applies: each component of the first vector is to be crossed into each component of the second vector. You should use that law to expand Eq. 3-22; check your result against the expansion given in Appendix E. You can also evaluate a cross product by setting up and evaluating a determinant, as shown in Appendix E.

We can use a cross product of unit vectors to see if any given rectangular coordinate system is a right-handed coordinate system: if **i** × **j** = **k**, it is. Check the system shown in Fig. 3-14.

CHECKPOINT 5: Vectors **C** and **D** have magnitudes of 3 units and 4 units, respectively. What is the angle between **C** and **D** if the magnitude of the vector product **C** × **D** equals (a) zero and (b) 12 units?

---

*You must use the smaller of the two angles between the vectors because $\sin \phi$ and $\sin(360° - \phi)$ differ in algebraic sign.

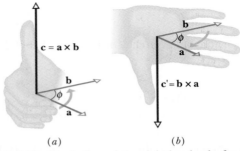

(a)   (b)

FIGURE 3-20 Illustration of the right-hand rule for vector products. (a) Swing vector **a** into vector **b** with the fingers of your right hand. Your thumb shows the direction of vector **c** = **a** × **b**. (b) Showing that (**a** × **b**) = −(**b** × **a**).

SAMPLE PROBLEM 3-7

Vector **a** lies in the xy plane in Fig. 3-21. It has a magnitude of 18 units and points in a direction 250° from the direction of increasing x. Vector **b** has a magnitude of 12 units and points along the direction of increasing z.

(a) What is the scalar product of these two vectors?

SOLUTION: The angle ϕ between these two vectors is 90° so that, from Eq. 3-15,

$$\mathbf{a} \cdot \mathbf{b} = ab \cos \phi = (18)(12)(\cos 90°) = 0. \quad \text{(Answer)}$$

The scalar product of any two vectors that are at right angles to each other is zero. This is consistent with the fact that neither of these vectors has a component along the direction of the other vector.

(b) What is the vector product **c** = **a** × **b**?

SOLUTION: The magnitude of the vector product is, from Eq. 3-20,

$$ab \sin \phi = (18)(12)(\sin 90°) = 216. \quad \text{(Answer)}$$

The direction of **c** is perpendicular to the plane formed by **a** and **b**. It must then be perpendicular to **b**, which means that it is perpendicular to the z axis. This in turn means that **c** must lie in the xy plane. The right-hand rule of Fig. 3-20 shows that **c** points as shown in Fig. 3-21. Because **c** is also perpendicular to **a**, the direction of **c** makes an angle of 250° − 90° = 160° with the direction of increasing x.

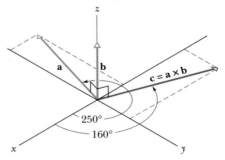

FIGURE 3-21 Sample Problem 3-7. A problem in vector multiplication.

SAMPLE PROBLEM 3-8

If $\mathbf{c} = \mathbf{a} \times \mathbf{b}$, where $\mathbf{a} = 3\mathbf{i} - 4\mathbf{j}$ and $\mathbf{b} = -2\mathbf{i} + 3\mathbf{k}$, what is $\mathbf{c}$?

SOLUTION: From Eq. 3-22 we have

$$\mathbf{c} = (3\mathbf{i} - 4\mathbf{j}) \times (-2\mathbf{i} + 3\mathbf{k}),$$

which, by the distributive law, becomes

$$\mathbf{c} = -(3\mathbf{i} \times 2\mathbf{i}) + (3\mathbf{i} \times 3\mathbf{k}) + (4\mathbf{j} \times 2\mathbf{i}) - (4\mathbf{j} \times 3\mathbf{k}).$$

We next evaluate each term with Eq. 3-20, determining the direction with the right-hand rule. (For the first term, $\phi = 0°$; for the other terms, $\phi = 90°$.) We find

$$\mathbf{c} = 0 - 9\mathbf{j} - 8\mathbf{k} - 12\mathbf{i} = -12\mathbf{i} - 9\mathbf{j} - 8\mathbf{k}. \quad \text{(Answer)}$$

The vector $\mathbf{c}$ is perpendicular to both $\mathbf{a}$ and $\mathbf{b}$, a fact that you can check by showing $\mathbf{c} \cdot \mathbf{a} = 0$ and $\mathbf{c} \cdot \mathbf{b} = 0$; that is, there is no component of $\mathbf{c}$ along the direction of either $\mathbf{a}$ or $\mathbf{b}$.

PROBLEM SOLVING TACTICS

TACTIC 5: *Common Errors with Cross Products*

Several errors are common in finding a cross product. (1) Failure to arrange vectors tail to tail is tempting when an illustration presents them head to tail: you must mentally shift (or better, redraw) one vector to the proper arrangement without changing its orientation. (2) Failing to use the right hand in applying the right-hand rule is easy when the right hand is occupied with a calculator or pencil. (3) Failure to sweep the first vector of the product into the second vector can occur when the orientations of the vectors require an awkward twisting of your hand to apply the right-hand rule. And sometimes that happens when you try to make the sweep mentally rather than actually using your hand. (4) Failure to work with a right-handed coordinate system results when you forget how to draw such a system (see Fig. 3-14).

REVIEW & SUMMARY

Scalars and Vectors

Scalars, such as temperature, have magnitude only. They are specified by a number with a unit (82°F) and obey the rules of arithmetic and ordinary algebra. *Vectors,* such as displacement, have both magnitude and direction (5 m, north) and obey the special rules of vector algebra.

Adding Vectors Geometrically

Two vectors $\mathbf{a}$ and $\mathbf{b}$ may be added geometrically by drawing them to a common scale and placing them head to tail. The vector connecting the tail of the first to the head of the second is the sum vector $\mathbf{s}$, as Fig. 3-3 shows. To subtract $\mathbf{b}$ from $\mathbf{a}$, reverse the direction of $\mathbf{b}$ to get $-\mathbf{b}$; then add $-\mathbf{b}$ to $\mathbf{a}$: see Fig. 3-7. Vector addition is commutative and obeys the associative law.

Components of a Vector

The (scalar) *components* a_x and a_y of any two-dimensional vector $\mathbf{a}$ are found by dropping perpendicular lines from the ends of $\mathbf{a}$ onto the coordinate axes. The components are given by

$$a_x = a \cos \theta \quad \text{and} \quad a_y = a \sin \theta, \quad (3-5)$$

where θ is the angle from the positive direction of the x axis to the direction of $\mathbf{a}$. The algebraic sign of a component indicates its direction along the associated axis. Given the components, we can reconstruct the vector from

$$a = \sqrt{a_x^2 + a_y^2} \quad \text{and} \quad \tan \theta = \frac{a_y}{a_x}. \quad (3-6)$$

Unit-Vector Notation

Often we find it useful to introduce *unit vectors* $\mathbf{i}$, $\mathbf{j}$, and $\mathbf{k}$, whose magnitudes are unity and whose directions are those of the x, y,

and z axes, respectively, in a right-handed coordinate system. We can write a vector $\mathbf{a}$ in terms of unit vectors as

$$\mathbf{a} = a_x\mathbf{i} + a_y\mathbf{j} + a_z\mathbf{k}, \quad (3-7)$$

in which $a_x\mathbf{i}$, $a_y\mathbf{j}$, and $a_z\mathbf{k}$ are the **vector components** and a_x, a_y, and a_z are the **scalar components** of $\mathbf{a}$.

Adding Vectors in Component Form

To add vectors in component form, we use the rules

$$r_x = a_x + b_x; \quad r_y = a_y + b_y; \quad r_z = a_z + b_z. \quad (3\text{-}10 \text{ to } 3\text{-}12)$$

Here $\mathbf{a}$ and $\mathbf{b}$ are the vectors to be added, and $\mathbf{r}$ is the vector sum.

Vectors and Physical Laws

Any physical situation involving vectors can be described using many possible coordinate systems. We usually choose the one that simplifies the work. However, the relation between the vector quantities does not depend on our choice. The laws of physics are also independent of our choice of coordinates.

Scalar Times Vector

The product of a scalar s and a vector $\mathbf{v}$ is a new vector whose magnitude is sv and whose direction is the same as that of $\mathbf{v}$ if s is positive, and opposite that of $\mathbf{v}$ if s is negative. To divide $\mathbf{v}$ by s, multiply $\mathbf{v}$ by $(1/s)$.

The Scalar Product

The scalar (or **dot**) product of two vectors $\mathbf{a}$ and $\mathbf{b}$ is written $\mathbf{a} \cdot \mathbf{b}$ and is the *scalar* quantity given by

$$\mathbf{a} \cdot \mathbf{b} = ab \cos \phi, \quad (3-15)$$

in which ϕ is the angle between the directions of **a** and **b**. The scalar product may be positive, zero, or negative, depending on the value of ϕ. Figure 3-19b indicates that a scalar product is the product of the magnitude of one vector and the component of the second vector along the direction of the first vector. In unit-vector notation, we have

$$\mathbf{a}\cdot\mathbf{b} = (a_x\mathbf{i} + a_y\mathbf{j} + a_z\mathbf{k})\cdot(b_x\mathbf{i} + b_y\mathbf{j} + b_z\mathbf{k}), \quad (3\text{-}17)$$

which obeys the distributive law. Note that $\mathbf{a}\cdot\mathbf{b} = \mathbf{b}\cdot\mathbf{a}$.

The Vector Product

The vector (or **cross**) product of two vectors **a** and **b** is written

$\mathbf{a} \times \mathbf{b}$ and is a *vector* **c** whose magnitude c is given by

$$c = ab \sin \phi, \quad (3\text{-}20)$$

in which ϕ is the smaller of the angles between the directions of **a** and **b**. The direction of **c** is perpendicular to the plane defined by **a** and **b** and is given by a right-hand rule, as shown in Fig. 3-20. Note that $\mathbf{a} \times \mathbf{b} = -(\mathbf{b} \times \mathbf{a})$. In unit-vector notation we have

$$\mathbf{a} \times \mathbf{b} = (a_x\mathbf{i} + a_y\mathbf{j} + a_z\mathbf{k}) \times (b_x\mathbf{i} + b_y\mathbf{j} + b_z\mathbf{k}), \quad (3\text{-}22)$$

to which we may apply the distributive law.

QUESTIONS

1. Displacement vector **N** points from coordinates (5 m, 3 m) to coordinates (7 m, 6 m) in the xy plane. Which of the following displacement vectors are equivalent to **N**: vector **A**, which points from (−6 m, −5 m) to (−4 m, −2 m); vector **B**, which points from (−6 m, 1 m) to (−4 m, 4 m); vector **C**, which points from (−8 m, −6 m) to (−10 m, −9 m)?

2. If $\mathbf{d} = \mathbf{a} + \mathbf{b} + (-\mathbf{c})$, does (a) $\mathbf{a} + (-\mathbf{d}) = \mathbf{c} + (-\mathbf{b})$, (b) $\mathbf{a} = (-\mathbf{b}) + \mathbf{d} + \mathbf{c}$, and (c) $\mathbf{c} + (-\mathbf{d}) = \mathbf{a} + \mathbf{b}$?

3. Equation 3-2 shows that the addition of two vectors **a** and **b** is commutative. Does that mean subtraction is commutative, so that $\mathbf{a} - \mathbf{b} = \mathbf{b} - \mathbf{a}$?

4. In a game within a three-dimensional maze, you must move your game piece from *start,* at xyz coordinates (0, 0, 0), to *finish,* at coordinates (−2 cm, 4 cm, −4 cm). The game piece can undergo only the displacements (in centimeters) given below. If, along the way, the game piece lands at coordinates (−5 cm, −1 cm, −1 cm) or (5 cm, 2 cm, −1 cm), you lose the game. Which vectors and in what sequence will get your game piece to *finish*?

$$\mathbf{p} = -7\mathbf{i} + 2\mathbf{j} - 3\mathbf{k} \qquad \mathbf{r} = 2\mathbf{i} - 3\mathbf{j} + 2\mathbf{k}$$
$$\mathbf{q} = 2\mathbf{i} - \mathbf{j} + 4\mathbf{k} \qquad \mathbf{s} = 3\mathbf{i} + 5\mathbf{j} - 3\mathbf{k}.$$

5. Describe two vectors **a** and **b** such that
(a) $\mathbf{a} + \mathbf{b} = \mathbf{c}$ and $a + b = c$;
(b) $\mathbf{a} + \mathbf{b} = \mathbf{a} - \mathbf{b}$;
(c) $\mathbf{a} + \mathbf{b} = \mathbf{c}$ and $a^2 + b^2 = c^2$.

6. (a) Can a vector have zero magnitude if one of its components is not zero? (b) Can two vectors having different magnitudes be combined to give a vector sum of zero? Can three vectors give a vector sum of zero if they (c) are in the same plane, and (d) are not in the same plane?

7. Can the magnitude of the difference between two vectors ever be greater than (a) the magnitude of one of the vectors, (b) the magnitudes of both vectors, and (c) the magnitude of their sum?

8. Can the sum of the magnitudes of two vectors ever be equal to the magnitude of the sum of the same two vectors? If no, why not? If yes, when?

9. Which of the arrangements of axes in Fig. 3-22 can be labeled

"right-handed coordinate system"? As usual, each axis label indicates the positive side of the axis.

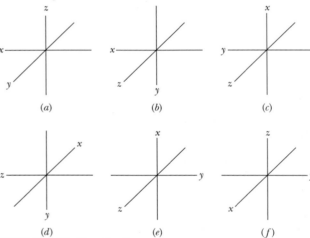

FIGURE 3-22   Question 9.

10. The x and y components of four vectors, **a**, **b**, **c**, and **d** are given below. For which vectors will your calculator give you the correct angle θ when you use it to find θ with Eq. 3-6? Answer first by examining Fig. 3-13, and then check your answers with your calculator.

$$a_x = 3 \qquad a_y = 3 \qquad c_x = -3 \qquad c_y = -3$$
$$b_x = -3 \qquad b_y = 3 \qquad d_x = 3 \qquad d_y = -3.$$

11. In Fig. 3-23, what are the signs of the x and y components of (a) **A**, and (b) **D**, where $\mathbf{D} = \mathbf{A} - \mathbf{B}$?

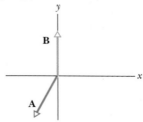

FIGURE 3-23   Question 11.

12. Figure 3-24 shows vector **A** and four choices for vector **B** that differ only in orientation. Rank the choices according to the absolute value of **A**·**B** that they give, greatest first.

FIGURE 3-24
Question 12.

13. Figure 3-25 shows vector **A** and four other vectors that have the same magnitude but differ in orientation. (a) Which of those other four vectors have the same dot product with **A**? (b) Which have a negative dot product with **A**?

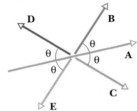

FIGURE 3-25 Question 13.

14. If **A** = 2**i** + 4**j** and **B** = 7**k,** what is **A**·**B**?

15. If **a**·**b** = **a**·**c**, must **b** equal **c**?

16. In Fig. 3-26, what is the direction of (a) **E** × **F**, (b) **F** × **E**, and (c) **G** × **E**, where **G** is in the *xy* plane? (d) Does the answer to (c) change if **G** is shifted parallel to the *z* axis without any change in its direction?

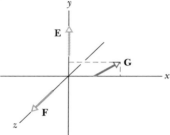

FIGURE 3-26 Question 16.

17. If **A** = 2**i** + 4**j**, what is **A** × **B** when (a) **B** = 8**i** + 16**j** and (b) **B** = − 8**i** − 16**j**? (This question can be answered without computation.)

<hr>

EXERCISES **&** PROBLEMS

SECTION 3-2 Adding Vectors: Graphical Method

1E. Consider two displacements, one of magnitude 3 m and another of magnitude 4 m. Show how the displacement vectors may be combined to get a resultant displacement of magnitude (a) 7 m, (b) 1 m, and (c) 5 m.

2E. A woman walks 250 m in the direction 30° east of north, then 175 m directly east. (a) Using graphical methods, find her final displacement from the starting point. (b) Compare the magnitude of her displacement with the distance she walked.

3E. A person walks in the following pattern: 3.1 km north, then 2.4 km west, and finally 5.2 km south. (a) Construct the vector diagram that represents this motion. (b) How far and in what direction would a bird fly in a straight line from the same starting point to the same final point?

4E. A car is driven east for a distance of 50 km, then north for 30 km, and then in a direction 30° east of north for 25 km. Draw the vector diagram and determine the total displacement of the car from its starting point.

5P. Vector **a** has a magnitude of 5.0 units and is directed east. Vector **b** is directed 35° west of north and has a magnitude of 4.0 units. Construct vector diagrams for calculating **a** + **b** and **b** − **a**. Estimate the magnitudes and directions of **a** + **b** and **b** − **a** from your diagrams.

6P. A bank in downtown Boston is robbed (see the map in Fig. 3-27). To elude police, the robbers escape by helicopter, making three successive flights described by the following displacements: 20 miles, 45° south of east; 33 miles, 26° north of west; 16

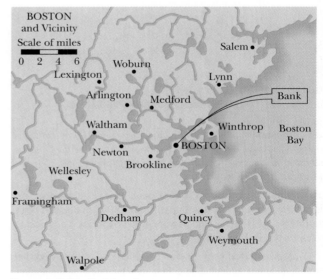

FIGURE 3-27 Problem 6.

miles, 18° east of south. At the end of the third flight they are captured. In what town are they apprehended? (Use the graphical method to add these displacements on the map.)

7P. Three vectors **a**, **b**, and **c**, each having a magnitude of 50 units, lie in the xy plane and make angles of 30°, 195°, and 315° with the positive x axis, respectively. Find graphically the magnitudes and directions of the vectors (a) **a** + **b** + **c**, (b) **a** − **b** + **c**, and (c) a vector **d** such that (**a** + **b**) − (**c** + **d**) = 0.

SECTION 3-3 Vectors and Their Components

8E. (a) Express the following angles in radians: 20.0°, 50.0°, 100°. (b) Convert the following angles to degrees: 0.330 rad, 2.10 rad, 7.70 rad.

9E. What are the x and y components of a vector **a** in the xy plane if its direction is 250° counterclockwise from the positive x axis and its magnitude is 7.3 units?

10E. The x component of a certain vector is −25.0 units and the y component is +40.0 units. (a) What is the magnitude of the vector? (b) What is the angle between the direction of the vector and the positive direction of x?

11E. A displacement vector **r** in the xy plane is 15 m long and directed as shown in Fig. 3-28. Determine the x and y components of the vector.

FIGURE 3-28
Exercise 11.

12E. A heavy piece of machinery is raised by sliding it 12.5 m along a plank oriented at 20.0° to the horizontal, as shown in Fig. 3-29. (a) How high above its original position is it raised? (b) How far is it moved horizontally?

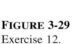

FIGURE 3-29
Exercise 12.

13E. The minute hand of a wall clock measures 10 cm from axis to tip. What is the displacement vector of its tip (a) from a quarter after the hour to half past, (b) in the next half hour, and (c) in the next hour?

14E. A ship sets out to sail to a point 120 km due north. An unexpected storm blows the ship to a point 100 km due east of its starting point. How far, and in what direction, must it now sail to reach its original destination?

15P. A person desires to reach a point that is 3.40 km from her present location and in a direction that is 35.0° north of east. However, she must travel along streets that are oriented either

north–south or east–west. What is the minimum distance she could travel to reach her destination?

16P. Rock *faults* are ruptures along which opposite faces of rock have slid past each other. In Fig. 3-30, points A and B coincided before the rock in the foreground slid down to the right. The net displacement AB is along the plane of the fault. The horizontal component of AB is the *strike-slip* AC. The component of AB that is directly down the plane of the fault is the *dip-slip* AD. (a) What is the net displacement AB if the strike-slip is 22.0 m and the dip-slip is 17.0 m? (b) If the plane of the fault is inclined 52.0° to the horizontal, what is the vertical component of AB?

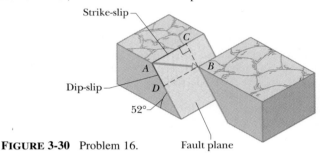

FIGURE 3-30 Problem 16.

17P. A wheel with a radius of 45.0 cm rolls without slipping along a horizontal floor (Fig. 3-31). At time t_1, the dot P painted on the rim of the wheel is at the point of contact between the wheel and the floor. At a later time t_2, the wheel has rolled through one-half of a revolution. What is the displacement of P during this interval?

FIGURE 3-31 Problem 17.

18P. A room has dimensions 10.0 ft (height) × 12.0 ft × 14.0 ft. A fly starting at one corner flies around, ending up at the diagonally opposite corner. (a) What is the magnitude of its displacement? (b) Could the length of its path be less than this distance? Greater than this distance? Equal to this distance? (c) Choose a suitable coordinate system and find the components of the displacement vector in that system. (d) If the fly walks rather than flies, what is the length of the shortest path it can take? (This can be answered without calculus.)

SECTION 3-5 Adding Vectors by Components

19E. Find the vector components of the sum **r** of the vector displacements **c** and **d** whose components in meters along three perpendicular directions are $c_x = 7.4$, $c_y = -3.8$, $c_z = -6.1$; $d_x = 4.4$, $d_y = -2.0$, $d_z = 3.3$.

20E. (a) What is the sum in unit-vector notation of the two vectors **a** = 4.0**i** + 3.0**j** and **b** = −13**i** + 7.0**j**? (b) What are the magnitude and direction of **a** + **b**?

21E. Calculate the x and y components, magnitudes, and directions of (a) $\mathbf{a} + \mathbf{b}$ and (b) $\mathbf{b} - \mathbf{a}$ if $\mathbf{a} = 3.0\mathbf{i} + 4.0\mathbf{j}$ and $\mathbf{b} = 5.0\mathbf{i} - 2.0\mathbf{j}$.

22E. Two vectors are given by $\mathbf{a} = 4\mathbf{i} - 3\mathbf{j} + \mathbf{k}$ and $\mathbf{b} = -\mathbf{i} + \mathbf{j} + 4\mathbf{k}$. Find (a) $\mathbf{a} + \mathbf{b}$, (b) $\mathbf{a} - \mathbf{b}$, and (c) a vector $\mathbf{c}$ such that $\mathbf{a} - \mathbf{b} + \mathbf{c} = 0$.

23E. Given two vectors $\mathbf{a} = 4.0\mathbf{i} - 3.0\mathbf{j}$ and $\mathbf{b} = 6.0\mathbf{i} + 8.0\mathbf{j}$, find the magnitudes and directions of (a) $\mathbf{a}$, (b) $\mathbf{b}$, (c) $\mathbf{a} + \mathbf{b}$, (d) $\mathbf{b} - \mathbf{a}$, and (e) $\mathbf{a} - \mathbf{b}$. How do the orientations of the last two compare?

24P. If $\mathbf{a} - \mathbf{b} = 2\mathbf{c}$, $\mathbf{a} + \mathbf{b} = 4\mathbf{c}$, and $\mathbf{c} = 3\mathbf{i} + 4\mathbf{j}$, then what are $\mathbf{a}$ and $\mathbf{b}$?

25P. If vector $\mathbf{B}$ is added to vector $\mathbf{A}$, the result is $6.0\mathbf{i} + 1.0\mathbf{j}$. If $\mathbf{B}$ is subtracted from $\mathbf{A}$, the result is $-4.0\mathbf{i} + 7.0\mathbf{j}$. What is the magnitude of $\mathbf{A}$?

26P. A vector $\mathbf{B}$, when added to the vector $\mathbf{C} = 3.0\mathbf{i} + 4.0\mathbf{j}$, yields a resultant vector that is in the positive y direction and has a magnitude equal to that of $\mathbf{C}$. What is the magnitude of $\mathbf{B}$?

27P. Two vectors $\mathbf{a}$ and $\mathbf{b}$ have equal magnitudes of 10.0 units. They are oriented as shown in Fig. 3-32 and their vector sum is $\mathbf{r}$. Find (a) the x and y components of $\mathbf{r}$, (b) the magnitude of $\mathbf{r}$, and (c) the angle $\mathbf{r}$ makes with the positive x axis.

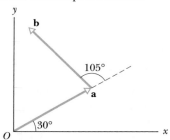

FIGURE 3-32 Problem 27.

28P. A golfer takes three putts to get the ball into the hole. The first putt displaces the ball 12 ft north, the second 6.0 ft southeast, and the third 3.0 ft southwest. What displacement was needed to get the ball into the hole on the first putt?

29P. A radar station detects an airplane approaching directly from the east. At first observation, the range to the plane is 1200 ft at 40° above the horizon. The airplane is tracked for another 123° in the vertical east–west plane, the range at final contact being 2580 ft. See Fig. 3-33. Find the displacement of the airplane during the period of observation.

FIGURE 3-33 Problem 29.

30P. (a) A man leaves his front door, walks 1000 m east, 2000 m north, and then takes a penny from his pocket and drops it from a cliff 500 m high. Set up a coordinate system and write

down an expression, using unit vectors, for the displacement of the penny from the house to its landing point. (b) The man then returns to his front door, following a different path. What is his resultant displacement for the round trip?

31P. A particle undergoes three successive displacements in a plane, as follows: 4.00 m southwest, 5.00 m east, and 6.00 m in a direction 60.0° north of east. Choose the y axis pointing north and the x axis pointing east and find (a) the components of each displacement, (b) the components of the resultant displacement, (c) the magnitude and direction of the resultant displacement, and (d) the displacement that would be required to bring the particle back to the starting point.

32P. Prove that two vectors must have equal magnitudes if their sum is perpendicular to their difference.

33P. Two vectors of lengths a and b make an angle θ with each other when placed tail to tail. Prove, by taking components along two perpendicular axes, that the length of their sum is

$$r = \sqrt{a^2 + b^2 + 2ab\cos\theta}.$$

34P. (a) Using unit vectors, write expressions for the body diagonals (the straight lines from one corner to another through the center) of a cube in terms of its edges, which have length a. (b) Determine the angles that the body diagonals make with the adjacent edges. (c) Determine the length of the body diagonals.

SECTION 3-6 Vectors and the Laws of Physics

35E. A vector $\mathbf{a}$ with a magnitude of 17.0 m is directed 56.0° counterclockwise from the $+x$ axis, as shown in Fig. 3-34. (a) What are the components a_x and a_y of the vector? (b) A second coordinate system is inclined by 18.0° with respect to the first. What are the components a'_x and a'_y in this primed coordinate system?

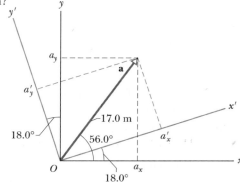

FIGURE 3-34 Exercise 35.

SECTION 3-7 Multiplying Vectors

36E. A vector $\mathbf{d}$ has a magnitude of 2.5 m and points north. What are the magnitudes and directions of the vectors (a) $4.0\mathbf{d}$ and (b) $-3.0\mathbf{d}$?

37E. Consider $\mathbf{a}$ in the positive direction of x, $\mathbf{b}$ in the positive direction of y, and a scalar d. What is the direction of $\mathbf{b}/d$ if d is (a) positive and (b) negative? What is the magnitude of (c) $\mathbf{a} \cdot \mathbf{b}$ and

(d) $\mathbf{a} \cdot \mathbf{b}/d$? What is the direction of (e) $\mathbf{a} \times \mathbf{b}$ and (f) $\mathbf{b} \times \mathbf{a}$? (g) What are the magnitudes of the cross products in (e) and (f)? (h) What are the magnitude and direction of $\mathbf{a} \times \mathbf{b}/d$?

38E. In a right-handed coordinate system show that

$$\mathbf{i} \cdot \mathbf{i} = \mathbf{j} \cdot \mathbf{j} = \mathbf{k} \cdot \mathbf{k} = 1 \quad \text{and} \quad \mathbf{i} \cdot \mathbf{j} = \mathbf{j} \cdot \mathbf{k} = \mathbf{k} \cdot \mathbf{i} = 0.$$

If the coordinate system is rectangular but not right-handed, do the results change?

39E. In a right-handed coordinate system show that

$$\mathbf{i} \times \mathbf{i} = \mathbf{j} \times \mathbf{j} = \mathbf{k} \times \mathbf{k} = 0$$

and $\quad\quad \mathbf{i} \times \mathbf{j} = \mathbf{k}; \quad \mathbf{k} \times \mathbf{i} = \mathbf{j}; \quad \mathbf{j} \times \mathbf{k} = \mathbf{i}.$

If the coordinate system is rectangular but not right-handed, do the results change?

40E. Show for any vector $\mathbf{a}$ that $\mathbf{a} \cdot \mathbf{a} = a^2$ and that $\mathbf{a} \times \mathbf{a} = 0$.

41E. Find (a) "north cross west," (b) "down dot south," (c) "east cross up," (d) "west dot west," and (e) "south cross south." Let each vector have unit magnitude.

42E. A vector $\mathbf{a}$ of magnitude 10 units and another vector $\mathbf{b}$ of magnitude 6.0 units point in directions differing by 60°. Find (a) the scalar product of the two vectors and (b) the magnitude of the vector product $\mathbf{a} \times \mathbf{b}$.

43E. Two vectors, $\mathbf{r}$ and $\mathbf{s}$, lie in the xy plane. Their magnitudes are 4.50 and 7.30 units, respectively, and their directions are 320° and 85.0°, respectively, as measured counterclockwise from the positive x axis. What are the values of (a) $\mathbf{r} \cdot \mathbf{s}$ and (b) $\mathbf{r} \times \mathbf{s}$?

44E. For the vectors in Fig. 3-35, calculate (a) $\mathbf{a} \cdot \mathbf{b}$, (b) $\mathbf{a} \cdot \mathbf{c}$, and (c) $\mathbf{b} \cdot \mathbf{c}$.

FIGURE 3-35   Exercises 44 and 45.

45E. For the vectors in Fig. 3-35, calculate (a) $\mathbf{a} \times \mathbf{b}$, (b) $\mathbf{a} \times \mathbf{c}$, and (c) $\mathbf{b} \times \mathbf{c}$.

46P. *Scalar Product in Unit-Vector Notation.* Let two vectors be represented in terms of their components as

$$\mathbf{a} = a_x\mathbf{i} + a_y\mathbf{j} + a_z\mathbf{k} \quad \text{and} \quad \mathbf{b} = b_x\mathbf{i} + b_y\mathbf{j} + b_z\mathbf{k}.$$

Show that

$$\mathbf{a} \cdot \mathbf{b} = a_xb_x + a_yb_y + a_zb_z.$$

47P. (a) Determine the components and magnitude of $\mathbf{r} = \mathbf{a} - \mathbf{b} + \mathbf{c}$ if $\mathbf{a} = 5.0\mathbf{i} + 4.0\mathbf{j} - 6.0\mathbf{k}$, $\mathbf{b} = -2.0\mathbf{i} + 2.0\mathbf{j} + 3.0\mathbf{k}$, and $\mathbf{c} = 4.0\mathbf{i} + 3.0\mathbf{j} + 2.0\mathbf{k}$. (b) Calculate the angle between $\mathbf{r}$ and the positive z axis.

48P. Use the definition of scalar product, $\mathbf{a} \cdot \mathbf{b} = ab \cos \theta$, and the fact that $\mathbf{a} \cdot \mathbf{b} = a_xb_x + a_yb_y + a_zb_z$ (see Problem 46) to calculate the angle between the two vectors given by $\mathbf{a} = 3.0\mathbf{i} + 3.0\mathbf{j} + 3.0\mathbf{k}$ and $\mathbf{b} = 2.0\mathbf{i} + 1.0\mathbf{j} + 3.0\mathbf{k}$.

49P. *Vector Product in Unit-Vector Notation.* Show that for the vectors $\mathbf{a}$ and $\mathbf{b}$ of Problem 46,

$$\mathbf{a} \times \mathbf{b} = \mathbf{i}(a_yb_z - a_zb_y) + \mathbf{j}(a_zb_x - a_xb_z) + \mathbf{k}(a_xb_y - a_yb_x).$$

50P. Two vectors are given by $\mathbf{a} = 3.0\mathbf{i} + 5.0\mathbf{j}$ and $\mathbf{b} = 2.0\mathbf{i} + 4.0\mathbf{j}$. Find (a) $\mathbf{a} \times \mathbf{b}$, (b) $\mathbf{a} \cdot \mathbf{b}$, and (c) $(\mathbf{a} + \mathbf{b}) \cdot \mathbf{b}$.

51P. Two vectors $\mathbf{a}$ and $\mathbf{b}$ have the components, in arbitrary units, $a_x = 3.2$, $a_y = 1.6$, $b_x = 0.50$, $b_y = 4.5$. (a) Find the angle between the directions of $\mathbf{a}$ and $\mathbf{b}$. (b) Find the components of a vector $\mathbf{c}$ that is perpendicular to $\mathbf{a}$, is in the xy plane, and has a magnitude of 5.0 units.

52P. Vector $\mathbf{a}$ lies in the yz plane 63° from the $+y$ axis, has a positive z component, and has magnitude 3.20 units. Vector $\mathbf{b}$ lies in the xz plane 48° from the $+x$ axis, has a positive z component, and has magnitude 1.40 units. Find (a) $\mathbf{a} \cdot \mathbf{b}$, (b) $\mathbf{a} \times \mathbf{b}$, and (c) the angle between $\mathbf{a}$ and $\mathbf{b}$.

53P. Three vectors are given by $\mathbf{a} = 3.0\mathbf{i} + 3.0\mathbf{j} - 2.0\mathbf{k}$, $\mathbf{b} = -1.0\mathbf{i} - 4.0\mathbf{j} + 2.0\mathbf{k}$, and $\mathbf{c} = 2.0\mathbf{i} + 2.0\mathbf{j} + 1.0\mathbf{k}$. Find (a) $\mathbf{a} \cdot (\mathbf{b} \times \mathbf{c})$, (b) $\mathbf{a} \cdot (\mathbf{b} + \mathbf{c})$, and (c) $\mathbf{a} \times (\mathbf{b} + \mathbf{c})$.

54P. Find the angles between the body diagonals of a cube with edge length a. See Problem 34.

55P. Show that the area of the triangle contained between the vectors $\mathbf{a}$ and $\mathbf{b}$ and the red line in Fig. 3-36 is $\frac{1}{2}|\mathbf{a} \times \mathbf{b}|$.

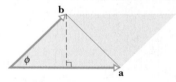

FIGURE 3-36   Problem 55.

56P. (a) Show that $\mathbf{a} \cdot (\mathbf{b} \times \mathbf{a})$ is zero for all vectors $\mathbf{a}$ and $\mathbf{b}$. (b) What is the value of $\mathbf{a} \times (\mathbf{b} \times \mathbf{a})$ if there is an angle ϕ between the directions of $\mathbf{a}$ and $\mathbf{b}$?

57P. Show that $\mathbf{a} \cdot (\mathbf{b} \times \mathbf{c})$ is equal in magnitude to the volume of the parallelepiped formed on the three vectors $\mathbf{a}$, $\mathbf{b}$, and $\mathbf{c}$ as shown in Fig. 3-37.

FIGURE 3-37   Problem 57.

58P. The three vectors shown in Fig. 3-38 have magnitudes $a = 3.00$, $b = 4.00$, and $c = 10.0$. (a) Calculate the x and y components of these vectors. (b) Find the numbers p and q such that $\mathbf{c} = p\mathbf{a} + q\mathbf{b}$.

FIGURE 3-38   Problem 58.

4
Motion in Two and Three Dimensions

In 1922, one of the Zacchinis, a famous family of circus performers, was the first human cannonball to be shot across an arena and into a net. To increase the excitement, the family gradually increased the height and distance of the flight until, in 1939 or 1940, Emanuel Zacchini soared over three Ferris wheels and through a horizontal distance of 225 feet. How could he know where to place the net? And how could he be certain he would clear the Ferris wheels?

4-1 MOVING IN TWO OR THREE DIMENSIONS

This chapter extends the material of the preceding two chapters to two and three dimensions. Many of the ideas of Chapter 2, such as position, velocity, and acceleration, are used here, but they are now a little more complex because of the extra dimensions. To keep the notation manageable, we use the vector algebra of Chapter 3. As you read this chapter, you might want to thumb back to those previous chapters to refresh your memory.

4-2 POSITION AND DISPLACEMENT

One general way of locating a particle-like object is with a **position vector r**, which is a vector that extends from a reference point (usually the origin of a coordinate system) to the object. In the unit-vector notation of Section 3-4, **r** can be written

$$\mathbf{r} = x\mathbf{i} + y\mathbf{j} + z\mathbf{k}, \qquad (4\text{-}1)$$

where $x\mathbf{i}$, $y\mathbf{j}$, and $z\mathbf{k}$ are the vector components of **r** and the coefficients x, y, and z are its scalar components. (This notation is slightly different from the notation we used in Chapter 3. Take a minute to convince yourself that the two are comparable.)

The coefficients x, y, and z give the object's location along the axes and relative to the origin. That is, the object has the rectangular coordinates (x, y, z). For instance, Fig. 4-1 shows an object P whose position vector is

$$\mathbf{r} = -3\mathbf{i} + 2\mathbf{j} + 5\mathbf{k}$$

and whose rectangular coordinates are $(-3, 2, 5)$. Along the x axis P is 3 units from the origin, in the $-\mathbf{i}$ direction. Along the y axis it is 2 units from the origin, in the $+\mathbf{j}$

direction. And along the z axis it is 5 units from the origin, in the $+\mathbf{k}$ direction.

As an object moves, its position vector changes in such a way that the vector always extends to the object from the origin. If the object has position vector $\mathbf{r}_1$ at time t_1 and position vector $\mathbf{r}_2$ at a later time $t_1 + \Delta t$, then its *displacement* $\Delta\mathbf{r}$ during the time interval Δt is

$$\Delta\mathbf{r} = \mathbf{r}_2 - \mathbf{r}_1. \qquad (4\text{-}2)$$

Using the unit-vector notation of Eq. 4-1, we can rewrite this displacement as

$$\Delta\mathbf{r} = (x_2\mathbf{i} + y_2\mathbf{j} + z_2\mathbf{k}) - (x_1\mathbf{i} + y_1\mathbf{j} + z_1\mathbf{k})$$

or

$$\Delta\mathbf{r} = (x_2 - x_1)\mathbf{i} + (y_2 - y_1)\mathbf{j} + (z_2 - z_1)\mathbf{k}, \qquad (4\text{-}3)$$

where coordinates (x_1, y_1, z_1) correspond to position vector $\mathbf{r}_1$ and coordinates (x_2, y_2, z_2) correspond to position vector $\mathbf{r}_2$. We can also rewrite the displacement by substituting Δx for $(x_2 - x_1)$, Δy for $(y_2 - y_1)$, and Δz for $(z_2 - z_1)$.

SAMPLE PROBLEM 4-1

The position vector for a particle is initially

$$\mathbf{r}_1 = -3\mathbf{i} + 2\mathbf{j} + 5\mathbf{k}$$

and then later is

$$\mathbf{r}_2 = 9\mathbf{i} + 2\mathbf{j} + 8\mathbf{k}$$

(see Fig. 4-2). What is the displacement from $\mathbf{r}_1$ to $\mathbf{r}_2$?

SOLUTION: Recall from Chapter 3 that we add (or subtract) two vectors in unit-vector notation by combining the components, axis by axis. So Eq. 4-2 becomes

$$\Delta\mathbf{r} = (9\mathbf{i} + 2\mathbf{j} + 8\mathbf{k}) - (-3\mathbf{i} + 2\mathbf{j} + 5\mathbf{k})$$
$$= 12\mathbf{i} + 3\mathbf{k}. \qquad \text{(Answer)}$$

The displacement vector is parallel to the xz plane, because it lacks any y component, a fact that is easier to pick out in the numerical result than in Fig. 4-2.

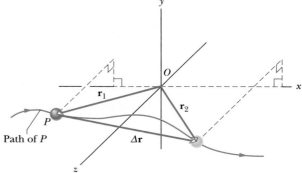

FIGURE 4-2 Sample Problem 4-1. The displacement $\Delta\mathbf{r} = \mathbf{r}_2 - \mathbf{r}_1$ extends from the head of $\mathbf{r}_1$ to the head of $\mathbf{r}_2$.

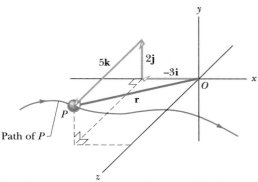

FIGURE 4-1 The position vector **r** for object P is the vector sum of its vector components.

CHECKPOINT 1: (a) If a bat flies from xyz coordinates $(-2$ m, 4 m, -3 m) to coordinates $(6$ m, -2 m, -3 m), what is its displacement $\Delta\mathbf{r}$ in unit-vector notation? (b) Is $\Delta\mathbf{r}$ parallel to one of the three coordinate planes? If so, which plane?

4-3 VELOCITY AND AVERAGE VELOCITY

If a particle moves through a displacement $\Delta\mathbf{r}$ in a time interval Δt, then its *average velocity* is

$$\overline{\mathbf{v}} = \frac{\Delta\mathbf{r}}{\Delta t}, \tag{4-4}$$

which can be written in expanded form as

$$\overline{\mathbf{v}} = \frac{\Delta x\mathbf{i} + \Delta y\mathbf{j} + \Delta z\mathbf{k}}{\Delta t}$$
$$= \frac{\Delta x}{\Delta t}\mathbf{i} + \frac{\Delta y}{\Delta t}\mathbf{j} + \frac{\Delta z}{\Delta t}\mathbf{k}. \tag{4-5}$$

The (instantaneous) *velocity* $\mathbf{v}$ is the value that $\overline{\mathbf{v}}$ approaches in the limit as we shrink Δt to 0. It can be written as the derivative

$$\mathbf{v} = \frac{d\mathbf{r}}{dt}. \tag{4-6}$$

Substituting for $\mathbf{r}$ from Eq. 4-1 yields

$$\mathbf{v} = \frac{d}{dt}(x\mathbf{i} + y\mathbf{j} + z\mathbf{k}) = \frac{dx}{dt}\mathbf{i} + \frac{dy}{dt}\mathbf{j} + \frac{dz}{dt}\mathbf{k},$$

which can be rewritten as

$$\mathbf{v} = v_x\mathbf{i} + v_y\mathbf{j} + v_z\mathbf{k}, \tag{4-7}$$

where

$$v_x = \frac{dx}{dt}, \quad v_y = \frac{dy}{dt}, \quad \text{and} \quad v_z = \frac{dz}{dt} \tag{4-8}$$

are the scalar components of $\mathbf{v}$.

Figure 4-3 shows the path of a particle P that is restricted to the xy plane. As the particle travels to the right along the curve, its position vector sweeps to the right. At t_1 the position vector is $\mathbf{r}_1$, and at an arbitrary later time $t_1 + \Delta t$ the position vector is $\mathbf{r}_2$. The particle's displacement during Δt is $\Delta\mathbf{r}$. The particle's average velocity $\overline{\mathbf{v}}$ during Δt is, by Eq. 4-4, in the same direction as $\Delta\mathbf{r}$.

Three things happen as we shrink interval Δt toward zero: (1) the vector $\mathbf{r}_2$ in Fig. 4-3 moves toward $\mathbf{r}_1$ so that

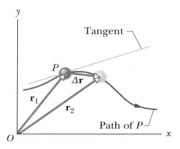

FIGURE 4-3 The position of particle P along its path, both at time t_1 and at a later time $t_1 + \Delta t$. The vector $\Delta\mathbf{r}$ is the displacement of the particle during Δt. The tangent to the particle's path at t_1 is shown.

$\Delta\mathbf{r}$ shrinks toward zero; (2) the direction of $\Delta\mathbf{r}$ (and thus the direction of $\overline{\mathbf{v}}$) approaches the direction of the tangent line in Fig. 4-3; and (3) the average velocity $\overline{\mathbf{v}}$ approaches the instantaneous velocity $\mathbf{v}$.

In the limit as $\Delta t \to 0$, we have $\overline{\mathbf{v}} \to \mathbf{v}$ and, most important here, $\overline{\mathbf{v}}$ takes on the direction of the tangent line. Hence $\mathbf{v}$ has that direction as well. That is:

> The instantaneous velocity $\mathbf{v}$ of a particle is always tangent to the path of the particle.

This is shown in Fig. 4-4, where both $\mathbf{v}$ and its scalar x and y components are included. The result is the same in three dimensions: $\mathbf{v}$ is always tangent to the particle's path.

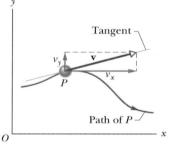

FIGURE 4-4 The velocity $\mathbf{v}$ of particle P along with the scalar components of $\mathbf{v}$. Note that $\mathbf{v}$ lies along the tangent to the path.

CHECKPOINT 2: The figure shows a circular path taken by a particle. If the instantaneous velocity of the particle is $\mathbf{v} = (2 \text{ m/s})\mathbf{i} - (2 \text{ m/s})\mathbf{j}$, through which quadrant is the particle moving when it is traveling (a) clockwise and (b) counterclockwise around the circle?

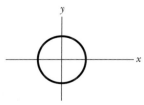

4-4 ACCELERATION AND AVERAGE ACCELERATION

When a particle's velocity changes from $\mathbf{v}_1$ to $\mathbf{v}_2$ in a time period Δt, its average acceleration $\bar{\mathbf{a}}$ during Δt is

$$\bar{\mathbf{a}} = \frac{\mathbf{v}_2 - \mathbf{v}_1}{\Delta t} = \frac{\Delta \mathbf{v}}{\Delta t}. \qquad (4\text{-}9)$$

If we shrink Δt to 0, then in the limit $\bar{\mathbf{a}}$ approaches the (instantaneous) *acceleration* $\mathbf{a}$; that is,

$$\mathbf{a} = \frac{d\mathbf{v}}{dt}. \qquad (4\text{-}10)$$

If the velocity changes in *either* magnitude *or* direction (or both), there is an acceleration.

Substituting $\mathbf{v}$ from Eq. 4-7 into Eq. 4-10 yields

$$\mathbf{a} = \frac{d}{dt}(v_x\mathbf{i} + v_y\mathbf{j} + v_z\mathbf{k})$$

$$= \frac{dv_x}{dt}\mathbf{i} + \frac{dv_y}{dt}\mathbf{j} + \frac{dv_z}{dt}\mathbf{k}$$

or

$$\mathbf{a} = a_x\mathbf{i} + a_y\mathbf{j} + a_z\mathbf{k}, \qquad (4\text{-}11)$$

in which the three scalar components of the acceleration vector are given by

$$a_x = \frac{dv_x}{dt}, \quad a_y = \frac{dv_y}{dt}, \quad \text{and} \quad a_z = \frac{dv_z}{dt}. \qquad (4\text{-}12)$$

Figure 4-5 shows an acceleration vector $\mathbf{a}$ and its scalar components for a particle P moving in two dimensions.

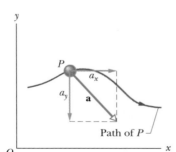

FIGURE 4-5 The acceleration $\mathbf{a}$ of particle P along with the scalar components of $\mathbf{a}$.

SAMPLE PROBLEM 4-2

A rabbit runs across a parking lot on which a set of coordinate axes has, strangely enough, been drawn. The rabbit's path is such that the components of its position with respect to an origin of coordinates are given as functions of time by

$$x = -0.31t^2 + 7.2t + 28$$

and

$$y = 0.22t^2 - 9.1t + 30,$$

with t in seconds and x and y in meters. The rabbit's position vector $\mathbf{r}$ is

$$\mathbf{r}(t) = x(t)\mathbf{i} + y(t)\mathbf{j}.$$

(a) What are the magnitude and direction of the rabbit's position vector at $t = 15$ s?

SOLUTION: At $t = 15$ s, the components of $\mathbf{r}$ are

$$x = (-0.31)(15)^2 + (7.2)(15) + 28 = 66 \text{ m}$$

and

$$y = (0.22)(15)^2 - (9.1)(15) + 30 = -57 \text{ m}.$$

The components and $\mathbf{r}$ itself are shown in Fig. 4-6a. The magnitude of $\mathbf{r}$ is

$$r = \sqrt{x^2 + y^2} = \sqrt{(66 \text{ m})^2 + (-57 \text{ m})^2}$$
$$= 87 \text{ m}. \qquad \text{(Answer)}$$

The angle θ between $\mathbf{r}$ and the direction of increasing x is

$$\theta = \tan^{-1}\frac{y}{x} = \tan^{-1}\left(\frac{-57 \text{ m}}{66 \text{ m}}\right) = -41°. \quad \text{(Answer)}$$

(Although $\theta = 139°$ has the same tangent as $-41°$, study of the signs of the components of $\mathbf{r}$ rules out $139°$.)

(b) Also calculate the position of the rabbit at $t = 0, 5, 10, 20,$ and 25 s and sketch the rabbit's path.

SOLUTION: Proceeding as in (a) leads to the following values of r and θ.

t (s)	x (m)	y (m)	r (m)	θ
0	28	30	41	+47°
5	56	−10	57	−10°
10	69	−39	79	−29°
15	66	−57	87	−41°
20	48	−64	80	−53°
25	14	−60	62	−77°

Figure 4-6b shows a plot of the rabbit's path, drawn with the x and y values.

SAMPLE PROBLEM 4-3

In Sample Problem 4-2, find the magnitude and direction of the rabbit's velocity vector at $t = 15$ s.

SOLUTION: The velocity component along the x axis (see Eq. 4-8) is

$$v_x = \frac{dx}{dt} = \frac{d}{dt}(-0.31t^2 + 7.2t + 28) = -0.62t + 7.2.$$

At $t = 15$ s, this becomes

$$v_x = (-0.62)(15) + 7.2 = -2.1 \text{ m/s}.$$

Similarly,

$$v_y = \frac{dy}{dt} = \frac{d}{dt}(0.22t^2 - 9.1t + 30) = 0.44t - 9.1.$$

At $t = 15$ s, this becomes

$$v_y = (0.44)(15) - 9.1 = -2.5 \text{ m/s}.$$

The vector **v** and its components are shown in Fig. 4-6c.
The magnitude and direction of **v** are

$$v = \sqrt{v_x^2 + v_y^2} = \sqrt{(-2.1 \text{ m/s})^2 + (-2.5 \text{ m/s})^2}$$
$$= 3.3 \text{ m/s} \qquad \text{(Answer)}$$

and

$$\theta = \tan^{-1}\frac{v_y}{v_x} = \tan^{-1}\left(\frac{-2.5 \text{ m/s}}{-2.1 \text{ m/s}}\right)$$
$$= \tan^{-1} 1.19 = -130°. \qquad \text{(Answer)}$$

(Although 50° has the same tangent, inspection of the signs of
the velocity components indicates that the desired angle is in
the third quadrant, given by $50° - 180° = -130°$.) The velocity vector in Fig. 4-6c is tangent to the path of the rabbit
and points in the direction in which the rabbit is running at
$t = 15$ s.

FIGURE 4-6 Sample Problems 4-2, 4-3,
and 4-4. (a) The vector **r** and its components at $t = 15$ s. The magnitude of **r** is
87 m. (b) The path of a rabbit across a
parking lot, showing the rabbit's position
at the indicated times. (c) The velocity **v**
of the rabbit at $t = 15$ s. Note that **v** is
tangent to the path at the position of the
rabbit at $t = 15$ s. (d) The acceleration **a**
of the rabbit at $t = 15$ s. The rabbit happens to have this same acceleration for all
points of its path.

(a)

(b)

(c)

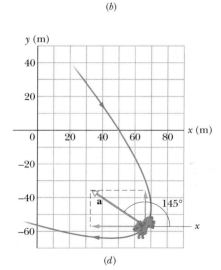

(d)

SAMPLE PROBLEM 4-4

In Sample Problem 4-2, determine the magnitude and direction of the rabbit's acceleration vector **a** at $t = 15$ s.

SOLUTION: The acceleration components (see Eq. 4-12) are

$$a_x = \frac{dv_x}{dt} = \frac{d}{dt}(-0.62t + 7.2) = -0.62 \text{ m/s}^2$$

and

$$a_y = \frac{dv_y}{dt} = \frac{d}{dt}(0.44t - 9.1) = 0.44 \text{ m/s}^2.$$

We see that the acceleration does not vary with time; it is a constant. In fact, we have differentiated the time variable completely away. The components of **a**, along with the vector itself, are shown in Fig. 4-6d.

The magnitude and direction of **a** are

$$a = \sqrt{a_x^2 + a_y^2} = \sqrt{(-0.62 \text{ m/s}^2)^2 + (0.44 \text{ m/s}^2)^2}$$

$$= 0.76 \text{ m/s}^2 \qquad \text{(Answer)}$$

and

$$\theta = \tan^{-1}\frac{a_y}{a_x} = \tan^{-1}\left(\frac{0.44 \text{ m/s}^2}{-0.62 \text{ m/s}^2}\right)$$

$$= 145°. \qquad \text{(Answer)}$$

The acceleration vector has the same magnitude and direction for all parts of the rabbit's path. Perhaps a strong southeast wind was blowing across the parking lot, causing the rabbit's acceleration toward the northwest.

CHECKPOINT **3:** Here are four descriptions of the position (in meters) of a hockey puck as it moves in the xy plane:

(1) $x = -3t^2 + 4t - 2$ and $y = 6t^2 - 4t$
(2) $x = -3t^3 - 4t$ and $y = -5t^2 + 6$
(3) $\mathbf{r} = 2t^2\mathbf{i} - (4t + 3)\mathbf{j}$
(4) $\mathbf{r} = (4t^3 - 2t)\mathbf{i} + 3\mathbf{j}$

For each description, determine whether the x and y components of the puck's acceleration are constant, and whether the acceleration **a** is constant.

SAMPLE PROBLEM 4-5

A particle with velocity $\mathbf{v_0} = -2.0\mathbf{i} + 4.0\mathbf{j}$ (in meters per second) at $t = 0$ undergoes a constant acceleration **a** of magnitude $a = 3.0$ m/s^2 at an angle $\theta = 130°$ from the positive direction of the x axis. What is the particle's velocity **v** at $t = 2.0$ s, in unit-vector notation and as a magnitude and direction (with respect to the positive direction of the x axis)?

SOLUTION: Since **a** is constant, Eq. 2-11 ($v = v_0 + at$) applies; it should, however, be used separately to find v_x and v_y

(the x and y components of velocity **v**) because they vary independently of each other. So we write

$$v_x = v_{0x} + a_x t \quad \text{and} \quad v_y = v_{0y} + a_y t.$$

Here v_{0x} ($= -2.0$ m/s) and v_{0y} ($= 4.0$ m/s) are the x and y components of $\mathbf{v_0}$, and a_x and a_y are the x and y components of **a**. To find a_x and a_y, we resolve **a** with Eqs. 3-5:

$$a_x = a \cos \theta = (3.0 \text{ m/s}^2)(\cos 130°) = -1.93 \text{ m/s}^2,$$

$$a_y = a \sin \theta = (3.0 \text{ m/s}^2)(\sin 130°) = +2.30 \text{ m/s}^2.$$

When these values are inserted into the equations for v_x and v_y, we find that

$$v_x = -2.0 \text{ m/s} + (-1.93 \text{ m/s}^2)(2.0 \text{ s}) = -5.9 \text{ m/s},$$

$$v_y = 4.0 \text{ m/s} + (2.30 \text{ m/s}^2)(2.0 \text{ s}) = 8.6 \text{ m/s}.$$

So at $t = 2.0$ s, we have, from Eq. 4-7,

$$\mathbf{v} = (-5.9 \text{ m/s})\mathbf{i} + (8.6 \text{ m/s})\mathbf{j}. \qquad \text{(Answer)}$$

The magnitude of **v** is

$$v = \sqrt{(-5.9 \text{ m/s})^2 + (8.6 \text{ m/s})^2}$$

$$= 10 \text{ m/s}. \qquad \text{(Answer)}$$

The angle of **v** is

$$\theta = \tan^{-1}\left(\frac{8.6 \text{ m/s}}{-5.9 \text{ m/s}}\right) = 124° \approx 120°. \qquad \text{(Answer)}$$

Check the last line with your calculator. Does 124° appear on the display, or does $-55.5°$ appear? Now sketch the vector **v** with its components to see which angle is reasonable. To see why some calculators give a mathematically possible but unreasonable result here, reread Tactic 3 in Chapter 3.

PROBLEM SOLVING TACTICS

TACTIC 1: *Trig Functions and Angles*

In Sample Problem 4-3, we computed $\theta = \tan^{-1} 1.19$ and had to find θ. Your calculator might tell you $\theta = 50°$. However, Fig. 3-13c shows that $\theta = 230°$ ($= 50° + 180°$) has the same tangent. Inspection of the signs of the velocity components v_x and v_y in Fig. 4-6c tells us that this latter angle is the correct one. Some calculators can choose the correct angle for you.

There is still another decision to make. We can stick with 230° or we can relabel it as $-130°$. They are exactly the same angle, as is pointed out in Tactic 1 of Chapter 3. Here, we chose $\theta = -130°$.

TACTIC 2: *Drawing Vectors*

The vectors in Fig. 4-6 were oriented in the following way. (1) Choose the point at which you wish the tail of the vector to be. (2) From that point, draw a line in the direction of increasing x. (3) Using a protractor, mark off the appropriate angle θ, counterclockwise from this line if θ is positive or clockwise if θ is negative.

The vector **r** in Fig. 4-6a should be drawn to the same scale as the two axes because it is a length. The velocity vector **v** in Fig. 4-6c and the acceleration vector **a** in Fig. 4-6d, however, have no established scale and you may make them as long or as short as you wish.

It makes no sense to ask whether, for example, a velocity vector should be longer or shorter than a displacement vector. They are different physical quantities, expressed in different units, and they have no common scale.

CHECKPOINT 4: If a marble's position is given by $\mathbf{r} = (4t^3 - 2t)\mathbf{i} + 3\mathbf{j}$, with **r** in meters and t in seconds, what must be the units of the coefficients 4, -2, and 3?

4-5 PROJECTILE MOTION

We next consider a particle that moves in a vertical plane during free fall, with its only acceleration being that of the free-fall acceleration **g**, which is downward. Such a **projectile** might be a golf ball (as in Fig. 4-7), a baseball, or any of a variety of other objects. Throughout, we shall assume that the air has no effect on the motion of the projectile.

Figure 4-8, which is analyzed in the next section, shows the path followed by a projectile under such ideal

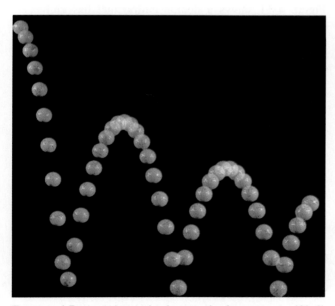

FIGURE 4-7 A stroboscopic photograph of an orange golf ball bouncing off a hard surface. Between impacts, the ball has projectile motion.

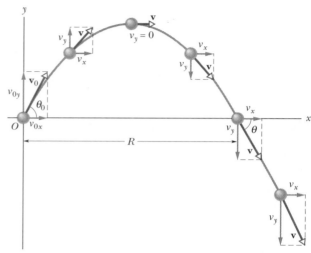

FIGURE 4-8 The path of a projectile that is launched at $x_0 = 0$ and $y_0 = 0$, with an initial velocity $\mathbf{v}_0$. The initial velocity and the velocities at various points along its path are shown, along with their components. Note that the horizontal velocity component remains constant but the vertical velocity component changes continuously. The *range R* is the horizontal distance the projectile has traveled *when it returns to its launch height.*

conditions. The projectile is launched with some initial velocity $\mathbf{v}_0$, which can be written

$$\mathbf{v}_0 = v_{0x}\mathbf{i} + v_{0y}\mathbf{j}. \qquad (4\text{-}13)$$

The components v_{0x} and v_{0y} can then be found if we know the angle θ_0 between $\mathbf{v}_0$ and the positive x direction:

$$v_{0x} = v_0 \cos \theta_0 \quad \text{and} \quad v_{0y} = v_0 \sin \theta_0. \qquad (4\text{-}14)$$

During its two-dimensional motion, the projectile's position vector **r** and velocity vector **v** change continuously, but its acceleration vector **a** is constant and *always* directed vertically downward. (The projectile has *no* horizontal acceleration.) As shown in Fig. 4-9, the angle between the direction of the velocity vector and that of the acceleration vector is not constant but varies during the motion.

Projectile motion, like that in Figs. 4-7 through 4-9, looks complicated, but we have the following simplifying feature (known from experiment):

The horizontal motion and the vertical motion are in dependent of each other; that is, neither motion affects the other.

This feature allows us to break up a problem involving two-dimensional motion into two separate and easier one-dimensional problems, one for the horizontal motion and

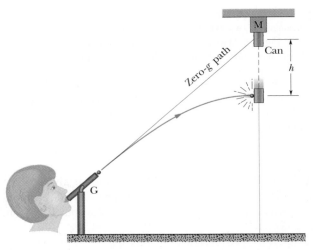

FIGURE 4-9 The velocity and acceleration vectors of a projectile for various motions. Note that the acceleration and velocity vectors do not have any fixed directional relation to each other.

one for the vertical motion. Here are two experiments that show that the horizontal motion and the vertical motion are independent.

Two Golf Balls

Figure 4-10 is a stroboscopic photograph of two golf balls, one simply dropped and the other fired horizontally by a spring mechanism. They have the same vertical motion, each ball falling through the same vertical distance in the

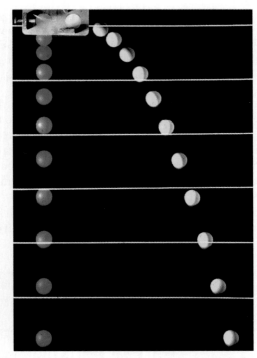

FIGURE 4-10 One ball is released from rest at the same instant that another ball is shot horizontally to the right. Their vertical motions are identical.

FIGURE 4-11 The projectile ball always hits the falling can. Each falls a distance h from where it would be were there no free-fall acceleration.

same interval of time. *The fact that one ball is moving horizontally while it is falling has no effect on its vertical motion.* To push the experiment to a limit, we say that if you were to fire a rifle horizontally and at the same time drop a bullet, in the absence of air resistance the two bullets would reach level ground at the same time.

A Great Student Rouser

Figure 4-11 shows a demonstration that has enlivened many a physics lecture. It involves a blow gun G, using a ball as a projectile. The target is a can suspended from a magnet M, and the tube of the blow gun is aimed directly at the can. The experiment is arranged so that the magnet releases the can just as the ball leaves the blow gun.

If g (the magnitude of the free-fall acceleration) were zero, the ball would follow the straight line shown in Fig. 4-11 and the can would float in place after the magnet released it. The ball would certainly hit the can.

However, g is *not* zero. The ball *still* hits the can! As Fig. 4-11 shows, during the time of flight of the ball, both ball and can fall the same distance h from their zero-g locations. The harder the demonstrator blows, the greater the ball's initial speed, the shorter the time of flight, and the smaller the value of h.

4-6 PROJECTILE MOTION ANALYZED

Now we are ready to analyze projectile motion, horizontally and vertically:

FIGURE 4-12 The vertical component of this skateboarder's velocity is changing, but not the horizontal component, which matches the skateboard's velocity. As a result, the skateboard stays underneath him, allowing him to land on it.

The Horizontal Motion

Because there is *no acceleration* in the horizontal direction, the horizontal component v_{0x} of the projectile's initial velocity remains unchanged throughout the motion, as demonstrated in Fig. 4-12. The horizontal displacement $x - x_0$ from an initial position x_0 is given by Eq. 2-15 with $a = 0$, which we write as

$$x - x_0 = v_{0x}t.$$

Because $v_{0x} = v_0 \cos \theta_0$, this becomes

$$x - x_0 = (v_0 \cos \theta_0)t. \qquad (4\text{-}15)$$

The Vertical Motion

The vertical motion is the motion we discussed in Section 2-8 for a particle in free fall. Equations 2-21 to 2-25 apply. Equation 2-22, for example, becomes

$$\begin{aligned} y - y_0 &= v_{0y}t - \tfrac{1}{2}gt^2 \\ &= (v_0 \sin \theta_0)t - \tfrac{1}{2}gt^2, \end{aligned} \qquad (4\text{-}16)$$

where the initial vertical velocity component v_{0y} is replaced with the equivalent $v_0 \sin \theta_0$. Equations 2-21 and 2-23 are also useful in analyzing projectile motion.

Adapted to our purpose, they are

$$v_y = v_0 \sin \theta_0 - gt \qquad (4\text{-}17)$$

and

$$v_y^2 = (v_0 \sin \theta_0)^2 - 2g(y - y_0). \qquad (4\text{-}18)$$

As is illustrated in Fig. 4-8 and Eq. 4-17, the vertical velocity component behaves just as for a ball thrown vertically upward. It is directed upward initially, its magnitude steadily decreasing to zero, *which marks the maximum height of the path*. The vertical component then reverses direction, and its magnitude becomes larger with time.

The Equation of the Path

We can find the equation of the path (the **trajectory**) of the projectile by eliminating t between Eqs. 4-15 and 4-16. Solving Eq. 4-15 for t and substituting into Eq. 4-16, we obtain, after a little rearrangement,

$$y = (\tan \theta_0)x - \frac{gx^2}{2(v_0 \cos \theta_0)^2} \qquad \text{(trajectory)}. \quad (4\text{-}19)$$

This is the equation of the path shown in Fig. 4-8. In deriving it, for simplicity we let $x_0 = 0$ and $y_0 = 0$ in Eqs. 4-15 and 4-16, respectively. Because g, θ_0, and v_0 are constants, Eq. 4-19 is of the form $y = ax + bx^2$, in which a and b are constants. This is the equation of a parabola, so the path is *parabolic*.

The Horizontal Range

The *horizontal range R* of the projectile, as Fig. 4-8 shows, is the *horizontal* distance the projectile has traveled when it returns to its initial (launch) height. To find range R, let us put $x - x_0 = R$ in Eq. 4-15 and $y - y_0 = 0$ in Eq. 4-16, obtaining

$$x - x_0 = (v_0 \cos \theta_0)t = R$$

and

$$y - y_0 = (v_0 \sin \theta_0)t - \tfrac{1}{2}gt^2 = 0.$$

Eliminating t between these two equations yields

$$R = \frac{2v_0^2}{g} \sin \theta_0 \cos \theta_0.$$

Using the identity $\sin 2\theta_0 = 2 \sin \theta_0 \cos \theta_0$ (see Appendix E), we obtain

$$R = \frac{v_0^2}{g} \sin 2\theta_0. \qquad (4\text{-}20)$$

Note that R has its maximum value when $\sin 2\theta_0 = 1$, which corresponds to $2\theta_0 = 90°$ or $\theta_0 = 45°$.

The horizontal range R is maximum for a launch angle of 45°.

The Effects of the Air

We have assumed that the air through which the projectile moves has no effect on its motion, a reasonable assumption at low speeds. However, for greater speeds, the disagreement between our calculations and the actual motion of the projectile can be large because the air resists (or opposes) the motion. Figure 4-13, for example, shows two paths for a fly ball that leaves the bat at an angle of 60° with the horizontal and an initial speed of 100 mi/h (data adapted from "The Trajectory of a Fly Ball," by Peter J. Brancazio, *The Physics Teacher,* January 1985). Path I (the baseball player's fly ball) is a calculated path that approximates normal conditions of play, in air. Path II (the physics professor's fly ball) is the path that the ball would follow in a vacuum. Table 4-1 gives some data for the two cases. We shall discuss details of the effect of the air on motion in Chapter 6.

CHECKPOINT **5:** A fly ball is hit to the outfield. During its flight (ignore the effects of air), what happens to its (a) horizontal and (b) vertical components of velocity? What are the (c) horizontal and (d) vertical components of its acceleration during its ascent and its descent, and at the topmost point of its flight?

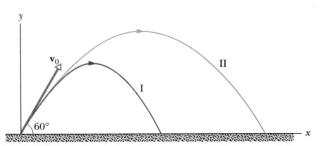

FIGURE 4-13 (I) The path of a fly ball, calculated (using a computer) by taking air resistance into account. (II) The path the ball would follow in a vacuum, calculated by the methods of this chapter. Also see Table 4-1.

TABLE 4-1 **TWO FLY BALLS**[a]

	PATH I (AIR)	PATH II (VACUUM)
Range	323 ft	581 ft
Maximum height	174 ft	252 ft
Time of flight	6.6 s	7.9 s

[a]See Fig. 4-13. The launch angle is 60° and the launch speed is 100 mi/h.

SAMPLE PROBLEM 4-6

A rescue plane is flying at a constant elevation of 1200 m with a speed of 430 km/h toward a point directly over a person struggling in the water (see Fig. 4-14). At what angle of sight ϕ should the pilot release a rescue capsule if it is to strike (very close to) the person in the water?

SOLUTION: The initial velocity of the capsule is the same as the velocity of the plane at the moment of release. That is, the initial capsule velocity $\mathbf{v}_0$ is horizontal and has a magnitude of 430 km/h. We know the vertical distance the capsule falls, so we can find its time of flight with Eq. 4-16,

$$y - y_0 = (v_0 \sin \theta_0)t - \tfrac{1}{2}gt^2.$$

Putting $y - y_0 = -1200$ m (we use the minus sign because the person is below the origin) and $\theta_0 = 0$, we obtain

$$-1200 \text{ m} = 0 - \tfrac{1}{2}(9.8 \text{ m/s}^2)t^2.$$

Solving for t yields

$$t = \sqrt{\frac{(2)(1200 \text{ m})}{9.8 \text{ m/s}^2}} = 15.65 \text{ s}.$$

The horizontal distance covered by the capsule (and by the plane) in that time is given by Eq. 4-15:

$$x - x_0 = (v_0 \cos \theta_0)t$$
$$= (430 \text{ km/h})(\cos 0°)(15.65 \text{ s})(1 \text{ h}/3600 \text{ s})$$
$$= 1.869 \text{ km} = 1869 \text{ m}.$$

If $x_0 = 0$, then $x = 1869$ m. The angle of sight is then (see Fig. 4-14)

$$\phi = \tan^{-1} \frac{x}{h} = \tan^{-1} \frac{1869 \text{ m}}{1200 \text{ m}} = 57°. \quad \text{(Answer)}$$

Because the plane and the capsule have the same horizontal velocity, the plane remains vertically over the capsule while the capsule is in flight.

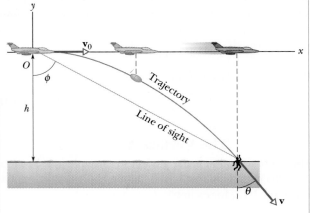

FIGURE 4-14 Sample Problem 4-6. A plane drops a rescue capsule and then continues in level flight. While the capsule is falling, its horizontal velocity component remains equal to the velocity of the plane. The capsule hits the water with velocity $\mathbf{v}$, at angle θ from the vertical.

SAMPLE PROBLEM 4-7

A movie stuntman is to run across a rooftop and jump horizontally off it, to land on the roof of the next building (Fig. 4-15). Before he attempts the jump, he wisely asks you to determine whether it is possible. Can he make the jump if his maximum rooftop speed is 4.5 m/s?

SOLUTION: The fall of 4.8 m will take a time t, which can be obtained from Eq. 4-16. Letting $y - y_0 = -4.8$ m (note the sign) and $\theta_0 = 0$, you rearrange Eq. 4-16 to obtain

$$t = \sqrt{-\frac{2(y - y_0)}{g}} = \sqrt{-\frac{(2)(-4.8 \text{ m})}{9.8 \text{ m/s}^2}}$$

$$= 0.990 \text{ s}.$$

You now ask: "How far would he move horizontally in this time?" The answer, from Eq. 4-15, is

$$x - x_0 = (v_0 \cos \theta_0)t$$

$$= (4.5 \text{ m/s})(\cos 0°)(0.990 \text{ s}) = 4.5 \text{ m}.$$

To reach the next building, the stuntman has to move 6.2 m horizontally. Your advice: "Don't jump."

FIGURE 4-15 Sample Problem 4-7. Can the stuntman make the jump?

SAMPLE PROBLEM 4-8

Figure 4-16 shows a pirate ship, moored 560 m from a fort defending the harbor entrance of an island. The harbor defense cannon, located at sea level, has a muzzle velocity of 82 m/s.

(a) To what angle must the cannon be elevated to hit the pirate ship?

SOLUTION: Solving Eq. 4-20 for $2\theta_0$ yields

$$2\theta_0 = \sin^{-1}\frac{gR}{v_0^2} = \sin^{-1}\frac{(9.8 \text{ m/s}^2)(560 \text{ m})}{(82 \text{ m/s})^2}$$

$$= \sin^{-1} 0.816.$$

There are two angles whose sine is 0.816, namely, 54.7° and 125.3°. Thus we find

$$\theta_0 = \tfrac{1}{2}(54.7°) \approx 27° \qquad \text{(Answer)}$$

and

$$\theta_0 = \tfrac{1}{2}(125.3°) \approx 63°. \qquad \text{(Answer)}$$

The commandant of the fort can elevate the cannon to either of these two angles and (if only there were no intervening air!) hit the pirate ship.

(b) What are the times of flight for the two elevation angles calculated above?

SOLUTION: Solving Eq. 4-15 for t gives, for $\theta_0 = 27°$,

$$t = \frac{x - x_0}{v_0 \cos \theta_0} = \frac{560 \text{ m}}{(82 \text{ m/s})(\cos 27°)}$$

$$= 7.7 \text{ s}. \qquad \text{(Answer)}$$

Repeating the calculation for $\theta_0 = 63°$ yields $t = 15$ s. It is reasonable that the time of flight for the higher elevation angle should be larger.

(c) How far should the pirate ship be from the fort if it is to be beyond range of the cannon?

SOLUTION: We have seen that maximum range corresponds to an elevation angle θ_0 of 45°. Thus, from Eq. 4-20 with $\theta_0 = 45°$,

$$R = \frac{v_0^2}{g} \sin 2\theta_0 = \frac{(82 \text{ m/s})^2}{9.8 \text{ m/s}^2} \sin(2 \times 45°)$$

$$= 690 \text{ m}. \qquad \text{(Answer)}$$

As the pirate ship sails away, the two elevation angles at which the ship can be hit draw closer together, eventually merging at $\theta_0 = 45°$ when the ship is 690 m away. Beyond that distance the ship is safe.

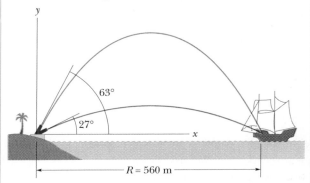

FIGURE 4-16 Sample Problem 4-8. At this range, the harbor defense cannon can hit the pirate ship at two elevation angles of the cannon.

SAMPLE PROBLEM 4-9

Figure 4-17 illustrates the flight of Emanuel Zacchini over three Ferris wheels, located as shown and each 18 m high. Zacchini is launched with speed $v_0 = 26.5$ m/s, at an angle $\theta_0 = 53°$ from the horizontal and with an initial height of 3.0 m above the ground. The net in which he is to land is at the same height.

(a) Does he clear the first Ferris wheel?

SOLUTION: We place the origin on the cannon muzzle so that $x_0 = 0$ and $y_0 = 0$. To find his height y when $x = 23$ m, we use Eq. 4-19:

$$y = (\tan \theta_0)x - \frac{gx^2}{2(v_0 \cos \theta_0)^2}$$

$$= (\tan 53°)(23 \text{ m}) - \frac{(9.8 \text{ m/s}^2)(23 \text{ m})^2}{2(26.5 \text{ m/s})^2(\cos 53°)^2}$$

$$= 20.3 \text{ m}. \hspace{2cm} \text{(Answer)}$$

Since he begins 3.0 m off the ground, he clears the Ferris wheel by about 5.3 m.

(b) If he reaches his maximum height when he is over the middle Ferris wheel, what is his clearance above it?

SOLUTION: At maximum height, $v_y = 0$ and Eq. 4-18 becomes

$$v_y^2 = (v_0 \sin \theta_0)^2 - 2gy = 0.$$

Solving for y gives

$$y = \frac{(v_0 \sin \theta_0)^2}{2g} = \frac{(26.5 \text{ m/s})^2(\sin 53°)^2}{(2)(9.8 \text{ m/s}^2)} = 22.9 \text{ m},$$

which means that he clears the middle Ferris wheel by 7.9 m.

(c) What is his time of flight t?

SOLUTION: Of the several ways to find t, we could use Eq. 4-16 and the fact that $y = 0$ when he lands. Then we have

$$y = (v_0 \sin \theta_0)t - \tfrac{1}{2}gt^2 = 0,$$

or $\hspace{1cm} t = \dfrac{2v_0 \sin \theta_0}{g} = \dfrac{(2)(26.5 \text{ m/s})(\sin 53°)}{9.8 \text{ m/s}^2}$

$$= 4.3 \text{ s}. \hspace{2cm} \text{(Answer)}$$

(d) How far from the cannon should the center of the net be positioned?

SOLUTION: One way to answer is with Eq. 4-15, with $x_0 = 0$:

$$x = (v_0 \cos \theta_0)t$$
$$= (26.5 \text{ m/s})(\cos 53°)(4.3 \text{ s})$$
$$= 69 \text{ m}, \hspace{2cm} \text{(Answer)}$$

which is the range R of the flight.

We can now answer the questions that opened this chapter: How could Zacchini know where to place the net? And how could he be certain he would clear the Ferris wheels? He (or someone) did the calculations as we have here. Although he could not take into account the complicated effects of the air on his flight, Zacchini knew that the air would slow him and thus decrease his range from the calculated value. So he used a wide net and biased it toward the cannon. He was then relatively safe whether the effects of the air in a particular flight happened to slow him considerably or very little. Still, the variability of this factor of air effects must have played on his imagination before each flight.

The circus performer still faced a subtle danger: even for shorter flights, his propulsion in the muzzle of the cannon was

so severe that he underwent a momentary blackout. If he landed during the blackout, he could break his neck. So, he had trained himself to awake quickly. Indeed, not waking up in time presents the only real danger to a human cannonball in a circus of today, where a flight is much shorter and safer than Zacchini's.

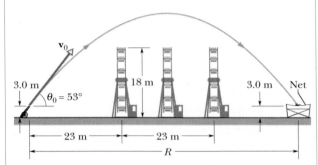

FIGURE 14-17   Sample Problem 4-9. The flight of a human cannonball over three Ferris wheels, and the desired placement of the net.

PROBLEM SOLVING TACTICS

TACTIC 3: *Numbers Versus Algebra*

One way to avoid numerical rounding errors is to solve problems algebraically, substituting numbers only in the final step. That is easy to do in Sample Problems 4-6 to 4-9, and that is the way experienced problem solvers operate. In these early chapters, however, we prefer to solve most problems in pieces, to give you a firmer numerical grasp of what is going on. Later we shall stick more to the algebra.

4-7 UNIFORM CIRCULAR MOTION

A particle is in **uniform circular motion** if it travels around a circle or circular arc at constant speed. Although the speed does not vary, *the particle is accelerating*. That fact may be surprising because we usually think of acceleration as an increase or decrease in speed. But actually **v** is a vector, not a scalar. If **v** changes, even only in direction, there is an acceleration, and that is what happens in uniform circular motion.

We use Fig. 4-18 to find the magnitude and direction of the acceleration. That figure represents a particle in uniform circular motion with speed v in a circle of radius r. Velocity vectors are drawn for two points, p and q, that are symmetric with respect to the y axis. These vectors, $\mathbf{v}_p$ and $\mathbf{v}_q$, have the same magnitude v but—because they point in different directions—they are different vectors. Their x and y components are

$$v_{px} = +v\cos\theta, \qquad v_{py} = +v\sin\theta$$

and $\qquad v_{qx} = +v\cos\theta, \qquad v_{qy} = -v\sin\theta.$

The time required for the particle to move from p to q at constant speed v is

$$\Delta t = \frac{\text{arc}(pq)}{v} = \frac{r(2\theta)}{v}, \qquad (4\text{-}21)$$

in which arc(pq) is the length of the arc from p to q.

We now have enough information to calculate the components of the average acceleration $\bar{\mathbf{a}}$ experienced by the particle as it moves from p to q in Fig. 4-18. For the x component, we have

$$\bar{a}_x = \frac{v_{qx} - v_{px}}{\Delta t} = \frac{v\cos\theta - v\cos\theta}{\Delta t} = 0.$$

This result is not surprising because it is clear from symmetry in Fig. 4-18 that the x component of velocity has the same value at q and at p.

For the y component of the average acceleration, we find, making use of Eq. 4-21,

$$\bar{a}_y = \frac{v_{qy} - v_{py}}{\Delta t} = \frac{-v\sin\theta - v\sin\theta}{\Delta t}$$

$$= -\frac{2v\sin\theta}{2r\theta/v} = -\left(\frac{v^2}{r}\right)\left(\frac{\sin\theta}{\theta}\right).$$

The minus sign tells us that this acceleration component points vertically downward in Fig. 4-18.

Now let us allow the angle θ in Fig. 4-18 to shrink, approaching zero as a limit. This requires that points p and q in that figure approach their midpoint, shown as point P at the top of the circle. The average acceleration $\bar{\mathbf{a}}$, whose components we have just found, then approaches the instantaneous acceleration $\mathbf{a}$ at point P.

The *direction* of this instantaneous acceleration vector at point P in Fig. 4-18 is downward, toward the center of the circle at O, because the direction of the average acceleration does not change as θ becomes smaller. To find the *magnitude* a of the instantaneous acceleration vector, we need only the mathematical fact that as θ shrinks, the ratio $(\sin\theta)/\theta$ approaches unity. From the relation given above for $\bar{a}_y$, we then have

$$a = \frac{v^2}{r} \qquad \text{(centripetal acceleration).} \qquad (4\text{-}22)$$

We conclude:

When a particle moves at constant speed v in a circle (or a circular arc) of radius r, the acceleration of the particle is directed toward the center of the circle and has a constant magnitude v^2/r.

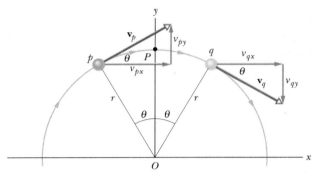

FIGURE 4-18 A particle moves in uniform circular motion at constant speed v in a circle of radius r. Its velocities $\mathbf{v}_p$ and $\mathbf{v}_q$ at points p and q, equidistant from the y axis, are shown, along with its velocity components at those points. The instantaneous acceleration of the particle at any point is directed toward the center of the circle and has a magnitude v^2/r.

In addition, during this acceleration at constant speed, the particle travels the circumference of the circle (a distance of $2\pi r$) in time

$$T = \frac{2\pi r}{v} \qquad \text{(period).} \qquad (4\text{-}23)$$

T is called the *period of revolution*, or simply the *period*, of the motion. It is, in general, the time for a particle to go around a closed path exactly once.

Figure 4-19 shows the relation between the velocity and acceleration vectors at various stages during uniform circular motion. Both vectors have constant magnitude as the motion progresses, but their directions change continuously. The velocity is always tangent to the circle in the direction of motion; the acceleration is always directed radially inward. Because of this, the acceleration associated with uniform circular motion is called a **centripetal** (meaning "center seeking") **acceleration**.

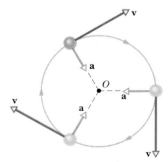

FIGURE 4-19 Velocity and acceleration vectors for a particle in uniform circular motion. Both have constant magnitude but vary continuously in direction.

The acceleration resulting from a change in the direction of a velocity is just as real as one resulting from a change in the magnitude of a velocity. In Fig. 2-8, for example, we saw Colonel John P. Stapp while his rocket sled was braked to rest. He was experiencing a velocity that was constant in direction but changing rapidly in magnitude. On the other hand, an astronaut whirling in a human centrifuge experiences a velocity that is constant in magnitude but changing rapidly in direction. The accelerations that these two people feel are indistinguishable.

CHECKPOINT 6: An object moves in a circular path in the xy plane, with the center at the origin. When the object is at $x = -2$ m, its velocity is $-(4$ m/s$)\mathbf{j}$. Give the object's (a) velocity and (b) centripetal acceleration when it is at $y = 2$ m.

SAMPLE PROBLEM 4-10

"Top gun" pilots have long worried about taking a turn too tightly. As a pilot's body undergoes centripetal acceleration, with the head toward the center of curvature, the blood pressure in the brain decreases, leading to loss of brain function.

There are several warning signs to signal a pilot to ease up: when the centripetal acceleration is $2g$ or $3g$, the pilot feels heavy. At about $4g$, the pilot's vision switches to black and white and narrows down to "tunnel vision." If that acceleration is sustained or increased, vision ceases and soon after, the pilot is unconscious—a condition known as g-LOC for "g-induced loss of consciousness."

What is the centripetal acceleration, in g units, of a pilot flying an F-22 at speed $v = 1600$ mi/h (716 m/s) through a circular arc with radius of curvature $r = 3.60$ mi (5.80 km)?

SOLUTION: Substituting the given data into Eq. 4-22, we have

$$a = \frac{v^2}{r} = \frac{(716 \text{ m/s})^2}{5800 \text{ m}}$$

$$= 88.39 \text{ m/s}^2 = 9.0g. \qquad \text{(Answer)}$$

If an unwary pilot caught in a dogfight puts the aircraft into such a tight turn, the pilot goes into g-LOC almost immediately, with no warning signs to signal the danger.

SAMPLE PROBLEM 4-11

A satellite is in circular Earth orbit, at an altitude $h = 200$ km above the Earth's surface. There the free-fall acceleration g is 9.20 m/s². What is the orbital speed v of the satellite?

SOLUTION: We have uniform circular motion around the Earth. The satellite's centripetal acceleration is then the free-fall acceleration g. So we can find v from Eq. 4-22, with $a = g$ and with $r = R_E + h$, where R_E is the Earth's radius (see inside front cover or Appendix C):

$$g = \frac{v^2}{R_E + h}.$$

Solving for v gives

$$\begin{aligned} v &= \sqrt{g(R_E + h)} \\ &= \sqrt{(9.20 \text{ m/s}^2)(6.37 \times 10^6 \text{ m} + 200 \times 10^3 \text{ m})} \\ &= 7770 \text{ m/s} = 7.77 \text{ km/s.} \qquad \text{(Answer)} \end{aligned}$$

You can show that this is equivalent to 17,400 mi/h and that the satellite would take 1.47 h to complete one orbital revolution; that is, the period T of the motion is 1.47 h.

4-8 RELATIVE MOTION IN ONE DIMENSION

Suppose you see a duck flying north at, say, 20 mi/h. To another duck flying alongside, the first duck is at rest. In other words, the velocity of a particle depends on the **reference frame** of whoever is doing the measuring. For our purposes, a reference frame is the physical object to which you attach your coordinate system.

The reference frame that seems most natural to us in our daily comings and goings is the ground beneath our feet. When a traffic officer tells you that you have been driving at 70 mi/h, the unspoken qualification, "in a coordinate system attached to the ground," is always taken for granted by each of you.

If you are traveling in an airplane or a spaceship, the reference frame of the Earth may not be the most convenient one. You are free to choose any reference frame that you wish. Once having made it, however, you must always be aware of your choice and be careful to make all your measurements with respect to that reference frame.

Suppose that Alex (frame A) is parked by the side of a highway, watching car P (the "particle") speed past. Barbara (frame B), driving along the highway at constant speed, is also watching car P. Suppose that, as in Fig. 4-20, they both measure the position of the car at a given moment. From the figure we see that

$$x_{PA} = x_{PB} + x_{BA}. \qquad (4\text{-}24)$$

The terms in Eq. 4-24 are scalars and may be of either sign. The equation is read: "The coordinate of P as measured by A *is equal to* the coordinate of P as measured by B *plus* the coordinate of B as measured by A." Note how this reading is supported by the sequence of the subscripts.

Taking the time derivative of Eq. 4-24, we obtain

FIGURE 4-20 Alex (frame A) and Barbara (frame B) watch car P. All motion is along the common x axis of the two frames. The vector $\mathbf{v}_{BA}$ shows the relative separation velocity of the two frames. The three position measurements shown are all made at the same instant of time.

$$\frac{d}{dt}(x_{PA}) = \frac{d}{dt}(x_{PB}) + \frac{d}{dt}(x_{BA}),$$

or (because $v = dx/dt$)

$$v_{PA} = v_{PB} + v_{BA}. \qquad (4\text{-}25)$$

This scalar equation is the relation between the velocities of the same object (car P) as measured in the two frames; those measured velocities are different. In words, Eq. 4-25 says: "The velocity of P as measured by A *is equal to* the velocity of P as measured by B *plus* the velocity of B as measured by A." The term v_{BA} is the separation velocity of frame B with respect to frame A; see Fig. 4-20.

We consider only frames that move at constant velocity; such frames are called **inertial reference frames.** In our example, this means that Barbara (frame B) will drive always at constant speed with respect to Alex (frame A). Car P (the moving particle), however, may speed up, slow down, come to rest, or reverse direction.

The time derivative of the velocity equation (Eq. 4-25) yields the acceleration equation,

$$a_{PA} = a_{PB}. \qquad (4\text{-}26)$$

Note that because v_{BA} is a constant, its time derivative is zero. Equation 4-26 tells us:

Observers on different inertial frames of reference (their velocity of separation is constant) will measure the same acceleration for a moving particle.

CHECKPOINT **7:** The table gives velocities (km/h) for Barbara and car P of Fig. 4-20 for three situations. For each, what is the missing value and how is the distance between Barbara and car P changing?

SITUATION	v_{BA}	v_{PA}	v_{PB}
1	+50	+50	
2	+30		+40
3		+60	−20

SAMPLE PROBLEM 4-12

Alex, parked by the side of an east–west road, is watching car P, which is moving in a westerly direction. Barbara, driving east at a speed $v_{BA} = 52$ km/h, watches the same car. Take the easterly direction as positive.

(a) If Alex measures a speed of 78 km/h for car P, what velocity will Barbara measure?

SOLUTION: Equation 4-25 may be rearranged to yield

$$v_{PB} = v_{PA} - v_{BA}.$$

We have $v_{PA} = -78$ km/h, the minus sign telling us that car P is moving west, in the negative direction. We also have $v_{BA} = 52$ km/h, so that

$$v_{PB} = (-78 \text{ km/h}) - (52 \text{ km/h})$$
$$= -130 \text{ km/h}. \qquad \text{(Answer)}$$

If car P were connected to Barbara's car by a string wound up on a spool, the string would be unwinding at this speed as the two cars separated.

(b) If Alex sees car P brake to a halt in 10 s, what acceleration (assumed constant) will he measure for it?

SOLUTION: From Eq. 2-11 ($v = v_0 + at$) we have

$$a = \frac{v - v_0}{t} = \frac{0 - (-78 \text{ km/h})}{10 \text{ s}}$$
$$= \left(\frac{78 \text{ km/h}}{10 \text{ s}}\right)\left(\frac{1 \text{ m/s}}{3.6 \text{ km/h}}\right)$$
$$= 2.2 \text{ m/s}^2. \qquad \text{(Answer)}$$

(c) What acceleration would Barbara measure for the braking car?

SOLUTION: Barbara sees the initial speed of the car as -130 km/h, as we calculated in (a). Although the car has braked to rest, it is at rest only in Alex's reference frame. To Barbara, car P is not at rest at all but is receding from her at 52 km/h so that its final velocity in her reference frame is -52 km/h. Thus, from the relation $v = v_0 + at$,

$$a = \frac{v - v_0}{t} = \frac{(-52 \text{ km/h}) - (-130 \text{ km/h})}{10 \text{ s}}$$
$$= 2.2 \text{ m/s}^2. \qquad \text{(Answer)}$$

This is exactly the same acceleration that Alex measured, which reassures us that we made no mistakes.

4-9 RELATIVE MOTION IN TWO DIMENSIONS

Now we move from the scalar world of relative motion in one dimension to the vector world of relative motion in two (and, by extension, in three) dimensions. Figure 4-21 shows reference frames A and B, now two-dimensional. Our two observers are again watching a moving particle P. We once more assume that the two frames are separating at a constant velocity $\mathbf{v}_{BA}$ (the frames are inertial) and, for simplicity, we further assume that their x and y axes remain parallel to each other as they do so.

Let observers on frames A and B each measure the position of particle P at a certain instant. From the vector triangle in Fig. 4-21, we have the vector equation

$$\mathbf{r}_{PA} = \mathbf{r}_{PB} + \mathbf{r}_{BA}. \tag{4-27}$$

This relation is the vector equivalent of the scalar Eq. 4-24.

If we take the time derivative of Eq. 4-27, we find a connection between the (vector) velocities of the particle as measured by our two observers; namely,

$$\mathbf{v}_{PA} = \mathbf{v}_{PB} + \mathbf{v}_{BA}. \tag{4-28}$$

This is the vector equivalent of the scalar Eq. 4-25. Note that the order of the subscripts is the same as in that equation, and that again $\mathbf{v}_{BA}$ is the constant relative velocity of frame B as observed by the observer on frame A.

If we take the time derivative of Eq. 4-28, we obtain a connection between the two measured accelerations; namely,

$$\mathbf{a}_{PA} = \mathbf{a}_{PB}. \tag{4-29}$$

It remains true for motion in three dimensions that all observers on inertial reference frames will measure the same acceleration for a moving particle.

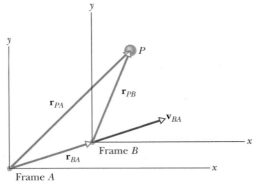

FIGURE 4-21 Reference frames in two dimensions. The vectors $\mathbf{r}_{PA}$ and $\mathbf{r}_{PB}$ show the positions of particle P in frames A and B, respectively. Vector $\mathbf{r}_{BA}$ shows the position of frame B with respect to frame A. Vector $\mathbf{v}_{BA}$ shows the (constant) separation velocity of the two frames.

SAMPLE PROBLEM 4-13

A bat detects an insect (lunch) while the two are flying with velocities $\mathbf{v}_{BG}$ and $\mathbf{v}_{IG}$, respectively, measured with respect to the ground. See Fig. 4-22a. What is the velocity $\mathbf{v}_{IB}$ of the insect with respect to the bat, in unit-vector notation?

SOLUTION: From Fig. 4-22a, the velocities of the insect and the bat, relative to the ground, are given by

$$\mathbf{v}_{IG} = (5.0 \text{ m/s})(\cos 50°)\mathbf{i} + (5.0 \text{ m/s})(\sin 50°)\mathbf{j}$$

and

$$\mathbf{v}_{BG} = (4.0 \text{ m/s})(\cos 150°)\mathbf{i} + (4.0 \text{ m/s})(\sin 150°)\mathbf{j},$$

where in each term, the angle is relative to the positive direction of the x axis. Now here is the key idea: the velocity $\mathbf{v}_{IB}$ of the *insect relative to the bat* is given by the vector sum of the velocity $\mathbf{v}_{IG}$ of the *insect relative to the ground* and the velocity $\mathbf{v}_{GB}$ of the *ground relative to the bat*; that is,

$$\mathbf{v}_{IB} = \mathbf{v}_{IG} + \mathbf{v}_{GB},$$

as shown in Fig. 4-22b. (Note that on the right-hand side of the equation, the two inner subscripts are the same. Also note that the two outer ones are the same as those on the left-hand side and appear in the same order.)

By definition the vector $\mathbf{v}_{GB}$ is in the direction opposite that of the vector $\mathbf{v}_{BG}$. Hence $\mathbf{v}_{GB} = -\mathbf{v}_{BG}$, and so

$$\mathbf{v}_{IB} = \mathbf{v}_{IG} + (-\mathbf{v}_{BG}).$$

Substituting the unit-vector expressions for $\mathbf{v}_{IG}$ and $\mathbf{v}_{BG}$ into this expression, we find

$$\mathbf{v}_{IB} = (5.0 \text{ m/s})(\cos 50°)\mathbf{i} + (5.0 \text{ m/s})(\sin 50°)\mathbf{j}$$
$$- (4.0 \text{ m/s})(\cos 150°)\mathbf{i} - (4.0 \text{ m/s})(\sin 150°)\mathbf{j}$$
$$= 3.21\mathbf{i} + 3.83\mathbf{j} + 3.46\mathbf{i} - 2.0\mathbf{j}$$
$$\approx (6.7 \text{ m/s})\mathbf{i} + (1.8 \text{ m/s})\mathbf{j}. \qquad \text{(Answer)}$$

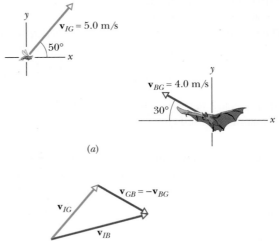

FIGURE 4-22 Sample Problem 4-13. (*a*) A bat detects an insect. (*b*) Velocity vectors of insect and bat.

SAMPLE PROBLEM 4-14

The compass in a plane indicates that the plane is headed (pointed) due east; its airspeed indicator reads 215 km/h. (Airspeed is speed relative to the air.) A steady wind of 65.0 km/h is blowing due north.

(a) What is the velocity of the plane with respect to the ground?

SOLUTION: The moving "particle" in this problem is the plane P. There are two reference frames, the ground (G) and the air mass (M). By a simple change of notation, we can rewrite Eq. 4-28 as

$$\mathbf{v}_{PG} = \mathbf{v}_{PM} + \mathbf{v}_{MG}. \qquad (4\text{-}30)$$

Figure 4-23a shows these vectors, which form a right triangle. The terms in Eq. 4-30 are, in sequence, the velocity of the plane with respect to the ground, the velocity of the plane with respect to the air, and the velocity of the air with respect to the ground (that is, the wind velocity). Note that the orientation of the plane in Fig. 4-23a is consistent with a due east reading on its compass. The plane is pointed due east but may not actually be moving in that direction.

The magnitude of the plane's velocity relative to the ground is found from the triangle in Fig. 4-23a:

$$\begin{aligned}
v_{PG} &= \sqrt{v_{PM}^2 + v_{MG}^2} \\
&= \sqrt{(215 \text{ km/h})^2 + (65.0 \text{ km/h})^2} \\
&= 225 \text{ km/h}. \qquad \text{(Answer)}
\end{aligned}$$

The angle α in Fig. 4-23a follows from

$$\begin{aligned}
\alpha &= \tan^{-1} \frac{v_{MG}}{v_{PM}} = \tan^{-1} \frac{65.0 \text{ km/h}}{215 \text{ km/h}} \\
&= 16.8°. \qquad \text{(Answer)}
\end{aligned}$$

Thus, with respect to the ground, the plane is flying at 225 km/h in a direction 16.8° north of east. Note that its speed relative to the ground (the "ground speed") is greater than its airspeed.

(b) If the pilot wishes to fly due east, what must be the heading? That is, what must the compass read?

SOLUTION: In this case the pilot must head into the wind somewhat, so that the wind velocity will help ensure that the velocity of the plane with respect to the ground points east. The wind velocity is unchanged from the preceding situation (a), and the vector diagram representing the new situation is as shown in Fig. 4-23b. Note that the three vectors still form a right triangle, as they did in Fig. 4-23a, and Eq. 4-30 still holds.

The pilot's ground speed is now, from Fig. 4-23b,

$$\begin{aligned}
v_{PG} &= \sqrt{v_{PM}^2 - v_{MG}^2} \\
&= \sqrt{(215 \text{ km/h})^2 - (65.0 \text{ km/h})^2} = 205 \text{ km/h}.
\end{aligned}$$

As the orientation of the plane in Fig. 4-23b indicates, the pilot must head into the wind by an angle θ given by

$$\theta = \sin^{-1} \frac{v_{MG}}{v_{PM}} = \sin^{-1} \frac{65.0 \text{ km/h}}{215 \text{ km/h}} = 17.6°. \quad \text{(Answer)}$$

Note that the ground speed is now less than the airspeed because the plane is headed into the wind.

(a)

(b)

FIGURE 4-23 Sample Problem 4-14. (a) An airplane, heading due east, is blown to the north. (b) To travel due east, the plane must head into the wind.

4-10 RELATIVE MOTION AT HIGH SPEEDS (OPTIONAL)

An orbiting satellite has a speed of 17,000 mi/h. Before you call that a high speed, you must answer this question: "High compared to what?" Nature has given us a standard; it is the speed of light c, where

$$c = 299{,}792{,}458 \text{ m/s} \qquad \text{(speed of light)}. \qquad (4\text{-}31)$$

As we shall examine later, no entity—be it particle or wave—can move faster than the speed of light, no matter what reference frame is used for observation. By this standard, all tangible large-scale objects—no matter how high their speeds seem by ordinary standards—are very slow indeed. The speed of the orbiting satellite, for example, is only 0.0025% of the speed of light. Subatomic particles such as electrons or protons, however, can acquire speeds very close to (but never equal to or greater than) the speed of light. Experiment shows, for example, that an electron accelerated through 10 million volts acquires a speed of $0.9988c$; if you accelerate it through 20 million volts, its speed increases, but only to $0.9997c$. The speed of light is a barrier that objects can approach but never reach. (Alas, the faster-than-light speeds utilized in science fiction, such as during "warp drive" in *Star Trek,* when the speed is $c2^n$, where n is the warp number, are only fictional.)

How can we be sure that the kinematics we have examined so far, which was developed by studying very slow, ordinary objects, also holds for very fast objects, such as high-speed electrons or protons? The answer, which we can find only from experiment, is that the kinematics for slow objects does *not* hold at speeds that approach the speed of light. Einstein's **theory of special relativity,** however, agrees with experiment at *all* speeds.

At "slow" speeds—all speeds that can be acquired by ordinary objects—the kinematic equations of Einstein's theory reduce to those of the kinematics we have examined. The failure of the "slow kinematics" is gradual, its predictions agreeing less and less well with experiment as the speed increases. Here is an example: Eq. 4-25,

$$v_{PA} = v_{PB} + v_{BA} \qquad \text{(slow speeds)},$$

relates the speed of particle P as seen by an observer in frame B to that seen by an observer in frame A. The corresponding equation in Einstein's theory is

$$v_{PA} = \frac{v_{PB} + v_{BA}}{1 + v_{PB}v_{BA}/c^2} \qquad \text{(all speeds)}. \qquad (4\text{-}32)$$

If $v_{PB} \ll c$ and $v_{BA} \ll c$ (which is always the case for ordinary objects), then the denominator in Eq. 4-32 is very close to unity and Eq. 4-32 reduces to Eq. 4-25, as we know it must.

The speed of light c is the central constant of Einstein's theory, and every relativistic equation contains it. A way to test the validity of such equations is to allow c to become infinitely large. Under those conditions, *all* speeds would be "slow" and "slow kinematics" would never fail. Putting $c \rightarrow \infty$ in Eq. 4-32 does indeed reduce that equation to Eq. 4-25.

SAMPLE PROBLEM 4-15

(Slow speeds) For the case where $v_{PB} = v_{BA} = 0.0001c$ ($= 67,000$ mi/h!), what do Eqs. 4-25 and 4-32 predict for v_{PA}?

SOLUTION: From Eq. 4-25,

$$v_{PA} = v_{PB} + v_{BA} = 0.0001c + 0.0001c$$
$$= 0.0002c. \qquad \text{(Answer)}$$

From Eq. 4-32,

$$v_{PA} = \frac{v_{PB} + v_{BA}}{1 + v_{PB}v_{BA}/c^2} = \frac{0.0001c + 0.0001c}{1 + (0.0001c)^2/c^2}$$
$$= \frac{0.0002c}{1.00000001} \approx 0.0002c. \qquad \text{(Answer)}$$

Conclusion: At any speed acquired by ordinary tangible objects, Eqs. 4-25 and 4-32 yield essentially the same answer. For such speeds, we can use Eq. 4-25 ("slow kinematics") without a second thought.

SAMPLE PROBLEM 4-16

(High speeds) For $v_{PB} = v_{BA} = 0.65c$, what do Eqs. 4-25 and 4-32 predict for v_{PA}?

SOLUTION: From Eq. 4-25,

$$v_{PA} = v_{PB} + v_{BA} = 0.65c + 0.65c$$
$$= 1.30c. \qquad \text{(Answer)}$$

From Eq. 4-32,

$$v_{PA} = \frac{v_{PB} + v_{BA}}{1 + v_{PB}v_{BA}/c^2} = \frac{0.65c + 0.65c}{1 + (0.65c)(0.65c)/c^2}$$
$$= \frac{1.30c}{1.423} = 0.91c. \qquad \text{(Answer)}$$

Conclusion: At high speeds, "slow kinematics" and special relativity predict very different results. "Slow kinematics" involves no upper limit on speed and can easily (as in this case) predict a speed greater than the speed of light. Special relativity, on the other hand, *never* predicts a speed that exceeds c, no matter how high the combining speeds. Experiment agrees with special relativity on all counts.

REVIEW & SUMMARY

Position Vector

The location of a particle relative to the origin of a coordinate system is given by a *position vector* $\mathbf{r}$, which in unit-vector notation is

$$\mathbf{r} = x\mathbf{i} + y\mathbf{j} + z\mathbf{k}. \qquad (4\text{-}1)$$

Here $x\mathbf{i}$, $y\mathbf{j}$, and $z\mathbf{k}$ are the *vector components* of position vector $\mathbf{r}$, and x, y, and z are its *scalar components*. A position vector is described by a magnitude and one or two angles for orientation, or by its vector or scalar components.

Displacement

If a particle moves so that its position vector changes from $\mathbf{r}_1$ to $\mathbf{r}_2$, then the particle's *displacement* $\Delta \mathbf{r}$ is

$$\Delta \mathbf{r} = \mathbf{r}_2 - \mathbf{r}_1. \qquad (4\text{-}2)$$

The displacement can also be written as

$$\Delta \mathbf{r} = (x_2 - x_1)\mathbf{i} + (y_2 - y_1)\mathbf{j} + (z_2 - z_1)\mathbf{k}, \qquad (4\text{-}3)$$

where coordinates (x_1, y_1, z_1) correspond to position vector $\mathbf{r}_1$ and coordinates (x_2, y_2, z_2) correspond to position vector $\mathbf{r}_2$.

Average Velocity

If a particle undergoes a displacement $\Delta \mathbf{r}$ in time Δt, its *average velocity* $\bar{\mathbf{v}}$ for that time interval is

$$\bar{\mathbf{v}} = \frac{\Delta \mathbf{r}}{\Delta t}. \tag{4-4}$$

Velocity

As Δt in Eq. 4-4 is shrunk to 0, $\bar{\mathbf{v}}$ reaches a limit called the (instantaneous) *velocity:*

$$\mathbf{v} = \frac{d\mathbf{r}}{dt}, \tag{4-6}$$

which can be rewritten in unit-vector notation as

$$\mathbf{v} = v_x\mathbf{i} + v_y\mathbf{j} + v_z\mathbf{k}, \tag{4-7}$$

where $v_x = dx/dt$, $v_y = dy/dt$, and $v_z = dz/dt$. When the position of a moving particle is plotted on a coordinate system, $\mathbf{v}$ is always tangent to the curve representing the particle's path.

Average Acceleration

If a particle's velocity changes from $\mathbf{v}_1$ to $\mathbf{v}_2$ in time interval Δt, its *average acceleration* during Δt is

$$\bar{\mathbf{a}} = \frac{\mathbf{v}_2 - \mathbf{v}_1}{\Delta t} = \frac{\Delta \mathbf{v}}{\Delta t}. \tag{4-9}$$

Acceleration

As Δt in Eq. 4-9 is shrunk to 0, $\bar{\mathbf{a}}$ reaches a limiting value called the (instantaneous) *acceleration,* which is

$$\mathbf{a} = \frac{d\mathbf{v}}{dt}. \tag{4-10}$$

In unit-vector notation,

$$\mathbf{a} = a_x\mathbf{i} + a_y\mathbf{j} + a_z\mathbf{k}, \tag{4-11}$$

where $a_x = dv_x/dt$, $a_y = dv_y/dt$, and $a_z = dv_z/dt$.

When $\mathbf{a}$ is constant, the components of $\mathbf{a}$, $\mathbf{v}$, and $\mathbf{r}$ along any axis can be treated as in the one-dimensional motion of Chapter 2.

Projectile Motion

Projectile motion is the motion of a particle that is launched with an initial velocity $\mathbf{v}_0$ and then has the free-fall acceleration $\mathbf{g}$. If $\mathbf{v}_0$ is expressed as a magnitude (the speed v_0) and an angle θ_0, the equations of motion along the horizontal x axis and vertical y axis are

$$x - x_0 = (v_0 \cos \theta_0)t, \tag{4-15}$$

$$y - y_0 = (v_0 \sin \theta_0)t - \tfrac{1}{2}gt^2, \tag{4-16}$$

$$v_y = v_0 \sin \theta_0 - gt, \tag{4-17}$$

$$v_y^2 = (v_0 \sin \theta_0)^2 - 2g(y - y_0). \tag{4-18}$$

The path of a particle in projectile motion is parabolic and is given by

$$y = (\tan \theta_0)x - \frac{gx^2}{2(v_0 \cos \theta_0)^2}, \tag{4-19}$$

where the origin has been chosen so that x_0 and y_0 are both zero. The particle's **range** R, which is the horizontal distance from the launch point to the point at which the particle returns to the launch height, is

$$R = \frac{v_0^2}{g} \sin 2\theta_0. \tag{4-20}$$

Uniform Circular Motion

If a particle travels along a circle or circular arc with radius r at constant speed v, it is in *uniform circular motion* and has an acceleration $\mathbf{a}$ of magnitude

$$a = \frac{v^2}{r}. \tag{4-22}$$

The direction of $\mathbf{a}$ is toward the center of the circle or circular arc, and $\mathbf{a}$ is said to be centripetal. The time for the particle to complete a circle is

$$T = \frac{2\pi r}{v}. \tag{4-23}$$

T is called the *period of revolution,* or simply the *period,* of the motion.

Relative Motion

When two frames of reference A and B are moving relative to each other at constant velocity, they are said to be **inertial reference frames.** The velocity of a moving particle as measured by an observer in frame A, in general, differs from that measured from frame B. The two measured velocities are related by

$$\mathbf{v}_{PA} = \mathbf{v}_{PB} + \mathbf{v}_{BA}, \tag{4-28}$$

in which $\mathbf{v}_{BA}$ is the velocity of B with respect to A. Both observers measure the same acceleration for the particle; that is,

$$\mathbf{a}_{PA} = \mathbf{a}_{PB}. \tag{4-29}$$

If speeds near the speed of light are involved, Eq. 4-25 ($v_{PA} = v_{PB} + v_{BA}$) must be replaced with an equation derived using the **special theory of relativity.** For one-dimensional motion, the correct result is

$$v_{PA} = \frac{v_{PB} + v_{BA}}{1 + v_{PB}v_{BA}/c^2}, \tag{4-32}$$

which becomes identical to Eq. 4-25 if the speeds are all negligible compared to the speed of light c.

QUESTIONS

1. Here are four descriptions for the velocity of a hockey puck in the xy plane, all in meters per second:
(1) $v_x = -3t^2 + 4t - 2$ and $v_y = 6t - 4$
(2) $v_x = -3$ and $v_y = -5t^2 + 6$
(3) $\mathbf{v} = 2t^2\mathbf{i} - (4t + 3)\mathbf{j}$
(4) $\mathbf{v} = -2t\mathbf{i} + 3\mathbf{j}$
For each description, determine whether the x and y components of the acceleration are constant, and whether the acceleration vector $\mathbf{a}$ is constant. In description 4, if $\mathbf{v}$ is in meters per second and t is in seconds, what must be the units of the coefficients -2 and 3?

2. Figure 4-24 shows three situations in which identical projectiles are launched from the ground (at the same levels) at identical initial speeds and angles. The projectiles do not land on the same terrain, however. Rank the situations according to the final speeds of the projectiles just before they land, greatest first.

(a)   (b)   (c)

FIGURE 4-24 Question 2.

3. At what point in the path of a projectile is the speed (a) a minimum and (b) a maximum?

4. At a certain instant, a fly ball has velocity $\mathbf{v} = 25\mathbf{i} - 4.9\mathbf{j}$ (x is horizontal, y is upward, and $\mathbf{v}$ is in meters per second). Has the ball passed the highest point of its trajectory?

5. You are to launch a rocket, from just above the ground, with one of the following initial velocity vectors: (1) $\mathbf{v}_0 = 20\mathbf{i} + 70\mathbf{j}$, (2) $\mathbf{v}_0 = -20\mathbf{i} + 70\mathbf{j}$, (3) $\mathbf{v}_0 = 20\mathbf{i} - 70\mathbf{j}$, (4) $\mathbf{v}_0 = -20\mathbf{i} - 70\mathbf{j}$. In your coordinate system, x runs along level ground and y increases upward. (a) Rank the vectors according to the launch speed of the projectile, greatest first. (b) Rank the vectors according to the time of flight of the projectile, greatest first.

6. A snowball is thrown from ground level (by someone in a hole) with initial speed v_0 at an angle of 45° relative to the (level) ground, on which the snowball later lands. If the launch angle is increased, do (a) the range and (b) the flight time increase, decrease, or stay the same?

7. A mud ball is launched 2 m above the ground with initial velocity $\mathbf{v}_0 = (2\mathbf{i} + 4\mathbf{j})$ m/s. What is its velocity just before it lands on a surface that is 2 m above the ground?

8. An airplane flying horizontally at a constant speed of 350 km/h over level ground releases a bundle of food supplies. Ignore the effect of the air on the bundle. What are the bundle's initial (a) vertical and (b) horizontal components of velocity? (c) What is its horizontal component of velocity just before hitting the ground? (d) If the airplane's speed were, instead, 450 km/h, would the time of fall be larger, smaller, or the same?

9. Figure 4-25 shows three paths for a kicked football. Ignoring the effects of air on the flight, rank the paths according to (a) time of flight, (b) initial vertical velocity component, (c) initial horizontal velocity component, and (d) initial speed. Place the greatest first in each part.

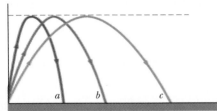

FIGURE 4-25 Question 9.

10. Figure 4-26 shows the velocity and acceleration of a particle at a particular instant in three situations. In which situation, and at that instant, is (a) the speed increasing, (b) the speed decreasing, (c) the speed not changing, (d) $\mathbf{v} \cdot \mathbf{a}$ positive, (e) $\mathbf{v} \cdot \mathbf{a}$ negative, and (f) $\mathbf{v} \cdot \mathbf{a} = 0$?

(1)   (2)   (3)

FIGURE 4-26 Question 10.

11. You are driving directly behind a pickup truck, going at the same speed as the truck. A crate falls from the bed of the truck to the road. (a) Will your car hit the crate before the crate hits the road if you neither brake nor swerve? (b) During the fall, is the horizontal speed of the crate more than, less than, or the same as that of the truck?

12. (a) Is it possible to be accelerating while traveling at constant speed? Is it possible to round a curve with (b) zero acceleration and (c) a constant magnitude of acceleration?

13. While riding in a moving car, you toss an egg directly upward. Does the egg tend to land behind you, in front of you, or back in your hands if the car is (a) traveling at a constant speed, (b) increasing in speed, and (c) decreasing in speed?

14. A passenger in an elevator drops a coin while the elevator descends at constant speed. According to (a) the passenger and (b) someone waiting at the next floor, is the acceleration of the coin equal to, less than, or more than g?

15. A pickpocket standing on the rear observation platform of a train moving with constant velocity drops a wallet over the rear guardrail. Describe the path of the wallet as seen by (a) the pickpocket, (b) her accomplice waiting alongside the track at the drop site, and (c) a police officer on another train moving at constant speed in the direction opposite that of the pickpocket's train, on a parallel track.

16. When the Germans shelled Paris from 70 mi away with the WWI long-range artillery piece nicknamed ''Big Bertha,'' the shells were fired at an angle greater than 45°; the Germans had discovered that a greater angle gave their gun a greater range, possibly even twice as long as that with a 45° angle. Does that result mean that the density of the air at high altitudes increases with altitude or decreases with altitude?

17. Figure 4-27 shows one of four star cruisers that are in a race. As each cruiser passes the starting line, a shuttle craft leaves the cruiser and races toward the finish line. You are the judge and are stationary relative to the start and finish lines. The speeds v_c of the four cruisers relative to you and the speeds v_s of the shuttles relative to their corresponding cruisers are, respectively: (1) 0.70c, 0.40c; (2) 0.40c, 0.70c; (3) 0.20c, 0.90c; (4) 0.50c, 0.60c. Without written calculation, determine (a) which shuttle wins the race and (b) which is last.

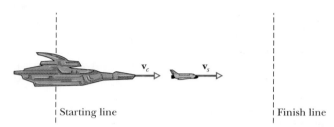

FIGURE 4-27 Question 17.

EXERCISES & PROBLEMS

SECTION 4-2 Position and Displacement

1E. A watermelon has the following coordinates: $x = -5.0$ m, $y = 8.0$ m, and $z = 0$ m. Find its position vector (a) in unit-vector notation and (b) in terms of the vector's magnitude and orientation. (c) Sketch the vector on a right-handed coordinate system. If the watermelon is moved to the xyz coordinates (3.00 m, 0 m, 0 m), what is its displacement (d) in unit-vector notation and (e) in terms of a magnitude and an orientation?

2E. The position vector for an electron is $\mathbf{r} = 5.0\mathbf{i} - 3.0\mathbf{j} + 2.0\mathbf{k}$, in meters. (a) Find the magnitude of $\mathbf{r}$. (b) Sketch the vector on a right-handed coordinate system.

3E. The position vector for a proton is initially $\mathbf{r} = 5.0\mathbf{i} - 6.0\mathbf{j} + 2.0\mathbf{k}$ and then later is $\mathbf{r} = -2.0\mathbf{i} + 6.0\mathbf{j} + 2.0\mathbf{k}$, all in meters. (a) What is the proton's displacement vector, and (b) to what plane is it parallel?

4E. A positron undergoes a displacement $\Delta\mathbf{r} = 2.0\mathbf{i} - 3.0\mathbf{j} + 6.0\mathbf{k}$, ending with the position vector $\mathbf{r} = 3.0\mathbf{j} - 4.0\mathbf{k}$, in meters. What was the positron's former position vector?

SECTION 4-3 Velocity and Average Velocity

5E. A plane flies 300 mi east from city A to city B in 45.0 min and then 600 mi south from city B to city C in 1.50 h. (a) What displacement vector represents the total trip? What are (b) the average velocity (a vector) and (c) the average speed for the trip?

6E. A train moving at a constant speed of 60.0 km/h moves east for 40.0 min, then in a direction 50.0° east of north for 20.0 min, and finally west for 50.0 min. What is the average velocity of the train during this run?

7E. In 3.50 h, a balloon drifts 21.5 km north, 9.70 km east, and 2.88 km in upward elevation from its release point on the ground. Find (a) the magnitude of its average velocity and (b) the angle its average velocity makes with the horizontal.

8E. An ion's position vector is initially $\mathbf{r} = 5.0\mathbf{i} - 6.0\mathbf{j} + 2.0\mathbf{k}$, and 10 s later it is $\mathbf{r} = -2.0\mathbf{i} + 8.0\mathbf{j} - 2.0\mathbf{k}$, all in meters. What was its average velocity during the 10 s?

9E. The position of an electron is given by $\mathbf{r} = 3.0t\mathbf{i} - 4.0t^2\mathbf{j} + 2.0\mathbf{k}$ (where t is in seconds and the coefficients have the proper units for $\mathbf{r}$ to be in meters). (a) What is $\mathbf{v}(t)$ for the electron? (b) In unit-vector notation, what is $\mathbf{v}$ at $t = 2.0$ s? (c) What are the magnitude and direction of $\mathbf{v}$ just then?

SECTION 4-4 Acceleration and Average Acceleration

10E. A proton initially has $\mathbf{v} = 4.0\mathbf{i} - 2.0\mathbf{j} + 3.0\mathbf{k}$ and then 4.0 s later has $\mathbf{v} = -2.0\mathbf{i} - 2.0\mathbf{j} + 5.0\mathbf{k}$ (in meters per second). (a) In unit-vector notation, what is the average acceleration $\bar{\mathbf{a}}$ over the 4.0 s? (b) What are the magnitude and orientation of $\bar{\mathbf{a}}$?

11E. A particle moves so that its position as a function of time in SI units is $\mathbf{r} = \mathbf{i} + 4t^2\mathbf{j} + t\mathbf{k}$. Write expressions for (a) its velocity and (b) its acceleration as functions of time.

12E. The position $\mathbf{r}$ of a particle moving in an xy plane is given by $\mathbf{r} = (2.00t^3 - 5.00t)\mathbf{i} + (6.00 - 7.00t^4)\mathbf{j}$. Here $\mathbf{r}$ is in meters and t in seconds. Calculate (a) $\mathbf{r}$, (b) $\mathbf{v}$, and (c) $\mathbf{a}$ when $t = 2.00$ s. (d) What is the orientation of a line that is tangent to the particle's path at $t = 2.00$ s?

13E. An ice boat sails across the surface of a frozen lake with constant acceleration produced by the wind. At a certain instant the boat's velocity is $6.30\mathbf{i} - 8.42\mathbf{j}$ in meters per second. Three seconds later, because of a wind shift, the boat is instantaneously at rest. What is its average acceleration during this 3 s interval?

14P. A particle starts from the origin at $t = 0$ with a velocity of $8.0\mathbf{j}$ m/s and moves in the xy plane with a constant acceleration of $(4.0\mathbf{i} + 2.0\mathbf{j})$ m/s². (a) At the instant the x coordinate of the particle is 29 m, what is its y coordinate? (b) What is the speed of the particle at this time?

15P. A particle leaves the origin with an initial velocity $\mathbf{v} =$

3.00i, in meters per second. It experiences a constant acceleration **a** = −1.00**i** − 0.500**j**, in meters per second squared. (a) What is the velocity of the particle when it reaches its maximum x coordinate? (b) Where is the particle at this time?

16P. The velocity **v** of a particle moving in the xy plane is given by **v** = $(6.0t − 4.0t^2)$**i** + 8.0**j**. Here **v** is in meters per second and t (> 0) is in seconds. (a) What is the acceleration when $t = 3.0$ s? (b) When (if ever) is the acceleration zero? (c) When (if ever) is the velocity zero? (d) When (if ever) does the speed equal 10 m/s?

17P. A particle A moves along the line $y = 30$ m with a constant velocity **v** ($v = 3.0$ m/s) directed parallel to the positive x axis (Fig. 4-28). A second particle B starts at the origin with zero speed and constant acceleration **a** ($a = 0.40$ m/s²) at the same instant that particle A passes the y axis. What angle θ between **a** and the positive y axis would result in a collision between these two particles? (If your computation involves an equation with a term such as t^4, substitute $u = t^2$ and then consider solving the resulting quadratic equation to get u.)

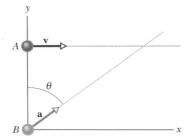

FIGURE 4-28
Problem 17.

SECTION 4-6 Projectile Motion Analyzed

In some of these problems, exclusion of the effects of the air is unwarranted but helps simplify the calculations.

18E. A dart is thrown horizontally toward the bull's-eye, point P on the dart board of Fig. 4-29, with an initial speed of 10 m/s. It hits at point Q on the rim, vertically below P, 0.19 s later. (a) What is the distance PQ? (b) How far away from the dart board did the dart thrower stand?

FIGURE 4-29
Exercise 18.

19E. A rifle is aimed horizontally at a target 100 ft away. The bullet hits the target 0.75 in. below the aiming point. (a) What is the bullet's time of flight? (b) What is its muzzle velocity?

20E. Electrons, like all other forms of matter, can undergo free fall. (a) If an electron is projected horizontally with a speed of 3.0 × 10⁶ m/s, how far will it fall in traversing 1.0 m of horizontal distance? (b) Does the answer increase or decrease if the initial speed is increased?

21E. In a cathode-ray tube, a beam of electrons is projected horizontally with a speed of 1.0 × 10⁹ cm/s into the region between a pair of horizontal plates 2.0 cm square. An electric field between the plates causes a constant downward acceleration of the electrons of magnitude 1.0 × 10¹⁷ cm/s². Find (a) the time required for an electron to pass through the plates, (b) the vertical displacement of the beam in passing through the plates (it does not run into a plate), and (c) the velocity of the beam as it emerges from the plates.

22E. A ball rolls horizontally off the edge of a tabletop that is 4.0 ft high. It strikes the floor at a point 5.0 ft horizontally away from the edge of the table. (a) How long was the ball in the air? (b) What was its speed at the instant it left the table?

23E. A projectile is fired horizontally from a gun that is 45.0 m above flat ground. The muzzle velocity is 250 m/s. (a) How long does the projectile remain in the air? (b) At what horizontal distance from the firing point does it strike the ground? (c) What is the magnitude of the vertical component of its velocity as it strikes the ground?

24E. A baseball leaves a pitcher's hand horizontally at a speed of 100 mi/h. The distance to the batter is 60 ft. (a) How long does it take for the ball to travel the first 30 ft horizontally? The second 30 ft? (b) How far does the ball fall under gravity during the first 30 ft of its horizontal travel? (c) During the second 30 ft? (d) Why aren't the quantities in (b) and (c) equal? (Ignore the effect of air resistance.)

25E. A projectile is launched with an initial speed of 30 m/s at an angle of 60° above the horizontal. Calculate the magnitude and direction of its velocity (a) 2.0 s and (b) 5.0 s after launch.

26E. A stone is catapulted rightward with an initial velocity of 20.0 m/s at an angle of 40.0° above level ground. Find its horizontal and vertical displacements (a) 1.10 s, (b) 1.80 s, and (c) 5.00 s after launch.

27E. You throw a ball from a cliff with an initial velocity of 15.0 m/s at an angle of 20.0° below the horizontal. Find (a) its horizontal displacement and (b) its vertical displacement 2.30 s later.

28E. You throw a ball with a speed of 25.0 m/s at an angle of 40.0° above the horizontal directly toward a wall (Fig. 4-30). The wall is 22.0 m from the release point of the ball. (a) How long does the ball take to reach the wall? (b) How far above the release point does the ball hit the wall? (c) What are the horizontal and vertical components of its velocity as it hits the wall? (d) When it hits, has it passed the highest point on its trajectory?

FIGURE 4-30
Exercise 28.

29E. (a) Prove that for a projectile fired from level ground at an angle θ_0 above the horizontal, the ratio of the maximum height H to the range R is given by $H/R = \frac{1}{4}\tan\theta_0$. See Fig. 4-31. (b) For what angle θ_0 does $H = R$?

30E. A projectile is fired from level ground at an angle θ_0 above the horizontal. (a) Show that the elevation angle ϕ of the highest point as seen from the launch point is related to θ_0, the elevation angle of projection, by $\tan\phi = \frac{1}{2}\tan\theta_0$. See Fig. 4-31 and Exercise 29. (b) Calculate ϕ for $\theta_0 = 45°$.

FIGURE 4-31 Exercises 29 and 30.

31E. A stone is projected at a cliff of height h with an initial speed of 42.0 m/s directed 60.0° above the horizontal, as shown in Fig. 4-32. The stone strikes at A, 5.50 s after launching. Find (a) the height h of the cliff, (b) the speed of the stone just before impact at A, and (c) the maximum height H reached above the ground.

FIGURE 4-32 Exercise 31.

32P. The launching speed of a certain projectile is five times the speed it has at its maximum height. Calculate the elevation angle at launching.

33P. In Sample Problem 4-6, find (a) the speed of the capsule when it hits the water and (b) the angle θ shown in Fig. 4-14.

34P. In the 1991 World Track and Field Championships in Tokyo, Mike Powell (Fig. 4-33) jumped 8.95 m, breaking the 23-year long-jump record set by Bob Beamon by a full 5 cm. Assume that Powell's speed on takeoff was 9.5 m/s (about equal to that of a sprinter) and that $g = 9.80$ in Tokyo. How close did Powell come to the maximum possible range in the absence of air resistance?

35P. A rifle with a muzzle velocity of 1500 ft/s shoots a bullet at a target 150 ft away. How high above the target must the rifle barrel be pointed to ensure that the bullet will hit the target?

36P. Show that the maximum height reached by a projectile is $y_{\max} = (v_0 \sin\theta_0)^2/2g$.

37P. A ball is shot from the ground into the air. At a height of 9.1 m, the velocity is observed to be $\mathbf{v} = 7.6\mathbf{i} + 6.1\mathbf{j}$ in meters per second ($\mathbf{i}$ horizontal, $\mathbf{j}$ upward). (a) To what maximum height will the ball rise? (b) What will be the total horizontal distance traveled by the ball? (c) What is the velocity of the ball (magnitude and direction) the instant before it hits the ground?

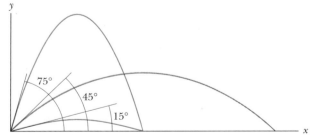

FIGURE 4-33 Problem 34. Mike Powell's jump.

38P. In a detective story, a body is found 15 ft from the base of a building and 80 ft below an open window. Would you guess the death to be accidental. Explain your answer.

39P. In Galileo's *Two New Sciences,* the author states that "for elevations [angles of projection] which exceed or fall short of 45° by equal amounts, the ranges are equal. . . ." Prove this statement. (See Fig. 4-34.)

FIGURE 4-34 Problem 39.

40P. The range of a projectile depends not only on v_0 and θ_0 but also on the value g of the free-fall acceleration, which varies from place to place. In 1936, Jesse Owens established a world's running broad jump record of 8.09 m at the Olympic Games at Berlin ($g = 9.8128$ m/s^2). Assuming the same values of v_0 and θ_0, by how much would his record have differed if he had competed instead in 1956 at Melbourne ($g = 9.7999$ m/s^2)?

41P. A third baseman wishes to throw to first base, 127 ft distant. His best throwing speed is 85 mi/h. (a) If he throws the ball horizontally 3.0 ft above the ground, how far from first base will it hit the ground? (b) At what upward angle must the third baseman throw the ball if the first baseman is to catch it 3.0 ft above the ground? (c) What will be the time of flight in that case?

42P. During volcanic eruptions, chunks of solid rock can be blasted out of the volcano; these projectiles are called *volcanic bombs*. Figure 4-35 shows a cross section of Mt. Fuji, in Japan. (a) At what initial speed would a bomb have to be ejected, at 35° to the horizontal, from the vent at *A* in order to fall at the foot of the volcano at *B*? Ignore, for the moment, the effects of air on the bomb's travel. (b) What would be the time of flight? (c) Would the effect of the air increase or decrease your answer in (a)?

FIGURE 4-35   Problem 42.

43P. At what initial speed must the basketball player throw the ball, at 55° above the horizontal, to make the foul shot, as shown in Fig. 4-36?

FIGURE 4-36   Problem 43.

44P. A football player punts the football so that it will have a "hang time" (time of flight) of 4.5 s and land 50 yd away. If the ball leaves the player's foot 5.0 ft above the ground, what initial velocity (magnitude and direction) must the ball have?

45P. A golfer tees off from the top of a rise, giving the golf ball an initial velocity of 43 m/s at an angle of 30° above the horizontal. The ball strikes the fairway a horizontal distance of 180 m from the tee. Assume the fairway is level. (a) How high is the rise above the fairway? (b) What is the speed of the ball as it strikes the fairway?

46P. A projectile is fired with an initial speed $v_0 = 30.0$ m/s from the level ground at a target on the ground a distance $R = 20.0$ m away (Fig. 4-37). Find the two projection angles.

FIGURE 4-37   Problem 46.

47P. What is the maximum vertical height to which a baseball player can throw a ball if his maximum throwing *range* is 60 m?

48P. A certain airplane has a speed of 180 mi/h and is diving at an angle of 30.0° below the horizontal when a radar decoy is released. (See Fig. 4-38.) The horizontal distance between the release point and the point where the decoy strikes the ground is 2300 ft. (a) How high was the plane when the decoy was released? (b) How long was the decoy in the air?

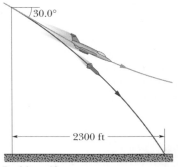

FIGURE 4-38   Problem 48.

49P. A football is kicked off with an initial speed of 64 ft/s at a projection angle of 45°. A receiver 60 yd away in the direction of the kick starts running to meet the ball at that instant. What must be his average speed if he is to catch the ball just before it hits the ground? Neglect air resistance.

50P. A ball rolls horizontally off the top of a stairway with a speed of 5.0 ft/s. The steps are 8.0 in. high and 8.0 in. wide. Which step will the ball hit first?

51P. An airplane, diving at an angle of 53.0° with the vertical, releases a projectile at an altitude of 730 m. The projectile hits the ground 5.00 s after being released. (a) What is the speed of the aircraft? (b) How far did the projectile travel horizontally during its flight? (c) What were the horizontal and vertical components of its velocity just before striking the ground?

52P. A ball is thrown horizontally from a height of 20 m and hits the ground with a speed that is three times its initial speed. What was the initial speed?

53P. (a) During a tennis match, a player serves at 23.6 m/s, the ball leaving the racquet horizontally 2.37 m above the court surface. By how much does the ball clear the net, which is 12 m away and 0.90 m high? (b) Suppose the player serves the ball as before except that the ball leaves the racquet at 5.00° below the horizontal. Does the ball clear the net now?

54P. In Sample Problem 4-8, suppose that a second identical harbor defense cannon is emplaced 30 m above sea level, rather that at sea level. How much longer is the horizontal distance from launch to impact of the second cannon than that of the first, which was found to be 690 m, if the elevation angle of fire is 45°?

55P. A batter hits a pitched ball whose center is 4.0 ft above the ground so that its angle of projection is 45° and its *range* is 350 ft. The ball will be a home run if it clears a 24 ft high fence that is 320 ft from home plate. Will the ball clear the fence? If so, by how much?

56P*. A football kicker can give the ball an initial speed of 25 m/s. Within what two elevation angles must he kick the ball to score a field goal from a point 50 m in front of goalposts whose horizontal bar is 3.44 m above the ground? (You might want to use $\sin^2 \theta + \cos^2 \theta = 1$ to get a relation between $\tan^2 \theta$ and $1/\cos^2 \theta$, and then solve the resulting quadratic equation.)

SECTION 4-7 Uniform Circular Motion

57E. In one model of the hydrogen atom, an electron orbits a proton in a circle of radius 5.28×10^{-11} m with a speed of 2.18×10^6 m/s. (a) What is the acceleration of the electron in this model? (b) What is the period of the motion?

58E. (a) What is the acceleration of a sprinter running at 10 m/s when rounding a bend with a turn radius of 25 m? (b) In what direction does the acceleration vector point?

59E. A magnetic field can force a charged particle to move in a circular path. Suppose that an electron experiences a radial acceleration of 3.0×10^{14} m/s² in a particular magnetic field. (a) What is the speed of the electron if the radius of its circular path is 15 cm? (b) What is the period of the motion?

60E. A sprinter runs at 9.2 m/s around a circular track with a centripetal acceleration of 3.8 m/s². (a) What is the track radius? (b) What is the period of the motion?

61E. An Earth satellite moves in a circular orbit 640 km above the Earth's surface. The period of the motion is 98.0 min. (a) What is the speed of the satellite? (b) What is the free-fall acceleration at the orbit height?

62E. Suppose a space probe can withstand the stresses of a 20*g* acceleration. (a) What is the minimum turning radius of such a craft moving at a speed of one-tenth the speed of light? (b) How long would it take to complete a 90° turn at this speed?

63E. A rotating fan completes 1200 revolutions every minute. Consider a point on the tip of a blade, at a radius of 0.15 m. (a) Through what distance does the point move in one revolution? (b) What is the speed of the point? (c) What is its acceleration? (d) What is the period of the motion?

64E. The fast train known as the TGV (Train à Grande Vitesse) that runs south from Paris, France, has a scheduled average speed of 216 km/h. (a) If the train goes around a curve at that speed and the acceleration experienced by the passengers is to be limited to 0.050*g*, what is the smallest radius of curvature for the track that can be tolerated? (b) If there is a curve with a 1.00 km radius, to what speed must the train be slowed to keep the acceleration below the limit?

65E. When a large star becomes a *supernova,* its core may be compressed so tightly that it becomes a *neutron star,* with a radius

of about 20 km (about the size of the San Francisco area). If a neutron star rotates once every second, (a) what is the speed of a particle on the star's equator and (b) what is the particle's centripetal acceleration in meters per second squared and in *g* units? (c) If the neutron star rotates even faster, what happens to the answers to (a) and (b)?

66E. An astronaut is rotated in a horizontal centrifuge at a radius of 5.0 m. (a) What is the astronaut's speed if the centripetal acceleration is 7.0*g*? (b) How many revolutions per minute are required to produce this acceleration? (c) What is the period of the motion?

67P. (a) What is the centripetal acceleration of an object on the Earth's equator owing to the rotation of the Earth? (b) What would the period of rotation of the Earth have to be for objects on the equator to have a centripetal acceleration equal to 9.8 m/s²?

68P. A carnival Ferris wheel has a 15 m radius and completes five turns about its horizontal axis every minute. (a) What is the period of the motion? (b) What is the centripetal acceleration of a passenger at the highest point? (c) What is the centripetal acceleration at the lowest point?

69P. Calculate the acceleration of a person at latitude 40° owing to the rotation of the Earth. (See Fig. 4-39.)

FIGURE 4-39 Problem 69.

70P. A particle *P* travels with constant speed on a circle of radius $r = 3.00$ m (Fig. 4-40) and completes one revolution in 20.0 s. The particle passes through *O* at $t = 0$. Find the magnitude and direction of each of the following vectors. (a) With respect to *O*, find the particle's position vector at $t = 5.00$ s, 7.50 s, and 10.0 s. For the 5.00 s interval from the end of the fifth second to the end of the tenth second, find the particle's (b) displacement and (c) average velocity. Find its (d) velocity and (e) acceleration at the beginning and end of that 5.00 s interval.

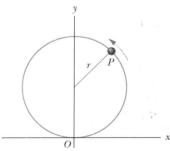

FIGURE 4-40 Problem 70.

71P. A boy whirls a stone in a horizontal circle 2.0 m above the ground by means of a string 1.5 m long. The string breaks, and the stone flies off horizontally and strikes the ground 10 m away. What was the centripetal acceleration of the stone while in circular motion?

SECTION 4-8 Relative Motion in One Dimension

72E. A boat is traveling upstream at 14 km/h with respect to the water of a river. The water itself is flowing at 9 km/h with respect to the ground. (a) What is the velocity of the boat with respect to the ground? (b) A child on the boat walks from front to rear at 6 km/h with respect to the boat. What is the child's velocity with respect to the ground?

73E. A person walks up a stalled 15 m long escalator in 90 s. When standing on the same escalator, now moving, the person is carried up in 60 s. How much time would it take that person to walk up the moving escalator? Does the answer depend on the length of the escalator?

74E. A transcontinental flight of 2700 mi is scheduled to take 50 min longer westward than eastward. The airspeed of the airplane is 600 mi/h, and the jet stream it will fly through is presumed to be moving either due east or due west. What assumptions about the jet stream wind velocity are made in preparing the schedule?

75E. A cameraman on a pickup truck is traveling westward at 40 mi/h while he videotapes a cheetah that is moving westward 30 mi/h faster than the truck. Suddenly, the cheetah stops, turns, and then runs at 60 mi/h eastward, as measured by a suddenly nervous crew member who stands alongside the cheetah's path. The change in the animal's velocity took 2.0 s. What was its acceleration from the perspective of the cameraman? From the perspective of the nervous crew member?

76P. The airport terminal in Geneva, Switzerland, has a "moving sidewalk" to speed passengers through a long corridor. Peter does not use the moving sidewalk; he takes 150 s to walk through the corridor. Paul, who simply stands on the moving sidewalk, covers the same distance in 70 s. Mary boards the sidewalk and walks along it. How long does Mary take to move through the corridor? Assume that Peter and Mary walk at the same speed.

SECTION 4-9 Relative Motion in Two Dimensions

77E. In rugby (Fig. 4-41) a player can legally pass the ball to a teammate as long as the pass is not "forward" (it must not have a velocity component parallel to the length of the field and directed toward the other team's goal). Suppose a player runs parallel to the field's length with a speed of 4.0 m/s while he passes the ball with a speed of 6.0 m/s relative to himself. What is the smallest angle from the forward direction that keeps the pass legal?

78E. Two highways intersect as shown in Fig. 4-42. At the instant shown, a police car P is 800 m from the intersection and moving at 80 km/h. Motorist M is 600 m from the intersection and moving at 60 km/h. (a) In unit-vector notation, what is the velocity of the motorist with respect to the police car? (b) For the

FIGURE 4-41 Exercise 77.

instant shown in Fig. 4-42, how does the direction of the velocity found in (a) compare to the line of sight between the two cars? (c) If the cars maintain their velocities, do the answers to (a) and (b) change as the cars move nearer the intersection?

FIGURE 4-42 Exercise 78.

79E. Snow is falling vertically at a constant speed of 8.0 m/s. At what angle from the vertical do the snowflakes appear to be falling as viewed by the driver of a car traveling on a straight road with a speed of 50 km/h?

80E. In a large department store, a shopper is standing on the "up" escalator, which is traveling at an angle of 40° above the horizontal and at a speed of 0.75 m/s. He passes his daughter, who is standing on the identical, adjacent "down" escalator. (See Fig. 4-43.) Find the velocity of the shopper relative to his daughter in unit-vector notation.

81P. A helicopter is flying in a straight line over a level field at a constant speed of 6.2 m/s and at a constant altitude of 9.5 m. A package is ejected horizontally from the helicopter with an initial velocity of 12 m/s relative to the helicopter, and in a direction opposite the helicopter's motion. (a) Find the initial speed of the package relative to the ground. (b) What is the horizontal distance between the helicopter and the package at the instant the package strikes the ground? (c) What angle does the velocity vector of the

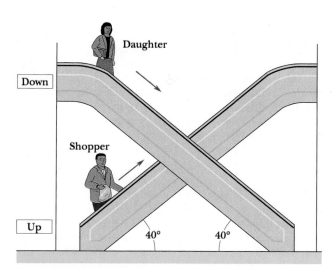

FIGURE 4-43 Exercise 80.

package make with the ground at the instant before impact, as seen from the ground?

82P. A train travels due south at 30 m/s (relative to the ground) in a rain that is blown toward the south by the wind. The path of each raindrop makes an angle of 70° with the vertical, as measured by an observer stationary on the Earth. An observer on the train, however, sees the drops fall perfectly vertically. Determine the speed of the raindrops relative to the Earth.

83P. A light plane attains an airspeed of 500 km/h. The pilot sets out for a destination 800 km to the north but discovers that the plane must be headed 20.0° east of north to fly there directly. The plane arrives in 2.00 h. What was the wind velocity vector?

84P. Two ships, *A* and *B*, leave port at the same time. Ship *A* travels northwest at 24 knots and ship *B* travels at 28 knots in a direction 40° west of south. (1 knot = 1 nautical mile per hour; see Appendix D.) (a) What are the magnitude and direction of the velocity of ship *A* relative to *B*? (b) After what time will they be 160 nautical miles apart? (c) What will be the bearing (direction) of *B* relative to *A* at that time?

85P. The New Hampshire State Police use aircraft to enforce highway speed limits. Suppose that one of the airplanes has a speed of 135 mi/h in still air. It is flying straight north so that it is at all times directly above a north–south highway. A ground observer tells the pilot by radio that a 70.0 mi/h wind is blowing, but neglects to give the wind direction. The pilot observes that in spite of the wind the plane can travel 135 mi along the highway in 1.00 h. In other words, the ground speed is the same as if there were no wind. (a) What is the direction of the wind? (b) What is the heading of the plane, that is, the angle between an axis along its length and the highway?

86P. A wooden boxcar is moving along a straight railroad track at speed v_1. A sniper fires a bullet (initial speed v_2) at it from a high-powered rifle. The bullet passes through both walls of the car, its entrance and exit holes being exactly opposite each other as viewed from within the car. From what direction, relative to the track, was the bullet fired? Assume that the bullet was not

deflected upon entering the car, but that its speed decreased by 20%. Take $v_1 = 85$ km/h and $v_2 = 650$ m/s. (Why don't you need to know the width of the boxcar?)

87P. A woman can row a boat at 4.0 mi/h in still water. (a) If she is crossing a river where the current is 2.0 mi/h, in what direction must her boat be headed if she wants to reach a point directly opposite her starting point? (b) If the river is 4.0 mi wide, how long will it take her to cross the river? (c) Suppose that instead of crossing the river she rows 2.0 mi *down* the river and then back to her starting point. How long will she take? (d) How long will she take to row 2.0 mi *up* the river and then back to her starting point? (e) In what direction should she head the boat if she wants to cross in the shortest possible time, and what is that time?

SECTION 4-10 Relative Motion at High Speeds

88E. As we head toward the center of our galaxy, we detect a burst of light traveling past us at speed *c* relative to us, toward that center. An observer on a second ship, traveling at a speed of 0.98*c* relative to us, also detects the light. What is the speed of the light as measured by the observer on the second ship if that ship is traveling (a) in the same direction as us and (b) in the opposite direction?

89E. An electron moves at speed 0.42*c* with respect to observer *B*. Observer *B* moves at speed 0.63*c* with respect to observer *A*, in the same direction as the electron. What does observer *A* measure for the speed of the electron?

90P. While traveling in a ship toward the star Betelgeuse, we detect a burst of protons traveling past us toward that star, with a velocity of 0.9800*c* relative to us. Detectors on a second ship, also traveling toward Betelgeuse along our line of travel, measure the velocity of the protons to be −0.9800*c* relative to them. What is the relative speed between us and the second ship?

91P. Galaxy Alpha is observed to be receding from us with a speed of 0.35*c*. Galaxy Beta, located in precisely the opposite direction, is also found to be receding from us at this same speed. What recessional speed would an observer on Galaxy Alpha find (a) for our galaxy and (b) for Galaxy Beta?

ELECTRONIC COMPUTATION

92. If the launch site of a projectile is above the landing place, a launch angle of 45° may not produce the greatest horizontal distance. Suppose a shot putter releases the shot from a point that is a distance *h* above a horizontal playing field. The initial speed of the shot is v_0 and the launch angle is θ. (a) Show that the horizontal distance from the putter's feet to the landing point is given by

$$d = \frac{v_0 \cos \theta}{g} [v_0 \sin \theta + \sqrt{v_0^2 \sin^2 \theta + 2gh}].$$

(b) Set up a program to calculate *d* for a series of values of θ, given v_0 and *h*. (c) Take $v_0 = 9.0$ m/s and $h = 2.1$ m and find to the nearest half degree the launch angle that produces the greatest horizontal distance. Also find that distance. (d) Does the result depend on the initial speed of the shot? Try $v_0 = 5.0$ m/s and $v_0 = 15$ m/s. (The last value is much greater than the speed generated by a champion shot putter.)

93. Immediately after launch, all projectiles move away from the launch site but later some move closer before moving further away again. It depends on the launch angle. Set up a program to compute the distance from the launch site to the projectile every 0.5 s from the time of launch to a few seconds after it is below the launch site. Take the initial speed to be 100 m/s and run the program for launch angles from 5° to 90°, with an interval of 5°. For each launch angle, search the list of distances to find when the projectile is moving away from and when it is moving toward the launch site. Classify the motions according to whether or not the projectile moves toward the launch site during any time interval and, if it does, estimate the interval. (See the article "Projectiles: Are They Coming or Going?" by James S. Walker in *The Physics Teacher*, May 1995.)

94. A batter strikes a baseball at a point 1.00 m above home plate, giving the ball an initial velocity of magnitude v_0, at angle θ above the horizontal. The ball is to barely clear a 2.40 m tall fence at a distance $R = 110$ m from home plate. (a) Find the minimum and maximum values of θ that allow this clearance for $v_0 = 35.0$ m/s. (*Hint:* rather than explicitly solving for θ in the equations of motion, find two expressions of θ, plot them, and then find their intersections.) (b) Find the minimum value of v_0 that allows the clearance for $\theta = 40.0°$. (c) Now find the minimum value of v_0 considering all values of θ and give the corresponding angle. (d) Repeat (c) for Fenway Park, with a 12.2 m

tall "wall" at 96.0 m down the foul line from home plate.

95. A golfer chips balls toward a vertical wall 20.0 m straight ahead, trying to hit a 30.0 cm diameter red circle painted on the wall. The target is centered about a point 1.20 m above the point where the wall intersects the horizontal ground. On one try, the ball leaves the ground with a speed of 15.0 m/s and at an angle of 35.0° above the horizontal. (a) How long does the ball take to reach the wall? (b) Does the ball hit the red circle? (c) What is the speed of the ball just before it hits? (d) Has the ball passed the highest point of its trajectory when it hits?

96. At one instant a bicyclist is 40.0 m due east of a park's flagpole, going due south with a speed of 10.0 m/s. Then 30.0 s later, the cyclist is 40.0 m due north of the flagpole, going due east with a speed of 10.0 m/s. Find (a) the displacement, (b) the average velocity, and (c) the average acceleration of the cyclist during the 30.0 s interval. (d) Compute $(\mathbf{v}_f - \mathbf{v}_i)/2$, where $\mathbf{v}_f$ and $\mathbf{v}_i$ are the velocities at the end and at the beginning of the 30.0 s interval, respectively.

97. At one instant a butterfly has the position vector $\mathbf{D}_i = (2.00 \text{ m})\mathbf{i} + (3.00 \text{ m})\mathbf{j} + (1.00 \text{ m})\mathbf{k}$ relative to the base of a bird-bath. Then 40 s later, the butterfly has the position vector $\mathbf{D}_f = (3.00 \text{ m})\mathbf{i} + (1.00 \text{ m})\mathbf{j} + (2.00 \text{ m})\mathbf{k}$. Find (a) the displacement (in unit-vector notation), (b) the magnitude of the displacement, (c) the average velocity, and (d) the average speed of the butterfly during the 40.0 s interval.

5
Force and Motion—I

On April 4, 1974, John Massis of Belgium managed to move two passenger cars belonging to New York's Long Island Railroad. He did so by clamping his teeth down on a bit that was attached to the cars with a rope and then leaning backward while pressing his feet against the railway ties. The cars weighed about 80 tons. Did Massis have to pull with superhuman force to accelerate them?

5-1 WHAT CAUSES AN ACCELERATION?

If you see the velocity of a particle-like body change in either magnitude or direction, you know that something must have *caused* that change (that acceleration). Indeed, out of common experience, you know that the change in velocity must be due to an interaction between the body and something in its surroundings. For example, if you see a hockey puck that is sliding across an ice rink suddenly stop or suddenly change direction, you will suspect that the puck bumped into a slight hill on the ice surface.

An interaction that causes an acceleration of a body is called a **force**, which is, loosely speaking, a push or pull. For example, the bump on the hockey puck by the hill is a push on the puck, causing an acceleration. The relationship between a force and the acceleration it causes was first understood by Isaac Newton (1642–1727) and is the subject of this chapter. The study of that relationship, as Newton presented it, is called *Newtonian mechanics*. We will focus on its three primary laws of motion.

Newtonian mechanics does not apply to all situations. As we discussed in Chapter 4, if the speeds of the interacting bodies are an appreciable fraction of the speed of light, we must replace Newtonian mechanics with Einstein's special theory of relativity, which holds at any speed, including those near the speed of light. If the interacting bodies are on the scale of atomic structure (for example, they might be electrons within an atom), we must replace Newtonian mechanics with quantum mechanics. Physicists now view Newtonian mechanics as a special case of these two more comprehensive theories. Still, it is a very important special case because it applies to the motion of objects ranging in size from the very small (almost on the scale of atomic structure) to astronomical (objects such as galaxies and clusters of galaxies). Let us now examine the first law of motion in Newtonian mechanics.

5-2 NEWTON'S FIRST LAW

Before Newton formulated his mechanics, it was thought that some influence, a "force," was needed to keep a body moving at constant velocity. Similarly, a body was thought to be in its "natural state" when it was at rest. For it to move with constant velocity, it seemingly had to be propelled in some way, by a push or a pull. Otherwise, it would "naturally" stop moving.

These ideas were reasonable. If you send a book sliding across a wooden floor, it does indeed slow and then stop. If you want to make it move across the floor with constant velocity, you have to continuously pull or push it.

Slide the book over the ice of a skating rink, however,

and it goes a lot farther. You can imagine longer and more slippery surfaces, over which the book would slide farther and farther. In the limit you can think of a long, extremely slippery surface (said to be a **frictionless surface**), over which the book would hardly slow. (We can in fact come close to this situation in the laboratory, by sending a book sliding over a horizontal air table, across which it moves on a film of air.)

We are led to conclude that you do *not* need a force to keep a body moving with constant velocity. And that leads us to the first of Newton's three laws of motion:

> **Newton's First Law:** Consider a body on which no force acts. If the body is at rest, it will remain at rest. If the body is moving with constant velocity, it will continue to do so.

This law fits in nicely with what we discussed in Section 4-8 about reference frames: a body on which no force acts may be stationary in one frame and moving at constant velocity with respect to another. (*Rest* and *moving with constant velocity* are not all that different.)

Newton's first law can be interpreted as a statement about reference frames, in that it defines the kinds of reference frames in which the laws of Newtonian mechanics hold: frames with constant separation velocity. From this point of view the first law is expressed as follows:

> **Newton's First Law:** If no force acts on a body, we can always find a reference frame in which that body has no acceleration.

Newton's first law is sometimes called the *law of inertia,* and the reference frames that it defines are called *inertial reference frames* or just *inertial frames.*

Figure 5-1 shows how you can test a particular frame to see whether it is an inertial frame. With the railroad car at rest, mark the position of the stationary pendulum bob on the table. With the car in motion, the bob remains over the mark *only* if the car is moving in a straight line at

FIGURE 5-1 Testing a railroad car to see whether it is an inertial reference frame.

constant speed. The car is then an inertial frame. If the car is gaining or losing speed or is rounding a corner, the bob moves from its mark, and the car is then a noninertial reference frame.

5-3 FORCE

A force causes the acceleration of a body. We now wish to define the unit of force carefully, in terms of the acceleration that it gives to a standard reference body. As the standard body, we use (or rather we imagine that we use) the standard kilogram of Fig. 1-6. This body has been assigned, exactly and by definition, a mass of 1 kg.

We put the standard body on a horizontal frictionless table and pull the body to the right (Fig. 5-2) so that by trial and error, it eventually experiences a measured acceleration of 1 m/s². We then declare, as a matter of definition, that the force we are exerting on the standard body has a magnitude of 1 newton (abbreviated N).

We can exert a 2 N force on our standard body by pulling it so that its measured acceleration is 2 m/s², and so on. Thus in general, if our standard body of 1 kg mass has an acceleration a, we know that force F must be acting on it and that the magnitude of the force (in newtons) is equal to the magnitude of the acceleration (in meters per second per second).

Thus, a force is measured by the acceleration it produces. But acceleration is a vector quantity, with both magnitude and direction. Is force also a vector quantity? We can easily assign a direction to a force (just assign the direction of the acceleration), but that is not sufficient. We must prove by experiment that forces are vector quantities. Actually, that has been done: forces are indeed vector quantities; they have magnitudes and directions and they combine according to the vector rules of Chapter 3.

Henceforth, we shall use boldface letters, most often **F**, to represent forces. And we shall use the symbol Σ**F** for the vector sum of several forces, which we call the **resultant force** or the **net force**. As with other vectors, a force or a net force can have components along coordinate axes. Finally, we note that Newton's first law holds not only when there is no force on a body, but also when the net force is equal to zero.

FIGURE 5-2 A force **F** on the standard kilogram gives that body an acceleration **a**.

CHECKPOINT **1:** In the figure, two perpendicular forces **F**₁ and **F**₂ are combined in six different ways. Which ways may be used to correctly determine the net force Σ**F**?

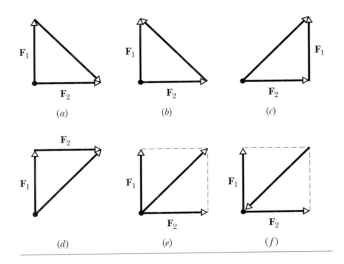

5-4 MASS

Everyday experience tells us that a given force produces different magnitudes of acceleration for different bodies. Put a baseball and a bowling ball on the floor and give both the same sharp kick. Even if you don't actually do this, you know the result: the baseball receives a noticeably larger acceleration than the bowling ball. The two accelerations differ because the mass of the baseball differs from the mass of the bowling ball. But what, exactly, is mass?

We can explain how to measure mass by imagining a series of experiments. In the first experiment we exert a force on our standard body, whose mass m_0 is defined to be 1.0 kg. Suppose that the standard body accelerates at 1.0 m/s². We can then say the force on that body is 1.0 N.

We next apply that same force (we would need some way of being certain it is the same force) to a second body, body X, whose mass is not known. Suppose we find that this body X accelerates at 0.25 m/s². We know that a *less massive* baseball receives a *greater acceleration* than a more massive bowling ball when the same force (kick) is applied to both. Let us then make the following conjecture: the ratio of the masses of two bodies is equal to the inverse of the ratio of their accelerations when the same force is applied to both. For body X and the standard body, this tells us that

$$\frac{m_X}{m_0} = \frac{a_0}{a_X}.$$

Solving for m_X yields

$$m_X = m_0 \frac{a_0}{a_X} = (1.0 \text{ kg}) \frac{1.0 \text{ m/s}^2}{0.25 \text{ m/s}^2} = 4.0 \text{ kg}.$$

Our conjecture will be useful, of course, only if it continues to hold when we change the applied force to other values. For example, if we apply an 8.0 N force to the standard body, we obtain an acceleration of 8.0 m/s². And when the 8.0 N force is applied to body X, we obtain an acceleration of 2.0 m/s². Our conjecture then gives us

$$m_X = m_0 \frac{a_0}{a_X} = (1.0 \text{ kg}) \frac{8.0 \text{ m/s}^2}{2.0 \text{ m/s}^2} = 4.0 \text{ kg},$$

consistent with our first experiment. Many experiments yielding similar results indicate that our conjecture provides a consistent and reliable means of assigning a mass to any given body.

Our measurement experiments indicate that mass is an *intrinsic* characteristic of a body—that is, a characteristic that automatically comes with the existence of the body. They also indicate that mass is a scalar quantity. However, the nagging question remains: What, exactly, is mass?

Since the word *mass* is used in everyday English, we should have some intuitive understanding of it, maybe something that we can physically sense. Is it a body's size, weight, or density? The answer is no, although those characteristics are sometimes confused with mass. We can say only that *the mass of a body is the characteristic that relates a force on the body to the resulting acceleration.* Mass has no more familiar definition than that; you can have a physical sensation of mass only when you attempt to accelerate a body, as in the kicking of a baseball or a bowling ball.

5-5 NEWTON'S SECOND LAW

All the definitions, experiments, and observations that we have described so far can be summarized in a simple vector equation, which is called Newton's second law of motion:

$$\sum \mathbf{F} = m\mathbf{a} \qquad \text{(Newton's second law).} \qquad (5\text{-}1)$$

In using Eq. 5-1, we must first be quite certain what body we are applying it to. Then $\sum \mathbf{F}$ in Eq. 5-1 is the vector sum, or net force, of *all* the forces that act *on* that body. Only forces that act *on* the body are to be included, not forces acting on other bodies that might be involved in a given problem. Finally, $\sum \mathbf{F}$ includes only *external* forces, that is, forces exerted on the body by other bodies. We do not include internal forces, in which one part of the body exerts a force on another part.

Like other vector equations, Eq. 5-1 is equivalent to three scalar equations:

$$\sum F_x = ma_x, \qquad \sum F_y = ma_y, \qquad \sum F_z = ma_z. \qquad (5\text{-}2)$$

These equations relate the three components of the net force acting on a body to the three components of the acceleration of that body.

You should note that Newton's second law includes the formal statement of Newton's first law as a special case. That is, if no force acts on a body, Eq. 5-1 tells us that the body will not be accelerated.

For SI units, Eqs. 5-2 tell us that

$$1 \text{ N} = (1 \text{ kg})(1 \text{ m/s}^2) = 1 \text{ kg} \cdot \text{m/s}^2, \qquad (5\text{-}3)$$

consistent with our discussion in Section 5-3. Although we shall use SI units almost exclusively from now on, other systems of units are still in use. Chief among these are the British system and the CGS (centimeter–gram–second) system. Table 5-1 shows the units in which Eqs. 5-1 and 5-2 are expressed in these systems. (See also Appendix D.)

To solve problems with Newton's second law, we often draw a **free-body diagram,** representing the body by a dot and each external force (or the net force $\sum \mathbf{F}$) that acts on the body by a vector with its tail on the dot. (Instead of the dot, we can sketch the body.) A set of coordinate axes is included, and sometimes a vector representing the acceleration of the body is also included.

We then should start the solution with the vector relation of Eq. 5-1. However, we shall usually quickly switch our attention to one or more of the scalar relations of Eqs. 5-2 and work along one axis at a time. The first relation of Eqs. 5-2 tells us that the sum of all the force components along the x axis causes the x component a_x of the body's acceleration but causes no acceleration in the y and z directions. Or, turned around, it says that the acceleration component a_x is caused only by the sum of the force components along the x axis. Similarly, the acceleration component a_y along the y axis is caused only by the sum of the force components along the y axis; and the acceleration component a_z along the z axis is caused only by the sum of the force components along the z axis. In general we have:

> The acceleration component along a given axis is caused only by the sum of the force components along that *same* axis and not by force components along some other axis.

In Sample Problem 5-1 we shall be concerned with a single force that acts along an x axis. So, we shall not need to find several force components. However, in Sample Problem 5-2, three forces act on a body and two make

TABLE 5-1 UNITS IN NEWTON'S SECOND LAW (Eqs. 5-1 and 5-2)

SYSTEM	FORCE	MASS	ACCELERATION
SI	newton (N)	kilogram (kg)	m/s^2
CGS[a]	dyne	gram (g)	cm/s^2
British[b]	pound (lb)	slug	ft/s^2

[a]1 dyne $= 1$ g·cm/s².   [b]1 lb $= 1$ slug·ft/s².

nonzero angles with an x axis and a y axis. In this two-dimensional situation, we must find the force components along the x axis and the y axis, and then use, separately, the first two relations of Eqs. 5-2.

Sample Problem 5-2 is also an example of a general type of problem: the body does not accelerate ($\mathbf{a} = 0$) in spite of the external forces on it. In such a situation, Eq. 5-1 tells us that $\Sigma\mathbf{F} = 0$. That is, the net force is zero, the forces acting on the body *balance* each other, and the body is said to be in *equilibrium*.

However, if we switch our attention to Eqs. 5-2, we see something that is more useful to problem solving. The lack of acceleration means that $a_x = 0$ and thus that $\Sigma F_x = 0$. That is, the x components of the forces also balance each other. If we replace ΣF_x with the actual x components of the forces involved, we then have an algebraic relation we can use in the solution. Similarly, $a_y = 0$ and thus $\Sigma F_y = 0$; after replacing ΣF_y with the actual y components of the forces involved, we have another algebraic relation.

In some problems you might find that the force components along one axis balance, but those along a second axis do not. So, there must be acceleration along that second axis only.

How will you know what to do in any given situation? Experience helps, and that is why this chapter contains many Sample Problems.

CHECKPOINT **2:** The figure shows two horizontal forces moving a block along a frictionless floor. Assume that a third horizontal force $\mathbf{F}_3$ also acts on the block. What are the magnitude and direction of $\mathbf{F}_3$ when the block is (a) stationary and (b) moving to the left with a constant speed of 5 m/s?

3 N   5 N

SAMPLE PROBLEM 5-1

A student (with cleated boots) pushes a loaded sled whose mass m is 240 kg for a distance d of 2.3 m over the frictionless surface of a frozen lake. He exerts a constant horizontal force

F, with magnitude $F = 130$ N, as he does so (see Fig. 5-3a).

(a) If the sled starts from rest, what is its final velocity?

SOLUTION: Figure 5-3b is a free-body diagram for the situation. We lay out a horizontal x axis, we take the direction of increasing x to be to the right, and we treat the sled as a particle, represented by a dot. We assume that the x component F_x of the force **F** exerted by the student is the only horizontal force acting on the sled. We can then find the magnitude of the acceleration a_x of the sled from Newton's second law:

$$a_x = \frac{F_x}{m} = \frac{130 \text{ N}}{240 \text{ kg}} = 0.542 \text{ m/s}^2.$$

Because the acceleration is constant, we can use Eq. 2-16, $v^2 = v_0^2 + 2a(x - x_0)$, to find the final velocity. Putting $v_0 = 0$ and $x - x_0 = d$, and identifying a_x as a, we solve for v:

$$v = \sqrt{2ad}$$
$$= \sqrt{(2)(0.542 \text{ m/s}^2)(2.3 \text{ m})} = 1.6 \text{ m/s}. \qquad \text{(Answer)}$$

The force, the acceleration, the displacement, and thus the final velocity of the sled are all positive, which means that they all point to the right in Fig. 5-3b.

(b) The student now wants to reverse the direction of the velocity of the sled in 4.5 s. With what constant force must he push on the sled to do so?

SOLUTION: Let us first find the constant acceleration required to reverse the sled's velocity in 4.5 s, using Eq. 2-11 ($v = v_0 + at$). Solving for a gives

(a)

(b)

(c)

FIGURE 5-3 Sample Problem 5-1. (a) A student pushes a loaded sled over a frictionless surface. (b) A free-body diagram for part (a) in the problem, showing the net force acting on the sled and the acceleration the net force produces. (c) A free-body diagram for part (b). The student now pushes in the opposite direction on the sled, reversing its acceleration.

$$a = \frac{v - v_0}{t} = \frac{(-1.6 \text{ m/s}) - (1.6 \text{ m/s})}{4.5 \text{ s}}$$

$$= -0.711 \text{ m/s}^2.$$

This is larger in magnitude than the acceleration in (a), namely 0.542 m/s², so it stands to reason that the student must push with a greater force this time. We find this greater force from Eqs. 5-2, with a_x being a:

$$F_x = ma_x = (240 \text{ kg})(-0.711 \text{ m/s}^2)$$

$$= -171 \text{ N.} \qquad \text{(Answer)}$$

The minus sign shows that the student must push the sled in the direction of decreasing x, that is, to the left in Fig. 5-3c, the free-body diagram for this situation.

SAMPLE PROBLEM 5-2

In a two-dimensional tug-of-war, Alex, Betty, and Charles pull on an automobile tire, at angles as shown in Fig. 5-4a, which is an overhead view. The tire remains stationary in spite of the three pulls. Alex pulls with force $\mathbf{F}_A$ of magnitude 220 N, and Charles pulls with force $\mathbf{F}_C$ of magnitude 170 N. The direction of $\mathbf{F}_C$ is not given. What is the magnitude of Betty's force $\mathbf{F}_B$?

SOLUTION: Figure 5-4b is a free-body diagram for the tire. Because the acceleration of the tire is zero, Eq. 5-1 tells us that the net force on the tire must also be zero:

$$\Sigma \mathbf{F} = \mathbf{F}_A + \mathbf{F}_B + \mathbf{F}_C = m\mathbf{a} = 0.$$

This vector relation is equivalent to the first two scalar relations of Eqs. 5-2. Along the x axis we have

$$\Sigma F_x = F_{Ax} + F_{Bx} + F_{Cx} = 0, \qquad (5\text{-}4)$$

and along the y axis we have

$$\Sigma F_y = F_{Ay} + F_{By} + F_{Cy} = 0. \qquad (5\text{-}5)$$

Using the given data and the angles in Fig. 5-4b, we now substitute into Eqs. 5-4 and 5-5, using signs to indicate the directions of the vector components. Equation 5-4 becomes

$$\Sigma F_x = -F_A \cos 47.0° + 0 + F_C \cos \phi = 0.$$

Substituting known values yields

$$-(220 \text{ N})(\cos 47.0°) + 0 + (170 \text{ N})(\cos \phi) = 0,$$

which gives us

$$\phi = \cos^{-1} \frac{(220 \text{ N})(\cos 47.0°)}{170 \text{ N}} = 28.0°.$$

Similarly, Eq. 5-5 becomes

$$\Sigma F_y = F_A \sin 47.0° - F_B + F_C \sin \phi = 0,$$

in which we have substituted $-F_B$ for F_{By} because Betty's pull is entirely along the negative direction of y. Substituting known values here yields

$$(220 \text{ N})(\sin 47.0°) - F_B + (170 \text{ N})(\sin \phi) = 0.$$

We next substitute 28.0° for ϕ and solve the equation for F_B, finding that

$$F_B = (220 \text{ N})(\sin 47.0°) + (170 \text{ N})(\sin 28.0°)$$

$$= 241 \text{ N.} \qquad \text{(Answer)}$$

Note that we first had to solve the equations along the x axis to use $\phi = 28.0°$ in the equations along the y axis. If we had started with the equations along the y axis, we would have just gotten "stuck." So, if you get stuck on one axis, try another.

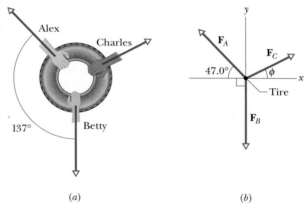

(a) $\qquad$ (b)

FIGURE 5-4 Sample Problem 5-2. (a) An overhead view of three people pulling on a tire. (b) A free-body diagram for the tire.

SAMPLE PROBLEM 5-3

Figure 5-5a shows an overhead view of a 2 kg cookie tin being accelerated at 8 m/s² across a frictionless surface by three horizontal forces. Forces $\mathbf{F}_1$ and $\mathbf{F}_2$ have magnitudes of 10 N and 12 N, respectively. Figure 5-5b is an incomplete free-body diagram for the situation, with the acceleration $\mathbf{a}$ included. What is the third force $\mathbf{F}_3$ in unit vector notation?

SOLUTION: The net force of the three horizontal forces causes the horizontal acceleration, and Eq. 5-1 gives us

$$\Sigma \mathbf{F} = \mathbf{F}_1 + \mathbf{F}_2 + \mathbf{F}_3 = m\mathbf{a}.$$

From Eqs. 5-2, we have, along the x axis,

$$\Sigma F_x = F_{1x} + F_{2x} + F_{3x} = ma_x, \qquad (5\text{-}6)$$

and along the y axis

$$\Sigma F_y = F_{1y} + F_{2y} + F_{3y} = ma_y. \qquad (5\text{-}7)$$

By rewriting Eq. 5-6 in terms of magnitudes and angles, and including signs to indicate directions, we obtain

$$-F_1 \cos 60° + 0 + F_{3x} = ma \sin 30°.$$

Substitution of the given data yields

$$-(10 \text{ N}) \cos 60° + 0 + F_{3x} = (2 \text{ kg})(8 \text{ m/s}^2) \sin 30°,$$

which tells us that

$$F_{3x} = (10 \text{ N}) \cos 60° + (2 \text{ kg})(8 \text{ m/s}^2) \sin 30°$$
$$= 13 \text{ N}.$$

Similarly, Eq. 5-7 becomes

$$-F_1 \sin 60° + F_2 + F_{3y} = -ma \cos 30°$$

and then

$$-(10 \text{ N}) \sin 60° + 12 \text{ N} + F_{3y} = -(2 \text{ kg})(8 \text{ m/s}^2) \cos 30°,$$

which tells us that

$$F_{3y} = -17.2 \text{ N} \approx 17 \text{ N}.$$

Thus, the third force is

$$\mathbf{F}_3 = (13 \text{ N})\mathbf{i} - (17 \text{ N})\mathbf{j}. \qquad \text{(Answer)}$$

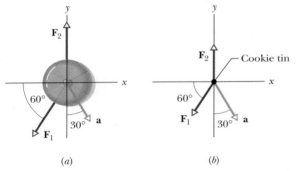

(a)   (b)

FIGURE 5-5 Sample Problem 5-3. (*a*) An overhead view of a cookie tin that is being accelerated by three horizontal forces, two of which are shown. (*b*) A free-body diagram for the cookie tin.

CHECKPOINT 3: The figure shows overhead views of four situations in which two forces accelerate the same block across a frictionless floor. Rank the situations according to the magnitudes of (a) the net force on the block and (b) the acceleration of the block, greatest first.

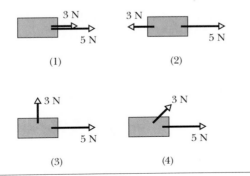

(1)   (2)

(3)   (4)

5-6 SOME PARTICULAR FORCES

Weight

The **weight W** of a body is a force that pulls the body directly toward a nearby astronomical body; in everyday circumstances that astronomical body is the Earth. The force is primarily due to an attraction—called a **gravitational attraction**—between the two bodies; we discuss it in detail in Chapter 14. For now we consider only situations in which a body with mass m is located at a point where the free-fall acceleration has magnitude g. Then the magnitude W of the weight (force) vector acting on the body is

$$W = mg. \tag{5-8}$$

The *weight vector* itself can be written either as

$$\mathbf{W} = -mg\mathbf{j} = -W\mathbf{j} \tag{5-9}$$

(where $+\mathbf{j}$ points upward, away from the Earth), or as

$$\mathbf{W} = m\mathbf{g}, \tag{5-10}$$

where $\mathbf{g}$ represents the free-fall acceleration vector. In many cases, the choice of notation is up to you, but you need to understand what you mean by it and not get tripped up by, say, writing Eq. 5-8 when you mean Eq. 5-10.

Since weight is a force, its SI unit is the newton. *It is not mass,* and its magnitude at any given location depends on the value of g there. A bowling ball might weigh 71 N on the Earth, but only 12 N on the Moon, where the free-fall acceleration is different. The ball's mass, 7.2 kg, is the same in either place, because mass is an intrinsic property of the ball alone. (If you want to lose weight, climb a mountain. Not only will the exercise reduce your mass, but the increased elevation means you are farther from the center of the Earth, and that means the value of g is less. So your weight will be less.)

Normally we assume that weight is measured from an inertial frame. If it is, instead, measured from a noninertial frame (an example comes up in Sample Problems 5-11*b* and *c*), the measurement gives an **apparent weight** instead of the actual weight.

We can *weigh* a body by placing it on one of the pans of an equal-arm balance (Fig. 5-6) and then adding reference bodies (whose masses are known) on the other pan until we strike a balance. The masses on the pans then match, and we know the mass m of the body. If we know the value of g for the location of the balance, we can find the weight of the body with Eq. 5-8.

We can also weigh a body with a spring scale (Fig. 5-7). The body stretches a spring, moving a pointer along a scale that has been calibrated and marked in either mass or weight units. (Most bathroom scales in the United States work this way and read in pounds.) If the scale is marked in mass units, it is accurate only where the value of g is the same as where the scale was calibrated.

The Normal Force

When a body is pressed against a surface, the body experiences a force that is perpendicular to the surface. The force is called the **normal force N,** the name coming from the mathematical term *normal,* meaning "perpendicular."

If a body rests on a horizontal surface as in Fig. 5-8*a*, $\mathbf{N}$ is directed upward and the body's weight $\mathbf{W} = m\mathbf{g}$ is

$$\mathbf{W}_L = -m_L g\mathbf{j} = m_L\mathbf{g} \qquad \mathbf{W}_R = -m_R g\mathbf{j} = m_R\mathbf{g}$$

FIGURE 5-6 An equal-arm balance. When the device is in balance, the masses on the left (*L*) and right (*R*) pans are equal.

FIGURE 5-7 A spring scale. The reading is proportional to the *weight* of the object placed on the pan, and the scale gives that weight if marked in weight units. If, instead, it is marked in mass units, the reading is accurate only if the free-fall acceleration g is the same as where the scale was calibrated.

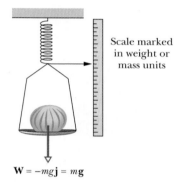

$$\mathbf{W} = -mg\mathbf{j} = m\mathbf{g}$$

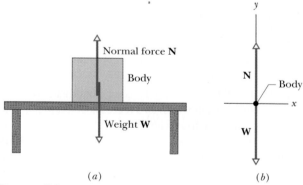

FIGURE 5-8 (*a*) A body resting on a tabletop experiences a normal force $\mathbf{N}$ perpendicular to the tabletop. (*b*) The corresponding free-body diagram.

directed downward. For this *particular* arrangement, we find the magnitude of $\mathbf{N}$ from the second of Eqs. 5-2:

$$\sum F_y = N - mg = ma_y, \tag{5-11}$$

and so, with $a_y = 0$,

$$N = mg. \tag{5-12}$$

CHECKPOINT **4**: In Fig. 5-8, is the magnitude of the normal force **N** greater than, less than, or equal to the weight magnitude *mg* if the body and table are in an elevator that is moving upward (a) at a constant speed and (b) at an increasing speed?

Friction

If we slide or attempt to slide a body over a surface, the motion is resisted by a bonding between the body and surface. (We discuss this more in the next chapter.) The resistance is considered to be a single force **f**, called the **frictional force,** or simply **friction**. This force is directed along the surface, opposite the direction of the intended motion (Fig. 5-9). Sometimes, to simplify a situation, friction is assumed to be negligible, and the surface is said to be *frictionless.*

Tension

When a cord (or a rope, cable, or other such object) is attached to a body and pulled taut, the cord is said to be under **tension**. It pulls on the body with a force **T**, whose direction is away from the body and along the cord at the point of attachment (Fig. 5-10a).

A cord is often said to be *massless* (meaning its mass is negligible compared to the body's mass) and unstretchable. The cord exists only as a connection between two

FIGURE 5-9 A frictional force **f** opposes the attempted slide of a body over a surface.

FIGURE 5-10 (a) The cord, pulled taut, is under tension. It pulls on the body and the hand with force **T**, even if it runs around a massless, frictionless pulley as in (b) and (c).

bodies. It pulls on both bodies with the same magnitude *T*, even if the bodies and the cord are accelerating and even if the cord runs around a *massless, frictionless pulley* (Figs. 5-10b and c). Such a pulley has negligible mass compared to the bodies and has negligible friction on its axle opposing its rotation.

CHECKPOINT **5**: The body that is suspended by a rope in Fig. 5-10c has a weight of 75 N. Is *T* equal to, greater than, or less than 75 N when the body is moving upward (a) at constant speed, (b) at increasing speed, and (c) at decreasing speed?

SAMPLE PROBLEM 5-4

Let us return to John Massis and the railroad cars, and assume that Massis pulled (with his teeth) on his end of the rope with a constant force that was 2.5 times his body weight, at an angle θ of 30° from the horizontal. His mass *m* was 80 kg. The weight *W* of the passenger cars was 7.0×10^5 N (about 80 tons), and he moved them 1.0 m along the rails. Assume that the rolling wheels encountered no retarding force from the rails. What was the train's speed at the end of the pull?

SOLUTION: Figure 5-11 is a free-body diagram for the cars, which are represented by a dot. The x axis runs along the rails. From Eqs. 5-2, we have

$$\sum F_x = T\cos\theta = Ma_x, \qquad (5\text{-}13)$$

in which M is the mass of the cars. We can find T and M from the given data.

With our assumptions, the pull from Massis is

$$T = 2.5mg = (2.5)(80\ \text{kg})(9.8\ \text{m/s}^2) = 1960\ \text{N}$$

(or 440 lb), which is about what a good middle-weight weight lifter can lift—and far from a superhuman force.

The weight W of the cars is, by Eq. 5-8,

$$W = Mg,$$

so their mass M must be

$$M = \frac{W}{g} = \frac{7.0 \times 10^5\ \text{N}}{9.8\ \text{m/s}^2} = 7.143 \times 10^4\ \text{kg}.$$

Now, from Eq. 5-13, we find their acceleration to be

$$a_x = \frac{T\cos\theta}{M} = \frac{(1960\ \text{N})(\cos 30°)}{7.143 \times 10^4\ \text{kg}}$$
$$= 2.376 \times 10^{-2}\ \text{m/s}^2.$$

To solve for the speed of the cars at the end of the pull, we use Eq. 2-16, with subscripts for the x axis and with $v_0 = 0$ and $x - x_0 = 1.0$ m:

$$v_x^2 = v_{0x}^2 + 2a_x(x - x_0),$$

or

$$v_x = \sqrt{0 + (2)(2.376 \times 10^{-2}\ \text{m/s}^2)(1.0\ \text{m})}$$
$$= 0.22\ \text{m/s}. \qquad \text{(Answer)}$$

Massis would have done better if the rope had been attached higher on the car, so that it was horizontal. Can you see why?

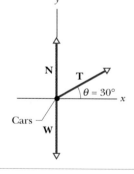

FIGURE 5-11   Sample Problem 5-4. Free-body diagram for the passenger cars pulled by Massis. The vectors are not drawn to scale; the tension in the rope is *much* smaller than the weight and normal force.

5-7 NEWTON'S THIRD LAW

Forces come in pairs. If a hammer exerts a force on a nail, the nail exerts a force of equal magnitude but opposite direction on the hammer. If you lean against a brick wall, the wall pushes back on you.

Let body A in Fig. 5-12 exert a force $\mathbf{F}_{BA}$ on body B; experiment shows that body B then exerts a force $\mathbf{F}_{AB}$ on body A. These two forces are equal in magnitude and op-

FIGURE 5-12   Newton's third law. Body A exerts a force $\mathbf{F}_{BA}$ on body B, while body B exerts a force $\mathbf{F}_{AB}$ on body A, where $\mathbf{F}_{AB} = -\mathbf{F}_{BA}$.

positely directed. That is,

$$\mathbf{F}_{AB} = -\mathbf{F}_{BA} \qquad \text{(Newton's third law).} \qquad (5\text{-}14)$$

Note the order of the subscripts. $\mathbf{F}_{AB}$, for example, is the force exerted *on* body A *by* body B. Equation 5-14 holds regardless of whether the bodies move or remain stationary.

Equation 5-14 sums up Newton's third law of motion. Commonly, one of these forces (it does not matter which) is called the **action force.** The other member of the pair is then called the **reaction force.** Every time you find a force, a good question is: Where is its reaction force?

The words "To every action there is always an equal and opposite reaction" have become enshrined in the popular language and mean various things to various speakers. In physics, however, these words mean Eq. 5-14 and nothing else. In particular, cause and effect are not involved; either force can be the action force.

You may think: "If every force has an associated force that is equal in magnitude and opposite in direction, why don't they cancel each other? How can anything ever get moving?" The answer is simple. As Fig. 5-12 shows:

The forces of an action–reaction pair *always* act on *different* bodies; thus they do not combine to give a net force and cannot cancel each other.

Two forces that act on the *same* body are *not* an action–reaction pair, even though they may be equal in magnitude

FIGURE 5-13   A satellite in Earth orbit. The forces shown are an action–reaction pair. Note that they act on different bodies.

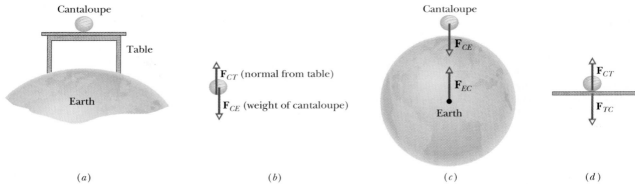

FIGURE 5-14 (a) A cantaloupe rests on a tabletop that rests on the Earth. (b) The forces *on the cantaloupe*, are $\mathbf{F}_{CT}$ and $\mathbf{F}_{CE}$. The cantaloupe is stationary because these two forces balance. (c) The action–reaction pair for the cantaloupe–Earth forces. (d) The action–reaction pair for the cantaloupe–table forces.

and opposite in direction. Let's identify the action–reaction pairs in two examples.

An Orbiting Satellite

Figure 5-13 shows an orbiting satellite. The only force that acts on it is $\mathbf{F}_{SE}$, the force exerted *on* the satellite *by* the gravitational pull of the Earth. Where is the corresponding reaction force? It is $\mathbf{F}_{ES}$, the force acting on the Earth due to the gravitational pull of the satellite; the pull is taken to be at the center of the Earth.

You may think that the tiny satellite cannot exert much of a gravitational pull on the Earth but it does, exactly as Newton's third law requires. That is, considering magnitudes only, $F_{ES} = F_{SE}$. The force $\mathbf{F}_{ES}$ causes the Earth to accelerate, but, because of the Earth's large mass, its acceleration is too small to be detected.

A Cantaloupe Resting on a Table

Figure 5-14a shows a cantaloupe at rest on a table.* The Earth pulls downward on the cantaloupe with a force $\mathbf{F}_{CE}$, the cantaloupe's weight. The cantaloupe does not accelerate because this force is canceled by an equal and opposite normal force $\mathbf{F}_{CT}$ exerted on the cantaloupe by the table. (See Fig. 5-14b.) However, $\mathbf{F}_{CE}$ and $\mathbf{F}_{CT}$ do *not* form an action–reaction pair *because they act on the same body, the cantaloupe.*

The reaction force to $\mathbf{F}_{CE}$ is $\mathbf{F}_{EC}$, the (gravitational) force with which the cantaloupe attracts the Earth. This action–reaction pair is shown in Fig. 5-14c.

The reaction force to $\mathbf{F}_{CT}$ is $\mathbf{F}_{TC}$, the force exerted on the table by the cantaloupe. This action–reaction pair is shown in Fig. 5-14d. The action–reaction pairs in this

problem, and the bodies on which they act, are then

first pair: $\mathbf{F}_{CE} = -\mathbf{F}_{EC}$ (cantaloupe and Earth)

and

second pair: $\mathbf{F}_{CT} = -\mathbf{F}_{TC}$ (cantaloupe and table).

CHECKPOINT 6: Suppose that the cantaloupe and table of Fig. 5-14 are in an elevator cab that begins to accelerate upward. (a) Do the magnitudes of forces $\mathbf{F}_{TC}$ and $\mathbf{F}_{CT}$ increase, decrease, or stay the same? (b) Are those two forces still equal in magnitude but opposite in direction? (c) Do the magnitudes of forces $\mathbf{F}_{CE}$ and $\mathbf{F}_{EC}$ increase, decrease, or stay the same? (d) Are those two forces still equal in magnitude but opposite in direction?

5-8 APPLYING NEWTON'S LAWS

The rest of this chapter consists of sample problems. You should pore over them, learning not just their particular answers but, instead, the procedures for attacking a problem. Especially important is knowing how to translate a sketch of a situation into a free-body diagram with appropriate axes, so that Newton's laws can be applied. We begin with Sample Problem 5-5, which is worked out in exhaustive detail, using a question-and-answer format.

SAMPLE PROBLEM 5-5

Figure 5-15 shows a block (the *sliding block*) whose mass M is 3.3 kg. It is free to move along a horizontal frictionless surface such as an air table. The sliding block is connected by a cord that extends around a massless, frictionless pulley to a second block (the *hanging block*), whose mass m is 2.1 kg. The hang-

---

*We ignore small complications caused by the rotation of the Earth.

ing block falls and the sliding block accelerates to the right. Find (a) the acceleration of the sliding block, (b) the acceleration of the hanging block, and (c) the tension in the cord.

Q *What is this problem all about?*

You are given two massive objects, the sliding block and the hanging block. It might not occur to you, but you are also given the Earth, which pulls on each of these objects; without the Earth, nothing would happen. A total of five forces act on the blocks, as shown in Fig. 5-16:

1. The cord pulls to the right on the sliding block with a force of magnitude *T*.

2. The cord pulls upward on the hanging block with a force of the same magnitude *T*. This upward force keeps the hanging block from falling freely, which it would otherwise do. We assume that the cord has the same tension throughout its length; the pulley just serves to change the direction of this force, without changing its magnitude.

3. The Earth pulls down on the sliding block with a force *M***g**, the weight of the sliding block.

4. The Earth pulls down on the hanging block with a force *m***g**, the weight of the hanging block.

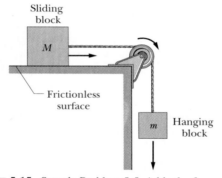

FIGURE 5-15 Sample Problem 5-5. A block of mass *M* on a horizontal frictionless surface is connected to a block of mass *m* by a cord that wraps over a pulley. Both cord and pulley are massless. The pulley is frictionless. The arrows indicate the motion when the system is released from rest.

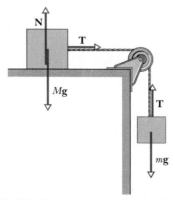

FIGURE 5-16 The forces acting on the two blocks.

5. The table pushes up on the sliding block with a normal force **N**.

There is another thing that you should note. We assume that the cord does not stretch, so that if the hanging block falls 1 mm in a certain time, the sliding block moves 1 mm to the right in that same interval. The blocks move together and their accelerations have the same magnitude *a*.

Q *How do I classify this problem? Should it suggest a particular law of physics to me?*

Yes, it should. Forces, masses, and accelerations are involved, and that should suggest Newton's second law of motion, $\Sigma\mathbf{F} = m\mathbf{a}$.

Q *If I apply that law to this problem, to what body should I apply it?*

We focus on two bodies in this problem, the sliding block and the hanging block. Although they are extended objects, we can treat each block as a particle because every small part of it (every atom, say) moves in exactly the same way. Apply Newton's second law separately to each block.

Q *What about the pulley?*

We cannot represent the pulley as a particle because different parts of it move in different ways. When we discuss rotation, we shall deal with pulleys in detail. Meanwhile, we get around the problem by using a pulley whose mass is negligible compared with the masses of the two blocks.

Q *OK. Now how do I apply $\Sigma\mathbf{F} = m\mathbf{a}$ to the sliding block?*

Represent the sliding block as a particle of mass *M* and draw *all* the forces that act *on* it, as in Fig. 5-17. This is the block's *free-body diagram*. There are three forces. Next, locate a set of axes. It makes sense to draw the *x* axis parallel to the table, in the direction in which the block moves.

Q *Thanks, but you still haven't told me how to apply $\Sigma\mathbf{F} = m\mathbf{a}$ to the sliding block. All you have done is explain how to draw a free-body diagram.*

Right you are. The expression $\Sigma\mathbf{F} = m\mathbf{a}$ is a vector equation and you can write it as three scalar equations:

$$\sum F_x = Ma_x, \quad \sum F_y = Ma_y, \quad \sum F_z = Ma_z, \quad (5\text{-}15)$$

in which ΣF_x, ΣF_y, and ΣF_z are the components of the net force. Because the sliding block does not move vertically, we know there is no net force in the *y* direction: the weight $\mathbf{W} = M\mathbf{g}$ of the sliding block is balanced by the upward-acting normal force **N** on the block. No force acts in the *z* direction, which is perpendicular to the page. We can, however, apply the first of Eqs. 5-15.

In the *x* direction, there is only one force component, so $\Sigma F_x = Ma_x$ becomes

$$T = Ma. \quad (5\text{-}16)$$

This equation contains two unknowns, *T* and *a*, so we cannot yet solve it. Recall, however, that we have not said anything about the hanging block.

Q *I agree. How do I apply $\Sigma\mathbf{F} = m\mathbf{a}$ to the hanging block?*

Draw a free-body diagram for the block, as in Fig. 5-18. This time, we use the second of Eqs. 5-15, finding

$$\sum F_y = T - mg = -ma, \qquad (5\text{-}17)$$

where the minus sign on the right side of the equation indicates that the hanging block accelerates downward, in the negative direction of the y axis. Equation 5-17 yields

$$mg - T = ma. \qquad (5\text{-}18)$$

This contains the same two unknowns as Eq. 5-16 does. If you add these equations, T will cancel out. Solving for a then yields

$$a = \frac{m}{M + m} g. \qquad (5\text{-}19)$$

Substituting this result into Eq. 5-16 yields

$$T = \frac{Mm}{M + m} g. \qquad (5\text{-}20)$$

Putting in the numbers gives, for these two quantities,

$$a = \frac{m}{M + m} g = \frac{2.1 \text{ kg}}{3.3 \text{ kg} + 2.1 \text{ kg}} (9.8 \text{ m/s}^2)$$

$$= 3.8 \text{ m/s}^2 \qquad \text{(Answer)}$$

and

$$T = \frac{Mm}{M + m} g = \frac{(3.3 \text{ kg})(2.1 \text{ kg})}{3.3 \text{ kg} + 2.1 \text{ kg}} (9.8 \text{ m/s}^2)$$

$$= 13 \text{ N}. \qquad \text{(Answer)}$$

Q *The problem is now solved, right?*

That's a fair question, but we are here not only to solve problems but also to learn physics. This problem is not really finished until we have studied the results to see if they make sense. This is often a much more confidence-building experience than simply getting the right answer.

Look first at Eq. 5-19. Note that it is dimensionally correct and also that the acceleration a will always be less than g. This is as it must be, because the hanging block is not in free fall. The cord pulls upward on it.

Look now at Eq. 5-20, which we rewrite in the form

$$T = \frac{M}{M + m} mg. \qquad (5\text{-}21)$$

In this form, it is easier to see that this equation is also dimensionally correct, because both T and mg are forces. Equation 5-21 also lets us see that the tension in the cord is always less than mg, the weight of the hanging block. That is a comforting thought because, if T were *greater* than mg, the hanging block would accelerate upward.

We can also check the results by studying special cases, in which we can guess what the answers must be. A simple example is to put g = 0, as if the experiment were carried out in interstellar space. We know that in that case, the blocks would not move from rest and there would be no tension in the cord. Do the formulas predict this? Yes, they do. If you put g = 0 in Eqs. 5-19 and 5-20, you find a = 0 and T = 0. Two more special cases that you might try are M = 0 and m → ∞.

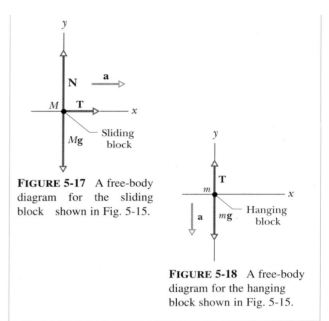

FIGURE 5-17 A free-body diagram for the sliding block shown in Fig. 5-15.

FIGURE 5-18 A free-body diagram for the hanging block shown in Fig. 5-15.

SAMPLE PROBLEM 5-5—ANOTHER WAY

The acceleration a of the blocks of Fig. 5-15 can be found in two lines of algebra if we (a) employ an unconventional axis, call it u, that runs through *both* blocks and along the cord as shown in Fig. 5-19a, and then (b) mentally straighten out the u axis as in Fig. 5-19b and treat the blocks as being portions of a single composite body with mass M + m. A free-body diagram for the two-block system is shown in Fig. 5-19c.

SOLUTION: Note that there is only one force acting on the composite body along the u axis, and that is the force mg in the positive direction of the axis. The tension **T** of Fig. 5-16 is now internal to the composite body and so does not enter into Newton's second law. The force exerted by the pulley on the cord is perpendicular to the u axis; so it too does not enter.

Using Eqs. 5-2 as a guide, we write a component equation for the acceleration along the u axis:

$$\sum F_u = (M + m)a_u,$$

where the mass of the body is M + m. The acceleration of the composite body along the u axis (and of the individual blocks, since they are connected) has magnitude a. The only force on the composite body along the u axis has a magnitude of mg. So our equation becomes

$$mg = (M + m)a,$$

or

$$a = \frac{m}{M + m} g, \qquad (5\text{-}22)$$

which matches Eq. 5-19.

To find T, we apply Newton's second law to either block, obtaining either Eq. 5-16 or Eq. 5-18. We then substitute for a from Eq. 5-22 and solve for T, getting Eq. 5-20.

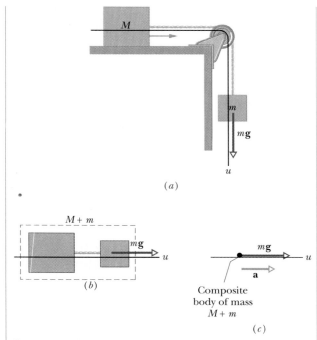

(a)

(b)

Composite
body of mass
$M + m$

(c)

FIGURE 5-19 (a) An "axis" u runs through the blocks-and-cord system of Fig. 5-15. (b) The blocks are rearranged to straighten u and then are treated as a single body with mass $M + m$. (c) The associated free-body diagram, considering only forces along u. There is one such force.

SAMPLE PROBLEM 5-6

A block whose mass M is 33 kg is pushed across a frictionless surface by a stick whose mass m is 3.2 kg, as in Fig. 5-20a. The block is moved (from rest) a distance $d = 77$ cm in 1.7 s at constant acceleration.

(a) Identify all horizontal action–reaction force pairs in this problem.

SOLUTION: As the exploded view of Fig. 5-20b shows, there are two action–reaction pairs:

first pair: $\mathbf{F}_{HS} = -\mathbf{F}_{SH}$ (hand and stick)

second pair: $\mathbf{F}_{SB} = -\mathbf{F}_{BS}$ (stick and block).

The force $\mathbf{F}_{HS}$ on the hand from the stick is the force that you would feel if the hand in Fig. 5-20 were yours.

(b) What force must the hand apply to the stick?

SOLUTION: This is the force that accelerates the block and stick. To find it, we must first find the constant acceleration a, using Eq. 2-15:

$$x - x_0 = v_0 t + \tfrac{1}{2} a t^2.$$

Setting $v_0 = 0$ and $x - x_0 = d$, and solving for a, give

$$a = \frac{2d}{t^2} = \frac{(2)(0.77 \text{ m})}{(1.7 \text{ s})^2} = 0.533 \text{ m/s}^2.$$

To find the force that the hand exerts, we apply Newton's second law to a system consisting of the stick and the block taken together. Thus,

$$F_{SH} = (M + m)a = (33 \text{ kg} + 3.2 \text{ kg})(0.533 \text{ m/s}^2)$$
$$= 19.3 \text{ N} \approx 19 \text{ N}. \qquad \text{(Answer)}$$

(c) With what force does the stick push on the block?

SOLUTION: To find this force, we apply Newton's second law to the block alone:

$$F_{BS} = Ma = (33 \text{ kg})(0.533 \text{ m/s}^2)$$
$$= 17.6 \text{ N} \approx 18 \text{ N}. \qquad \text{(Answer)}$$

(d) What is the net force on the stick?

SOLUTION: We can find the magnitude F of this force in two ways. First, using results from (b) and (c), we have

$$F = F_{SH} - F_{SB} = 19.3 \text{ N} - 17.6 \text{ N}$$
$$= 1.7 \text{ N}. \qquad \text{(Answer)}$$

Note that we have used Newton's third law here, in assuming that $\mathbf{F}_{SB}$, the force on the stick from the block, has the same magnitude (17.6 N before rounding) as $\mathbf{F}_{BS}$.

The second way to arrive at an answer is to apply Newton's second law to the stick directly. We have

$$F = ma = (3.2 \text{ kg})(0.533 \text{ m/s}^2) = 1.7 \text{ N}, \qquad \text{(Answer)}$$

in agreement with our first result. This is as it must be because the two methods are algebraically identical; check it out.

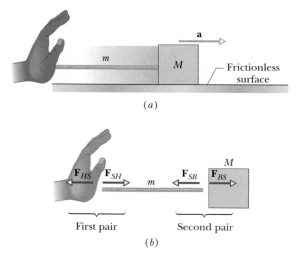

(a)

First pair   Second pair

(b)

FIGURE 5-20 Sample Problem 5-6. (a) A block of mass M is pushed over a frictionless surface by a stick of mass m. (b) An exploded view, showing the action–reaction pairs between the hand and the stick (first pair) and between the stick and the block (second pair).

SAMPLE PROBLEM 5-7

Figure 5-21a shows a block of mass $m = 15$ kg hanging from three cords. What are the tensions in the cords?

SOLUTION: In the free-body diagram for the block (Fig. 5-21b), tension $\mathbf{T}_C$ from cord C pulls upward while the block's weight $m\mathbf{g}$ is directed downward. Since the system is at rest, Newton's second law for the block yields

$$\sum \mathbf{F} = \mathbf{T}_C + m\mathbf{g} = 0.$$

Because forces $\mathbf{T}_C$ and $m\mathbf{g}$ are only vertical, this equation gives us one scalar equation:

$$\sum F_y = T_C - mg = 0.$$

Substituting known values, we find

$$T_C = mg = (15 \text{ kg})(9.8 \text{ m/s}^2)$$
$$= 147 \text{ N} \approx 150 \text{ N}. \qquad \text{(Answer)}$$

The clue to our next step is to realize that the knot where the three cords join is the only point at which all three forces act, and it is this knot to which we should apply Newton's second law. Figure 5-21c is the free-body diagram for the knot. Since the knot is not accelerated, the net force acting on it must be zero. Thus

$$\sum \mathbf{F} = \mathbf{T}_A + \mathbf{T}_B + \mathbf{T}_C = 0.$$

This vector equation is equivalent to the two scalar equations

$$\sum F_y = T_A \sin 28° + T_B \sin 47° - T_C = 0 \quad (5\text{-}23)$$

and

$$\sum F_x = -T_A \cos 28° + T_B \cos 47° = 0. \quad (5\text{-}24)$$

Note carefully that, when we write the x component of $\mathbf{T}_A$ as $T_A \cos 28°$, we must include a minus sign to show that it extends in the negative direction of the x axis.

Substituting numerical values into Eqs. 5-23 and 5-24 leads to

$$T_A(0.469) + T_B(0.731) = 147 \text{ N} \quad (5\text{-}25)$$

and

$$T_B(0.682) = T_A(0.883). \quad (5\text{-}26)$$

From Eq. 5-26, we have

$$T_B = \frac{0.883}{0.682} T_A = 1.29 T_A.$$

Substituting this into Eq. 5-25 and solving for T_A, we obtain

$$T_A = \frac{147 \text{ N}}{0.469 + (1.29)(0.731)}$$
$$= 104 \text{ N} \approx 100 \text{ N}. \qquad \text{(Answer)}$$

Finally, T_B is found from

$$T_B = 1.29 T_A = (1.29)(104 \text{ N})$$
$$= 134 \text{ N} \approx 130 \text{ N}. \qquad \text{(Answer)}$$

(a)

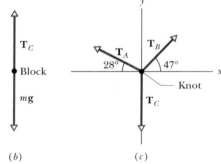

(b)   (c)

FIGURE 5-21 Sample Problem 5-7. (a) A block of mass m hangs from three cords. (b) A free-body diagram for the block. (c) A free-body diagram for the knot at the intersection of the three cords.

SAMPLE PROBLEM 5-8

Figure 5-22a shows a block of mass $m = 15$ kg held by a cord on a frictionless inclined plane. What is the tension in the cord if $\theta = 27°$? What force does the plane exert on the block?

SOLUTION: Figure 5-22b is the free-body diagram for the block. The following forces act on it: (1) a normal force $\mathbf{N}$, exerted outward on the block by the plane on which it rests, (2) the tension $\mathbf{T}$ in the cord, and (3) the weight $\mathbf{W}$ ($= m\mathbf{g}$) of the block. Because the acceleration of the block is zero, the net force acting on the block must also be zero by Newton's second law:

$$\sum \mathbf{F} = \mathbf{T} + \mathbf{N} + m\mathbf{g} = 0. \quad (5\text{-}27)$$

We choose a coordinate system with the x axis parallel to the plane. With this choice, not one but two forces ($\mathbf{N}$ and $\mathbf{T}$) line up with coordinate axes (a bonus). Note that the angle between the weight vector and the negative direction of the y axis equals the slant angle of the plane. The x and y components of that vector are found with the triangle of Fig. 5-22c. The component versions of Eq. 5-27 are

$$\sum F_x = T - mg \sin \theta = 0$$

and

$$\sum F_y = N - mg \cos \theta = 0.$$

Thus

$$T = mg \sin \theta$$
$$= (15 \text{ kg})(9.8 \text{ m/s}^2)(\sin 27°)$$
$$= 67 \text{ N} \qquad \text{(Answer)}$$

and

$$N = mg \cos \theta$$
$$= (15 \text{ kg})(9.8 \text{ m/s}^2)(\cos 27°)$$
$$= 131 \text{ N} \approx 130 \text{ N}. \qquad \text{(Answer)}$$

(a)

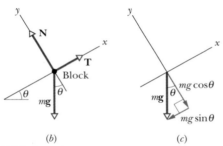

(b)   (c)

FIGURE 5-22 Sample Problems 5-8 and 5-9. (a) A block of mass m rests on a smooth plane, held there by a cord. (b) A free-body diagram for the block. Note how the coordinate axes are placed. (c) Finding the x and y components of $m\mathbf{g}$.

$\mathbb{C}$HECKPOINT 7: In the figure, horizontal force $\mathbf{F}$ is applied to the block. (a) Is the component of $\mathbf{F}$ that is perpendicular to the ramp $F \cos \theta$ or $F \sin \theta$? (b) Does the presence of $\mathbf{F}$ increase or decrease the magnitude of the normal force on the block from the ramp?

SAMPLE PROBLEM 5-9

Suppose you cut the cord holding the block on the plane in Fig. 5-22a. With what acceleration will the block move?

SOLUTION: Cutting the cord removes the tension $\mathbf{T}$ in Fig. 5-22b. The two remaining forces do not cancel and, indeed, they cannot, because they do not act along the same line. Applying Newton's second law to the x components of the forces $\mathbf{N}$ and $m\mathbf{g}$ in Fig. 5-22b now yields

$$\sum F_x = 0 - mg \sin \theta = ma,$$

so

$$a = -g \sin \theta. \qquad (5\text{-}28)$$

Note that the normal force $\mathbf{N}$ plays no role in producing the acceleration because its x component is zero.

Equation 5-28 yields

$$a = -(9.8 \text{ m/s}^2)(\sin 27°) = -4.4 \text{ m/s}^2. \qquad \text{(Answer)}$$

The minus sign indicates that the acceleration is in the direction of decreasing x, that is, down the plane.

Equation 5-28 reveals that the acceleration of the block is independent of its mass, just as the acceleration of a freely falling body is independent of the mass of the falling body. Indeed, Eq. 5-28 shows that an inclined plane can be used to "dilute" gravitation—to "slow down" free fall. For $\theta = 90°$, Eq. 5-28 yields $a = -g$; for $\theta = 0°$, it yields $a = 0$. Both are expected results.

SAMPLE PROBLEM 5-10

Figure 5-23a shows two blocks connected by a cord that passes over a massless, frictionless pulley (the arrangement is known as *Atwood's machine*). Let $m = 1.3$ kg and $M = 2.8$ kg. Find the tension in the cord and the (common) magnitude of the acceleration of the two blocks.

SOLUTION: Figures 5-23b and 5-23c are free-body diagrams for the blocks. We are given $M > m$, so we expect M to fall and m to rise. That information allows us to assign the proper algebraic signs to the accelerations of the blocks.

Before we begin the calculations, we note that the tension in the cord must be less than the weight of block M (otherwise, that block would not fall from rest) and greater than the weight of block m (otherwise, that block would not rise). The vectors in the two free-body diagrams of Fig. 5-23 are drawn to represent these facts.

Applying Newton's second law to the block with mass m, which has an acceleration of magnitude a in the positive direction of the y axis, we find

$$T - mg = ma. \qquad (5\text{-}29)$$

For the block with mass M, which has an acceleration of magnitude a in the negative direction of the y axis, we have

$$T - Mg = -Ma, \qquad (5\text{-}30)$$

or

$$-T + Mg = Ma. \qquad (5\text{-}31)$$

By adding Eqs. 5-29 and 5-31 (or eliminating T through substitution), we obtain

$$a = \frac{M - m}{M + m}\, g. \qquad (5\text{-}32)$$

Substituting this result in either Eq. 5-29 or Eq. 5-31 and solving for T, we obtain

$$T = \frac{2mM}{M + m}\, g. \qquad (5\text{-}33)$$

Inserting the given data, we obtain

$$a = \frac{M - m}{M + m}\, g = \frac{2.8\ \text{kg} - 1.3\ \text{kg}}{2.8\ \text{kg} + 1.3\ \text{kg}}\,(9.8\ \text{m/s}^2)$$

$$= 3.6\ \text{m/s}^2 \qquad \text{(Answer)}$$

and

$$T = \frac{2Mm}{M + m}\, g = \frac{(2)(2.8\ \text{kg})(1.3\ \text{kg})}{2.8\ \text{kg} + 1.3\ \text{kg}}\,(9.8\ \text{m/s}^2)$$

$$= 17\ \text{N}. \qquad \text{(Answer)}$$

You can show that the weights of the two blocks are 13 N ($= mg$) and 27 N ($= Mg$). So the tension ($= 17$ N) does indeed lie between these two values.

FIGURE 5-23 Sample Problem 5-10. (*a*) A block of mass M and one of smaller mass m are connected by a cord that passes over a pulley. The directions in which the system will accelerate from rest are shown by the arrows. Free-body diagrams for (*b*) block m and (*c*) block M.

(*b*)   (*c*)

SAMPLE PROBLEM 5-10—ANOTHER WAY

Just as we redid Sample Problem 5-5 with an unconventional axis u, we can redo Sample Problem 5-10 with u.

SOLUTION: Run the u axis through the system as shown in Fig. 5-24*a*. Straighten out the axis as in Fig. 5-24*b*, and treat the blocks as a single body with a mass of $M + m$. Then draw a free-body diagram as in Fig. 5-24*c*. Note that along the u

axis there are two forces on the composite of the blocks: $m\mathbf{g}$ in the negative direction of the u axis and $M\mathbf{g}$ in the positive direction. (The force exerted on the cord by the pulley is perpendicular to the u axis and so doesn't enter our calculation.) The two forces along the u axis give the composite body (and each block) an acceleration $\mathbf{a}$. Newton's second law in component form for motion along u is

$$\sum F_u = Mg - mg = (M + m)a, \qquad (5\text{-}34)$$

which yields

$$a = \frac{M - m}{M + m}\, g,$$

as previously. To get T we apply Newton's second law to either block, using a conventional axis y. For the block with mass m, we obtain Eq. 5-29. With the above result for a substituted into Eq. 5-29, we obtain Eq. 5-33.

FIGURE 5-24 (*a*) An "axis" u runs through the system of Fig. 5-23. (*b*) The blocks are rearranged to straighten u and then treated as a single body with mass $M + m$. (*c*) The associated free-body diagram, considering only forces along u. There are two such forces.

SAMPLE PROBLEM 5-11

A passenger of mass $m = 72.2$ kg stands on a platform scale in an elevator cab (Fig. 5-25). What is the scale reading for the acceleration values given in the figure?

SOLUTION: We consider this problem from the point of view of an observer in an (inertial) reference frame fixed with respect to the Earth. Let this observer apply Newton's second law to the accelerating passenger. Figure 5-25*a*–*c* presents free-body diagrams for the passenger, treated as a particle (some particle!), for various accelerations of the cab.

Regardless of the acceleration of the cab, the Earth pulls downward on the passenger with a force having magnitude mg, in which $g = 9.80$ m/s^2 is the free-fall acceleration in the reference frame of the Earth. The elevator pushes upward on the scale platform. The scale platform, in turn, pushes upward on the passenger with a normal force whose magnitude N equals the reading of the scale. The weight that the accelerating passenger would judge himself to have is what he reads on the scale. This quantity is often called the *apparent weight;* the term *weight* (or *true weight*) is reserved for the quantity mg.

Newton's second law yields

$$N - mg = ma,$$

or

$$N = m(g + a). \tag{5-35}$$

(a) What does the scale read if the cab is at rest or moving with constant speed? (See Fig. 5-25*a*.)

SOLUTION: Here $a = 0$ and we have

$$N = m(g + a) = (72.2 \text{ kg})(9.80 \text{ m/s}^2 + 0)$$
$$= 708 \text{ N.} \qquad \text{(Answer)}$$

This is the weight of the passenger.

(b) What does the scale read if the cab has an upward acceleration of magnitude 3.20 m/s²? (See Fig. 5-25*b*.)

SOLUTION: An upward acceleration means that the cab is either moving up with increasing speed or down with decreasing speed. In either case, the cab is a noninertial frame and Eq. 5-35 yields

$$N = m(g + a) = (72.2 \text{ kg})(9.80 \text{ m/s}^2 + 3.20 \text{ m/s}^2)$$
$$= 939 \text{ N.} \qquad \text{(Answer)}$$

The passenger presses down on the scale with a greater force than if he were at rest. The passenger—reading the scale— might conclude that he has gained 231 N!

(c) What does the scale read if the cab has a downward acceleration of magnitude 3.20 m/s²? (See Fig. 5-25*c*.)

SOLUTION: A downward acceleration means that the cab is either moving up with decreasing speed or down with increas-

ing speed. Again the cab is a noninertial frame. Equation 5-35 now yields

$$N = m(g + a) = (72.2 \text{ kg})(9.80 \text{ m/s}^2 - 3.20 \text{ m/s}^2)$$
$$= 477 \text{ N.} \qquad \text{(Answer)}$$

The passenger presses down on the scale with less force than if the cab were at rest. He seems to have lost 231 N.

Reading of weight or apparent weight

FIGURE 5-25 Sample Problem 5-11. A passenger with mass m in an elevator, standing on a spring scale which indicates his weight or apparent weight. The free-body diagrams are for the cases in which (*a*) the acceleration of the elevator cab is zero, (*b*) $a = +3.20$ m/s², and (*c*) $a = -3.20$ m/s².

Cʜᴇᴄᴋᴘᴏɪɴᴛ **8:** What does the scale read if the cable breaks, so that the cab falls freely? That is, what is the apparent weight of the passenger in free fall?

REVIEW & SUMMARY

Newtonian Mechanics

The velocity of a particle or a particle-like body can change— that is, the particle can accelerate—when the particle is acted on by one or more **forces**—pushes or pulls—exerted by other objects. *Newtonian mechanics* relates accelerations and forces.

Force

The magnitudes of forces are defined in terms of the acceleration they give the standard kilogram. A force that accelerates that standard body by exactly 1 m/s² is defined to have a magnitude of 1 N. The direction of the force is the direction of the acceleration. Forces are experimentally found to be vector quantities, so they are combined according to the rules of vector algebra. The **net force** on a body is the vector sum of all the forces acting on it.

Mass

The **mass** of a body is the characteristic of that body that relates the body's acceleration to the force (or net force) causing the acceleration. Masses are scalar quantities.

Newton's First Law

If there is no net force on a body, the body must remain at rest if it is initially at rest, or move in a straight line at constant speed if it is in motion. For such a body, there are reference frames, called *inertial frames,* from which the body's acceleration **a** will be measured as being zero.

Newton's Second Law

The net force $\Sigma\mathbf{F}$ on a body with mass m is related to the body's

15. Figure 5-35 shows four choices for the direction of a force of magnitude F to be applied to a block on an inclined plane. The directions are either horizontal or vertical. (For choices a and b, the force is not enough to lift the block off the plane.) Rank the choices according to the magnitude of the normal force on the block from the plane, greatest first.

FIGURE 5-35
Question 15.

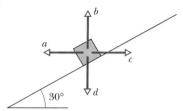

EXERCISES & PROBLEMS

SECTION 5-3 Force

1E. If the 1 kg standard body has an acceleration of 2.00 m/s² at 20° to the positive direction of the x axis, then (a) what are the x and y components of the net force on it, and (b) what is the net force in unit-vector notation?

2E. If the 1 kg standard body is accelerated by $\mathbf{F}_1 = (3.0\ \text{N})\mathbf{i} + (4.0\ \text{N})\mathbf{j}$ and $\mathbf{F}_2 = (-2.0\ \text{N})\mathbf{i} + (-6.0\ \text{N})\mathbf{j}$, then (a) what is the net force in unit-vector notation, and what are the magnitude and direction of (b) the net force and (c) the acceleration?

3P. Suppose that the 1 kg standard body accelerates at 4.00 m/s² at 160° from the positive direction of the x axis, owing to two forces, one of which is $\mathbf{F}_1 = (2.50\ \text{N})\mathbf{i} + (4.60\ \text{N})\mathbf{j}$. What is the other force (a) in unit-vector notation and (b) as a magnitude and direction?

SECTION 5-5 Newton's Second Law

4E. Two horizontal forces act on a 2.0 kg chopping block that can slide over a frictionless kitchen counter, which lies in an xy plane. One force is $\mathbf{F}_1 = (3.0\ \text{N})\mathbf{i} + (4.0\ \text{N})\mathbf{j}$. Find the acceleration of the chopping block in unit-vector notation when the other force is (a) $\mathbf{F}_2 = (-3.0\ \text{N})\mathbf{i} + (-4.0\ \text{N})\mathbf{j}$, (b) $\mathbf{F}_2 = (-3.0\ \text{N})\mathbf{i} + (4.0\ \text{N})\mathbf{j}$, and (c) $\mathbf{F}_2 = (3.0\ \text{N})\mathbf{i} + (-4.0\ \text{N})\mathbf{j}$.

5E. While two forces act on it, a particle is to move continuously with $\mathbf{v} = (3\ \text{m/s})\mathbf{i} - (4\ \text{m/s})\mathbf{j}$. One of the forces is $\mathbf{F}_1 = (2\ \text{N})\mathbf{i} + (-6\ \text{N})\mathbf{j}$. What is the other force?

6E. Three forces act on a particle that moves with an unchanging velocity of $\mathbf{v} = (2\ \text{m/s})\mathbf{i} - (7\ \text{m/s})\mathbf{j}$. Two of the forces are $\mathbf{F}_1 = (2\ \text{N})\mathbf{i} + (3\ \text{N})\mathbf{j} + (-2\ \text{N})\mathbf{k}$ and $\mathbf{F}_2 = (-5\ \text{N})\mathbf{i} + (8\ \text{N})\mathbf{j} + (-2\ \text{N})\mathbf{k}$. What is the third force?

7E. There are two horizontal forces on the 2.0 kg box in the overhead view of Fig. 5-36 but only one is shown. The box moves strictly along the x axis. For each of the following values for the acceleration a_x of the box, find the second force: (a) 10 m/s², (b) 20 m/s², (c) 0, (d) −10 m/s², and (e) −20 m/s².

FIGURE 5-36 Exercise 7.

8E. There are two forces on the 2.0 kg box in the overhead view of Fig. 5-37 but only one is shown. The figure also shows the acceleration of the box. Find the second force (a) in unit-vector notation and (b) as a magnitude and direction.

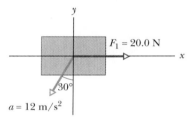

FIGURE 5-37 Exercise 8.

9E. Five forces pull on the 4.0 kg box in Fig. 5-38. Find the box's acceleration (a) in unit-vector notation and (b) as a magnitude and direction.

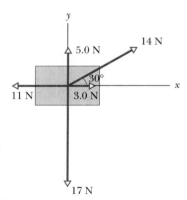

FIGURE 5-38 Exercise 9.

10P. Three astronauts, propelled by jet backpacks, push and guide a 120 kg asteroid toward a processing dock, exerting the forces shown in Fig. 5-39. What is the asteroid's acceleration (a) in unit-vector notation and (b) as a magnitude and direction?

FIGURE 5-39 Problem 10.

11P. Figure 5-40 is an overhead view of a 12 kg tire that is to be pulled by three ropes. One force (F_1, with magnitude 50 N) is indicated. Orient the other two forces F_2 and F_3 so that the magnitude of the resulting acceleration of the tire is least, and find that magnitude if (a) $F_2 = 30$ N, $F_3 = 20$ N; (b) $F_2 = 30$ N, $F_3 = 10$ N; and (c) $F_2 = F_3 = 30$ N.

FIGURE 5-40 Problem 11.

SECTION 5-6 Some Particular Forces

12E. What are the mass and weight of (a) a 1400 lb snowmobile and (b) a 421 kg heat pump?

13E. What are the weight in newtons and the mass in kilograms of (a) a 5.0 lb bag of sugar, (b) a 240 lb fullback, and (c) a 1.8 ton automobile? (1 ton = 2000 lb.)

14E. A space traveler whose mass is 75 kg leaves the Earth. Compute his weight (a) on Earth, (b) on Mars, where $g = 3.8$ m/s², and (c) in interplanetary space, where $g = 0$. (d) What is his mass at each of these locations?

15E. A certain particle has a weight of 22 N at a point where the free-fall acceleration is 9.8 m/s². (a) What are the weight and mass of the particle at a point where the free-fall acceleration is 4.9 m/s²? (b) What are the weight and mass of the particle if it is moved to a point in space where the free-fall acceleration is zero?

16E. A weight-conscious penguin with a mass of 15.0 kg rests on a bathroom scale (Fig. 5-41). What are (a) the penguin's weight **W** and (b) the normal force **N** on the penguin? (c) What is the reading on the scale, assuming it is calibrated in weight units?

FIGURE 5-41 Exercise 16.

17E. Figure 5-42a shows a crude mobile hanging from a ceiling, with two metal pieces strung by cords of negligible mass. The masses of the pieces are given. What is the tension in (a) the bottom cord and (b) the top cord? Figure 5-42b shows a similar mobile consisting of three metal pieces. The masses of the highest and lowest pieces are given. The tension in the top cord is 199 N. What is the tension in (c) the lowest cord and (d) the middle cord?

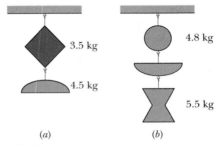

FIGURE 5-42 Exercise 17.

18E. (a) An 11.0 kg salami is supported by a cord that runs to a spring scale, which is supported by another cord from the ceiling (Fig. 5-43a). What is the reading on the scale? (b) In Fig. 5-43b the salami is supported by a cord that runs around a pulley and to a scale. The opposite end of the scale is attached by a cord to a wall. What is the reading on the scale? (c) In Fig. 5-43c the wall has been replaced with a second 11.0 kg salami on the left, and the assembly is stationary. What is the reading on the scale now?

FIGURE 5-43 Exercise 18.

SECTION 5-8 Applying Newton's Laws

19E. When an airplane is in level flight, its weight is balanced by a vertical "lift," which is a force exerted by the air. How large is the lift on an airplane in such flight if the mass of the airplane is 1.20×10^3 kg?

20E. What is the magnitude of the net force acting on a 3800 lb automobile accelerating at 12 ft/s²?

21E. An experimental rocket sled can be accelerated at a constant rate from rest to 1600 km/h in 1.8 s. What is the magnitude of the required average force if the sled has a mass of 500 kg?

22E. A car traveling at 53 km/h hits a bridge abutment. A passenger in the car moves forward a distance of 65 cm (with respect to the road) while being brought to rest by an inflated air bag. What magnitude of force (assumed constant) acts on the passenger's upper torso, which has a mass of 41 kg?

23E. If a nucleus captures a stray neutron, it must bring the neutron to a stop within the diameter of the nucleus by means of the *strong force*. That force, which "glues" the nucleus together, is essentially zero outside the nucleus. Suppose that a stray neutron with an initial speed of 1.4×10^7 m/s is just barely captured by a nucleus with diameter $d = 1.0 \times 10^{-14}$ m. Assuming that the force on the neutron is constant, find the magnitude of that force. The neutron's mass is 1.67×10^{-27} kg.

24E. In a modified tug-of-war game, two people pull in opposite directions, not on a rope, but on a 25 kg sled resting on (frictionless) ice. If the participants exert forces of 90 N and 92 N, what will be the magnitude of the sled's acceleration?

25E. A 450 lb motorcycle accelerates from 0 to 55 mi/h in 6.0 s. (a) What is the magnitude of the motorcycle's constant acceleration? (b) What is the magnitude of the net force causing the acceleration?

26E. Refer to Fig. 5-15 and suppose the two masses are $m = 2.0$ kg and $M = 4.0$ kg. (a) Decide without any calculations which of them should be hanging if the magnitude of the acceleration is to be largest. What then are (b) the magnitude of the acceleration and (c) the associated tension in the cord?

27E. Refer to Fig. 5-22. Let the mass of the block be 8.5 kg and the angle θ equal 30°. Find (a) the tension in the cord and (b) the normal force acting on the block. (c) If the cord is cut, find the magnitude of the block's acceleration.

28E. A jet airplane starts from rest on the runway and accelerates for takeoff at 2.3 m/s². It has two jet engines, each of which exerts a force (thrust) on the airplane of 1.4×10^5 N. What is the weight of the airplane?

29E. *Sunjamming.* A "sun yacht" is a spacecraft with a large sail that is pushed by sunlight. Although such a push is tiny in everyday circumstances, it can be large enough to send the spacecraft outward from the Sun on a cost-free but slow trip. Suppose that the spacecraft has a mass of 900 kg and receives a push of 20 N. (a) What is the magnitude of the resulting acceleration? If the craft starts from rest, (b) how far will it travel in 1 day and (c) how fast will it then be moving?

30E. The tension at which a fishing line snaps is commonly called the line's "strength." What minimum strength is needed for a line that is to stop a 19 lb salmon in 4.4 in. if the fish is initially drifting at 9.2 ft/s? Assume a constant deceleration.

31E. In a laboratory experiment, an initially stationary electron (mass = 9.11×10^{-31} kg) undergoes a constant acceleration through 1.5 cm, reaching a speed of 6.0×10^6 m/s at the end of that distance. (a) What is the magnitude of the force accelerating the electron? (b) What is the weight of the electron?

32E. An electron is projected horizontally at a speed of 1.2×10^7 m/s into an electric field that exerts a constant vertical force of 4.5×10^{-16} N on it. The mass of the electron is 9.11×10^{-31} kg. Determine the vertical distance the electron is deflected during the time it has moved 30 mm horizontally.

33E. A car that weighs 1.30×10^4 N is initially moving at a speed of 40 km/h when the brakes are applied and the car is brought to a stop in 15 m. Assuming that the force that stops the car is constant, find (a) the magnitude of that force and (b) the time required for the change in speed. If the initial speed is, instead, twice as great, and the car experiences the same force during the braking, how are (c) the stopping distance and (d) the stopping time changed? (There could be a lesson here about the danger of driving at high speeds.)

34E. Compute the initial upward acceleration of a rocket of mass 1.3×10^4 kg if the initial upward force produced by its engine (the thrust) is 2.6×10^5 N. Do not neglect the weight of the rocket.

35E. A rocket and its payload have a total mass of 5.0×10^4 kg. How large is the force produced by the engine (the thrust) when (a) the rocket is "hovering" over the launchpad just after ignition, and (b) when the rocket is accelerating upward at 20 m/s²?

36P. A 40 kg girl and an 8.4 kg sled are on the surface of a frozen lake, 15 m apart. By means of a rope, the girl exerts a horizontal 5.2 N force on the sled, pulling it toward her. (a) What is the acceleration of the sled? (b) What is the acceleration of the girl? (c) How far from the girl's initial position do they meet, assuming that no frictional forces act?

37P. A 160 lb firefighter slides down a vertical pole with an acceleration of 10 ft/s², directed downward. What are the magnitudes and directions of the vertical forces (a) exerted by the pole on the firefighter and (b) exerted by the firefighter on the pole?

38P. A sphere of mass 3.0×10^{-4} kg is suspended from a cord. A steady horizontal breeze pushes the sphere so that the cord makes an angle of 37° with the vertical when at rest. Find (a) the magnitude of that push and (b) the tension in the cord.

39P. A worker drags a crate across a factory floor by pulling on a rope tied to the crate (Fig. 5-44). The worker exerts a force of 450 N on the rope, which is inclined at 38° to the horizontal, and the floor exerts a horizontal force of 125 N that opposes the motion. Calculate the acceleration of the crate (a) if its mass is 310 kg and (b) if its weight is 310 N.

FIGURE 5-44 Problem 39.

40P. Two blocks are in contact on a frictionless table. A horizontal force is applied to one block, as shown in Fig. 5-45. (a) If

$m_1 = 2.3$ kg, $m_2 = 1.2$ kg, and $F = 3.2$ N, find the force between the two blocks. (b) Show that if a force of the same magnitude F is applied to m_2 but in the opposite direction, the force between the blocks is 2.1 N, which is not the same value calculated in (a). Explain the difference.

FIGURE 5-45   Problem 40.

41P. You pull a short refrigerator with a constant force **F** across a greased (frictionless) floor, either with **F** horizontal (case 1) or with **F** tilted upward at an angle θ (case 2). (a) What is the ratio of the refrigerator's speed in case 2 to its speed in case 1 if you pull for a certain time t? (b) What is this ratio if you pull for a certain distance d?

42P. For sport, a 12-kg armadillo runs onto a large pond of level, frictionless ice with an initial velocity of 5.0 m/s along the positive direction of the x axis. Take its initial position on the ice as being the origin. It slips over the ice while being pushed by a wind with a force of 17 N in the positive direction of the y axis. In unit-vector notation, what are the animal's (a) velocity and (b) position vectors when it has slid for 3.0 s?

43P. An elevator and its load have a combined mass of 1600 kg. Find the tension in the supporting cable when the elevator, originally moving downward at 12 m/s, is brought to rest with constant acceleration in a distance of 42 m.

44P. An object is hung from a spring balance attached to the ceiling of an elevator. The balance reads 65 N when the elevator is standing still. What is the reading when the elevator is moving upward (a) with a constant speed of 7.6 m/s and (b) with a speed of 7.6 m/s while decelerating at a rate of 2.4 m/s²?

45P. A 1400 kg jet engine is fastened to the fuselage of a passenger jet by just three bolts (this is the usual practice). Assume that each bolt supports one-third of the load. (a) Calculate the force on each bolt as the plane waits in line for clearance to take off. (b) During flight, the plane encounters turbulence, which suddenly imparts an upward vertical acceleration of 2.6 m/s² to the plane. Calculate the force on each bolt now.

46P. In Fig. 5-46 a 15,000 kg helicopter is lifting a 4500 kg truck with an upward acceleration of 1.4 m/s². Calculate (a) the force the air exerts on the helicopter blades and (b) the tension in the upper supporting cable.

FIGURE 5-46   Problem 46.

47P. An 80 kg man jumps down to a concrete patio from a window ledge only 0.50 m above the ground. He neglects to bend his knees on landing, so that his motion is arrested in a distance of 2.0 cm. (a) What is the average acceleration of the man from the time his feet first touch the patio to the time he is brought fully to rest? (b) With what force does this jump jar his bone structure?

48P. Three blocks are connected, as shown in Fig. 5-47, on a horizontal frictionless table and pulled to the right with a force $T_3 = 65.0$ N. If $m_1 = 12.0$ kg, $m_2 = 24.0$ kg, and $m_3 = 31.0$ kg, calculate (a) the acceleration of the system and (b) the tensions T_1 and T_2 in the interconnecting cords.

FIGURE 5-47   Problem 48.

49P. Figure 5-48 shows four penguins that are being playfully pulled along very slippery (frictionless) ice by a curator. The masses of three penguins and the tension in two of the cords are given. Find the penguin mass that is not given.

FIGURE 5-48   Problem 49.

50P. An elevator weighing 6240 lb is pulled upward by a cable with an acceleration of 4.00 ft/s². (a) Calculate the tension in the cable. (b) What is the tension when the elevator is decelerating at the rate of 4.00 ft/s² but is still moving upward?

51P. An 80-kg person is parachuting and experiencing a downward acceleration of 2.5 m/s². The mass of the parachute is 5.0 kg. (a) What upward force is exerted on the open parachute by the air? (b) What downward force is exerted by the person on the parachute?

52P. An 85-kg man lowers himself to the ground from a height of 10.0 m by holding onto a rope that runs over a frictionless pulley to a 65-kg sandbag. (a) With what speed does the man hit the ground if he started from rest? (b) Is there anything he could do to reduce the speed with which he hits the ground?

53P. A new 26-ton Navy jet (Fig. 5-49) requires an airspeed of 280 ft/s for liftoff (1 ton = 2000 lb). The engine develops a maximum force of 24,000 lb, but that is insufficient for reaching takeoff speed in the 300 ft available on an aircraft carrier. What minimum force (assumed constant) is needed from the catapult that is used to help launch the jet? Assume that the catapult and the jet's engine each exert a constant force over the 300 ft distance used for takeoff.

FIGURE 5-49 Problem 53.

54P. Imagine a landing craft approaching the surface of Callisto, one of Jupiter's moons. If the engine provides an upward force (thrust) of 3260 N, the craft descends at constant speed; if the engine provides only 2200 N, the craft accelerates downward at 0.39 m/s². (a) What is the weight of the landing craft in the vicinity of Callisto's surface? (b) What is the mass of the craft? (c) What is the free-fall acceleration near the surface of Callisto?

55P. A 52 kg circus performer is to slide down a rope that will snap if the tension exceeds 425 N. (a) What happens if the performer hangs stationary on the rope? (b) At what magnitude of acceleration does the performer just avoid breaking the rope?

56P. A chain consisting of five links, each of mass 0.100 kg, is lifted vertically with a constant acceleration of 2.50 m/s², as shown in Fig. 5-50. Find (a) the forces acting between adjacent links, (b) the force **F** exerted on the top link by the person lifting the chain, and (c) the *net* force accelerating each link.

FIGURE 5-50 Problem 56.

57P. A 1.0 kg mass on a 37° incline is connected to a 3.0 kg mass on a horizontal surface (Fig. 5-51). The surfaces and the pulley are frictionless. If $F = 12$ N, what is the tension in the connecting cord?

FIGURE 5-51 Problem 57.

58P. A block of mass $m_1 = 3.70$ kg on a frictionless inclined plane of angle 30.0° is connected by a cord over a massless, frictionless pulley to a second block of mass $m_2 = 2.30$ kg hanging vertically (Fig. 5-52). What are (a) the magnitude of the acceleration of each block and (b) the direction of the acceleration of m_2? (c) What is the tension in the cord?

FIGURE 5-52 Problem 58.

59P. You need to lower a bundle of old roofing material weighing 100 lb to the ground with a rope that will snap if the tension in it exceeds 87 lb. (a) How can you avoid snapping the rope during the descent? (b) If the descent is 20 ft and you just barely avoid snapping the rope, with what speed will the bundle hit the ground?

60P. A block is projected up a frictionless inclined plane with initial speed v_0. The angle of incline is θ. (a) How far up the plane does it go? (b) How long does it take to get there? (c) What is its speed when it gets back to the bottom? Find numerical answers for $\theta = 32.0°$ and $v_0 = 3.50$ m/s.

61P. A spaceship lifts off vertically from the moon, where the free-fall acceleration is 1.6 m/s². If the spaceship has an upward acceleration of 1.0 m/s² as it lifts off, what is the force of the spaceship on an astronaut who weighs 735 N on Earth?

62P. A lamp hangs vertically from a cord in a descending elevator that decelerates at 2.4 m/s². (a) If the tension in the cord is 89 N, what is the lamp's mass? (b) What is the cord's tension when the elevator ascends with an upward acceleration of 2.4 m/s²?

63P. A 100 kg crate is pushed at constant speed up the frictionless 30.0° ramp shown in Fig. 5-53. (a) What horizontal force **F** is required? (b) What force is exerted by the ramp on the crate?

FIGURE 5-53 Problem 63.

64P. A 10 kg monkey climbs up a massless rope that runs over a frictionless tree limb and back down to a 15 kg package on the ground (Fig. 5-54). (a) What is the magnitude of the least acceleration the monkey must have if it is to lift the package off the ground? If, after the package has been lifted, the monkey stops its climb and holds onto the rope, what are (b) the monkey's acceleration and (c) the tension in the rope?

FIGURE 5-54   Problem 64.

65P. Figure 5-55 shows a section of an alpine cable-car system. The maximum permissible mass of each car with occupants is 2800 kg. The cars, riding on a support cable, are pulled by a second cable attached to each pylon. What is the difference in tension between adjacent sections of pull cable if the cars are at the maximum permissible mass and are being accelerated up the 35° incline at 0.81 m/s²?

FIGURE 5-55   Problem 65.

66P. An interstellar ship has a mass of 1.20×10^6 kg and is initially at rest relative to a star system. (a) What constant acceleration is needed to bring the ship up to a speed of $0.10c$ (where c is the speed of light) relative to the star system in 3.0 days? (You need not consider Einstein's special relativity.) (b) What is that acceleration in g units? (c) What force is required for the acceleration? (d) If the engines are shut down when $0.10c$ is reached, how long does the ship take (start to finish) to journey 5.0 light-months, the distance that light travels in 5.0 months?

67P. The elevator in Fig. 5-56 consists of a 1150 kg cage (*A*), a 1400 kg counterweight (*B*), a driving mechanism (*C*), a cable, and two pulleys. In operation, mechanism *C* grabs the cable, either forcing it along or retarding its motion. This process means that tension T_1 in the cable on one side of *C* differs from tension T_2 on the other side. Suppose that the upward acceleration of *A* and the downward acceleration of *B* have the magnitude $a = 2.0$ m/s². Neglecting the pulleys and the mass of the cable, find (a) T_1, (b) T_2, and (c) the force on the cable produced by *C*.

FIGURE 5-56   Problem 67.

68P. A 5.00 kg block is pulled along a horizontal frictionless floor by a cord that exerts a force $F = 12.0$ N at an angle $\theta = 25.0°$ above the horizontal, as shown in Fig. 5-57. (a) What is the acceleration of the block? (b) The force F is slowly increased. What is its value just before the block is lifted (completely) off the floor? (c) What is the acceleration of the block just before it is lifted (completely) off the floor?

FIGURE 5-57   Problem 68.

69P. In earlier days, horses pulled barges down canals in the manner shown in Fig. 5-58. Suppose that the horse pulls on the rope with a force of 7900 N at an angle of 18° to the direction of motion of the barge, which is headed straight along the canal. The mass of the barge is 9500 kg, and its acceleration is 0.12 m/s². Calculate the force exerted by the water on the barge.

FIGURE 5-58   Problem 69.

70P. A hot-air balloon of mass M is descending vertically with downward acceleration a (Fig. 5-59). How much mass must be thrown out to give the balloon an upward acceleration a (same magnitude but opposite direction)? Assume that the upward force from the air (the lift) does not change because of the mass (the ballast) that is lost.

Ballast

FIGURE 5-59 Problem 70.

71P. A certain force gives mass m_1 an acceleration of 12.0 m/s² and mass m_2 an acceleration of 3.30 m/s². What acceleration would the force give to an object with a mass of (a) $m_2 - m_1$ and (b) $m_2 + m_1$?

72P. A rocket with mass 3000 kg is fired from the ground at an angle of elevation of 60°. The motor creates a force (thrust) on the rocket of 6.0×10^4 N at a constant angle of 60° to the horizontal for 50 s and then cuts out. As a rough approximation, ignore the mass of fuel consumed and neglect forces from the air. Calculate (a) the altitude of the rocket at motor cutout and (b) the total horizontal distance from firing point to eventual impact with the ground (assumed to be level).

73P. A block of mass M is pulled along a horizontal frictionless surface by a rope of mass m, as shown in Fig. 5-60. A horizontal force **F** is applied to one end of the rope. (a) Show that the rope *must* sag, even if only by an imperceptible amount. Then, assuming that the sag is negligible, find (b) the acceleration of rope and block, (c) the force that the rope exerts on the block, and (d) the tension in the rope at its midpoint.

FIGURE 5-60 Problem 73.

74P. Figure 5-61 shows a man sitting in a bosun's chair that dangles from a massless rope, which runs over a massless, frictionless pulley and back down to the man's hand. The combined mass of the man and chair is 95.0 kg. (a) With what force must the man pull on the rope for him to rise at constant speed? (b) What force is needed for an upward acceleration of 1.30 m/s²? (c) Suppose, instead, that the rope on the right is held by a person on the ground. Repeat (a) and (b) for this new situation. (d) In each of the four cases, what is the force exerted on the ceiling by the pulley system?

Electronic Computation

75. Fig. 5-62 shows two blocks on a frictionless plane, with inclination angle θ, and a third block. Blocks m_1 and m_2 are con-

FIGURE 5-61 Problem 74.

nected by a cord with tension T_1; blocks m_2 and m_3 are connected by a cord (around a frictionless, massless pulley) with tension T_2. If $\theta = 20°$, $m_1 = 2.00$ kg, $m_2 = 1.00$ kg, and $m_3 = 3.00$ kg, find T_1, T_2, and the acceleration of the blocks. (*Hint:* Use Newton's second law for each block to write three simultaneous equations, which can then be solved electronically.)

FIGURE 5-62 Problem 75.

76. A 2.00 kg object is subjected to three forces that give it an acceleration $\mathbf{a} = -(8.00 \text{ m/s}^2)\mathbf{i} + (6.00 \text{ m/s}^2)\mathbf{j}$. If two of the three forces are $\mathbf{F}_1 = (30.0 \text{ N})\mathbf{i} + (16.0 \text{ N})\mathbf{j}$ and $\mathbf{F}_2 = -(12.0 \text{ N})\mathbf{i} + (8.00 \text{ N})\mathbf{j}$, find the third force.

6
Force and Motion—II

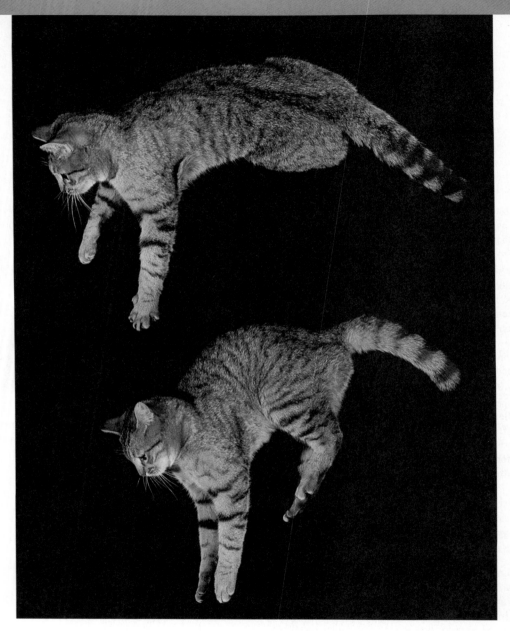

Cats, who enjoy sleeping on window sills, are often kept in apartment buildings. When a cat accidentally falls out of a window and onto a sidewalk, the extent of injury (such as the number of fractured bones or the certainty of death) <u>decreases</u> with height if the fall is more than seven or eight floors. (There is even a record of a cat who fell 32 floors and suffered only slight damage to its thorax and one tooth.) How can that be?

6-1 FRICTION

Frictional forces are unavoidable in our daily lives. Left to act alone, they would stop every moving object and bring to a halt every rotating shaft. In an automobile, about 20% of the gasoline is used to counteract friction in the engine and in the drive train. On the other hand, if friction were totally absent, we could not get an automobile to go anywhere, and we could not walk or ride a bicycle. We could not hold a pencil and, if we could, it would not write. Nails and screws would be useless, woven cloth would fall apart, and knots would come undone.

Here we deal with the frictional forces that exist between dry solid surfaces, moving across each other at slow speeds. Consider two simple experiments:

1. *First experiment.* Send a book sliding across a tabletop. A frictional force, exerted by the tabletop on the bottom of the sliding book, slows the book and eventually stops it. If you want to make the book move across the table with constant velocity, you must push or pull it with a steady force of equal magnitude and opposite direction to that of the frictional force, which opposes the motion.

2. *Second experiment.* A heavy crate is resting on the floor of a warehouse. You push on it horizontally with a steady force but it does not move. That is because the force that you apply is balanced by a frictional force, exerted horizontally on the bottom of the crate by the floor and opposite the direction of your push. Remarkably, this frictional force automatically adjusts itself, in both magnitude and direction, to cancel exactly whatever force you decide to apply. Of course, if you can push hard enough, you will be able to move the crate (see the first experiment).

Figure 6-1 shows a similar situation in detail. In Fig. 6.1a, a block rests on a tabletop, its weight **W** balanced by an equal but opposite normal force **N**. In Fig. 6-1b, you exert a force **F** on the block, attempting to pull it to the left. In response, a frictional force **f**$_s$ arises, pointing to the right, exactly matching the force that you have applied. The force **f**$_s$ is called the **static frictional force.**

Figures 6-1c and 6-1d show that as you increase your applied force, the static frictional force **f**$_s$ increases also and the block remains at rest. When the applied force reaches a certain value, however, the block "breaks away" from its intimate contact with the tabletop and accelerates leftward (Fig. 6-1e). The frictional force that then opposes the motion is called the **kinetic frictional force f**$_k$.

Usually, the kinetic frictional force, which acts when there is motion, is less than the maximum value of the static frictional force, which acts when there is no motion. So, if you wish the block to move across the surface with a

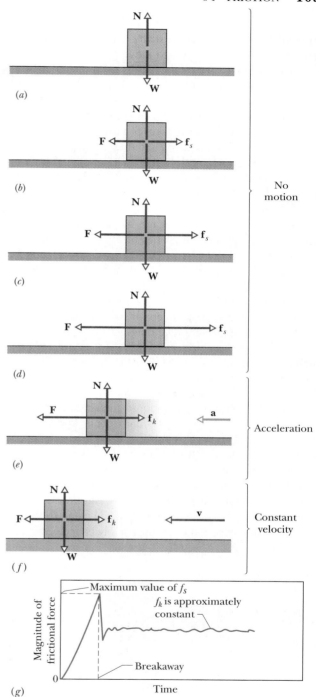

FIGURE 6-1 (a) The forces on a stationary block. (b–d) An external force **F**, applied to the block, is counterbalanced by an equal but opposite static frictional force **f**$_s$. As **F** is increased, **f**$_s$ also increases, until **f**$_s$ reaches a certain maximum value. (e) The block then "breaks away," accelerating suddenly to the left. (f) If the block is now to move with constant velocity, the applied force **F** must be reduced from the maximum value it had just before the block broke away. (g) Some experimental results for the sequence (a) through (f).

constant speed, you must usually decrease the applied force once the block begins to move, as in Fig. 6-1*f*. For example, Fig. 6-1*g* shows the results of an experiment in which the force on a block was slowly increased until breakaway occurred. Note the reduced force needed to keep the block moving at constant speed after breakaway.

Basically, the frictional force is a force acting between the surface atoms of one body and those of another. If two highly polished and carefully cleaned metal surfaces are brought together in a very good vacuum, they cannot be made to slide over each other. Instead, they *cold-weld* together instantly, forming a single piece of metal. If a machinist's specially polished gage blocks are brought together in air, they stick almost as firmly to each other and can be separated only by means of a wrenching motion. Usually, however, such intimate atom-to-atom contact is not possible. Even a highly polished metal surface is far from being flat on the atomic scale. Moreover, the surfaces of everyday objects have layers of oxides and other contaminants that reduce cold-welding.

When two such surfaces are placed together, only the high points touch each other. (It is like having the Alps of Switzerland turned over and placed down on the Alps of Austria.) The actual *micro*scopic area of contact is much less than the apparent *macro*scopic contact area, perhaps by a factor of 10^4. Nonetheless, many contact points cold-weld together. These welds produce static friction when an applied force attempts to slide the surfaces.

When the surfaces are pulled across each other, there is first a tearing of welds (breakaway) and then a continuous tearing apart and re-forming of welds as additional chance contacts are made (Fig. 6-2). The kinetic friction $\mathbf{f}_k$ is the vector sum of forces at those many chance contacts. Often, the motion is "jerky," because the two surfaces

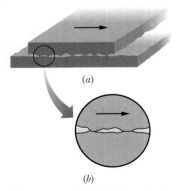

(a)

(b)

FIGURE 6-2 The mechanism of sliding friction. (*a*) The upper surface is sliding to the right over the lower surface in this enlarged view. (*b*) A detail, showing two spots where cold-welding has occurred. Force is required to break these welds and maintain the motion.

briefly stick together and then slip. Such repetitive *stick-and-slip* can produce squeaking or squealing, as when tires skid on dry pavement, fingernails scratch along a chalkboard, a rusty hinge is opened, and a bow is drawn across a violin string.

6-2 PROPERTIES OF FRICTION

Experiment shows that when a dry and unlubricated body is pressed against a surface in the same condition, and an applied force $\mathbf{F}$ attempts to slide the body along the surface, the resulting frictional force has three properties:

PROPERTY 1. If the body does not move, then the static frictional force $\mathbf{f}_s$ and the component of $\mathbf{F}$ that is parallel to the surface are equal in magnitude, and $\mathbf{f}_s$ is directed opposite that component of $\mathbf{F}$.

PROPERTY 2. The magnitude of $\mathbf{f}_s$ has a maximum value $f_{s,\max}$ that is given by

$$f_{s,\max} = \mu_s N, \qquad (6\text{-}1)$$

where μ_s is the **coefficient of static friction** and N is the magnitude of the normal force. If the magnitude of the component of $\mathbf{F}$ that is parallel to the surface exceeds $f_{s,\max}$, then the body begins to slide along the surface.

PROPERTY 3. If the body begins to slide along the surface, the magnitude of the frictional force rapidly decreases to a value f_k given by

$$f_k = \mu_k N, \qquad (6\text{-}2)$$

where μ_k is the **coefficient of kinetic friction.** Thereafter during the sliding, the kinetic frictional force $\mathbf{f}_k$ has a magnitude given by Eq. 6-2.

Properties 1 and 2 are worded in terms of a single applied force $\mathbf{F}$, but they also hold for the resultant of several applied forces acting on the body. Equations 6-1 and 6-2 are *not* vector equations; the direction of $\mathbf{f}_s$ or $\mathbf{f}_k$ is always parallel to the surface and opposite the intended motion, and $\mathbf{N}$ is perpendicular to the surface.

The coefficients μ_s and μ_k are dimensionless and must be determined experimentally. Their values depend on certain properties of both the body and the surface; hence, they are usually referred to with the preposition "between," as in "the value of μ_s *between* a sled and asphalt is 0.5." We assume that the value of μ_k does not depend on the speed at which the body slides along the surface.

CHECKPOINT **1:** A block lies on a floor. (a) What is the magnitude of the frictional force on it from the floor? (b) If a horizontal force of 5 N is now applied to the block, but the block does not move, what is the magnitude of the frictional force on it? (c) If the maximum value $f_{s,max}$ of the static frictional force on the block is 10 N, will the block move if the horizontally applied force is 8 N? (d) If the force is 12 N?

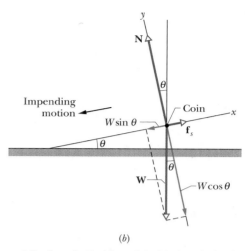

(b)

FIGURE 6-3 Sample Problem 6-1. (a) A coin is about to slide from rest down the cover of a book. (b) A free-body diagram for the coin, showing the three forces (drawn to scale) that act on it. The weight **W** is shown resolved into its components along the x and the y axes, whose orientations are chosen to simplify the problem.

SAMPLE PROBLEM 6-1

Figure 6-3a shows a coin resting on a book that has been tilted at an angle θ with the horizontal. By trial and error you find that when θ is increased to 13°, the coin begins to slide down the book. What is the coefficient of static friction μ_s between the coin and the book?

SOLUTION: Figure 6-3b is a free-body diagram for the coin when it is on the verge of sliding. The forces on the coin are the normal force **N**, pushing outward from the plane of the book, the weight **W** of the coin, and the frictional force $\mathbf{f}_s$, which points up the plane because the impending motion is down the plane. Since the coin is in equilibrium, the net force on it must be zero. From Newton's second law, we have

$$\sum \mathbf{F} = \mathbf{f}_s + \mathbf{W} + \mathbf{N} = 0. \qquad (6\text{-}3)$$

For the x components, this vector equation gives us

$$\sum F_x = f_s - W \sin \theta = 0, \quad \text{or} \quad f_s = W \sin \theta. \qquad (6\text{-}4)$$

For the y components, we have

$$\sum F_y = N - W \cos \theta = 0, \quad \text{or} \quad N = W \cos \theta. \qquad (6\text{-}5)$$

When the coin is on the verge of sliding (*and only then*) the magnitude of the static frictional force acting on it has the maximum value $\mu_s N$. Substituting $\mu_s N$ into Eq. 6-4 and dividing by Eq. 6-5, we obtain

$$\frac{f_s}{N} = \frac{\mu_s N}{N} = \frac{W \sin \theta}{W \cos \theta} = \tan \theta,$$

or

$$\mu_s = \tan \theta = \tan 13° = 0.23. \quad \text{(Answer)} \qquad (6\text{-}6)$$

You can easily measure μ_s for a coin and this text; you do not need a protractor. You can measure the two lengths shown in Fig. 6-3a with a ruler, and their ratio h/d is $\tan \theta$.

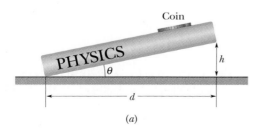

(a)

SAMPLE PROBLEM 6-2

If a car's wheels are "locked" (kept from rolling) during emergency braking, the car slides along the road. Ripped-off bits of tire and small melted sections of road form the "skid marks" that reveal the cold-welding during the slide. The record for the longest skid marks on a public road was reportedly set in 1960 by a Jaguar on the M1 highway in England— the marks were 290 m (about 950 ft) long! Assuming that $\mu_k = 0.60$, how fast was the car going when the wheels became locked?

SOLUTION: Figure 6-4a depicts the car's travel; Fig. 6-4b is the car's free-body diagram during deceleration, showing the car's weight **W**, the normal force **N**, and the kinetic frictional force $\mathbf{f}_k$ acting on the car. We can use Eq. 2-16,

$$v^2 = v_0^2 + 2a_x(x - x_0),$$

with $v = 0$ and $x - x_0 = d$, to find the car's initial speed v_0. Substituting these values and rearranging yield

$$v_0 = \sqrt{-2a_x d}. \qquad (6\text{-}7)$$

To find a_x, we apply Newton's second law along the x axis. If we ignore the effects of the air on the car, the only force with a component along the x axis is $\mathbf{f}_k$, and that component is $-f_k$. So we have

$$-f_k = ma_x, \quad \text{or} \quad a_x = -\frac{f_k}{m} = -\frac{\mu_k N}{m}, \qquad (6\text{-}8)$$

where m is the car's mass, and Eq. 6-2 tells us that $f_k = \mu_k N$.

The normal force **N** has magnitude $N = W = mg$. Substituting this result into Eq. 6-8 yields

$$a_x = -\frac{\mu_k mg}{m} = -\mu_k g. \qquad (6\text{-}9)$$

Substituting Eq. 6-9 into Eq. 6-7, we find

$$v_0 = \sqrt{2\mu_k g d} = \sqrt{(2)(0.60)(9.8 \text{ m/s}^2)(290 \text{ m})}$$
$$= 58 \text{ m/s} = 210 \text{ km/h} \qquad \text{(Answer)}$$

(about 130 mi/h). In obtaining the answer, we implicitly assumed that $v = 0$ at the far end of the skid marks. Actually, the marks ended only because the Jaguar left the road after 290 m. So v_0 was at least 210 km/h, and possibly much more.

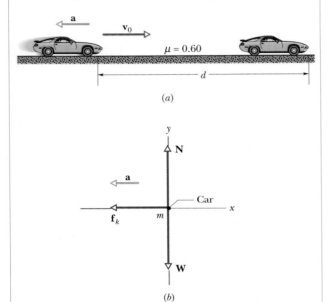

(a)

(b)

FIGURE 6-4 Sample Problem 6-2. (a) A car, sliding to the right and finally stopping after a displacement d. (b) A free-body diagram for the decelerating car. The acceleration vector points to the left, in the direction of the frictional force $\mathbf{f}_k$.

SAMPLE PROBLEM 6-3

A woman pulls a loaded sled of mass $m = 75$ kg along a horizontal surface at constant velocity. The coefficient of kinetic friction μ_k between the runners and the snow is 0.10, and the angle ϕ in Fig. 6-5 is 42°.

(a) What is the tension T in the rope?

SOLUTION: Figure 6-5b is the free-body diagram for the sled. Applying Newton's second law in the horizontal direction yields

$$T \cos \phi - f_k = ma_x = 0, \qquad (6\text{-}10)$$

where a_x is zero because the velocity is constant. In the vertical direction, we have

$$T \sin \phi + N - mg = ma_y = 0, \qquad (6\text{-}11)$$

in which mg is the weight of the sled. From Eq. 6-2,

$$f_k = \mu_k N. \qquad (6\text{-}12)$$

These three equations contain T, N, and f_k as unknowns. Eliminating N and f_k will allow us to find the remaining variable T. We start by adding Eqs. 6-10 and 6-12, obtaining

$$T \cos \phi = \mu_k N,$$

or

$$N = \frac{T \cos \phi}{\mu_k}. \qquad (6\text{-}13)$$

Substituting this into Eq. 6-11 and solving for T, we obtain

$$T = \frac{\mu_k mg}{\cos \phi + \mu_k \sin \phi} \qquad (6\text{-}14)$$
$$= \frac{(0.10)(75 \text{ kg})(9.8 \text{ m/s}^2)}{\cos 42° + (0.10)(\sin 42°)}$$
$$= 91 \text{ N}, \qquad \text{(Answer)}$$

which is considerably less than the weight of the sled.

(b) What is the normal force with which the snow pushes vertically upward on the sled?

SOLUTION: We can find N by substituting $T = 91$ N and given data into either Eq. 6-11 or 6-13. With Eq. 6-11 we find

$$N = mg - T \sin \phi$$
$$= (75 \text{ kg}) (9.8 \text{ m/s}^2) - (91 \text{ N}) \sin 42°$$
$$= 670 \text{ N}. \qquad \text{(Answer)}$$

(a)

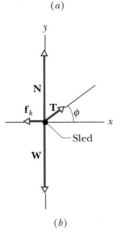

(b)

FIGURE 6-5 Sample Problem 6-3. (a) A woman exerts a force $\mathbf{T}$ on a sled, pulling it at a constant velocity. (b) A free-body diagram for the sled and its load.

CHECKPOINT 2: In the figure, horizontal force $\mathbf{F}_1$ of magnitude 10 N is applied to a box on a floor, but the box does not slide. Then, as the magnitude of vertically applied force $\mathbf{F}_2$ is increased from zero, but before the box begins to slide, do the following quantities increase, decrease, or stay the same: (a) the magnitude of the frictional force on the box; (b) the magnitude of the normal force on the box from the floor; (c) the maximum value $f_{s,\text{max}}$ of the static frictional force on the box?

SAMPLE PROBLEM 6-4

In Fig. 6-6a, a crate of dilled pickles with mass $m_1 = 14$ kg moves along a plane that makes an angle of $\theta = 30°$ with the horizontal. That crate is connected to a crate of pickled dills with mass $m_2 = 14$ kg by a taut, massless cord that runs around a frictionless, massless pulley. The hanging crate of dills descends with constant velocity.

(a) What are the magnitude and direction of the frictional force exerted on m_1 by the plane?

SOLUTION: The fact that m_2 descends indicates that m_1 moves up the plane, and so a *kinetic* frictional force $\mathbf{f}_k$ must point down the plane.

We cannot use Eq. 6-2 to find the magnitude of $\mathbf{f}_k$, because we do not know the coefficient of kinetic friction μ_k between m_1 and the plane. However, we can use the techniques of Chapter 5. To start, we draw the free-body diagrams for m_1 and m_2 in Figs. 6-6b and 6-6c, where $\mathbf{T}$ is the pull from the tension in the cord and the weight vectors are $\mathbf{W}_1 = m_1\mathbf{g}$ and $\mathbf{W}_2 = m_2\mathbf{g}$.

With $\mathbf{W}_1$ resolved into x and y components we have, from Newton's second law applied to the x axis in Fig. 6-6b,

$$\sum F_x = T - f_k - m_1 g \sin \theta = m_1 a_x = 0, \qquad (6\text{-}15)$$

where $a_x = 0$ because m_1 must move at constant velocity. Next, for m_2 we apply Newton's second law along the y axis in Fig. 6-6c and use the fact that m_2 moves at constant velocity. We find

$$\sum F_y = T - m_2 g = m_2 a_y = 0,$$

or

$$T = m_2 g. \qquad (6\text{-}16)$$

We then substitute T from Eq. 6-16 into Eq. 6-15 and solve for f_k, obtaining

$$f_k = m_2 g - m_1 g \sin \theta$$
$$= (14 \text{ kg})(9.8 \text{ m/s}^2) - (14 \text{ kg})(9.8 \text{ m/s}^2)(\sin 30°)$$
$$= 68.6 \text{ N} \approx 69 \text{ N.} \qquad \text{(Answer)}$$

(b) What is μ_k?

SOLUTION: We can use Eq. 6-2 to find μ_k, but first we need the magnitude of the normal force $\mathbf{N}$ acting on m_1. To find N, we apply Newton's second law for m_1 along the y axis in Fig. 6-6b:

$$\sum F_y = N - m_1 g \cos \theta = m_1 a_y = 0,$$

or

$$N = m_1 g \cos \theta.$$

From Eq. 6-2 we now have

$$\mu_k = \frac{f_k}{N} = \frac{f_k}{m_1 g \cos \theta}$$

$$= \frac{68.6 \text{ N}}{(14 \text{ kg})(9.8 \text{ m/s}^2)(\cos 30°)} = 0.58. \qquad \text{(Answer)}$$

(a)

(b)

(c)

FIGURE 6-6 Sample Problem 6-4. (a) Mass m_1 moves up the plane while mass m_2 descends at constant velocity. (b) A free-body diagram for mass m_1. (c) A free-body diagram for mass m_2.

6-3 THE DRAG FORCE AND TERMINAL SPEED

A **fluid** is anything that can flow—generally either a gas or a liquid. When there is a relative velocity between a fluid and a body (either because the body moves through the fluid or because the fluid moves past the body), the body experiences a **drag force D** that opposes the relative motion and points in the direction in which the fluid flows relative to the body.

Here we examine only cases in which air is the fluid, the body is blunt (like a baseball) rather than slender (like a javelin), and the relative motion is fast enough so that the air becomes turbulent (breaks up into swirls) behind the body. In such cases, the magnitude of the drag force **D** is related to the relative speed v by an experimentally determined **drag coefficient** C according to

$$D = \tfrac{1}{2}C\rho Av^2, \qquad (6\text{-}17)$$

where ρ is the air density (mass per volume) and A is the **effective cross-sectional area** of the body (the area of a cross section taken perpendicular to the velocity **v**). The drag coefficient C (typical values range from 0.4 to 1.0) is not truly a constant for a given body, because if v varies significantly, the value of C can vary as well. Here, we ignore such complications.

Downhill speed skiers know well that drag depends on A and v^2. To reach high speeds a skier must reduce D as much as possible by, for example, riding the skis in the "egg position" (Fig. 6-7) to minimize A.

Equation 6-17 indicates that when a blunt object falls from rest through air, D gradually increases from zero as the speed of the body increases. As suggested in Fig. 6-8, if the body falls far enough, D eventually equals the body's weight $W\ (= mg)$, and the net vertical force on the body is then zero. By Newton's second law, the acceleration must also be zero then, and so the body's speed no longer increases. The body then falls at a constant **terminal speed** v_t, which we find by setting $D = mg$ in Eq. 6-17, obtaining

$$\tfrac{1}{2}C\rho Av_t^2 = mg,$$

which gives

$$v_t = \sqrt{\frac{2mg}{C\rho A}}. \qquad (6\text{-}18)$$

FIGURE 6-7 This skier crouches in an "egg position" so as to minimize her effective cross-sectional area and thus the air drag acting on her.

FIGURE 6-8 The forces that act on a body falling through air: (a) the body when it has just begun to fall, (b) the free-body diagram just then, and (c) the free-body diagram a little later, after a drag force has developed. (d) The drag force has increased until it balances the weight of the body. The body now falls at its constant terminal speed.

Table 6-1 gives values of v_t for some common objects.

According to calculations* based on Eq. 6-17, a cat must fall about six floors to reach terminal speed. Until it does so, $W > D$ and the cat accelerates downward because of the net downward force. Recall from Chapter 2 that your body is an accelerometer, not a speedometer. Because the cat too senses the acceleration, it is frightened and keeps its feet underneath its body, its head tucked in, and its spine bent upward, making A small, v_t large, and injury on landing likely.

However, if the cat does reach v_t, the acceleration vanishes and the cat relaxes somewhat, stretching its legs

TABLE 6-1 SOME TERMINAL SPEEDS IN AIR

OBJECT	TERMINAL SPEED (m/s)	95% DISTANCE[a] (m)
16 lb Shot	145	2500
Sky diver (typical)	60	430
Baseball	42	210
Tennis ball	31	115
Basketball	20	47
Ping-Pong ball	9	10
Raindrop (radius = 1.5 mm)	7	6
Parachutist (typical)	5	3

[a]This is the distance through which the body must fall from rest to reach 95% of its terminal speed.

Source: Adapted from Peter J. Brancazio, *Sport Science,* Simon & Schuster, New York, 1984.

*W. O. Whitney and C. J. Mehlhaff, "High-rise syndrome in cats," *The Journal of the American Veterinary Medical Association,* Vol. 191, pages 1399–1403 (1987).

and neck horizontally outward and straightening its spine (it then resembles a flying squirrel). These actions increase A and D, and the cat begins to slow because now $D > W$ —the net force is upward—until a new, smaller v_t is reached. The decrease in v_t reduces the possibility of serious injury on landing. Just before the end of the fall, when it sees it is nearing the ground, the cat pulls its legs back beneath its body to prepare for the landing.

SAMPLE PROBLEM 6-5

If a falling cat reaches a first terminal speed of 60 mi/h while it is tucked in and then stretches out, doubling A, how fast is it falling when it reaches a new terminal speed?

SOLUTION: We let v_{to} and v_{tn} represent the original and new terminal speeds, and A_o and A_n the original and new areas. We then use Eq. 6-18 to set up a ratio of speeds:

$$\frac{v_{tn}}{v_{to}} = \frac{\sqrt{2mg/C\rho A_n}}{\sqrt{2mg/C\rho A_o}} = \sqrt{\frac{A_o}{A_n}} = \sqrt{\frac{A_o}{2A_o}} = \sqrt{0.5} \approx 0.7,$$

which means that $v_{tn} \approx 0.7 v_{to}$, or about 40 mi/h.

In April 1987, during a jump, parachutist Gregory Robertson noticed that fellow parachutist Debbie Williams had been knocked unconscious in a collision with a third sky diver and was unable to open her parachute. Robertson, who was well above Williams at the time and who had not yet opened his parachute for the 13,500-ft plunge, reoriented his body head-down so as to minimize A and maximize his downward speed. Reaching an estimated v_t of 200 mi/h, he caught up with Williams and then went into a horizontal "spread eagle" (as in Fig. 6-9) to increase D so that he could grab her. He opened her parachute and then, after releasing her, his own, with a scant 10 s before impact. Williams received extensive internal injuries due to her lack of control on landing but survived.

FIGURE 6-9 A parachutist in a horizontal "spread eagle" maximizes the air drag.

SAMPLE PROBLEM 6-6

A raindrop with radius $R = 1.5$ mm falls from a cloud that is at height $h = 1200$ m above the ground. The drag coefficient C for the drop is 0.60. Assume that the drop is spherical throughout its fall. The density of water ρ_w is 1000 kg/m³, and the density of air ρ_a is 1.2 kg/m³.

(a) What is the terminal speed of the drop?

SOLUTION: The volume of a sphere is $\frac{4}{3}\pi R^3$, and its effective area A is that of a circle with radius R. So, for the drop:
$$m = \tfrac{4}{3}\pi R^3 \rho_w \quad \text{and} \quad A = \pi R^2.$$

Then, from Eq. 6-18, we find

$$v_t = \sqrt{\frac{2mg}{C\rho_a A}} = \sqrt{\frac{8\pi R^3 \rho_w g}{3C\rho_a \pi R^2}} = \sqrt{\frac{8R\rho_w g}{3C\rho_a}}$$
$$= \sqrt{\frac{(8)(1.5 \times 10^{-3}\text{ m})(1000\text{ kg/m}^3)(9.8\text{ m/s}^2)}{(3)(0.60)(1.2\text{ kg/m}^3)}}$$
$$= 7.4 \text{ m/s} \ (= 17 \text{ mi/h}). \qquad \text{(Answer)}$$

Note that the height of the cloud does not enter into the calculation. The raindrop (see Table 6-1) reaches terminal speed after falling just a few meters.

(b) What would have been the speed just before impact if there had been no drag force?

SOLUTION: From Eq. 2-23, with $h = -(y - y_0)$ and $v_0 = 0$, we find
$$v = \sqrt{2gh} = \sqrt{(2)(9.8\text{ m/s}^2)(1200\text{ m})}$$
$$= 150 \text{ m/s} \ (= 340 \text{ mi/h}). \qquad \text{(Answer)}$$

Under these conditions, Shakespeare would scarcely have written, "it droppeth as the gentle rain from heaven, upon the place beneath."

CHECKPOINT **3**: Near the ground, is the speed of large raindrops greater than, less then, or the same as that of small raindrops, assuming that both large and small raindrops are spherical?

6-4 UNIFORM CIRCULAR MOTION

Recall that when a body moves in a circle (or a circular arc) at constant speed v, it is said to be in uniform circular motion. Also recall that the body has a centripetal acceleration (directed toward the center of the circle), of constant magnitude given by Eq. 4-22:

$$a = \frac{v^2}{r} \qquad \text{(centripetal acceleration)}, \qquad (6\text{-}19)$$

where r is the radius of the circle.

This centripetal acceleration is caused by a **centripetal force** that acts on the body and that is directed toward the center of the circle. The magnitude F of this force is constant and is given by Newton's second law as

$$F = ma = \frac{mv^2}{r} \quad \text{(centripetal force).} \quad (6\text{-}20)$$

If this force is not present, the body cannot undergo uniform circular motion. Both the centripetal acceleration and the centripetal force are vector quantities whose magnitudes are constant but whose directions are always changing so as to point toward the center of the circle.

If the body in uniform circular motion is, say, a hockey puck whirled around on the end of a string as in Fig. 6-10, the centripetal force is provided by the tension in the string. For the Moon in its (nearly) uniform circular motion around the Earth, the centripetal force is the gravitational attraction of the Earth. Thus a centripetal force is not a new kind of force; it can be a tension force, a gravitational force, or any other force.

Let us compare two familiar examples of uniform circular motion:

1. *Rounding a curve in a car.* You are sitting in the center of the rear seat of a car moving at high speed over flat road. When the driver suddenly turns left, rounding a corner in a circular arc, you are slid across the seat toward the right and jammed against the interior wall of the car. What is going on?

While the car is moving in the circular arc, it is in uniform circular motion. The centripetal force that causes the motion is a frictional force exerted by the roadway on the tires. This force points radially inward and—spread over the four tires—has a magnitude given by Eq. 6-20.

You would have been in uniform circular motion in the center of the seat if the frictional force exerted on you

by the seat had been great enough. However, it was not, and so you slid across the seat. Viewed from the reference frame of the ground, you actually continued moving in a straight line while the seat slid beneath you, until you met the car's wall. Its push on you was a centripetal force, and you then joined the car's uniform circular motion.

2. *Orbiting the Earth.* This time you are a passenger in the space shuttle *Atlantis,* orbiting the Earth and experiencing "weightlessness." What is going on in this case?

The centripetal force that keeps you and the shuttle in uniform circular motion is the gravitational attraction of the Earth for you and the shuttle. This force is directed radially inward, toward the center of the Earth, and has a magnitude given by Eq. 6-20.

In both car and shuttle you are in uniform circular motion, acted on by a centripetal force. Yet your experiences in the two situations are quite different. In the car, jammed up against the wall, you are aware of being compressed by the wall. In the orbiting shuttle, on the other hand, you are floating around with no sensation of any force acting on you. Why this great difference?

The difference is due to the nature of the two centripetal forces. In the car, the centripetal force is a *contact force,* exerted by the wall externally on the part of your body touching the wall. In the shuttle, the centripetal force is a *volume force,* due to the Earth's gravitational pull on every atom of your body and on every atom of the shuttle, in proportion to the mass of that atom. Thus there is no compression on any one part of your body, nor even any sensation of a force acting on you.

SAMPLE PROBLEM 6-7

Igor is a cosmonaut-engineer in the spacecraft *Vostok II,* orbiting the Earth at an altitude h of 520 km with a speed v of 7.6 km/s. Igor's mass m is 79 kg.

(a) What is his acceleration?

SOLUTION: Igor is in uniform circular motion in a circle of radius $R_E + h$, where R_E is the radius of the Earth. His centripetal acceleration is given by Eq. 6-19:

$$a = \frac{v^2}{r} = \frac{v^2}{R_E + h}$$

$$= \frac{(7.6 \times 10^3 \text{ m/s})^2}{6.37 \times 10^6 \text{ m} + 0.52 \times 10^6 \text{ m}}$$

$$= 8.38 \text{ m/s}^2 \approx 8.4 \text{ m/s}^2, \quad \text{(Answer)}$$

which is the value of the free-fall acceleration at Igor's altitude. If he were lifted to that altitude and released, instead of being put into orbit there, he would fall toward the Earth's center, starting out with that value for the acceleration. The

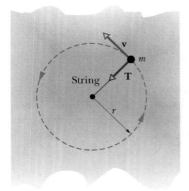

FIGURE 6-10 A hockey puck of mass m moves with constant speed v in a circular path on a horizontal frictionless surface. The centripetal force acting on the puck is **T**, the pull from the string.

difference in the two situations is that when he orbits the Earth, he always has a ''sideways'' motion as well: as he falls, he also moves to the side, so that he ends up moving along a curved path around the Earth.

(b) What (centripetal) gravitational force does the Earth exert on Igor?

SOLUTION: The centripetal force is

$$F = ma = (79 \text{ kg})(8.38 \text{ m/s}^2)$$
$$= 660 \text{ N} \approx 150 \text{ lb.} \qquad \text{(Answer)}$$

If Igor were to stand on a scale placed on the top of a tower with height $h = 520$ km, the scale would read 660 N or 150 lb. In orbit, the scale (if Igor could ''stand'' on it) would read zero because he and the scale are in free fall together, and therefore his feet do not actually press against it.

PROBLEM SOLVING TACTICS

TACTIC 1: *Looking Things Up*

In Sample Problem 6-7, we had to know the radius of the Earth, which was not given in the problem statement. You need to become familiar with sources of this kind of information, starting with this book. Many useful data are given in the inside covers, in the various appendixes, and in tables. The *Handbook of Chemistry and Physics,* updated every year by the publisher, CRC Press, is an invaluable resource.

For practice, see if you can track down the density of iron, the series expansion of e^x, the number of centimeters in a mile, the mean distance of Saturn from the Sun, the mass of the proton, the speed of light, and the atomic number of samarium. You can find them all in this book.

CHECKPOINT **4:** When you ride in a Ferris wheel at constant speed, what are the directions of your acceleration $\mathbf{a}$ and the normal force $\mathbf{N}$ on you (from the seat) as you pass through (a) the highest point and (b) the lowest point of the ride?

SAMPLE PROBLEM 6-8

In a 1901 circus performance, Allo ''Dare Devil'' Diavolo introduced the stunt of riding a bicycle in a loop-the-loop (Fig. 6-11a). Assuming that the loop is a circle with radius $R = 2.7$ m, what is the least speed v Diavolo could have at the top of the loop if he was to remain in contact with it there?

SOLUTION: Figure 6-11b is a free-body diagram for Diavolo and the bicycle (taken together as being a single particle) at the top of the loop, showing the normal force $\mathbf{N}$ exerted on them by the loop, and their weight $\mathbf{W} = m\mathbf{g}$. Their accelera-

tion $\mathbf{a}$ points downward, toward the center of the loop and, according to Eq. 6-19, has magnitude $a = v^2/R$. Applying Newton's second law along the y axis, we have

$$\sum F = -N - mg = -ma = -m\frac{v^2}{R}.$$

If he is on the verge of losing contact with the loop, $N = 0$, and so we then have

$$mg = m\frac{v^2}{R},$$

or
$$v = \sqrt{gR} = \sqrt{(9.8 \text{ m/s}^2)(2.7 \text{ m})}$$
$$= 5.1 \text{ m/s.} \qquad \text{(Answer)}$$

To avoid losing contact, Diavolo made certain that his speed at the top of the loop was greater than 5.1 m/s. Then $N > 0$ and, more important to him, he and the bicycle exerted a force of magnitude N on the track.

FIGURE 6-11 Sample Problem 6-8. (*a*) Contemporary advertisement for Diavolo and (*b*) free-body diagram for the performer at the top of the loop.

SAMPLE PROBLEM 6-9

Figure 6-12a shows a *conical pendulum.* Its bob, whose mass m is 1.5 kg, whirls around in a horizontal circle at constant speed v at the end of a cord whose length L, measured to the center of the bob, is 1.7 m. The cord makes an angle θ of 37° with the vertical. As the bob swings around in a circle, the cord sweeps out the surface of a cone. Find the period τ of the pendulum bob.

SOLUTION: Figure 6-12b, the free-body diagram for the bob, shows the forces on the bob: the pull $\mathbf{T}$ from the cord due to the cord's tension, and the bob's weight $\mathbf{W}$ (= $m\mathbf{g}$). We place the origin of axes at the center of the bob, as shown. Instead of the usual x axis (which is stationary), we use a radial axis r that always points from the bob toward the center of the circle.

The y and r components of **T** are $T \cos \theta$ and $T \sin \theta$, respectively. Since $a_y = 0$, Newton's second law gives

$$T \cos \theta - mg = ma_y = 0, \quad \text{or} \quad T \cos \theta = mg. \quad (6\text{-}21)$$

There must be a net force along the r axis to provide the centripetal acceleration for the bob. The only force component in that direction is $T \sin \theta$. So along the r axis, Eq. 6-20 gives

$$T \sin \theta = ma_r = \frac{mv^2}{R}, \quad (6\text{-}22)$$

where R is the radius of the bob's circular path. Dividing Eq. 6-22 by Eq. 6-21 and solving for v, we obtain

$$v = \sqrt{\frac{gR \sin \theta}{\cos \theta}}.$$

We can substitute $2\pi R/\tau$ (the distance around the circle divided by the period) for the speed v of the bob. Doing so and solving for τ, we obtain

$$\tau = 2\pi \sqrt{\frac{R \cos \theta}{g \sin \theta}}. \quad (6\text{-}23)$$

Now, from Fig. 6-12a we see that $R = L \sin \theta$. Making this substitution in Eq. 6-23 yields

$$\tau = 2\pi \sqrt{\frac{L \cos \theta}{g}} \quad (6\text{-}24)$$

$$= 2\pi \sqrt{\frac{(1.7 \text{ m})(\cos 37°)}{9.8 \text{ m/s}^2}} = 2.3 \text{ s}. \quad \text{(Answer)}$$

From Eq. 6-24 we see that the period τ does not depend on the mass of the bob, but only on $L \cos \theta$, the vertical distance of the bob from its point of support. Thus, as shown in Fig. 6-12c, if several conical pendulums of different lengths are swung from the same support *with the same period*, their bobs will all lie in the same horizontal plane.

(a)

(b)

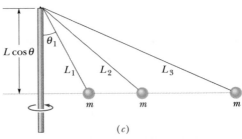

(c)

FIGURE 6-12 Sample Problem 6-9. (*a*) A conical pendulum, its cord making an angle θ with the vertical. (*b*) A free-body diagram for the pendulum bob. The axes point in the vertical and radial directions. The resultant force (and thus the acceleration) points radially inward toward the center of the circle. (*c*) Three pendulums, of different lengths, are whirled around by a rotating shaft; their bobs circulate in the same horizontal plane, as Eq. 6-24 predicts.

SAMPLE PROBLEM 6-10

Figure 6-13a represents a car of mass $m = 1600$ kg traveling at a constant speed $v = 20$ m/s along a flat, circular road of radius $R = 190$ m. What is the minimum value of μ_s between the tires of the car and the road that will prevent the car from slipping?

SOLUTION: The centripetal force that causes the car to travel in a circle is the radial frictional force $\mathbf{f}_s$ exerted on the tires by the road. (Although the car is moving, it is not sliding radially, and so the frictional force is $\mathbf{f}_s$ and not $\mathbf{f}_k$.)

Figure 6-13b, the free-body diagram for the car, shows the forces on the car: $\mathbf{f}_s$, **N**, and $\mathbf{W} = m\mathbf{g}$. Since the car is not accelerating vertically, $a_y = 0$, and Newton's second law leads to the now familiar result $N = W = mg$.

In the radial direction, however, there must be a net force ΣF_r to give the car a centripetal acceleration $\mathbf{a}_r$. (Otherwise the car would move off the road along a straight line.) According to Eq. 6-20, $\Sigma F_r = mv^2/R$. Because the only radial force we have is f_s, it must be true that

$$f_s = \frac{mv^2}{R}. \quad (6\text{-}25)$$

Next, recall that a body is on the verge of slipping when f_s reaches its maximum value, $\mu_s N$. Since the problem concerns just such a critical situation, we set $f_s = \mu_s N$ in Eq. 6-25 and then substitute mg for N. We get

$$\mu_s mg = \frac{mv^2}{R},$$

or

$$\mu_s = \frac{v^2}{gR} \quad (6\text{-}26)$$

$$= \frac{(20 \text{ m/s})^2}{(9.8 \text{ m/s}^2)(190 \text{ m})} = 0.21. \quad \text{(Answer)}$$

If $\mu_s \geq 0.21$, the car will be held in a circle by $\mathbf{f}_s$. But if $\mu_s < 0.21$, the car will slide out of the circle.

Note two additional features of Eq. 6-26. First, the value

of μ_s depends on the square of v. That means that much more friction is required as the turning speed increases. You may have noted this effect if you have ever taken a flat turn too fast and suddenly felt the tires slip. Second, the mass m dropped out in our derivation of Eq. 6-26. That means Eq. 6-26 holds for a vehicle of any mass, from a kiddy car to a bicycle to a heavy truck.

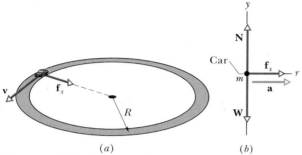

	(a)		(b)

FIGURE 6-13 Sample Problem 6-10. (a) A car moves around a flat curved road at constant speed. The frictional force $\mathbf{f}_s$ provides the necessary centripetal force. (b) A free-body diagram (not to scale) for the car, in the vertical plane.

$$N_r = N \sin \theta = \frac{mv^2}{R}. \qquad (6\text{-}28)$$

Dividing Eq. 6-28 by Eq. 6-27 gives

$$\tan \theta = \frac{v^2}{gR}.$$

Thus we have

$$\theta = \tan^{-1} \frac{v^2}{gR} \qquad (6\text{-}29)$$

$$= \tan^{-1} \frac{(20 \text{ m/s})^2}{(9.8 \text{ m/s}^2)(190 \text{ m})} = 12°. \quad \text{(Answer)}$$

Equations 6-26 and 6-29 tell us that the critical coefficient of friction for an unbanked road is the same as the tangent of the bank angle for a banked road. The road must produce a certain centripetal force one way or the other—either with friction or by being banked.

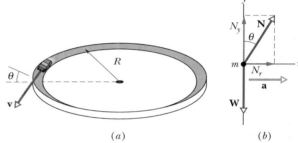

	(a)		(b)

FIGURE 6-14 Sample Problem 6-11. (a) A car moves around a curved banked road at constant speed. The bank angle is exaggerated for clarity. (b) A free-body diagram for the car, assuming that friction between tires and road is zero. The radially inward component of the normal force provides the necessary centripetal force. The resulting acceleration also points radially inward.

CHECKPOINT 5: In Fig. 6-13, suppose the car is on the verge of sliding when the radius of the circle is R_1. (a) If we double the car's speed, what is the least radius that would now keep the car from sliding? (b) If we also double the weight of the car (say by adding sandbags), what is the least radius that would now keep the car from sliding?

SAMPLE PROBLEM 6-11

You cannot always count on friction to get your car around a curve, especially if the road is icy or wet. That is why highway curves are banked. As in Sample Problem 6-10, suppose that a car of mass m moves at a constant speed v of 20 m/s around a curve, now banked, whose radius R is 190 m (Fig. 6-14a). What bank angle θ makes reliance on friction unnecessary?

SOLUTION: The centripetal acceleration and the required centripetal force ΣF_r are the same as in the preceding sample problem. The effect of banking is to tilt the normal force $\mathbf{N}$ toward the center of curvature of the road, so that now it has an inward radial component N_r that can supply the required centripetal force.

Since there is no acceleration in the vertical direction,

$$N_y = N \cos \theta = W = mg. \qquad (6\text{-}27)$$

In the radial direction the only force component is N_r (we assume a frictional force is unnecessary). So, by Eq. 6-20,

SAMPLE PROBLEM 6-12

Even some seasoned roller coaster riders blanch at the thought of riding the Rotor, which is essentially a large hollow cylinder that is rotated rapidly around its central axis (Fig. 6-15). Before the ride begins, a rider enters the cylinder through a door on the side and stands on a floor, up against a canvas-covered wall. The door is closed, and as the cylinder begins to turn, the rider, wall, and floor move in unison. When the rider's speed reaches some predetermined value, the floor abruptly and alarmingly falls away. The rider does not fall with it but instead is pinned to the wall while the cylinder rotates, as if an unseen (and somewhat unfriendly) agent is pressing the body to the wall. Later, the floor is eased back to the rider's feet, the cylinder slows, and the rider sinks a few centimeters to regain footing on the floor. (Some riders consider all this to be fun.)

Suppose that the coefficient of static friction μ_s between the rider's clothing and the canvas is 0.40 and that the cylinder's radius R is 2.1 m.

(a) What minimum speed v must the cylinder and rider have if the rider is not to fall when the floor drops?

SOLUTION: The rider will not be pulled down by her weight **W** provided the magnitudes of **W** and the frictional force **f**$_s$, which is exerted upward on her by the wall, are equal. At the minimum required speed, she is on the verge of slipping, which means that f_s must be at its maximum value of $\mu_s N$. So in this critical situation

$$\mu_s N = mg, \qquad (6\text{-}30)$$

where m is her mass.

The normal force **N** is, as usual, perpendicular to the surface against which the body (here, the woman) is pressed, but note that now it points horizontally toward the central axis. That force accounts for the centripetal force that provides the woman's centripetal acceleration a_r and keeps her moving in a circle. Thus, by Eq. 6-20,

$$N = \frac{mv^2}{R}. \qquad (6\text{-}31)$$

Substituting this expression for N in Eq. 6-30 and solving for v, we have

$$v = \sqrt{\frac{gR}{\mu_s}} = \sqrt{\frac{(9.8 \text{ m/s}^2)(2.1 \text{ m})}{0.40}}$$

$$= 7.17 \text{ m/s} \approx 7.2 \text{ m/s}. \qquad \text{(Answer)}$$

Note that the result is independent of the rider's mass; it holds for anyone riding the Rotor, from a child to a sumo wrestler.

(b) If the rider's mass is 49 kg, what is the magnitude of the centripetal force on her?

SOLUTION: According to Eq. 6-31,

$$N = \frac{mv^2}{R} = \frac{(49 \text{ kg})(7.17 \text{ m/s})^2}{2.1 \text{ m}}$$

$$\approx 1200 \text{ N}. \qquad \text{(Answer)}$$

FIGURE 6-15 Sample Problem 6-12. A Rotor in an amusement park, showing the forces on a rider. The centripetal force is the normal force with which the wall pushes inward on the rider.

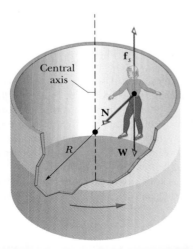

Although this force is directed toward the central axis, the rider has an overwhelming sensation that the force pinning her against the wall is directed radially outward. Her sensation stems from the fact that she is in a noninertial frame (she and it are accelerating). As measured from such frames, forces can be illusionary. The illusion is part of the Rotor's attraction.

CHECKPOINT **6:** If the Rotor of Sample Problem 6-12 initially moves at the minimum required speed for the rider not to fall and then its speed is increased in steps, do the following increase, decrease, or remain the same: (a) the magnitude of **f**$_s$; (b) the magnitude of **N**; (c) the value of $f_{s,\text{max}}$?

6-5 THE FORCES OF NATURE

We have used **F** as a generic symbol for force. We have also used other symbols: **W** for the weight of a body, **T** for the pull from a cord under tension, **f** for a frictional force, **N** for a normal force, and **D** for the drag force exerted, for example, by air on a sky diver. At the fundamental level, all these forces fall into two types: (1) the **gravitational force,** of which weight is our only example, and (2) the **electromagnetic force,** which includes—without exception—all the others. The electromagnetic force is the combination of electrical forces and magnetic forces. The force that makes an electrically charged balloon stick to a wall and the force with which a magnet picks up an iron nail are other examples of it. In fact, aside from the gravitational force, *all* forces that we can experience directly as a push or pull are electromagnetic in nature. That is, all such forces, including frictional forces, normal forces, contact forces, and tension forces, involve, fundamentally, electromagnetic forces exerted by one atom on another. The tension in a taut rope, for example, exists only because the atoms of the rope attract one another.

Only two other fundamental forces are known, and they both act over such short distances that we cannot ex-

Banked tracks are needed for turns that are taken so quickly that friction alone cannot provide enough centripetal force.

TABLE 6-2 THE QUEST FOR THE SUPERFORCE—A PROGRESS REPORT

DATE	RESEARCHER	ACHIEVEMENT
1687	Newton	Showed that the same laws apply to astronomical bodies and to objects on Earth. Unified celestial and terrestrial mechanics.
1820 1830s	Oersted Faraday	Showed, by brilliant experiments, that the then separate sciences of electricity and magnetism are intimately linked.
1873	Maxwell	Unified the sciences of electricity, magnetism, and optics into the single subject of electromagnetism.
1979	Glashow, Salam, Weinberg	Received the Nobel prize for showing that the weak force and the electromagnetic force could be viewed as different aspects of a single *electroweak force.* This combination of forces reduced the number of fundamental forces from four to three.
1984	Rubbia, van der Meer	Received the Nobel prize for verifying experimentally the predictions of the theory of the electroweak force.

Work in Progress

Grand unification theories (GUTs), seek to unify the electroweak force and the strong force.
Supersymmetry theories: seek to unify all forces, including the gravitational force, within a single framework.
Superstring theories: interpret pointlike particles, such as electrons, as being unimaginably tiny, closed loops. Strangely, extra dimensions beyond the familiar four dimensions of spacetime appear to be required.

perience them directly through our senses. They are the **weak force,** which is involved in certain kinds of radioactive decay, and the **strong force,** which binds together the quarks that make up protons and neutrons and is the ''glue'' that holds together an atomic nucleus.

Physicists have long believed that nature has an underlying simplicity and that the number of fundamental forces can be reduced. Einstein spent most of his working life trying to interpret these forces as different aspects of a single *superforce.* He failed, but in the 1960s and 1970s, other physicists showed that the weak force and the electromagnetic force are different aspects of a single **electroweak force.** The quest for further reduction continues today, at the very forefront of physics. Table 6-2 lists the progress that has been made toward **unification** (as the goal is called) and gives some hints about the future.

REVIEW & SUMMARY

Friction

When a force **F** attempts to slide a body along a surface, a **frictional force** is exerted on the body by the surface. The frictional force is parallel to the surface and directed so as to oppose the sliding. It is due to bonding between the body and the surface.

If the body does not slide, the frictional force is a **static frictional force** $\mathbf{f}_s$. If there is sliding, the frictional force is a **kinetic frictional force** $\mathbf{f}_k$.

Properties of Friction

Property 1. If the body does not move, then the static frictional force $\mathbf{f}_s$ and the component of **F** that is parallel to the surface are equal in magnitude, and $\mathbf{f}_s$ is directed opposite that component. If that parallel component increases, $\mathbf{f}_s$ also increases.

Property 2. The magnitude of $\mathbf{f}_s$ has a maximum value $f_{s,\max}$ that is given by

$$f_{s,\max} = \mu_s N, \qquad (6\text{-}1)$$

where μ_s is the **coefficient of static friction** and N is the magnitude of the normal force. If the component of **F** that is parallel to the surface exceeds $f_{s,\max}$, then the body slides on the surface.

Property 3. If the body begins to slide along the surface, the magnitude of the frictional force rapidly decreases to a constant value f_k given by

$$f_k = \mu_k N, \qquad (6\text{-}2)$$

where μ_k is the **coefficient of kinetic friction.**

Drag Force

When there is a relative velocity between air (or some other fluid) and a body, the body experiences a **drag force D** that opposes the relative motion and points in the direction in which the fluid flows relative to the body. The magnitude of **D** is related to the relative speed v by an experimentally determined **drag coefficient** C according to

$$D = \tfrac{1}{2} C\rho A v^2, \qquad (6\text{-}17)$$

where ρ is the fluid density (mass per volume) and A is the **effective cross-sectional area** of the body (the area of a cross section taken perpendicular to the relative velocity **v**).

Terminal Speed

When a blunt object falls far enough through air, the magnitudes of the drag force and the object's weight are equal. The body then falls at a constant **terminal speed** v_t given by

$$v_t = \sqrt{\frac{2mg}{C\rho A}}, \qquad (6\text{-}18)$$

where m is the body's mass.

Uniform Circular Motion

If a particle moves in a circle or a circular arc with radius r at constant speed v, it is said to be in **uniform circular motion.** It then has a **centripetal acceleration** with a magnitude given by

$$a = \frac{v^2}{r}, \qquad (6\text{-}19)$$

which is due to a **centripetal force** with a magnitude given by

$$F = \frac{mv^2}{r}, \qquad (6\text{-}20)$$

where m is the particle's mass. The vectors **a** and **F** point toward the center of curvature of the particle's path.

Fundamental Forces

The myriad examples of forces can be reduced to three fundamental types: **gravitational, electroweak** (a combination of the historic grouping of **electric** and **magnetic** forces with the **weak** force), and **strong.** Only the gravitational, electric, and magnetic forces are readily apparent in the everyday world. Physicists hope to reduce the list of three fundamental forces to a single force, the elusive *superforce* that would include all others.

QUESTIONS

1. Figure 6-16 shows four blocks arranged on a board. The board will be lifted by its right end (like the book in Fig. 6-3a) until the blocks begin to slide down it. The blocks are made of the same material and their masses are:

block 1, 5 kg   block 3, 10 kg

block 2, 10 kg   block 4, 5 kg

In which order, left to right, should the blocks be placed for them to begin sliding at the smallest possible angle between the board and the horizontal?

FIGURE 6-16   Question 1.

2. In Fig. 6-17, horizontal force $\mathbf{F}_1$ of magnitude 10 N is applied to a box on a floor, but the box does not slide. Then, as the magnitude of vertical force $\mathbf{F}_2$ is increased from zero, do the following quantities increase, decrease, or stay the same: (a) the magnitude of the frictional force $\mathbf{f}_s$ on the box; (b) the magnitude of the normal force $\mathbf{N}$ on the box from the floor; (c) the maximum value $f_{s,\max}$ of the static frictional force on the box? (d) Does the box eventually slide?

FIGURE 6-17   Question 2.

3. If you press an apple crate against a wall so hard that the crate cannot slide down the wall, what is the direction of (a) the static frictional force $\mathbf{f}_s$ on the crate from the wall and (b) the normal force $\mathbf{N}$ on the crate from the wall? If you increase your push, what happens to (c) f_s, (d) N, and (e) $f_{s,\max}$?

4. A box is on a ramp that is at angle θ to the horizontal. As θ is increased from zero, and before the box slips, do the following increase, decrease, or remain the same: (a) the weight component of the box along the ramp, (b) the magnitude of the static frictional force on the box from the ramp, (c) the weight component of the box perpendicular to the ramp, (d) the normal force on the box from the ramp, and (e) the maximum value $f_{s,\max}$ of the static frictional force?

5. In Fig. 6-18, a block is held stationary on a ramp by the frictional force on it from the ramp. A force **F**, directed up the ramp, is then applied to the block and gradually increased in magnitude from zero. During the increase, what happens to the direction and magnitude of the frictional force on the block?

FIGURE 6-18   Question 5.

6. Reconsider Question 5 but with the force **F** now directed down the ramp. As the magnitude of **F** is increased from zero, what happens to the direction and magnitude of the frictional force on the block?

7. In Fig. 6-19, if the angle θ of force **F** on the stationary box is increased, do the following quantities increase, decrease, or remain the same: (a) F_x; (b) $\mathbf{f}_s$; (c) **N**; (d) $f_{s,\max}$?

FIGURE 6-19   Question 7.

8. Repeat Question 7 for force **F** angled upward instead of downward.

9. In the conical pendulum of Sample Problem 6-9, what (a) period and (b) speed are associated with $\theta = 90°$?

10. A particle is made to move around three circular arcs, with the following speeds and radii of curvature:

ARC	SPEED	RADIUS
1	$2v_0$	r_0
2	$3v_0$	$3r_0$
3	$2v_0$	$4r_0$

Rank the arcs according to the magnitude of the centripetal force acting on the particle, greatest first.

11. Figure 6-20 shows an overhead view of a amusement park ride that travels at constant speed through five circular arcs of radii R_0, $2R_0$, and $3R_0$. Rank the arcs according to the magnitude of the centripetal force acting on a rider while traveling in the arcs, greatest first.

FIGURE 6-20   Question 11.

12. Figure 6-21 shows a section of a circular space station that rotates about its center so as to give an apparent weight to the crew. One of the crew is shown at the outer wall of the station, which has velocity $\mathbf{v}_s$. (a) If the astronaut moves to a point closer to the center of the station (say by taking an elevator), does his apparent weight increase, decrease, or remain the same? (b) If,

instead, the astronaut runs along the outer wall in the direction opposite $\mathbf{v}_s$ (with a speed less than the magnitude of $\mathbf{v}_s$), does his apparent weight increase, decrease, or remain the same?

FIGURE 6-21   Question 12.

13. Figure 6-22 shows overhead views of two stones that travel in circles over a frictionless surface. Each stone is tied to a cord whose opposite end is anchored at the center of the circle. Is the tension in the longer cord greater than, less than, or the same as that in the shorter cord if the stones travel (a) at the same speed and (b) with the same period of motion?

FIGURE 6-22   Question 13.

14. A coin lies on a turntable whose speed can be gradually increased from zero in steps. What happens to the magnitude of the frictional force on the coin from the turntable as the speed is increased to a large value?

EXERCISES & PROBLEMS

SECTION 6-2 Properties of Friction

1E. A bedroom bureau with a mass of 45 kg, including drawers and clothing, rests on the floor. (a) If the coefficient of static friction between the bureau and the floor is 0.45, what is the minimum horizontal force a person must apply to start the bureau moving? (b) If the drawers and clothing, with 17 kg mass, are removed before the bureau is pushed, what is the new minimum magnitude?

2E. A baseball player with mass $m = 79$ kg, sliding into second base, is retarded by a force of friction $f = 470$ N. What is the coefficient of kinetic friction μ_k between the player and the ground?

3E. The coefficient of static friction between Teflon and scrambled eggs is about 0.04. What is the smallest angle from the horizontal that will cause the eggs to slide across the bottom of a Teflon-coated skillet?

4E. A 100 N force, directed at an angle θ above the horizontal, is applied to a 25.0 kg chair sitting on the floor. (a) For each of the following angles θ, calculate the magnitude of the normal force of the floor on the chair and the horizontal component of the applied force: (i) 0°, (ii) 30.0°, (iii) 60.0°. (b) Take the coefficient of static friction between the chair and the floor to be 0.420 and, for each of the values of θ, decide if the chair remains at rest or slides.

5E. In Nevada and southern California, stones leave trails in the hard-baked desert floor, as if they had been migrating (Fig. 6-23). For years curiosity mounted about the unseen motion that caused the trails. The answer finally came in the 1970s: when an occasional storm hits the desert, a thin layer of mud may form over a still-firm base, greatly reducing the coefficient of friction between the stones and ground. If a strong wind accompanies the storm, it pushes the stones, leaving trails that are later baked hard by the Sun. Suppose a stone's mass is 300 kg (about the greatest stone mass that has left a trail) and the coefficient of static friction is reduced to 0.15. Of what magnitude is the force from a horizontal gust that is needed to move the stone?

FIGURE 6-23 Exercise 5.

6E. What is the greatest acceleration that can be generated by a runner if the coefficient of static friction between shoes and track is 0.95? (Only one foot is on the track during the acceleration.)

7E. A worker pushes horizontally on a 35 kg crate with a 110 N force. The coefficient of static friction between the crate and the floor is 0.37. (a) What is the frictional force exerted on the crate by the floor? (b) What is the maximum magnitude $f_{s,\max}$ of the static frictional force under the circumstances? (c) Does the crate move? (d) Suppose, next, that a second worker pulls directly upward on the crate to help out. What is the least pull she can exert that will allow the first worker's 110 N push to move the crate? (e) If, instead, the second worker pulls horizontally to help out, what is the least pull she can exert to get the crate moving?

8E. A person pushes horizontally with a force of 220 N on a 55 kg crate to move it across a level floor. The coefficient of kinetic friction is 0.35. (a) What is the magnitude of the frictional force? (b) What is the acceleration of the crate?

9E. A trunk with a weight of 220 N rests on the floor. The coefficient of static friction between the trunk and the floor is 0.41, while the coefficient of kinetic friction is 0.32. (a) What is the minimum magnitude for a horizontal force with which a person must push on the trunk to start it moving? (b) Once the trunk is moving, what magnitude of horizontal force must the person apply to keep it moving with constant velocity? (c) If the person continued to push with the force used to start the motion, what would be the acceleration of the trunk?

10E. A filing cabinet with a weight of 556 N rests on the floor. The coefficient of static friction between it and the floor is 0.68, and the coefficient of kinetic friction is 0.56. In four different attempts to move it, it is pushed with horizontal forces of (a) 222 N, (b) 334 N, (c) 445 N, and (d) 556 N. For each attempt, determine whether the cabinet moves, and calculate the magnitude of the frictional force the floor exerts on it. The cabinet is initially at rest for each attempt.

11E. A horizontal force F of 12 N pushes a block weighing 5.0 N against a vertical wall (Fig. 6-24). The coefficient of static

FIGURE 6-24 Exercise 11.

friction between the wall and the block is 0.60, and the coefficient of kinetic friction is 0.40. Assume that the block is not moving initially. (a) Will the block start moving? (b) In unit-vector notation, what is the force exerted on the block by the wall?

12E. A 49 kg rock climber is climbing a "chimney" between two rock slabs as shown in Fig. 6-25. The static coefficient of friction between her shoes and the rock is 1.2; between her back and the rock it is 0.80. She has reduced her push against the rock until her back and her shoes are on the verge of slipping. (a) What is her push against the rock? (b) What fraction of her weight is supported by the frictional force on her shoes?

FIGURE 6-25 Exercise 12.

13E. A house is built on the top of a hill with a nearby 45° slope (Fig. 6-26). An engineering study indicates that the slope angle should be reduced because the top layers of soil along the slope might slip past the lower layers. If the static coefficient of friction between two such layers is 0.5, what is the least angle ϕ through which the present slope should be reduced to prevent slippage?

FIGURE 6-26 Exercise 13.

14E. The coefficient of kinetic friction in Fig. 6-27 is 0.20. What is the acceleration of the block if (a) it is sliding down the slope and (b) it has been given an upward shove and is still sliding up the slope?

FIGURE 6-27 Exercise 14.

15E. A 110 g hockey puck slides on the ice for 15 m before it stops. (a) If its initial speed was 6.0 m/s, what was the magnitude of the frictional force on the puck during the sliding? (b) What was the coefficient of friction between the puck and the ice?

16P. A student, crazed with final exams, uses a force **P** of magnitude 80 N to push a 5.0 kg block across the ceiling of his room, as shown in Fig. 6-28. If the coefficient of kinetic friction between the block and surface is 0.40, what is the magnitude of the acceleration of the block?

FIGURE 6-28 Problem 16.

17P. A student wants to determine the coefficients of static friction and kinetic friction between a box and a plank. She places the box on the plank and gradually raises one end of the plank. When the angle of inclination with the horizontal reaches 30°, the box starts to slip, and it slides 2.5 m down the plank in 4.0 s. What are the coefficients of friction?

18P. A worker wishes to pile a cone of sand onto a circular area in his yard. The radius of the circle is R, and no sand is to spill onto the surrounding area (Fig. 6-29). If μ_s is the static coefficient of friction between each layer of sand along the slope and the sand beneath it (along which it might slip), show that the greatest volume of sand that can be stored in this manner is $\pi \mu_s R^3/3$. (The volume of a cone is $Ah/3$, where A is the base area and h is the cone's height.)

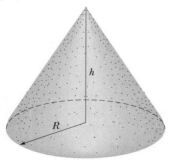

FIGURE 6-29 Problem 18.

19P. A ski that is placed on snow will stick to the snow. However, when the ski is moved along the snow, the rubbing warms and partially melts the snow, reducing the coefficient of friction and promoting sliding. Waxing the ski makes it water repellent

and reduces friction with the resulting layer of water. A magazine reports that a new type of plastic ski is especially water repellent and that, on a gentle 200 m slope in the Alps, a skier reduced his top-to-bottom time from 61 s with standard skis to 42 s with the new skis. (a) Determine the magnitudes of his average acceleration with each pair of skis. (b) Assuming a 3.0° slope, compute the coefficient of kinetic friction for each case.

20P. An 11 kg block of steel is at rest on a horizontal table. The coefficient of static friction between block and table is 0.52. (a) What is the magnitude of the horizontal force that will just start the block moving? (b) What is the magnitude of a force acting upward 60° from the horizontal that will just start the block moving? (c) If the force acts down at 60° from the horizontal, how large can its magnitude be without causing the block to move?

21P. A railroad flatcar is loaded with crates having a coefficient of static friction of 0.25 with the floor. If the train is moving at 48 km/h, in how short a distance can the train be stopped at constant deceleration without causing the crates to slide?

22P. A block slides down an inclined plane of slope angle θ with constant velocity. It is then projected up the same plane with an initial speed v_0. (a) How far up the incline will it move before coming to rest? (b) Will it slide down again? Give an argument to back your answer.

23P. A 68 kg crate is dragged across a floor by pulling on a rope inclined 15° above the horizontal. (a) If the coefficient of static friction is 0.50, what minimum tension in the rope is required to start the crate moving? (b) If $\mu_k = 0.35$, what is the magnitude of the initial acceleration of the crate?

24P. A pig slides down a 35° incline (Fig. 6-30) in twice the time it would take to slide down a frictionless 35° incline. What is the coefficient of kinetic friction between the pig and the incline?

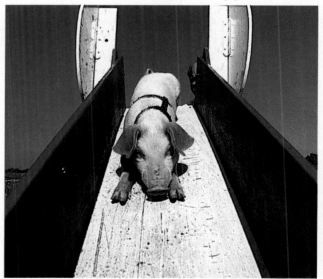

FIGURE 6-30 Problem 24.

25P. In Fig. 6-31, A and B are blocks with weights of 44 N and 22 N, respectively. (a) Determine the minimum weight (block C) that must be placed on A to keep it from sliding, if μ_s between A

and the table is 0.20. (b) Block *C* suddenly is lifted off *A*. What is the acceleration of block *A*, if μ_k between *A* and the table is 0.15?

FIGURE 6-31   Problem 25.

26P.   A 3.5 kg block is pushed along a horizontal floor by a force $F = 15$ N that makes an angle $\theta = 40°$ with the horizontal (Fig. 6-32). The coefficient of kinetic friction between the block and floor is 0.25. Calculate (a) the magnitude of the frictional force exerted on the block and (b) the acceleration of the block.

FIGURE 6-32   Problem 26.

27P.   In Fig. 6-33 a fastidious worker pushes directly along the handle of a mop with a force **F**. The handle is at an angle θ with the vertical, and μ_s and μ_k are the coefficients of static and kinetic friction between the head of the mop and the floor. Ignore the mass of the handle and assume that all the mop's mass *m* is in its head. (a) If the mop head moves along the floor with a constant velocity, then what is *F*? (b) Show that if θ is less than a certain value θ_0, then **F** (still directed along the handle) is unable to move the mop head. Find θ_0.

FIGURE 6-33   Problem 27.

28P.   A 5.0 kg block on an inclined plane is acted on by a horizontal force **F** with magnitude 50 N (Fig. 6-34). The coefficient of kinetic friction between block and plane is 0.30. The coefficient of static friction is not given (but you might still know something about it). (a) What is the acceleration of the block if it is moving up the plane? (b) With the horizontal force still acting, how far up the plane will the block go if it has an initial upward speed of 4.0 m/s? (c) What happens to the block after it reaches the highest point? Give an argument to back your answer.

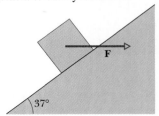

FIGURE 6-34   Problem 28.

29P.   Figure 6-35 shows the cross section of a road cut into the side of a mountain. The solid line *AA'* represents a weak bedding plane along which sliding is possible. Block *B* directly above the highway is separated from uphill rock by a large crack (called a *joint*), so that only the force of friction between the block and the bedding plane prevents sliding. The mass of the block is 1.8×10^7 kg, the *dip angle* θ of the failure plane is 24°, and the coefficient of static friction between block and plane is 0.63. (a) Show that the block will not slide. (b) Water seeps into the joint and expands upon freezing, exerting on the block a force **F** parallel to *AA'*. What minimum value of *F* will trigger a slide?

FIGURE 6-35   Problem 29.

30P.   A block weighing 80 N rests on a plane inclined at 20° to the horizontal (Fig. 6-36). The coefficient of static friction is 0.25, and the coefficient of kinetic friction is 0.15. (a) What is the minimum magnitude of the force **F**, parallel to the plane, that will prevent the block from slipping down the plane? (b) What is the minimum magnitude *F* that will start the block moving up the plane? (c) What value of *F* is required to move the block up the plane at constant velocity?

FIGURE 6-36   Problem 30.

31P.   Block *B* in Fig. 6-37 weighs 711 N. The coefficient of static friction between block and horizontal surface is 0.25. Find the maximum weight of block *A* for which the system will be stationary.

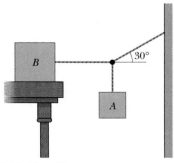

FIGURE 6-37   Problem 31.

32P. Body A in Fig. 6-38 weighs 102 N, and body B weighs 32 N. The coefficients of friction between A and the incline are $\mu_s = 0.56$ and $\mu_k = 0.25$. Angle θ is 40°. Find the acceleration of the system if (a) A is initially at rest, (b) A is moving up the incline, and (c) A is moving down the incline.

33P. Two blocks are connected over a pulley as shown in Fig. 6-38. The mass of block A is 10 kg and the coefficient of kinetic friction is 0.20. Angle θ is 30°. Block A slides down the incline at constant speed. What is the mass of block B?

FIGURE 6-38 Problems 32 and 33.

34P. Block m_1 in Fig. 6-39 has a mass of 4.0 kg and m_2 has a mass of 2.0 kg. The coefficient of kinetic friction between m_2 and the horizontal plane is 0.50. The inclined plane is frictionless. Find (a) the tension in the cord and (b) the acceleration of the blocks.

FIGURE 6-39 Problem 34.

35P. Two blocks, of weights 8.0 lb and 16 lb, are connected by a massless string and slide down a 30° inclined plane. The coefficient of kinetic friction between the 8.0 lb block and the plane is 0.10; that between the 16 lb block and the plane is 0.20. Assuming that the 8.0 lb block leads, find (a) the acceleration of the blocks and (b) the tension in the string. (c) Describe the motion if the blocks are reversed.

36P. Two masses, $m_1 = 1.65$ kg and $m_2 = 3.30$ kg, attached by a massless rod parallel to the inclined plane on which both slide (Fig. 6-40), travel down along the plane with m_1 trailing m_2. The angle of incline is $\theta = 30°$. The coefficient of kinetic friction between m_1 and the incline is $\mu_1 = 0.226$; that between m_2 and the incline is $\mu_2 = 0.113$. Compute (a) the tension in the rod and (b) the common acceleration of the two masses. (c) How would the answers to (a) and (b) change if m_2 trailed m_1?

FIGURE 6-40 Problem 36.

37P. A 4.0 kg block is put on top of a 5.0 kg block. To cause the top block to slip on the bottom one, while the bottom one is held fixed, a horizontal force of at least 12 N must be applied to the top block. The assembly of blocks is now placed on a horizontal, frictionless table (Fig. 6-41). Find (a) the magnitude of the maximum horizontal force **F** that can be applied to the lower block so that the blocks will move together and (b) the resulting acceleration of the blocks.

FIGURE 6-41 Problem 37.

38P. The two blocks (with $m = 16$ kg and $M = 88$ kg) shown in Fig. 6-42 are not attached. The coefficient of static friction between the blocks is $\mu_s = 0.38$, but the surface beneath M is frictionless. What is the minimum magnitude of the horizontal force **F** required to hold m against M?

FIGURE 6-42 Problem 38.

39P. A 40 kg slab rests on a frictionless floor. A 10 kg block rests on top of the slab (Fig. 6-43). The coefficient of static friction μ_s between the block and the slab is 0.60, whereas the kinetic coefficient μ_k is 0.40. The 10 kg block is pulled by a horizontal force with a magnitude of 100 N. What are the resulting acceleration magnitudes of (a) the block and (b) the slab?

FIGURE 6-43 Problem 39.

40P. A crate slides down an inclined right-angled trough as in Fig. 6-44. The coefficient of kinetic friction between the crate and the trough is μ_k. What is the acceleration of the crate in terms of μ_k, θ, and g?

FIGURE 6-44 Problem 40.

41P. A locomotive accelerates a 25-car train along a level track. Every car has a mass of 50 metric tons and is subject to a frictional force $f = 250v$, where the speed v is in meters per second and the force f is in newtons. At the instant when the speed of the train is 30 km/h, the acceleration is 0.20 m/s². (a) What is the tension in the coupling between the first car and the locomotive? (b) If this tension is the maximum force the locomotive can exert on the train, what is the steepest grade up which the locomotive can pull the train at 30 km/h?

42P. An initially stationary box of sand is to be pulled across a floor by means of a cord in which the tension should not exceed 1100 N. The coefficient of static friction between the box and floor is 0.35. (a) What should be the angle between the cord and the horizontal in order to pull the greatest possible amount of sand, and (b) what is the weight of the sand and box in that situation?

43P*. A 1000 kg boat is traveling at 90 km/h when its engine is shut off. The magnitude of the frictional force $\mathbf{f}_k$ between boat and water is proportional to the speed v of the boat: $f_k = 70v$, where v is in meters per second and f_k is in newtons. Find the time required for the boat to slow down to 45 km/h.

SECTION 6-3 The Drag Force and Terminal Speed

44E. Calculate the drag force on a missile 53 cm in diameter cruising with a speed of 250 m/s at low altitude, where the density of air is 1.2 kg/m³. Assume $C = 0.75$.

45E. The terminal speed of a sky diver in the spread-eagle position is 160 km/h. In the nosedive position, the terminal speed is 310 km/h. Assuming that C does not change from one position to the other, find the ratio of the effective cross-sectional area A in the slower position to that in the faster position.

46E. Calculate the ratio of the drag force on a passenger jet flying with a speed of 1000 km/h at an altitude of 10 km to the drag force on a prop-driven transport flying at half the speed and half the altitude of the jet. At 10 km the density of air is 0.38 kg/m³ and at 5.0 km it is 0.67 kg/m³. Assume that the airplanes have the same effective cross-sectional area and the same drag coefficient C.

47P. From the data in Table 6-1, deduce the diameter of the 16 lb shot. Assume that $C = 0.49$ and the density of air is 1.2 kg/m³.

SECTION 6-4 Uniform Circular Motion

48E. If the coefficient of static friction for tires on a road is 0.25, at what maximum speed can a car round a level curve of 47.5 m radius without slipping?

49E. What is the smallest radius of an unbanked curve around which a bicyclist can travel if her speed is 18 mi/h and the coefficient of static friction between the tires and the road is 0.32?

50E. During an Olympic bobsled run, a European team takes a turn of radius 25 ft at a speed of 60 mi/h. How many g's do the riders experience during the turn?

51E. A car weighing 10.7 kN and traveling at 13.4 m/s attempts to round an unbanked curve with a radius of 61.0 m. (a) What force of friction is required to keep the car on its circular path? (b) If the coefficient of static friction between the tires and road is 0.35, is the attempt at taking the curve successful?

52E. A circular curve of highway is designed for traffic moving at 60 km/h. (a) If the radius of the curve is 150 m, what is the

correct angle of banking of the road? (b) If the curve were not banked, what would be the minimum coefficient of friction between tires and road that would keep traffic from skidding at this speed?

53E. A banked circular highway curve is designed for traffic moving at 60 km/h. The radius of the curve is 200 m. Traffic is moving along the highway at 40 km/h on a rainy day. What is the minimum coefficient of friction between tires and road that will allow cars to negotiate the turn without sliding off the road?

54E. A child places a picnic basket on the outer rim of a merry-go-round that has a radius of 4.6 m and revolves once every 30 s. (a) What is the speed of a point on that rim? (b) How large must the coefficient of static friction between the basket and the merry-go-round be for the basket to stay on the ride?

55E. A conical pendulum is formed by attaching a 50 g mass to a 1.2 m string. The mass swings around a horizontal circle of radius 25 cm. (a) What is the speed of the mass? (b) What is the acceleration of the mass? (c) What is the tension in the string?

56E. In the Bohr model of the hydrogen atom, the electron revolves in a circular orbit around the nucleus. If the radius is 5.3×10^{-11} m and the electron circles 6.6×10^{15} times per second, find (a) the speed of the electron, (b) the acceleration (magnitude and direction) of the electron, and (c) the centripetal force acting on the electron. (This force is the result of the attraction between the positively charged nucleus and the negatively charged electron.) The electron's mass is 9.11×10^{-31} kg.

57E. A mass m on a frictionless table is attached to a hanging mass M by a cord through a hole in the table (Fig. 6-45). Find the speed with which m must move in order for M to stay at rest.

FIGURE 6-45   Exercise 57.

58E. A stuntman drives a car over the top of a hill, the cross section of which can be approximated by a circle of radius 250 m, as in Fig. 6-46. What is the greatest speed at which he can drive without the car leaving the road at the top of the hill?

250 m

FIGURE 6-46   Exercise 58.

59P. A small coin is placed on a flat, horizontal turntable. The turntable is observed to make three revolutions in 3.14 s. (a) What is the speed of the coin when it rides without slipping at a distance of 5.0 cm from the center of the turntable? (b) What is the acceleration (magnitude and direction) of the coin? (c) What is the magnitude of the frictional force acting on the coin if the coin has a mass of 2.0 g? (d) What is the coefficient of static friction between the coin and the turntable if the coin is observed to slide off the turntable when it is more than 10 cm from the center of the turntable?

60P. A small object is placed 10 cm from the center of a phonograph turntable. It remains in place when the table rotates at $33\frac{1}{3}$ rev/min but slides off when the table rotates at 45 rev/min. Between what limits must the coefficient of static friction between the object and the surface of the turntable lie?

61P. A bicyclist travels in a circle of radius 25.0 m at a constant speed of 9.00 m/s. The combined mass of the bicycle and rider is 85.0 kg. Calculate the magnitudes of (a) the force of friction exerted by the road on the bicycle and (b) the total force exerted by the road on the bicycle.

62P. A 150 lb student on a steadily rotating Ferris wheel has an apparent weight of 125 lb at the highest point. (a) What is the student's apparent weight at the lowest point? (b) What is the student's apparent weight at the highest point if the wheel's speed is doubled?

63P. A car is rounding a flat curve of radius $R = 220$ m at the curve's maximum design speed $v = 94.0$ km/h. What *total* force does a passenger with mass $m = 85.0$ kg exert on the seat cushion?

64P. A stone tied to the end of a string is whirled around in a vertical circle of radius R. Find the critical speed below which the string would become slack at the highest point.

65P. A certain string can withstand a maximum tension of 9.0 lb without breaking. A child ties a 0.82 lb stone to one end and, holding the other end, whirls the stone in a vertical circle of radius 3.0 ft, slowly increasing the speed until the string breaks. (a) Where is the stone on its path when the string breaks? (b) What is the speed of the stone as the string breaks?

66P. An airplane is flying in a horizontal circle at a speed of 480 km/h. If the wings of the plane are tilted 40° to the horizontal, what is the radius of the circle in which the plane is flying? (See Fig. 6-47.) Assume that the required force is provided entirely by an "aerodynamic lift" that is perpendicular to the wing surface.

40°

FIGURE 6-47 Problem 66.

67P. A frigate bird is soaring in a horizontal circular path. Its bank angle (relative to the horizontal) is estimated to be 25° and the bird takes 13 s to complete one circle. (a) How fast is the bird flying? (b) What is the radius of the circle?

68P. A model airplane of mass 0.75 kg is flying at constant speed in a horizontal circle at one end of a 30 m cord and at a height of 18 m. The other end of the cord is tethered to the ground. The airplane circles 4.4 times per minute and has its wings horizontal so that the air is pushing vertically upward. (a) What is the acceleration of the plane? (b) What is the tension in the cord? (c) What is the total upward force (lift) on the plane's wings?

69P. An old streetcar rounds a corner on unbanked tracks. If the radius of the tracks is 30 ft and the car's speed is 10 mi/h, what angle with the vertical will be made by the loosely hanging hand straps?

70P. As shown in Fig. 6-48, a 1.34 kg ball is connected by means of two massless strings to a vertical, rotating rod. The strings are tied to the rod, are taut, and form two sides of an equilateral triangle. The tension in the upper string is 35 N. (a) Draw the free-body diagram for the ball. (b) What is the tension in the lower string? (c) What is the net force on the ball at the instant shown in Fig. 6-48? (d) What is the speed of the ball?

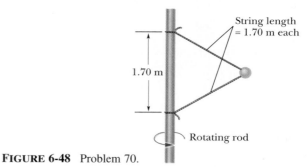

String length
= 1.70 m each

1.70 m

Rotating rod

FIGURE 6-48 Problem 70.

71P. Assume that the standard kilogram mass would weigh exactly 9.80 N at sea level on the Earth's equator if the Earth did not rotate. Then take into account the fact that the Earth does rotate, so that this mass moves in a circle of radius 6.40×10^6 m (the Earth's radius) at a constant speed of 465 m/s. (a) Determine the centripetal force needed to keep the standard mass moving in its circular path. (b) Determine the force exerted by the standard mass on a spring balance from which it is suspended at the equator (that force is its "apparent weight").

72P. Suppose that the space station in Question 12 has a radius of 500 m. (a) If a crew member weighs 600 N on Earth, what must be the speed v_s of the outer wall of the station if that crew member is to have an apparent weight of 300 N when standing near the outer wall (as in Fig. 6-21)? (b) What is the apparent weight if the crew member sprints along the outer wall at 10 m/s (relative to the outer wall) in the same direction as $\mathbf{v}_s$?

7
Kinetic Energy and Work

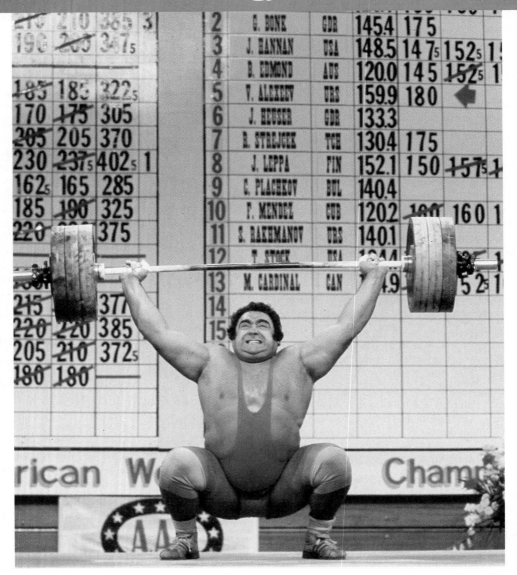

In the weight-lifting competition of the 1976 Olympics, Vasili Alexeev astounded the world by lifting a record-breaking 562 lb (2500 N) from the floor to over his head (about 2 m). In 1957 Paul Anderson stooped beneath a reinforced wood platform, placed his hands on a short stool to brace himself, and then pushed upward on the platform with his back, lifting the platform and its load about a centimeter. On the platform were auto parts and a safe filled with lead; the composite weight of the load was 6270 lb (27,900 N)! Who, Alexeev or Anderson, did more work on the objects they lifted?

7-1 KINETIC ENERGY

We begin this chapter with a definition: **energy** is a scalar quantity that is associated with a state of one or more objects. The term *state* here has its common meaning: it is the condition of an object. In this chapter we shall focus on one form of energy, **kinetic energy** K, which is associated with the *state of motion* of an object. The faster the object moves, the greater is its kinetic energy. And when the object is stationary, its kinetic energy is zero.

For an object of mass m and whose speed v is well below the speed of light, we define kinetic energy as

$$K = \tfrac{1}{2}mv^2 \qquad \text{(kinetic energy).} \qquad (7\text{-}1)$$

Kinetic energy can never be negative because m and v^2 can never be negative.

The SI unit of kinetic energy (and every other type of energy) is the **joule** (J), named for James Prescott Joule, an English scientist of the 1800s. It is derived directly from the units for mass and velocity:

$$1 \text{ joule} = 1 \text{ J} = 1 \text{ kg} \cdot \text{m}^2/\text{s}^2. \qquad (7\text{-}2)$$

A convenient unit of energy for dealing with atoms or with subatomic particles is the **electron-volt** (eV):

$$1 \text{ electron-volt} = 1 \text{ eV} = 1.60 \times 10^{-19} \text{ J}. \qquad (7\text{-}3)$$

Three common multiples of this unit are the kiloelectron-volt (1 keV = 10^3 eV), the megaelectron-volt (1 MeV = 10^6 eV), and the gigaelectron-volt (1 GeV = 10^9 eV).

or $\qquad\qquad v = 40.8 \text{ m/s}$

(about 90 mi/h). To find the mass m of each locomotive, we divide its weight by g:

$$m = \frac{1.2 \times 10^6 \text{ N}}{9.8 \text{ m/s}^2} = 1.22 \times 10^5 \text{ kg}.$$

Now, using Eq. 7-1, we find the total kinetic energy of both locomotives just before the collision as:

$$K = 2(\tfrac{1}{2}mv^2) = (1.22 \times 10^5 \text{ kg})(40.8 \text{ m/s})^2$$
$$= 2.0 \times 10^8 \text{ J}. \qquad \text{(Answer)}$$

This is equivalent to a detonation of about 100 lb of TNT.

FIGURE 7-1 Sample Problem 7-1. The aftermath of an 1896 crash of two locomotives.

SAMPLE PROBLEM 7-1

In 1896 in Waco, Texas, William Crush of the "Katy" railroad parked two locomotives at opposite ends of a 6.4 km track, fired them up, tied their throttles open, and then allowed them to crash head-on at full speed (Fig. 7-1), in front of 30,000 spectators. Hundreds of people were hurt by flying debris; several were killed. Assuming the weight of each locomotive was 1.2×10^6 N and its acceleration prior to the collision was a constant 0.26 m/s², what was the total kinetic energy of the two locomotives just before the collision?

SOLUTION: To find the kinetic energy of each locomotive, we need its mass and its speed just before the collision. To find the speed v, we use Eq. 2-16,

$$v^2 = v_0^2 + 2a(x - x_0).$$

With $v_0 = 0$ and $x - x_0 = 3.2 \times 10^3$ m (half the initial separation), this yields

$$v^2 = 0 + 2(0.26 \text{ m/s}^2)(3.2 \times 10^3 \text{ m}),$$

7-2 WORK

The energy of an object changes if an exchange of energy occurs between the object and its environment. Such a transfer can occur due to a force or due to an exchange of heat. We discuss the exchange of heat in Chapter 19. Here, we discuss the transfer of energy via a force, a process known as doing **work.**

If you accelerate an object to a greater speed by applying a force to the object, you increase the kinetic energy K ($= \tfrac{1}{2}mv^2$) of the object. Similarly, if you decelerate the object to a lesser speed by applying a force, you decrease the kinetic energy of the object. We account for these changes in kinetic energy by saying that your force has transferred energy *to* the object from yourself or *from* the object to yourself.

In such a transfer of energy via a force, *work W* is said to be *done on the object by the force*. More formally, we define work as follows:

Work W is energy transferred to or from an object by means of a force acting on the object. Energy transferred to the object is positive work, and energy transferred from the object is negative work.

"Work," then, is transferred energy; "doing work" is the act of transferring the energy. Work has the same units as energy and is a scalar quantity.

The term *transfer* can be misleading; it does not mean that anything material flows into or out of the object. That is, the transfer is not like a flow of water. Rather it is like the electronic transfer of money between two bank accounts: the number in one account goes up while the number in the other account goes down, with nothing material passing between the two accounts.

Note that we are not concerned here with the common meaning of the word "work," which implies that *any* physical or mental labor is work. For example, if you push hard against a wall, you tire owing to the continuously repeated contractions of your muscles that are required and you are, in the common sense, working. But such effort does not cause an energy transfer to or from the wall and thus is not work done on the wall as defined here.

To avoid confusion in this chapter, we shall use the symbol W only for work and shall represent weight with $m\mathbf{g}$ or mg.

7-3 WORK AND KINETIC ENERGY

Let us now relate the work done on an object by a force and the corresponding change in the kinetic energy of the object. If the force changes the speed of the object, it also changes the kinetic energy of the object. If the kinetic energy is the only type of energy of the object being changed by the force, then the change ΔK in kinetic energy is equal to the work W done by the force:

$$\Delta K = K_f - K_i = W \qquad \begin{array}{l}\text{(work–kinetic}\\ \text{energy theorem)}.\end{array} \qquad (7\text{-}4)$$

Here K_i is the initial kinetic energy ($= \frac{1}{2}mv_0^2$) and K_f is the kinetic energy ($+\frac{1}{2}mv^2$) after the work—the energy transfer—is done.

The right-hand equality of Eq. 7-4 can also be written

$$K_f = K_i + W. \qquad (7\text{-}5)$$

Equations 7-4 and 7-5 are equivalent statements of the **work–kinetic energy theorem.**

If the object's energy other than kinetic energy is being changed by the force, then Eqs. 7-4 and 7-5 do not apply. A kinetic frictional force, for example, can change

both the kinetic energy and the *thermal energy* of an object. (Thermal energy is associated with the random motions of atoms and molecules within an object.) Equations 7-4 and 7-5 then do not apply.

Some forces cause a transfer of energy within the object itself, and again Eqs. 7-4 and 7-5 do not apply. For example, suppose you push yourself away from a wall while ice skating; then the push (a force) transfers energy internally from a biological type of energy in your muscles to kinetic energy of your body as a whole. We shall consider kinetic frictional forces in Chapter 8 and internal energy transfers in Chapter 9.

CHECKPOINT 1: A particle moves along an x axis. Does the kinetic energy of the particle increase, decrease, or remain the same if the particle's velocity changes (a) from -3 m/s to -2 m/s and (b) from -2 m/s to 2 m/s? (c) In each situation, is the work done on the particle positive, negative, or zero?

Work Done by a Single Force

We would now like to relate the change ΔK in kinetic energy of an object to the magnitude F of the force causing the change. We start with a particle; the only type of energy this simplest type of object can have is kinetic energy. In Fig. 7-2, a particle moves along an x axis on a horizontal frictionless floor while a constant force $\mathbf{F}$ acts on it at a constant angle ϕ to the particle's path. Because the horizontal force component $F \cos \phi$ gives the particle an acceleration a_x along the path, the force changes the particle's velocity from its initial value $\mathbf{v}_0$. Thus the force also changes the particle's kinetic energy.

Suppose the force acts on the particle through a displacement $\mathbf{d}$, giving it a velocity $\mathbf{v}$ whose magnitude is, by Eq. 2-16,

$$v^2 = v_0^2 + 2a_x d. \qquad (7\text{-}6)$$

Multiplying both sides of Eq. 7-6 by the mass m of the particle and rearranging yield

$$\tfrac{1}{2}mv^2 - \tfrac{1}{2}mv_0^2 = ma_x d. \qquad (7\text{-}7)$$

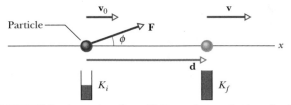

FIGURE 7-2 A constant force $\mathbf{F}$ directed at angle ϕ to the displacement $\mathbf{d}$ of a particle accelerates the particle along that path, changing the velocity of the particle from $\mathbf{v}_0$ to $\mathbf{v}$. A "kinetic energy gauge" indicates the resulting change in the kinetic energy of the particle, from the value K_i to K_f.

The left side of Eq. 7-7 is the difference between the initial kinetic energy K_i $(= \frac{1}{2}mv_0^2)$ of the particle and the final kinetic energy K_f $(= \frac{1}{2}mv^2)$. This difference is the change ΔK in the kinetic energy of the particle due to the force **F**. Making that substitution, and substituting $F \cos \phi$ for the product ma_x according to Newton's second law, we get

$$\Delta K = Fd \cos \phi. \qquad (7\text{-}8)$$

Now by comparing Eq. 7-8 with Eq. 7-4, we see that the right side of Eq. 7-8 gives the work W done on the particle by force **F**. Thus, we may write

$$W = Fd \cos \phi \quad \begin{array}{l}\text{(work done by a}\\\text{constant force).}\end{array} \qquad (7\text{-}9)$$

If the angle ϕ is less than 90°, then the work W is positive, which means that energy is transferred *to* the particle and the kinetic energy of the particle *increases*. If ϕ is greater than 90° (up to 180°), then the work W is negative, which means that energy is transferred *from* the particle and the kinetic energy of the particle *decreases*.

From Eq. 7-9 we see that in addition to the joule, another SI unit of work is the newton-meter (N·m). The corresponding unit in the British system is the foot-pound (ft·lb). Thus, we can extend Eq. 7-2 to

$$1 \text{ J} = 1 \text{ kg·m}^2/\text{s}^2 = 1 \text{ N·m} = 0.738 \text{ ft·lb}. \quad (7\text{-}10)$$

The right side of Eq. 7-9 is equivalent to the scalar (or dot) product $\mathbf{F} \cdot \mathbf{d}$. So in vector form Eq. 7-9 is

$$W = \mathbf{F} \cdot \mathbf{d} \quad \begin{array}{l}\text{(work by a}\\\text{constant force).}\end{array} \qquad (7\text{-}11)$$

(This is our first application of a scalar product; you may wish to review its discussion in Section 3-7.) Equation 7-11 is especially useful when **F** and **d** are given in unit-vector notation.

As derived, Eqs. 7-9 and 7-11 give the work done by a constant force that changes the kinetic energy of a particle. However, in certain cases we can extend them to objects that are clearly not particles:

If a force acting alone on an object changes only the kinetic energy of the object (and no other energy of the object), then the work done by the force is given by Eqs. 7-9 and 7-11.

Such objects are said to be *particle-like*.

For example, Fig. 7-3 shows a student propelling a bed in an intramural bed race. If his force on the bed is constant, do Eqs. 7-9 and 7-11 give the work done by the force? We can easily see that most but not all of the trans-

FIGURE 7-3 A bed race. We can approximate the bed as being a particle for the purpose of calculating the work done on the bed by the force applied by the student.

ferred energy goes into the kinetic energy of the bed and penguin-rider moving along the street. Some small amount goes into the rotation of the wheels. However, if we choose to neglect that small amount, then Eqs. 7-9 and Eq. 7-11 do give the work done by the force applied by the student, and the bed and rider are particle-like.

CHECKPOINT **2:** The figure shows four situations in which a force acts on a box while the box slides rightward a distance d across a frictionless floor. The magnitudes of the forces are identical; their orientations are as shown. Rank the situations according to the work done on the box during the displacement, from most positive to most negative.

(a)   (b)   (c)   (d)

Work Done by Multiple Forces

If several forces act on a particle, we can replace **F** in Eq. 7-11 with the net force $\Sigma\mathbf{F}$, where

$$\Sigma\mathbf{F} = \mathbf{F}_1 + \mathbf{F}_2 + \mathbf{F}_3 + \cdots, \qquad (7\text{-}12)$$

where $\mathbf{F}_j$ are the individual forces. Then

$$W = \left(\Sigma\mathbf{F}\right) \cdot \mathbf{d} \qquad (7\text{-}13)$$

is the work done by the net force during a displacement **d** of the particle. With Eq. 7-12, we can rewrite Eq. 7-13 as

$$\begin{aligned}W &= \mathbf{F}_1 \cdot \mathbf{d} + \mathbf{F}_2 \cdot \mathbf{d} + \mathbf{F}_3 \cdot \mathbf{d} + \cdots\\ &= W_1 + W_2 + W_3 + \cdots \quad \text{(total work)}.\end{aligned} \quad (7\text{-}14)$$

This equation tells us that the total work done on the particle is the sum of the work done by all the forces acting on the particle.

We can now rewrite Eq. 7-4 (the work–kinetic energy theorem) as

$$\Delta K = K_f - K_i = W_1 + W_2 + W_3 + \cdots. \quad (7\text{-}15)$$

This tells us that the change ΔK in the kinetic energy of a particle is equal to the total work done by all the forces acting on the particle.

SAMPLE PROBLEM 7-2

Figure 7-4a shows two industrial spies sliding an initially stationary 225 kg floor safe a distance of 8.50 m along a straight line toward their truck. The push $\mathbf{F}_1$ of Spy 001 is 12.0 N, directed at an angle of 30° downward from the horizontal; the pull $\mathbf{F}_2$ of Spy 002 is 10.0 N, directed at 40° above the horizontal. The floor and safe make frictionless contact.

(a) What is the total work done on the safe by forces $\mathbf{F}_1$ and $\mathbf{F}_2$ during the 8.50 m displacement $\mathbf{d}$?

SOLUTION: Figure 7-4b is a free-body diagram for the safe, considered to be a particle. We can find the total work done on the safe by finding the work done by each force and then adding the results. From Eq. 7-9, the work done by $\mathbf{F}_1$ is

$$W_1 = F_1 d \cos \phi_1 = (12.0 \text{ N})(8.50 \text{ m})(\cos 30°)$$
$$= 88.33 \text{ J},$$

and the work done by $\mathbf{F}_2$ is

$$W_2 = F_2 d \cos \phi_2 = (10.0 \text{ N})(8.50 \text{ m})(\cos 40°)$$
$$= 65.11 \text{ J}.$$

So, from Eq. 7-14, the total work W is

$$W = W_1 + W_2 = 88.33 \text{ J} + 65.11 \text{ J}$$
$$= 153.4 \text{ J} \approx 153 \text{ J}. \qquad \text{(Answer)}$$

Thus during the 8.50 m displacement, the spies transfer 153 J of energy to the kinetic energy of the safe.

(b) During the displacement, what is the work W_g done on the safe by its weight $m\mathbf{g}$ and what is the work W_N done on the safe by the normal force $\mathbf{N}$ due to the floor?

SOLUTION: Both these forces are perpendicular to the displacement. Thus Eq. 7-9 tells us that

$$W_g = mgd \cos 90° = mgd(0) = 0 \qquad \text{(Answer)}$$

and

$$W_N = Nd \cos 90° = Nd(0) = 0. \qquad \text{(Answer)}$$

These forces do not transfer any energy to or from the safe.

(c) The safe is initially stationary. What is its speed v at the end of the 8.50 m displacement?

SOLUTION: The speed of the safe changes because its kinetic energy is changed when energy is transferred to it by the forces. We relate the speed to the work done by combining Eqs. 7-4 and 7-1:

$$W = K_f - K_i = \tfrac{1}{2}mv^2 - \tfrac{1}{2}mv_0^2.$$

The initial speed v_0 is zero, and we now know that the work done is 153.4 J. Solving for v and then substituting the known data, we find that

$$v = \sqrt{\frac{2W}{m}} = \sqrt{\frac{2(153.4 \text{ J})}{225 \text{ kg}}}$$
$$= 1.17 \text{ m/s}. \qquad \text{(Answer)}$$

FIGURE 7-4   Sample Problem 7-2. (a) Two spies move a floor safe. (b) A free-body diagram for the safe, with the displacement $\mathbf{d}$ of the safe included.

SAMPLE PROBLEM 7-3

A runaway crate of prunes slides over a floor toward you. To slow the crate you push against it with a force $\mathbf{F} = (2.0 \text{ N})\mathbf{i} + (-6.0 \text{ N})\mathbf{j}$ while running backward (Fig. 7-5). During your pushing, the crate goes through a displacement $\mathbf{d} = (-3.0 \text{ m})\mathbf{i}$.

(a) How much work has your force done on the crate during the displacement?

SOLUTION: From Eq. 7-11, the work is

$$W = \mathbf{F} \cdot \mathbf{d} = [(2.0 \text{ N})\mathbf{i} + (-6.0 \text{ N})\mathbf{j}] \cdot [(-3.0 \text{ m})\mathbf{i}].$$

Of the possible unit-vector dot products, only $\mathbf{i} \cdot \mathbf{i}$, $\mathbf{j} \cdot \mathbf{j}$, and $\mathbf{k} \cdot \mathbf{k}$ are nonzero (see Section 3-7). Here we have

$$W = (2.0)(-3.0 \text{ m})\mathbf{i} \cdot \mathbf{i} + (-6.0 \text{ N})(-3.0 \text{ m})\mathbf{j} \cdot \mathbf{i}$$
$$= (-6.0 \text{ J})(1) + 0 = -6.0 \text{ J}. \qquad \text{(Answer)}$$

Thus the force transfers 6.0 J of energy from the kinetic energy of the crate.

(b) If the crate has a kinetic energy of 10 J at the beginning of the displacement $\mathbf{d}$, what is its kinetic energy at the end of the displacement?

SOLUTION: Using Eq. 7-5 with $K_i = 10$ J and $W = -6$ J, we find

$$K_f = K_i + W = 10 \text{ J} + (-6.0 \text{ J}) = 4.0 \text{ J}. \qquad \text{(Answer)}$$

FIGURE 7-5 Sample Problem 7-3. Slowing a runaway prune crate with force **F** applied during displacement **d**.

7-4 WORK DONE BY WEIGHT

We next examine the work done on an object by a particular type of force, namely, its weight. Figure 7-6 shows a particle-like tomato of mass m that is thrown upward with initial speed v_0 and thus with initial kinetic energy $K_i = \frac{1}{2}mv_0^2$. As the tomato rises, it slows because a constant downward force, its weight mg, acts on it.

Because the tomato slows, its kinetic energy decreases. We know from experimental observation that if the weight mg is the only force acting on the tomato (air drag is somehow eliminated), then kinetic energy is the only energy of the tomato that changes. Thus, to find the work W_g done on the tomato by its weight, we substitute mg for F in Eq. 7-9, finding

$$W_g = mgd \cos \phi \quad \text{(work done by weight).} \quad (7\text{-}16)$$

For such a rising object, the weight mg is directed opposite the displacement **d**, as indicated in Fig. 7-6. Then $\phi = 180°$ and

$$W_g = mgd \cos (180°) = mgd(-1) = -mgd. \quad (7\text{-}17)$$

The minus sign tells us that during the object's rise, the weight of the object transfers energy in the amount mgd from the kinetic energy of the object. This is consistent with the slowing of the object as it rises.

After the object has reached its maximum height and

FIGURE 7-6 A particle-like tomato of mass m thrown upward slows from velocity **v₀** to velocity **v** during displacement **d** because its weight mg acts on it. A kinetic energy gauge indicates the resulting change in the kinetic energy of the object, from K_i ($= \frac{1}{2}mv_0^2$) to K_f ($= \frac{1}{2}mv^2$).

is falling back down, the angle ϕ between the weight mg and the displacement **d** is zero. Thus,

$$W_g = mgd \cos (0°) = mgd(+1) = +mgd. \quad (7\text{-}18)$$

The plus sign tells us that the weight now transfers energy in the amount mgd to the kinetic energy of the object. This is consistent with the speeding up of the object as it falls. (Actually, as we shall see in Chapter 8, energy transfers associated with lifting and lowering an object involve not just the object, but the full object-Earth system. Without the Earth, of course, "lifting" would be meaningless.)

Work Done in Lifting and Lowering an Object

Now suppose we lift a particle-like object by applying a force **F** to it. During the upward displacement, our applied force does positive work W_a on the object while the object's weight also does negative work W_g on it. That is, our force tends to transfer energy to the object while its weight tends to transfer energy from it. By Eq. 7-15, the change ΔK in the kinetic energy of the object due to these two energy transfers is

$$\Delta K = K_f - K_i = W_a + W_g, \quad (7\text{-}19)$$

in which K_f is the kinetic energy at the end of the displacement and K_i is that at the start of the displacement. This equation also applies if we lower the object; but then the weight tends to transfer energy *to* the object while our force tends to transfer energy *from* it.

In one common situation the object is stationary before and after the lift—for example, when you lift a book from the floor to a shelf. Then K_f and K_i are both zero, and Eq. 7-19 reduces to

$$W_a + W_g = 0$$

or

$$W_a = -W_g. \quad (7\text{-}20)$$

Note that we get the same result if K_f and K_i are not zero but are still equal. Either way, the result means that the work done by the applied force is the negative of the work done by the weight. That is, the applied force transfers the same amount of energy to the object as its weight transfers from the object. Using Eq. 7-16, we can rewrite Eq. 7-20 as

$$W_a = -mgd \cos \phi \quad \text{(work in lifting and lowering; } K_f = K_i), \quad (7\text{-}21)$$

with ϕ being the angle between mg and **d**. If the displacement is vertically upward (Fig. 7-7a), then $\phi = 180°$ and the work done by our force equals mgd. If the displacement is vertically downward (Fig. 7-7b), then $\phi = 0°$ and the work done by the applied force equals $-mgd$.

Equations 7-20 and 7-21 apply to any situation in

(a)   (b)

FIGURE 7-7 (*a*) An applied force **F** lifts an object. The displacement **d** of the object makes an angle $\phi = 180°$ with the weight vector $m\mathbf{g}$ of the object. The applied force does positive work on the object. (*b*) An applied force **F** lowers an object. The displacement **d** of the object makes an angle $\phi = 0°$ with weight vector $m\mathbf{g}$. The applied force does negative work on the object.

which an object is lifted or lowered, with the object stationary before and after the lift. They are independent of the magnitude of the force used. For example, when Alexeev lifted the record-breaking weight of 2500 N, his force on the weights varied considerably during the lift. Still, because the weights were stationary before and after the lift, the work he did is given by Eqs. 7-20 and 7-21, where, in Eq. 7-21, $m\mathbf{g}$ is the weight he lifted and d is the distance he lifted that weight.

SAMPLE PROBLEM 7-4

Let us return to the weight-lifting feats of Vasili Alexeev and Paul Anderson.

(a) When Alexeev lifted a weight of 2500 N a distance of 2.0 m, how much work was done on the weights by their weight $m\mathbf{g}$?

SOLUTION: The magnitude of the weight vector $m\mathbf{g}$ is mg. The angle ϕ between that vector and the displacement vector **d** is 180°. From Eq. 7-16, the work done by $m\mathbf{g}$ is

$$W_g = mgd \cos \phi = (2500 \text{ N})(2.0 \text{ m})(\cos 180°)$$
$$= -5000 \text{ J.} \qquad \text{(Answer)}$$

(b) How much work was done by Alexeev's force during the lift?

SOLUTION: Because the weights were stationary at the start and end of the lift, we can use Eq. 7-20, finding

$$W_{VA} = -W_g = +5000 \text{ J.} \qquad \text{(Answer)}$$

(c) While Alexeev held the weights stationary above his head, how much work was done by his force on the weights?

SOLUTION: When he supported the weights, they were stationary. Thus their displacement $d = 0$ and, by Eq. 7-9, the

Using a harness, Paul Anderson lifts 30 people having a combined weight of about 2400 lb.

work done on the weights was zero (even though supporting the weights was a very tiring task).

(d) How much work was done by the force Paul Anderson applied to lift a weight of 27,900 N a distance of 1.0 cm?

SOLUTION: From Eq. 7-21, with $mg = 27{,}900$ N and $d = 1.0$ cm, we find

$$W_{PA} = -mgd \cos \phi = -mgd \cos 180°$$
$$= -(27{,}900 \text{ N})(0.010 \text{ m})(-1) = 280 \text{ J.} \qquad \text{(Answer)}$$

Anderson's lift required a tremendous upward force but only a small energy transfer of 280 J, owing to the short displacement involved.

SAMPLE PROBLEM 7-5

An initially stationary 15.0 kg crate of cheese is pulled, via a cable, a distance $L = 5.70$ m up a frictionless ramp, to a height h of 2.50 m, where it stops (Fig. 7-8a).

(a) How much work is done on the crate by its weight $m\mathbf{g}$ during the lift?

SOLUTION: We calculate this work with Eq. 7-16, using L for the magnitude of the displacement. The angle between $m\mathbf{g}$ and the displacement is $\theta + 90°$ (see the free-body diagram of Fig. 7-8b). We have

$$W_g = mg \, L \cos (\theta + 90°) = -mgL \sin \theta.$$

From Fig. 7-8a, we see that $L \sin \theta$ is the height h moved by the crate. So we have

$$W_g = -mgh. \qquad (7\text{-}22)$$

This means that the work done by the weight depends on the

vertical displacement of the crate and not on the horizontal displacement. (We shall return to this point in Chapter 8.) Inserting the given data in Eq. 7-22 yields

$$W_g = -(15.0 \text{ kg})(9.8 \text{ m/s}^2)(2.50 \text{ m})$$
$$= -368 \text{ J}. \qquad \text{(Answer)}$$

(b) How much work is done on the crate by the force **T** applied by the cable, which pulls the crate up the ramp?

SOLUTION: Because the crate is stationary before and after the lift, the change ΔK in its kinetic energy must be zero. Then from Eq. 7-15,

$$\Delta K = W_1 + W_2 + W_3 + \cdots, \qquad (7\text{-}23)$$

we know that the total work done by all the forces acting on it must be zero. Besides the weight of the crate, there are only two other forces acting on the crate: the normal force **N** due to the ramp and the force **T** from the cable. Because **N** is perpendicular to the displacement of the crate along the ramp, the normal force does zero work on the crate. So, with W_T representing the work done by the force **T**, Eq. 7-23 becomes

$$0 = W_g + W_T.$$

Substituting -368 J for W_g, we find

$$W_T = 368 \text{ J}. \qquad \text{(Answer)}$$

(a)

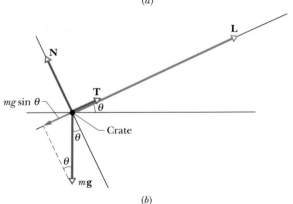

(b)

FIGURE 7-8 Sample Problem 7-5. (a) A crate is pulled up a frictionless ramp by a force parallel to the ramp. (b) A free-body diagram for the crate, showing all the forces that act on it. Its displacement **L** is also shown.

CHECKPOINT **3**: Suppose we raise the crate in Sample Problem 7-5 by the same height h but with a longer ramp. (a) Is the work done by force **T** now greater, smaller, or the same as before? (b) Is the magnitude of **T** needed to move the crate now greater, smaller, or the same as before?

SAMPLE PROBLEM 7-6

A 500 kg elevator cab is descending with speed $v_i = 4.0$ m/s when the cable that controls it begins to slip, allowing it to fall with constant acceleration $\mathbf{a} = \mathbf{g}/5$ (Fig. 7-9a).

(a) During its fall through a distance $d = 12$ m, what is the work W_1 done on the cab by its weight $m\mathbf{g}$?

SOLUTION: The cab's free-body diagram during the 12-m fall is shown in Fig. 7-9b. Note that the angle between the cab's displacement **d** and its weight $m\mathbf{g}$ is 0°. With Eq. 7-16 we find

$$W_1 = mgd \cos 0° = (500 \text{ kg})(9.8 \text{ m/s}^2)(12 \text{ m})(1)$$
$$= 5.88 \times 10^4 \text{ J} \approx 5.9 \times 10^4 \text{ J}. \qquad \text{(Answer)}$$

(b) During the 12 m fall, what is the work W_2 done on the cab by the upward pull **T** exerted by the elevator cable?

SOLUTION: The situation here differs from those in Sample Problems 7-4 and 7-5, because the kinetic energy of the cab at the beginning of the displacement does not equal that at the end of the displacement. So Eqs. 7-20 and 7-21 do not apply; the work done by force **T** on the cab *is not* the negative of the

(a)   (b)

FIGURE 7-9 Sample Problem 7-6. An elevator cab, descending with speed v_i, suddenly begins to accelerate downward. (a) It moves through a displacement **d** with constant acceleration $\mathbf{a} = \mathbf{g}/5$. (b) A free-body diagram for the cab, with the displacement included.

work done by the weight of the cab. We must calculate the work W_2 done by **T** with Eq. 7-9 ($W = Fd \cos \phi$). To do so, we first must find the magnitude of **T**. Applying Newton's second law to the cab yields

$$\sum F = T - mg = ma.$$

Since **a** has magnitude $g/5$ and is directed downward, we get

$$T = m(g + a) = m(g - g/5)$$
$$= (500 \text{ kg})(\tfrac{4}{5})(9.8 \text{ m/s}^2) = 3920 \text{ N}.$$

We can now use Eq. 7-9 to find the work done by **T**. The angle between **T** and the cab's displacement **d** is 180°. The magnitude of **T** is 3920 N. And the magnitude of the displacement is 12 m. So

$$W_2 = Td \cos 180° = (3920 \text{ N})(12 \text{ m})(-1)$$
$$= -4.70 \times 10^4 \text{ J}. \qquad \text{(Answer)}$$

(c) What is the total work W done on the cab in the 12 m fall?

SOLUTION: The total work is the algebraic sum of the work done by the two forces (by Eq. 7-14):

$$W = W_1 + W_2 = 5.88 \times 10^4 \text{ J} - 4.70 \times 10^4 \text{ J}$$
$$= 1.18 \times 10^4 \text{ J} \approx 1.2 \times 10^4 \text{ J}. \qquad \text{(Answer)}$$

Thus, during the 12 m fall, a net energy of 1.2×10^4 J is transferred to the cab.

We can also find W a different way. We first find the net force on the cab, using Newton's second law:

$$\sum F = ma = (500 \text{ kg}) \left(-\frac{9.8 \text{ m/s}^2}{5} \right) = -980 \text{ N}.$$

We next find the work done on the cab by that net force, which acts downward and thus at an angle of 0° with **d**:

$$W = (980 \text{ N})(12 \text{ m}) \cos 0°$$
$$= 1.18 \times 10^4 \text{ J} \approx 1.2 \times 10^4 \text{ J}. \quad \text{(Answer)}$$

(d) What is the cab's kinetic energy at the end of the 12 m fall?

SOLUTION: The kinetic energy K_i at the start of the fall, when the speed is $v_i = 4.0$ m/s, is

$$K_i = \tfrac{1}{2}mv_i^2 = \tfrac{1}{2}(500 \text{ kg})(4.0 \text{ m/s})^2 = 4000 \text{ J}.$$

The kinetic energy K_f at the end of the fall is given by Eq. 7-5,

$$K_f = K_i + W = 4000 \text{ J} + 1.18 \times 10^4 \text{ J}$$
$$= 1.58 \times 10^4 \text{ J} \approx 1.6 \times 10^4 \text{ J}. \qquad \text{(Answer)}$$

(e) What is the speed v_f of the cab at the end of the 12 m fall?

SOLUTION: From Eq. 7-1, we have

$$K_f = \tfrac{1}{2}mv_f^2,$$

which we solve for v_f:

$$v_f = \sqrt{\frac{2K_f}{m}} = \sqrt{\frac{(2)(1.58 \times 10^4 \text{ J})}{500 \text{ kg}}}$$
$$= 7.9 \text{ m/s}. \qquad \text{(Answer)}$$

7-5 WORK DONE BY A VARIABLE FORCE

One-Dimensional Analysis

Let us return to the situation of Fig. 7-2 but now consider the force to be directed along the x axis and the force magnitude to vary with position x. Thus, as the particle moves, the magnitude of the force doing work on it changes. Only the magnitude of this **variable force** changes, not its direction. Moreover, its magnitude changes with the position of the particle, but not over time.

Figure 7-10a shows a plot of such a one-dimensional variable force. How do we find the work done on the particle by this force as the particle moves from an initial point x_i to a final point x_f? We cannot use Eq. 7-9, because it applies only for a constant force **F**. To develop a new approach, let us divide the total displacement of the particle into a number of intervals of width Δx. We choose Δx small enough to permit us to take the force $F(x)$ as being reasonably constant over that interval. We let $\overline{F_j(x)}$ be the average value of $F(x)$ within the jth interval.

The increment (small amount) of work ΔW_j done by the force in the jth interval is now given by Eq. 7-9 and is

$$\Delta W_j = \overline{F_j(x)} \, \Delta x. \qquad (7\text{-}24)$$

On the graph of Fig. 7-10b, $\overline{F_j(x)}$ is the height of the jth strip, and Δx is its width; ΔW_j is then equal in magnitude to the area of the strip.

To approximate the total work W done by the force as the particle moves from x_i to x_f, we add the areas of all the strips between x_i and x_f in Fig. 7-10b. That is,

$$W = \sum \Delta W_j = \sum \overline{F_j(x)} \, \Delta x. \qquad (7\text{-}25)$$

Equation 7-25 is an approximation because the broken "skyline" formed by the tops of the rectangular strips in Fig. 7-10b only approximates the actual curve.

We can make the approximation better by reducing the strip width Δx and using more strips, as in Fig. 7-10c. In the limit, we let the strip width approach zero; the number of strips then becomes infinitely large and we have, as an exact result,

$$W = \lim_{\Delta x \to 0} \sum \overline{F_j(x)} \, \Delta x. \qquad (7\text{-}26)$$

This limit is exactly what we mean by the integral of the function $F(x)$ between the limits x_i and x_f. Thus Eq. 7-26 becomes

$$W = \int_{x_i}^{x_f} F(x) \, dx \qquad \text{(work: variable force).} \qquad (7\text{-}27)$$

If we know the function $F(x)$, we can substitute it into

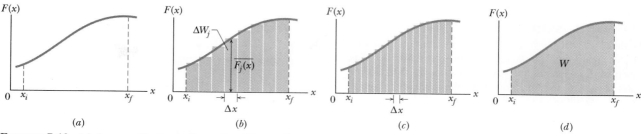

FIGURE 7-10 (a) A generalized one-dimensional force plotted against the displacement x of a particle on which it acts. The particle moves from x_i to x_f. (b) Same as (a) but with the area under the curve divided into narrow strips. (c) Same as (b) but with the area divided into narrower strips. (d) The limiting case. The work done by the force is given by Eq. 7-27 and is represented geometrically by the shaded area between the curve and the x axis and between x_i and x_f.

Eq. 7-27, introduce the proper limits of integration, carry out the integration, and thus find the work. (Appendix E contains a list of common integrals.) Geometrically, the work is equal to the area between the $F(x)$ curve and the x axis, between the limits x_i and x_f (shaded in Fig. 7-10d).

Three-Dimensional Analysis

Consider now a particle that is acted on by a three-dimensional force

$$\mathbf{F} = F_x\mathbf{i} + F_y\mathbf{j} + F_z\mathbf{k}, \qquad (7\text{-}28)$$

and let some or all of the components F_x, F_y, and F_z depend on the position of the particle, which means that they are functions of that position. Furthermore, let the particle move through an incremental displacement

$$d\mathbf{r} = dx\mathbf{i} + dy\mathbf{j} + dz\mathbf{k}. \qquad (7\text{-}29)$$

The increment of work dW done on the particle by $\mathbf{F}$ during the displacement $d\mathbf{r}$ is, by Eq. 7-11,

$$dW = \mathbf{F} \cdot d\mathbf{r} = F_x\,dx + F_y\,dy + F_z\,dz. \qquad (7\text{-}30)$$

The work W done by $\mathbf{F}$ while the particle moves from an initial position r_i with coordinates (x_i, y_i, z_i) to a final position r_f with coordinates (x_f, y_f, z_f) is then

$$W = \int_{r_i}^{r_f} dW = \int_{x_i}^{x_f} F_x\,dx + \int_{y_i}^{y_f} F_y\,dy + \int_{z_i}^{z_f} F_z\,dz. \qquad (7\text{-}31)$$

If $\mathbf{F}$ has only an x component, then the y and z terms in Eq. 7-31 are zero and the equation reduces to Eq. 7-27.

Work–Kinetic Energy Theorem with a Variable Force

Equation 7-27 gives the work done by a variable force on a particle in a one-dimensional situation. Let us now make certain that the work calculated with Eq. 7-27 is indeed

equal to the change in kinetic energy of the particle, as the work–kinetic energy theorem states.

Consider a particle of mass m, moving along the x axis and acted on by a net force $F(x)$ that points along that axis. The work done on the particle by this force as the particle moves from an initial position x_i to a final position x_f is given by Eq. 7-27 as

$$W = \int_{x_i}^{x_f} F(x)\,dx = \int_{x_i}^{x_f} ma\,dx, \qquad (7\text{-}32)$$

in which we use Newton's second law to replace $F(x)$ with ma. We can write the quantity $ma\,dx$ in Eq. 7-32 as

$$ma\,dx = m\frac{dv}{dt}\,dx. \qquad (7\text{-}33)$$

From the "chain rule" of calculus, we have

$$\frac{dv}{dt} = \frac{dv}{dx}\frac{dx}{dt} = \frac{dv}{dx}v, \qquad (7\text{-}34)$$

and Eq. 7-33 becomes

$$ma\,dx = m\frac{dv}{dx}v\,dx = mv\,dv. \qquad (7\text{-}35)$$

Substituting Eq. 7-35 into Eq. 7-32 yields

$$W = \int_{v_i}^{v_f} mv\,dv = m\int_{v_i}^{v_f} v\,dv$$
$$= \tfrac{1}{2}mv_f^2 - \tfrac{1}{2}mv_i^2. \qquad (7\text{-}36)$$

Note that when we change the variable from x to v we are required to express the limits on the integral in terms of the new variable. Note also that because the mass m is a constant, we are able to move it outside the integral.

Recognizing the terms on the right of Eq. 7-36 as kinetic energies allows us to write this equation as

$$W = K_f - K_i = \Delta K,$$

which is the work–kinetic energy theorem.

SAMPLE PROBLEM 7-7

Force $\mathbf{F} = (3x\text{ N})\mathbf{i} + (4\text{ N})\mathbf{j}$, with x in meters, acts on a particle, changing only the kinetic energy of the particle. How much work is done on the particle as it moves from coordinates (2 m, 3 m) to (3 m, 0 m)? Does the speed of the particle increase, decrease, or remain the same?

SOLUTION: From Eq. 7-31 we have

$$W = \int_2^3 3x\,dx + \int_3^0 4\,dy = 3\int_2^3 x\,dx + 4\int_3^0 dy.$$

Using the list of integrals in Appendix E, we obtain

$$W = 3\left[\tfrac{1}{2}x^2\right]_2^3 + 4\left[y\right]_3^0$$
$$= \tfrac{3}{2}[3^2 - 2^2] + 4[0 - 3]$$
$$= -4.5\text{ J} \approx -5\text{ J}. \qquad \text{(Answer)}$$

The negative result tells us that energy is transferred from the particle by force $\mathbf{F}$. Because the kinetic energy of the particle decreases, its speed must decrease.

7-6 WORK DONE BY A SPRING FORCE

We next want to examine the work done on a particle-like object by a particular type of variable force, namely, a **spring force**—the force exerted by a spring. Many forces in nature have the same mathematical form as the spring force. So, by examining this one force, you can gain an understanding of many others.

The Force Exerted by a Spring

Figure 7-11a shows a spring in its **relaxed state,** that is, neither compressed nor extended. One end is fixed, and a particle-like object, say, a block, is attached to the other, free end. In Fig. 7-11b, we stretch the spring by pulling the block to the right. In reaction, the spring pulls on the block toward the left, in the direction that will restore the relaxed state. (A spring's force is sometimes said to be a *restoring force.*) In Fig. 7-11c, we compress the spring by pushing the block to the left. The spring now pushes on the block toward the right, again so as to restore the relaxed state.

To a good approximation for many springs, the force $\mathbf{F}$ exerted by the spring is proportional to the displacement $\mathbf{d}$ of the free end from its position when the spring is in the relaxed state. That is, the *spring force* is given by

$$\mathbf{F} = -k\mathbf{d} \qquad \text{(Hooke's law),} \qquad (7\text{-}37)$$

which is known as **Hooke's law** after Robert Hooke, an English scientist of the late 1600s. The minus sign in Eq. 7-37 indicates that the spring force is always opposite in direction from the displacement of its free end. The con-

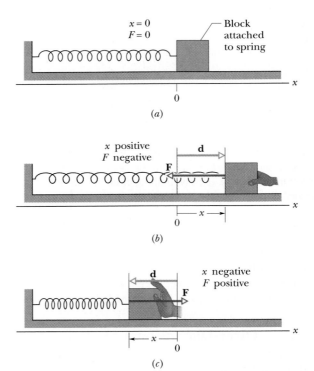

FIGURE 7-11   (a) A spring in its relaxed state. The origin of an x axis has been placed at the end of the spring that is attached to a block. (b) The block is displaced by $\mathbf{d}$, and the spring is stretched by an amount x. Note the restoring force $\mathbf{F}$ exerted by the spring. (c) The spring is compressed by an amount x. Again, note the restoring force.

stant k is called the **spring constant** (or **force constant**) and is a measure of the stiffness of the spring. The larger k is, the stiffer the spring; that is, the stronger the spring will pull or push for a given displacement. The SI unit for k is the newton per meter.

In Fig. 7-11 an x axis has been placed parallel to the length of the spring, with the origin ($x = 0$) at the position of the free end when the spring is in its relaxed state. For this common arrangement, Eq. 7-37 becomes

$$F = -kx \qquad \text{(Hooke's law).} \qquad (7\text{-}38)$$

Note that a spring force is a variable force because it depends on the position of the free end: F can be symbolized as $F(x)$, as in Section 7-5. Also note that Hooke's law is a linear relationship between the force magnitude F and the position x of the free end.

Work Done by a Spring Force

Suppose we give the block in Fig. 7-11a an abrupt rightward jerk to provide the block with kinetic energy. And let us assume that the contact between block and floor is frictionless, that the spring has negligible mass compared to the block (the spring is *massless*), and that the spring obeys Hooke's law exactly (it is *ideal*).

As the block moves rightward, the spring force **F** slows it and thus decreases its kinetic energy. Experimentally we find that when our assumptions about the spring and floor hold, the only energy of the block that is changed by **F** is the kinetic energy. Thus the work–kinetic energy theorem applies to this situation. To find the work, we must use Eq. 7-27 because the spring force is a variable force, with F given by Eq. 7-38. So, as the block moves from a position x_i to x_f, the work done on the block by the spring force is

$$W_s = \int_{x_i}^{x_f} F\,dx = \int_{x_i}^{x_f} (-kx)\,dx = -k\int_{x_i}^{x_f} x\,dx$$
$$= \left(-\tfrac{1}{2}k\right)\left[x^2\right]_{x_i}^{x_f} = \left(-\tfrac{1}{2}k\right)(x_f^2 - x_i^2). \qquad (7\text{-}39)$$

Multiplied out this yields

$$W_s = \tfrac{1}{2}kx_i^2 - \tfrac{1}{2}kx_f^2 \qquad \begin{array}{l}\text{(work by a}\\ \text{spring force).}\end{array} \qquad (7\text{-}40)$$

This work W_s done by the spring force can have a positive or negative value, depending on whether the *net* transfer of energy is to or from the block as the block moves from x_i to x_f. If $x_i = 0$ and if we call the final position x, then Eq. 7-40 becomes

$$W_s = -\tfrac{1}{2}kx^2 \qquad \begin{array}{l}\text{(work by a}\\ \text{spring force).}\end{array} \qquad (7\text{-}41)$$

Now suppose that we displace the block along x while continuing to apply a force $\mathbf{F}_a$ to it. During the displacement, our applied force does work W_a on the block while the spring force does work W_s. By Eq. 7-15, the change ΔK in the kinetic energy of the block due to these two energy transfers is

$$\Delta K = K_f - K_i = W_a + W_s, \qquad (7\text{-}42)$$

in which K_f is the kinetic energy at the end of the displacement and K_i is that at the start of the displacement. If the block is stationary before and after the displacement, then K_f and K_i are both zero and Eq. 7-42 reduces to

$$W_a + W_s = 0,$$
or $$W_a = -W_s. \qquad (7\text{-}43)$$

The result means that the work done on the block by the applied force is the negative of the work done on the block by the spring force.

Note that the length of the spring does not appear explicitly in the expressions for the spring force (Eqs. 7-37 and 7-38) and for the work done by the spring force (Eqs. 7-40 and 7-41). The length of the spring is one of several factors that determine the spring constant k; thus, the length is in those equations implicitly.

SAMPLE PROBLEM 7-8

You apply a 4.9 N force F_a to a block attached to the free end of a spring to keep the spring stretched from its relaxed length by 12 mm, as in Fig. 7-11*b*.

(a) What is the spring constant of the spring?

SOLUTION: The stretched spring pulls with a force of -4.9 N. From Eq. 7-38, with $x = 12$ mm, we have

$$k = -\frac{F}{x} = -\frac{-4.9 \text{ N}}{12 \times 10^{-3} \text{ m}}$$
$$= 408 \text{ N/m} \approx 410 \text{ N/m}. \qquad \text{(Answer)}$$

Note that we do not need to know the length of the spring to find k. The plot of Eq. 7-38 in Fig. 7-12 refers to this spring. The slope of the line is -410 N/m.

(b) What force does the spring exert on the block if you stretch the spring by 17 mm?

SOLUTION: From Eq. 7-38 we have

$$F = -kx = -(408 \text{ N/m})(17 \times 10^{-3} \text{ m})$$
$$= -6.9 \text{ N}. \qquad \text{(Answer)}$$

The dot on the curve of Fig. 7-12 represents this force and the corresponding displacement. Note that x is positive and F is negative, as required by Eq. 7-38.

(c) How much work does the spring force do on the block as the spring is stretched 17 mm as in (b)?

SOLUTION: Because the spring is initially in its relaxed state, we can use Eq. 7-41:

$$W_s = -\tfrac{1}{2}kx^2 = -(\tfrac{1}{2})(408 \text{ N/m})(17 \times 10^{-3} \text{ m})^2$$
$$= -5.9 \times 10^{-2} \text{ J} = -59 \text{ mJ}. \qquad \text{(Answer)}$$

The shaded area in Fig. 7-12 represents this work. The work is negative because the spring force and the displacement of the block are in opposite directions. Note that the amount of work done by the spring force would be the same if the spring had been compressed (rather than stretched) by 17 mm.

(d) With the spring initially stretched by 17 mm, you allow the block to return to $x = 0$ (the spring returns to its relaxed state); you then compress the spring by 12 mm. How much work does the spring force do on the block during this total displacement of the block?

SOLUTION: For this situation, we have $x_i = +17$ mm (the spring is initially stretched) and $x_f = -12$ mm (the spring is finally compressed). Equation 7-40 becomes

$$W_s = \tfrac{1}{2}kx_i^2 - \tfrac{1}{2}kx_f^2 = \tfrac{1}{2}k(x_i^2 - x_f^2)$$
$$= \tfrac{1}{2}(408 \text{ N/m})[(17 \times 10^{-3} \text{ m})^2 - (-12 \times 10^{-3} \text{ m})^2]$$
$$= 0.030 \text{ J} = 30 \text{ mJ}. \qquad \text{(Answer)}$$

This work done on the block by the spring force is positive because the spring force does more positive work as the block moves from $x_i = +17$ mm to $x = 0$ than it does negative work as the block moves from $x = 0$ to $x_f = -12$ mm.

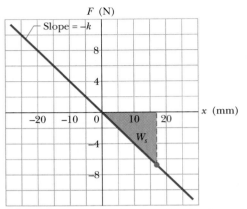

FIGURE 7-12 The force–distance plot for the spring of Sample Problem 7-8. The spring obeys Hooke's law (Eqs. 7-37 and 7-38) and has a spring constant $k = 410$ N/m. For the significance of the dot and the shaded area marked W_s, see Sample Problem 7-8(b) and 7-8(c), respectively.

CHECKPOINT **4:** For three situations, the initial and final positions, respectively, along the x axis for the block in Fig. 7-11 are (a) -3 cm, 2 cm; (b) 2 cm, 3 cm; and (c) -2 cm, 2 cm. In each situation is the work done by the spring force on the block positive, negative, or zero?

SAMPLE PROBLEM 7-9

A block whose mass m is 5.7 kg slides on a horizontal frictionless tabletop with a constant speed v of 1.2 m/s. It is brought momentarily to rest as it compresses a spring in its path (Fig. 7-13). By what distance d is the spring compressed? The spring constant k is 1500 N/m.

SOLUTION: From Eq. 7-41, the work done *by* the spring force *on* the block as the spring is compressed a distance d from its rest state is given by

$$W_s = -\tfrac{1}{2}kd^2.$$

The change in the kinetic energy of the block as it is stopped is

$$\Delta K = K_f - K_i = 0 - \tfrac{1}{2}mv^2.$$

The work–kinetic energy theorem (Eq. 7-4) requires that these two quantities be equal. Setting them so and solving for d, we obtain

$$d = v\sqrt{\frac{m}{k}} = (1.2 \text{ m/s})\sqrt{\frac{5.7 \text{ kg}}{1500 \text{ N/m}}}$$

$$= 7.4 \times 10^{-2} \text{ m} = 7.4 \text{ cm}. \qquad \text{(Answer)}$$

FIGURE 7-13 Sample Problem 7-9. A block moves toward a spring with velocity **v**. When it is momentarily stopped by the spring, it will have compressed the spring by a distance d.

PROBLEM SOLVING TACTICS

TACTIC 1: *Derivatives and Integrals; Slopes and Areas*

If you know a function $y = F(x)$, you can find the value of its derivative (for any value of x) or of its integral (between any two values of x) from the rules of calculus. If you do not know the function analytically but have a plot of it, you can find the values of both its derivative and its integral graphically. Finding a derivative graphically is shown in Tactic 5 of Chapter 2. Here we evaluate an integral graphically.

Figure 7-14 is a plot of a particular force function $F(x)$. Let us calculate graphically the work W done by this force as the particle on which it acts moves from $x_i = 2.0$ cm to $x_f = 5.0$ cm. According to Eq. 7-27, the work is

$$W = \int_{x_i}^{x_f} F(x)\, dx,$$

which is equal to the (total) shaded area shown under the curve between the two points.

You can approximate this area with a rectangle formed by drawing a horizontal line across Fig. 7-14. Draw it at a level such that the areas marked "1" and "2" appear to be equal. A line at $F = 44$ N is about right, and the area of the equivalent rectangle ($= W$) is then

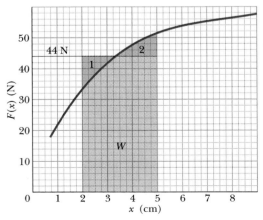

FIGURE 7-14 The graph of a one-dimensional force $F(x)$. The shaded area under the curve (which represents the work done by F) is approximated by a rectangle formed by excluding area 2 and including (approximately equal) area 1.

$$W = \text{height} \times \text{base} = (44 \text{ N})(5.0 \text{ cm} - 2.0 \text{ cm})$$
$$= 132 \text{ N} \cdot \text{cm} \approx 1.3 \text{ N} \cdot \text{m} = 1.3 \text{ J}.$$

You can also find the area by counting the small squares underneath the curve. The shaded area contains about 260 squares, and each square represents $(2 \text{ N})(0.25 \text{ cm}) = 0.5$ N·cm. The work is then

$$W = (260 \text{ squares})\left(\frac{0.5 \text{ N} \cdot \text{cm}}{1 \text{ square}}\right) = 130 \text{ N} \cdot \text{cm}$$

$$= 1.3 \text{ J},$$

just as above. *Remember:* On a two-dimensional graph, a derivative is represented by a slope, and an integral by an area.

7-7 POWER

A contractor wishes to lift a load of bricks from the sidewalk to the top of a building by means of a winch. We can now calculate how much work the force applied by the winch must do on the load to make the lift. The contractor, however, is much more interested in the *rate* at which that work is done. Will the job take 5 minutes (acceptable) or a week (unacceptable)?

The rate at which work is done by a force is said to be the **power** due to the force. If an amount of work W is done in an amount of time Δt by a force, the **average power** due to the force during that time interval is

$$\bar{P} = \frac{W}{\Delta t} \qquad \text{(average power).} \qquad (7\text{-}44)$$

The **instantaneous power** P is the instantaneous rate of doing work, which we can write as

$$P = \frac{dW}{dt} \qquad \text{(instantaneous power).} \qquad (7\text{-}45)$$

The SI unit of power is the joule per second. This unit is used so often that it has a special name, the **watt** (W), after James Watt, who greatly improved the rate at which steam engines could do work. In the British system, the unit of power is the foot-pound per second. Often the horsepower is used. Some relations among these units are

$$1 \text{ watt} = 1 \text{ W} = 1 \text{ J/s} = 0.738 \text{ ft} \cdot \text{lb/s} \qquad (7\text{-}46)$$

and

$$1 \text{ horsepower} = 1 \text{ hp} = 550 \text{ ft} \cdot \text{lb/s}$$
$$= 746 \text{ W}. \qquad (7\text{-}47)$$

Inspection of Eq. 7-44 shows that work can be expressed as power multiplied by time, as in the common unit, the kilowatt-hour. Thus

$$1 \text{ kilowatt-hour} = 1 \text{ kW} \cdot \text{h} = (10^3 \text{ W})(3600 \text{ s})$$
$$= 3.60 \times 10^6 \text{ J} = 3.60 \text{ MJ}. \qquad (7\text{-}48)$$

Perhaps because of our utility bills, the watt and the kilowatt-hour have become identified as electrical units. They can be used equally well as units for other examples of power and energy. Thus if you pick up this book from the floor and put it on a tabletop, you are free to report the work that you have done as $4 \times 10^{-6} \text{ kW} \cdot \text{h}$ (or more conveniently as $4 \text{ mW} \cdot \text{h}$).

We can also express the rate at which a force does work on a particle (or particle-like object) in terms of that force and the body's velocity. For a particle moving along a straight line (say, the x axis) and acted on by a constant force $\mathbf{F}$ directed at some angle ϕ to that line, Eq. 7-45 becomes

$$P = \frac{dW}{dt} = \frac{F \cos \phi \, dx}{dt} = F \cos \phi \left(\frac{dx}{dt}\right),$$

or

$$P = Fv \cos \phi. \qquad (7\text{-}49)$$

Reorganizing the right side of Eq. 7-49 as the dot product $\mathbf{F} \cdot \mathbf{v}$, we may also write Eq. 7-49 as

$$P = \mathbf{F} \cdot \mathbf{v} \qquad \text{(instantaneous power).} \qquad (7\text{-}50)$$

For example, the truck in Fig. 7-15 exerts a force $\mathbf{F}$ on the trailing load, which has velocity $\mathbf{v}$ at some instant. The instantaneous power due to $\mathbf{F}$ is the rate at which $\mathbf{F}$ transfers energy to the load at that instant and is given by Eqs. 7-49 and 7-50. Saying that this power is "the power of the truck" is often acceptable, but we should keep in mind what is meant: power is the rate at which the applied *force* transfers energy.

FIGURE 7-15 The power due to the truck's applied force on the trailing load is the rate at which that force does work on the load.

SAMPLE PROBLEM 7-10

Figure 7-16 shows forces $\mathbf{F}_1$ and $\mathbf{F}_2$ acting on a box as the box slides rightward across a frictionless floor. Force $\mathbf{F}_1$ is horizontal, with magnitude 2.0 N; force $\mathbf{F}_2$ is angled upward by 60° to the floor and has magnitude 4.0 N. The speed v of the box at a certain instant is 3.0 m/s.

(a) What is the power due to each force acting on the box at that instant, and what is the net power? Is the net power changing at that instant?

SOLUTION: We use Eq. 7-49 to find the power due to each force. For force $\mathbf{F}_1$, at angle $\phi_1 = 180°$ to velocity $\mathbf{v}$, we have

$$P_1 = F_1 v \cos \phi_1 = (2.0 \text{ N})(3.0 \text{ m/s}) \cos 180°$$
$$= -6.0 \text{ W.} \qquad \text{(Answer)}$$

This result tells us that force $\mathbf{F}_1$ is transferring energy from the box at the rate of 6.0 J/s.

For force $\mathbf{F}_2$, at angle $\phi_2 = 60°$ to velocity $\mathbf{v}$, we have

$$P_2 = F_2 v \cos \phi_2 = (4.0 \text{ N})(3.0 \text{ m/s}) \cos 60°$$
$$= 6.0 \text{ W.} \qquad \text{(Answer)}$$

This result tells us that force $\mathbf{F}_2$ is transferring energy to the box at the rate of 6.0 J/s.

The net power is the sum of the individual powers:

$$P_{\text{net}} = P_1 + P_2$$
$$= -6.0 \text{ W} + 6.0 \text{ W} = 0, \qquad \text{(Answer)}$$

which tells us that the net rate of transfer of energy to or from the box is zero. Thus, the kinetic energy ($K = \frac{1}{2}mv^2$) of the box is not changing, and so the speed of the box will remain at 3.0 m/s. With neither the forces $\mathbf{F}_1$ and $\mathbf{F}_2$ nor the velocity $\mathbf{v}$ changing, we see from Eq. 7-50 that P_1 and P_2 are not changing and thus neither is P_{net}.

(b) If the magnitude of $\mathbf{F}_2$ is, instead, 6.0 N, what now is the power due to each force acting on the box at the given instant, and what is the net power? Is the net power changing?

SOLUTION: For force $\mathbf{F}_2$, we now have

$$P_2 = F_2 v \cos \theta_2 = (6.0 \text{ N})(3.0 \text{ m/s}) \cos 60°$$
$$= 9.0 \text{ W.} \qquad \text{(Answer)}$$

The power of force $\mathbf{F}_1$ is still

$$P_1 = -6.0 \text{ W.} \qquad \text{(Answer)}$$

So, the net power is now

$$P_{\text{net}} = P_1 + P_2 = -6.0 \text{ W} + 9.0 \text{ W}$$
$$= 3.0 \text{ W}, \qquad \text{(Answer)}$$

which tells us that the net rate of transfer of energy to the box has a positive value. Thus, the kinetic energy of the box is increasing, and so also is the speed of the box. With the speed increasing, we see from Eq. 7-50 that the values of P_1 and P_2, and thus also of P_{net}, will be changing. Hence, this net power of 3.0 W is the net power only at the instant the speed is the given 3.0 m/s.

FIGURE 7-16 Sample Problem 7-10. Two forces $\mathbf{F}_1$ and $\mathbf{F}_2$ act on a box that slides rightward across a frictionless floor. The velocity of the box is $\mathbf{v}$.

CHECKPOINT 5: A block moves with uniform circular motion because a cord tied to the block is anchored at the center of a circle. Is the power of the force exerted on the block by the cord positive, negative, or zero?

7-8 KINETIC ENERGY AT HIGH SPEEDS (OPTIONAL)

In Section 4-10 we saw that for particles moving at speeds near the speed of light, Newtonian mechanics fails and must be replaced by Einstein's theory of special relativity. One consequence is that we can no longer use the expression $K = \frac{1}{2}mv^2$ for the kinetic energy of a particle. Instead, we must use

$$K = mc^2 \left(\frac{1}{\sqrt{1 - (v/c)^2}} - 1 \right), \qquad (7\text{-}51)$$

in which c is the speed of light.

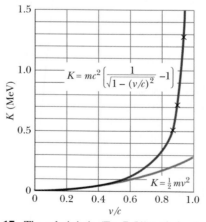

FIGURE 7-17 The relativistic (Eq. 7-51) and classical (Eq. 7-1) equations for the kinetic energy of an electron, plotted as a function of v/c, where v is the speed of the electron and c is the speed of light. Note that the two curves blend together at low speeds and diverge widely at high speeds. Experimental data (at the $\times$ marks) show that at high speeds the relativistic curve agrees with experiment but the classical curve does not.

Figure 7-17 shows that these two formulas, which look so different, do indeed give widely different results at high speeds. Experiment shows that—beyond any doubt—the relativistic expression (Eq. 7-51) is correct and the classical expression (Eq. 7-1) is not. At low speeds, however, the two formulas merge, yielding the same result. In particular, both formulas yield $K = 0$ for $v = 0$.

All relativistic formulas must reduce to their classical counterparts at low speeds. To see how this comes about for Eq. 7-51, let us write that equation in the form

$$K = mc^2[(1 - \beta^2)^{-1/2} - 1]. \qquad (7\text{-}52)$$

Here, for convenience, we have substituted the dimensionless *speed parameter* β for the speed ratio v/c.

At very low speeds, $v \ll c$ and therefore $\beta \ll 1$. At low speeds, we can then expand the quantity $(1 - \beta^2)^{-1/2}$ by the binomial theorem, obtaining (see Tactic 2, below)

$$(1 - \beta^2)^{-1/2} = 1 + \tfrac{1}{2}\beta^2 + \cdots. \qquad (7\text{-}53)$$

Substituting Eq. 7-53 into Eq. 7-52 leads to

$$K = mc^2[(1 + \tfrac{1}{2}\beta^2 + \cdots) - 1]. \qquad (7\text{-}54)$$

For very small β, the terms represented by the dots in Eq. 7-54 decrease rapidly in size. We can then, with little error, replace the entire sum in parentheses with its first two terms, obtaining

$$K \approx (mc^2)[(1 + \tfrac{1}{2}\beta^2) - 1]$$

or, with $\beta = v/c$,

$$K \approx (mc^2)(\tfrac{1}{2}\beta^2) = \tfrac{1}{2}mv^2,$$

which is exactly what we set out to show.

PROBLEM SOLVING TACTICS

TACTIC 2: *Approximations with the Binomial Theorem*
Often we have a quantity written in the form $(a + b)^n$ and we want to find its approximate value for the case in which $b \ll a$. To do so, it is simplest to recast the quantity in the form $(1 + x)^n$, where x is dimensionless and is much less than unity. Thus we can put

$$(a + b)^n = a^n(1 + b/a)^n = (a^n)(1 + x)^n.$$

This is of the desired form, with $x (= b/a)$ being dimensionless. We can then evaluate $(1 + x)^n$ with the binomial theorem, keeping only as many terms as are appropriate to the problem. (That decision takes some experience.)

The binomial theorem (given in Appendix E) is

$$(1 + x)^n = 1 + \frac{n}{1!}x + \frac{n(n - 1)}{2!}x^2 + \cdots. \qquad (7\text{-}55)$$

The exclamation marks in Eq. 7-55 identify factorials, or products of all the whole numbers from the given number

down to 1. As an example, $4! = 4 \times 3 \times 2 \times 1 = 24$; you probably have a key for factorials on your calculator.

We applied the binomial theorem in Eq. 7-53, letting $x = -\beta^2$ and $n = -\tfrac{1}{2}$.

As an exercise, evaluate $(1 + 0.045)^{-2.3}$ both on your calculator and by expansion, using Eq. 7-55 with $x = 0.045$ and $n = -2.3$. Check the various terms in the binomial sum to see how rapidly they decrease.

7-9 REFERENCE FRAMES

We apply Newton's laws of mechanics only in inertial reference frames. Recall that these frames move at constant velocity.

For some physical quantities, observers in different inertial reference frames will measure the exact same values. In Newtonian mechanics, these *invariant* quantities (as they are called) are force, mass, acceleration, and time. As an example, if an observer in one inertial frame finds that a certain particle has a mass of 3.15 kg, observers in all other inertial frames will measure the same mass for the particle. For other physical quantities, such as the displacement and velocity of a particle, observers in different inertial frames will measure different values; these quantities are *not* invariant.

If the displacement of a particle depends on the reference frame of the observer, so must the work done on the particle because work ($W = \mathbf{F} \cdot \mathbf{d}$) is defined in terms of displacement. If the displacement of a particle during a given interval is measured to be $+2.47$ m in one reference frame, it might be measured to be zero in another and -3.64 m in a third. Because the force $\mathbf{F}$ does not change (it is invariant), work that is positive in one reference frame can be zero in another and negative in a third.

What about the kinetic energy of the particle? If its velocity depends on the reference frame of the observer, then so must its kinetic energy because kinetic energy ($K = \tfrac{1}{2}mv^2$) is defined in terms of velocity. Does that mean that the work–kinetic energy theorem is in trouble?

From Galileo to Einstein, physicists have come to believe in the **principle of invariance:**

The laws of physics must have the same form in all inertial reference frames.

That is, even though some *physical quantities* have different values in different reference frames, the *laws of physics* must hold true in all frames. Behind this formal statement of invariance is a feeling that—in some deep sense—if different observers look at the same event, they must see nature working in the same way.

Among the laws to which this invariance principle applies is the work–kinetic energy theorem. Thus, even though different observers watching the same moving particle might measure different values for work and for kinetic energy, they would all find that the work–kinetic energy theorem holds in their respective frames. Let us look at a simple example.

In Fig. 7-18, Sally rides up in an elevator cab at constant speed v, holding a book. Steve, standing on a facing balcony, observes her as the cab rises through a height h. From these two references frames, what are the work–kinetic energy relations that apply to the book?

1. *Sally's report.* "My reference frame is the elevator cab. I am exerting an upward force on the book, but this force does no work because the book is not moving in my reference frame. The weight of the book, acting downward, does no work, for the same reason. Thus the total work done on the book by all the forces that act on it is zero. According to the work–kinetic energy theorem, the kinetic energy of the book should not change. That is what I observe; the kinetic energy of the book is zero in my reference frame and remains so. Everything fits together."

2. *Steve's report.* "My reference frame is the balcony. I see that Sally exerts a force **F** on her book. In my frame, the point of application of **F** is moving and the work done by **F** as the book moves upward through a height h is $+mgh$. I also know that the book's weight does work on the book, in the amount $-mgh$. Thus the total work done on the book during its rise is zero. According to the work–kinetic energy theorem, the kinetic energy of the book

FIGURE 7-18 Sally rides up in an elevator, holding a book. Steve watches her. Both check the work–kinetic energy theorem in their respective reference frames, as it applies to the motion of the book.

should not change. That is what I observe. In my frame, the kinetic energy is a constant $\frac{1}{2}mv^2$. Everything fits."

Although Steve and Sally do not agree about the displacement of the book and its kinetic energy, they do agree that the work–kinetic energy theorem holds in their respective reference frames.

It does not matter what (inertial) reference frame you pick in which to solve a problem as long as you (1) make sure you know what that frame is and (2) use that frame consistently throughout the problem.

REVIEW & SUMMARY

Kinetic Energy

The **kinetic energy** K associated with the motion of a particle of mass m and speed v, where v is well below the speed of light, is

$$K = \tfrac{1}{2}mv^2 \qquad \text{(kinetic energy).} \qquad (7\text{-}1)$$

Work

Work W is energy transferred to or from an object via a force acting on the object. Energy transferred to the object is positive work, and that transferred from the object is negative work.

Work and Kinetic Energy

If the only type of energy of an object being changed by work W is kinetic energy, then we can write the change ΔK in kinetic energy of the object as

$$\Delta K = K_f - K_i = W \qquad \begin{array}{l}\text{(work–kinetic} \\ \text{energy theorem),}\end{array} \qquad (7\text{-}4)$$

in which K_i is the initial kinetic energy of the object and K_f is the kinetic energy after the work is done. Equation 7-4 rearranged gives us

$$K_f = K_i + W. \qquad (7\text{-}5)$$

Work Done by a Constant Force

The work done on a particle by a constant force **F** during displacement **d** of the particle is

$$W = Fd \cos \phi = \mathbf{F} \cdot \mathbf{d} \qquad \begin{array}{l}\text{(work,} \\ \text{constant force),}\end{array} \qquad (7\text{-}9, \ 7\text{-}11)$$

in which ϕ is the constant angle between **F** and **d**. When several forces $\mathbf{F}_1, \mathbf{F}_2, \mathbf{F}_3, \ldots$ act on a particle, the work done by the net force $\Sigma\mathbf{F}$ is

$$W = \left(\sum \mathbf{F}\right) \cdot \mathbf{d} \qquad \begin{array}{l}\text{(work by constant} \\ \text{net force),}\end{array} \qquad (7\text{-}13)$$

This work is equal to the total work done by all the forces:

$$W = \mathbf{F}_1 \cdot \mathbf{d} + \mathbf{F}_2 \cdot \mathbf{d} + \mathbf{F}_3 \cdot \mathbf{d} + \cdots$$
$$= W_1 + W_2 + W_3 + \cdots \quad \text{(total work).} \quad (7\text{-}14)$$

Substituting this total work into the work–kinetic energy theorem, we have

$$\Delta K = K_f - K_i = W_1 + W_2 + W_3 + \cdots, \quad (7\text{-}15)$$

which gives us the change in the kinetic energy of the particle due to the total work.

Work Done by Weight

The work W_g done by the weight $m\mathbf{g}$ of a particle-like object during a displacement $\mathbf{d}$ of the object is given by

$$W_g = mgd \cos \phi, \quad (7\text{-}16)$$

in which ϕ is the angle between $m\mathbf{g}$ and $\mathbf{d}$.

Work Done in Lifting and Lowering an Object

The work W_a done by an applied force during a lifting or lowering of a particle-like object is related to the work W_g done by the object's weight and the change ΔK in the object's kinetic energy by

$$\Delta K = K_f - K_i = W_a + W_g. \quad (7\text{-}19)$$

If the kinetic energy at the beginning of a lift equals that at the end of the lift, then Eq. 7-19 reduces to

$$W_a = -W_g, \quad (7\text{-}20)$$

which tells us that the applied force transfers as much energy to the object as the weight of the object transfers from the object.

Work Done by a Variable Force

When the force $\mathbf{F}$ on a particle-like object depends on the position of the object, the work done by $\mathbf{F}$ on the object while the object moves from an initial position r_i with coordinates (x_i, y_i, z_i) to a final position r_f with coordinates (x_f, y_f, z_f) is

$$W = \int_{x_i}^{x_f} F_x \, dx + \int_{y_i}^{y_f} F_y \, dy + \int_{z_i}^{z_f} F_z \, dz. \quad (7\text{-}31)$$

If $\mathbf{F}$ has only an x component, then Eq. 7-31 reduces to

$$W = \int_{x_i}^{x_f} F(x) \, dx. \quad (7\text{-}27)$$

Spring Force

The force $\mathbf{F}$ exerted by a spring is

$$\mathbf{F} = -k\mathbf{d} \quad \text{(Hooke's law),} \quad (7\text{-}37)$$

where $\mathbf{d}$ is the displacement of its free end from the position when the spring is in its **relaxed state** (neither compressed or extended)

and k is the **spring constant** (a measure of the spring's stiffness). If an x axis lies along the spring, with the origin at the location of the spring's free end when the spring is in its relaxed state, Eq. 7-37 can be written as

$$F = -kx \quad \text{(Hooke's law).} \quad (7\text{-}38)$$

Work Done by a Spring Force

If an object is attached to the spring's free end, the work W_s done on the object by the spring force when the object is moved from an initial position x_i to a final position x_f is

$$W_s = \tfrac{1}{2}kx_i^2 - \tfrac{1}{2}kx_f^2. \quad (7\text{-}40)$$

If $x_i = 0$ and $x_f = x$, then Eq. 7-40 becomes

$$W_s = -\tfrac{1}{2}kx^2. \quad (7\text{-}41)$$

Power

The **power** due to a force is the *rate* at which that force does work on an object. If the force does work W during a time interval Δt, the *average power* due to the force over that time interval is

$$\overline{P} = \frac{W}{\Delta t}. \quad (7\text{-}44)$$

Instantaneous power is the instantaneous rate of doing work:

$$P = \frac{dW}{dt}. \quad (7\text{-}45)$$

If a force $\mathbf{F}$ is at an angle ϕ to the direction of travel of the object, the instantaneous power is

$$P = Fv \cos \phi = \mathbf{F} \cdot \mathbf{v}, \quad (7\text{-}49, \, 7\text{-}50)$$

in which $\mathbf{v}$ is the instantaneous velocity of the object.

Relativistic Kinetic Energy

The kinetic energy of objects moving at speeds near the speed of light c must be calculated using the relativistic expression

$$K = mc^2 \left(\frac{1}{\sqrt{1 - (v/c)^2}} - 1 \right). \quad (7\text{-}51)$$

This equation reduces to Eq. 7-1 when v is much less than c.

The Principle of Invariance

Some quantities (such as mass, force, acceleration, and time in Newtonian mechanics) are *invariant;* that is, they have the same numerical values when measured in different inertial reference frames. Others (for example, velocity, kinetic energy, and work) have different values in different frames. However, the *laws* of physics have the same *form* in all inertial reference frames. This is called the **principle of invariance.**

QUESTIONS

1. Rank the following velocities according to the kinetic energy a particle will have with each velocity, greatest first: (a) $\mathbf{v} = 4\mathbf{i} + 3\mathbf{j}$, (b) $\mathbf{v} = -4\mathbf{i} + 3\mathbf{j}$, (c) $\mathbf{v} = -3\mathbf{i} + 4\mathbf{j}$, (d) $\mathbf{v} = 3\mathbf{i} - 4\mathbf{j}$, (e) $\mathbf{v} = 5\mathbf{i}$, and (f) $v = 5$ m/s at 30° to the horizontal.

2. For $t > 0$, is the kinetic energy of a particle increasing, decreasing, or remaining the same if the particle's velocity is given by (a) $\mathbf{v} = 3\mathbf{i}$ and (b) $v = -4t$?

3. For $t > 0$, is the kinetic energy of a particle increasing, decreasing, or remaining the same if the particle's position is given by (a) $x = 4t^2 - 2$, (b) $x = -3t + 14$, (c) $\mathbf{r} = 2\mathbf{i} - 3t\mathbf{j}$, and (d) $\mathbf{r} = (2t^2 - 3)\mathbf{i} + (4t - 2)\mathbf{j}$?

4. A force $\mathbf{F}$ that is directed along an x axis moves a particle 5 m along that axis. Can we use Eqs. 7-9 and 7-11 to find the work done if the magnitude (in newtons) of $\mathbf{F}$ is given by (a) $F = 3$; (b) $F = 2x$; (c) $F = 2t$?

5. Is positive or negative work done by a constant force $\mathbf{F}$ on a particle during a straight-line displacement $\mathbf{d}$ if (a) the angle between $\mathbf{F}$ and $\mathbf{d}$ is 30°; (b) the angle is 100°; (c) $\mathbf{F} = 2\mathbf{i} - 3\mathbf{j}$ and $\mathbf{d} = -4\mathbf{i}$?

6. If a particle-like object of mass 5 kg travels in uniform circular motion at speed 4 m/s, how much work is done on the particle by the centripetal force acting on it (a) at any instant and (b) during one complete revolution?

7. Figure 7-19 shows six situations in which two forces act simultaneously on a box after the box has been sent sliding over a frictionless surface, either to the left or to the right. The forces are either 1 N or 2 N in magnitude, as indicated by the vector lengths. For each situation, is the work done on the box by the net force during the indicated displacement $\mathbf{d}$ positive, negative, or zero?

FIGURE 7-19 Question 7.

8. In overhead views, Fig. 7-20 shows three situations in which a box is acted on by two forces ($\mathbf{F}_1$ and $\mathbf{F}_2$) of equal magnitude. As the box moves, the forces maintain their orientation relative to the velocity $\mathbf{v}$. For each situation, is the total work done on the box positive, negative, or zero?

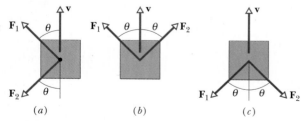

FIGURE 7-20 Question 8.

9. In Fig. 7-21, a greased pig has a choice of three frictionless slides along which to slide to the ground. Rank the slides according to how much work the pig's weight does on the pig during the descent, greatest first.

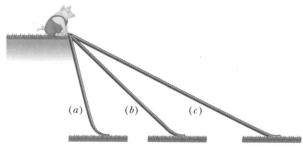

FIGURE 7-21 Question 9.

10. You lift an armadillo to a shelf. Does the work done by your force on the armadillo depend on (a) the mass of the armadillo, (b) the weight of the armadillo, (c) the height of the shelf, (d) the time you take, or (e) whether you move the armadillo sideways or directly upward?

11. Figure 7-22 shows a bundle of magazines that is lifted by a cord through distance d. The table shows six pairs of values of the initial speed v_0 and final speed v (in meters per second) of the bundle at the beginning and end of distance d. Rank the pairs according to the work done by the cord's force on the bundle over distance d, most positive first, most negative last.

	a	b	c	d	e	f
$v =$	0	2	2	2	0	1
$v_0 =$	0	2	0	1	2	2

FIGURE 7-22 Question 11.

12. Figure 7-23 shows four graphs (drawn to the same scale) of the values of a variable force $\mathbf{F}$ (directed along an x axis) versus the position x of a particle on which the force acts. Rank the graphs according to the work done by $\mathbf{F}$ on the particle from $x = 0$ to x_1, from most positive work to most negative work.

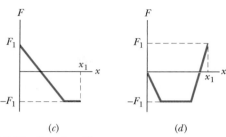

(c)   (d)

FIGURE 7-23   Question 12.

13. Figure 7-24 shows the values of a force **F**, directed along an x axis, that will act on a particle at the corresponding values of x. If the particle begins at rest at x = 0, what is the particle's coordinate when it has (a) its greatest kinetic energy, (b) its greatest speed, and (c) zero speed? (d) What is its direction of travel after it reaches x = 6 m?

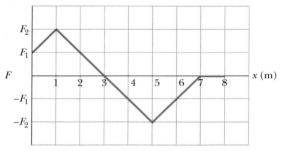

FIGURE 7-24   Question 13.

14. A block is attached to a relaxed spring as in Fig. 7-25a. The spring constant k of the spring is such that for a certain rightward displacement **d** of the block, the spring force acting on the block has magnitude F_1 and has done work W_1 on the block. A second, identical spring is then attached on the opposite side of the block,

as shown in Fig. 7-25b; both springs are in their relaxed state in the figure. If the block is again displaced by **d**, (a) what is the magnitude of the net force on it from both springs and (b) how much work has been done on it by the spring forces?

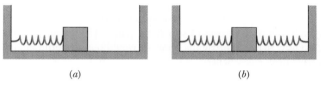

(a)   (b)

FIGURE 7-25   Question 14.

15. Spring A is stiffer than spring B, that is, $k_A > k_B$. The spring force of which spring does more work if the springs are compressed (a) the same distance and (b) by the same applied force?

16. Figure 7-26 shows three arrangements of a block attached to identical springs that are in their relaxed state when the block is centered as shown. Rank the arrangements according to the magnitude of the net force on the block, largest first, when the block is displaced by distance d (a) to the right and (b) to the left. Rank the arrangements according to the work done on the block by the spring forces, greatest first, when the block is displaced by d (c) to the right and (d) to the left.

(1)   (2)   (3)

FIGURE 7-26   Question 16.

17. If you cut a spring in half, the spring constant of either half is what multiple of the spring constant of the original spring? (*Hint:* Consider the amount by which each half stretches for a given value of force.)

EXERCISES & PROBLEMS

SECTION 7-1   Kinetic Energy

1E. If a Saturn V rocket with an Apollo spacecraft attached has a combined mass of 2.9×10^5 kg and is to reach a speed of 11.2 km/s, how much kinetic energy will it then have?

2E. If an electron (mass $m = 9.11 \times 10^{-31}$ kg) in copper near the lowest possible temperature has a kinetic energy of 6.7×10^{-19} J, what is the speed of the electron?

3E. Calculate the kinetic energies of the following objects moving at the given speeds: (a) a 110 kg football linebacker running at 8.1 m/s; (b) a 4.2 g bullet at 950 m/s; (c) the aircraft carrier *Nimitz*, 91,400 tons at 32 knots.

4E. On August 10, 1972, a large meteorite skipped across the atmosphere above western United States and Canada, much like a stone skipped across water. The accompanying fireball was so bright that it could be seen in the daytime sky (Fig. 7-27). The meteorite's mass was about 4×10^6 kg; its speed was about 15

FIGURE 7-27   Exercise 4 A large meteorite skips across the atmosphere in the sky above the mountains (upper right).

km/s. Had it entered the atmosphere vertically, it would have hit the Earth's surface with about the same speed. (a) Calculate the meteorite's loss of kinetic energy (in joules) that would have been associated with the vertical impact. (b) Express the energy as a multiple of the explosive energy of 1 megaton of TNT, which is 4.2×10^{15} J. (c) The energy associated with the atomic bomb explosion over Hiroshima was equivalent to 13 kilotons of TNT. To how many "Hiroshima bombs" would the meteorite impact have been equivalent?

5E. An explosion at ground level leaves a crater with a diameter that is proportional to the energy of the explosion raised to the $\frac{1}{3}$ power; an explosion of 1 megaton of TNT leaves a crater with a 1 km diameter. Below Lake Huron in Michigan there appears to be an ancient impact crater with a 50 km diameter. What was the kinetic energy associated with that impact, in terms of (a) megatons of TNT and (b) Hiroshima bomb equivalents (see Exercise 4)? (Such meteorite or comet impacts may have significantly altered the Earth's climate and contributed to the extinction of the dinosaurs and other life-forms.)

6P. A proton (mass $m = 1.67 \times 10^{-27}$ kg) is being accelerated along a straight line at 3.6×10^{15} m/s^2 in a machine called a linear accelerator. If the proton has an initial speed of 2.4×10^7 m/s and travels 3.5 cm, what then is (a) its speed and (b) the increase in its kinetic energy in electron-volts?

7P. A man racing his son has half the kinetic energy of the son, who has half the mass of the father. The man speeds up by 1.0 m/s and then has the same kinetic energy as the son. What were the original speeds of man and son?

8P. What is the kinetic energy associated with the Earth's orbiting of the Sun? (See Appendix C for numerical data.)

SECTION 7-3 Work and Kinetic Energy

9E. A 102 kg object moves in a horizontal straight line with an initial speed of 53 m/s. If it is stopped along that line with a deceleration of 2.0 m/s^2, (a) what magnitude of force is required, (b) what distance does it travel while decelerating, and (c) what work is done by the decelerating force on the object? (d) Answer questions (a) to (c) for a deceleration of 4.0 m/s^2.

10E. To pull a 50 kg crate across a frictionless floor, a worker applies a force of 210 N, directed 20° above the horizontal. As the crate moves 3.0 m, what work is done on the crate by (a) the worker's force, (b) the weight of the crate, and (c) the normal force exerted by the floor on the crate? (d) What is the total work done on the crate?

11E. A floating ice block is pushed through a displacement $\mathbf{d} = (15\text{ m})\mathbf{i} - (12\text{ m})\mathbf{j}$ along a straight embankment by rushing water, which exerts a force $\mathbf{F} = (210\text{ N})\mathbf{i} - (150\text{ N})\mathbf{j}$ on the block. How much work does the force do on the block during the displacement?

12E. A particle moves along a straight path through a displacement $\mathbf{d} = (8\text{ m})\mathbf{i} + c\mathbf{j}$ while a force $\mathbf{F} = (2\text{ N})\mathbf{i} - (4\text{ N})\mathbf{j}$ acts on it. (Other forces also act on the particle.) What is the value of c if the work done by $\mathbf{F}$ on the particle is (a) zero, (b) positive, and (c) negative?

13E. An initially stationary proton is accelerated in a cyclotron to a final speed of 3.0×10^6 m/s. How much work, in electron-volts, is done on the proton by the (electric) force accelerating it? (Although the speed is 1% the speed of light, relativity theory is not required here.)

14E. A single force acts on a body that moves along a straight line. Figure 7-28 shows the velocity versus time for the body. Find the sign (plus or minus) of the work done by the force on the body in each of the intervals AB, BC, CD, and DE.

FIGURE 7-28 Exercise 14.

15E. A firehose 12 m long (Fig. 7-29) is uncoiled by pulling the nozzle end horizontally along a frictionless surface at the steady speed of 2.3 m/s. The mass of 1.0 m of the hose is 0.25 kg. How much work has been done on the hose by the applied force when the entire hose is moving?

FIGURE 7-29 Exercise 15.

16P. Figure 7-30 shows three forces applied to a greased trunk that moves leftward by 3.00 m over a frictionless floor. The force magnitudes are $F_1 = 5.00$ N, $F_2 = 9.00$ N, and $F_3 = 3.00$ N. During the displacement, (a) what is the net work done on the trunk by the three forces and (b) does the kinetic energy of the trunk increase or decrease?

FIGURE 7-30 Problem 16.

17P. A force acts on a 3.0 kg particle-like object in such a way that the position of the object as a function of time is given by $x = 3.0t - 4.0t^2 + 1.0t^3$, with x in meters and t in seconds. Find the work done on the object by the force from $t = 0$ to $t = 4.0$ s. (*Hint:* What are the speeds at those times?)

18P. Figure 7-31 shows an overhead view of three horizontal forces acting on a cargo canister that is initially stationary but which now moves across a frictionless floor. The force magnitudes are $F_1 = 3.00$ N, $F_2 = 4.00$ N, and $F_3 = 10.0$ N. What is the net work done on the canister by the three forces during the first 4.00 m of displacement?

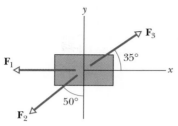

FIGURE 7-31 Problem 18.

19P. A constant force of magnitude 10 N makes an angle of 150° (measured counterclockwise) with the direction of increasing x as the force acts on a 2.0 kg object. How much work is done on the object by the force as the object moves from the origin to the point with position vector $(2.0 \text{ m})\mathbf{i} - (4.0 \text{ m})\mathbf{j}$?

SECTION 7-4 Work Done by Weight

20E. (a) In 1975 the roof of Montreal's Velodrome, with a weight of 41,000 tons, was lifted by 4.0 in. so that it could be centered. How much work was done on the roof by the forces making the lift? (b) In 1960 Mrs. Maxwell Rogers of Tampa, Florida, reportedly raised one end of a car that had fallen onto her son when a jack failed. If her panic lift effectively raised 900 lb of the 3600 lb car by 2 in., how much work was done by her force acting on the car?

21E. To push a 25.0 kg crate up a frictionless incline, angled at 25.0° to the horizontal, a worker exerts a force of 209 N, parallel to the incline. As the crate slides 1.50 m, how much work is done on the crate by (a) the worker's applied force, (b) the weight of the crate, and (c) the normal force exerted by the incline on the crate? (d) What is the total work done on the crate?

22E. A 45 kg block of ice slides down a frictionless incline 1.5 m long and 0.91 m high. A worker pushes up against the ice, parallel to the incline, so that the block slides down at constant speed. (a) Find the force exerted by the worker. How much work is done on the block by (b) the worker's force, (c) the weight of the block, (d) the normal force exerted by the surface of the incline on the block, and (e) the resultant force on the block?

23E. In Fig. 7-32, a cord runs around two massless, frictionless pulleys; a canister with mass $m = 20$ kg hangs from one pulley; and you exert a force $\mathbf{F}$ on the free end of the cord. (a) What must

FIGURE 7-32 Exercise 23.

be the magnitude of $\mathbf{F}$ if you are to lift the canister at a constant speed? (b) To lift the canister by 2.0 cm, how far must you pull the free end of the cord? During that lift, what is the work done on the canister by (c) your force (via the cord) and (d) the weight $m\mathbf{g}$ of the canister? (*Hint:* When a cord loops around a pulley as shown, it pulls on the pulley with a net force that is twice the tension in the cord.)

24P. A helicopter lifts a 72 kg astronaut 15 m vertically from the ocean by means of a cable. The acceleration of the astronaut is $g/10$. How much work is done on the astronaut by (a) the force from the helicopter and (b) her weight? What are (c) the kinetic energy and (d) the speed of the astronaut just before she reaches the helicopter?

25P. A cord is used to vertically lower an initially stationary block of mass M at a constant downward acceleration of $g/4$. When the block has fallen a distance d, find (a) the work done by the cord's force on the block, (b) the work done by the weight of the block, (c) the kinetic energy of the block, and (d) the speed of the block.

26P. A cave rescue team lifts an injured spelunker directly upward and out of a sinkhole by means of a motor-driven cable. The lift is performed in three stages, each requiring a vertical distance of 10.0 m: (1) the initially stationary spelunker is accelerated to a speed of 5.00 m/s; (2) he is then lifted at the constant speed of 5.00 m/s; (3) finally he is decelerated to zero speed. How much work is done on the 80.0 kg rescuee by the force lifting him during each stage?

SECTION 7-5 Work Done by a Variable Force

27E. A 5.0 kg block moves in a straight line on a horizontal frictionless surface under the influence of a force that varies with position as shown in Fig. 7-33. How much work is done by the force as the block moves from the origin to $x = 8.0$ m?

FIGURE 7-33 Exercise 27.

28E. A 10 kg mass moves along an x axis. Its acceleration as a function of its position is shown in Fig. 7-34. What is the net work performed on the mass by the force causing the acceleration as the mass moves from $x = 0$ to $x = 8.0$ m?

FIGURE 7-34 Exercise 28.

29P. (a) Estimate the work done by the force represented by the graph of Fig. 7-35 in displacing a particle from $x = 1$ m to $x = 3$ m. Refine your method to see how close you can come to the exact answer of 6 J. (b) The curve is given by $F = a/x^2$, with $a = 9$ N·m². Calculate the work using integration.

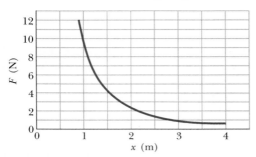

FIGURE 7-35 Problem 29.

30P. The force on a particle is $F = F_0(x/x_0 - 1)$. Find the work done by the force in moving the particle from $x = 0$ to $x = 2x_0$ by (a) plotting $F(x)$ and measuring the work from the graph and (b) integrating $F(x)$.

31P. What work is done by a force $\mathbf{F} = (2x \text{ N})\mathbf{i} + (3 \text{ N})\mathbf{j}$, with x in meters, that moves a particle from a position $\mathbf{r}_i = (2 \text{ m})\mathbf{i} + (3 \text{ m})\mathbf{j}$ to a position $\mathbf{r}_f = -(4 \text{ m})\mathbf{i} - (3 \text{ m})\mathbf{j}$?

32P. A force $\mathbf{F}$ in the direction of increasing x acts on an object moving along the x axis. If the magnitude of the force is $F = 10e^{-x/2.0}$ N, where x is in meters, find the work done by $\mathbf{F}$ as the object moves from $x = 0$ to $x = 2.0$ m (a) by plotting $F(x)$ and finding the area under the curve and (b) by integrating to find the work analytically.

33P. The only force acting on a 2.0 kg body as it moves along the x axis varies as shown in Fig. 7-36. The velocity of the body at $x = 0$ is 4.0 m/s. (a) What is the kinetic energy of the body at $x = 3.0$ m? (b) At what value of x will the body have a kinetic energy of 8.0 J? (c) What is the maximum kinetic energy attained by the body between $x = 0$ and $x = 5.0$ m?

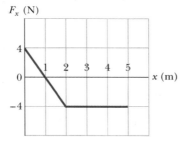

FIGURE 7-36 Problem 33.

34P. A 230 kg crate hangs from the end of a 12.0 m rope. You push horizontally on the crate with a varying force $\mathbf{F}$ to move it 4.00 m to the side (Fig. 7-37). (a) What is the magnitude of $\mathbf{F}$ when the crate is in this final position? During the crate's displacement, what are (b) the total work done on it, (c) the work done by the weight of the crate, and (d) the work done by the pull on the crate from the rope? (e) Knowing that the crate is motion-

less before and after its displacement, use the answers to (b), (c), and (d) to find the work your force $\mathbf{F}$ does on the crate. (f) Why is the work of your force not equal to the product of the horizontal displacement and the answer to (a)?

FIGURE 7-37 Problem 34.

SECTION 7-6 Work Done by a Spring Force

35E. A spring with a spring constant of 15 N/cm has a cage attached to one end (Fig. 7-38). (a) How much work does the spring force do on the cage if the spring is stretched from its relaxed length by 7.6 mm? (b) How much additional work is done by the spring force if it is stretched by an additional 7.6 mm?

FIGURE 7-38 Exercise 35.

36E. During spring semester at MIT, residents of the parallel buildings of the East Campus dorms battle one another with large catapults that are made with surgical hose mounted on a window frame. A balloon filled with dyed water is placed in a pouch attached to the hose, which is then stretched through the width of the room. Assume that the stretching of the hose obeys Hooke's law and has a spring constant of 100 N/m. If the hose is stretched by 5.00 m and then released, how much work does the force from the hose do on the balloon in the pouch by the time the hose reaches its relaxed length?

37P. The only force acting on a 2.0 kg body as it moves along the positive x axis has an x component $F_x = -6x$ N, where x is in meters. The velocity of the body at $x = 3.0$ m is 8.0 m/s. (a) What is the velocity of the body at $x = 4.0$ m? (b) At what positive value of x will the body have a velocity of 5.0 m/s?

38P. A spring with a pointer attached is hanging next to a scale marked in millimeters. Three different weights are hung from the spring, in turn, as shown in Fig. 7-39. (a) If all weight is removed from the spring, which mark on the scale will the pointer indicate? (b) What is the weight W?

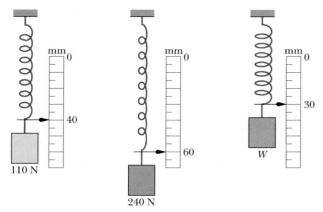

FIGURE 7-39   Problem 38.

39P. In Fig. 7-40, two identical springs, each with a relaxed length of 50 cm and a spring constant of 500 N/m, are connected by a short cord of length 10 cm. The upper spring is attached to the ceiling; a box that weighs 100 N hangs from the lower spring. Two additional cords, each 85 cm long, are also tied to the assembly; they are limp. (a) If the short cord is cut, so that the box then hangs from the springs and the two longer cords, does the box move up or down? (b) How far does the box move? (c) How much total work do the two spring forces (one directly, the other via a cord) do on the box during that move?

FIGURE 7-40   Problem 39.

40P. A 250 g block is dropped onto a vertical spring with spring constant $k = 2.5$ N/cm (Fig. 7-41). The block becomes attached to the spring, and the spring compresses 12 cm before momentarily stopping. While the spring is being compressed, what work is

FIGURE 7-41   Problem 40.

done on the block (a) by its weight and (b) by the spring force? (c) What is the speed of the block just before it hits the spring? (Assume that friction is negligible.) (d) If the speed at impact is doubled, what is the maximum compression of the spring?

SECTION 7-7   Power

41E. The loaded cab of an elevator has a mass of 3.0×10^3 kg and moves 210 m up the shaft in 23 s at constant speed. At what average rate does the force from the cable do work on the cab?

42E. If a ski lift raises 100 passengers averaging 150 lb a height of 500 ft in 60 s, at constant speed, what average power is required of the force making the lift?

43E. An elevator cab in the New York Marriott Marquis has a mass of 4500 kg and can carry a maximum load of 1800 kg. If the cab is moving upward at full load at 3.80 m/s, what power is required of the force moving the cab to maintain that speed?

44E. (a) At a certain instant, a particle experiences a force $\mathbf{F} = (4.0 \text{ N})\mathbf{i} - (2.0 \text{ N})\mathbf{j} + (9.0 \text{ N})\mathbf{k}$ while having a velocity $\mathbf{v} = -(2.0 \text{ m/s})\mathbf{i} + (4.0 \text{ m/s})\mathbf{k}$. What is the instantaneous rate at which the force does work on the particle? (b) At some other time, the velocity consists of only a $\mathbf{j}$ component. If the force is unchanged, and the instantaneous power is -12 W, what is the velocity of the particle just then?

45P. A 100 kg block is pulled at a constant speed of 5.0 m/s across a horizontal floor by an applied force of 122 N directed 37° above the horizontal. What is the rate at which the force does work on the block?

46P. A horse pulls a cart with a force of 40 lb at an angle of 30° above the horizontal and moves along at a speed of 6.0 mi/h. (a) How much work does the force do in 10 min? (b) What is the average power (in horsepower) of the force?

47P. An initially stationary 2.0 kg object accelerates horizontally and uniformly to a speed of 10 m/s in 3.0 s. (a) In that 3.0 s interval, how much work is done on the object by the force accelerating it? What is the instantaneous power due to that force (b) at the end of the interval and (c) at the end of the first half of the interval?

48P. A force of 5.0 N acts on a 15 kg body initially at rest. Compute (a) the work done by the force in the first, second, and third seconds and (b) the instantaneous power due to the force at the end of the third second.

49P. A fully loaded, slow-moving freight elevator has a cab with a total mass of 1200 kg, which is required to travel upward 54 m in 3.0 min. The elevator's counterweight has a mass of only 950 kg. So, the elevator motor must help pull the cab upward. What average power (in horsepower) is required of the force that motor exerts on the cab via the cable?

50P. The force (but not the power) required to tow a boat at constant velocity is proportional to the speed. If a speed of 2.5 mi/h requires 10 hp, how much horsepower does a speed of 7.5 mi/h require?

51P. A 0.30 kg mass sliding on a horizontal frictionless surface is attached to one end of a horizontal spring (with $k = 500$ N/m) whose other end is fixed. The mass has a kinetic energy of 10 J as it passes through its equilibrium position (the point at which the spring force is zero). (a) At what rate is the spring doing work on the mass as the mass passes through its equilibrium position? (b) At what rate is the spring doing work on the mass when the spring is compressed 0.10 m and the mass is moving away from the equilibrium position?

SECTION 7-8 Kinetic Energy at High Speeds

52E. An electron moves through 5.1 cm in 0.25 ns. (a) What is its speed in terms of the speed of light? (b) What is its kinetic energy in electron-volts? (c) What percentage error do you make if you use the classical formula to compute the kinetic energy?

53E. The work–kinetic energy theorem applies to particles at all speeds. How much work, in electron-volts, must be done to increase the speed of an electron from rest to (a) $0.500c$, (b) $0.990c$, and (c) $0.999c$?

54P. An electron has a speed of $0.999c$. (a) What is its kinetic energy? (b) If its speed is increased by 0.05%, by what percentage does its kinetic energy increase?

Electronic Computation

55P. A force $\mathbf{F} = (3.00 \text{ N})\mathbf{i} + (7.00 \text{ N})\mathbf{j} + (7.00 \text{ N})\mathbf{k}$ acts on a 2.00 kg object which moves from an initial position of $\mathbf{d}_i = (3.00 \text{ m})\mathbf{i} - (2.00 \text{ m})\mathbf{j} + (5.00 \text{ m})\mathbf{k}$ to a final position of $\mathbf{d}_f = -(5.00 \text{ m})\mathbf{i} + (4.00 \text{ m})\mathbf{j} + (7.00 \text{ m})\mathbf{k}$ in 4.00 s. Find (a) the work done on the object by the force in the 4.00 s interval, (b) the average power due to the force during the 4.00 s interval, and (c) the angle between vectors $\mathbf{d}_i$ and $\mathbf{d}_f$.

8
Potential Energy and Conservation of Energy

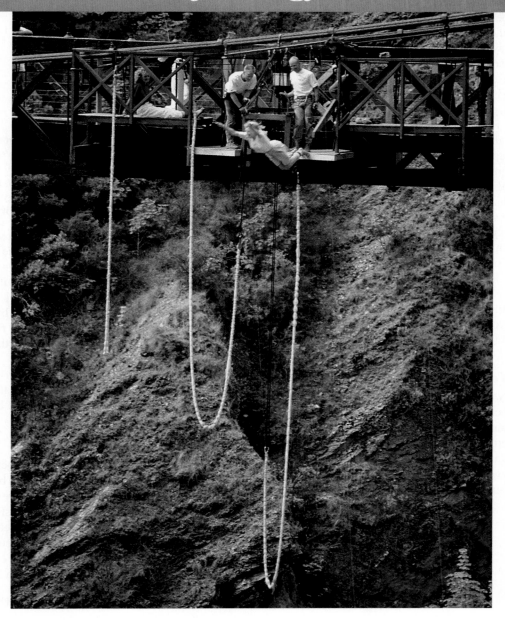

In the dangerous "sport" of bungee-jumping, a participant attaches a specially designed elastic cord to the ankles and leaps from a tall structure. How can we tell just how far down a jumper will go? The answer is, of course, important to the jumper, who is keenly aware of what lies below.❓

8-1 POTENTIAL ENERGY

In this chapter we continue the discussion of energy that we began in Chapter 7. To do so, we define a second form of energy: **potential energy** U is energy that can be associated with the configuration (or arrangement) of a system of objects that exert a force on one another. If the configuration of the system changes, then the potential energy of the system also changes.

One type of potential energy is the **gravitational potential energy** that is associated with the state of separation between objects, which attract one another via the gravitational force. For example, when Vasili Alexeev lifted 562 lb above his head in the 1976 Olympics, he increased the separation between the weights and the Earth. His effort changed the gravitational potential energy of the weights–Earth system by changing the configuration of the system, that is, by shifting the relative locations of the Earth and the weights (Fig. 8-1).

Another type of potential energy is **elastic potential energy,** which is associated with the state of compression or extension of an elastic (springlike) object. The force involved here is the spring force. If you compress or extend a spring, you change the relative locations of the coils within the spring. A spring force resists that change, and the result of your effort is an increase in the elastic potential energy of the spring.

Work and Potential Energy

In Chapter 7 we discussed the relation between work and a change in kinetic energy. Here we discuss the relation between work and a change in potential energy.

Let us throw a tomato upward (Fig. 8-2). We already know that as the tomato rises, the work W_g done on the tomato by its weight (the gravitational force acting on it) is negative because the weight transfers energy *from* the ki-

(a)   (b)

FIGURE 8-1 When Alexeev lifted the weights above his head, he increased the separation between the weights and the Earth and thus changed the configuration of the weights–Earth system from that in (a) to that in (b).

Negative work done   Positive work done

FIGURE 8-2 A tomato is thrown upward. As it rises, its weight does negative work on it, decreasing its kinetic energy by transferring energy to the gravitational potential energy of the tomato–Earth system. As the tomato descends, its weight does positive work on it, increasing its kinetic energy by transferring energy from the gravitational potential energy of the system.

netic energy of the tomato. We can now finish the story by saying that this energy is transferred by the weight *to* the gravitational potential energy of the tomato–Earth system.

The tomato slows, stops, and then begins to fall back down because of its weight. During the fall, the transfer is reversed: the work W_g done on the tomato by its weight is now positive—the weight transfers energy *from* the gravitational potential energy of the tomato–Earth system *to* the kinetic energy of the tomato.

For either rise or fall, the change ΔU in the gravitational potential energy is defined to equal the negative of the work done on the tomato by its weight. Using the general symbol W for work, we write this as

$$\Delta U = -W. \tag{8-1}$$

The same relationship applies to a block–spring system, as in Fig. 8-3. If we abruptly shove the block to send it rightward, the spring force does negative work on the block during the rightward motion, transferring energy from the kinetic energy of the block to the elastic potential energy of the spring. The block slows, stops, and then begins to move in the opposite direction because of the spring force. The transfer of energy is then reversed.

Conservative and Nonconservative Forces

Let us list the key elements of the two situations we just discussed:

1. The *system* consists of two or more objects.

2. A *force* acts between a particle-like object (tomato or block) in the system and the rest of the system.

3. When the system configuration changes, the force does *work* (call it W_1) on the particle-like object, transferring

FIGURE 8-3 A block, attached to a spring and initially at rest at $x = 0$, is set in motion toward the right. (a) As the block moves rightward (as indicated by the arrow), the spring force does negative work on it, decreasing its kinetic energy by transferring energy to the elastic potential energy of the spring. The block momentarily stops when it has zero kinetic energy. (b) Then, as the block moves back toward $x = 0$, the spring force does positive work on it, increasing its kinetic energy by transferring energy from the elastic potential energy.

energy between the kinetic energy K of the object and some other form of energy of the system.

4. When the configuration change is reversed, the force reverses the energy transfer, doing work W_2 in the process.

If $W_1 = -W_2$ is always true, then the other form of energy is a potential energy, and the force is said to be a **conservative force.** As you might suspect, the gravitational force and the spring force are both conservative (since otherwise we could not have spoken of gravitational potential energy and elastic potential energy, as above.)

Not all forces are conservative. For an example, let us send a block sliding across a floor that is not frictionless. During the sliding, a kinetic frictional force from the floor does negative work on the block, slowing the block by transferring energy from its kinetic energy to thermal energy of the block–floor system. We know from experiment that this energy transfer cannot be reversed (thermal energy cannot be transferred back to kinetic energy of the block by the kinetic frictional force). So, although we have a system, a force that acts between parts of the system, and a transfer of energy by the force, the force is not conservative (it is **nonconservative**). Thus the thermal energy of the block–floor system is not a potential energy.

When only conservative forces act on a particle-like object, we can greatly simplify otherwise difficult problems involving motion of the object. The next section, in which we develop a test for identifying conservative forces, provides one means for simplifying such problems.

8-2 PATH INDEPENDENCE OF CONSERVATIVE FORCES

The primary test for determining whether a force is conservative or nonconservative is this: let the force act (alone) on a particle that moves along a *closed path* begin-

ning at some initial position and eventually returning to that position (so that the particle makes a *round trip* beginning and ending at the initial position). The force is conservative only if the total energy it transfers to and from the particle during the round trip is zero. In other words:

The net work done by a conservative force on a particle moving around any closed path is zero.

We know from experiment that the gravitational force passes this test. An example is the tossed tomato of Fig. 8-2. The tomato leaves the launch point with a speed v_0 and kinetic energy $\frac{1}{2}mv_0^2$. The gravitational force acting on the tomato slows it, stops it, and then makes it fall back down. When the tomato returns to the launch point, it again has speed v_0 and kinetic energy $\frac{1}{2}mv_0^2$. Thus the gravitational force transfers as much energy *from* the tomato during the ascent as it transfers *to* the tomato during the descent back to the launch point. The net work done on the tomato by the gravitational force during the round trip is zero.

Figure 8-4a shows an arbitrary round trip for a particle acted upon by a single force. The particle moves from an initial point a to point b along path 1 and then back to point a along path 2. The force does work on the particle as the particle moves along each path. Without worrying about where positive work is done and where negative work is done, let us just represent the work done from a to b along path 1 as $W_{ab,1}$ and the work done from b back to a along path 2 as $W_{ba,2}$. If the force is conservative, then the net work done during the round trip must be zero:

$$W_{ab,1} + W_{ba,2} = 0,$$

and thus

$$W_{ab,1} = -W_{ba,2}. \tag{8-2}$$

In words, the work done along the outward path must be the negative of the work done along the path back.

Let us now consider the work $W_{ab,2}$ done on the particle by the force as the particle moves from a to b along path 2, as indicated in Fig. 8-4b. If the force is conservative, that

FIGURE 8-4 (a) A particle, acted on by a conservative force, moves in a round trip from point a to point b along path 1 and then back to point a along path 2. (b) The particle can move from point a to point b by following either path 1 or path 2.

work is the negative of $W_{ba,2}$:

$$W_{ab,2} = -W_{ba,2}. \tag{8-3}$$

Substituting $W_{ab,2}$ for $-W_{ba,2}$ in Eq. 8-2, we obtain

$$W_{ab,1} = W_{ab,2}. \tag{8-4}$$

This simple equation is a powerful one, because it allows us to simplify difficult problems when only a conservative force is involved. It tells us that the work done on a particle by a conservative force as the particle moves from some initial point a to some other point b does not depend on *which* path is taken.

> The work done by a conservative force on a particle moving between two points does not depend on the path taken by the particle.

Suppose you need to calculate the work done by a conservative force along a given path between two points, and the calculation is difficult. You can find the work by using another path between those two points for which the calculation is easier. Sample Problem 8-1 gives an example.

SAMPLE PROBLEM 8-1

Figure 8-5*a* shows a 2.0 kg block of slippery cheese that slides along a frictionless track from point a to point b. The cheese travels through a total distance of 2.0 m along the track, and a net vertical distance of 0.80 m. How much work is done on the cheese by its weight during the slide?

SOLUTION: From Eq. 7-16, we know that the work done by a weight force $m\mathbf{g}$, over a displacement $\mathbf{d}$ at angle ϕ to $m\mathbf{g}$, is given by $W = mgd \cos \phi$. But using that equation here appears to be impossible because we do not know the actual shape of the track. And even if we did, the calculation would be difficult because the angle ϕ is not constant over the path taken by the cheese.

However, because weight is a conservative force, we can find the work by choosing another path between a and b—one that makes the calculation easy. Let us choose the dashed path in Fig. 8-5*b*; it consists of two straight segments. Along the horizontal segment, the angle ϕ is a constant 90°. Even though we do not know the displacement along that horizontal segment, Eq. 7-16 tells us that the work W_h done there is

$$W_h = mgd \cos 90° = 0.$$

Along the vertical segment, the displacement d is 0.80 m and, with $m\mathbf{g}$ and $\mathbf{d}$ both downward, the angle ϕ is a constant 0°. So Eq. 7-16 gives us, for the work W_v done along the vertical part of the dashed path,

$$W_v = mgd \cos 0°$$
$$= (2.0 \text{ kg})(9.8 \text{ m/s}^2)(0.80 \text{ m})(1) = 15.7 \text{ J}.$$

The total work done on the cheese by its weight as the cheese moves from point a to point b along the dashed path is then

$$W = W_h + W_v = 0 + 15.7 \text{ J} \approx 16 \text{ J}. \quad \text{(Answer)}$$

This is also the work done as the cheese moves along the track from a to b.

FIGURE 8-5 Sample Problem 8-1. (*a*) A block of cheese slides along a frictionless track from point a to point b. (*b*) Finding the work done on the cheese by its weight is easier along the dashed path than along the actual path taken by the cheese; the result is the same for both paths.

CHECKPOINT **1:** The figure shows three paths connecting points a and b. A single force $\mathbf{F}$ does the indicated work on a particle moving along each path in the indicated direction. On the basis of this information, is force $\mathbf{F}$ conservative?

8-3 DETERMINING POTENTIAL ENERGY VALUES

Consider now a particle-like object (say, a melon) that is part of a system (say, a melon–Earth system) in which a conservative force $\mathbf{F}$ acts. When that force does work W on the object, the change ΔU in the potential energy associated with the system is the negative of the work done. We wrote this fact as Eq. 8-1 ($\Delta U = -W$). For the most general case, in which the force may vary with position, we may write the work W as in Eq. 7-27:

$$W = \int_{x_i}^{x_f} F(x) \, dx. \tag{8-5}$$

This equation gives the work done by the force when the object moves from point x_i to point x_f, changing the configuration of the system. (Because F is a conservative force, the work is the same for all paths between those two points.)

Substituting Eq. 8-5 into Eq. 8-1, we find that the change in potential energy due to the change in configuration is

$$\Delta U = -\int_{x_i}^{x_f} F(x) \, dx. \qquad (8\text{-}6)$$

Gravitational Potential Energy

Now consider a particle with mass m moving vertically along a y axis. As the particle moves from point y_i to point y_f, its weight $m\mathbf{g}$ does work on it. To find the corresponding change in the gravitational potential energy of the particle–Earth system, we change the integration in Eq. 8-6 to be along the y axis and we substitute $-mg$ for force F. We then have

$$\Delta U = -\int_{y_i}^{y_f} (-mg) \, dy = mg \int_{y_i}^{y_f} dy = mg \left[y \right]_{y_i}^{y_f},$$

which yields

$$\Delta U = mg(y_f - y_i) = mg \, \Delta y. \qquad (8\text{-}7)$$

Only a change ΔU in gravitational potential energy (or any other type of potential energy) is physically important. However, to simplify a calculation or a discussion, we can say that a certain gravitational potential energy U is associated with any given configuration of the system, with the particle at a given position y. To do so, we rewrite Eq. 8-7 as

$$U - U_i = mg(y - y_i). \qquad (8\text{-}8)$$

Then we take U_i to be the gravitational potential energy of the system when it is in a **reference configuration,** with the particle at a **reference point** y_i. Usually, we set $U_i = 0$ and $y_i = 0$. Doing this changes Eq. 8-8 to

$$U - 0 = mg(y - 0),$$

which gives us

$$U = mgy \qquad \text{(gravitational potential energy).} \qquad (8\text{-}9)$$

Equation 8-9 tells us that the gravitational potential energy associated with a particle–Earth system depends on the vertical position y (or height) of the particle relative to the reference position of $y = 0$, not on the horizontal position.

Elastic Potential Energy

We next consider the block–spring system shown in Fig. 8-3, with the block moving on the end of a spring of spring constant k. As the block moves from point x_i to point x_f, the spring force $F = -kx$ does work on the block. To find the

corresponding change in the elastic potential energy of the block–spring system, we substitute $-kx$ for $F(x)$ in Eq. 8-6. We then have

$$\Delta U = -\int_{x_i}^{x_f} (-kx) \, dx = k \int_{x_i}^{x_f} x \, dx = \tfrac{1}{2}k \left[x^2 \right]_{x_i}^{x_f},$$

and

$$\Delta U = \tfrac{1}{2}kx_f^2 - \tfrac{1}{2}kx_i^2. \qquad (8\text{-}10)$$

To associate a potential energy U with any given configuration of the system, with the block at position x, we set the reference point for the block as $x_i = 0$, which is always at the equilibrium position of the block. And we set the corresponding elastic potential energy of the system as $U_i = 0$. Thus Eq. 8-10 becomes

$$U - 0 = \tfrac{1}{2}kx^2 - 0,$$

which gives us

$$U(x) = \tfrac{1}{2}kx^2 \qquad \text{(elastic potential energy).} \qquad (8\text{-}11)$$

PROBLEM SOLVING TACTICS

TACTIC 1: *Using the Term "Potential Energy"*

A potential energy is associated with a system as a whole. However, you might see statements that associate it with only part of the system. For example, you might read, "An apple hanging in a tree has a gravitational potential energy of 30 J." Such statements are often acceptable, but you should always keep in mind that the potential energy is actually associated with a system—here the apple–Earth system. Also keep in mind that assigning a particular potential energy value, such as 30 J here, to an object or even a system makes sense *only* if the reference potential energy value is known.

SAMPLE PROBLEM 8-2

A 2.0 kg sloth clings to a limb that is 5.0 m above the ground (Fig. 8-6).

(a) What is the gravitational potential energy U of the sloth–Earth system if we take the reference point $y = 0$ to be (1) at the ground, (2) at the bottom of a balcony that is 3.0 m above the ground, (3) at the limb, and (4) 1.0 m above the limb? Take the gravitational potential energy to be zero at $y = 0$.

SOLUTION: With Eq. 8-9, we can calculate U for each choice of $y = 0$. For example, for (1) the sloth is initially at $y = 5.0$ m, and

$$U = mgy = (2.0 \text{ kg})(9.8 \text{ m/s}^2)(5.0 \text{ m})$$

$$= 98 \text{ J} \qquad \text{(Answer)}$$

For the other choices, the values of U are

(2) $U = mgy = mg(2.0 \text{ m}) = 39 \text{ J},$

(3) $U = mgy = mg(0) = 0 \text{ J},$

(4) $U = mgy = mg(-1.0 \text{ m}) = -19.6 \text{ J}$
$$\approx -20 \text{ J.} \qquad \text{(Answer)}$$

(b) The sloth drops to the ground. For each choice of reference point, what is the change in the potential energy of the sloth–Earth system due to the fall?

SOLUTION: For all four situations, we have $\Delta y = -5.0$ m. So for (1) to (4), Eq. 8-7 tells us that

$$\Delta U = mg \, \Delta y = (2.0 \text{ kg})(9.8 \text{ m/s}^2)(-5.0 \text{ m})$$
$$= -98 \text{ J.} \qquad \text{(Answer)}$$

Thus, although the value of U depends on the choice of where we let $y = 0$, the *change* in potential energy does not. Remember, only the change ΔU in potential energy is physically important—not the value of U, which depends on arbitrary choices about the reference configuration.

FIGURE 8-6 Sample Problem 8-2. Four choices of reference point $y = 0$. Each y axis is marked in units of meters.

CHECKPOINT 2: A particle is to move along the x axis from $x = 0$ to x_1 while a conservative force, directed along the x axis, acts on the particle. The figure shows three situations in which the magnitude of that force varies with x. The force has the same maximum magnitude F_1 in all three situations. Rank the situations according to the change in the associated potential energy during the particle's motion, most positive first, most negative last.

(1)   (2)   (3)

8-4 CONSERVATION OF MECHANICAL ENERGY

The **mechanical energy** E of a system is the sum of its potential energy U and the kinetic energy K of the objects within it:

$$E = K + U \quad \text{(mechanical energy).} \qquad (8\text{-}12)$$

In this section, we examine what happens to this mechanical energy when a conservative force acts within the system. (Later we shall consider the effects of nonconservative forces.)

When a conservative force does work W on an object within the system, it transfers energy between kinetic en-

In olden days, a native Alaskan would be tossed via a blanket to be able to see farther over the flat terrain. Nowadays, it is done just for fun. During the ascent of the child in the photograph, energy is transferred from kinetic energy to gravitational potential energy. The maximum height is reached when that transfer is complete. Then the transfer is reversed during the fall.

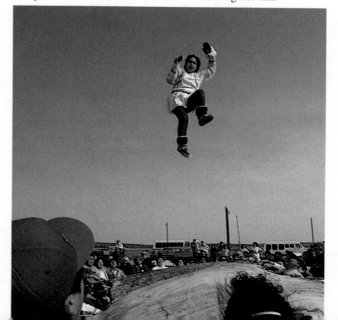

ergy K of the object and potential energy U of the system. From Eq. 7-4, the change ΔK in kinetic energy is

$$\Delta K = W \qquad (8\text{-}13)$$

and from Eq. 8-1, the change ΔU in potential energy is

$$\Delta U = -W. \qquad (8\text{-}14)$$

Combining Eqs. 8-13 and Eq. 8-14, we find that

$$\Delta K = -\Delta U. \qquad (8\text{-}15)$$

In other words, one of these forms of energy increases exactly as much as the other decreases.

We can rewrite Eq. 8-15 as

$$K_2 - K_1 = -(U_2 - U_1), \qquad (8\text{-}16)$$

where the subscripts 1 and 2 refer to two different instants during the particle's motion, and thus to two different configurations of the system. Rearranging Eq. 8-16 yields

$$K_2 + U_2 = K_1 + U_1 \qquad \begin{array}{l}\text{(conservation of}\\ \text{mechanical energy).}\end{array} \qquad (8\text{-}17)$$

The left side of Eq. 8-17 is the mechanical energy of the system at one instant; the right side is the mechanical energy at another instant and in a different configuration. The equation tells us that these two energies are equal:

When only a conservative force acts within a system, the kinetic energy and potential energy can change. However, their sum, the mechanical energy E of the system, does not change.

This result is called the **principle of conservation of mechanical energy.** (Now you can see where *conservative* forces got their name.) With the aid of Eq. 8-15, we can write this principle in one more form, as

$$\Delta E = \Delta K + \Delta U = 0. \qquad (8\text{-}18)$$

The principle of conservation of mechanical energy allows us to solve problems that would be quite difficult to solve using only Newton's laws:

When the mechanical energy of a system is conserved, we can relate the total of kinetic energy and potential energy at one instant to that at another instant *without considering the intermediate motion* and *without finding the work done by the forces involved.*

Figure 8-7 shows an example in which the principle of conservation of mechanical energy can be applied: as a

pendulum swings, the energy of the pendulum–Earth system is transferred back and forth between kinetic energy K and gravitational potential energy U, with the sum $K + U$ being constant. If we are given the gravitational potential energy when the pendulum bob is at its highest point (Fig. 8-7c), we can find the kinetic energy of the bob at the lowest point (Fig. 8-7e) by using Eq. 8-17.

For example, let us choose the lowest point as the reference point and set the corresponding gravitational potential energy $U_2 = 0$. Suppose then that the potential energy at the highest point is $U_1 = 20$ J relative to the reference point. Because the bob momentarily stops at its highest point, the kinetic energy there is $K_1 = 0$. Substituting these values into Eq. 8-17 gives us the kinetic energy K_2 at the lowest point:

$$K_2 + 0 = 0 + 20\text{ J} \quad \text{or} \quad K_2 = 20\text{ J}.$$

Note that we get this result without considering the motion between the highest and lowest points (such as in Fig. 8-7d) and without finding the work done by any forces involved in the motion.

CHECKPOINT 3: The figure shows four situations— one in which an initially stationary block is dropped and three in which the block is allowed to slide down frictionless ramps. (a) Rank the situations according to the kinetic energy of the block at point B, greatest first. (b) Rank them according to the speed of the block at point B, greatest first.

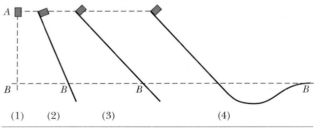

(1)   (2)   (3)   (4)

SAMPLE PROBLEM 8-3

In Fig. 8-8, a child of mass m is released from rest at the top of a water slide, at height $h = 8.5$ m above the bottom of the slide. Assuming that the slide is frictionless because of the water on it, find the child's speed at the bottom of the slide.

SOLUTION: If we use only the physics of Chapters 2 through 6, this problem is impossible to solve because we are given no information about the shape of the slide. However, with the physics of Chapters 7 and 8, it is easy. We first note that, in the absence of friction, the only force exerted on the child by the slide is a normal force, which is always perpendicular to the surface of the slide. Since the child always moves *along* the slide, this force is always perpendicular to

FIGURE 8-7 A pendulum, with its mass concentrated in a bob at the lower end, swings back and forth. One full cycle of the motion is shown. During the cycle the values of the potential and kinetic energies of the pendulum–Earth system vary as the bob rises and falls, but the mechanical energy E of the system remains constant. The energy E can be described as continuously shifting between the kinetic and potential forms. In stages (a) and (e), all the energy is kinetic energy. The bob then has its greatest speed and is at its lowest point. In stages (c) and (g), all the energy is potential energy. The bob then has zero speed and is at its highest point. In stages (b), (d), (f), and (h), half the energy is kinetic energy and half is potential energy. If the swinging involved a frictional force at the point where the pendulum is attached to the ceiling, or a drag force due to the air, then E would not be conserved, and eventually the pendulum would stop.

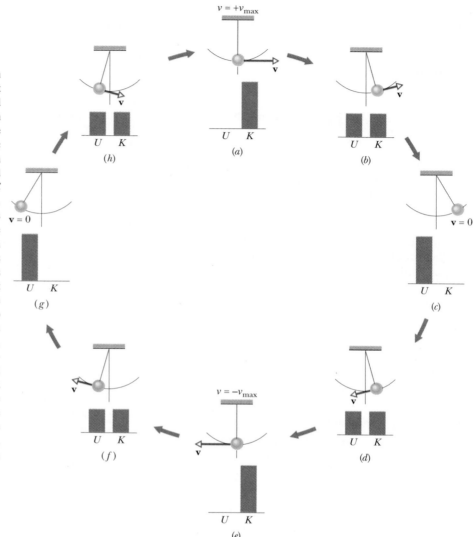

the child's displacement and thus cannot do work on the child. The only force that does work is the weight mg of the child, a conservative force. We therefore have a child–Earth system for which the mechanical energy E is conserved throughout the child's motion.

Thus, the mechanical energy E_b at the bottom of the slide

FIGURE 8-8 Sample Problem 8-3. A child slides down a water slide as she descends a height h.

and the mechanical energy E_t at the top are equal: $E_b = E_t$. Writing this conservation of energy equation in the form of Eq. 8-17 gives us

$$K_b + U_b = K_t + U_t,$$

or $\frac{1}{2}mv_b^2 + mgy_b = \frac{1}{2}mv_t^2 + mgy_t.$

Dividing by m and rearranging yield

$$v_b^2 = v_t^2 + 2g(y_t - y_b).$$

Putting $v_t = 0$ and $y_t - y_b = h$ leads to

$$v_b = \sqrt{2gh} = \sqrt{(2)(9.8 \text{ m/s}^2)(8.5 \text{ m})}$$
$$= 13 \text{ m/s.} \qquad \text{(Answer)}$$

This is the same speed that the child would reach if she fell 8.5 m. On an actual slide, some frictional forces would act and the child would not be moving quite so fast.

This problem is hard to solve directly with Newton's laws. Using conservation of mechanical energy makes the solution much easier. However, if you were asked to find the time taken for the child to reach the bottom of the slide, energy methods would be of no use; you would need to know the shape of the slide and you would have a difficult problem.

SAMPLE PROBLEM 8-4

The spring of a spring gun is compressed a distance d of 3.2 cm from its relaxed state, and a ball of mass $m = 12$ g is put in the barrel. With what speed will the ball leave the barrel when the gun is fired? The spring constant k is 7.5 N/cm. Assume no friction and a horizontal gun barrel. Also assume that the ball leaves the spring and the spring stops when the spring reaches its relaxed length.

SOLUTION: Let E_i be the mechanical energy of the gun–ball system in the initial state (before the gun has been fired), and E_f be the system's mechanical energy in the final state (as the ball leaves the barrel). Initially, the mechanical energy is the spring's potential energy $U_i = \frac{1}{2}kd^2$. In the final state, the mechanical energy is the ball's kinetic energy $K_f = \frac{1}{2}mv^2$. Because mechanical energy is conserved, we have

$$E_i = E_f,$$

$$U_i + K_i = U_f + K_f,$$

and $$\tfrac{1}{2}kd^2 + 0 = 0 + \tfrac{1}{2}mv^2.$$

Solving for v yields

$$v = d\sqrt{\frac{k}{m}} = (0.032 \text{ m}) \sqrt{\frac{750 \text{ N/m}}{12 \times 10^{-3} \text{ kg}}}$$

$$= 8.0 \text{ m/s.} \qquad \text{(Answer)}$$

SAMPLE PROBLEM 8-5

A 61.0 kg bungee-cord jumper is on a bridge 45.0 m above a river. In its relaxed state, the elastic bungee cord has length $L = 25.0$ m. Assume that the cord obeys Hooke's law, with a spring constant of 160 N/m.

(a) If the jumper stops before reaching the water, what is the height h of her feet above the water at her lowest point?

SOLUTION: As shown in Fig. 8-9, let d be the amount by which the cord has stretched when she momentarily stops at her lowest point. As a result of her jump, the change ΔU_g in her gravitational potential energy is

$$\Delta U_g = mg\,\Delta y = -mg(L + d),$$

where m is her mass. The change in the elastic potential energy U_e of the cord is

$$\Delta U_e = \tfrac{1}{2}kd^2.$$

Her kinetic energy K is zero both initially and when she stops.

From Eq. 8-18 and the above, we have for the cord–jumper–Earth system,

$$\Delta K + \Delta U_e + \Delta U_g = 0$$

$$0 + \tfrac{1}{2}kd^2 - mg(L + d) = 0$$

$$\tfrac{1}{2}kd^2 - mgL - mgd = 0.$$

Inserting the given data, we obtain

$$\tfrac{1}{2}(160 \text{ N/m})d^2 - (61.0 \text{ kg})(9.8 \text{ m/s}^2)(25.0 \text{ m})$$
$$- (61.0 \text{ kg})(9.8 \text{ m/s}^2)d = 0,$$

which we solve for d with the quadratic formula, getting

$$d = 17.9 \text{ m.}$$

(The quadratic formula also gives a negative value for d, which has no meaning here.) Her feet are then a distance of $(L + d) = 42.9$ m below their initial height. Thus

$$h = 45.0 \text{ m} - 42.9 \text{ m} = 2.1 \text{ m.} \qquad \text{(Answer)}$$

If she happens to be especially tall, the jumper could be in for a dunking.

(b) What is the net force on her at her lowest point (in particular, is it zero)?

SOLUTION: Her weight $m\mathbf{g}$ acts downward and has magnitude $mg = 597.8$ N. The upward force exerted on her by the cord at the moment of stopping is given by Hooke's law, $F = -kx$, where x is the displacement of the free end of the cord. Here, the displacement is downward, so $x = -d$ and

$$F = -kx = -(160 \text{ N/m})(-17.9 \text{ m}) = 2864 \text{ N.}$$

The net force on the jumper is then

$$2864 \text{ N} - 597.8 \text{ N} \approx 2270 \text{ N.} \qquad \text{(Answer)}$$

FIGURE 8-9 Sample Problem 8-5. A bungee-cord jumper at the lowest point of the jump.

Thus at the lowest point, where she momentarily stops, there is a net upward force of 2270 N (almost four times her weight) acting on her. This force will yank her back upward.

PROBLEM SOLVING TACTICS

TACTIC 2: *Conservation of Mechanical Energy*

Asking the following questions will help you to solve problems involving the conservation of mechanical energy.

For what system is mechanical energy conserved? You should be able to separate your system from its environment. Imagine drawing a closed surface such that whatever is inside is your system and whatever is outside is the environment of that system. In Sample Problem 8-3 the system is the *child + Earth*. In Sample Problem 8-4, it is the *ball + gun*. In Sample Problem 8-5, it is the *cord + jumper + Earth*.

Is friction or drag present? If friction or drag is present, mechanical energy is not conserved.

Is your system isolated? Conservation of mechanical energy applies only to isolated systems. That means that no *external forces* (forces exerted by objects outside the system) should do work on the objects in the system. In Sample Problem 8-4, if you chose the ball alone as your system, it is not isolated because the spring force does work on it. Thus the mechanical energy of the ball is not conserved.

What are the initial and final states of your system? The system changes from some initial state (or configuration) to some final state. You apply the principle of conservation of mechanical energy by saying that E has the same value in both these states. Be very clear about what these two states are.

8-5 READING A POTENTIAL ENERGY CURVE

Once again consider a particle that is part of a system in which a conservative force acts. This time suppose that the particle is constrained to move along an x axis while the conservative force does work on it. We can learn a lot about the motion of the particle from a plot of the system's potential energy $U(x)$. However, before we discuss such plots, we need one more relationship.

Finding the Force Analytically

Equation 8-6 tells us how to find the potential energy $U(x)$ in a one-dimensional situation if we know the force $F(x)$. But now we want to go the other way. That is, we know the potential energy function $U(x)$ and want to find the force.

For one-dimensional motion, the work W done by a force that acts on a particle as the particle moves through a distance Δx is $F(x)\,\Delta x$. We can then write Eq. 8-1 as

$$\Delta U(x) = -W = -F(x)\,\Delta x.$$

Solving for $F(x)$ and passing to the differential limit yield

$$F(x) = -\frac{dU(x)}{dx} \qquad \begin{array}{c}\text{(one-dimensional}\\ \text{motion).}\end{array} \qquad (8\text{-}19)$$

We can check this result by putting $U(x) = \frac{1}{2}kx^2$, which is the elastic potential energy function for a spring force. Equation 8-19 then yields, as expected, $F(x) = -kx$, which is Hooke's law. Similarly, we can substitute $U(x) = mgx$, which is the gravitational potential energy function for a particle of mass m at height x above the Earth's surface. Equation 8-19 then yields $F = -mg$, which is the weight that acts on the particle.

The Potential Energy Curve

Figure 8-10a is a plot of a potential energy function $U(x)$ for a system in which a particle is in one-dimensional motion while a conservative force $F(x)$ does work on it. We can easily find $F(x)$ by (graphically) taking the slope of the $U(x)$ curve at various points, as Eq. 8-19 tells us to do. Figure 8-10b is a plot of $F(x)$ found in this way.

Turning Points

In the absence of a nonconservative force, the mechanical energy E of the system has a constant value given by

$$U(x) + K(x) = E. \qquad (8\text{-}20)$$

Here $K(x)$ is the *kinetic energy function* of the particle (this $K(x)$ gives the kinetic energy as a function of the particle's location x). We may rewrite Eq. 8-20 as

$$K(x) = E - U(x). \qquad (8\text{-}21)$$

Suppose that E (which has a constant value, remember) happens to be 5.0 J. It would be represented in Fig. 8-10 by a horizontal line that runs through the value 5.0 J on the energy axis. (It is shown in Fig. 8-10a.)

Equation 8-21 tells us how to determine the kinetic energy K for any location x of the particle: on the $U(x)$ curve, find U for that location x and then subtract U from E. For example, if the particle is at any point to the right of x_5, then $K = 1.0$ J. The value of K is greatest (5.0 J) when the particle is at x_2, and least (0 J) when the particle is at x_1.

Since K can never be negative (because v^2 is always positive), the particle can never move to the left of x_1, where $E - U$ is negative. As the particle moves toward x_1 from x_2, K decreases (the particle slows) until $K = 0$ at x_1 (the particle stops).

Note that when the particle reaches x_1, the force on the particle, given by Eq. 8-19, is positive (because the slope dU/dx is negative). This means that the particle does not remain at x_1 but instead begins to move to the right, opposite its earlier motion. Hence x_1 is a **turning point,** a place where $K = 0$ and the particle changes direction. There is

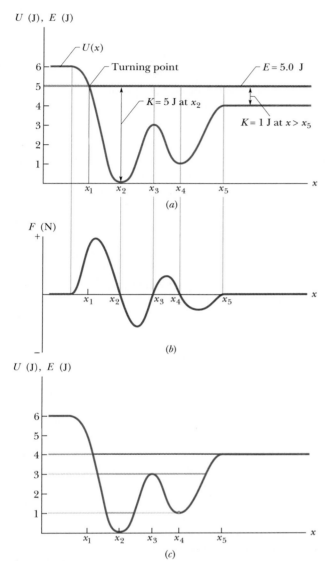

FIGURE 8-10 (*a*) A plot of $U(x)$, the potential energy function of a system containing a particle confined to move along the x axis. There is no friction, so mechanical energy is conserved. (*b*) A plot of the force $F(x)$ acting on the particle, derived from the potential energy plot by taking its slope at various points. (*c*) The $U(x)$ plot of (*a*) with three possible values of E shown.

no turning point (where $K = 0$) on the right side of the graph. When the particle heads toward increasing values of x, it will continue indefinitely.

Equilibrium Points

Figure 8-10*c* shows three additional values for E superposed on the plot of the potential energy $U(x)$. Let us see how they would change the situation. If $E = 4.0$ J (purple line), the turning point shifts from x_1 to a point between x_1 and x_2. Also, at any point to the right of x_5, the system's

mechanical energy is equal to its potential energy; thus, the particle has no kinetic energy and (by Eq. 8-19) no force acts on it, and so it must be stationary there. A particle at such a position is said to be in **neutral equilibrium.** (A marble placed on a horizontal tabletop is in that state.)

If $E = 3.0$ J (pink line), there are two turning points: one is between x_1 and x_2, and the other is between x_4 and x_5. In addition, x_3 is a point at which $K = 0$. If the particle is located exactly there, the force on it is also zero, and the particle remains stationary. However, if it is displaced even slightly in either direction, a nonzero force pushes it farther in the same direction, and the particle continues to move. A particle at such a position is said to be in **unstable equilibrium.** (A marble balanced on top of a bowling ball is an example.)

Next consider the particle's behavior if $E = 1.0$ J (green line). If we place it at x_4, it is stuck there. It cannot move left or right on its own because to do so would require a negative kinetic energy. If we push it slightly left or right, a restoring force appears that moves it back to x_4. A particle at such a position is said to be in **stable equilibrium.** (A marble placed at the bottom of a hemispherical bowl is an example.) If we place the particle in the cuplike *potential well* centered at x_2, it can still move left and right somewhat, but only partway to x_1 or x_3.

CHECKPOINT **4:** The figure gives the potential energy function $U(x)$ for a system in which a particle is in one-dimensional motion. (a) Rank regions *AB*, *BC*, and *CD* according to the magnitude of the force on the particle, greatest first. (b) What is the direction of the force when the particle is in region *AB*?

8-6 WORK DONE BY NONCONSERVATIVE FORCES

When only a conservative force does work on an object within a system, the mechanical energy E of the system stays constant. However, when a nonconservative force does work on an object, the mechanical energy of the system changes. In that situation we *cannot* use Eqs. 8-17 and 8-18. What, then, can we say about the energy changes? We answer this question by examining nonconservative forces of two types: an applied force (applied by, say, you) and a kinetic frictional force.

Work Done by an Applied Force

Suppose you push a bowling ball directly upward. During the upward motion and before the ball leaves your hand, your applied force does work W_{app} on the ball while the weight of the ball does work W_g on the ball. From observation we find that the ball can be treated as a particle. Thus, we can relate the work done on the ball to the change ΔK in its kinetic energy by using the work–kinetic energy theorem (Eq. 7-15):

$$W_{app} + W_g = \Delta K. \tag{8-22}$$

Because the weight of the ball is a conservative force, we can use Eq. 8-1 to relate W_g to the corresponding change ΔU in the gravitational potential energy of the ball–Earth system:

$$W_g = -\Delta U. \tag{8-23}$$

Substituting for W_g in Eq. 8-22, we get

$$W_{app} = \Delta K + \Delta U. \tag{8-24}$$

The right side of this equation is the change ΔE in the mechanical energy of the ball–Earth system, so

$$W_{app} = \Delta E. \tag{8-25}$$

Equation 8-25 says that the work done by your applied force changes the mechanical energy of the ball–Earth system. Specifically, your force transfers energy to the kinetic energy of the ball (you make it move faster) and to the gravitational potential energy of ball–Earth system (you increase the separation between the ball and the Earth).

Previously we defined work as being energy transferred to or from an object via a force. Now we can expand that definition to include situations in which the object is part of a system:

Work is energy transferred to or from a system via a force.

In some situations, the system consists of just an object, as in Chapter 7; in other situations, the system consists of two (or more) objects, as in the ball–Earth system here.

Work Done by a Kinetic Frictional Force

Figure 8-11 shows a block of mass m and initial speed v_0 sliding across a floor that is not frictionless. A kinetic frictional force $\mathbf{f}_k$ stops the block during displacement $\mathbf{d}$. From experiment we know that this force reduces the kinetic energy K of the block by transferring energy to thermal energy E_{int} of the block and the portion of the floor along which the block slides. Thermal energy is said to be an

FIGURE 8-11   A block of mass m and initial speed v_0 slides across a floor. A kinetic frictional force $\mathbf{f}_k$ does work on the block, stopping it during displacement $\mathbf{d}$.

internal energy of an object because it involves random motions of the atoms and molecules within the object; hence, we represent it with the symbol E_{int}.

Because we also know from experiment that the transfer to thermal energy is not reversible, we say that the block's kinetic energy is **dissipated** by force $\mathbf{f}_k$. The work W_f done on the block by $\mathbf{f}_k$ is not the entire amount of dissipated energy, but only the part that is transferred *from* the block to the floor. The rest of the dissipated energy remains within the block as thermal energy.

As an example, suppose we find experimentally that the change in kinetic energy of the block is $\Delta K = -100$ J. Then 100 J is dissipated by $\mathbf{f}_k$, and the change in thermal energy of the block and floor is $\Delta E_{int} = +100$ J. Suppose we also find experimentally that 40 J is transferred (internally) to thermal energy of the block and 60 J is transferred (externally) to thermal energy of the floor. Then the work W_f done on the block by $\mathbf{f}_k$ is $W_f = -60$ J.

The situation in Fig. 8-11 is comparable to that in Fig. 7-2 (Section 7-3), so Eq. 7-8 applies here with two changes. The force magnitude is now f_k instead of F, and the angle ϕ between the displacement and the force is 180°. With those changes, Eq. 7-8 becomes

$$\Delta K = f_k d \cos 180° = -f_k d. \tag{8-26}$$

Thus the product $-f_k d$ tells us the amount by which $\mathbf{f}_k$ decreases the kinetic energy K of the block in Fig. 8-11. If a potential energy were also involved (say, the block were sliding along a ramp), the product $-f_k d$ would tell us the amount by which $\mathbf{f}_k$ decreases (by dissipating) the mechanical energy E of the system:

$$\Delta E = -f_k d \qquad \begin{array}{c}\text{(dissipated mechanical}\\ \text{energy).}\end{array} \tag{8-27}$$

Only a portion of this dissipated energy is transferred from the block to the floor, and only that portion is equal to the work W_f done by $\mathbf{f}_k$. *Without additional information, we cannot determine W_f.*

SAMPLE PROBLEM 8-6

Figure 8-12 shows a disabled robot of mass $m = 40$ kg being dragged by a cable up the 30° inclined wall inside a volcano crater. The applied force $\mathbf{F}$ exerted on the robot by the cable has a magnitude of 380 N. The kinetic frictional force $\mathbf{f}_k$ acting on the robot has a magnitude of 140 N. The robot moves through a displacement $\mathbf{d}$ of magnitude 0.50 m along the crater wall.

(a) How much of the mechanical energy of the robot–Earth system is dissipated by the kinetic frictional force $\mathbf{f}_k$ during displacement $\mathbf{d}$?

SOLUTION: From Eq. 8-27 and the given data, we have

$$\Delta E = -f_k d = -(140 \text{ N})(0.50 \text{ m}) = -70 \text{ J}. \quad \text{(Answer)}$$

This result means that during the displacement, 70 J is transferred to thermal energy of the robot and the wall along which it slides.

(b) What is the work W_g done on the robot by its weight during the displacement?

SOLUTION: From Eqs. 8-1 and 8-7, we know that

$$W_g = -\Delta U = -mg\,\Delta y. \quad \text{(8-28)}$$

In Fig. 8-12, the change Δy in the height of the robot is $h = d\sin 30°$. Substituting this and given data into Eq. 8-28, we find

$$\begin{aligned} W_g &= -mgh = -mgd\sin 30° \\ &= -(40 \text{ kg})(9.8 \text{ m/s}^2)(0.50 \text{ m})(0.5) \\ &= -98 \text{ J}. \quad \text{(Answer)} \end{aligned}$$

This result means that during the displacement, 98 J is transferred to gravitational potential energy of the robot–Earth system.

(c) What is the work W_{app} done by the applied force $\mathbf{F}$?

SOLUTION: We use Eq. 7-9 ($W = Fd\cos\phi$) to find the work W_{app} done by $\mathbf{F}$ in Fig. 8-12. The angle ϕ between $\mathbf{F}$ and $\mathbf{d}$ is 0°. Substituting this and given data into Eq. 7-9, we find

$$\begin{aligned} W_{app} &= Fd\cos\phi = (380 \text{ N})(0.50 \text{ m})\cos 0° \\ &= 190 \text{ J}. \quad \text{(Answer)} \end{aligned}$$

This result means that during the displacement, 190 J is transferred by the applied force $\mathbf{F}$ to the robot–Earth system.

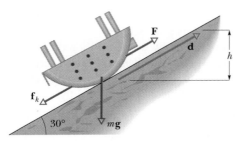

FIGURE 8-12 Sample Problem 8-6. A disabled robot is dragged up the wall inside a volcano crater, through a displacement $\mathbf{d}$ and vertical distance h, by force $\mathbf{F}$. A kinetic frictional force $\mathbf{f}_k$ opposes the motion.

CHECKPOINT 5: The figure shows three choices for the orientation of a plane that is not frictionless and for the direction in which a block slides along the plane. The block begins with the same speed in all three choices and slides until the kinetic frictional force has stopped it. Rank the choices according to the amount of mechanical energy that is dissipated, greatest first.

(1)   (2)   (3)

8-7 CONSERVATION OF ENERGY

Picture a block sliding over a frictionless floor. We can consider the block to be an isolated system in the sense that no external force transfers energy into or out of the system. So, a conservation principle applies to our system (albeit a simple one): the energy of the block is conserved.

Now suppose the block slides over a floor that is not frictionless. If we again take the block to be our system, we must say that the system is no longer isolated: Now an external force—the kinetic frictional force—acts on the block, and part of the energy transferred by that force is transferred to the floor. So now the energy of the system (the block) is not conserved.

However, we can expand the system to include both the block and the floor. For this new system, the kinetic frictional force is no longer an external force, and so it no longer transfers energy from the system, only within the system. So again we have an isolated system within which energy is conserved. Expanding what we mean by *the system* allows us to write a new conservation principle, one that helps us account for the energy transfers.

To find this new conservation principle, we use Eq. 8-27 to write the amount of mechanical energy dissipated by $\mathbf{f}_k$. With the block sliding across the floor, the only mechanical energy involved is kinetic energy, so we write Eq. 8-27 as

$$\Delta K = -f_k d, \quad \text{(8-29)}$$

in which ΔK is the change (decrease) in kinetic energy. This entire amount of energy is transferred to thermal energy of the block and floor. Thus, the change ΔE_{int} in thermal energy is

$$\Delta E_{int} = -\Delta K, \quad \text{(8-30)}$$

which gives us

$$\Delta K + \Delta E_{int} = 0. \quad \text{(8-31)}$$

So, although the mechanical energy of the block is not conserved, the sum of the mechanical energy of the block and the thermal energy of the block and floor *is* conserved. That sum is called the **total energy** E_{tot} of the block–floor system, and our new conservation principle is called the **law of conservation of energy.** Using Eq. 8-31, we can write this law as

$$\Delta E_{tot} = \Delta K + \Delta E_{int} = 0. \qquad (8\text{-}32)$$

If, in addition, conservative forces act on parts of our system, we can again expand the meaning of *the system* to include any transfers to and from potential energy by those forces. Then the total energy of this new, expanded isolated system does not change and our conservation law is written as

$$\Delta E_{tot} = \Delta K + \Delta U + \Delta E_{int} = 0 \qquad \begin{array}{l}\text{(isolated system,}\\ \text{conservation of}\\ \text{total energy).}\end{array}$$
$$(8\text{-}33)$$

If we discover other forms of energy, we can always include them in our conservation law by rewriting Eq. 8-33 as

$$\Delta E_{tot} = \Delta K + \Delta U + \Delta E_{int}$$
$$+ \left(\begin{array}{l}\text{changes in other}\\ \text{forms of energy}\end{array}\right) = 0. \qquad (8\text{-}34)$$

Note that this law of conservation of energy is *not* something we derived. Instead, it is a law based on countless experiments. Scientists and engineers have never found an exception to it. The law is extremely powerful in analyzing difficult situations because it tells us something that does

FIGURE 8-13 To descend, the rock climber must transfer energy from the gravitational potential energy of a system consisting of her, her gear, and the Earth. She has wrapped the rope around metal rings so that the rope rubs against the rings. This allows most of the transferred energy to go to the thermal energy of the rope and bars rather than to her kinetic energy.

not change when various types of energies *are* changing. In words, it says:

> In an isolated system, energy can be transferred from one type to another, but the total energy of the system remains constant.

In less formal wording, it says: energy cannot magically appear or disappear.

For example, in Fig. 8-13, we can consider the Earth and the climber and her gear to be one system. As the climber rappels down the rock face, changing the configuration of the system, she needs to control the transfer of energy from the gravitational potential energy of the system (it cannot just magically disappear). Some of it is transferred to her kinetic energy, but she obviously does not want much transferred to that form of energy. So she has wrapped the rope around metal rings. The resulting frictional force between the rope and the rings dissipates much of the mechanical energy of the system in a controlled way by slowly transferring energy to thermal energy of the rope and the rings.

If a system is not isolated and external forces transfer energy to or from the system, then Eqs. 8-33 and 8-34 do not apply. Instead we account for those transfers by writing

$$W = \Delta E_{tot} = \Delta K + \Delta U + \Delta E_{int}, \qquad (8\text{-}35)$$

in which W is the work done on the system by the external forces. For example, in Fig. 8-13, if we consider the rope to be external to the system, then the frictional force exerted by the rope on the metal rings of the system does work W on the system, transferring energy from the system to thermal energy of the rope while, within the system, the values of K, U, and E_{int} change.

Power

Now that you have seen how energy can be transferred from one form to another and have been given a hint that additional forms of energy exist, we can expand the definition of power given in Section 7-7. There it is the rate at which work is done by a force. In a more general sense, power P is the rate at which energy is transferred by a force from one form to another. If an amount of energy ΔE is transferred in an amount of time Δt, the **average power** due to the force is

$$\overline{P} = \frac{\Delta E}{\Delta t}. \qquad (8\text{-}36)$$

Similarly, the **instantaneous power** due to the force is

$$P = \frac{dE}{dt}. \qquad (8\text{-}37)$$

SAMPLE PROBLEM 8-7

In Fig. 8-14, a circus beagle of mass 6.0 kg runs onto the left end of a curved ramp with speed $v_0 = 7.8$ m/s at height $y_0 = 8.5$ m above the floor. It then slides to the right and comes to a momentary stop when it reaches a height $y = 11.1$ m from the floor. What is the increase in thermal energy of the beagle and ramp due to the sliding?

SOLUTION: The isolated system to which Eq. 8-33 applies is the beagle–Earth–ramp system, because the only forces acting on the beagle are its weight $m\mathbf{g}$ (due to the Earth) and frictional and normal forces (due to the ramp). The weight is, of course, a conservative force with which we associate a change ΔU in gravitational potential energy. The frictional force dissipates that potential energy and the beagle's kinetic energy during the slide, increasing the thermal energy of the beagle and ramp by an amount ΔE_{int}. The normal force causes no energy transfers because it is always perpendicular to the path of the beagle.

At the stopping point, the speed of the beagle is zero, and thus so is its kinetic energy. Applying Eq. 8-33 to the beagle–Earth–ramp system, we find that

$$\Delta K + \Delta U + \Delta E_{int} = 0$$

and

$$(0 - \tfrac{1}{2}mv_0^2) + mg(y - y_0) + \Delta E_{int} = 0.$$

Solving for ΔE_{int}, we get

$$\Delta E_{int} = \tfrac{1}{2}mv_0^2 - mg(y - y_0)$$
$$= \tfrac{1}{2}(6.0 \text{ kg})(7.8 \text{ m/s})^2$$
$$\quad -(6.0 \text{ kg})(9.8 \text{ m/s}^2)(11.1 \text{ m} - 8.5 \text{ m})$$
$$\approx 30 \text{ J.} \qquad \text{(Answer)}$$

FIGURE 8-14 Sample Problem 8-7. A beagle slides along a curved ramp, starting with speed v_0 at height y_0, and reaching a height y where it comes to a momentary stop.

SAMPLE PROBLEM 8-8

A steel ball whose mass m is 5.2 g is fired vertically downward from a height h_1 of 18 m with an initial speed v_0 of 14 m/s (Fig. 8-15a). It buries itself in sand to a depth h_2 of 21 cm.

(a) What is the change in the mechanical energy of the ball?

SOLUTION: At the stopping depth h_2, the speed of the ball is zero, and thus so is its kinetic energy. The change in the mechanical energy of the ball is given by

$$\Delta E = \Delta K + \Delta U, \qquad (8\text{-}38)$$

or, with Eq. 8-8 ($\Delta U = mg \, \Delta y$),

$$\Delta E = (0 - \tfrac{1}{2}mv_0^2) - mg(h_1 + h_2),$$

where $-(h_1 + h_2)$ is the total *downward* displacement of the ball. Inserting the given data, we find

$$\Delta E = -\tfrac{1}{2}(5.2 \times 10^{-3} \text{ kg})(14 \text{ m/s})^2$$
$$\quad -(5.2 \times 10^{-3} \text{ kg})(9.8 \text{ m/s}^2)(18 \text{ m} + 0.21 \text{ m})$$
$$= -1.437 \text{ J} \approx -1.4 \text{ J.} \qquad \text{(Answer)}$$

(Note that here we assigned the potential energy to the ball alone rather than, more properly, to the ball–Earth system—recall Problem Solving Tactic 1.)

(b) What is the change in the internal energy of the ball–Earth–sand system?

SOLUTION: This system is an isolated system because once the ball has been fired, the only forces that act on it are its weight $m\mathbf{g}$ (due to the Earth) and the average upward force $\mathbf{F}$ exerted on it by the sand (Fig. 8-15b). So Eq. 8-33 applies. Substituting Eq. 8-38 into Eq. 8-33, we find that for the ball–Earth–sand system

$$\Delta E + \Delta E_{int} = 0,$$

or

$$\Delta E_{int} = -\Delta E = -(-1.437 \text{ J}) \approx 1.4 \text{ J.} \qquad \text{(Answer)}$$

As the ball moves through the sand, $\mathbf{F}$ dissipates all the ball's mechanical energy, transferring that energy to thermal energy of the ball and sand.

(c) What is the magnitude of the average force $\mathbf{F}$ exerted on the ball by the sand?

SOLUTION: The mechanical energy of the ball is conserved until it reaches the sand. Then, as the ball moves through a distance h_2 in the sand, its mechanical energy is changed by ΔE. So Eq. 8-27 ($-f_k d = \Delta E$) can be rewritten as

$$-Fh_2 = \Delta E.$$

Solving this for F, we find

$$F = \frac{\Delta E}{-h_2} = \frac{-1.437 \text{ J}}{-0.21 \text{ m}} = 6.84 \text{ N} \approx 6.8 \text{ N.} \qquad \text{(Answer)}$$

We could also find F by first using the techniques of Chapter 2 to find the ball's speed at the surface of the sand and its average deceleration within the sand. Then, using Newton's second law, we would have F. Obviously, more algebraic steps would be required.

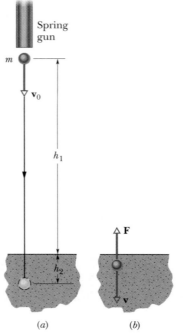

FIGURE 8-15   Sample Problem 8-8. (*a*) A ball is fired downward, coming to rest in sand. Mechanical energy is conserved over path h_1 but not over path h_2, where a nonconservative force **F** acts on the ball. (*b*) Force **F**, due to the sand, transfers energy primarily to thermal energy of the ball and the sand through which the ball travels.

8-8 MASS AND ENERGY (OPTIONAL)

The science of chemistry was built up by assuming that in chemical reactions, energy and mass are each conserved. In 1905, Einstein showed that as a consequence of his theory of special relativity, mass can be considered to be another form of energy. Thus the law of conservation of energy is really the law of conservation of mass–energy.

In a chemical reaction, the amount of mass that is transferred into other forms of energy (or vice versa) is such a tiny fraction of the total mass involved that there is no hope of measuring the mass change with even the best laboratory balances. Mass and energy truly *seem* to be separately conserved. In a nuclear reaction, however, the energy released is often about a million times greater than in a chemical reaction, and the change in mass can easily be measured. Taking mass–energy transfers into account where nuclear reactions are involved becomes a matter of necessary laboratory routine.

Mass and energy are related by what is certainly the best-known equation in physics (see Fig. 8-16), namely,

FIGURE 8-16   In 1979, students of Shenandoah Junior High School, in Miami, Florida, honored Albert Einstein on the 100th anniversary of his birth by spelling out his famous formula with their bodies.

$$E = mc^2, \qquad (8\text{-}39)$$

in which E is the energy equivalent (called the **mass energy**) of mass m, and c is the speed of light. (If you continue your study of physics beyond this book, you will see more refined discussions of the relation of mass and energy. You might even encounter disagreements about just what that relation is and means.)

Table 8-1 shows the energy equivalents of the masses of a few objects. The amount of energy lying dormant in ordinary objects is enormous. The energy equivalent of the mass of a penny, for example, would cost over a million dollars to purchase from your local utility company. The mass equivalents of some energies are equally striking. The entire annual U.S. electrical energy production, for example, corresponds to a mass of only a few hundred kilograms of matter (stones, potatoes, anything!).

In applying Eq. 8-39 to nuclear or chemical reactions between particles, we can rewrite it as

$$Q = -\Delta m \, c^2, \qquad (8\text{-}40)$$

TABLE 8-1   THE ENERGY EQUIVALENTS OF A FEW OBJECTS

OBJECT	MASS (kg)	ENERGY EQUIVALENT
Electron	9.11×10^{-31}	8.2×10^{-14} J ($= 511$ keV)
Proton	1.67×10^{-27}	1.5×10^{-10} J ($= 938$ MeV)
Uranium atom	4.0×10^{-25}	3.6×10^{-8} J ($= 225$ GeV)
Dust particle	1×10^{-13}	1×10^{4} J ($= 2$ kcal)
Penny	3.1×10^{-3}	2.8×10^{14} J ($= 78$ GW·h)

in which Q (called simply the Q *of the reaction*) is the energy released (positive value) or absorbed (negative value) in the reaction, and Δm is the corresponding decrease or increase in the mass of the particles as a result of the reaction. If the reaction is nuclear fission (a nucleus splits into smaller nuclei), less than 0.1% of the mass initially present is transformed into other forms of energy. If the reaction is a chemical one, the percentage is much less, typically a million times less.

In practice, SI units are rarely used with Eq. 8-40, because they are too large to be convenient. Masses are usually measured in atomic mass units (abbreviated u; see Section 1-6), where

$$1 \text{ u} = 1.66 \times 10^{-27} \text{ kg}, \qquad (8\text{-}41)$$

and energies are usually measured in electron-volts or multiples thereof. Equation 7-3 tells us that

$$1 \text{ eV} = 1.60 \times 10^{-19} \text{ J}. \qquad (8\text{-}42)$$

In the units of Eqs. 8-41 and 8-42, the multiplying constant c^2 has the values

$$c^2 = 9.315 \times 10^8 \text{ eV/u} = 9.315 \times 10^5 \text{ keV/u}$$
$$= 931.5 \text{ MeV/u}. \qquad (8\text{-}43)$$

SAMPLE PROBLEM 8-9

In the *nuclear fission* reaction

$$n + {}^{235}U \rightarrow {}^{140}Ce + {}^{94}Zr + n + n,$$

a neutron (n) combines with a uranium nucleus (^{235}U), making the nucleus unstable and causing it to split (to fission) into two smaller nuclei (^{140}Ce and ^{94}Zr) and to release two neutrons. The masses involved are

mass(^{235}U) = 235.04 u   mass(^{94}Zr) = 93.91 u

mass(^{140}Ce) = 139.91 u   mass(n) = 1.00867 u

(a) What is the fractional change in the mass of the two interacting particles?

SOLUTION: To find the mass change Δm, we subtract the mass of the reacting particles from the mass of the product particles:

$$\Delta m = (139.91 + 93.91 + 2 \times 1.00867)$$
$$- (235.04 + 1.00867)$$
$$= -0.211 \text{ u}.$$

The mass of the reacting particles is

$$M = 235.04 \text{ u} + 1.00867 \text{ u} = 236.05 \text{ u},$$

so the fractional decrease in mass is

$$\frac{|\Delta m|}{M} = \frac{0.211 \text{ u}}{236.05 \text{ u}}$$
$$= 0.00089, \text{ or about } 0.1\%. \qquad \text{(Answer)}$$

Although this is small, it is easily measurable.

(b) How much energy is released during each fission reaction?

SOLUTION: From Eq. 8-40 we have

$$Q = -\Delta m \, c^2 = -(-0.211 \text{ u})(931.5 \text{ MeV/u})$$
$$= 197 \text{ MeV}, \qquad \text{(Answer)}$$

where the value for c^2 is taken from Eq. 8-43. The energy release of 197 MeV per reaction is much larger than the energy releases of a few electron-volts per reaction that are typical of chemical reactions.

SAMPLE PROBLEM 8-10

The nucleus of an atom of deuterium (or heavy hydrogen) is called a **deuteron.** It is composed of a proton and a neutron. If a deuteron is torn apart, how much energy is involved in the process? Is energy absorbed or released by the reaction? The masses involved are

deuteron: $m_d = 2.01355$ u

proton: $m_p = 1.00728$ u

neutron: $m_n = 1.00867$ u $\Big\}$ 2.01595 u

SOLUTION: Since the total mass of the separated proton and neutron is greater than the deuteron mass, energy must be added to the deuteron to cause the reaction. The increase in mass due to the reaction is

$$\Delta m = (m_p + m_n) - m_d$$
$$= (1.00728 + 1.00867) - (2.01355) = 0.00240 \text{ u}.$$

The corresponding energy is then, from Eq. 8-40,

$$Q = -\Delta m \, c^2 = -(0.00240 \text{ u})(931.5 \text{ MeV/u})$$
$$= -2.24 \text{ MeV}. \qquad \text{(Answer)}$$

The minus sign means that energy must be added to the deuteron. The quantity 2.24 MeV is called the *binding energy* of the deuteron. The binding energy of any nucleus is the amount of energy that must be supplied to separate the nucleus into its basic units of protons and neutrons. Conversely, it is the amount of energy that would be released if the nucleus could somehow be constructed from those basic units.

CHECKPOINT **6:** If we could somehow bring together six protons and six neutrons to form the nucleus of a ^{12}C atom, would the total mass of those 12 particles decrease or increase?

8-9 QUANTIZED ENERGY

So far, we have assumed that the energy within a system can take on *any* value. That seems a reasonable assumption for everyday systems involving springs and weights and such. However, it is not true for such microscopic systems as an atom and a *quantum dot* (a laboratory-produced confinement of electrons that resembles an atom). For these microscopic systems, the energy internal to the system is **quantized;** that is, the energy is restricted to certain values. When the system has a particular energy value, it is said to be in the **quantum state** that is characterized by that value.

Figure 8-17 is a typical diagram of the energy values (or *energy levels*) for an atom: energy is plotted vertically and each of the five lowest possible energy values E_0, E_1, . . . , E_4 is represented by a horizontal line (hence the term *level*). The atom cannot have any intermediate value of energy. The quantum state associated with the lowest energy, labeled E_0 in Fig. 8-17 and assigned the arbitrary value of zero, is called the *ground state* of the atom. The quantum states with greater energies are called *excited states* of the atom: the state with energy E_1 is the first excited state, the state with energy E_2 is the second excited state, and so on.

The atom tends to be in its ground state so as to have the lowest allowed energy (much as a ball tends to roll downhill). The atom can move to an excited state, with greater energy, only if an external source provides energy equal to the energy difference between the ground state and the excited state. The atom can gain energy in a collision with another atom or a free electron. Or it can gain energy by absorbing light. However, in either process the atom must gain enough energy to put it into one of the higher energy levels. Thus, not only the energy values of an atom are quantized, but also the amounts of energy that an atom can gain (or lose) as it makes *quantum jumps* (or transitions) from one energy level to another are quantized.

The upward arrow in Fig. 8-17 represents a quantum jump from the ground state to the second excited state, which (according to the diagram) requires that the atom gain 2.5 eV of energy. If it gains the energy by absorbing light, that light ceases to exist and its energy is fully transferred to the energy of the atom. Thus, that light must have an energy of 2.5 eV if it is to be absorbed by the atom. If the light has slightly more or slightly less energy, it cannot be absorbed.

If an atom is to absorb light, the energy of that light must equal the energy difference between the initial energy level of the atom and a higher level.

When the atom reaches an excited state, it does not stay there but quickly *de-excites* by decreasing its energy, either in a collision or by emitting light. In the latter process, the emitted light is actually created by the atom (the

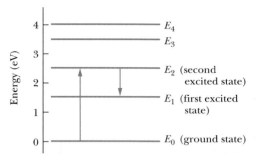

FIGURE 8-17 An energy level diagram for an atom showing the quantized energy values (or levels). Each energy level corresponds to a quantum state of the atom. The lowest energy, marked E_0, is associated with the ground state. The upward arrow represents a quantum jump of the atom from the ground state to the second excited state, which has energy E_2. The downward arrow represents a quantum jump from there to the first excited state, which has energy E_1.

light did not exist prior to de-excitation). In either process, the atom must lose enough energy to reach one of its lower energy levels.

If an atom emits light, the energy of that light equals the energy difference between the initial energy level of the atom and a lower level.

As an example, the downward arrow in Fig. 8-17 represents a quantum jump from the second excited state to the first excited state. The light that is emitted in the jump has an energy of 1.0 eV. The atom must next jump to the ground state with a second emission of light having an energy of 1.5 eV. Instead of this two-jump process, the atom could have jumped from the second excited state directly to the ground state with a single emission of light having an energy of 2.5 eV.

CHECKPOINT **7:** An atom that is being monitored emits light four times without being re-excited. Three of the emissions are detected at energies of 1.1 eV, 1.4 eV, and 2.8 eV. The sequence of the emissions is not known. Here are the energy levels of the atom:

$E_8 = 7.7$ eV	$E_5 = 4.8$ eV	$E_2 = 2.7$ eV
$E_7 = 6.6$	$E_4 = 4.2$	$E_1 = 1.3$
$E_6 = 5.5$	$E_3 = 3.9$	$E_0 = 0$

(a) What was the initial state of the atom and (b) what was the energy of the fourth emission?

REVIEW & SUMMARY

Conservative Forces

A force is a **conservative force** if the net work it does on a particle moving along a closed path from an initial point and then back to that point is zero. Or, equivalently, it is conservative if its work on a particle moving between two points does not depend on the path taken by the particle. The gravitational force (weight) and the spring force are conservative forces; the kinetic frictional force is a **nonconservative force.**

Potential Energy

A **potential energy** is energy that is associated with the configuration of a system in which a conservative force acts. When the conservative force does work W on a particle within the system, the change ΔU in the potential energy of the system is

$$\Delta U = -W. \tag{8-1}$$

If the particle moves from point x_i to point x_f, the change in the potential energy of the system is

$$\Delta U = -\int_{x_i}^{x_f} F(x)\, dx. \tag{8-6}$$

Gravitational Potential Energy

The potential energy associated with a system consisting of the Earth and a nearby particle is the **gravitational potential energy.** If the particle moves from height y_i to height y_f, the change in the gravitational potential energy of the particle–Earth system is

$$\Delta U = mg(y_f - y_i) = mg\,\Delta y. \tag{8-7}$$

If the **reference position** of the particle is set as $y_i = 0$ and the corresponding gravitational potential energy of the system is set as $U_i = 0$, then the gravitational potential energy U when the particle is at any position y is

$$U = mgy. \tag{8-9}$$

Elastic Potential Energy

Elastic potential energy is the energy associated with the state of compression or extension of an elastic object. For a spring that exerts a spring force $F = -kx$, the elastic potential energy is

$$U(x) = \tfrac{1}{2}kx^2. \tag{8-11}$$

The reference configuration is with the spring at its relaxed length, with $x = 0$ and $U = 0$.

Mechanical Energy

The **mechanical energy** E of a system is the sum of its kinetic energy K and its potential energy U:

$$E = K + U. \tag{8-12}$$

If only a conservative force within the system does work, then the mechanical energy E of the system cannot change. This **conservation of mechanical energy** is written as

$$K_2 + U_2 = K_1 + U_1, \tag{8-17}$$

in which the subscripts refer to different instants during an energy transfer process. This conservation can also be written as

$$\Delta E = \Delta K + \Delta U = 0. \tag{8-18}$$

Potential Energy Curves

If we know the **potential energy function** $U(x)$ for a system in which a force F acts on a particle, we can find the force as

$$F(x) = -\frac{dU(x)}{dx}. \tag{8-19}$$

If $U(x)$ is given on a graph, then at any value of x, the force F is the negative of the slope of the curve there and the kinetic energy of the particle is given by

$$K(x) = E - U(x), \tag{8-21}$$

where E is the mechanical energy of the system. A **turning point** is a point x where the particle reverses its motion (there, $K = 0$). The particle is in **equilibrium** at points where the slope of the $U(x)$ curve is zero (there, $F(x) = 0$).

Work by Nonconservative Forces

If a nonconservative applied force $\mathbf{F}$ does work on particle that is part of a system having a potential energy, then the work W_{app} done on the system by $\mathbf{F}$ is equal to the change ΔE in the mechanical energy of the system:

$$W_{app} = \Delta K + \Delta U = \Delta E. \tag{8-24, 8-25}$$

If a kinetic frictional force $\mathbf{f}_k$ does work on an object, the change ΔE in the total mechanical energy of the object and any system containing it is given by

$$\Delta E = -f_k d, \tag{8-27}$$

in which d is the displacement of the object during the work. The mechanical energy lost by this transfer is said to be **dissipated** by $\mathbf{f}_k$. A portion of the dissipated energy is transferred from the object; that amount is equal to the work done by $\mathbf{f}_k$.

Principle of Conservation of Energy

In an isolated system, energy may be transferred from one type to another, but the total energy E_{tot} of the system always remains constant. This conservation law is written as

$$\Delta E_{tot} = \Delta K + \Delta U + \Delta E_{int} + \left(\begin{array}{c} \text{changes in other} \\ \text{forms of energy} \end{array}\right) = 0. \tag{8-34}$$

Here ΔE_{int} is the change in the internal energy of the bodies within the system.

If, instead, the system is not isolated, then an external force can change the total energy of the system by doing work W. In that case

$$W = \Delta E_{tot} = \Delta K + \Delta U + \Delta E_{int}. \tag{8-35}$$

Power

The **power** of a force is the *rate* at which that force transfers energy. If an amount of energy ΔE is transferred by a force in an amount of time Δt, the **average power** of the force is

$$\overline{P} = \frac{\Delta E}{\Delta t}. \qquad (8\text{-}36)$$

The **instantaneous power** of a force is

$$P = \frac{dE}{dt}. \qquad (8\text{-}37)$$

Mass and Energy

The **mass energy** E of an object is the energy equivalent of its mass m:

$$E = mc^2, \qquad (8\text{-}39)$$

in which c is the speed of light. To convert a mass given in atomic mass units (u) to an energy equivalent in megaelectron-volts, multiply by 931.5 MeV/u. In a nuclear or chemical reaction, the energy released or absorbed is the Q of the reaction, given as

$$Q = -\Delta m\, c^2, \qquad (8\text{-}40)$$

in which Δm is the corresponding decrease or increase in mass. (Q is positive for an energy release, corresponding to a mass decrease.)

Quantized Energy

The energy within a microscopic system, such as an atom, is **quantized** (restricted to certain values) and the system is said to be in a **quantum state** that is characterized by its energy value. The system cannot have any other values of energy. The lowest energy corresponds to the **ground state** of the system. Greater values of energy correspond to **excited states** of the system. If the system gains or loses energy via light, the energy of the light must equal the difference between two of the allowed energy values of the system.

QUESTIONS

1. When a particle moves from f to i and from j to i along the paths shown in Fig. 8-18, and in the indicated directions, a conservative force **F** does the indicated amounts of work on it. How much work is done by **F** when the particle moves from f to j?

FIGURE 8-18 Question 1.

2. Figure 8-19 shows one direct path and four indirect paths from point i to point f. Along the direct path and three of the indirect paths, only a conservative force F_c acts on a certain particle. Along the fourth indirect path, both F_c and a nonconservative force F_{nc} act on the particle. The work (in joules) done on a particle in going from i to f is indicated along each straight line segment of the indirect paths. (a) How much work is done on the particle in moving from i to f along the direct path? (b) How much work is done on the particle by F_{nc} along the one path where it acts?

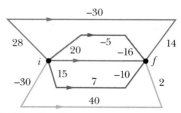

FIGURE 8-19 Question 2.

3. A spring is initially stretched by 3.0 cm from its relaxed length. Here are four choices: the initial stretch is changed to (a) a stretch of 2.0 cm, (b) a compression of 2.0 cm, (c) a compression of 4.0 cm, and (d) a stretch of 4.0 cm. Rank the choices according to the change in the elastic potential energy of the spring, most positive first, most negative last.

4. In Fig. 8-20, a brave skater slides down three slopes of frictionless ice whose vertical heights d are identical. Rank the slopes according to (a) the work done on the skater by her weight during the descent on each slope and (b) the change in her kinetic energy produced along the slope, greatest first.

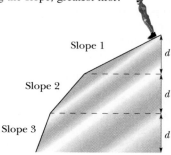

FIGURE 8-20 Question 4.

5. A coconut is thrown from a cliff edge toward a wide, flat valley, with initial speed 8 m/s. Rank the following choices for the launch direction according to (a) the initial kinetic energy of the coconut and (b) its kinetic energy just before hitting the valley bottom, greatest first: (1) **v** almost vertically upward, (2) **v** angled upward by 45°, (3) **v** horizontal, (4) **v** angled downward by 45°, and (5) **v** almost vertically downward.

6. Figure 8-21 shows three plums that are launched from the same level with the same speed. One moves straight upward, one is launched at a small angle to the vertical, and one is launched along a frictionless incline. Rank the plums according to their speed when they reach the level of the dashed line, greatest first.

FIGURE 8-21 Question 6. (1) (2) (3)

7. In Fig. 8-22, a horizontally moving block can take three frictionless routes, differing in elevation, to reach the dashed finish line. Rank the routes according to (a) the speed of the block at the finish line and (b) the travel time of the block to the finish line, greatest first.

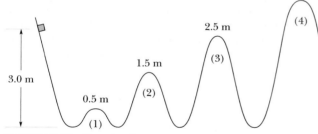

FIGURE 8-22 Question 7. (3)

8. In Fig. 8-23, a small, initially stationary block is released on a frictionless ramp at a height of 3.0 m. Hill heights along the ramp are as shown. The hills have identical circular tops (assume that the block does not fly off any hill). (a) Which hill is the first the block cannot cross? (b) What does it do after failing to cross that hill? On which hilltop is (c) the centripetal acceleration of the block greatest? (d) On which hilltop is the normal force on the block least?

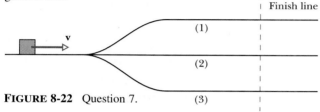

FIGURE 8-23 Question 8.

9. Figure 8-24 shows two arrangements of the same two blocks on a frictionless plane; the blocks are connected with a taut cord that runs over a massless, frictionless pulley. In each arrangement, the hanging block descends when the blocks are released. Consider the total kinetic energy of the two blocks when that block has descended by a certain distance d. Is the total kinetic energy in arrangement (a) more than, less than, or the same as that in arrangement (b)?

(a) (b)

FIGURE 8-24 Question 9.

10. In Fig. 8-25, an initially stationary block is released at time t_1 to slide down a frictionless ramp to a massless spring, which it reaches at time t_2 and then compresses until the maximum compression is reached at time t_3. From t_1 to t_3, what happens to (a) the kinetic energy of the block, (b) the gravitational potential energy of the block–Earth system, (c) the elastic potential energy of the spring, (d) the mechanical energy of the block–Earth system, (e) the mechanical energy of the spring, and (f) the mechanical energy of the block–Earth–spring system?

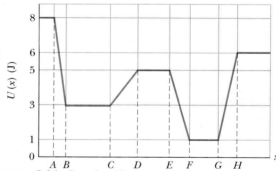

FIGURE 8-25
Question 10.

11. In Checkpoint 4, what values must the mechanical energy E and kinetic energy K of the particle not exceed if the particle is to be (a) trapped in the potential well and (b) able to move to the left of point D but not to the right of it.

12. Figure 8-26 gives the potential energy function of a particle. (a) Rank regions AB, BC, CD, and DE according to the magnitude of the force on the particle, greatest first. What value must the energy E of the particle not exceed if the particle is to be (b) trapped in the potential well at the left, (c) trapped in the potential well at the right, and (d) able to move between the two potential wells but not to the right of point H? For the situation of (d), in what region—BC, DE, or FG—will the particle have (e) the greatest kinetic energy and (f) the least speed?

FIGURE 8-26 Question 12.

13. In Fig. 8-27, a block slides from A to C along a frictionless ramp, and then it passes through horizontal region CD, where a frictional force acts on it. Is the block's kinetic energy increasing, decreasing, or constant in (a) region AB, (b) region BC, and (c) region CD? (d) Is the block's mechanical energy increasing, decreasing, or constant in those regions?

FIGURE 8-27
Question 13.

EXERCISES & PROBLEMS

SECTION 8-3   Determining Potential Energy Values

1E. The only force acting on a particle is conservative force **F**. If the particle is at point A, the potential energy of the system associated with **F** and the particle is 40 J. If the particle moves from point A to point B, the work done on the particle by **F** is +25 J. What is the potential energy of the system with the particle at B?

2E. What is the spring constant of a spring that stores 25 J of elastic potential energy when compressed by 7.5 cm from its relaxed length?

3E. You drop a 2.00 kg textbook to a friend who stands on the ground 10.0 m below the textbook with outstretched hands 1.50 m above the ground (Fig. 8-28). (a) How much work is done on the textbook by its weight as it drops to your friend's hands? (b) What is the change in the gravitational potential energy of the textbook–Earth system during the drop? If the gravitational potential energy of that system is taken to be zero at ground level, what is its potential energy when the textbook (c) is released and (d) reaches the hands?

10.0 m

1.50 m

FIGURE 8-28   Exercises 3 and 12.

4E. In Fig. 8-29, a 2.00 g ice flake is released from the edge of a hemispherical bowl whose radius r is 22.0 cm. The flake–bowl contact is frictionless. (a) How much work is done on the flake by its weight during the flake's descent to the bottom of the bowl? (b) What is the change in the potential energy of the flake–Earth system during that descent? (c) If that potential energy is taken to be zero at the bottom of the bowl, what is its value when the flake is released? (d) If, instead, the potential energy is taken to be zero at the release point, what is its value when the flake reaches the bottom of the bowl?

Ice flake

r

FIGURE 8-29   Exercises 4 and 13.

5E. In Fig. 8-30, a frictionless roller coaster of mass m tops the first hill with speed v_0. How much work does its weight do on it from that point to (a) point A, (b) point B, and (c) point C? If the gravitational potential energy of the coaster–Earth system is taken to be zero at point C, what is its value when the coaster is at (d) point B and (e) point A?

First hill

v_0

A

B

h h $h/2$ C

FIGURE 8-30   Exercises 5 and 14.

6E. Figure 8-31 shows a ball with mass m attached to the end of a thin rod with length L and negligible mass. The other end of the rod is pivoted so that the ball can move in a vertical circle. The rod is held in the horizontal position as shown and then given enough of a downward push to cause the ball to swing down and around and just reach the vertically upward position, with zero speed there. How much work is done on the ball by its weight from the initial point to (a) the lowest point, (b) the highest point, and (c) the point on the right at which the ball is level with the initial point? (d) If the gravitational potential energy of the ball–Earth system is taken to be zero at the initial point, what is its value when the ball reaches those three other points, respectively?

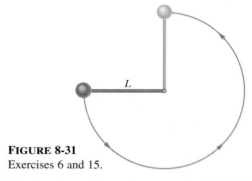

L

FIGURE 8-31
Exercises 6 and 15.

7P. A spring with a spring constant of 3200 N/m is initially stretched until the elastic potential energy is 1.44 J. ($U = 0$ for no stretch.) What is the change in the elastic potential energy if the initial stretch is changed to (a) a stretch of 2.0 cm, (b) a compression of 2.0 cm, and (c) a compression of 4.0 cm?

8P. A 1.50 kg snowball is fired from a cliff 12.5 m high with an initial velocity of 14.0 m/s, directed 41.0° above the horizontal. (a) How much work is done on the snowball by its weight during its flight to the ground below the cliff? (b) What is the change in the gravitational potential energy of the snowball–Earth system during the flight? (c) If that gravitational potential energy is taken to be zero at the height of the cliff, what is its value when the snowball reaches the ground?

9P. Figure 8-32 shows a thin rod, of length L and negligible mass, that can pivot about one end to rotate in a vertical circle. A heavy ball of mass m is attached to the other end. The rod is pulled aside through an angle θ and released. As the ball descends to its lowest point, (a) how much work does its weight do on it and (b) what is the change in the gravitational potential energy of the ball–Earth system? (c) If the gravitational potential energy is taken to be zero at the lowest point, what is its value just as the ball is released?

FIGURE 8-32 Problems 9 and 20.

10P. In Fig. 8-33, a small block of mass m can slide along the frictionless loop-the-loop. The block is released from rest at point P, at height $h = 5R$ above the bottom of the loop. How much work does the weight of the block do on the block as the block travels from point P to (a) point Q and (b) the top of the loop? If the gravitational potential energy of the block–Earth system is taken to be zero at the bottom of the loop, what is that potential energy when the block is (c) at point P, (d) at point Q, and (e) at the top of the loop?

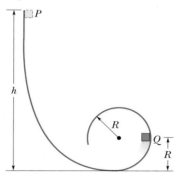

FIGURE 8-33 Problems 10 and 39.

11P. A force $\mathbf{F}$ that is directed along an x axis acts on a particle as the particle moves from $x = 1.0$ m to $x = 4.0$ m and then back to $x = 1.0$ m. What is the net work done on the particle by $\mathbf{F}$ for the round trip if the values of the force during the outward and the return trips are (a) 3.0 N and -3.0 N; (b) 5.0 N and 5.0 N; (c) $2.0x$ and $-2.0x$; (d) $3.0x^2$ and $3.0x^2$? Here, x is in meters and F is in newtons. (e) In which situation(s) is the force conservative?

SECTION 8-4 Conservation of Mechanical Energy

12E. (a) In Exercise 3, what is the speed of the textbook when it reaches the hands? (b) If we substituted a second textbook with twice the mass, what would its speed be?

13E. (a) In Exercise 4, what is the speed of the flake when it reaches the bottom of the bowl? (b) If we substituted a second flake with twice the mass, what would its speed be?

14E. In Exercise 5, what is the speed of the coaster at (a) point A, (b) point B, and (c) point C? (d) How high will it go on the last hill, which is too high for it to cross? (e) If we substitute a second coaster with twice the mass, what then are the answers (a) through (d)?

15E. (a) In Exercise 6, what initial speed is given the ball? What is its speed at (b) the lowest point and (c) the point on the right at which the ball is level with the initial point?

16E. A 70.0 kg man jumping from a window lands in an elevated fire rescue net 11.0 m below the window. He momentarily stops when he has stretched the net by 1.50 m. Assuming that mechanical energy is conserved during this process and that the net functions like an ideal spring, find the elastic potential energy of the net when it is stretched by 1.50 m.

17E. In Fig. 8-34, a runaway truck with failed brakes is moving downgrade at 80 mi/h just before the driver has the truck travel up an emergency escape ramp with an inclination of 15°. What minimum length L must the ramp have if the truck is to stop (momentarily) along it? Why are real escape ramps often covered with a thick layer of sand or gravel?

FIGURE 8-34 Exercise 17.

18E. A volcanic ash flow is moving across horizontal ground when it encounters a 10° upslope. The front of the flow then travels 920 m on the upslope before stopping. Assume that the gases entrapped in the flow lift the flow and thus make the frictional force from the ground negligible; assume also that mechanical energy of the front of the flow is conserved. What was the initial speed of the front of the flow?

19P. A 1.50 kg water balloon is shot straight up with an initial speed of 3.00 m/s. (a) What is the kinetic energy of the balloon just as it is launched? (b) How much work does the weight of the balloon do on the balloon during the balloon's full ascent? (c) What is the change in the gravitational potential energy of the balloon–Earth system during the full ascent? (d) If the gravitational potential energy is taken to be zero at the launch point, what is its value when the balloon reaches its maximum height? (e) If, instead, the gravitational potential energy is taken to be zero at the maximum height, what is its value at the launch point? (f) What is the maximum height of the balloon?

20P. In Problem 9, what is the speed of the ball at the lowest point if $L = 2.00$ m and $\theta = 30.0°$?

21P. (a) In Problem 8, using energy techniques, rather than techniques of Chapter 4, find the speed of the snowball as it reaches the ground below the cliff. (b) What is that speed if, instead, the launch angle is 41.0° *below* the horizontal?

22P. Figure 8-35 shows an 8.00 kg stone resting on a spring. The spring is compressed 10.0 cm by the stone. (a) What is the spring constant? (b) The stone is pushed down an additional 30.0 cm and released. What is the elastic potential energy of the compressed spring just before that release? (c) What is the change in the gravitational potential energy of the stone–Earth system when the stone moves from the release point to its maximum height? (d) What is that maximum height, measured from the release point?

FIGURE 8-35 Problem 22.

23P. A 5.0 g marble is fired vertically upward using a spring gun. The spring must be compressed 8.0 cm if the marble is to just reach a target 20 m above the marble's position on the compressed spring. (a) What is the change in the gravitational potential energy of the marble–Earth system during the 20 m ascent? (b) What is the change in the elastic potential energy of the spring during its launch of the marble? (c) What is the spring constant of the spring?

24P. Figure 8-36a applies to the spring in a cork gun (Fig. 8-36b): it shows the spring force as a function of the stretch or compression of the spring. The spring is compressed by 5.5 cm and used to propel a 3.8 g cork from the gun. (a) What is the speed of the cork if it is released as the spring passes through its relaxed position? (b) Suppose, instead, that the cork sticks to the spring and stretches it 1.5 cm before separation occurs. What now is the speed of the cork at the time of release?

FIGURE 8-36 Problem 24.

25P. A 2.00 kg block is placed against a spring on a frictionless 30.0° incline (Fig. 8-37). The spring, whose spring constant is 19.6 N/cm, is compressed 20.0 cm and then released. (a) What is the elastic potential energy of the compressed spring? (b) What is the change in the gravitational potential energy of the block–Earth system as the block moves from the release point to its

highest point on the incline? (c) How far along the incline is the highest point from the release point?

FIGURE 8-37 Problem 25.

26P. In Fig. 8-38, a 12 kg block is released from rest on an incline angled at $\theta = 30°$. Below the block is a spring that can be compressed 2.0 cm by a force of 270 N. The block momentarily stops when it compresses the spring by 5.5 cm. (a) How far has the block moved down the incline to this stopping point? (b) What is the speed of the block just as it touches the spring?

FIGURE 8-38 Problem 26.

27P. A 0.55 kg projectile is launched from the edge of a cliff with an initial kinetic energy of 1550 J and at its highest point is 140 m above the launch point. (a) What is the horizontal component of its velocity? (b) What was the vertical component of its velocity just after launch? (c) At one instant during its flight the vertical component of its velocity is 65 m/s. At that time, how far is it above or below the launch point?

28P. A 50 g ball is thrown from a window with an initial velocity of 8.0 m/s at an angle of 30° above the horizontal. Using energy methods, determine (a) the kinetic energy of the ball at the top of its flight and (b) its speed when it is 3.0 m below the window. Does the answer to (b) depend on either (c) the mass of the ball or (d) the initial angle?

29P. The spring of a child's spring gun has a spring constant of 4.0 lb/in. When the gun is inclined upward by 30° to the horizontal, a 2.0 oz ball is shot to a height of 6.0 ft above the muzzle of the gun. (a) What was the muzzle speed of the ball? (b) By how much must the spring have been compressed initially?

30P. In Fig. 8-39, the pulley is massless, and both it and the inclined plane are frictionless. If the masses are released from rest with the connecting cord taut, what is their total kinetic energy when the 2.0 kg mass has fallen 25 cm?

FIGURE 8-39 Problem 30.

31P. A 1.50 kg snowball is shot upward at an angle of 34.0° to the horizontal with an initial speed of 20.0 m/s. (a) What is its initial kinetic energy? (b) By how much does the gravitational potential energy of the snowball–Earth system change as the snowball moves from the launch point to the point of maximum height? (c) What is that maximum height?

32P. A pendulum consists of a 2.0 kg stone swinging on a 4.0 m string of negligible mass. The stone has a speed of 8.0 m/s when it passes its lowest point. (a) What is the speed when the string is at 60° to the vertical? (b) What is the greatest angle with the vertical that the string will reach during the stone's motion? (c) If the potential energy of the pendulum–Earth system is taken to be zero at the stone's lowest point, what is the total mechanical energy of the system?

33P. The string in Fig. 8-40 is $L = 120$ cm long, and the distance d to the fixed peg at point P is 75.0 cm. When the initially stationary ball is released with the string horizontal as shown, it will swing along the dashed arc. What is its speed when it reaches (a) its lowest point and (b) its highest point after the string catches on the peg?

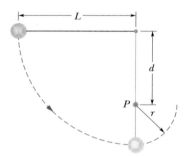

FIGURE 8-40 Problems 33 and 41.

34P. A 2.0 kg block is dropped from a height of 40 cm onto a spring of spring constant $k = 1960$ N/m (Fig. 8-41). Find the maximum distance the spring is compressed.

FIGURE 8-41 Problem 34.

35P. Figure 8-42 shows a pendulum of length L. Its bob (which effectively has all the mass) has speed v_0 when the cord makes an angle θ_0 with the vertical. (a) Derive an expression for the speed of the bob when it is in its lowest position. What is the least value that v_0 can have if the pendulum is to swing down and then up (b) to a horizontal position, and (c) to a vertical position with the cord remaining straight?

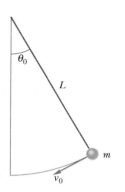

FIGURE 8-42 Problem 35.

36P. Two children are playing a game in which they try to hit a small box on the floor with a marble fired from a spring-loaded gun that is mounted on a table. The target box is 2.20 m horizontally from the edge of the table; see Fig. 8-43. Bobby compresses the spring 1.10 cm, but the center of the marble falls 27.0 cm short of the center of the box. How far should Rhoda compress the spring to score a direct hit?

FIGURE 8-43 Problem 36.

37P. The magnitude of the gravitational force between a particle of mass m_1 and one of mass m_2 is given by

$$F(x) = G\frac{m_1 m_2}{x^2},$$

where G is a constant and x is the distance between the particles. (a) What is the corresponding potential energy function $U(x)$? Assume that $U(x) \to 0$ as $x \to \infty$. (b) How much work is required to increase the separation of the particles from $x = x_1$ to $x = x_1 + d$?

38P. A 20 kg object is acted on by a conservative force given by $F = -3.0x - 5.0x^2$, with F in newtons and x in meters. Take the potential energy associated with the force to be zero when the object is at $x = 0$. (a) What is the potential energy of the system associated with the force when the object is at $x = 2.0$ m? (b) If the object has a velocity of 4.0 m/s in the negative direction of the x axis when it is at $x = 5.0$ m, what is its speed when it passes through the origin? (c) What are the answers to (a) and (b) if the potential energy of the system is taken to be -8.0 J when the object is at $x = 0$?

39P. (a) In Problem 10, what is the *net* force acting on the block when it reaches point Q? (b) At what height h should the block be released from rest so that it is on the verge of losing contact with the track at the top of the loop? (*On the verge of losing contact* means that the normal force on the block from the track has just then become zero.)

40P. Tarzan, who weighs 688 N, swings from a cliff at the end of a convenient vine that is 18 m long (Fig. 8-44). From the top of the cliff to the bottom of the swing, he descends by 3.2 m. The vine will break if the force on it exceeds 950 N. (a) Does the vine break? (b) If no, what is the greatest force on it during the swing? If yes, at what angle does it break?

FIGURE 8-44   Problem 40.

41P.   In Fig. 8-40 show that, if the ball is to swing completely around the fixed peg, then $d > 3L/5$. (*Hint:* The ball must still be moving at the top of its swing. Do you see why?)

42P.   An effectively massless rigid rod of length L has a ball with mass m attached to its end, forming a pendulum. The pendulum is inverted, with the rod straight up, and then released. What are (a) the ball's speed at the lowest point and (b) the tension in the rod at that point? (c) The same pendulum is next put in a horizontal position and released from rest. At what angle from the vertical do the magnitudes of the tension in the rod and the weight of the ball match?

43P*.   A chain is held on a frictionless table with one-fourth of its length hanging over the edge, as shown in Fig. 8-45. If the chain has length L and mass m, how much work is required to pull the hanging part back onto the table?

FIGURE 8-45   Problem 43.

44P*.   A 3.20 kg block starts at rest and slides a distance d down a frictionless 30.0° incline, where it runs into a spring (Fig. 8-46). The block slides an additional 21.0 cm before it is brought to rest momentarily by compressing the spring, whose spring constant k is 431 N/m. (a) What is the value of d? (b) What is the distance between the point of first contact and the point where the block's speed is greatest?

FIGURE 8-46   Problem 44.

45P*.   A boy is seated on the top of a hemispherical mound of ice (Fig. 8-47). He is given a very small push and starts sliding down the ice. Show that he leaves the ice at a point whose height is $2R/3$ if the ice is frictionless. (*Hint:* The normal force vanishes as he leaves the ice.)

FIGURE 8-47   Problem 45.

SECTION 8-5   Reading a Potential Energy Curve

46E.   A conservative force $F(x)$ acts on a particle that moves along the x axis. Figure 8-48 shows how the potential energy $U(x)$ associated with $F(x)$ varies with the particle's position. (a) Plot $F(x)$, using the same x scale as in Fig. 8-48. (b) The mechanical energy E of the system is 4.0 J. Plot the particle's kinetic energy $K(x)$ on Fig. 8-48.

FIGURE 8-48   Exercise 46.

47P.   The potential energy of a diatomic molecule (a two-atom system like H_2 or O_2) is given by

$$U = \frac{A}{r^{12}} - \frac{B}{r^6},$$

with r being the separation of the two atoms of the molecule and A and B being positive constants. This potential energy is associated with the force that binds the two atoms together. (a) Find the *equilibrium separation,* that is, the distance between the atoms at which the force on each atom is zero. Is the force repulsive (the atoms are pushed apart) or attractive (they are pulled together) if their separation is (b) smaller and (c) larger than the equilibrium separation?

48P.   A conservative force $F(x)$ acts on a 2.0 kg particle that

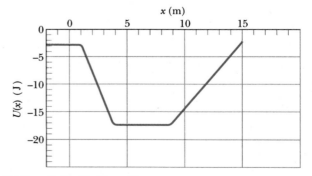

FIGURE 8-49   Problem 48.

moves along the x axis. The potential energy $U(x)$ associated with $F(x)$ is graphed in Fig. 8-49. When the particle is at $x = 2.0$ m, its velocity is -1.5 m/s. (a) What are the magnitude and direction of $F(x)$ at this position? (b) Between what limits of x does the particle move? (c) What is its speed at $x = 7.0$ m?

49P. Figure 8-50a shows a molecule consisting of two atoms of masses m and M (with $m \ll M$) and separation r. Figure 8-50b shows the potential energy $U(r)$ of the molecule as a function of r. Describe the motion of the atoms (a) if the total mechanical energy E of the two-atom system is greater than zero (as is E_1), and (b) if E is less than zero (as is E_2). For $E_1 = 1 \times 10^{-19}$ J and $r = 0.3$ nm, find (c) the potential energy of the system, (d) the total kinetic energy of the atoms, and (e) the force (magnitude and direction) acting on each atom. For what values of r is the force (f) repulsive, (g) attractive, and (h) zero?

(a)

(b)

FIGURE 8-50 Problem 49.

SECTION 8-6 Work Done by Nonconservative Forces

50E. A collie drags its bed box across a floor by applying a horizontal force of 8.0 N. The kinetic frictional force acting on the box has magnitude 5.0 N. As the box is dragged through 0.70 m along the way, (a) what is the work done by the collie's applied force and (b) how much mechanical energy is dissipated by the frictional force?

51E. The temperature of a plastic cube is monitored while the cube is pushed 3.0 m across a floor at constant speed by a horizontal force of 15 N. The monitoring reveals that the thermal energy of the cube increases by 20 J. How much work is done on the cube by the kinetic frictional force acting on it?

52P. A worker pushed a 27 kg block 9.2 m along a level floor at constant speed with a force directed 32° below the horizontal. (a) If the coefficient of kinetic friction is 0.20, how much work was done by the worker's force? (b) How much energy was dissipated by the frictional force?

53P. A 50 kg trunk is pushed 6.0 m at constant speed up a 30° incline by a constant horizontal force. The coefficient of kinetic friction between the trunk and the incline is 0.20. Calculate the work done by (a) the applied force and (b) the weight of the

trunk. (c) How much energy was dissipated by the frictional force acting on the trunk?

54P. A 3.57 kg block is drawn at constant speed 4.06 m along a horizontal floor by a rope exerting a 7.68 N force at an angle of 15.0° above the horizontal. Compute (a) the work done by the rope's force and (b) the coefficient of kinetic friction between block and floor. (c) How much energy is dissipated by the frictional force?

55P. A 1400 kg block of granite is pulled up an incline at a constant speed of 1.34 m/s by a cable and winch (Fig. 8-51). The coefficient of kinetic friction between the block and the incline is 0.40. What is the power due to the force applied to the block by the cable?

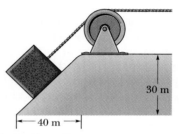

FIGURE 8-51 Problem 55.

SECTION 8-7 Conservation of Energy

56E. If a 70 kg baseball player steals home by sliding into the plate with an initial speed of 10 m/s, (a) how much kinetic energy is dissipated by the frictional force stopping him and (b) what is the change in the thermal energy of his body and the ground along which he slides?

57E. A 75 g Frisbee is thrown from a point 1.1 m above the ground with a speed of 12 m/s. When it has reached a height of 2.1 m, its speed is 10.5 m/s. (a) How much work was done on the Frisbee by its weight? (b) How much of the Frisbee's mechanical energy was dissipated by air drag?

58E. An outfielder throws a baseball with an initial speed of 120 ft/s. Just before an infielder catches the ball at the same level, its speed is 110 ft/s. How much of the ball's mechanical energy is dissipated by air drag? The weight of a baseball is 9.0 oz.

59E. A 0.63 kg ball is thrown up with an initial speed of 14 m/s and reaches a maximum height of 8.1 m. How much energy is dissipated by the air drag acting on the ball during the ascent?

60E. A 9.4 kg projectile is fired vertically upward. Air drag dissipates 68 kJ during its ascent. How much higher would it have gone were air drag negligible?

61E. A river descends 15 m through rapids. The speed of the water is 3.2 m/s upon entering the rapids and 13 m/s upon leaving. What percentage of the gravitational potential energy of the water–Earth system is transferred to kinetic energy during the descent? (*Hint:* Consider the descent of, say, 10 kg of water.)

62E. Approximately 5.5×10^6 kg of water falls 50 m over Niagara Falls each second. (a) What is the decrease in the gravitational potential energy of the water–Earth system each second?

(b) If all this energy could be converted to electrical energy by an electric generating plant (it cannot be), at what rate would electrical energy by supplied? (The mass of 1 m³ of water is 1000 kg.) (c) If the electrical energy were sold at 1 cent/kW·h, what would be the yearly cost?

63E. Each second, 1200 m³ of water passes over a waterfall 100 m high. Three-fourths of the kinetic energy gained by the water in falling is transferred to electrical energy by a hydroelectric generator. At what rate does the generator produce electrical energy? (The mass of 1 m³ of water is 1000 kg.)

64E. The area of the continental United States is about 8 × 10⁶ km², and the average elevation of its land surface is about 500 m (above sea level). The average yearly rainfall is 75 cm. Two-thirds of this rainwater returns to the atmosphere by evaporation, but the rest eventually flows into the ocean. If the associated decrease in gravitational potential energy of the water–Earth system could be fully converted to electrical energy, what would be the average power of the conversion? (The mass of 1 m³ of water is 1000 kg.)

65E. A 68 kg sky diver falls at a constant terminal speed of 59 m/s. (a) At what rate is the gravitational potential energy of the Earth–sky diver system being reduced? (b) At what rate is mechanical energy being dissipated?

66E. A 25 kg bear slides, from rest, 12 m down a lodgepole pine tree, moving with a speed of 5.6 m/s just before hitting the ground. (a) What change occurs in the gravitational potential energy of the bear–Earth system during the slide? (b) What is the kinetic energy of the bear just before hitting the ground? (c) What is the average frictional force that acts on the bear?

67E. During a rockslide, a 520 kg rock slides from rest down a hillside that is 500 m long and 300 m high. The coefficient of kinetic friction between the rock and the hill surface is 0.25. (a) If the gravitational potential energy U of the rock–Earth system is set to zero at the bottom of the hill, what is the value of U just before the slide? (b) How much mechanical energy is dissipated by frictional forces during the slide? (c) What is the kinetic energy of the rock as it reaches the bottom of the hill? (d) What is its speed then?

68E. A 30 g bullet, with a horizontal velocity of 500 m/s, stops 12 cm within a solid wall. (a) What is the change in its mechanical energy? (b) What is the magnitude of the average force from the wall stopping it?

69P. As Fig. 8-52 shows, a 3.5 kg block is accelerated by a compressed spring whose spring constant is 640 N/m. After leaving the spring at the spring's relaxed length, the block travels over a horizontal surface, with a coefficient of kinetic friction of 0.25, for a distance of 7.8 m before stopping. (a) How much mechanical energy was dissipated by the frictional force in stopping the block? (b) What was the maximum kinetic energy of the block? (c) Through what distance was the spring compressed before the block began to move?

FIGURE 8-52 Problem 69.

70P. In Fig. 8-53, a block is moved down an incline a distance of 5.0 m from point A to point B by a force **F** that is parallel to the incline and has a magnitude of 2.0 N. The magnitude of the frictional force acting on the block is 10 N. If the kinetic energy of the block increases by 35 J between A and B, how much work is done on the block by its weight as the block moves from A to B?

FIGURE 8-53 Problem 70.

71P. Fasten one end of a vertical spring to a ceiling, attach a cabbage to the other end, and then slowly lower the cabbage until the upward force exerted by the spring on the cabbage balances the weight of the cabbage. Show that the loss of gravitational potential energy of the cabbage–Earth system equals twice the gain in the spring's potential energy. Why are these two quantities not equal?

72P. You push a 2.0 kg block against a horizontal spring, compressing the spring by 15 cm. When you release the block, the spring forces it to slide across a tabletop. It stops 75 cm from where you released it. The spring constant is 200 N/m. What is the coefficient of kinetic friction between the block and the table?

73P. A moving 2.5 kg block (Fig. 8-54) collides with a horizontal spring whose spring constant is 320 N/m. The block compresses the spring a maximum distance of 7.5 cm from its rest position. The coefficient of kinetic friction between the block and the horizontal surface is 0.25. (a) How much work is done by the spring in bringing the block to rest? (b) How much mechanical energy is dissipated by the force of friction while the block is being brought to rest by the spring? (c) What was the speed of the block when it hit the spring?

FIGURE 8-54 Problem 73.

74P. Two snowy peaks are 850 m and 750 m above the valley between them. A ski run extends down from the top of the higher peak and then back up to the top of the lower one, with a total length of 3.2 km and an average slope of 30° (Fig. 8-55). (a) A skier starts from rest on the higher peak. At what speed will he arrive at the top of the lower peak if he just coasts without using the poles? Ignore friction. (b) Approximately what coefficient of kinetic friction between snow and skis would make him stop just at the top of the lower peak?

FIGURE 8-55 Problem 74.

75P. A factory worker accidentally releases a 400 lb crate that was being held at rest at the top of a 12 ft long ramp inclined at 39° to the horizontal. The coefficient of kinetic friction between the crate and the ramp, and between the crate and the horizontal factory floor, is 0.28. (a) How fast is the crate moving as it reaches the bottom of the ramp? (b) How far will it subsequently slide across the factory floor? (Assume that the crate's kinetic energy does not change as it moves from the ramp onto the floor.) (c) Why don't the answers to (a) and (b) depend on the mass of the crate?

76P. Two blocks are connected by a string, as shown in Fig. 8-56. They are released from rest. Show that, after they have moved a distance L, their common speed is given by

$$v = \sqrt{\frac{2(m_2 - \mu m_1)gL}{m_1 + m_2}},$$

in which μ is the coefficient of kinetic friction between the upper block and the surface. Assume that the pulley is massless and frictionless.

FIGURE 8-56 Problem 76.

77P. A 4.0 kg bundle starts up a 30° incline with 128 J of kinetic energy. How far will it slide up the plane if the coefficient of friction is 0.30?

78P. A cookie jar is moving up a 40° incline. At a point 1.8 ft from the bottom of the incline (measured along the incline), it has a speed of 4.5 ft/s. The coefficient of kinetic friction between jar and incline is 0.15. (a) How much farther up the incline will the jar move? (b) How fast will it be going when it slides back to the bottom of the incline?

79P. A certain spring is found *not* to conform to Hooke's law. The force (in newtons) it exerts when stretched a distance x (in meters) is found to have magnitude $52.8x + 38.4x^2$ in the direction opposing the stretch. (a) Compute the work required to stretch the spring from $x = 0.500$ m to $x = 1.00$ m. (b) With one end of the spring fixed, a particle of mass 2.17 kg is attached to the other end of the spring when it is extended by an amount $x = 1.00$ m. If the particle is then released from rest, compute its speed at the instant the spring has returned to the configuration in which the extension is $x = 0.500$ m. (c) Is the force exerted by the spring conservative or nonconservative? Explain.

80P. A girl whose weight is 267 N slides down a 6.1 m playground slide that makes an angle of 20° with the horizontal. The coefficient of kinetic friction is 0.10. (a) Find the work done on her by her weight. (b) Find the amount of energy dissipated by the frictional force. (c) If she starts at the top with a speed of 0.457 m/s, what is her speed at the bottom?

81P. In Fig. 8-57, a block slides along a track from one level to a higher level, by moving through an intermediate valley. The track is frictionless until the block reaches the higher level. There a frictional force stops the block in a distance d. The block's initial speed v_0 is 6.0 m/s; the height difference h is 1.1 m; and the coefficient of kinetic friction μ is 0.60. Find d.

FIGURE 8-57 Problem 81.

82P. In the hydrogen atom, the magnitude of the force of attraction between the positively charged nucleus (a proton) and the negatively charged electron is given by

$$F = k\frac{e^2}{r^2},$$

where e is the magnitude of the charge of the electron and the proton, k is a constant, and r is the separation between electron and nucleus. Assume that the nucleus is fixed in place. Imagine that the electron, which is initially moving in a circle of radius r_1 about the nucleus, suddenly "jumps" into a circular orbit of smaller radius r_2 (Fig. 8-58). (a) Calculate the change in kinetic energy of the electron, using Newton's second law. (b) Using the relation between force and potential energy, calculate the change in potential energy of the atom. (c) By how much has the atom's total energy decreased in this process? (The total energy of the atom must decrease to provide the energy of the light that the atom emits because of the electron's jump.)

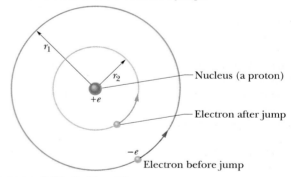

FIGURE 8-58 Problem 82.

83P. A stone with weight w is thrown vertically upward into the air from ground level with initial speed v_0. If a constant force f due to air drag acts on the stone throughout its flight, (a) show that the maximum height reached by the stone is

$$h = \frac{v_0^2}{2g(1 + f/w)}.$$

(b) Show that the speed of the stone just before impact with the ground is

$$v = v_0 \left(\frac{w - f}{w + f}\right)^{1/2}.$$

84P. A playground slide is in the form of an arc of a circle with a maximum height of 4.0 m, with a radius of 12 m, and with the ground tangent to the circle (Fig. 8-59). A 25 kg child starts from rest at the top of the slide and is observed to have a speed of 6.2 m/s at the bottom. (a) What is the length of the slide? (b) What average frictional force acts on the child over this distance? If, instead of the ground, a vertical line through the *top of the slide* is tangent to the circle, what are (c) the length of the slide and (d) the average frictional force on the child?

FIGURE 8-59   Problem 84.

85P. A particle can slide along a track with elevated ends and a flat central part, as shown in Fig. 8-60. The flat part has length L. The curved portions of the track are frictionless, but for the flat part the coefficient of kinetic friction is $\mu_k = 0.20$. The particle is released from rest at point A, which is a height $h = L/2$ above the flat part of the track. Where does the particle finally come to rest?

FIGURE 8-60   Problem 85.

86P. A massless rigid rod of length L has a ball of mass m attached to one end (Fig. 8-61). The other end is pivoted in such a way that the ball will move in a vertical circle. The system is launched downward from the horizontal position A with initial speed v_0. The ball just reaches point D and then stops. (a) Derive an expression for v_0 in terms of L, m, and g. (b) What is the tension in the rod when the ball is at B? (c) A little grit is placed on the pivot, after which the ball just reaches C when launched from A with the same speed as before. How much mechanical energy is dissipated by friction during this motion? (d) How much total mechanical energy has been dissipated by friction when the ball finally comes to rest at B after several oscillations?

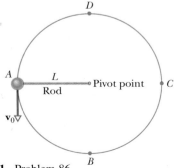

FIGURE 8-61   Problem 86.

87P. The cable of the 4000 lb elevator in Fig. 8-62 snaps when the elevator is at rest at the first floor, where the bottom is a distance $d = 12$ ft above a cushioning spring whose spring constant is $k = 10,000$ lb/ft. A safety device clamps the elevator against guide rails so that a constant frictional force of 1000 lb opposes the motion of the elevator. (a) Find the speed of the elevator just before it hits the spring. (b) Find the maximum distance x that the spring is compressed. (c) Find the distance that the elevator will bounce back up the shaft. (d) Using conservation of energy, find the approximate total distance that the elevator will move before coming to rest. Why is the answer not exact?

FIGURE 8-62   Problem 87.

88P. A metal tool is sharpened by being held against the rim of a wheel on a grinding machine by a force of 180 N. The wheel has a radius of 20 cm and rotates at 2.5 rev/s. The coefficient of kinetic friction between the wheel and the tool is 0.32. At what rate is energy being transferred from the motor driving the wheel to the thermal energy of the wheel and tool and to the kinetic energy of the material thrown from the tool?

89P*. At a factory, a 300 kg crate is dropped vertically from a packing machine onto a conveyor belt moving at 1.20 m/s (Fig. 8-63). (A motor maintains the belt's constant speed.) The coefficient of kinetic friction between belt and crate is 0.400. After a short time, slipping between the belt and the crate ceases and the crate then moves along with the belt. For the period of time during which the crate is being brought to rest relative to the belt, calculate, for a coordinate system at rest in the factory, (a) the kinetic energy supplied to the crate, (b) the magnitude of the kinetic frictional force acting on the crate, and (c) the energy supplied by the motor. (d) Explain why the answers to (a) and (c) are different.

FIGURE 8-63   Problem 89.

SECTION 8-8   Mass and Energy

90E. (a) How much energy in joules is represented by a mass of 102 g? (b) For how many years would this supply the energy needs of a one-family home consuming energy at the average rate of 1.00 kW?

91E. The magnitude M of an earthquake on the so-called Richter scale is related to the released energy E in joules by the equation

$$\log E = 5.24 + 1.44M.$$

(a) The 1906 San Francisco earthquake was of magnitude 8.2 (Fig. 8-64). How much energy was released? (b) How much mass is equivalent to this amount of energy?

92E. The United States generated about 2.31×10^{12} kW·h of electrical energy in 1983. What is the mass equivalent of this electrical energy?

93P. What is the minimum energy that is required to break a nucleus of ^{12}C (of mass 11.99671 u) into three nuclei of ^{4}He (of mass 4.00151 u each)?

94P. The nucleus of a gold atom contains 79 protons and 118 neutrons and has a mass of 196.9232 u. What is its binding energy? See Sample Problem 8-10 for additional data.

95P. In the nuclear fusion reaction $d + t \rightarrow {}^{4}He + n$, a deuteron (d) combines with the nucleus of a tritium atom (t, with one proton and two neutrons) to produce the nucleus of a helium atom (with two protons and two neutrons) and a free neutron (n). The masses involved are

$$\text{d: } 2.01355 \text{ u} \qquad {}^{4}\text{He: } 4.00151 \text{ u}$$
$$\text{t: } 3.01550 \text{ u} \qquad \text{n: } 1.00867 \text{ u}$$

(a) Does the reaction release or absorb energy? (b) How much energy is absorbed or released?

SECTION 8-9   Quantized Energy

96E. Here are the lowest five energy levels, in electron-volts, of atoms of three types:

TYPE A	TYPE B	TYPE C
3.8	3.2	3.1
3.0	2.7	2.9
2.4	2.0	2.2
1.4	1.2	1.5
0	0	0

Two emissions of light from atoms are detected at the energies of (a) 1.4 eV and (b) 1.5 eV. For each emission, identify all transitions that could have produced the light.

97E. An atom that is being monitored emits light five times without being re-excited. The energies detected for four of those emissions are 0.7 eV, 0.8 eV, 0.9 eV, and 2.0 eV. The energy information for the other emission was lost by the computer controlling the detectors, as was information about the sequence of detection. The lowest 12 energy levels of the atom, in electron-volts, are

6.5	4.1	2.6
5.3	3.8	2.0
4.9	3.4	1.5
4.5	2.9	0

(a) From which energy level did the atom begin its quantum jumping downward in energy? (b) What value of energy was lost by the computer?

ELECTRONIC COMPUTATION

98. A nerfball™ has a mass of 9.8 g and a terminal velocity in air of 7.3 m/s. It is thrown straight up with an initial speed of 15 m/s. Use numerical integration to calculate the coordinate and velocity of the ball every 0.1 s from the time it is thrown to the time it returns to its initial position. For each of those times calculate the potential energy, kinetic energy, and total mechanical energy. On the way up how much of the mechanical energy of the ball-Earth system is dissipated? How much is dissipated on the way down? Take the potential energy to be zero initially.

99. A 700 g block is released from rest at height h_0 above a vertical spring with negligible mass and a spring constant $k = 400$ N/m. The block sticks to the spring and momentarily comes to rest after compressing the spring 19.0 cm. How much work is done (a) by the block on the spring and (b) by the spring on the block? (c) What was the value of h_0? (d) If the block was released from a height $2h_0$ above the spring, what would be the maximum compression of the spring?

100. To make a pendulum, a 300 g mass is tied to the end of a string 1.4 m long. The mass is pulled to one side until the string makes an angle of 30.0° with the vertical; then (with the string taut) the mass is released from rest. Find (a) the speed of the mass when the string makes an angle of 20.0° with the vertical and (b) the maximum speed of the mass. (c) What is the angle between the string and the vertical when the speed of the mass is one third its maximum value?

FIGURE 8-64   Exercise 91. Destruction on Nob Hill in San Francisco due to the earthquake of 1906. Over 400 km of the San Andreas fault line ruptured during the earthquake.

9
Systems of Particles

If you leap forward, chances are that your head and torso will follow a parabolic path, like a baseball thrown in from the outfield. But when a skilled ballet dancer leaps across the stage in a <u>grand jeté</u> as shown in the photograph, the path taken by her head and torso is nearly horizontal during much of the jump. She seems to be floating across the stage. The audience may not know about gravitational attraction, but they still sense that something unusual has happened. How does the ballerina seemingly "turn off" gravity?

9-1 A SPECIAL POINT

Physicists love to look at something complicated and find in it something simple and familiar. Here is an example. If you flip a baseball bat into the air, its motion as it turns is clearly more complicated than that of, say, a nonspinning tossed ball (Fig. 9-1a), which moves like a particle. Every part of the bat moves in a different way from every other part, so you cannot represent the bat as a particle that is tossed into the air; instead, it is a system of particles.

However, if you look closely, you will find that one special point of the bat moves in a simple parabolic path, just as a particle would if tossed into the air (Fig. 9-1b). In fact, that special point moves as though (1) the bat's total mass were concentrated there and (2) the weight of the bat acted only there. This special point is said to be the **center of mass** of the bat. In general:

> The center of mass of a body or a system of bodies is the point that moves as though all of the mass were concentrated there and all external forces were applied there.

The center of mass of a baseball bat lies along the central axis. You can locate it by balancing the bat horizontally on an outstretched finger: the center of mass is on the bat's axis just above your finger.

9-2 THE CENTER OF MASS

We shall now spend some time determining how to find the center of mass in various systems. We start with a system of a few particles, and then we consider a system of a great many particles (as in a baseball bat).

Systems of Particles

Figure 9-2a shows two particles of masses m_1 and m_2 separated by a distance d. We have arbitrarily chosen the origin of the x axis to coincide with m_1. We *define* the position of the center of mass of this two-particle system to be

$$x_{cm} = \frac{m_2}{m_1 + m_2} d. \qquad (9\text{-}1)$$

Suppose, as an example, that $m_2 = 0$. Then there is only one particle (m_1), and the center of mass must lie at the position of that particle; Eq. 9-1 dutifully reduces to $x_{cm} = 0$. If $m_1 = 0$, there is again only one particle (m_2), and we have, as we expect, $x_{cm} = d$. If $m_1 = m_2$, the masses of the particles are equal and the center of mass

(a)

(b)

FIGURE 9-1 (a) A ball tossed into the air follows a parabolic path. (b) The center of mass (the black dot) of a baseball bat that is flipped into the air does also, but all other points of the bat follow more complicated curved paths.

FIGURE 9-2 (a) Two particles of masses m_1 and m_2 are separated by a distance d. The dot labeled cm shows the position of the center of mass, calculated from Eq. 9-1. (b) The same as (a) except that the origin is located farther from the particles. The position of the center of mass is calculated from Eq. 9-2. The relative location of the center of mass (with respect to the particles) is the same in both cases.

should be halfway between them; Eq. 9-1 reduces to $x_{cm} = \frac{1}{2}d$, again as we expect. Finally, Eq. 9-1 tells us that if neither m_1 nor m_2 is zero, x_{cm} can have only values that lie between zero and d; that is, the center of mass must lie somewhere between the two particles.

Figure 9-2b shows a more generalized situation, in which the coordinate system has been shifted leftward. The position of the center of mass is now defined as

$$x_{cm} = \frac{m_1 x_1 + m_2 x_2}{m_1 + m_2}. \qquad (9\text{-}2)$$

Note that if we put $x_1 = 0$, then x_2 becomes d and Eq. 9-2 reduces to Eq. 9-1, as it must. Note also that in spite of the shift of the coordinate system, the center of mass is still the same distance from each particle.

We can rewrite Eq. 9-2 as

$$x_{cm} = \frac{m_1 x_1 + m_2 x_2}{M}, \qquad (9\text{-}3)$$

in which M is the total mass of the system. Here, $M = m_1 + m_2$. We can extend this equation to a more general situation in which n particles are strung out along the x axis. Then the total mass is $M = m_1 + m_2 + \cdots + m_n$,

and the location of the center of mass is

$$x_{cm} = \frac{m_1 x_1 + m_2 x_2 + m_3 x_3 + \cdots + m_n x_n}{M}$$

$$= \frac{1}{M} \sum_{i=1}^{n} m_i x_i. \qquad (9\text{-}4)$$

Here the subscript i is a running number, or index, that takes on all integer values from 1 to n. It identifies the various particles, their masses, and their x coordinates.

If the particles are distributed in three dimensions, the center of mass must be identified by three coordinates. By extension of Eq. 9-4, they are

$$x_{cm} = \frac{1}{M} \sum_{i=1}^{n} m_i x_i,$$

$$y_{cm} = \frac{1}{M} \sum_{i=1}^{n} m_i y_i, \qquad (9\text{-}5)$$

$$z_{cm} = \frac{1}{M} \sum_{i=1}^{n} m_i z_i.$$

We can also define the center of mass with the language of vectors. The position of a particle whose coordinates are x_i, y_i, and z_i is given by a position vector:

$$\mathbf{r}_i = x_i \mathbf{i} + y_i \mathbf{j} + z_i \mathbf{k}. \qquad (9\text{-}6)$$

Here the index identifies the particle, and $\mathbf{i}$, $\mathbf{j}$, and $\mathbf{k}$ are unit vectors pointing, respectively, in the direction of the x, y, and z axes. Similarly, the position of the center of mass of a system of particles is given by a position vector:

$$\mathbf{r}_{cm} = x_{cm} \mathbf{i} + y_{cm} \mathbf{j} + z_{cm} \mathbf{k}. \qquad (9\text{-}7)$$

The three scalar equations of Eq. 9-5 can now be replaced by a single vector equation,

$$\mathbf{r}_{cm} = \frac{1}{M} \sum_{i=1}^{n} m_i \mathbf{r}_i, \qquad (9\text{-}8)$$

where again M is the total mass of the system. You can check that this equation is correct by substituting Eqs. 9-6 and 9-7 into it, and then separating out the x, y, and z components. The scalar relations of Eq. 9-5 result.

Rigid Bodies

An ordinary object, such as a baseball bat, contains so many particles (atoms) that we can best treat it as a continuous distribution of matter. The "particles" then become differential mass elements dm, the sums of Eq. 9-5 become integrals, and the coordinates of the center of mass are defined as

$$x_{cm} = \frac{1}{M} \int x \, dm,$$

$$y_{cm} = \frac{1}{M} \int y \, dm, \qquad (9\text{-}9)$$

$$z_{cm} = \frac{1}{M} \int z \, dm,$$

where M is now the mass of the object. The integrals are to be evaluated over all the mass elements in the object. In practice, however, we rewrite them in terms of the coordinates of the mass elements. If the object has uniform density ρ (mass per volume), then we can write

$$\rho = \frac{dm}{dV} = \frac{M}{V}, \qquad (9\text{-}10)$$

where dV is the volume occupied by a mass element dm, and V is the total volume of the object (ρ is ''rho''). We next substitute dm from Eq. 9-10 into Eq. 9-9, finding

$$x_{cm} = \frac{1}{V} \int x \, dV,$$

$$y_{cm} = \frac{1}{V} \int y \, dV, \qquad (9\text{-}11)$$

$$z_{cm} = \frac{1}{V} \int z \, dV.$$

These integrals are evaluated over the volume of the object. An example is given in Sample Problem 9-4.

Many objects have a point, a line, or a plane of symmetry. The center of mass of such an object then lies at that point, on that line, or in that plane. For example, the center of mass of a homogeneous sphere (which has a point of symmetry) is at the center of the sphere. The center of mass of a homogeneous cone (whose axis is a line of symmetry) lies on the axis of the cone. The center of mass of a banana (which has a plane of symmetry that splits it into two equal parts) lies somewhere in that plane.

The center of mass of an object need not lie within the object. There is no dough at the center of mass of a doughnut, and no iron at the center of mass of a horseshoe.

SAMPLE PROBLEM 9-1

Figure 9-3 shows three particles of masses $m_1 = 1.2$ kg, $m_2 = 2.5$ kg, and $m_3 = 3.4$ kg located at the corners of an equilateral triangle of edge $a = 140$ cm. Where is the center of mass?

SOLUTION: We choose our x and y coordinate axes so that one of the particles is located at the origin and the x axis coincides with one of the sides of the triangle. The three particles then have the following coordinates:

PARTICLE	MASS (kg)	x (cm)	y (cm)
m_1	1.2	0	0
m_2	2.5	140	0
m_3	3.4	70	121

Because of our wise choice of coordinate axes, three of the coordinates in the table are zero, simplifying the calculations. The total mass M of the system is 7.1 kg.

From Eq. 9-5, the coordinates of the center of mass are

$$x_{cm} = \frac{1}{M} \sum_{i=1}^{3} m_i x_i = \frac{m_1 x_1 + m_2 x_2 + m_3 x_3}{M}$$

$$= \frac{(1.2 \text{ kg})(0) + (2.5 \text{ kg})(140 \text{ cm}) + (3.4 \text{ kg})(70 \text{ cm})}{7.1 \text{ kg}}$$

$$= 83 \text{ cm} \qquad \text{(Answer)}$$

and

$$y_{cm} = \frac{1}{M} \sum_{i=1}^{3} m_i y_i = \frac{m_1 y_1 + m_2 y_2 + m_3 y_3}{M}$$

$$= \frac{(1.2 \text{ kg})(0) + (2.5 \text{ kg})(0) + (3.4 \text{ kg})(121 \text{ cm})}{7.1 \text{ kg}}$$

$$= 58 \text{ cm}. \qquad \text{(Answer)}$$

In Fig. 9-3, the center of mass is located by the position vector $\mathbf{r}_{cm}$, with components x_{cm} and y_{cm}.

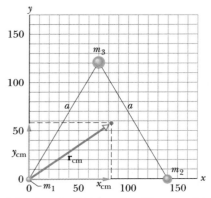

FIGURE 9-3 Sample Problem 9-1. Three particles having different masses form an equilateral triangle of side a. The center of mass is located by the position vector $\mathbf{r}_{cm}$.

SAMPLE PROBLEM 9-2

Find the center of mass of the uniform triangular plate that is shown in each part of Fig. 9-4.

SOLUTION: Figure 9-4a shows the plate divided into thin slats, parallel to one side of the triangle. From symmetry, the center of mass of a thin, uniform slat is at its midpoint. The center of mass of the triangular plate must then lie somewhere along the line that connects the midpoints of all the slats. That bisecting line also connects the upper vertex with the midpoint of the opposite side. The plate would balance if it were placed on a knife-edge coinciding with this line of symmetry.

In Figs. 9-4b and 9-4c, we subdivide the plate into slats parallel to the other two sides of the triangle. Again, the center of mass must lie somewhere along each of the bisecting lines shown. Hence the center of mass of the plate must lie at the intersection of these three symmetry lines, as Fig. 9-4d shows. It is the only point that the three lines have in common.

You can check this conclusion experimentally by taking advantage of the (correct) intuitive notion that an object suspended from a point will orient itself so that its center of mass lies vertically below that point. Suspend the triangle from each vertex in turn, and draw a line vertically downward from the suspension point, as in Fig. 9-4e. The center of mass of the triangle will be at the intersection of the three lines.

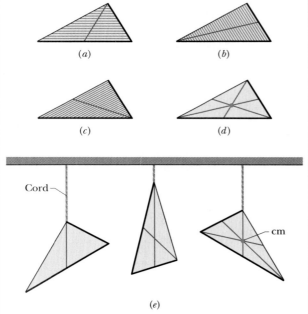

FIGURE 9-4 Sample Problem 9-2. In (a), (b), and (c), the triangular plate is divided into thin slats, parallel to a side. The center of mass must lie along the bisecting lines shown. (d) The dot, the only point common to all three lines, is the position of the center of mass. (e) Finding the center of mass by suspending the triangle from each vertex in turn.

$\mathcal{C}$HECKPOINT 1: The figure shows a uniform square plate from which four identical squares at the corners will be removed. (a) Where is the center of mass of the plate originally? Where is it after the removal of (b) square 1; (c) squares 1 and 2; (d) squares 1 and 3; (e) squares 1, 2, and 3; (f) all four squares? Answer in terms of quadrants, axes, or points (without calculation, of course).

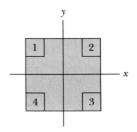

SAMPLE PROBLEM 9-3

Figure 9-5a shows a uniform circular metal plate of radius $2R$ from which a disk of radius R has been removed. Let us call this plate with a hole object X. Its center of mass is shown as a dot on the x axis. Locate this point.

SOLUTION: Figure 9-5b shows object X before the disk was removed. Call the disk object D, and the original composite plate object C. From symmetry, the center of mass of object C is at the origin of a coordinate system placed as shown.

In finding the center of mass of a composite object, we can assume that the masses of its components are concentrated at their individual centers of mass. Thus object C can be treated as equivalent to two point masses, representing objects X and D. Figure 9-5c shows the positions of the centers of mass of these three objects.

The position of the center of mass of object C is given, from Eq. 9-2, as

$$x_C = \frac{m_D x_D + m_X x_X}{m_D + m_X},$$

in which x_D and x_X are the positions of the centers of mass of objects D and X, respectively. Noting that $x_C = 0$ and solving for x_X, we obtain

$$x_X = -\frac{x_D m_D}{m_X}. \tag{9-12}$$

If ρ is the density (mass per volume) of the plate material and t is the uniform thickness of the plate, we have

$$m_D = \pi R^2 \rho t \quad \text{and} \quad m_X = \pi(2R)^2 \rho t - \pi R^2 \rho t.$$

With these substitutions and with $x_D = -R$, Eq. 9-12 becomes

$$x_X = -\frac{(-R)(\pi R^2 \rho t)}{\pi(2R)^2 \rho t - \pi R^2 \rho t} = \tfrac{1}{3}R. \quad \text{(Answer)}$$

Note that the uniform density and the uniform thickness of the plate cancel out and thus do not determine x_X.

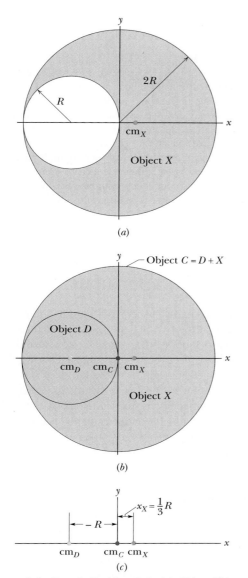

FIGURE 9-5 Sample Problem 9-3. (*a*) Object X is a metal disk of radius $2R$ with a hole of radius R cut in it; its center of mass is at point cm$_X$. (*b*) Object D is a metal disk that fills the hole in object X; its center of mass is at point cm$_D$, at $x = -R$. Object C is the composite object made up of objects X and D; its center of mass is at the origin of coordinates. (*c*) The centers of mass of the three objects.

SAMPLE PROBLEM 9-4

Silbury Hill (Fig. 9-6*a*), a mound on the plains near Stonehenge, was built 4600 years ago for unknown reasons, perhaps as a burial site. It is an incomplete right-circular cone (see Fig. 9-6*b*), with a flattened top of radius $r_2 = 16$ m, a base radius $r_1 = 88$ m, a height $h = 40$ m, and a volume V of $4.09 \times$

(*a*)

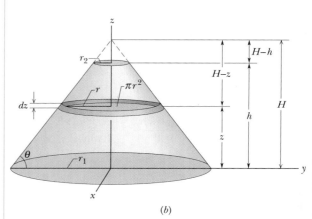

(*b*)

FIGURE 9-6 Sample Problem 9-4. (*a*) Silbury Hill in England, built by Neolithic people, required an estimated 1.8×10^7 hours of labor. (*b*) An incomplete, right-circular cone that resembles Silbury Hill. A ''wafer'' of radius r and thickness dz is shown at height z from the base.

10^5 m³. The sides of the cone make an angle $\theta = 30°$ with the horizontal.

(a) Where is the center of mass of the mound?

SOLUTION: Because of the circular symmetry of the mound, the center of mass lies on the central vertical axis of the cone, at height z_{cm} above the base. To find z_{cm}, we use the last part of Eq. 9-11. We can simplify the integral by using the symmetry of the mound. To do this, we consider a thin, horizontal ''wafer,'' as shown in Fig. 9-6*b*. The wafer has radius r, thickness dz, and horizontal area πr^2 and is at height z from the base. Its volume dV is

$$dV = \pi r^2 \, dz. \tag{9-13}$$

The mound consists of a stack of such wafers, with radii ranging from r_1 at the bottom of the stack to r_2 at the top. If the cone were complete, it would have a height that we call H in

Fig. 9-6b. The radius r of each wafer is related to H by

$$\tan \theta = \frac{H}{r_1} = \frac{H - z}{r},$$

or
$$r = (H - z)\frac{r_1}{H}. \tag{9-14}$$

Substituting Eqs. 9-13 and 9-14 into the last part of Eq. 9-11, we have

$$z_{cm} = \frac{1}{V}\int z\, dV = \frac{\pi r_1^2}{VH^2}\int_0^h z(H - z)^2\, dz$$

$$= \frac{\pi r_1^2}{VH^2}\int_0^h (z^3 - 2z^2H + zH^2)\, dz$$

$$= \frac{\pi r_1^2}{VH^2}\left[\frac{z^4}{4} - \frac{2z^3H}{3} + \frac{z^2H^2}{2}\right]_0^h$$

$$= \frac{\pi r_1^2 h^4}{VH^2}\left[\frac{1}{4} - \frac{2H}{3h} + \frac{H^2}{2h^2}\right].$$

Substituting known values, we find

$$z_{cm} = \frac{\pi(88\text{ m})^2(40\text{ m})^4}{(4.09 \times 10^5\text{ m}^3)(50.8\text{ m})^2}$$

$$\times \left[\frac{1}{4} - \frac{2(50.8\text{ m})}{3(40\text{ m})} + \frac{(50.8\text{ m})^2}{2(40\text{ m})^2}\right]$$

$$= 12.37\text{ m} \approx 12\text{ m}. \qquad \text{(Answer)}$$

(b) If Silbury Hill has density $\rho = 1.5 \times 10^3$ kg/m³, then how much work was required to lift the dirt from the level of the base to build the mound?

SOLUTION: To find the work dW required to lift a mass element dm to height z, we use Eq. 7-21, with $\phi = 180°$:

$$dW = -dm\, gz\cos 180° = dm\, gz.$$

With Eq. 9-10, we substitute $\rho\, dV$ for dm, finding

$$dW = \rho gz\, dV.$$

To find the total work required to lift all the mass of Silbury Hill into place, we sum, via integration, the work dW associated with each volume element dV:

$$W = \int dW = \int \rho gz\, dV = \rho g\int z\, dV.$$

Using Eq. 9-11, we replace the integral with Vz_{cm}, finding

$$W = \rho V g z_{cm}. \tag{9-15}$$

This tells us that the work required to lift all the mass of Silbury Hill into place is the same as if all the mass were lifted to (and somehow concentrated at) the center of mass of the hill. Substituting the known data in Eq. 9-15 yields

$$W = (1.5 \times 10^3\text{ kg/m}^3)(4.09 \times 10^5\text{ m}^3)$$

$$\times (9.8\text{ m/s}^2)(12.37\text{ m})$$

$$= 7.4 \times 10^{10}\text{ J}. \qquad \text{(Answer)}$$

PROBLEM SOLVING TACTICS

TACTIC 1: *Center-of-Mass Problems*

Sample Problems 9-1 to 9-3 provide three strategies for simplifying center-of-mass problems. (1) Make full use of the symmetry of the object, be it point, line, or plane. (2) If the object can be divided into several parts, treat each of these parts as a particle, located at its own center of mass. (3) Choose your axes wisely: if your system is a group of particles, choose one of the particles as your origin. If your system is a body with a line of symmetry, there is your x axis! The choice of origin is completely arbitrary; the location of the center of mass is the same regardless of the origin from which it is measured.

9-3 NEWTON'S SECOND LAW FOR A SYSTEM OF PARTICLES

If you roll a cue ball at a second billiard ball that is at rest, you expect that the two-ball system will continue to have some forward motion after impact. You would be surprised, for example, if both balls came back toward you or if both moved to the right or to the left.

What continues to move forward, its steady motion completely unaffected by the collision, is the center of mass of the two balls. If you focus on this point—which is always halfway between these bodies of identical mass—you can easily convince yourself by trial at a pool table that this is so. No matter whether the collision is glancing, head on, or somewhere in between, the center of mass moves majestically forward, as if the collision had never occurred. Let us look into this in more detail.

To do so, we replace the pair of billiard balls with an assemblage of n particles of (possibly) different masses. We are interested not in the individual motions of these particles but *only* in the motion of their center of mass. Although the center of mass is just a point, it moves like a particle whose mass is equal to the total mass of the system; we can assign a position, a velocity, and an acceleration to it. We state (and shall prove below) that the (vector) equation that governs the motion of the center of mass of such a system of particles is

$$\sum \mathbf{F}_{ext} = M\mathbf{a}_{cm} \qquad \text{(system of particles).} \tag{9-16}$$

Equation 9-16 is Newton's second law governing the motion of the center of mass of a system of particles. Remarkably, it retains the same form ($\Sigma\mathbf{F} = m\mathbf{a}$) that holds for the motion of a single particle. In using Eq. 9-16, the three quantities that appear in it must be evaluated with some care:

1. $\Sigma \mathbf{F}_{ext}$ is the *vector* sum of *all* the *external forces* that act on the system. Forces exerted by one part of the system on another are called *internal forces,* and you must be careful not to include them when using Eq. 9-16.

2. M is the *total mass* of the system. We assume that no mass enters or leaves the system as it moves, so that M remains constant. The system is said to be **closed.**

3. $\mathbf{a}_{cm}$ is the acceleration of the *center of mass* of the system. Equation 9-16 gives no information about the acceleration of any other point of the system.

Equation 9-16, like all other vector equations, is equivalent to three equations involving the components of $\Sigma \mathbf{F}_{ext}$ and $\mathbf{a}_{cm}$ along the three coordinate axes. These equations are

$$\sum F_{ext,x} = Ma_{cm,x},$$
$$\sum F_{ext,y} = Ma_{cm,y}, \qquad (9\text{-}17)$$
$$\sum F_{ext,z} = Ma_{cm,z}.$$

Now we can go back and examine the behavior of the billiard balls. Once the cue ball has begun to roll, no net external force acts on the (two-ball) system. And thus, because $\Sigma \mathbf{F}_{ext} = 0$, Eq. 9-16 tells us that $\mathbf{a}_{cm} = 0$ also. Because acceleration is the rate of change of velocity, we conclude that the velocity of the center of mass of the system of two balls does not change. When the two balls collide, the forces that come into play are *internal* forces, exerted by one ball on the other. Such forces do not contribute to $\Sigma \mathbf{F}_{ext}$, which remains zero. Thus the center of mass of the system, which was moving forward before the collision, must continue to move forward after the collision, with the same speed and in the same direction.

Equation 9-16 applies not only to a system of particles but also to a solid body, such as the bat of Fig. 9-1*b*. In that case, M in Eq. 9-16 is the mass of the bat and $\Sigma \mathbf{F}_{ext}$ is the weight $M\mathbf{g}$ of the bat. Equation 9-16 then tells us that $\mathbf{a}_{cm} = \mathbf{g}$. In other words, the center of mass of the bat moves as if the bat were a single particle of mass M, with force $M\mathbf{g}$ acting on it.

Figure 9-7 shows another interesting case. Suppose that at a fireworks display, a rocket is launched on a parabolic path. At a certain point, it explodes into fragments. If the explosion had not occurred, the rocket would have continued along the trajectory shown in the figure. The forces of the explosion are *internal* to the system (the rocket or its fragments); that is, they are forces exerted by one part of the system on another part. If we ignore air resistance, the total *external* force $\Sigma \mathbf{F}_{ext}$ acting on the system is the weight $M\mathbf{g}$ of the system, regardless of whether the rocket explodes. Thus, from Eq. 9-16, the acceleration

FIGURE 9-7 A fireworks rocket explodes in flight. In the absence of air drag, the center of mass of the fragments would continue to follow the original parabolic path, until fragments began to hit the ground.

of the center of mass of the fragments (while they are in flight) remains equal to $\mathbf{g}$, and the center of mass of the fragments follows the same parabolic trajectory that the unexploded rocket would have followed.

When a ballet dancer leaps across the stage in a grand jeté, she raises her arms and stretches her legs out horizontally as soon as her feet leave the stage (Fig. 9-8). These actions shift her center of mass upward through her body. Although the shifting center of mass faithfully follows a parabolic path across the stage, its movement relative to the body decreases the height that would be attained by the head and torso in a normal jump. The result is that the head and torso follow a nearly horizontal path.

Proof of Equation 9-16

Now let us prove this important equation. From Eq. 9-8 we have, for a system of n particles,

$$M\mathbf{r}_{cm} = m_1\mathbf{r}_1 + m_2\mathbf{r}_2 + m_3\mathbf{r}_3 + \cdots + m_n\mathbf{r}_n, \quad (9\text{-}18)$$

in which M is the system's total mass and $\mathbf{r}_{cm}$ is the vector locating the position of the system's center of mass.

Differentiating Eq. 9-18 with respect to time gives

$$M\mathbf{v}_{cm} = m_1\mathbf{v}_1 + m_2\mathbf{v}_2 + m_3\mathbf{v}_3 + \cdots + m_n\mathbf{v}_n. \quad (9\text{-}19)$$

Here $\mathbf{v}_i$ ($= d\mathbf{r}_i/dt$) is the velocity of the ith particle, and $\mathbf{v}_{cm}$ ($= d\mathbf{r}_{cm}/dt$) is the velocity of the center of mass.

Differentiating Eq. 9-19 with respect to time leads to

$$M\mathbf{a}_{cm} = m_1\mathbf{a}_1 + m_2\mathbf{a}_2 + m_3\mathbf{a}_3 + \cdots + m_n\mathbf{a}_n. \quad (9\text{-}20)$$

Here $\mathbf{a}_i$ ($= d\mathbf{v}_i/dt$) is the acceleration of the ith particle, and $\mathbf{a}_{cm}$ ($= d\mathbf{v}_{cm}/dt$) is the acceleration of the center of mass. Although the center of mass is just a geometrical point, it has a position, a velocity, and an acceleration, as if it were a particle.

From Newton's second law, $m_i\mathbf{a}_i$ is equal to the result-

Path of head

Path of center of mass

FIGURE 9-8 A grand jeté. (Adapted from *The Physics of Dance,* by Kenneth Laws, Schirmer Books, 1984.)

ant force $\mathbf{F}_i$ that acts on the *i*th particle. Thus we can rewrite Eq. 9-20 as

$$M\mathbf{a}_{cm} = \mathbf{F}_1 + \mathbf{F}_2 + \mathbf{F}_3 + \cdots + \mathbf{F}_n. \quad (9\text{-}21)$$

Among the forces that contribute to the right side of Eq. 9-21 will be forces that the particles of the system exert on each other (internal forces) and forces exerted on the particles from outside the system (external forces). By Newton's third law, the internal forces form action–reaction pairs and cancel out in the sum that appears on the right side of Eq. 9-21. What remains is the vector sum of all the *external* forces that act on the system. Equation 9-21 then reduces to Eq. 9-16, the relation that we set out to prove.

CHECKPOINT **2:** Two skaters on frictionless ice hold opposite ends of a pole of negligible mass. An axis runs along the pole, and the origin of the axis is at the center of mass of the two-skater system. One skater, Fred, weighs twice a much as the other skater, Ethel. Where do the skaters meet if (a) Fred pulls hand over hand along the pole so as to draw himself to Ethel, (b) Ethel pulls hand over hand to draw herself to Fred, and (c) both skaters pull hand over hand?

SAMPLE PROBLEM 9-5

Figure 9-9*a* shows a system of three particles, each particle acted on by a different external force and all initially at rest. What is the acceleration of the center of mass of this system?

SOLUTION: The position of the center of mass, calculated by the method of Sample Problem 9-1, is marked by a dot in the figure. As Fig. 9-9*b* suggests, we treat this point as if it were a real particle, assigning to it a mass M equal to the total mass of the system (16 kg) and assuming that all external forces are applied at that point.

The *x* component of the net external force $\Sigma\mathbf{F}_{ext}$ acting on the center of mass is

$$\sum F_{ext,x} = 14 \text{ N} - 6.0 \text{ N} + (12 \text{ N})(\cos 45°)$$
$$= 16.5 \text{ N},$$

and the *y* component is

$$\sum F_{ext,y} = (12 \text{ N})(\sin 45°) = 8.49 \text{ N}.$$

The net external force thus has the magnitude

$$\sum F_{ext} = \sqrt{(16.5 \text{ N})^2 + (8.49 \text{ N})^2} = 18.6 \text{ N}$$

and makes an angle with the *x* axis given by

$$\theta = \tan^{-1}\frac{8.49 \text{ N}}{16.5 \text{ N}} = \tan^{-1} 0.515$$
$$= 27°. \qquad \text{(Answer)}$$

This is also the direction of the acceleration $\mathbf{a}_{cm}$ of the center of mass. From Eq. 9-16, the magnitude of $\mathbf{a}_{cm}$ is given by

$$a_{cm} = \frac{\sum F_{ext}}{M} = \frac{18.6 \text{ N}}{16 \text{ kg}} = 1.16 \text{ m/s}^2$$
$$\approx 1.2 \text{ m/s}^2. \qquad \text{(Answer)}$$

The three particles of Fig. 9-9*a* and their center of mass move with (different) constant accelerations. Since the particles start from rest, each will move, with ever-increasing speed, along a straight line in the direction of the force acting *on it*. The center of mass will move in the directon of $\mathbf{a}_{cm}$.

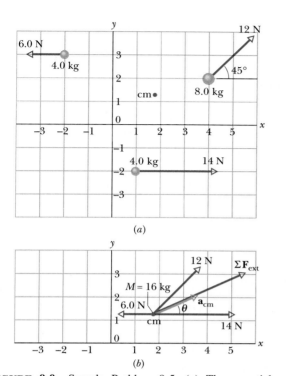

FIGURE 9-9 Sample Problem 9-5. (*a*) Three particles, placed at rest in the positions shown, are acted on by the external forces shown. The center of mass of the system is marked. (*b*) The forces are now transferred to the center of mass of the system, which behaves like a particle whose mass *M* is equal to the total mass of the system. The net force and the acceleration of the center of mass are shown.

9-4 LINEAR MOMENTUM

Momentum is another word that has several meanings in everyday language but only a single precise meaning in physics. The **linear momentum** of a particle is a vector **p**, defined as

$$\mathbf{p} = m\mathbf{v} \qquad \begin{array}{c}\text{(linear momentum} \\ \text{of a particle),}\end{array} \qquad (9\text{-}22)$$

in which *m* is the mass of the particle and **v** is its velocity. (The adjective *linear* is often dropped, but it serves to distinguish **p** from *angular* momentum, which is introduced in Chapter 12 and which is associated with rotation.) Note that since *m* is always a positive scalar quantity, Eq. 9-22 tells us that **p** and **v** are in the same direction. Equation 9-22 also tells us that the SI unit for momentum is the kilogram-meter per second.

Newton actually expressed his second law of motion in terms of momentum:

The rate of change of the momentum of a particle is proportional to the net force acting on the particle and is in the direction of that force.

In equation form this becomes

$$\sum \mathbf{F} = \frac{d\mathbf{p}}{dt}. \qquad (9\text{-}23)$$

Substituting for **p** from Eq. 9-22 gives

$$\sum \mathbf{F} = \frac{d\mathbf{p}}{dt} = \frac{d}{dt}(m\mathbf{v}) = m\frac{d\mathbf{v}}{dt} = m\mathbf{a}.$$

Thus the relations $\sum \mathbf{F} = d\mathbf{p}/dt$ and $\sum \mathbf{F} = m\mathbf{a}$ are completely equivalent expressions of Newton's second law of motion as it applies to the motion of single particles in classical mechanics.

Momentum at Very High Speeds

For particles moving with speeds that are near the speed of light, Newtonian mechanics predicts results that do not agree with experiment. In such cases, we must use Einstein's theory of special relativity. In relativity, the formulation $\mathbf{F} = d\mathbf{p}/dt$ is correct, *provided* we define the momentum of a particle not as *m***v** but as

$$\mathbf{p} = \frac{m\mathbf{v}}{\sqrt{1 - (v/c)^2}}, \qquad (9\text{-}24)$$

in which *c*, the speed of light, is a sure indicator of a relativistic equation.

The speeds of common macroscopic objects such as baseballs, bullets, or space probes are so much lower than the speed of light that the quantity $(v/c)^2$ in Eq. 9-24 is very close to zero. Under these conditions, Eq. 9-24 reduces to Eq. 9-22 and Einstein's relativity theory reduces to Newtonian mechanics. For electrons and other subatomic particles, however, speeds very close to that of light are easily obtained and the definition in Eq. 9-24 *must* be used, often as a matter of routine engineering practice.

9-5 THE LINEAR MOMENTUM OF A SYSTEM OF PARTICLES

Consider now a system of *n* particles, each with its own mass, velocity, and linear momentum. The particles may interact with each other, and external forces may act on them as well. The system as a whole has a total linear momentum **P**, which is defined to be the vector sum of the individual particles' linear momenta. Thus

TABLE 9-1   SOME DEFINITIONS AND LAWS IN CLASSICAL MECHANICS

LAW OR DEFINITION	SINGLE PARTICLE	SYSTEM OF PARTICLES
Newton's second law	$\Sigma \mathbf{F} = m\mathbf{a}$	$\Sigma \mathbf{F}_{ext} = M\mathbf{a}_{cm}$ (9-16)
Linear momentum	$\mathbf{p} = m\mathbf{v}$ (9-22)	$\mathbf{P} = M\mathbf{v}_{cm}$ (9-26)
Newton's second law	$\Sigma \mathbf{F} = \dfrac{d\mathbf{p}}{dt}$ (9-23)	$\Sigma \mathbf{F}_{ext} = \dfrac{d\mathbf{P}}{dt}$ (9-28)
Work–kinetic energy theorem	$W = \Delta K$	

$$\mathbf{P} = \mathbf{p}_1 + \mathbf{p}_2 + \mathbf{p}_3 + \cdots + \mathbf{p}_n$$
$$= m_1\mathbf{v}_1 + m_2\mathbf{v}_2 + m_3\mathbf{v}_3 + \cdots + m_n\mathbf{v}_n. \quad (9\text{-}25)$$

If we compare this equation with Eq. 9-19, we see that

$$\mathbf{P} = M\mathbf{v}_{cm} \qquad \begin{array}{l}\text{(linear momentum,}\\ \text{system of particles),}\end{array} \quad (9\text{-}26)$$

which gives us another way to define the linear momentum of a system of particles:

> The linear momentum of a system of particles is equal to the product of the total mass M of the system and the velocity of the center of mass.

If we take the time derivative of Eq. 9-26, we find

$$\frac{d\mathbf{P}}{dt} = M\frac{d\mathbf{v}_{cm}}{dt} = M\mathbf{a}_{cm}. \quad (9\text{-}27)$$

Comparing Eqs. 9-16 and 9-27 allows us to write Newton's second law for a system of particles in the equivalent form

$$\sum \mathbf{F}_{ext} = \frac{d\mathbf{P}}{dt}. \quad (9\text{-}28)$$

This equation is the generalization of the single-particle equation $\Sigma \mathbf{F} = d\mathbf{p}/dt$ to a system of many particles. Table 9-1 displays the important relations that we have derived for single particles and for systems of particles.

$\mathbf{C}$HECKPOINT 3: The figure gives the linear momentum versus time for a particle moving along an axis. A force directed along the axis acts on the particle. (a) Rank the four regions indicated according to the magnitude of the force, greatest first. (b) In which region is the particle slowing?

SAMPLE PROBLEM 9-6

Figure 9-10a shows a 2.0 kg toy race car before and after taking a turn on a track. Its speed is 0.50 m/s before the turn and 0.40 m/s after the turn. What is the change $\Delta \mathbf{P}$ in the linear momentum of the car?

SOLUTION: Before we can use Eq. 9-26 to find the car's linear momenta before and after the turn, we need its velocity $\mathbf{v}_i$ before the turn and its velocity $\mathbf{v}_f$ after the turn. Using the coordinate system of Fig. 9-10a, we write $\mathbf{v}_i$ and $\mathbf{v}_f$ as

$$\mathbf{v}_i = -(0.50 \text{ m/s})\mathbf{j} \quad \text{and} \quad \mathbf{v}_f = (0.40 \text{ m/s})\mathbf{i}.$$

Equation 9-26 then gives us the linear momentum $\mathbf{P}_i$ before the turn and the linear momentum $\mathbf{P}_f$ after the turn:

$$\mathbf{P}_i = M\mathbf{v}_i = (2.0 \text{ kg})(-0.50 \text{ m/s})\mathbf{j} = (-1.0 \text{ kg}\cdot\text{m/s})\mathbf{j}$$

and

$$\mathbf{P}_f = M\mathbf{v}_f = (2.0 \text{ kg})(0.40 \text{ m/s})\mathbf{i} = (0.80 \text{ kg}\cdot\text{m/s})\mathbf{i}.$$

Because these linear momenta are not along the same axis, we *cannot* find the change in linear momentum $\Delta \mathbf{P}$ by merely subtracting the magnitude of $\mathbf{P}_i$ from the magnitude of $\mathbf{P}_f$. Instead, we write the change in linear momentum as

$$\Delta \mathbf{P} = \mathbf{P}_f - \mathbf{P}_i \quad (9\text{-}29)$$

and then as

$$\Delta \mathbf{P} = (0.80 \text{ kg}\cdot\text{m/s})\mathbf{i} - (-1.0 \text{ kg}\cdot\text{m/s})\mathbf{j}$$
$$= (0.8\mathbf{i} + 1.0\mathbf{j}) \text{ kg}\cdot\text{m/s}. \qquad \text{(Answer)}$$

Figure 9-10b shows $\Delta \mathbf{P}$, $\mathbf{P}_f$, and $-\mathbf{P}_i$. Note that we subtract $\mathbf{P}_i$ from $\mathbf{P}_f$ by adding $-\mathbf{P}_i$ to $\mathbf{P}_f$.

(a)   (b)

FIGURE 9-10   Sample Problem 9-6. (a) A toy car takes a turn. (b) The change $\Delta \mathbf{P}$ in the car's linear momentum is the vector sum of its final linear momentum $\mathbf{P}_f$ and the negative of its initial linear momentum $\mathbf{P}_i$.

9-6 CONSERVATION
OF LINEAR MOMENTUM

Suppose that the sum of the external forces acting on a system of particles is zero (the system is isolated) and that no particles leave or enter the system (the system is closed). Putting $\Sigma\mathbf{F}_{ext} = 0$ in Eq. 9-28 then yields $d\mathbf{P}/dt = 0$, or

$$\mathbf{P} = \text{constant} \qquad \text{(closed, isolated system).} \qquad (9\text{-}30)$$

This important result, called the **law of conservation of linear momentum,** can also be written as

$$\mathbf{P}_i = \mathbf{P}_f \qquad \text{(closed, isolated system),} \qquad (9\text{-}31)$$

where the subscripts refer to the values of $\mathbf{P}$ at some initial time i and some later time f. Equations 9-30 and 9-31 tell us that if no net external force acts on a system of particles, the total linear momentum of the system remains constant.

Like the law of conservation of energy that you met in Chapter 8, the law of conservation of linear momentum is a more general law than Newtonian mechanics itself. It holds in the subatomic realm, where Newton's laws fail. It holds for the highest particle speeds, where Einstein's relativity theory prevails, that is, where Eq. 9-24, rather than Eq. 9-22, must be used to find the linear momentum.

From Eq. 9-26 ($\mathbf{P} = M\mathbf{v}_{cm}$) we see that if $\mathbf{P}$ is constant, then $\mathbf{v}_{cm}$, the velocity of the center of mass, is also constant. This in turn means that $\mathbf{a}_{cm}$, the acceleration of the center of mass, must be zero, which is consistent with Eq. 9-16 ($\Sigma\mathbf{F}_{ext} = M\mathbf{a}_{cm}$) when $\Sigma\mathbf{F}_{ext} = 0$. Thus the law of conservation of linear momentum is consistent with both forms of Newton's second law for systems of particles, as displayed in Table 9-1.

Equations 9-30 and 9-31 are vector equations and, as such, both are equivalent to three equations corresponding to the conservation of linear momentum in three mutually perpendicular directions. Depending on the forces acting on a system, linear momentum might be conserved in one or two directions but not in all directions:

> If a component of the net *external* force on a closed system is zero along an axis, then the component of the linear momentum of the system along that axis cannot change.

As an example, suppose that you toss a grapefruit across a room. During its flight, the only external force acting on the grapefruit (which we take as the system) is its weight $m\mathbf{g}$, which is directed vertically downward. Thus, the vertical component of the linear momentum of the grapefruit changes, but since no horizontal external force acts on the grapefruit, the horizontal component of the linear momentum cannot change.

Note we focus on external forces acting on a closed system. Although internal forces can change the linear momentum of portions of the system, they cannot change the total linear momentum of the system itself.

CHECKPOINT **4:** An initially stationary device lying on a frictionless floor explodes into two pieces, which then slide across the floor. One piece slides in the positive direction of an x axis. (a) What is the sum of the momenta of the two pieces after the explosion? (b) Can the second piece move at an angle to the x axis? (c) What is the direction of the momentum of the second piece?

SAMPLE PROBLEM 9-7

A box with mass $m = 6.0$ kg slides with speed $v = 4.0$ m/s across a frictionless floor in the positive direction of an x axis. It suddenly explodes into two pieces: one piece, with mass $m_1 = 2.0$ kg, moves in the positive direction of the x axis with speed $v_1 = 8.0$ m/s. What is the velocity of the second piece, with mass m_2?

SOLUTION: Our reference frame will be that of the floor. Our system, which initially consists of the original box and then of the two pieces, is closed but is not isolated, since the box and pieces have weight and each experiences a normal force from the floor. However, those forces are all vertical and thus cannot change the horizontal component of the momentum of the system. Neither can the forces produced by the explosion, because those forces are internal to the system. Thus, the horizontal component of the momentum of the system is conserved, and we can apply Eq. 9-31 along the x axis.

The initial momentum of the system was that of the box:

$$\mathbf{P}_i = m\mathbf{v}.$$

Similarly, we can write the final momenta of the two pieces as

$$\mathbf{P}_{f1} = m_1\mathbf{v}_1 \quad \text{and} \quad \mathbf{P}_{f2} = m_2\mathbf{v}_2.$$

The final total momentum $\mathbf{P}_f$ of the system is the vector sum of the momenta of the two pieces:

$$\mathbf{P}_f = \mathbf{P}_{f1} + \mathbf{P}_{f2} = m_1\mathbf{v}_1 + m_2\mathbf{v}_2.$$

Since all the velocities and momenta in this problem are vectors along a single axis, we can represent them by their magnitudes, along with an implied plus sign or an explicit minus sign. Doing so while applying Eq. 9-31, we now obtain

$$P_i = P_f$$

or

$$mv = m_1v_1 + m_2v_2.$$

Inserting the given data, and using the fact that the mass of the second piece is $m_2 = m - m_1 = 4.0$ kg, we find

$$(6.0 \text{ kg})(4.0 \text{ m/s}) = (2.0 \text{ kg})(8.0 \text{ m/s}) + (4.0 \text{ kg})v_2$$

and thus

$$v_2 = 2.0 \text{ m/s.} \qquad \text{(Answer)}$$

Since the result is positive, the second piece moves in the positive direction of the x axis.

SAMPLE PROBLEM 9-8

As Fig. 9-11 shows, a cannon whose mass M is 1300 kg fires a ball whose mass m is 72 kg in a horizontal direction with a velocity **v** relative to the cannon, which recoils (freely) with velocity **V** relative to the Earth. The magnitude of **v** is 55 m/s.

(a) What is **V**?

SOLUTION: We choose the cannon plus the ball as our system. By doing so, we ensure that the forces involved in the firing of the cannon are internal to the system, and we do not have to deal with them. The external forces acting on the system have no components in the horizontal direction. Thus the horizontal component of the total linear momentum of the system must remain unchanged as the cannon is fired.

We choose the Earth as a reference frame. Because all the velocities here are horizontal, they are positive if they point rightward in Fig. 9-11 and negative if they point leftward. Although **V** is shown pointing leftward, we actually do not yet know its direction.

The velocity of the ball relative to the Earth is v_E, represented by v_E. The ball's velocity relative to the cannon is the difference between its velocity relative to the Earth and the cannon's velocity relative to the Earth. We represent this as

$$v = v_E - V,$$

and thus

$$v_E = v + V. \qquad (9\text{-}32)$$

Before the cannon is fired, the system has an initial linear momentum P_i of zero. While the ball is in flight, the system has a horizontal linear momentum P_f which, with the aid of Eq. 9-32, is given by

$$P_f = MV + mv_E = MV + m(v + V),$$

in which the first term on the right is the linear momentum of the cannon and the second term that of the ball.

Conservation of linear momentum in the horizontal direction requires that $P_i = P_f$, or

$$0 = MV + m(v + V).$$

Solving for V yields

$$V = -\frac{mv}{M + m} = -\frac{(72 \text{ kg})(55 \text{ m/s})}{1300 \text{ kg} + 72 \text{ kg}}$$

$$= -2.9 \text{ m/s.} \qquad \text{(Answer)}$$

The minus sign tells us that the cannon recoils to the left in Fig. 9-11, as we know it does.

(b) What is v_E?

SOLUTION: From Eq. 9-32, we find

$$v_E = v + V = 55 \text{ m/s} + (-2.9 \text{ m/s})$$

$$= 52 \text{ m/s.} \qquad \text{(Answer)}$$

Because of the recoil, the ball is moving a little slower relative to the Earth than it would were there no recoil.

Note the importance in this problem of choosing the system (*cannon + ball*) wisely and of being clear about the reference frame (Earth or recoiling cannon) to which the various measurements are referenced.

FIGURE 9-11   Sample Problem 9-8. A cannon of mass M fires a ball of mass m. The ball has velocity v_E relative to the Earth and velocity **v** relative to the cannon. The recoiling cannon has velocity **V** relative to the Earth.

SAMPLE PROBLEM 9-9

Imagine a spaceship and cargo module, of total mass M, traveling in deep space with velocity $v_i = 2100$ km/h relative to the Sun (Fig. 9-12a). With a small explosion, the ship ejects the cargo module, of mass $0.20M$ (Fig. 9-12b). The ship then travels 500 km/h faster than the module; that is, the relative speed v_{rel} between the module and the ship is 500 km/h. What then is the velocity v_f of the ship relative to the Sun?

SOLUTION: Because the ship–module system is closed and isolated, its total linear momentum is conserved; that is,

$$P_i = P_f, \qquad (9\text{-}33)$$

where the subscripts i and f refer to values before and after the ejection, respectively. Before the ejection, we have

$$P_i = Mv_i. \qquad (9\text{-}34)$$

Let U be the velocity of the ejected module relative to the Sun. The total linear momentum of the system after the ejection is then

$$P_f = (0.20M)U + (0.80M)v_f, \qquad (9\text{-}35)$$

where the first term on the right is the linear momentum of the module and the second term is that of the ship.

The relative speed v_{rel} between the module and the ship is the difference in their velocities

$$v_{rel} = v_f - U,$$

from which we write

$$U = v_f - v_{rel}.$$

Substituting this expression for U into Eq. 9-35, and then substituting Eqs. 9-34 and 9-35 into Eq. 9-33, we find

$$Mv_i = 0.20M(v_f - v_{rel}) + 0.80Mv_f,$$

which gives us

$$v_f = v_i + 0.20v_{rel},$$

or

$$v_f = 2100 \text{ km/h} + (0.20)(500 \text{ km/h})$$
$$= 2200 \text{ km/h}. \qquad \text{(Answer)}$$

FIGURE 9-12 Sample Problem 9-9. (*a*) A spaceship, with a cargo module, moving at velocity $\mathbf{v}_i$. (*b*) The spaceship has ejected the cargo module. The spaceship now moves at velocity $\mathbf{v}_f$ and the cargo module moves at velocity $\mathbf{U}$.

$\mathbb{C}$HECKPOINT **5:** The table gives velocities of the ship and module in Sample Problem 9-9 (after ejection and relative to the Sun), and the relative speed between the ship and the module for three situations. What are the missing values?

	VELOCITIES (km/h)		RELATIVE SPEED (km/h)
	MODULE	SHIP	
(a)	1500	2000	
(b)		3000	400
(c)	1000		600

SAMPLE PROBLEM 9-10

Figure 9-13 shows two blocks that are connected by an ideal spring and are free to slide on a frictionless horizontal surface. Block 1 has mass m_1 and block 2 has mass m_2. The blocks are pulled in opposite directions (stretching the spring) and then released from rest.

(a) What is the ratio v_1/v_2 of the velocity of block 1 to the velocity of block 2 as the separation between the blocks decreases?

SOLUTION: We take the two blocks and the spring as our system, and the horizontal surface on which they slide as our reference frame. The initial momentum P_i of the system before the blocks are released is zero. The momentum of the system at any instant as the blocks move toward each other is

$$P_f = m_1v_1 + m_2v_2.$$

Conservation of momentum requires that $P_i = P_f$, or

$$0 = m_1v_1 + m_2v_2. \qquad (9\text{-}36)$$

Thus we have

$$\frac{v_1}{v_2} = -\frac{m_2}{m_1}. \qquad (9\text{-}37)$$

The minus sign tells us that the two velocities always have opposite signs. Equation 9-37 holds at every instant after release, no matter what the actual speeds of the blocks are.

(b) What is the ratio K_1/K_2 of the kinetic energies of the blocks as their separation decreases?

SOLUTION: We can write the ratio K_1/K_2 as

$$\frac{K_1}{K_2} = \frac{\frac{1}{2}m_1v_1^2}{\frac{1}{2}m_2v_2^2} = \frac{m_1}{m_2}\left(\frac{v_1}{v_2}\right)^2.$$

Substituting for v_1/v_2 from Eq. 9-37 and simplifying, we find

$$\frac{K_1}{K_2} = \frac{m_2}{m_1}. \qquad (9\text{-}38)$$

As the blocks move toward each other and the stretch of the spring decreases, energy is transferred from the elastic potential energy of the spring to the kinetic energies K_1 and K_2 of the blocks. Although K_1 and K_2 then increase, Eq. 9-38 tells us that their ratio does not change but is preset by the ratio of the masses. After the spring reaches its rest length and begins to be compressed by the blocks, the energy transfer is reversed, but Eq. 9-38 still holds.

Equations 9-36 through 9-38 apply to other situations in which two bodies attract (or repel) each other. For example, they apply to a stone falling toward the Earth. The stone corresponds to, say, block 1 in Fig. 9-13 and the Earth to block 2. The gravitational force between the stone and the Earth corresponds to the mutual force between the blocks that is provided by the spring in Fig. 9-13. Our reference frame is the frame in which the center of mass of the stone–Earth system is stationary. In this frame, Eq. 9-36 tells us that the magnitudes of the linear momenta of the stone and the Earth remain equal throughout the fall. Equations 9-37 and 9-38 tell us that, because $m_2 \gg m_1$, the stone has much greater speed and kinetic energy than the Earth during the fall.

FIGURE 9-13 Sample Problem 9-10. Two blocks, resting on a frictionless surface and connected by a spring, have been pulled apart and then released from rest. The vector sum of their linear momenta remains zero during their subsequent motions. The system boundary is shown.

SAMPLE PROBLEM 9-11

A firecracker placed inside a coconut of mass M, initially at rest on a frictionless floor, blows the fruit into three pieces and sends them sliding across the floor. An overhead view is shown in Fig. 9-14a. Piece C, with mass 0.30M, has final speed $v_{fC} = 5.0$ m/s.

(a) What is the speed of piece B, with mass 0.20M?

SOLUTION: We superimpose an xy coordinate system as shown in Fig. 9-14b, with the direction of decreasing x coinciding with the direction of $\mathbf{v}_{fA}$. The x axis makes an angle of 80° with $\mathbf{v}_{fC}$ and an angle of 50° with $\mathbf{v}_{fB}$.

The linear momentum of the coconut (and its pieces) is conserved along both the x and y axes, because the forces involved in the explosion are internal forces, and because no external force acts on the coconut along the x or y axis. Along the y axis the conservation of linear momentum is written as

$$P_{iy} = P_{fy}, \qquad (9\text{-}39)$$

where the subscript i refers to the initial value (before the explosion), and the subscript y refers to the y component of the vector $\mathbf{P}_i$ or $\mathbf{P}_f$.

The component P_{iy} of the initial linear momentum is zero, because the coconut is initially at rest. To get an expression for P_{fy}, we find the y component of the final momentum of each piece:

$$p_{fA,y} = 0,$$

$$p_{fB,y} = -0.20Mv_{fB,y} = -0.20Mv_{fB} \sin 50°,$$

$$p_{fC,y} = 0.30Mv_{fC,y} = 0.30Mv_{fC} \sin 80°.$$

(Note that $p_{fA,y} = 0$ because of our choice of axes.) Equation 9-39 can now be written as

$$P_{iy} = P_{fy} = p_{fA,y} + p_{fB,y} + p_{fC,y}.$$

Then, with $v_{fC} = 5.0$ m/s, we have

$$0 = 0 - 0.20Mv_{fB} \sin 50° + (0.30M)(5.0 \text{ m/s}) \sin 80°,$$

from which we find

$$v_{fB} = 9.64 \text{ m/s} \approx 9.6 \text{ m/s}. \qquad \text{(Answer)}$$

(b) What is the speed of piece A?

SOLUTION: Because linear momentum is also conserved along the x axis, we have

$$P_{ix} = P_{fx}, \qquad (9\text{-}40)$$

where $P_{ix} = 0$ because the coconut is initially at rest. To get P_{fx}, we find the x components of the final momenta, using the fact that piece A must have a mass of 0.50M:

$$p_{fA,x} = -0.50Mv_{fA},$$

$$p_{fB,x} = 0.20Mv_{fB,x} = 0.20Mv_{fB} \cos 50°,$$

$$p_{fC,x} = 0.30Mv_{fC,x} = 0.30Mv_{fC} \cos 80°.$$

Equation 9-40 can now be written as

$$P_{ix} = P_{fx} = p_{fA,x} + p_{fB,x} + p_{fC,x}.$$

Then, with $v_{fC} = 5.0$ m/s and $v_{fB} = 9.64$ m/s, we have

$$0 = -0.50Mv_{fA} + 0.20M(9.64 \text{ m/s}) \cos 50°$$

$$+ 0.30M(5.0 \text{ m/s}) \cos 80°,$$

from which we find

$$v_{fA} = 3.0 \text{ m/s}. \qquad \text{(Answer)}$$

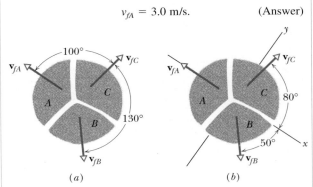

FIGURE 9-14   Sample Problem 9-11. Three pieces of an exploded coconut move off in three directions along a frictionless floor. (a) An overhead view of the event. (b) The same with a two-dimensional axis system imposed.

CHECKPOINT 6: Suppose that the exploding coconut in Sample Problem 9-11 is accelerating in the negative direction of y in Fig. 9-14 (say it is on a ramp that slants downward in that direction). Is linear momentum conserved along (a) the x axis (as stated in Eq. 9-40) and (b) the y axis (as stated in Eq. 9-39)?

PROBLEM SOLVING TACTICS

TACTIC 2: *Conservation of Linear Momentum*

This is a good time to reread Tactic 2 of Chapter 8, which deals with problems involving the conservation of mechanical energy. The same broad suggestions hold here for problems involving the conservation of linear momentum.

First, make sure that you have chosen a closed, isolated system. *Closed* means that no matter (no particles) passes through the system boundary in any direction. *Isolated* means that the net external force acting on the system is zero. If the system is *not* isolated or *not* closed, then Eqs. 9-30 and 9-31 do not hold.

Remember that linear momentum is a vector quantity so that each component can be conserved separately, provided only that the corresponding component of the net external force is zero. In Sample Problem 9-8, no net external force acts horizontally on the *cannon + ball* system, so the horizontal component of linear momentum is conserved. However, the net external force acting vertically on this system is not zero; weight acts on the cannonball while it is in flight.

Thus the vertical component of linear momentum for this system is not conserved.

Select two appropriate states of the system (which you may choose to call the initial state and the final state) and write expressions for the linear momentum of the system in each of these two states. In writing these expressions, make sure that you know what inertial reference frame you are using, and make sure also that you include the entire system, not missing any part of it and not including objects that do not belong to your system. In Sample Problem 9-8 we must be particularly clear about the reference frame (Earth or recoiling cannon) to which the various measurements are referenced.

Finally, set your expressions for P_i and P_f equal to each other and solve for what is requested.

9-7 SYSTEMS WITH VARYING MASS: A ROCKET (OPTIONAL)

In the systems we have dealt with so far, we have assumed that the total mass of the system remains constant. Sometimes, as in a rocket (Fig. 9-15), it does not. Most of the mass of a rocket on its launching pad is fuel, all of which will eventually be burned and ejected from the nozzle of the rocket engine.

We handle the variation of the mass of the rocket as the rocket accelerates by applying Newton's second law, not to the rocket alone but to the rocket and its ejected

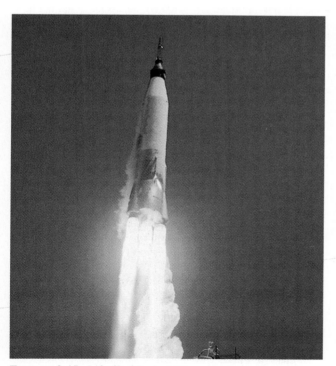

FIGURE 9-15 Liftoff of Project Mercury spacecraft.

combustion products taken together. The mass of *this* system does *not* change as the rocket accelerates.

Finding the Acceleration

Assume that you are in an inertial reference frame, watching a rocket accelerate through deep space with no gravitational or atmospheric drag forces acting on it. For this one-dimensional motion, let M be the mass of the rocket and v its velocity at an arbitrary time t (see Fig. 9-16a).

Figure 9-16b shows how things stand a time interval dt later. The rocket now has velocity $v + dv$ and mass $M + dM$, where the change in mass dM is a *negative quantity*. The exhaust products released by the rocket during interval dt have mass $-dM$ and velocity U relative to our inertial reference frame.

Our system consists of the rocket and the exhaust products released during interval dt. The system is closed and isolated, so the linear momentum of the system must be conserved during dt; that is,

$$P_i = P_f, \qquad (9\text{-}41)$$

where the subscripts i and f indicate the values at the beginning and end of time interval dt. We can rewrite Eq. 9-41 as

$$Mv = -dM\,U + (M + dM)(v + dv), \qquad (9\text{-}42)$$

where the first term on the right is the linear momentum of the exhaust products released during interval dt and the second term is the linear momentum of the rocket at the end of interval dt.

FIGURE 9-16 (*a*) An accelerating rocket of mass M at time t, as seen from an inertial reference frame. (*b*) The same but at time $t + dt$. The exhaust products released during interval dt are shown.

We can simplify Eq. 9-42 by using the speed u of the exhaust products relative to the rocket. We find this relative speed u by subtracting the velocity U of the exhaust products from the velocity $v + dv$ of the rocket at the end of interval dt:

$$u = (v + dv) - U.$$

This gives us

$$U = v + dv - u. \tag{9-43}$$

Substituting this result for U into Eq. 9-42 yields, with a little algebra,

$$-dM\, u = M\, dv. \tag{9-44}$$

Dividing each side by dt yields

$$-\frac{dM}{dt}\, u = M\,\frac{dv}{dt}. \tag{9-45}$$

We replace dM/dt (the rate at which the rocket loses mass) by $-R$, where R is the (positive) rate of fuel consumption. And we recognize that dv/dt is the acceleration of the rocket. With these changes, Eq. 9-45 becomes

$$Ru = Ma \quad \text{(first rocket equation).} \tag{9-46}$$

Equation 9-46 holds at any instant, with the mass M, the fuel consumption rate R, and the acceleration a evaluated at that instant.

The left side of Eq. 9-46 has the dimensions of a force ($\text{kg} \cdot \text{m/s}^2 = \text{N}$) and depends only on design characteristics of the rocket engine, namely, the rate R at which it consumes fuel mass and the speed u with which that mass is ejected relative to the rocket. We call this term Ru the **thrust** of the rocket engine and represent it with T. Newton's second law emerges clearly if we write Eq. 9-46 as $T = Ma$, in which a is the acceleration of the rocket at the time that its mass is M.

Finding the Velocity

How will the velocity of a rocket change as it consumes its fuel? From Eq. 9-44 we have

$$dv = -u\,\frac{dM}{M}.$$

Integrating leads to

$$\int_{v_i}^{v_f} dv = -u \int_{M_i}^{M_f} \frac{dM}{M},$$

in which M_i is the initial mass of the rocket and M_f its final mass. Evaluating the integral then gives

$$v_f - v_i = u \ln \frac{M_i}{M_f} \quad \begin{array}{l}\text{(second rocket}\\ \text{equation)}\end{array} \tag{9-47}$$

for the increase in the speed of the rocket during the change in mass from M_i to M_f.* We see here the advantage of multistage rockets, in which M_f is reduced by discarding successive stages when their fuel is depleted. An ideal rocket would reach its destination with only its payload remaining.

SAMPLE PROBLEM 9-12

A rocket whose initial mass M_i is 850 kg consumes fuel at the rate $R = 2.3$ kg/s. The speed u of the exhaust gases relative to the rocket engine is 2800 m/s.

(a) What thrust does the rocket engine provide?

SOLUTION: The thrust is

$$T = Ru = (2.3 \text{ kg/s})(2800 \text{ m/s})$$
$$= 6440 \text{ N} \approx 6400 \text{ N.} \quad \text{(Answer)}$$

(b) What is the initial acceleration of the rocket?

SOLUTION: From Newton's second law, we have

$$a = \frac{T}{M_i} = \frac{6440 \text{ N}}{850 \text{ kg}} = 7.6 \text{ m/s}^2. \quad \text{(Answer)}$$

To be launched from the Earth's surface, a rocket must have an initial acceleration greater than $g = 9.8$ m/s². Put another way, the thrust T of the rocket engine must exceed the initial weight of the rocket, which here is $M_i g = 850$ kg $\times$ 9.8 m/s² = 8300 N. Because the acceleration or thrust requirement is not met (here $T = 6400$ N), our rocket could not be launched from the Earth's surface. It would have to be launched into space by another, more powerful, rocket.

(c) Suppose, instead, that the rocket is launched from a spacecraft already in deep space, where we can neglect any gravitational force acting on it. The mass M_f of the rocket when its fuel is exhausted is 180 kg. What is its speed relative to the spacecraft at that time? Assume that the spacecraft is so massive that the launch does not alter its speed.

SOLUTION: The rocket begins with a speed of $v_i = 0$ relative to the spacecraft. From Eq. 9-47 we have

$$v_f = u \ln \frac{M_i}{M_f}$$
$$= (2800 \text{ m/s}) \ln \frac{850 \text{ kg}}{180 \text{ kg}}$$
$$= (2800 \text{ m/s}) \ln 4.72 \approx 4300 \text{ m/s.} \quad \text{(Answer)}$$

Note that the ultimate speed of the rocket can exceed the exhaust speed u.

---

*The symbol "ln" in Eq. 9-47 means the *natural logarithm*, which is the logarithm taken to the base e (= 2.718 . . .); punch "ln," not "log," on your calculator.

9-8 EXTERNAL FORCES AND INTERNAL ENERGY CHANGES

As the ice skater in Fig. 9-17a pushes herself away from a railing, the railing exerts a force $\mathbf{F}_{ext}$ on her at angle ϕ to the horizontal. This force accelerates her from an initial velocity of zero to some final velocity with which she leaves the railing (Fig. 9-17b). Thus the force increases her kinetic energy. This example differs in two ways from earlier examples in which an external force changes an object's kinetic energy:

1. Previously, each part of an object moved rigidly in the same direction; here the skater's arm does not move like the rest of her body.

2. Previously, the external force transferred energy between the object and its environment; here the external force transfers energy from an internal biochemical energy of the skater's muscles to the kinetic energy of her body as a whole.

   In spite of these two differences, we can relate the external force to the change in kinetic energy much as we did for a particle in Section 7-3. To do so, we treat the skater as a particle located at her center of mass, with all of her mass M concentrated there. Then we treat the external force $\mathbf{F}_{ext}$ as acting on the center of mass (Fig. 9-17c). The horizontal force component $F_{ext} \cos \phi$ gives the center of mass a horizontal acceleration $a_{cm,x}$, changing the velocity of the center of mass from an initial value $\mathbf{v}_{cm,0}$.

   Suppose that with displacement $\mathbf{d}_{cm}$ of the center of mass, the velocity of the center of mass becomes $\mathbf{v}_{cm}$. Then the magnitude of $\mathbf{v}_{cm}$ is, by Eq. 2-16,

$$v_{cm}^2 = v_{cm,0}^2 + 2a_{cm,x}d_{cm}. \qquad (9\text{-}48)$$

Multiplying both sides of Eq. 9-48 by mass M and rearranging yield

$$\tfrac{1}{2}Mv_{cm}^2 - \tfrac{1}{2}Mv_{cm,0}^2 = Ma_{cm,x}d_{cm}. \qquad (9\text{-}49)$$

The left side of Eq. 9-49 is the difference between the initial kinetic energy $K_{cm,i}$ of the center of mass and the final kinetic energy $K_{cm,f}$. This difference is the change ΔK_{cm} in the kinetic energy of the center of mass due to the force $\mathbf{F}_{ext}$. Making that substitution, and substituting the product $F_{ext} \cos \phi$ for the product $Ma_{cm,x}$ according to Newton's second law, we get

$$\Delta K_{cm} = F_{ext}d_{cm} \cos \phi. \qquad (9\text{-}50)$$

Because the skater is an isolated system in the sense that no energy is transferred to or from her, the change

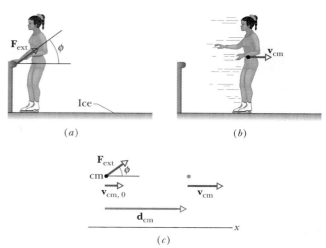

FIGURE 9-17   (a) As a skater pushes herself away from a railing, the railing exerts a force $\mathbf{F}_{ext}$ on her. (b) After the skater leaves the railing, her center of mass has velocity $\mathbf{v}_{cm}$. (c) External force $\mathbf{F}_{ext}$ is taken to act at the skater's center of mass; the vector is at angle ϕ with a horizontal x axis. When the center of mass goes through displacement $\mathbf{d}_{cm}$, its velocity is changed from $\mathbf{v}_{cm,0}$ to $\mathbf{v}_{cm}$ by the horizontal component of $\mathbf{F}_{ext}$.

ΔK_{cm} in her kinetic energy must be accompanied by a change ΔE_{int} in an internal energy. (Let us assume that the only internal energy changed is muscular biochemical energy.) From the principle of the conservation of energy, we write

$$\Delta K_{cm} + \Delta E_{int} = 0,$$

or

$$\Delta E_{int} = -\Delta K_{cm}. \qquad (9\text{-}51)$$

Substituting Eq. 9-50 into Eq. 9-51 yields the change in her internal energy:

$$\Delta E_{int} = -F_{ext}d_{cm} \cos \phi. \qquad (9\text{-}52)$$

Taken together, Eqs. 9-51 and 9-52 say that the external force transfers an amount of energy equal to $F_{ext}d_{cm} \cos \phi$ from internal energy to kinetic energy.

   If the height of the skater's center of mass changes while the external force is applied, a change ΔU_{cm} in the gravitational potential energy of the skater–Earth system occurs. Then the isolated system to which the principle of conservation of energy applies is the skater–Earth system, and that principle is written as

$$\Delta K_{cm} + \Delta U_{cm} + \Delta E_{int} = 0. \qquad (9\text{-}53)$$

Substituting Eq. 9-52 into Eq. 9-53, we find

$$\Delta K_{cm} + \Delta U_{cm} - F_{ext}d_{cm} \cos \phi = 0,$$

FIGURE 9-18   A car, initially at rest, accelerates to the right. The road exerts four frictional forces (two of them shown) on the bottom surfaces of the tires. Taken together, these four forces makeup the net external force $\mathbf{F}_{ext}$ acting on the car.

or

$$\Delta E = \Delta K_{cm} + \Delta U_{cm} = F_{ext}d_{cm} \cos \phi, \quad (9\text{-}54)$$

where ΔE is the change in the mechanical energy of the skater–Earth system. In words, Eq. 9-54 tells us that the external force transfers an amount of energy equal to the product $F_{ext}d_{cm} \cos \phi$ from internal energy to the mechanical energy of the system, increasing either the kinetic energy of the skater or the height of the skater, or both.

If the external force $\mathbf{F}_{ext}$ is not constant in magnitude, we can replace the symbol F_{ext} in Eqs. 9-52 and 9-54 with the symbol $\overline{F}_{ext}$ for the average magnitude.

Although we have used an ice skater to derive Eqs. 9-52 and 9-54, the equations hold for other objects on which an external force causes a change ΔE_{int} in an internal energy of the object. For example, consider a car that is increasing in speed. During the acceleration, the engine causes the tires to push backward on the road's surface. This push produces frictional forces that act on each tire in the forward direction (Fig. 9-18). The net external force $\mathbf{F}_{ext}$, which is the sum of these frictional forces, gives the car's center of mass an acceleration $\mathbf{a}_{cm}$. It also transfers energy from an internal energy (released by the combustion of fuel) to the kinetic energy of the car. If $\mathbf{F}_{ext}$ is constant, then for a given displacement $\mathbf{d}_{cm}$ of the car's center of mass along level road, we can relate the change ΔK_{cm} in the kinetic energy to the external force $\mathbf{F}_{ext}$ with Eq. 9-54, setting $\Delta U_{cm} = 0$ and $\phi = 0°$.

If the driver applies the brakes, Eq. 9-54 still applies. Now the net external force $\mathbf{F}_{ext}$ due to the frictional forces is toward the rear, and $\phi = 180°$. And now energy is transferred from kinetic energy of the car's center of mass to thermal energy of the brakes.

SAMPLE PROBLEM 9-13

When a click beetle is upside down on its back, it jumps upward by suddenly arching its back, transferring energy stored in a muscle to kinetic energy of its center of mass. This launching mechanism produces an audible click, giving the beetle its name. Videotape of a certain jump shows that the center of mass of a click beetle of mass $m = 4.0 \times 10^{-6}$ kg moved upward by displacement $\mathbf{d}_{cm} = 0.77$ mm during the launch and then to a maximum height of $h = 0.30$ m. What was the average magnitude of the external force $\mathbf{F}_{ext}$ on the beetle's back from the floor during the launch?

SOLUTION: The beetle–Earth system is an isolated system in the sense that no energy is transferred to or from it. Thus we can apply the principle of conservation of energy to the system during the time interval T from the beginning of the launch to the moment the center of mass reaches maximum height h. During interval T, the changes in energy are the following: (1) the change ΔK_{cm} in the kinetic energy of the center of mass is zero (the beetle is stationary both initially and at maximum height h); (2) the change ΔU_{cm} in the gravitational potential energy of the beetle–Earth system is equal to mgh; (3) during the launch, the external force $\mathbf{F}_{ext}$ on the beetle causes a change ΔE_{int} in an internal energy of the beetle.

We write the conservation of energy during time interval T as

$$\Delta K_{cm} + \Delta U_{cm} + \Delta E_{int} = 0.$$

Substituting for ΔK_{cm} and ΔU_{cm}, we have

$$0 + mgh + \Delta E_{int} = 0,$$

or

$$\Delta E_{int} = -mgh. \quad (9\text{-}55)$$

Substituting Eq. 9-52 into Eq. 9-55 yields

$$\overline{F}_{ext}d_{cm} \cos \phi = mgh,$$

or

$$\overline{F}_{ext} = \frac{mgh}{d_{cm} \cos \phi}. \quad (9\text{-}56)$$

The angle ϕ between upward force $\mathbf{F}_{ext}$ and upward displacement $\mathbf{d}_{cm}$ is $0°$. Substituting this and given data into Eq. 9-56, we find

$$\overline{F}_{ext} = \frac{(4.0 \times 10^{-6} \text{ kg})(9.8 \text{ m/s}^2)(0.30 \text{ m})}{(7.7 \times 10^{-4} \text{ m})(\cos 0°)}$$

$$= 1.5 \times 10^{-2} \text{ N}. \quad \text{(Answer)}$$

This force magnitude may seem small, but to the click beetle it is enormous because, as you can show, it gives the beetle an acceleration of over $380g$ during the launch.

REVIEW & SUMMARY

Center of Mass

The **center of mass** of a system of particles is defined to be the point whose coordinates are given by

$$x_{cm} = \frac{1}{M} \sum_{i=1}^{n} m_i x_i, \quad y_{cm} = \frac{1}{M} \sum_{i=1}^{n} m_i y_i, \quad z_{cm} = \frac{1}{M} \sum_{i=1}^{n} m_i z_i, \tag{9-5}$$

or

$$\mathbf{r}_{cm} = \frac{1}{M} \sum_{i=1}^{n} m_i \mathbf{r}_i, \tag{9-8}$$

where M is the total mass of the system. If the mass is continuously distributed, the center of mass is given by

$$x_{cm} = \frac{1}{M} \int x \, dm, \quad y_{cm} = \frac{1}{M} \int y \, dm, \quad z_{cm} = \frac{1}{M} \int z \, dm. \tag{9-9}$$

If the density (mass per volume) is uniform, then Eq. 9-9 can be rewritten as

$$x_{cm} = \frac{1}{V} \int x \, dV, \quad y_{cm} = \frac{1}{V} \int y \, dV, \quad z_{cm} = \frac{1}{V} \int z \, dV, \tag{9-11}$$

where V is the volume occupied by M.

Newton's Second Law for a System of Particles

The motion of the center of mass of any system of particles is governed by **Newton's second law for a system of particles**, which is written as

$$\sum \mathbf{F}_{ext} = M \mathbf{a}_{cm}. \tag{9-16}$$

Here $\sum \mathbf{F}_{ext}$ is the resultant of the *external* forces acting on the system, M is the total mass of the system, and $\mathbf{a}_{cm}$ is the acceleration of the system's center of mass.

Linear Momentum and Newton's Second Law

For a single particle, we define a vector quantity $\mathbf{p}$ called the **linear momentum** as

$$\mathbf{p} = m\mathbf{v}, \tag{9-22}$$

and write Newton's second law in terms of momentum:

$$\sum \mathbf{F} = \frac{d\mathbf{p}}{dt}. \tag{9-23}$$

For a system of particles these relations become

$$\mathbf{P} = M\mathbf{v}_{cm} \quad \text{and} \quad \sum \mathbf{F}_{ext} = \frac{d\mathbf{P}}{dt}. \tag{9-26, 9-28}$$

Relativistic Momentum

A more complete (relativistic) definition of linear momentum is

$$\mathbf{p} = \frac{m\mathbf{v}}{\sqrt{1 - (v/c)^2}}. \tag{9-24}$$

Equation 9-24 must be used whenever a particle's speed is near the speed of light c. It is valid under all circumstances and is equivalent to Eq. 9-22 whenever $v \ll c$.

Conservation of Linear Momentum

If a system is isolated so that no net *external* force acts on the system, the linear momentum $\mathbf{P}$ of the system remains constant:

$$\mathbf{P} = \text{constant} \quad \text{(closed, isolated system)}, \tag{9-30}$$

which can also be written as

$$\mathbf{P}_i = \mathbf{P}_f \quad \text{(closed, isolated system)}, \tag{9-31}$$

where the subscripts refer to the values of $\mathbf{P}$ at some initial time and at a later time. Equations 9-30 and 9-31 are equivalent statements of the **law of conservation of linear momentum.**

Variable-Mass Systems

For a system of varying mass, we redefine the system, enlarging its boundaries until it encompasses a larger system whose mass *does* remain constant; then we apply the law of conservation of linear momentum. For a rocket, this means that the system includes both the rocket and its exhaust gases. Such analysis shows that in the absence of external forces a rocket accelerates at an instantaneous rate given by

$$Ru = Ma \quad \text{(first rocket equation)}, \tag{9-46}$$

in which M is the rocket's instantaneous mass (including unexpended fuel), R is the fuel consumption rate, and u is the fuel's exhaust speed relative to the rocket. The term Ru is the **thrust** of the rocket engine. For a rocket with constant R and u, whose speed changes from v_i to v_f when its mass changes from M_i to M_f,

$$v_f - v_i = u \ln \frac{M_i}{M_f} \quad \text{(second rocket equation)}. \tag{9-47}$$

External Forces and Internal Energy Changes

An external force $\mathbf{F}_{ext}$ acting on an object can cause a change ΔE_{int} in an internal energy of the object:

$$\Delta E_{int} = -F_{ext} d_{cm} \cos \phi. \tag{9-52}$$

Here $\mathbf{d}_{cm}$ is the displacement of the object's center of mass and ϕ is the angle between $\mathbf{d}_{cm}$ and $\mathbf{F}_{ext}$. If the external force transfers energy from an internal energy to the kinetic energy of the object's center of mass, the change ΔK_{cm} in that kinetic energy is

$$\Delta K_{cm} = F_{ext} d_{cm} \cos \phi. \tag{9-50}$$

If the transfer is to the mechanical energy of a system containing the object and having potential energy U, then the change ΔE in the mechanical energy is

$$\Delta E = \Delta K_{cm} + \Delta U_{cm} = F_{ext} d_{cm} \cos \phi. \tag{9-54}$$

QUESTIONS

1. An amateur sculptor cuts the shape of a bird from a single sheet of metal of uniform thickness (Fig. 9-19). Of the numbered points, which is most likely to be the center of mass of the "sculpture"?

FIGURE 9-19   Question 1.

2. Figure 9-20 shows four situations in which a uniform square metal plate has had a section removed. The origin of the x and y axes is at the center of the original plate, and in each situation the center of mass of the removed section was at the origin. In each situation, where is the center of mass of the remaining section of the plate? Give quadrant, line, or point.

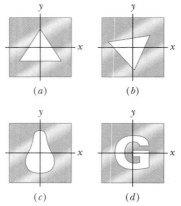

FIGURE 9-20   Question 2.

3. In Fig. 9-21, a penguin stands at the left edge of a uniform sled of length L, which lies on frictionless ice. The sled and penguin have equal masses. (a) Where is the center of mass of the sled? (b) How far and in what direction is the center of the sled from the center of mass of sled–penguin system? The penguin then waddles to the right edge of the sled and the sled slides on the ice. (c) Does the center of mass of the sled–penguin system move leftward, rightward, or not at all? (d) Now how far and in what direction is the center of the sled from the center of mass of the sled–penguin system? (e) How far does the penguin move relative to the sled? Relative to the center of mass of the sled–penguin system, how far does (f) the center of the sled move and (g) the penguin move? (Warmup for Problem 23)

FIGURE 9-21   Questions 3 and 4.

4. In Question 3 and Fig. 9-21, suppose that the sled and penguin are initially moving rightward at speed v_0. (a) As the penguin waddles to the right edge of the sled, is the speed v of the sled less than, greater than, or equal to v_0? (b) If the penguin then waddles back to the left edge, during that motion, is the speed v of the sled less than, greater than, or equal to v_0?

5. Figure 9-22 shows an overhead view of four particles of equal mass sliding over a frictionless surface at constant velocity. The directions of the velocities are indicated; their magnitudes are equal. Consider pairing the particles. Which pairs form a system with a center of mass that (a) is stationary, (b) is stationary and at the origin, and (c) passes through the origin?

FIGURE 9-22   Question 5.

6. If you drop two watermelons side by side from a bridge, what is the acceleration of the center of mass of the two melon system? (b) If you delay dropping one of the melons, what is the acceleration of the two-melon system while both are falling?

7. Figure 9-23 shows an overhead view of three particles on which external forces act. The magnitudes and directions of the forces on two of the particles are indicated. What are the magnitude and direction of the force acting on the third particle if the center of mass of the three-particle system is (a) stationary, (b) moving at a constant velocity rightward, (c) accelerating rightward?

FIGURE 9-23   Question 7.

8. A container sliding along an x axis on a frictionless surface explodes into three pieces. The pieces then move along the x axis in the directions indicated in Figure 9-24. The following table gives four sets of magnitudes (in kg·m/s) for the momenta $\mathbf{p}_1$, $\mathbf{p}_2$, and $\mathbf{p}_3$ of the pieces. Rank the sets according to the initial speed of the container, greatest first.

	p_1	p_2	p_3		p_1	p_2	p_3
(a)	10	2	6	(b)	10	6	2
(c)	2	10	6	(d)	6	2	10

FIGURE 9-24   Question 8.

9. Consider a box, like that in Sample Problem 9-7, which explodes into two pieces while moving with a constant positive velocity along an x axis. If one piece, with mass m_1, ends up with positive velocity $\mathbf{v}_1$, then the second piece, with mass m_2, could end up with (a) a positive velocity $\mathbf{v}_2$ (Fig. 9-25a) or (b) a negative velocity $\mathbf{v}_2$ (Fig. 9-25b), or (c) zero velocity (Fig. 9-25c). Rank those three possible results for the second piece according to the corresponding magnitude of $\mathbf{v}_1$, greatest first.

(a)　　　　　(b)　　　　　(c)

FIGURE 9-25   Question 9.

10. Figure 9-26 shows overhead views of three two-dimensional explosions in which a stationary grapefruit is blown by a fire-cracker into three pieces, seven pieces, and nine pieces. The pieces then slide over a frictionless floor. For each situation, Fig. 9-26 shows the momentum vectors of all but one piece; that piece has momentum $\mathbf{P}'$. The numbers next to the vectors are the magnitudes of the momenta (in kilogram-meter per second). Rank the three situations according to the magnitudes of (a) P'_x, (b) P'_y, and (c) $\mathbf{P}'$, greatest first.

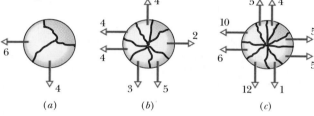

(a)　　　　　(b)　　　　　(c)

FIGURE 9-26   Question 10.

11. Figure 9-27 shows an overhead view showing six particles that have emerged from a region in which a two-dimensional explosion took place. The explosion was of an object that had been stationary on a frictionless floor. The directions of the momenta of the particles are indicated by the vectors; the numbers give the magnitudes of the momenta (in kg·m/s). (a) Will more particles be emerging from the explosion region? (b) If so, give (b) their net momentum and (c) their direction of travel.

FIGURE 9-27   Question 11.

12. For three pairs of particles, here are the masses of the particles and their initial separation d:

	m_1	m_2	INITIAL SEPARATION d
Pair 1	$2m$	$8m$	1.0 m
Pair 2	$3m$	$6m$	2.0 m
Pair 3	$4m$	$9m$	0.5 m

The particles in each pair attract each other. They are released from rest at the given initial separation d. When the separation has decreased to $d/2$, the speed of m_1 is v_1 and the speed of m_2 is v_2. Without written calculation, rank the pairs according to the ratio v_1/v_2, greatest first. (*Hint:* See Sample Problem 9-10.)

EXERCISES & PROBLEMS

SECTION 9-2   The Center of Mass

1E. (a) How far is the center of mass of the Earth–Moon system from the center of the Earth? (Appendix C gives the masses of the Earth and the Moon and the distance between the two.) (b) Express the answer to (a) as a fraction of the Earth's radius.

2E.   The distance between the centers of the carbon (C) and oxygen (O) atoms in a carbon monoxide (CO) gas molecule is 1.131×10^{-10} m. Locate the center of mass of a CO molecule relative to the carbon atom. (Find the masses of C and O in Appendix D.)

3E. (a) What are the coordinates of the center of mass of the three-particle system shown in Fig. 9-28? (b) What happens to the center of mass as the mass of the topmost particle is gradually increased?

FIGURE 9-28   Exercise 3.

4E. Three thin rods each of length L are arranged in an inverted U, as shown in Fig. 9-29. The two rods on the arms of the U each have mass M; the third rod has mass $3M$. Where is the center of mass of the assembly?

FIGURE 9-29 Exercise 4.

5E. A uniform square plate 6 m on a side has had a square piece 2 m on a side cut out of it (Fig. 9-30). The center of that piece is at $x = 2$ m, $y = 0$. The center of the square plate is at $x = y = 0$. Find the coordinates of the center of mass of the remaining piece.

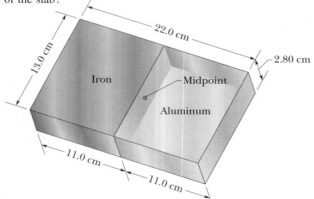

FIGURE 9-30 Exercise 5.

6P. Show that the ratio of the distances of two particles from the center of mass of the two-particle system is the inverse ratio of their masses.

7P. Figure 9-31 shows the dimensions of a composite slab; half the slab is made of aluminum (density = 2.70 g/cm³) and half is made of iron (density = 7.85 g/cm³). Where is the center of mass of the slab?

FIGURE 9-31 Problem 7.

8P. In the ammonia (NH₃) molecule (see Fig. 9-32), the three hydrogen (H) atoms form an equilateral triangle; the center of the triangle is 9.40×10^{-11} m from each hydrogen atom. The nitrogen (N) atom is at the apex of a pyramid, with the three hydrogen atoms forming the base. The nitrogen-to-hydrogen distance is 10.14×10^{-11} m, and the nitrogen-to-hydrogen atomic mass

ratio is 13.9. Locate the center of mass of the molecule relative to the nitrogen atom.

FIGURE 9-32 Problem 8.

9P. A cubical box, open at the top, with edge length 40 cm, is constructed from metal plate of uniform density and negligible thickness (Fig. 9-33). Find the coordinates of the center of mass of the box.

FIGURE 9-33 Problem 9.

10P. The Great Pyramid of Cheops at El Gizeh, Egypt (Fig. 9-34a), had height $H = 147$ m before its topmost stone fell. Its base is a square with edge length $L = 230$ m (see Fig. 9-34b). Its volume V is equal to $L^2H/3$. Assuming that it has uniform density $\rho = 1.8 \times 10^3$ kg/m³, find (a) the original height of its center of mass above the base and (b) the work required to lift the blocks into place from the base level.

(a)

(b)

FIGURE 9-34
Problem 10.

11P*. A right cylindrical can with mass M, height H, and uniform density is initially filled with pop of mass m (Fig. 9-35). We punch small holes in the top and bottom to drain the pop; we then consider the height h of the center of mass of the can and any pop within it. What is h (a) initially, and (b) when all the pop has drained? (c) What happens to h during the draining of the pop? (d) If x is the height of the remaining pop at any given instant, find x (in terms of M, H, and m) when the center of mass reaches its lowest point.

FIGURE 9-35 Problem 11.

SECTION 9-3 Newton's Second Law for a System of Particles

12E. Two skaters, one with mass 65 kg and the other with mass 40 kg, stand on an ice rink holding a pole with a length of 10 m and a negligible mass. Starting from the ends of the pole, the skaters pull themselves along the pole until they meet. How far will the 40 kg skater move?

13E. An old Chrysler with mass 2400 kg is moving along a straight stretch of road at 80 km/h. It is followed by a Ford with mass 1600 kg moving at 60 km/h. How fast is the center of mass of the two cars moving?

14E. A man of mass m clings to a rope ladder suspended below a balloon of mass M; see Fig. 9-36. The balloon is stationary with respect to the ground. (a) If the man begins to climb the ladder at speed v (with respect to the ladder), in what direction and with what speed (with respect to the Earth) will the balloon move? (b) What is the state of the motion after the man stops climbing?

FIGURE 9-36 Exercise 14.

15E. Two particles P and Q are initially at rest 1.0 m apart. P has a mass of 0.10 kg and Q a mass of 0.30 kg. P and Q attract each other with a constant force of 1.0×10^{-2} N. No external forces act on the system. (a) Describe the motion of the center of mass. (b) At what distance from P's original position do the particles collide?

16E. A cannon and a supply of cannonballs are inside a sealed railroad car of length L, as in Fig. 9-37. The cannon fires to the right; the car recoils to the left. Fired cannonballs remain in the car after hitting the far wall and landing on the floor there. (a) After all the cannonballs have been fired, what is the greatest distance the car can have moved from its original position? (b) Under what circumstance would the car move that distance? (c) What is the speed of the car just after all the cannonballs have been fired?

FIGURE 9-37 Exercise 16.

17P. At the instant a 3.0 kg particle has a velocity of 6.0 m/s in the direction of decreasing y, a 4.0 kg particle has a velocity of 7.0 m/s in the direction of increasing x. What is the speed of the center of mass of the two-particle system?

18P. A stone is dropped at $t = 0$. A second stone, with twice the mass of the first, is dropped from the same point at $t = 100$ ms. (a) Where is the center of mass of the two stones at $t = 300$ ms? Neither stone has yet reached the ground. (b) How fast is the center of mass of the two-stone system moving at that time?

19P. A 1000 kg automobile is at rest at a traffic signal. At the instant the light turns green, the automobile starts to move with a constant acceleration of 4.0 m/s². At the same instant a 2000 kg truck, traveling at a constant speed of 8.0 m/s, overtakes and passes the automobile. (a) How far is the center of mass of the automobile–truck system from the traffic light at $t = 3.0$ s? (b) What is the speed of the center of mass of the automobile–truck system then?

20P. Ricardo, mass 80 kg, and Carmelita, who is lighter, are enjoying Lake Merced at dusk in a 30 kg canoe. When the canoe is at rest in the placid water, they exchange seats, which are 3.0 m apart and symmetrically located with respect to the canoe's center. Ricardo notices that the canoe moved 40 cm relative to a submerged log during the exchange and calculates Carmelita's mass, which she has not told him. What is it?

21P. A shell is fired from a gun with a muzzle velocity of 20 m/s, at an angle of 60° with the horizontal. At the top of the trajectory, the shell explodes into two fragments of equal mass (Fig. 9-38). One fragment, whose speed immediately after the explosion is zero, falls vertically. How far from the gun does the other fragment land, assuming that the terrain is level and that the air drag is negligible?

FIGURE 9-38 Problem 21.

22P. Two identical containers of sugar are connected by a massless cord that passes over a massless, frictionless pulley with a diameter of 50 mm (Fig. 9-39). The two containers are at the same level. Each originally has a mass of 500 g. (a) Locate the horizontal position of their center of mass. (b) Now 20 g of sugar is transferred from one container to the other, but the containers are prevented from moving. Locate the new horizontal position of their center of mass. (c) The two containers are now released. In what direction does the center of mass move? (d) What is its acceleration?

FIGURE 9-39 Problem 22.

23P. A dog, weighing 10.0 lb, is standing on a flatboat so that he is 20.0 ft from the shore (to the left in Fig. 9-40a). He walks 8.00 ft on the boat toward shore and then halts. The boat weighs 40.0 lb, and one can assume there is no friction between it and the water. How far is the dog then from the shore? (*Hint:* See Fig. 9-40b. The dog moves leftward; the boat moves rightward; but does the center of mass of the *boat* + *dog* system move?)

FIGURE 9-40 Problem 23.

SECTION 9-5 The Linear Momentum of a System of Particles

24E. Suppose that your mass is 80 kg. How fast would you have to run to have the same linear momentum as a 1600 kg car moving at 1.2 km/h?

25E. How fast must an 816 kg Geo travel (a) to have the same

linear momentum as a 2650 kg Cadillac going 16 km/h and (b) to have the same kinetic energy?

26E. What is the linear momentum of an electron of speed $0.99c$ ($= 2.97 \times 10^8$ m/s)?

27E. The linear momentum of a particle moving at 1.5×10^8 m/s is measured to be 2.9×10^{-19} kg·m/s. By finding its mass, determine whether the particle is an electron or a proton.

28E. A 0.70 kg mass is moving horizontally with a speed of 5.0 m/s when it strikes a vertical wall. The mass rebounds with a speed of 2.0 m/s. What is the magnitude of the change in linear momentum of the mass?

29P. A 0.165 kg cue ball with an initial speed of 2.00 m/s bounces off the rail in a game of pool, as shown from overhead in Fig. 9-41. For x and y axes located as shown, the bounce reverses the y component of the ball's velocity but does not alter the x component. (a) What is θ in Fig. 9-41? (b) What is the change in the ball's linear momentum in unit-vector notation? (The fact that the ball rolls is not relevant to either question.)

FIGURE 9-41 Problem 29.

30P. A 2100 kg truck traveling north at 41 km/h turns east and accelerates to 51 km/h. (a) What is the change in the kinetic energy of the truck? (b) What are the magnitude and direction of the change in the linear momentum of the truck?

31P. An object is tracked by a radar station and found to have a position vector given by $\mathbf{r} = (3500 - 160t)\mathbf{i} + 2700\mathbf{j} + 300\mathbf{k}$, with $\mathbf{r}$ in meters and t in seconds. The radar station's x axis points east, its y axis north, and its z axis vertically up. If the object is a 250 kg meteorological missile, what are (a) its linear momentum, (b) its direction of motion, and (c) the net force on it?

32P. A 50 g ball is thrown from ground level into the air with an initial speed of 16 m/s at an angle of 30° above the horizontal. (a) What are the values of the kinetic energy of the ball initially and just before it hits the ground? (b) Find the corresponding values of the linear momentum (magnitude and direction). (c) Show that the change in linear momentum is just equal to the weight of the ball multiplied by the time of flight.

33P. A particle with mass m has linear momentum $\mathbf{p}$ of magnitude mc. What is its speed in terms of c, the speed of light?

SECTION 9-6 Conservation of Linear Momentum

34E. A 200 lb man standing on a surface of negligible friction kicks forward a 0.15 lb stone lying at his feet, giving it a speed of 13 ft/s. What velocity does the man acquire as a result?

35E. Two blocks of masses 1.0 kg and 3.0 kg are connected by a spring and rest on a frictionless surface. They are given velocities toward each other such that the 1.0 kg block travels initially at 1.7 m/s toward the center of mass, which remains at rest. What is the initial velocity of the other block?

36E. A space vehicle is traveling at 4300 km/h relative to the Earth when the exhausted rocket motor is disengaged and sent backward with a speed of 82 km/h relative to the command module. The mass of the motor is four times the mass of the module. What is the speed of the command module relative to Earth after the separation?

37E. A 75 kg man is riding on a 39 kg cart traveling at a speed of 2.3 m/s. He jumps off with zero horizontal speed. What is the resulting change in the speed of the cart?

38E. A railroad flatcar of weight W can roll without friction along a straight horizontal track. Initially, a man of weight w is standing on the car, which is moving to the right with speed v_0; see Fig. 9-42. What is the change in velocity of the car if the man runs to the left so that his speed relative to the car is v_{rel}?

FIGURE 9-42 Exercise 38.

39P. The last stage of a rocket is traveling at a speed of 7600 m/s. This last stage is made up of two parts that are clamped together, namely, a rocket case with a mass of 290.0 kg and a payload capsule with a mass of 150.0 kg. When the clamp is released, a compressed spring causes the two parts to separate with a relative speed of 910.0 m/s. (a) What are the speeds of the two parts after they have separated? Assume that all velocities are along the same line. (b) Find the total kinetic energy of the two parts before and after they separate; account for any difference.

40P. A radioactive nucleus, initially at rest, decays by emitting an electron and a neutrino perpendicular to each other. (A *neutrino* is one of the fundamental particles of physics.) The linear momentum of the electron is 1.2×10^{-22} kg·m/s and that of the neutrino is 6.4×10^{-23} kg·m/s. (a) Find the direction and magnitude of the linear momentum of the nucleus as it recoils from the decay. (b) The mass of the residual nucleus is 5.8×10^{-26} kg. What is its kinetic energy as it recoils?

41E. An electron (mass $m_1 = 9.11 \times 10^{-31}$ kg) and a proton (mass $m_2 = 1.67 \times 10^{-27}$ kg) attract each other via an electrical force. Suppose that an electron and a proton are released from rest with an initial separation $d = 3.0 \times 10^{-6}$ m. When their separation has decreased to 1.0×10^{-6} m, what is the ratio of (a) the electron's linear momentum to the proton's linear momentum, (b) the electron's speed to the proton's speed, and (c) the electron's kinetic energy to the proton's kinetic energy? (d) As the separation continues to decrease, do the answers to (a) through (c) increase, decrease, or remain the same?

42P. A 4.0 kg mass sliding on a frictionless surface explodes into two 2.0 kg masses. After the explosion the velocities of the 2.0 kg masses are 3.0 m/s, due north, and 5.0 m/s, 30° north of east. What was the original speed of the 4.0 kg mass?

43P. A vessel at rest explodes, breaking into three pieces. Two pieces, having equal mass, fly off perpendicular to one another with the same speed of 30 m/s. The third piece has three times the mass of each other piece. What are the direction and magnitude of its velocity immediately after the explosion?

44P. A 2140 kg railroad flatcar, which can move with negligible friction, is motionless next to a platform. A 242 kg sumo wrestler runs at 5.3 m/s along the platform (parallel to the track) and then jumps onto the flatcar. What is the speed of the flatcar if he then (a) stands on it, (b) runs at 5.3 m/s relative to it in his original direction, and (c) turns and runs at 5.3 m/s relative to the flatcar opposite his original direction?

45P. A rocket sled with a mass of 2900 kg moves at 250 m/s on a set of rails. At a certain point, a scoop on the sled dips into a trough of water located between the tracks and scoops water into an empty tank on the sled. By applying the principle of conservation of linear momentum, determine the speed of the sled after 920 kg of water has been scooped up. Ignore any retarding force on the scoop.

46P. A body of mass 8 kg is traveling at 2 m/s with no external force acting. At a certain instant an internal explosion occurs, splitting the body into two chunks of 4 kg mass each. The explosion gives the chunks an additional 16 J of kinetic energy. Neither chunk leaves the line of original motion. Determine the speed and direction of motion of each of the chunks after the explosion.

47P. You are on an iceboat on frictionless, flat ice; you and the boat have a combined mass M. Along with you are two stones with masses m_1 and m_2 such that $M = 6.00m_1 = 12.0m_2$. To get the boat moving, you throw the stones rearward, either in succession or together, but in each case with a certain speed v_{rel} relative to the boat. What is the resulting speed of the boat if you throw the stones (a) simultaneously, (b) m_1 and then m_2, and (c) m_2 and then m_1?

48P. A 1400 kg cannon, which fires a 70.0 kg shell with a speed of 556 m/s relative to the muzzle, is set at an elevation angle of 39.0° above the horizontal. The cannon is mounted on frictionless rails, so that it recoils freely. (a) What is the speed of the shell with respect to the Earth? (b) At what angle with the ground is the shell projected? (*Hint:* The horizontal component of the linear momentum of the system remains unchanged as the gun is fired.)

SECTION 9-7 Systems with Varying Mass: A Rocket

49E. A rocket at rest in space, where there is virtually no gravitational force, has a mass of 2.55×10^5 kg, of which 1.81×10^5 kg is fuel. The engine consumes fuel at the rate of 480 kg/s, and the exhaust speed is 3.27 km/s. The engine is fired for 250 s. (a) Find the thrust of the rocket engine. (b) What is the mass of the rocket after the engine burn? (c) What is the final speed attained?

50E. A rocket is moving away from the solar system at a speed of 6.0×10^3 m/s. It fires its engine, which ejects exhaust with a velocity of 3.0×10^3 m/s relative to the rocket. The mass of the rocket at this time is 4.0×10^4 kg, and its acceleration is 2.0 m/s². (a) What is the thrust of the engine? (b) At what rate, in kilograms per second, was exhaust ejected during the firing?

51E. A 6090 kg space probe, moving nose-first toward Jupiter at 105 m/s relative to the Sun, fires its rocket engine, ejecting 80.0 kg of exhaust at a speed of 253 m/s relative to the space probe. What is the final velocity of the probe?

52E. Consider a rocket at rest in deep space. For the rocket's speed to be (a) equal to the exhaust speed and (b) equal to twice the exhaust speed after firing of the rocket engine, what must be the rocket's *mass ratio* (ratio of initial to final mass)?

53E. During a lunar mission, it is necessary to increase the speed of a spacecraft by 2.2 m/s when it is moving at 400 m/s. The exhaust speed of rocket engine is 1000 m/s. What fraction of the initial mass of the spacecraft must be burned and ejected for the increase?

54E. A railroad car moves at a constant speed of 3.20 m/s under a grain elevator. Grain drops into it at the rate of 540 kg/min. What force must be applied to the railroad car, in the absence of friction, to keep it moving at constant speed?

55P. A single-stage rocket, at rest in a certain inertial reference frame, has mass M when the rocket engine is ignited. Show that when the mass has decreased to $0.368M$, the gases streaming out of the rocket engine at that time will be at rest in the original reference frame.

56P. A 6100 kg rocket is set for vertical firing from the Earth's surface. If the exhaust speed is 1200 m/s, how much gas must be ejected each second if the thrust (a) is to equal the weight of the rocket and (b) is to give the rocket an initial upward acceleration of 21 m/s²?

57P. Two long barges are moving in the same direction in still water, one with a speed of 10 km/h and the other with a speed of 20 km/h. While they are passing each other, coal is shoveled from the slower to the faster one at a rate of 1000 kg/min; see Fig. 9-43. How much additional force must be provided by the driving engines of each barge if neither is to change speed? Assume that the shoveling is always perfectly sideways and that the frictional forces between the barges and the water do not depend on the weight of the barges.

FIGURE 9-43   Problem 57.

58P. A jet airplane is traveling 180 m/s. Each second, the engine takes in 68 m³ of air, which has a mass of 70 kg. The air is used to burn 2.9 kg of fuel each second. The energy is used to compress the products of combustion and to eject them at the rear of the plane at 490 m/s relative to the plane. Find (a) the thrust of the jet engine and (b) its power in watts.

SECTION 9-8   External Forces and Internal Energy Changes

59E. The summit of Mount Everest is 8850 m above sea level. (a) How much energy would a 90 kg climber expend against the gravitational force (his weight) in climbing to the summit from sea level? (b) How many candy bars, at 300 kcal per bar, would supply an energy equivalent to this? Your answer should suggest that work done against the gravitational force is a very small part of the energy expended in climbing a mountain.

60E. A 51 kg boy climbs, with constant speed, a vertical rope 6.0 m long in 10 s. (a) What is the increase in the boy–Earth gravitational potential energy? (b) What is the boy's average power during the climb?

61E. A 55 kg woman runs up a flight of stairs with a height of 4.5 m in 3.5 s. What average power is required of her?

62E. A sprinter who weighs 670 N runs the first 7.0 m of a race in 1.6 s, starting from rest and accelerating uniformly. What are the sprinter's (a) speed and (b) kinetic energy at the end of that 1.6 s? (c) What average power does the sprinter generate during the 1.6 s interval?

63E. The luxury liner *Queen Elizabeth 2* has a diesel-electric powerplant with a maximum power of 92 MW at a cruising speed of 32.5 knots. What forward force is exerted on the ship at this highest attainable speed? (1 knot = 6076 ft/h.)

64E. What power, in horsepower, is required of the engine of a 1600 kg car moving at 25.1 m/s on a level road if the forces of resistance total 703 N?

65E. A swimmer moves through the water at an average speed of 0.22 m/s. The average drag force opposing this motion is 110 N. What average power is required of the swimmer?

66E. The energy required for a person to run is about 335 J/m, regardless of speed. What is a runner's average power during (a) a 100 m dash (time = 10 s) and (b) a marathon (distance = 26.2 mi; time = 2 h 10 min)?

67E. An automobile with passengers has weight 16,400 N and is moving at 113 km/h when the driver brakes to a stop. The road exerts a frictional force of 8230 N on the wheels. Find the stopping distance.

68E. You crouch from a standing position, lowering your center of mass 18 cm in the process. Then you jump vertically. The average force exerted on you by the floor while you jump is three times your weight. What is your upward speed as you pass through your standing position in leaving the floor?

69E. A 55-kg woman leaps vertically from a crouching position in which her center of mass is 40 cm above the ground. As her feet leave the floor her center of mass is 90 cm above the ground;

it rises to 120 cm at the top of her leap. (a) What average force was exerted on her by the ground during the jump? (b) What maximum speed does she attain?

70P. A 110 kg ice hockey player skates at 3.0 m/s toward a railing at the edge of the ice and then stops himself by grasping the railing with his outstretched arms. During this stopping process his center of mass moves 30 cm toward the railing. (a) What is the change in the kinetic energy of his center of mass during this process? (b) What average force must he exert on the railing?

71P. A 1500 kg automobile starts from rest on a horizontal road and gains a speed of 72 km/h in 30 s. (a) What is the kinetic energy of the auto at the end of the 30 s? (b) What is the average power required of the car during the 30 s interval? (c) What is the instantaneous power at the end of the 30 s interval, assuming that the acceleration was constant?

72P. While a 1710 kg automobile is moving at a constant speed of 15.0 m/s, the engine supplies 16.0 kW of power to overcome friction, wind resistance, and so on. (a) What is the effective retarding force associated with all the frictional forces combined? (b) What power must the engine supply if the car is to move up an 8.00% grade (8.00 m vertically for each 100 m horizontally) at 15.0 m/s? (c) On what downgrade, expressed as a percentage, would the car coast at 15.0 m/s?

73P. If a locomotive with a power capability of 1.5 MW can accelerate a train from a speed of 10 m/s to 25 m/s in 6.0 min, (a) calculate the mass of the train. Find (b) the speed of the train and (c) the force accelerating the train as functions of time (in seconds) during the 6.0 min interval. (d) Find the distance moved by the train during the interval.

74P. Resistance to the motion of an automobile consists of road friction, which is almost independent of speed, and air drag, which is proportional to speed squared. For a car with a weight of 12,000 N, the total resistant force F is given by $F = 300 + 1.8v^2$, where F is in newtons and v is in meters per second. Calculate the power required to accelerate the car at 0.92 m/s^2 when the speed is 80 km/h.

75P*. A dragster with mass m races another dragster through a distance d from an initial dead stop. Assume that its engine provides a constant instantaneous power P for the entire race and that the dragster can be treated as a particle. Find the elapsed time for the race.

Electronic Computation

76. The following gives the masses of three objects and, for a certain instant, the (xy) coordinates and the velocities of the objects. At that instant, what are (a) the position and (b) the velocity of the center of mass of the three particle system, and (c) what is the net linear momentum of the system?

OBJECT	MASS (kg)	COORDINATES (m)	VELOCITY (m/s)
1	4.00	(0, 0)	$1.50i - 2.50j$
2	3.00	(7.00, 3.00)	0
3	5.00	(3.00, 2.00)	$2.00i - 1.00j$

77. A 2.00 kg block is released from rest over the side of a very tall building ($y = 0$) at time $t = 0$. At time $t = 1.00$ s, a 3.00 kg block is released from rest at the same point. The first block hits the ground at $t = 5.00$ s. Plot, for the time interval $t = 0$ to $t = 6.00$ s, (a) the position and (b) the speed of the center of mass of the two block system.

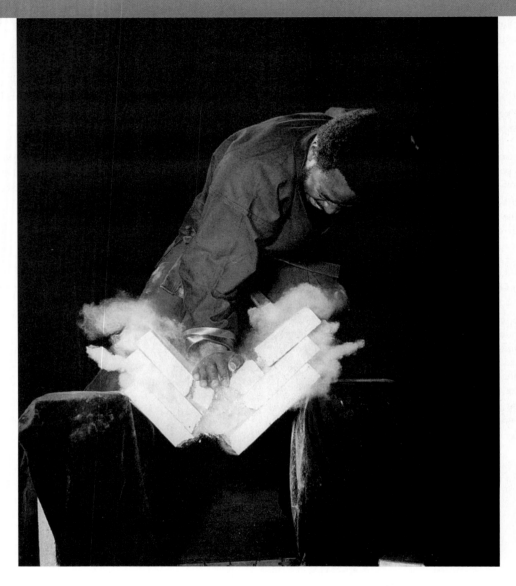

Ronald McNair, a physicist and one of the astronauts killed in the explosion of the Challenger *space shuttle, held a black belt in karate. Here he breaks several concrete slabs with one blow. In such karate demonstrations, a pine board or a concrete "patio block" is typically used. When struck, the board or block bends, storing energy like a stretched spring does, until a critical energy is reached. Then the object breaks. Surprisingly, the energy necessary to break the block is about one-third of that for the board, yet the board is considerably easier to break. Why?*

10-1 WHAT IS A COLLISION?

In everyday language, a *collision* occurs when objects crash into each other. Although we will refine that definition, it conveys the meaning well enough and covers com-

(a)

(b)

(c)

FIGURE 10-1 Collisions range widely in scale. (*a*) Meteor Crater in Arizona is about 1200 m wide and 200 m deep. (*b*) An alpha particle coming in from the left (whose trail is colored yellow in this false-color photograph) bounces off a nitrogen nucleus that had been stationary and that now moves toward the bottom right (red trail). (*c*) In a tennis match, the ball is in contact with the racquet for about 4 ms in each collision (for a cumulative time of only 1 s per set).

mon collisions, such as those between billiard balls, a hammer and a nail, and—too commonly—automobiles. Figure 10-1*a* shows the lasting result of a collision (an impressive crash) that occurred about 20,000 years ago. Collisions range from the microscopic scale of subatomic particles (Fig. 10-1*b*) to the astronomic scale of colliding stars and colliding galaxies. Even when they are on a normal scale, they are often too brief to be visible, although they involve significant distortion of the *colliding bodies* (Fig. 10-1*c*).

We shall use the following more formal definition of collision:

> A **collision** is an isolated event in which two or more bodies (the colliding bodies) exert relatively strong forces on each other for a relatively short time.

We must be able to distinguish times that are *before, during,* and *after* a collision, as suggested in Fig. 10-2. In the illustration, a system boundary surrounds the two colliding bodies. The forces that the bodies exert on each other are internal to the system.

Note that our formal definition of collision does not require the "crash" of our informal definition. When a space probe swings around a large planet to pick up speed (a *slingshot* encounter), that too is a collision. The probe and planet do not actually "touch," but a collision does not require contact, and a collision force does not have to be a force of contact; it can just as easily be a gravitational force, as in this case.

Many physicists today spend their time playing what we can call "the collision game." A principal goal of this game is to find out as much as possible about the forces that act during a collision, knowing the state of the particles both before and after the collision. Virtually all our knowledge of the subatomic world—electrons, protons, neutrons, muons, quarks, and the like—comes from experiments involving collisions. The rules of the game are the laws of conservation of momentum and of energy.

System boundary

Before   During   After

FIGURE 10-2 A flowchart showing the system in which a collision occurs.

10-2 IMPULSE AND LINEAR MOMENTUM

Single Collision

Figure 10-3 shows the equal but opposite forces, $\mathbf{F}(t)$ and $-\mathbf{F}(t)$, that act during a simple head-on collision between two particle-like bodies of different masses. These forces will change the linear momentum of both bodies; the amount of the change will depend not only on the average values of the forces, but also on the time Δt during which they act. To see this quantitatively, let us apply Newton's second law in the form $\mathbf{F} = d\mathbf{p}/dt$ to body R on the right in Fig. 10-3. We have

$$d\mathbf{p} = \mathbf{F}(t)\,dt, \tag{10-1}$$

in which $\mathbf{F}(t)$ is the time-varying force, displayed as the curve in Fig. 10-4a. Let us integrate Eq. 10-1 over the collision interval Δt, that is, from an initial time t_i (just before the collision) to a final time t_f (just after the collision). We obtain

$$\int_{\mathbf{p}_i}^{\mathbf{p}_f} d\mathbf{p} = \int_{t_i}^{t_f} \mathbf{F}(t)\,dt. \tag{10-2}$$

The left side of this equation is $\mathbf{p}_f - \mathbf{p}_i$, the change in linear momentum of the body R. The right side, which is a measure of both the strength and the duration of the collision force, is called the **impulse J** of the collision. Thus

$$\mathbf{J} = \int_{t_i}^{t_f} \mathbf{F}(t)\,dt \qquad \text{(impulse defined).} \tag{10-3}$$

Equation 10-3 tells us that the impulse is equal to the area under the $F(t)$ curve of Fig. 10-4a.

From Eqs. 10-2 and 10-3 we see that the change in the linear momentum of each body in a collision is equal to the impulse that acts on that body:

$$\mathbf{p}_f - \mathbf{p}_i = \Delta\mathbf{p} = \mathbf{J} \qquad \begin{array}{l}\text{(impulse–linear} \\ \text{momentum theorem).}\end{array} \tag{10-4}$$

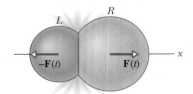

FIGURE 10-3 Two particle-like bodies L and R collide with each other. During the collision, body L exerts force $\mathbf{F}(t)$ on body R, and body R exerts force $-\mathbf{F}(t)$ on body L. Forces $\mathbf{F}(t)$ and $-\mathbf{F}(t)$ are an action–reaction pair. Their magnitudes vary with time during the collision, but at any given instant these magnitudes are equal.

(Furthermore, from the conservation of linear momentum, we know that $\Delta\mathbf{p}$ for body R in Fig. 10-3 equals $-\Delta\mathbf{p}$ for body L.) Equation 10-4 can also be written in component form as

$$p_{fx} - p_{ix} = \Delta p_x = J_x, \tag{10-5}$$

$$p_{fy} - p_{iy} = \Delta p_y = J_y, \tag{10-6}$$

and

$$p_{fz} - p_{iz} = \Delta p_z = J_z. \tag{10-7}$$

Impulse and linear momentum are both vectors, and they have the same units and dimensions. The **impulse–linear momentum theorem** of Eq. 10-4, like the work–kinetic energy theorem, is not a new and independent theorem but is a direct consequence of Newton's second law. Both theorems are special forms of this law, useful for special purposes.

If $\overline{F}$ is the average magnitude of the force in Fig. 10-4a, we can write the magnitude of the impulse as

$$J = \overline{F}\,\Delta t, \tag{10-8}$$

where Δt is the duration of the collision. The value of $\overline{F}$ is chosen so that the area within the rectangle of Fig. 10-4b is equal to the area under the $F(t)$ curve of Fig. 10-4a.

CHECKPOINT 1: A paratrooper whose chute fails to open lands in snow; he is hurt slightly. Had he landed on bare ground, the stopping time would have been 10 times shorter and the collision lethal. Does the presence of the snow increase, decrease, or leave unchanged the values of (a) the paratrooper's change in momentum, (b) the impulse stopping the paratrooper, and (c) the force stopping the paratrooper?

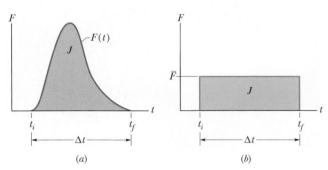

(a)   (b)

FIGURE 10-4 (*a*) The curve shows the magnitude of the time-varying force $F(t)$ that acts on body R during the collision of Fig. 10-3. The area under the curve for $F(t)$ is equal to the magnitude of the impulse $\mathbf{J}$ on body R in the collision. (*b*) The height of the rectangle represents the average force $\overline{F}$ acting on body R over the time interval Δt. The area within the rectangle is equal to the area under the curve in (*a*) and thus is also equal to the magnitude of the impulse $\mathbf{J}$ in the collision.

Series of Collisions

In Fig. 10-5, a steady stream of bodies, with identical linear momenta mv, collides with body R, which is fixed in place. In each collision of this one-dimensional situation, the impulse J acting on body R and the change Δp in linear momentum of the incident body have the same magnitude but opposite directions; that is, $J = -\Delta p$. If n bodies collide with R in time interval Δt, then by Eq. 10-4 the total impulse J acting on body R during Δt is

$$J = -n\,\Delta p. \qquad (10\text{-}9)$$

By rearranging Eq. 10-8 and substituting Eq. 10-9, we find the average force $\overline{F}$ acting on body R during the collisions:

$$\overline{F} = \frac{J}{\Delta t} = -\frac{n}{\Delta t}\,\Delta p = -\frac{n}{\Delta t}\,m\,\Delta v. \qquad (10\text{-}10)$$

This equation gives us $\overline{F}$ in terms of $n/\Delta t$, the rate at which the bodies collide with body R, and Δv, the change in the velocity of those bodies.

If the colliding bodies stop upon impact, then in Eq. 10-10 we substitute

$$\Delta v = v_f - v_i = 0 - v = -v, \qquad (10\text{-}11)$$

where v_i $(= v)$ and v_f $(= 0)$ are the velocities before and after the collision, respectively. If, instead, the colliding bodies bounce directly backward from body R with no change in speed, then $v_f = -v$ and we substitute

$$\Delta v = v_f - v_i = -v - v = -2v. \qquad (10\text{-}12)$$

In time interval Δt, an amount of mass $\Delta m = nm$ collides with body R. With this result, we can rewrite Eq. 10-10 as

$$\overline{F} = -\frac{\Delta m}{\Delta t}\,\Delta v. \qquad (10\text{-}13)$$

This equation gives the average force $\overline{F}$ in terms of $\Delta m/\Delta t$, the rate at which mass collides with body R. Here again we can substitute Eq. 10-11 or 10-12 for Δv, depending on what the colliding bodies do.

FIGURE 10-5 A steady stream of bodies, with identical linear momenta, collides with body R, which is fixed in place. The average force $\overline{\mathbf{F}}$ on body R is to the right and has a magnitude that depends on the rate at which the bodies collide or, equivalently, the rate at which mass collides.

A pitched 140 g baseball, in horizontal flight with a speed v_i of 39 m/s, is struck by a batter. After leaving the bat, the ball travels in the opposite direction with speed v_f, also 39 m/s.

(a) What impulse J acts on the ball while it is in contact with the bat?

SOLUTION: We can calculate the impulse from the change it produces in the ball's linear momentum, using Eq. 10-4 for one-dimensional motion. Let us choose the direction in which the bat is moving to be positive. From Eq. 10-4 we have

$$J = p_f - p_i = mv_f - mv_i$$
$$= (0.14\ \text{kg})(39\ \text{m/s}) - (0.14\ \text{kg})(-39\ \text{m/s})$$
$$= 10.9\ \text{kg} \cdot \text{m/s} \approx 11\ \text{kg} \cdot \text{m/s}. \qquad \text{(Answer)}$$

With our sign convention, the initial velocity of the ball is negative and the final velocity is positive. The impulse turned out to be positive, which tells us that the direction of the impulse vector acting on the ball is the direction in which the bat was swinging, which makes sense.

(b) The impact time Δt for the baseball–bat collision is 1.2 ms. What average force acts on the baseball?

SOLUTION: From Eq. 10-8 we have

$$\overline{F} = \frac{J}{\Delta t} = \frac{10.9\ \text{kg} \cdot \text{m/s}}{0.0012\ \text{s}}$$
$$= 9100\ \text{N}, \qquad \text{(Answer)}$$

which is about a ton. The *maximum* force will be larger than this. The sign of the average force exerted on the ball is positive, which means that the direction of the force vector is the same as that of the impulse vector.

(c) What is the average acceleration a of the baseball?

SOLUTION: We find this with

$$\overline{a} = \frac{\overline{F}}{m} = \frac{9100\ \text{N}}{0.14\ \text{kg}} = 6.5 \times 10^4\ \text{m/s}^2, \qquad \text{(Answer)}$$

or about $6600g$.

In defining a collision, we assumed that no net external force acts on the colliding bodies. That is not true in this case, because the weight $m\mathbf{g}$ of the ball always acts on the ball, whether the ball is in flight or in contact with the bat. However, this force, with a magnitude of 1.4 N, is negligible compared to the average force exerted by the bat, which has a magnitude of 9100 N. We are quite safe in treating the collision as "isolated."

As in Sample Problem 10-1, the baseball approaches the bat horizontally at a speed v_i of 39 m/s, but now the collision is not head-on, and the ball leaves the bat with a speed v_f of 45 m/s at an upward angle of 30° from the horizontal (Fig. 10-6). What is the average force $\overline{\mathbf{F}}$ exerted on the ball if the collision lasts 1.2 ms?

SOLUTION: We find the components J_x and J_y of the impulse from Eqs. 10-5 and 10-6:

$$J_x = p_{fx} - p_{ix} = m(v_{fx} - v_{ix})$$
$$= (0.14 \text{ kg})[(45 \text{ m/s})(\cos 30°) - (-39 \text{ m/s})]$$
$$= 10.92 \text{ kg} \cdot \text{m/s},$$

and

$$J_y = p_{fy} - p_{iy} = m(v_{fy} - v_{iy})$$
$$= (0.14 \text{ kg})[(45 \text{ m/s})(\sin 30°) - 0]$$
$$= 3.150 \text{ kg} \cdot \text{m/s}.$$

The magnitude of the impulse $\mathbf{J}$ is given by

$$J = \sqrt{J_x^2 + J_y^2}$$
$$= \sqrt{(10.92 \text{ kg} \cdot \text{m/s})^2 + (3.150 \text{ kg} \cdot \text{m/s})^2}$$
$$= 11.37 \text{ kg} \cdot \text{m/s}.$$

From Eq. 10-8, the magnitude $\overline{F}$ of the average force is

$$\overline{F} = \frac{J}{\Delta t} = \frac{11.37 \text{ kg} \cdot \text{m/s}}{0.0012 \text{ s}}$$
$$= 9475 \text{ N} \approx 9500 \text{ N}. \qquad \text{(Answer)}$$

The impulse $\mathbf{J}$ is angled upward from the horizontal by θ, where θ is given by

$$\tan \theta = \frac{J_y}{J_x} = \frac{3.150 \text{ kg} \cdot \text{m/s}}{10.92 \text{ kg} \cdot \text{m/s}} = 0.288,$$

or

$$\theta = 16°. \qquad \text{(Answer)}$$

The average force $\overline{\mathbf{F}}$ is in the same direction as $\mathbf{J}$. Note that, unlike Sample Problem 10-1, the direction of $\overline{\mathbf{F}}$ and $\mathbf{J}$ in this two-dimensional situation is different from the direction taken by the ball as it leaves the bat.

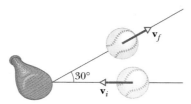

FIGURE 10-6 Sample Problem 10-2. A bat collides with a pitched baseball, sending the ball off at an angle of 30° from the horizontal.

CHECKPOINT **2:** The figure shows an overhead view of a ball bouncing from a wall without any change in its speed. Consider the change $\Delta\mathbf{p}$ in the ball's linear momentum. (a) Is Δp_x positive, negative, or zero? (b) Is Δp_y positive, negative, or zero? (c) What is the direction of $\Delta\mathbf{p}$?

10-3 ELASTIC COLLISIONS IN ONE DIMENSION

Stationary Target

Consider a simple head-on collision of two bodies of masses m_1 and m_2 as illustrated in Fig. 10-7. For convenience, we take one of the bodies to be stationary, with velocity $\mathbf{v}_{2i} = 0$ before the collision. That body will be the "target," and the other body will be the "projectile,"* with velocity $\mathbf{v}_{1i}$ before the collision. We assume that this two-body system is closed (no mass enters or leaves it) and isolated (no net external force acts on it). Let us also make another, special assumption: the kinetic energy of the system is the same before and after the collision. The collision is then said to be of a special type called **elastic collisions.**

> In an elastic collision, the kinetic energy of each colliding body can change, but the total kinetic energy of the system does not change.

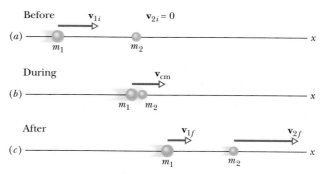

FIGURE 10-7 Two bodies undergo an elastic collision. One of them (the target body with mass m_2) is initially at rest before the collision. The velocities are shown (a) before, (b) during, and (c) after the collision. (The velocity during the collision is the velocity of the center of mass of the two bodies, which are momentarily touching.) The velocities are drawn to scale for the case in which $m_1 = 3m_2$.

---

*If the target body is moving with respect to our laboratory frame, we can always find another inertial reference frame in which the target body is initially stationary.

Note carefully that the linear momentum of a closed, isolated system is *always* conserved in a collision, regardless of whether the collision is elastic, because the forces involved in the collision are all internal forces.

> In a closed, isolated system, the linear momentum of each colliding body can change, but the net linear momentum cannot change, regardless of whether the collision is elastic.

The conservations of linear momentum and of kinetic energy for the collision of Fig. 10-7 give us

$$m_1 v_{1i} = m_1 v_{1f} + m_2 v_{2f} \quad \text{(linear momentum)} \quad (10\text{-}14)$$

and

$$\tfrac{1}{2} m_1 v_{1i}^2 = \tfrac{1}{2} m_1 v_{1f}^2 + \tfrac{1}{2} m_2 v_{2f}^2 \quad \text{(kinetic energy)}. \quad (10\text{-}15)$$

In each of these equations, the subscript i identifies the initial velocities and the subscript f the final velocities of the bodies. If we know the masses of the bodies and if we also know v_{1i}, the initial velocity of body 1, the only unknown quantities are v_{1f} and v_{2f}, the final velocities of the two bodies. With two equations at our disposal, we should be able to find these two unknowns.

To do so, we rewrite Eq. 10-14 as

$$m_1(v_{1i} - v_{1f}) = m_2 v_{2f} \quad (10\text{-}16)$$

and Eq. 10-15 as*

$$m_1(v_{1i} - v_{1f})(v_{1i} + v_{1f}) = m_2 v_{2f}^2. \quad (10\text{-}17)$$

After dividing Eq. 10-17 by Eq. 10-16 and doing some more algebra, we obtain

$$v_{1f} = \frac{m_1 - m_2}{m_1 + m_2} v_{1i} \quad (10\text{-}18)$$

and

$$v_{2f} = \frac{2m_1}{m_1 + m_2} v_{1i}. \quad (10\text{-}19)$$

We note from Eq. 10-19 that v_{2f} is always positive (the target body with mass m_2 always moves forward). From Eq. 10-18 we see that v_{1f} may be of either sign (the projectile body with mass m_1 moves forward if $m_1 > m_2$ but rebounds if $m_1 < m_2$).

CHECKPOINT **3:** What is the final linear momentum of the target in Fig. 10-7 if the initial linear momen-

tum of the projectile is $6 \text{ kg} \cdot \text{m/s}$ and the final linear momentum of the projectile is (a) $2 \text{ kg} \cdot \text{m/s}$ and (b) $-2 \text{ kg} \cdot \text{m/s}$? (c) What is the final kinetic energy of the target if the initial and final kinetic energies of the projectile are, respectively, 5 J and 2 J?

Let us look at a few special situations.

1. *Equal masses.* If $m_1 = m_2$, Eqs. 10-18 and 10-19 reduce to

$$v_{1f} = 0 \quad \text{and} \quad v_{2f} = v_{1i},$$

which we might call a pool player's result. It predicts that after a head-on collision of bodies with equal masses, body 1 (initially moving) stops dead in its tracks and body 2 (initially at rest) takes off with the initial speed of body 1. In head-on collisions, bodies of equal mass simply exchange velocities. This is true even if the target particle (body 2) is not initially at rest.

2. *A massive target.* In terms of Fig. 10-7, a massive target means that $m_2 \gg m_1$. For example, we might fire a golf ball at a cannonball. Equations 10-18 and 10-19 then reduce to

$$v_{1f} \approx -v_{1i} \quad \text{and} \quad v_{2f} \approx \left(\frac{2m_1}{m_2}\right) v_{1i}. \quad (10\text{-}20)$$

This tells us that body 1 (the golf ball) simply bounces back in the same direction from which it came, its speed essentially unchanged. Body 2 (the cannonball) moves forward at a low speed, because the quantity in parentheses in Eq. 10-20 is much less than unity. All this is what we should expect.

3. *A massive projectile.* This is the opposite case; that is, $m_1 \gg m_2$. This time, we fire a cannonball at a golf ball. Equations 10-18 and 10-19 reduce to

$$v_{1f} \approx v_{1i} \quad \text{and} \quad v_{2f} \approx 2v_{1i}. \quad (10\text{-}21)$$

Equation 10-21 tells us that body 1 (the cannonball) simply keeps on going, scarcely slowed by the collision. Body 2 (the golf ball) charges ahead at twice the speed of the cannonball.

You may wonder: "Why twice the speed?" As a starting point in thinking about the matter, recall the collision described by Eq. 10-20, in which the velocity of the incident light body (the golf ball) changed from $+v$ to $-v$, a velocity *change* of $2v$. The same *change* in velocity (from zero to $2v$) occurs in this example also.

4. *Motion of the center of mass.* The center of mass of two colliding bodies continues to move, totally uninfluenced by the collision. This follows from conservation

---

*In this step, we use the identity $a^2 - b^2 = (a - b)(a + b)$. It reduces the amount of algebra needed to solve the simultaneous equations, Eqs. 10-16 and 10-17.

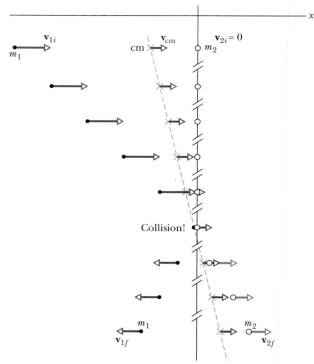

FIGURE 10-8 Some freeze-frames of two bodies undergoing an elastic collision. Body 2 is initially at rest, and $m_2 = 3m_1$. The velocity of the center of mass is also shown. Note that it is unaffected by the collision.

of linear momentum and from Eq. 9-26,

$$P = Mv_{cm} = (m_1 + m_2)v_{cm}, \quad (10\text{-}22)$$

which relates the linear momentum P of the system of two bodies to v_{cm}, the velocity of their center of mass. Because the momentum P is unchanged by the collision, v_{cm} must also be unchanged. So the center of mass continues to move in the same direction and at the same speed. From Eq. 10-22, the velocity of the center of mass for the collision shown in Fig. 10-7 (target initially at rest) is

$$v_{cm} = \frac{P}{m_1 + m_2} = \frac{m_1}{m_1 + m_2} v_{1i}. \quad (10\text{-}23)$$

Figure 10-8, a series of freeze-frames of a typical elastic collision, shows that the center of mass does indeed move steadily forward, unaffected in any way by the collision.

Moving Target

Now that we have examined the elastic collision of a projectile and a stationary target, let us examine the situation in which both bodies are moving before they undergo an elastic collision.

FIGURE 10-9 Two bodies headed for an elastic collision.

For the situation of Fig. 10-9, the conservation of linear momentum is written as

$$m_1 v_{1i} + m_2 v_{2i} = m_1 v_{1f} + m_2 v_{2f}, \quad (10\text{-}24)$$

and the conservation of kinetic energy is written as

$$\tfrac{1}{2} m_1 v_{1i}^2 + \tfrac{1}{2} m_2 v_{2i}^2 = \tfrac{1}{2} m_1 v_{1f}^2 + \tfrac{1}{2} m_2 v_{2f}^2. \quad (10\text{-}25)$$

To solve these simultaneous equations for v_{1f} and v_{2f}, we first rewrite Eq. 10-24 as

$$m_1(v_{1i} - v_{1f}) = -m_2(v_{2i} - v_{2f}), \quad (10\text{-}26)$$

and Eq. 10-25 as

$$m_1(v_{1i} - v_{1f})(v_{1i} + v_{1f})$$
$$= -m_2(v_{2i} - v_{2f})(v_{2i} + v_{2f}). \quad (10\text{-}27)$$

After dividing Eq. 10-27 by Eq. 10-26 and doing some more algebra, we obtain

$$v_{1f} = \frac{m_1 - m_2}{m_1 + m_2} v_{1i} + \frac{2m_2}{m_1 + m_2} v_{2i} \quad (10\text{-}28)$$

and

$$v_{2f} = \frac{2m_1}{m_1 + m_2} v_{1i} + \frac{m_2 - m_1}{m_1 + m_2} v_{2i}. \quad (10\text{-}29)$$

Note that the assignment of subscripts 1 and 2 to the bodies is arbitrary. If we exchange those subscripts in Fig. 10-9 and in Eqs. 10-28 and 10-29, we end up with the same set of equations. Note also that if we set $v_{2i} = 0$, body 2 becomes a stationary target, and Eqs. 10-28 and 10-29 reduce to Eqs. 10-18 and 10-19, respectively.

From Eq. 10-22, the constant velocity v_{cm} of the center of mass of the two-body system of Fig. 10-9 is

$$v_{cm} = \frac{P}{m_1 + m_2} = \frac{m_1 v_{1i} + m_2 v_{2i}}{m_1 + m_2}. \quad (10\text{-}30)$$

Because the two-body system is closed and isolated, its linear momentum P is unchanged by the collision. Thus, the velocity v_{cm} of its center of mass is also unchanged.

CHECKPOINT 4: The initial momenta of body 1 and body 2 in Fig. 10-9 are, respectively, 10 kg·m/s and -8 kg·m/s. What is the final momentum of body 2 if the final momentum of body 1 is (a) 2 kg·m/s, and (b) -2 kg·m/s?

SAMPLE PROBLEM 10-3

Two metal spheres, suspended by vertical cords, initially just touch, as shown in Fig. 10-10. Sphere 1, with mass $m_1 = 30$ g, is pulled to the left to height $h_1 = 8.0$ cm, and then released. After swinging down, it undergoes an elastic collision with sphere 2, whose mass $m_2 = 75$ g.

(a) What is the velocity v_{1f} of sphere 1 just after the collision?

SOLUTION: We let v_{1i} represent the speed of sphere 1 just before the collision. When the sphere begins its descent, its kinetic energy is zero and the gravitational potential energy is m_1gh_1. Just before the collision, sphere 1 has kinetic energy $\frac{1}{2}m_1v_{1i}^2$ and the gravitational potential energy is zero. For the descent, the conservation of mechanical energy gives us

$$\tfrac{1}{2}m_1v_{1i}^2 = m_1gh_1,$$

which we solve for the speed v_{1i} of sphere 1 just before the collision:

$$v_{1i} = \sqrt{2gh_1} = \sqrt{(2)(9.8 \text{ m/s}^2)(0.080 \text{ m})} = 1.252 \text{ m/s}.$$

Although sphere 1 swings down in a two-dimensional arc, its velocity is horizontal when it collides with sphere 2. Thus the collision is one-dimensional, and we can represent the velocity of sphere 1 just before that collision as v_{1i}.

To find the velocity v_{1f} of sphere 1 just after the collision, we use Eq. 10-18:

$$v_{1f} = \frac{m_1 - m_2}{m_1 + m_2} v_{1i} = \frac{0.030 \text{ kg} - 0.075 \text{ kg}}{0.030 \text{ kg} + 0.075 \text{ kg}} (1.252 \text{ m/s})$$

$$= -0.537 \text{ m/s} \approx -0.54 \text{ m/s}. \qquad \text{(Answer)}$$

The minus sign tells us that sphere 1 moves to the left just after the collision.

(b) To what height h_1' does sphere 1 swing to the left after the collision?

SOLUTION: When sphere 1 begins its upward swing, its kinetic energy is $\frac{1}{2}m_1v_{1f}^2$ and the gravitational potential energy is zero. When it momentarily stops at height h_1', its kinetic energy is zero and the gravitational potential energy is m_1gh_1'. Conserving mechanical energy for the upward swing, we find

$$m_1gh_1' = \tfrac{1}{2}m_1v_{1f}^2,$$

or $\qquad h_1' = \dfrac{v_{1f}^2}{2g} = \dfrac{(-0.537 \text{ m/s})^2}{(2)(9.8 \text{ m/s}^2)}$

$$= 0.0147 \text{ m} \approx 1.5 \text{ cm}. \qquad \text{(Answer)}$$

(c) What is the velocity v_{2f} of sphere 2 just after the collision?

SOLUTION: From Eq. 10-19, we have

$$v_{2f} = \frac{2m_1}{m_1 + m_2} v_{1i} = \frac{(2)(0.030 \text{ kg})}{0.030 \text{ kg} + 0.075 \text{ kg}} (1.252 \text{ m/s})$$

$$= 0.715 \text{ m/s} \approx 0.72 \text{ m/s}. \qquad \text{(Answer)}$$

(d) To what height h_2 does sphere 2 swing after the collision?

SOLUTION: Sphere 2 begins its ascent with kinetic energy $\frac{1}{2}m_2v_{2f}^2$. When it momentarily stops at height h_2, the gravitational potential energy is m_2gh_2. Conservation of mechanical energy for the ascent gives us

$$m_2gh_2 = \tfrac{1}{2}m_2v_{2f}^2,$$

or $\qquad h_2 = \dfrac{v_{2f}^2}{2g} = \dfrac{(0.715 \text{ m/s})^2}{(2)(9.8 \text{ m/s}^2)}$

$$= 0.0261 \text{ m} \approx 2.6 \text{ cm}. \qquad \text{(Answer)}$$

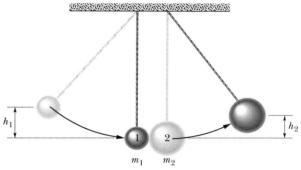

FIGURE 10-10   Sample Problem 10-3. Two metal spheres suspended by cords just touch when they are at rest. Sphere 1, with mass m_1, is pulled to the left to height h_1 and then released. The subsequent elastic collision with sphere 2 sends sphere 2 to height h_2.

SAMPLE PROBLEM 10-4

In a nuclear reactor, newly produced fast neutrons must be slowed down before they can participate effectively in the chain-reaction process. This is done by allowing them to collide with the nuclei of atoms in a *moderator*.

(a) By what fraction is the kinetic energy of a neutron (of mass m_1) reduced in a head-on elastic collision with a nucleus of mass m_2, initially at rest?

SOLUTION: The initial and final kinetic energies of the neutron are

$$K_i = \tfrac{1}{2}m_1v_{1i}^2 \quad \text{and} \quad K_f = \tfrac{1}{2}m_1v_{1f}^2.$$

The fraction we seek is then

$$\text{frac} = \frac{K_i - K_f}{K_i} = \frac{v_{1i}^2 - v_{1f}^2}{v_{1i}^2} = 1 - \frac{v_{1f}^2}{v_{1i}^2}. \qquad (10\text{-}31)$$

For such a collision we have, from Eq. 10-18,

$$\frac{v_{1f}}{v_{1i}} = \frac{m_1 - m_2}{m_1 + m_2}. \qquad (10\text{-}32)$$

Substituting Eq. 10-32 into Eq. 10-31 yields, after a little algebra,

$$\text{frac} = \frac{4m_1m_2}{(m_1 + m_2)^2}. \quad \text{(Answer)} \quad (10\text{-}33)$$

(b) Evaluate the fraction for lead, carbon, and hydrogen. The ratios of the mass of a nucleus to the mass of a neutron ($= m_2/m_1$) for these nuclei are 206 for lead, 12 for carbon, and about 1 for hydrogen.

SOLUTION: The following values of the fraction can be calculated with Eq. 10-33: for lead ($m_2 = 206m_1$),

$$\text{frac} = \frac{(4)(206)}{(1 + 206)^2} = 0.019 \text{ or } 1.9\%; \quad \text{(Answer)}$$

for carbon ($m_2 = 12m_1$),

$$\text{frac} = \frac{(4)(12)}{(1 + 12)^2} = 0.28 \text{ or } 28\%; \quad \text{(Answer)}$$

and for hydrogen ($m_2 \approx m_1$),

$$\text{frac} = \frac{(4)(1)}{(1 + 1)^2} = 1 \text{ or } 100\%. \quad \text{(Answer)}$$

These results partially explain why water, which contains lots of hydrogen, is a much better moderator of neutrons than lead.

10-4 INELASTIC COLLISIONS IN ONE DIMENSION

An **inelastic collision** is one in which the kinetic energy of the system of colliding bodies is not conserved. If you drop a Superball onto a hard floor, it loses very little of its kinetic energy on impact and rebounds to almost its original height. If the ball did regain the original height, its collision with the floor would be elastic. However, the small loss of kinetic energy in the collision lowers the rebound height, and therefore the collision is somewhat inelastic.

A dropped golf ball will lose more of its kinetic energy and will rebound to only 60% of its original height. This collision is noticeably inelastic. If you drop a ball of wet putty onto the floor, it sticks to the floor and does not rebound at all. Because the putty sticks and does not rebound, this collision is said to be a **completely inelastic collision.**

The kinetic energy that is lost in any inelastic collision is transferred to some other form of energy, perhaps thermal energy. Nonetheless, the linear momentum of the system is always conserved (provided the system is isolated and closed). Since kinetic energy and linear momentum both involve the speeds of the colliding bodies, the conservation of linear momentum limits just how much kinetic energy is lost by a system in an inelastic collision. When the bodies stick together in a completely inelastic collision, the amount of kinetic energy that is lost is the maximum

FIGURE 10-11 A completely inelastic collision between two bodies. Before the collision, the body of mass m_2 is at rest. Afterward, the bodies stick together, which is the criterion for a *completely* inelastic collision. The velocities are drawn to scale for the case in which $m_1 = 3m_2$.

allowed by the conservation of linear momentum. In some situations, this maximum loss is all the kinetic energy of the system.

In this section, we restrict ourselves to completely inelastic collisions. Figure 10-11 shows a one-dimensional inelastic collision in which one body is initially stationary. The law of conservation of linear momentum holds, so

$$m_1v = (m_1 + m_2)V, \quad (10\text{-}34)$$

or

$$V = v\,\frac{m_1}{m_1 + m_2}, \quad (10\text{-}35)$$

where V represents the final velocity of the stuck-together bodies. Equation 10-35 tells us that the final speed is always less than that of the incoming body.

Figure 10-12 (compare it with Fig. 10-8) shows that the motion of the center of mass in a completely inelastic collision is unaffected by the collision. Here, even though the bodies stick together, the kinetic energy associated with the motion of the center of mass is still present. Kinetic energy cannot vanish *totally* in an inelastic collision unless the reference frame is fixed with respect to the center of mass of the colliding bodies. In such cases, all motion relative to the reference frame ceases when the bodies stick together. If the target body m_2 happens to be exceedingly massive, the center of mass of the system essentially coincides with that of the target. This is the situation in which a putty ball is dropped onto the floor, the target body being the Earth. As the ball hits, all the kinetic energy is dissipated, being transferred to some other form of energy.

If both bodies are moving prior to a completely inelastic collision, we replace Eq. 10-34 with

$$m_1v_1 + m_2v_2 = (m_1 + m_2)V, \quad (10\text{-}36)$$

where m_1v_1 is the initial linear momentum of one body, and m_2v_2 is that of the other body. Here again, all the kinetic energy is dissipated only if the reference frame is fixed with respect to the center of mass of the bodies. Figure

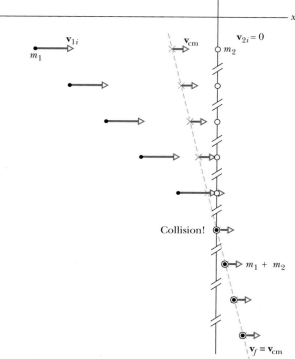

FIGURE 10-12   Some freeze-frames of two bodies undergoing a completely inelastic collision. Body 2 is initially at rest, and $m_2 = 3m_1$. The bodies stick together after the collision and move forward together. The velocity of the center of mass is also shown. Note that it is unaffected by the collision and that it is equal to the final velocity of the stuck-together bodies.

FIGURE 10-13   Two cars after an almost head-on, almost completely inelastic collision.

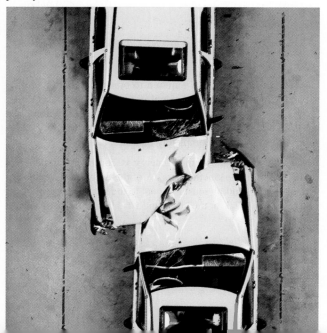

10-13 shows an example: identical cars driven at identical speeds underwent an almost head-on and almost completely inelastic collision. Their center of mass was stationary relative to the Earth. So a bystander, who would have been stationary relative to the Earth, would have seen these cars stop dead because of the collision, rather than move afterward in one direction or another.

CHECKPOINT 5: Body 1 and body 2 are in a completely inelastic one-dimensional collision. What is their final momentum if their initial momenta are, respectively, (a) 10 kg·m/s and 0; (b) 10 kg·m/s and 4 kg·m/s; (c) 10 kg·m/s and −4 kg·m/s?

SAMPLE PROBLEM 10-5

The *ballistic pendulum* was used to measure the speeds of bullets before electronic timing devices were developed. The version shown in Fig. 10-14 consists of a large block of wood of mass $M = 5.4$ kg, hanging from two long cords. A bullet of mass $m = 9.5$ g is fired into the block, coming quickly to rest. The *block + bullet* then swing upward, their center of mass rising a vertical distance $h = 6.3$ cm before the pendulum comes momentarily to rest at the end of its arc.

(a) What was the speed v of the bullet just prior to the collision?

SOLUTION: Just after the collision, the *bullet + block* have speed V. Applying the conservation of linear momentum to

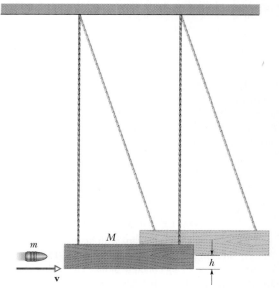

FIGURE 10-14   Sample Problem 10-5. A ballistic pendulum, used to measure the speeds of bullets.

the collision, we have

$$mv = (M + m)V.$$

Since the bullet and block stick together, the collision is completely inelastic, and kinetic energy is not conserved during it. However, *after* the collision, mechanical energy *is* conserved, because no force then acts to dissipate that energy. So the kinetic energy of the system when the block is at the bottom of its arc must equal the potential energy of the system when the block is at the top:

$$\tfrac{1}{2}(M + m)V^2 = (M + m)gh.$$

Eliminating V between these two equations leads to

$$v = \frac{M + m}{m}\sqrt{2gh}$$

$$= \left(\frac{5.4 \text{ kg} + 0.0095 \text{ kg}}{0.0095 \text{ kg}}\right)\sqrt{(2)(9.8 \text{ m/s}^2)(0.063 \text{ m})}$$

$$= 630 \text{ m/s}. \hspace{2cm} \text{(Answer)}$$

The ballistic pendulum is a kind of "transformer," exchanging the high speed of a light object (the bullet) for the low — and thus more easily measurable — speed of a massive object (the block).

(b) What is the initial kinetic energy of the bullet? How much of this energy remains as mechanical energy of the swinging pendulum?

SOLUTION: The kinetic energy of the bullet is

$$K_b = \tfrac{1}{2}mv^2 = (\tfrac{1}{2})(0.0095 \text{ kg})(630 \text{ m/s})^2$$

$$= 1900 \text{ J}. \hspace{2cm} \text{(Answer)}$$

The mechanical energy of the swinging pendulum is equal to the gravitational potential energy when the block is at the top of its swing:

$$E = (M + m)gh$$

$$= (5.4 \text{ kg} + 0.0095 \text{ kg})(9.8 \text{ m/s}^2)(0.063 \text{ m})$$

$$= 3.3 \text{ J}. \hspace{2cm} \text{(Answer)}$$

Thus only 3.3/1900, or 0.2%, of the original kinetic energy of the bullet is transferred to mechanical energy of the pendulum. The rest is transferred to thermal energy of the block and bullet, or goes into the breaking of wood fibers as the bullet bores into the wood.

SAMPLE PROBLEM 10-6

A karate expert strikes downward with his fist (of mass $m_1 = 0.70$ kg), breaking a 0.14 kg board (Fig. 10-15a). He then does the same to a 3.2 kg concrete block. The spring constants k for bending are 4.1×10^4 N/m for the board and 2.6×10^6 N/m for the block. Breaking occurs at a deflection d of 16 mm for the board and 1.1 mm for the block (Fig. 10-15c).*

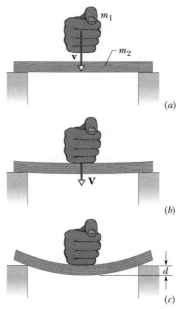

FIGURE 10-15 Sample Problem 10-6. (*a*) A karate expert strikes at a flat object with speed v. (*b*) Fist and object undergo a completely inelastic collision, and bending begins. The *fist + object* then have speed V. (*c*) The object breaks when its center has been deflected by an amount d.

(a) Just before the board and block break, what is the energy stored in each?

SOLUTION: We treat the bending as the compression of a spring for which Hooke's law applies. The stored energy is then, from Eq. 8-11, $U = \tfrac{1}{2}kd^2$. For the board,

$$U = \tfrac{1}{2}(4.1 \times 10^4 \text{ N/m})(0.016 \text{ m})^2$$

$$= 5.248 \text{ J} \approx 5.2 \text{ J}. \hspace{1cm} \text{(Answer)}$$

For the block,

$$U = \tfrac{1}{2}(2.6 \times 10^6 \text{ N/m})(0.0011 \text{ m})^2$$

$$= 1.573 \text{ J} \approx 1.6 \text{ J}. \hspace{1cm} \text{(Answer)}$$

(b) What fist speed v is required to break the board and the block? Assume that mechanical energy is conserved during the bending, that the fist and struck object stop just before the break, and that the fist–object collision at the onset of bending (Fig. 10-15b) is completely inelastic.

SOLUTION: If mechanical energy is conserved during the bending, then the kinetic energy K of the fist–object system at the onset of bending must be equal to the stored energy U just at breaking: 5.2 J for the board and 1.6 J for the block. The fist

*The data are taken from "The Physics of Karate," by S. R. Wilk, R. E. McNair, and M. S. Feld, *American Journal of Physics,* September 1983.

speed required to break the object is the speed required to produce that kinetic energy K. To find it, we must first find the speed V of *fist* + *object* at the onset of bending. We write the equality between K and U as

$$K = \tfrac{1}{2}(m_1 + m_2)V^2 = U,$$

from which we find

$$V = \sqrt{\frac{2U}{m_1 + m_2}},$$

where V is the speed of *fist* + *object* at the onset of bending, $m_1 = 0.70$ kg, and m_2 is 0.14 kg for the board or 3.2 kg for the block. For the board we have

$$V = \sqrt{\frac{2(5.248 \text{ J})}{0.70 \text{ kg} + 0.14 \text{ kg}}} = 3.534 \text{ m/s}.$$

For the block we have

$$V = \sqrt{\frac{2(1.573 \text{ J})}{0.70 \text{ kg} + 3.2 \text{ kg}}} = 0.8981 \text{ m/s}.$$

Now let the fist have speed v just before hitting the board or block. Then Eq. 10-35 holds for the collision. By rearranging that equation, we get

$$v = \left(\frac{m_1 + m_2}{m_1}\right)V.$$

For the board we find

$$v = \left(\frac{0.70 \text{ kg} + 0.14 \text{ kg}}{0.70 \text{ kg}}\right)(3.534 \text{ m/s})$$

$$\approx 4.2 \text{ m/s}, \qquad \text{(Answer)}$$

and for the block we find

$$v = \left(\frac{0.70 \text{ kg} + 3.2 \text{ kg}}{0.70 \text{ kg}}\right)(0.8981 \text{ m/s})$$

$$\approx 5.0 \text{ m/s}. \qquad \text{(Answer)}$$

The fist speed must be about 20% faster for the fist to break the block, because the larger mass of the block makes the transfer of energy to the block more difficult.

10-5 COLLISIONS IN TWO DIMENSIONS

Here we consider a *glancing collision* (it is not head-on) between a projectile body and a target body at rest. Figure 10-16 shows a typical situation. After the collision, the two bodies fly off at angles θ_1 and θ_2, as the figure shows.

From the conservation of linear momentum (a vector relation), we can write two scalar equations:

$$m_1 v_{1i} = m_1 v_{1f}\cos\theta_1 + m_2 v_{2f}\cos\theta_2 \quad \text{(x component)}$$
$$(10\text{-}37)$$

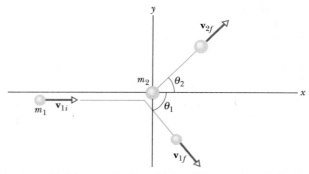

FIGURE 10-16 An elastic collision between two bodies in which the collision is not head-on. The body with mass m_2 (the target) is initially at rest.

and

$$0 = -m_1 v_{1f}\sin\theta_1 + m_2 v_{2f}\sin\theta_2 \quad \text{(y component)}.$$
$$(10\text{-}38)$$

If the collision is elastic, kinetic energy is also conserved:

$$\tfrac{1}{2}m_1 v_{1i}^2 = \tfrac{1}{2}m_1 v_{1f}^2 + \tfrac{1}{2}m_2 v_{2f}^2 \quad \text{(kinetic energy)}. \quad (10\text{-}39)$$

These three equations contain seven variables: two masses, m_1 and m_2; three speeds, v_{1i}, v_{1f}, and v_{2f}; and two angles, θ_1 and θ_2. If we know any four of these quantities, we can solve the three equations for the remaining three quantities. Often the known quantities are the two masses, the initial speed, and one of the angles. The unknowns to be solved for are then the two final speeds and the remaining angle.

CHECKPOINT **6:** In Fig. 10-16, suppose that the projectile has an initial momentum of 6 kg·m/s, a final x component of momentum of 4 kg·m/s, and a final y component of momentum of -3 kg·m/s. For the target, what then are (a) the final x component of momentum and (b) the final y component of momentum?

SAMPLE PROBLEM 10-7

Two particles of equal masses have an elastic collision, the target particle being initially at rest. Show that (unless the collision is head-on) the two particles will always move off perpendicular to each other after the collision.

SOLUTION: You might be tempted to jump into the problem and solve it in a straightforward way, by applying Eqs. 10-37, 10-38, and 10-39. There is a neater way.

Figure 10-17a shows the situation before and after the collision, each particle with its linear momentum vector attached. Because linear momentum is conserved in the collision, these vectors must form a closed triangle, as Fig. 10-17b shows. (The vector $m\mathbf{v}_{1i}$ must be the vector sum of $m\mathbf{v}_{1f}$ and

mv_{2f}.) Because the masses of the particles are equal, the closed linear momentum triangle of Fig. 10-17b is also a closed velocity triangle, because dividing by the scalar m does not change the relation of the vectors. That is, we may draw the velocity vectors as in Fig. 10-17c, because

$$\mathbf{v}_{1i} = \mathbf{v}_{1f} + \mathbf{v}_{2f}. \qquad (10\text{-}40)$$

Because kinetic energy is conserved, Eq. 10-39 holds. With the equal terms $\frac{1}{2}m$ canceled out, this equation becomes

$$v_{1i}^2 = v_{1f}^2 + v_{2f}^2. \qquad (10\text{-}41)$$

Equation 10-41 applies to the lengths of the sides in the triangle of Fig. 10-17c. For it to hold, the triangle must be a right triangle (and Eq. 10-41 is then the Pythagorean theorem). Therefore the angle ϕ between the vectors $\mathbf{v}_{1f}$ and $\mathbf{v}_{2f}$ in Fig. 10-17 must be 90°, which is what we set out to prove.

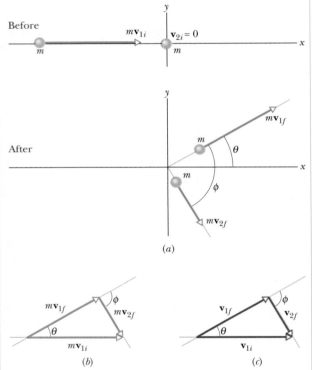

(a)

(b)   (c)

FIGURE 10-17 Sample Problem 10-7. A neat proof that in an elastic collision between two particles of the same mass, the particles fly off at 90° to each other afterward. For this to hold true, the target particle must be initially at rest and the collision must not be head-on.

SAMPLE PROBLEM 10-8

Two skaters collide and embrace, in a completely inelastic collision. That is, they stick together after impact, as suggested by Fig. 10-18, where the origin is placed at the point of collision. Alfred, whose mass m_A is 83 kg, is originally moving east with speed $v_A = 6.2$ km/h. Barbara, whose mass m_B is 55 kg, is originally moving north with speed $v_B = 7.8$ km/h.

(a) What is the velocity $\mathbf{V}$ of the couple after impact?

SOLUTION: Linear momentum is conserved during the collision. We can write, for the linear momentum components in the x and y directions,

$$m_A v_A = MV \cos \theta \qquad (x \text{ component}) \qquad (10\text{-}42)$$

and

$$m_B v_B = MV \sin \theta \qquad (y \text{ component}), \qquad (10\text{-}43)$$

in which $M = m_A + m_B$. Dividing Eq. 10-43 by Eq. 10-42 yields

$$\tan \theta = \frac{m_B v_B}{m_A v_A} = \frac{(55 \text{ kg})(7.8 \text{ km/h})}{(83 \text{ kg})(6.2 \text{ km/h})} = 0.834.$$

Thus

$$\theta = \tan^{-1} 0.834 = 39.8° \approx 40°. \qquad (\text{Answer})$$

From Eq. 10-43 we then have

$$V = \frac{m_B v_B}{M \sin \theta} = \frac{(55 \text{ kg})(7.8 \text{ km/h})}{(83 \text{ kg} + 55 \text{ kg})(\sin 39.8°)}$$

$$= 4.86 \text{ km/h} \approx 4.9 \text{ km/h}. \qquad (\text{Answer})$$

(b) What is the velocity of the center of mass of the two skaters before and after the collision?

SOLUTION: We can answer this without further calculation. After the collision, the velocity of the center of mass is the same as the velocity that we calculated in part (a), namely, 4.9 km/h at 40° north of east. Because the velocity of the center of mass is not changed by the collision, the same value must prevail before the collision.

(c) What is the fractional change in the kinetic energy of the skaters because of the collision?

SOLUTION: The initial kinetic energy is

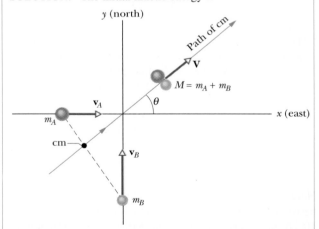

FIGURE 10-18 Sample Problem 10-8. Two skaters, Alfred (A) and Barbara (B), represented with spheres in this simplified overhead view, have a completely inelastic collision. Afterward, they move off together at angle θ, with speed V. The path of their center of mass is shown. The position of the center of mass for the indicated positions of the skaters before the collision is also shown.

$$K_i = \tfrac{1}{2}m_A v_A^2 + \tfrac{1}{2}m_B v_B^2$$
$$= (\tfrac{1}{2})(83 \text{ kg})(6.2 \text{ km/h})^2 + (\tfrac{1}{2})(55 \text{ kg})(7.8 \text{ km/h})^2$$
$$= 3270 \text{ kg} \cdot \text{km}^2/\text{h}^2.$$

The final kinetic energy is

$$K_f = \tfrac{1}{2}MV^2$$
$$= (\tfrac{1}{2})(83 \text{ kg} + 55 \text{ kg})(4.86 \text{ km/h})^2$$
$$= 1630 \text{ kg} \cdot \text{km}^2/\text{h}^2.$$

The fractional change is then

$$\text{frac} = \frac{K_f - K_i}{K_i}$$
$$= \frac{1630 \text{ kg} \cdot \text{km}^2/\text{h}^2 - 3270 \text{ kg} \cdot \text{km}^2/\text{h}^2}{3270 \text{ kg} \cdot \text{km}^2/\text{h}^2}$$
$$= -0.50. \qquad \text{(Answer)}$$

Thus 50% of the initial kinetic energy is lost as a result of the collision.

CHECKPOINT **7:** In Sample Problem 10-8 and Fig. 10-18, does θ increase, decrease, or remain the same if we increase (a) the speed of Barbara and (b) the mass of Barbara?

PROBLEM SOLVING TACTICS

TACTIC 1: *Switch to SI Units?*

More often than not, it is wise to express all physical quantities in their basic SI units; thus all speeds in meters per second, all masses in kilograms, and so on. Sometimes, however, it is not necessary to do this. In Sample Problem 10-8a, for example, the units cancel out when we calculate the angle θ. In Sample Problem 10-8c, they cancel when we calculate the dimensionless quantity frac. In this latter case, for example, there is no need to change the kinetic energy units to joules, the basic SI energy unit; we can leave them in the units kg·km²/h² because we can look ahead and see that they are going to cancel when we compute the ratio frac.

10-6 REACTIONS
AND DECAY PROCESSES (OPTIONAL)

Here we consider certain collisions (called **nuclear reactions**) in which the identity and even the number of interacting nuclear particles may change because of the collisions. We consider too the *spontaneous decay* of unstable particles, in which one particle is transformed to two other particles. For both kinds of event, there is a clear distinction between times "before the event" and times "after the event," and both linear momentum and *total* energy are conserved. In short, we can treat these events by the methods we have already developed for collisions.

SAMPLE PROBLEM 10-9

A radioactive nucleus of uranium-235 decays spontaneously to thorium-231 by emitting an alpha particle (the nucleus of a helium atom, symbolized as α, helium-4, or ^{4}He):

$$^{235}\text{U} \rightarrow \alpha + {}^{231}\text{Th}.$$

The alpha particle ($m_\alpha = 4.00$ u) has a kinetic energy K_α of 4.60 MeV. What is the kinetic energy of the thorium-231 nucleus ($m_{\text{Th}} = 231$ u)?

SOLUTION: The ^{235}U nucleus is initially at rest in the laboratory. After decay, the alpha particle flies off and the ^{231}Th moves in the opposite direction, with kinetic energies K_α and K_{Th}, respectively. Applying the law of conservation of linear momentum leads to

$$0 = m_{\text{Th}} v_{\text{Th}} + m_\alpha v_\alpha,$$

which we can recast as

$$m_{\text{Th}} v_{\text{Th}} = -m_\alpha v_\alpha. \qquad (10\text{-}44)$$

Squaring both sides of Eq. 10-44 yields

$$m_{\text{TH}}^2 \, v_{\text{Th}}^2 = m_\alpha^2 \, v_\alpha^2. \qquad (10\text{-}45)$$

Because $K = \tfrac{1}{2}mv^2$, we can write Eq. 10-45 as

$$m_{\text{Th}} K_{\text{Th}} = m_\alpha K_\alpha.$$

Thus

$$K_{\text{Th}} = K_\alpha \frac{m_\alpha}{m_{\text{Th}}} = (4.60 \text{ MeV}) \frac{4.00 \text{ u}}{231 \text{ u}}$$
$$= 7.97 \times 10^{-2} \text{ MeV} = 79.7 \text{ keV}. \quad \text{(Answer)}$$

We see that, of the total amount of kinetic energy made available during the decay (4.60 MeV + 0.0797 MeV = 4.68 MeV), the heavy thorium-231 nucleus receives only about 0.0797/4.68, or 1.7%.

SAMPLE PROBLEM 10-10

A nuclear reaction of great importance for the generation of energy by nuclear fusion is the so-called d-d reaction, one form of which is

$$d + d \rightarrow t + p. \qquad (10\text{-}46)$$

The particles represented by these letters are all isotopes of hydrogen, with the following properties:

SYMBOLS		NAME	MASS
p	^{1}H	Proton	$m_p = 1.00783$ u
d	^{2}H	Deuteron	$m_d = 2.01410$ u
t	^{3}H	Triton	$m_t = 3.01605$ u

(a) How much energy appears because of the mass change Δm that occurs in this reaction?

SOLUTION: From Eq. 8-40, the Q of a reaction (the energy released or absorbed in the reaction) is $Q = -\Delta m\, c^2$. Here the change Δm in mass is $\Delta m = m_p + m_t - 2m_d$. So, here Q is

$$Q = -\Delta m\, c^2 = (2m_d - m_p - m_t)c^2$$
$$= (2 \times 2.01410\text{ u} - 1.00783\text{ u} - 3.01605\text{ u})$$
$$\times (931.5\text{ MeV/u})$$
$$= (0.00432\text{ u})(931.5\text{ MeV/u})$$
$$= 4.02\text{ MeV}. \qquad \text{(Answer)}$$

We have used the value of 931.5 MeV/u for c^2 (Eq. 8-43).

A positive value for Q (as in this case) means that the reaction is **exothermic**; in this reaction mass energy is transferred to kinetic energy of the products. Only $0.00432/(2 \times 2.01410)$, or about 0.1%, of the mass energy originally present has been so transferred. A negative Q signals an **endothermic** reaction, in which the transfer is from kinetic energy to mass energy. And $Q = 0$ means an *elastic encounter*, with no mass change and kinetic energy conserved.

(b) A deuteron of kinetic energy $K_d = 1.50$ MeV strikes a stationary deuteron, initiating the reaction of Eq. 10-46. A proton is observed to move off at an angle of 90° to the incident direction, with a kinetic energy of 3.39 MeV; see Fig. 10-19. What is the kinetic energy of the triton?

SOLUTION: The energy Q released due to the decrease in mass appears as an increase in the kinetic energies of the particles. Thus we can write

$$Q = \Delta K = K_p + K_t - K_d,$$

where Q is the energy computed for this reaction in (a). Solving for K_t gives us

$$K_t = Q + K_d - K_p$$
$$= 4.02\text{ MeV} + 1.50\text{ MeV} - 3.39\text{ MeV}$$

$$= 2.13\text{ MeV}. \qquad \text{(Answer)}$$

(c) At what angle ϕ with the incident direction (see Fig. 10-19) does the triton emerge?

SOLUTION: We have not yet made use of the fact that linear momentum is conserved in the reaction of Eq. 10-46. The law of conservation of linear momentum yields two scalar equations:

$$m_d v_d = m_t v_t \cos\phi \qquad (x\text{ component}) \qquad (10\text{-}47)$$

and

$$0 = m_p v_p + m_t v_t \sin\phi \qquad (y\text{ component}). \qquad (10\text{-}48)$$

From Eq. 10-48 we have

$$\sin\phi = -\frac{m_p v_p}{m_t v_t}. \qquad (10\text{-}49)$$

Using the relation $K = \frac{1}{2}mv^2$, we can rewrite each linear momentum mv as $\sqrt{2mK}$ and thus recast Eq. 10-49 as

$$\phi = \sin^{-1}\left(-\sqrt{\frac{m_p K_p}{m_t K_t}}\right)$$

$$= \sin^{-1}\left(-\sqrt{\frac{(1.01\text{ u})(3.39\text{ MeV})}{(3.02\text{ u})(2.13\text{ MeV})}}\right)$$

$$= \sin^{-1}(-0.730) = -46.9°. \qquad \text{(Answer)}$$

Before After

FIGURE 10-19 Sample Problem 10-10. A moving deuteron (d) strikes a stationary deuteron, producing a proton (p) and a triton (t).

REVIEW & SUMMARY

Collisions

In a **collision**, two bodies exert strong forces on each other for a relatively short time. These forces are internal to the two-body system and are significantly larger than any external force during the collision. The laws of conservation of energy and of linear momentum, applied immediately before and after a collision, allow us to predict the outcome of the collision and to understand the interactions between the colliding bodies.

Impulse and Linear Momentum

Applying Newton's second law in momentum form to a particle-like body involved in a collision leads to the **impulse–linear momentum theorem:**

$$\mathbf{p}_f - \mathbf{p}_i = \Delta\mathbf{p} = \mathbf{J}, \qquad (10\text{-}4)$$

where $\mathbf{p}_f - \mathbf{p}_i = \Delta\mathbf{p}$ is the change in the body's linear momentum, and $\mathbf{J}$ is the **impulse** due to the force $\mathbf{F}(t)$ exerted on the body by the other body in the collision:

$$\mathbf{J} = \int_{t_i}^{t_f} \mathbf{F}(t)\, dt. \qquad (10\text{-}3)$$

If $\bar{F}$ is the average of $\mathbf{F}(t)$ during the collision and Δt is the duration of the collision, then for one-dimensional motion

$$J = \bar{F}\,\Delta t. \qquad (10\text{-}8)$$

When a steady stream of bodies, each with mass m and speed v, collides with a body fixed in position, the average force on the fixed body is

$$\bar{F} = -\frac{n}{\Delta t} \Delta p = -\frac{n}{\Delta t} m \Delta v, \qquad (10\text{-}10)$$

where $n/\Delta t$ is the rate at which the bodies collide with the fixed body, and Δv is the change in velocity of each colliding body. This average force can also be written as

$$\bar{F} = -\frac{\Delta m}{\Delta t} \Delta v, \qquad (10\text{-}13)$$

where $\Delta m/\Delta t$ is the rate at which mass collides with the fixed body. In Eqs. 10-10 and 10-13, $\Delta v = -v$ if the bodies stop upon impact or $\Delta v = -2v$ if they bounce directly backward with no change in speed.

Elastic Collision — One Dimension

An *elastic collision* is one in which the kinetic energy of a system of two colliding bodies is conserved. For a one-dimensional situation in which one body (the target) is stationary and the other body (the projectile) is initially moving, we get the following relations from conservation of kinetic energy and linear momentum:

$$v_{1f} = \frac{m_1 - m_2}{m_1 + m_2} v_{1i} \qquad (10\text{-}18)$$

and

$$v_{2f} = \frac{2m_1}{m_1 + m_2} v_{1i}. \qquad (10\text{-}19)$$

Here subscripts i and f refer to the velocities immediately before and after the collision, respectively. If both bodies are moving prior to the collision, their velocities immediately after the collision are given by

$$v_{1f} = \frac{m_1 - m_2}{m_1 + m_2} v_{1i} + \frac{2m_2}{m_1 + m_2} v_{2i} \qquad (10\text{-}28)$$

and

$$v_{2f} = \frac{2m_1}{m_1 + m_2} v_{1i} + \frac{m_2 - m_1}{m_1 + m_2} v_{2i}. \qquad (10\text{-}29)$$

Inelastic Collision — One Dimension

An *inelastic collision* is one in which the kinetic energy of a system of two colliding bodies is not conserved. The total linear momentum of the system must, however, still be conserved. If the colliding bodies stick together, the collision is **completely inelastic,** and the reduction in kinetic energy is maximum (but not necessarily to zero). For one-dimensional motion, with one body initially stationary and the other having initial velocity v, the velocity V of the stuck-together bodies is found by applying conservation of linear momentum to the system:

$$m_1 v = (m_1 + m_2) V. \qquad (10\text{-}34)$$

If both bodies are moving prior to the collision, conservation of linear momentum is written as

$$m_1 v_1 + m_2 v_2 = (m_1 + m_2) V. \qquad (10\text{-}36)$$

Motion of the Center of Mass

The center of mass of a closed, isolated system of colliding bodies is unaffected by the collision, whether the collision is elastic or inelastic. For a one-dimensional collision, the constant velocity v_{cm} of the center of mass is given by

$$v_{cm} = \frac{P}{m_1 + m_2} = \frac{m_1 v_{1i} + m_2 v_{2i}}{m_1 + m_2} \qquad (10\text{-}30)$$

before and after the collision.

Collisions in Two Dimensions

Collisions in two dimensions are governed by the conservation of vector linear momentum, a condition that leads to two-component equations. These determine the final motion if the collision is completely inelastic. Otherwise, the laws of conservation of linear momentum and of energy generally lead to equations that cannot be solved completely unless other data, such as the final direction of one of the velocities, are available.

Nuclear Reactions and Decay

In a *nuclear reaction* or *decay* of nuclear particles, linear momentum and *total* energy are conserved. If the mass of a system of such particles changes by Δm, the mass energy of the system changes by $\Delta m\, c^2$. The Q of the reaction or decay is defined as

$$Q = -\Delta m\, c^2.$$

This process is said to be **exothermic,** and Q is a positive quantity, if mass energy is transferred to kinetic energy of particles in the system. It is said to be **endothermic,** and Q is a negative quantity, if kinetic energy of particles in the system is transferred to mass energy.

QUESTIONS

1. Figure 10-20 shows three graphs of force magnitude versus time for a body involved in a collision. Rank the graphs according to the magnitude of the impulse on the body, greatest first.

FIGURE 10-20 Question 1.

2. Figure 10-21 shows an overhead view of a golf ball bouncing off a tree trunk with no change in its speed. During the collision the force $\mathbf{F}$ on the ball by the trunk causes a change $\Delta \mathbf{p}$ in the linear momentum of the ball. If angle θ is increased (and assuming that the duration of the collision is unchanged), do the following increase, decrease, or remain the same: (a) Δp_x, (b) Δp_y, (c) magnitude of $\Delta \mathbf{p}$, (d) F_x, (e) F_y, and (f) magnitude of $\mathbf{F}$?

3. The following table gives, for three situations, the masses (kilograms) and velocities (meters per second) for the two parti-

FIGURE 10-21
Question 2.

cles in Fig. 10-9. For which situations is the center of mass of the two-particle system stationary?

SITUATION	m_1	v_1	m_2	v_2
a	2	3	4	−3
b	6	2	3	−4
c	4	3	4	−3

4. (a) In Sample Problem 10-3, does the rebound height of sphere 1 increase, decrease, or remain the same if we increase the mass of sphere 2? If, instead, we set $m_2 = m_1$, (b) what is the rebound height of sphere 1, and (c) what is the height h_2 reached by sphere 2?

5. Two bodies undergo an elastic one-dimensional collision along an x axis. Figure 10-22 is a graph of position versus time for those bodies and for their center of mass. (a) Were both bodies initially moving, or was one initially stationary? Which line segment corresponds to the motion of the center of mass (b) before the collision and (c) after the collision? (d) Is the mass of the body that was moving faster before the collision greater than, less than, or equal to that of the other body?

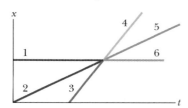

FIGURE 10-22 Question 5.

6. One body catches up with a second body and they then undergo a one-dimensional collision. Figure 10-23 is a graph of position versus time for those bodies and for their center of mass. Which line segment corresponds to (a) the faster body before the collision, (b) the center of mass before the collision, (c) the center of mass after the collision, and (d) the initially faster body after the collision? (e) Is the mass of the initially faster body greater than, less than, or equal to that of the other body?

FIGURE 10-23 Question 6.

7. Figure 10-24 shows, for four situations, three identical blocks that undergo elastic collisions on a frictionless surface. In situations 1 and 2, two of the blocks are glued together. In all four situations, the initially moving blocks have the same velocity **v**. Rank the situations according to (a) the total linear momentum of the blocks after the collisions and (b) the speed of the rightmost block after the collisions, greatest first.

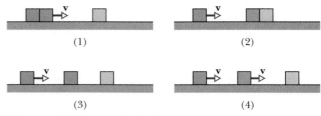

FIGURE 10-24 Question 7.

8. Figure 10-25 shows seven identical blocks on a frictionless floor. Initially, blocks a and b are moving rightward and block g is moving leftward, each with speed $v = 3$ m/s. The other blocks are stationary. A series of elastic collisions occurs. After the last collision, what are the speeds and directions of motion of each of the seven blocks?

FIGURE 10-25 Question 8.

9. In Fig. 10-26, blocks A and B have linear momenta with directions as shown and with magnitudes of 9 kg · m/s and 4 kg · m/s, respectively. (a) What is the direction of motion of the center of mass of the two-block system over the frictionless floor? (b) If the blocks stick together during their collision, in what direction will they move? (c) If, instead, block A ends up moving to the left, will the magnitude of its momentum then be smaller than, more than, or the same as that of block B?

FIGURE 10-26 Question 9.

10. Figure 10-27 shows four graphs of position versus time for two bodies and their center of mass. The two bodies undergo a completely inelastic, one-dimensional collision while moving along an x axis. In graph 1, are (a) the two bodies and (b) the center of mass moving in the positive or negative direction of the x axis? (c) Which graphs correspond to a physically impossible situation?

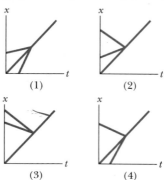

FIGURE 10-27
Question 10.

11. Body Q, with linear momentum $\mathbf{p}_Q = (2\mathbf{i} - 3\mathbf{j})$ kg·m/s, collides with and also sticks to body R, with linear momentum $\mathbf{p}_R = (8\mathbf{i} + 3\mathbf{j})$ kg·m/s. In what direction do the bodies move after the collision?

12. Drop, in succession, a baseball and a basketball from about shoulder height above a hard floor, and note how high each rebounds. Then align the baseball above the basketball (with a small separation as in Fig. 10-28a) and drop them simultaneously. (Be prepared to duck, and guard your face.) (a) Is the rebound height of the basketball now higher or lower than before (Fig. 10-28b)? (b) Is the rebound height of the baseball less than or greater than the sum of the individual baseball and basketball rebound heights? (See also Problem 37.)

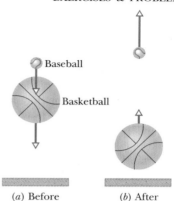

(a) Before   (b) After

FIGURE 10-28   Question 12 and Problem 37.

EXERCISES & PROBLEMS

SECTION 10-2 Impulse and Linear Momentum

1E. The linear momentum of a 1500 kg car increased by 9.0×10^3 kg·m/s in 12 s. (a) What is the magnitude of the constant force that accelerated the car? (b) By how much did the speed of the car increase?

2E. A cue stick strikes a stationary pool ball, exerting an average force of 50 N over a time of 10 ms. If the ball has mass 0.20 kg, what speed does it have after impact?

3E. The National Transportation Safety Board is testing the crash-worthiness of a new car. The 2300 kg vehicle, moving at 15 m/s, is allowed to collide with a bridge abutment, being brought to rest in a time of 0.56 s. What force, assumed constant, acted on the car during impact?

4E. A ball of mass m and speed v strikes a wall perpendicularly and rebounds in the opposite direction with the same speed. (a) If the time of collision is Δt, what is the average force exerted by the ball on the wall? (b) Evaluate this average force numerically for a rubber ball with mass 140 g moving at 7.8 m/s; the duration of the collision is 3.8 ms.

5E. A 150 g (weight $\approx$ 5.3 oz) baseball pitched at a speed of 40 m/s ($\approx$ 130 ft/s) is hit straight back to the pitcher at a speed of 60 m/s ($\approx$ 200 ft/s). What average force was exerted by the bat if it was in contact with the ball for 5.0 ms?

6E. Until he was in his seventies, Henri LaMothe excited audiences by belly-flopping from a height of 40 ft into 12 in. of water (Fig. 10-29). Assuming that he stops just as he reaches the bottom of the water, what is the average force on him from the water? Assume his weight is 160 lb.

7E. In February 1955, a paratrooper fell 1200 ft from an airplane without being able to open his chute but happened to land in snow, suffering only minor injuries. Assume that his speed at impact was 56 m/s (terminal speed), that his mass (including gear) was 85 kg, and that the force on him from the snow was at the survivable limit of 1.2×10^5 N. What is the minimum depth of snow that would have stopped him safely?

8E. A force that averages 1200 N is applied to a 0.40 kg steel ball moving at 14 m/s in a collision lasting 27 ms. If the force is in a direction opposite the initial velocity of the ball, find the final speed and direction of the ball.

9E. A 1.2 kg medicine ball drops vertically onto a floor, hitting with a speed of 25 m/s. It rebounds with an initial speed of 10 m/s. (a) What impulse acts on the ball during the contact? (b) If the ball is in contact with the floor for 0.020 s, what is the average force exerted on the floor?

10E. A golfer hits a golf ball, giving it an initial velocity of magnitude 50 m/s directed 30° above the horizontal. Assuming that the mass of the ball is 46 g and the club and ball are in contact

FIGURE 10-29   Exercise 6.

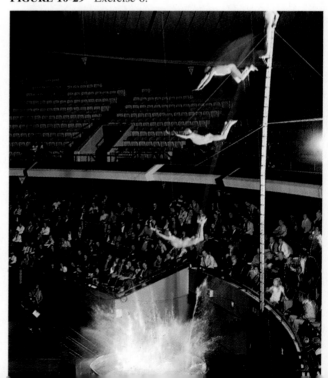

for 1.7 ms, find (a) the impulse on the ball, (b) the impulse on the club, (c) the average force exerted on the ball by the club, and (d) the work done on the ball.

11P. A 1400 kg car moving at 5.3 m/s is initially traveling north in the positive y direction. After completing a 90° right-hand turn to the positive x direction in 4.6 s, the inattentive operator drives into a tree, which stops the car in 350 ms. In unit-vector notation, what is the impulse on the car (a) during the turn and (b) during the collision? What is the magnitude of the average force that acts on the car (c) during the turn and (d) during the collision? (e) What is the angle between the average force in (c) and the positive x direction?

12P. The force on a 10 kg object increases uniformly from zero to 50 N in 4.0 s. What is the object's speed at the end of the 4.0 s interval if it started from rest?

13P. A pellet gun fires ten 2.0 g pellets per second with a speed of 500 m/s. The pellets are stopped by a rigid wall. (a) What is the momentum of each pellet? (b) What is the kinetic energy of each pellet? (c) What is the average force exerted by the stream of pellets on the wall? (d) If each pellet is in contact with the wall for 0.6 ms, what is the average force exerted on the wall by each pellet during contact? Why is this average force so different from the average force calculated in (c)?

14P. A movie set machine gun fires 50 g bullets at a speed of 1000 m/s. An actor, holding the machine gun in his hands, can exert an average force of 180 N against the gun. Determine the maximum number of bullets he can fire per minute while still holding the gun steady.

15P. It is well known that bullets and other missiles fired at Superman simply bounce off his chest (Fig. 10-30). Suppose that a gangster sprays Superman's chest with 3 g bullets at the rate of 100 bullets/min, the speed of each bullet being 500 m/s. Suppose too that the bullets rebound straight back with no change in speed. What is the average force exerted by the stream of bullets on Superman's chest?

16P. During a violent thunderstorm, hail of diameter 1.0 cm falls at a speed of 25 m/s. There are estimated to be 120 hailstones per cubic meter of air. Ignore the bounce of the hail on impact. (a) What is the mass of each hailstone (density = 0.92 g/cm³)? (b) What average force is exerted by hail on a 10 m × 20 m flat roof during the storm?

17P. A stream of water impinges on a stationary "dished" turbine blade, as shown in Fig. 10-31. The speed of the water is v, both before and after it strikes the curved surface of the blade, and the mass of water striking the blade per unit time is constant at the value μ. Find the force exerted by the water on the blade.

FIGURE 10-31   Problem 17.

18P. A stream of water from a hose is sprayed on a wall. If the speed of the water is 5.0 m/s and the hose sprays 300 cm³/s, what is the average force exerted on the wall by the stream of water? Assume that the water does not spatter back appreciably. Each cubic centimeter of water has a mass of 1.0 g.

19P. Figure 10-32 shows an approximate plot of force versus time during the collision of a 58 g tennis ball with a wall. The initial velocity of the ball is 34 m/s perpendicular to the wall; it rebounds with the same speed, also perpendicular to the wall. What is F_{max}, the maximum value of the contact force during the collision?

FIGURE 10-32   Problem 19.

20P. A ball having a mass of 150 g strikes a wall with a speed of 5.2 m/s and rebounds with only 50% of its initial kinetic energy.

FIGURE 10-30   Problem 15.

(a) What is the speed of the ball immediately after rebounding? (b) What was the magnitude of the impulse of the ball on the wall? (c) If the ball was in contact with the wall for 7.6 ms, what was the magnitude of the average force exerted by the wall on the ball during this time interval?

21P. In the overhead view of Fig. 10-33, a 300 g ball with a speed v of 6.0 m/s strikes a wall at an angle θ of 30° and then rebounds with the same speed and angle. It is in contact with the wall for 10 ms. (a) What is the impulse on the ball? (b) What is the average force exerted by the ball on the wall?

FIGURE 10-33 Problem 21.

22P. A 2500 kg unmanned space probe is moving in a straight line at a constant speed of 300 m/s. Control rockets on the space probe execute a burn in which a thrust of 3000 N acts for 65.0 s. (a) What is the change in linear momentum (magnitude only) of the probe if the thrust is backward, forward, or directly sideways? (b) What is the change in kinetic energy under the same three conditions? Assume that the mass of the ejected fuel is negligible compared to the mass of the space probe.

23P. A force exerts an impulse J on an object of mass m, changing its speed from v to u. The force and the object's motion are along the same straight line. Show that the work done by the force is $\frac{1}{2}J(u + v)$.

24P. A spacecraft is separated into two parts by detonating the explosive bolts that hold them together. The masses of the parts are 1200 kg and 1800 kg; the magnitude of the impulse on each part is 300 N·s. With what relative speed do the two parts separate because of the detonation?

25P. A stationary croquet ball with mass 0.50 kg is struck by a mallet, receiving the impulse shown in Fig. 10-34. What is the ball's velocity just after the force has become zero?

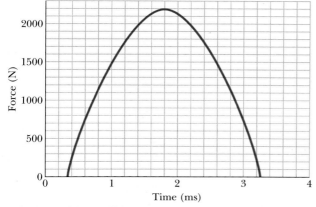

FIGURE 10-34 Problem 25.

26P. A soccer player kicks a soccer ball of mass 0.45 kg that is initially at rest. The player's foot is in contact with the ball for 3.0×10^{-3} s, and the force of the kick is given by

$$F(t) = [(6.0 \times 10^6)t - (2.0 \times 10^9)t^2] \text{ N},$$

for $0 \le t \le 3.0 \times 10^{-3}$ s, where t is in seconds. Find the magnitudes of the following: (a) the impulse imparted to the ball, (b) the average force exerted by the player's foot on the ball during the period of contact, (c) the maximum force exerted by the player's foot on the ball during the period of contact, and (d) the ball's velocity immediately after it loses contact with the player's foot.

27P. Spacecraft *Voyager 2* (of mass m and speed v relative to the Sun) approaches the planet Jupiter (of mass M and speed V relative to the Sun) as shown in Fig. 10-35. The spacecraft rounds the planet and departs in the opposite direction. What is its speed, relative to the Sun, after this slingshot encounter, which can be analyzed as a collision? Assume $v = 12$ km/s and $V = 13$ km/s (the orbital speed of Jupiter). The mass of Jupiter is very much greater than the mass of the spacecraft; $M \gg m$.

FIGURE 10-35 Problem 27.

SECTION 10-3 Elastic Collisions in One Dimension

28E. The blocks in Fig. 10-36 slide without friction. (a) What is the velocity **v** of the 1.6 kg block after the collision? (b) Is the collision elastic? (c) Suppose the initial velocity of the 2.4 kg block is the reverse of what is shown. Can the velocity **v** of the 1.6 kg block after the collision be in the direction shown?

FIGURE 10-36 Exercise 28.

29E. A hovering fly is approached by an enraged elephant charging at 2.1 m/s. Assuming that the collision is elastic, at what speed does the fly rebound? Note that the projectile (the elephant) is much more massive than the stationary target (the fly).

30E. An electron collides elastically with a hydrogen atom initially at rest. (All motions are along the same straight line.) What percentage of the electron's initial kinetic energy is transferred to the hydrogen atom? The mass of the hydrogen atom is 1840 times the mass of the electron.

31E. A cart with mass 340 g moving on a frictionless linear air track at an initial speed of 1.2 m/s strikes a second cart of unknown mass at rest. The collision between the carts is elastic.

After the collision, the first cart continues in its original direction at 0.66 m/s. (a) What is the mass of the second cart? (b) What is its speed after impact? (c) What is the speed of the two-cart center of mass?

32E. An alpha particle (mass 4 u) experiences an elastic head-on collision with a gold nucleus (mass 197 u) that is originally at rest. What percentage of its original kinetic energy does the alpha particle lose?

33E. A body of mass 2.0 kg makes an elastic collision with another body at rest and continues to move in the original direction but with one-fourth of its original speed. (a) What is the mass of the struck body? (b) What is the speed of the two-body center of mass if the initial speed of the 2.0 kg body was 4.0 m/s?

34P. A steel ball of mass 0.500 kg is fastened to a cord 70.0 cm long and fixed at the far end, and is released when the cord is horizontal (Fig. 10-37). At the bottom of its path, the ball strikes a 2.50 kg steel block initially at rest on a frictionless surface. The collision is elastic. Find (a) the speed of the ball and (b) the speed of the block, both just after the collision.

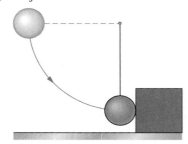

FIGURE 10-37 Problem 34.

35P. A platform scale is calibrated to indicate the mass in kilograms of an object placed on it. Particles fall from a height of 3.5 m and collide with the platform of the scale. The collisions are elastic; the particles rebound upward with the same speed they had before hitting the pan. If each particle has a mass of 110 g and collisions occur at the rate of 42 s^{-1}, what is the average scale reading?

36P. Two titanium spheres approach each other head-on with the same speed and collide elastically. After the collision, one of the spheres, whose mass is 300 g, remains at rest. (a) What is the mass of the other sphere? (b) What is the speed of the two-sphere center of mass if the initial speed of each sphere was 2.0 m/s?

37P. A ball of mass m is aligned above a ball of mass M (with a slight separation, as in Fig. 10-28a), and the two are dropped simultaneously from height h. (Assume the radius of each ball is negligible compared to h.) (a) If M rebounds elastically from the floor and then m rebounds elastically from M, what ratio m/M results in M stopping upon its collision with m? (The answer is approximately the mass ratio of a baseball to a basketball, as in Question 12.) (b) What height does m then reach?

38P. A block of mass m_1 is at rest on a long frictionless table, one end of which terminates at a wall. A block of mass m_2 is placed between the first block and the wall and set in motion to the left, toward m_1, with constant speed v_{2i}, as in Fig. 10-38. Assuming that all collisions are elastic, find the value of m_2 (in

terms of m_1) for which both blocks move with the same velocity after m_2 has collided once with m_1 and once with the wall. Assume the wall to have infinite mass.

FIGURE 10-38 Problem 38.

39P. A target glider, whose mass m_2 is 350 g, is at rest on an air track, a distance d = 53 cm from the end of the track. A projectile glider, whose mass m_1 is 590 g, approaches the target glider with velocity v_{1i} = −75 cm/s and collides elastically with it (Fig. 10-39). The target glider rebounds elastically from a short spring at the end of the track and meets the projectile glider for a second time. How far from the end of the track does this second collision occur?

FIGURE 10-39 Problem 39.

SECTION 10-4   Inelastic Collisions in One Dimension

40E. Meteor Crater in Arizona (Fig. 10-1a) is thought to have been formed by the impact of a meteor with the Earth some 20,000 years ago. The mass of the meteor is estimated at 5 × 10^{10} kg, and its speed at 7200 m/s. What speed would such a meteor impart to the Earth in a head-on collision?

41E. A 6.0 kg box sled is coasting across frictionless ice at a speed of 9.0 m/s when a 12 kg package is dropped into it from above. What is the new speed of the sled?

42E. A 5.20 g bullet moving at 672 m/s strikes a 700 g wooden block at rest on a frictionless surface. The bullet emerges with its speed reduced to 428 m/s. (a) What is the resulting speed of the block? (b) What is the speed of the bullet–block center of mass?

43E. Two 2.0 kg masses, A and B, collide. The velocities before the collision are $\mathbf{v}_A$ = 15$\mathbf{i}$ + 30$\mathbf{j}$ and $\mathbf{v}_B$ = −10$\mathbf{i}$ + 5.0$\mathbf{j}$. After the collision, $\mathbf{v}'_A$ = −5.0$\mathbf{i}$ + 20$\mathbf{j}$. All speeds are given in meters per second. (a) What is the final velocity of B? (b) How much kinetic energy was gained or lost in the collision?

44E. A bullet of mass 10 g strikes a ballistic pendulum of mass 2.0 kg. The center of mass of the pendulum rises a vertical distance of 12 cm. Assuming that the bullet remains embedded in the pendulum, calculate the bullet's initial speed.

45E. A bullet of mass 4.5 g is fired horizontally into a 2.4 kg wooden block at rest on a horizontal surface. The coefficient of kinetic friction between block and surface is 0.20. The bullet comes to rest in the block, which moves 1.8 m. (a) What is the speed of the block immediately after the bullet comes to rest within it? (b) At what speed is the bullet fired?

46E. A 5.0 kg block with a speed of 3.0 m/s collides with a 10 kg block that has a speed of 2.0 m/s in the same direction. After the collision, the 10 kg block is observed to be traveling in the original direction with a speed of 2.5 m/s. (a) What is the speed of the 5.0 kg block immediately after the collision? (b) By how much does the total kinetic energy of the system of two blocks change because of the collision? (c) Suppose, instead, that the 10 kg block ends up with a speed of 4.0 m/s. What then is the change in the total kinetic energy? (d) Account for the result you obtained in (c).

47P. Two cars A and B slide on an icy road as they attempt to stop at a traffic light. The mass of A is 1100 kg, and the mass of B is 1400 kg. The coefficient of kinetic friction between the locked wheels of either car and the road is 0.13. Car A succeeds in coming to rest at the light, but car B cannot stop and rear-ends car A. After the collision, A comes to rest 8.2 m ahead of the impact point and B 6.1 m ahead; see Fig. 10-40. Both drivers had their brakes locked throughout the incident. (a) From the distance each car moved after the collision, find the speed of each car immediately after impact. (b) Use conservation of linear momentum to find the speed at which car B struck car A. On what grounds can the use of linear momentum conservation be criticized here?

FIGURE 10-40 Problem 47.

48P. A 3000 kg weight falls vertically through 6.0 m and then collides with a 500 kg pile, driving it 3.0 cm into bedrock. Assuming that the weight–pile collision is completely inelastic, find the magnitude of the average force on the pile by the bedrock during the 3.0 cm descent.

49P. Two particles, one having twice the mass of the other, are held together with a compressed spring between them. The energy stored in the spring is 60 J. How much kinetic energy does each particle have after the two are released? Assume that all the stored energy is transferred to the particles and that neither particle is attached to the spring after the release.

50P. A 3.50 g bullet is fired horizontally at two blocks resting on a smooth tabletop, as shown in Fig. 10-41a. The bullet passes

FIGURE 10-41 Problem 50.

through the first block, with mass 1.20 kg, and embeds itself in the second, with mass 1.80 kg. Speeds of 0.630 m/s and 1.40 m/s, respectively, are thereby imparted to the blocks, as shown in Fig. 10-41b. Neglecting the mass removed from the first block by the bullet, find (a) the speed of the bullet immediately after it emerges from the first block and (b) the bullet's original speed.

51P. An object with mass m and speed v explodes into two pieces, one three times as massive as the other; the explosion takes place in gravity-free space. The less massive piece comes to rest. How much kinetic energy was added to the system in the explosion?

52P. A box is put on a scale that is marked in units of mass and adjusted to read zero when the box is empty. A stream of marbles is then poured into the box from a height h above its bottom at a rate of R (marbles per second). Each marble has mass m. If the collisions between the marbles and the box are completely inelastic, find the scale reading at time t after the marbles begin to fill the box. Determine a numerical answer when $R = 100$ s^{-1}, $h = 7.60$ m, $m = 4.50$ g, and $t = 10.0$ s.

53P. A 35-ton railroad freight car collides with a stationary caboose car. They couple together, and 27% of the initial kinetic energy is dissipated as heat, sound, vibrations, and so on. Find the weight of the caboose.

54P. A ball of mass m is projected with speed v_i into the barrel of a spring gun of mass M initially at rest on a frictionless surface; see Fig. 10-42. The ball sticks in the barrel at the point of maximum compression of the spring. No mechanical energy is dissipated by friction. (a) What is the speed of the spring gun after the ball comes to rest in the barrel? (b) What fraction of the initial kinetic energy of the ball is stored in the spring?

FIGURE 10-42 Problem 54.

55P. A block of mass $m_1 = 2.0$ kg slides along a frictionless table with a speed of 10 m/s. Directly in front of it, and moving in the same direction, is a block of mass $m_2 = 5.0$ kg moving at 3.0 m/s. A massless spring with spring constant $k = 1120$ N/m is attached to the near side of m_2, as shown in Fig. 10-43. When the blocks collide, what is the maximum compression of the spring? (*Hint:* At the moment of maximum compression of the spring, the two blocks move as one. Find the velocity by noting that the collision is completely inelastic to this point.)

FIGURE 10-43 Problem 55.

56P. A 1.0 kg block at rest on a horizontal frictionless surface is connected to an unstretched spring ($k = 200$ N/m) whose other end is fixed (Fig. 10-44). A 2.0 kg block whose speed is 4.0 m/s collides with the 1.0 kg block. If the two blocks stick together after the one-dimensional collision, what maximum compression of the spring occurs when the blocks momentarily stop?

FIGURE 10-44   Problem 56.

57P. Two 22.7 kg ice sleds are placed a short distance apart, one directly behind the other, as shown in Fig. 10-45. A 3.63 kg cat, standing on one sled, jumps across to the other and immediately back to the first. Both jumps are made at a speed of 3.05 m/s relative to the ice. Find the final speeds of the two sleds.

FIGURE 10-45   Problem 57.

58P. The bumper of a 1200 kg car is designed so that it can just absorb all the energy when the car runs head-on into a solid wall at 5.00 km/h. The car is involved in a collision in which it runs at 70.0 km/h into the rear of a 900 kg car moving at 60.0 km/h in the same direction. The 900 kg car is accelerated to 70.0 km/h as a result of the collision. (a) What is the speed of the 1200 kg car immediately after impact? (b) What is the ratio of the kinetic energy absorbed in the collision to that which can be absorbed by the bumper of the 1200 kg car?

59P. A railroad freight car weighing 32 tons and traveling at 5.0 ft/s overtakes one weighing 24 tons and traveling at 3.0 ft/s in the same direction. If the cars couple together, find (a) the speed of the cars after collision and (b) the loss of kinetic energy during collision. (c) If instead, as is very unlikely, the collision is elastic, find the speeds of the cars after collision.

SECTION 10-5   Collisions in Two Dimensions

60E. An alpha particle collides with an oxygen nucleus, initially at rest. The alpha particle is scattered at an angle of 64.0° above its initial direction of motion, and the oxygen nucleus recoils at an angle of 51.0° below this initial direction. The final speed of the nucleus is 1.20×10^5 m/s. Find (a) the final speed and (b) the initial speed of the alpha particle. (The mass of an alpha particle is 4.0 u; the mass of an oxygen nucleus is 16 u.)

61E. A proton (atomic mass 1 u) with a speed of 500 m/s collides elastically with another proton at rest. The projectile proton is scattered 60° from its initial direction. (a) What is the direction of the velocity of the target proton after the collision? (b) What are the speeds of the two protons after the collision?

62E. A certain nucleus, at rest, spontaneously disintegrates into three particles. Two of them are detected; their masses and velocities are as shown in Fig. 10-46. (a) In unit-vector notation, what is the linear momentum of the third particle, which is known to have a mass of 11.7×10^{-27} kg? (b) How much kinetic energy appears in the disintegration process?

FIGURE 10-46   Exercise 62.

63E. In a game of pool, the cue ball strikes another ball initially at rest. After the collision, the cue ball moves at 3.50 m/s along a line making an angle of 22.0° with its original direction of motion, and the second ball has a speed of 2.00 m/s. Find (a) the angle between the direction of motion of the second ball and the original direction of motion of the cue ball and (b) the original speed of the cue ball. (c) Is kinetic energy conserved?

64E. Two vehicles A and B are traveling west and south, respectively, toward the same intersection, where they collide and lock together. Before the collision, A (total weight 2700 lb) is moving with a speed of 40 mi/h and B (total weight 3600 lb) has a speed of 60 mi/h. Find the magnitude and direction of the velocity of the (interlocked) vehicles immediately after the collision.

65E. In a game of billiards, the cue ball is given an initial speed V and strikes the pack of 15 stationary balls. All 16 balls then engage in numerous ball–ball and ball–cushion collisions. Some time later, it is observed that (by some accident) all 16 balls have the same speed v. Assuming that all collisions are elastic and ignoring the rotational aspect of the balls' motion, calculate v in terms of V.

66P. A 20.0 kg body is moving in the direction of the positive x axis with a speed of 200 m/s when, owing to an internal explosion, it breaks into three parts. One part, whose mass is 10.0 kg, moves away from the point of explosion with a speed of 100 m/s in the direction of the positive y axis. A second fragment, with a mass of 4.00 kg, moves in the direction of the negative x axis with a speed of 500 m/s. (a) What is the velocity of the third (6.00 kg) fragment? (b) How much energy was released in the explosion? Ignore effects due to gravity.

67P. Two balls A and B, having different but unknown masses, collide. A is initially at rest, and B has speed v. After collision, B has speed $v/2$ and moves perpendicularly to its original motion. (a) Find the direction in which ball A moves after collision. (b) Can you determine the speed of A from the information given? Explain.

68P. Show that if a neutron is scattered through 90° in an elastic collision with a deuteron that is initially at rest, the neutron loses two-thirds of its initial kinetic energy to the deuteron. (The mass of a neutron is 1.0 u; the mass of a deuteron is 2.0 u.)

69P. After a completely inelastic collision, two objects of the same mass and same initial speed are found to move away together at half their initial speed. Find the angle between the initial velocities of the objects.

70P. Two pendulums, both of length l, are initially situated as in Fig. 10-47. The left pendulum is released and strikes the other. Assume that the collision is completely inelastic, and neglect the mass of the strings and any frictional effects. How high does the center of mass of the pendulum system rise after the collision?

FIGURE 10-47   Problem 70.

71P. A billiard ball moving at a speed of 2.2 m/s strikes an identical stationary ball a glancing blow. After the collision, one ball is found to be moving at a speed of 1.1 m/s in a direction making a 60° angle with the original line of motion. (a) Find the velocity of the other ball. (b) Can the collision be inelastic, given these data?

72P. A ball with an initial speed of 10 m/s collides elastically with two identical stationary balls whose centers are on a line perpendicular to the initial velocity and that are initially in contact with each other (Fig. 10-48). The first ball is aimed directly at the contact point, and all motion is frictionless. Find the velocities of all three balls after the collision. (*Hint:* With friction absent, each impulse is directed along the line connecting the centers of the colliding balls, normal to the colliding surfaces.)

FIGURE 10-48   Problem 72.

73P. A barge with mass 1.50×10^5 kg is proceeding down river at 6.2 m/s in heavy fog when it collides broadside with a barge heading directly across the river (see Fig. 10-49). The second

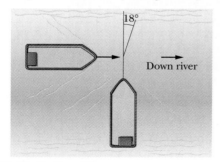

FIGURE 10-49   Problem 73.

barge has mass 2.78×10^5 kg and was moving at 4.3 m/s. Immediately after impact, the second barge finds its course deflected by 18° in the downriver direction and its speed increased to 5.1 m/s. The river current was practically zero at the time of the

accident. (a) What are the speed and direction of motion of the first barge immediately after the collision? (b) How much kinetic energy is lost in the collision?

SECTION 10-6   Reactions and Decay Processes

74E. The precise masses in the reaction

$$p + {}^{19}F \rightarrow \alpha + {}^{16}O$$

have been determined to be

$$m_p = 1.007825 \text{ u}, \qquad m_\alpha = 4.002603 \text{ u},$$
$$m_F = 18.998405 \text{ u}, \qquad m_O = 15.994915 \text{ u}.$$

Calculate the Q of the reaction from these data.

75E. A particle called Σ^- (sigma minus) is initially at rest and decays spontaneously into two other particles according to

$$\Sigma^- \rightarrow \pi^- + n.$$

The masses are

$$m_\Sigma = 2340.5 m_e, \quad m_\pi = 273.2 m_e, \quad m_n = 1838.65 m_e,$$

where m_e (9.11×10^{-31} kg) is the electron mass. (a) How much energy is transferred to kinetic energy in this process? (b) How do the linear momenta of the decay products (π^- and n) compare? (c) Which product gets the larger share of the kinetic energy?

76P*. An alpha particle with kinetic energy 7.70 MeV strikes an ^{14}N nucleus at rest. An ^{17}O nucleus and a proton are produced; the proton is emitted at 90° to the direction of the incident alpha particle and has a kinetic energy of 4.44 MeV. The masses of the various particles are: alpha particle, 4.00260 u; ^{14}N, 14.00307 u; proton, 1.007825 u; and ^{17}O, 16.99914 u. (a) What is the kinetic energy of the oxygen nucleus? (b) What is the Q of the reaction?

77P*. Consider the alpha decay of radium (Ra) to radon (Rn), according to the reaction

$$^{226}Ra \rightarrow \alpha + {}^{222}Rn.$$

The masses of the various nuclei are: ^{226}Ra, 226.0254 u; alpha, 4.0026 u; ^{222}Rn, 222.0175 u. (a) Calculate the Q of the reaction. (b) What value of Q would be obtained if the accurate masses given above were rounded off to three significant figures? What is the kinetic energy of (c) the alpha particle and (d) the radon nucleus? (For this calculation the rounded-off values of the masses *can* be used. Why?)

ELECTRONIC COMPUTATION

78. A 6.00 kg model rocket is traveling horizontally and due south with a speed of 20.0 m/s when it explodes into two pieces. The velocity of one piece, with a mass of 2.00 kg, is

$$\mathbf{v}_1 = (-12.0 \text{ m/s})\mathbf{i} + (30.0 \text{ m/s})\mathbf{j} - (15.0 \text{ m/s})\mathbf{k},$$

with $\mathbf{i}$ pointing due east, $\mathbf{j}$ pointing due north, and $\mathbf{k}$ pointing vertically upward. (a) What is the linear momentum of the other piece, in unit-vector notation? (b) What is the kinetic energy of the other piece? (c) How much kinetic energy is produced by the explosion?

11
Rotation

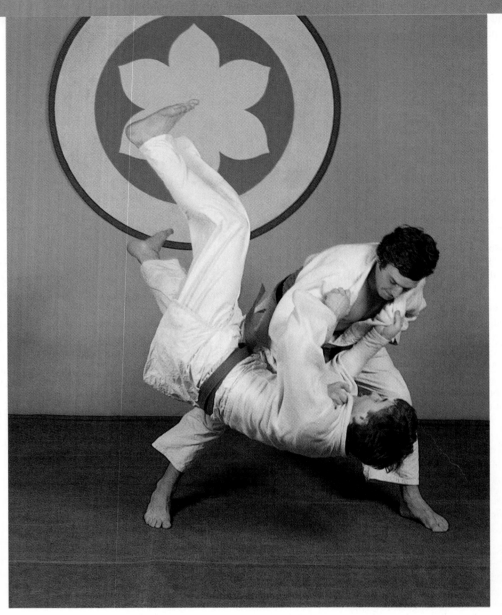

In judo, a weaker and smaller fighter who understands physics can defeat a stronger and larger fighter who does not. This fact is demonstrated by the basic "hip throw," in which a fighter rotates the fighter's opponent around his hip and—if the throw is successful—onto the mat. Without the proper use of physics, the throw requires considerable strength and can easily fail. What is the advantage offered by physics?

11-1 TRANSLATION AND ROTATION

The graceful movement of a figure skater can be used to illustrate, in an aesthetically pleasing way, two kinds of pure, or unmixed, motion. Figure 11-1a shows a skater gliding across the ice in a straight line with constant speed. Her motion is one of pure **translation.** Figure 11-1b shows her spinning at a constant rate about a fixed vertical axis, in a motion of pure **rotation.** This second kind of motion is our focus in this chapter.

Translation is motion along a straight line, the motion we have discussed almost exclusively so far. Rotation is the motion of wheels, gears, motors, the hands of clocks, the rotors of jet engines, and the blades of helicopters. It is the motion of hurricanes, planets, stars, and galaxies.

FIGURE 11-1 Figure skater Kristi Yamaguchi in motion of (a) pure translation along a fixed direction and (b) pure rotation about a fixed axis.

(a)

(b)

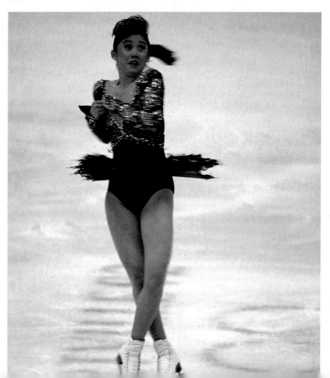

11-2 THE ROTATIONAL VARIABLES

In this chapter, we deal with the rotation of a *rigid* body about a *fixed* axis. The first of these restrictions means that we shall not examine the rotation of such objects as the Sun, because the Sun—a ball of gas—is not a rigid body. Our second restriction rules out objects like a bowling ball rolling down a bowling lane. Such a ball is in *rolling* motion, rotating about a *moving* axis.

Figure 11-2 shows a rigid body of arbitrary shape in pure *rotation around a fixed axis,* called the **axis of rotation** or the **rotation axis.** Every point of the body moves in a circle whose center lies on the axis of rotation, and every point moves through the same angle during a particular time interval. (In pure translation, every point of the body moves in a straight line, and every point moves through the same *linear distance* during a particular time interval. Comparisons between linear and angular motion will be a constant part of what follows.)

We deal now—one at a time—with the angular equivalents of the linear quantities position, displacement, velocity, and acceleration.

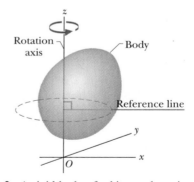

FIGURE 11-2 A rigid body of arbitrary shape in pure rotation about the *z* axis of a coordinate system. The position of the *reference line* with respect to the rigid body is arbitrary, but it is perpendicular to the rotation axis. It is fixed in the body and rotates with the body.

Angular Position

Figure 11-2 also shows a reference line, fixed in the body, perpendicular to the rotation axis, and rotating with the body. We can describe the motion of the rotating body by specifying the **angular position** of this line, that is, the angle of the line relative to a fixed direction. In Fig. 11-3, the angular position θ is measured relative to the positive direction of the x axis, and θ is given by

$$\theta = \frac{s}{r} \quad \text{(radian measure)}. \qquad (11\text{-}1)$$

Here s is the length of arc (or the arc distance) along a circle and between the x axis and the reference line, and r is the radius of that circle.

An angle defined in this way is measured in **radians** (rad) rather than in revolutions (rev) or degrees. The radian, being the ratio of two lengths, is a pure number and thus has no dimension. Because the circumference of a circle of radius r is $2\pi r$, there are 2π radians in a complete circle:

$$1 \text{ rev} = 360° = \frac{2\pi r}{r} = 2\pi \text{ rad}, \qquad (11\text{-}2)$$

and thus

$$1 \text{ rad} = 57.3° = 0.159 \text{ rev}. \qquad (11\text{-}3)$$

We do *not* reset θ to zero with each complete rotation of the reference line about the rotation axis. If the reference line completes two revolutions, then the angular position θ is $\theta = 4\pi$ rad.

For pure translational motion along the x direction, we know all there is to know about a moving body if we know $x(t)$, its position as a function of time. Similarly, for pure rotation, we know all there is to know about a rotating body if we know $\theta(t)$, the angular position of the body's reference line as a function of time.

FIGURE 11-3 The rotating rigid body of Fig. 11-2 in cross section, viewed from above. The plane of the cross section is perpendicular to the rotation axis, which now extends out of the page, toward you. In this position of the body, the reference line makes an angle θ with the x axis.

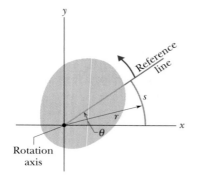

Angular Displacement

If the body of Fig. 11-3 rotates about the rotation axis as in Fig. 11-4, changing the angular position of the reference line from θ_1 to θ_2, the body undergoes an **angular displacement** $\Delta\theta$ given by

$$\Delta\theta = \theta_2 - \theta_1. \qquad (11\text{-}4)$$

This definition of angular displacement holds not only for the rigid body as a whole but also for *every particle within that body*.

If a body is in translational motion along an x axis, its displacement Δx is either positive or negative, depending on whether the body is moving in the direction of increasing x or decreasing x. Similarly, the angular displacement $\Delta\theta$ of a rotating body can be either positive or negative, depending on whether the body is rotating in the direction of increasing θ (counterclockwise, as in Figs. 11-3 and 11-4) or decreasing θ (clockwise).

CHECKPOINT 1: The disk in the figure can rotate about its central axis like a merry-go-round. Which of the following pairs of values for its initial and final angular positions, respectively, give a negative angular displacement: (a) -3 rad, $+5$ rad, (b) -3 rad, -7 rad, (c) 7 rad, -3 rad?

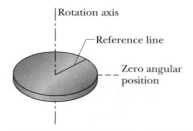

FIGURE 11-4 The reference line of the rigid body of Figs. 11-2 and 11-3 is at angular position θ_1 at time t_1 and at angular position θ_2 at a later time t_2. The quantity $\Delta\theta$ ($= \theta_2 - \theta_1$) is the angular displacement that occurs during the interval Δt ($= t_2 - t_1$). The body itself is not shown.

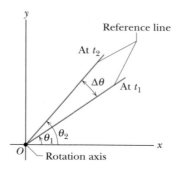

Angular Velocity

Suppose (see Fig. 11-4) that our rotating body is at angular position θ_1 at time t_1 and at angular position θ_2 at time t_2. We define the **average angular velocity** of the body in the time interval Δt from t_1 to t_2 to be

$$\overline{\omega} = \frac{\theta_2 - \theta_1}{t_2 - t_1} = \frac{\Delta\theta}{\Delta t}, \tag{11-5}$$

in which $\Delta\theta$ is the angular displacement that occurs during Δt (ω is the lowercase Greek letter omega).

The **(instantaneous) angular velocity** ω, with which we shall be most concerned, is the limit of the ratio in Eq. 11-5 as Δt is made to approach zero. Thus

$$\omega = \lim_{\Delta t \to 0} \frac{\Delta\theta}{\Delta t} = \frac{d\theta}{dt}. \tag{11-6}$$

If we know $\theta(t)$, we can find the angular velocity ω by differentiation.

Equations 11-5 and 11-6 hold not only for the rotating rigid body as a whole but also for *every particle of that body.* The unit of angular velocity is commonly the radian per second (rad/s) or the revolution per second (rev/s).

If a particle moves in translation along an x axis, its linear velocity v can be either positive or negative, depending on whether the particle is moving in the direction of increasing x or decreasing x. Similarly, the angular velocity ω of a rotating rigid body can be either positive or negative, depending on whether the body is rotating in the direction of increasing θ (counterclockwise) or of decreasing θ (clockwise). The magnitude of an angular velocity is called the **angular speed,** which is also represented with ω.

Angular Acceleration

If the angular velocity of a rotating body is not constant, then the body has an angular acceleration. Let ω_2 and ω_1 be the angular velocities at times t_2 and t_1, respectively. The **average angular acceleration** of the rotating body in the interval from t_1 to t_2 is defined as

$$\overline{\alpha} = \frac{\omega_2 - \omega_1}{t_2 - t_1} = \frac{\Delta\omega}{\Delta t}, \tag{11-7}$$

in which $\Delta\omega$ is the change in the angular velocity that occurs during the time interval Δt. The **(instantaneous) angular acceleration** α, with which we shall be most concerned, is the limit of this quantity as Δt is made to approach zero. Thus

$$\alpha = \lim_{\Delta t \to 0} \frac{\Delta\omega}{\Delta t} = \frac{d\omega}{dt}. \tag{11-8}$$

Equations 11-7 and 11-8 hold not only for the rotating rigid body as a whole but also for *every particle of that body.* The unit of angular acceleration is commonly the radian per second-squared (rad/s²) or the revolution per second-squared (rev/s²).

SAMPLE PROBLEM 11-1

The angular position of a reference line on a spinning wheel is given by

$$\theta = t^3 - 27t + 4,$$

where t is in seconds and θ is in radians.

(a) Find $\omega(t)$ and $\alpha(t)$.

SOLUTION: To get $\omega(t)$, we differentiate $\theta(t)$ with respect to t:

$$\omega = \frac{d\theta(t)}{dt} = 3t^2 - 27. \qquad \text{(Answer)}$$

To get $\alpha(t)$, we differentiate $\omega(t)$ with respect to t:

$$\alpha = \frac{d\omega(t)}{dt} = \frac{d(3t^2 - 27)}{dt} = 6t. \qquad \text{(Answer)}$$

(b) Do we ever find $\omega = 0$?

SOLUTION: Setting $\omega(t) = 0$ yields

$$0 = 3t^2 - 27,$$

which we solve, finding

$$t = \pm 3 \text{ s}. \qquad \text{(Answer)}$$

That is, the angular velocity is momentarily zero 3 s before and 3 s after our clock reads zero.

(c) Describe the wheel's motion for $t \geq 0$.

SOLUTION: To answer, we examine the expressions for $\theta(t)$, $\omega(t)$, and $\alpha(t)$.

At $t = 0$, the reference line on the wheel is at $\theta = 4$ rad, and the wheel is rotating with an angular velocity of -27 rad/s (that is, *clockwise* at an angular speed of 27 rad/s) and an angular acceleration of zero.

For $0 < t < 3$ s, the wheel continues to rotate clockwise, but at decreasing angular speed, because it now has a positive (counterclockwise) angular acceleration. (Check $\omega(t)$ and $\alpha(t)$ for, say, $t = 2$ s.)

At $t = 3$ s, the wheel stops momentarily ($\omega = 0$) and has rotated as far clockwise as it will ever get (the reference line is now at $\theta = -50$ rad).

For $t > 3$ s, the wheel's angular acceleration continues to increase. Its angular velocity, which is now also counterclockwise, increases rapidly because the signs of ω and α are the same.

SAMPLE PROBLEM 11-2

A child's top is spun with angular acceleration

$$\alpha = 5t^3 - 4t,$$

where the coefficients are in units compatible with seconds and radians. At $t = 0$, the top has angular velocity 5 rad/s, and a reference line on it is at angular position $\theta = 2$ rad.

(a) Obtain an expression for the angular velocity $\omega(t)$ of the top.

SOLUTION: From Eq. 11-8 we have

$$d\omega = \alpha \, dt,$$

which we integrate to get

$$\omega = \int \alpha \, dt = \int (5t^3 - 4t) \, dt$$

$$= \tfrac{5}{4}t^4 - \tfrac{4}{2}t^2 + C.$$

To evaluate the constant of integration C, we note that $\omega = 5$ rad/s at $t = 0$. Substituting these values in our expression for ω yields

$$5 \text{ rad/s} = 0 - 0 + C,$$

so $C = 5$ rad/s. Then

$$\omega = \tfrac{5}{4}t^4 - 2t^2 + 5. \qquad \text{(Answer)}$$

(b) Obtain an expression for the angular position $\theta(t)$ of the top.

SOLUTION: From Eq. 11-6 we have

$$d\theta = \omega \, dt,$$

which we integrate to get

$$\theta = \int \omega \, dt = \int (\tfrac{5}{4}t^4 - 2t^2 + 5) \, dt,$$

$$= \tfrac{1}{4}t^5 - \tfrac{2}{3}t^3 + 5t + C'$$

$$= \tfrac{1}{4}t^5 - \tfrac{2}{3}t^3 + 5t + 2, \qquad \text{(Answer)}$$

where C' has been evaluated by noting that $\theta = 2$ rad at $t = 0$.

11-3 ARE ANGULAR QUANTITIES VECTORS?

We can describe the position, velocity, and acceleration of a single particle by means of vectors. If the particle is confined to a straight line, however, we do not really need the power of vectors. Such a particle has only two directions available to it, and we can designate these with plus and minus signs.

In the same way, a rigid body rotating about a fixed

axis can rotate only clockwise or counterclockwise about this axis, and again we can select between them by means of plus and minus signs. The question arises: "Can we treat the angular displacement, velocity, and acceleration of a rotating body as vectors?" The answer is a qualified "yes" (see caution below, in connection with angular displacements).

Consider the angular velocity. Figure 11-5a shows a phonograph record rotating about a fixed spindle. The record has a fixed rotation rate $\omega \, (= 33\tfrac{1}{3}$ rev/min) and a fixed direction of rotation (clockwise as viewed from above). By convention, we represent its angular velocity as a vector $\boldsymbol{\omega}$ pointing along the axis of rotation, as in Fig. 11-5b. We choose the length of this vector according to some convenient scale, for example, with 1 cm corresponding to 10 rev/min.

We establish a direction for the vector $\boldsymbol{\omega}$ by using a **right-hand rule,** as Fig. 11-5c shows. Curl your right hand about the rotating record, your fingers pointing *in the direction of rotation.* Your extended thumb will then point in the direction of the angular velocity vector. If the record were to rotate in the opposite sense, the right-hand rule would tell you that the angular velocity vector then points in the opposite direction.

It is not easy to get used to representing angular quantities as vectors. We instinctively expect that something should be moving *along* the direction of a vector. That is not the case here. Instead, something (the rigid body) is rotating *around* the direction of the vector. In the world of pure rotation, a vector defines an axis of rotation, not a direction in which something moves. Nonetheless, the

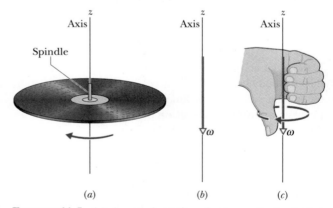

FIGURE 11-5 (a) A record rotating about a vertical axis that coincides with the axis of the spindle. (b) The angular velocity of the rotating record can be represented by the vector $\boldsymbol{\omega}$, lying along the axis and pointing down, as shown. (c) We establish the direction of the angular velocity vector as downward by using a right-hand rule. When the fingers of the right hand curl around the record and point the way it is moving, the extended thumb points in the direction of $\boldsymbol{\omega}$.

vector also defines the motion. Furthermore, it obeys all the rules for vector manipulation discussed in Chapter 3. The angular acceleration $\boldsymbol{\alpha}$ is another vector, and it too obeys those rules.

In this chapter we consider only rotations that are about a fixed axis. For such situations, we need not consider vectors—we can represent angular velocity with ω and angular acceleration with α, and we can indicate a direction with an implied plus sign for counterclockwise or an explicit minus sign for clockwise. In more complicated situations, however, we would use vectors $\boldsymbol{\omega}$ and $\boldsymbol{\alpha}$.

Now for the caution. Angular *displacements* (unless they are very small) *cannot* be treated as vectors. Why not? We can certainly give them both magnitude and direction, just as we did for the angular velocity vector in Fig. 11-5. However, that is (as the mathematicians say) a necessary condition but not sufficient. To be represented as a vector, a quantity must *also* obey the rules of vector addition, one of which says that if you add two vectors, the order in which you add them does not matter. Angular displacements fail this test.

To see this, place a book flat on a table, as in Fig. 11-6*a*. Now give the book two successive 90° angular displacements, *first* about the (horizontal) *x* axis and *then* about the (vertical) *y* axis, using the right-hand rule as a guide to positive rotation in each case.

Now, with a book in the same initial position (Fig. 11-6*b*), carry out these two angular displacements in the reverse order (that is, *first* about the *y* axis and *then* about the *x* axis). As the diagrams show, the book ends up in a very different orientation.

Thus the same two operations produce different results, depending on the order in which you carry them out. Addition of angular displacements is thus not commutative, so angular displacements are not vector quantities. With some practice, you should be able to show that the final positions of the book are much closer together if you use displacements much smaller than 90°. In the limiting case of differential angular displacements (such as $d\theta$ in Eq. 11-6), angular displacements *can* be treated as vectors.

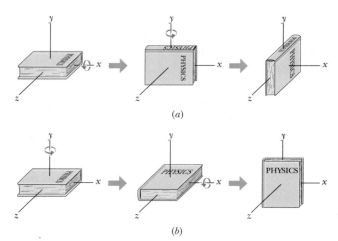

FIGURE 11-6 (*a*) From its initial position on the left, the book is given two successive 90° rotations, first about the (horizontal) *x* axis and then about the (vertical) *y* axis. (*b*) The book is given the same rotations, but in the reverse order. If angular displacement were truly a vector quantity, the order of these displacements would not matter. It clearly does matter, so (large) angular displacements are not vector quantities, even though we can assign magnitude and direction to them.

11-4 ROTATION WITH CONSTANT ANGULAR ACCELERATION

In pure translation, motion with a *constant linear acceleration* (for example, that of a falling body) is an important special case. In Table 2-1, we displayed a series of equations that hold for such motion.

In pure rotation, the case of *constant angular acceleration* is also important, and a parallel set of equations holds for this case also. We shall not derive them here, but simply write them from the corresponding linear equations, substituting equivalent angular quantities for the linear ones. This is done in Table 11-1, which displays both sets of equations (Eqs. 2-11 and 2-15 to 2-18; 11-9 to 11-13). For simplicity, we let $x_0 = 0$ and $\theta_0 = 0$ in these equa-

TABLE 11-1 EQUATIONS OF MOTION FOR CONSTANT LINEAR AND FOR CONSTANT ANGULAR ACCELERATION

EQUATION NUMBER	LINEAR FORMULA	MISSING VARIABLE		ANGULAR FORMULA	EQUATION NUMBER
(2-11)	$v = v_0 + at$	x	θ	$\omega = \omega_0 + \alpha t$	(11-9)
(2-15)	$x = v_0 t + \frac{1}{2}at^2$	v	ω	$\theta = \omega_0 t + \frac{1}{2}\alpha t^2$	(11-10)
(2-16)	$v^2 = v_0^2 + 2ax$	t	t	$\omega^2 = \omega_0^2 + 2\alpha\theta$	(11-11)
(2-17)	$x = \frac{1}{2}(v_0 + v)t$	a	α	$\theta = \frac{1}{2}(\omega_0 + \omega)t$	(11-12)
(2-18)	$x = vt - \frac{1}{2}at^2$	v_0	ω_0	$\theta = \omega t - \frac{1}{2}\alpha t^2$	(11-13)

tions. With those *initial conditions,* a linear displacement $\Delta x\ (= x - x_0)$ is equal to x, and an angular displacement $\Delta\theta\ (= \theta - \theta_0)$ is equal to θ.

$\mathbb{C}$HECKPOINT **2:** In four situations, a rotating body has angular position $\theta(t)$ given by (a) $\theta = 3t - 4$, (b) $\theta = -5t^3 + 4t^2 + 6$, (c) $\theta = 2/t^2 - 4/t$, and (d) $\theta = 5t^2 - 3$. To which situations do the angular equations of Table 11-1 apply?

SAMPLE PROBLEM 11-3

A grindstone (Fig. 11-7) has a constant angular acceleration $\alpha = 0.35$ rad/s². It starts from rest (that is, $\omega_0 = 0$) with an arbitrary reference line horizontal, at angular position $\theta_0 = 0$.

(a) What is the angular displacement θ of the reference line (hence of the wheel) at $t = 18$ s?

SOLUTION: From Eq. 11-10 of Table 11-1 $(\theta = \omega_0 t + \frac{1}{2}\alpha t^2)$, we obtain:

$$\theta = (0)(18\ \text{s}) + (\tfrac{1}{2})(0.35\ \text{rad/s}^2)(18\ \text{s})^2$$
$$= 56.7\ \text{rad} \approx 57\ \text{rad} \approx 3200° \approx 9.0\ \text{rev}. \quad \text{(Answer)}$$

(b) What is the wheel's angular velocity at $t = 18$ s?

SOLUTION: From Eq. 11-9 of Table 11-1 $(\omega = \omega_0 + \alpha t)$ we now get:

$$\omega = 0 + (0.35\ \text{rad/s}^2)(18\ \text{s})$$
$$= 6.3\ \text{rad/s} = 360°/\text{s} = 1.0\ \text{rev/s}. \quad \text{(Answer)}$$

FIGURE 11-7 Sample Problems 11-3 and 11-4. A grindstone. At $t = 0$ the reference line (which we imagine to be marked on the stone) is horizontal.

SAMPLE PROBLEM 11-4

For the grindstone of Sample Problem 11-3, let us assume the same angular acceleration ($\alpha = 0.35$ rad/s²), but let us now assume that the wheel does not start from rest but has an initial angular velocity ω_0 of -4.6 rad/s; that is, the angular acceleration acts initially to slow the wheel (because the signs of α and ω_0 are opposite).

(a) At what time t will the grindstone momentarily stop?

SOLUTION: Solving Eq. 11-9 $(\omega = \omega_0 + \alpha t)$ for t yields

$$t = \frac{\omega - \omega_0}{\alpha} = \frac{0 - (-4.6\ \text{rad/s})}{0.35\ \text{rad/s}^2} = 13\ \text{s}. \quad \text{(Answer)}$$

(b) At what time will the grindstone have rotated such that its angular displacement is five revolutions in the positive direction of rotation? (The angular displacement of the reference line will then be $\theta = 5$ rev.)

SOLUTION: The wheel is initially rotating in the negative (clockwise) direction with $\omega_0 = -4.6$ rad/s, but its angular acceleration α is positive (counterclockwise). This initial opposition of the signs of angular velocity and angular acceleration means that the wheel slows in its rotation in the negative direction, stops, and then reverses to rotate in the positive direction. After the reference line comes back through its initial orientation of $\theta = 0$, the wheel must turn an additional five revolutions to reach the angular displacement we want. This is all "taken care of" if we use Eq. 11-10:

$$\theta = \omega_0 t + \tfrac{1}{2}\alpha t^2.$$

Substituting known values and setting $\theta = 5$ rev $= 10\pi$ rad give us

$$10\pi\ \text{rad} = (-4.6\ \text{rad/s})t + (\tfrac{1}{2})(0.35\ \text{rad/s}^2)t^2.$$

Note that for t in seconds, the units in this equation are consistent. Dropping units (for convenience) and rearranging give

$$t^2 - 26.3t - 180 = 0. \quad (11\text{-}14)$$

Solving this quadratic equation for t and discarding the negative root, we obtain

$$t = 32\ \text{s}. \quad \text{(Answer)}$$

PROBLEM SOLVING TACTICS

TACTIC 1: *Unexpected Answers*

Do not hasten to throw away one root of a quadratic equation as meaningless. Often, as in Sample Problem 11-4, a discarded root has physical meaning.

The two solutions to Eq. 11-14 are $t = 32$ s and $t = -5.6$ s. We chose the first (positive) solution and ignored the second as perhaps meaningless. But is it? A negative time in this problem simply means a time before $t = 0$, that is, a time before we started to pay attention to what was going on.

Figure 11-8 is a plot of the angular position θ of the reference line on the grindstone of Sample Problem 11-4 as a function of time, for both positive and negative times. It is a plot of Eq. 11-10 $(\theta = \omega_0 t + \frac{1}{2}\alpha t^2)$, using as constants $\omega_0 = -4.6$ rad/s and $\alpha = +0.35$ rad/s². Point a corresponds to $t = 0$, at which time we arbitrarily took the angular position of the reference line to be zero. The wheel is moving in the direction of decreasing θ at that time, and continues to do so until coming to rest at point b at $t = 13$ s. It then reverses, with the reference line returning to its $\theta = 0$ position at point c and

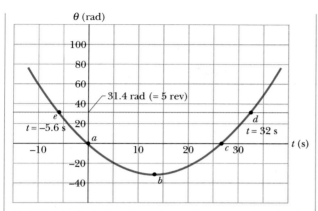

FIGURE 11-8 A plot of angular position versus time for the grindstone of Sample Problem 11-4. Negative times (that is, times before $t = 0$) have been included. The two roots of Eq. 11-14 are indicated by points d and e.

going on for five additional revolutions (= 31.4 rad) to point d. This latter point (with $t = 32$ s) represents the root that we accepted as the answer to our problem.

Note, however, that the reference line was at this same angular position at $t = -5.6$ s, before the "official start" of the problem. This root (point e) is just as valid as the root at point d. More important, by asking if this negative root can have any physical meaning, you learn a little more about the rotation of the wheel.

SAMPLE PROBLEM 11-5

During an analysis of a helicopter engine, you determine that the rotor's velocity changes from 320 rev/min to 225 rev/min in 1.50 min as the rotor is slowing to a stop.

(a) What is the average angular acceleration of the rotor blades during this interval?

SOLUTION: From Eq. 11-7,

$$\bar{\alpha} = \frac{\omega - \omega_0}{\Delta t} = \frac{225 \text{ rev/min} - 320 \text{ rev/min}}{1.50 \text{ min}}$$
$$= -63.3 \text{ rev/min}^2. \qquad \text{(Answer)}$$

(b) Assume the angular acceleration α is constant at this average value. How long will the rotor blades take to stop from their initial angular velocity of 320 rev/min?

SOLUTION: Solving Eq. 11-9 ($\omega = \omega_0 + \alpha t$) for t gives

$$t = \frac{\omega - \omega_0}{\alpha} = \frac{0 - 320 \text{ rev/min}}{-63.3 \text{ rev/min}^2}$$
$$= 5.1 \text{ min}. \qquad \text{(Answer)}$$

(c) How many revolutions will the rotor blades make in coming to rest from their initial angular velocity of 320 rev/min?

SOLUTION: Solving Eq. 11-11 ($\omega^2 = \omega_0^2 + 2\alpha\theta$) for θ gives us

$$\theta = \frac{\omega^2 - \omega_0^2}{2\alpha} = \frac{0 - (320 \text{ rev/min})^2}{(2)(-63.3 \text{ rev/min}^2)}$$
$$= 809 \text{ rev}. \qquad \text{(Answer)}$$

11-5 RELATING THE LINEAR AND ANGULAR VARIABLES

In Section 4-7, we discussed uniform circular motion, in which a particle travels at constant linear speed v along a circle and around an axis of rotation. When a rigid body, such as a merry-go-round, turns around an axis of rotation, each particle in the body moves in its own circle around that axis. Since the body is rigid, all the particles make one revolution in the same amount of time; that is, they all have the same angular speed ω.

However, the farther a particle is from the axis, the greater the circumference of its circle is, and so the faster its linear speed v must be. You can notice this on a merry-go-round. You turn with the same angular speed ω regardless of your distance from the center, but your linear speed v increases noticeably if you move to the outside edge of the merry-go-round.

We often need to relate the linear variables s, v, and a for a particular point in a rotating body to the angular variables θ, ω, and α for that body. The two sets of variables are related by r, the *perpendicular distance* of the point from the rotation axis. This perpendicular distance is the distance between the point and the rotation axis, measured along a perpendicular to the axis. It is also the radius r of the circle traveled by the point around the axis of rotation.

The Position

If a reference line on a rigid body is rotated through an angle θ, a point within the body is moved a distance s along a circular arc, where s is given by Eq. 11-1:

$$s = \theta r \qquad \text{(radian measure).} \qquad (11\text{-}15)$$

This is the first of our linear–angular relations. The angle θ must be measured in radians because Eq. 11-15 is itself the definition of angular measure in radians.

The Speed

Differentiating Eq. 11-15 with respect to time—with r held constant—leads to

$$\frac{ds}{dt} = \frac{d\theta}{dt} r.$$

But ds/dt is the linear speed (the magnitude of the linear velocity) of the point in question and $d\theta/dt$ is the angular speed ω of the rotating body, so

$$v = \omega r \qquad \text{(radian measure).} \qquad (11\text{-}16)$$

Again, the angular speed ω must be expressed in radian measure. Equation 11-16 tells us that since all points within the rigid body have the same angular speed ω, points with greater radius r have greater linear speed v. Figure 11-9a shows that the linear velocity is always tangent to the circular path of the point in question.

If the angular speed ω of the rigid body is constant, then Eq. 11-16 tells us that the linear speed v for any point within it is also constant. Thus each point within the body undergoes uniform circular motion. The period of revolution T for the motion of each point and for the rigid body itself is given by Eq. 4-23:

$$T = \frac{2\pi r}{v}. \qquad (11\text{-}17)$$

This equation tells us that the time for one revolution is the distance $2\pi r$ traveled in one revolution divided by the speed at which that distance is traveled. Substituting for v from Eq. 11-16 and canceling r, we find also that

$$T = \frac{2\pi}{\omega} \qquad \text{(radian measure).} \qquad (11\text{-}18)$$

This equivalent equation says that the time for one revolution is the angle 2π rad traversed in one revolution divided by the angular speed at which that angle is traversed.

The Acceleration

Differentiating Eq. 11-16 with respect to time—again with r held constant—leads to

$$\frac{dv}{dt} = \frac{d\omega}{dt} r. \qquad (11\text{-}19)$$

Here we run up against a complication. In Eq. 11-19, dv/dt represents only the part of the linear acceleration that is responsible for changes in the *magnitude* v of the linear velocity $\mathbf{v}$. Like $\mathbf{v}$, that part of the linear acceleration is tangent to the path of the point in question. We call it the *tangential component* a_t of the linear acceleration of the point, and we write

$$a_t = \alpha r \qquad \text{(radian measure),} \qquad (11\text{-}20)$$

where $\alpha = d\omega/dt$.

In addition, as Eq. 4-22 tells us, a particle (or point) moving in a circular path has a *radial component* of linear acceleration, $a_r = v^2/r$ (radially inward), that is responsi-

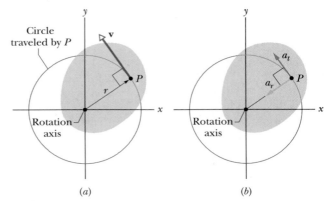

FIGURE 11-9 The rotating rigid body of Fig. 11-2, shown in cross section. Every point of the body (such as P) moves in a circle around the rotation axis. (a) The linear velocity $\mathbf{v}$ of every point is tangent to the circle in which the point moves. (b) The linear acceleration $\mathbf{a}$ of the point has (in general) two components: a tangential component a_t and a radial component a_r.

ble for changes in the *direction* of the linear velocity $\mathbf{v}$. By substituting for v from Eq. 11-16, we can write this component as

$$a_r = \frac{v^2}{r} = \omega^2 r \qquad \text{(radian measure).} \qquad (11\text{-}21)$$

Thus, as Fig. 11-9b shows, the linear acceleration of a point on a rotating rigid body has, in general, two components. The radially inward component a_r (given by Eq. 11-21) is present unless the angular velocity of the body is zero. The tangential component a_t (given by Eq. 11-20) is present unless the angular acceleration is zero.

CHECKPOINT **3:** A cockroach rides the rim of a rotating merry-go-round. If the angular speed of this system (*merry-go-round + cockroach*) is constant, does the cockroach have (a) radial acceleration and (b) tangential acceleration? If the angular speed is decreasing, does the cockroach have (c) radial and (d) tangential acceleration?

SAMPLE PROBLEM 11-6

Figure 11-10 shows a centrifuge used to accustom astronaut trainees to high accelerations. The radius r of the circle traveled by an astronaut is 15 m.

(a) At what constant angular velocity must the centrifuge rotate if the astronaut is to be subject to a linear acceleration that is $11g$?

SOLUTION: Because the angular velocity is constant, the

angular acceleration α ($= d\omega/dt$) is zero and so is the tangential component of the linear acceleration (see Eq. 11-20). This leaves only the radial component. From Eq. 11-21 ($a_r = \omega^2 r$), with $a_r = 11g$, we have

$$\omega = \sqrt{\frac{a_r}{r}} = \sqrt{\frac{(11)(9.8 \text{ m/s}^2)}{15 \text{ m}}}$$

$$= 2.68 \text{ rad/s} \approx 26 \text{ rev/min}. \qquad \text{(Answer)}$$

(b) What is the tangential acceleration of the astronaut if the centrifuge accelerates uniformly from rest to the angular velocity of (a) in 120 s?

SOLUTION: Since the angular acceleration is constant during the speed-up of the centrifuge, Eq. 11-9 applies and we get

$$\alpha = \frac{\omega - \omega_0}{t} = \frac{2.68 \text{ rad/s} - 0}{120 \text{ s}} = 0.0223 \text{ rad/s}^2.$$

With Eq. 11-20, we then find

$$a_t = \alpha r = (0.0223 \text{ rad/s}^2)(15 \text{ m})$$

$$= 0.33 \text{ m/s}^2. \qquad \text{(Answer)}$$

Although the final radial acceleration a_r ($= 11g$) is large (and alarming), the tangential acceleration a_t ($= 0.034g$) during the speed-up is not.

FIGURE 11-10 Sample Problem 11-6. A centrifuge in Cologne, Germany, is used to accustom astronauts to the large acceleration experienced during a liftoff.

PROBLEM SOLVING TACTICS

TACTIC 2: *Units for Angular Variables*
In Eq. 11-1 ($\theta = s/r$), we committed ourselves to the use of radian measure for all angular variables whenever we are using equations that contain both angular and linear variables. That is, we must express angular displacements in radians, angular velocities in rad/s and rad/min, and angular accelerations in rad/s^2 and rad/min^2. Equations 11-15, 11-16, 11-18, 11-20, and 11-21 are marked to emphasize this. The only exceptions to this rule are equations that involve *only* angular variables, such as the angular equations listed in Table 11-1. Here you are free to use any unit you wish for the angular variables. That is, you may use radians, degrees, or revolutions, as long as you use them consistently.

In equations where radian measure must be used, you need not keep track of the unit "radian" (rad) algebraically, as you must do for other units. You can add or delete it at will, to suit the context. In Sample Problem 11-6a the unit was added to the answer; in Sample Problem 11-6b it was omitted from the answer.

11-6 KINETIC ENERGY OF ROTATION

The rapidly rotating blade of a table saw certainly has kinetic energy. How can we express it? We cannot use the familiar formula $K = \frac{1}{2}mv^2$ directly because it applies only to particles, and we would not know what to use for v.

Instead, we shall treat the table saw (and any other rotating rigid body) as a collection of particles —all with different speeds. We can then add up the kinetic energies of these particles to find the kinetic energy of the body as a whole. In this way we obtain, for the kinetic energy of a rotating body,

$$K = \tfrac{1}{2}m_1 v_1^2 + \tfrac{1}{2}m_2 v_2^2 + \tfrac{1}{2}m_3 v_3^2 + \cdots$$
$$= \sum \tfrac{1}{2} m_i v_i^2, \qquad (11\text{-}22)$$

in which m_i is the mass of the ith particle and v_i is its speed. The sum is taken over all the particles in the body.

The problem with Eq. 11-22 is that v_i is not the same for all particles. We solve this problem by substituting for v from Eq. 11-16 ($v = \omega r$), so that we have

$$K = \sum \tfrac{1}{2} m_i (\omega r_i)^2 = \tfrac{1}{2} \left(\sum m_i r_i^2 \right) \omega^2, \quad (11\text{-}23)$$

in which ω *is* the same for all particles.

The quantity in parentheses on the right side of Eq. 11-23 tells us how the mass of the rotating body is distributed about its axis of rotation. We call that quantity the **rotational inertia** (or **moment of inertia**) I of the body with respect to the axis of rotation. It is a constant for a particular rigid body and for a particular rotation axis. (That axis must always be specified if the value of I is to be meaningful.)

We may now write

$$I = \sum m_i r_i^2 \qquad \text{(rotational inertia)} \quad (11\text{-}24)$$

and substitute into Eq. 11-23, obtaining

$$K = \tfrac{1}{2} I \omega^2 \qquad \text{(radian measure)} \quad (11\text{-}25)$$

as the expression we seek. Because we have used the relation $v = \omega r$ in deriving Eq. 11-25, ω must be expressed in radian measure. The SI unit for I is the kilogram–square meter $(kg \cdot m^2)$.

Equation 11-25, which gives the kinetic energy of a rigid body in pure rotation, is the angular equivalent of the formula $K = \frac{1}{2}Mv_{cm}^2$, which gives the kinetic energy of a rigid body in pure translation. In each case, there is a factor of $\frac{1}{2}$. Where mass M (which can be called the *translational inertia*) appears in one formula, I (the *rotational inertia*) appears in the other. Finally, each equation contains as a factor the square of a speed—translational or rotational as appropriate. The kinetic energies of translation and of rotation are not different kinds of energy. They are both kinetic energy, expressed in ways that are appropriate to the motion at hand.

The rotational inertia of a rotating body depends not only on its mass but also on how that mass is distributed with respect to the rotation axis. Figure 11-11 suggests a convincing way to develop a physical feeling for rotational inertia. Figure 11-11a shows the exterior of either of two plastic rods that outwardly appear to be identical. The dimensions and the weights of the two rods are the same, and both rods balance at their midpoints. If you grasp each rod at its center and move it back and forth rapidly in translational motion, you still cannot tell them apart.

However, a truly striking difference appears if, with a twisting wrist motion, you twist the rods (like a baton) rapidly in back-and-forth angular motion. One rod twists quite easily; the other does not. As Figs. 11-11b and 11-11c show, the "easy" rod has two internal weights near its

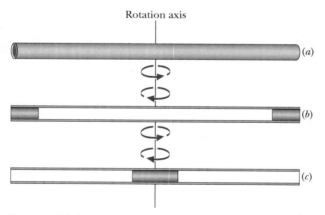

FIGURE 11-11 (a) Two plastic rods like this seem identical until you try to twist them back and forth rapidly about their midpoints. Rod (c) twists readily; rod (b) does not. Although both rods have the same mass, the rotational inertia of rod (b) about an axis through its midpoint is considerably greater than that of rod (c) because of the different internal distributions of mass.

center and the "hard" rod has one at each end. Although the masses of the rods are equal, their rotational inertias about a central axis are quite different because their mass distributions are different.

C$_{HECKPOINT}$ **4:** The figure shows three masses that rotate about a vertical axis. The perpendicular distance between the axis and the center of each mass is given. Rank the three masses according to their rotational inertia about that axis, greatest first.

Rotation axis — 1 m ● 36 kg | 2 m ● 9 kg | 3 m ● 4 kg

11-7 CALCULATING THE ROTATIONAL INERTIA

If a rigid body is made up of discrete particles, we can calculate its rotational inertia from Eq. 11-24. If the body is continuous, we can replace the sum in Eq. 11-24 with an integral, and the definition of rotational inertia becomes

$$I = \int r^2\, dm \qquad \text{(rotational inertia, continuous).} \qquad (11\text{-}26)$$

In the sample problems that follow this section, we calculate I for bodies of both kinds. In general, the rotational inertia of any rigid body with respect to a rotation axis depends on (1) the shape of the body, (2) the perpendicular distance from the axis to the body's center of mass, and (3) the orientation of the body with respect to the axis.

Table 11-2 gives the rotational inertias of several common bodies, about various axes. Note how the distribution of mass relative to the rotational axis affects the value of the rotational inertia I. For example, the rod in (f) has more mass far from the rotational axis than does the equally long rod in (e). So, if the rods have the same mass M, the rod in (f) has the greater rotational inertia.

The Parallel-Axis Theorem

If you know the rotational inertia of a body about any axis that passes through its center of mass, you can find its rotational inertia about any other axis parallel to that axis with the **parallel-axis theorem:**

$$I = I_{cm} + Mh^2 \qquad \text{(parallel-axis theorem).} \qquad (11\text{-}27)$$

Here M is the mass of the body and h is the perpendicular

TABLE 11-2 SOME ROTATIONAL INERTIAS

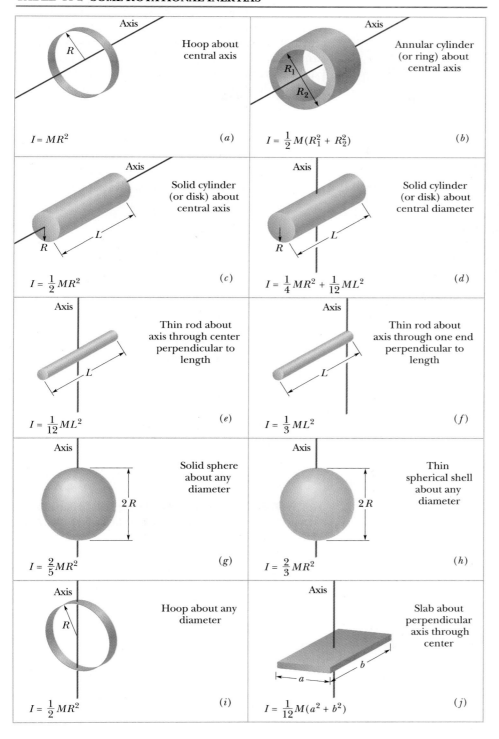

Hoop about
central axis

$I = MR^2$ (a)

Annular cylinder
(or ring) about
central axis

$I = \frac{1}{2}M(R_1^2 + R_2^2)$ (b)

Solid cylinder
(or disk) about
central axis

$I = \frac{1}{2}MR^2$ (c)

Solid cylinder
(or disk) about
central diameter

$I = \frac{1}{4}MR^2 + \frac{1}{12}ML^2$ (d)

Thin rod about
axis through center
perpendicular to
length

$I = \frac{1}{12}ML^2$ (e)

Thin rod about
axis through one end
perpendicular to
length

$I = \frac{1}{3}ML^2$ (f)

Solid sphere
about any
diameter

$I = \frac{2}{5}MR^2$ (g)

Thin
spherical shell
about any
diameter

$I = \frac{2}{3}MR^2$ (h)

Hoop about any
diameter

$I = \frac{1}{2}MR^2$ (i)

Slab about
perpendicular
axis through
center

$I = \frac{1}{12}M(a^2 + b^2)$ (j)

distance between the two (parallel) axes. In words, this theorem can be stated as follows:

> The rotational inertia of a body about any axis is equal to the rotational inertia (= Mh^2) it would have about that axis if all its mass were concentrated at its center of mass *plus* its rotational inertia (= I_{cm}) about a parallel axis through its center of mass.

Proof of the Parallel-Axis Theorem

Let O be the center of mass of the arbitrarily shaped body shown in cross section in Fig. 11-12. Place the origin of coordinates at O. Consider an axis through O perpendicular to the plane of the figure, and another axis through point P parallel to the first axis. Let the coordinates of P be a and b.

Let dm be a mass element with coordinates x and y. The rotational inertia of the body about the axis through P is then, from Eq. 11-26,

$$I = \int r^2 \, dm = \int [(x-a)^2 + (y-b)^2] \, dm,$$

which we can rearrange as

$$I = \int (x^2 + y^2) \, dm - 2a \int x \, dm$$

$$- 2b \int y \, dm + \int (a^2 + b^2) \, dm. \quad (11\text{-}28)$$

From the definition of the center of mass (Eq. 9-9), the middle two integrals of Eq. 11-28 give the coordinates of the center of mass (multiplied by a constant) and thus must

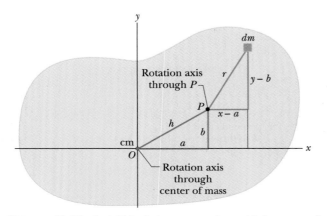

FIGURE 11-12 A rigid body in cross section, with its center of mass at O. The parallel-axis theorem (Eq. 11-27) relates the rotational inertia of the body about an axis through O to that about a parallel axis through a point such as P, a distance h from the body's center of mass. Both axes are perpendicular to the plane of the figure.

each be zero. Because $x^2 + y^2$ is equal to R^2, where R is the distance from O to dm, the first integral is simply I_{cm}, the rotational inertia of the body about an axis through its center of mass. Inspection of Fig. 11-12 shows that the last term in Eq. 11-28 is Mh^2, where M is the total mass. Thus Eq. 11-28 reduces to Eq. 11-27, which is the relation that we set out to prove.

$\mathbb{C}$HECKPOINT **5:** The figure shows a booklike object (one side is longer than the other) and four choices of rotation axes, all perpendicular to the face of the object. Rank the choices according to the rotational inertia of the object about the axis, greatest first.

(1)   (2)   (3) (4)

SAMPLE PROBLEM 11-7

Figure 11-13 shows a rigid body consisting of two particles of mass m connected by a rod of length L and negligible mass.

(a) What is the rotational inertia of this body about an axis through its center, perpendicular to the rod (see Fig. 11-13a)?

SOLUTION: From Eq. 11-24 we have

$$I = \sum m_i r_i^2 = (m)(\tfrac{1}{2}L)^2 + (m)(\tfrac{1}{2}L)^2$$
$$= \tfrac{1}{2}mL^2. \qquad \text{(Answer)}$$

(b) What is the rotational inertia of the body about an axis through one end of the rod and parallel to the first axis, as in Fig. 11-13b?

(a)

(b)

FIGURE 11-13 Sample Problem 11-7. A rigid body that consists of two particles of mass m which are joined by a rod of negligible mass.

SOLUTION: We can use the parallel-axis theorem of Eq. 11-27. We have just calculated I_{cm} in (a), and the distance h between the parallel axes is half the length of the rod. Thus, from Eq. 11-27,

$$I = I_{cm} + Mh^2 = \tfrac{1}{12}mL^2 + (2m)(\tfrac{1}{2}L)^2$$
$$= mL^2. \hspace{2cm} \text{(Answer)}$$

We can check this result by direct calculation, using Eq. 11-24:

$$I = \sum m_i r_i^2 = (m)(0)^2 + (m)(L)^2$$
$$= mL^2. \hspace{2cm} \text{(Answer)}$$

SAMPLE PROBLEM 11-8

Figure 11-14 shows a thin, uniform rod of mass M and length L.

(a) What is its rotational inertia about an axis perpendicular to the rod, through its center of mass?

SOLUTION: We place the rod on an x axis, with its center of mass at the origin, and we choose as a mass element a slice dx of the rod. The center of the slice is a distance x from the rotation axis. The mass per unit length of the rod is M/L, so the mass dm of the element dx is

$$dm = \left(\frac{M}{L}\right) dx.$$

From Eq. 11-26 we have

$$I = \int r^2 \, dm = \int_{x=-L/2}^{x=+L/2} x^2 \left(\frac{M}{L}\right) dx$$
$$= \frac{M}{3L}\left[x^3\right]_{-L/2}^{+L/2} = \frac{M}{3L}\left[\left(\frac{L}{2}\right)^3 - \left(-\frac{L}{2}\right)^3\right]$$
$$= \tfrac{1}{12}ML^2. \hspace{2cm} \text{(Answer)}$$

This agrees with the result given in Table 11-2(e).

(b) What is the rotational inertia of the rod about an axis perpendicular to the rod through one end?

SOLUTION: We combine the result in (a) with the parallel-axis theorem (Eq. 11-27), obtaining

$$I = I_{cm} + Mh^2$$
$$= \tfrac{1}{12}ML^2 + (M)(\tfrac{1}{2}L)^2 = \tfrac{1}{3}ML^2, \hspace{1cm} \text{(Answer)}$$

which agrees with the result given in Table 11-2(f).

FIGURE 11-14 Sample Problem 11-8. A uniform rod of length L and mass M.

SAMPLE PROBLEM 11-9

A hydrogen chloride molecule consists of a hydrogen atom whose mass m_H is 1.01 u and a chlorine atom whose mass m_{Cl} is 35.0 u. The centers of the two atoms are a distance $d = 1.27 \times 10^{-10}$ m $= 127$ pm apart (Fig. 11-15). What is the rotational inertia of the molecule about an axis perpendicular to the line joining the two atoms and passing through the center of mass of the molecule?

SOLUTION: Let x be the distance from the center of mass of the molecule to the chlorine atom. Then from Fig. 11-15 and Eq. 9-3, we see that

$$0 = \frac{-m_{Cl}x + m_H(d-x)}{m_{Cl} + m_H},$$

or

$$m_{Cl}x = m_H(d-x).$$

This yields

$$x = \frac{m_H}{m_{Cl} + m_H} d. \hspace{2cm} \text{(11-29)}$$

From Eq. 11-24, the rotational inertia about an axis through the center of mass is

$$I = \sum m_i r_i^2 = m_H(d-x)^2 + m_{Cl}x^2.$$

Substituting for x from Eq. 11-29 leads, after some algebra, to

$$I = d^2 \frac{m_H m_{Cl}}{m_{Cl} + m_H} = (127 \text{ pm})^2 \frac{(1.01 \text{ u})(35.0 \text{ u})}{35.0 \text{ u} + 1.01 \text{ u}}$$
$$= 15{,}800 \text{ u} \cdot \text{pm}^2. \hspace{2cm} \text{(Answer)}$$

These units are convenient enough when dealing with the rotational inertia of molecules. If we use the angstrom as a length unit (1 Å $= 10^{-10}$ m), the answer above becomes $I = 1.58$ u $\cdot$ Å^2, an even more convenient unit.

FIGURE 11-15 Sample Problem 11-9. A hydrogen chloride molecule, shown schematically. A rotation axis extends through its center of mass and is perpendicular to the line joining the two atomic centers.

SAMPLE PROBLEM 11-10

With modern technology, it is possible to construct a flywheel that stores enough energy to run an automobile. The energy is stored as rotational kinetic energy when the flywheel is initially made to spin by a machine. The stored energy is then gradually transferred to the automobile by a gear system as the automobile is being driven. Suppose that such a wheel is a solid cylinder whose mass M is 75 kg and whose radius R is

25 cm. If the wheel is spun at 85,000 rev/min, how much rotational kinetic energy can it store?

SOLUTION: The rotational inertia of the cylindrical wheel follows from Table 11-2(c):

$$I = \tfrac{1}{2}MR^2 = (\tfrac{1}{2})(75 \text{ kg})(0.25 \text{ m})^2 = 2.34 \text{ kg} \cdot \text{m}^2.$$

The angular velocity of the wheel is

$$\omega = (85{,}000 \text{ rev/min})(2\pi \text{ rad/rev})(1 \text{ min/60 s})$$
$$= 8900 \text{ rad/s}.$$

From Eq. 11-25, the kinetic energy of rotation is then

$$K = \tfrac{1}{2}I\omega^2 = (\tfrac{1}{2})(2.34 \text{ kg} \cdot \text{m}^2)(8900 \text{ rad/s})^2$$
$$= 9.3 \times 10^7 \text{ J} = 26 \text{ kW} \cdot \text{h}. \qquad \text{(Answer)}$$

This amount of energy, used with the expected efficiency, would take a small car about 200 mi.

11-8   TORQUE

A doorknob is located as far as possible from the door's hinge line for a good reason. If you want to open a heavy door you must certainly apply a force; that alone, however, is not enough. Where you apply that force and in what direction you push are also important. If you apply your force nearer to the hinge line than the knob, or at any angle other than 90° to the plane of the door, you must use a greater force to move the door than if you apply the force at the knob and perpendicular to the door's plane.

Figure 11-16a shows a cross section of a body that is free to rotate about an axis passing through O and perpendicular to the cross section. A force $\mathbf{F}$ is applied at point P, whose position relative to O is defined by a position vector $\mathbf{r}$. Vectors $\mathbf{F}$ and $\mathbf{r}$ make an angle ϕ with each other. (For simplicity, we consider only forces that have no compo-

nent parallel to the rotation axis; thus, $\mathbf{F}$ is in the plane of the page.)

To determine how $\mathbf{F}$ results in a rotation of the body around the rotation axis, we resolve $\mathbf{F}$ into two components (Fig. 11-16b). One component, called the *radial component* F_r, points along $\mathbf{r}$. This component does not cause rotation, because it acts along a line that extends through O. (If you pull on a door parallel to the plane of the door, you do not rotate the door.) The other component of $\mathbf{F}$, called the *tangential component* F_t, is perpendicular to $\mathbf{r}$ and has magnitude $F_t = F \sin \phi$. This component *does* cause rotation. (If you pull on a door perpendicular to its plane, you rotate the door.)

The ability of $\mathbf{F}$ to rotate the body depends not only on the magnitude of its tangential component F_t, but also on just how far from O it is applied. To include both these factors, we define a quantity called **torque** τ as the product of the two factors and write it as

$$\tau = (r)(F \sin \phi). \qquad (11\text{-}30)$$

Two equivalent ways of computing the torque are

$$\tau = (r)(F \sin \phi) = rF_t \qquad (11\text{-}31)$$

and $\qquad \tau = (r \sin \phi)(F) = r_\perp F, \qquad (11\text{-}32)$

where $r_\perp$ is the perpendicular distance between the rotation axis at O and an extended line running through the vector $\mathbf{F}$ (Fig. 11-16c). This extended line is called the **line of action** of $\mathbf{F}$, and $r_\perp$ is called the **moment arm** of $\mathbf{F}$. Figure 11-16b shows that we can describe r, the magnitude of $\mathbf{r}$, as being the moment arm of the force component F_t.

Torque, which comes from the Latin word meaning "to twist," may be loosely identified as the turning or twisting action of the force $\mathbf{F}$. When you apply a force to an object—such as a screwdriver or pipe wrench—with the

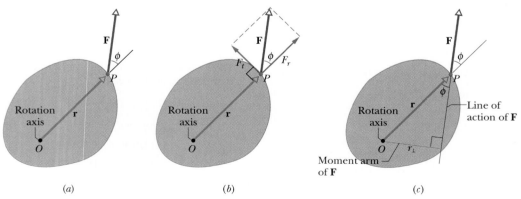

(a)   (b)   (c)

FIGURE 11-16   (a) A force $\mathbf{F}$ acts at point P on a rigid body that is free to rotate about an axis through O; the axis is perpendicular to the plane of the cross section shown here. (b) The torque exerted by this force is (r) $(F \sin \phi)$. We can also write it as rF_t, where F_t is the tangential component of $\mathbf{F}$. (c) The torque can also be written as $r_\perp F$, where $r_\perp$ is the moment arm of $\mathbf{F}$.

purpose of turning that object, you are applying a torque. The SI unit of torque is the newton-meter (N·m).* A torque τ is positive if it tends to rotate the body counter-clockwise, in the direction of increasing θ, as in Fig. 11-16. It is negative if it tends to rotate the body clockwise.

The definition of torque in Eq. 11-30 can be rewritten as a *vector cross product:*

$$\tau = \mathbf{r} \times \mathbf{F}. \qquad (11\text{-}33)$$

Thus torque τ is a vector that is directed perpendicular to the plane containing $\mathbf{r}$ and $\mathbf{F}$. The magnitude of τ is given by Eqs. 11-30, 11-31, and 11-32, and its direction is given by the right-hand rule. We shall make use of Eq. 11-33 in Chapter 12.

CHECKPOINT **6:** The figure shows an overhead view of a meter stick that can pivot about the dot at the position marked 20 (for 20 cm). All five horizontal forces on the stick have the same magnitude. Rank those forces according to the magnitude of the torque that they produce, greatest first.

11-9 NEWTON'S SECOND LAW FOR ROTATION

Figure 11-17 shows a simple case of rotation about a fixed axis. The rotating rigid body consists of a single particle of mass m fastened to the end of a massless rod of length r. A force $\mathbf{F}$ acts as shown, causing the particle to move in a circle about the axis. The particle has a tangential component of acceleration a_t governed by Newton's second law:

$$F_t = ma_t.$$

The torque acting on the particle is, from Eq. 11-31,

$$\tau = F_t r = ma_t r.$$

From Eq. 11-20 ($a_t = \alpha r$) we can write this as

$$\tau = m(\alpha r)r = (mr^2)\alpha. \qquad (11\text{-}34)$$

The quantity in parentheses on the right side of Eq. 11-34 is the rotational inertia of the particle about the rotation axis

---

*The newton-meter is also the unit of work. Torque and work, however, are quite different quantities and must not be confused. Work is often expressed in joules (1 J = 1 N·m), but torque never is.

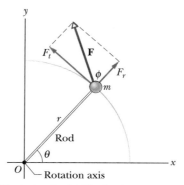

FIGURE 11-17 A simple rigid body, free to rotate about an axis through O, consists of a particle of mass m fastened to the end of a rod of length r and negligible mass. An applied force $\mathbf{F}$ causes the body to rotate.

(see Eq. 11-24). So, Eq. 11-34 reduces to

$$\tau = I\alpha \qquad \text{(radian measure).} \qquad (11\text{-}35)$$

For the situation in which more than one force is applied to the particle, we can extend Eq. 11-35 as

$$\sum \tau = I\alpha \qquad \text{(radian measure),} \qquad (11\text{-}36)$$

where $\Sigma\tau$ is the net torque (the sum of all external torques) acting on the particle. Equation 11-36 is the angular (or rotational) form of Newton's second law.

Although we derived Eq. 11-35 for the special case of a single particle rotating about a fixed axis, it holds for any rigid body rotating about a fixed axis, because any such body can be analyzed as an assembly of single particles.

CHECKPOINT **7:** The figure shows an overhead view of a meter stick that can pivot about the point indicated, which is to the left of the stick's midpoint. Two horizontal forces, $\mathbf{F}_1$ and $\mathbf{F}_2$, are applied to the stick. Only $\mathbf{F}_1$ is shown. $\mathbf{F}_2$ is perpendicular to the stick and is applied at the right end. If the stick is not to turn, (a) what should be the direction of $\mathbf{F}_2$, and (b) should F_2 be greater than, less than, or equal to F_1?

SAMPLE PROBLEM 11-11

Figure 11-18a shows a uniform disk, whose mass M is 2.5 kg and whose radius R is 20 cm, mounted on a fixed horizontal axle. A block whose mass m is 1.2 kg hangs from a massless cord that is wrapped around the rim of the disk. Find the acceleration of the falling block (assuming that it does fall), the angular acceleration of the disk, and the tension in the cord. The cord does not slip, and there is no friction at the axle.

SOLUTION: Figure 11-18b is a free-body diagram for the block. We assume the block accelerates downward, so the magnitude mg of its weight must exceed the tension T in the cord. From Newton's second law, we have

$$T - mg = ma. \qquad (11\text{-}37)$$

Figure 11-18c is a free-body diagram for the disk. The torque acting on the disk is $-TR$, negative because it rotates the disk clockwise. From Table 11-2(c), we know that the rotational inertia I of the disk is $\frac{1}{2}MR^2$. (Two other forces also act on the disk, its weight Mg and the normal force $\mathbf{N}$ exerted on the disk by its support. Since, however, both these forces act at the axis of the disk, they exert no torque on the disk.) Applying Newton's second law in angular form ($\tau = I\alpha$) to the disk, we obtain

$$-TR = \tfrac{1}{2}MR^2\alpha.$$

Because the cord does not slip, we assume that the linear acceleration a of the block and the (tangential) linear acceleration a_t of the rim of the disk are equal. Then, by Eq. 11-20, $\alpha = a/R$, and the equation above reduces to

$$T = -\tfrac{1}{2}Ma. \qquad (11\text{-}38)$$

Combining Eqs. 11-37 and 11-38 leads to

$$a = -g\,\frac{2m}{M + 2m} = -(9.8 \text{ m/s}^2)\,\frac{(2)(1.2 \text{ kg})}{2.5 \text{ kg} + (2)(1.2 \text{ kg})}$$

$$= -4.8 \text{ m/s}^2. \qquad \text{(Answer)}$$

We then use Eq. 11-38 to find T:

$$T = -\tfrac{1}{2}Ma = -\tfrac{1}{2}(2.5 \text{ kg})(-4.8 \text{ m/s}^2)$$

$$= 6.0 \text{ N}. \qquad \text{(Answer)}$$

As we should expect, the acceleration of the falling block is

less than g, and the tension in the cord ($= 6.0$ N) is less than the weight of the hanging block ($= mg = 11.8$ N). We see also that the acceleration of the block and the tension depend on the mass of the disk but not on its radius. As a check, we note that the formulas derived above predict $a = -g$ and $T = 0$ for the case of a massless disk ($M = 0$). This is what we would expect; the block simply falls as a free body, trailing the string behind it.

From Eq. 11-20, the angular acceleration of the disk is

$$\alpha = \frac{a}{R} = \frac{-4.8 \text{ m/s}^2}{0.20 \text{ m}} = -24 \text{ rad/s}^2. \qquad \text{(Answer)}$$

SAMPLE PROBLEM 11-12

To throw an 80 kg opponent with a basic judo hip throw, you intend to pull his uniform with a force $\mathbf{F}$ and a moment arm $d_1 = 0.30$ m from a pivot point (rotation axis) on your right hip, about which you wish to rotate him with an angular acceleration of -6.0 rad/s^2, that is, with a clockwise acceleration (Fig. 11-19). Assume that his rotational inertia I relative to the pivot point is 15 kg·m^2.

(a) What must the magnitude of $\mathbf{F}$ be if you initially bend your opponent forward to bring his center of mass to your hip (Fig. 11-19a)?

SOLUTION: If your opponent's center of mass is at the rotation axis, his weight vector produces no torque about that axis. The only torque on him then is due to your pull $\mathbf{F}$. From Eqs. 11-31 and 11-35, we have, for the clockwise torque,

$$\tau = -d_1F = I\alpha,$$

which gives

$$F = \frac{-I\alpha}{d_1} = \frac{-(15 \text{ kg·m}^2)(-6.0 \text{ rad/s}^2)}{0.30 \text{ m}}$$

$$= 300 \text{ N}. \qquad \text{(Answer)}$$

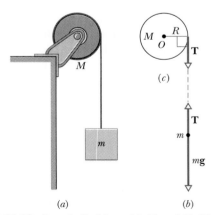

(a)   (b)

FIGURE 11-18   Sample Problems 11-11 and 11-13. (a) The falling block causes the disk to rotate. (b) A free-body diagram for the block. (c) A free-body diagram for the disk.

FIGURE 11-19   Sample Problem 11-12. A judo hip throw (a) correctly executed and (b) incorrectly executed.

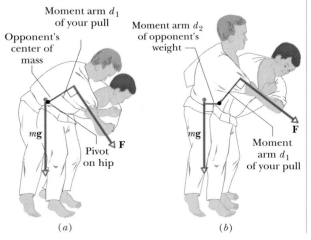

(a)   (b)

(b) What must the magnitude of **F** be if he remains upright and his weight mg has a moment arm $d_2 = 0.12$ m from the pivot point (Fig. 11-19b)?

SOLUTION: In this situation, your opponent's weight mg provides a positive (counterclockwise) torque that counters your torque. From Eqs. 11-31 and 11-36, we have

$$\sum \tau = -d_1 F + d_2 mg = I\alpha,$$

which gives

$$F = -\frac{I\alpha}{d_1} + \frac{d_2 mg}{d_1}.$$

From (a), we know that the first term on the right is equal to 300 N. Substituting this and the given data, we have

$$F = 300 \text{ N} + \frac{(0.12 \text{ m})(80 \text{ kg})(9.8 \text{ m/s}^2)}{0.30 \text{ m}}$$

$$= 613.6 \text{ N} \approx 610 \text{ N}. \qquad \text{(Answer)}$$

The results indicate that you will have to pull much harder if you do not initially bend your opponent to bring his center of mass to your hip. A good judo fighter knows this lesson from physics. (An analysis of judo and aikido is given in "The Amateur Scientist," *Scientific American*, July 1980.)

11-10 WORK AND ROTATIONAL KINETIC ENERGY

Let us again consider the situation of Fig. 11-17, in which force **F** rotates a rigid body consisting of a single particle of mass m fastened to the end of a massless rod. During the rotation, force **F** does work on the body. Let us assume that the only energy of the body that is changed by **F** is the kinetic energy. Then we can apply the work–kinetic energy theorem of Eq. 7-4:

$$\Delta K = K_f - K_i = W. \qquad (11\text{-}39)$$

Using $K = \frac{1}{2}mv^2$ and Eq. 11-16 ($v = \omega r$), we can rewrite Eq. 11-39 as

$$\Delta K = \frac{1}{2}mr^2\omega_f^2 - \frac{1}{2}mr^2\omega_i^2 = W. \qquad (11\text{-}40)$$

From Eq. 11-24, the rotational inertia for a body with one particle is $I = mr^2$. Substituting this into Eq. 11-40 yields

$$\Delta K = \frac{1}{2}I\omega_f^2 - \frac{1}{2}I\omega_i^2 = W. \qquad (11\text{-}41)$$

Equation 11-41 says that the work W done by **F** on the body of Fig. 11-17 changes the rotational kinetic energy K ($= \frac{1}{2}I\omega^2$) of the body. Equation 11-41 is the angular equivalent of the work–kinetic energy theorem for translational motion. We derived it for a rigid body with one particle, but it holds for any rigid body rotated about a fixed axis.

We next relate the work W done on the body in Fig. 11-17 to the torque τ on the body due to force **F**. If the particle in Fig. 11-17 were to move a differential distance ds along its circular path, the body would rotate through differential angle $d\theta$, with $ds = r\,d\theta$. We would use Eq. 7-11 to show the work dW done on the body by force **F** during this motion:

$$dW = \mathbf{F} \cdot d\mathbf{s} = F_t\,ds = F_t r\,d\theta. \qquad (11\text{-}42)$$

(Recall that F_t is the component of **F** along the particle's path.) From Eq. 11-31, we see that the product $F_t r$ is equal to the torque τ. So, we can rewrite Eq. 11-42 as

$$dW = \tau\,d\theta. \qquad (11\text{-}43)$$

The work done during a finite angular displacement from θ_i to θ_f is then

$$W = \int_{\theta_i}^{\theta_f} \tau\,d\theta. \qquad (11\text{-}44)$$

TABLE 11-3 **SOME CORRESPONDING RELATIONS FOR TRANSLATIONAL AND ROTATIONAL MOTION**

PURE TRANSLATION (FIXED DIRECTION)		PURE ROTATION (FIXED AXIS)	
Position	x	Angular position	θ
Velocity	$v = dx/dt$	Angular velocity	$\omega = d\theta/dt$
Acceleration	$a = dv/dt$	Angular acceleration	$\alpha = d\omega/dt$
Mass	m	Rotational inertia	I
Newton's second law	$F = ma$	Newton's second law	$\tau = I\alpha$
Work	$W = \int F\,dx$	Work	$W = \int \tau\,d\theta$
Kinetic energy	$K = \frac{1}{2}mv^2$	Kinetic energy	$K = \frac{1}{2}I\omega^2$
Power	$P = Fv$	Power	$P = \tau\omega$
Work–kinetic energy theorem	$W = \Delta K$	Work–kinetic energy theorem	$W = \Delta K$

Equation 11-44, which holds for any rigid body rotating about a fixed axis, is the rotational equivalent of Eq. 7-27, which is

$$W = \int_{x_i}^{x_f} F \, dx.$$

We can find the power P for rotational motion from Eq. 11-43:

$$P = \frac{dW}{dt} = \tau \frac{d\theta}{dt} = \tau\omega. \qquad (11\text{-}45)$$

This is the rotational analog of $P = Fv$ (from Eq. 7-49), which gives the rate at which a force F does work on a particle moving with speed v, with $\mathbf{F}$ and $\mathbf{v}$ parallel.

Table 11-3 summarizes the equations that apply to the rotation of a rigid body about a fixed axis and the equivalent relations for translational motion.

SAMPLE PROBLEM 11-13

(a) In the arrangement of Fig. 11-18, through what angle does the disk rotate in 2.5 s, starting from rest.

SOLUTION: From Eq. 11-10 ($\theta = \omega_0 t + \frac{1}{2}\alpha t^2$), we have, putting $\omega_0 = 0$ and using the value of α calculated in Sample Problem 11-11,

$$\theta = 0 + (\tfrac{1}{2})(-24 \text{ rad/s}^2)(2.5 \text{ s})^2$$
$$= -75 \text{ rad}. \qquad \text{(Answer)}$$

(b) What is the angular velocity of the disk at $t = 2.5$ s?

SOLUTION: We can find this with Eq. 11-9 ($\omega = \omega_0 + \alpha t$). Putting $\omega_0 = 0$ and using the value of α calculated in Sample Problem 11-11, we obtain

$$\omega = 0 + (-24 \text{ rad/s}^2)(2.5 \text{ s})$$
$$= -60 \text{ rad/s}. \qquad \text{(Answer)}$$

(c) What is the kinetic energy K of the disk at $t = 2.5$ s?

SOLUTION: From Eq. 11-25, the kinetic energy of the disk is $\frac{1}{2}I\omega^2$, in which $I = \frac{1}{2}MR^2$. Thus, using the value of ω found in (b), we have

$$K = \tfrac{1}{2}I\omega^2 = \tfrac{1}{2}(\tfrac{1}{2}MR^2)\omega^2$$
$$= (\tfrac{1}{4})(2.5 \text{ kg})(0.20 \text{ m})^2(-60 \text{ rad/s})^2$$
$$= 90 \text{ J}. \qquad \text{(Answer)}$$

Another approach to this problem is to calculate the change in the rotational kinetic energy using the work–kinetic energy theorem. To do so, we must calculate the work W done on the disk by the torque that acts on it. This torque, exerted by the tension T in the cord, is constant. Equation 11-44 then gives us,

$$W = \int_{\theta_i}^{\theta_f} \tau \, d\theta = \tau \int_{\theta_i}^{\theta_f} d\theta = \tau(\theta_f - \theta_i).$$

For the torque τ we use $-TR$, in which T is the tension in the cord ($= 6.0$ N; see Sample Problem 11-11) and R ($= 0.20$ m) is the radius of the disk. The quantity $\theta_f - \theta_i$ is just the angular displacement that we calculated in (a). Thus

$$W = \tau(\theta_f - \theta_i) = -TR(\theta_f - \theta_i)$$
$$= -(6.0 \text{ N})(0.20 \text{ m})(-75 \text{ rad})$$
$$= 90 \text{ J}. \qquad \text{(Answer)}$$

Because K_i is zero (the system starts from rest), Eq. 11-41 tells us that this answer is equal to K.

SAMPLE PROBLEM 11-14

A rigid sculpture, consisting of a thin hoop (of mass m and radius $R = 0.15$ m) and two thin rods (each of mass m and length $L = 2.0R$), is arranged as shown in Fig. 11-20. The sculpture can pivot around a horizontal axis in the plane of the hoop, passing through its center.

(a) In terms of m and R, what is the sculpture's rotational inertia I about the rotation axis?

SOLUTION: From Table 11-2(i), the hoop has rotational inertia $I_{\text{hoop}} = \frac{1}{2}mR^2$ about its diameter. From Table 11-2(e), rod A has rotational inertia $I_{\text{cm},A} = mL^2/12$ about an axis through its center of mass and parallel to the rotation axis. To find the rotational inertia I_A of rod A about the rotation axis, we use Eq. 11-27, the parallel-axis theorem:

$$I_A = I_{\text{cm},A} + mh_{\text{cm},A}^2 = \frac{mL^2}{12} + m\left(R + \frac{L}{2}\right)^2$$
$$= 4.33mR^2,$$

where we have used the fact that $L = 2.0R$ and where $h_{\text{cm},A}$ ($= R + L/2$) is the perpendicular distance between the center of rod A and the rotation axis.

We treat rod B similarly: it has zero rotational inertia $I_{\text{cm},B}$ about an axis along its length, and its rotational inertia I_B about the rotation axis is

$$I_B = I_{\text{cm},B} + mh_{\text{cm},B}^2 = 0 + mR^2 = mR^2.$$

FIGURE 11-20 Sample Problem 11-14. A rigid sculpture consisting of a hoop and two rods can rotate around a horizontal axis.

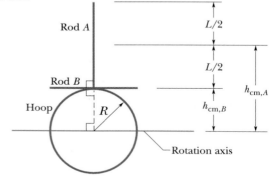

Here $h_{cm, B} (= R)$ is the perpendicular distance between rod B and the rotation axis. Thus the rotational inertia I of the sculpture about the rotation axis is

$$I = I_{hoop} + I_A + I_B = \tfrac{1}{2}mR^2 + 4.33mR^2 + mR^2$$
$$= 5.83mR^2 \approx 5.8mR^2. \qquad \text{(Answer)}$$

(b) Starting from rest, the sculpture rotates around the rotation axis from the initial upright orientation of Fig. 11-20. What is its angular speed ω about the axis when it is inverted?

SOLUTION: Before the sculpture moves, its center of mass is a distance y_{cm} above the rotation axis, with y_{cm} given by Eq. 9-5:

$$y_{cm} = \frac{m(0) + mR + m(R + L/2)}{3m} = R,$$

where $3m$ is the mass of the sculpture. As the sculpture rotates, its center of mass descends to the same distance *below* the rotation axis. Thus the center of mass undergoes a vertical displacement of $\Delta y_{cm} = -2R$. During the descent, gravita-

tional potential energy U of the sculpture is transferred to kinetic energy K of its rotation. The change ΔU in potential energy is the product of the magnitude of the sculpture's weight $(3mg)$ and the vertical displacement Δy_{cm}. Thus

$$\Delta U = 3mg \, \Delta y_{cm} = 3mg(-2R) = -6mgR.$$

From Eq. 11-25, the associated change in kinetic energy is

$$\Delta K = \tfrac{1}{2}I\omega^2 - 0 = \tfrac{1}{2}I\omega^2.$$

So by conservation of mechanical energy

$$\Delta K + \Delta U = 0,$$

and

$$\tfrac{1}{2}I\omega^2 - 6mgR = 0.$$

Substituting $I = 5.83mR^2$ and solving for ω give

$$\omega = \sqrt{\frac{12g}{5.83R}} = \sqrt{\frac{(12)(9.8 \text{ m/s}^2)}{(5.83)(0.15 \text{ m})}}$$
$$= 12 \text{ rad/s}. \qquad \text{(Answer)}$$

REVIEW & SUMMARY

Angular Position

To describe the rotation of a rigid body about a fixed axis, called the **rotation axis,** we assume a **reference line** is fixed in the body, perpendicular to that axis and rotating with the body. We measure the **angular position** θ of this line relative to a fixed direction. When θ is measured in **radians,**

$$\theta = \frac{s}{r} \qquad \text{(radian measure),} \qquad (11\text{-}1)$$

where s is the arc length of a circular path of radius r and angle θ. Radian measure is related to angle measure in revolutions and degrees by

$$1 \text{ rev} = 360° = 2\pi \text{ rad.} \qquad (11\text{-}2)$$

Angular Displacement

A body that rotates about a rotation axis, changing its angular position from θ_1 to θ_2, undergoes an **angular displacement**

$$\Delta\theta = \theta_2 - \theta_1, \qquad (11\text{-}4)$$

where $\Delta\theta$ is positive for counterclockwise rotation and negative for clockwise rotation.

Angular Velocity and Speed

If a body rotates through an angular displacement $\Delta\theta$ in a time interval Δt, its **average angular velocity** $\bar{\omega}$ is

$$\bar{\omega} = \frac{\Delta\theta}{\Delta t}. \qquad (11\text{-}5)$$

The **(instantaneous) angular velocity** ω of the body is

$$\omega = \frac{d\theta}{dt}. \qquad (11\text{-}6)$$

Both $\bar{\omega}$ and ω are vectors, with directions given by the **right-hand rule** of Fig. 11-5. They are positive for counterclockwise rotation and negative for clockwise rotation. The magnitude of the body's angular velocity is the **angular speed.**

Angular Acceleration

If the angular velocity of a body changes from ω_1 to ω_2 in a time interval $\Delta t = t_2 - t_1$, the **average angular acceleration** $\bar{\alpha}$ of the body is

$$\bar{\alpha} = \frac{\omega_2 - \omega_1}{t_2 - t_1} = \frac{\Delta\omega}{\Delta t}. \qquad (11\text{-}7)$$

The **(instantaneous) angular acceleration** α of a body is

$$\alpha = \frac{d\omega}{dt}. \qquad (11\text{-}8)$$

Both $\bar{\alpha}$ and α are vectors.

The Kinematic Equations for Constant Angular Acceleration

Constant angular acceleration ($\alpha = $ constant) is an important special case of rotational motion. The appropriate kinematic equations, given in Table 11-1, are

$$\omega = \omega_0 + \alpha t, \qquad (11\text{-}9) \qquad \theta = \tfrac{1}{2}(\omega_0 + \omega)t, \qquad (11\text{-}12)$$
$$\theta = \omega_0 t + \tfrac{1}{2}\alpha t^2, \qquad (11\text{-}10) \qquad \theta = \omega t - \tfrac{1}{2}\alpha t^2. \qquad (11\text{-}13)$$
$$\omega^2 = \omega_0^2 + 2\alpha\theta, \qquad (11\text{-}11)$$

Linear and Angular Variables Related

A point in a rigid rotating body, at a *perpendicular distance r* from the rotation axis, moves in a circle with radius r. If the body rotates through an angle θ, the point moves along an arc with length s given by

$$s = \theta r \quad \text{(radian measure),} \quad (11\text{-}15)$$

where θ is in radians.

The linear velocity $\mathbf{v}$ of the point is tangent to the circle; the point's linear speed v is given by

$$v = \omega r \quad \text{(radian measure),} \quad (11\text{-}16)$$

where ω is the angular speed (in radians per second) of the body.

The linear acceleration $\mathbf{a}$ of the point has both *tangential* and *radial* components. The tangential component is

$$a_t = \alpha r \quad \text{(radian measure),} \quad (11\text{-}20)$$

where α is the magnitude of the angular acceleration (in radians per second-squared) of the body. The radial component of $\mathbf{a}$ is

$$a_r = \frac{v^2}{r} = \omega^2 r \quad \text{(radian measure).} \quad (11\text{-}21)$$

If the point moves in uniform circular motion, the period T of the motion for the point and the body is

$$T = \frac{2\pi r}{v} = \frac{2\pi}{\omega} \quad \text{(radian measure).} \quad (11\text{-}17, \ 11\text{-}18)$$

Rotational Kinetic Energy and Rotational Inertia

The kinetic energy K of a rigid body rotating about a fixed axis is given by

$$K = \tfrac{1}{2}I\omega^2 \quad \text{(radian measure),} \quad (11\text{-}25)$$

in which I is the **rotational inertia** of the body, defined as

$$I = \sum m_i r_i^2 \quad (11\text{-}24)$$

for a system of discrete particles and as

$$I = \int r^2 \, dm \quad (11\text{-}26)$$

for a body with continuously distributed mass. The r and r_i in these expressions represent the perpendicular distance from the axis of rotation to each mass element in the body.

The Parallel-Axis Theorem

The *parallel-axis theorem* relates the rotational inertia I of a body about any axis to that of the same body about a parallel axis through the center of mass:

$$I = I_{\text{cm}} + Mh^2. \quad (11\text{-}27)$$

Here h is the perpendicular distance between the two axes.

Torque

Torque is a turning or twisting action on a body about a rotation axis due to a force $\mathbf{F}$. If $\mathbf{F}$ is exerted at a point given by the position vector $\mathbf{r}$ relative to the axis, then the torque τ (a vector quantity) is

$$\tau = \mathbf{r} \times \mathbf{F}. \quad (11\text{-}33)$$

The magnitude of the torque is

$$\tau = rF_t = r_\perp F = rF \sin \phi, \quad (11\text{-}31, \ 11\text{-}32, \ 11\text{-}30)$$

where F_t is the component of $\mathbf{F}$ perpendicular to $\mathbf{r}$, and ϕ is the angle between $\mathbf{r}$ and $\mathbf{F}$. The quantity $r_\perp$ is the perpendicular distance between the rotation axis and an extended line running through the $\mathbf{F}$ vector. This line is called the **line of action** of $\mathbf{F}$, and $r_\perp$ is called the **moment arm** of $\mathbf{F}$. Similarly, r is the moment arm of F_t.

The SI unit of torque is the newton-meter ($\text{N} \cdot \text{m}$). A torque τ is positive if it tends to rotate the body counterclockwise and negative if it tends to rotate the body in the clockwise direction.

Newton's Second Law in Angular Form

The rotational analog of Newton's second law is

$$\sum \tau = I\alpha, \quad (11\text{-}36)$$

where $\Sigma\tau$ is the net torque acting on a particle or rigid body, I is the rotational inertia of the particle or body about the rotation axis, and α is the resulting angular acceleration about that axis.

Work and Rotational Kinetic Energy

The equations used for calculating work and power in rotational motion are analogs of the corresponding equations governing translational motion and are

$$W = \int_{\theta_i}^{\theta_f} \tau \, d\theta \quad (11\text{-}44)$$

and

$$P = \frac{dW}{dt} = \tau\frac{d\theta}{dt} = \tau\omega. \quad (11\text{-}45)$$

The form of the work–kinetic energy theorem used for rotating bodies is

$$\Delta K = \tfrac{1}{2}I\omega_f^2 - \tfrac{1}{2}I\omega_i^2 = W. \quad (11\text{-}41)$$

QUESTIONS

1. Figure 11-21b is a graph of the angular position of the rotating disk of Fig. 11-21a. Is the angular velocity of the disk positive, negative, or zero at (a) $t = 1$ s, (b) $t = 2$ s, and (c) $t = 3$ s? (d) Is the angular acceleration positive or negative?

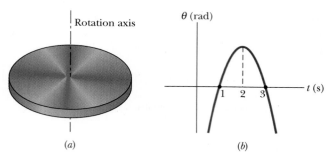

FIGURE 11-21   Question 1.

2. Figure 11-22 is a graph of the angular velocity of the rotating disk of Fig. 11-21*a*. What are the (a) initial and (b) final directions of rotation? (c) Does the disk momentarily stop? (d) Is the angular acceleration positive or negative? (e) Is the angular acceleration constant or varying?

FIGURE 11-22
Question 2.

3. At *t* = 0, the rotating disk of Fig. 11-21*a* is at angular position $\theta_0 = -2$ rad. Here are the signs of the disk's initial angular velocity and constant angular acceleration, respectively, for four situations: (1) +, +; (2) +, −; (3) −, +; (4) −, −. For which situations will the disk (a) undergo a momentary stop, (b) definitely pass through angular position $\theta = 0$ (given enough time), and (c) definitely not pass through that angular position?

4. For the rotating disk of Fig. 11-21*a*, here are the initial and final angular velocities, respectively, in four situations: (a) 2 rad/s, 3 rad/s; (b) −2 rad/s, 3 rad/s; (c) −2 rad/s, −3 rad/s; (d) 2 rad/s, −3 rad/s. The magnitude of the angular acceleration of the disk has the same constant value in all four situations. Rank the situations according to the magnitude of the disk's displacement, greatest first, during the change from the initial to the final angular velocity.

5. For which of the following expressions for $\omega(t)$ of a rotating object do the angular equations of Table 11-1 apply? (a) $\omega = 3$; (b) $\omega = 4t^2 + 2t − 6$; (c) $\omega = 3t − 4$; (d) $\omega = 5t^2 − 3$.

6. Figure 11-23 gives the angular velocity versus time for the rotating disk of Fig. 11-21*a*. For a point on the rim of the disk, rank the four instants *a*, *b*, *c*, and *d* according to the magnitude of (a) the tangential acceleration and (b) the radial acceleration, greatest first.

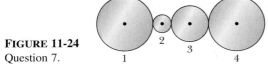

FIGURE 11-23
Question 6.

7. Figure 11-24 shows four gears that rotate together because of friction between them. Gear 2 has radius *R*; gear 3 has radius 2*R*; and gears 1 and 4 have radius 3*R*. Gear 2 is forced to rotate by a motor. Rank the four gears according to (a) the linear speed of their rims and (b) their angular speed of rotation, greatest first.

8. Circular disks *A* and *B* have the same weight and thickness, but the density of *A* is greater than that of *B*. Is the rotational inertia of *A* about its central axis greater than, less than, or the same as that of *B*?

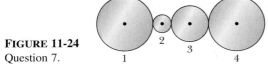

FIGURE 11-24
Question 7.

9. Figure 11-25 shows three uniform disks, along with their radii *R* and masses *M*. Rank the disks according to their rotational inertias about their central axes, greatest first.

R:	1 m	2 m	3 m
M:	26 kg	7 kg	3 kg

FIGURE 11-25
Question 9.

(a)   (b)   (c)

10. Five solids with identical masses are shown in cross section in Fig. 11-26. The cross sections have equal widths at the widest part and equal heights (but not necessarily equal thicknesses). (a) Which one has the greatest rotational inertia about an axis through its center of mass and perpendicular to its cross section? (b) Which has the least?

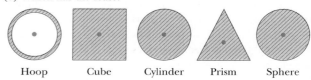

Hoop   Cube   Cylinder   Prism   Sphere

FIGURE 11-26   Question 10.

11. Figure 11-27*a* shows a meter stick, half wood and half steel, that is pivoted at the wood end at *O*. A force **F** is applied to the steel end at *a*. In Fig. 11-27*b*, the stick is reversed and pivoted at the steel end at *O'*, and the same force is applied at the wood end at *a'*. Is the resulting angular acceleration of Fig. 11-27*a* greater than, less than, or the same as that of Fig. 11-27*b*?

(a)   (b)

FIGURE 11-27   Question 11.

12. Figure 11-28*a* shows an overhead view of a horizontal bar that can pivot about the point indicated. Two horizontal forces act on the bar, but the bar is stationary. If the angle between the bar and force $\mathbf{F}_2$ is now decreased from the initial 90°, and the bar is still not to turn, should the magnitude of $\mathbf{F}_2$ be made larger, be made smaller, or left the same?

13. Figure 11-28*b* shows an overhead view of a horizontal bar that is rotated about the pivot point by two horizontal forces, $\mathbf{F}_1$ and $\mathbf{F}_2$, at opposite ends of the bar. The direction of $\mathbf{F}_2$ is at angle ϕ to the bar. Rank the following values of ϕ according to the magnitude of the angular acceleration of the bar, greatest first: 90°, 70°, and 110°.

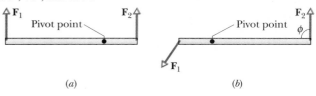

(a)   (b)

FIGURE 11-28   Questions 12 and 13.

14. A force is applied to the rim of the disk in Fig. 11-21a so as to change its angular velocity. Its initial and final angular velocities, respectively, for four situations are: (a) −2 rad/s, 5 rad/s; (b) 2 rad/s, 5 rad/s; (c) −2 rad/s, −5 rad/s; and (d) 2 rad/s, −5 rad/s. Rank the situations according to the work done by the torque due to the force, greatest first.

15. Hold your right arm downward, palm toward your thigh. Keeping your wrist rigid, (1) lift the arm until it is horizontal and forward, (2) move it horizontally until it is pointed toward the right, and (3) then bring it down to your side. Your palm faces forward. If you start over, but reverse the steps, why does your palm *not* face forward?

EXERCISES & PROBLEMS

SECTION 11-2 The Rotational Variables

1E. (a) What angle in radians is subtended by an arc that has length 1.80 m and is part of a circle of radius 1.20 m? (b) Express the same angle in degrees. (c) The angle between two radii of a circle is 0.620 rad. What arc length is subtended if the radius is 2.40 m?

2E. During a time interval t the flywheel of a generator turns through the angle $\theta = at + bt^3 - ct^4$, where a, b, and c are constants. Write expressions for the wheel's (a) angular velocity and (b) angular acceleration.

3E. What is the angular speed of (a) the second hand, (b) the minute hand, and (c) the hour hand of a watch? Answer in radians per second.

4E. Our Sun is 2.3×10^4 ly (light-years) from the center of our Milky Way galaxy and is moving in a circle around that center at a speed of 250 km/s. (a) How long does it take the Sun to make one revolution about the galactic center? (b) How many revolutions has the Sun completed since it was formed about 4.5×10^9 years ago?

5E. The angular position of a point on the rim of a rotating wheel is given by $\theta = 4.0t - 3.0t^2 + t^3$, where θ is in radians if t is given in seconds. (a) What are the angular velocities at $t = 2.0$ s and $t = 4.0$ s? (b) What is the average angular acceleration for the time interval that begins at $t = 2.0$ s and ends at $t = 4.0$ s? (c) What are the instantaneous angular accelerations at the beginning and end of this time interval?

6E. The angular position of a point on a rotating wheel is given by $\theta = 2 + 4t^2 + 2t^3$, where θ is in radians and t is in seconds. At $t = 0$, what are (a) the angular position and (b) the angular velocity? (c) What is the angular velocity at $t = 4.0$ s? (d) Calculate the angular acceleration at $t = 2.0$ s. (e) Is the angular acceleration constant?

7P. A wheel rotates with an angular acceleration α given by $\alpha = 4at^3 - 3bt^2$, where t is the time and a and b are constants. If the wheel has an initial angular speed ω_0, write equations for (a) the angular speed and (b) the angular displacement of the wheel as functions of time.

8P. A good baseball pitcher can throw a baseball toward home plate at 85 mi/h with a spin of 1800 rev/min. How many revolutions does the baseball make on its way to home plate? For simplicity, assume that the 60 ft trajectory is a straight line.

9P. A diver makes 2.5 complete revolutions on the way from a 10 m high platform to the water. Assuming zero initial vertical velocity, find the diver's average angular velocity during a dive.

10P. The wheel in Fig. 11-29 has eight spokes and a radius of 30 cm. It is mounted on a fixed axle and is spinning at 2.5 rev/s. You want to shoot a 20 cm long arrow parallel to this axle and through the wheel without hitting any of the spokes. Assume that the arrow and the spokes are very thin. (a) What minimum speed must the arrow have? (b) Does it matter where between the axle and rim of the wheel you aim? If so, where is the best location?

FIGURE 11-29
Problem 10.

SECTION 11-4 Rotation with Constant Angular Acceleration

11E. The angular speed of an automobile engine is increased from 1200 rev/min to 3000 rev/min in 12 s. (a) What is its angular acceleration in rev/min², assuming it to be uniform? (b) How many revolutions does the engine make during this time?

12E. A phonograph turntable rotating at $33\frac{1}{3}$ rev/min slows down and stops in 30 s after the motor is turned off. (a) Find its (uniform) angular acceleration in units of rev/min². (b) How many revolutions did it make in this time?

13E. A disk, initially rotating at 120 rad/s, is slowed down with a constant angular acceleration of magnitude 4.0 rad/s². (a) How much time elapses before the disk stops? (b) Through what angle does the disk rotate in coming to rest?

14E. A pulley wheel that is 8.0 cm in diameter has a 5.6 m long cord wrapped around its periphery. Starting from rest, the wheel is given a constant angular acceleration of 1.5 rad/s². (a) Through what angle must the wheel turn for the cord to unwind? (b) How long does the unwinding take?

15E. A heavy flywheel rotating on its central axis is slowing down because of friction in its bearings. At the end of the first minute of slowing, its angular velocity is 0.90 of its initial angular velocity of 250 rev/min. Assuming constant frictional forces, find its angular velocity at the end of the second minute.

16E. The flywheel of an engine is rotating at 25.0 rad/s. When the engine is turned off, the flywheel decelerates at a constant rate and comes to rest after 20.0 s. Calculate (a) the angular acceleration (in rad/s^2) of the flywheel, (b) the angle (in rad) through which the flywheel rotates in coming to rest, and (c) the number of revolutions made by the flywheel in coming to rest.

17E. Starting from rest, a disk rotates about its central axis with constant angular acceleration. In 5.0 s, it has rotated 25 rad. (a) What was the angular acceleration during this time? (b) What was the average angular velocity? (c) What is the instantaneous angular velocity of the disk at the end of the 5.0 s? (d) Assuming that the acceleration does not change, through what additional angle will the disk turn during the next 5.0 s?

18P. Starting from rest at $t = 0$, a wheel undergoes a constant angular acceleration. When $t = 2.0$ s, the angular velocity of the wheel is 5.0 rad/s. The acceleration continues until $t = 20$ s, when it abruptly ceases. Through what angle does the wheel rotate in the interval $t = 0$ to $t = 40$ s?

19P. A wheel, starting from rest, rotates with a constant angular acceleration of 2.00 rad/s^2. During a certain 3.00 s interval, it turns through 90.0 rad. (a) How long had the wheel been turning before the start of the 3.00 s interval? (b) What was the angular velocity of the wheel at the start of the 3.00 s interval?

20P. A wheel has a constant angular acceleration of 3.0 rad/s^2. In a 4.0 s interval, it turns through an angle of 120 rad. Assuming that the wheel started from rest, how long had it been in motion at the start of this 4.0 s interval?

21P. A flywheel completes 40 rev as it slows from an angular speed of 1.5 rad/s to a complete stop. (a) Assuming uniform acceleration, what is the time required for it to come to rest? (b) What is the angular acceleration? (c) How much time is required for it to complete the first 20 of the 40 revolutions?

22P. At $t = 0$, a flywheel has an angular velocity of 4.7 rad/s, an angular acceleration of -0.25 rad/s^2, and a reference line at $\theta_0 = 0$. (a) Through what maximum angle θ_{max} will the reference line turn in the positive direction? At what times t will the line be at (b) $\theta = \frac{1}{2}\theta_{max}$ and (c) $\theta = -10.5$ rad (consider both positive and negative values of t)? (d) Graph θ versus t, and indicate the answers to (a), (b), and (c) on the graph.

23P. A disk rotates about its central axis starting from rest and accelerates with constant angular acceleration. At one time it is rotating at 10 rev/s. After 60 more complete revolutions, its angular speed is 15 rev/s. Calculate (a) the angular acceleration, (b) the time required to complete the 60 revolutions mentioned, (c) the time required to attain the 10 rev/s angular speed, and (d) the number of revolutions from rest until the time the disk attained the 10 rev/s angular speed.

24P. A wheel turns through 90 revolutions in 15 s, its angular speed at the end of the period being 10 rev/s. (a) What was its angular speed at the beginning of the 15 s interval, assuming constant angular acceleration? (b) How much time had elapsed between the time the wheel was at rest and the beginning of the 15 s interval?

SECTION 11-5 Relating The Linear and Angular Variables

25E. What is the acceleration of a point on the rim of a 12 in. ($= 30$ cm) diameter record rotating at $33\frac{1}{3}$ rev/min?

26E. A phonograph record on a turntable rotates at $33\frac{1}{3}$ rev/min. (a) What is the angular speed in rad/s? What is the linear speed of a point on the record at the needle at (b) the beginning and (c) the end of the recording? The distances of the needle from the turntable axis are 5.9 in. and 2.9 in. at these two positions.

27E. What is the angular speed of a car traveling at 50 km/h and rounding a circular turn of radius 110 m?

28E. A flywheel with a diameter of 1.20 m is rotating at 200 rev/min. (a) What is the angular velocity of the flywheel in rad/s? (b) What is the linear velocity of a point on the rim of the flywheel? (c) What constant angular acceleration (in rev/min^2) will increase the wheel's angular speed to 1000 rev/min in 60 s? (d) How many revolutions does the wheel make during this 60 s acceleration?

29E. A point on the rim of a 0.75 m diameter grinding wheel changes speed uniformly from 12 m/s to 25 m/s in 6.2 s. What is the average angular acceleration of the wheel during this interval?

30E. The Earth's orbit about the Sun is almost a circle. With respect to the Sun, what are the Earth's (a) angular speed, (b) linear speed, and (c) acceleration?

31E. At 7:14 A.M. on June 30, 1908, a huge explosion occurred above remote central Siberia at latitude 61° N and longitude 102° E; the fireball thus created was the brightest flash seen by anyone before the advent of nuclear weapons. The *Tunguska Event,* which according to one chance witness "covered an enormous part of the sky," was probably the explosion of a *stony asteroid* about 140 m wide. (a) Considering only the Earth's rotation, determine how much later the asteroid would have had to arrive to put the explosion above Helsinki at longitude 25° E? This would have obliterated the city. (b) If the asteroid had, instead, been a *metallic asteroid,* it could have reached the Earth's surface. How much later would such an asteroid have had to arrive to put the impact in the Atlantic Ocean at longitude 20° W? (The resulting tsunamis would have wiped out coastal civilization on both sides of the Atlantic.)

32E. An astronaut is being tested in a centrifuge. The centrifuge has a radius of 10 m and, in starting, rotates according to $\theta = 0.30t^2$, where t in seconds gives θ in radians. When $t = 5.0$ s, what are the astronaut's (a) angular velocity, (b) linear speed, (c) tangential acceleration (magnitude only), and (d) radial acceleration (magnitude only)?

33E. What are (a) the angular speed, (b) the radial acceleration, and (c) the tangential acceleration of a spaceship negotiating a circular turn of radius 3220 km at a speed of 29,000 km/h?

34E. A coin of mass M is placed a distance R from the center of a phonograph turntable. The coefficient of static friction is μ_s. The angular speed of the turntable is slowly increased to a value ω_0, at which time the coin slides off. (a) Find ω_0 in terms of the quantities M, R, g, and μ_s. (b) Make a sketch showing the approximate path of the coin as it flies off the turntable.

35P. The flywheel of a steam engine runs with a constant angular speed of 150 rev/min. When steam is shut off, the friction of the bearings and of the air brings the wheel to rest in 2.2 h. (a) What is the constant angular acceleration, in rev/min², of the wheel during the slowdown? (b) How many rotations does the wheel make before coming to rest? (c) At the instant the flywheel is turning at 75 rev/min, what is the tangential component of the linear acceleration of a flywheel particle that is 50 cm from the axis of rotation? (d) What is the magnitude of the net linear acceleration of the particle in (c)?

36P. A gyroscope flywheel of radius 2.83 cm is accelerated from rest at 14.2 rad/s² until its angular speed is 2760 rev/min. (a) What is the tangential acceleration of a point on the rim of the flywheel? (b) What is the radial acceleration of this point when the flywheel is spinning at full speed? (c) Through what distance does a point on the rim move during the acceleration?

37P. If an airplane propeller rotates at 2000 rev/min while the airplane flies at a speed of 480 km/h relative to the ground, what is the speed of a point on the tip of the propeller, at radius 1.5 m, as seen by (a) the pilot and (b) an observer on the ground? The plane's velocity is parallel to the propeller's axis of rotation.

38P. An early method of measuring the speed of light makes use of a rotating slotted wheel. A beam of light passes through a slot at the outside edge of the wheel, as in Fig. 11-30, travels to a distant mirror, and returns to the wheel just in time to pass through the next slot in the wheel. One such slotted wheel has a radius of 5.0 cm and 500 slots at its edge. Measurements taken when the mirror was $l = 500$ m from the wheel indicated a speed of light of 3.0×10^5 km/s. (a) What was the (constant) angular speed of the wheel? (b) What was the linear speed of a point on the edge of the wheel?

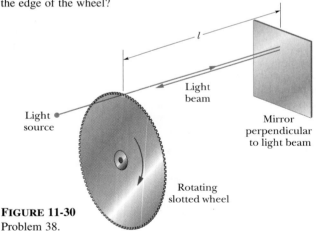

FIGURE 11-30
Problem 38.

39P. A car starts from rest and moves around a circular track of radius 30.0 m. Its speed increases at the constant rate of 0.500 m/s². (a) What is the magnitude of its *net* linear acceleration 15.0 s later? (b) What angle does this net acceleration vector make with the car's velocity at this time?

40P. (a) What is the angular speed about the polar axis of a point on the Earth's surface at a latitude of 40° N? (b) What is the linear speed of the point? (c) What are the corresponding values for a point at the equator?

41P. In Fig. 11-31, wheel A of radius $r_A = 10$ cm is coupled by belt B to wheel C of radius $r_C = 25$ cm. Wheel A increases its angular speed from rest at a uniform rate of 1.6 rad/s². Find the time for wheel C to reach a rotational speed of 100 rev/min, assuming the belt does not slip. (*Hint:* If the belt does not slip, the linear speeds at the rims of the two wheels must be equal.)

FIGURE 11-31 Problem 41.

42P. In Fig. 11-32, four pulleys are connected by two belts. Pulley A (radius 15 cm) is the drive pulley, and it rotates at 10 rad/s. Pulley B (radius 10 cm) is connected by belt 1 to pulley A. Pulley B' (radius 5 cm) is concentric with pulley B and is rigidly attached to it. Pulley C (radius 25 cm) is connected by belt 2 to pulley B'. Calculate (a) the linear speed of a point on belt 1, (b) the angular speed of pulley B, (c) the angular speed of pulley B', (d) the linear speed of a point on belt 2, and (e) the angular speed of pulley C. (See hint to Problem 41.)

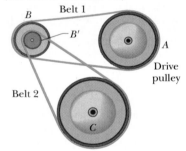

FIGURE 11-32 Problem 42.

43P. A phonograph turntable is rotating at $33\frac{1}{3}$ rev/min. A small object is on the turntable 6.0 cm from the axis of rotation. (a) Calculate the acceleration of the object, assuming that the object does not slip. (b) What is the minimum value of the coefficient of static friction between the object and the turntable? (c) Suppose that the turntable achieved its angular speed by starting from rest and undergoing a constant angular acceleration for 0.25 s. Calculate the minimum coefficient of static friction required for the object not to slip during the acceleration period.

44P. A pulsar is a rapidly rotating neutron star that emits radio pulses with precise synchronization, one such pulse for each rotation of the star. The *period T* of rotation is found by measuring the time between pulses. At present, the pulsar in the central region of the Crab nebula (see Fig. 11-33) has a period of rotation of $T = 0.033$ s, and this period is observed to be increasing at the rate of 1.26×10^{-5} s/y. (a) What is the value of the angular acceleration in rad/s²? (b) If its angular acceleration is constant, how many years from now will the pulsar stop rotating? (c) The pulsar originated in a supernova explosion seen in the year A.D. 1054. What was T for the pulsar when it was born? (Assume constant angular acceleration since then.)

cylinder axis (Fig. 11-34). Calculate the satellite's (a) rotational inertia about the rotation axis and (b) rotational kinetic energy.

FIGURE 11-34 Exercise 49.

50E. Two particles, each with mass m, are fastened to each other, and to a rotation axis at O, by two thin rods, each with length l and mass M as shown in Fig. 11-35. The combination rotates around the rotation axis with angular velocity ω. Obtain algebraic expressions for (a) the rotational inertia of the combination about O and (b) the kinetic energy of rotation about O.

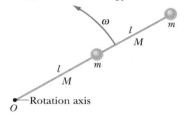

FIGURE 11-35 Exercise 50.

51E. Each of the three helicopter rotor blades shown in Fig 11-36 is 5.20 m long and has a mass of 240 kg. The rotor is rotating at 350 rev/min. (a) What is the rotational inertia of the rotor assembly about the axis of rotation? (Each blade can be considered to be a thin rod.) (b) What is the kinetic energy of rotation?

FIGURE 11-36 Exercise 51 and Problem 85.

52E. Assume the Earth to be a sphere of uniform density. Calculate (a) its rotational inertia and (b) its rotational kinetic energy. (c) Suppose that this energy could be harnessed for our use. How long could the Earth supply 1.0 kW of power to each of the 6.4×10^9 persons on Earth?

FIGURE 11-33 Problem 44. The Crab nebula resulted from a star whose explosion was seen in A.D. 1054. In addition to the gaseous debris seen here, the explosion left a spinning neutron star at its center. The star has a diameter of only 30 km.

SECTION 11-6 Kinetic Energy of Rotation

45E. Calculate the rotational inertia of a wheel that has a kinetic energy of 24,400 J when rotating at 602 rev/min.

46P. The oxygen molecule, O_2, has a total mass of 5.30×10^{-26} kg and a rotational inertia of 1.94×10^{-46} kg·m² about an axis through the center of the line joining the atoms and perpendicular to that line. Suppose that such a molecule in a gas has a speed of 500 m/s and that its rotational kinetic energy is two-thirds of its translational kinetic energy. Find its angular velocity.

SECTION 11-7 Calculating the Rotational Inertia

47E. Calculate the kinetic energies of two uniform solid cylinders, each rotating about its central axis. They have the same mass, 1.25 kg, and rotate with the same angular velocity, 235 rad/s, but the first has a radius of 0.25 m and the second a radius of 0.75 m.

48E. A molecule has a rotational inertia of 14,000 u·pm² and is spinning at an angular speed of 4.3×10^{12} rad/s. (a) Express the rotational inertia in kg·m². (b) Calculate the rotational kinetic energy in electron-volts.

49E. A communications satellite is a solid cylinder with mass 1210 kg, diameter 1.21 m, and length 1.75 m. Prior to launching from the shuttle cargo bay, it is set spinning at 1.52 rev/s about the

53E. Calculate the rotational inertia of a meter stick, with mass 0.56 kg, about an axis perpendicular to the stick and located at the 20 cm mark. (Treat the stick as a thin rod.)

54P. Show that the axis about which a given rigid body has its smallest rotational inertia must pass through its center of mass.

55P. Derive the expression for the rotational inertia of a hoop of mass M and radius R about its central axis; see Table 11-2(a).

56P. Figure 11-37 shows a uniform solid block of mass M and edge lengths a, b, and c. Calculate its rotational inertia about an axis through one corner and perpendicular to the large faces.

FIGURE 11-37   Problem 56.

57P. The masses and coordinates of four particles are as follows: 50 g, $x = 2.0$ cm, $y = 2.0$ cm; 25 g, $x = 0$, $y = 4.0$ cm; 25 g, $x = -3.0$ cm, $y = -3.0$ cm; 30 g, $x = -2.0$ cm, $y = 4.0$ cm. What is the rotational inertia of this collection with respect to the (a) x, (b) y, and (c) z axes? (d) If the answers to (a) and (b) are A and B, respectively, then what is the answer to (c) in terms of A and B?

58P. (a) Show that the rotational inertia of a solid cylinder of mass M and radius R about its central axis is equal to the rotational inertia of a thin hoop of mass M and radius $R/\sqrt{2}$ about its central axis. (b) Show that the rotational inertia I of any given body of mass M about any given axis is equal to the rotational inertia of an *equivalent hoop* about that axis, if the hoop has the same mass M and a radius k given by

$$k = \sqrt{\frac{I}{M}}.$$

The radius k of the equivalent hoop is called the *radius of gyration* of the given body.

59P. Delivery trucks that operate by making use of energy stored in a rotating flywheel have been used in Europe. The trucks are charged by using an electric motor to get the flywheel up to its top speed of 200π rad/s. One such flywheel is a solid, homogeneous cylinder with a mass of 500 kg and a radius of 1.0 m. (a) What is the kinetic energy of the flywheel after charging? (b) If the truck operates with an average power requirement of 8.0 kW, for how many minutes can it operate between chargings?

SECTION 11-8 Torque

60E. The length of a bicycle pedal arm is 0.152 m, and a downward force of 111 N is applied by the foot. What is the magnitude of the torque about the pivot point when the arm makes an angle of (a) 30°, (b) 90°, and (c) 180° with the vertical?

61E. A small 0.75 kg ball is attached to one end of a 1.25 m long massless rod, and the other end of the rod is hung from a pivot. When the resulting pendulum is 30° from the vertical, what is the magnitude of the torque about the pivot?

62E. A bicyclist of mass 70 kg puts all his weight on each downward-moving pedal as he pedals up a steep road. Take the diameter of the circle in which the pedals rotate to be 0.40 m, and determine the magnitude of the maximum torque he exerts.

63P. The body in Fig. 11-38 is pivoted at O, and two forces act on it as shown. (a) Find an expression for the magnitude of the net torque on the body about the pivot. (b) If $r_1 = 1.30$ m, $r_2 = 2.15$ m, $F_1 = 4.20$ N, $F_2 = 4.90$ N, $\theta_1 = 75.0°$, and $\theta_2 = 60.0°$, what is the net torque about the pivot?

FIGURE 11-38   Problem 63.

64P. The body in Fig. 11-39 is pivoted at O. Three forces act on it in the directions shown on the figure: $F_A = 10$ N at point A, 8.0 m from O; $F_B = 16$ N at point B, 4.0 m from O; and $F_C = 19$ N at point C, 3.0 m from O. What is the net torque about O?

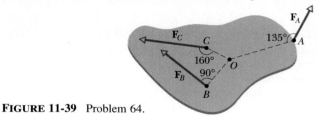

FIGURE 11-39   Problem 64.

SECTION 11-9 Newton's Second Law for Rotation

65E. When a torque of 32.0 N·m is applied to a certain wheel, the wheel acquires an angular acceleration of 25.0 rad/s². What is the rotational inertia of the wheel?

66E. Launching herself from a board, a diver changed her angular velocity from zero to 6.20 rad/s in 220 ms. Her rotational inertia is 12.0 kg·m². (a) What was her angular acceleration during the launch? (b) What external torque acted on the diver during the launch?

67E. A cylinder having a mass of 2.0 kg can rotate about its central axis through point O. Forces are applied as in Fig. 11-40:

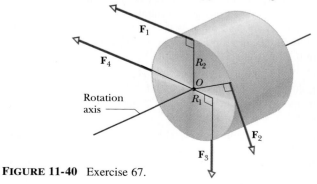

FIGURE 11-40   Exercise 67.

$F_1 = 6.0$ N, $F_2 = 4.0$ N, $F_3 = 2.0$ N, and $F_4 = 5.0$ N. Also, $R_1 = 5.0$ cm and $R_2 = 12$ cm. Find the magnitude and direction of the angular acceleration of the cylinder. (During the rotation, the forces maintain their same angles relative to the cylinder).

68E. A small object with mass 1.30 kg is mounted on one end of a rod 0.780 m long and of negligible mass. The system rotates in a horizontal circle about the other end of the rod at 5010 rev/min. (a) Calculate the rotational inertia of the system about the axis of rotation. (b) There is an air drag of 2.30×10^{-2} N on the object, directed opposite its motion. What torque must be applied to the system to keep it rotating at constant speed?

69E. A thin spherical shell has a radius of 1.90 m. An applied torque of 960 N·m imparts to the shell an angular acceleration equal to 6.20 rad/s² about an axis through the center of the shell. (a) What is the rotational inertia of the shell about the axis of rotation? (b) Calculate the mass of the shell.

70P. Figure 11-41 shows the massive shield door at a neutron test facility at Lawrence Livermore Laboratory; this is the world's heaviest hinged door. The door has a mass of 44,000 kg, a rotational inertia about an axis through its hinges of 8.7×10^4 kg·m², and a (front) face width of 2.4 m. Neglecting friction, what steady force, applied at its outer edge and perpendicular to the plane of the door, can move it from rest through an angle of 90° in 30 s? Assume no friction acts on the hinges.

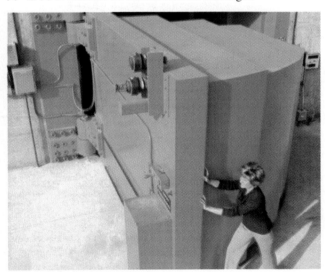

FIGURE 11-41 Problem 70.

71P. A pulley, with a rotational inertia of 1.0×10^{-3} kg·m² about its axle and a radius of 10 cm, is acted on by a force applied tangentially at its rim. The force magnitude varies in time as $F = 0.50t + 0.30t^2$, where F is in newtons if t is given in seconds. The pulley is initially at rest. At $t = 3.0$ s what are (a) its angular acceleration and (b) its angular velocity?

72P. A wheel of radius 0.20 m is mounted on a frictionless horizontal axis. A massless cord is wrapped around the wheel and attached to a 2.0 kg object that slides on a frictionless surface inclined at an angle of 20° with the horizontal, as shown in Fig. 11-42. The object accelerates down the incline at 2.0 m/s². What is the rotational inertia of the wheel about its axis of rotation?

FIGURE 11-42 Problem 72.

73P. Two uniform solid spheres have the same mass, 1.65 kg, but one has a radius of 0.226 m while the other has a radius of 0.854 m. (a) For each of the spheres, find the torque required to bring the sphere from rest to an angular velocity of 317 rad/s in 15.5 s. Each sphere rotates about an axis through its center. (b) For each sphere, what force applied tangentially at the equator would provide the needed torque?

74P. In an Atwood's machine (Fig. 5-23), one block has a mass of 500 g, and the other a mass of 460 g. The pulley, which is mounted in horizontal frictionless bearings, has a radius of 5.00 cm. When released from rest, the heavier block is observed to fall 75.0 cm in 5.00 s (without the cord slipping on the pulley). (a) What is the acceleration of each block? What is the tension in the part of the cord that supports (b) the heavier block and (c) the lighter block? (d) What is the angular acceleration of the pulley? (e) What is its rotational inertia?

75P. Figure 11-43 shows two blocks, each of mass m, suspended from the ends of a rigid weightless rod of length $l_1 + l_2$, with $l_1 = 20$ cm and $l_2 = 80$ cm. The rod is held horizontally on the fulcrum shown in the figure and then released. Calculate the accelerations of the two blocks as they start to move.

FIGURE 11-43 Problem 75.

76P. Two identical blocks, each of mass M, are connected by a massless string over a pulley of radius R and rotational inertia I (Fig. 11-44). The string does not slip on the pulley; it is not known whether there is friction between the table and the sliding block; the pulley's axis is frictionless. When this system is released, it is found that the pulley turns through an angle θ in time t

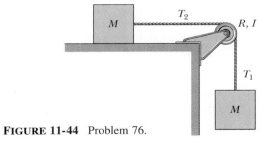

FIGURE 11-44 Problem 76.

and the acceleration of the blocks is constant. (a) What is the angular acceleration of the pulley? (b) What is the acceleration of the two blocks? (c) What are the tensions in the upper and lower sections of the string? Express the answers in terms of M, I, R, θ, g, and t.

SECTION 11-10 Work and Rotational Kinetic Energy

77E. (a) If $R = 12$ cm, $M = 400$ g, and $m = 50$ g in Fig. 11-18, find the speed of the block after it has descended 50 cm starting from rest. Solve the problem using energy conservation principles. (b) Repeat (a) with $R = 5.0$ cm.

78E. An automobile crankshaft develops 100 hp ($= 74.6$ kW) when rotating at a speed of 1800 rev/min. What torque does the crankshaft deliver?

79E. A 32.0 kg wheel, essentially a thin hoop with radius 1.20 m, is rotating at 280 rev/min. It must be brought to a stop in 15.0 s. (a) How much work must be done to stop it? (b) What is the required power?

80E. A thin rod of length l and mass m is suspended freely from one end. It is pulled to one side and then allowed to swing like a pendulum, passing through its lowest position with an angular speed ω. (a) Calculate its kinetic energy as it passes through its lowest position. (b) How high does its center of mass rise above its lowest position? Neglect friction and air resistance.

81P. Calculate (a) the torque, (b) the energy, and (c) the average power required to accelerate the Earth in 1 day from rest to its present angular speed about its axis.

82P. A meter stick is held vertically with one end on the floor and is then allowed to fall. Find the speed of the other end when it hits the floor, assuming that the end on the floor does not slip. (*Hint:* Consider the stick to be a thin rod and use the conservation of energy.)

83P. A rigid body is made of three identical thin rods, each with length l, fastened together in the form of a letter H (Fig. 11-45). The body is free to rotate about a horizontal axis that runs along the length of one of the legs of the H. The body is allowed to fall from rest from a position in which the plane of the H is horizontal. What is the angular speed of the body when the plane of the H is vertical?

FIGURE 11-45 Problem 83.

84P. A uniform cylinder of radius 10 cm and mass 20 kg is mounted so as to rotate freely about a horizontal axis that is parallel to and 5.0 cm from the central longitudinal axis of the cylinder. (a) What is the rotational inertia of the cylinder about the axis of rotation? (b) If the cylinder is released from rest with its central longitudinal axis at the same height as the axis about which the cylinder rotates, what is the angular speed of the cylinder as it passes through its lowest position? (*Hint:* Use the principle of conservation of energy.)

85P. A uniform helicopter rotor blade (see Fig. 11-36) is 7.80 m long and has a mass of 110 kg. (a) What force is exerted on the bolt attaching the blade to the rotor axle when the rotor is turning at 320 rev/min? (*Hint:* For this calculation the blade can be considered to be a point mass at the center of mass. Why?) (b) Calculate the torque that must be applied to the rotor to bring it to full speed from rest in 6.7 s. Ignore air resistance. (The blade cannot be considered to be a point mass for this calculation. Why not? Assume the mass distribution of a uniform, thin rod.) (c) How much work did the torque do on the blade in order for the blade to reach a speed of 320 rev/min?

86P. A uniform spherical shell of mass M and radius R rotates about a vertical axis on frictionless bearings (Fig. 11-46). A massless cord passes around the equator of the shell, over a pulley of rotational inertia I and radius r, and is attached to a small object of mass m. There is no friction on the pulley's axle; the cord does not slip on the pulley. What is the speed of the object after it has fallen a distance h from rest? Use work–energy considerations.

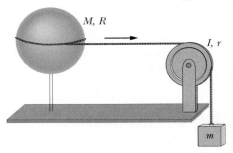

FIGURE 11-46 Problem 86.

87P. A tall, cylinder-shaped chimney falls over when its base is ruptured. Treating the chimney as a thin rod with height h, express the (a) radial and (b) tangential components of the linear acceleration of the top of the chimney as a function of the angle θ made by the chimney with the vertical. (c) At what angle θ does the linear acceleration equal g?

88P. Attached to each end of a thin steel rod of length 1.20 m and mass 6.40 kg is a small ball of mass 1.06 kg. The rod is constrained to rotate in a horizontal plane about a vertical axis through its midpoint. At a certain instant, it is observed to be rotating with an angular velocity of 39.0 rev/s. Because of friction, it comes to rest 32.0 s later. Assuming a constant frictional torque, compute (a) the angular acceleration, (b) the retarding torque exerted by the friction, (c) the total mechanical energy dissipated by the friction, and (d) the number of revolutions executed during the 32.0 s. (e) Now suppose that the frictional torque is known not to be constant. Which, if any, of the quantities (a), (b), (c), or (d) can still be computed without additional information? If such a quantity exists, give its value.

89P. A uniform rod of mass 1.5 kg is 2.0 m long (Fig. 11-47). The rod can pivot about a horizontal, frictionless pin through one end. It is released from rest at an angle of 40° above the horizontal. (a) What is the angular acceleration of the rod at the instant it is released? (The rotational inertia of the rod about the pin is 2.0 kg·m².) (b) Use the principle of conservation of energy to determine the angular speed of the rod as it passes through the horizontal position.

FIGURE 11-47 Problem 89.

90P. The rigid body shown in Fig. 11-48 consists of three particles connected by massless rods. It is to be rotated about an axis perpendicular to its plane through point P. If $M = 0.40$ kg, $a = 30$ cm, and $b = 50$ cm, how much work is required to take the body from rest to an angular speed of 5.0 rad/s?

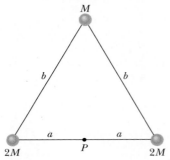

FIGURE 11-48 Problem 90.

91P*. A car is fitted with an energy-conserving flywheel, which in operation is geared to the driveshaft so that the flywheel rotates at 240 rev/s when the car is traveling at 80 km/h. The total mass of the car is 800 kg; the flywheel weighs 200 N and is a uniform disk 1.1 m in diameter. The car descends a 1500-m-long, 5° slope, starting from rest, with the flywheel engaged and no power generated from the engine. Neglecting friction and the rotational inertia of the wheels, find (a) the speed of the car at the bottom of the slope, (b) the angular acceleration of the flywheel at the bottom of the slope, and (c) the rate at which energy is stored in the rotation of the flywheel by the driveshaft as the car reaches the bottom of the slope.

ELECTRONIC COMPUTATION

92. In a certain machine two disks are free to rotate independently on parallel axles. The first has a radius of 7.0 cm and a rotational inertia of 1.5 kg·m². The second has a radius of 15 cm and a rotational inertia of 3.5 kg·m². While the first disk is at rest and the second is rotating at 175 rad/s, the axles are moved closer together so that the rims of the disks touch. Each disk exerts a normal force of 150 N on the other. (a) Take the coefficient of kinetic friction between the rims to be 0.25 and generate a table listing the values of the angular velocities for every 2 s during the first 80 s after the disks touch. Plot these values as functions of time. Also plot the total rotational kinetic energy of the disks. Is it conserved? (b) What influence does the frictional force have on the motions of the disks? Repeat the calculations for a coefficient of kinetic friction of 0.50. Is the time the disks take to reach their final angular velocities for the first μ_k value the same as for the second value? Are the final angular velocities the same for the two values? Is the final rotational kinetic energy the same?

93. Five particle-like objects, positioned in the xy plane according to the following table, form a rigidly connected body. What is the rotational inertia of the body about (a) the x axis, (b) the y axis, and (c) the z axis? (d) What is the center of mass of the body?

Object	1	2	3	4	5
Mass (grams)	500	400	300	600	450
x (cm)	15	−13	17	−4.0	−5.0
y (cm)	20	13	−6.0	−7.0	9.0

94. In Fig. 11-49, two blocks, of mass $m_1 = 400$ g and $m_2 = 600$ g, are connected by a massless cord that is wrapped around a uniform disk of mass $M = 500$ g and radius $R = 12.0$ cm. The disk can rotate without friction about a fixed horizontal axis through its center; the cord cannot slip on the disk. The system is released from rest. Find (a) the magnitude of the acceleration of the blocks, (b) the tension T_1 in the cord at the left, and (c) the tension T_2 in the cord at the right.

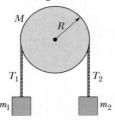

FIGURE 11-49 Problem 94.

12
Rolling, Torque, and Angular Momentum

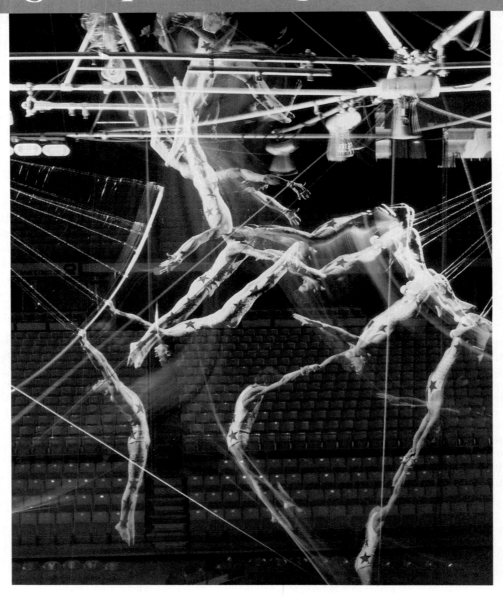

In 1897, a European "aerialist" made the first triple somersault during the flight from a swinging trapeze to the hands of a partner. For the next 85 years aerialists attempted to complete a <u>quadruple</u> somersault, but not until 1982 was it done before an audience: Miguel Vazquez of the Ringling Bros. and Barnum & Bailey Circus rotated his body in four complete circles in midair before his brother Juan caught him. Both were stunned by their success. Why was the feat so difficult, and what feature of physics made it (finally) possible?

FIGURE 12-1 A time exposure photograph of a rolling disk. Small lights have been attached to the disk, one at its center and one at its edge. The latter traces out a curve called a *cycloid*.

12-1 ROLLING

When a bicycle moves along a straight track, the center of each wheel moves forward in pure translation. A point on the rim of the wheel, however, traces out a more complex path, as Fig. 12-1 shows. In what follows, we analyze the motion of a rolling wheel first by viewing it as a combination of pure translation and pure rotation, and then by viewing it as rotation alone.

Rolling as Rotation and Translation Combined

Imagine that you are watching the wheel of a bicycle, which passes you at constant speed while rolling smoothly, without slipping, along a street. As shown in Fig. 12-2, the center of mass O of the wheel moves forward at constant speed v_{cm}. The point P at which the wheel contacts the street also moves forward at speed v_{cm}, so that it always remains directly below O.

During a time interval t, you see both O and P move forward by a distance s. The bicycle rider sees the wheel rotate through an angle θ about the center of the wheel, with the point of the wheel that was touching the street at the beginning of t moving through arc length s. Equation

FIGURE 12-2 The center of mass O of a rolling wheel moves a distance s at velocity $\mathbf{v}_{cm}$, while the wheel rotates through angle θ. The contact point P between the wheel and the surface over which the wheel rolls also moves a distance s.

11-15 relates the arc length s to the rotation angle θ:

$$s = R\theta, \tag{12-1}$$

where R is the radius of the wheel. The linear speed v_{cm} of the center of the wheel (the center of mass of this uniform wheel) is ds/dt, and the angular speed ω of the wheel about its center is $d\theta/dt$. So, differentiating Eq. 12-1 with respect to time gives us

$$v_{cm} = \omega R \quad \text{(rolling motion).} \tag{12-2}$$

(Note that Eq. 12-2 holds only if the wheel rolls *smoothly;* that is, it does not slip over the street.)

Figure 12-3 shows that the rolling motion of a wheel is a combination of purely translational and purely rotational motions. Figure 12-3a shows the purely rotational motion (as if the rotation axis through the center were stationary): every point on the wheel rotates about the center with angular speed ω. (This is the type of motion we considered in Chapter 11.) Every point on the outside edge of the wheel has linear speed v_{cm} given by Eq. 12-2. Figure 12-3b shows the purely translational motion (as if the wheel did not rotate at all): every point on the wheel moves to the right with speed v_{cm}.

The combination of Figs. 12-3a and 12-3b yields the actual rolling motion of the wheel, Fig. 12-3c. Note that in this combination of motions, the portion of the wheel at the bottom (at point P) is stationary and the portion at the top (at point T) is moving at speed $2v_{cm}$, faster than any other portion of the wheel. These results are demonstrated in Fig. 12-4, which is a time exposure of a rolling bicycle wheel. The blurring of the spokes near the top of the wheel com-

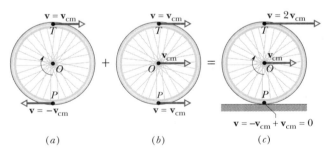

FIGURE 12-3 Rolling motion of a wheel as a combination of purely rotational motion and purely translational motion. (a) The purely rotational motion: all points on the wheel move with the same angular speed ω. Points on the outside edge of the wheel all move with the same linear speed $v = v_{cm}$. The linear velocities $\mathbf{v}$ of two such points, at top (T) and bottom (P) of the wheel, are shown. (b) The purely translational motion: all points on the wheel move to the right with the same linear velocity $\mathbf{v}_{cm}$ as the center of the wheel. (c) The rolling motion of the wheel is the combination of (a) and (b).

FIGURE 12-4 A photograph of a rolling bicycle wheel. The spokes near the top of the wheel are more blurred than those near the bottom of the wheel because they are moving faster, as Fig. 12-3c shows.

pared with the sharper images of the spokes near the bottom of the wheel shows that the wheel is moving faster near its top than near its bottom.

The motion of any round body rolling smoothly over a surface can be separated into purely rotational and purely translational motions, as in Figs. 12-3a and 12-3b.

Rolling as Pure Rotation

Figure 12-5 suggests another way to look at the rolling motion of a wheel, namely, as pure rotation about an axis that always extends through the point where the wheel contacts the street as the wheel moves. That is, we consider the rolling motion to be pure rotation about an axis passing through point P in Fig. 12-3c and perpendicular to the

Rotational axis at P

FIGURE 12-5 Rolling can be viewed as pure rotation, with angular speed ω, about an axis that always extends through P. The vectors show the instantaneous linear velocities of selected points on the rolling wheel. You can obtain the vectors by combining the translational and rotational motions as in Fig. 12-3.

plane of the figure. The vectors in Fig. 12-5 then represent the instantaneous velocities of points on the rolling wheel.

Question. What angular speed about this new axis will a stationary observer assign to a rolling bicycle wheel?
Answer. The same angular speed ω that the rider assigns to the wheel as she or he observes it in pure rotation about an axis through its center of mass.

To verify this answer, let us use it to calculate the linear speed of the top of the rolling wheel from the point of view of a stationary observer. If we call the wheel's radius R, the top is a distance $2R$ from the axis through P in Fig. 12-5, so the linear speed at the top should be (using Eq. 12-2)

$$v_{\text{top}} = (\omega)(2R) = 2(\omega R) = 2v_{\text{cm}},$$

in exact agreement with Fig. 12-3c. You can similarly verify the linear speeds shown for the portion of the wheel at points O and P in Fig. 12-3c.

CHECKPOINT 1: The rear wheel on a clown's bicycle has twice the radius of the front wheel. (a) Is the linear speed at the very top of the rear wheel greater than, less than, or the same as that of the front wheel? (b) Is the angular speed of the rear wheel greater than, less than, or the same as that of the front wheel?

The Kinetic Energy

Let us now calculate the kinetic energy of the rolling wheel as measured by the stationary observer. If we view the

As this hoop rolls, does the dog walk as fast as the pony?

rolling as pure rotation about an axis through P in Fig. 12-5, we have

$$K = \tfrac{1}{2}I_P\omega^2, \qquad (12\text{-}3)$$

in which ω is the angular speed of the wheel and I_P is the rotational inertia of the wheel about the axis through P. From the parallel-axis theorem (Eq. 11-27) we have

$$I_P = I_{cm} + MR^2, \qquad (12\text{-}4)$$

in which M is the mass of the wheel and I_{cm} is its rotational inertia about an axis through its center of mass. Substituting Eq. 12-4 into Eq. 12-3, we obtain

$$K = \tfrac{1}{2}I_{cm}\omega^2 + \tfrac{1}{2}MR^2\omega^2,$$

and using the relation $v_{cm} = \omega R$ (Eq. 12-2) yields

$$K = \tfrac{1}{2}I_{cm}\omega^2 + \tfrac{1}{2}Mv_{cm}^2. \qquad (12\text{-}5)$$

We can interpret the first of these terms ($\tfrac{1}{2}I_{cm}\omega^2$) as the kinetic energy associated with the rotation of the wheel about an axis through its center of mass (Fig. 12-3a), and the second term ($\tfrac{1}{2}Mv_{cm}^2$) as the kinetic energy associated with the translational motion of the wheel (Fig. 12-3b).

Friction and Rolling

If the wheel rolls at constant speed, as in Fig. 12-2, it has no tendency to slide at the point of contact P, and thus there is no frictional force acting on the wheel there. However, if a force acts on the wheel, changing the speed v_{cm} of the center of the wheel or the angular speed ω about the center, then there is a tendency for the wheel to slide at P, and a frictional force acts on the wheel at P to oppose that tendency. Until the wheel actually begins to slide, the frictional force is a *static* frictional force $\mathbf{f}_s$. If the wheel begins to slide, then the force is a *kinetic* frictional force $\mathbf{f}_k$.

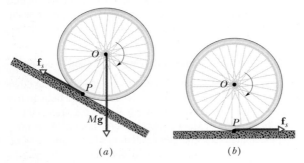

(a)   (b)

FIGURE 12-6 (*a*) A wheel rolls down an incline without sliding. A static frictional force $\mathbf{f}_s$ acts on the wheel at P, opposing the wheel's tendency to slide because of its weight $M\mathbf{g}$. (*b*) A wheel rolls horizontally without sliding while its angular speed is increased. A static frictional force $\mathbf{f}_s$ acts on the wheel at P, opposing the tendency to slide. If in (*a*) or (*b*) the wheel slides, the frictional force is a kinetic frictional force $\mathbf{f}_k$.

In Fig. 12-6a, a wheel rolls down an incline. Its weight $M\mathbf{g}$ acts on the wheel at the wheel's center of mass. Since that force $M\mathbf{g}$ lacks any moment arm relative to the center of the wheel, it cannot produce a torque about the center and cause the wheel to rotate about the center. But since $M\mathbf{g}$ tends to slide the wheel down the incline, a frictional force acts on the wheel at contact point P to oppose the tendency to slide. This force, which points up along the incline, *does* have a moment arm relative to the center; the moment arm is the radius of the wheel. Thus this frictional force produces a torque about the center and causes the wheel to rotate about the center.

In Fig. 12-6b, a wheel is being made to rotate faster while rolling along a flat surface, as on an accelerating bicycle. The increase in ω tends to slide the bottom of the wheel toward the left. A frictional force acts on the wheel toward the right at P to oppose the tendency to slide. (This frictional force is the external force acting on an accelerating bicycle–rider system that causes the acceleration.)

SAMPLE PROBLEM 12-1

A uniform solid cylindrical disk, whose mass M is 1.4 kg and whose radius R is 8.5 cm, rolls across a horizontal table at a speed v of 15 cm/s.

(a) What is the speed of the top of the rolling disk?

SOLUTION: When we speak of the speed of a rolling object, we always mean the speed of its center of mass. From Fig. 12-3c we see that the speed of the top of the disk is just twice this, or

$$v_{top} = 2v_{cm} = (2)(15 \text{ cm/s}) = 30 \text{ cm/s}. \quad \text{(Answer)}$$

(b) What is the angular speed ω of the rolling disk?

SOLUTION: From Eq. 12-2 we have

$$\omega = \frac{v_{cm}}{R} = \frac{15 \text{ cm/s}}{8.5 \text{ cm}}$$

$$= 1.8 \text{ rad/s} = 0.28 \text{ rev/s}. \quad \text{(Answer)}$$

This value applies whether the axis of rotation is taken to be an axis through point P in Fig. 12-5 or an axis through the center of mass.

(c) What is the kinetic energy K of the rolling disk?

SOLUTION: From Eq. 12-5 we have, putting $I_{cm} = \tfrac{1}{2}MR^2$ and using the relation $v_{cm} = \omega R$,

$$K = \tfrac{1}{2}I_{cm}\omega^2 + \tfrac{1}{2}Mv_{cm}^2$$

$$= (\tfrac{1}{2})(\tfrac{1}{2}MR^2)(v_{cm}/R)^2 + \tfrac{1}{2}Mv_{cm}^2 = \tfrac{3}{4}Mv_{cm}^2$$

$$= \tfrac{3}{4}(1.4 \text{ kg})(0.15 \text{ m/s})^2$$

$$= 0.024 \text{ J} = 24 \text{ mJ}. \quad \text{(Answer)}$$

TABLE 12-1 THE RELATIVE SPLITS BETWEEN ROTATIONAL AND TRANSLATIONAL ENERGIES FOR ROLLING OBJECTS

OBJECT	ROTATIONAL INERTIA I_{cm}	PERCENTAGE OF ENERGY IN	
		TRANSLATION	ROTATION
Hoop	$1MR^2$	50%	50%
Disk	$\frac{1}{2}MR^2$	67%	33%
Sphere	$\frac{2}{5}MR^2$	71%	29%
General^a	βMR^2	$100\dfrac{1}{1+\beta}$%	$100\dfrac{\beta}{1+\beta}$%

$^a\beta$ may be computed for any rolling object as I_{cm}/MR^2.

(d) What fraction of the kinetic energy is associated with the motion of translation and what fraction with the motion of rotation about an axis through the center of mass?

SOLUTION: The kinetic energy associated with translation is the second term of Eq. 12-5, or $\frac{1}{2}Mv_{cm}^2$. The fraction we seek is then, using the expression derived in (c),

$$\text{frac} = \frac{\frac{1}{2}Mv_{cm}^2}{\frac{3}{4}Mv_{cm}^2} = \frac{2}{3} \quad \text{or} \quad 67\%. \quad \text{(Answer)}$$

The remaining 33% is associated with rotation about an axis through the center of mass.

The relative split between translational and rotational energy depends on the rotational inertia of the rolling object. As Table 12-1 shows, the rolling object (the hoop) that has its mass farthest from the central axis of rotation (and so has the largest rotational inertia) has the largest share of its kinetic energy in rotational motion. The object (the sphere) that has its mass closest to the central axis of rotation (and so has the smallest rotational inertia) has the smallest share in that form.

The formulas at the bottom of Table 12-1 apply to the generic rolling object with **rotational inertia parameter** β. This parameter has the value 1 for a hoop, $\frac{1}{2}$ for a disk, and $\frac{2}{5}$ for a sphere.

SAMPLE PROBLEM 12-2

A bowling ball, whose radius R is 11 cm and whose mass M is 7.2 kg, rolls from rest down a ramp whose length L is 2.1 m. The ramp is inclined at an angle θ of 34° to the horizontal; see the sphere in Fig. 12-7. How fast is the ball moving when it reaches the bottom of the ramp? Assume the ball is uniform in density.

SOLUTION: The center of the ball falls a vertical distance $h = L \sin \theta$; so the decrease in gravitational potential energy is $MgL \sin \theta$. This loss of potential energy equals the gain in kinetic energy. Thus we can write (see Eq. 12-5)

$$MgL \sin \theta = \tfrac{1}{2}I_{cm}\omega^2 + \tfrac{1}{2}Mv_{cm}^2. \quad (12\text{-}6)$$

From Table 11-2(g) we see that, for a solid sphere, $I_{cm} = \frac{2}{5}MR^2$. We can also replace ω with its equal, v_{cm}/R. Substituting both these quantities into Eq. 12-6 yields

$$MgL \sin \theta = (\tfrac{1}{2})(\tfrac{2}{5})(MR^2)(v_{cm}/R)^2 + \tfrac{1}{2}Mv_{cm}^2.$$

Solving for v_{cm} yields

$$\begin{aligned} v_{cm} &= \sqrt{(\tfrac{10}{7})\, gL \sin \theta} \\ &= \sqrt{(\tfrac{10}{7})(9.8 \text{ m/s}^2)(2.1 \text{ m})(\sin 34°)} \\ &= 4.1 \text{ m/s}. \quad \text{(Answer)} \end{aligned}$$

Note that the answer does not depend on the mass or radius of the ball.

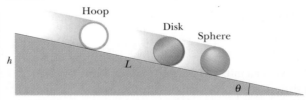

FIGURE 12-7 Sample Problems 12-2 and 12-3. A hoop, a disk, and a sphere roll from rest down a ramp of angle θ. Although released from rest at the same position and time, they arrive at the bottom in the order shown.

SAMPLE PROBLEM 12-3

Here we generalize the result of Sample Problem 12-2. A uniform hoop, disk, and sphere, having the same mass M and the same radius R, are released simultaneously from rest at the top of a ramp whose length L is 2.5 m and whose ramp angle θ is 12° (Fig. 12-7).

(a) Which object reaches the bottom first?

SOLUTION: Table 12-1 gives us the answer. The sphere puts the largest share of its kinetic energy (71%) into translational motion, so it wins the race. Next comes the disk and then the hoop.

(b) How fast are the objects moving at the ramp's bottom?

SOLUTION: The center of mass of each object rolling down the ramp falls the same vertical distance h. Like a body in free fall, the object loses potential energy in amount Mgh and thus gains this amount of kinetic energy. At the bottom of the ramp then, the total kinetic energies of all three objects are the same. How these kinetic energies are divided between the translational and rotational forms depends on each object's distribution of mass.

From Eq. 12-5 we can write (putting $\omega = v_{cm}/R$)

$$\begin{aligned} Mgh &= \tfrac{1}{2}I_{cm}\omega^2 + \tfrac{1}{2}Mv_{cm}^2 \\ &= \tfrac{1}{2}I_{cm}(v_{cm}^2/R^2) + \tfrac{1}{2}Mv_{cm}^2 \\ &= \tfrac{1}{2}(I_{cm}/R^2)v_{cm}^2 + \tfrac{1}{2}Mv_{cm}^2. \quad (12\text{-}7) \end{aligned}$$

Putting $h = L \sin \theta$ and solving for v_{cm}, we obtain

$$v_{cm} = \sqrt{\frac{2gL \sin \theta}{1 + I_{cm}/MR^2}}, \qquad \text{(Answer)} \qquad (12\text{-}8)$$

which is the symbolic answer to the question.

Note that the speed depends not on the mass or the radius of the rolling object, but only on the distribution of its mass about its central axis, which enters through the term I_{cm}/MR^2. A marble and a bowling ball will have the same speed at the bottom of the ramp and will thus roll down the ramp in the same time. A bowling ball will beat a disk of any mass or radius, and almost anything that rolls will beat a hoop.

For the rolling hoop (see the hoop listing in Table 12-1) we have $I_{cm}/MR^2 = 1$, so Eq. 12-8 yields

$$v_{cm} = \sqrt{\frac{2gL \sin \theta}{1 + I_{cm}/MR^2}}$$

$$= \sqrt{\frac{(2)(9.8 \text{ m/s}^2)(2.5 \text{ m})(\sin 12°)}{1 + 1}}$$

$$= 2.3 \text{ m/s.} \qquad \text{(Answer)}$$

From a similar calculation, we obtain $v_{cm} = 2.6$ m/s for the disk ($I_{cm}/MR^2 = \frac{1}{2}$) and 2.7 m/s for the sphere ($I_{cm}/MR^2 = \frac{2}{5}$). This supports our prediction of (a) that the win, place, and show sequence in this race will be sphere, disk, and hoop.

SAMPLE PROBLEM 12-4

Figure 12-8 shows a round uniform body of mass M and radius R rolling down a ramp at an angle θ. This time let us analyze the motion directly from Newton's laws, rather than by energy methods as we did in Sample Problem 12-3.

(a) What is the linear acceleration of the rolling body?

SOLUTION: Figure 12-8 also shows the forces that act on the body: its weight $M\mathbf{g}$, a normal force $\mathbf{N}$, and a static frictional force $\mathbf{f}_s$. The weight can be considered to act at the center of mass, which is at the center of this uniform body. The normal force and the frictional force act on the portion of the body in contact with the ramp at point P. The weight and normal force have zero moment arms about an axis through the center of the body. So they cannot cause the body to rotate about that center. Clockwise rotation of the body results from a negative torque due to the frictional force; that force has a moment arm of R about the center of the body.

We now apply the linear form of Newton's second law ($\Sigma F = Ma$) along the ramp, taking the positive direction to be up the ramp. We obtain

$$\Sigma F = f_s - Mg \sin \theta = Ma. \qquad (12\text{-}9)$$

This equation has two unknowns, f_s and a. To obtain another equation in the same two unknowns, we next apply the angular form of Newton's second law ($\Sigma \tau = I\alpha$) about the rotation axis through the center of mass. (Although we derived the

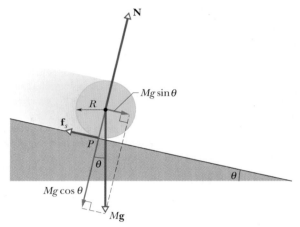

FIGURE 12-8 Sample Problem 12-4. A round uniform body of radius R rolls down a ramp. The forces that act on it are its weight $M\mathbf{g}$, a normal force $\mathbf{N}$, and a frictional force $\mathbf{f}_s$ pointing up the ramp. (For clarity, $\mathbf{N}$ has been shifted along its line of action until its tail is at the center of the body.)

relation $\Sigma \tau = I\alpha$ in Chapter 11 for an axis fixed in an inertial frame, it holds also for a rotation axis through the center of mass of this or another accelerating body, provided the axis does not change direction.) We get

$$\Sigma \tau = -f_s R = I_{cm}\alpha = \frac{I_{cm}a}{R}, \qquad (12\text{-}10)$$

where we have used the relation $\alpha = a/R$ (Eq. 11-20).

Solving Eq. 12-10 for the frictional force f_s gives

$$f_s = -\frac{I_{cm}a}{R^2}, \qquad (12\text{-}11)$$

where the minus sign reminds us that the frictional force $\mathbf{f}_s$ acts in the direction opposite that of the acceleration $\mathbf{a}$. Substituting Eq. 12-11 into Eq. 12-9 and solving for a, we have

$$a = -\frac{g \sin \theta}{1 + I_{cm}/MR^2}. \qquad \text{(Answer)} \qquad (12\text{-}12)$$

We could have, instead, found a second equation by summing torques and applying Newton's law in angular form about an axis through the *point of contact P*. This time, $\Sigma \tau$ would consist only of the torque due to the force component $Mg \sin \theta$, acting at the center of the body with moment arm R:

$$\Sigma \tau = -(Mg \sin \theta)(R) = I_P\alpha = \frac{I_P a}{R}, \qquad (12\text{-}13)$$

where I_P is the rotational inertia about an axis through P. To find I_P, we would use the parallel-axis theorem:

$$I_P = I_{cm} + MR^2. \qquad (12\text{-}14)$$

Substituting I_P from Eq. 12-14 in Eq. 12-13 and solving for a, we would again obtain Eq. 12-12.

(b) What is the frictional force f_s?

SOLUTION: Substituting Eq. 12-12 into Eq. 12-11 yields

$$f_s = Mg \frac{\sin \theta}{1 + MR^2/I_{cm}}. \quad \text{(Answer)} \quad (12\text{-}15)$$

Study of Eq. 12-15 shows that the frictional force is less than $Mg \sin \theta$, the component of the weight that acts parallel to the ramp. This is necessarily true if the object is to accelerate down the ramp.

Table 12-1 shows that if the rolling body is a solid disk, $I_{cm}/MR^2 = \frac{1}{2}$. The acceleration and the frictional force then follow from Eqs. 12-12 and 12-15 and are

$$a = -\tfrac{2}{3}g \sin \theta \quad \text{and} \quad f_s = \tfrac{1}{3}Mg \sin \theta.$$

(c) What is the speed of the rolling body at the bottom of the ramp if the ramp has length L?

SOLUTION: The motion is one of constant acceleration, so we can use the relation

$$v^2 = v_0^2 + 2a(x - x_0). \quad (12\text{-}16)$$

Putting $x - x_0 = -L$ and $v_0 = 0$ and introducing a from Eq. 12-12, we obtain Eq. 12-8—the result that we derived by energy methods.

CHECKPOINT 2: Disks A and B are identical and roll across a floor with equal speeds. Then disk A rolls up an incline, reaching a maximum height h, and disk B moves up an incline that is identical except that it is frictionless. Is the maximum height reached by disk B greater than, less than, or equal to h?

12-2 THE YO-YO

A yo-yo is a physics lab that you can fit in your pocket. If a yo-yo rolls down its string for a distance h, it loses potential energy in amount mgh but gains kinetic energy in both translational ($\frac{1}{2}mv_{cm}^2$) and rotational ($\frac{1}{2}I_{cm}\omega^2$) forms. When it is climbing back up, it loses kinetic energy and regains potential energy.

In a modern yo-yo the string is not tied to the axle but is looped around it. When the yo-yo "hits" the bottom of its string, an upward force on the axle from the string stops the descent, removing the small remaining translational kinetic energy. The yo-yo then spins, axle inside loop, with only rotational kinetic energy. The yo-yo keeps spinning ("sleeping") until you "wake it" by jerking on the string, causing the string to catch on the axle and the yo-yo to climb back up. The rotational kinetic energy of the yo-yo at the bottom of its string (and thus the sleeping time) can be considerably increased by throwing the yo-yo downward so that it starts down the string with initial speeds v_{cm} and ω instead of rolling down from rest.

FIGURE 12-9 (a) A yo-yo, shown in cross section. The string, of assumed negligible thickness, is wound around an axle of radius R_0. (b) A free-body diagram for the falling yo-yo. Only the axle is shown. In a real yo-yo, the thickness of the string cannot be neglected. It changes the effective radius of the yo-yo axle, which then varies with the amount of wound-up string.

Let us analyze the motion of the yo-yo directly with Newton's second law. Figure 12-9a shows an idealized yo-yo, in which the thickness of the string can be neglected. Figure 12-9b is a free-body diagram, in which only the yo-yo axle is shown. We seek an expression for the linear acceleration a of the yo-yo. Applying Newton's second law in its linear form ($\Sigma F = ma$) yields

$$\Sigma F = T - Mg = Ma. \quad (12\text{-}17)$$

Here M is the mass of the yo-yo, and T is the tension in the yo-yo's string.

Applying Newton's second law in angular form ($\Sigma\tau = I\alpha$) about an axis through the center of mass yields

$$\Sigma\tau = TR_0 = I\alpha, \quad (12\text{-}18)$$

where R_0 is the radius of the yo-yo axle and I is the rotational inertia of the yo-yo about its central axis. The linear acceleration a of the yo-yo is downward (and thus *negative*). From the perspective of Fig. 12-9, the angular acceleration α of the yo-yo is counterclockwise (and thus *positive*) because from that perspective the torque given by Eq. 12-18 is counterclockwise. So we relate α to a with the relation $a = -\alpha R_0$. Solving this relation for α ($= -a/R_0$) and substituting into Eq. 12-18, we find

$$TR_0 = -\frac{Ia}{R_0}.$$

After eliminating T between this and Eq. 12-17, we solve for a, obtaining

$$a = -g \frac{1}{1 + I/MR_0^2}. \qquad (12\text{-}19)$$

Thus an ideal yo-yo rolls down its string with constant acceleration. For a small acceleration, you need a light yo-yo with a large rotational inertia and a small axle radius.

SAMPLE PROBLEM 12-5

A yo-yo is constructed of two brass disks whose thickness b is 8.5 mm and whose radius R is 3.5 cm, joined by a short axle whose radius R_0 is 3.2 mm.

(a) What is the yo-yo's rotational inertia about its central axis? Neglect the rotational inertia of the axle. The density ρ of brass is 8400 kg/m³.

SOLUTION: The rotational inertia I of a disk about its central axis is $\frac{1}{2}MR^2$. In this problem, we can treat the two disks together, as a single disk. We first find its mass from its density and its volume V:

$$M = V\rho = (2)(\pi R^2)(b)(\rho)$$
$$= (2)(\pi)(0.035 \text{ m})^2(0.0085 \text{ m})(8400 \text{ kg/m}^3)$$
$$= 0.550 \text{ kg.}$$

The rotational inertia is then

$$I = \tfrac{1}{2}MR^2 = (\tfrac{1}{2})(0.550 \text{ kg})(0.035 \text{ m})^2$$
$$= 3.4 \times 10^{-4} \text{ kg} \cdot \text{m}^2. \qquad \text{(Answer)}$$

(b) A string of length $l = 1.1$ m and of negligible thickness is wound on the axle. What is the linear acceleration of the yo-yo as it rolls down the string from rest?

SOLUTION: From Eq. 12-19,

$$a = -g \frac{1}{1 + I/MR_0^2}$$
$$= -\frac{9.8 \text{ m/s}^2}{1 + \dfrac{3.4 \times 10^{-4} \text{ kg} \cdot \text{m}^2}{(0.550 \text{ kg})(0.0032 \text{ m})^2}}$$
$$= -0.16 \text{ m/s}^2. \qquad \text{(Answer)}$$

The acceleration points downward and has this value whether the yo-yo is rolling down the string or climbing it.

Note that the quantity I/MR_0^2 in Eq. 12-19 is simply the rotational inertia parameter β that was introduced in Table 12-1. For this yo-yo, we have $\beta = 60$, a much greater value than that for any of the objects listed in that table. The acceleration of our yo-yo is small, corresponding to that of a hoop rolling down a 1.9° ramp.

(c) What is the tension in the string of the yo-yo?

SOLUTION: We can find this by substituting a from Eq. 12-19 into Eq. 12-17 and solving for T. We find

$$T = \frac{Mg}{1 + MR_0^2/I}, \qquad (12\text{-}20)$$

which tells us, as it must, that the tension in the string is smaller than the weight of the yo-yo. Numerically, we have

$$T = \frac{(0.550 \text{ kg})(9.8 \text{ m/s}^2)}{1 + (0.550 \text{ kg})(0.0032 \text{ m})^2/(3.4 \times 10^{-4} \text{ kg} \cdot \text{m}^2)}$$
$$= 5.3 \text{ N.} \qquad \text{(Answer)}$$

This is the tension whether the yo-yo is falling down or climbing up the string.

12-3 TORQUE REVISITED

In Chapter 11 we defined torque τ for a rigid body that can rotate around a fixed axis, with each particle in the body forced to move in a path that is a circle about that axis. We now expand the definition of torque to apply it to an individual particle that moves along any path relative to a fixed *point* (rather than a fixed axis). The path need no longer be a circle.

Figure 12-10a shows such a particle at point P in the xy plane. A single force $\mathbf{F}$ in that plane acts on the particle, and the particle's position relative to the origin O is given by position vector $\mathbf{r}$. The torque τ acting on the particle relative to the fixed point O is a vector quantity defined as

$$\tau = \mathbf{r} \times \mathbf{F} \qquad \text{(torque defined).} \qquad (12\text{-}21)$$

We can evaluate the vector (or cross) product in this definition of τ by using the rules for such products given in Section 3-7. To find the direction of τ, we slide the vector $\mathbf{F}$ (without changing its direction) until its tail is at origin O, so that the two vectors in the vector product are tail to tail as in Fig. 12-10b. We then use the right-hand rule for vector products in Fig. 3-20a, sweeping the fingers of the right hand from $\mathbf{r}$ (the first vector in the product) into $\mathbf{F}$ (the second vector). The outstretched right thumb then gives the direction of τ. In Fig. 12-10b, τ points along the positive direction of the z axis.

To find the magnitude of τ, we use the general result of Eq. 3-20 ($c = ab \sin \phi$), finding

$$\tau = rF \sin \phi, \qquad (12\text{-}22)$$

where ϕ is the angle between $\mathbf{r}$ and $\mathbf{F}$ when the vectors are tail to tail. From Fig. 12-10b, we see that Eq. 12-22 can be rewritten as

$$\tau = rF_\perp, \qquad (12\text{-}23)$$

where $F_\perp$ ($= F \sin \phi$) is the component of $\mathbf{F}$ perpendicular to $\mathbf{r}$. From Fig. 12-10c, we see that Eq. 12-22 can also be

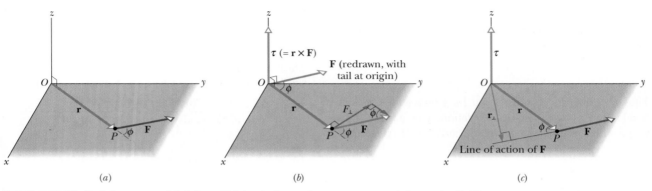

FIGURE 12-10 Defining torque. (*a*) A force **F**, lying in the *xy* plane, acts on a particle at point *P*. (*b*) This force produces a torque τ ($= \mathbf{r} \times \mathbf{F}$) on the particle with respect to the origin *O*. By the right-hand rule for vector (cross) products, the torque vector points in the direction of increasing *z*. Its magnitude is given by $rF_\perp$ in (*b*) and by $r_\perp F$ in (*c*).

rewritten as

$$\tau = r_\perp F, \tag{12-24}$$

where $r_\perp$ ($= r \sin \phi$) is the moment arm of **F** (the perpendicular distance between *O* and the line of action of **F**).

SAMPLE PROBLEM 12-6

In Fig. 12-11*a*, three forces, each of magnitude 2.0 N, act on a particle. The particle is in the *xz* plane at point *P* given by position vector **r**, where $r = 3.0$ m and $\theta = 30°$. Force $\mathbf{F}_1$ is parallel to the *x* axis, force $\mathbf{F}_2$ is parallel to the *z* axis, and force $\mathbf{F}_3$ is parallel to the *y* axis. What is the torque, with respect to the origin *O*, due to each force?

SOLUTION: Figures 12-11*b* and 12-11*c* are direct views of the *xz* plane, redrawn with vectors $\mathbf{F}_1$ and $\mathbf{F}_2$ shifted so their tails are at the origin, to better show the angles between those vectors and vector **r**. The angle between **r** and $\mathbf{F}_3$ is 90°. Applying Eq. 12-22 for each force, we find the magnitudes of the torques to be

$$\tau_1 = rF_1 \sin \phi_1 = (3.0 \text{ m})(2.0 \text{ N})(\sin 150°)$$
$$= 3.0 \text{ N} \cdot \text{m},$$

$$\tau_2 = rF_2 \sin \phi_2 = (3.0 \text{ m})(2.0 \text{ N})(\sin 120°)$$
$$= 5.2 \text{ N} \cdot \text{m},$$

and $$\tau_3 = rF_3 \sin \phi_3 = (3.0 \text{ m})(2.0 \text{ N})(\sin 90°)$$
$$= 6.0 \text{ N} \cdot \text{m}. \tag{Answer}$$

To find the directions of these torques, we use the right-hand rule, placing the fingers of the right hand so as to rotate **r** into **F** through the *smaller* of the two angles they make. In Fig. 12-11*d*, we represent $\mathbf{F}_3$ with a circled cross $\otimes$, suggesting the tail of an arrow. (Were it in the opposite direction, $\mathbf{F}_3$ would be represented by a circled dot $\odot$, suggesting the tip of an arrow.) Rotating **r** into $\mathbf{F}_3$ with the fingers of the right hand yields the direction of τ_3 (in the direction of the thumb). All three of the torque vectors are shown in Fig. 12-11*e*.

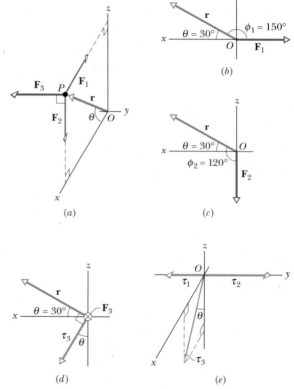

FIGURE 12-11 Sample Problem 12-6. (*a*) A particle at point *P* is acted on by three forces, each parallel to a coordinate axis. The angle ϕ (used in finding torque) is shown (*b*) for $\mathbf{F}_1$ and (*c*) for $\mathbf{F}_2$. (*d*) Torque τ_3 is perpendicular to both **r** and $\mathbf{F}_3$ (force $\mathbf{F}_3$ is directed into the plane of the figure). (*e*) The torques (relative to the origin *O*) acting on the particle.

PROBLEM SOLVING TACTICS

TACTIC 1: *Vector Products and Torques*

Equation 12-21 for torques is our first application of the vector (or cross) product. You might want to review Section 3-7, where the rules for the vector prouct are given. In that section, Problem Solving Tactic 5 lists many common errors in finding the direction of the vector product.

Keep in mind that a torque is calculated *with respect to* (or *about*) a point, which must be known if the value of the torque is to be meaningful. Changing the point can change the torque in both magnitude and direction. For example, in Sample Problem 12-6, the torques due to the three forces are calculated about the origin O. You can show that the torques due to the same three forces are each zero if they are calculated about point P (at the position of the particle).

CHECKPOINT **3:** The position vector **r** of a particle points along the positive direction of the z axis. If the torque on the particle is (a) zero, (b) in the negative direction of x, and (c) in the negative direction of y, in what direction is the force producing the torque?

12-4 ANGULAR MOMENTUM

Like all other linear quantities, linear momentum has its angular counterpart. Figure 12-12 shows a particle with linear momentum **p** ($= m\mathbf{v}$) located at point P in the xy plane. The **angular momentum** ℓ of this particle with respect to the origin O is a vector quantity defined as

$$\ell = \mathbf{r} \times \mathbf{p} = m(\mathbf{r} \times \mathbf{v}) \quad \text{(angular momentum defined)}, \quad (12\text{-}25)$$

where **r** is the position vector of the particle with respect to O. As the particle moves relative to O, in the direction of its momentum **p** ($= m\mathbf{v}$), position vector **r** rotates around O. Note carefully that to have angular momentum about O, the particle does *not* itself have to rotate around O. Comparison of Eqs. 12-21 and 12-25 shows that angular momentum bears the same relation to linear momentum that torque does to force. The SI unit of angular momentum is the kilogram-meter-squared per second (kg·m²/s), equivalent to the joule-second (J·s).

To find the direction of the angular momentum vector ℓ in Fig. 12-12, we slide the vector **p** until its tail is at origin O. Then we use the right-hand rule for vector products, sweeping the fingers from **r** into **p**. The outstretched thumb then shows that vector ℓ points in the positive direction of the z axis in Fig. 12-12. This positive direction is consistent with the counterclockwise rotation of the particle's position vector **r** about the z axis, as the particle continues to move. (A negative direction of ℓ would be consistent with a clockwise rotation of **r** about the z axis.)

To find the magnitude of ℓ, we use the general result of Eq. 3-20, to write

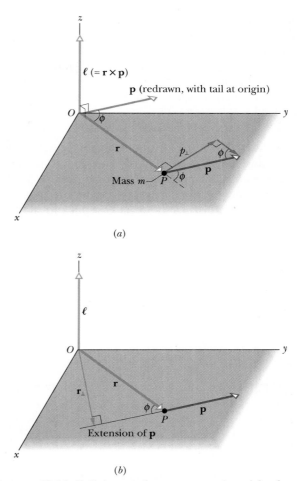

FIGURE 12-12 Defining angular momentum. A particle of mass m at point P has linear momentum **p** ($= m\mathbf{v}$), assumed to lie in the xy plane. The particle has angular momentum ℓ ($= \mathbf{r} \times \mathbf{p}$) with respect to the origin O. By the right-hand rule, the angular momentum vector points in the direction of increasing z. (a) The magnitude of ℓ is given by $\ell = rp_\perp = rmv_\perp$. (b) The magnitude of ℓ is also given by $\ell = r_\perp p = r_\perp mv$.

$$\ell = rmv \sin \phi, \quad (12\text{-}26)$$

where ϕ is the angle between **r** and **p** when these two vectors are tail to tail. From Fig. 12-12a, we see that Eq. 12-26 can be rewritten as

$$\ell = rp_\perp = rmv_\perp, \quad (12\text{-}27)$$

where $p_\perp$ is the component of **p** perpendicular to **r** and $p_\perp = mv_\perp$. From Fig. 12-12b, we see that Eq. 12-26 can also be rewritten as

$$\ell = r_\perp p = r_\perp mv, \quad (12\text{-}28)$$

where $r_\perp$ is the perpendicular distance between O and an extension of **p**.

Just as is true for torque, angular momentum has

meaning only with respect to a specified origin. Moreover, if the particle in Fig. 12-12 did not lie in the xy plane, or if the linear momentum $\mathbf{p}$ of the particle did not also lie in that plane, the angular momentum ℓ would not be parallel to the z axis. The direction of the angular momentum vector is always perpendicular to the plane formed by the vectors $\mathbf{r}$ and $\mathbf{p}$.

CHECKPOINT **4:** In part a of the figure, particles 1 and 2 move around point O in opposite directions, in circles with radii 2 m and 4 m. In part b, particles 3 and 4 travel along straight lines, in the same direction, with their closest distances to point O being 2 m and 4 m. Particle 5 moves directly away from O. All five particles have the same mass and the same constant speed. (a) Rank the particles according to the magnitudes of their angular momentum about point O, greatest first. (b) Which particles have negative angular momentum about point O?

(a)   (b)

SAMPLE PROBLEM 12-7

Figure 12-13 shows an overhead view of two particles moving at constant momentum along horizontal paths. Particle 1, with momentum magnitude $p_1 = 5.0$ kg·m/s, has position vector $\mathbf{r}_1$ and will pass 2.0 m from point O. Particle 2, with momentum magnitude $p_2 = 2.0$ kg·m/s, has position vector $\mathbf{r}_2$ and will pass 4.0 m from point O. What is the net angular momentum $\mathbf{L}$ about point O of the two-particle system?

SOLUTION: To find $\mathbf{L}$, we first must find the individual angular momentum vectors ℓ_1 and ℓ_2 of the two particles. To find the magnitude of ℓ_1, we use Eq. 12-28, substituting $r_{\perp 1} = 2.0$ m and $p_1 = 5.0$ kg·m/s:

$$\ell_1 = r_{\perp 1} p_1 = (2.0 \text{ m})(5.0 \text{ kg·m/s})$$
$$= 10 \text{ kg·m}^2/\text{s}.$$

To find the direction of vector ℓ_1, we use Eq. 12-25 and the right-hand rule for vector products. For $\mathbf{r}_1 \times \mathbf{p}_1$, the vector product is out of the page, perpendicular to the plane of Fig. 12-13. This is a positive direction, consistent with the counterclockwise rotation of the particle's position vector $\mathbf{r}_1$ around O as particle 1 moves. Thus the angular momentum vector for particle 1 is

$$\ell_1 = +10 \text{ kg·m}^2/\text{s}.$$

Similarly, the magnitude of ℓ_2 is

$$\ell_2 = r_{\perp 2} p_2 = (4.0 \text{ m})(2.0 \text{ kg·m/s})$$
$$= 8.0 \text{ kg·m}^2/\text{s},$$

and the vector product $\mathbf{r}_2 \times \mathbf{p}_2$ is into the page, which is a negative direction, consistent with the clockwise rotation of $\mathbf{r}_2$ around O as particle 2 moves. Thus, the angular momentum vector for particle 2 is

$$\ell_2 = -8.0 \text{ kg·m}^2/\text{s}.$$

The net angular momentum for the two-particle system is then

$$L = \ell_1 + \ell_2$$
$$= +10 \text{ kg·m}^2/\text{s} + (-8.0 \text{ kg·m}^2/\text{s})$$
$$= +2.0 \text{ kg·m}^2/\text{s}. \qquad \text{(Answer)}$$

The plus sign means that the system's net angular momentum about point O is out of the page.

FIGURE 12-13   Sample Problem 12-7. Two particles pass near point O.

12-5 NEWTON'S SECOND LAW IN ANGULAR FORM

Newton's second law written in the form

$$\sum \mathbf{F} = \frac{d\mathbf{p}}{dt} \qquad \text{(single particle)} \qquad (12\text{-}29)$$

expresses the close relation between force and linear momentum for a single particle. We have seen enough of the parallelism between linear and angular quantities to be pretty sure that there is also a close relation between torque and angular momentum. Guided by Eq. 12-29, we can even guess that it must be

$$\sum \boldsymbol{\tau} = \frac{d\ell}{dt} \qquad \text{(single particle)} \qquad (12\text{-}30)$$

Equation 12-30 is indeed an angular form of Newton's second law for a single particle:

The (vector) sum of all the torques acting on a particle is equal to the time rate of change of the angular momentum of that particle.

Equation 12-30 has no meaning unless the torques $\boldsymbol{\tau}$ and the angular momentum $\boldsymbol{\ell}$ are defined with respect to the same origin.

Proof of Equation 12-30

We start with Eq. 12-25, the definition of the angular momentum of a particle:

$$\boldsymbol{\ell} = m(\mathbf{r} \times \mathbf{v}),$$

where $\mathbf{r}$ is the position vector of the particle and $\mathbf{v}$ is the velocity of the particle. Differentiating* each side with respect to time t yields

$$\frac{d\boldsymbol{\ell}}{dt} = m\left(\mathbf{r} \times \frac{d\mathbf{v}}{dt} + \frac{d\mathbf{r}}{dt} \times \mathbf{v}\right). \qquad (12\text{-}31)$$

But $d\mathbf{v}/dt$ is the acceleration $\mathbf{a}$ of the particle, and $d\mathbf{r}/dt$ is its velocity $\mathbf{v}$. Thus we can rewrite Eq. 12-31 as

$$\frac{d\boldsymbol{\ell}}{dt} = m(\mathbf{r} \times \mathbf{a} + \mathbf{v} \times \mathbf{v}).$$

Now $\mathbf{v} \times \mathbf{v} = 0$ (the vector product of any vector with itself is zero because the angle between the two vectors is necessarily zero). This leads to

$$\frac{d\boldsymbol{\ell}}{dt} = m(\mathbf{r} \times \mathbf{a}) = \mathbf{r} \times m\mathbf{a}.$$

We now use Newton's second law ($\Sigma\mathbf{F} = m\mathbf{a}$) to replace $m\mathbf{a}$ with its equal, the vector sum of the forces that act on the particle, obtaining

$$\frac{d\boldsymbol{\ell}}{dt} = \mathbf{r} \times \left(\sum \mathbf{F}\right) = \sum(\mathbf{r} \times \mathbf{F}). \qquad (12\text{-}32)$$

Finally, Eq. 12-21 shows us that $\mathbf{r} \times \mathbf{F}$ is the torque associated with the force $\mathbf{F}$, so Eq. 12-32 becomes

$$\sum \boldsymbol{\tau} = \frac{d\boldsymbol{\ell}}{dt}.$$

This is Eq. 12-30, the relation that we set out to prove.

CHECKPOINT 5: The figure shows the position vector $\mathbf{r}$ of a particle at a certain instant, and four choices for the direction of a force that is to accelerate the particle. All four choices are in the xy plane. (a) Rank the choices according to the magnitude of the time rate of change ($d\boldsymbol{\ell}/dt$) they produce in the angular momentum of the particle about point O, greatest first. (b) Which

---

*In differentiating a vector product, be sure not to change the order of the two quantities (here $\mathbf{r}$ and $\mathbf{v}$) that form that product. (See Eq. 3-21.)

choice results in a negative rate of change about O?

SAMPLE PROBLEM 12-8

A penguin of mass m falls from rest at point A, a horizontal distance d from the origin O in Fig. 12-14.

(a) Find an expression for the angular momentum of the falling penguin about O.

SOLUTION: The angular momentum is given by Eq. 12-25 ($\boldsymbol{\ell} = \mathbf{r} \times \mathbf{p}$); its magnitude is (from Eq. 12-26)

$$\ell = rmv \sin \phi.$$

Here, $r \sin \phi = d$ no matter how far the penguin falls, and $v = gt$. Thus ℓ has magnitude

$$\ell = mgtd. \qquad \text{(Answer)} \qquad (12\text{-}33)$$

The right-hand rule shows that the angular momentum vector $\boldsymbol{\ell}$ is directed into the plane of Fig. 12-14, in the direction of decreasing z. We represent $\boldsymbol{\ell}$ with a circled cross $\otimes$ at the origin. The vector $\boldsymbol{\ell}$ changes with time in magnitude only; its direction remains unchanged.

(b) What torque does the weight $m\mathbf{g}$ acting on the penguin exert about the origin O?

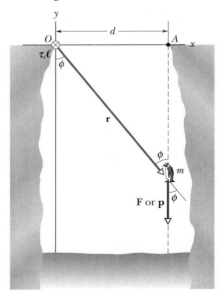

FIGURE 12-14 Sample Problem 12-8. A penguin of mass m falls vertically from point A. The torque $\boldsymbol{\tau}$ and the angular momentum $\boldsymbol{\ell}$ of the falling penguin with respect to the origin O are directed into the plane of the figure at O.

SOLUTION: The torque is given by Eq. 12-21 ($\boldsymbol{\tau} = \mathbf{r} \times \mathbf{F}$); its magnitude is (from Eq. 12-22)

$$\tau = rF \sin \phi.$$

Again $r \sin \phi = d$, and $F = mg$. Thus

$$\tau = mgd = \text{a constant.} \quad \text{(Answer)} \quad (12\text{-}34)$$

Note that the magnitude of the torque is simply the product of the force magnitude mg and the moment arm d. The right-hand rule shows that the torque vector $\boldsymbol{\tau}$ is directed into the plane of Fig. 12-14, in the direction of decreasing z, hence parallel to $\boldsymbol{\ell}$. (Note that we can also derive Eq. 12-34 by differentiating Eq. 12-33 with respect to t and then substituting the result into Eq. 12-30.)

We see that $\boldsymbol{\tau}$ and $\boldsymbol{\ell}$ depend very much (through d) on the location of the origin. If the penguin falls from the origin, we have $d = 0$ and thus no torque or angular momentum.

12-6 THE ANGULAR MOMENTUM OF A SYSTEM OF PARTICLES

Now we turn our attention to the motion of a system of particles with respect to an origin. Note that "a system of particles" includes a rigid body as a special case. The total angular momentum $\mathbf{L}$ of a system of particles is the (vector) sum of the angular momenta $\boldsymbol{\ell}$ of the particles:

$$\mathbf{L} = \boldsymbol{\ell}_1 + \boldsymbol{\ell}_2 + \boldsymbol{\ell}_3 + \cdots + \boldsymbol{\ell}_n = \sum_{i=1}^{n} \boldsymbol{\ell}_i, \quad (12\text{-}35)$$

in which $i (= 1, 2, 3, \ldots)$ labels the particles.

With time, the angular momenta of individual particles may change, either because of interactions within the system (between the individual particles) or because of influences that may act on the system from the outside. We can find the change in $\mathbf{L}$ as these changes take place by taking the time derivative of Eq. 12-35. Thus

$$\frac{d\mathbf{L}}{dt} = \sum_{i=1}^{n} \frac{d\boldsymbol{\ell}_i}{dt}. \quad (12\text{-}36)$$

From Eq. 12-30, $d\boldsymbol{\ell}_i/dt$ is just $\Sigma\boldsymbol{\tau}_i$, the (vector) sum of the torques that act on the ith particle.

Some torques are *internal,* associated with forces that the particles within the system exert on one another; other torques are *external,* associated with forces that act from outside the system. The internal forces, because of Newton's law of action and reaction, cancel in pairs. So, to add the torques, we need consider only those associated with external forces. Equation 12-36 then becomes

$$\sum \boldsymbol{\tau}_{\text{ext}} = \frac{d\mathbf{L}}{dt} \quad \text{(system of particles).} \quad (12\text{-}37)$$

Equation 12-37 is Newton's second law for rotation in angular form, expressed for a system of particles; it is analogous to $\Sigma\mathbf{F}_{\text{ext}} = d\mathbf{P}/dt$ (Eq. 9-28). In words, Eq. 12-37 tells us that the (vector) sum of the *external torques* acting on a system of particles is equal to the time rate of change of the *angular momentum* of that system. Equation 12-37 has meaning only if the torque and angular momentum vectors are referred to the same origin. In an inertial reference frame, Eq. 12-37 can be applied with respect to any point. In an accelerating frame (such as a wheel rolling down a ramp), Eq. 12-37 can be applied *only* with respect to the *center of mass* of the system.

12-7 THE ANGULAR MOMENTUM OF A RIGID BODY ROTATING ABOUT A FIXED AXIS

We next evaluate the angular momentum of a system of particles that form a rigid body which rotates about a fixed axis. Figure 12-15a shows such a body. The fixed axis of rotation is the z axis, and the body rotates about it with constant angular speed ω. We wish to find the angular momentum of the body about the axis of rotation.

We can find the angular momentum by summing the z components of the angular momenta of the mass elements in the body. In Fig. 12-15a, a typical mass element Δm_i of the body moves around the z axis in a circular path. The position of the mass element is located relative to the origin

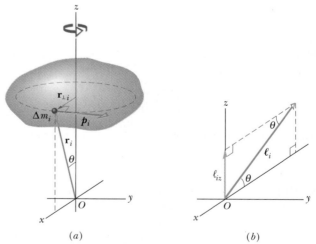

(a)   (b)

FIGURE 12-15 (a) A rigid body rotates about the z axis with angular speed ω. A mass element Δm_i within the body moves about the z axis in a circle with radius $r_{\perp i}$. The mass element has linear momentum $\mathbf{p}_i$, and it is located relative to the origin O by position vector $\mathbf{r}_i$. Here Δm_i is shown when $r_{\perp i}$ is parallel to the x axis. (b) The angular momentum $\boldsymbol{\ell}_i$, with respect to O, of the mass element in (a). The z component ℓ_{iz} of $\boldsymbol{\ell}_i$ is also shown.

O by position vector $\mathbf{r}_i$. The radius of the mass element's circular path is $r_{\perp i}$, the perpendicular distance between the element and the z axis.

The magnitude of the angular momentum ℓ_i of this mass element, with respect to O, is given by Eq. 12-26:

$$\ell_i = (r_i)(p_i)(\sin 90°) = (r_i)(\Delta m_i \, v_i),$$

where p_i and v_i are the linear momentum and linear speed of the mass element, and 90° is the angle between $\mathbf{r}_i$ and $\mathbf{p}_i$. The angular momentum vector ℓ_i for the mass element in Fig. 12-15a is shown in Fig. 12-15b; its direction must be perpendicular to those of $\mathbf{r}_i$ and $\mathbf{p}_i$.

We are interested in the component of ℓ_i that is parallel to the rotation axis, here the z axis. That z component is

$$\ell_{iz} = \ell_i \sin \theta = (r_i \sin \theta)(\Delta m_i \, v_i) = r_{\perp i} \, \Delta m_i \, v_i.$$

The z component of the angular momentum for the rotating rigid body as a whole is found by adding up the contributions of all the mass elements that make up the body. Thus, because $v = \omega r_\perp$, we may write

$$L_z = \sum_{i=1}^{n} \ell_{iz} = \sum_{i=1}^{n} \Delta m_i \, v_i r_{\perp i} = \sum_{i=1}^{n} \Delta m_i (\omega r_{\perp i}) r_{\perp i}$$
$$= \omega \left(\sum_{i=1}^{n} \Delta m_i \, r_{\perp i}^2 \right). \qquad (12\text{-}38)$$

We can remove ω from the summation here because it is a constant: it has the same value for all points of the rotating rigid body.

The quantity $\Sigma \Delta m_i \, r_{\perp i}^2$ in Eq. 12-38 is the rotational inertia I of the body about the fixed axis (see Eq. 11-24). Thus Eq. 12-38 reduces to

$$L = I\omega \qquad \text{(rigid body, fixed axis).} \qquad (12\text{-}39)$$

We have dropped the subscript z, but you must remember that the angular momentum defined by Eq. 12-39 is the angular momentum about the rotation axis. Also, I in that equation is the rotational inertia about that same axis.

Table 12-2, which supplements Table 11-3, extends our list of corresponding linear and angular relations.

SAMPLE PROBLEM 12-9

A tomahawk expert knows how to throw the instrument so that it completes an integer number of full revolutions about its center of mass during its flight, to bury its edge in the target (Fig. 12-16). Suppose that for a flight of horizontal distance $d = 5.90$ m, with a horizontal component of velocity $v_x = 20.0$ m/s, a tomahawk rotates 1.00 rev. Suppose also that the rotational inertia I of the tomahawk about its center of mass is 1.95×10^{-3} kg·m².

(a) What is the magnitude of the tomahawk's angular momentum about the center of mass during the tomahawk's flight?

SOLUTION: The tomahawk rotates at a constant angular speed ω about an axis that is through its center of mass. So, we can calculate the angular momentum of the tomahawk about that axis with Eq. 12-39 ($L = I\omega$) if we first find ω. From Eq. 11-5, we know that ω is related to an angle of rotation $\Delta\theta$ and the time interval Δt_1 for the rotation by

$$\omega = \frac{\Delta\theta}{\Delta t_1}.$$

Replacing Δt_1 with d/v_x and substituting $\Delta\theta = 1.00$ rev $= 2\pi$ rad, we have

$$\omega = \frac{v_x \, \Delta\theta}{d} = \frac{(20.0 \text{ m/s})(2\pi \text{ rad})}{5.90 \text{ m}} = 21.3 \text{ rad/s}.$$

Substituting this and the given value of I in Eq. 12-39, we find

$$L = I\omega = (1.95 \times 10^{-3} \text{ kg·m}^2)(21.3 \text{ rad/s})$$
$$= 4.15 \times 10^{-2} \text{ kg·m}^2/\text{s}. \qquad \text{(Answer)}$$

(b) The launch required a time interval $\Delta t_2 = 0.150$ s. About

---

TABLE 12-2 **MORE CORRESPONDING RELATIONS FOR TRANSLATIONAL AND ROTATIONAL MOTION**[a]

TRANSLATIONAL		ROTATIONAL	
Force	$\mathbf{F}$	Torque	$\boldsymbol{\tau} \, (= \mathbf{r} \times \mathbf{F})$
Linear momentum	$\mathbf{p}$	Angular momentum	$\boldsymbol{\ell} \, (= \mathbf{r} \times \mathbf{p})$
Linear momentum[b]	$\mathbf{P} \, (= \Sigma \mathbf{p}_i)$	Angular momentum[b]	$\mathbf{L} \, (= \Sigma \boldsymbol{\ell}_i)$
Linear momentum[b]	$\mathbf{P} = M\mathbf{v}_{\text{cm}}$	Angular momentum[c]	$L = I\omega$
Newton's second law[b]	$\Sigma \mathbf{F}_{\text{ext}} = \dfrac{d\mathbf{P}}{dt}$	Newton's second law[b]	$\Sigma \boldsymbol{\tau}_{\text{ext}} = \dfrac{d\mathbf{L}}{dt}$
Conservation law[d]	$\mathbf{P} = $ a constant	Conservation law[d]	$\mathbf{L} = $ a constant

[a] See also Table 11-3.   [b] For systems of particles, including rigid bodies.

[c] For a rigid body about a fixed axis, with L being the component along that axis.

[d] For an isolated system.

the tomahawk's center of mass and from the perspective of Fig. 12-16, what was the average torque applied by the expert to the tomahawk during the launch?

SOLUTION: From Eq. 12-37, we can relate the average torque $\bar{\tau}$ to the change in angular momentum ΔL during the launching time interval Δt_2 with

$$\bar{\tau} = \frac{\Delta L}{\Delta t_2} = \frac{L_f - L_i}{\Delta t_2}.$$

Here, the initial and final angular momenta for the launching period are $L_i = 0$ and $L_f = -4.15 \times 10^{-2}$ kg·m²/s (negative because the tomahawk rotates clockwise in Fig. 12-16). Substituting these values and the given value for Δt_2 yields

$$\bar{\tau} = \frac{-4.15 \times 10^{-2} \text{ kg·m}^2/\text{s}}{0.150 \text{ s}} = -0.277 \text{ N·m.} \quad \text{(Answer)}$$

FIGURE 12-16 Sample Problem 12-9. Once thrown, a tomahawk rotates around an axis through its center of mass. (The parabolic path of the center of mass is not shown.)

CHECKPOINT 6: In the figure, a disk, a hoop, and a solid sphere are made to spin about fixed central axes (like a top) by means of strings wrapped around them, with the strings producing the same constant tangential force **F** on all three objects. The three objects have the same mass and radius, and they are initially stationary. Rank the objects according to (a) their angular momentum about their central axes and (b) their angular speed, greatest first, when the strings have been pulled for a certain time t.

Disk   Hoop   Sphere

12-8 CONSERVATION OF ANGULAR MOMENTUM

So far we have discussed two powerful conservation laws, the conservation of energy and the conservation of linear momentum. Now we meet a third law of this type, the conservation of angular momentum. We start from Eq. 12-37 ($\Sigma \tau_{\text{ext}} = d\mathbf{L}/dt$), which is Newton's second law in an-

gular form. If no net external torque acts on the system, this equation becomes $d\mathbf{L}/dt = 0$, or

$$\mathbf{L} = \text{a constant} \quad \text{(isolated system).} \quad (12\text{-}40)$$

This result, called the **law of conservation of angular momentum,** can also be written as

$$\mathbf{L}_i = \mathbf{L}_f \quad \text{(isolated system),} \quad (12\text{-}41)$$

where the subscripts refer to the values of **L** at some initial time i and later time f. Equations 12-40 and 12-41 tell us:

> If the net external torque acting on a system is zero, the angular momentum **L** of the system remains constant, no matter what changes take place within the system.

Equations 12-40 and 12-41 are vector equations; as such, they are equivalent to three scalar equations corresponding to the conservation of angular momentum in three mutually perpendicular directions. Depending on the torques acting on a system, the angular momentum of the system might be conserved in only one or two directions but not in all directions:

> If any component of the net *external* torque on a system is zero, then that component of the angular momentum of the system along that axis cannot change, no matter what changes take place within the system.

We can apply this law to the isolated body in Fig. 12-15, which rotates around the z axis. Suppose that the initially rigid body somehow redistributes its mass relative to that rotation axis, changing its rotational inertia. Equations 12-40 and 12-41 state that the angular momentum of the body cannot change. Substituting Eq. 12-39 (for the angular momentum along the rotational axis) into Eq. 12-41, we write this conservation law as

$$I_i \omega_i = I_f \omega_f. \quad (12\text{-}42)$$

Here the subscripts refer to the values of the rotational inertia I and angular speed ω before and after the redistribution of mass.

Like the other two conservation laws that we have discussed, Eqs. 12-40 and 12-41 hold beyond the limitations of Newtonian mechanics. They hold for particles whose speeds approach that of light (where the theory of relativity reigns), and they remain true in the world of subatomic particles (where quantum mechanics reigns). No exceptions to the law of conservation of angular momentum have ever been found.

We now discuss four examples involving this law.

1. *The spinning volunteer.* Figure 12-17 shows a student seated on a stool that can rotate freely about a vertical axis. The student, who has been set into rotation at a modest initial angular speed ω_i, holds two dumbbells in his outstretched hands. His angular momentum vector **L** lies along the vertical axis, pointing upward.

The instructor now asks the student to pull in his arms; this action reduces his rotational inertia from its initial value I_i to a smaller value I_f, because he moves mass closer to the rotation axis. His rate of rotation increases markedly, from ω_i to ω_f. If the student wishes to slow down, he has only to extend his arms once more.

No net external torque acts on the system consisting of the student, stool, and dumbbells. Thus the angular momentum of that system about the rotation axis must remain constant, no matter how the student maneuvers the weights. In Fig. 12-17a, the student's angular speed ω_i is relatively low and his rotational inertia I_i relatively high. According to Eq. 12-42, his angular speed in Fig. 12-17b must be greater to compensate for the decreased rotational inertia due to mass being closer to the rotational axis.

2. *The springboard diver.* Figure 12-18 shows a diver doing a forward one-and-a-half-somersault dive. As you should expect, her center of mass follows a parabolic path. She leaves the springboard with a definite angular momentum **L** about an axis through her center of mass, represented by a vector pointing into the plane of Fig. 12-18, perpendicular to the page. When she is in the air, the diver

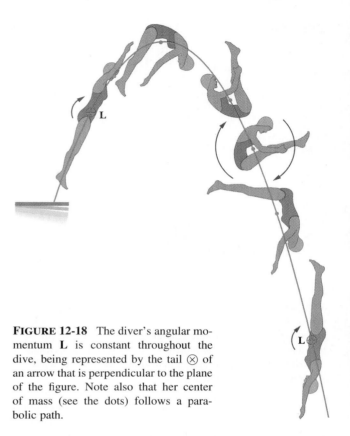

FIGURE 12-18 The diver's angular momentum **L** is constant throughout the dive, being represented by the tail $\otimes$ of an arrow that is perpendicular to the plane of the figure. Note also that her center of mass (see the dots) follows a parabolic path.

forms an isolated system and her angular momentum cannot change. By pulling her arms and legs into the closed *tuck position,* she can considerably reduce her rotational inertia about the same axis and thus, according to Eq. 12-42, considerably increase her angular speed. Pulling out of the tuck position (into the *open layout position*) at the end of the dive increases her rotational inertia and thus slows her rotation rate so she can enter the water with little splash. Even in a more complicated dive involving both twisting and somersaulting, the angular momentum of the diver must be conserved, in both magnitude *and* direction, throughout the dive.

3. *Spacecraft orientation.* Figure 12-19, which represents a spacecraft with a rigidly mounted flywheel, suggests a scheme (albeit crude) for orientation control. The *spacecraft + flywheel* form an isolated system. So, if the total angular momentum **L** of the system is zero because neither spacecraft nor flywheel is turning, it must remain zero (as long as the system remains isolated).

To change the orientation of the spacecraft, the flywheel is started up (Fig. 12-19a). The spacecraft will start to rotate in the opposite sense to maintain the system's angular momentum at zero. When the flywheel is then brought to rest, the spacecraft will also stop rotating but will have changed its orientation (Fig. 12-19b). Throughout, the angular momentum of the system *spacecraft + flywheel* never differs from zero.

Interestingly, the spacecraft *Voyager 2*, on its 1986

Rotation axis

(a)   (b)

FIGURE 12-17 (a) The student has a relatively large rotational inertia and a relatively small angular speed. (b) By decreasing his rotational inertia, the student automatically increases his angular speed. The angular momentum **L** of the rotating system remains unchanged.

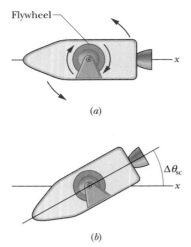

Flywheel

(a)

$\Delta\theta_{sc}$

x

(b)

FIGURE 12-19 (a) An idealized spacecraft containing a flywheel. If the flywheel is made to rotate clockwise as shown, the spacecraft itself will rotate counterclockwise because the total angular momentum must remain zero. (b) When the flywheel is braked to rest, the spacecraft will also stop rotating but will have reoriented its axis by the angle $\Delta\theta_{sc}$.

flyby of the planet Uranus, was set into unwanted rotation by this flywheel effect every time its tape recorder was turned on at high speed. The ground staff at the Jet Propulsion Laboratory had to program the on-board computer to turn on counteracting thruster jets every time the tape recorder was turned on or off.

4. The incredible shrinking star. When the nuclear fire in the core of a star burns low, the star may eventually begin to collapse, building up pressure in its interior. The collapse may go so far as to reduce the radius of the star from something like that of the Sun to the incredibly small value of a few kilometers. The star then becomes a *neutron star*—its material has been compressed to an incredibly dense gas of neutrons.

During this shrinking process, the star is an isolated system and its angular momentum **L** cannot change. Because its rotational inertia is greatly reduced, its angular speed is correspondingly greatly increased, to as much as 600–800 revolutions per *second*. For comparison, the Sun, a typical star, rotates at about one revolution per month.

C**HECKPOINT 7:** A rhinoceros beetle rides the rim of a small disk that rotates like a merry-go-round. If the beetle crawls toward the center of the disk, do the following (each relative to the central axis) increase, decrease, or remain the same: (a) the rotational inertia of the beetle–disk system, (b) the angular momentum of the system, and (c) the angular speed of the beetle and disk?

SAMPLE PROBLEM 12-10

Figure 12-20a shows a student, again sitting on a stool that can rotate freely about a vertical axis. The student, initially at rest, is holding a bicycle wheel whose rim is loaded with lead and whose rotational inertia I about its central axis is 1.2 kg·m². The wheel is rotating at an angular speed ω_i of 3.9 rev/s; as seen from overhead, the rotation is counterclockwise. The axis of the wheel is vertical, and the angular momentum $\mathbf{L}_i$ of the wheel points vertically upward. The student now inverts the wheel (Fig. 12-20b); as a result, the student and stool rotate about the stool axis. With what angular speed and direction does the student then rotate? (The rotational inertia I_0 of the *student + stool + wheel* system about the stool axis is 6.8 kg·m².)

SOLUTION: There is no net torque acting on the *student + stool + wheel* to change the angular momentum of that system about any vertical axis. The initial angular momentum $\mathbf{L}_i$ of the system is that of the bicycle wheel alone. After the wheel has been inverted, the system must still have a net angular momentum of the same magnitude *and* direction.

After the inversion, the angular momentum of the wheel is $-\mathbf{L}_i$. In addition, the *student + stool* must acquire some angular momentum; call it **L**. Then, as shown in Fig. 12-20c, we have

$$L_i = L + (-L_i),$$

or

$$L = 2L_i = I_0\omega,$$

in which ω is the angular speed acquired by the student after the wheel's inversion. This yields

$$\omega = \frac{2L_i}{I_0} = \frac{2I\omega_i}{I_0}$$

(a)   (b)

FIGURE 12-20 Sample Problem 12-10. (a) A student holds a bicycle wheel rotating around the vertical. (b) The student inverts the wheel, setting himself into rotation. (c) The net angular momentum of the system must remain the same in spite of the inversion.

$$\mathbf{L}_i = \mathbf{L} + (-\mathbf{L}_i)$$

Initial   Final

(c)

$$= \frac{(2)(1.2 \text{ kg} \cdot \text{m}^2)(3.9 \text{ rev/s})}{6.8 \text{ kg} \cdot \text{m}^2}$$

$$= 1.4 \text{ rev/s}. \qquad \text{(Answer)}$$

This positive result tells us that the student rotates counterclockwise about the stool axis as seen from overhead. If the student wishes to stop rotating, he has only to invert the wheel once more.

In inverting the wheel, the student will become well aware of the need to apply a torque. However, this torque is internal to the *student + stool + wheel* system and so cannot change the total angular momentum of this system.

We could, however, decide to adopt as our system *student + stool* alone; the wheel then would be external to our system. From this point of view, as the student exerts a torque on the wheel, the wheel exerts a reaction torque on him and this torque is now an external torque. It is the action of this external torque that changes the angular momentum of the *student + stool* system, setting it spinning. Whether a torque is internal or external depends only on how you choose to define your system.

SAMPLE PROBLEM 12-11

During a jump to his partner, an aerialist is to make a triple somersault lasting a time $t = 1.87$ s. For the first and last quarter-revolution, he is in the extended orientation shown in Fig. 12-21, with rotational inertia $I_1 = 19.9 \text{ kg} \cdot \text{m}^2$ around his center of mass. During the rest of the flight he is in a moderate tuck, with rotational inertia $I_2 = 5.50 \text{ kg} \cdot \text{m}^2$.

(a) What must be his initial angular speed ω_1 around his center of mass?

SOLUTION: He rotates in the extended position for two quarter-turns, or a total angle $\theta_1 = 0.500$ rev, in total time t_1; he is in the tuck for an angle θ_2 during a time t_2. These two times are given by

$$t_1 = \frac{\theta_1}{\omega_1}, \qquad \text{and} \qquad t_2 = \frac{\theta_2}{\omega_2}, \qquad (12\text{-}43)$$

where ω_2 is his angular speed in the tuck. We can find an expression for ω_2 by noting that his angular momentum is conserved throughout the flight. Then, from Eq. 12-42,

$$I_2\omega_2 = I_1\omega_1,$$

from which

$$\omega_2 = \frac{I_1}{I_2}\,\omega_1. \qquad (12\text{-}44)$$

His total flight time is

$$t = t_1 + t_2,$$

which, with substitutions from Eqs. 12-43 and 12-44, becomes

$$t = \frac{\theta_1}{\omega_1} + \frac{\theta_2 I_2}{\omega_1 I_1} = \frac{1}{\omega_1}\left(\theta_1 + \theta_2\frac{I_2}{I_1}\right). \qquad (12\text{-}45)$$

Inserting the given data, we obtain

$$1.87 \text{ s} = \frac{1}{\omega_1}\left(0.500 \text{ rev} + 2.50 \text{ rev}\,\frac{5.50 \text{ kg} \cdot \text{m}^2}{19.9 \text{ kg} \cdot \text{m}^2}\right),$$

where, for the triple somersault, he must spend $\theta = 2.5$ revolutions in a tuck. From this equation we find

$$\omega_1 = 0.6369 \text{ rev/s} \approx 0.637 \text{ rev/s}. \qquad \text{(Answer)}$$

(b) If he now attempts a quadruple somersault, with the same ω_1 and t, by using a tighter tuck, what must his rotational inertia I_2 be during the tuck?

SOLUTION: The angle of rotation θ_2 during the tuck is now 3.50 rev, and Eq. 12-45 becomes

$$1.87 \text{ s} = \frac{1}{0.6369 \text{ rev/s}}$$
$$\times \left(0.500 \text{ rev} + 3.50 \text{ rev}\,\frac{I_2}{19.9 \text{ kg} \cdot \text{m}^2}\right),$$

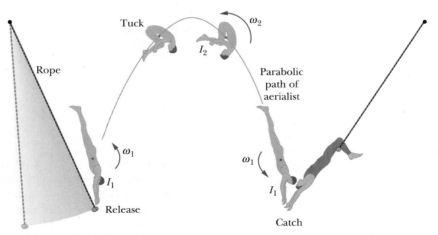

Tuck ω_2

I_2

Rope

Parabolic path of aerialist

ω_1

I_1

ω_1

I_1

Release

Catch

FIGURE 12-21 Sample Problem 12-11. The triple somersault.

from which we find

$$I_2 = 3.929 \text{ kg} \cdot \text{m}^2 \approx 3.93 \text{ kg} \cdot \text{m}^2. \quad \text{(Answer)}$$

This smaller value for I_2 allows faster turning in the tuck and is about the smallest value possible for an aerialist. To make a four-and-a-half somersault, an aerialist would have to increase either the time of flight or the initial angular speed, but either change would make the catch by his partner more difficult. (Can you see why?)

(c) For the quadruple somersault, what is the rotation period T during the tuck?

SOLUTION: We first find the angular speed ω_2 during the tuck from Eq. 12-44:

$$\omega_2 = \frac{I_1}{I_2}\omega_1 = \frac{19.9 \text{ kg} \cdot \text{m}^2}{3.929 \text{ kg} \cdot \text{m}^2} \, 0.6369 \text{ rev/s}$$

$$= 3.226 \text{ rev/s}.$$

Then we find the time T for one rotation from

$$T = \frac{1 \text{ rev}}{\omega_2} = \frac{1 \text{ rev}}{3.226 \text{ rev/s}} = 0.310 \text{ s}. \quad \text{(Answer)}$$

One reason the quadruple somersault is so difficult is that rotation occurs too quickly for the aerialist to see his surroundings clearly or to "fine-tune" the angular speed by adjusting his rotational inertia during flight.

SAMPLE PROBLEM 12-12

Four thin rods, each with mass M and length $d = 1.0$ m, are rigidly connected in the form of a plus sign; the entire assembly rotates in a horizontal plane around a vertical axle at the center, with initial (clockwise) angular velocity $\omega_i = -2.0$ rad/s (see Fig. 12-22). A mud ball with mass m and initial speed $v_i = 12$ m/s is thrown at, and sticks to, the end of one rod. Let $M = 3m$. What is the final angular velocity ω_f of the *plus sign* + *mud ball* system if the initial path of the mud ball is each of the four paths shown in Fig. 12-22: path 1 (contact is made when the ball's velocity is perpendicular to the rod), path 2 (radial contact), path 3 (perpendicular contact), and path 4 (contact is made at 60° to the perpendicular)?

SOLUTION: The total angular momentum L of the system about the axle is conserved during the collision:

$$L_f = L_i, \quad (12\text{-}46)$$

where the subscripts f and i represent final and initial values. Let I_+ represent the rotational inertia of the plus sign about the axle. From Table 11-2(f), we have, for the four rods,

$$I_+ = 4\left(\frac{Md^2}{3}\right).$$

The rotational inertia of the mud ball about the axle as the ball rotates on the plus sign is $I_{mb} = md^2$. Let ℓ_i represent the

initial (before contact) angular momentum of the mud ball about the axle, and ω_f represent the final angular velocity of the system. Using $L = I\omega$, we may rewrite Eq. 12-46 as

$$I_+\omega_f + I_{mb}\omega_f = I_+\omega_i + \ell_i,$$

and then as

$$(\tfrac{4}{3}Md^2)\omega_f + (md^2)\omega_f = (\tfrac{4}{3}Md^2)\omega_i + \ell_i. \quad (12\text{-}47)$$

Substituting $M = 3m$ and $\omega_i = -2.0$ rad/s and solving for ω_f, we find that

$$\omega_f = \frac{1}{5md^2}\left(4md^2\,(-2 \text{ rad/s}) + \ell_i\right). \quad (12\text{-}48)$$

We evaluate the magnitude of ℓ_i for paths 1 and 3 with Eq. 12-28, where $r_\perp = d$ and $v = v_i$. For path 2, we use Eq. 12-28 with $r_\perp = 0$. For path 4 we use Eq. 12-27, where $r = d$ and $v_\perp = v_i \cos 60°$. To determine the sign of ℓ_i for an approaching mud ball, we draw a position vector from the axle of the plus sign to the mud ball. Then we note how the position vector rotates around the axle as the mud ball continues to move toward the plus sign. If the position vector rotates clockwise, ℓ_i is negative; if it rotates counterclockwise, ℓ_i is positive. The results are:

path 1: $\ell_i = -mdv_i$;   path 2: $\ell_i = 0$;

path 3: $\ell_i = mdv_i$;   path 4: $\ell_i = mdv_i \cos 60°$.

The speed v_i is given as 12 m/s. Substituting these values in turn into Eq. 12-48, we find ω_f to be:

path 1: -4.0 rad/s;   path 2: -1.6 rad/s;

path 3: 0.80 rad/s;   path 4: -0.40 rad/s.   (Answer)

FIGURE 12-22 Sample Problem 12-12. An overhead view of four rigidly connected rods rotating freely around a central axle, and four paths a mud ball can take to stick onto one of the rods.

12-9 QUANTIZED ANGULAR MOMENTUM

A physical quantity is said to be **quantized** if it can have only certain discrete values, and all intermediate values are prohibited. We have met two examples so far, both on a microscopic level: the quantization of mass (Section 2-9) and the quantization of energy (Section 8-9). Angular momentum, on a microscopic level, is a third example.

Fundamental particles such as electrons and protons always have a certain intrinsic angular momentum called **spin angular momentum** S, as if they always spin like a top at a certain rate. (However, they do not spin. Spin angular momentum is a more abstract quantity than the angular momentum of, say, a merry-go-round, which involves actual rotation.) Spin angular momentum S is quantized; its value (along any given direction) is given by

$$S = m_s \frac{h}{2\pi}, \tag{12-49}$$

where h is Planck's constant (6.63×10^{-34} J·s) and m_s is a quantum number. For electrons, protons, positrons, and antiprotons, the values of m_s can be only $+\frac{1}{2}$ and $-\frac{1}{2}$. If $m_s = +\frac{1}{2}$, the particle is said to be *spin up* and it has spin $S = +\frac{1}{2}(h/2\pi)$. If $m_s = -\frac{1}{2}$, the particle is said to be *spin down* and it has spin $S = -\frac{1}{2}(h/2\pi)$.

When fundamental particles collide or transform into other fundamental particles, the angular momentum of the system of those particles must be conserved. For example, it must be conserved when a proton (p) and an antiproton ($\bar{\text{p}}$) collide and *annihilate* each other. The result can be the appearance of four positive pions (π^+) and four negative pions (π^-), which are also fundamental particles:

$$\text{p} + \bar{\text{p}} \rightarrow 4\pi^+ + 4\pi^-.$$

For any pion, $m_s = 0$. So, by Eq. 12-49, for any pion, $S = 0$. Thus, after the p–$\bar{\text{p}}$ annihilation, the system has a total spin angular momentum of zero. By conservation of angular momentum, the total spin angular momentum of the colliding proton and antiproton must also be zero if the annihilation is to occur by the process above. That is, either the proton or the antiproton must be spin up and the other must be spin down.

REVIEW & SUMMARY

Rolling Bodies

For a wheel of radius R that rolls without slipping,

$$v_{cm} = \omega R, \tag{12-2}$$

where v_{cm} is the linear speed of the wheel's center and ω is the angular speed of the wheel about its center. The wheel may also be viewed as rotating instantaneously about the point P of the "road" that is in contact with the wheel. The angular speed of the wheel about this point is the same as the angular speed of the wheel about its center. The rolling wheel has kinetic energy

$$K = \tfrac{1}{2}I_{cm}\omega^2 + \tfrac{1}{2}Mv_{cm}^2, \tag{12-5}$$

where I_{cm} is the rotational moment of the wheel about its center.

Torque as a Vector

In three dimensions, *torque* $\boldsymbol{\tau}$ is a vector quantity defined relative to a fixed point (usually an origin); it is

$$\boldsymbol{\tau} = \mathbf{r} \times \mathbf{F}, \tag{12-21}$$

where $\mathbf{F}$ is a force applied to a particle and $\mathbf{r}$ is a position vector locating the particle relative to the fixed point (or origin). The magnitude of $\boldsymbol{\tau}$ is given by

$$\tau = rF \sin \phi = rF_\perp = r_\perp F, \quad \text{(12-22, 12-23, 12-24)}$$

where ϕ is the angle between $\mathbf{F}$ and $\mathbf{r}$, $F_\perp$ is the component of $\mathbf{F}$ perpendicular to $\mathbf{r}$, and $r_\perp$ is the moment arm of $\mathbf{F}$. The direction of $\boldsymbol{\tau}$ is given by the right-hand rule for cross products.

Angular Momentum of a Particle

The **angular momentum** $\boldsymbol{\ell}$ of a particle with linear momentum $\mathbf{p}$, mass m, and linear velocity $\mathbf{v}$ is a vector quantity defined relative to a fixed point (usually an origin); it is

$$\boldsymbol{\ell} = \mathbf{r} \times \mathbf{p} = m(\mathbf{r} \times \mathbf{v}). \tag{12-25}$$

The magnitude of $\boldsymbol{\ell}$ is given by

$$\ell = rmv \sin \phi \tag{12-26}$$
$$= rp_\perp = rmv_\perp \tag{12-27}$$
$$= r_\perp p = r_\perp mv, \tag{12-28}$$

where ϕ is the angle between $\mathbf{r}$ and $\mathbf{p}$, $p_\perp$ and $v_\perp$ are the components of $\mathbf{p}$ and $\mathbf{v}$ perpendicular to $\mathbf{r}$, and $r_\perp$ is the perpendicular distance between the fixed point and an extension of $\mathbf{p}$. The direction of $\boldsymbol{\ell}$ is given by the right-hand rule.

Newton's Second Law in Angular Form

Newton's second law for a particle can be written in vector angular form as

$$\sum \boldsymbol{\tau} = \frac{d\boldsymbol{\ell}}{dt}, \tag{12-30}$$

where $\sum \boldsymbol{\tau}$ is the net torque acting on the particle, and $\boldsymbol{\ell}$ is the angular momentum of the particle.

Angular Momentum of a System of Particles

The angular momentum $\mathbf{L}$ of a system of particles is the vector sum of the angular momenta of the individual particles:

$$\mathbf{L} = \boldsymbol{\ell}_1 + \boldsymbol{\ell}_2 + \cdots + \boldsymbol{\ell}_n = \sum_{i=1}^{n} \boldsymbol{\ell}_i. \tag{12-35}$$

The time rate of change of this angular momentum is equal to the sum of the external torques on the system (the torques due to interactions of the particles of the system with particles external to the system):

$$\sum \boldsymbol{\tau}_{ext} = \frac{d\mathbf{L}}{dt} \quad \text{(system of particles).} \tag{12-37}$$

Angular Momentum of a Rigid Body

For a rigid body rotating about a fixed axis, the component of its angular momentum parallel to the rotation axis is

$$L = I\omega \qquad \text{(rigid body, fixed axis)}. \qquad (12\text{-}39)$$

Conservation of Angular Momentum

The angular momentum **L** of a system remains constant if the net external torque acting on the system is zero:

$$\mathbf{L} = \text{a constant} \qquad \text{(isolated system)} \qquad (12\text{-}40)$$

or

$$\mathbf{L}_i = \mathbf{L}_f \qquad \text{(isolated system)}. \qquad (12\text{-}41)$$

This is the **law of conservation of angular momentum.** It is one of the fundamental conservation laws of nature, having been veri-fied even in situations (involving high-speed particles or sub-atomic dimensions) in which Newton's laws are not applicable.

Quantized Angular Momentum

The spin angular momentum S of fundamental particles is quantized; its value (along any given direction) is

$$S = m_s \frac{h}{2\pi}, \qquad (12\text{-}49)$$

where h is Planck's constant (6.63×10^{-34} J·s) and m_s is a quantum number. *Spin up* and *spin down* correspond to $m_s = +\frac{1}{2}$ and $m_s = -\frac{1}{2}$, respectively. The angular momentum of a system of fundamental particles is conserved during a collision or a transformation.

QUESTIONS

1. In Fig. 12-23, a block slides down a frictionless ramp and a sphere rolls without sliding down a ramp of the same angle θ. The block and sphere have the same mass, start from rest at point A, and descend through point B. (a) In that descent, is the work done by the block's weight on the block greater than, less than, or the same as the work done by the sphere's weight on the sphere? At B, which object has more (b) translational kinetic energy and (c) speed down the ramp?

FIGURE 12-23 Question 1.

2. A cannonball and a marble roll from rest down an incline without sliding. (a) Does the cannonball take more, less, or the same time as the marble to reach the bottom? (b) Is the fraction of the cannonball's kinetic energy that is associated with translation more than, less than, or the same as that of the marble?

3. A cannonball rolls down an incline without sliding. If the roll is now repeated with an incline that is less steep but of the same height as the first incline, are (a) the ball's time to reach the bottom and (b) its translational kinetic energy at the bottom greater than, less than, or the same as previously?

4. A solid brass cylinder and a solid wood cylinder have the same radius and mass (the wood cylinder is longer). Released together from rest, they roll down an incline. (a) Which cylinder reaches the bottom first, or do they tie? (b) The wood cylinder is then shortened to match the length of the brass cylinder, and the brass cylinder is drilled out along its long axis to match the mass of the wood cylinder. Which cylinder now wins the race?

5. In Fig. 12-24, a woman rolls a cylindrical drum, by means of a board on top, through the distance $L/2$, which is half the board's length. The drum rolls smoothly, and the board does not slide over the drum. (a) What length of board has rolled over the top of the drum? (b) How far has the woman walked?

FIGURE 12-24 Question 5.

6. The position vector **r** of a particle relative to a certain point has a magnitude of 3 m, and the force **F** on the particle has a magnitude of 4 N. What is the angle between the directions of **r** and **F** if the magnitude of the associated torque equals (a) zero and (b) 12 N·m?

7. Figure 12-25 shows a particle moving at constant velocity **v** and five points with their xy coordinates. Rank the points according to the magnitude of the angular momentum of the particle measured about them, greatest first.

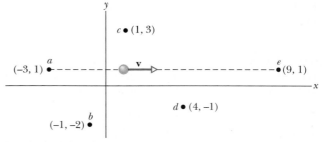

FIGURE 12-25 Question 7.

8. (a) In Checkpoint 4, what is the torque on particles 1 and 2 about point O due to the centripetal forces that cause those particles to circle at constant speed? (b) As particles 3, 4, and 5 move from the left to the right of point O, do their individual angular momenta increase, decrease, or stay the same?

9. Figure 12-26 shows three particles of the same mass and the same constant speed moving as indicated by the velocity vectors. Points a, b, c, and d form a square, with point e at the center. Rank the points according to the magnitude of the net angular momentum of the three-particle system when measured about the points, greatest first.

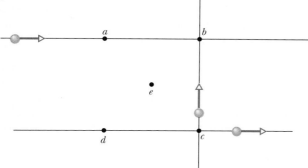

FIGURE 12-26   Question 9.

10. The following gives the angular momentum $\ell(t)$ of a particle in four situations: (1) $\ell = 3t + 4$; (2) $\ell = -6t^2$; (3) $\ell = 2$; (4) $\ell = 4/t$. In which situation is the net torque on the particle (a) zero, (b) positive and constant, (c) negative and increasing in magnitude ($t > 0$), and (d) negative and decreasing in magnitude ($t > 0$)?

11. A bola, which consists of three heavy balls connected to a common point by identical lengths of sturdy string, is readied for launch by holding one of the balls overhead and rotating the wrist, causing the other two balls to rotate in a horizontal circle about the hand. The bola is then released, and its configuration rapidly changes from that shown in the overhead view of Fig. 12-27a to that of Fig. 12-27b. During that change do its (a) angular momentum about its center and (b) angular speed increase, decrease, or stay the same?

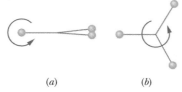

(a)   (b)

FIGURE 12-27   Question 11.

12. A rhinoceros beetle rides the rim of a disk rotating like a merry-go-round counterclockwise. If it then walks along the rim in the direction of the rotation, will the following increase, decrease, or remain the same: (a) the angular momentum of the beetle–disk system, (b) the angular momentum and angular velocity of the beetle, and (c) the angular momentum and angular velocity of the disk? (d) What are your answers if the beetle walks in the direction opposite the rotation?

13. In Fig. 12-28, a disk is spinning freely at the bottom of an axle and with an angular momentum (about its center) of 50 units, counterclockwise. Four more spinning disks are to be dropped down the axle to land on and couple (via friction) to the first disk. Their angular momenta are: (1) 20 units clockwise, (2) 10 units counterclockwise, (3) 10 units clockwise, and (4) 60 units clockwise. (a) What is the final angular momentum of the system? (b) In what order should the five disks be added so that at one stage the disks that are then coupled on the axle are stationary?

FIGURE 12-28   Question 13.

14. Figure 12-29 shows an overhead view of a rectangular slab that can spin like a merry-go-round about its center at O. Also shown are seven paths along which wads of bubble gum can be thrown (all with the same speed and mass) to stick onto the stationary slab. (a) Rank the paths according to the angular speed that the slab (and gum) will have after the gum sticks, greatest first. (b) For which paths will the angular momentum of the slab (and gum) about O be negative?

FIGURE 12-29   Question 14.

15. What happens to the initially stationary yo-yo in Fig. 12-30 if you pull it via its string with (a) force $\mathbf{F}_2$ (the line of action passes through the point of contact on the table, as indicated), (b) force $\mathbf{F}_1$ (the line of action passes above the point of contact), and (c) force $\mathbf{F}_3$ (the line of action passes to the right of the point of contact)?

FIGURE 12-30   Question 15.

EXERCISES & PROBLEMS

SECTION 12-1 Rolling

1E. A thin-walled pipe rolls along the floor. What is the ratio of its translational kinetic energy to its rotational kinetic energy about an axis parallel to its length and through its center of mass?

2E. A 140 kg hoop rolls along a horizontal floor so that its center of mass has a speed of 0.150 m/s. How much work must be done on the hoop to stop it?

3E. An automobile traveling 80.0 km/h has tires of 75.0 cm diameter. (a) What is the angular speed of the tires about the axle? (b) If the car is brought to a stop uniformly in 30.0 turns of the tires (without skidding), what is the angular acceleration of the wheels? (c) How far does the car move during the braking?

4E. A 1000 kg car has four 10 kg wheels. When the car is moving, what fraction of the total kinetic energy of the car is due to rotation of the wheels about their axles? Assume that the wheels have the same rotational inertia as uniform disks of the same mass and size. Why do you not need the radius of the wheels?

5E. A wheel of radius 0.250 m, which is moving initially at 43.0 m/s, rolls to a stop in 225 m. Calculate (a) its linear acceleration and (b) its angular acceleration. (c) The wheel's rotational inertia is 0.155 kg · m² about its central axis. Calculate the torque exerted by the friction on the wheel, about the central axis.

6E. An automobile has a total mass of 1700 kg. It accelerates from rest to 40 km/h in 10 s. Assume each wheel is a uniform 32 kg disk. Find, for the end of the 10 s interval, (a) the rotational kinetic energy of each wheel about its axle, (b) the total kinetic energy of each wheel, and (c) the total kinetic energy of the automobile.

7E. A uniform sphere rolls down an incline. (a) What must be the incline angle if the linear acceleration of the center of the sphere is to be $0.10g$? (b) For this angle, what would be the acceleration of a frictionless block sliding down the incline?

8E. A solid sphere of weight 8.00 lb rolls up an incline with an inclination angle of 30.0°. At the bottom of the incline the center of mass of the sphere has a translational speed of 16.0 ft/s. (a) What is the kinetic energy of the sphere at the bottom of the incline? (b) How far does the sphere travel up the incline? (c) Does the answer to (b) depend on the weight of the sphere?

9P. A constant horizontal force of 10 N is applied to a wheel of mass 10 kg and radius 0.30 m as shown in Fig. 12-31. The wheel

FIGURE 12-31 Problem 9.

rolls without slipping on the horizontal surface, and the acceleration of its center of mass is 0.60 m/s². (a) What are the magnitude and direction of the frictional force on the wheel? (b) What is the rotational inertia of the wheel about an axis through its center of mass and perpendicular to the plane of the wheel?

10P. Consider a 66 cm diameter tire on a car traveling at 80 km/h on a level road in the direction of increasing x. As seen by a passenger in the car, what are the linear velocity and the magnitude of the linear acceleration of (a) the center of the wheel, (b) a point at the top of the tire, and (c) a point at the bottom of the tire? (d) Repeat (a) to (c), in the same order, for a stationary observer alongside the road.

11P. A body of radius R and mass m is rolling smoothly with speed v on a horizontal surface. It then rolls up a hill to a maximum height h. (a) If $h = 3v^2/4g$, what is the body's rotational inertia about the rotational axis through its center of mass? (b) What might the body be?

12P. A homogeneous sphere starts from rest at the upper end of the track shown in Fig. 12-32 and rolls without slipping until it rolls off the right-hand end. If $H = 6.0$ m and $h = 2.0$ m and the track is horizontal at the right-hand end, how far horizontally from point A does the sphere land on the floor?

FIGURE 12-32 Problem 12.

13P. A small sphere, with radius r and mass m, rolls without slipping on the inside of a large fixed hemisphere with radius R and a vertical axis of symmetry. It starts at the top from rest. (a) What is its kinetic energy at the bottom? (b) What fraction of its kinetic energy at the bottom is associated with rotation about an axis through its center of mass? (c) What normal force does the small sphere exert on the hemisphere at the bottom if $r \ll R$?

14P. A solid cylinder of radius 10 cm and mass 12 kg starts from rest and rolls without slipping a distance of 6.0 m down a house roof that is inclined at 30°. (See Fig. 12-33.) (a) What is the angular speed of the cylinder about its center as it leaves the house roof? (b) The outside wall of the house is 5.0 m high. How far from the edge of the roof does the cylinder hit the level ground?

FIGURE 12-33   Problem 14.

15P. A small solid marble of mass m and radius r will roll without slipping along the loop-the-loop track shown in Fig. 12-34 if it is released from rest somewhere on the straight section of track. (a) From what minimum height h above the bottom of the track must the marble be released to ensure that it does not leave the track at the top of the loop? (The radius of the loop-the-loop is R; assume $R \gg r$.) (b) If the marble is released from height $6R$ above the bottom of the track, what is the horizontal component of the force acting on it at point Q?

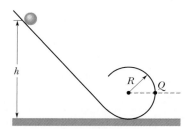

FIGURE 12-34   Problem 15.

16P. A bowler throws a bowling ball of radius $R = 11$ cm down a lane. The ball slides on the lane, with initial speed $v_{cm,0} = 8.5$ m/s and initial angular speed $\omega_0 = 0$. The coefficient of kinetic friction between the ball and the lane is 0.21. The kinetic frictional force $\mathbf{f}_k$ acting on the ball (Fig. 12-35) causes a linear acceleration of the ball while producing a torque that causes an angular acceleration of the ball. When speed v_{cm} has decreased enough and angular speed ω has increased enough, the ball stops sliding and then rolls smoothly. (a) What then is v_{cm} in terms of ω? During the sliding, what are the ball's (b) linear acceleration and (c) angular acceleration? (d) How long does the ball slide? (e) How far does the ball slide? (f) What is the speed of the ball when smooth rolling begins?

FIGURE 12-35   Problem 16.

SECTION 12-2   The Yo-Yo

17E. A yo-yo has a rotational inertia of 950 g·cm² and a mass of 120 g. Its axle radius is 3.2 mm, and its string is 120 cm long. The yo-yo rolls from rest down to the end of the string. (a) What is its linear acceleration? (b) How long does it take to reach the end of the string? As it reaches the end of the string, what are its (c) linear speed, (d) translational kinetic energy, (e) rotational kinetic energy, and (f) angular speed?

18P. Suppose that the yo-yo in Exercise 17, instead of rolling from rest, is thrown so that its initial speed down the string is 1.3 m/s. (a) How long does the yo-yo take to reach the end of the string? As it reaches the end of the string, what are its (b) total kinetic energy, (c) linear speed, (d) translational kinetic energy, (e) angular speed, and (f) rotational kinetic energy?

SECTION 12-3   Torque Revisited

19E. Given that $\mathbf{r} = x\mathbf{i} + y\mathbf{j} + z\mathbf{k}$ and $\mathbf{F} = F_x\mathbf{i} + F_y\mathbf{j} + F_z\mathbf{k}$, show that the torque $\boldsymbol{\tau} = \mathbf{r} \times \mathbf{F}$ is given by

$$\boldsymbol{\tau} = (yF_z - zF_y)\mathbf{i} + (zF_x - xF_z)\mathbf{j} + (xF_y - yF_x)\mathbf{k}.$$

20E. Show that, if $\mathbf{r}$ and $\mathbf{F}$ lie in a given plane, the torque $\boldsymbol{\tau} = \mathbf{r} \times \mathbf{F}$ has no component in that plane.

21E. What are the magnitude and direction of the torque about the origin on a plum located at coordinates $(-2.0$ m, 0, 4.0 m$)$ due to force $\mathbf{F}$ whose only component is (a) $F_x = 6.0$ N, (b) $F_x = -6.0$ N, (c) $F_z = 6.0$ N, and (d) $F_z = -6.0$ N?

22E. What are the magnitude and direction of the torque about the origin on a particle located at coordinates $(0, -4.0$ m, 3.0 m$)$ due to (a) force $\mathbf{F}_1$ with components $F_{1x} = 2.0$ N and $F_{1y} = F_{1z} = 0$, and (b) force $\mathbf{F}_2$ with components $F_{2x} = 0$, $F_{2y} = 2.0$ N, and $F_{2z} = 4.0$ N?

23P. Force $\mathbf{F} = (2.0$ N$)\mathbf{i} - (3.0$ N$)\mathbf{k}$ acts on a pebble with position vector $\mathbf{r} = (0.50$ m$)\mathbf{j} - (2.0$ m$)\mathbf{k}$, relative to the origin. What is the resulting torque acting on the pebble about (a) the origin and (b) a point with coordinates (2.0 m, 0, -3.0 m)?

24P. What is the torque about the origin on a jar of jalapeño peppers located at coordinates (3.0 m, -2.0 m, 4.0 m) due to (a) force $\mathbf{F}_1 = (3.0$ N$)\mathbf{i} - (4.0$ N$)\mathbf{j} + (5.0$ N$)\mathbf{k}$, (b) force $\mathbf{F}_2 = (-3.0$ N$)\mathbf{i} - (4.0$ N$)\mathbf{j} - (5.0$ N$)\mathbf{k}$, and (c) the vector sum of $\mathbf{F}_1$ and $\mathbf{F}_2$? (d) Repeat (c) for a point with coordinates (3.0 m, 2.0 m, 4.0 m) instead of the origin.

25P. What is the net torque about the origin on a flea located at coordinates $(0, -4.0$ m, 5.0 m$)$ when forces $\mathbf{F}_1 = (3.0$ N$)\mathbf{k}$ and $\mathbf{F}_2 = (-2.0$ N$)\mathbf{j}$ act on the flea?

26P. Force $\mathbf{F} = (-8.0$ N$)\mathbf{i} + (6.0$ N$)\mathbf{j}$ acts on a particle with position vector $\mathbf{r} = (3.0$ m$)\mathbf{i} + (4.0$ m$)\mathbf{j}$. What are (a) the torque on the particle about the origin and (b) the angle between the directions of $\mathbf{r}$ and $\mathbf{F}$?

SECTION 12-4   Angular Momentum

27E. Two objects are moving as shown in Fig. 12-36. What is their total angular momentum about point O?

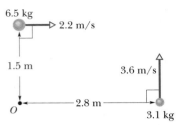

FIGURE 12-36   Exercise 27.

28E. A 1200 kg airplane is flying in a straight line at 80 m/s, 1.3 km above the ground. What is the magnitude of its angular momentum with respect to a point on the ground directly under the path of the plane?

29E. A particle P with mass 2.0 kg has position vector $\mathbf{r}$ (r = 3.0 m) and velocity $\mathbf{v}$ (v = 4.0 m/s) as shown in Fig. 12-37. It is

FIGURE 12-37
Exercise 29.

acted on by force $\mathbf{F}$ (F = 2.0 N). All three vectors lie in the xy plane. About the origin, what are (a) the angular momentum of the particle and (b) the torque acting on the particle?

30E. If we are given r, p, and ϕ, we can calculate the angular momentum of a particle from Eq. 12-26. Sometimes, however, we are given the components (x, y, z) of $\mathbf{r}$ and (v_x, v_y, v_z) of $\mathbf{v}$ instead. (a) Show that the components of $\boldsymbol{\ell}$ along the x, y, and z axes are then given by $\ell_x = m(yv_z - zv_y)$, $\ell_y = m(zv_x - xv_z)$, and $\ell_z = m(xv_y - yv_x)$. (b) Show that if the particle moves only in the xy plane, the angular momentum vector has only a z component.

31E. At a certain time, the position vector in meters of a 0.25 kg object is $\mathbf{r} = 2.0\mathbf{i} - 2.0\mathbf{k}$. At that instant, its velocity in meters per second is $\mathbf{v} = -5.0\mathbf{i} + 5.0\mathbf{k}$, and the force in newtons acting on it is $\mathbf{F} = 4.0\mathbf{j}$. (a) What is the angular momentum of the object about the origin? (b) What torque acts on it? (*Hint:* See Exercises 19 and 30.)

32P. What is the magnitude of the angular momentum, about the Earth's center, of an 84 kg person on the equator due to the rotation of the Earth?

33P. Two particles, each of mass m and speed v, travel in opposite directions along parallel lines separated by a distance d. (a) In terms of m, v, and d, find an expression for the magnitude L of the angular momentum of the two-particle system around a point

midway between the two lines. (b) Does the expression change if we change the point about which L is calculated? (c) Now reverse the direction of travel for one of the particles and repeat (a) and (b).

34P. A 2.0 kg object moves in a plane with velocity components v_x = 30 m/s and v_y = 60 m/s as it passes through the point (x, y) = $(3.0, -4.0)$ m. (a) What is its angular momentum relative to the origin at this moment? (b) What is its angular momentum relative to the point $(-2.0, -2.0)$ m at this same moment?

35P. (a) Use the data given in the appendices to compute the total of the magnitudes of the angular momenta of all the planets owing to their revolution about the Sun. (b) What fraction of this total is associated with the planet Jupiter?

SECTION 12-5   Newton's Second Law in Angular Form

36E. A 3.0 kg particle is at x = 3.0 m, y = 8.0 m with a velocity of $\mathbf{v}$ = $(5.0$ m/s$)\mathbf{i} - (6.0$ m/s$)\mathbf{j}$. It is acted on by a 7.0 N force in the negative x direction. (a) What is the angular momentum of the particle about the origin? (b) What torque about the origin acts on the particle? (c) At what rate is the angular momentum of the particle changing with time?

37E. A particle is acted on by two torques about the origin: $\boldsymbol{\tau}_1$ has a magnitude of 2.0 N·m and points in the direction of increasing x, and $\boldsymbol{\tau}_2$ has a magnitude of 4.0 N·m and points in the direction of decreasing y. What are the magnitude and direction of $d\boldsymbol{\ell}/dt$, where $\boldsymbol{\ell}$ is the angular momentum of the particle about the origin?

38E. What torque about the origin acts on a particle moving in the xy plane if the particle has the following values of angular momentum about the origin:

(a) -4.0 kg·m²/s, (c) $-4.0\sqrt{t}$ kg·m²/s,
(b) $-4.0t^2$ kg·m²/s, (d) $-4.0/t^2$ kg·m²/s?

39E. A 3.0 kg toy car on the x axis has velocity $\mathbf{v}$ = $-2.0t^3$ m/s along that axis. About the origin and for $t > 0$, what are (a) the car's angular momentum and (b) the torque acting on the car? (c) Repeat (a) and (b) for a point with coordinates $(2.0$ m, 5.0 m, $0)$ instead of the origin. (d) Repeat (a) and (b) for a point with coordinates $(2.0$ m, -5.0 m, $0)$ instead of the origin.

40P. At t = 0, a 2.0 kg particle has position vector $\mathbf{r}$ = $(4.0$ m$)\mathbf{i} - (2.0$ m$)\mathbf{j}$ relative to the origin. Its velocity is given by $\mathbf{v}$ = $(-6.0t^4$ m/s$)\mathbf{i} + (3.0$ m/s$)\mathbf{j}$. About the origin and for $t > 0$, what are (a) the particle's angular momentum and (b) the torque acting on the particle? (c) Repeat (a) and (b) for a point with coordinates $(-2.0$ m, -3.0 m, $0)$ instead of the origin.

41P. A projectile of mass m is fired from the ground with an initial speed v_0 and an initial angle θ_0 above the horizontal. (a) Find an expression for the magnitude of its angular momentum about the firing point as a function of time. (b) Find the rate at which the angular momentum changes with time. (c) Evaluate the magnitude of $\mathbf{r} \times \mathbf{F}$ directly and compare the result with (b). Why should the results be identical?

SECTION 12-7 The Angular Momentum of a Rigid Body Rotating About a Fixed Axis

42E. A sanding disk with rotational inertia 1.2×10^{-3} kg·m² is attached to an electric drill whose motor delivers a torque of 16 N·m. Find (a) the angular momentum of the disk about its central axis and (b) the angular speed of the disk 33 ms after the motor is turned on.

43E. The angular momentum of a flywheel having a rotational inertia of 0.140 kg·m² about its axis decreases from 3.00 to 0.800 kg·m²/s in 1.50 s. (a) What is the average torque acting on the flywheel about its central axis during this period? (b) Assuming a uniform angular acceleration, through what angle will the flywheel have turned? (c) How much work was done on the wheel? (d) What is the average power of the flywheel?

44E. Three particles, each of mass m, are fastened to each other and to a rotation axis at O by three massless strings, each with length l as shown in Fig. 12-38. The combination rotates around the rotational axis with angular velocity ω in such a way that the particles remain in a straight line. In terms of m, l, and ω, and relative to point O, what are (a) the rotational inertia of the combination, (b) the angular momentum of the middle particle, and (c) the total angular momentum of the three particles?

FIGURE 12-38 Exercise 44.

45E. A uniform rod rotates in a horizontal plane about a vertical axis through one end. The rod is 6.00 m long, weighs 10.0 N, and rotates at 240 rev/min clockwise when seen from above. Calculate (a) the rotational inertia of the rod about the axis of rotation and (b) the angular momentum of the rod about that axis.

46P. Figure 12-39 shows a rigid structure consisting of a circular hoop, of radius R and mass m, and a square made of four thin bars, each of length R and mass m. The rigid structure rotates at a constant speed about a vertical axis with a period of rotation of 2.5 s. Assuming $R = 0.50$ m and $m = 2.0$ kg, calculate (a) the structure's rotational inertia about the axis of rotation and (b) its angular momentum about that axis.

FIGURE 12-39 Problem 46.

47P. Wheels A and B in Fig. 12-40 are connected by a belt that does not slip. The radius of wheel B is three times the radius of

wheel A. What would be the ratio of the rotational inertias I_A/I_B if both wheels had (a) the same angular momenta about their central axes and (b) the same rotational kinetic energies?

FIGURE 12-40 Problem 47.

48P. An impulsive force $F(t)$ acts for a short time Δt on a rotating rigid body with rotational inertia I. Show that

$$\int \tau \, dt = \overline{F}R \, \Delta t = I(\omega_f - \omega_i),$$

where R is the moment arm of the force, $\overline{F}$ is the average value of the force during the time it acts on the body, and ω_i and ω_f are the angular velocities of the body just before and just after the force acts. (The quantity $\int \tau \, dt = \overline{F}R \, \Delta t$ is called the *angular impulse*, in analogy with $\overline{F} \, \Delta t$, the linear impulse.)

49P*. Two cylinders having radii R_1 and R_2 and rotational inertias I_1 and I_2 about the central axis are supported by axes perpendicular to the plane of Fig. 12-41. The large cylinder is initially

FIGURE 12-41 Problem 49.

rotating with angular velocity ω_0. The small cylinder is moved to the right until it touches the large cylinder and is caused to rotate by the frictional force between the two. Eventually, slipping ceases, and the two cylinders rotate at constant rates in opposite directions. Find the final angular velocity ω_2 of the small cylinder in terms of I_1, I_2, R_1, R_2, and ω_0. (*Hint:* Neither angular momentum nor kinetic energy is conserved. Apply the angular impulse equation of Problem 48.)

SECTION 12-8 Conservation of Angular Momentum

50E. The rotor of an electric motor has rotational inertia $I_m = 2.0 \times 10^{-3}$ kg·m² about its central axis. The motor is used to change the orientation of the space probe in which it is mounted. The motor axis is mounted parallel to the axis of the probe, which has rotational inertia $I_p = 12$ kg·m² about its axis. Calculate the number of revolutions of the rotor required to turn the probe through 30° about its axis.

51E. A man stands on a platform that is rotating (without friction) with an angular speed of 1.2 rev/s; his arms are outstretched and he holds a weight in each hand. The rotational inertia of the system of man, weights, and platform about the central axis is 6.0 kg·m². If by moving the weights the man decreases the rotational inertia of the system to 2.0 kg·m², (a) what is the resulting angular speed of the platform and (b) what is the ratio of the new kinetic energy of the system to the original kinetic energy? (c) What provided the added kinetic energy?

52E. Two disks are mounted on low-friction bearings on the same axle and can be brought together so that they couple and rotate as one unit. (a) The first disk, with rotational inertia 3.3 kg·m² about its central axis, is set spinning at 450 rev/min. The second disk, with rotational inertia 6.6 kg·m² about its central axis, is set spinning at 900 rev/min in the same direction as the first. They then couple together. What is their angular speed after coupling? (b) If instead the second disk is set spinning at 900 rev/min in the direction opposite the first disk's rotation, what is the angular speed after coupling?

53E. A wheel is rotating freely with an angular speed of 800 rev/min on a shaft whose rotational inertia is negligible. A second wheel, initially at rest and with twice the rotational inertia of the first, is suddenly coupled to the same shaft. (a) What is the angular speed of the resultant combination of the shaft and two wheels? (b) What fraction of the original rotational kinetic energy is lost?

54E. The rotational inertia of a collapsing spinning star changes to one-third its initial value. What is the ratio of the new rotational kinetic energy to the initial rotational kinetic energy?

55E. Suppose that the Sun runs out of nuclear fuel and suddenly collapses to form a white dwarf star, with a diameter equal to that of the Earth. Assuming no mass loss, what would then be the Sun's new rotation period, which currently is about 25 days? Assume that the Sun and the white dwarf are uniform, solid spheres.

56E. In a playground, there is a small merry-go-round of radius 1.20 m and mass 180 kg. Its radius of gyration (see Problem 58 of Chapter 11) is 91.0 cm. A child of mass 44.0 kg runs at a speed of 3.00 m/s along a path that is tangent to the rim of the initially stationary merry-go-round and then jumps on. Neglect friction between the bearings and the shaft of the merry-go-round. Calculate (a) the rotational inertia of the merry-go-round about its axis of rotation, (b) the angular momentum of the child, while running, about the axis of rotation of the merry-go-round, and (c) the angular speed of the merry-go-round and child after the child has jumped on.

57E. A horizontal platform in the shape of a circular disk rotates on a frictionless bearing about a vertical axle through the center of the disk. The platform has a mass of 150 kg, a radius of 2.0 m, and a rotational inertia of 300 kg·m² about the axis of rotation. A 60 kg student walks slowly from the rim of the platform toward the center. If the angular speed of the system is 1.5 rad/s when the student starts at the rim, what is the angular speed when she is 0.50 m from the center?

58E. With center and spokes of negligible mass, a certain bicycle wheel has a thin rim of radius 1.14 ft and weight 8.36 lb; it can turn on its axle with negligible friction. A man holds the wheel above his head with the axle vertical while he stands on a turntable free to rotate without friction; the wheel rotates clockwise, as seen from above, with an angular speed of 57.7 rad/s, and the

turntable is initially at rest. The rotational inertia of *wheel + man + turntable* about the common axis of rotation is 1.54 slug·ft². The man's free hand suddenly stops the rotation of the wheel (relative to the turntable). Determine the resulting angular velocity (magnitude and direction) of the system.

59P. Two skaters, each of mass 50 kg, approach each other along parallel paths separated by 3.0 m. They have equal and opposite velocities of 1.4 m/s. The first skater carries one end of a long pole with negligible mass, and the second skater grabs the other end of it as she passes; see Fig. 12-42. Assume frictionless ice. (a) Describe quantitatively the motion of the skaters after they have become connected by the pole. (b) By pulling on the pole, the skaters reduce their separation to 1.0 m. What is their angular speed then? (c) Calculate the kinetic energy of the system in (a) and (b). (d) What is the source of the added kinetic energy?

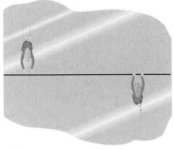

FIGURE 12-42 Problem 59.

60P. Two children, each with mass M, sit on opposite ends of a narrow board with length L and mass M (the same as the mass of each child). The board is pivoted at its center and is free to rotate in a horizontal circle without friction. (Treat it as a thin rod.) (a) What is the rotational inertia of the board plus the children about a vertical axis through the center of the board? (b) What is the angular momentum of the system if it is rotating with angular speed ω_0 in a clockwise direction as seen from above? What is the direction of the angular momentum? (c) While the system is rotating, the children pull themselves toward the center of the board until they are half as far from the center as before. What is the resulting angular speed in terms of ω_0? (d) What is the change in kinetic energy of the system as a result of the children changing their positions? (What is the source of the added kinetic energy?)

61P. A toy train track is mounted on a large wheel that is free to turn with negligible friction about a vertical axis (Fig. 12-43). A toy train of mass m is placed on the track and, with the system initially at rest, the electrical power is turned on. The train reaches a steady speed v with respect to the track. What is the angular

FIGURE 12-43 Problem 61.

velocity ω of the wheel, if its mass is M and its radius R? (Treat the wheel as a hoop, and neglect the mass of the spokes and hub.)

62P. A cockroach of mass m runs counterclockwise around the rim of a lazy Susan (a circular dish mounted on a vertical axle) of radius R and rotational inertia I and having frictionless bearings. The cockroach's speed (relative to the Earth) is v, whereas the lazy Susan turns clockwise with angular speed ω_0. The cockroach finds a bread crumb on the rim and, of course, stops. (a) What is the angular speed of the lazy Susan after the cockroach stops? (b) Is mechanical energy conserved?

63P. A girl of mass M stands on the rim of a frictionless merry-go-round of radius R and rotational inertia I that is not moving. She throws a rock of mass m horizontally in a direction that is tangent to the outer edge of the merry-go-round. The speed of the rock, relative to the ground, is v. Afterwards what are (a) the angular speed of the merry-go-round and (b) the linear speed of the girl?

64P. A phonograph record of mass 0.10 kg and radius 0.10 m rotates about a vertical axis through its center with an angular speed of 4.7 rad/s. The rotational inertia of the record about its axis of rotation is 5.0×10^{-4} kg·m². A wad of putty of mass 0.020 kg drops vertically onto the record from above and sticks to the edge of the record. What is the angular speed of the record immediately after the putty sticks to it?

65P. A uniform thin rod of length 0.50 m and mass 4.0 kg can rotate in a horizontal plane about a vertical axis through its center. The rod is at rest when a 3.0-g bullet traveling in the horizontal plane of the rod is fired into one end of the rod. As viewed from above, the direction of the bullet's velocity makes an angle of 60° with the rod (Fig. 12-44). If the bullet lodges in the rod and the angular velocity of the rod is 10 rad/s immediately after the collision, what is the magnitude of the bullet's velocity just before impact?

FIGURE 12-44 Problem 65.

66P. A cockroach of mass m lies on the rim of a uniform disk of mass $10.0m$ that can rotate freely about its center like a merry-go-round. Initially the cockroach and disk rotate together with an angular velocity of ω_0. Then the cockroach walks half way to the center of the disk. (a) What is the change $\Delta\omega$ in the angular velocity of the cockroach-disk system? (b) What is the ratio K/K_0 of the new kinetic energy of the system to its initial kinetic energy? (c) What accounts for the change in the kinetic energy?

67P. A uniform disk of mass $10m$ and radius $3.0r$ can rotate freely about its fixed center like a merry-go-round. A smaller uniform disk of mass m and radius r lies on top of the larger disk, concentric with it. Initially the two disks rotate together with an angular velocity of 20 rad/s. Then a slight disturbance causes the smaller disk to slide outward across the larger disk, until the outer edge of the smaller disk catches on the outer edge of the larger

disk. Afterwards, the two disks again rotate together (without further sliding). (a) What then is their angular velocity about the center of the larger disk? (b) What is the ratio K/K_0 of the new kinetic energy of the two-disk system to the system's initial kinetic energy?

68P. If the Earth's polar ice caps melted and the water returned to the oceans, the oceans would be made deeper by about 30 m. What effect would this have on the Earth's rotation? Make an estimate of the resulting change in the length of the day. (Concern has been expressed that warming of the atmosphere resulting from industrial pollution could cause the ice caps to melt.)

69P*. The particle of mass m in Fig. 12-45 slides down the frictionless surface and collides with the uniform vertical rod, sticking to it. The rod pivots about O through the angle θ before momentarily coming to rest. Find θ in terms of the other parameters given in the figure.

FIGURE 12-45 Problem 69.

70P*. Two 2.00 kg balls are attached to the ends of a thin rod of negligible mass, 50.0 cm long. The rod is free to rotate in a vertical plane without friction about a horizontal axis through its center. While the rod is horizontal (Fig. 12-46), a 50.0 g putty wad drops onto one of the balls with a speed of 3.00 m/s and sticks to it. (a) What is the angular speed of the system just after the putty wad hits? (b) What is the ratio of the kinetic energy of the entire system after the collision to that of the putty wad just before? (c) Through what angle will the system rotate until it momentarily stops?

FIGURE 12-46 Problem 70.

SECTION 12-9 Quantized Angular Momentum

71E. What is the value of spin angular momentum S (along any given direction) for an electron?

72E. A positive pion (π^+) can spontaneously decay to (suddenly transform to) a positive muon (μ^+) and a neutrino (ν):

$$\pi^+ \rightarrow \mu^+ + \nu.$$

If the neutrino has $m_s = +\frac{1}{2}$, what are the values of (a) m_s and (b) S for the positive muon?

73E. Two protons colliding at high speeds can transform to three protons and one antiproton:

$$p + p \rightarrow p + p + p + \bar{p}.$$

If the two colliding protons are both spin up, what are the spin orientations (up or down) of the transformation products?

ELECTRONIC COMPUTATION

74. When a bowling ball is bowled, it usually skids for a short distance, then rolls without slipping. While it is skidding, a force of kinetic friction acts to slow the translational motion and increase the angular speed. Skidding stops when $\omega R = v$, where ω is the angular speed, R is the radius, and v is the speed of the center of mass. The net force acting on the ball has magnitude $f = \mu_k N$, where μ_k is the coefficient of kinetic friction and N is the magnitude of the normal force. Since the alley is horizontal, $N = mg$, where m is the mass of the ball. The net torque on the ball has magnitude $\tau = fR$, where R is the radius of the ball. A bowling ball has a mass of 7.25 kg, a radius of 10.9 cm, and a rotational inertia that is nearly the same as that of a uniform sphere. Set up a program to compute the angular speed and translational speed every 0.1 s from the time the ball is bowled to the time it stops skidding. Take the coefficient of kinetic friction to be 0.35 and the initial translational speed to be 20 m/s. For each of the initial conditions given below, plot a graph of the speed as a function of time. Find the time when the ball stops skidding and also the linear speed of the ball at that time. Except for a slight decrease due to air drag, this is the speed with which the ball hits the pins. (a) The ball is not rotating initially. (b) The ball is released with an angular speed of 150 rad/s, spinning in the same direction as when it stops skidding. (c) The ball is released with

the same angular speed but in the "wrong" direction. (d) How should a bowling ball be released to have the greatest linear speed as it hits the pins?

75. A projectile with a mass of 0.15 kg is fired with an initial speed of 50 m/s at an angle of 35° above the horizontal. (a) Neglect air resistance and use a numerical integration program to calculate its angular momentum about the firing point every 0.1 s from the time it is fired until it again reaches the height from which it was fired. Use $\tau = \mathbf{r} \times \mathbf{F}$, where $\mathbf{r}$ is the position vector of the projectile and $\mathbf{F}$ is the gravitational force on the projectile, to calculate the torque τ on the projectile (about the firing point) at each of those times. Plot the magnitude of the angular momentum as a function of the square of the time and verify that the graph is a straight line. Plot the magnitude of the torque as a function of time and verify that the graph is a straight line. Are these graphs consistent with Newton's second law for rotation, $\tau = dL/dt$? (b) Now include air resistance. Take the terminal speed to be $v_t = 75$ m/s. The acceleration can be written $\mathbf{a} = -g\mathbf{j} - bv\mathbf{v}$, where $b = g/v_t^2$. Does the magnitude of the angular momentum increase in proportion to the square of the time? Is the magnitude of the torque proportional to the time? At any position of the projectile, is the magnitude of the torque greater or less than it is in the absence of drag? Give an argument based on the directions of the forces acting on the projectile to show that your numerical result is plausible.

76. At the instant the displacement of a 2.00 kg object relative to the origin is $\mathbf{d} = (2.00 \text{ m})\mathbf{i} + (4.00 \text{ m})\mathbf{j} - (3.00 \text{ m})\mathbf{k}$, its velocity is $\mathbf{v} = -(6.00 \text{ m/s})\mathbf{i} + (3.00 \text{ m/s})\mathbf{j} + (3.00 \text{ m/s})\mathbf{k}$, and it is subject to a force $\mathbf{F} = (6.00 \text{ N})\mathbf{i} - (8.00 \text{ N})\mathbf{j} + (4.0 \text{ N})\mathbf{k}$. Find (a) the acceleration of the object, (b) the angular momentum of the object about the origin, (c) the torque about the origin acting on the object, and (d) the angle between the velocity of the object and the force acting on the object.

13
Equilibrium and Elasticity

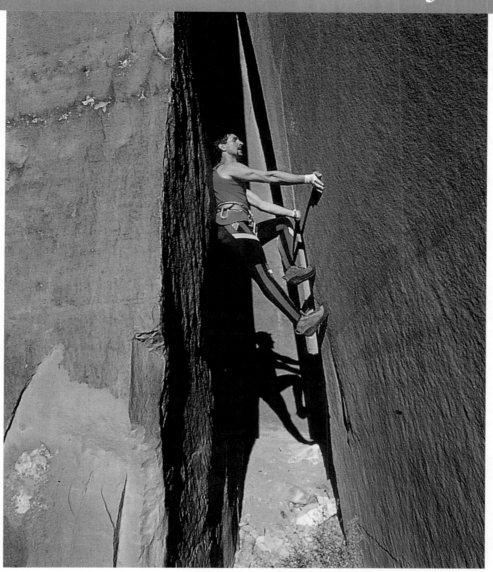

Rock climbing may be the ultimate physics exam. Failure can mean death, and even "partial credit" can mean severe injury. For example, in a long chimney climb, in which your shoulders are pressed against one wall of a wide, vertical fissure and your feet are pressed against the opposite wall, you need to rest occasionally or you will fall due to exhaustion. Here the exam consists of a single question: What can you do to relax your push on the walls in order to rest? If you relax without considering the physics, the walls will not hold you up. So, what is the answer to this life-and-death, one-question exam?

13-1 EQUILIBRIUM

Consider these objects: (1) a book resting on a table, (2) a hockey puck sliding across a frictionless surface with constant velocity, (3) the rotating blades of a ceiling fan, and (4) the wheel of a bicycle that is traveling along a straight path at constant speed. For each of these four objects:

1. The linear momentum **P** of its center of mass is constant.

2. Its angular momentum **L** about its center of mass, or about any other point, is also constant.

We say that such objects are in **equilibrium.** The two requirements for equilibrium are then

$$\mathbf{P} = \text{a constant} \quad \text{and} \quad \mathbf{L} = \text{a constant.} \quad (13\text{-}1)$$

Our concern in this chapter is with situations in which the constants in Eq. 13-1 are in fact zero. That is, we are concerned largely with objects that are not moving in any way—either in translation or in rotation—in the reference frame from which we observe them. Such objects are in **static equilibrium.** Of the four objects mentioned at the beginning of this section, only one—the book resting on the table—is in static equilibrium.

The balancing rock of Fig. 13-1 is another example of an object that, for the present at least, is in static equilib-

FIGURE 13-1 A balancing rock near Petrified Forest National Park in Arizona. Although its perch seems precarious, the rock is in static equilibrium.

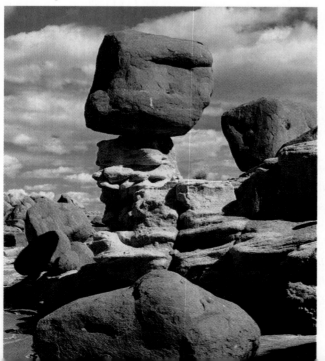

rium. It shares this property with countless other structures, such as cathedrals, houses, filing cabinets, and taco stands, that remain stationary over time.

As we discussed in Section 8-5, if a body returns to a state of static equilibrium after having been displaced from it by a force, the body is said to be in *stable* static equilibrium. A marble placed at the bottom of a hemispherical bowl is an example. If, however, a small force can displace the body and end the equilibrium, the body is in *unstable* static equilibrium.

For example, suppose we balance a domino as in Fig. 13-2a, with the domino's center of mass vertically above a supporting edge. The torque about the supporting edge due to the domino's weight **W** is zero, because the line of action of **W** is through that edge. Since the weight then cannot cause the domino to rotate about the edge, the domino is in equilibrium. Of course, even a slight force on it due to some chance disturbance ends the equilibrium: as the weight's line of action moves to one side of the supporting edge (as in Fig. 13-2b), the torque due to the weight increases the domino's rotation. Thus the domino in Fig. 13-2a is in unstable static equilibrium.

The domino in Fig. 13-2c is not quite as unstable. To topple this domino, a force would have to rotate it through and beyond the balance position of Fig. 13-2a, in which the center of mass is above a supporting edge. So, a slight force will not topple this domino, but a vigorous flick of the finger against the domino certainly will. (If we arrange a chain of such upright dominos, a finger flick against the first can cause the whole chain to fall.)

(a)   (b)   (c)   (d)

FIGURE 13-2 (a) A domino balanced on one edge, with its center of mass vertically above that edge. The domino's weight **W** is directed through the supporting edge. (b) If the domino is rotated even slightly from the balanced orientation, its weight **W** causes a torque that increases the rotation. (c) A domino upright on a narrow side is somewhat more stable than the domino in (a). (d) A square block is even more stable.

FIGURE 13-3 A construction worker balanced above New York City is in static equilibrium but is more stable parallel to the beam than perpendicular to it.

The child's square block in Fig. 13-2*d* is even more stable because its center of mass would have to be moved even farther to get it to pass above a supporting edge. A flick of the finger may not topple the block. (This is why you never see a chain of toppling square blocks.) The worker in Fig. 13-3 is like both the domino and the square block: parallel to the beam, his stance is wide and he is stable; perpendicular to the beam, his stance is narrow and he is unstable (and at the mercy of a chance gust of wind).

The analysis of static equilibrium is very important in engineering practice. The design engineer must isolate and identify all the external forces and torques that may act on a structure and, by good design and wise choice of materials, make sure that the structure will tolerate these loads. Such analysis is necessary to make sure, for example, that bridges do not collapse under their traffic and wind loads, and that the landing gear of aircraft will survive the shock of rough landings.

13-2 THE REQUIREMENTS OF EQUILIBRIUM

The translational motion of a body is governed by Newton's second law in its linear momentum form, given by Eq. 9-28 as

$$\sum \mathbf{F}_{\text{ext}} = \frac{d\mathbf{P}}{dt}. \qquad (13\text{-}2)$$

If the body is in translational equilibrium, that is, if **P** is a constant, then $d\mathbf{P}/dt = 0$ and we must have

$$\sum \mathbf{F}_{\text{ext}} = 0 \qquad \text{(balance of forces)}. \quad (13\text{-}3)$$

The rotational motion of a body is governed by Newton's second law in its angular momentum form, given by Eq. 12-37 as

$$\sum \boldsymbol{\tau}_{\text{ext}} = \frac{d\mathbf{L}}{dt}. \qquad (13\text{-}4)$$

If the body is in rotational equilibrium, that is, if **L** is a constant, then $d\mathbf{L}/dt = 0$ and we must have

$$\sum \boldsymbol{\tau}_{\text{ext}} = 0 \qquad \text{(balance of torques)}. \quad (13\text{-}5)$$

Thus the two requirements for a body to be in equilibrium are as follows:

---

1. The vector sum of all the external forces that act on the body must be zero.

2. The vector sum of all the external torques that act on the body, measured about *any* possible point, must also be zero.

These requirements obviously hold for *static* equilibrium as well as the more general equilibrium in which **P** and **L** are constant but not zero.

Equations 13-3 and 13-5, as vector equations, are each equivalent to three independent scalar equations, one for each direction of the coordinate axes:

Balance of forces	Balance of torques	
$\sum F_x = 0$	$\sum \tau_x = 0$	
$\sum F_y = 0$	$\sum \tau_y = 0$	(13-6)
$\sum F_z = 0$	$\sum \tau_z = 0$	

For convenience, we have dropped the subscript ext.

We shall simplify matters by considering only situations in which the forces that act on the body lie in the *xy* plane. This means that the only torques that can act on the body must tend to cause rotation around an axis parallel to the *z* axis. With this assumption, we eliminate one force equation and two torque equations from Eq. 13-6, leaving

$$\sum F_x = 0 \quad \text{(balance of forces),} \quad (13\text{-}7)$$

$$\sum F_y = 0 \quad \text{(balance of forces),} \quad (13\text{-}8)$$

$$\sum \tau_z = 0 \quad \text{(balance of torques).} \quad (13\text{-}9)$$

Here, F_x and F_y are, respectively, the x and the y components of the external forces that act on the body, and τ_z represents the torques that these forces exert either about the z axis or about *any* axis parallel to it.

A hockey puck sliding at constant velocity over ice satisfies Eqs. 13-7, 13-8, and 13-9 and is thus in equilibrium *but not in static equilibrium*. For static equilibrium, the linear momentum **P** of the puck must be not only constant but also zero; the puck must be resting on the ice. Thus there is another requirement for static equilibrium:

---

3. The linear momentum **P** of the body must be zero.

CHECKPOINT 1: The figure gives six overhead views of a uniform rod on which two or more forces act perpendicularly to the rod. If the magnitudes of the forces are adjusted properly (but kept nonzero), in which situations can the rod be in static equilibrium?

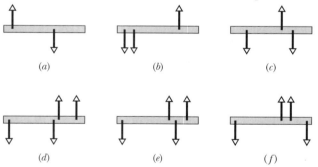

13-3 THE CENTER OF GRAVITY

The weight **W** of an extended body is the vector sum of the gravitational forces acting on the individual elements (the atoms) of the body. Instead of considering all those individual elements, we can say that a single force **W** effectively acts at a single point called the **center of gravity** (cg) of the body. By *effectively* we mean that if the forces on the individual elements were somehow turned off and force **W** at the center of gravity turned on, the net force and the net torque (about any point) acting on the body would not change.

Michel Menin walks a taut wire 10,335 ft above French farmland, adjusting the position of a heavy pole to keep the center of mass of the Menin–pole system over the wire in spite of wind gusts.

Up until now, we have assumed that the weight **W** acts at the center of mass (cm) of the body, which is equivalent to assuming that the center of gravity is at the center of mass. Here we show that this assumption is valid *provided* the acceleration **g** due to the gravitational force is constant over the body.

Figure 13-4a shows an extended body, of mass M, and one of its elements, of mass m_i. Each such element has weight $m_i\mathbf{g}_i$, where $\mathbf{g}_i$ is the acceleration due to the gravitational force at the location of the element. In Fig. 13-4a, each weight $m_i\mathbf{g}_i$ produces a torque τ_i on the element about the origin O. Using Eq. 11-32 ($\tau = r_\perp F$), we can write torque τ_i as

$$\tau_i = x_i m_i g_i, \quad (13\text{-}10)$$

FIGURE 13-4 (a) An element of mass m_i in an extended body has weight $m_i\mathbf{g}_i$, with moment arm x_i about the origin O of a coordinate system. (b) The weight **W** of a body is said to act at the center of gravity (cg) of the body. The moment arm of **W** about O is equal to x_{cg}.

where x_i is the moment arm $r_\perp$ of force $m_i\mathbf{g}_i$. The net torque on all the elements of the body is then

$$\tau_{\text{net}} = \sum \tau_i = \sum x_i m_i g_i. \qquad (13\text{-}11)$$

Figure 13-4b shows the body's weight $\mathbf{W}$ acting at the body's center of gravity. From Eq. 11-32, the torque about O due to $\mathbf{W}$ is

$$\tau = x_{\text{cg}}W, \qquad (13\text{-}12)$$

where x_{cg} is the moment arm of $\mathbf{W}$. The body's weight $\mathbf{W}$ is equal to the sum of the weights $m_i\mathbf{g}_i$ of its elements. So we can substitute $\sum m_i g_i$ for W in Eq. 13-12 to write

$$\tau = x_{\text{cg}} \sum m_i g_i. \qquad (13\text{-}13)$$

By definition, the torque due to weight $\mathbf{W}$ acting at the center of gravity is equal to the net torque due to the weights of all the individual elements of the body. So, τ in Eq. 13-13 is equal to τ_{net} in Eq. 13-11, and we can write

$$x_{\text{cg}} \sum m_i g_i = \sum x_i m_i g_i \qquad (13\text{-}14)$$

If the accelerations g_i are all equal, we can cancel them from the two sides of Eq. 13-14. Substituting the body's mass M for $\sum m_i$ on the left side of Eq. 13-14 and then rearranging yield

$$x_{\text{cg}} = \frac{1}{M} \sum x_i m_i. \qquad (13\text{-}15)$$

With Eq. 9-4 we see that the right side of Eq. 13-15 gives the coordinate x_{cm} for the center of mass. We can now write

$$x_{\text{cg}} = x_{\text{cm}}. \qquad (13\text{-}16)$$

Thus, the body's center of gravity and its center of mass have the same x coordinate.

We can generalize this result for three dimensions by using vector notation. The generalized result is: If the gravitational acceleration is constant over a body, the body's center of gravity is at its center of mass. For the rest of this book, we shall assume that these points coincide.

From Eq. 13-12, we see that the torque due to the weight of a body is zero only if the moment arm x_{cg} is zero. This means that if the body is suspended from some arbitrary point S about which it can rotate, it rotates (due to the torque $\tau = x_{\text{cg}}W$ about S) until x_{cg} is zero. Its center of gravity is then vertically below the suspension point, as in Figs. 13-5a and b, and the body is in equilibrium. If the body is suspended at its center of gravity, as in Fig. 13-5c, then for any orientation of the body, x_{cg} is zero and the body is in equilibrium.

13-4 SOME EXAMPLES OF STATIC EQUILIBRIUM

In this section we examine six sample problems involving static equilibrium. In each, we select a system of one or more objects to which we apply the equations of equilibrium (Eqs. 13-7, 13-8, and 13-9). The forces involved in the equilibrium are all in the xy plane, which means that the torques involved are parallel to the z axis. Thus, in applying Eq. 13-9, the balance of torques, we select an axis parallel to the z axis about which to calculate the torques. Although Eq. 13-9 is satisfied for *any* such choice of axis, we shall see that certain choices simplify the application of Eq. 13-9 by eliminating one or more unknown force terms.

SAMPLE PROBLEM 13-1

A uniform beam of length L whose mass m is 1.8 kg rests with its ends on two digital scales, as in Fig. 13-6a. A uniform block whose mass M is 2.7 kg rests on the beam, its center a distance $L/4$ from the beam's left end. What do the scales read?

We choose as our system the beam and the block, taken together. Figure 13-6b is a free-body diagram for this system, showing all the forces that act on it. The scales push upward at the ends of the beam with forces $\mathbf{F}_l$ and $\mathbf{F}_r$. The magnitudes of these two forces are the scale readings that we seek. The weight of the beam, $m\mathbf{g}$, acts downward at the beam's center of mass. Similarly, $M\mathbf{g}$, the weight of the block, acts downward at *its* center of mass. In Fig. 13-6b, the block is represented by a dot within the boundary of the beam, and the vector $M\mathbf{g}$ is drawn with its tail on that dot. (In drawing Fig. 13-6b from 13-6a, the vector $M\mathbf{g}$ is shifted in the direction it points, that is, along its line of action. The shift does not alter $M\mathbf{g}$, or a torque due to $M\mathbf{g}$ about any axis perpendicular to the figure.)

Our system is in static equilibrium so that the balance of forces equations (Eqs. 13-7 and 13-8) and the balance of torques equation (Eq. 13-9) apply. We solve this problem in two equivalent ways.

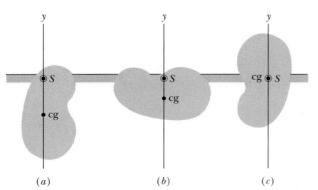

(a) $\qquad\qquad$ (b) $\qquad\qquad$ (c)

FIGURE 13-5 A body free to rotate about a suspension point S will rotate until its center of gravity is vertically below S as in (a) and (b), unless the center of gravity is at S as in (c).

FIRST SOLUTION: The forces have no x components, so Eq. 13-7, which is $\Sigma F_x = 0$, provides no information. Equation 13-8 gives, for the magnitudes of the y components,

$$\sum F_y = F_l + F_r - Mg - mg = 0. \qquad (13\text{-}17)$$

We have two unknown forces (F_l and F_r), but we cannot find them separately because we have only one equation. Fortunately, we have another equation at hand, namely, Eq. 13-9, the balance of torques equation.

We can apply Eq. 13-9 to *any* axis perpendicular to the plane of Fig. 13-6. Let us choose an axis through the left end of the beam. We take as positive those torques that—acting alone—would produce a counterclockwise rotation of the beam about our axis. We then have, from Eq. 13-9,

$$\sum \tau_z = (F_l)(0) - (Mg)(L/4) - (mg)(L/2) + (F_r)(L)$$
$$= 0,$$

or

$$F_r = (g/4)(2m + M)$$
$$= (\tfrac{1}{4})(9.8 \text{ m/s}^2)(2 \times 1.8 \text{ kg} + 2.7 \text{ kg})$$
$$= 15 \text{ N}. \qquad \text{(Answer)} \quad (13\text{-}18)$$

Note how choosing an axis that passes through the application point of one of the unknown forces, F_l, eliminates that force

from Eq. 13-9 and allows us to solve directly for the other force. *Such a choice can help simplify a problem.*

If we solve Eq. 13-17 for F_l and substitute known quantities, we find

$$F_l = (M + m)g - F_r$$
$$= (2.7 \text{ kg} + 1.8 \text{ kg})(9.8 \text{ m/s}^2) - 15 \text{ N}$$
$$= 29 \text{ N}. \qquad \text{(Answer)}$$

SECOND SOLUTION: As a check, let us solve this problem in a different way, by applying the balance of torques equation about two different axes. Choosing first an axis through the left end of the beam, as we did above, we find Eq. 13-18 and the solution $F_r = 15$ N.

For an axis passing through the right end of the beam, Eq. 13-9 yields

$$\sum \tau_z = -(F_l)(L) + (Mg)(3L/4) + (mg)(L/2) + (F_r)(0)$$
$$= 0.$$

Solving for F_l, we find

$$F_l = (g/4)(2m + 3M)$$
$$= (\tfrac{1}{4})(9.8 \text{ m/s}^2)(2 \times 1.8 \text{ kg} + 3 \times 2.7 \text{ kg})$$
$$= 29 \text{ N}, \qquad \text{(Answer)}$$

in agreement with our earlier result. Note that the length of the beam enters this problem not explicitly but only as it affects the mass of the beam.

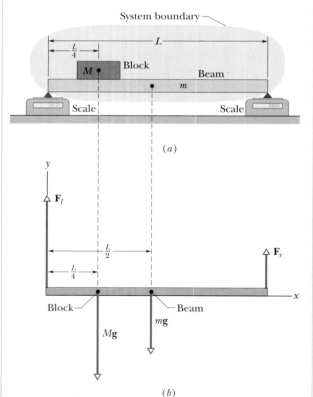

(a)

(b)

FIGURE 13-6 Sample Problem 13-1. (*a*) A beam of mass m supports a block of mass M. The system boundary is marked. (*b*) A free-body diagram, showing the forces that act on the system *beam + block*.

CHECKPOINT **2:** The figure gives an overhead view of a uniform rod in static equilibrium. (a) Can you find the magnitudes of unknown forces $\mathbf{F}_1$ and $\mathbf{F}_2$ by balancing the forces? (b) If you wish to find the magnitude of force $\mathbf{F}_2$ by using a single equation, where should you place a rotational axis? (c) The magnitude of $\mathbf{F}_2$ turns out to be 65 N. What then is the magnitude of $\mathbf{F}_1$?

SAMPLE PROBLEM 13-2

A bowler holds a bowling ball whose mass M is 7.2 kg in the palm of his hand. As Fig. 13-7*a* shows, his upper arm is vertical and his lower arm is horizontal. What forces must the biceps muscle and the bony structure of the upper arm exert on the lower arm? The forearm and hand together have a mass m

of 1.8 kg; the needed dimensions are as shown in Fig. 13-7a.

SOLUTION: Our system is the lower arm and the bowling ball, taken together. Figure 13-7b shows a free-body diagram of the system. (The ball is represented by a dot within the boundary of the lower arm; the ball's weight vector Mg has its tail on that dot. In drawing Fig. 13-7b from 13-7a, the vector Mg is shifted along its line of action; the shift does not alter Mg, or a torque due to Mg about any axis perpendicular to the figure.) The unknown forces are T, the force exerted by the biceps muscle, and F, the force exerted by the upper arm on the lower arm. The forces are all vertical.

From Eq. 13-8, which is $\Sigma F_y = 0$, we find

$$\sum F_y = T - F - mg - Mg = 0. \qquad (13\text{-}19)$$

Applying Eq. 13-9 about an axis through O at the elbow, and taking torques that would cause counterclockwise rotations as positive, we obtain

$$\sum \tau_z = (F)(0) + (T)(d) - (mg)(D) - (Mg)(L)$$
$$= 0. \qquad (13\text{-}20)$$

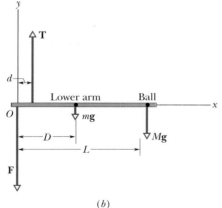

FIGURE 13-7 Sample Problem 13-2. (a) A hand holds a bowling ball. The system boundary is marked. (b) A free-body diagram of the *lower arm + ball* system, showing the forces that act. The vectors are not to scale; the powerful forces exerted by the biceps muscle and the elbow joint are many times larger than either weight.

By choosing our axis to pass through point O, we have eliminated the variable F from Eq. 13-20. Solved for T, it yields

$$T = g \frac{mD + ML}{d}$$

$$= (9.8 \text{ m/s}^2) \frac{(1.8 \text{ kg})(15 \text{ cm}) + (7.2 \text{ kg})(33 \text{ cm})}{4.0 \text{ cm}}$$

$$= 648 \text{ N} \approx 650 \text{ N}. \qquad \text{(Answer)}$$

Thus the biceps muscle must pull up on the forearm with a force that is about nine times larger than the weight of the bowling ball—holding the ball as in Fig. 13-7a is difficult.

If we now solve Eq. 13-19 for F and substitute known quantities into it, we find

$$F = T - g(M + m)$$
$$= 648 \text{ N} - (9.8 \text{ m/s}^2)(7.2 \text{ kg} + 1.8 \text{ kg})$$
$$= 560 \text{ N}, \qquad \text{(Answer)}$$

which is about eight times the weight of the bowling ball.

SAMPLE PROBLEM 13-3

A ladder whose length L is 12 m and whose mass m is 45 kg rests against a wall. Its upper end is a distance h of 9.3 m above the ground, as in Fig. 13-8a. The center of mass of the ladder is one-third of the way up the ladder. A firefighter whose mass M is 72 kg climbs the ladder until her center of mass is halfway up. Assume that the wall, but not the ground, is frictionless. What forces are exerted on the ladder by the wall and by the ground?

SOLUTION: Figure 13-8b shows a free-body diagram of the *firefighter + ladder* system. (The firefighter is represented by a dot within the boundary of the ladder; her weight vector Mg has its tail on that dot. In drawing Fig. 13-8b from 13-8a, the vector Mg is shifted along its line of action; the shift does not alter Mg, or a torque due to Mg about any axis perpendicular to the figure.) The wall exerts a horizontal force F_w on the ladder; it can exert no vertical force because the wall–ladder contact is assumed to be frictionless. The ground exerts a force F_g on the ladder, with horizontal component F_{gx} (due to friction) and vertical component F_{gy} (the usual normal force). We choose coordinate axes as shown, with the origin O at the point where the ladder meets the ground. The distance a from the wall to the foot of the ladder is readily found from

$$a = \sqrt{L^2 - h^2} = \sqrt{(12 \text{ m})^2 - (9.3 \text{ m})^2} = 7.58 \text{ m}.$$

From Eqs. 13-7 and 13-8, the balance of forces equations, we have for the system, respectively,

$$\sum F_x = F_w - F_{gx} = 0 \qquad (13\text{-}21)$$

and

$$\sum F_y = F_{gy} - Mg - mg = 0. \qquad (13\text{-}22)$$

Equation 13-22 yields

$$F_{gy} = g(M + m) = (9.8 \text{ m/s}^2)(72 \text{ kg} + 45 \text{ kg})$$
$$= 1146.6 \text{ N} \approx 1100 \text{ N}. \quad \text{(Answer)}$$

We next balance the torques acting on the system, choosing an axis through O, perpendicular to the plane of the figure. The moment arms about O for $\mathbf{F}_w$, $M\mathbf{g}$, $m\mathbf{g}$, $\mathbf{F}_{gx}$ and $\mathbf{F}_{gy}$ are h, $a/2$, $a/3$, zero, and zero, respectively. The zero moment arms for $\mathbf{F}_{gx}$ and $\mathbf{F}_{gy}$ mean that these forces produce zero torque about O. From Eq. 13-9, the balance of torques equation, we then have

$$\sum \tau_z = -(F_w)(h) + (Mg)(a/2) + (mg)(a/3)$$
$$= 0. \quad \text{(13-23)}$$

Solving Eq. 13-23 for F_w, we find that

$$F_w = \frac{ga(M/2 + m/3)}{h}$$
$$= \frac{(9.8 \text{ m/s}^2)(7.58 \text{ m})(72/2 \text{ kg} + 45/3 \text{ kg})}{9.3 \text{ m}}$$
$$= 407 \text{ N} \approx 410 \text{ N}. \quad \text{(Answer)}$$

From Eq. 13-21 we then have

$$F_{gx} = F_w = 410 \text{ N}. \quad \text{(Answer)}$$

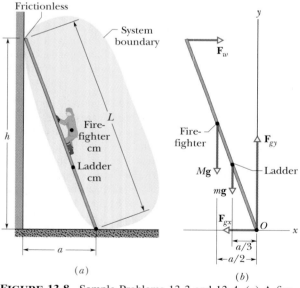

(a)

(b)

FIGURE 13-8 Sample Problems 13-3 and 13-4. (*a*) A firefighter climbs halfway up a ladder that is leaning against a frictionless wall. The ground is not frictionless. (*b*) A free-body diagram, showing the forces that act on the firefighter–ladder system. The origin O of a coordinate system is chosen at the point of application of the unknown force $\mathbf{F}_g$ (whose vector components $\mathbf{F}_{gx}$ and $\mathbf{F}_{gy}$ are shown). This choice simplifies the solution for the other unknown force, $\mathbf{F}_w$.

In Sample Problem 13-3, let the coefficient of static friction μ_s between the ladder and the ground be 0.53. How far up the ladder can the firefighter go before the ladder starts to slip?

SOLUTION: The forces that act have the same labels as in Fig. 13-8. Let q be the fraction of the ladder's length the firefighter climbs before slipping occurs (she is then a horizontal distance qa from O). At the onset of slipping, we have

$$F_{gx} = \mu_s F_{gy}, \quad \text{(13-24)}$$

in which F_{gx} is the static frictional force (usually symbolized as f_s) and F_{gy} is the normal force (usually symbolized as N).

If we apply Eq. 13-9, the balance of torques equation, about an axis through O we have, also at the onset of slipping,

$$\sum \tau_z = -(F_w)(h) + (Mg)(qa) + (mg)(a/3) = 0,$$

or

$$F_w = \frac{ga}{h} (\tfrac{1}{3}m + Mq). \quad \text{(13-25)}$$

This equation shows us that as the firefighter climbs the ladder, that is, as q increases, the force F_w exerted by the wall must increase if equilibrium is to be maintained. To find q at the onset of slipping, we must first find F_w.

Equation 13-7, the balance of forces equation for the x direction, gives

$$\sum F_x = F_w - F_{gx} = 0.$$

If we combine this equation with Eq. 13-24 we find, for the ladder at the onset of slipping,

$$F_w = F_{gx} = \mu_s F_{gy}. \quad \text{(13-26)}$$

From Eq. 13-8, the balance of forces equation for the y direction, we have

$$\sum F_y = F_{gy} - Mg - mg,$$

or

$$F_{gy} = (M + m)g. \quad \text{(13-27)}$$

Combining Eqs. 13-26 and 13-27, we find

$$F_w = \mu_s g(M + m). \quad \text{(13-28)}$$

If, finally, we combine Eqs. 13-25 and 13-28 and solve for q, we have

$$q = \frac{\mu_s h}{a} \frac{(M + m)}{M} - \frac{m}{3M} \quad \text{(13-29)}$$

$$= \frac{(0.53)(9.3 \text{ m})}{7.6 \text{ m}} \frac{(72 \text{ kg} + 45 \text{ kg})}{72 \text{ kg}} - \frac{45 \text{ kg}}{(3)(72 \text{ kg})}$$

$$= 0.85. \quad \text{(Answer)}$$

The firefighter can climb 85% of the way up the ladder before it starts to slip.

You can show from Eq. 13-29 that the firefighter can climb all the way up the ladder (which corresponds to $q = 1$) without slipping if $\mu_s > 0.61$. On the other hand, the ladder will slip when weight is put on the first rung (which corresponds to $q \approx 0$) if $\mu_s < 0.11$.

CHECKPOINT 3: In the figure, a stationary 5 kg rod AC is held against a wall by a rope and friction between rod and wall. The uniform rod is 1 m long, and angle $\theta = 30°$. (a) If you are to find the magnitude of the force **T** on the rod from the rope with a single equation, at what labeled point should a rotational axis be placed? With that choice of axis and counterclockwise torques positive, what is the sign of (b) the torque τ_w due to the rod's weight and (c) the torque τ_r due to the pull on the rod by the rope? (d) Is τ_r greater than, less than, or equal to τ_w?

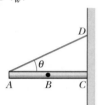

SAMPLE PROBLEM 13-5

Figure 13-9a shows a safe, whose mass M is 430 kg, hanging by a rope from a boom whose dimensions a and b are 1.9 m and 2.5 m, respectively. The boom's uniform beam has a mass m of 85 kg; the mass of the horizontal cable is negligible.

(a) Find the tension T in the cable.

SOLUTION: Figure 13-9b is a free-body diagram of the beam, which we take as our system. The forces acting on it are the tension **T** exerted by the cable, the weight Mg of the safe, the weight mg of the beam itself (acting at the center of the beam), and the force components F_h and F_v exerted on the beam by the hinge that fastens the beam to the wall.

   Let us apply Eq. 13-9, the balance of torques equation, to an axis through the hinge perpendicular to the plane of the figure. Taking torques that would cause counterclockwise rotations as positive, we have

$$\sum \tau_z = (T)(a) - (Mg)(b) - (mg)(\tfrac{1}{2}b) = 0.$$

By our wise choice of an axis, we have eliminated the unknown forces F_h and F_v from this equation (since they create no torque about that axis), leaving only the unknown force T. Solving for T yields

$$T = \frac{gb(M + \tfrac{1}{2}m)}{a}$$

$$= \frac{(9.8 \text{ m/s}^2)(2.5 \text{ m})(430 \text{ kg} + 85/2 \text{ kg})}{1.9 \text{ m}}$$

$$= 6090 \text{ N} \approx 6100 \text{ N}. \qquad \text{(Answer)}$$

(b) Find the force components F_h and F_v exerted on the beam by the hinge.

SOLUTION: We now apply the balance of forces equations. From Eq. 13-7 we have

$$\sum F_x = F_h - T = 0,$$

and so $\qquad F_h = T = 6090 \text{ N} \approx 6100 \text{ N}. \qquad$ (Answer)

From Eq. 13-8 we have

$$\sum F_y = F_v - mg - Mg = 0,$$

and so

$$F_v = g(m + M) = (9.8 \text{ m/s}^2)(85 \text{ kg} + 430 \text{ kg})$$

$$= 5047 \text{ N} \approx 5000 \text{ N}. \qquad \text{(Answer)}$$

(c) What is the magnitude F of the net force exerted by the hinge on the beam?

SOLUTION: From the figure we see that

$$F = \sqrt{F_h^2 + F_v^2}$$

$$= \sqrt{(6090 \text{ N})^2 + (5047 \text{ N})^2} \approx 7900 \text{ N}. \quad \text{(Answer)}$$

Note that F is substantially greater than either the combined weights of the safe and the beam, 5000 N, or the tension in the horizontal wire, 6100 N.

(a)

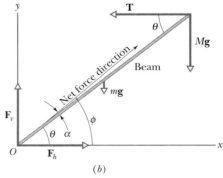

(b)

FIGURE 13-9 Sample Problem 13-5. (a) A heavy safe is hung from a boom consisting of a horizontal steel cable and a uniform beam. (b) A free-body diagram for the beam. Note that the net force acting on the beam due to $\mathbf{F}_v$ and $\mathbf{F}_h$ does not point directly along the beam axis.

(d) What is the angle α between the direction of the net force **F** exerted on the beam by the hinge and the center line of the beam?

SOLUTION: From the figure we see that

$$\theta = \tan^{-1}\frac{a}{b} = \tan^{-1}\frac{1.9 \text{ m}}{2.5 \text{ m}} = 37.2°,$$

and

$$\phi = \tan^{-1}\frac{F_v}{F_h} = \tan^{-1}\frac{5047 \text{ N}}{6090 \text{ N}} = 39.6°.$$

So

$$\alpha = \phi - \theta = 39.6° - 37.2° = 2.4°. \quad \text{(Answer)}$$

If the weight of the beam were small enough to neglect, you would find $\alpha = 0$; that is, the hinge force would point directly along the beam axis.

SAMPLE PROBLEM 13-6

In Fig. 13-10, a rock climber with mass $m = 55$ kg rests during a "chimney climb," pressing only with her shoulders and feet against the walls of a fissure of width $w = 1.0$ m. Her center of mass is a horizontal distance $d = 0.20$ m from the wall against which her shoulders are pressed. The coefficient of static friction between her shoes and the wall is $\mu_1 = 1.1$, and between her shoulders and the wall it is $\mu_2 = 0.70$.

(a) What minimum horizontal push must the climber exert on the walls to keep from falling?

SOLUTION: Her horizontal push against the wall is the same at her shoulders and at her feet. That push must be equal in magnitude to the normal force **N** on her due to the wall at each of those places. Thus there is no net horizontal force on her and Eq. 13-7, $\Sigma F_x = 0$, is satisfied.

As her weight mg acts to slide her down the walls, static frictional forces $\mathbf{f}_1$ and $\mathbf{f}_2$ automatically act upward at feet and shoulders, respectively, to counter the tendency to slide. As long as she does not fall, Eq. 13-8 ($\Sigma F_y = 0$) is satisfied and we have

$$f_1 + f_2 = mg. \quad (13\text{-}30)$$

Let us assume that the climber initially pushes hard against the walls, and then slowly relaxes her push. As she decreases her push, the magnitude N of the normal force decreases, and so do the products $\mu_1 N$ and $\mu_2 N$ that limit the static friction at her feet and shoulders, respectively (see Eq. 6-1).

Suppose N is reduced until the limit $\mu_1 N$ equals the magnitude f_1 of the frictional force at her feet and the limit $\mu_2 N$ equals the magnitude f_2 of the frictional force at her shoulders. She is then on the verge of slipping at both places. If she were to relax her push any more, the resulting decrease in the limits $\mu_1 N$ and $\mu_2 N$ would leave $f_1 + f_2$ less than mg, and she would fall. Thus when she is on the verge of slipping, Eq. 13-30 becomes

$$\mu_1 N + \mu_2 N = mg, \quad (13\text{-}31)$$

which yields

$$N = \frac{mg}{\mu_1 + \mu_2} = \frac{(55 \text{ kg})(9.8 \text{ m/s}^2)}{1.1 + 0.70} = 299 \text{ N} \approx 300 \text{ N}.$$

Thus her minimum horizontal push must be about 300 N.

(b) For that push, what must be the vertical distance h between her feet and shoulders if she is to be stable?

SOLUTION: To satisfy Eq. 13-9, $\Sigma \tau = 0$, the forces acting on her must not create a net torque about *any* rotation axis. Let us consider a rotational axis perpendicular to the page at her shoulders. The net torque about that axis is given by

$$\sum \tau = -f_1 w + Nh + mgd = 0. \quad (13\text{-}32)$$

Solving this for h, setting $f_1 = \mu_1 N$, and substituting $N = 299$ N and other known values, we find

$$h = \frac{f_1 w - mgd}{N} = \frac{\mu_1 N w - mgd}{N} = \mu_1 w - \frac{mgd}{N}$$

$$= (1.1)(1.0 \text{ m}) - \frac{(55 \text{ kg})(9.8 \text{ m/s}^2)(0.20 \text{ m})}{299 \text{ N}}$$

$$= 0.739 \text{ m} \approx 0.74 \text{ m}. \quad \text{(Answer)}$$

We would find the same required value of h if we chose any other rotation axis perpendicular to the page, such as one at her feet.

(c) What are the magnitudes of the frictional forces supporting the climber?

SOLUTION: For $N = 299$ N, we have

$$f_1 = \mu_1 N = (1.1)(299 \text{ N})$$

$$= 328.9 \text{ N} \approx 330 \text{ N}, \quad \text{(Answer)}$$

and from Eq. 13-30 we then have

$$f_2 = mg - f_1 = (55 \text{ kg})(9.8 \text{ m/s}^2) - 328.9 \text{ N}$$

$$= 210.1 \text{ N} \approx 210 \text{ N}. \quad \text{(Answer)}$$

(d) Is she stable if she exerts the same force (299 N) on the walls when her feet are higher, with $h = 0.37$ m?

SOLUTION: Again $N = 299$ N, and the net torque on the climber about any rotation axis must be zero. For the axis at her shoulders, we obtain f_1 from Eq. 13-32:

$$f_1 = \frac{Nh + mgd}{w}$$

$$= \frac{(299 \text{ N})(0.37 \text{ m}) + (55 \text{ kg})(9.8 \text{ m/s}^2)(0.20 \text{ m})}{1.0 \text{ m}}$$

$$= 218 \text{ N}.$$

This is less than the limit $\mu_1 N \,(= 329 \text{ N})$ and thus is possible to attain.

Next we use Eq. 13-30 to find the value of f_2 that will ensure that $\Sigma F_y = 0$:

$$f_2 = mg - f_1 = (55 \text{ kg})(9.8 \text{ m/s}^2) - 218 \text{ N} = 321 \text{ N}.$$

This exceeds the limit $\mu_2 N \,(= 209 \text{ N})$ and thus is impossible to attain with a push of 299 N. The only way the climber can avoid slipping when $h = 0.37$ m (or for any value of h less than 0.74 m) is by pushing harder than 299 N on the walls so as to increase the limit $\mu_2 N$.

Similarly, if $h > 0.74$ m, she must also exert a force greater than 299 N on the walls to be stable. Here, then, is the advantage of knowing the physics before you climb a chimney. When you need to rest, you will avoid the (dire) error of novice climbers who place their feet too high or too low. Instead, you will know that there is a "best" distance between shoulders and feet, requiring the least push, and giving you a good chance to rest.

FIGURE 13-10 Sample Problem 13-6. The forces on a climber resting in a rock chimney. The push of the climber on the chimney walls gives rise to the normal forces **N** (which are equal in magnitude) and the frictional forces $\mathbf{f}_1$ and $\mathbf{f}_2$.

tem of Fig. 13-8 was chosen to include the firefighter so that those unknown forces are *internal* to the system and thus need not be found in order to solve Sample Problems 13-3 and 13-4.

3. Draw a *free-body diagram* of the system. Show all the forces that act on the system, labeling them clearly and making sure that their points of application and lines of action are correctly shown.
4. Draw in the *x and y axes* of a coordinate system. Choose them so that at least one axis is parallel to one or more unknown force. Resolve into components the forces that do not lie along one of the axes. In all our sample problems it made sense to choose the *x* axis horizontal and the *y* axis vertical.
5. Write the two *balance of forces equations,* using symbols throughout.
6. Choose one or more rotational axes perpendicular to the plane of the figure and write the *balance of torques equation* for each axis. If you choose an axis that passes through the line of action of an unknown force, the equation will be simplified because that force will not appear in it.
7. *Solve* your equations *algebraically* for the unknowns. Some students feel more confident in substituting numbers with units in the independent equations at this stage, especially if the algebra is particularly involved. However, experienced problem solvers prefer the algebraic approach, which reveals the dependence of solutions on the various variables.
8. Finally, *substitute numbers* with units in your algebraic solutions, obtaining numerical values for the unknowns.
9. Look at your answer—does it make sense? Is it obviously too large or too small? Is the sign correct? Are the units appropriate?

13-5 INDETERMINATE STRUCTURES

For the problems of this chapter, we have only three independent equations at our disposal, usually two balance of forces equations and one balance of torques equation about a given axis. Thus if a problem has more than three unknowns, we cannot solve it.

It is easy to find such problems. In Sample Problems 13-3 and 13-4, for example, we could have assumed that there is friction between the wall and the top of the ladder. Then there would have been a vertical frictional force acting where the ladder touches the wall, making a total of four unknown forces. With only three equations, we could not have solved this problem.

Consider also an unsymmetrically loaded car. What are the forces—all different—on the four tires? Again, we cannot find them because we have only three independent equations with which to work. Similarly, we can solve an equilibrium problem for a table with three legs but not one with four legs. Problems like these, in which there are more unknowns than equations, are called **indeterminate.**

And yet, solutions to indeterminate problems exist in the real world. If you rest the tires of the car on four platform scales, each scale will register a definite reading, the sum of the readings being the weight of the car. What is eluding us in our efforts to solve equations for the individual scale readings?

The problem is that we have assumed—without making a great point of it—that the bodies to which we apply the equations of static equilibrium are perfectly rigid. By this we mean that they do not deform when forces are applied to them. Strictly, there are no such bodies. The tires of the car, for example, deform easily under load until the car settles into a position of static equilibrium.

We have all had experience with a wobbly restaurant table, which we usually level by putting folded paper under one of the legs. If a big enough elephant sat on such a table, however, you may be sure that if the table did not collapse, it would deform just like the tires of a car. Its legs would all touch the floor, the forces acting upward on the table legs would all assume definite (and different) values as in Fig. 13-11, and the table would no longer wobble. But how do we find the values of those forces acting on the legs?

To solve such indeterminate equilibrium problems, we must supplement equilibrium equations with some knowledge of *elasticity,* the branch of physics and engineering that describes how real bodies deform when forces are applied to them. The next section provides an introduction to this subject.

FIGURE 13-11 The table is an indeterminate structure. The four forces on the table legs are different in magnitude and cannot be found from the laws of static equilibrium alone.

CHECKPOINT **4:** A horizontal uniform bar of weight 10 N is to hang from a ceiling by two wires that exert upward forces F_1 and F_2. The figure shows four arrangements for the wires. Which arrangements, if any, are indeterminate (so that we cannot solve for numerical values of F_1 and F_2)?

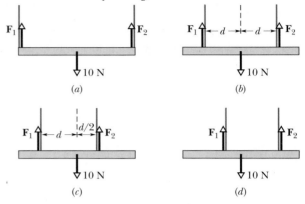

13-6 ELASTICITY

When a large number of atoms come together to form a metallic solid, such as an iron nail, they settle into equilibrium positions in a three-dimensional *lattice,* a repetitive arrangement in which each atom has a well-defined equilibrium distance from its nearest neighbors. The atoms are held together by interatomic forces that are represented by springs in Fig. 13-12.* The lattice is remarkably rigid, which is another way of saying that the "interatomic springs" are extremely stiff. It is for this reason that we perceive many ordinary objects such as metal ladders, tables, and spoons as perfectly rigid. Of course, some ordinary objects, such as garden hoses or rubber gloves, do not strike us as rigid at all. The atoms that make up these objects *do not* form a rigid lattice like that of Fig. 13-12 but are aligned in long flexible molecular chains, each chain being only loosely bound to its neighbors.

All real "rigid" bodies are to some extent **elastic,** which means that we can change their dimensions slightly by pulling, pushing, twisting, or compressing them. To get a feeling for the orders of magnitude involved, consider a vertical steel rod 1 m long and 1 cm in diameter. If you hang a subcompact car from the end of such a rod, the rod

---

*Ordinary metal objects, such as an iron nail, are made up of grains of iron, each grain formed as a more or less perfect lattice, such as that of Fig. 13-12. The forces between the grains are much weaker than the forces that hold the lattice together, so that rupture usually occurs along grain boundaries.

FIGURE 13-12 The atoms of a metallic solid are distributed on a repetitive three-dimensional lattice. The springs represent interatomic forces.

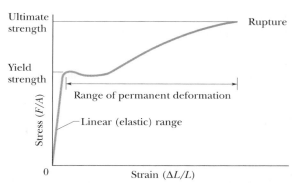

FIGURE 13-14 A stress-strain curve for a steel test specimen such as that of Fig. 13-15. The specimen deforms permanently when the stress is equal to the *yield strength* of the material. It ruptures when the stress is equal to the *ultimate strength* of the material.

will stretch, but only by about 0.5 mm, or 0.05%. Furthermore, the rod will return to its original length when the car is removed.

If you hang two cars from the rod, the rod will be permanently stretched and will not recover its original length when you remove the load. If you hang three cars from the rod, the rod will break. Just before rupture, the elongation of the rod will be less than 0.2%. Although deformations of this size seem small, they are important in engineering practice. (Whether a wing under load will stay on an airplane is obviously important.)

Figure 13-13 shows three ways that a solid might change its dimensions when forces act on it. In Fig. 13-13*a*, a cylinder is stretched. In Fig. 13-13*b*, a cylinder is deformed by a force perpendicular to its axis, much as one might deform a pack of cards or a book. In Fig. 13-13*c*, a

solid object, placed in a fluid under high pressure, is compressed uniformly on all sides. What the three deformation types have in common is that a **stress,** or deforming force per unit area, produces a **strain,** or unit deformation. In Fig. 13-13, *tensile stress* (associated with stretching) is in (*a*), *shearing stress* is in (*b*), and *hydraulic stress* is in (*c*).

The stresses and the strains take different forms in the three cases of Fig. 13-13, but—over the range of engineering usefulness—stress and strain are proportional to each other. The constant of proportionality is called a **modulus of elasticity,** so that

$$\text{stress} = \text{modulus} \times \text{strain}. \qquad (13\text{-}33)$$

Figure 13-14 shows the relation between stress and strain for a steel test cylinder such as that of Fig. 13-15. In a

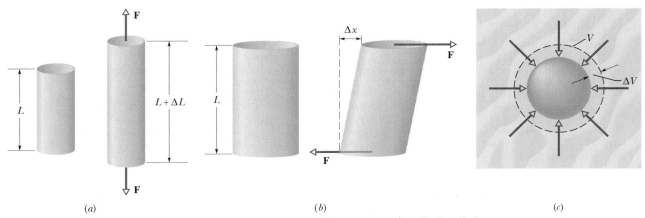

FIGURE 13-13 (*a*) A cylinder subject to *tensile stress* stretches by an amount ΔL. (*b*) A cylinder subject to *shearing stress* deforms by an amount Δx, somewhat like a pack of playing cards would. (*c*) A solid sphere subject to uniform *hydraulic stress* from a fluid shrinks in volume by an amount ΔV. All the deformations shown are greatly exaggerated.

FIGURE 13-15 A test specimen, used to determine a stress–strain curve such as that of Fig. 13-14. The change that occurs in a certain length L is measured in a tensile stress–strain test.

standard test, the tensile stress on a test cylinder is slowly increased from zero to the point where the cylinder fractures, and the strain is carefully measured and graphed. For a substantial range of applied stresses, the stress–strain relation is linear, and the specimen recovers its original dimensions when the stress is removed; it is here that Eq. 13-33 applies. If the stress is increased beyond the **yield strength** S_y of the specimen, the specimen becomes permanently deformed. If the stress continues to increase, the specimen eventually ruptures, at a stress called the **ultimate strength** S_u.

Tension and Compression

For simple tension or compression, the stress is defined as F/A, the force divided by the area over which it acts (the force is perpendicular to the area, as you can see in Fig. 13-13a). The strain, or unit deformation, is then the dimensionless quantity $\Delta L/L$, the fractional (or sometimes percentage) change in the length of the specimen. If the specimen is a long rod and the stress does not exceed the yield strength, then not only the entire rod but also every section of it experiences the same strain when a given stress is applied. Because the strain is dimensionless, the modulus in Eq. 13-33 has the same dimensions as the stress, namely, force per unit area.

The modulus for tensile and compressive stresses is called the **Young's modulus** and is represented in engi-

FIGURE 13-16 A strain gauge of overall dimensions 9.8 mm by 4.6 mm. The gauge is fastened with adhesive to the object whose strain is to be measured; it experiences the same strain as the object. The electrical resistance of the gauge varies with the strain, permitting strains up to about 3% to be measured.

neering practice by the symbol E. Equation 13-33 becomes

$$\frac{F}{A} = E \frac{\Delta L}{L}. \tag{13-34}$$

The strain $\Delta L/L$ in a specimen can often be measured conveniently with a *strain gauge* (Fig. 13-16). These simple and useful devices, which can be attached directly to operating machinery with adhesives, are based on the principle that the electrical properties of the gauge are dependent on the strain it undergoes.

Although the Young's modulus for an object may be almost the same for tension and compression, the object's ultimate strength may well be different for the two types of stress. Concrete, for example, is very strong in compression but is so weak in tension that it is almost never used in that manner. Table 13-1 shows the Young's modulus and other elastic properties for some materials of engineering interest.

TABLE 13-1
SOME ELASTIC PROPERTIES OF SELECTED MATERIALS OF ENGINEERING INTEREST

MATERIAL	DENSITY ρ (kg/m³)	YOUNG'S MODULUS E (10^9 N/m²)	ULTIMATE STRENGTH S_u (10^6 N/m²)	YIELD STRENGTH S_y (10^6 N/m²)
Steel[a]	7860	200	400	250
Aluminum	2710	70	110	95
Glass	2190	65	50[b]	—
Concrete[c]	2320	30	40[b]	—
Wood[d]	525	13	50[b]	—
Bone	1900	9[b]	170[b]	—
Polystyrene	1050	3	48	—

[a] Structural steel (ASTM-A36).   [b] In compression.   [c] High strength.   [d] Douglas fir.

Shearing

In the case of shearing, the stress is also a force per unit area, but the force vector lies in the plane of the area rather than at right angles to it. The strain is the dimensionless ratio $\Delta x/L$, with the quantities defined as shown in Fig. 13-13b. The corresponding modulus, which is given the symbol G in engineering practice, is called the **shear modulus.** For shearing, Eq. 13-33 is written as

$$\frac{F}{A} = G\frac{\Delta x}{L}. \tag{13-35}$$

Shearing stresses play a critical role in the buckling of shafts that rotate under load and in bone fractures caused by bending.

Hydraulic Stress

In Fig. 13-13c, the stress is the fluid pressure p on the object, which, as you will see in Chapter 15, is a force per unit area. The strain is $\Delta V/V$, where V is the original volume of the specimen and ΔV is the absolute value of the change in volume. The corresponding modulus, with symbol B, is called the **bulk modulus** of the material. The object is said to be under *hydraulic compression,* and the pressure can be called the *hydraulic stress.* For this situation, we write Eq. 13-33 as

$$p = B\frac{\Delta V}{V}. \tag{13-36}$$

The bulk modulus of water is 2.2×10^9 N/m², and that of steel is 16×10^{10} N/m². The pressure at the bottom of the Pacific Ocean, at its average depth of about 4000 m, is 4.0×10^7 N/m². The fractional compression $\Delta V/V$ of a volume of water due to this pressure is 1.8%; that for a steel object is only about 0.025%. In general, solids — with their rigid atomic lattices — are less compressible than liquids, in which the atoms or molecules are less tightly coupled to their neighbors.

SAMPLE PROBLEM 13-7

A structural steel rod has a radius R of 9.5 mm and a length L of 81 cm. A force F of 6.2×10^4 N (about 7 tons) stretches it along its length.

(a) What is the stress in the rod?

SOLUTION: From its definition, the stress is

$$\text{stress} = \frac{F}{A} = \frac{F}{\pi R^2} = \frac{6.2 \times 10^4 \text{ N}}{(\pi)(9.5 \times 10^{-3} \text{ m})^2}$$

$$= 2.2 \times 10^8 \text{ N/m}^2. \quad \text{(Answer)}$$

The yield strength for structural steel is 2.5×10^8 N/m², so that this rod is dangerously close to its yield strength.

(b) What is the elongation of the rod under this load? What is the strain?

SOLUTION: From Eq. 13-34, using the result we have just calculated and the value of E for steel (Table 13-1), we obtain

$$\Delta L = \frac{(F/A)L}{E} = \frac{(2.2 \times 10^8 \text{ N/m}^2)(0.81 \text{ m})}{2.0 \times 10^{11} \text{ N/m}^2}$$

$$= 8.9 \times 10^{-4} \text{ m} = 0.89 \text{ mm}. \quad \text{(Answer)}$$

Thus the strain is

$$\frac{\Delta L}{L} = \frac{8.9 \times 10^{-4} \text{ m}}{0.81 \text{ m}}$$

$$= 1.1 \times 10^{-3} = 0.11\%. \quad \text{(Answer)}$$

SAMPLE PROBLEM 13-8

The femur, which is the principal bone of the thigh, has a minimum diameter in an adult male of about 2.8 cm, corresponding to a cross-sectional area A of 6×10^{-4} m². At what compressive load would it break?

SOLUTION: From Table 13-1 we see that the ultimate strength S_u for bone in compression is 170×10^6 N/m². The compressive force at fracture is the force F that produces the stress S_u; so

$$F = S_u A = (170 \times 10^6 \text{ N/m}^2)(6 \times 10^{-4} \text{ m}^2)$$

$$= 1.0 \times 10^5 \text{ N}. \quad \text{(Answer)}$$

This is 23,000 lb or 11 tons. Although this is a large force, it can be encountered during, for example, an unskillful parachute landing on hard ground. The force need not be sustained to break the bone; a few milliseconds will do it.

SAMPLE PROBLEM 13-9

A table has three legs that are 1.00 m in length and a fourth leg that is longer by a distance $d = 0.50$ mm, so that the table wobbles slightly. A heavy steel cylinder whose mass M is 290 kg is placed upright on the table (whose mass is much less than M) so that all four legs compress and the table no longer wobbles. The legs are wooden cylinders whose cross-sectional area A is 1.0 cm². The Young's modulus E for the wood is 1.3×10^{10} N/m². Assume that the tabletop remains level and that the legs do not buckle. With what force does the floor push upward on each leg?

SOLUTION: We take the table plus steel cylinder as our system. The situation is like Fig. 13-11, except we now have a steel cylinder on the table. If the tabletop remains level, each of the three short legs must be compressed by the same amount (call it ΔL_3) and thus by the same force F_3. The single long leg must be compressed by a larger amount ΔL_4, by a larger force F_4, and we must have

$$\Delta L_4 = \Delta L_3 + d. \qquad (13\text{-}37)$$

We can rewrite Eq. 13-34 as $\Delta L = FL/EA$. We then use this relationship to substitute for L_4 and L_3 in Eq. 13-37, letting L represent, in turn, the initial length of the three short legs (1.0 m) and the approximate initial length of the long leg (also 1.0 m). Equation 13-37 becomes

$$F_4 L = F_3 L + dAE. \qquad (13\text{-}38)$$

From Eq. 13-8, the balance of forces in the vertical direction, we have for our system

$$\sum F_y = 3F_3 + F_4 - Mg = 0. \qquad (13\text{-}39)$$

If we solve Eqs. 13-38 and 13-39 for the unknown force F_3, we find

$$F_3 = \frac{Mg}{4} - \frac{dAE}{4L}$$

$$= \frac{(290 \text{ kg})(9.8 \text{ m/s}^2)}{4}$$

$$- \frac{(5.0 \times 10^{-4} \text{ m})(10^{-4} \text{ m}^2)(1.3 \times 10^{10} \text{ N/m}^2)}{(4)(1.00 \text{ m})}$$

$$= 711 \text{ N} - 163 \text{ N} = 548 \text{ N} \approx 550 \text{ N}. \qquad \text{(Answer)}$$

From Eq. 13-39 we then find

$$F_4 = Mg - 3F_3 = (290 \text{ kg})(9.8 \text{ m/s}^2) - 3(548 \text{ N})$$

$$\approx 1200 \text{ N}. \qquad \text{(Answer)}$$

You can show that to reach their equilibrium configuration, the three short legs were each compressed by 0.42 mm and the single long leg by 0.92 mm, the difference being 0.50 mm.

CHECKPOINT 5: The figure shows a horizontal block that is suspended by two wires, A and B, which are identical except for their original lengths. The center of mass of the block is closer to wire B than to wire A. (a) Measuring torques about the block's center of mass, state whether the torque due to wire A is greater than, less than, or equal to the torque due to wire B. (b) Which wire exerts more force on the block? (c) If the wires are now equal in length, which one was originally shorter?

REVIEW & SUMMARY

Static Equilibrium

A rigid body at rest is said to be in **static equilibrium.** For such a body, the vector sum of the external forces acting on it is zero:

$$\sum \mathbf{F}_{ext} = 0 \qquad \text{(balance of forces).} \qquad (13\text{-}3)$$

If all the forces lie in the xy plane, this vector equation is equivalent to two scalar component equations:

$$\sum F_x = 0 \quad \text{and} \quad \sum F_y = 0 \qquad \begin{array}{c}\text{(balance of}\\ \text{forces).}\end{array} \qquad (13\text{-}7, 13\text{-}8)$$

Static equilibrium also implies that the vector sum of the external torques acting on the body about *any* point is zero, or

$$\sum \tau_{ext} = 0 \qquad \text{(balance of torques).} \qquad (13\text{-}5)$$

If the forces lie in the xy plane, all torque vectors are parallel to the z axis, and Eq. 13-5 is equivalent to the single scalar component equation

$$\sum \tau_z = 0 \qquad \text{(balance of torques).} \qquad (13\text{-}9)$$

Center of Gravity

The gravitational force acts individually on all the individual elements of a body. The net effect of all the individual actions may be found by imagining an equivalent total gravitational force $M\mathbf{g}$

acting at a specific point called the **center of gravity.** If the gravitational acceleration $\mathbf{g}$ is the same for all the particles in a body, the center of gravity is at the center of mass.

Elastic Moduli

Three **elastic moduli** are used to describe the elastic behavior (deformations) of objects as they respond to the forces that act on them. The **strain** (fractional change in length) is linearly related to the applied **stress** (force per unit area) by the modulus in each case. The general relation is

$$\text{stress} = \text{modulus} \times \text{strain}. \qquad (13\text{-}33)$$

Tension and Compression

When an object is under tension or compression, Eq. 13-33 is written as

$$\frac{F}{A} = E \frac{\Delta L}{L}, \qquad (13\text{-}34)$$

where $\Delta L/L$ is the strain of the object, F is the magnitude of the applied force $\mathbf{F}$ causing the strain, A is the cross-sectional area over which $\mathbf{F}$ is applied (perpendicular to A, as in Fig. 13-13a), and E is the **Young's modulus** for the object. The stress is F/A.

Shearing

When an object is under a shearing stress, Eq. 13-33 is written as

$$\frac{F}{A} = G \frac{\Delta x}{L}, \tag{13-35}$$

where $\Delta x/L$ is the strain of the object, Δx is the displacement of one end of the object in the direction of the applied force **F** (as in Fig. 13-13b), and G is the **shear modulus** of the object. The stress is F/A.

Hydraulic Stress

When an object undergoes *hydraulic compression* due to a stress exerted by a surrounding fluid, Eq. 13-33 is written as

$$p = B \frac{\Delta V}{V}, \tag{13-36}$$

where p is the pressure (*hydraulic stress*) on the object due to the fluid, $\Delta V/V$ (the strain) is the absolute value of the fractional change in the object's volume due to that pressure, and B is the **bulk modulus** of the object.

QUESTIONS

1. Figure 13-17 shows an overhead view of a uniform stick on which four forces act. Suppose we choose a rotational axis through point O, calculate the torques about that axis due to the forces, and find that these torques balance. Will the torques balance if, instead, the rotational axis is chosen to be at (a) point A, (b) point B, or (c) point C? (d) Suppose, instead, that we find that the torques about point O do not balance. Is there another point about which the torques will balance?

FIGURE 13-17 Question 1.

2. Figure 13-18 shows four overhead views of rotating uniform disks that are sliding across a frictionless floor. Three forces, of magnitude F, $2F$, or $3F$, act on each disk; each force is applied either at the rim, at the center, or halfway between rim and center. The force vectors rotate along with the disks, and, in the "snapshots" of Fig. 13-18, point left or right. Which disks are in equilibrium?

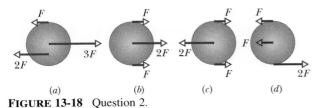

(a)	(b)	(c)	(d)

FIGURE 13-18 Question 2.

3. Figure 13-19 shows overhead views of two structures on which three forces act. The directions of the forces are as indicated. If the magnitudes of the forces are adjusted properly (but kept nonzero), which structure can be in static equilibrium?

FIGURE 13-19 Question 3. (a) (b)

4. Figure 13-20 shows a mobile of toy penguins hanging from a ceiling. Each crossbar is horizontal, has negligible mass, and extends three times as far to the right of the wire supporting it as to the left. Penguin 1 has mass $m_1 = 48$ kg. What are the masses of the other penguins?

FIGURE 13-20 Question 4.

5. Figure 13-21 shows an overhead view of a metal square lying on a frictionless floor. Three forces, which are drawn to scale, act at the corners of the square. (a) Is the first requirement of equilibrium (in Eq. 13-1) satisfied? (b) Is the second requirement of equilibrium satisfied? (c) If the answer to either (a) or (b) is no, could a fourth force acting on the square then satisfy both requirements of equilibrium?

FIGURE 13-21 Question 5.

6. (a) How many different free-standing towers can you build with three of the small (four-projection) Lego blocks? (Two such blocks are stackable one directly above the other, or one displaced half a block length left or right of the other. Mirror-image arrangements count as one.) How many such towers are in (b) stable equilibrium and (c) unstable equilibrium (center of mass almost over an edge)? (d) Which arrangement is most stable (hardest to topple)? Why?

7. Figure 13-22 shows four arrangements in which a painting is suspended from a wall with two identical lengths of wire. The wires in Fig. 13-22b and c all make the same angle with the horizontal. Rank the four arrangements according to the tension in the wires, largest first.

(a)   (b)   (c)   (d)

FIGURE 13-22 Question 7.

8. A ladder leans against a frictionless wall but is prevented from falling because of friction between it and the ground. Suppose you shift the base of the ladder toward the wall. Tell whether the following become larger, smaller, or stay the same: (a) the normal force on the ladder from the ground, (b) the force on the ladder from the wall, (c) the static frictional force on the ladder from the ground, and (d) the maximum value $f_{s,max}$ of the static frictional force.

9. A physical therapist gone wild has constructed the (stationary) assembly of massless pulleys and cords seen in Fig. 13-23. One long cord wraps around all the pulleys and shorter cords suspend pulleys from the ceiling or weights from the pulleys. Except for one, the weights (in newtons) are indicated. (a) What is that last weight? (*Hint:* When a cord loops halfway around a pulley as here, it pulls on the pulley with a net force that is twice the tension in the cord.) (b) What is the tension in the short cord labeled with T?

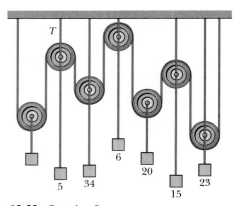

FIGURE 13-23 Question 9.

10. Three piñatas hang from the (stationary) assembly of massless pulleys and cords seen in Fig. 13-24. One long cord runs from the ceiling at the right to the lower pulley at the left. Several shorter cords suspend pulleys from the ceiling or piñatas from the pulleys. The weights (in newtons) of two piñatas are given.

(a) What is the weight of the third piñata? (*Hint:* When a cord loops around a pulley, it pulls on the pulley with a net force that is twice the tension in the cord.) (b) What is the tension in the short cord labeled with T?

FIGURE 13-24 Question 10.

11. (a) In Checkpoint 3, to express τ_r in terms of T, should you use sin θ or cos θ? (b) If angle θ is decreased (by shortening the rope but still keeping the rod horizontal), does the torque τ_r required for equilibrium become larger, smaller, or stay the same? (c) Does the corresponding force T become larger, smaller, or stay the same?

12. The table gives the areas of three surfaces and the magnitude of a force that is applied perpendicular to the surface and uniformly across it. Rank the surfaces according to the stress on them, greatest first.

	AREA	FORCE
Surface A	$0.5A_0$	$2F_0$
Surface B	$2A_0$	$4F_0$
Surface C	$3A_0$	$6F_0$

13. The table gives the initial lengths of three rods and the changes in their length when forces are applied to their ends to put them under strain. Rank the rods according to their strain, greatest first.

	INITIAL LENGTH	CHANGE IN LENGTH
Rod A	$2L_0$	ΔL_0
Rod B	$4L_0$	$2\Delta L_0$
Rod C	$10L_0$	$4\Delta L_0$

EXERCISES & PROBLEMS

SECTION 13-4 Some Examples of Static Equilibrium

1E. An eight-member family, whose weights in pounds are indicated in Fig. 13-25, is balanced on a seesaw. What is the number of the person who causes the largest torque, about the rotation axis at *fulcrum f*, directed (a) out of the page and (b) into the page?

FIGURE 13-25   Exercise 1.

2E. A certain nut is known to require forces of 40 N exerted on its shell from both sides to crack it. What force components $F_\perp$, perpendicular to the handles, will be required when the nut is placed in the nutcracker shown in Fig. 13-26?

FIGURE 13-26   Exercise 2.

3E. The leaning Tower of Pisa (see Fig. 13-27) is 55 m high and 7.0 m in diameter. The top of the tower is displaced 4.5 m from the vertical. Treat the tower as a uniform, circular cylinder. (a)

FIGURE 13-27   Exercise 3. The leaning Tower of Pisa (the photograph is not distorted).

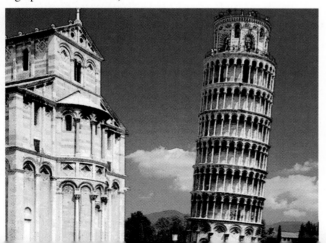

What additional displacement, measured at the top, will bring the tower to the verge of toppling? (b) What angle with the vertical will the tower make at that moment?

4E. A particle is acted on by forces given, in newtons, by $F_1 = 10i - 4j$ and $F_2 = 17i + 2j$. (a) What force F_3 balances these forces? (b) What direction does F_3 have relative to the *x* axis?

5E. A bow is drawn until the tension in the string is equal to the force exerted by the archer. What is the angle between the two parts of the string?

6E. In Fig. 13-28, a man is trying to get his car out of mud on the shoulder of a road. He ties one end of a rope tightly around the front bumper and the other end tightly around a utility pole 60 ft away. He then pushes sideways on the rope at its midpoint with a force of 125 lb, displacing the center of the rope 1.0 ft from its previous position, and the car barely moves. What force does the rope exert on the car? (The rope stretches somewhat.)

FIGURE 13-28   Exercise 6.

7E. A rope, assumed massless, is stretched horizontally between two supports that are 3.44 m apart. When an object of weight 3160 N is hung at the center of the rope, the rope is observed to sag by 35.0 cm. What is the tension in the rope?

8E. The system in Fig. 13-29 is in equilibrium, but it begins to slip if any additional mass is added to the 5.0 kg object. What is the coefficient of static friction between the 10 kg block and the plane on which it rests?

FIGURE 13-29   Exercise 8.

9E. A scaffold of mass 60 kg and length 5.0 m is supported in a horizontal position by one vertical cable at each end. A window washer of mass 80 kg stands at a point 1.5 m from one end. What is the tension in the cable (a) closest to the window washer and (b) farthest away from the window washer?

10E. A beam is carried by three men, one man at one end and the other two supporting the beam between them on a crosspiece placed so that the load is equally divided among the three men.

Where is the crosspiece placed? (Neglect the mass of the crosspiece.)

11E. A uniform cubical crate is 0.750 m on each side and weighs 500 N. It rests on the floor with one edge against a very small, fixed obstruction. At what height above the floor must a horizontal force of 350 N be applied to the crate to just tip it?

12E. A uniform sphere of weight W and radius r is held in place by a rope attached to a frictionless wall a distance L above the center of the sphere, as in Fig. 13-30. Find (a) the tension in the rope and (b) the force exerted on the sphere by the wall.

FIGURE 13-30 Exercise 12.

13E. An automobile with a mass of 1360 kg has 3.05 m between the front and rear axles. Its center of gravity is located 1.78 m behind the front axle. Determine (a) the force exerted on each of the front wheels (assumed the same) and (b) the force exerted on each of the back wheels (assumed the same) by the level ground.

14E. A 160 lb man is walking across a level bridge and stops one-fourth of the way from one end. The bridge is uniform and weighs 600 lb. What are the vertical forces exerted on the bridge by its supports at (a) the far end and (b) the near end?

15E. A diver of weight 580 N stands at the end of a 4.5 m diving board of negligible weight. The board is attached to two pedestals 1.5 m apart, as shown in Fig. 13-31. What are the magnitude and direction of the force on the board from (a) the left pedestal and (b) the right pedestal? (c) Which pedestal is being stretched, and which compressed?

FIGURE 13-31 Exercise 15.

16E. A meter stick balances horizontally on a knife-edge at the 50.0 cm mark. With two nickels stacked over the 12.0 cm mark, the stick is found to balance at the 45.5 cm mark. A nickel has a mass of 5.0 g. What is the mass of the meter stick?

17E. A 75 kg window cleaner uses a 10 kg ladder that is 5.0 m long. He places one end on the ground 2.5 m from a wall, rests the upper end against a cracked window, and climbs the ladder. He climbs 3.0 m up the ladder when the window breaks. Neglecting friction between the ladder and window and assuming that the base of the ladder did not slip, find (a) the force exerted on the

window by the ladder just before the window breaks and (b) the magnitude and direction of the force exerted on the ladder by the ground just before the window breaks.

18E. Figure 13-32 shows the anatomical structures in the lower leg and foot that are involved in standing tiptoe with the heel raised off the floor so that the foot effectively contacts the floor at only one point, shown as P in the figure. Calculate, in terms of a person's weight W, the forces that must be exerted on the foot by (a) the calf muscle (at A) and (b) the lower-leg bones (at B) when the person stands tiptoe on one foot. Assume that $a = 5.0$ cm and $b = 15$ cm.

FIGURE 13-32 Exercise 18.

19P. By means of a turnbuckle G, bar AB of the square frame $ABCD$ in Fig. 13-33 is put in tension, as if its ends A and B were subject to the horizontal, outward forces **T** shown. Determine the forces on the other bars; identify those bars that are in tension and those that are being compressed. The diagonals AC and BD pass each other freely at E. Symmetry considerations can lead to considerable simplification in this and similar problems.

FIGURE 13-33 Problem 19.

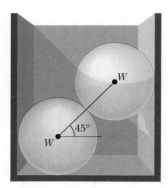

FIGURE 13-34 Problem 20.

20P. Two identical, uniform, frictionless spheres, each of weight W, rest in a rigid rectangular container as shown in Fig. 13-34. Find, in terms of W, the forces acting on the spheres due to (a) the container surfaces and (b) one another, if the line of centers of the spheres makes an angle of 45° with the horizontal.

21P. An 1800 lb construction bucket is suspended by a cable A that is attached at O to two other cables B and C, making angles of 51° and 66° with the horizontal (Fig. 13-35). Find the tension in (a) cable A, (b) cable B, and (c) cable C. (*Hint:* To avoid solving two equations in two unknowns, position the axes as shown in the figure.)

FIGURE 13-35 Problem 21.

22P. The force **F** in Fig. 13-36 is just sufficient to hold the 14 lb block and weightless pulleys in equilibrium. There is no appreciable friction. Calculate the tension T in the upper cable.

FIGURE 13-36 Problem 22.

23P. The system in Fig. 13-37 is in equilibrium with the string in the center exactly horizontal. Find (a) tension T_1, (b) tension T_2, (c) tension T_3, and (d) angle θ.

FIGURE 13-37 Problem 23.

24P. A balance is made up of a rigid, massless rod supported at and free to rotate about a point not at the center of the rod. It is balanced by unequal weights placed in the pans at each end of the rod. When an unknown mass m is placed in the left-hand pan, it is balanced by a mass m_1 placed in the right-hand pan; and when the mass m is placed in the right-hand pan, it is balanced by a mass m_2 in the left-hand pan. Show that $m = \sqrt{m_1 m_2}$.

25P. A 15 kg weight is being lifted by the pulley system shown in Fig. 13-38. The upper arm is vertical, whereas the forearm makes an angle of 30° with the horizontal. What forces are being exerted on the forearm by (a) the triceps muscle and (b) the upper-arm bone (the humerus)? The forearm and hand together have a mass of 2.0 kg with a center of mass 15 cm (measured along the arm) from the point where the forearm and upper-arm bones are in contact. The triceps muscle pulls vertically upward at a point 2.5 cm behind the contact point.

FIGURE 13-38 Problem 25.

26P. A 50.0 kg uniform square sign, 2.00 m on a side, is hung from a 3.00 m rod of negligible mass. A cable is attached to the end of the rod and to a point on the wall 4.00 m above the point where the rod is fixed to the wall (Fig. 13-39). (a) What is the tension in the cable? What are the (b) horizontal and (c) vertical components of the force exerted by the wall on the rod?

FIGURE 13-39 Problem 26.

27P. In Fig. 13-40, a 55 kg rock climber is in a lie-back climb along a fissure, with hands pulling on one side of the fissure and feet pressed against the opposite side. The fissure has width $w = 0.20$ m, and the center of mass of the climber is a horizontal distance $d = 0.40$ m from the fissure. The coefficient of static friction between hands and rock is $\mu_1 = 0.40$, and between boots and rock it is $\mu_2 = 1.2$. (a) What is the least horizontal pull by the hands and push by the feet that will keep him stable? (b) For the horizontal pull of (a), what must be the vertical distance h between hands and feet? (c) If the climber encounters wet rock, so that μ_1 and μ_2 are reduced, what happens to the answers to (a) and (b), respectively?

FIGURE 13-40 Problem 27.

28P. Forces $\mathbf{F}_1$, $\mathbf{F}_2$, and $\mathbf{F}_3$ act on the structure of Fig. 13-41, shown in an overhead view. We wish to put the structure in equilibrium by applying a force, at a point such as P, whose vector components are $\mathbf{F}_h$ and $\mathbf{F}_v$. We are given that $a = 2.0$ m, $b = 3.0$ m, $c = 1.0$ m, $F_1 = 20$ N, $F_2 = 10$ N, and $F_3 = 5.0$ N. Find (a) F_h, (b) F_v, and (c) d.

FIGURE 13-41 Problem 28.

29P. In Fig. 13-42, what magnitude of force $\mathbf{F}$ applied horizontally at the axle of the wheel is necessary to raise the wheel over an obstacle of height h? Take r as the radius of the wheel and W as its weight.

FIGURE 13-42 Problem 29.

30P. A trap door in a ceiling is 0.91 m square, has a mass of 11 kg, and is hinged along one side with a catch at the opposite side. If the center of gravity of the door is 10 cm toward the hinged side from the door's center, what forces must (a) the catch and (b) the hinge sustain?

31P. Four identical uniform bricks, each of length L, are put on top of one another (Fig. 13-43) in such a way that part of each extends beyond the one beneath. Find, in terms of L, the maximum values of (a) a_1, (b) a_2, (c) a_3, (d) a_4, and (e) h, such that the stack is in equilibrium.

FIGURE 13-43 Problem 31.

32P. One end of a uniform beam that weighs 50.0 lb and is 3.00 ft long is attached to a wall with a hinge. The other end is supported by a wire (see Fig. 13-44). (a) Find the tension in the wire. What are the (b) horizontal and (c) vertical components of the force of the hinge on the beam?

FIGURE 13-44 Problem 32.

33P. The system in Fig. 13-45 is in equilibrium. A mass of 225 kg hangs from the end of the uniform strut whose mass is 45.0 kg. Find (a) the tension T in the cable and the (b) horizontal and (c) vertical force components exerted on the strut by the hinge.

FIGURE 13-45 Problem 33.

34P. A door 2.1 m high and 0.91 m wide has a mass of 27 kg. A hinge 0.30 m from the top and another 0.30 m from the bottom each support half the door's weight. Assume that the center of gravity is at the geometrical center of the door and determine the (a) vertical and (b) horizontal force components exerted by each hinge on the door.

35P. A nonuniform bar of weight W is suspended at rest in a horizontal position by two massless cords as shown in Fig. 13-46. One cord makes the angle $\theta = 36.9°$ with the vertical; the other makes the angle $\phi = 53.1°$ with the vertical. If the length L of the bar is 6.10 m, compute the distance x from the left-hand end of the bar to its center of gravity.

FIGURE 13-46 Problem 35.

36P. In Fig. 13-47, a thin horizontal bar AB of negligible weight and length L is pinned to a vertical wall at A and supported at B by a thin wire BC that makes an angle θ with the horizontal. A weight W can be moved anywhere along the bar; its position is defined by the distance x from the wall to its center of mass. As a function of x, find (a) the tension in the wire, and the (b) horizontal and (c) vertical components of the force exerted on the bar by the pin at A.

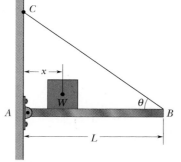

FIGURE 13-47 Problems 36 and 37.

37P. In Fig. 13-47, suppose the length L of the uniform bar is 3.0 m and its weight is 200 N. Also, let $W = 300$ N and $\theta = 30°$. The wire can withstand a maximum tension of 500 N. (a) What is the maximum possible distance x before the wire breaks? With W placed at this maximum x, what are the (b) horizontal and (c) vertical components of the force exerted on the bar by the pin at A?

38P. Two uniform beams, A and B, are attached to a wall with hinges and then loosely bolted together as in Fig. 13-48. Find the

FIGURE 13-48 Problem 38.

horizontal and vertical components of the force on (a) beam A due to its hinge, (b) beam A due to the bolt, (c) beam B due to its hinge, and (d) beam B due to the bolt.

39P. Four identical, uniform bricks of length L are stacked on a table in two ways, as shown in Fig. 13-49 (compare with Problem 31). We seek to maximize the overhang distance h in both arrangements. Find the optimum distances a_1, a_2, b_1, and b_2, and calculate h for the two arrangements. (See "The Amateur Scientist," *Scientific American,* June 1985, for a discussion and an even better version of arrangement (b).)

FIGURE 13-49 Problem 39.

40P. A uniform plank, with length $L = 20$ ft and weight $W = 100$ lb, rests on the ground and against a frictionless roller at the top of a wall of height $h = 10$ ft (see Fig. 13-50). The plank remains in equilibrium for any value of $\theta \geq 70°$ but slips if $\theta < 70°$. Find the coefficient of static friction between the plank and the ground.

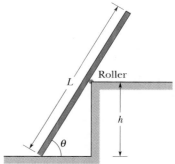

FIGURE 13-50 Problem 40.

41P. For the stepladder shown in Fig. 13-51, sides AC and CE are each 8.0 ft long and hinged at C. Bar BD is a tie-rod 2.5 ft long, halfway up. A man weighing 192 lb climbs 6.0 ft along the ladder. Assuming that the floor is frictionless and neglecting the weight of the ladder, find (a) the tension in the tie-rod and the forces exerted on the ladder by the floor at (b) *A* and (c) *E*. (*Hint:* It will help to isolate parts of the ladder in applying the equilibrium conditions.)

FIGURE 13-51 Problem 41.

42P. A uniform cube of side length *L* rests on a horizontal floor. The coefficient of static friction between cube and floor is μ. A horizontal pull *P* is applied perpendicular to one of the vertical faces of the cube, at a distance *h* above the floor on the vertical midline of the cube face. As *P* is slowly increased, the cube will either (a) begin to slide or (b) begin to tip. What is the condition on μ for (a) to occur? For (b)? (*Hint:* At the onset of tipping, where is the normal force located?)

43P. A cubical box is filled with sand and weighs 890 N. We wish to ''roll'' the box by pushing horizontally on one of the upper edges. (a) What minimum force is required? (b) What minimum coefficient of static friction between box and floor is required? (c) Is there a more efficient way to roll the box? If so, find the smallest possible force that would have to be applied directly to the box to roll it. (*Hint:* See the hint for Problem 42.)

44P. A crate, in the form of a cube with edge lengths of 4.0 ft, contains a piece of machinery whose design is such that the center of gravity of the crate and its contents is located 1.0 ft above its geometrical center. The crate rests on a ramp that makes an angle θ with the horizontal. As θ is increased from zero, an angle will be reached at which the crate will either start to slide down the ramp or tip over. Which event will occur (a) when the coefficient of static friction between ramp and crate is 0.60 and (b) when it is 0.70? In each case, give the angle at which the event occurs. (*Hint:* See the hint for Problem 42.)

45P*. A car on a horizontal road makes an emergency stop by applying the brakes so that all four wheels lock and skid along the road. The coefficient of kinetic friction between tires and road is 0.40. The separation between the front and rear axles is 4.2 m, and the center of mass of the car is located 1.8 m behind the front axle and 0.75 m above the road; see Fig. 13-52. The car weighs 11 kN. Calculate (a) the braking deceleration of the car, (b) the normal force on each wheel, and (c) the braking force on each wheel. (*Hint:* Although the car is not in translational equilibrium, it *is* in rotational equilibrium.)

FIGURE 13-52 Problem 45.

SECTION 13-6 Elasticity

46E. Figure 13-53 shows the stress–strain curve for quartzite. What are (a) the Young's modulus and (b) the approximate yield strength for this material?

FIGURE 13-53 Exercise 46.

47E. After a fall, a 95 kg rock climber finds himself dangling from the end of a rope that had been 15 m long and 9.6 mm in diameter but which has stretched by 2.8 cm. For the rope, calculate (a) the strain, (b) the stress, and (c) the Young's modulus.

48E. A mine elevator is supported by a single steel cable 2.5 cm in diameter. The total mass of the elevator cage plus occupants is 670 kg. By how much does the cable stretch when the elevator is (a) at the surface, 12 m below the elevator motor, and (b) at the bottom of the shaft, which is 350 m deep? (Neglect the mass of the cable.)

49E. Suppose the (square) beam in Fig. 13-9*a* is of Douglas fir. What must be its thickness to keep the compressive stress on it to $\frac{1}{6}$ of its ultimate strength? (See Sample Problem 13-5.)

50E. A horizontal aluminum rod 4.8 cm in diameter projects 5.3 cm from a wall. A 1200 kg object is suspended from the end of the rod. The shear modulus of aluminum is 3.0×10^{10} N/m². Neglecting the rod's weight, find (a) the shear stress on the rod and (b) the vertical deflection of the end of the rod.

51E. A solid copper cube has an edge length of 85.5 cm. How much pressure must be applied to the cube to reduce the edge length to 85.0 cm? The bulk modulus of copper is 1.4×10^{11} N/m².

52P. A tunnel 150 m long, 7.2 m high, and 5.8 m wide (with a flat roof) is to be constructed 60 m beneath the ground. (See Fig. 13-54.) The tunnel roof is to be supported entirely by square steel columns, each with a cross-sectional area of 960 cm². The density of the ground material is 2.8 g/cm³. (a) What is the total weight that the columns must support? (b) How many columns are needed to keep the compressive stress on each column at one-half its ultimate strength?

FIGURE 13-54 Problem 52.

53P. A rectangular slab of slate rests on a 26° incline; see Fig. 13-55. The slab has dimensions 43 m long, 2.5 m thick, and 12 m wide. Its density is 3.2 g/cm³. The coefficient of static friction between the slab and the underlying rock is 0.39. (a) Calculate the component of the slab's weight acting parallel to the incline. (b) Calculate the static frictional force on the slab. By comparing (a) and (b), you can see that the slab is in danger of sliding and is prevented from doing so only by chance protrusions between the slab and the underlying rock. (c) To stabilize the slab, bolts are driven perpendicular to the incline. If each bolt has a cross-sectional area of 6.4 cm² and will snap under a shearing stress of 3.6 × 10⁸ N/m², what is the minimum number of bolts needed? Assume that the bolts do not affect the normal force.

FIGURE 13-55 Problem 53.

54P. In Fig. 13-56, a lead brick rests horizontally on cylinders A and B. The areas of the top faces of the cylinders are related by $A_A = 2A_B$; the Young's moduli of the cylinders are related by

$E_A = 2E_B$. The cylinders had identical lengths before the brick was placed on them. What fraction of the brick's weight is supported (a) by cylinder A and (b) by cylinder B? The horizontal distances between the center of mass of the brick and the center-lines of the cylinders are d_A for cylinder A and d_B for cylinder B. (c) What is the ratio d_A/d_B?

FIGURE 13-56 Problem 54.

55P. In Fig. 13-57, a 103 kg uniform log hangs by two steel wires, A and B, both of radius 1.20 mm. Initially, wire A was 2.50 m long and 2.00 mm shorter than wire B. The log is now horizontal. What forces are exerted on it by (a) wire A and (b) wire B? (c) What is the ratio d_A/d_B?

FIGURE 13-57 Problem 55.

56P. Figure 13-58 is an overhead view of a rigid rod that turns about a vertical axle until the identical rubber stoppers A and B are forced against rigid walls at distances r_A and r_B from the axle. Initially the stoppers touch the walls, without being compressed. Then force F is applied perpendicular to the rod at a distance R from the axle. Find expressions for the forces compressing (a) stopper A and (b) stopper B?

FIGURE 13-58 Problem 56.

14
Gravitation

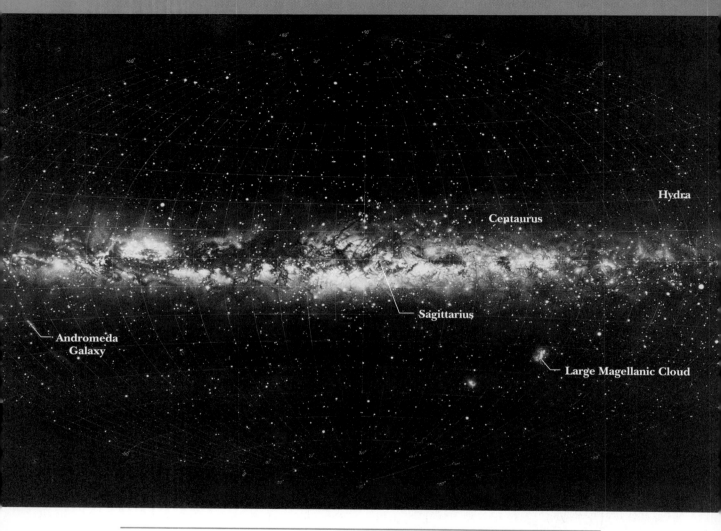

Hydra

Centaurus

Sagittarius

Andromeda
Galaxy

Large Magellanic Cloud

The Milky Way galaxy is a disk-shaped collection of dust, planets, and billions of stars, including our Sun and solar system. The force that binds it or any other galaxy together is the same force that holds Earth's moon in orbit and you on Earth—the gravitational force. That force is also responsible for one of nature's strangest objects, the black hole, a star that has completely collapsed onto itself. The gravitational force near a black hole is so strong that not even light can escape it. But if that is the case, how can a black hole be detected?

14-1 THE WORLD AND THE GRAVITATIONAL FORCE

The drawing that opens this chapter shows our view of the Milky Way galaxy: we are near the edge of the disk of the galaxy, about 26,000 light-years (2.5×10^{20} m) from its center, which in the drawing lies in the star collection known as Sagittarius. Our galaxy is a member of the Local Group of galaxies, which includes the Andromeda galaxy (Fig. 14-1) at a distance of 2.3×10^6 light-years, and several closer dwarf galaxies, such as the Large Magellanic Cloud shown in the opening drawing.

The Local Group is part of the Local Supercluster of galaxies. Measurements taken during and since the 1980s suggest that the Local Supercluster and the supercluster consisting of the clusters Hydra and Centaurus are all moving toward an exceptionally massive region called the Great Attractor. This region appears to be about 150 million light-years away, on the opposite side of the Milky Way from us, past the clusters Hydra and Centaurus.

The force that binds together these progressively larger structures, from star to galaxy to supercluster, and may be drawing them all toward the Great Attractor, is the gravitational force. That force not only holds you on Earth but also reaches out across intergalactic space.

FIGURE 14-1 The Andromeda galaxy. Located 2.3×10^6 light-years from us, and faintly visible to the naked eye, it is very similar to our home galaxy, the Milky Way.

14-2 NEWTON'S LAW OF GRAVITATION

Physicists like to examine seemingly unrelated phenomena to show that a relationship can be found if they are examined closely enough. This search for unification has been going on for centuries. In 1665, the 23-year-old Isaac Newton made a basic contribution to physics when he showed that the force that holds the Moon in its orbit is the same force that makes an apple fall. We take this so much for granted now that it is not easy for us to comprehend the ancient belief that the motions of earthbound bodies and heavenly bodies were different in kind and were governed by different laws.

Newton concluded that not only does Earth attract an apple and the Moon but every body in the universe attracts every other body; this tendency of bodies to move toward each other is called **gravitation.** Newton's conclusion takes a little getting used to, because the familiar attraction of Earth for earthbound bodies is so great that it overwhelms the attraction that earthbound bodies have for each other. For example, Earth attracts an apple with a force of several ounces. You also attract a nearby apple (and it attracts you), but the force of attraction is less than the weight of a speck of dust.

Quantitatively, Newton proposed a *force law* that we call **Newton's law of gravitation:** every particle attracts any other particle with a **gravitational force** whose magnitude is given by

$$F = G \frac{m_1 m_2}{r^2}$$

(Newton's law of gravitation).   (14-1)

Here m_1 and m_2 are the masses of the particles, r is the distance between them, and G is the **gravitational constant,** whose value is now known to be

$$G = 6.67 \times 10^{-11} \ \text{N} \cdot \text{m}^2/\text{kg}^2$$
$$= 6.67 \times 10^{-11} \ \text{m}^3/\text{kg} \cdot \text{s}^2. \quad (14\text{-}2)$$

As Fig. 14-2 shows, a particle m_2 attracts a particle m_1 with a gravitational force **F** that is directed toward particle m_2.

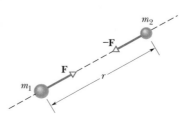

FIGURE 14-2 Two particles, of masses m_1 and m_2 and with a separation of r, attract each other according to Newton's law of gravitation, Eq. 14-1. The forces of attraction, **F** and $-$**F**, are equal in magnitude and are in opposite directions.

And particle m_1 attracts particle m_2 with a gravitational force $-\mathbf{F}$ that is directed toward m_1. The forces $\mathbf{F}$ and $-\mathbf{F}$ form an action–reaction pair and are opposite in direction but equal in magnitude. They depend on the separation of the two particles, but not on their location: the particles could be in a deep cave or in deep space. And $\mathbf{F}$ and $-\mathbf{F}$ are not altered by the presence of other bodies, even if those bodies lie between the two particles we are considering.

The strength of the gravitational force, that is, how strongly two given masses at a given separation attract each other, depends on the value of the gravitational constant G. If G—by some miracle—were suddenly multiplied by a factor of 10, you would be crushed to the floor by Earth's attraction. If it were divided by this factor, Earth's attraction would be weak enough for you to jump over a tall building.

Although Newton's law of gravitation applies strictly to particles, we can also apply it to real objects as long as the sizes of the objects are small compared to the distance between them. The Moon and Earth are far enough apart so that, to a good approximation, we can treat them both as particles. But what about an apple and Earth? From the point of view of the apple, the broad and level Earth, stretching out to the horizon beneath the apple, certainly does not look like a particle.

Newton solved the apple–Earth problem by proving an important theorem called the *shell theorem:*

A uniform spherical shell of matter attracts a particle that is outside the shell as if all the shell's mass were concentrated at its center.

Earth can be thought of as a nest of such shells, one within another, and each attracting a particle outside Earth's surface as if the mass of that shell were at the center of the shell. Thus, from the apple's point of view, Earth *does* behave like a particle, located at the center of Earth and having a mass equal to that of the planet.

Suppose, as in Fig. 14-3, that Earth pulls down on an apple with a force of 0.80 N. The apple must then pull up on Earth with a force of 0.80 N, which we take to act at the

FIGURE 14-3 The apple pulls up on Earth just as hard as Earth pulls down on the apple.

center of Earth. Although the forces are matched in magnitude, they produce different accelerations when the apple is released. For the apple, the acceleration is about 9.8 m/s², the familiar acceleration of a falling body near Earth's surface. For Earth, the acceleration measured in a reference frame attached to the center of mass of the apple–Earth system is only about 1×10^{-25} m/s².

$\mathbf{C}$HECKPOINT **1:** A particle is to be placed, in turn, outside four objects, each of mass m: (1) a large uniform solid sphere, (2) a large uniform spherical shell, (3) a small uniform solid sphere, and (4) a small uniform shell. In each situation, the distance between the particle and the center of the object is d. Rank the objects according to the magnitude of the gravitational force they exert on the particle, greatest first.

14-3 GRAVITATION AND THE PRINCIPLE OF SUPERPOSITION

Given a group of particles, we find the net (or resultant) gravitational force exerted on any one of them by using the **principle of superposition.** This is a general principle that says a net effect is the sum of the individual effects. Here, the principle means that we first compute the gravitational force that acts on our selected particle due to each of the other particles, in turn. We then find the net force by adding these forces vectorially, as usual.

For n interacting particles, we can write the principle of superposition for gravitational forces as

$$\mathbf{F}_1 = \mathbf{F}_{12} + \mathbf{F}_{13} + \mathbf{F}_{14} + \mathbf{F}_{15} + \cdots + \mathbf{F}_{1n}. \quad (14\text{-}3)$$

Here $\mathbf{F}_1$ is the net force on particle 1 and, for example, $\mathbf{F}_{13}$ is the force exerted on particle 1 by particle 3. We can express this equation more compactly as a vector sum:

$$\mathbf{F}_1 = \sum_{i=2}^{n} \mathbf{F}_{1i}. \quad (14\text{-}4)$$

What about the gravitational force exerted on a particle by a real extended object? The force is found by dividing the object into units small enough to treat as particles and then using Eq. 14-4 to find the vector sum of the forces exerted on the particle by all the units. In the limiting case, we can divide the extended object into differential units of mass dm, each of which exerts only a differential force $d\mathbf{F}$ on the particle. In this limit, the sum of Eq. 14-4 becomes an integral and we have

$$\mathbf{F}_1 = \int d\mathbf{F}, \quad (14\text{-}5)$$

in which the integral is taken over the entire extended object. If the object is a uniform sphere or a spherical shell, we can avoid the integration of Eq. 14-5 by assuming that the object's mass is concentrated at the object's center and using Eq. 14-1.

SAMPLE PROBLEM 14-1

Figure 14-4a shows an arrangement of five particles, with masses $m_1 = 8.0$ kg, $m_2 = m_3 = m_4 = m_5 = 2.0$ kg, and with $a = 2.0$ cm and $\theta = 30°$. What is the net gravitational force $\mathbf{F}_1$ acting on m_1 due to the other particles?

SOLUTION: From Eq. 14-4 we know that $\mathbf{F}_1$ is the vector sum of forces $\mathbf{F}_{12}$, $\mathbf{F}_{13}$, $\mathbf{F}_{14}$, and $\mathbf{F}_{15}$, which are the gravitational forces acting on m_1 due to the other particles. Because m_2 and m_4 are equal and because both are a distance $r = 2a$ from m_1, we have from Eq. 14-1

$$F_{12} = F_{14} = \frac{Gm_1m_2}{(2a)^2}. \qquad (14\text{-}6)$$

Similarly, since m_3 and m_5 are equal and are both a distance $r = a$ from m_1, we have

$$F_{13} = F_{15} = \frac{Gm_1m_3}{a^2}. \qquad (14\text{-}7)$$

Figure 14-4b is a free-body diagram for m_1. It and Eq. 14-6 show that $\mathbf{F}_{12}$ and $\mathbf{F}_{14}$ are equal in magnitude but opposite in direction; thus those forces cancel. Inspection of Fig. 14-4b and Eq. 14-7 reveals that the x components of $\mathbf{F}_{13}$ and $\mathbf{F}_{15}$ also cancel, and that their y components are identical in magnitude and both point toward increasing y. Thus $\mathbf{F}_1$ points toward increasing y, and its magnitude is twice the y component of $\mathbf{F}_{13}$:

$$F_1 = 2F_{13} \cos \theta = 2\frac{Gm_1m_3}{a^2} \cos \theta$$

$$= 2\frac{(6.67 \times 10^{-11} \text{ m}^3/\text{kg}\cdot\text{s}^2)(8.0 \text{ kg})(2.0 \text{ kg})}{(0.020 \text{ m})^2} \cos 30°$$

$$= 4.6 \times 10^{-6} \text{ N}. \qquad \text{(Answer)}$$

Note that the presence of m_5 along the line between m_1 and m_4 does not alter the gravitational force exerted by m_4 on m_1.

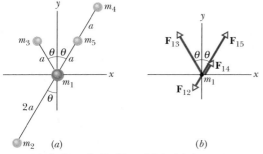

FIGURE 14-4 Sample Problem 14-1. (a) An arrangement of five particles. (b) The forces acting on the particle of mass m_1 due to the other four particles.

CHECKPOINT **2:** The figure shows four arrangements of three particles of equal masses. (a) Rank the arrangements according to the magnitude of the net gravitational force on the particle labeled m, greatest first. (b) In arrangement 2, is the direction of the net force closer to the line of length d or to the line of length D?

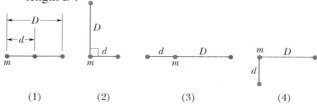

SAMPLE PROBLEM 14-2

In Fig. 14-5, a particle of mass $m_1 = 0.67$ kg is a distance $d = 23$ cm from one end of a uniform rod with length $L = 3.0$ m and mass $M = 5.0$ kg. What is the magnitude of the gravitational force $\mathbf{F}_1$ on the particle due to the rod?

SOLUTION: We consider a differential mass dm of the rod, located a distance r from m_1 and occupying a length dr along the rod. From Eq. 14-1, we may write the differential gravitational force dF_1 on m_1 due to dm as

$$dF_1 = \frac{Gm_1}{r^2} dm. \qquad (14\text{-}8)$$

The direction of this force is to the right in Fig. 14-5. In fact, because m_1 is located on a line through the length of the rod, each differential force dF_1 due to each differential mass dm in the rod is directed to the right. Thus, we can find the magnitude F_1 of the net force on m_1 simply by summing the magnitudes of these differential forces. We do so by integrating Eq. 14-8 along the length of the rod. (If m_1 were not on the central axis of the rod, the differential force vectors would point in different directions and the net force would have to be found by vector summation.)

The right-hand side of Eq. 14-8 contains two variables, r and dm. Before we can integrate, we must eliminate dm from that equation; we do so by using the fact that the rod is uniform in density to write

$$\frac{dm}{dr} = \frac{M}{L}. \qquad (14\text{-}9)$$

FIGURE 14-5 Sample Problem 14-2. A particle of mass m_1 is a distance d from one end of a rod of length L. A differential mass dm of the rod is a distance r from m_1.

This allows us to substitute $dm = (M/L)\,dr$ in Eq. 14-8. Then we integrate as indicated in Eq. 14-5 to get

$$F_1 = \int dF_1 = \int_d^{L+d} \frac{Gm_1}{r^2} \frac{M}{L}\,dr = \frac{Gm_1M}{L}\int_d^{L+d}\frac{dr}{r^2}$$

$$= -\frac{Gm_1M}{L}\left[\frac{1}{r}\right]_d^{L+d} = -\frac{Gm_1M}{L}\left[\frac{1}{L+d} - \frac{1}{d}\right]$$

$$= \frac{Gm_1M}{d(L+d)}$$

$$= \frac{(6.67 \times 10^{-11}\ \text{m}^3/\text{kg}\cdot\text{s}^2)(0.67\ \text{kg})(5.0\ \text{kg})}{(0.23\ \text{m})(3.0\ \text{m} + 0.23\ \text{m})}$$

$$= 3.0 \times 10^{-10}\ \text{N}. \qquad\qquad \text{(Answer)}$$

PROBLEM SOLVING TACTICS

TACTIC 1: *Drawing Gravitational Force Vectors*
When you are given a diagram of particles, such as Fig. 14-4a, and asked to find the net gravitational force on one of them, you should usually draw a free-body diagram showing only the particle of concern and the forces *it alone* experiences, as in Fig. 14-4b. If, instead, you choose to superimpose the vectors on the given diagram, be sure to draw the vectors with either their tails (preferably) or their heads on the particle experiencing those forces. If you draw the vectors elsewhere, you invite confusion. And confusion is guaranteed if you draw the vectors on the particles *causing* the forces on the particle of concern.

TACTIC 2: *Simplifying a Sum of Forces with Symmetry*
In Sample Problem 14-1 we used the symmetry of the situation to reduce the time and amount of calculation involved in the solution. By realizing that m_2 and m_4 are positioned symmetrically about m_1, and thus that $\mathbf{F}_{12}$ and $\mathbf{F}_{14}$ cancel, we avoided calculating either force. And by realizing that the x components of $\mathbf{F}_{13}$ and $\mathbf{F}_{15}$ cancel and that their y components are identical and add, we saved even more effort.

$\mathbf{C}$HECKPOINT **3:** In the figure, what is the direction of the net gravitational force on the particle of mass m_1 due to the other particles, each of mass m, that are arranged symmetrically relative to the y axis?

14-4 GRAVITATION NEAR EARTH'S SURFACE

Let us ignore the rotation of Earth for a moment and assume the planet to be a nonrotating uniform sphere of mass M. The magnitude of the gravitational force acting on a particle of mass m, located outside Earth a distance r from Earth's center, is then given by Eq. 14-1 as

$$F = G\frac{Mm}{r^2}. \qquad (14\text{-}10)$$

If the particle is released, it will fall toward the center of Earth, as a result of the gravitational force F, with an acceleration we shall call the **gravitational acceleration** a_g. Newton's second law tells us that F and a_g are related by

$$F = ma_g. \qquad (14\text{-}11)$$

Now, substituting F from Eq. 14-10 into Eq. 14-11 and solving for a_g, we find

$$a_g = \frac{GM}{r^2}. \qquad (14\text{-}12)$$

Table 14-1 shows values of a_g computed for various altitudes above Earth's surface.

The gravitational acceleration a_g computed with Eq. 14-12 is not the same as the free-fall acceleration g that we would measure for the falling particle (and that we have approximated as 9.80 m/s² near Earth's surface). The two accelerations differ for three reasons: (1) Earth is not uniform, (2) it is not a perfect sphere, and (3) it rotates. Moreover, because g differs from a_g, the weight mg of the particle differs from the gravitational force on the particle as given by Eq. 14-10 for the same three reasons. Let us now examine those reasons.

1. *Earth is not uniform.* The density of Earth varies radially as shown in Fig. 14-6, and the density of the crust (or outer section) of Earth varies from region to region over Earth's surface. Thus g varies from region to region.

2. *Earth is not a sphere.* Earth is approximately an ellipsoid, flattened at the poles and bulging at the equator. Its equatorial radius is greater than its polar radius by 21 km.

TABLE 14-1   **VARIATION OF a_g WITH ALTITUDE**

ALTITUDE (km)	a_g (m/s²)	ALTITUDE EXAMPLE
0	9.83	Mean Earth surface
8.8	9.80	Mt. Everest
36.6	9.71	Highest manned balloon
400	8.70	Space shuttle orbit
35,700	0.225	Communications satellite

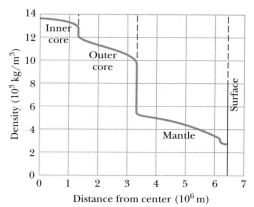

FIGURE 14-6 The density of Earth as a function of distance from the center. The limits of the solid inner core, the largely liquid outer core, and the solid mantle are shown, but the crust of Earth is too thin to show clearly on this plot.

Thus a point at the poles is closer to the dense core of Earth than is a point on the equator. This is one reason the free-fall acceleration g increases as one proceeds, at sea level, from the equator toward the poles.

3. **Earth is rotating.** The rotation axis runs through the north and south poles of Earth. An object located on Earth's surface anywhere except at those poles must rotate in a circle about the rotation axis and thus must have a centripetal acceleration that points toward the center of the circle. This centripetal acceleration requires a centripetal force that also points toward that center.

To see how Earth's rotation causes the free-fall acceleration g to differ from the gravitational acceleration a_g, and causes weight to differ from the gravitational force given by Eq. 14-10, we analyze a simple situation in which a crate of mass m rests on a scale at the equator. Figure 14-7a shows this situation as viewed from a point in space above the north pole.

Figure 14-7b is a free-body diagram for the crate. The centripetal acceleration **a** of the crate points toward the center of the circle it travels, which is coincident with the center of Earth (assumed spherical). Earth exerts a gravitational force on the crate with magnitude ma_g (Eq. 14-11). The scale exerts an upward normal force **N** on the crate. Applying Newton's second law to the crate, with the positive direction being radially inward toward Earth's center, we find

$$\sum F = ma_g - N = ma. \qquad (14\text{-}13)$$

The magnitude of **N** is the reading on the scale and thus is equal to the weight mg of the crate. Substituting mg for N in Eq. 14-13 yields

$$ma_g - mg = ma, \qquad (14\text{-}14)$$

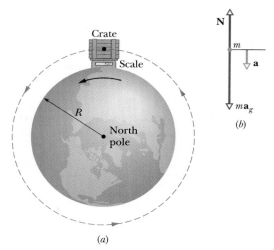

FIGURE 14-7 (a) A crate resting on a platform scale at Earth's equator as Earth rotates. The view is along Earth's rotational axis, looking down on the north pole. (b) A free-body diagram for the crate. The crate is in uniform circular motion and is thus accelerated toward the center of Earth. The gravitational force on it has magnitude ma_g. The normal force **N** from the scale has magnitude mg, where g is the free-fall acceleration.

which shows that the weight mg of the crate differs from the magnitude ma_g of the gravitational force acting on the crate. Dividing Eq. 14-14 by m shows also that g is different from a_g by an amount equal to the centripetal acceleration a.

For the centripetal acceleration a we can substitute $\omega^2 R$, where ω is the angular speed of Earth's rotation and R is the radius of the circular path taken by the crate. (R is approximately the radius of Earth.) For ω we can substitute $2\pi/T$, where $T = 24$ h is approximately the time for one rotation of Earth. When we make these substitutions and divide by m, Eq. 14-14 gives us

$$a_g - g = \omega^2 R = \left(\frac{2\pi}{T}\right)^2 R \qquad (14\text{-}15)$$

$$= 0.034 \text{ m/s}^2.$$

Thus the free-fall acceleration g (9.78 m/s^2 $\approx$ 9.8 m/s^2) measured on the equator of the real, rotating planet is slightly less than the gravitational acceleration a_g due strictly to the gravitational force. The difference between g and a_g becomes progressively smaller as the crate is moved to progressively greater latitudes, because the crate then travels in a smaller circle, making R in Eq. 14-15 smaller. So, often we can approximate the free-fall acceleration g with the gravitational acceleration a_g. And we can approximate the weight mg of an object with the gravitational force (given by Eq. 14-10).

SAMPLE PROBLEM 14-3

Consider a pulsar, a collapsed star of extremely high density, with a mass M equal to that of the Sun (1.98×10^{30} kg), a radius R of only 12 km, and a rotational period T of 0.041 s. At its equator, by what percentage does the free-fall acceleration g differ from the gravitational acceleration a_g?

SOLUTION: To find a_g on the surface of the pulsar, we use Eq. 14-12, with R replacing r and with M being the mass of the pulsar. Substituting the known values gives

$$a_g = \frac{GM}{R^2} = \frac{(6.67 \times 10^{-11} \text{ m}^3/\text{kg} \cdot \text{s}^2)(1.98 \times 10^{30} \text{ kg})}{(12{,}000 \text{ m})^2}$$

$$= 9.2 \times 10^{11} \text{ m/s}^2.$$

Dividing Eq. 14-15 (which applies to any rotating body) by a_g and substituting the known data, we now find

$$\frac{a_g - g}{a_g} = \left(\frac{2\pi}{T}\right)^2 \frac{R}{a_g} = \left(\frac{2\pi}{0.041 \text{ s}}\right)^2 \frac{12{,}000 \text{ m}}{9.2 \times 10^{11} \text{ m/s}^2}$$

$$= 3.1 \times 10^{-4} = 0.031\%. \qquad \text{(Answer)}$$

Even though a pulsar rotates extremely rapidly, its rotation reduces the free-fall acceleration from the gravitational acceleration only slightly, because its radius is so small.

SAMPLE PROBLEM 14-4

(a) An astronaut whose height h is 1.70 m floats feet "down" in an orbiting space shuttle at a distance $r = 6.77 \times 10^6$ m from the center of Earth. What is the difference in the gravitational acceleration at her feet and at her head?

SOLUTION: Equation 14-12 tells us the gravitational acceleration at any distance r from the center of Earth:

$$a_g = \frac{GM_E}{r^2}, \qquad (14\text{-}16)$$

where M_E is Earth's mass. We cannot simply apply Eq. 14-16 twice, first with, say, $r = 6.77 \times 10^6$ m for the feet and then with $r = 6.77 \times 10^6$ m + 1.70 m for the head. If we do, a calculator will give us the same value for a_g twice, and thus a difference of zero, because h is so small compared to r. Instead, we differentiate Eq. 14-16 with respect to r, obtaining

$$da_g = -2\frac{GM_E}{r^3} \, dr, \qquad (14\text{-}17)$$

where da_g is the differential change in the gravitational acceleration due to a differential change dr in r. For the astronaut, $dr = h$ and $r = 6.77 \times 10^6$ m. Substituting data into Eq. 14-17, we find

$$da_g = -2 \frac{(6.67 \times 10^{-11} \text{ m}^3/\text{kg} \cdot \text{s}^2)(5.98 \times 10^{24} \text{ kg})}{(6.77 \times 10^6 \text{ m})^3} (1.70 \text{ m})$$

$$= -4.37 \times 10^{-6} \text{ m/s}^2. \qquad \text{(Answer)}$$

This result means that the gravitational acceleration of the astronaut's feet toward Earth is slightly greater than the gravitational acceleration of her head toward Earth. This difference in acceleration tends to stretch her body, but the difference is so small that the stretching is unnoticeable.

(b) If the astronaut is now feet "down" at the same orbital radius r of 6.77×10^6 m about a black hole of mass $M_h = 1.99 \times 10^{31}$ kg (which is 10 times our Sun's mass), what is the difference in the gravitational acceleration at her feet and head? The black hole has a surface (called the *horizon* of the black hole) of radius $R_h = 2.95 \times 10^4$ m. Nothing, not even light, can escape from the surface or anywhere inside it. Note that the astronaut is (wisely) well outside the surface (at $r = 229R_h$).

SOLUTION: We can again use Eq. 14-17 if we substitute $M_h = 1.99 \times 10^{31}$ kg for M_E. We find

$$da_g = -2\frac{(6.67 \times 10^{-11} \text{ m}^3/\text{kg} \cdot \text{s}^2)(1.99 \times 10^{31} \text{ kg})}{(6.77 \times 10^6 \text{ m})^3} (1.70 \text{ m})$$

$$= -14.5 \text{ m/s}^2. \qquad \text{(Answer)}$$

This means that the gravitational acceleration of the astronaut's feet toward the black hole is noticeably larger than that of her head. The resulting tendency to stretch her body would be bearable but quite painful. If she drifted closer to the black hole, the stretching tendency would increase drastically.

14-5 GRAVITATION INSIDE EARTH

Newton's shell theorem can also be applied to a situation in which a particle is located *inside* a uniform shell, to show the following:

> A uniform shell of matter exerts no *net* gravitational force on a particle located inside it.

If the density of Earth were uniform, the gravitational force acting on a particle would be a maximum at Earth's surface. The force would, as we expect, decrease as the particle moved outward. If the particle were to move inward, perhaps down a deep mine shaft, the gravitational force would change for two reasons. (1) It would tend to increase because the particle would be moving closer to the center of Earth. (2) It would tend to decrease because the shell of material lying outside the particle's radial position would not exert any net force on the particle.

For a uniform Earth, the second influence prevails and the force on the particle steadily decreases to zero as the particle approaches the center of Earth. However, for the real (nonuniform) Earth, the force on the particle actually increases as the particle begins to descend. The force reaches a maximum at a certain depth; only then does it begin to decrease as the particle descends further.

SAMPLE PROBLEM 14-5

Suppose a tunnel runs through Earth from pole to pole, as in Fig. 14-8. Assume that Earth is a nonrotating, uniform sphere. Find the gravitational force on a particle of mass m dropped into the tunnel when it reaches a distance r from Earth's center.

SOLUTION: The force that acts on the particle is associated only with the mass M' of Earth that lies within a sphere of radius r. The portion of Earth that lies outside this sphere does not exert any net force on the particle.

Mass M' is given by

$$M' = \rho V' = \rho \, \frac{4\pi r^3}{3}, \qquad (14\text{-}18)$$

in which V' is the volume occupied by M' (which lies within the dashed line in Fig. 14-8), and ρ is the assumed uniform density of Earth.

The force acting on the particle is then, using Eqs. 14-1 and 14-18,

$$F = -\frac{GmM'}{r^2} = -\frac{Gm\rho 4\pi r^3}{3r^2} = -\left(\frac{4\pi mG\rho}{3}\right) r$$

$$= -Kr, \qquad \text{(Answer)} \qquad (14\text{-}19)$$

in which K, a constant, is equal to $4\pi mG\rho/3$. We have inserted a minus sign to indicate that the force $\mathbf{F}$ and the displacement $\mathbf{r}$ are in opposite directions, the former being toward the center of Earth and the latter away from that point. Thus Eq. 14-19 tells us that the force acting on the particle is proportional to the displacement of the particle but oppositely directed, precisely the situation for Hooke's law.

FIGURE 14-8 Sample Problem 14-5. A particle is dropped into a tunnel drilled through Earth.

14-6 GRAVITATIONAL POTENTIAL ENERGY

In Section 8-3, we discussed the gravitational potential energy of a particle–Earth system. We were careful to keep the particle near Earth's surface, so that we could regard the gravitational force as constant. We then chose some reference configuration of the system as having a gravitational potential energy of zero. Often, in this configuration the particle was on Earth's surface. For particles not on Earth's surface, the gravitational potential energy decreased when the separation between the particle and Earth decreased.

Here, we broaden our view and consider the gravitational potential energy U of two particles, of masses m and M, separated by a distance r. We again choose a reference configuration with U equal to zero. However, to simplify the equations, the separation distance r in the reference configuration is now large enough to be approximated as *infinite*. As before, the gravitational potential energy decreases when the separation decreases. Since $U = 0$ for $r = \infty$, the potential energy is negative for any finite separation and becomes progressively more negative as the particles move closer together.

With these facts in mind and as we shall justify below, we take the gravitational potential energy of the two-particle system to be

$$U = -\frac{GMm}{r} \qquad \begin{array}{l}\text{(gravitational}\\ \text{potential energy).}\end{array} \qquad (14\text{-}20)$$

Note that $U(r)$ approaches zero as r approaches infinity and that for any finite value of r, the value of $U(r)$ is negative.

The potential energy given by Eq. 14-20 is a property of the system of two particles rather than of either particle alone. There is no way to divide this energy and say that so much belongs to one particle and so much to the other. Nevertheless, if $M \gg m$, as is true for Earth and a baseball, we often speak of "the potential energy of the baseball." We can get away with this because, when a baseball moves in the vicinity of Earth, changes in the potential energy of the baseball–Earth system appear almost entirely as changes in the kinetic energy of the baseball, since changes in the kinetic energy of Earth are too small to be measured. Similarly, in Section 14-8 we shall speak of "the potential energy of an artificial satellite" orbiting Earth, because the satellite's mass is so much smaller than Earth's mass. When we speak of the potential energy of bodies of comparable mass, however, we have to be careful to treat them as a system.

If our system contains more than two particles, we consider each pair of particles in turn, calculate the gravi-

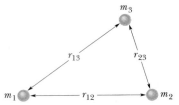

FIGURE 14-9 Three particles exert gravitational forces on one another. The gravitational potential energy of the system is the sum of the gravitational potential energies of each of the three possible pairs of the particles.

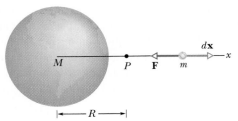

FIGURE 14-11 A baseball of mass m falls toward Earth from infinity, along a radial line (an x axis) passing through point P at a distance R from the center of Earth.

tational potential energy of that pair with Eq. 14-20 as if the other particles were not there, and then algebraically sum the results. Applying Eq. 14-20 to each of the three pairs of Fig. 14-9, for example, gives the potential energy of the system as

$$U = -\left(\frac{Gm_1m_2}{r_{12}} + \frac{Gm_1m_3}{r_{13}} + \frac{Gm_2m_3}{r_{23}}\right). \quad (14\text{-}21)$$

A *globular cluster* (Fig. 14-10) in the constellation Sagittarius is a good example of a naturally occurring system of particles. It contains about 70,000 stars, which can be paired in 2.5×10^9 different ways. Contemplation of this structure suggests the enormous amount of gravitational potential energy stored in the universe.

Proof of Eq. 14-20

Let a baseball, starting from rest at a great (infinite) distance from Earth, fall toward point P, as in Fig. 14-11. The potential energy of the baseball–Earth system is initially

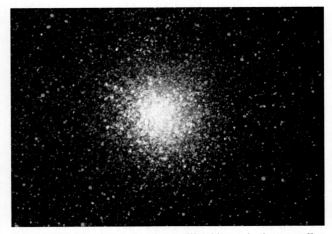

FIGURE 14-10 A globular cluster, like this one in the constellation Sagittarius, contains tens of thousands of stars in an almost spherical collection. There are many such clusters in our Milky Way galaxy, some of which are visible with a small telescope.

zero. When the baseball reaches P, the potential energy is the negative of the work W done by the gravitational force as the baseball moves to P from its distant position. Thus, from Eqs. 8-1 and 8-6 (generalized to vector form), we have

$$U = -W = -\int_{\infty}^{R} \mathbf{F}(x) \cdot d\mathbf{x}. \quad (14\text{-}22)$$

The limits on the integral are the initial distance of the baseball, which we have taken to be infinitely great, and its final distance R.

The vector $\mathbf{F}(x)$ in Eq. 14-22 points radially inward toward the center of Earth in Fig. 14-11, and the vector $d\mathbf{x}$ points radially outward, the angle ϕ between them being $180°$. Thus

$$\mathbf{F}(x) \cdot d\mathbf{x} = F(x)(\cos 180°)(dx)$$
$$= -F(x)\, dx. \quad (14\text{-}23)$$

For $F(x)$ in Eq. 14-23, we now substitute from Newton's law of gravitation (Eq. 14-1) written as a function of x rather than r, obtaining

$$\mathbf{F}(x) \cdot d\mathbf{x} = -\frac{GMm}{x^2}\, dx.$$

Putting that result into Eq. 14-22, we obtain

$$U = \int_{\infty}^{R} \left(\frac{GMm}{x^2}\right) dx = -\left[\frac{GMm}{x}\right]_{\infty}^{R} = -\frac{GMm}{R},$$

which, with a change in symbol from R to r, is Eq. 14-20.

Equation 14-20 holds no matter what path the baseball takes in moving in toward Earth. Consider a path made up of small steps, as in Fig. 14-12. No work is done along circular steps like AB, because along them the (radial) gravitational force is perpendicular to the displacement. The total work done along all the radial steps, such as BC, is the same as the work done in moving along a single radial line, as in Fig. 14-11. Thus the work done on a particle by the gravitational force when the particle moves between any two points is independent of the path that the

FIGURE 14-12 The work done by the gravitational force as a baseball moves from A to E is independent of the path followed.

particle follows and depends only on the particle's initial and final positions. The work may be computed simply as the negative of the difference in potential energy between the two points:

$$W = -\Delta U = -(U_f - U_i), \qquad (14\text{-}24)$$

where U_f and U_i are the potential energies associated with the final and initial positions. This is what we mean when we say, as we did in Chapter 8, that the gravitational force is *conservative*. And, as discussed there, if the work *did* depend on the path, as it does for frictional forces, there would be no potential energy associated with the gravitational force.

Potential Energy and Force

In the proof of Eq. 14-20, we derived the potential energy function U from the force function F. We should be able to go the other way, that is, to start from the potential energy function and derive the force function. Guided by Eq. 8-19, we can write

$$F = -\frac{dU}{dr} = -\frac{d}{dr}\left(-\frac{GMm}{r}\right)$$

$$= -\frac{GMm}{r^2}. \qquad (14\text{-}25)$$

This is just Newton's law of gravitation (Eq. 14-1). The minus sign indicates that the force on mass m points radially inward, toward mass M.

Escape Speed

If you fire a projectile upward, usually it will slow, stop momentarily, and return to Earth. There is, however, a certain minimum initial speed that will cause it to move upward forever, theoretically coming to rest only at infinity. This initial speed is called the **escape speed.**

Consider a projectile of mass m, leaving the surface of a planet (or some other astronomical body or system) with escape speed v. It has a kinetic energy K given by $\frac{1}{2}mv^2$ and a potential energy U given by Eq. 14-20:

$$U = -\frac{GMm}{R},$$

in which M is the mass of the planet and R its radius.

When the projectile reaches infinity, it stops and thus has no kinetic energy. It also has no potential energy because this is our zero-potential-energy configuration. Its total energy at infinity is therefore zero. From the principle of conservation of energy, its total energy at the planet's surface must also have been zero, so

$$K + U = \tfrac{1}{2}mv^2 + \left(-\frac{GMm}{R}\right) = 0.$$

This yields

$$v = \sqrt{\frac{2GM}{R}}. \qquad (14\text{-}26)$$

The escape speed does not depend on the direction in which a projectile is fired. However, attaining that speed is easier if the projectile is fired in the direction the launch site is moving as Earth rotates about its axis. For example, rockets are launched eastward at Cape Canaveral to take advantage of the Cape's eastward speed of 1500 km/h.

Equation 14-26 can be applied to find the escape speed of a projectile from any astronomical body, provided we substitute the mass of the body for M and the radius of the body for R. Table 14-2 shows escape speeds from some astronomical bodies.

CHECKPOINT 4: You move a ball of mass m away from a sphere of mass M. (a) Does the gravitational potential energy of the ball–sphere system increase or decrease? (b) Is positive or negative work done by the gravitational force between the ball and the sphere?

TABLE 14-2 **SOME ESCAPE SPEEDS**

BODY	MASS (kg)	RADIUS (m)	ESCAPE SPEED (km/s)
Ceres[a]	1.17×10^{21}	3.8×10^5	0.64
Earth's moon	7.36×10^{22}	1.74×10^6	2.38
Earth	5.98×10^{24}	6.37×10^6	11.2
Jupiter	1.90×10^{27}	7.15×10^7	59.5
Sun	1.99×10^{30}	6.96×10^8	618
Sirius B[b]	2×10^{30}	1×10^7	5200
Neutron star[c]	2×10^{30}	1×10^4	2×10^5

[a] The most massive of the asteroids.

[b] A *white dwarf* (a star in a final stage of evolution) that is a companion of the bright star Sirius.

[c] The collapsed core of a star that remains after that star has exploded in a *supernova* event.

SAMPLE PROBLEM 14-6

An asteroid, headed directly toward Earth, has a speed of 12 km/s relative to the planet when it is at a distance of 10 Earth radii from Earth's center. Ignoring the effects of the terrestrial atmosphere on the asteroid, find the asteroid's speed when it reaches Earth's surface.

SOLUTION: Because the mass of an asteroid is much less than that of Earth, we can assign the gravitational potential energy of the asteroid–Earth system to the asteroid alone, and we can neglect any change in the speed of Earth relative to the asteroid during the asteroid's fall. Because we are to ignore the effects of the atmosphere on the asteroid, the mechanical energy of the asteroid is conserved during its fall; that is,

$$K_f + U_f = K_i + U_i,$$

where K and U are the kinetic energy and gravitational potential energy of the asteroid, and the subscripts i and f refer to initial values (at a distance of 10 Earth radii) and final values (at a distance of one Earth radius).

Let m represent the mass of the asteroid, M the mass of Earth ($= 5.98 \times 10^{24}$ kg), and R the radius of Earth ($= 6.37 \times 10^6$ m). Using Eq. 14-20 for the gravitational potential energy and $\frac{1}{2}mv^2$ for K, we have

$$\tfrac{1}{2}mv_f^2 - \frac{GMm}{R} = \tfrac{1}{2}mv_i^2 - \frac{GMm}{10R}.$$

Rearranging and substituting known values, we find

$$v_f^2 = v_i^2 + \frac{2GM}{R}\left(1 - \frac{1}{10}\right)$$

$$= (12 \times 10^3 \text{ m/s})^2$$

$$+ \frac{2(6.67 \times 10^{-11} \text{ m}^3/\text{kg} \cdot \text{s}^2)(5.98 \times 10^{24} \text{ kg})}{6.37 \times 10^6 \text{ m}} 0.9$$

$$= 2.567 \times 10^8 \text{ m}^2/\text{s}^2,$$

and thus

$$v_f = 1.60 \times 10^4 \text{ m/s} = 16 \text{ km/s}. \qquad \text{(Answer)}$$

At this speed, the asteroid would not have to be particularly large to do considerable damage. As an example, if it were only 5 m across, the impact could release about as much energy as the nuclear explosion at Hiroshima. Alarmingly, about 500 million asteroids of this size are near Earth's orbit, and in 1994 one of them apparently penetrated Earth's atmosphere and exploded at an altitude of 20 km near a remote South Pacific island (setting off nuclear-explosion warnings on six military satellites). An asteroid 500 m across (there may be a million of them near Earth's orbit) could end modern civilization and almost eliminate humans worldwide.

14-7 PLANETS AND SATELLITES: KEPLER'S LAWS

The motions of the planets, as they seemingly wander against the background of the stars, have been a puzzle since the dawn of history. The "loop-the-loop" motion of Mars, shown in Fig. 14-13, was particularly baffling. Johannes Kepler (1571–1630), after a lifetime of study, worked out the empirical laws that govern these motions. Tycho Brahe (1546–1601), the last of the great astronomers to make observations without the help of a telescope, compiled the extensive data from which Kepler was able to derive the three laws of planetary motion that now bear his name. Later, Newton (1642–1727) showed that his law of gravitation leads to Kepler's empirical laws.

We discuss each of Kepler's laws in turn. Although we apply the laws here to planets orbiting the Sun, they hold equally well for satellites, either natural or artificial, orbiting Earth or any other massive central body.

1. THE LAW OF ORBITS: All planets move in elliptical orbits, with the Sun at one focus.

Figure 14-14 shows a planet, of mass m, moving in such an orbit around the Sun, whose mass is M. We assume that $M \gg m$, so that the center of mass of the planet–Sun system is virtually at the center of the Sun.

The orbit in Fig. 14-14 is described by giving its **semimajor axis** a and its **eccentricity** e, the latter defined

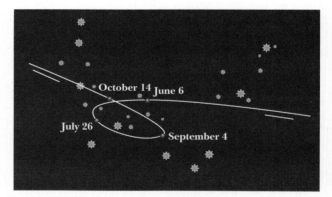

FIGURE 14-13 The path of the planet Mars as it moved against a background of the constellation Capricorn during 1971. Its position on four selected days is marked. Both Mars and Earth are moving in orbits around the Sun so that we see the position of Mars relative to us; this sometimes results in an apparent loop in the path of Mars.

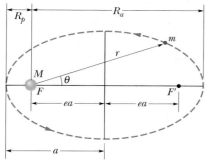

FIGURE 14-14 A planet of mass m moving in an elliptical orbit around the Sun. The Sun, of mass M, is at one focus F of the ellipse. The other, or "empty," focus is F'. Each focus is a distance ea from the center, e being the eccentricity of the ellipse. The semimajor axis a of the ellipse, the perihelion (nearest the Sun) distance R_p, and the aphelion (farthest from the Sun) distance R_a are also shown.

so that ea is the distance from the center of the ellipse to either focus F or F'. *An eccentricity of zero corresponds to a circle,* in which the two foci merge to a single central point. The eccentricities of the planetary orbits are not large, so—sketched on paper—the orbits look circular. The eccentricity of the ellipse of Fig. 14-14, which has been exaggerated for clarity, is 0.74. The eccentricity of Earth's orbit is only 0.0167.

---

2. THE LAW OF AREAS: A line that connects a planet to the Sun sweeps out equal areas in equal times.

Qualitatively, this second law tells us that the planet will move most slowly when it is farthest from the Sun and most rapidly when it is nearest to the Sun. As it turns out, Kepler's second law is totally equivalent to the law of conservation of angular momentum. Let us prove it.

The area of the shaded wedge in Fig. 14-15a closely approximates the area swept out in time Δt by a line connecting the Sun and the planet, which are separated by a distance r. The area ΔA of the wedge is approximately the area of a triangle with base $r\,\Delta\theta$ and height r. Thus $\Delta A \approx \frac{1}{2}r^2\,\Delta\theta$. This expression for ΔA becomes more exact as Δt (hence $\Delta\theta$) approaches zero. The instantaneous rate at which area is being swept out is then

$$\frac{dA}{dt} = \frac{r^2}{2}\frac{d\theta}{dt} = \frac{r^2\omega}{2}, \qquad (14\text{-}27)$$

in which ω is the angular speed of the rotating line connecting Sun and planet.

Figure 14-15b shows the linear momentum $\mathbf{p}$ of the planet, along with its components. From Eq. 12-27, the magnitude of the angular momentum $\mathbf{L}$ of the planet about the Sun is given by the product of r and the component of $\mathbf{p}$ perpendicular to r, or

$$L = rp_\perp = (r)(mv_\perp) = (r)(m\omega r)$$
$$= mr^2\omega, \qquad (14\text{-}28)$$

where we have replaced $v_\perp$ with its equivalent ωr (Eq. 11-16). Eliminating $r^2\omega$ between Eqs. 14-27 and 14-28 leads to

$$\frac{dA}{dt} = \frac{L}{2m}. \qquad (14\text{-}29)$$

If dA/dt is constant, as Kepler said it is, then Eq. 14-29 means that L must also be constant—angular momentum is conserved. So Kepler's second law is indeed equivalent to the law of conservation of angular momentum.

---

3. THE LAW OF PERIODS: The square of the period of any planet is proportional to the cube of the semimajor axis of its orbit.

To see this, consider a circular orbit with radius r (the radius is equivalent to the semimajor axis of an ellipse).

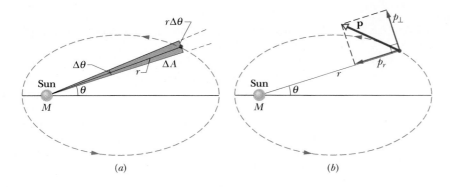

FIGURE 14-15 (a) In time Δt, the line r connecting the planet to the Sun (of mass M) sweeps through an angle $\Delta\theta$, sweeping out an area ΔA. (b) The linear momentum $\mathbf{p}$ of the planet and its components.

FIGURE 14-16 A planet of mass m moving around the Sun in a circular orbit of radius r.

Applying Newton's second law, $F = ma$, to the orbiting planet in Fig. 14-16 yields

$$\frac{GMm}{r^2} = (m)(\omega^2 r). \qquad (14\text{-}30)$$

Here we have substituted from Eq. 14-1 for the force F and used Eq. 11-21 to substitute $\omega^2 r$ for the centripetal acceleration. If we replace ω with $2\pi/T$ (Eq. 11-18), where T is the period of the motion, we obtain Kepler's third law:

$$T^2 = \left(\frac{4\pi^2}{GM}\right) r^3 \qquad \text{(law of periods).} \qquad (14\text{-}31)$$

The quantity in parentheses is a constant, its value depending only on the mass of the central body.

Equation 14-31 holds also for elliptical orbits, provided we replace r with a, the semimajor axis of the ellipse.

On February 7, 1984, at a height of 102 km above Hawaii and with a speed of about 29,000 km/h, Bruce McCandless stepped (untethered) into space from a space shuttle and became the first human satellite.

TABLE 14-3
KEPLER'S LAW OF PERIODS FOR THE SOLAR SYSTEM

PLANET	SEMIMAJOR AXIS a (10^{10} m)	PERIOD T (y)	T^2/a^3 (10^{-34} y^2/m^3)
Mercury	5.79	0.241	2.99
Venus	10.8	0.615	3.00
Earth	15.0	1.00	2.96
Mars	22.8	1.88	2.98
Jupiter	77.8	11.9	3.01
Saturn	143	29.5	2.98
Uranus	287	84.0	2.98
Neptune	450	165	2.99
Pluto	590	248	2.99

This law predicts that the ratio T^2/a^3 has essentially the same value for every planetary orbit around a given massive body. Table 14-3 shows how well it holds for the orbits of the planets of the solar system.

CHECKPOINT 5: Satellite 1 is in a certain circular orbit about a planet, while satellite 2 is in a larger circular orbit. Which satellite has (a) the longer period and (b) the greater speed?

SAMPLE PROBLEM 14-7

A satellite in circular orbit at an altitude h of 230 km above Earth's surface has a period T of 89 min. What mass of Earth follows from these data?

SOLUTION: We apply Kepler's law of periods to the satellite–Earth system. Solving Eq. 14-31 for M yields

$$M = \frac{4\pi^2 r^3}{GT^2}. \qquad (14\text{-}32)$$

The radius r of the satellite orbit is

$$r = R + h = 6.37 \times 10^6 \text{ m} + 230 \times 10^3 \text{ m}$$
$$= 6.60 \times 10^6 \text{ m},$$

in which R is the radius of Earth. Substituting this value and the period T into Eq. 14-32 yields

$$M = \frac{(4\pi^2)(6.60 \times 10^6 \text{ m})^3}{(6.67 \times 10^{-11} \text{ m}^3/\text{kg} \cdot \text{s}^2)(89 \times 60 \text{ s})^2}$$
$$= 6.0 \times 10^{24} \text{ kg.} \qquad \text{(Answer)}$$

In the same way, we could find the mass of our Sun from the period and radius of Earth's orbit (assumed circular), and the mass of Jupiter from the period and orbital radius of any one of its moons (without knowing the mass of that moon).

SAMPLE PROBLEM 14-8

Comet Halley orbits about the Sun with a period of 76 years and, in 1986, had a distance of closest approach to the Sun, its *perihelion distance* R_p, of 8.9×10^{10} m. Table 14-3 shows that this is between the orbits of Mercury and Venus.

(a) What is the comet's farthest distance from the Sun, its *aphelion distance* R_a?

SOLUTION: We can find the semimajor axis of the orbit of comet Halley from Eq. 14-31. Substituting a for r in that equation and solving for a yield

$$a = \left(\frac{GMT^2}{4\pi^2} \right)^{1/3}. \qquad (14\text{-}33)$$

If we substitute the mass M of the Sun, 1.99×10^{30} kg, and the period T of the comet, 76 years or 2.4×10^9 s, into Eq. 14-33 we find that $a = 2.7 \times 10^{12}$ m.

Study of Fig. 14-14 shows that $R_a + R_p = 2a$, or

$$R_a = 2a - R_p$$
$$= (2)(2.7 \times 10^{12} \text{ m}) - 8.9 \times 10^{10} \text{ m}$$
$$= 5.3 \times 10^{12} \text{ m}. \qquad \text{(Answer)}$$

Table 14-3 shows that this is just a little less than the semimajor axis of the orbit of Pluto.

(b) What is the eccentricity of the orbit of comet Halley?

SOLUTION: Figure 14-14 shows that $ea = a - R_p$, or

$$e = \frac{a - R_p}{a} = 1 - \frac{R_p}{a}$$
$$= 1 - \frac{8.9 \times 10^{10} \text{ m}}{2.7 \times 10^{12} \text{ m}} = 0.97. \qquad \text{(Answer)}$$

This cometary orbit, with an eccentricity approaching unity, is a long thin ellipse.

SAMPLE PROBLEM 14-9

Observations of the light from a certain star indicate that it is part of a binary (two-star) system. This visible star has orbital speed $v = 270$ km/s, orbital period $T = 1.70$ days, and approximate mass $m_1 = 6M_s$, where M_s is the Sun's mass, 1.99×10^{30} kg. Assuming that the visible star and its companion object, which is dark and unseen, are in circular orbits (see Fig. 14-17), determine the approximate mass m_2 of the dark object.

SOLUTION: As in the two particle systems of Section 9-2, the center of mass of this two-star system lies along a line connecting their centers. As with the freely rotating bodies and systems of Chapter 12, this two-star system rotates about its center of mass. In Fig. 14-17, the center of mass is labeled point O; the visible star and dark object orbit O with orbital

radii r_1 and r_2, respectively, and have a separation of $r = r_1 + r_2$. From Eq. 14-1, the gravitational force on the visible star from the dark object is

$$F = \frac{Gm_1m_2}{r^2}.$$

Applying Newton's second law, $F = ma$, to the visible star then yields

$$\frac{Gm_1m_2}{r^2} = m_1a = (m_1)(\omega^2 r_1), \qquad (14\text{-}34)$$

where ω is the angular speed of the visible star and $\omega^2 r_1$ is its centripetal acceleration toward O.

We can obtain another equation in these same variables by locating the center of mass O. Because O is a distance r_1 from the visible star, we can use Eq. 9-1 to write

$$r_1 = \frac{m_2 r}{m_1 + m_2}.$$

This yields

$$r = r_1 \frac{m_1 + m_2}{m_2}. \qquad (14\text{-}35)$$

Now substituting Eq. 14-35 for r and $2\pi/T$ for ω into Eq. 14-34 gives us, after some algebraic manipulation,

$$\frac{m_2^3}{(m_1 + m_2)^2} = \frac{4\pi^2}{GT^2} r_1^3. \qquad (14\text{-}36)$$

We still have two unknowns, m_2 and r_1. We can, however, find r_1 from the orbital motion of the visible star: the orbital period T is equal to the circumference of the orbit $(2\pi r_1)$ divided by the speed v of the planet. This gives us

$$T = \frac{2\pi r_1}{v},$$

or

$$r_1 = \frac{vT}{2\pi}. \qquad (14\text{-}37)$$

If we substitute $m_1 = 6M_s$ and Eq. 14-37 for r_1, then Eq. 14-36 becomes

$$\frac{m_2^3}{(6M_s + m_2)^2} = \frac{v^3 T}{2\pi G}$$
$$= \frac{(2.7 \times 10^5 \text{ m/s})^3 (1.70 \text{ days})(86,400 \text{ s/day})}{(2\pi)(6.67 \times 10^{-11} \text{ N} \cdot \text{m}^2/\text{kg}^2)}$$
$$= 6.90 \times 10^{30} \text{ kg,}$$

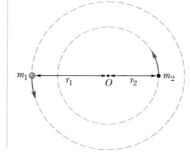

FIGURE 14-17 Sample Problem 14-9. A visible star with mass m_1 and a dark, unseen object with mass m_2 orbit around the center of mass of the two-star system at O.

or
$$\frac{m_2^3}{(6M_s + m_2)^2} = 3.47 M_s. \qquad (14\text{-}38)$$

We can solve this cubic equation for m_2, or, since we are working with approximate masses anyway, we can substitute integer multiples of M_s for m_2 until we find

$$m_2 \approx 9M_s. \qquad \text{(Answer)}$$

The data here approximate those for the binary system LMC X-3 in the Large Magellanic Cloud (shown in the figure that begins this chapter). From other data, the dark object is known to be especially compact: it may be a star that collapsed under its own gravitational pull to become a neutron star or a black hole. Since a neutron star cannot have a mass larger than about $2M_s$, the result $m_2 \approx 9M_s$ strongly suggests that the dark object is a black hole.

Thus, one can detect the presence of a black hole if it is part of a binary system with a visible star whose mass, orbital speed, and orbital period can be measured.

14-8 SATELLITES: ORBITS AND ENERGY (OPTIONAL)

As a satellite orbits Earth on its elliptical path, both its speed, which fixes its kinetic energy K, and its distance from the center of Earth, which fixes its gravitational potential energy U, fluctuate with fixed periods. However, the mechanical energy E of the satellite remains constant. (Since the satellite's mass is so much smaller than Earth's mass, we assign U and E of the Earth–satellite system to the satellite alone.)

The potential energy is given by Eq. 14-20 and is

$$U = -\frac{GMm}{r}$$

(with $U = 0$ for infinite separation). Here r is the radius of the orbit, assumed for the time being to be circular.

To find the kinetic energy of a satellite in a circular orbit, we write Newton's second law, $F = ma$, as

$$\frac{GMm}{r^2} = m\frac{v^2}{r}, \qquad (14\text{-}39)$$

where v^2/r is the centripetal acceleration of the satellite. Then from Eq. 14-39, the kinetic energy is

$$K = \tfrac{1}{2}mv^2 = \frac{GMm}{2r}, \qquad (14\text{-}40)$$

which shows us that for a satellite in a circular orbit,

$$K = -\frac{U}{2} \quad \text{(circular orbit)}. \qquad (14\text{-}41)$$

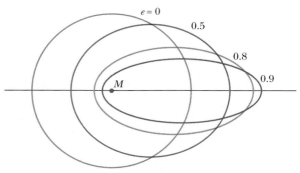

FIGURE 14-18 Four orbits about an object of mass M. All four orbits have the same semimajor axis a and thus correspond to the same total mechanical energy E. Their eccentricities e are marked.

The total mechanical energy of the orbiting satellite is

$$E = K + U = \frac{GMm}{2r} - \frac{GMm}{r}$$

or
$$E = -\frac{GMm}{2r} \quad \text{(circular orbit)}. \qquad (14\text{-}42)$$

This tells us that for a satellite in a circular orbit, the total energy E is the negative of the kinetic energy K:

$$E = -K \quad \text{(circular orbit)}. \qquad (14\text{-}43)$$

For a satellite in an elliptical orbit of semimajor axis a, we can substitute a for r in Eq. 14-42 to find the mechanical energy as

$$E = -\frac{GMm}{2a} \quad \text{(elliptical orbit)}. \qquad (14\text{-}44)$$

Equation 14-44 tells us that the total energy of an orbiting satellite depends only on the semimajor axis of its orbit and not on its eccentricity e. For example, four orbits with the same semimajor axis are shown in Fig. 14-18; the same satellite would have the same total mechanical energy E in all four orbits. Figure 14-19 shows the variation of K, U, and E with r for a satellite moving in a circular orbit about a massive central body.

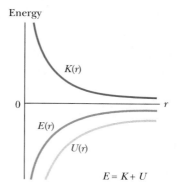

FIGURE 14-19 The variation of kinetic energy K, potential energy U, and total energy E with radius r for a satellite in a circular orbit. For any value of r, the values of U and E are negative, the value of K is positive, and $E = -K$. As r approaches infinity, all three energy curves approach a value of zero.

CHECKPOINT **6:** In the figure, a space shuttle is initially in a circular orbit of radius r about Earth. At point P, the pilot briefly fires a forward-pointing thruster to decrease the shuttle's kinetic energy K and mechanical energy E. (a) Which of the dashed elliptical orbits shown in the figure will the shuttle then take? (b) Is the orbital period T of the shuttle (the time to return to P) then greater than, less than, or the same as in the circular orbit?

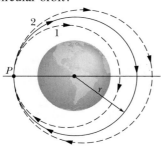

SAMPLE PROBLEM 14-10

A playful astronaut releases a bowling ball, of mass $m = 7.20$ kg, into a circular orbit about Earth at an altitude h of 350 km.

(a) What is the mechanical energy E of the ball in its orbit?

SOLUTION: The orbital radius r is given by

$$r = R + h = 6370 \text{ km} + 350 \text{ km} = 6.72 \times 10^6 \text{ m},$$

in which R is the radius of Earth. From Eq. 14-42, the mechanical energy is

$$E = -\frac{GMm}{2r}$$

$$= -\frac{(6.67 \times 10^{-11} \text{ N} \cdot \text{m}^2/\text{kg}^2)(5.98 \times 10^{24} \text{ kg})(7.20 \text{ kg})}{(2)(6.72 \times 10^6 \text{ m})}$$

$$= -2.14 \times 10^8 \text{ J} = -214 \text{ MJ}. \qquad \text{(Answer)}$$

(b) What was the mechanical energy E_0 of the ball on the launchpad at Cape Canaveral? From there to the orbit, what was the change ΔE in the ball's mechanical energy?

SOLUTION: On the launchpad, the ball has some kinetic energy because of the rotation of Earth, but we can show that this is small enough to neglect. Thus the total energy E_0 is equal to the potential energy U_0, which is given by Eq. 14-20:

$$E_0 = U_0 = -\frac{GMm}{R}$$

$$= -\frac{(6.67 \times 10^{-11} \text{ N} \cdot \text{m}^2/\text{kg}^2)(5.98 \times 10^{24} \text{ kg})(7.20 \text{ kg})}{6.37 \times 10^6 \text{ m}}$$

$$= -4.51 \times 10^8 \text{ J} = -451 \text{ MJ}. \qquad \text{(Answer)}$$

You might be tempted to say that the potential energy of the ball on Earth's surface is zero. Recall, however, that our zero-potential-energy configuration is one in which the ball is removed to a great distance from Earth. You also might be tempted to use Eq. 14-42 to find E_0, but recall that Eq. 14-42 applies to a satellite *in orbit*.

The *increase* in the mechanical energy of the ball from launchpad to orbit was

$$\Delta E = E - E_0 = (-214 \text{ MJ}) - (-451 \text{ MJ})$$

$$= 237 \text{ MJ}. \qquad \text{(Answer)}$$

You can buy this amount of energy from your utility company for a few dollars.

14-9 EINSTEIN AND GRAVITATION

Principle of Equivalence

Einstein once said: "I was . . . in the patent office at Bern when all of a sudden a thought occurred to me: 'If a person falls freely, he will not feel his own weight.' I was startled. This simple thought made a deep impression on me. It impelled me toward a theory of gravitation."

Thus Einstein tells us how he began to form his **general theory of relativity.** The fundamental postulate of this theory about gravitation (the gravitating of objects toward each other) is called the **principle of equivalence,** which says that gravitation and acceleration are equivalent. If a physicist were locked up in a small box as in Fig. 14-20, he would not be able to tell whether the box was at

(a)   (b)

FIGURE 14-20 (a) A physicist in a box resting on Earth sees a cantaloupe falling with acceleration $a = 9.8$ m/s². (b) If he and the box accelerate in deep space at 9.8 m/s², the cantaloupe has the same acceleration relative to him. It is not possible, by doing experiments within the box, for the physicist to tell which situation he is in. For example, the platform scale on which he stands reads the same weight in both situations.

rest on Earth (and subject only to Earth's gravitational force), as in Fig. 14-20a, or accelerating through interstellar space at 9.8 m/s² (and subject only to the force producing that acceleration), as in Fig. 14-20b. In both situations he would feel the same and would read the same value for his weight on a scale. Moreover, if he watched an object fall past him, the object would have the same acceleration relative to him in both situations.

Curvature of Space

We have thus far explained gravitation as due to a force between masses. Einstein showed that, instead, gravitation is due to a curvature (or shape) of space that is caused by the masses. (As is discussed later in this book, space and time are entangled; so, the curvature of which Einstein spoke is really a curvature of *spacetime,* the combined four dimensions of our universe.)

Picturing how space (such as vacuum) can have curvature is difficult. An analogy might help: Suppose that

from orbit we watch a race in which two boats begin on the equator with a separation of 20 km and head due south (Fig. 14-21a). To the sailors, the boats travel along flat, parallel paths. However, with time the boats draw together until, nearer the south pole, they touch. The sailors in the boats can interpret this drawing together in terms of a force acting on the boats. However, we can see that the boats

(a)

(b)

FIGURE 14-22 (a) Light from a distant quasar follows curved paths around a galaxy or a large black hole because the mass of the galaxy or black hole has curved the adjacent space. If the light is detected, it appears to have originated along the backward extensions of the final paths (dashed lines). (b) The Einstein ring known as MG1131+0456 on the computer screen of a telescope. The source of the light (actually, radio waves, which is a form of invisible light) is far behind the large, unseen galaxy that produces the ring; a portion of the source appears as the two bright spots seen along the ring.

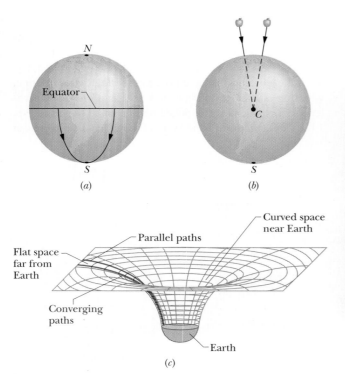

FIGURE 14-21 (a) Two objects moving along lines of longitude toward the south pole converge because of the curvature of Earth's surface. (b) Two objects falling freely near Earth move along lines that converge toward the center of Earth because of the curvature of space near Earth. (c) Far from Earth (and other masses), space is flat and parallel paths remain parallel. Close to Earth, the parallel paths begin to converge because space is curved by Earth's mass.

draw together simply because of the curvature of Earth's surface. We can see this because we are viewing the race from "outside" that surface.

Figure 14-21b shows a similar race: two horizontally separated apples are dropped from the same height above Earth. Although the apples may appear to travel along parallel paths, they actually move toward each other because they both fall toward the Earth's center. We can interpret the motion of the apples in terms of the gravitational force on the apples by Earth. But we can also interpret the motion in terms of a curvature of the space near Earth, due to the presence of Earth's mass. This time we cannot see the curvature because we cannot get "outside" the curved space, as we got "outside" the curved Earth in the boat example. But we can depict the curvature with a drawing like Fig. 14–21c; there the apples would move along a surface that curves toward Earth because of Earth's mass.

When light passes near Earth, its path bends slightly because of the curvature of space there, an effect called *gravitational lensing*. When it passes a more massive structure, like a galaxy or a black hole having large mass, its path can be bent more. If such a massive structure is between us and a quasar (an extremely bright, extremely distant source of light), the light from the quasar can bend around the massive structure and toward us (Fig. 14-22a). Then, because the light seems to be coming from slightly different directions in the sky, we see the same quasar in those different directions. In some situations, the quasars we see blend together to form a giant luminous arc, which is called an *Einstein ring* (Fig. 14-22b).

Is gravitation best attributed to the curvature of space-time due to the presence of masses or to a force between masses? Or should it be attributed to the actions of a type of fundamental particle called a *graviton,* as conjectured in some modern physics theories? We do not know.

REVIEW & SUMMARY

The Law of Gravitation

Any particle in the universe attracts any other particle with a **gravitational force** whose magnitude is

$$F = G\frac{m_1 m_2}{r^2} \quad \text{(Newton's law of gravitation)}, \quad (14\text{-}1)$$

where m_1 and m_2 are the masses of the particles and r is their separation. The **gravitational constant** G is a universal constant whose value is 6.67×10^{-11} N·m²/kg².

The Gravitational Behavior of Uniform Spherical Shells

Equation 14-1 holds only for particles. The gravitational force between extended bodies must generally be found by adding (integrating) the individual forces on individual particles within the bodies. However, if either of the bodies is a uniform spherical shell or a spherically symmetric solid, the net gravitational force it exerts on an *external* object may be computed as if all the mass of the shell or body were located at its center.

Superposition

Gravitational forces obey the **principle of superposition;** that is, the total force $\mathbf{F}_1$ on a particle labeled as particle 1 is the sum of the forces exerted on it by all other particles taken one at a time:

$$\mathbf{F}_1 = \sum_{i=2}^{n} \mathbf{F}_{1i}, \quad (14\text{-}4)$$

in which the sum is a vector sum of the forces $\mathbf{F}_{1i}$ exerted on particle 1 by particles 2, 3, $\cdots$, n. The gravitational force $\mathbf{F}_1$ exerted on a particle by an extended body is found by dividing the body into units of differential mass dm, each of which exerts a differential force $d\mathbf{F}$ on the particle, and then integrating to find the sum of those forces:

$$\mathbf{F}_1 = \int d\mathbf{F}. \quad (14\text{-}5)$$

Gravitational Acceleration

The *gravitational acceleration* a_g of a particle (of mass m) is due solely to the gravitational force acting on it. When the particle is at distance r from the center of a uniform, spherical body of mass M, the magnitude of the gravitational force on the particle is given by Eq. 14-1. In addition, by Newton's second law,

$$F = ma_g, \quad (14\text{-}11)$$

and thus a_g is given by

$$a_g = \frac{GM}{r^2}. \quad (14\text{-}12)$$

Free-Fall Acceleration and Weight

The actual free-fall acceleration $\mathbf{g}$ of a particle near Earth differs slightly from the gravitational acceleration $\mathbf{a}_g$, and the weight mg of the particle differs from the gravitational force (Eq. 14-1) acting on the particle, because Earth is not uniform or spherical and because Earth rotates.

Gravitation Within a Spherical Shell

A uniform shell of matter exerts no gravitational force on a particle located inside it. This means that if a particle is located inside a uniform solid sphere of matter at distance r from its center, the gravitational force exerted on the particle is due only to the mass

M' that lies within a sphere of radius r. This mass M' is given by

$$M' = \rho \frac{4\pi r^3}{3}, \qquad (14\text{-}18)$$

where ρ is the density of the sphere.

Gravitational Potential Energy

The gravitational potential energy $U(r)$ of a system of two particles, with masses M and m and separated by a distance r, is the negative of the work that would be done by the gravitational force of either particle acting on the other as the separation between the particles changes from infinite (very large) to r. This energy is

$$U = -\frac{GMm}{r} \qquad \begin{array}{l}\text{(gravitational} \\ \text{potential energy).}\end{array} \qquad (14\text{-}20)$$

Potential Energy of a System

If a system contains more than two particles, the total gravitational potential energy is the sum of terms representing the potential energies of all the pairs; as an example, we have for three particles, of masses m_1, m_2, and m_3,

$$U = -\left(\frac{Gm_1 m_2}{r_{12}} + \frac{Gm_1 m_3}{r_{13}} + \frac{Gm_2 m_3}{r_{23}} \right). \qquad (14\text{-}21)$$

Escape Speed

An object will escape the gravitational pull of an astronomical body of mass M and radius R if the object's speed near the body's surface is at least equal to the **escape speed,** given by

$$v = \sqrt{\frac{2GM}{R}}. \qquad (14\text{-}26)$$

Kepler's Laws

Gravitational attraction holds the solar system together and makes possible orbiting Earth satellites, both natural and artificial. Such motions are governed by Kepler's three laws of planetary motion, all of which are direct consequences of Newton's laws of motion and gravitation:

1. **The law of orbits.** All planets move in elliptical orbits with the Sun at one focus.

2. **The law of areas.** A line joining any planet to the Sun sweeps out equal areas in equal times (a statement equivalent to conservation of angular momentum).

3. **The law of periods.** The square of the period T of any planet about the Sun is proportional to the cube of the semimajor axis a of the orbit. For circular orbits with radius r, the semimajor axis a is replaced by r and the law is written as

$$T^2 = \left(\frac{4\pi^2}{GM} \right) r^3 \qquad \text{(law of periods),} \qquad (14\text{-}31)$$

where M is the mass of the attracting body—the Sun in the case of the solar system. This result is generally valid for elliptical planetary orbits, when the semimajor axis a is inserted in place of the circular radius r.

Energy in Planetary Motion

When a planet or satellite with mass m moves in a circular orbit with radius r, its potential energy U and kinetic energy K are given by

$$U = -\frac{GMm}{r} \quad \text{and} \quad K = \frac{GMm}{2r}. \qquad (14\text{-}20, 14\text{-}40)$$

The mechanical energy $E = K + U$ is

$$E = -\frac{GMm}{2r}. \qquad (14\text{-}42)$$

For an elliptical orbit of semimajor axis a,

$$E = -\frac{GMm}{2a}. \qquad (14\text{-}44)$$

Einstein's View of Gravitation

Einstein pointed out that gravitation and acceleration are equivalent. This **principle of equivalence** led him to a theory of gravitation (the **general theory of relativity**) that explains gravitational effects in terms of a curvature of space.

QUESTIONS

1. In Fig. 14-23, two particles, of masses m and $2m$, are fixed in place on an axis. (a) Where on the axis can a third particle of mass $3m$ be placed (other than at infinity) so that the net gravitational force on it from the first two particles is zero: to the left of the first two particles, to their right, between them but closer to the more massive particle, or between them but closer to the less massive particle? (b) Does the answer change if the third particle has, instead, a mass of $16m$? (c) Is there a point off the axis at which the net force on the third particle would be zero?

2. In Fig. 14-24, a central particle is surrounded by two circular rings of particles, at radii r and R, with $R > r$. All the particles have mass m. What are the magnitude and direction of the net gravitational force on the central particle due to the particles in the rings?

FIGURE 14-23
Question 1.

FIGURE 14-24
Question 2.

3. In Fig. 14-25, a central particle of mass M is surrounded by a square array of other particles, separated by either distance d or $d/2$ along the perimeter of the square. What are the magnitude and direction of the net gravitational force on the central particle due to the other particles?

FIGURE 14-25
Question 3.

4. Figure 14-26 shows a particle of mass m that is moved from an infinite distance to the center of a ring of mass M, along the central axis of the ring. For the trip, how does the magnitude of the gravitational force on the particle due to the ring change?

FIGURE 14-26
Question 4.

5. Figure 14-27 shows four arrangements of a particle of mass m and one or more uniform rods of mass M and length L, each a distance d from the particle. Rank the arrangements according to the magnitude of the net gravitational force on the particle from the rods, greatest first.

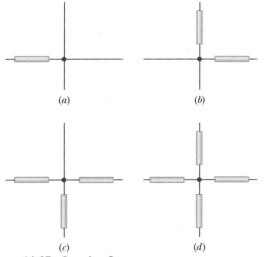

FIGURE 14-27 Question 5.

6. Rank the four systems of equal-mass particles in Checkpoint 2 according to the absolute value of the gravitational potential energy of the system, greatest first.

7. In Fig. 14-28, a particle of mass m (not shown) is to be moved from an infinite distance to one of the three possible locations a, b, and c. Two other particles, of masses m and $2m$, are fixed in place. Rank the three possible locations according to the work done by the net gravitational force on the moving particle due to the fixed particles, greatest first.

FIGURE 14-28 Question 7.

8. In Fig. 14-29, a particle of mass m is initially at point A, at distance d from the center of one uniform sphere and distance $4d$ from the center of another uniform sphere, both of mass $M \gg m$. State whether, if you moved the particle to point D, the following would be positive, negative, or zero: (a) the change in the gravitational potential energy of the particle, (b) the work done by the net gravitational force on the particle, (c) the work done by your force. (d) What are the answers if, instead, the move were from point B to point C?

FIGURE 14-29 Question 8.

9. Reconsider the situation of Question 8. Would the work done by you be positive, negative, or zero if you moved the particle (a) from A to B, (b) from A to C, (c) from B to D? (d) Rank those moves according to the absolute value of the work done by your force, greatest first.

10. The following are the masses and radii of three planets: planet A, $2M$ and R; planet B, $3M$ and $2R$; planet C, $4M$ and $2R$. Rank the planets according to the escape speeds from their surfaces, greatest first.

11. Figure 14-30 shows six paths by which a rocket orbiting a moon might move from point a to point b. Rank the paths according to (a) the corresponding change in the gravitational potential energy of the rocket–moon system and (b) the net work done on the rocket by the gravitational force from the moon, greatest first.

FIGURE 14-30
Question 11.

12. Which of the orbital paths (for, say, a spy satellite) in Fig. 14-31 do not require continuous adjustments by booster rockets? Orbit 1 is at latitude 60°; orbit 2 is in an equatorial plane; orbit 3 is about the center of Earth, between latitudes 60° N and 60° S.

FIGURE 14-31
Question 12.

13. A satellite, with speed v_1 and mass m, is in a circular orbit about a planet of mass M_1. Another satellite, with speed v_2 and mass $2m$, is in a circular orbit of the same radius about a planet of mass M_2. Is M_2 greater than, less than, or equal to M_1 (a) if the satellites have the same period and (b) if $v_2 > v_1$?

EXERCISES & PROBLEMS

SECTION 14-2 Newton's Law of Gravitation

1E. What must the separation be between a 5.2 kg particle and a 2.4 kg particle in order for their gravitational attraction to be 2.3×10^{-12} N?

2E. Some believe that the positions of the planets at the time of birth influence the newborn. Others deride this belief and claim that the gravitational force exerted on a baby by the obstetrician is greater than that exerted by the planets. To check this claim, calculate and compare the gravitational force exerted on a 3 kg baby (a) by a 70 kg obstetrician who is 1 m away and roughly approximated as a point mass, (b) by the massive planet Jupiter ($m = 2 \times 10^{27}$ kg) at its closest approach to Earth ($= 6 \times 10^{11}$ m), and (c) by Jupiter at its greatest distance from Earth ($= 9 \times 10^{11}$ m). (d) Is the claim correct?

3E. The Sun and Earth each exert a gravitational force on the Moon. What is the ratio F_{Sun}/F_{Earth} of these two forces? (The average Sun–Moon distance is equal to the Sun–Earth distance.)

4E. One of the *Echo* satellites consisted of an inflated spherical aluminum balloon 30 m in diameter and of mass 20 kg. Suppose a meteor having a mass of 7.0 kg passes within 3.0 m of the surface of the satellite. What is the gravitational force on the meteor from the satellite at the closest approach?

5P. A mass M is split into two parts, m and $M - m$, which are then separated by a certain distance. What ratio m/M maximizes the gravitational force between the parts?

SECTION 14-3 Gravitation and the Principle of Superposition

6E. How far from Earth must a space probe be along a line toward the Sun so that the Sun's gravitational pull on the probe balances Earth's pull?

7E. A spaceship is on a straight-line path between Earth and its moon. At what distance from Earth is the net gravitational force on the spaceship zero?

8P. What is the percent change in the acceleration of Earth toward the Sun when the alignment of Earth, Sun, and Moon changes from an eclipse of the Sun (with the Moon between Earth and Sun) to an eclipse of the Moon (Earth between Moon and Sun)?

9P. Four spheres, with masses $m_1 = 400$ kg, $m_2 = 350$ kg, $m_3 = 2000$ kg, and $m_4 = 500$ kg, have (x, y) coordinates of $(0, 50$ cm$)$, $(0, 0)$, $(-80$ cm, $0)$, and $(40$ cm, $0)$, respectively. What is the net gravitational force $\mathbf{F}_2$ on m_2 due to the other masses?

10P. In Fig. 14-32a, four spheres form the corners of a square whose side is 2.0 cm long. What are the magnitude and direction of the net gravitational force from them on a central sphere with mass $m_5 = 250$ kg?

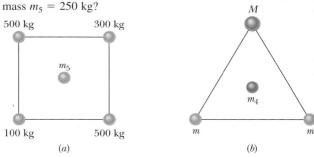

FIGURE 14-32   Problems 10 and 11.

11P. In Fig. 14-32b, two spheres of mass m and a third sphere of mass M form an equilateral triangle, and a fourth sphere of mass m_4 is at the center of the triangle. The net gravitational force on m_4 from the three other spheres is zero; what is M in terms of m?

12P. Two spheres with masses $m_1 = 800$ kg and $m_2 = 600$ kg are separated by 0.25 m. What is the net gravitational force (in both magnitude and direction) from them on a 2.0 kg sphere located 0.20 m from m_1 and 0.15 m from m_2?

13P. The masses and coordinates of three spheres are as follows: 20 kg, $x = 0.50$ m, $y = 1.0$ m; 40 kg, $x = -1.0$ m, $y = -1.0$ m; 60 kg, $x = 0$ m, $y = -0.50$ m. What is the magnitude of the gravitational force on a 20 kg sphere located at the origin due to the other spheres?

14P. Figure 14-33 shows two uniform rods of the same length L and mass M, lying on an axis with separation d. Using the result of Sample Problem 14-2, set up an integral (with integration limits and ready for integration) to find the gravitational force of one rod on the other.

FIGURE 14-33   Problem 14.

15P. Figure 14-34 shows a spherical hollow inside a lead sphere of radius R; the surface of the hollow passes through the center of the sphere and "touches" the right side of the sphere. The mass of the sphere before hollowing was M. With what gravitational force does the hollowed-out lead sphere attract a small sphere of mass m that lies at a distance d from the center of the lead sphere, on the straight line connecting the centers of the spheres and of the hollow?

FIGURE 14-34 Problem 15.

SECTION 14-4 Gravitation Near Earth's Surface

16E. Calculate the gravitational acceleration on the surface of the Moon from values of the mass and radius of the Moon found in Appendix C.

17E. At what altitude above Earth's surface would the gravitational acceleration be 4.9 m/s²?

18E. You weigh 120 lb at sidewalk level outside the World Trade Center in New York City. Suppose that you ride from this level to the top of one of its 1350 ft towers. Ignoring Earth's rotation, how much less would you weigh there (because you are slightly farther from the center of Earth)?

19E. A typical neutron star may have a mass equal to that of the Sun but a radius of only 10 km. (a) What is the gravitational acceleration at the surface of such a star? (b) How fast would an object be moving if it fell from rest through a distance of 1.0 m on such a star? (Assume the star does not rotate.)

20E. An object lying on Earth's equator is accelerated (a) toward the center of Earth because Earth rotates, (b) toward the Sun because Earth revolves around the Sun in an almost circular orbit, and (c) toward the center of our galaxy because the Sun moves about the galactic center. For the latter, the period is 2.5×10^8 y and the radius is 2.2×10^{20} m. Calculate these three accelerations as multiples of $g = 9.8$ m/s².

21E. (a) What will an object weigh on the Moon's surface if it weighs 100 N on Earth's surface? (b) How many Earth radii must this same object be from the center of Earth if it is to weigh the same as it does on the Moon?

22P. If g is to be determined by dropping an object through a distance of exactly 10 m, what percent error in the measurement of the time of fall results in a 0.1% error in the value of g?

23P. The fastest possible rate of rotation of a planet is that for which the gravitational force on material at the equator just barely provides the centripetal force needed for the rotation. (Why?) (a) Show that the corresponding shortest period of rotation is

$$T = \sqrt{\frac{3\pi}{G\rho}},$$

where ρ is the uniform density of the spherical planet. (b) Calculate the rotation period assuming a density of 3.0 g/cm³, typical of many planets, satellites, and asteroids. No astronomical object has ever been found to be spinning with a period shorter than that determined by this analysis.

24P. In Fig. 14-35, identical masses m hang from strings on a balance at the surface of Earth. The strings have negligible mass

and differ in length by h. Assume that Earth is spherical, with density $\rho = 5.5$ g/cm³. (a) Show that the difference ΔW in the weights, due to one mass being closer to Earth than the other, is $8\pi G\rho mh/3$. (b) Find the difference in length that will give a ratio $\Delta W/W = 1 \times 10^{-6}$, where W is either weight.

FIGURE 14-35 Problem 24.

25P. A body is suspended from a spring scale in a ship sailing along the equator with a speed v. (a) Show that the scale reading will be very close to $W_0(1 \pm 2\omega v/g)$, where ω is the angular speed of Earth and W_0 is the scale reading when the ship is at rest. (b) Explain the $\pm$ sign.

26P. Certain neutron stars (extremely dense stars) are believed to be rotating at about 1 rev/s. If such a star has a radius of 20 km, what must be its minimum mass so that material on its surface remains in place during the rapid rotation?

27P. The radius R_h and mass M_h of a black hole are related by $R_h = 2GM_h/c^2$, where c is the speed of light. Assume that the gravitational acceleration a_g of an object at a distance $r_o = 1.001R_h$ from the center of a black hole is given by Eq. 14-12 (it is, for large black holes). (a) Find an expression for a_g at r_o in terms of M_h. (b) Does a_g at r_o increase or decrease with an increase of M_h? (c) What is a_g at r_o for a very large black hole whose mass is 1.55×10^{12} times the solar mass of 1.99×10^{30} kg? (d) If the astronaut of Sample Problem 14-4 is at r_o with her feet toward this black hole, what is the difference in gravitational acceleration at head and feet? (e) Is the tendency to stretch the astronaut severe?

SECTION 14-5 Gravitation Inside Earth

28E. Two concentric shells of uniform density having masses M_1 and M_2 are situated as shown in Fig. 14-36. Find the force on a particle of mass m when the particle is located at (a) $r = a$, (b) $r = b$, and (c) $r = c$. The distance r is measured from the center of the shells.

FIGURE 14-36
Exercise 28.

29E. With what speed would mail pass through the center of Earth if it were dropped down the tunnel of Sample Problem 14-5?

30E. Show that, at the bottom of a vertical mine shaft dug to depth D, the measured value of a_g will be

$$a_g = a_{gs}\left(1 - \frac{D}{R}\right),$$

a_{gs} being the surface value. Assume that Earth is a uniform sphere of radius R.

31P. A solid sphere of uniform density has a mass of 1.0×10^4 kg and a radius of 1.0 m. What is the gravitational force due to the sphere on a particle of mass m located at a distance of (a) 1.5 m and (b) 0.50 m from the center of the sphere? (c) Write a general expression for the gravitational force on m at a distance $r \leq 1.0$ m from the center of the sphere.

32P. A uniform solid sphere of radius R produces a gravitational acceleration of a_g on its surface. At what two distances from the center of the sphere is the gravitational acceleration $a_g/3$? (*Hint:* Consider distances both inside and outside the sphere.)

33P. Figure 14-37 shows, not to scale, a cross section through the interior of Earth. Rather than being uniform throughout, Earth is divided into three zones: an outer *crust,* a *mantle,* and an inner *core.* The dimensions of these zones and the masses contained within them are shown on the figure. Earth has a total mass of 5.98×10^{24} kg and a radius of 6370 km. Ignore rotation and assume that Earth is spherical. (a) Calculate a_g at the surface. (b) Suppose that a bore hole (the *Mohole*) is driven to the crust–mantle interface at a depth of 25 km; what would be the value of a_g at the bottom of the hole? (c) Suppose that Earth were a uniform sphere with the same total mass and size. What would be the value of a_g at a depth of 25 km? (See Exercise 30.) (Precise measurements of a_g are sensitive probes of the interior structure of Earth, although results can be clouded by local density variations.)

25 km

6345 km

Core, 1.93×10^{24} kg

Mantle, 4.01×10^{24} kg

Crust, 3.94×10^{22} kg

3490 km

FIGURE 14-37 Problem 33. Not to scale.

SECTION 14-6 Gravitational Potential Energy

34E. (a) What is the gravitational potential energy of the two-particle system in Exercise 1? If you triple the separation between the particles, how much work is done (b) by the gravitational force between the particles and (c) by you?

35E. (a) In Problem 9, remove m_1 and calculate the gravitational potential energy of the remaining three-particle system. (b) If m_1 is then put back in place, is the potential energy of the four-particle system more or less than that of the system in (a)? (c) In (a), is the work done by you to remove m_1 positive or negative? (d) In (b), is the work done by you to replace m_1 positive or negative?

36E. In Problem 5, what ratio m/M gives the least gravitational potential energy for the system?

37E. The mean diameters of Mars and Earth are 6.9×10^3 km and 1.3×10^4 km, respectively. The mass of Mars is 0.11 times Earth's mass. (a) What is the ratio of the mean density of Mars to that of Earth? (b) What is the value of g on Mars? (c) What is the escape speed on Mars?

38E. A spaceship is idling at the fringes of our galaxy, 80,000 light-years from the galactic center. What is the ship's escape speed from the galaxy? The mass of the galaxy is 1.4×10^{11} times that of our sun. Assume, for simplicity, that the matter forming the galaxy is distributed in a uniform sphere.

39E. Calculate the amount of energy required to escape from (a) Earth's moon and (b) Jupiter relative to that required to escape from Earth.

40E. Show that the escape speed from the Sun at Earth's distance from the Sun is $\sqrt{2}$ times the speed of the Earth in its orbit, assumed to be a circle. (This is a specific case of a general result for circular orbits: $v_{esc} = \sqrt{2}v_{orb}$.)

41E. A particle of comet dust with mass m is a distance R from Earth's center and a distance r from the Moon's center. If Earth's mass is M_E and the Moon's mass is M_m, what is the total gravitational potential energy of the particle–Earth system and the particle–Moon system?

42E. Upon "burning out," a large star can collapse under its own gravitational force to become a black hole. The star's surface then has a radius R_s such that the removal of a mass m from the surface to infinity would require work equal to the mass energy mc^2 of that mass. If M_s is the star's mass, show with Newton's law of gravitation that $R_s = GM_s/c^2$. (The correct value for R_s is actually twice this. To get it, we would have to apply Einstein's general theory of relativity rather than Newton's law.)

43P. The three spheres in Fig. 14-38, with masses $m_1 = 800$ g, $m_2 = 100$ g, and $m_3 = 200$ g, have their centers on a common line, with $L = 12$ cm and $d = 4.0$ cm. You move the middle sphere until its center-to-center separation from m_3 is $d = 4.0$ cm. How much work is done on m_2 (a) by you and (b) by the net gravitational force on m_2 due to m_1 and m_3?

L

d

d

m_1

m_2

m_3

FIGURE 14-38 Problem 43.

44P. A rocket is accelerated to a speed of $v = 2\sqrt{gR_e}$ near Earth's surface (where Earth's radius is R_e) and then coasts upward. (a) Show that it will escape from Earth. (b) Show that very far from Earth its speed is $v = \sqrt{2gR_e}$.

45P. (a) What is the escape speed on a spherical asteroid whose radius is 500 km and whose gravitational acceleration at the surface is 3.0 m/s²? (b) How far from the surface will a particle go if it leaves the asteroid's surface with a radial speed of 1000 m/s? (c) With what speed will an object hit the asteroid if it is dropped from 1000 km above the surface?

46P. Zero, a hypothetical planet, has a mass of 5.0×10^{23} kg, a radius of 3.0×10^6 m, and no atmosphere. A 10 kg space probe is to be launched vertically from its surface. (a) If the probe is launched with an initial energy of 5.0×10^7 J, what will be its kinetic energy when it is 4.0×10^6 m from the center of Zero? (b) If the probe is to achieve a maximum distance of 8.0×10^6 m from the center of Zero, with what initial kinetic energy must it be launched from the surface of Zero?

47P. In a double-star system, two stars of mass 3.0×10^{30} kg each rotate about the system's center of mass at a radius of 1.0×10^{11} m. (a) What is their common angular speed? (b) If a meteoroid passes through this center of mass perpendicular to the orbital plane of the stars, what value must its speed exceed at that point if it is to escape to "infinity" from the star system?

48P. Two neutron stars are separated by a distance of 10^{10} m. They each have a mass of 10^{30} kg and a radius of 10^5 m. They are initially at rest with respect to each other. (a) How fast are they moving when their separation has decreased to one-half its initial value? (b) How fast are they moving just before they collide?

49P. A projectile is fired vertically from Earth's surface with an initial speed of 10 km/s. Neglecting air drag, how far above the surface of Earth will it go?

50P. A sphere of matter, of mass M and radius a, has a concentric cavity of radius b, as shown in cross section in Fig. 14-39. (a) Sketch a curve of the gravitational force F exerted by the sphere on a particle of mass m, located a distance r from the center of the sphere, as a function of r in the range $0 \le r \le \infty$. Consider $r = 0$, b, a, and ∞ in particular. (b) Sketch the corresponding curve for the potential energy $U(r)$ of the system.

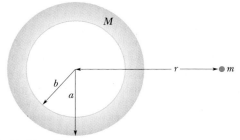

FIGURE 14-39 Problem 50.

51P. A 20 kg mass is located at the origin, and a 10 kg mass is located on the x axis at $x = 0.80$ m. The 10 kg mass is released from rest while the 20 kg mass is held in place at the origin. (a) What is the gravitational potential energy of the two-mass system immediately after the 10 kg mass is released? (b) What is the

kinetic energy of the 10 kg mass after it has moved 0.20 m toward the 20 kg mass?

52P*. Several planets (Jupiter, Saturn, Uranus) possess nearly circular surrounding rings, perhaps composed of material that failed to form a satellite. In addition, many galaxies contain ring-like structures. Consider a homogeneous ring of mass M and radius R. (a) What gravitational attraction does it exert on a particle of mass m located a distance x from the center of the ring along its axis? See Fig. 14-40. (b) Suppose the particle falls from rest as a result of the attraction of the ring of matter. Find an expression for the speed with which it passes through the center of the ring.

FIGURE 14-40 Problem 52.

53P*. The gravitational force between two particles with masses m and M, initially at rest at great separation, pulls them together. Show that <u>at any instant the speed of either particle relative to the other is $\sqrt{2G(M + m)/d}$</u>, where d is their separation at that instant. (*Hint:* Use the laws of conservation of energy and conservation of linear momentum.)

SECTION 14-7 Planets and Satellites: Kepler's Laws

54E. The mean distance of Mars from the Sun is 1.52 times that of Earth from the Sun. From Kepler's law of periods, calculate the number of years required for Mars to make one revolution about the Sun; compare your answer with the value given in Appendix C.

55E. The planet Mars has a satellite, Phobos, which travels in an orbit of radius 9.4×10^6 m with a period of 7 h 39 min. Calculate the mass of Mars from this information.

56E. Determine the mass of Earth from the period T and the radius r of the Moon's orbit about Earth: $T = 27.3$ days and $r = 3.82 \times 10^5$ km. Assume the Moon orbits the center of Earth rather than the center of mass of the Earth–Moon system.

57E. Our Sun, with mass 2.0×10^{30} kg, revolves about the center of the Milky Way galaxy, which is 2.2×10^{20} m away, once every 2.5×10^8 years. Assuming that each of the stars in the galaxy has a mass equal to that of our Sun, that the stars are distributed uniformly in a sphere about the galactic center, and that our Sun is essentially at the edge of that sphere, estimate roughly the number of stars in the galaxy.

58E. A satellite is placed in a circular orbit with a radius equal to one-half the radius of the Moon's orbit. What is its period of revolution in lunar months? (A lunar month is the period of revolution of the Moon.)

59E. (a) What linear speed must an Earth satellite have to be in a circular orbit at an altitude of 160 km? (b) What is the period of revolution?

60E. Most asteroids revolve around the Sun between Mars and Jupiter. However, several "Apollo asteroids" with diameters of about 30 km move in orbits that cross the orbit of Earth. The orbit of one of these asteroids is shown to scale in Fig. 14-41. By taking measurements directly from the figure, calculate the asteroid's period of revolution in years.

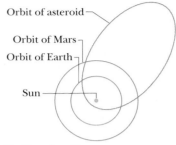

Orbit of asteroid
Orbit of Mars
Orbit of Earth
Sun

FIGURE 14-41 Exercise 60.

61E. A satellite, moving in an elliptical orbit, is 360 km above Earth's surface at its farthest point and 180 km above at its closest point. Calculate (a) the semimajor axis and (b) the eccentricity of the orbit. (*Hint:* See Sample Problem 14-8.)

62E. The Sun's center is at one focus of Earth's orbit. How far from this focus is the other focus? Express your answer in terms of the solar radius, 6.96×10^8 m. The eccentricity of Earth's orbit is 0.0167, and the semimajor axis may be taken to be 1.50×10^{11} m. See Fig. 14-14.

63E. (a) Using Kepler's third law (Eq. 14-31), express the gravitational constant G in terms of the astronomical unit as a length unit, the solar mass as a mass unit, and the year as a time unit. (One astronomical unit = 1 AU = 1.496×10^{11} m. One solar mass = $1M_s = 1.99 \times 10^{30}$ kg. One year = 1 y = 3.156×10^7 s.) (b) What form does Kepler's third law take in these units?

64E. A satellite hovers over a certain spot on the equator of (rotating) Earth. What is the altitude of its orbit (called a *geosynchronous orbit*)?

65E. A comet that was seen in April 574 by Chinese astronomers on a day known by them as the Woo Woo day was spotted again in May 1994. Assume the time between observations is the period of the Woo Woo day comet and take its eccentricity as 0.11. What are (a) the semimajor axis of the comet's orbit and (b) its greatest distance from the Sun in terms of the mean orbital radius R_P of Pluto.

66E. In 1993 the spacecraft *Galileo* sent home an image (Fig. 14-42) of asteroid 243 Ida and an orbiting tiny moon (now known as Dactyl), the first confirmed example of an asteroid–moon system. In the image, the moon, which is 1.5 km wide, is 100 km from the center of the asteroid, which is 55 km long. The shape of the moon's orbit is not well known; assume it is circular with a period of 27 h. (a) What is the mass of the asteroid? (b) The volume of the asteroid, measured from the *Galileo* images, is 14,100 km³. What is the density of the asteroid?

67P. Assume that the satellite of Exercise 64 is in orbit at the longitude of Chicago. You are in Chicago (latitude 47.5°) and

FIGURE 14-42 Exercise 66. A picture from the spacecraft *Galileo* shows a tiny moon orbiting asteroid 243 Ida.

want to pick up the signals the satellite broadcasts. In what direction should you point your antenna?

68P. In 1610, Galileo used his telescope to discover four prominent moons around Jupiter. Their mean orbital radii a and periods T are

NAME	a (10^8 m)	T (days)
Io	4.22	1.77
Europa	6.71	3.55
Ganymede	10.7	7.16
Callisto	18.8	16.7

(a) Plot log a (y axis) against log T (x axis) and show that you get a straight line. (b) Measure the slope of the line and compare it with the value that you expect from Kepler's third law. (c) Find the mass of Jupiter from the intercept of this line with the y axis.

69P. Show how, guided by Kepler's third law (Eq. 14-31), Newton could deduce that the force holding the Moon in its orbit, assumed circular, depends on the inverse square of the distance from the center of Earth.

70P. In a certain binary-star system, each star has the same mass as our Sun, and they revolve about their center of mass. The distance between them is the same as the distance between Earth and the Sun. What is their period of revolution in years?

71P. A certain triple-star system consists of two stars, each of mass m, revolving about a central star of mass M in the same circular orbit of radius r (Fig. 14-43). The two stars are always at opposite ends of a diameter of the circular orbit. Derive an expression for the period of revolution of the stars.

72P. (a) What is the escape speed from the Sun for an object in Earth's orbit (of orbital radius R) but far from Earth? (b) If an object already has a speed equal to Earth's orbital speed, what additional speed must it be given to escape as in (a)? (c) Suppose an object is launched from Earth in the direction of Earth's orbital

FIGURE 14-43 Problem 71.

motion. What speed must it be given during the launch so that when it is far from Earth, but still at a distance of about R from the Sun, it can escape from the Sun? (This is the speed required for an Earth-launched object to escape from the Sun.)

73P*. Three identical stars of mass M are located at the vertices of an equilateral triangle with side L. At what speed must they move if they all revolve under the influence of one another's gravitational force in a circular orbit circumscribing the triangle while still preserving the equilateral triangle?

74P*. A satellite in a circular orbit had been intended to hover over a certain spot on Earth's surface. Through error, however, the satellite's orbital radius was made 1.0 km too large for this to happen. At what rate and in what direction does the point directly below the satellite move across Earth's surface?

SECTION 14-8 Satellites: Orbits and Energy

75E. An asteroid, whose mass is 2.0×10^{-4} times the mass of Earth, revolves in a circular orbit around the Sun at a distance that is twice Earth's distance from the Sun. (a) Calculate the period of revolution of the asteroid in years. (b) What is the ratio of the kinetic energy of the asteroid to that of Earth?

76E. Consider two satellites, A and B, of equal mass m, moving in the same circular orbit of radius r around Earth, of mass M_E, but in opposite senses of rotation and therefore on a collision course (see Fig. 14-44). (a) In terms of G, M_E, m, and r, find the total mechanical energy $E_A + E_B$ of the two-satellite-plus-Earth system before collision. (b) If the collision is completely inelastic so that the wreckage remains as one piece of tangled material (mass $= 2m$), find the total mechanical energy immediately after collision. (c) Describe the subsequent motion of the wreckage.

FIGURE 14-44 Exercise 76.

77P. Two Earth satellites, A and B, each of mass m, are to be launched into circular orbits about Earth's center. Satellite A is to orbit at an altitude of 4000 mi. Satellite B is to orbit at an altitude of 12,000 mi. The radius of Earth R_E is 4000 mi. (a) What is the

ratio of the potential energy of satellite B to that of satellite A, in orbit? (b) What is the ratio of the kinetic energy of satellite B to that of satellite A, in orbit? (c) Which satellite has the greater total energy if each has a mass of 14.6 kg? By how much?

78P. Use the conservation of mechanical energy and Eq. 14-44 to show that if an object is in an elliptical orbit about a planet, then its distance r from the planet and speed v are related by

$$v^2 = GM \left(\frac{2}{r} - \frac{1}{a} \right).$$

79P. Use the result of Problem 78 and data contained in Sample Problem 14-8 to calculate (a) the speed v_p of comet Halley at perihelion and (b) its speed v_a at aphelion. (c) Using conservation of angular momentum relative to the Sun, find the ratio of the comet's perihelion distance R_p to its aphelion distance R_a in terms of v_p and v_a.

80P. (a) Does it take more energy to get a satellite up to 1000 mi above Earth than to put it in circular orbit once it is there? (Take Earth's radius to be 4000 mi.) (b) What about 2000 mi? (c) What about 3000 mi?

81P. One way to attack a satellite in Earth orbit is to launch a swarm of pellets in the same orbit as the satellite but in the opposite direction. Suppose a satellite in a circular orbit 500 km above Earth's surface collides with a pellet having mass 4.0 g. (a) What is the kinetic energy of the pellet in the reference frame of the satellite? (b) What is the ratio of this kinetic energy to the kinetic energy of a 4.0 g bullet from a modern army rifle with a muzzle velocity of 950 m/s?

82P. Consider a satellite in a circular orbit about Earth. State how the following properties of the satellite depend on the radius r of its orbit: (a) period, (b) kinetic energy, (c) angular momentum, and (d) speed.

83P. What are (a) the speed and (b) the period of a 220 kg satellite in an approximately circular orbit 640 km above the surface of Earth? Suppose the satellite loses mechanical energy at the average rate of 1.4×10^5 J per orbital revolution. Adopting the reasonable approximation that the trajectory is a "circle of slowly diminishing radius," determine the satellite's (c) altitude, (d) speed, and (e) period at the end of its 1500th revolution. (f) What is the magnitude of the average retarding force? (g) Is angular momentum around Earth's center conserved for the satellite or the satellite–Earth system?

84P. The orbit of Earth about the Sun is *almost* circular: the closest and farthest distances are 1.47×10^8 km and 1.52×10^8 km, respectively. Determine the corresponding variations in (a) total energy, (b) potential energy, (c) kinetic energy, and (d) orbital speed. (*Hint:* Use the laws of conservation of energy and conservation of angular momentum.)

85P. In a shuttle craft of mass $m = 2000$ kg, Captain Janeway orbits a planet of mass $M = 5.98 \times 10^{24}$ kg, in a circular orbit of radius $r = 6.80 \times 10^6$ m. What are (a) the period of the orbit and (b) the speed of the shuttle craft? Janeway briefly fires a forward-pointing thruster, reducing her speed by 1.00%. Just then, what are (c) the speed, (d) the kinetic energy, (e) the gravitational potential energy, and (f) the mechanical energy of the shuttle craft? (g) What is the semimajor axis of the elliptical orbit now taken by the craft? (h) What is the difference between the period of the original circular orbit and that of the new elliptical orbit, and which orbit has the smaller period?

SECTION 14-9 Einstein and Gravitation

86E. In Fig. 14-20b, the scale on which the 60 kg physicist stands reads 220 N. How long will the cantaloupe take to reach the floor if the physicist drops it from rest (relative to himself), 2.1 m from the floor?

87P. Figure 14-45 shows the walls of a duct in a ship in deep space; the ship has an acceleration $\mathbf{a} = (2.5 \text{ m/s}^2)\mathbf{i}$. An electron is sent across the duct's width of 3.0 cm, from the origin shown and with an initial velocity of $\mathbf{v}_0 = (0.40 \text{ m/s})\mathbf{j}$. In unit vector notation and relative to the ship, what are (a) the displacement of the electron at the end of its flight and (b) its velocity just before impact with the far wall?

FIGURE 14-45 Problem 87.

Electronic Computation

88. A 6000 kg planetary probe is in a circular solar orbit with a radius of 108×10^6 km (the orbital radius of Venus). The space agency wants to put the probe into the same orbit as Earth, a circular orbit with a radius of 150×10^6 km. The first step is to increase the probe's speed so that the probe is in an elliptical orbit with a perihelion distance equal to the radius of its initial orbit and an aphelion distance equal to the radius of its final orbit. The mass of the Sun is 1.99×10^{30} kg. (a) Calculate the increases in speed and energy that are required. Plot the elliptical orbit. (In polar coordinates, with the origin at the center, the equation for an ellipse is

$$r^2 = \frac{r_p r_a (r_p + r_a)^2}{(r_p + r_a)^2 \sin^2 \theta + 4 r_p r_a \cos^2 \theta},$$

where r_p is the perihelion distance and r_a is the aphelion distance.) (b) When the aphelion position is reached, the speed of the probe is again changed to put it into the desired circular orbit. What changes in the speed and energy are required?

89. In your calculator, make a list of the periods T and a separate list for the semimajor axes a for the planets in Table 14-3. Multiply the list for T by an appropriate factor so that the unit of T is seconds. (a) Store values of T^2 and a^3 in two new lists. Have the calculator do a linear regression fit of T^2 versus a^3. From the parameters of the fit and using the known value of G, determine the mass of the Sun. (b) Store values of $\log T$ in a list and $\log a$ in another list. Have the calculator plot $\log T$ versus $\log a$ and then do a linear regression fit of the plot. From the parameters of this fit and using the known value of G, again determine the mass of the Sun.

15
Fluids

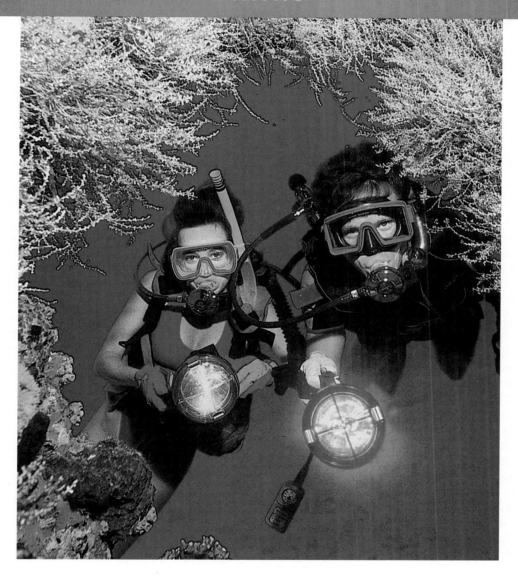

The force exerted by water on the body of a descending diver increases noticeably, even for a relatively shallow descent to the bottom of a swimming pool. However, in 1975, using scuba gear with a special gas mixture for breathing, William Rhodes emerged from a chamber that had been lowered 1000 ft into the Gulf of Mexico, and he then swam to a record depth of 1148 ft. Strangely, a novice scuba diver practicing in a swimming pool might be in more danger from the force exerted by the water than was Rhodes. And occasionally, novice scuba divers die because they have neglected that danger. What is this potentially lethal risk?

15-1 FLUIDS AND THE WORLD AROUND US

Fluids—which include both liquids and gases—play a central role in our daily lives. We breathe and drink them, and a rather vital fluid circulates in the human cardiovascular system. There is the fluid ocean and the fluid atmosphere.

In a car, there are fluids in the tires, the gas tank, the radiator, the combustion chambers of the engine, the exhaust manifold, the battery, the air conditioning system, the windshield wiper reservoir, the lubrication system, and the hydraulic system. (*Hydraulic* means operated via a liquid.) The next time you see a large piece of earthmoving machinery, count the hydraulic cylinders that permit the machine to do its work. Large jet planes have scores of them.

We use the kinetic energy of a moving fluid in windmills, and the potential energy of another fluid in hydroelectric power plants. Given time, fluids carve the landscape. We often travel great distances just to watch fluids move. Perhaps it is time to see what physics can tell us about fluids.

15-2 WHAT IS A FLUID?

A **fluid,** in contrast to a solid, is a substance that can flow. Fluids conform to the boundaries of any container in which we put them. They do so because a fluid cannot sustain a force that is tangential to its surface. (In the more formal language of Section 13-6, a fluid is a substance that flows because it cannot withstand a shearing stress. It can, however, exert a force in the direction perpendicular to its surface.) Some materials, such as pitch, take a long time to conform to the boundaries of a container, but they do so eventually; thus we classify them as fluids.

You may wonder why we lump liquids and gases together and call them fluids. After all (you may say), liquid water is as different from steam as it is from ice. Actually, it is not. Ice, like other crystalline solids, has its constituent atoms organized in a fairly rigid three-dimensional array called a crystalline lattice. In neither steam nor liquid water, however, is there any such orderly long-range arrangement.

15-3 DENSITY AND PRESSURE

When we discuss rigid bodies, we are concerned with particular lumps of matter, such as wooden blocks, baseballs, or metal rods. Physical quantities that we find useful, and in whose terms we express Newton's laws, are *mass* and *force*. We might speak, for example, of a 3.6 kg block acted on by a 25 N force.

With fluids, we are more interested in properties that vary from point to point in the extended substance than with properties of specific lumps of that substance. It is more useful to speak of **density** and **pressure** than of mass and force.

Density

To find the density ρ (rho) of a fluid at any point, we isolate a small volume element ΔV around that point and measure the mass Δm of the fluid contained within that element. The **density** is then

$$\rho = \frac{\Delta m}{\Delta V}. \qquad (15\text{-}1)$$

In theory, the density at any point in a fluid is the limit of this ratio as the volume element ΔV at that point is made smaller and smaller. In practice, we assume that a fluid sample is large compared to atomic dimensions and thus "smooth" (with uniform density), rather than "lumpy" with atoms. This assumption allows us to write Eq. 15-1 in the form $\rho = m/V$, where m and V are the mass and volume of the sample.

Density is a scalar property; its SI unit is the kilogram per cubic meter. Table 15-1 shows the densities of some

TABLE 15-1 **SOME DENSITIES**

MATERIAL OR OBJECT	DENSITY (kg/m³)
Interstellar space	10^{-20}
Best laboratory vacuum	10^{-17}
Air: 20°C and 1 atm	1.21
20°C and 50 atm	60.5
Styrofoam	1×10^2
Water: 20°C and 1 atm	0.998×10^3
20°C and 50 atm	1.000×10^3
Seawater: 20°C and 1 atm	1.024×10^3
Whole blood	1.060×10^3
Ice	0.917×10^3
Iron	7.9×10^3
Mercury	13.6×10^3
Earth: average	5.5×10^3
core	9.5×10^3
crust	2.8×10^3
Sun: average	1.4×10^3
core	1.6×10^5
White dwarf star (core)	10^{10}
Uranium nucleus	3×10^{17}
Neutron star (core)	10^{18}
Black hole (1 solar mass)	10^{19}

substances and the average densities of some objects. Note that the density of a gas (see Air in the table) varies considerably with pressure, but the density of a liquid (see Water) does not. That is, gases are readily *compressible* but liquids are not.

Pressure

Let a small pressure-sensing device be suspended inside a fluid-filled vessel, as in Fig. 15-1a. The sensor (Fig. 15-1b) consists of a piston of area ΔA riding in a close-fitting cylinder and resting against a spring. A readout arrangement allows us to record the amount by which the (calibrated) spring is compressed by the surrounding fluid, thus indicating the magnitude ΔF of the force that acts on the piston. We define the **pressure** exerted by the fluid on the piston as

$$p = \frac{\Delta F}{\Delta A}. \tag{15-2}$$

In theory, the pressure at any point in the fluid is the limit of this ratio as the area ΔA of the piston, centered on that point, is made smaller and smaller. However, if the force is uniform over a flat area A, we can write Eq. 15-2 as $p = F/A$. (When we say a force is uniform over an area, we mean that the same force would be measured at every point of the area.)

We find by experiment that at a given point in a fluid at rest, the pressure p defined by Eq. 15-2 has the same value no matter how the pressure sensor is oriented. Pressure is a scalar, having no directional properties. It is true that the force acting on the piston of our pressure sensor is a vector, but Eq. 15-2 involves only the *magnitude* of that force, a scalar quantity.

The SI unit of pressure is the newton per square meter, which is given a special name, the **pascal** (Pa). In metric countries, tire pressure gauges are calibrated in kilopascals. The pascal is related to some other common (non-SI) pressure units as follows:

$$1 \text{ atm} = 1.01 \times 10^5 \text{ Pa} = 760 \text{ torr} = 14.7 \text{ lb/in.}^2.$$

The *atmosphere* (atm) is, as the name suggests, the approximate average pressure of the atmosphere at sea level. The *torr* (named for Evangelista Torricelli, who invented the mercury barometer in 1674) was formerly called the *millimeter of mercury* (mm Hg). The pound per square inch is often abbreviated psi. Table 15-2 shows some pressures.

TABLE 15-2 SOME PRESSURES

	PRESSURE (Pa)
Center of the Sun	2×10^{16}
Center of Earth	4×10^{11}
Highest sustained laboratory pressure	1.5×10^{10}
Deepest ocean trench (bottom)	1.1×10^8
Spike heels on a dance floor	1×10^6
Automobile tire[a]	2×10^5
Atmosphere at sea level	1.0×10^5
Normal blood pressure[a,b]	1.6×10^4
Best laboratory vacuum	10^{-12}

[a] Pressure in excess of atmospheric pressure.

[b] The systolic pressure, corresponding to 120 torr on the physician's pressure gauge.

SAMPLE PROBLEM 15-1

A living room has floor dimensions of 3.5 m and 4.2 m and a height of 2.4 m.

(a) What does the air in the room weigh?

SOLUTION: If V is the volume of the room and ρ is the density of air at 1 atm (see Table 15-1), then the weight W of the air is

$$W = mg = \rho V g$$
$$= (1.21 \text{ kg/m}^3)(3.5 \text{ m} \times 4.2 \text{ m} \times 2.4 \text{ m})(9.8 \text{ m/s}^2)$$
$$= 418 \text{ N} \approx 420 \text{ N}. \quad \text{(Answer)}$$

This is about 94 lb.

(b) What force does the atmosphere exert on the floor of the room?

SOLUTION: The force is

$$F = pA = (1.0 \text{ atm})\left(\frac{1.01 \times 10^5 \text{ N/m}^2}{1 \text{ atm}}\right)(3.5 \text{ m})(4.2 \text{ m})$$
$$= 1.5 \times 10^6 \text{ N}. \quad \text{(Answer)}$$

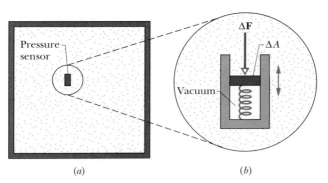

FIGURE 15-1 (a) A fluid-filled vessel containing a small pressure sensor, shown in (b). The pressure is measured by the relative position of the movable piston in the sensor.

This force ($\approx$ 170 tons) is the weight of a column of air covering the floor and extending all the way to the top of the atmosphere. It is equal to the force that would be exerted on the floor if (in the absence of the atmosphere) the room were filled with mercury to a depth of 30 in. Why doesn't this enormous force break the floor?

15-4 FLUIDS AT REST

Figure 15-2a shows a tank of water—or other liquid—open to the atmosphere. As every diver knows, the pressure *increases* with depth below the air–water interface. The diver's depth gauge, in fact, is a pressure sensor much like that of Fig. 15-1b. As every mountaineer knows, the pressure *decreases* with altitude as one ascends into the atmosphere. The pressures encountered by the diver and the mountaineer are usually called *hydrostatic pressures,* because they are due to fluids that are static (at rest).

Let us look first at the increase in pressure with depth below the water's surface. We set up a vertical y axis, its origin being at the air–water interface and the direction of increasing y being up. Consider a water sample contained in an imaginary right circular cylinder of horizontal base area A, and let y_1 and y_2 (both of which are *negative* numbers) be the depths below the surface of the upper and the lower cylinder faces, respectively.

Figure 15-2b shows a free-body diagram for the water in the cylinder. The water sample is in equilibrium, its weight (downward) being exactly balanced by the difference between the force of magnitude $F_2 = p_2A$ acting upward on its lower face and the force of magnitude $F_1 = p_1A$ acting downward on its upper face. Thus

$$F_2 = F_1 + W. \tag{15-3}$$

The volume V of the cylinder is $A(y_1 - y_2)$. Thus the mass m of the water in the cylinder is $\rho A(y_1 - y_2)$, in which ρ is the uniform density of water. The weight W is then $mg = \rho A g(y_1 - y_2)$. Substituting this for W in Eq. 15-3 and substituting for F_1 and F_2 yield

$$p_2A = p_1A + \rho Ag(y_1 - y_2),$$

or

$$p_2 = p_1 + \rho g(y_1 - y_2). \tag{15-4}$$

This equation can be used to find pressures both in a liquid and in the atmosphere. For the former, suppose we seek the pressure p at a depth h below the liquid surface. Then we choose level 1 to be the surface, level 2 to be a distance h below it (as in Fig. 15-3), and p_0 to represent the atmospheric pressure on the surface. We then substitute

$$y_1 = 0, \quad p_1 = p_0 \quad \text{and} \quad y_2 = -h, \quad p_2 = p$$

into Eq. 15-4, which becomes

$$p = p_0 + \rho gh \quad \text{(pressure at depth } h\text{)}. \tag{15-5}$$

Note that the pressure at a given depth depends on that depth but not on any horizontal dimension. Thus Eq. 15-5 holds no matter what the shape of the containing vessel. If the bottom surface of the vessel is at depth h, then Eq. 15-5 gives the pressure p on that surface.

In Eq. 15-5, p is said to be the total pressure or **absolute pressure** at level 2. To see why, note in Fig. 15-3 that the pressure p at level 2 consists of two contributions: (1) p_0, the pressure due to the atmosphere, which bears down on the liquid, and (2) ρgh, the pressure due to the liquid above level 2, which bears down on level 2. In general, the difference between an absolute pressure and an atmospheric pressure is called the **gauge pressure.** So, for the situation of Fig. 15-3, the gauge pressure is ρgh.

Equation 15-4 also holds above the liquid surface: it gives the atmospheric pressure at a given distance above level 1 in terms of the atmospheric pressure p_1 at level 1

(a)

(b)

FIGURE 15-2   (a) A tank of water in which a sample of water is contained in an imaginary cylinder of horizontal base area A. (b) A free-body diagram of the water sample. The water in the sample is in static equilibrium, its weight being balanced by the net upward buoyant force that acts on it.

FIGURE 15-3   The pressure p increases with depth h below the water surface according to Eq. 15-5.

(*assuming* that the atmospheric density is uniform over that distance). For example, to find the atmospheric pressure at a distance d above level 1 in Fig. 15-3, we substitute

$$y_1 = 0, \quad p_1 = p_0 \quad \text{and} \quad y_2 = d, \quad p_2 = p.$$

Then with $\rho = \rho_{air}$, we obtain

$$p = p_0 - \rho_{air}gd.$$

CHECKPOINT 1: The figure shows four containers of olive oil. Rank them according to the pressure at depth h, greatest first.

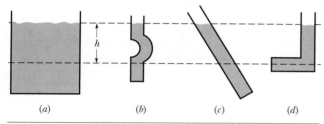

(*a*)   (*b*)   (*c*)   (*d*)

SAMPLE PROBLEM 15-2

(a) An enterprising diver reasons that if a typical snorkel tube, which is 20 cm long, works, a tube 6.0 m long should also work. If he foolishly uses such a tube (Fig. 15-4), what is the pressure difference Δp between the external pressure on him and the air pressure in his lungs? Why is he in danger?

SOLUTION: First consider the diver at depth $L = 6.0$ m without the snorkel tube. The external pressure on him is given by Eq. 15-5 as

$$p = p_0 + \rho g L.$$

His body adjusts to that pressure by contracting slightly until the internal pressures are in equilibrium with the external pressure. In particular, the average blood pressure increases, and the average air pressure in his lungs increases until it equals the external pressure p.

If he then foolishly uses the 6.0 m tube to breathe, the pressurized air in his lungs will be expelled upward through the tube to the atmosphere, and the air pressure in his lungs will rapidly drop to atmospheric pressure p_0. Assuming he is in fresh water of density 1000 kg/m³, the pressure difference Δp between the lower pressure within his lungs and the higher pressure on his chest will then be

$$\Delta p = p - p_0 = \rho g L$$
$$= (1000 \text{ kg/m}^3)(9.8 \text{ m/s}^2)(6.0 \text{ m})$$
$$= 5.9 \times 10^4 \text{ Pa.} \qquad \text{(Answer)}$$

This pressure difference, about 0.6 atm, is sufficient to collapse the lungs and force the still-pressurized blood into them, a process known as lung squeeze.

(b) A novice scuba diver practicing in a swimming pool takes enough air from his tank to fully expand his lungs before abandoning the tank at depth L and swimming to the surface. He ignores instructions and fails to exhale during his ascent. When he reaches the surface, the pressure difference between the external pressure on him and the air in his lungs is 70 torr. From what depth did he start? What potentially lethal danger does he face?

SOLUTION: When he fills his lungs at depth L, the external pressure on him (and the air pressure within his lungs) is again given by Eq. 15-5 as

$$p = p_0 + \rho g L.$$

As he ascends, the external pressure on him decreases, until it is atmospheric pressure p_0 at the surface. His blood pressure also decreases, until it is normal. But because he does not exhale, the air pressure in his lungs remains at the value it had at depth L. So, at the surface, the pressure difference between the higher pressure in his lungs and the lower pressure on his chest is given by

$$\Delta p = p - p_0 = \rho g L,$$

from which we find

$$L = \frac{\Delta p}{\rho g}$$
$$= \frac{70 \text{ torr}}{(1000 \text{ kg/m}^3)(9.8 \text{ m/s}^2)} \left(\frac{1.01 \times 10^5 \text{ Pa}}{760 \text{ torr}} \right)$$
$$= 0.95 \text{ m.} \qquad \text{(Answer)}$$

The pressure difference of 70 torr (about 9% of atmospheric pressure) is sufficient to rupture the diver's lungs and force air from them into the depressurized blood, which then carries the air to the heart, killing the diver. If the diver follows instructions and gradually exhales as he ascends, he allows the pressure in his lungs to equalize with the external pressure, and then there is no danger.

FIGURE 15-4 Sample Problem 15-2. DON'T TRY THIS with a tube that is longer than a standard snorkel tube or the attempt to breathe through the tube could kill you. The reason is that, because the external (water) pressure on your chest can be so much greater than the internal (air) pressure, you might not be able to expand your lungs to inhale.

SAMPLE PROBLEM 15-3

The U-tube in Fig. 15-5 contains two liquids in static equilibrium: water of density ρ_w is in the right arm, and oil of unknown density ρ_x is in the left. Measurement gives $l = 135$ mm and $d = 12.3$ mm. What is the density of the oil?

SOLUTION: If the pressure at the oil–water interface in the left arm is p_{int}, then the pressure in the right arm at the level of that interface must also be p_{int}, because the left and right arms are connected by water below the level of the interface. In the right arm, the interface is a distance l below the free surface of the *water* and we have, from Eq. 15-5,

$$p_{int} = p_0 + \rho_w gl \quad \text{(right arm).}$$

In the left arm, the interface is a distance $l + d$ below the free surface of the *oil* and we have, again from Eq. 15-5,

$$p_{int} = p_0 + \rho_x g(l + d) \quad \text{(left arm).}$$

Equating these two expressions and solving for the unknown density yield

$$\rho_x = \rho_w \frac{l}{l + d} = (1000 \text{ kg/m}^3)\frac{135 \text{ mm}}{135 \text{ mm} + 12.3 \text{ mm}}$$

$$= 916 \text{ kg/m}^3. \qquad \text{(Answer)}$$

Note that the answer does not depend on the atmospheric pressure p_0 or the free-fall acceleration g.

FIGURE 15-5 Sample Problem 15-3. The oil in the left arm stands higher than the water in the right arm because the oil is less dense than the water. Both fluid columns produce the same pressure P_{int} at the level of the interface.

15-5 MEASURING PRESSURE

The Mercury Barometer

Figure 15-6a shows a very basic *mercury barometer,* a device used to measure the pressure of the atmosphere. The long glass tube is filled with mercury and inverted with its open end in a dish of mercury, as the figure shows. The space above the mercury column contains only mercury vapor, whose pressure is so small at ordinary temperatures that it can be neglected.

We can use Eq. 15-4 to find the atmospheric pressure p_0 in terms of the height h of the mercury column. We choose level 1 of Fig. 15-2 to be that of the air–mercury

FIGURE 15-6 (*a*) A mercury barometer. (*b*) Another mercury barometer. The distance h is the same in both cases.

interface and level 2 to be that of the top of the mercury column, as labeled in Fig. 15-6a. We then substitute

$$y_1 = 0, \quad p_1 = p_0 \quad \text{and} \quad y_2 = h, \quad p_2 = 0$$

into Eq. 15-4, finding that

$$p_0 = \rho gh, \qquad (15\text{-}6)$$

where ρ is the density of the mercury.

For a given pressure, the height h of the mercury column does not depend on the cross-sectional area of the vertical tube. The fanciful mercury barometer of Fig. 15-6b gives the same reading as that of Fig. 15-6a; all that counts is the vertical distance h between the mercury levels.

Equation 15-6 shows that, for a given pressure, the height of the column of mercury depends on the value of g at the location of the barometer and on the density of mercury, which varies with temperature. The column height (in millimeters) is numerically equal to the pressure (in torr) *only* if the barometer is at a place where g has its accepted standard value of 9.80665 m/s² *and* the temperature of the mercury is 0°C. If these conditions do not prevail (and they rarely do), small corrections must be made before the height of the mercury column can be transformed into a pressure.

The Open-Tube Manometer

An *open-tube manometer* (Fig. 15-7) measures the gauge pressure of a gas. It consists of a U-tube containing a liquid, with one end of the tube connected to the vessel whose gauge pressure we wish to measure and the other end open to the atmosphere. We can use Eq. 15-4 to find the gauge pressure in terms of the height h shown in Fig. 15-7. Let us choose levels 1 and 2 as shown in Fig. 15-7. We then

FIGURE 15-7 An open-tube manometer, connected so that the gauge pressure of the gas in the tank on the left is read. The right arm of the U-tube is open to the atmosphere.

substitute

$$y_1 = 0, \quad p_1 = p_0 \quad \text{and} \quad y_2 = -h, \quad p_2 = p$$

into Eq. 15-4, finding that

$$p_g = p - p_0 = \rho g h. \tag{15-7}$$

The gauge pressure p_g is directly proportional to h.

The gauge pressure can be positive or negative, depending on whether $p > p_0$ or $p < p_0$. In inflated tires or the human circulatory system, the (absolute) pressure is greater than atmospheric pressure, so the gauge pressure is a positive quantity, sometimes called the *overpressure*. If you suck on a straw to pull fluid up the straw, the (absolute) pressure in your lungs is actually less than atmospheric pressure. The gauge pressure in your lungs is then a negative quantity.

SAMPLE PROBLEM 15-4

The column in a mercury barometer has a measured height h of 740.35 mm. The temperature is $-5.0°C$, at which temperature the density of mercury is 1.3608×10^4 kg/m³. The free-fall acceleration g at the site of the barometer is 9.7835 m/s². What is the atmospheric pressure in pascals and in torr?

SOLUTION: From Eq. 15-6 we have

$$p_0 = \rho g h$$

$$= (1.3608 \times 10^4 \text{ kg/m}^3)(9.7835 \text{ m/s}^2)(0.74035 \text{ m})$$

$$= 9.8566 \times 10^4 \text{ Pa.} \quad \text{(Answer)}$$

Barometer readings are usually expressed in torr, where 1 torr is the pressure exerted by a column of mercury 1 mm high at a place where g has an accepted standard value of 9.80665 m/s² and at a temperature (0.0°C) at which mercury has a density of 1.35955×10^4 kg/m³. Thus from Eq. 15-6,

$$1 \text{ torr} = (1.35955 \times 10^4 \text{ kg/m}^3)(9.80665 \text{ m/s}^2)$$

$$\times (1 \times 10^{-3} \text{ m})$$

$$= 133.326 \text{ Pa.}$$

Applying this conversion factor yields, for the atmospheric pressure recorded on the barometer,

$$p_0 = 9.8566 \times 10^4 \text{ Pa} = 739.29 \text{ torr.} \quad \text{(Answer)}$$

Note that the pressure in torr (739.29 torr) is numerically close to—but otherwise differs significantly from—the height h of the mercury column expressed in mm (740.35 mm).

15-6 PASCAL'S PRINCIPLE

When you squeeze one end of a tube of toothpaste, you are watching **Pascal's principle** in action. This principle is also the basis for the Heimlich maneuver, in which a sharp pressure increase properly applied to the abdomen is transmitted to the throat, forcefully ejecting food lodged there. The principle was first stated clearly in 1652 by Blaise Pascal (for whom the unit of pressure is named):

> A change in the pressure applied to an enclosed incompressible fluid is transmitted undiminished to every portion of the fluid and to the walls of the containing vessel.

Demonstrating Pascal's Principle

Consider the case in which the fluid is an incompressible liquid contained in a tall cylinder, as in Fig. 15-8. The cylinder is fitted with a piston on which a container of lead shot rests. The atmosphere, container, and shot exert pressure p_{ext} on the piston and thus on the liquid. The pressure p at any point P in the liquid is then

$$p = p_{ext} + \rho g h. \tag{15-8}$$

Let us add a little more lead shot to the container to increase p_{ext} by an amount Δp_{ext}. The quantities ρ, g, and h in Eq. 15-8 are unchanged, so the pressure change at P is

$$\Delta p = \Delta p_{ext}. \tag{15-9}$$

FIGURE 15-8 Weights loaded onto the piston create a pressure p_{ext} at the top of the enclosed (incompressible) liquid. If p_{ext} is increased, by adding more weights, the pressure increases by the same amount at all points within the liquid.

This pressure change is independent of h, so it must hold for all points within the liquid, as Pascal's principle states.

Pascal's Principle and the Hydraulic Lever

Figure 15-9 shows how Pascal's principle can be made the basis of a hydraulic lever. In operation, let an external force of magnitude F_i be exerted downward on the left-hand (or input) piston, whose area is A_i. An incompressible liquid in the device then exerts an upward force of magnitude F_o on the righthand (or output) piston, whose area is A_o. To keep the system in equilibrium, an external load (not shown) must exert a downward force of magnitude F_o on the output piston. The force F_i applied on the left and the downward force F_o exerted by the load on the right produce a change Δp in the pressure of the liquid that is given by

$$\Delta p = \frac{F_i}{A_i} = \frac{F_o}{A_o},$$

so
$$F_o = F_i \frac{A_o}{A_i}. \qquad (15\text{-}10)$$

Equation 15-10 shows that the output force F_o exerted on the load must be larger than the input force F_i if $A_o > A_i$, as is the case in Fig. 15-9.

If we move the input piston downward a distance d_i, the output piston moves upward a distance d_o, such that the same volume V of the incompressible liquid is displaced at both pistons. Then

$$V = A_i d_i = A_o d_o,$$

which we can write as

$$d_o = d_i \frac{A_i}{A_o}. \qquad (15\text{-}11)$$

This shows that, if $A_o > A_i$ (as in Fig. 15-9), the output piston moves a smaller distance than the input piston moves.

From Eqs. 15-10 and 15-11 we can write the output work as

$$W = F_o d_o = \left(F_i \frac{A_o}{A_i} \right) \left(d_i \frac{A_i}{A_o} \right) = F_i d_i, \qquad (15\text{-}12)$$

which shows that the work W done *on* the input piston by the applied force is equal to the work W done *by* the output piston in lifting the load placed on it.

We see here that a given force exerted over a given distance can be transformed to a larger force exerted over a smaller distance. The product of force and distance remains unchanged so that the same work is done. However, there is often tremendous advantage in being able to exert the larger force. Most of us, for example, cannot lift an automobile and thus welcome the availability of a hydraulic jack, even though we have to pump the handle farther than the automobile rises. In this device, the displacement d_i is accomplished not in a single stroke but over a series of small strokes.

15-7 ARCHIMEDES' PRINCIPLE

Figure 15-10 shows a student in a swimming pool, manipulating a very thin plastic sack filled with water. She will find that it is in static equilibrium, tending neither to rise nor to sink. Yet the water in the sack has weight and—for that reason—should sink. The weight must be balanced by an upward force whose magnitude is equal to the weight of the water in the sack.

This upward **buoyant force F_b** is exerted on the water in the sack by the water that surrounds the sack. This buoyant force exists because—as you have already seen—the pressure in the water increases with depth below the surface, so the pressure near the bottom of the sack is greater than the pressure near the top.

Let us now remove the sack of water. Figure 15-11a shows the forces acting at the hole in the water formerly occupied by the sack. The buoyant force F_b, which points up, is the vector sum of all these forces.

FIGURE 15-9   A hydraulic arrangement, used to magnify a force F_i. The work done by F_i, however, is not magnified and is the same for both the input and output forces.

FIGURE 15-10   A thin-walled plastic sack of water is in static equilibrium in the pool. Its weight must be balanced by a net upward force exerted on the sack by the water surrounding it.

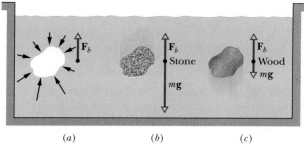

FIGURE 15-11 (*a*) The water surrounding the hole in the water exerts a resultant upward buoyant force on whatever fills the hole. (*b*) For a stone of the same volume as the hole, the weight exceeds the buoyant force. (*c*) For a lump of wood of the same volume, the weight is less than the buoyant force.

Let us fill the hole in Fig. 15-11*a* with a stone of exactly the same dimensions, as in Fig. 15-11*b*. *The same upward buoyant force that acted on the water-filled sack will act on the stone.* However, this force is too small to balance the weight of the stone, so the stone will sink. Even though the stone sinks, the water's buoyant force reduces its apparent weight, making the stone easier to lift as long as it is under water.

If we now fill the hole of Fig. 15-11*a* with a block of wood of the same dimensions, as in Fig. 15-11*c*, the same upward buoyant force acts on the wood. This time, however, the upward force is *greater* than the weight of the wood, and the wood will rise toward the surface. We summarize these facts with **Archimedes' principle**:

A body fully or partially immersed in a fluid is buoyed up by a force equal to the weight of the fluid that the body displaces.

It is this principle that explains floating. When, say, the piece of wood in Fig. 15-11*c* rises enough to break through the water's surface, it displaces less water than when it was submerged. According to Archimedes' principle, its buoyant force decreases. The wood continues to rise out of the water until the buoyant force acting on it has decreased to exactly the weight of the wood. The wood is then in static equilibrium; it is floating.

Recall that we say the weight of an object acts effectively at the object's center of mass. Similarly, we can say the buoyant force on a fully or partially immersed object acts effectively at a point called the **center of buoyancy.** This point is located where the center of mass of the displaced fluid would have been, had it not been displaced. If a uniformly dense object is fully submerged, then the center of buoyancy and the object's center of mass coincide. If the object is only partially submerged, those points are separated; then the buoyant force can produce a torque on the object about its center of mass.

CHECKPOINT 2: A penguin floats first in a fluid of density ρ_0, then in a fluid of density $0.95\rho_0$, and then in a fluid of density $1.1\rho_0$. (a) Rank the densities according to the buoyant force on the penguin, greatest first. (b) Rank the densities according to the amount of fluid displaced by the penguin, greatest first.

SAMPLE PROBLEM 15-5

The "tip of the iceberg" in popular speech has come to mean a small visible fraction of something that is mostly hidden. For real icebergs, what is this fraction?

In the late evening of August 21, 1986, something (possibly a volcanic tremor) disturbed Cameroon's Lake Nyos, which has a high concentration of dissolved carbon dioxide. The disturbance caused that gas to form bubbles. Being lighter than the surrounding fluid (the water), those bubbles were buoyed to the surface, where they released the carbon dioxide. The gas, being heavier than the surrounding fluid (now the air), rushed down the mountainside like a river, asphyxiating 1700 persons and the scores of animals seen here.

SOLUTION: The weight of an iceberg of total volume V_i is

$$W_i = \rho_i V_i g,$$

where $\rho_i = 917 \text{ kg/m}^3$ is the density of ice.

The weight of the displaced seawater, which is the buoyancy force F_b, is

$$W_w = F_b = \rho_w V_w g,$$

where $\rho_w = 1024 \text{ kg/m}^3$ is the density of seawater and V_w is the volume of the displaced water, that is, the submerged volume of the iceberg. For the floating iceberg, these two forces are equal, or

$$\rho_i V_i g = \rho_w V_w g.$$

From this equation, we find that the fraction we seek is

$$
\begin{aligned}
\text{frac} &= \frac{V_i - V_w}{V_i} = 1 - \frac{V_w}{V_i} = 1 - \frac{\rho_i}{\rho_w} \\
&= 1 - \frac{917 \text{ kg/m}^3}{1024 \text{ kg/m}^3} \\
&= 0.10 \text{ or } 10\%. \qquad \text{(Answer)}
\end{aligned}
$$

SAMPLE PROBLEM 15-6

A spherical, helium-filled balloon has a radius R of 12.0 m. The balloon, support cables, and basket have a mass m of 196 kg. What maximum load M can the balloon carry? Take $\rho_{He} = 0.160 \text{ kg/m}^3$ and $\rho_{air} = 1.25 \text{ kg/m}^3$; the volume of air displaced by the load is negligible.

SOLUTION: The weight of the displaced air, which is the buoyant force, and the weight of the helium in the balloon are

$$W_{air} = \rho_{air} V g \quad \text{and} \quad W_{He} = \rho_{He} V g,$$

in which $V\ (= 4\pi R^3/3)$ is the volume of the balloon.

At equilibrium, from Archimedes' principle,

$$W_{air} = W_{He} + mg + Mg$$

or

$$
\begin{aligned}
M &= (\tfrac{4}{3}\pi)(R^3)(\rho_{air} - \rho_{He}) - m \\
&= (\tfrac{4}{3}\pi)(12.0 \text{ m})^3(1.25 \text{ kg/m}^3 - 0.160 \text{ kg/m}^3) \\
&\quad - 196 \text{ kg} \\
&= 7690 \text{ kg}. \qquad \text{(Answer)}
\end{aligned}
$$

An object with this mass would weigh 17,000 lb at sea level.

15-8 IDEAL FLUIDS IN MOTION

The motion of *real fluids* is very complicated and not yet fully understood. Instead, we shall discuss the motion of an **ideal fluid,** which is simpler to handle mathematically and

yet provides useful results. Here are four assumptions that we make about our ideal fluid:

1. *Steady flow* In *steady* (or *laminar*) *flow* the velocity of the moving fluid at any fixed point does not change with time, either in magnitude or in direction. The gentle flow of water near the center of a quiet stream is steady; that in a chain of rapids is not. Figure 15-12 shows a transition from steady flow to *nonsteady* (or *turbulent*) flow for a rising stream of smoke. The speed of the smoke particles increases as they rise and, at a certain critical speed, the flow changes from steady to nonsteady.

2. *Incompressible flow* We assume, as we have already done for fluids at rest, that our ideal fluid is incompressible. That is, its density has a constant, uniform value.

3. *Nonviscous flow* Roughly speaking, the viscosity of a fluid is a measure of how resistive the fluid is to flow. For example, thick honey is more resistive to flow than water, and so honey is said to be more viscous than water. Viscosity is the fluid analog of friction between solids; both are mechanisms by which the kinetic energy of moving objects can be transferred to thermal energy. In the absence of friction, a block could glide at constant speed along a horizontal surface. In the same way, an object moving through a nonviscous fluid would experience no *viscous drag force,* that is, no resistive force due to viscosity. The British scientist Lord Rayleigh noted that in an ideal fluid a ship's propeller would not work but, on the other hand, a ship (once set into motion) would not need a propeller!

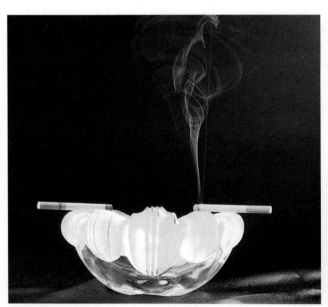

FIGURE 15-12 At a certain point, the rising flow of smoke and heated gas changes from steady to turbulent.

4. *Irrotational flow* Although it need not concern us further, we also assume that the flow is *irrotational*. To test for this property, let a tiny grain of dust move with the fluid. Although this test body may (or may not) move in a circular path, in irrotational flow the test body will not rotate about an axis through its own center of mass. For a loose analogy, the motion of a Ferris wheel is rotational; that of its passengers is irrotational.

15-9 STREAMLINES AND THE EQUATION OF CONTINUITY

Figure 15-13 shows streamlines traced out by dye injected into a moving fluid; Fig. 15-14 shows similar streamlines revealed by smoke. A **streamline** is the path traced out by a tiny fluid element, which we may call a fluid "particle." As the fluid particle moves, its velocity may change, both in magnitude and in direction. However, its velocity vector at any point is always tangent to the streamline at that point (Fig. 15-15). Streamlines never cross because, if they did, a fluid particle arriving at the intersection would have two velocities simultaneously, an impossibility.

In flows like that of Figs. 15-13 and 15-14, we can isolate a *tube of flow* whose boundary is made up of streamlines. Such a tube acts like a pipe because any fluid particle that enters it cannot escape through its walls; if a particle did escape, we would have a case of streamlines crossing each other.

Figure 15-16 shows two cross sections, of areas A_1 and A_2, along a thin tube of flow. Let us station ourselves at B and monitor the fluid, moving with speed v_1, for a short time interval Δt. During this interval, a fluid particle moves

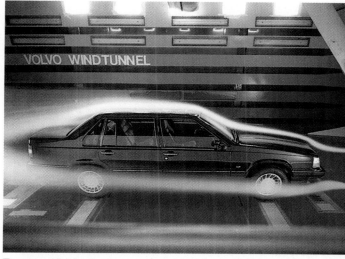

FIGURE 15-14 Smoke reveals streamlines in airflow past a car in a wind-tunnel test.

a small distance $v_1 \Delta t$, and a volume ΔV of fluid, given by

$$\Delta V = A_1 v_1 \Delta t,$$

passes through area A_1.

Assume that the fluid is incompressible and cannot be created or destroyed. Thus in this same time interval, the same volume of fluid must pass point C, farther down the tube of flow. If the speed there is v_2, this means that

$$\Delta V = A_1 v_1 \Delta t = A_2 v_2 \Delta t,$$

or

$$A_1 v_1 = A_2 v_2.$$

FIGURE 15-13 The steady flow of a fluid around a cylinder, as revealed by a dye tracer.

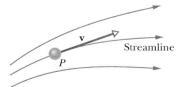

FIGURE 15-15 A fluid particle P traces out a streamline as it moves. The velocity vector of the particle is tangent to the streamline at every point.

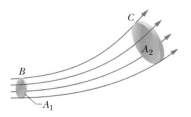

FIGURE 15-16 A tube of flow is defined by the streamlines that form its boundary. The flow rate of fluid must be the same for all cross sections of the tube of flow.

FIGURE 15-17 When a channel, such as a pipe, constricts, the streamlines draw closer together, signaling an increase in the fluid velocity. The arrow shows the direction of flow.

Thus, along the tube of flow,

$$R = Av = \text{a constant}, \tag{15-13}$$

in which R, whose SI unit is the cubic meter per second, is the **volume flow rate** of the fluid. Equation 15-13 is called the **equation of continuity** for fluid flow. It tells us that the flow is faster in the narrower parts of a tube of flow, where the streamlines are closer together, as in Fig. 15-17.

Equation 15-13 is actually an expression of the law of conservation of mass in a form useful in fluid mechanics. In fact, if we multiply R by the (assumed uniform) density of the fluid, we get the quantity $Av\rho$, which is the **mass flow rate**, whose SI unit is the kilogram per second. Then Eq. 15-13 effectively tells us that the mass that flows through point B in Fig. 15-16 each second must be equal to the mass that flows through point C each second.

CHECKPOINT 3: The figure shows a pipe and gives the volume flow rate (in cm³/s) and the direction of flow for all but one section. What are the volume flow rate and the direction of flow for that section?

SAMPLE PROBLEM 15-7

The cross-sectional area A_0 of the aorta (the major blood vessel emerging from the heart) of a normal resting person is 3 cm², and the speed v_0 of the blood is 30 cm/s. A typical capillary (diameter $\approx$ 6 μm) has a cross-sectional area A of 3×10^{-7} cm² and a flow speed v of 0.05 cm/s. How many capillaries does such a person have?

SOLUTION: All the blood that passes through the capillaries must have passed through the aorta so that, from Eq. 15-13,

$$A_0 v_0 = nAv,$$

where n is the number of capillaries. Solving for n yields

$$n = \frac{A_0 v_0}{Av} = \frac{(3 \text{ cm}^2)(30 \text{ cm/s})}{(3 \times 10^{-7} \text{ cm}^2)(0.05 \text{ cm/s})}$$

$$= 6 \times 10^9 \text{ or 6 billion.} \qquad \text{(Answer)}$$

You can easily show that the combined cross-sectional area of the capillaries is about 600 times the area of the aorta.

SAMPLE PROBLEM 15-8

Figure 15-18 shows how the stream of water emerging from a faucet "necks down" as it falls. The cross-sectional area A_0 is 1.2 cm², and A is 0.35 cm². The two levels are separated by a vertical distance $h = 45$ mm. At what rate does water flow from the tap?

SOLUTION: From the equation of continuity (Eq. 15-13) we have

$$A_0 v_0 = Av, \tag{15-14}$$

where v_0 and v are the water velocities at the corresponding levels. From Eq. 2-23 we can also write, because the water is falling freely with acceleration g,

$$v^2 = v_0^2 + 2gh. \tag{15-15}$$

Eliminating v between Eqs. 15-14 and 15-15 and solving for v_0, we obtain

$$v_0 = \sqrt{\frac{2ghA^2}{A_0^2 - A^2}}$$

$$= \sqrt{\frac{(2)(9.8 \text{ m/s}^2)(0.045 \text{ m})(0.35 \text{ cm}^2)^2}{(1.2 \text{ cm}^2)^2 - (0.35 \text{ cm}^2)^2}}$$

$$= 0.286 \text{ m/s} = 28.6 \text{ cm/s}.$$

The volume flow rate R is then

$$R = A_0 v_0 = (1.2 \text{ cm}^2)(28.6 \text{ cm/s})$$

$$= 34 \text{ cm}^3/\text{s}. \qquad \text{(Answer)}$$

FIGURE 15-18 Sample Problem 15-8. As water falls from a tap, its speed increases. Because the flow rate must be the same at all cross sections, the stream must "neck down."

15-10 BERNOULLI'S EQUATION

Figure 15-19 represents a tube of flow (or an actual pipe, for that matter) through which an ideal fluid is flowing at a steady rate. In a time interval Δt, suppose that a volume of fluid ΔV, colored purple in Fig. 15-19a, enters the tube at the left (or input) end and an identical volume, colored green in Fig. 15-19b, emerges at the right (or output) end. The emerging volume must be the same as the entering volume because the fluid is incompressible, with an assumed constant density ρ.

Let y_1, v_1, and p_1 be the elevation, speed, and pressure of the fluid entering at the left, and y_2, v_2, and p_2 be the corresponding quantities for the fluid emerging at the right. By applying the principle of conservation of energy to the fluid, we shall show that these quantities are related by

$$p_1 + \tfrac{1}{2}\rho v_1^2 + \rho g y_1 = p_2 + \tfrac{1}{2}\rho v_2^2 + \rho g y_2. \quad (15\text{-}16)$$

We can also write this equation as

$$p + \tfrac{1}{2}\rho v^2 + \rho g y = \text{a constant}. \quad (15\text{-}17)$$

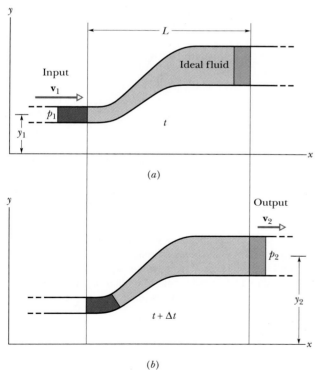

(a)

(b)

FIGURE 15-19 Fluid flows at a steady rate through a length L of a tube, from the input end at the left to the output end at the right. From time t in (a) to time $t + \Delta t$ in (b), the amount of fluid shown in purple enters the input end and the equal amount shown in green emerges from the output end.

Equations 15-16 and 15-17 are equivalent forms of **Bernoulli's equation,** after Daniel Bernoulli, who studied fluid flow in the 1700s.* Like the equation of continuity (Eq. 15-13), Bernoulli's equation is not a new principle but simply the reformulation of a familiar principle in a form more suitable to fluid mechanics. As a check, let us apply Bernoulli's equation to fluids at rest, by putting $v_1 = v_2 = 0$ in Eq. 15-16. The result is

$$p_2 = p_1 + \rho g(y_1 - y_2),$$

which is Eq. 15-4.

A major prediction of Bernoulli's equation emerges if we take y to be a constant ($y = 0$, say) so that the fluid does not change elevation as it flows. Equation 15-16 then becomes

$$p_1 + \tfrac{1}{2}\rho v_1^2 = p_2 + \tfrac{1}{2}\rho v_2^2, \quad (15\text{-}18)$$

which tells us that:

> If the speed of a fluid particle increases as it travels along a horizontal streamline, the pressure of the fluid must decrease, and conversely.

Put another way, where the streamlines are relatively close together (that is, where the velocity is relatively great), the pressure is relatively low, and conversely.

The link between a change in speed and a change in pressure makes sense if you consider a fluid particle. When the particle nears a narrow region, the higher pressure behind it accelerates it so that it then has a greater speed in the narrow region. And when it nears a wide region, the higher pressure ahead of it decelerates it so that it then has a lesser speed in the wide region.

Bernoulli's equation is strictly valid only to the extent that the fluid is ideal. If viscous forces are present, thermal energy will be involved. We take no account of this in the derivation that follows.

Proof of Bernoulli's Equation

Let us take as our system the entire volume of the (ideal) fluid shown in Fig. 15-19. We shall apply the principle of conservation of energy to this system as it moves from its initial state (Fig. 15-19a) to its final state (Fig. 15-19b). The fluid lying between the two vertical planes separated by a distance L in Fig. 15-19 does not change its properties during this process; we need be concerned only with changes that take place at the input and the output ends.

---

*For irrotational flow (which we assume), the constant in Eq. 15-17 has the same value for all points within the tube of flow; the points do not have to lie along the same streamline. Similarly, the points 1 and 2 in Eq. 15-16 can lie anywhere within the tube of flow.

We apply energy conservation in the form of the work–kinetic energy theorem,

$$W = \Delta K, \tag{15-19}$$

which tells us that the change in the kinetic energy of our system must equal the net work done on the system. The change in kinetic energy results from the change in speed between the ends of the pipe and is

$$\Delta K = \tfrac{1}{2}\Delta m\, v_2^2 - \tfrac{1}{2}\Delta m\, v_1^2$$
$$= \tfrac{1}{2}\rho\, \Delta V(v_2^2 - v_1^2), \tag{15-20}$$

in which $\Delta m\,(=\rho\,\Delta V)$ is the mass of the fluid that enters at the input end and leaves at the output end during a small time interval Δt.

The work done on the system arises from two sources. The work W_g done by the weight ($\Delta m\,\mathbf{g}$) of mass Δm during the vertical lift of the mass from the input level to the output level is

$$W_g = -\Delta m\, g(y_2 - y_1)$$
$$= -\rho g\, \Delta V(y_2 - y_1). \tag{15-21}$$

This work is negative because the upward displacement and the downward weight point in opposite directions.

Work W_p must also be done *on* the system (at the input end) to push the entering fluid into the tube and *by* the system (at the output end) to push forward the fluid ahead of the emerging fluid. In general, the work done by a force of magnitude F, acting on a fluid sample contained in a tube of area A to move the fluid through a distance Δx, is

$$F\,\Delta x = (pA)(\Delta x) = p(A\,\Delta x) = p\,\Delta V.$$

The work done on the system is then $p_1\,\Delta V$, and the work done by the system is $-p_2\,\Delta V$. Their sum W_p is

$$W_p = -p_2\,\Delta V + p_1\,\Delta V$$
$$= -(p_2 - p_1)\Delta V. \tag{15-22}$$

The work–kinetic energy theorem of Eq. 15-19 now becomes

$$W = W_g + W_p = \Delta K.$$

Substituting from Eqs. 15-20, 15-21, and 15-22 yields

$$-\rho g\,\Delta V(y_2 - y_1) - \Delta V(p_2 - p_1) = \tfrac{1}{2}\rho\,\Delta V(v_2^2 - v_1^2).$$

This, after a slight rearrangement, matches Eq. 15-16, which we set out to prove.

SAMPLE PROBLEM 15-9

Ethanol of density $\rho = 791$ kg/m³ flows smoothly through a horizontal pipe that tapers (as in Fig. 15-17) in cross-sectional area from $A_1 = 1.20 \times 10^{-3}$ m² to $A_2 = A_1/2$. The pressure difference Δp between the wide and narrow sections of pipe is 4120 Pa. What is the volume flow rate R of the ethanol?

SOLUTION: Rearranging Eq. 15-18 (Bernoulli's equation for level flow) yields

$$p_1 - p_2 = \tfrac{1}{2}\rho v_2^2 - \tfrac{1}{2}\rho v_1^2 = \tfrac{1}{2}\rho(v_2^2 - v_1^2), \tag{15-23}$$

where subscripts 1 and 2 refer to the wide and narrow sections of pipe, respectively. Equation 15-13 (the continuity equation) tells us that the flow is faster in the narrower section. Here that means $v_2 > v_1$. Equation 15-23 then tells us that $p_1 > p_2$.

Equation 15-13 also tells us the volume flow rate R is the same in the wide and narrow sections. So

$$R = v_1 A_1 = v_2 A_2.$$

These equations, with $A_2 = A_1/2$, give us

$$v_1 = \frac{R}{A_1} \quad \text{and} \quad v_2 = \frac{R}{A_2} = \frac{2R}{A_1}.$$

Substituting these expressions into Eq. 15-23, setting $p_1 - p_2 = \Delta p$, and rearranging give us

$$\Delta p = \tfrac{1}{2}\rho\left(\frac{4R^2}{A_1^2} - \frac{R^2}{A_1^2}\right) = \frac{3\rho R^2}{2A_1^2}.$$

Solving for R, we find

$$R = A_1\sqrt{\frac{2\Delta p}{3\rho}}$$
$$= 1.20 \times 10^{-3} \text{ m}^2 \sqrt{\frac{(2)(4120 \text{ Pa})}{(3)(791 \text{ kg/m}^3)}}$$
$$= 2.24 \times 10^{-3} \text{ m}^3/\text{s}. \tag{Answer}$$

SAMPLE PROBLEM 15-10

In the old West, a desperado fires a bullet into an open water tank (Fig. 15-20), creating a hole a distance h below the water surface. What is the speed v of the water emerging from the hole?

SOLUTION: This situation is essentially that of water moving (downward) with speed V through a wide pipe (the tank) of cross-sectional area A and then moving (horizontally) with speed v through a narrow pipe (the hole) of cross-sectional area a. From Eq. 15-13, we know that

$$R = av = AV$$

and thus that

$$V = \frac{a}{A}v.$$

Because $a \ll A$, we see that $V \ll v$.

We can also relate v to V (and to h) through the Bernoulli equation (Eq. 15-16). We take the level of the hole as our reference level for measuring elevations (and thus gravitational potential energy). Noting that the pressure at the top of the tank and at the bullet hole is the atmospheric pressure p_0 (because both places are exposed to the atmosphere), we write Eq. 15-16 as

$$p_0 + \tfrac{1}{2}\rho V^2 + \rho g h = p_0 + \tfrac{1}{2}\rho v^2 + \rho g(0). \quad (15\text{-}24)$$

(Here the top of the tank is represented by the left side of the equation, and the hole by the right side. The zero on the right

FIGURE 15-20 Sample Problem 15-10. Water pours through a hole in a water tank, at a distance h below the water surface. The water pressure at the surface and at the hole is atmospheric pressure p_0.

indicates that the hole is at our reference level.) Before we solve Eq. 15-24 for v, we can use our result that $V \ll v$ to simplify it: We assume that V^2, and thus the term $\tfrac{1}{2}\rho V^2$ in Eq. 15-24, is negligible compared to the other terms, and we drop it. Solving the remaining equation for v then yields

$$v = \sqrt{2gh}. \qquad \text{(Answer)}$$

This is the same speed that an object would have when falling a height h from rest.

CHECKPOINT **4:** Water flows smoothly through the pipe shown in the figure, descending in the process. Rank the four numbered sections of pipe according to (a) the volume flow rate R through them, (b) the flow speed v through them, and (c) the water pressure p within them, greatest first.

REVIEW & SUMMARY

Density

The **density** ρ of any material is defined as its mass per unit volume:

$$\rho = \frac{\Delta m}{\Delta V}. \qquad (15\text{-}1)$$

Usually, where a material sample is large compared with atomic dimensions, we can write Eq. 15-1 as $\rho = m/V$.

Fluid Pressure

A **fluid** is a substance that can flow; it conforms to the boundaries of its container because it cannot withstand shearing stress. It can, however, exert a force perpendicular to its surface. That force is described in terms of **pressure** p:

$$p = \frac{\Delta F}{\Delta A}, \qquad (15\text{-}2)$$

in which ΔF is the force acting on a surface element of area ΔA. If the force is uniform over a flat area, Eq. 15-2 can be written as $p = F/A$. The force resulting from fluid pressure at a particular point in a fluid has the same magnitude in all directions. *Gauge pressure* is the difference between the actual pressure (or *absolute pressure*) at a point and the atmospheric pressure.

Pressure Variation with Height and Depth

Pressure in a fluid at rest varies with vertical position y. For y measured positive upward,

$$p_2 = p_1 + \rho g(y_1 - y_2). \qquad (15\text{-}4)$$

The pressure is the same for all points at the same level. If h is the *depth* of a fluid sample below some reference level at which the pressure is p_0, Eq. 15-4 becomes

$$p = p_0 + \rho g h. \qquad (15\text{-}5)$$

Pascal's Principle

Pascal's principle, which can be derived from Eq. 15-4, states that a change in the pressure applied to an enclosed fluid is transmitted undiminished to every portion of the fluid and to the walls of the containing vessel.

Archimedes' Principle

The surface of an immersed object is acted on by forces associated with the fluid pressure. The vector sum of those forces (called the **buoyant force**) acts vertically upward, through the center of mass of the displaced fluid (the **center of buoyancy**). Archimedes' principle states that the magnitude of the buoyant force is equal to the weight of the fluid displaced by the object. When an object floats, its weight is equal to the buoyant force acting on it.

Flow of Ideal Fluids

An *ideal fluid* is incompressible and lacks viscosity, and its flow is steady and irrotational. A **streamline** is the path followed by an individual fluid particle. A *tube of flow* is a bundle of streamlines. The principle of conservation of mass shows that the flow within any tube of flow obeys the **equation of continuity:**

$$R = Av = \text{a constant,} \qquad (15\text{-}13)$$

in which R is the **volume flow rate,** A the cross-sectional area of the tube of flow at any point, and v the speed of the fluid at that point, assumed to be constant across A. The **mass flow rate** $Av\rho$ is also constant.

Bernoulli's Equation

Applying the principle of conservation of mechanical energy to the flow of an ideal fluid leads to **Bernoulli's equation:**

$$p + \tfrac{1}{2}\rho v^2 + \rho g y = \text{a constant} \qquad (15\text{-}17)$$

along any tube of flow.

QUESTIONS

1. Figure 15-21 shows a tank filled with water. Five horizontal floors and ceilings are indicated; all have the same area and are located at distances L, $2L$, or $3L$ below the top of the tank. Rank the floors and ceilings according to the force on them due to the water, greatest first.

FIGURE 15-21   Question 1.

2. Figure 15-22 shows four situations in which a red liquid and a gray liquid are in a U-tube. In one situation the liquids cannot be in static equilibrium. (a) Which situation is that? (b) For the other three situations, assume static equilibrium. For each, is the density of the red liquid greater than, less than, or equal to the density of the gray liquid?

    (1)   (2)   (3)   (4)

FIGURE 15-22   Question 2.

3. The containers of Fig. 15-23 have the same level of water and same base area, and are made of the same material. (a) Rank the containers (and contents) according to their weights on a scale, greatest first. (b) Rank the containers according to the pressure of the water on the base of the container. (c) Does Eq. 15-2 indicate that the answers to (a) and (b) are inconsistent? This apparent inconsistency is known as the *hydrostatic paradox.*

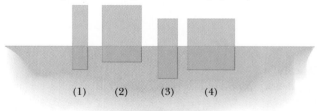

FIGURE 15-23   Question 3.   (1)   (2)   (3)

4. We fully submerge an irregular 3 kg lump of material in a certain fluid. The fluid that would have been in the space now occupied by the lump has a mass of 2 kg. (a) When we release the lump, does it move upward, move downward, or remain in place? (b) If we next fully submerge the lump in a less dense fluid and again release it, what does it do?

5. Figure 15-24 shows four solid objects floating in corn syrup. Rank the objects according to their density, greatest first.

    (1)   (2)   (3)   (4)

FIGURE 15-24   Question 5.

6. Figure 15-25 shows three identical open-top containers filled to the brim with water; toy ducks float in two of them. Rank the containers and contents according to their weight, greatest first.

    (a)   (b)   (c)

FIGURE 15-25   Question 6.

7. A boat with an anchor on board floats in a swimming pool that is somewhat wider than the boat. Does the water level in the pool move upward, move downward, or remain the same if the anchor is (a) dropped into the water or (b) thrown onto the surrounding ground? (c) Does the water level in the pool move upward, move downward, or remain the same if, instead, a cork is dropped from the boat into the water, where it floats?

8. Three gas bubbles of identical sizes are submerged in water: bubble 1 is filled with hydrogen, bubble 2 is filled with helium, and bubble 3 is filled with carbon dioxide. Rank the bubbles according to the magnitude of the buoyant force acting on them, greatest first.

9. A block of wood floats at a certain level in a pail of water in a stationary elevator. Does the block float higher, lower, or at the same level when the elevator cab (a) moves upward at constant speed, (b) moves downward at constant speed, (c) accelerates upward, and (d) accelerates downward with an acceleration magnitude less than 9.8 m/s²?

10. A container of water is suspended from a spring balance. Does the weight reading increase, decrease, or remain the same

when we, without spilling any water, (a) lower a heavy metal object via string into the water and (b) float a cork in the water?

11. Figure 15-26 shows two rectangular, uniformly dense blocks that are floating in water. Both blocks have been tipped by hand and then released. For each block, (a) does buoyancy cause a clockwise torque or a counterclockwise torque about the block's center of mass and (b) will the block right itself or tip over more?

FIGURE 15-26 Question 11.

EXERCISES & PROBLEMS

SECTION 15-3 Density and Pressure

1E. Convert a density of 1.0 g/cm³ to kg/m³.

2E. Three liquids that will not mix are poured into a cylindrical container. The volumes and densities of the liquids are 0.50 L, 2.6 g/cm³; 0.25 L, 1.0 g/cm³; and 0.40 L, 0.80 g/cm³. What is the force acting on the bottom of the container due to these liquids? One liter = 1 L = 1000 cm³. (Ignore the contribution due to the atmosphere.)

3E. Find the pressure increase in the fluid in a syringe when a nurse applies a force of 42 N to the syringe's circular piston, which has a radius of 1.1 cm.

4E. You inflate the front tires on your car to 28 psi. Later, you measure your blood pressure, obtaining a reading of 120/80, the readings being in mm Hg. In metric countries (which is to say, most of the rest of the world), these pressures are customarily reported in kilopascals (kPa). In kilopascals, what are (a) your tire pressure and (b) your blood pressure?

5E. An office window has dimensions 3.4 m by 2.1 m. As a result of the passage of a storm, the outside air pressure drops to 0.96 atm, but inside the pressure is held at 1.0 atm. What net force pushes out on the window?

6E. A fish maintains its depth in fresh water by adjusting the air content of porous bone or air sacs to make its average density the same as that of the water. Suppose that with its air sacs collapsed, a fish has a density of 1.08 g/cm³. To what fraction of its expanded body volume must the fish inflate the air sacs to reduce its density to that of water?

7P. An airtight box having a lid with an area of 12 in.² is partially evacuated. If a force of 108 lb is required to pull the lid off the box and the outside atmospheric pressure is 15 lb/in.², what is the air pressure in the box?

8P. In 1654 Otto von Guericke, inventor of the air pump, gave a demonstration before the noblemen of the Holy Roman Empire in which two teams of eight horses could not pull apart two evacu-

ated brass hemispheres. (a) Assuming that the hemispheres have thin walls, so that R in Fig. 15-27 may be considered both the inside and outside radius, show that the force F required to pull apart the hemispheres is $F = \pi R^2 \Delta p$, where Δp is the difference between the pressures outside and inside the sphere. (b) Taking R equal to 1.0 ft and the inside pressure as 0.10 atm, find the force the teams of horses would have had to exert to pull apart the hemispheres. (c) Why are two teams of horses used? Would not one team have proved the point just as well if the hemispheres were attached to a sturdy wall?

FIGURE 15-27 Problem 8.

SECTION 15-4 Fluids at Rest

9E. Calculate the hydrostatic difference in blood pressure between the brain and the foot in a person of height 1.83 m. The density of blood is 1.06×10^3 kg/m³.

10E. Find the pressure, in pascals, 150 m below the surface of the ocean. The density of seawater is 1.03 g/cm³, and the atmospheric pressure at sea level is 1.01×10^5 Pa.

11E. The sewer outlets of a house constructed on a slope are 8.2 m below street level. If the sewer is 2.1 m below street level, find the minimum pressure differential that must be created by the sewage pump to transfer waste of average density 900 kg/m³.

12E. Figure 15-28 displays the *phase diagram* of carbon, showing the ranges of temperature and pressure in which carbon will crystallize either as diamond or graphite. What is the minimum depth at which diamonds can form if the temperature at that depth is 1000°C and the subsurface rocks have density 3.1 g/cm³? Assume that, as in a fluid, the pressure is due to the weight of material lying above.

FIGURE 15-28 Exercise 12.

13E. The human lungs can operate against a pressure differential of up to about $\frac{1}{20}$ of an atmosphere. If a diver uses a snorkel for breathing, about how far below water level can she or he swim?

14E. A swimming pool has the dimensions 80 ft × 30 ft × 8.0 ft. (a) When it is filled with water, what is the force (resulting from the water alone) on the bottom, on the short sides, and on the long sides? (b) If you are concerned with the possibility that the concrete walls and floor will collapse, is it appropriate to take the atmospheric pressure into account? Why?

15E. (a) Find the total weight of water on top of a nuclear submarine at a depth of 200 m, assuming that its (horizontal cross-sectional) hull area is 3000 m². (b) What water pressure would a diver experience at this depth? Express your answer in atmospheres. Do you think that occupants of a damaged submarine at this depth could escape without special equipment? Assume the density of seawater is 1.03 g/cm³.

16E. Crew members attempt to escape from a damaged submarine 100 m below the surface. What force must they apply to a pop-out hatch, which is 1.2 m by 0.60 m, to push it out? Assume that the density of the ocean water is 1025 kg/m³.

17E. A simple open U-tube contains mercury. When 11.2 cm of water is poured into the right arm of the tube, how high above its initial level does the mercury rise in the left arm?

FIGURE 15-29 Exercise 18.

18E. A cylindrical barrel has a narrow tube fixed to the top, as shown with dimensions in Fig. 15-29. The vessel is filled with water to the top of the tube. Calculate the ratio of the hydrostatic force exerted on the bottom of the barrel to the weight of the water contained inside the barrel. Why is the ratio not equal to one? (Ignore the presence of the atmosphere.)

19P. Two identical cylindrical vessels with their bases at the same level each contain a liquid of density ρ. The area of either base is A, but in one vessel the liquid height is h_1, and in the other h_2. Find the work done by the gravitational force in equalizing the levels when the two vessels are connected.

20P. (a) Consider a container of fluid subject to a *vertical upward* acceleration of magnitude a. Show that the pressure variation with depth in the fluid is given by

$$p = \rho h(g + a),$$

where h is the depth and ρ is the density. (b) Show also that if the fluid as a whole undergoes a *vertical downward* acceleration of magnitude a, the pressure at a depth h is given by

$$p = \rho h(g - a).$$

(c) What is the pressure if the fluid is in free fall?

21P. In analyzing certain geological features, it is often appropriate to assume that the pressure at some horizontal *level of compensation,* deep inside Earth, is the same over a large region and is equal to the pressure exerted by the weight of the overlying material. That is, the pressure on the level of compensation is given by the fluid pressure formula. This model requires, for example, that mountains have *roots* (Fig. 15-30). Consider a mountain 6.0 km high. The continental rocks have a density of 2.9 g/cm³, and beneath the continent is the mantle, with a density of 3.3 g/cm³. Calculate the depth D of the root. (*Hint:* Set the pressure at points a and b equal; the depth y of the level of compensation will cancel out.)

FIGURE 15-30
Problem 21.

22P. In Fig. 15-31, the ocean is about to overrun the continent. Find the depth h of the ocean using the level-of-compensation method shown in Problem 21.

23P. Water stands at a depth D behind the vertical upstream face of a dam, as shown in Fig. 15-32. Let W be the width of the dam. (a) Find the resultant horizontal force exerted on the dam by the gauge pressure of the water and (b) the net torque owing to that gauge pressure about a line through O parallel to the width of the dam. (c) Find the moment arm of the resultant horizontal force about the line through O.

FIGURE 15-31
Problem 22.

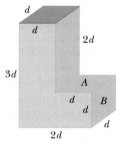

FIGURE 15-32
Problem 23.

24P. The L-shaped tank shown in Fig. 15-33 is filled with water and is open at the top. If $d = 5.0$ m, what are (a) the force on face A and (b) the force on face B due to the water?

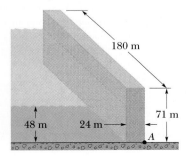

FIGURE 15-33
Problem 24.

25P. Figure 15-34 shows a dam and part of the freshwater reservoir backed up behind it. The dam is made of concrete of density 3.2 g/cm³ and has the dimensions shown on the figure. (a) The force exerted by the water pushes horizontally on the dam face, and this is resisted by the force of static friction between the dam and the bedrock foundation on which it rests. The coefficient of friction is 0.47. Calculate the factor of safety against sliding, that is, the ratio of the maximum possible friction force to the force exerted by the water. (b) The water also tries to rotate the dam about a line running along the base of the dam through point A; see Problem 23. The torque resulting from the weight of the dam acts in the opposite sense. Calculate the factor of safety against rotation, that is, the ratio of the torque owing to the weight of the dam to the torque exerted by the water.

FIGURE 15-34
Problem 25.

SECTION 15-5 Measuring Pressure

26E. Calculate the height of a column of water that gives a pressure of 1 atm at the bottom. Assume that $g = 9.80$ m/s².

27E. To suck lemonade of density 1000 kg/m³ up a straw to a maximum height of 4.0 cm, what minimum gauge pressure (in atmospheres) must you produce in your lungs?

28P. What would be the height of the atmosphere if the air density (a) were constant and (b) decreased linearly to zero with height? Assume a sea-level density of 1.3 kg/m³.

SECTION 15-6 Pascal's Principle

29E. A piston of small cross-sectional area a is used in a hydraulic press to exert a small force $\mathbf{f}$ on the enclosed liquid. A connecting pipe leads to a larger piston of cross-sectional area A (Fig. 15-35). (a) What force $\mathbf{F}$ will the larger piston sustain? (b) If the small piston has a diameter of 1.5 in. and the large piston one of 21 in., what weight on the small piston will support 2.0 tons on the large piston?

FIGURE 15-35
Exercises 29 and 30.

30E. In the hydraulic press of Exercise 29, through what distance must the large piston be moved to raise the small piston a distance of 3.5 ft?

SECTION 15-7 Archimedes' Principle

31E. A tin can has a total volume of 1200 cm³ and a mass of 130 g. How many grams of lead shot could it carry without sinking in water? The density of lead is 11.4 g/cm³.

32E. A boat floating in fresh water displaces 8000 lb of water. (a) How many pounds of water would this boat displace if it were floating in salt water of density 68.6 lb/ft³? (b) Would the volume of water displaced change? If so, by how much?

33E. About one-third of the body of a physicist swimming in the Dead Sea will be above the water line. Assuming that the human body density is 0.98 g/cm³, find the density of the water in the Dead Sea. (Why is it so much greater than 1.0 g/cm³?)

34E. An iron anchor appears 200 N lighter in water than in air. (a) What is the volume of the anchor? (b) How much does it weigh in air? The density of iron is 7870 kg/m³.

35E. An object hangs from a spring balance. The balance registers 30 N in air, 20 N when this object is immersed in water, and 24 N when the object is immersed in another liquid of unknown density. What is the density of that other liquid?

36E. A cubical object of dimensions $L = 2.00$ ft on a side and weight $W = 1000$ lb in a vacuum is suspended by a rope in an open tank of liquid of density $\rho = 2.00$ slugs/ft³ as in Fig. 15-36. (a) Find the total downward force exerted by the liquid and the atmosphere on the top of the object. (b) Find the total upward force on the bottom of the object. (c) Find the tension in the rope. (d) Calculate the buoyant force on the object using Archimedes' principle. What relation exists among all these quantities?

FIGURE 15-36
Exercise 36.

37E. A block of wood floats in water with two-thirds of its volume submerged. In oil the block floats with 0.90 of its volume submerged. Find the density of (a) the wood and (b) the oil.

38E. It has been proposed to move natural gas from the North Sea gas fields in huge dirigibles, using the gas itself to provide lift. Calculate the force required to tether such an airship to the ground for off-loading when it is fully loaded with 1.0×10^6 m³ of gas at a density of 0.80 kg/m³. (The weight of the airship is negligible by comparison.)

39E. A blimp is cruising slowly at low altitude, filled as usual with helium gas. Its maximum useful payload, including crew and cargo, is 1280 kg. How much more payload could the blimp carry if you replaced the helium with hydrogen? (Why not do it?) The volume of the helium-filled interior space is 5000 m³. The density of helium gas is 0.16 kg/m³, and the density of hydrogen is 0.081 kg/m³.

40E. A helium balloon is used to lift a 40 kg payload to an altitude of 27 km, where the air density is 0.035 kg/m³. The balloon has a mass of 15 kg, and the density of the gas in the balloon is 0.0051 kg/m³. What is the volume of the balloon? Neglect the volume of the payload.

41P. A hollow sphere of inner radius 8.0 cm and outer radius 9.0 cm floats half-submerged in a liquid of density 800 kg/m³. (a) What is the mass of the sphere? (b) Calculate the density of the material of which the sphere is made.

42P. A hollow spherical iron shell floats almost completely submerged in water. The outer diameter is 60.0 cm, and the density of iron is 7.87 g/cm³. Find the inner diameter.

43P. An iron casting containing a number of cavities weighs 6000 N in air and 4000 N in water. What is the volume of the cavities in the casting? The density of iron is 7.87 g/cm³.

44P. (a) What is the minimum area of the top surface of a slab of ice 0.30 m thick floating on fresh water that will hold up an automobile of mass 1100 kg? (b) Does it matter where the car is placed on the block of ice?

45P. Three children, each of weight 80 lb, make a log raft by lashing together logs of diameter 1.0 ft and length 6.0 ft. How many logs will be needed to keep them afloat? Take the density of wood to be 50 lb/ft³.

46P. Assume the density of brass weights to be 8.0 g/cm³ and that of air to be 0.0012 g/cm³. What percent error arises from neglecting the buoyancy of air in weighing an object of mass m and density ρ on a beam balance?

47P. A car has a total mass of 1800 kg. The volume of air space in the passenger compartment is 5.00 m³. The volume of the motor and front wheels is 0.750 m³, and the volume of the rear wheels, gas tank, and trunk is 0.800 m³; water cannot enter these areas. The car is parked on a hill; the handbrake cable snaps and the car rolls down the hill into a lake (Fig. 15-37). (a) At first, no water enters the passenger compartment. How much of the car, in cubic meters, is below the water surface with the car floating as shown? (b) As water slowly enters, the car sinks. How many cubic meters of water are in the car as it disappears below the water surface? (The car remains horizontal, owing to a heavy load in the trunk.)

FIGURE 15-37 Problem 47.

48P. A block of wood has a mass of 3.67 kg and a density of 600 kg/m³. It is to be loaded with lead so that it will float in water with 0.90 of its volume immersed. What mass of lead is needed (a) if the lead is on top of the wood and (b) if the lead is attached below the wood? The density of lead is 1.13×10^4 kg/m³.

49P. You place a glass beaker, partially filled with water, in a sink (Fig. 15-38). The beaker has a mass of 390 g and an interior volume of 500 cm³. You now start to fill the sink with water and you find, by experiment, that if the beaker is less than half full, it will float; but if it is more than half full, it remains on the bottom of the sink as the water rises to its rim. What is the density of the material of which the beaker is made?

FIGURE 15-38
Problem 49.

50P. What is the acceleration of a rising hot-air balloon if the ratio of the air density outside the balloon to that inside is 1.39? Neglect the mass of the balloon fabric and the basket.

51P. A metal rod of length 80 cm and mass 1.6 kg has a uniform cross-sectional area of 6.0 cm². Due to a nonuniform density, the center of mass of the rod is 20 cm from one end of the rod. The rod is suspended in a horizontal position in water by ropes attached to both ends as shown in Fig. 15-39. (a) What is the ten-

FIGURE 15-39
Problem 51.

sion in the rope closer to the center of mass? (b) What is the tension in the rope farther from the center of mass?

52P*. The tension in a string holding a solid block below the surface of a liquid (of density greater than the solid) is T_0 when the containing vessel (Fig. 15-40) is at rest. Show that when the vessel has an upward vertical acceleration of magnitude a, the tension T is equal to $T_0(1 + a/g)$.

FIGURE 15-40 Problem 52.

SECTION 15-9 Streamlines and the Equation of Continuity

53E. Figure 15-41 shows the merging of two streams to form a river. One stream has a width of 8.2 m, depth of 3.4 m, and current speed of 2.3 m/s. The other stream is 6.8 m wide and 3.2 m deep, and flows at 2.6 m/s. The width of the river is 10.5 m, and the current speed is 2.9 m/s. What is its depth?

FIGURE 15-41 Exercise 53.

54E. The water flowing through a 0.75 in. (inside diameter) pipe flows out through three 0.50 in. pipes. (a) If the flow rates in the three smaller pipes are 7.0, 5.0, and 3.0 gal/min, what is the flow rate in the 0.75 in. pipe? (b) What is the ratio of the speed of water in the 0.75 in. pipe to that in the pipe carrying 7.0 gal/min?

55E. A garden hose having an internal diameter of 0.75 in. is connected to a (stationary) lawn sprinkler that consists merely of an enclosure with 24 holes, each 0.050 in. in diameter. If the water in the hose has a speed of 3.0 ft/s, at what speed does it leave the sprinkler holes?

56P. Water is pumped steadily out of a flooded basement at a speed of 5.0 m/s through a uniform hose of radius 1.0 cm. The hose passes out through a window 3.0 m above the waterline. What is the power of the pump?

57P. A river 20 m wide and 4.0 m deep drains a 3000 km² land area in which the average precipitation is 48 cm/y. One-fourth of this rainfall returns to the atmosphere by evaporation, but the remainder ultimately drains into the river. What is the average speed of the river current?

SECTION 15-10 Bernoulli's Equation

58E. Water is moving with a speed of 5.0 m/s through a pipe with a cross-sectional area of 4.0 cm². The water gradually descends 10 m as the pipe increases in area to 8.0 cm². (a) What is the speed at the lower level? (b) If the pressure at the upper level is 1.5×10^5 Pa, what is the pressure at the lower level?

59E. Models of torpedoes are sometimes tested in a horizontal pipe of flowing water, much as a wind tunnel is used to test model airplanes. Consider a circular pipe of internal diameter 25.0 cm and a torpedo model, aligned along the axis of the pipe, with a diameter of 5.00 cm. The torpedo is to be tested with water flowing past it at 2.50 m/s. (a) With what speed must the water flow in the part of the pipe that is unconstricted by the model? (b) What will the pressure difference be between the constricted and unconstricted parts of the pipe?

60E. A water intake at a pump storage reservoir (Fig. 15-42) has a cross-sectional area of 8.00 ft². The water flows in at a speed of 1.33 ft/s. At the generator building 600 ft below the intake point, the cross-sectional area is smaller than at the intake and the water flows out at 31.0 ft/s. What is the difference in pressure, in pounds per square inch, between inlet and outlet?

FIGURE 15-42
Exercise 60.

61E. A water pipe having a 1.0 in. inside diameter carries water into the basement of a house at a speed of 3.0 ft/s and a pressure of 25 lb/in.². If the pipe tapers to 0.50 in. and rises to the second floor 25 ft above the input point, what are (a) the speed and (b) the water pressure at the second floor?

62E. How much work is done by pressure in forcing 1.4 m³ of water through a pipe having an internal diameter of 13 mm if the difference in pressure at the two ends of the pipe is 1.0 atm?

63E. In a horizontal oil pipeline that has a constant cross-sectional area, the pressure decrease between two points 1000 ft apart is 5.0 lb/in.². What is the energy loss per cubic foot of oil per foot of pipe?

64E. A tank of large area is filled with water to a depth $D = 1.0$ ft. A hole of cross section $A = 1.0$ in.² in the bottom of the tank allows water to drain out. (a) What is the rate at which water flows out in cubic feet per second? (b) At what distance below the bottom of the tank is the cross-sectional area of the stream equal to one-half the area of the hole?

65E. Suppose that two tanks, 1 and 2, each with a large opening at the top, contain different liquids. A small hole is made in the side of each tank at the same depth h below the liquid surface, but the hole in tank 1 has half the cross-sectional area of the hole in tank 2. (a) What is the ratio ρ_1/ρ_2 of the densities of the liquids if the mass flow rate is the same for the two holes? (b) What is the ratio of the volume flow rates from the two tanks? (c) To what height above the hole in the second tank should liquid be added or drained to equalize the volume flow rates?

66E. Air flows over the top of an airplane wing of area A with speed v_t and past the underside of the wing (also of area A) with speed v_u. Show that in this simplified situation Bernoulli's equation predicts that the magnitude L of the upward lift force on the wing will be

$$L = \tfrac{1}{2}\rho A(v_t^2 - v_u^2),$$

where ρ is the density of the air.

67E. If the speed of flow past the lower surface of an airplane wing is 110 m/s, what speed of flow over the upper surface will give a pressure difference of 900 Pa between upper and lower surfaces? Take the density of air to be 1.30×10^{-3} g/cm³, and see Exercise 66.

68E. An airplane has a wing area (each wing) of 10.0 m². At a certain airspeed, air flows over the upper wing surface at 48.0 m/s and over the lower wing surface at 40.0 m/s. What is the mass of the plane? Assume that the plane travels at constant velocity, that the air density is 1.20 kg/m², and that lift effects associated with the fuselage and tail assembly are small. Discuss the lift if the airplane, flying at the same airspeed, is (a) in level flight, (b) climbing at 15°, and (c) descending at 15°. (See Exercise 66.)

69P. Water flows through a horizontal pipe and is delivered into the atmosphere at a speed of 15 m/s as shown in Fig. 15-43. The diameters of the left and right sections of the pipe are 5.0 cm and 3.0 cm, respectively. (a) What volume of water is delivered into the atmosphere during a 10 min period? (b) What is the flow speed of the water in the left section of the pipe? (c) What is the gauge pressure in the left section of the pipe?

FIGURE 15-43 Problem 69.

70P. An opening of area 0.25 cm² in an otherwise closed beverage keg is 50 cm below the level of the liquid (of density 1.0 g/cm³) in the keg. What is the speed of the liquid flowing through the opening if the gauge pressure in the air space above the liquid is (a) zero and (b) 0.40 atm?

71P. In a hurricane, the air (density 1.2 kg/m³) is blowing over the roof of a house at a speed of 110 km/h. (a) What is the pressure difference between inside and outside that tends to lift the roof? (b) What would be the lifting force on a roof of area 90 m²?

72P. The windows in an office building are of dimensions 4.00 m by 5.00 m. On a stormy day, air is blowing at 30.0 m/s past a window on the top floor. Calculate the net force on the window. The air density is 1.23 kg/m³ there.

73P. A sniper fires a rifle bullet into a gasoline tank, making a hole 50 m below the surface of the gasoline. The tank was sealed

and is under 3.0 atm absolute pressure, as shown in Fig. 15-44. The stored gasoline has a density of 660 kg/m³. At what speed v does the gasoline begin to shoot out of the hole?

FIGURE 15-44
Problem 73.

74P. The fresh water behind a reservoir dam is 15 m deep. A horizontal pipe 4.0 cm in diameter passes through the dam 6.0 m below the water surface, as shown in Fig. 15-45. A plug secures the pipe opening. (a) Find the friction force between plug and pipe wall. (b) The plug is removed. What volume of water flows out of the pipe in 3.0 h?

FIGURE 15-45
Problem 74.

75P. A tank is filled with water to a height H. A hole is punched in one of the walls at a depth h below the water surface (Fig. 15-46). (a) Show that the distance x from the base of the tank to the point at which the resulting stream strikes the floor is given by $x = 2\sqrt{h(H - h)}$. (b) Could a hole be punched at another depth to produce a second stream that would have the same range? If so, at what depth? (c) At what depth should the hole be placed to make the emerging stream strike the ground at the maximum distance from the base of the tank?

FIGURE 15-46 Problem 75.

76P. A *siphon* is a device for removing liquid from a container. It operates as shown in Fig. 15-47. Tube ABC must initially be filled, but once this has been done, liquid will flow through the tube until the liquid surface in the container is level with the tube opening at A. The liquid has density ρ and negligible viscosity. (a) With what speed does the liquid emerge from the tube at C? (b) What is the pressure in the liquid at the topmost point B? (c) Theoretically, what is the greatest possible height h_1 that a siphon can lift water?

FIGURE 15-47 Problem 76.

77P. A *Venturi meter* is used to measure the flow speed of a fluid in a pipe. The meter is connected between two sections of the pipe (Fig. 15-48); the cross-sectional area A of the entrance and exit of the meter matches the pipe's cross-sectional area. Between the entrance and exit, the fluid flows from the pipe with speed v and then through a narrow "throat" of cross-sectional area a with speed V. A manometer connects the wider portion of the meter to the narrower portion. The change in the fluid's speed is accompanied by a change Δp in the fluid's pressure, which causes a height difference h of the liquid in the two arms of the manometer. (a) By applying Bernoulli's equation and the equation of continuity to points 1 and 2 in Fig. 15-48, show that

$$v = \sqrt{\frac{2a^2\,\Delta p}{\rho(A^2 - a^2)}},$$

where ρ is the density of the fluid. (b) Suppose that the fluid is water; that the cross-sectional diameters are 10 in.^{2} in the pipe, 5.0 in.^{2} in the throat; that the pressure is 8.0 lb/in.^{2} in the pipe and 6.0 lb/in.^{2} in the throat. What is the rate of water flow in cubic feet per second?

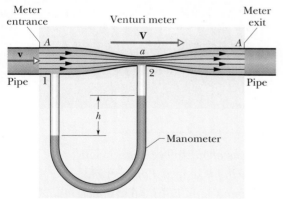

FIGURE 15-48 Problems 77 and 78.

78P. Consider the Venturi tube of Problem 77 and Fig. 15-48 without the manometer. Let A equal $5a$. Suppose that the pressure p_1 at A is 2.0 atm. (a) Compute the values of v at A and V at a that would make the pressure p_2 at a equal to zero. (b) Compute the corresponding volume flow rate if the diameter at A is 5.0 cm. The phenomenon that occurs at a when p_2 falls to nearly zero is known as cavitation. The water vaporizes into small bubbles.

79P. A pitot tube (Fig. 15-49) is used to determine the airspeed of an airplane. It consists of an outer tube with a number of small holes B (four are shown); the tube is connected to one arm of a U-tube. The other arm of the U-tube is connected to a hole, A, at the front end of the device, which points in the direction the plane is headed. At A the air becomes stagnant so that $v_A = 0$. At B, however, the speed of the air presumably equals the airspeed of the aircraft. (a) Use Bernoulli's equation to show that

$$v = \sqrt{\frac{2\rho g h}{\rho_{air}}},$$

where v is the airspeed of the plane and ρ is the density of the liquid in the U-tube. (b) The tube contains alcohol and indicates a level difference h of 26.0 cm. What is the plane's speed relative to the air? The density of the air is 1.03 kg/m^3 and that of alcohol is 810 kg/m^3.

FIGURE 15-49 Problems 79 and 80.

80P. A pitot tube (see Problem 79) on a high-altitude aircraft measures a differential pressure of 180 Pa. What is the airspeed if the density of the air is 0.031 kg/m^3?

16
Oscillations

Just as the third game of the 1989 World Series was about to begin near San Francisco, seismic waves from a magnitude 7.1 earthquake near Loma Prieta, 100 km distant, hit the area, causing extensive damage and killing 67 people. The photograph shows part of a 1.4 km stretch of the Nimitz Freeway, where dozens died when an upper deck collapsed onto a lower deck, trapping motorists. Obviously, the collapse was due to violent shaking by the seismic waves. But why was that particular stretch so severely damaged when the rest of the freeway, almost identical in construction, escaped collapse?

16-1 OSCILLATIONS

We are surrounded by oscillations—motions that repeat themselves. There are swinging chandeliers, boats bobbing at anchor, and the surging pistons in the engines of cars. There are oscillating guitar strings, drums, bells, diaphragms in telephones and speaker systems, and quartz crystals in wristwatches. Less evident are the oscillations of the air molecules that transmit the sensation of sound, the oscillations of the atoms in a solid that convey the sensation of temperature, and the oscillations of the electrons in the antennas of radio and TV transmitters.

Oscillations are not confined to material objects such as violin strings and electrons. Light, radio waves, x rays, and gamma rays are also oscillatory phenomena. You will study such oscillations in later chapters and will be helped greatly there by analogy with the mechanical oscillations that you are about to study here.

Oscillations in the real world are usually *damped;* that is, the motion dies out gradually, transferring mechanical energy to thermal energy by the action of frictional forces. Although we cannot totally eliminate such loss of mechanical energy, we can replenish the energy from some source. The children in Fig. 16-1, for example, know that by swinging their legs or torsos they can "pump" the swing and maintain or enhance the oscillations. In doing this, they transfer biochemical energy to mechanical energy of the oscillating systems.

FIGURE 16-1 A child soon learns how to maintain the oscillations of a swing by transferring energy into the swing's motion.

16-2 SIMPLE HARMONIC MOTION

Figure 16-2 shows a sequence of "snapshots" of a simple oscillating system, a particle moving repeatedly back and forth about the origin of the x axis. In this section we simply describe the motion. Later, we shall discuss how to attain such motion.

One important property of oscillatory motion is its **frequency,** or number of oscillations that are completed each second. The symbol for frequency is f, and its SI unit is the **hertz** (abbreviated Hz), where

$$1 \text{ hertz} = 1 \text{ Hz} = 1 \text{ oscillation per second}$$
$$= 1 \text{ s}^{-1}. \tag{16-1}$$

Related to the frequency is the **period** T of the motion, which is the time for one complete oscillation (or **cycle**). That is,

$$T = \frac{1}{f}. \tag{16-2}$$

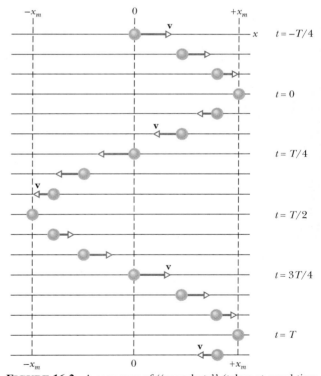

FIGURE 16-2 A sequence of "snapshots" (taken at equal time intervals) showing the position of a particle as it oscillates back and forth about the origin along the x axis, between the limits $+x_m$ and $-x_m$. The vector arrows are scaled to indicate the speed of the particle. The speed is maximum when the particle is at the origin and zero when it is at $\pm x_m$. If the time t is chosen to be zero when the particle is at $+x_m$, then the particle returns to $+x_m$ at $t = T$, where T is the period of the motion. The motion is then repeated.

Any motion that repeats itself at regular intervals is called **periodic motion** or **harmonic motion.** We are interested here in motion that repeats itself in a particular way, namely, like that in Fig. 16-2. It turns out that for such motion the displacement x of the particle from the origin is given as a function of time by

$$x(t) = x_m \cos(\omega t + \phi) \quad \text{(displacement)}, \quad (16\text{-}3)$$

in which x_m, ω, and ϕ are constants. This motion is called **simple harmonic motion** (SHM), a term that means that the periodic motion is a sinusoidal function of time.

The quantity x_m in Eq. 16-3, a positive constant whose value depends on how the motion was started, is called the **amplitude** of the motion; the subscript m stands for *maximum* because the amplitude is the magnitude of the maximum displacement of the particle in either direction. The cosine function in Eq. 16-3 varies between the limits ± 1, so the displacement $x(t)$ varies between the limits $\pm x_m$, as Fig. 16-2 shows.

(a)

(b)

(c)

FIGURE 16-3 In all three cases, the blue curve is obtained from Eq. 16-3 with $\phi = 0$. (a) The red curve differs from the blue curve *only* in that its amplitude x'_m is greater. (b) The red curve differs from the blue curve *only* in that its period is $T' = T/2$. (c) The red curve differs from the blue curve *only* in that $\phi = -\pi/4$ rad rather than zero.

The time-varying quantity $(\omega t + \phi)$ in Eq. 16-3 is called the **phase** of the motion, and the constant ϕ is called the **phase constant** (or **phase angle**). The value of ϕ depends on the displacement and velocity of the particle at $t = 0$. For the $x(t)$ plots of Fig. 16-3a, the phase constant ϕ is zero (compare the plots and Eq. 16-3 for $t = 0$).

It remains to interpret the constant ω. The displacement $x(t)$ must return to its initial value after one period T of the motion. That is, $x(t)$ must equal $x(t + T)$ for all t. To simplify our analysis, let us put $\phi = 0$ in Eq. 16-3. From that equation we then have

$$x_m \cos \omega t = x_m \cos [\omega(t + T)].$$

The cosine function first repeats itself when its argument (the phase) has increased by 2π rad, so that we must have, in the equation above,

$$\omega(t + T) = \omega t + 2\pi$$

or, $$\omega T = 2\pi.$$

Thus from Eq. 16-2,

$$\omega = \frac{2\pi}{T} = 2\pi f. \quad (16\text{-}4)$$

The quantity ω is called the **angular frequency** of the motion; its SI unit is the radian per second. (To be consistent, then, ϕ must be in radians.) Figure 16-3 compares $x(t)$ for two simple harmonic motions that differ either in amplitude, in period (and thus in frequency and angular frequency), or in phase constant.

CHECKPOINT 1: A particle undergoing simple harmonic oscillation of period T (like that in Fig. 16-2) is at $-x_m$ at time $t = 0$. Is it at $-x_m$, at $+x_m$, at 0, between $-x_m$ and 0, or between 0 and $+x_m$ when (a) $t = 2.00T$, (b) $t = 3.50T$, and (c) $t = 5.25T$?

The Velocity of SHM

By differentiating Eq. 16-3, we can find an expression for the velocity of a particle moving with simple harmonic motion. That is,

$$v(t) = \frac{dx}{dt} = \frac{d}{dt} [x_m \cos(\omega t + \phi)]$$

or

$$v(t) = -\omega x_m \sin(\omega t + \phi) \quad \text{(velocity)}. \quad (16\text{-}5)$$

Figure 16-4a is a plot of Eq. 16-3 with $\phi = 0$. Figure 16-4b shows Eq. 16-5, also with $\phi = 0$. Analogous to the amplitude x_m in Eq. 16-3, the positive quantity ωx_m in Eq.

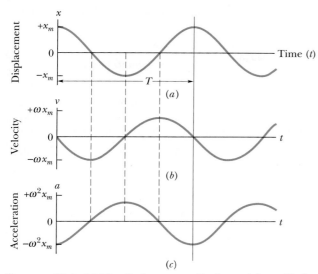

FIGURE 16-4 (a) The displacement $x(t)$ of a particle oscillating in SHM with phase angle ϕ equal to zero. The period T marks one complete oscillation. (b) The velocity $v(t)$ of the particle. (c) The acceleration $a(t)$ of the particle.

16-5 is called the **velocity amplitude** v_m. As you can see in Fig. 16-4b, the velocity of the oscillating particle varies between the limits $\pm v_m = \pm \omega x_m$. Note also in that figure that the curve of $v(t)$ is *shifted* (to the left) from the curve of $x(t)$ by one-quarter period: when the magnitude of the displacement is greatest (that is, $x(t) = x_m$), the magnitude of the velocity is least (that is, $v(t) = 0$). And when the magnitude of the displacement is least (that is, zero), the magnitude of the velocity is greatest (that is, $v_m = \omega x_m$).

The Acceleration of SHM

Knowing the velocity $v(t)$ for simple harmonic motion, we can find an expression for the acceleration of the oscillating particle by differentiating once more. Thus we have, from Eq. 16-5,

$$a(t) = \frac{dv}{dt} = \frac{d}{dt}[-\omega x_m \sin(\omega t + \phi)]$$

or

$$a(t) = -\omega^2 x_m \cos(\omega t + \phi) \quad \text{(acceleration)}. \quad (16\text{-}6)$$

Figure 16-4c is a plot of Eq. 16-6 for the case $\phi = 0$. The positive quantity $\omega^2 x_m$ in Eq. 16-6 is called the **acceleration amplitude** a_m. That is, the acceleration of the particle varies between the limits $\pm a_m = \pm \omega^2 x_m$, as Fig. 16-4c shows. Note also that the curve of $a(t)$ is shifted (to the left) by one-quarter period relative to the curve of $v(t)$.

We can combine Eqs. 16-3 and 16-6 to yield

$$a(t) = -\omega^2 x(t), \quad (16\text{-}7)$$

which is the hallmark of simple harmonic motion: the ac-celeration is proportional to the displacement but opposite in sign, and the two quantities are related by the square of the angular frequency. Thus, as Fig. 16-4 shows, when the displacement has its greatest positive value, the acceleration has its greatest negative value, and conversely. When the displacement is zero, the acceleration is also zero.

PROBLEM SOLVING TACTICS

TACTIC 1: *Phase Angles*
Note the effect of the phase angle ϕ on a plot of $x(t)$. When $\phi = 0$, $x(t)$ has a graph like that in Fig. 16-4a, a typical cosine curve. A *negative* value for ϕ shifts the curve *rightward* along the t axis (as in Fig. 16-3c), while a *positive* value shifts it *leftward*.

Two plots of SHM with different phase angles are said to have a *phase difference;* or, each is said to be *phase-shifted* from the other, or *out of phase* with the other. The curves in Fig. 16-3c, for example, have a phase difference of $\pi/4$ rad.

Because SHM repeats after each period T and the cosine function repeats after each 2π rad, one period T represents a phase difference of 2π rad. In Fig. 16-4, $x(t)$ is phase-shifted to the right from $v(t)$ by one-quarter period, or $-\pi/2$ rad; it is shifted to the right from $a(t)$ by one-half period, or $-\pi$ rad. A phase shift of 2π rad causes a curve of SHM to coincide with itself; that is, it looks unchanged.

16-3 THE FORCE LAW FOR SIMPLE HARMONIC MOTION

Once we know how the acceleration of a particle varies with time, we can use Newton's second law to learn what force must act on the particle to give it that acceleration. If we combine Newton's second law and Eq. 16-7, we find, for simple harmonic motion,

$$F = ma = -(m\omega^2)x. \quad (16\text{-}8)$$

This result—a force proportional to the displacement but opposite in sign—is familiar. It is Hooke's law,

$$F = -kx, \quad (16\text{-}9)$$

for a spring, the spring constant here being

$$k = m\omega^2. \quad (16\text{-}10)$$

We can in fact take Eq. 16-9 as an alternative definition of simple harmonic motion. It says:

Simple harmonic motion is the motion executed by a particle of mass m subject to a force that is proportional to the displacement of the particle but opposite in sign.

FIGURE 16-5 A linear simple harmonic oscillator. Like the particle of Fig. 16-2, the block moves in simple harmonic motion once it has been pulled to the side and released. Its displacement is then given by Eq. 16-3.

The block–spring system of Fig. 16-5 forms a **linear simple harmonic oscillator** (linear oscillator, for short), where "linear" indicates that F is proportional to x rather than to some other power of x. The angular frequency ω of the simple harmonic motion of the block is related to the spring constant k and the mass m of the block by Eq. 16-10, which yields

$$\omega = \sqrt{\frac{k}{m}} \quad \text{(angular frequency).} \quad (16\text{-}11)$$

By combining Eqs. 16-4 and 16-11, we can write, for the **period** of the linear oscillator of Fig. 16-5,

$$T = 2\pi\sqrt{\frac{m}{k}} \quad \text{(period).} \quad (16\text{-}12)$$

Equations 16-11 and 16-12 tell us that a large angular frequency (and thus a small period) goes with a stiff spring (large k) and a light block (small m).

Every oscillating system, be it the linear oscillator of Fig. 16-5, a diving board, or a violin string, has some element of "springiness" and some element of "inertia," or mass, and thus resembles a linear oscillator. In the linear oscillator of Fig. 16-5, these elements are located in separate parts of the system, the springiness being entirely in the spring, which we assume to be massless, and the inertia being entirely in the block, which we assume to be rigid. In a violin string, however, the two elements are both within the string itself, as you will see in Chapter 17.

CHECKPOINT **2:** Which of the following relationships between the force F on a particle and the particle's position x implies simple harmonic oscillation: (a) $F = -5x$, (b) $F = -400x^2$, (c) $F = 10x$, (d) $F = 3x^2$?

SAMPLE PROBLEM 16-1

A block whose mass m is 680 g is fastened to a spring whose spring constant k is 65 N/m. The block is pulled a distance $x = 11$ cm from its equilibrium position at $x = 0$ on a frictionless surface and released from rest at $t = 0$.

(a) What force does the spring exert on the block just before the block is released?

SOLUTION: From Hooke's law

$$F = -kx = -(65 \text{ N/m})(0.11 \text{ m})$$
$$= -7.2 \text{ N.} \quad \text{(Answer)}$$

The minus sign reminds us that the spring force acting on the block, which points back toward the origin, is opposite the displacement of the block, which points away from the origin.

(b) What are the angular frequency, the frequency, and the period of the resulting oscillation?

SOLUTION: From Eq. 16-11 we have

$$\omega = \sqrt{\frac{k}{m}} = \sqrt{\frac{65 \text{ N/m}}{0.68 \text{ kg}}} = 9.78 \text{ rad/s}$$
$$\approx 9.8 \text{ rad/s.} \quad \text{(Answer)}$$

The frequency follows from Eq. 16-4, which yields

$$f = \frac{\omega}{2\pi} = \frac{9.78 \text{ rad/s}}{2\pi} = 1.56 \text{ Hz} \approx 1.6 \text{ Hz.} \quad \text{(Answer)}$$

The period follows from Eq. 16-2, which yields

$$T = \frac{1}{f} = \frac{1}{1.56 \text{ Hz}} = 0.64 \text{ s} = 640 \text{ ms.} \quad \text{(Answer)}$$

(c) What is the amplitude of the oscillation?

SOLUTION: As we discussed in Section 8-4, the mechanical energy of a spring–block system like that in Fig. 16-5 is conserved because friction is not involved. Since the block is released from rest 11 cm from its equilibrium point, it has kinetic energy of zero whenever it is again 11 cm from that point. Thus its maximum displacement is 11 cm; that is,

$$x_m = 11 \text{ cm.} \quad \text{(Answer)}$$

(d) What is the maximum speed of the oscillating block?

SOLUTION: From Eq. 16-5 we see that the velocity amplitude is

$$v_m = \omega x_m = (9.78 \text{ rad/s})(0.11 \text{ m})$$
$$= 1.1 \text{ m/s.} \quad \text{(Answer)}$$

This maximum speed occurs when the oscillating block is rushing through the origin; compare Figs. 16-4a and 16-4b, where you can see that the speed is a maximum whenever $x = 0$.

(e) What is the magnitude of the maximum acceleration of the block?

SOLUTION: From Eq. 16-6 we see that the acceleration amplitude is

$$a_m = \omega^2 x_m = (9.78 \text{ rad/s})^2 (0.11 \text{ m})$$
$$= 11 \text{ m/s}^2. \quad \text{(Answer)}$$

This maximum acceleration occurs when the block is at the ends of its path. At those points, the force acting on the block has its maximum magnitude; compare Figs. 16-4a and 16-4c, where you can see that the magnitudes of the displacement and acceleration are maximum at the same times.

(f) What is the phase constant ϕ for the motion?

SOLUTION: At $t = 0$, the moment of release, the displacement of the block has its maximum value x_m and the velocity of the block is zero. If we put these *initial conditions,* as they are called, into Eqs. 16-3 and 16-5, we find

$$1 = \cos \phi \quad \text{and} \quad 0 = \sin \phi,$$

respectively. The smallest angle that satisfies both these requirements is

$$\phi = 0. \qquad \text{(Answer)}$$

(Any angle that is an integer multiple of 2π rad also satisfies these requirements.)

This equation has two solutions:

$$\phi = -25° \quad \text{and} \quad \phi = 155°.$$

You will see in part (c) how to choose between them.

(c) What is the amplitude x_m of the motion?

SOLUTION: From Eq. 16-13 we have, provisionally putting $\phi = 155°$,

$$x_m = \frac{x(0)}{\cos \phi} = \frac{-0.0850 \text{ m}}{\cos 155°} = 0.094 \text{ m}$$

$$= 9.4 \text{ cm.} \qquad \text{(Answer)}$$

If we had chosen $\phi = -25°$, we would have found $x_m = -9.4$ cm. However, the amplitude of the motion must always be a *positive* constant, so $-25°$ cannot be the correct phase constant. We must therefore have, in part (b),

$$\phi = 155°. \qquad \text{(Answer)}$$

SAMPLE PROBLEM 16-2

At $t = 0$, the displacement $x(0)$ of the block in a linear oscillator like that of Fig. 16-5 is -8.50 cm. Its velocity $v(0)$ then is -0.920 m/s, and its acceleration $a(0)$ is $+47.0$ m/s².

(a) What are the angular frequency ω and the frequency f of this system?

SOLUTION: If we put $t = 0$ in Eqs. 16-3, 16-5, and 16-6, we find

$$x(0) = x_m \cos \phi, \qquad (16\text{-}13)$$

$$v(0) = -\omega x_m \sin \phi, \qquad (16\text{-}14)$$

and

$$a(0) = -\omega^2 x_m \cos \phi. \qquad (16\text{-}15)$$

These three equations contain three unknowns, namely, x_m, ϕ, and ω. We should be able to find all three, but here we need only ω.

If we divide Eq. 16-15 by Eq. 16-13, the result is

$$\omega = \sqrt{-\frac{a(0)}{x(0)}} = \sqrt{-\frac{47.0 \text{ m/s}^2}{-0.0850 \text{ m}}}$$

$$= 23.5 \text{ rad/s.} \qquad \text{(Answer)}$$

The frequency f follows from Eq. 16-4 and is

$$f = \frac{\omega}{2\pi} = \frac{23.5 \text{ rad/s}}{2\pi} = 3.74 \text{ Hz.} \qquad \text{(Answer)}$$

(b) What is the phase constant ϕ?

SOLUTION: If we divide Eq. 16-14 by Eq. 16-13, we find

$$\frac{v(0)}{x(0)} = \frac{-\omega x_m \sin \phi}{x_m \cos \phi} = -\omega \tan \phi.$$

Solving for $\tan \phi$, we find

$$\tan \phi = -\frac{v(0)}{\omega x(0)} = -\frac{-0.920 \text{ m/s}}{(23.5 \text{ rad/s})(-0.0850 \text{ m})}$$

$$= -0.461.$$

SAMPLE PROBLEM 16-3

In Fig. 16-6a, a uniform bar with mass m lies symmetrically across two rapidly rotating, fixed rollers, A and B, with distance $L = 2.0$ cm between the bar's center of mass and each roller. The rollers, whose directions of rotation are shown in the figure, slip against the bar with coefficient of kinetic friction $\mu_k = 0.40$. Suppose the bar is displaced horizontally by a distance x, as in Fig. 16-6b, and then released. What is the angular frequency ω of the resulting horizontal simple harmonic (back and forth) motion of the bar?

SOLUTION: To find ω, we find an expression for the horizontal acceleration a of the bar as a function of x and compare it with Eq. 16-7. We do so by applying Newton's second law vertically and horizontally; then we use the law in angular form about the contact point between the bar and roller A.

The vertical forces acting on the bar are its weight $m\mathbf{g}$ and supporting forces $\mathbf{F}_A$ due to roller A and $\mathbf{F}_B$ due to roller B. Since there is no net vertical force acting on the bar, Newton's second law gives us

$$\sum F_y = F_A + F_B - mg = 0. \qquad (16\text{-}16)$$

The horizontal forces acting on the bar are the kinetic frictional forces $f_{kA} = \mu_k F_A$ (toward the right) due to roller A

FIGURE 16-6 Sample Problem 16-3. (*a*) A bar is in equilibrium on two rotating rollers, A and B, that slip beneath it. (*b*) The bar is displaced from equilibrium by a distance x and then released.

and $f_{kB} = -\mu_k F_B$ (toward the left) due to roller B. Horizontally, Newton's second law gives us

$$\mu_k F_A - \mu_k F_B = ma,$$

or

$$a = \frac{\mu_k F_A - \mu_k F_B}{m}. \qquad (16\text{-}17)$$

The bar experiences no net torque about an axis perpendicular to the plane of Fig. 16-6 through the contact point between the bar and roller A. So Newton's second law for torques about that axis gives us

$$\sum \tau = F_A(0) + F_B 2L - mg(L + x) + f_{kA}(0) + f_{kB}(0)$$
$$= 0, \qquad (16\text{-}18)$$

where the forces $\mathbf{F}_A$, $\mathbf{F}_B$, $m\mathbf{g}$, $\mathbf{f}_{kA}$, and $\mathbf{f}_{kB}$ have moment arms about that axis of 0, 2L, L + x, 0, and 0, respectively.

Solving Eqs. 16-16 and 16-18 for F_A and F_B, we find

$$F_A = \frac{mg(L - x)}{2L} \quad \text{and} \quad F_B = \frac{mg(L + x)}{2L}.$$

Substituting these results into Eq. 16-17 yields

$$a = -\frac{\mu_k g}{L} x. \qquad (16\text{-}19)$$

Comparison of Eq. 16-19 with Eq. 16-7 reveals that the bar must be undergoing simple harmonic motion with angular frequency ω given by

$$\omega^2 = \frac{\mu_k g}{L}.$$

Thus

$$\omega = \sqrt{\frac{\mu_k g}{L}} = \sqrt{\frac{(0.40)(9.8 \text{ m/s}^2)}{0.020 \text{ m}}}$$
$$= 14 \text{ rad/s.} \qquad \text{(Answer)}$$

PROBLEM SOLVING TACTICS

TACTIC 2: *Identifying SHM*

In Sample Problem 16-3 we were able to identify the motion as linear simple harmonic motion once you reached Eq. 16-19. In linear SHM the acceleration a and displacement x are related by an equation of the form

$$a = -(\text{a positive constant})x,$$

which says that the acceleration is proportional to the displacement from the equilibrium position but is in the opposite direction. Once you find such an expression, you can immediately compare it to Eq. 16-7, identify the positive constant as being equal to ω^2, and so quickly get an expression for the angular frequency of the motion. With Eq. 16-4 you then can find the period T and the frequency f.

As you will see in Sample Problem 16-8, the same technique can be used to identify angular SHM. In such motion, the angular acceleration α and angular displacement θ are re-

lated by an equation of the form

$$\alpha = -(\text{a positive constant})\theta,$$

which says that the angular acceleration is proportional to the angular displacement from the equilibrium position but is in the opposite direction. As for linear SHM, you can identify the positive constant with ω^2 to find ω, T, and f.

In some problems you might derive an expression for the force F as a function of displacement x. If the motion is linear SHM, the force and displacement are related by

$$F = -(\text{a positive constant})x,$$

which says that the force is proportional to the displacement but is in the opposite direction. Once you have found such an expression, you can immediately compare it to Eq. 16-9 and identify the positive constant as being k. If you know the mass that is involved, you can then use Eqs. 16-11, 16-12, and 16-4 to find the angular frequency ω, period T, and frequency f.

You can similarly identify angular SHM. In such motion the torque τ and angular displacement θ are related by

$$\tau = -(\text{a positive constant})\theta,$$

which says that the torque is proportional to the angular displacement from the equilibrium position but is in the opposite direction.

16-4 ENERGY IN SIMPLE HARMONIC MOTION

In Chapter 8 we saw that the energy of a linear oscillator shuttles back and forth between kinetic and potential forms, while their sum—the mechanical energy E of the oscillator—remains constant. We now consider this situation quantitatively.

The potential energy of a linear oscillator like that of Fig. 16-5 is associated entirely with the spring. Its value depends on how much the spring is stretched or compressed, that is, on $x(t)$. We can use Eqs. 8-11 and 16-3 to find

$$U(t) = \tfrac{1}{2}kx^2 = \tfrac{1}{2}kx_m^2 \cos^2(\omega t + \phi). \quad (16\text{-}20)$$

Note carefully that a function in the form $\cos^2 A$ (as here) is the same as $(\cos A)^2$ but is *not* the same as $\cos A^2$, which means $\cos(A^2)$.

The kinetic energy of the system is associated entirely with the block. Its value depends on how fast the block is moving, that is, on $v(t)$. We can use Eq. 16-5 to find

$$K(t) = \tfrac{1}{2}mv^2 = \tfrac{1}{2}m\omega^2 x_m^2 \sin^2(\omega t + \phi). \quad (16\text{-}21)$$

If we use Eq. 16-11 to substitute k/m for ω^2, we can write

Eq. 16-21 as

$$K(t) = \tfrac{1}{2}mv^2 = \tfrac{1}{2}kx_m^2 \sin^2(\omega t + \phi). \quad (16\text{-}22)$$

The mechanical energy follows from Eqs. 16-20 and 16-22 and is

$$E = U + K$$
$$= \tfrac{1}{2}kx_m^2 \cos^2(\omega t + \phi) + \tfrac{1}{2}kx_m^2 \sin^2(\omega t + \phi)$$
$$= \tfrac{1}{2}kx_m^2 [\cos^2(\omega t + \phi) + \sin^2(\omega t + \phi)].$$

For any angle α,

$$\cos^2 \alpha + \sin^2 \alpha = 1.$$

Thus the quantity in the square brackets above is unity and we have

$$E = U + K = \tfrac{1}{2}kx_m^2. \quad (16\text{-}23)$$

The mechanical energy of a linear oscillator is indeed a constant, independent of time. The potential energy and kinetic energy of the linear oscillator are shown as functions of time in Fig. 16-7a, and as functions of displacement in Fig. 16-7b.

You might now understand why an oscillating system normally contains an element of springiness and an ele-

(a)

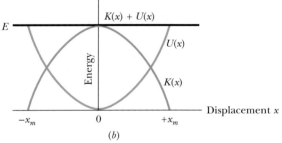

(b)

FIGURE 16-7 (a) The potential energy $U(t)$, kinetic energy $K(t)$, and mechanical energy E as functions of time, for a linear harmonic oscillator. Note that all energies are positive and that the potential energy and the kinetic energy peak twice during every period. (b) The potential energy $U(x)$, kinetic energy $K(x)$, and mechanical energy E as functions of position, for a linear harmonic oscillator with amplitude x_m. For $x = 0$ the energy is all kinetic, and for $x = \pm x_m$ it is all potential.

ment of inertia: it uses the former to store its potential energy and the latter to store its kinetic energy. The system of Sample Problem 16-3 is an exception, because no potential energy is involved. Instead, the energy transferred to the kinetic energy of the bar by the rollers actually varies over time.

$\mathcal{C}$HECKPOINT **3:** In Fig. 16-5, the block has a kinetic energy of 3 J and the spring has an elastic potential energy of 2 J when the block is at $x = +2.0$ cm. (a) What is the kinetic energy at $x = 0$? What are the elastic potential energies at (b) $x = -2.0$ cm and (c) $x = -x_m$?

SAMPLE PROBLEM 16-4

(a) What is the mechanical energy of the linear oscillator of Sample Problem 16-1?

SOLUTION: We find, substituting data from Sample Problem 16-1 into Eq. 16-23,

$$E = \tfrac{1}{2}kx_m^2 = (\tfrac{1}{2})(65 \text{ N/m})(0.11 \text{ m})^2$$
$$= 0.393 \text{ J} \approx 0.39 \text{ J}. \quad \text{(Answer)}$$

This value remains constant throughout the motion.

(b) What is the potential energy of this oscillator when the block is halfway to its end point, that is, when $x = \pm\tfrac{1}{2}x_m$?

SOLUTION: For any displacement, the potential energy of a spring–block system is given by $U = \tfrac{1}{2}kx^2$. Here,

$$U = \tfrac{1}{2}kx^2 = \tfrac{1}{2}k(\tfrac{1}{2}x_m)^2 = \tfrac{1}{4}(\tfrac{1}{2}kx_m^2)$$
$$= \tfrac{1}{4}E = (\tfrac{1}{4})(0.393 \text{ J}) = 0.098 \text{ J}. \quad \text{(Answer)}$$

(c) What is the kinetic energy of the oscillator when $x = \tfrac{1}{2}x_m$?

SOLUTION: We find this from

$$K = E - U$$
$$= 0.393 \text{ J} - 0.098 \text{ J} \approx 0.30 \text{ J}. \quad \text{(Answer)}$$

Thus, at this point during the oscillation, 25% of the energy is in potential form and 75% in kinetic form.

16-5 AN ANGULAR SIMPLE HARMONIC OSCILLATOR

Figure 16-8 shows an angular version of a simple harmonic oscillator; the element of springiness or elasticity is associated with the twisting of a suspension wire rather than the extension and compression of a spring as we previously had. The device is called a **torsion pendulum,** with *torsion* referring to the twisting.

FIGURE 16-8 An angular simple harmonic oscillator, or torsion pendulum, is an angular version of the linear simple harmonic oscillator of Fig. 16-5. The disk oscillates in a horizontal plane; the reference line oscillates with angular amplitude θ_m. The twist in the suspension wire stores potential energy as a spring does and provides the restoring torque.

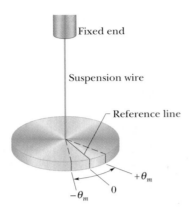

Fixed end

Suspension wire

Reference line

$+\theta_m$

0

$-\theta_m$

If we rotate the disk in Fig. 16-8 from its rest position (where the reference line is at 0) and release it, it will oscillate about that position in **angular simple harmonic motion.** Rotating the disk through an angle θ in either direction introduces a restoring torque given by

$$\tau = -\kappa\theta. \qquad (16\text{-}24)$$

Here κ (Greek *kappa*) is a constant, called the **torsion constant,** that depends on the length, diameter, and material of the suspension wire.

Comparison of Eq. 16-24 with Eq. 16-9 leads us to suspect that Eq. 16-24 is the angular form of Hooke's law, and that we can transform Eq. 16-12, which gives the period of linear SHM, into an equation for the period of angular SHM: we replace the spring constant k in Eq. 16-12 with its equivalent, the constant κ of Eq. 16-24, and we replace the mass m in Eq. 16-12 with *its* equivalent, the rotational inertia I of the oscillating disk. These replacements lead to

$$T = 2\pi\sqrt{\frac{I}{\kappa}} \qquad \text{(torsion pendulum)} \qquad (16\text{-}25)$$

for the period of the angular simple harmonic oscillator, or torsion pendulum.

SAMPLE PROBLEM 16-5

As Fig. 16-9a shows, a thin rod whose length L is 12.4 cm and whose mass m is 135 g is suspended at its midpoint from a long wire. Its period T_a of angular SHM is measured to be 2.53 s. An irregularly shaped object, which we call object X, is then hung from the same wire, as in Fig. 16-9b, and its period T_b is found to be 4.76 s.

(a) What is the rotational inertia of object X about its suspension axis?

SOLUTION: In Table 11-2(e), the rotational inertia of a thin rod about a perpendicular axis through its midpoint is given as $\frac{1}{12}mL^2$. Thus we have

$$I_a = \tfrac{1}{12}mL^2 = (\tfrac{1}{12})(0.135 \text{ kg})(0.124 \text{ m})^2$$
$$= 1.73 \times 10^{-4} \text{ kg}\cdot\text{m}^2.$$

Now let us write Eq. 16-25 twice, once for the rod and once for object X:

$$T_a = 2\pi\sqrt{\frac{I_a}{\kappa}} \quad \text{and} \quad T_b = 2\pi\sqrt{\frac{I_b}{\kappa}}.$$

Here the subscripts refer to Figs. 16-9a and 16-9b. The constant κ, which is a property of the wire, is the same for both figures; only the periods and the rotational inertias differ.

Let us square each of these equations, divide the second by the first, and solve the resulting equation for I_b. The result is

$$I_b = I_a\frac{T_b^2}{T_a^2} = (1.73 \times 10^{-4} \text{ kg}\cdot\text{m}^2)\frac{(4.76 \text{ s})^2}{(2.53 \text{ s})^2}$$
$$= 6.12 \times 10^{-4} \text{ kg}\cdot\text{m}^2. \qquad \text{(Answer)}$$

(b) What would be the period of oscillation if both objects were fastened together and hung from the wire, as in Fig. 16-9c?

SOLUTION: We can again write Eq. 16-25 twice, this time as

$$T_a = 2\pi\sqrt{\frac{I_a}{\kappa}} \quad \text{and} \quad T_c = 2\pi\sqrt{\frac{I_c}{\kappa}}.$$

After dividing again and putting $I_c = I_a + I_b$, we find

$$T_c = T_a\sqrt{\frac{I_c}{I_a}} = T_a\sqrt{\frac{I_a + I_b}{I_a}} = T_a\sqrt{1 + \frac{I_b}{I_a}}$$
$$= (2.53)\sqrt{1 + \frac{6.12 \times 10^{-4} \text{ kg}\cdot\text{m}^2}{1.73 \times 10^{-4} \text{ kg}\cdot\text{m}^2}}$$
$$= 5.39 \text{ s}. \qquad \text{(Answer)}$$

Suspension wire

Rod

L

Object X

(a)   (b)   (c)

FIGURE 16-9 Sample Problem 16-5. Three torsion pendulums consisting of a wire and (a) a rod, (b) an irregularly shaped object, and (c) the rod and the object rigidly connected.

16-6 PENDULUMS

We turn now to a class of simple harmonic oscillators in which the springiness is associated with the gravitational force rather than with the elastic properties of a twisted wire or a compressed or stretched spring.

The Simple Pendulum

If you hang an apple at the end of a long thread fixed at its upper end, and then set the apple swinging back and forth a small distance, you easily see that the apple's motion is periodic. Is it, in fact, simple harmonic motion? To idealize this situation, we consider a **simple pendulum,** which consists of a particle of mass m (called the *bob* of the pendulum) suspended from an unstretchable, massless string of length L, as in Fig. 16-10a. The bob is free to swing back and forth in the plane of the page, to the left and right of a vertical line through the point at which the upper end of the string is fixed.

The element of inertia in this pendulum is the mass m of the particle, and the element of springiness is in the gravitational attraction between the particle and Earth. Potential energy can be associated with the varying vertical distance between the swinging particle and Earth; we may view that varying distance as the varying length of a "gravitational spring."

The forces acting on the particle, shown in Fig. 16-10b, are its weight $m\mathbf{g}$ and the tension $\mathbf{T}$ in the string. We resolve $m\mathbf{g}$ into a radial component $mg \cos \theta$ and a component $mg \sin \theta$ that is tangent to the path taken by the particle. This tangential component is a restoring force, because it always acts opposite the displacement of the particle so as to bring the particle back toward its central location, the equilibrium position ($\theta = 0$), where it would be at rest were it not swinging. We write the restoring force as

$$F = -mg \sin \theta, \qquad (16\text{-}26)$$

where the minus sign indicates that F acts opposite the displacement.

If we assume that the angle θ in Fig. 16-10 is small, then $\sin \theta$ is very nearly equal to θ in radians. [For example, when $\theta = 5.00°$ ($= 0.0873$ rad), $\sin \theta = 0.0872$, a difference of only about 0.1%.] Also, the displacement s of the particle measured along its arc is equal to $L\theta$. Thus, for small θ, Eq. 16-26 becomes

$$F \approx -mg\theta = -mg \frac{s}{L} = -\left(\frac{mg}{L}\right)s. \quad (16\text{-}27)$$

A glance back at Eq. 16-9 shows that we again have Hooke's law, with the displacement now being arc length s instead of x. Thus *if a simple pendulum swings through a small angle,* it is a linear oscillator like the block–spring oscillator of Fig. 16-5; that is, it undergoes simple harmonic motion. Now the amplitude of the motion is measured as the **angular amplitude** θ_m, the maximum angle of swing. And the spring constant k is mg/L, the effective spring constant of the pendulum's gravitational spring.

By substituting mg/L for k in Eq. 16-12, we find, for the period of a simple pendulum,

$$T = 2\pi \sqrt{\frac{m}{k}} = 2\pi \sqrt{\frac{m}{mg/L}} \qquad (16\text{-}28)$$

or

$$T = 2\pi \sqrt{\frac{L}{g}} \qquad \text{(simple pendulum).} \quad (16\text{-}29)$$

Equation 16-29 holds only if the angular amplitude θ_m is small (which is what we assume in the exercises and problems for this chapter unless otherwise stated).

The element of inertia seems to be missing in Eq. 16-29 because the period is independent of the mass of the particle. This comes about because the element of springiness, which is the gravitational spring constant mg/L, is itself proportional to the mass of the particle, and the two masses cancel in Eq. 16-28. Figure 8-7 shows how energy shuttles back and forth between potential and kinetic forms during every oscillation of a simple pendulum.

The Physical Pendulum

Most pendulums in the real world are not even approximately "simple." Figure 16-11 shows a generalized **physical pendulum,** as we shall call realistic pendulums, with its weight $m\mathbf{g}$ acting at its center of mass C.

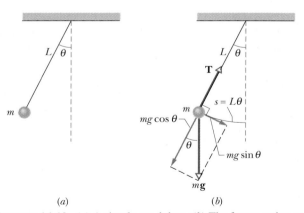

(a)   (b)

FIGURE 16-10 (a) A simple pendulum. (b) The forces acting on the bob are its weight $m\mathbf{g}$ and the tension $\mathbf{T}$ in the string. The tangential component $mg \sin \theta$ of the weight is a restoring force that brings the pendulum back to the central position.

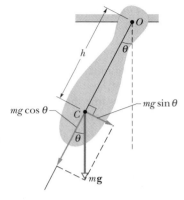

FIGURE 16-11 A physical pendulum. The restoring torque is $(mg \sin \theta)(h)$. When $\theta = 0$, the center of mass C hangs directly below the point of suspension O.

When the pendulum of Fig. 16-11 is displaced through an angle θ in either direction from its equilibrium position, a restoring torque appears. This torque acts about an axis through the suspension point O in Fig. 16-11 and has the magnitude

$$\tau = -(mg \sin \theta)(h). \qquad (16\text{-}30)$$

Here $mg \sin \theta$ is the tangential component of the weight $m\mathbf{g}$, and h (which is equal to OC) is the moment arm of this force component. The minus sign indicates that the torque is a restoring torque. That is, the torque always acts to reduce the angle θ to zero.

We once more decide to limit our interest to small amplitudes, so that $\sin \theta \approx \theta$. Then Eq. 16-30 becomes

$$\tau \approx -(mgh)\theta. \qquad (16\text{-}31)$$

Comparison with Eq. 16-24 shows that we again have Hooke's law in angular form and that the physical pendulum is an angular harmonic oscillator. Thus a physical pendulum undergoes simple harmonic motion *if* the angular amplitude θ_m of its motion is small. The term mgh of Eq. 16-31 is analogous to the torsion constant κ of Eq. 16-24. Substituting mgh for κ in Eq. 16-25, we find

$$T = 2\pi \sqrt{\frac{I}{mgh}} \quad \text{(physical pendulum)} \quad (16\text{-}32)$$

for the period of a physical pendulum when θ_m is small. Here I is the rotational inertia of the pendulum (about an axis through its point of support and perpendicular to its plane of swing), and h is the distance between the point of support and the center of mass of the swinging pendulum.

A physical pendulum will not swing if we hang it by its center of mass. Formally, this corresponds to putting $h = 0$ in Eq. 16-32. That equation then predicts $T \rightarrow \infty$, which implies that such a pendulum will never complete one swing.

Corresponding to any physical pendulum that oscillates about a given suspension point O with period T is a simple pendulum of length L_0 with the same period T. We can find L_0 with Eq. 16-29. The point along the physical pendulum at distance L_0 from point O is called the *center of oscillation* of the physical pendulum for the given suspension point.

The physical pendulum of Fig. 16-11 includes the simple pendulum as a special case. For the latter, h is the length L of the string and I is mL^2. Making these substitutions in Eq. 16-32 leads to

$$T = 2\pi \sqrt{\frac{I}{mgh}} = 2\pi \sqrt{\frac{mL^2}{mgL}} = 2\pi \sqrt{\frac{L}{g}},$$

which is exactly Eq. 16-29, the expression for the period of a simple pendulum.

Measuring *g*

We can use a physical pendulum to measure the free-fall acceleration g. (Countless thousands of such measurements have been made during geophysical prospecting.)

To analyze a simple case, take the pendulum to be a uniform rod of length L, suspended from one end. For such a pendulum, h in Eq. 16-32, the distance between the suspension point and the center of mass, is $\frac{1}{2}L$. Table 11-2(f) tells us that the rotational inertia of this pendulum about a perpendicular axis through one end is $\frac{1}{3}mL^2$. If we put $h = \frac{1}{2}L$ and $I = \frac{1}{3}mL^2$ in Eq. 16-32 and solve for g, we find

$$g = \frac{8\pi^2 L}{3T^2}. \qquad (16\text{-}33)$$

Thus by measuring L and the period T, we can find the value of g. (If precise measurements are to be made, a number of refinements are needed, such as swinging the pendulum in an evacuated chamber.)

CHECKPOINT **4:** Three physical pendulums, of masses m_0, $2m_0$, and $3m_0$, have the same shape and size and are suspended at the same point. Rank the masses according to the periods of the pendulums, greatest period first.

SAMPLE PROBLEM 16-6

A meter stick, suspended from one end, swings as a physical pendulum, as in Fig. 16-12a.

(a) What is its period of oscillation?

SOLUTION: From Table 11-2(f) we see that the rotational inertia of a rod or stick of length L about a perpendicular axis through one end is $\frac{1}{3}mL^2$. The distance h from the point of suspension to the center of mass, which is point C in Fig.

16-12a, is $\frac{1}{2}L$. If we substitute these two quantities into Eq. 16-32, we find

$$T = 2\pi \sqrt{\frac{I}{mgh}} = 2\pi \sqrt{\frac{\frac{1}{3}mL^2}{mg(\frac{1}{2}L)}} = 2\pi \sqrt{\frac{2L}{3g}} \quad (16\text{-}34)$$

$$= 2\pi \sqrt{\frac{(2)(1.00\text{ m})}{(3)(9.8\text{ m/s}^2)}} = 1.64\text{ s}. \quad \text{(Answer)}$$

(b) For the stick of Fig. 16-12a, what is the distance L_0 between the suspension point O of the stick and the center of oscillation of the stick?

SOLUTION: To locate the center of oscillation of the meter stick, we find the length L_0 of a simple pendulum (Fig. 16-12b) having the same period T as the meter stick. Setting Eqs. 16-29 and 16-34 equal yields

$$T = 2\pi \sqrt{\frac{L_0}{g}} = 2\pi \sqrt{\frac{2L}{3g}}.$$

You can see by inspection that

$$L_0 = \tfrac{2}{3}L = (\tfrac{2}{3})(100\text{ cm}) = 66.7\text{ cm}. \quad \text{(Answer)}$$

In Fig. 16-12a, point P marks this distance from suspension point O. Thus point P is the stick's center of oscillation for the given suspension point.

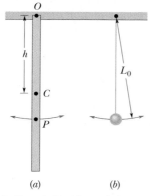

(a)   (b)

FIGURE 16-12 Sample Problem 16-6. (a) A meter stick suspended from one end as a physical pendulum. (b) A simple pendulum whose length L is chosen so that the periods of the two pendulums are equal. Point P on the pendulum of (a) marks the center of oscillation, a distance L_0 from the suspension point.

SAMPLE PROBLEM 16-7

A disk whose radius R is 12.5 cm is suspended, as a physical pendulum, from a point at distance h from its center C (Fig. 16-13). Its period T is 0.871 s when $h = R/2$. What is the free-fall acceleration g at the location of the pendulum?

SOLUTION: The rotational inertia I_{cm} of a disk about its central axis is $\frac{1}{2}mR^2$. From the parallel-axis theorem, the rotational inertia about an axis parallel to the central axis and through the point of suspension O, as in Fig. 16-13, is

$$I = I_{cm} + mh^2 = \tfrac{1}{2}mR^2 + m(\tfrac{1}{2}R)^2 = \tfrac{3}{4}mR^2.$$

If we put $I = \frac{3}{4}mR^2$ and $h = \frac{1}{2}R$ in Eq. 16-32, we find

$$T = 2\pi \sqrt{\frac{I}{mgh}} = 2\pi \sqrt{\frac{\frac{3}{4}mR^2}{mg(\frac{1}{2}R)}} = 2\pi \sqrt{\frac{3R}{2g}}.$$

Solving for g then gives us

$$g = \frac{6\pi^2 R}{T^2} = \frac{(6\pi^2)(0.125\text{ m})}{(0.871\text{ s})^2} = 9.76\text{ m/s}^2. \quad \text{(Answer)}$$

FIGURE 16-13 Sample Problem 16-7. A physical pendulum consisting of a uniform disk suspended from a point (O) that is halfway from the center of the disk to the rim.

SAMPLE PROBLEM 16-8

In Fig. 16-14, a penguin (obviously skilled in aquatic sports) dives from a uniform board that is hinged at the left and attached to a spring at the right. The board has length $L = 2.0$ m and mass $m = 12$ kg; the spring constant k is 1300 N/m. When the penguin dives, it leaves the board and spring oscillating with a small amplitude. Assume that the board is stiff enough not to bend, and find the period T of the oscillations.

SOLUTION: The spring applies a time-varying torque τ (about the hinge) to the board, which undergoes a time-varying angular acceleration α about the hinge. Since a spring is involved, we might guess that simple harmonic motion is also involved, but we will not draw that conclusion yet. Instead, we use Eqs. 11-30 and 11-35 to write

$$\tau = LF \sin 90° = I\alpha. \quad (16\text{-}35)$$

Here I is the rotational inertia of the board about the hinge, F is the force on the right end of the board from the spring, and 90° is the angle between the length of the board and the direction of that force.

We may treat the board as being a thin rod pivoted about one end, so that from Table 11-2(f), $I = mL^2/3$. The force F from the spring is $F = -kx$, where x is the vertical displacement of the right end of the board.

Substituting these expressions for I and F into the rightmost equality of Eq. 16-35 and evaluating sin 90°, we find

$$-Lkx = \frac{mL^2\alpha}{3}. \qquad (16\text{-}36)$$

We now have a mixture of linear displacement x (vertically) and rotational acceleration α (about the hinge). We can rewrite Eq. 16-36 in terms of just the rotational motion of the right end of the board about the hinge by using Eq. 11-15,

$$s = \theta r.$$

Here θ is the angle of the board's rotation about the hinge, $r = L$ is the radius of the rotation, and s is the arc length, which is approximated by the vertical displacement x (for small θ). We can rewrite this as $x = \theta L$ and then substitute for x in Eq. 16-36 to find

$$-Lk\theta L = \frac{mL^2\alpha}{3},$$

which becomes

$$\alpha = -\frac{3k}{m}\theta. \qquad (16\text{-}37)$$

Equation 16-37 is an angular version of Eq. 16-7. It tells us that the board does indeed undergo simple harmonic motion, with angular acceleration α and angular displacement θ. Comparing Eqs. 16-37 and 16-7 reveals that

$$\omega^2 = \frac{3k}{m}$$

and thus that $\omega = \sqrt{3k/m}$. Using Eq. 16-4, $\omega = 2\pi/T$, we then have

$$T = 2\pi\sqrt{\frac{m}{3k}} = 2\pi\sqrt{\frac{12\text{ kg}}{3(1300\text{ N/m})}}$$

$$= 0.35\text{ s.} \qquad \text{(Answer)}$$

Perhaps surprisingly, the period is independent of the board's length L.

FIGURE 16-14   Sample Problem 16-8. The dive by the penguin causes the board and spring to oscillate; the board pivots about the hinge at the left.

16-7 SIMPLE HARMONIC MOTION AND UNIFORM CIRCULAR MOTION

In 1610, Galileo, using his newly constructed telescope, discovered the four principal moons of Jupiter. Over weeks of observation, each moon seemed to him to be moving back and forth relative to the planet in what today we would call simple harmonic motion; the disk of the planet was the midpoint of the motion. The record of Galileo's observations, written in his own hand, is still available. A. P. French of MIT used Galileo's data to work out the position of the moon Callisto relative to Jupiter. In the results shown in Fig. 16-15, the circles are based on Galileo's observations and the curve is a best fit to the data. The curve strongly suggests Eq. 16-3, the displacement function for SHM. A period of about 16.8 days can be measured from the plot.

Actually, Callisto moves with essentially constant speed in an essentially circular orbit around Jupiter. Its true motion—far from being simple harmonic —is uniform circular motion. What Galileo saw—and what you can see with a good pair of binoculars and a little patience—is the projection of this uniform circular motion on a line in the plane of the motion. We are led by Galileo's remarkable observations to the conclusion that simple harmonic motion is uniform circular motion viewed edge-on. In more formal language:

> Simple harmonic motion is the projection of uniform circular motion on a diameter of the circle in which the latter motion occurs.

FIGURE 16-15   The angle between Jupiter and its moon Callisto as seen from Earth. The circles are based on Galileo's 1610 measurements. The curve is a best fit, strongly suggesting simple harmonic motion. At Jupiter's mean distance, 10 minutes of arc corresponds to about 2×10^6 km. (Adapted from A. P. French, *Newtonian Mechanics,* W. W. Norton & Company, New York, 1971, p. 288.)

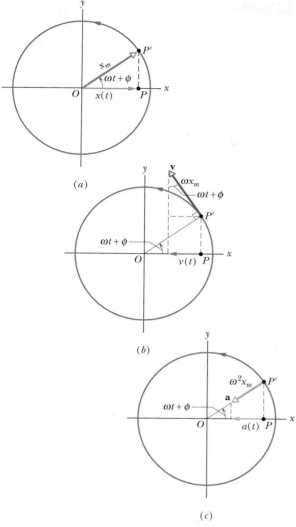

(a)

(b)

(c)

FIGURE 16-16 (a) A reference particle P' moving with uniform circular motion in a reference circle of radius x_m. Its projection P on the x axis executes simple harmonic motion. (b) The projection of the velocity $\mathbf{v}$ of the reference particle is the velocity of SHM. (c) The projection of the acceleration $\mathbf{a}$ of the reference particle is the acceleration of SHM.

Figure 16-16a gives an example. It shows a *reference particle P'* moving in uniform circular motion with (constant) angular speed ω in a *reference circle*. The radius x_m of the circle is the magnitude of the particle's position vector. At any time t, the angular position of the particle is $\omega t + \phi$, where ϕ is the angular position at $t = 0$.

The projection of particle P' onto the x axis is a point P, which we take to be a second particle. The projection of the position vector of particle P' onto the x axis gives the location $x(t)$ of P. Thus we find

$$x(t) = x_m \cos(\omega t + \phi),$$

which is precisely Eq. 16-3. Our conclusion is correct. If reference particle P' moves in uniform circular motion, its projection particle P moves in simple harmonic motion.

This relation throws new light on the angular frequency ω of simple harmonic motion, and you can see more clearly how the "angular" originates. The quantity ω is simply the constant angular speed with which the reference particle P' moves along its reference circle. The phase constant ϕ has a value determined by the position of reference particle P' along its reference circle at $t = 0$.

Figure 16-16b shows the velocity of the reference particle. The magnitude of the velocity vector is ωx_m, and its projection on the x axis is

$$v(t) = -\omega x_m \sin(\omega t + \phi),$$

which is exactly Eq. 16-5. The minus sign appears because the velocity component of P in Fig. 16-16b points to the left, in the direction of decreasing x.

Figure 16-16c shows the acceleration of the reference particle. The magnitude of the acceleration vector is $\omega^2 x_m$ and its projection on the x axis is

$$a(t) = -\omega^2 x_m \cos(\omega t + \phi),$$

which is exactly Eq. 16-6. Thus whether we look at the displacement, the velocity, or the acceleration, the projection of uniform circular motion is indeed simple harmonic motion.

16-8 DAMPED SIMPLE HARMONIC MOTION (OPTIONAL)

A pendulum will swing hardly at all under water, because the water exerts a drag force on the pendulum that quickly eliminates the motion. A pendulum swinging in air does better, but still the motion dies out because the air exerts a drag force on the pendulum (and friction acts at its support), transferring energy from the pendulum's motion.

When the motion of an oscillator is reduced by an external force, the oscillator and its motion are said to be **damped.** An idealized example of a damped oscillator is shown in Fig. 16-17: a block with mass m oscillates on a spring with spring constant k. From the mass, a rod extends to a vane (both assumed massless) that is submerged in a liquid. As the vane moves up and down, the liquid exerts an inhibiting drag force on it and thus on the entire oscillating system. With time, the mechanical energy of the block–spring system decreases, as energy is transferred to thermal energy of the liquid and vane.

Let us assume that the liquid exerts a **damping force** $\mathbf{F}_d$ that is proportional in magnitude to the velocity $\mathbf{v}$ of the

FIGURE 16-17 An idealized damped simple harmonic oscillator. A vane immersed in a liquid exerts a damping force on the oscillating block.

vane and block (an assumption that is accurate if the vane moves slowly). Then

$$F_d = -bv, \quad (16\text{-}38)$$

where b is a **damping constant** that depends on the characteristics of both the vane and the liquid and has the SI unit of kilogram per second. The minus sign indicates that F_d opposes the motion. The total force acting on the block is then

$$\sum F = -kx - bv,$$

or, if we set $v = dx/dt$,

$$\sum F = -kx - b\frac{dx}{dt}. \quad (16\text{-}39)$$

Substituting this total force into Newton's second law yields the differential equation

$$m\frac{d^2x}{dt^2} + b\frac{dx}{dt} + kx = 0,$$

whose solution is

$$x(t) = x_m e^{-bt/2m}\cos(\omega't + \phi), \quad (16\text{-}40)$$

where ω', the angular frequency of the damped oscillator, is given by

$$\omega' = \sqrt{\frac{k}{m} - \frac{b^2}{4m^2}}. \quad (16\text{-}41)$$

If $b = 0$ (there is no damping), then Eq. 16-41 reduces to Eq. 16-11 ($\omega = \sqrt{k/m}$) for the angular frequency of an undamped oscillator, and Eq. 16-40 reduces to Eq. 16-3 for

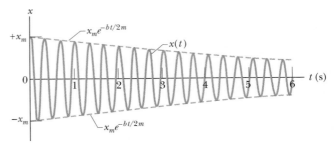

FIGURE 16-18 The displacement function $x(t)$ for the damped oscillator of Fig. 16-17. See Sample Problem 16-9. The amplitude, which is $x_m\,e^{-bt/2m}$, decreases exponentially with time.

the displacement of an undamped oscillator. If the damping constant is not zero but small (so that $b \ll \sqrt{km}$), then $\omega' \approx \omega$.

We can regard Eq. 16-40 as a cosine function whose amplitude, which is $x_m\,e^{-bt/2m}$, gradually decreases with time, as Fig. 16-18 suggests. For an undamped oscillator, the mechanical energy is constant and is given by Eq. 16-23: $E = \frac{1}{2}kx_m^2$. If the oscillator is damped, the mechanical energy is not constant but decreases with time. If the damping is small, we can find $E(t)$ by replacing x_m in Eq. 16-23 with $x_m\,e^{-bt/2m}$, the amplitude of the damped oscillations. Doing so, we find

$$E(t) \approx \tfrac{1}{2}kx_m^2\,e^{-bt/m}, \quad (16\text{-}42)$$

which tells us that the mechanical energy decreases exponentially with time.

SAMPLE PROBLEM 16-9

For the damped oscillator of Fig. 16-17, $m = 250$ g, $k = 85$ N/m, and $b = 70$ g/s.

(a) What is the period of the motion?

SOLUTION: Because $b \ll \sqrt{km} = 4.6$ kg/s, the period is approximately that of the undamped oscillator. From Eq. 16-12, we then have

$$T = 2\pi\sqrt{\frac{m}{k}} = 2\pi\sqrt{\frac{0.25\text{ kg}}{85\text{ N/m}}} = 0.34\text{ s.} \quad \text{(Answer)}$$

(b) How long does it take for the amplitude of the damped oscillations to drop to half its initial value?

SOLUTION: The amplitude at time t is displayed in Eq. 16-40 as $x_m\,e^{-bt/2m}$. It has the value x_m at $t = 0$. Thus we must find the value of t for which

$$x_m\,e^{-bt/2m} = \tfrac{1}{2}x_m.$$

Let us divide both sides by x_m and take the natural logarithm of the equation that remains. We find

$$\ln \tfrac{1}{2} = \ln(e^{-bt/2m}) = -bt/2m,$$

or

$$t = \frac{-2m \ln \tfrac{1}{2}}{b} = \frac{-(2)(0.25 \text{ kg})(\ln \tfrac{1}{2})}{0.070 \text{ kg/s}}$$

$$= 5.0 \text{ s.} \qquad \text{(Answer)}$$

Note, from the answer we found in (a), that this is about 15 periods of oscillation.

(c) How long does it take for the mechanical energy to drop to one-half its initial value?

SOLUTION: From Eq. 16-42 we see that the mechanical energy at time t is $\tfrac{1}{2}kx_m^2 \, e^{-bt/m}$. It has the value $\tfrac{1}{2}kx_m^2$ at $t = 0$. Thus we must find the value of t for which

$$\tfrac{1}{2}kx_m^2 \, e^{-bt/m} = \tfrac{1}{2}(\tfrac{1}{2}kx_m^2).$$

If we divide by $\tfrac{1}{2}kx_m^2$ and solve for t as we did above, we find

$$t = \frac{-m \ln \tfrac{1}{2}}{b} = \frac{-(0.25 \text{ kg})(\ln \tfrac{1}{2})}{0.070 \text{ kg/s}} = 2.5 \text{ s.} \quad \text{(Answer)}$$

This is exactly half the time calculated in (b), or about 7.5 periods of oscillation. Figure 16-18 was drawn to illustrate this sample problem.

CHECKPOINT **5:** Here are three sets of values for the spring constant, damping constant, and mass for the damped oscillator of Fig. 16-17. Rank the sets according to the time for the mechanical energy to decrease to one-fourth of its initial value, greatest first.

Set 1	$2k_0$	b_0	m_0
Set 2	k_0	$6b_0$	$4m_0$
Set 3	$3k_0$	$3b_0$	m_0

16-9 FORCED OSCILLATIONS AND RESONANCE

A person swinging passively in a swing is an example of *free oscillation*. If a kind friend pulls or pushes the swing periodically, as in Fig. 16-19, we have *forced,* or *driven, oscillations*. There are now *two* angular frequencies with which to deal: (1) the *natural* angular frequency ω of the system, which is the angular frequency at which it would oscillate if it were suddenly disturbed and then left to oscillate freely, and (2) the angular frequency ω_d of the external driving force.

We can use Fig. 16-17 to represent an idealized forced simple harmonic oscillator if we allow the structure marked "rigid support" to move up and down at a variable

FIGURE 16-19 Two frequencies are suggested in this painting by Nicholas Lancret: (1) the natural frequency at which the lady —left to herself—would swing, and (2) the frequency at which her friend tugs on the rope. If these two frequencies match, there is resonance.

angular frequency ω_d. A forced oscillator oscillates at the angular frequency ω_d of the driving force, and its displacement $x(t)$ is given by

$$x(t) = x_m \cos(\omega_d t + \phi), \qquad (16\text{-}43)$$

where x_m is the amplitude of the oscillations.

How large the displacement amplitude x_m is depends on a complicated function of ω_d and ω. The velocity amplitude v_m of the oscillations is easier to describe: it is greatest when

$$\omega_d = \omega \qquad \text{(resonance)}, \qquad (16\text{-}44)$$

a condition called **resonance.** Equation 16-44 is also *approximately* the condition at which the displacement amplitude x_m of the oscillations is greatest. Thus if you push a swing at its natural frequency, the displacement and velocity amplitudes will increase to large values, a fact that children learn quickly by trial and error. If you push at other frequencies, either higher or lower, the displacement and velocity amplitudes will be smaller.

Figure 16-20 shows how the displacement amplitude of an oscillator depends on the angular frequency ω_d of the driving force, for three values of the damping coefficient b. Note that for all three the amplitude is approximately greatest when $\omega_d/\omega = 1$, that is, when the resonance condition of Eq. 16-44 is satisfied. The curves of Fig. 16-20 show that the smaller the damping, the taller and narrower is the *resonance peak*.

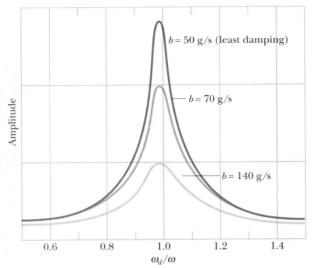

FIGURE 16-20 The displacement amplitude x_m of a forced oscillator varies as the angular frequency ω_d of the driving force is varied. The amplitude is greatest approximately at $\omega_d/\omega = 1$, the resonance condition. The curves here correspond to three values of the damping constant b.

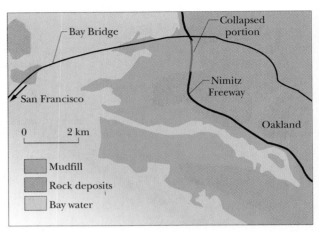

FIGURE 16-21 The geological structure in the Oakland portion of the San Francisco Bay area. The collapsed stretch of the Nimitz Freeway is indicated. (Adapted from "Sediment-Induced Amplification and the Collapse of the Nimitz Freeway," by S. E. Hough et al., *Nature*, April 26, 1990.)

All mechanical structures have one or more natural frequencies, and if a structure is subjected to a strong external driving force that matches one of these frequencies, the resulting oscillations of the structure may rupture it. Thus, for example, aircraft designers must make sure that none of the natural frequencies at which a wing can vibrate matches the angular frequency of the engines at cruising speed. It obviously would be dangerous for a wing to flap violently at certain engine speeds.

The collapse of a 1.4 km stretch of the Nimitz Freeway, shown in this chapter's opening photograph, is an example of the destruction that resonance can produce. Oscillations of the ground due to the earthquake's seismic waves had their greatest velocity amplitude at an angular frequency of about 9 rad/s, which almost exactly matched a natural angular frequency of the individual horizontal sections of the freeway. The collapse was confined to a particular 1.4 km stretch because that section of the freeway was built on a loosely structured mudfill whose velocity amplitude during the quake was *at least five times larger* than that of the rock deposits under the rest of the freeway (Fig. 16-21).

REVIEW & SUMMARY

Frequency

Every oscillatory, or periodic, motion has a *frequency f* (the number of oscillations per second) measured, in the SI system, in hertz:

$$1 \text{ hertz} = 1 \text{ Hz} = 1 \text{ oscillation per second}$$
$$= 1 \text{ s}^{-1}. \qquad (16\text{-}1)$$

Period

The *period T* is the time required for one complete oscillation, or **cycle.** It is related to the frequency by

$$T = \frac{1}{f}. \qquad (16\text{-}2)$$

Simple Harmonic Motion

In *simple harmonic motion* (SHM), the displacement $x(t)$ of a particle from its equilibrium position is described by the equation

$$x = x_m \cos(\omega t + \phi) \quad \text{(displacement)}, \qquad (16\text{-}3)$$

in which x_m is the **amplitude** of the displacement, the quantity $(\omega t + \phi)$ is the **phase** of the motion, and ϕ is the **phase constant.** The **angular frequency** ω is related to the period and frequency of the motion by

$$\omega = \frac{2\pi}{T} = 2\pi f \quad \text{(angular frequency)}. \qquad (16\text{-}4)$$

Differentiating Eq. 16-3 leads to equations for the particle's velocity and acceleration during SHM as functions of time:

$$v = -\omega x_m \sin(\omega t + \phi) \qquad \text{(velocity)} \qquad (16\text{-}5)$$

and $$a = -\omega^2 x_m \cos(\omega t + \phi) \qquad \text{(acceleration)}. \qquad (16\text{-}6)$$

In Eq. 16-5, the positive quantity ωx_m is the **velocity amplitude** v_m of the motion. In Eq. 16-6, the positive quantity $\omega^2 x_m$ is the **acceleration amplitude** a_m of the motion.

The Linear Oscillator

A particle with mass m that moves under the influence of a Hooke's law restoring force given by $F = -kx$ exhibits simple harmonic motion with

$$\omega = \sqrt{\frac{k}{m}} \qquad \text{(angular frequency)} \qquad (16\text{-}11)$$

and $$T = 2\pi \sqrt{\frac{m}{k}} \qquad \text{(period)}. \qquad (16\text{-}12)$$

Such a system is called a **linear simple harmonic oscillator.**

Energy

A particle in simple harmonic motion has, at any time, kinetic energy $K = \frac{1}{2}mv^2$ and potential energy $U = \frac{1}{2}kx^2$. If no friction is present, the mechanical energy $E = K + U$ remains constant even though K and U change.

Pendulums

Examples of devices that undergo simple harmonic motion are the **torsion pendulum** of Fig. 16-8, the **simple pendulum** of Fig. 16-10, and the **physical pendulum** of Fig. 16-11. Their periods of oscillation for small oscillations are, respectively,

$$T = 2\pi \sqrt{I/\kappa}, \qquad (16\text{-}25)$$

$$T = 2\pi \sqrt{L/g}, \qquad (16\text{-}29)$$

and $$T = 2\pi \sqrt{I/mgh}. \qquad (16\text{-}32)$$

In each case, the period involves an inertial term divided by a "springiness" term that measures the strength of the oscillator's restoring force.

Simple Harmonic Motion and Uniform Circular Motion

Simple harmonic motion is the projection of uniform circular motion onto the diameter of the circle in which the latter motion occurs. Figure 16-16 shows that all parameters of circular motion (position, velocity, and acceleration) project to the corresponding values for simple harmonic motion.

Damped Harmonic Motion

The mechanical energy E in a real oscillating system decreases during the oscillations because external forces, such as a drag force, inhibit the oscillations and transfer mechanical energy to thermal energy. The real oscillator and its motion are then said to be **damped.** If the **damping force** is given by $F_d = -bv$, where v is the velocity of the oscillator and b is a **damping constant,** then the displacement of the oscillator is given by

$$x(t) = x_m e^{-bt/2m} \cos(\omega' t + \phi), \qquad (16\text{-}40)$$

where ω', the angular frequency of the damped oscillator, is given by

$$\omega' = \sqrt{\frac{k}{m} - \frac{b^2}{4m^2}}. \qquad (16\text{-}41)$$

If the damping constant is small ($b \ll \sqrt{km}$), then $\omega' \approx \omega$, where ω is the angular frequency of the undamped oscillator. For small b, the mechanical energy E of the oscillator is given by

$$E(t) \approx \frac{1}{2}kx_m^2 \, e^{-bt/m}. \qquad (16\text{-}42)$$

Forced Oscillations and Resonance

If an external driving force with angular frequency ω_d acts on an oscillating system with *natural* angular frequency ω, the system oscillates with angular frequency ω_d. The velocity amplitude v_m of the system is greatest when

$$\omega_d = \omega, \qquad (16\text{-}44)$$

a condition called **resonance.** The amplitude x_m of the system is (approximately) greatest under the same condition.

QUESTIONS

1. Which of the following relationships between the acceleration a and the displacement x of a particle involve SHM: (a) $a = 0.5x$, (b) $a = 400x^2$, (c) $a = -20x$, (d) $a = -3x^2$?

2. The acceleration $a(t)$ of a particle undergoing SHM is graphed in Fig. 16-22. (a) Which of the labeled points corresponds to the particle being at $-x_m$? (b) At point 4, is the velocity of the particle positive, negative, or zero? (c) At point 5, is the particle at $-x_m$, at $+x_m$, at 0, between $-x_m$ and 0, or between 0 and $+x_m$?

3. A particle oscillates according to

$$x = x_m \cos(\omega t + \phi).$$

FIGURE 16-22 Question 2.

At $t = 0$, is the particle at $-x_m$, at $+x_m$, at 0, between $-x_m$ and 0, or between 0 and $+x_m$ if ϕ is (a) $\pi/2$, (b) $-\pi/3$, (c) $-3\pi/4$, and (d) $3\pi/4$?

4. Which of the following describe ϕ for the SHM of Fig. 16-23a: (a) $-\pi < \phi < -\pi/2$, (b) $\pi < \phi < 3\pi/2$, (c) $-3\pi/2 < \phi < -\pi$?

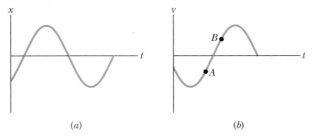

(a)   (b)

FIGURE 16-23   Questions 4 and 5.

5. The velocity $v(t)$ of a particle undergoing SHM is graphed in Fig. 16-23b. Is the particle momentarily stationary, headed toward $-x_m$, or headed toward $+x_m$ at (a) point A on the graph and (b) point B? Is the particle at $-x_m$, at $+x_m$, at 0, between $-x_m$ and 0, or between 0 and $+x_m$ when its velocity is represented by (c) point A and (d) point B? Is the speed of the particle increasing or decreasing at (e) point A and (f) point B?

6. What is the phase difference of the two linear oscillators, with identical masses and spring constants, in (a) Fig. 16-24a and (b) Fig. 16-24b? (c) What is the phase difference between the red oscillator in Fig. 16-24a and the green oscillator in Fig. 16-24b?

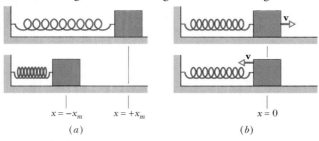

$x = -x_m$ $x = +x_m$ $x = 0$

(a)   (b)

FIGURE 16-24   Question 6.

7. A spring and a block are arranged to oscillate (1) on a frictionless horizontal surface as in Fig. 16-5, (2) on a frictionless slope of 45° (with the block at the lower end of the spring), and (3) vertically while hanging from a ceiling. Rank the arrangements according to (a) the rest length of the spring and (b) the frequency of oscillation, greatest first.

8. Three springs each hang from a ceiling with a stationary block attached to the lower end. The blocks (with $m_1 > m_2 > m_3$) stretch the springs by equal distances. Each block–spring system is then set into vertical SHM. Rank the masses according to the period of oscillation, greatest first.

9. Figure 16-25 shows three arrangements of a block attached to identical springs that are in the relaxed state when the block is centered. Rank the arrangements according to the frequency of oscillation of the block, greatest first.

(1)   (2)   (3)

FIGURE 16-25   Question 9.

10. A mass m is suspended from a spring with spring constant k. The spring is then cut in half and the same mass is suspended from one of the halves. Which, the original spring or the half-spring, gives a greater frequency of oscillation when the mass is put into vertical SHM?

11. When mass m_1 is hung from spring A and a smaller mass m_2 is hung from spring B, the springs are stretched by the same distance. If the two spring–mass systems are then put into vertical SHM with the same amplitude, which system will have more energy?

12. Do the following increase, decrease, or remain the same if the amplitude of a simple harmonic oscillator is doubled: (a) the period, (b) the spring constant, (c) the total energy, (d) the maximum speed, (e) the maximum acceleration?

13. Figure 16-26 shows three physical pendulums consisting of identical uniform spheres of the same mass that are rigidly connected by identical rods of negligible mass. Each pendulum is vertical and can pivot about suspension point O. Rank the pendulums according to period of oscillation, greatest first.

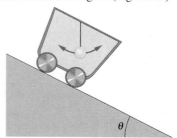

FIGURE 16-26   Question 13.   (a)   (b)   (c)

14. A pendulum suspended from the ceiling of an elevator cab has period T when the cab is stationary. Is the period larger, smaller, or the same when the cab moves (a) upward with constant speed, (b) downward with constant speed, (c) downward with constant upward acceleration, (d) upward with constant upward acceleration, (e) upward with constant downward acceleration $a = g$, and (f) downward with constant downward acceleration $a = g$?

15. A pendulum mounted in a cart has period T when the cart is stationary and on a horizontal plane. Is the period larger, smaller, or the same if the cart is on a plane inclined at angle θ (Fig. 16-27)

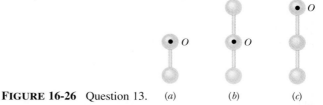

FIGURE 16-27
Question 15.

and (a) stationary, (b) moving down the plane with constant speed, (c) moving up the plane with constant speed, (d) moving up the plane with constant acceleration directed up the plane, (e) moving down the plane with constant acceleration directed up the plane, (f) moving down the plane with acceleration $a = g \sin \theta$ directed down the plane, and (g) moving up the plane with acceleration $a = g \sin \theta$ directed down the plane?

16. You are to build the oscillation transfer device shown in Fig. 16-28. It consists of two spring–block systems hanging from a flexible rod. When the spring of system 1 is stretched and then released, the resulting SHM of system 1 at frequency f_1 oscillates the rod. The rod then provides a driving force on system 2, at the same frequency f_1. For components you have to choose from four springs with spring constants k of 1600, 1500, 1400, and 1200

N/m, and four blocks with masses m of 800, 500, 400, and 200 kg. Without written calculation, determine which spring should go with which block in each of the two systems to maximize the amplitude of oscillations in system 2.

FIGURE 16-28 Question 16.

EXERCISES & PROBLEMS

SECTION 16-3 The Force Law for Simple Harmonic Motion

1E. An object undergoing simple harmonic motion takes 0.25 s to travel from one point of zero velocity to the next such point. The distance between those points is 36 cm. Calculate (a) the period, (b) the frequency, and (c) the amplitude of the motion.

2E. An oscillating mass–spring system takes 0.75 s to begin repeating its motion. Find (a) the period, (b) the frequency in hertz, and (c) the angular frequency in radians per second.

3E. A 4.00 kg block hangs from a spring, extending it 16.0 cm from its unstretched position. (a) What is the spring constant? (b) The block is removed and a 0.500 kg body is hung from the same spring. If the spring is then stretched and released, what is its period of oscillation?

4E. An oscillator consists of a block of mass 0.500 kg connected to a spring. When set into oscillation with amplitude 35.0 cm, it is observed to repeat its motion every 0.500 s. Find (a) the period, (b) the frequency, (c) the angular frequency, (d) the spring constant, (e) the maximum speed, and (f) the maximum force exerted on the block.

5E. The vibration frequencies of atoms in solids at normal temperatures are of the order of 10^{13} Hz. Imagine the atoms to be connected to one another by springs. Suppose that a single silver atom in a solid vibrates with this frequency and that all the other atoms are at rest. Compute the effective spring constant. One mole of silver (6.02×10^{23} atoms) has a mass of 108 g.

6E. What is the maximum acceleration of a platform that vibrates with an amplitude of 2.20 cm at a frequency of 6.60 Hz?

7E. A loudspeaker produces a musical sound by means of the oscillation of a diaphragm. If the amplitude of oscillation is limited to 1.0×10^{-3} mm, what frequencies will result in the acceleration of the diaphragm exceeding g?

8E. The scale of a spring balance which reads from 0 to 32.0 lb is 4.00 in. long. A package suspended from the balance is found to oscillate vertically with a frequency of 2.00 Hz. (a) What is the spring constant? (b) How much does the package weigh?

9E. A 20 N weight is hung from the bottom of a vertical spring, causing the spring to stretch 20 cm. (a) What is the spring constant? (b) This spring is now placed horizontally on a frictionless table. One end of it is held fixed, and the other end is attached to a 5.0 N weight. The weight is then moved (stretching the spring) and released from rest. What is the period of oscillation?

10E. A 50.0 g mass is attached to the bottom of a vertical spring and set vibrating. If the maximum speed of the mass is 15.0 cm/s and the period is 0.500 s, find (a) the spring constant of the spring, (b) the amplitude of the motion, and (c) the frequency of oscillation.

11E. A particle with a mass of 1.00×10^{-20} kg is vibrating with simple harmonic motion with a period of 1.00×10^{-5} s and a maximum speed of 1.00×10^3 m/s. Calculate (a) the angular frequency and (b) the maximum displacement of the particle.

12E. A small body of mass 0.12 kg is undergoing simple harmonic motion of amplitude 8.5 cm and period 0.20 s. (a) What is the maximum value of the force acting on it? (b) If the oscillations are produced by a spring, what is the spring constant?

13E. In an electric shaver, the blade moves back and forth over a distance of 2.0 mm. The motion is simple harmonic, with frequency 120 Hz. Find (a) the amplitude, (b) the maximum blade speed, and (c) the maximum blade acceleration.

14E. A loudspeaker diaphragm is vibrating in simple harmonic motion with a frequency of 440 Hz and a maximum displacement of 0.75 mm. What are (a) the angular frequency, (b) the maximum speed, and (c) the maximum acceleration?

15E. An automobile can be considered to be mounted on four identical springs as far as vertical oscillations are concerned. The springs of a certain car are adjusted so that the vibrations have a frequency of 3.00 Hz. (a) What is the spring constant of each spring if the mass of the car is 1450 kg and the weight is evenly distributed over the springs? (b) What will be the vibration frequency if five passengers, averaging 73.0 kg each, ride in the car? (Again, consider an even distribution of weight.)

16E. A body oscillates with simple harmonic motion according to the equation

$$x = (6.0 \text{ m}) \cos[(3\pi \text{ rad/s})t + \pi/3 \text{ rad}].$$

At $t = 2.0$ s, what are (a) the displacement, (b) the velocity, (c) the acceleration, and (d) the phase of the motion? Also, what are (e) the frequency and (f) the period of the motion?

17E. A particle executes linear SHM with frequency 0.25 Hz about the point $x = 0$. At $t = 0$, it has displacement $x = 0.37$ cm and zero velocity. For the motion, determine (a) the period, (b) the angular frequency, (c) the amplitude, (d) the displacement at time t, (e) the velocity at time t, (f) the maximum speed, (g) the maximum acceleration, (h) the displacement at $t = 3.0$ s, and (i) the speed at $t = 3.0$ s.

18E. The piston in the cylinder head of a locomotive has a stroke (twice the amplitude) of 0.76 m. If the piston moves with simple harmonic motion with an angular frequency of 180 rev/min, what is its maximum speed?

19P. Figure 16-29 shows an astronaut on a body-mass measuring device (BMMD). Designed for use on orbiting space vehicles, its purpose is to allow astronauts to measure their mass in the "weightless" conditions in Earth orbit. The BMMD is a spring-mounted chair; an astronaut measures his or her period of oscillation in the chair; the mass follows from the formula for the period of an oscillating block–spring system. (a) If M is the mass of the astronaut and m the effective mass of that part of the BMMD that also oscillates, show that

$$M = (k/4\pi^2)T^2 - m,$$

where T is the period of oscillation and k is the spring constant. (b) The spring constant was $k = 605.6$ N/m for the BMMD on Skylab Mission Two; the period of oscillation of the empty chair was 0.90149 s. Calculate the effective mass of the chair. (c) With an astronaut in the chair, the period of oscillation became 2.08832 s. Calculate the mass of the astronaut.

20P. A 2.00 kg block hangs from a spring. A 300 g body hung below the block stretches the spring 2.00 cm farther. (a) What is the spring constant? (b) If the 300 g body is removed and the block is set into oscillation, find the period of the motion.

21P. The end point of a spring vibrates with a period of 2.0 s when a mass m is attached to it. When this mass is increased by 2.0 kg, the period is found to be 3.0 s. Find the value of m.

22P. The end of one of the prongs of a tuning fork that executes simple harmonic motion of frequency 1000 Hz has an amplitude of 0.40 mm. Find (a) the maximum acceleration and (b) the maximum speed of the end of the prong. Find (c) the acceleration and (d) the speed of the end of the prong when the end has a displacement of 0.20 mm.

23P. A 0.10 kg block oscillates back and forth along a straight line on a frictionless horizontal surface. Its displacement from the origin is given by

$$x = (10 \text{ cm}) \cos[(10 \text{ rad/s})t + \pi/2 \text{ rad}].$$

(a) What is the oscillation frequency? (b) What is the maximum speed acquired by the block? At what value of x does this occur? (c) What is the maximum acceleration of the block? At what value of x does this occur? (d) What force, applied to the block, results in the given oscillation?

24P. At a certain harbor, the tides cause the ocean surface to rise and fall a distance d in simple harmonic motion, with a period of 12.5 h. How long does it take for the water to fall a distance $d/4$ from its maximum height?

25P. Two blocks ($m = 1.0$ kg and $M = 10$ kg) and a spring ($k = 200$ N/m) are arranged on a horizontal, frictionless surface as shown in Fig. 16-30. The coefficient of static friction between the two blocks is 0.40. What is the maximum possible amplitude of simple harmonic motion of the spring–blocks system if no slippage is to occur between the blocks?

FIGURE 16-30   Problem 25.

26P. A block is on a horizontal surface (a shake table) that is moving horizontally with simple harmonic motion of frequency 2.0 Hz. The coefficient of static friction between block and surface is 0.50. How great can the amplitude of the SHM be if the block is not to slip along the surface?

27P. A block is on a piston that is moving vertically with simple harmonic motion. (a) If the SHM has period 1.0 s, at what amplitude of motion will the block and piston separate? (b) If the piston has an amplitude of 5.0 cm, what is the maximum frequency for which the block and piston will be in contact continuously?

28P. An oscillator consists of a block attached to a spring ($k = 400$ N/m). At some time t, the position (measured from the system's equilibrium location), velocity, and acceleration of the

FIGURE 16-29   Problem 19.

block are $x = 0.100$ m, $v = -13.6$ m/s, and $a = -123$ m/s². Calculate (a) the frequency of oscillation, (b) the mass of the block, and (c) the amplitude of the motion.

29P. A simple harmonic oscillator consists of a block of mass 2.00 kg attached to a spring of spring constant 100 N/m. When $t = 1.00$ s, the position and velocity of the block are $x = 0.129$ m and $v = 3.415$ m/s. (a) What is the amplitude of the oscillations? What were the (b) position and (c) velocity of the mass at $t = 0$ s?

30P. A massless spring hangs from the ceiling with a small object attached to its lower end. The object is initially held at rest in a position y_i such that the spring is at its rest length. The object is then released from y_i and oscillates up and down, with its lowest position being 10 cm below y_i. (a) What is the frequency of the oscillation? (b) What is the speed of the object when it is 8.0 cm below the initial position? (c) An object of mass 300 g is attached to the first object, after which the system oscillates with half the original frequency. What is the mass of the first object? (d) Relative to y_i, where is the new equilibrium (rest) position with both objects attached to the spring?

31P. Two particles oscillate in simple harmonic motion along a common straight line segment of length A. Each particle has a period of 1.5 s, but they differ in phase by $\pi/6$ rad. (a) How far apart are they (in terms of A) 0.50 s after the lagging particle leaves one end of the path? (b) Are they then moving in the same direction, toward each other, or away from each other?

32P. Two particles execute simple harmonic motion of the same amplitude and frequency along the same straight line. They pass each other moving in opposite directions each time their displacement is half their amplitude. What is the phase difference between them?

33P. Two identical springs are attached to a block of mass m and to fixed supports as shown in Fig. 16-31. Show that the frequency of oscillation on the frictionless surface is

$$f = \frac{1}{2\pi}\sqrt{\frac{2k}{m}}.$$

FIGURE 16-31 Problems 33 and 34.

34P. Suppose that the two springs in Fig. 16-31 have different spring constants k_1 and k_2. Show that the frequency f of oscillation of the block is then given by

$$f = \sqrt{f_1^2 + f_2^2},$$

where f_1 and f_2 are the frequencies at which the block would oscillate if connected only to spring 1 or only to spring 2.

35P. Two springs are joined and connected to a mass m as shown in Fig. 16-32. The surface is frictionless. If the springs both have force constant k, show that the frequency of oscillation of m is

$$f = \frac{1}{2\pi}\sqrt{\frac{k}{2m}}.$$

FIGURE 16-32 Problem 35.

36P. A block weighing 14.0 N, which slides without friction on a 40.0° incline, is connected to the top of the incline by a massless spring of unstretched length 0.450 m and spring constant 120 N/m, as shown in Fig. 16-33. (a) How far from the top of the incline does the block stop? (b) If the block is pulled slightly down the incline and released, what is the period of the ensuing oscillations?

Frictionless

40.0°

FIGURE 16-33 Problem 36.

37P. A uniform spring whose unstretched length is L has a spring constant k. The spring is cut into two pieces of unstretched lengths L_1 and L_2, with $L_1 = nL_2$. (a) What are the corresponding spring constants k_1 and k_2 in terms of n and k? (b) If a block is attached to the original spring, as in Fig. 16-5, it oscillates with frequency f. If the spring is replaced with the piece L_1 or L_2, the corresponding frequency is f_1 or f_2. Find f_1 and f_2 in terms of f.

38P. Three 10,000 kg ore cars are held at rest on a 30° incline on a mine railway using a cable that is parallel to the incline (Fig. 16-34). The cable is observed to stretch 15 cm just before the coupling between the lower cars breaks, detaching the lowest car. Assuming that the cable obeys Hooke's law, find (a) the frequency and (b) the amplitude of the resulting oscillations of the remaining two cars.

Car that breaks free

30°

FIGURE 16-34 Problem 38.

39P. To alleviate the traffic congestion between two cities such as Boston and Washington, DC, engineers have proposed building a rail tunnel along a chord line connecting the cities (Fig. 16-35). A train, unpropelled by any engine and starting from rest, would fall through the first half of the tunnel and then move up

FIGURE 16-35 Problem 39.

FIGURE 16-36 Exercise 46.

the second half. Assuming Earth to be a uniform sphere and ig-noring air drag and friction, (a) show that the travel between cities is half a cycle of simple harmonic motion and (b) find the travel time between cities.

SECTION 16-4 Energy in Simple Harmonic Motion

40E. Find the mechanical energy of a block–spring system having a spring constant of 1.3 N/cm and an amplitude of 2.4 cm.

41E. An oscillating block–spring system has a mechanical energy of 1.00 J, an amplitude of 10.0 cm, and a maximum speed of 1.20 m/s. Find (a) the spring constant, (b) the mass of the block and (c) the frequency of oscillation.

42E. A 5.00 kg object on a horizontal frictionless surface is at-tached to a spring with spring constant 1000 N/m. The object is displaced from equilibrium 50.0 cm horizontally and given an initial velocity of 10.0 m/s back toward the equilibrium position. (a) What is the frequency of the motion? What are (b) the initial potential energy of the block–spring system, (c) the initial kinetic energy, and (d) the amplitude of the oscillation?

43E. A vertical spring stretches 9.6 cm when a 1.3 kg block is hung from its end. (a) Calculate the spring constant. This block is then displaced an additional 5.0 cm downward and released from rest. Find (b) the period, (c) the frequency, (d) the amplitude, and (e) the maximum speed of the resulting SHM.

44E. A (hypothetical) large slingshot is stretched 1.50 m to launch a 130 g projectile with speed sufficient to escape from Earth (11.2 km/s). Assume the elastic slingshot obeys Hooke's law. (a) What is the spring constant of the device, if all the elastic potential energy is converted to kinetic energy? (b) Assume that an average person can exert a force of 220 N. How many people would be required to stretch the slingshot?

45E. When the displacement in SHM is one-half the amplitude x_m, what fraction of the total energy is (a) kinetic energy and (b) potential energy? (c) At what displacement, in terms of the am-plitude, is the energy of the system half kinetic energy and half potential energy?

46E. A block of mass M, at rest on a horizontal frictionless table, is attached to a rigid support by a spring of constant k. A bullet of mass m and velocity v strikes the block as shown in Fig. 16-36. The bullet remains embedded in the block. Determine (a) the velocity of the block immediately after the collision and (b) the amplitude of the resulting simple harmonic motion.

47P. A 3.0 kg particle is in simple harmonic motion in one di-mension and moves according to the equation

$$x = (5.0 \text{ m})\cos[(\pi/3 \text{ rad/s})t - \pi/4 \text{ rad}].$$

(a) At what value of x is the potential energy of the particle equal to half the total energy? (b) How long does it take the particle to move to this position x from the equilibrium position?

48P. A 10 g particle is undergoing simple harmonic motion with an amplitude of 2.0×10^{-3} m. The maximum acceleration expe-rienced by the particle is 8.0×10^3 m/s²; the phase constant is $-\pi/3$ rad. (a) Write an equation for the force on the particle as a function of time. (b) What is the period of the motion? (c) What is the maximum speed of the particle? (d) What is the total mechan-ical energy of this simple harmonic oscillator?

49P. A massless spring with spring constant 19 N/m hangs ver-tically. A body of mass 0.20 kg is attached to its free end and then released. Assume that the spring was unstretched before the body was released. Find (a) how far below the initial position the body descends, and (b) the frequency and (c) the amplitude of the re-sulting motion, assumed to be simple harmonic.

50P. A 4.0 kg block is suspended from a spring with a spring constant of 500 N/m. A 50 g bullet is fired into the block from directly below with a speed of 150 m/s and is embedded in the block. (a) Find the amplitude of the resulting simple harmonic motion. (b) What fraction of the original kinetic energy of the bullet appears as mechanical energy in the harmonic oscillator?

51P*. A solid cylinder is attached to a horizontal massless spring so that it can roll without slipping along a horizontal sur-face (Fig. 16-37). The spring constant k is 3.0 N/m. If the system is released from rest at a position in which the spring is stretched by 0.25 m, find (a) the translational kinetic energy and (b) the rotational kinetic energy of the cylinder as it passes through the equilibrium position. (c) Show that under these conditions the center of mass of the cylinder executes simple harmonic motion with period

$$T = 2\pi\sqrt{\frac{3M}{2k}},$$

where M is the mass of the cylinder. (*Hint:* Find the time deriva-tive of the total mechanical energy.)

FIGURE 16-37 Problem 51.

SECTION 16-5 An Angular Simple Harmonic Oscillator

52E. A flat uniform circular disk has a mass of 3.00 kg and a radius of 70.0 cm. It is suspended in a horizontal plane by a vertical wire attached to its center. If the disk is rotated 2.50 rad about the wire, a torque of 0.0600 N·m is required to maintain the disk in this position. Calculate (a) the rotational inertia of the disk about the wire, (b) the torsion constant, and (c) the angular frequency of this torsion pendulum when it is set oscillating.

53P. A 95 kg solid sphere with a 15 cm radius is suspended by a vertical wire attached to the ceiling of a room. A torque of 0.20 N·m is required to twist the sphere through an angle of 0.85 rad. What is the period of oscillation when the sphere is released from this position?

54P. An engineer has an odd-shaped object of mass 10 kg and needs to find its rotational inertia about an axis through its center of mass. The object is supported with a wire along the desired axis. The wire has a torsion constant $\kappa = 0.50$ N·m. If this torsion pendulum oscillates through 20 complete cycles in 50 s, what is the rotational inertia of the odd-shaped object?

55P. The balance wheel of a watch oscillates with an angular amplitude of π rad and a period of 0.500 s. Find (a) the maximum angular speed of the wheel, (b) the angular speed of the wheel when its displacement is $\pi/2$ rad, and (c) the angular acceleration of the wheel when its displacement is $\pi/4$ rad.

SECTION 16-6 Pendulums

56E. What is the length of a simple pendulum whose period is 1.00 s at a point where $g = 32.2$ ft/s^2?

57E. A 2500 kg demolition ball swings from the end of a crane, as shown in Fig. 16-38. The length of the swinging segment of cable is 17 m. (a) Find the period of swing, assuming that the system can be treated as a simple pendulum. (b) Does the period depend on the ball's mass?

FIGURE 16-38 Exercise 57.

58E. What is the length of a simple pendulum that marks seconds by completing a full swing from left to right and then back again every 2.0 s?

59E. If a simple pendulum with length 1.50 m makes 72.0 oscillations in 180 s, what is the acceleration of gravity at its location?

60E. Two oscillating systems that you have studied are the block–spring and the simple pendulum. You can prove an interesting relation between them. Suppose that you hang a weight on the end of a spring, and when the weight is at rest, the spring is stretched a distance h as in Fig. 16-39. Show that the frequency of this block–spring system is the same as that of a simple pendulum whose length is h.

FIGURE 16-39 Exercise 60.

61E. A performer, seated on a trapeze, is swinging back and forth with a period of 8.85 s. If she stands up, thus raising the center of mass of the system *trapeze + performer* by 35.0 cm, what will be the new period of the trapeze? Treat *trapeze + performer* as a simple pendulum.

62E. A simple pendulum with length L is swinging freely with small angular amplitude. As the pendulum passes its central (or equilibrium) position, its cord is suddenly and rigidly clamped at its midpoint. In terms of the original period T of the pendulum, what will the new period be?

63E. A physical pendulum consists of a meter stick that is pivoted at a small hole drilled through the stick a distance x from the 50 cm mark. The period of oscillation is observed to be 2.5 s. Find the distance x.

64E. A pendulum is formed by pivoting a long thin rod of length L and mass m about a point on the rod that is a distance d above the center of the rod. (a) Find the period of this pendulum in terms of d, L, m, and g, assuming that it swings with small amplitude. What happens to the period if (b) d is decreased, (c) L is increased, or (d) m is increased?

65E. A physical pendulum consists of a uniform solid disk (of mass M and radius R) supported in a vertical plane by a pivot located a distance d from the center of the disk (Fig. 16-40). The disk is displaced by a small angle and released. Find an expression for the period of the resulting simple harmonic motion.

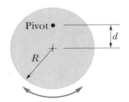

FIGURE 16-40 Exercise 65.

66E. A uniform circular disk whose radius R is 12.5 cm is suspended, as a physical pendulum, from a point on its rim. (a) What is its period of oscillation? (b) At what radial distance $r < R$ is there a point of suspension that gives the same period?

67E. A pendulum consists of a uniform disk with radius 10.0 cm and mass 500 g attached to a uniform rod with length 500 mm and mass 270 g (Fig. 16-41). (a) Calculate the rotational inertia of the pendulum about the pivot. (b) What is the distance between the pivot and the center of mass of the pendulum? (c) Calculate the period of oscillation.

FIGURE 16-41
Exercise 67.

500 mm

10.0 cm

68E. (a) If the physical pendulum of Sample Problem 16-6 is inverted and suspended at point P, what is its period of oscillation? (b) Is the period now greater than, less than, or equal to its previous value?

69E. In Sample Problem 16-6, we saw that a physical pendulum has a center of oscillation at distance $2L/3$ from its point of suspension. Show that the distance between the point of suspension and the center of oscillation for a physical pendulum of any form is I/mh, where I and h have the meanings assigned to them in Eq. 16-32, and m is the mass of the pendulum.

70E. A meter stick swinging from one end oscillates with a frequency f_0. What would be the frequency, in terms of f_0, if the bottom half of the stick were cut off?

71P. A stick with length L oscillates as a physical pendulum, pivoted about point O in Fig. 16-42. (a) Derive an expression for the period of the pendulum in terms of L and x, the distance from the point of support to the center of mass of the pendulum. (b) For what value of x/L is the period a minimum? (c) Show that if $L = 1.00$ m and $g = 9.80$ m/s², this minimum is 1.53 s.

FIGURE 16-42 Problem 71.

72P. The center of oscillation of a physical pendulum has this interesting property: if an impulsive force (assumed horizontal and in the plane of oscillation) acts at the center of oscillation, no reaction is felt at the point of support. Baseball players (and players of many other sports) know that unless the ball hits the bat at this point (called the "sweet spot" by athletes), the reaction due to the impact will sting their hands. To prove this property, let the stick in Fig. 16-12a simulate a baseball bat. Suppose that a horizontal force **F** (due to impact with the ball) acts toward the right at P, the center of oscillation. The batter is assumed to hold the bat at O, the point of support of the stick. (a) What acceleration does point O undergo as a result of **F**? (b) What angular acceleration is produced by **F** about the center of mass of the stick? (c) As a result of the angular acceleration in (b), what linear acceleration does point O undergo? (d) Considering the magnitudes and directions of the accelerations in (a) and (c), convince yourself that P is indeed the "sweet spot."

73P. The fact that g varies from place to place over Earth's surface drew attention when Jean Richer in 1672 took a pendulum clock from Paris to Cayenne, French Guiana, and found that it lost 2.5 min/day. If $g = 9.81$ m/s² in Paris, what is g in Cayenne?

74P. A scientist is making a precise measurement of g at a certain point in the Indian Ocean (on the equator) by timing the swings of a pendulum of accurately known construction. To provide a stable base, the measurements are conducted in a submerged submarine. It is observed that a slightly different result for g is obtained when the submarine is moving eastward than when it is moving westward, the speed in each case being 16 km/h. Account for this difference and calculate the fractional error $\Delta g/g$ in g for either travel direction.

75P. A long uniform rod of length L and mass m is free to rotate in a horizontal plane about a vertical axis through its center. A spring with force constant k is connected horizontally between one end of the rod and a fixed wall, as shown in the overhead view of Fig. 16-43. When the rod is in equilibrium, it is parallel to the wall. What is the period of the small oscillations that result when the rod is rotated slightly and released?

Wall

k

Rotation axis

FIGURE 16-43 Problem 75.

76P. What is the frequency of a simple pendulum 2.0 m long (a) in a room, (b) in an elevator accelerating upward at a rate of 2.0 m/s², and (c) in free fall?

77P. A simple pendulum of length L and mass m is suspended in a car that is traveling with constant speed v around a circle of radius R. If the pendulum undergoes small oscillations in a radial direction about its equilibrium position, what will its frequency of oscillation be?

78P. For a simple pendulum, find the angular amplitude θ_m at which the restoring torque required for simple harmonic motion deviates from the actual restoring torque by 1.0%. (See "Trigonometric Expansions" in Appendix E.)

79P. The bob on a simple pendulum of length R moves in an arc of a circle. (a) By considering that the acceleration of the bob as it moves through its equilibrium position is that for uniform circular motion (mv^2/R), show that the tension in the string at that position is $mg(1 + \theta_m^2)$ if the angular amplitude θ_m is small. (See "Trigonometric Expansions" in Appendix E.) (b) Is the tension at other positions of the bob larger, smaller, or the same?

80P. A wheel is free to rotate about its fixed axle. A spring is attached to one of its spokes a distance r from the axle, as shown in Fig. 16-44. (a) Assuming that the wheel is a hoop of mass m and radius R, obtain the angular frequency of small oscillations of this system in terms of m, R, r, and the spring constant k. How does the result change if (b) $r = R$ and (c) $r = 0$?

FIGURE 16-44 Problem 80.

81P. A 2.5 kg disk, 42 cm in diameter, is supported by a massless rod, 76 cm long, which is pivoted at its end, as in Fig. 16-45. (a) The massless torsion spring is initially not connected. What then is the period of oscillation? (b) The torsion spring is now connected so that, in equilibrium, the rod hangs vertically. What should be the torsional constant of the spring so that the new period of oscillation is 0.50 s shorter than before?

FIGURE 16-45 Problem 81.

82P. A physical pendulum has two possible pivot points A and B; point A has a fixed position, and B is adjustable along the length of the pendulum, as shown in Fig. 16-46. The period of the pendulum when suspended from A is found to be T. The pendulum is then reversed and suspended from B, which is moved until the pendulum again has period T. Show that the free-fall acceleration g is given by

$$g = \frac{4\pi^2 L}{T^2},$$

in which L is the distance between A and B for equal periods T. (Note that g can be measured in this way without knowing the rotational inertia of the pendulum or any of its dimensions except L.)

FIGURE 16-46 Problem 82.

83P*. A uniform rod with length L swings from a pivot as a physical pendulum. How far, in terms of L, should the pivot be from the center of mass to minimize the period?

SECTION 16-8 Damped Simple Harmonic Motion

84E. The amplitude of a lightly damped oscillator decreases by 3.0% during each cycle. What fraction of the energy of the oscillator is lost in each full oscillation?

85E. In Sample Problem 16-9, what is the ratio of the amplitude of the damped oscillations to the initial amplitude when 20 full oscillations have elapsed?

86E. For the system shown in Fig. 16-17, the block has a mass of 1.50 kg and the spring constant is 8.00 N/m. The damping force is given by $-b(dx/dt)$, where $b = 230$ g/s. Suppose that the block is initially pulled down a distance 12.0 cm and released. (a) Calculate the time required for the amplitude to fall to one-third of its initial value. (b) How many oscillations are made by the block in this time?

87P. A damped harmonic oscillator consists of a block ($m = 2.00$ kg), a spring ($k = 10.0$ N/m), and a damping force $F = -bv$. Initially, it oscillates with an amplitude of 25.0 cm; because of the damping, the amplitude falls to three-fourths of this initial value at the completion of four oscillations. (a) What is the value of b? (b) How much energy has been "lost" during these four oscillations?

88P. (a) In Eq. 16-39, find the ratio of the maximum damping force ($-b\, dx/dt$) to the maximum spring force ($-kx$) during the first oscillation for the data of Sample Problem 16-9. (b) Does this ratio change appreciably during later oscillations?

89P. Assume that you are examining the characteristics of the suspension system of a 2000 kg automobile. The suspension "sags" 10 cm when the weight of the entire automobile is placed on it. In addition, the amplitude of oscillation decreases by 50% during one complete oscillation. Estimate the values of k and b for the spring and shock absorber system of one wheel, assuming each wheel supports 500 kg.

SECTION 16-9 Forced Oscillations and Resonance

90E. For Eq. 16-43, suppose the amplitude x_m is given by

$$x_m = \frac{F_m}{[m^2(\omega_d^2 - \omega^2)^2 + b^2\omega_d^2]^{1/2}},$$

where F_m is the (constant) amplitude of the external oscillating force exerted on the spring by the rigid support in Fig. 16-17. At resonance, what are (a) the amplitude and (b) the velocity amplitude of the oscillating object?

91P. A 2200 lb car carrying four 180 lb people travels over a rough "washboard" dirt road with corrugations 13 ft apart which causes the car to bounce on its spring suspension.. The car bounces with maximum amplitude when its speed is 10 mi/h. The car now stops, and the four people get out. By how much does the car body rise on its suspension owing to this decrease in weight?

Electronic Computation

92. You see two carts connected by a spring and oscillating on a horizontal air track. You happen to know that the spring constant of the spring is $k = 50.0$ N/m. You use Newton's second law to determine that the period of oscillation is the same for each cart and is related to the masses of the carts by

$$T = 2\pi \sqrt{\frac{m_1 m_2}{k(m_1 + m_2)}}.$$

Using a position detector, you determine that the positions of the carts as functions of time can be expressed as $x_1(t) = 2.70[1 - \cos(18.0t)]$ and $x_2(t) = 10.70 + 1.29 \cos(18.0t)$, where the coordinates are in centimeters and the time is in seconds. (a) Use conservation of momentum to find the masses of the carts. (b) Generate a table of the values of x_1 and x_2 for every 0.01 s from $t = 0$ to $t = 35$ s. For each of these times, also calculate the position of the center of mass, the total momentum of the two-cart system, and the force of the spring on each cart. Verify that the center of mass does not move, that the total momentum is conserved, and that the forces on the carts have the same magnitude but opposite directions. (c) Use the table to find the equilibrium length of the spring.

93. A 2.0 kg block is attachd to the end of a spring with a spring constant of 350 N/m and forced to oscillate by an applied force $F = (15$ N$) \sin(\omega t)$, where $\omega = 35$ rad/s. The damping constant is $b = 15$ kg/s. At $t = 0$, the block is at rest with the spring at its rest length. (a) Use numerical integration to plot the displacement of the block for the first 1.0 s of its motion. Use the motion near the end of the 1.0 s interval to estimate the amplitude, period, and angular frequency. Repeat the calculation for (b) $\omega = \sqrt{k/m}$ and (c) $\omega = 20$ rad/s.

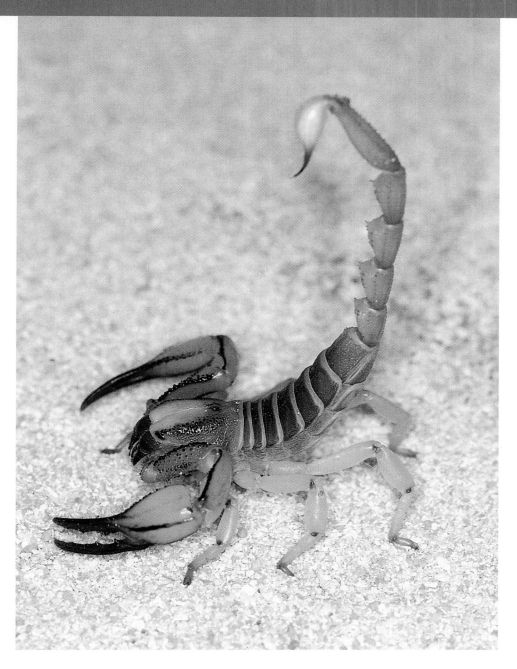

When a beetle moves along the sand within a few tens of centimeters of this sand scorpion, the scorpion immediately turns toward the beetle and dashes to it (for lunch). The scorpion can do this without seeing (it is nocturnal) or hearing the beetle. How can it so precisely locate its prey?

17-1 WAVES AND PARTICLES

Two ways to get in touch with a friend in a distant city are to write a letter and to use the telephone.

The first choice (the letter) involves the concept of "particle": a material object moves from one point to another, carrying with it information and energy. Most of the preceding chapters deal with particles or with systems of particles.

The second choice (the telephone) involves the concept of "wave," the subject of this chapter and the next. In a wave, information and energy move from one point to another but no material object makes that journey. In your telephone call, a sound wave carries your message from your vocal cords to the telephone. There, an electromagnetic wave takes over, passing along a copper wire or an optical fiber or through the atmosphere, possibly by way of a communications satellite. At the receiving end there is another sound wave, from a telephone to your friend's ear. Although the message is passed, nothing that you have touched reaches your friend. Leonardo da Vinci understood about waves when he wrote of water waves: "it often happens that the wave flees the place of its creation, while the water does not; like the waves made in a field of grain by the wind, where we see the waves running across the field while the grain remains in place."

Particle and *wave* are the two great concepts in classical physics, in the sense that we seem able to associate almost every branch of the subject with one or the other. The two concepts are quite different. The word *particle* suggests a tiny concentration of matter capable of transmitting energy. The word *wave* suggests just the opposite, namely, a broad distribution of energy, filling the space through which it passes. The job at hand is to put aside particles for a while and to learn something about waves.

17-2 TYPES OF WAVES

Waves are of three main types:

1. *Mechanical waves.* These waves are most familiar because we encounter them almost constantly; common examples include water waves, sound waves, and seismic waves. All these waves have certain central features: they are governed by Newton's laws, and they can exist only within a material medium, such as water, air, and rock.

2. *Electromagnetic waves.* These waves are less familiar, but you use them constantly; common examples include visible and ultraviolet light, radio and television waves, microwaves, x rays, and radar waves. These waves require no material medium to exist. Light waves from stars, for example, travel through the vacuum of space to reach us.

All electromagnetic waves travel through a vacuum at the same speed c, given by

$$c = 299,792,458 \text{ m/s} \quad \text{(speed of light).} \quad (17\text{-}1)$$

3. *Matter waves.* Although these waves are commonly used in modern technology, their type is probably very unfamiliar to you. Electrons, protons, and other fundamental particles, and even atoms and molecules, travel as waves. Because we commonly think of these things as constituting matter, these waves are called matter waves.

Much of what we discuss in this chapter applies to waves of all kinds. However, for specific examples we shall refer to mechanical waves.

17-3 TRANSVERSE AND LONGITUDINAL WAVES

A wave sent along a stretched, taut string is the simplest mechanical wave. If you give one end of a stretched string a single up-and-down jerk, a wave in the form of a single *pulse* travels along the string, as in Fig. 17-1a. This pulse and its motion can occur because the string is under ten-

(a)

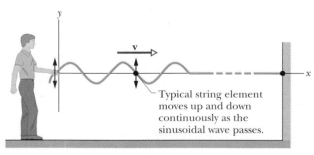

(b)

FIGURE 17-1 (a) Sending a single pulse down a long stretched string. (b) Sending a continuous sinusoidal wave down the string. Because the oscillations of any string element (represented here by a dot) are perpendicular to the direction in which the wave moves, the wave is a *transverse wave*.

sion. When you pull your end of the string upward, it begins to pull upward on the adjacent section of the string via tension between the two sections. As the adjacent section moves upward, it begins to pull the next section upward, and so on. Meanwhile, you have pulled down on your end of the string. So, as each section moves upward in turn, it begins to be pulled back downward by neighboring sections that are already on the way down. The net result is that a distortion in the string's shape (the pulse) moves along the string at some velocity **v**.

If you move your hand up and down in continuous simple harmonic motion, a continuous wave travels along the string at velocity **v**. Because the motion of your hand is a sinusoidal function of time, the wave has a sinusoidal shape at any given instant, as in Fig. 17-1b. That is, the wave has the shape of a sine curve or a cosine curve.

We consider here only an "ideal" string, in which no frictionlike forces within the string cause the wave to die out as it travels along the string. In addition, we assume that the string is so long that we need not consider a wave rebounding from the far end.

One way to study the waves of Fig. 17-1 is to monitor the **wave form** (shape of the wave) as it moves to the right. Alternatively, we can monitor the motion of an element of the string as the element oscillates up and down while the wave passes through it. We would find that the displacement of every such oscillating string element is *perpendicular* to the direction of travel of the wave, as indicated in Fig. 17-1. This motion is said to be **transverse,** and the wave is said to be a **transverse wave.**

Figure 17-2 shows how a sound wave can be produced by a piston in a long, air-filled pipe. If you suddenly move the piston rightward and then leftward, you can send a pulse of sound along the pipe. The rightward motion of the piston moves the elements of air next to it rightward, changing the air pressure there. The increased air pressure then pushes rightward on the elements of air somewhat

farther along the pipe. Once they have moved rightward, the elements move back leftward. Thus the motion of the air and the change in air pressure travel rightward along the pipe as a pulse.

If you push and pull on the piston in simple harmonic motion, as is being done in Fig. 17-2, a sinusoidal wave travels along the pipe. Because the motion of the elements of air is parallel to the direction of the wave's travel, the motion is said to be **longitudinal,** and the wave is said to be a **longitudinal wave.** In this chapter we concentrate on transverse waves, and string waves in particular; in Chapter 18 we shall concentrate on longitudinal waves, and sound waves in particular.

Both a transverse wave and a longitudinal wave are said to be **traveling waves** because the wave travels from one point to another, as from one end of the string to the other end in Fig. 17-1 or from one end of the pipe to the other end in Fig. 17-2. Note that it is the wave that moves between the two points and not the material (string or air) through which the wave moves.

The sand scorpion shown in the photograph opening this chapter uses waves of both transverse and longitudinal motion to locate its prey. When a beetle even slightly disturbs the sand, it sends pulses along the sand's surface (Fig. 17-3). One set of pulses is longitudinal, traveling with speed $v_l = 150$ m/s. A second set is transverse, traveling with speed $v_t = 50$ m/s.

The scorpion, with its eight legs spread roughly in a circle about 5 cm in diameter, intercepts the faster longitudinal pulses first and learns the direction of the beetle: it is in the direction of whichever leg is disturbed earliest by the pulses. The scorpion then senses the time interval Δt between that first interception and the interception of the slower transverse waves and uses it to determine the distance d to the beetle. This distance is given by

$$\Delta t = \frac{d}{v_t} - \frac{d}{v_l},$$

FIGURE 17-2 A sound wave is set up in an air-filled pipe by moving a piston back and forth. Because the oscillations of an element of the air (represented by the black dot) are parallel to the direction in which the wave travels, the wave is a *longitudinal wave.*

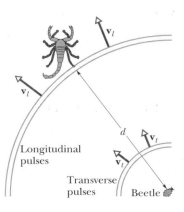

FIGURE 17-3 A beetle's motion sends fast longitudinal pulses and slower transverse pulses along the sand's surface. The sand scorpion first intercepts the longitudinal pulses; here, it is the rear-most right leg that senses the pulses earliest.

and it turns out to be

$$d = (75 \text{ m/s}) \, \Delta t.$$

For example, if $\Delta t = 4.0$ ms, then $d = 30$ cm, which gives the scorpion a perfect fix on the beetle.

17-4 WAVELENGTH AND FREQUENCY

To completely describe a wave on a string (and the motion of any element along its length), we need a function that gives the shape of the wave. This means that we need a relation in the form $y = h(x, t)$, in which y is the transverse displacement of any string element as a function h of the time t and the position x of the element along the string. In general, a sinusoidal shape like the wave in Fig. 17-1b can be described with h being either a sine function or a cosine function; both give the same general shape for the wave. In this chapter we use the sine function.

For a sinusoidal wave like that in Fig. 17-1b, traveling toward increasing values of x, the transverse displacement y of a string element at position x at time t is given by

$$y(x, t) = y_m \sin(kx - \omega t), \qquad (17\text{-}2)$$

(a)

(b)

FIGURE 17-4 (a) A snapshot of a string at $t = 0$ as the sinusoidal wave of Eq. 17-2 travels along it. A typical wavelength λ is shown: it is the horizontal distance between two successive points on the string (represented by dots) with the same displacement. The amplitude y_m is also shown: it is the maximum displacement of the string. (b) A graph showing the displacement of a string element at $x = 0$ as a function of time as the sinusoidal wave passes through the element. A typical period T is shown: it is the time between identical displacements of the string element (represented by dots).

Here y_m is the **amplitude** of the wave; the subscript m stands for maximum, because the amplitude is the magnitude of the maximum displacement of the string element in either direction parallel to the y axis. (Hence, y_m is always a positive quantity.) The quantities k and ω are constants whose meanings we are about to discuss. The quantity $kx - \omega t$ is called the **phase** of the wave.

You may wonder why, of the infinite variety of wave forms that are available, we pick the sinusoidal wave of Eq. 17-2 for detailed study. This is in fact a wise choice because, as you will see in Section 17-8, *all* wave forms—including the pulse of Fig. 17-1a—can be constructed by adding sinusoidal waves of carefully selected amplitudes and values of k. Understanding sinusoidal waves is thus the key to understanding waves of any shape.

Because Eq. 17-2 has two independent variables (x and t), we cannot represent the dependent variable y on a single two-dimensional graph. We would need a videotape to show it fully and in real time. We can, however, learn a lot from plots like those of Fig. 17-4.

Wavelength and Angular Wave Number

Figure 17-4a shows how the transverse displacement y of Eq. 17-2 varies with position x at an instant, arbitrarily called $t = 0$. That is, the figure is a "snapshot" of the wave at that instant. With $t = 0$, Eq. 17-2 becomes

$$y(x, 0) = y_m \sin kx \qquad (t = 0). \qquad (17\text{-}3)$$

Figure 17-4a is a plot of this equation; it shows the shape of the actual wave at time $t = 0$.

The **wavelength** λ of a wave is the distance (along the direction of the wave) between repetitions of the wave shape. A typical wavelength is marked in Fig. 17-4a. By definition, the displacement y is the same at both ends of this wavelength, that is, at $x = x_1$ and $x = x_1 + \lambda$. Thus, by Eq. 17-3,

$$y_m \sin kx_1 = y_m \sin k(x_1 + \lambda)$$
$$= y_m \sin(kx_1 + k\lambda). \qquad (17\text{-}4)$$

A sine function begins to repeat itself when its angle (or argument) is increased by 2π rad; so Eq. 17-4 will be accurate only if $k\lambda = 2\pi$, or if

$$k = \frac{2\pi}{\lambda} \qquad \text{(angular wave number).} \qquad (17\text{-}5)$$

We call k the **angular wave number** of the wave; its SI unit is the radian per meter. (Note that the symbol k here does *not* represent a spring constant as previously.)

Period, Angular Frequency, and Frequency

Figure 17-4b shows how the displacement y of Eq. 17-2 varies with time t at a fixed position, taken to be $x = 0$. If you were to monitor the string, you would see that the single element of the string at that position moves up and down in simple harmonic motion given by Eq. 17-2 with $x = 0$:

$$y(0, t) = y_m \sin(-\omega t)$$
$$= -y_m \sin \omega t \quad (x = 0). \quad (17\text{-}6)$$

Here we have made use of the fact that $\sin(-\alpha) = -\sin \alpha$, where α is any angle. Figure 17-4b is a plot of this equation; it *does not* show the shape of the wave.

We define the **period** of oscillation T of a wave to be the time interval between repetitions of the motion of an oscillating string element (at any position x). A typical period is marked in Fig. 17-4b. Applying Eq. 17-6 to both ends of this time interval and equating the results yield

$$-y_m \sin \omega t_1 = -y_m \sin \omega(t_1 + T)$$
$$= -y_m \sin(\omega t_1 + \omega T). \quad (17\text{-}7)$$

This can be true only if $\omega T = 2\pi$, or if

$$\omega = \frac{2\pi}{T} \quad \text{(angular frequency).} \quad (17\text{-}8)$$

We call ω the **angular frequency** of the wave; its SI unit is the radian per second.

The **frequency** f of the wave is defined as $1/T$ and is related to the angular frequency ω by

$$f = \frac{1}{T} = \frac{\omega}{2\pi} \quad \text{(frequency).} \quad (17\text{-}9)$$

Like the frequency of simple harmonic motion in Chapter 16, this frequency f is a number of oscillations per unit time—here, made by a string element as the wave moves through it. As in Chapter 16, f is usually measured in hertz or its multiples.

CHECKPOINT 1: The figure shows a snapshot of three waves traveling along a string. The phases for the waves are given by (a) $2x - 4t$, (b) $4x - 8t$, and (c) $8x - 16t$. Which phase corresponds to which wave in the figure?

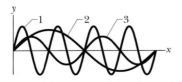

17-5 THE SPEED OF A TRAVELING WAVE

Figure 17-5 shows two snapshots of the wave of Eq. 17-2, taken a small time interval Δt apart. The wave is traveling in the direction of increasing x (to the right in Fig. 17-5), the entire wave pattern moving a distance Δx in that direction during the interval Δt. The ratio $\Delta x/\Delta t$ (or, in the differential limit, dx/dt) is the **wave speed** v. How can we find its value?

As the wave in Fig. 17-5 moves, each point of the moving wave form (such as point A) retains its displacement y. (Points on the string do not retain their displacement, but points on the wave *form* do.) For each such point, Eq. 17-2 tells us that the argument of the sine function must be a constant:

$$kx - \omega t = \text{a constant.} \quad (17\text{-}10)$$

Note that although this argument is constant, both x and t are changing. In fact, as t increases, x must also, to keep the argument constant. This confirms that the wave pattern is moving toward increasing x.

To find the wave speed v, we take the derivative of Eq. 17-10, getting

$$k\frac{dx}{dt} - \omega = 0$$

or

$$\frac{dx}{dt} = v = \frac{\omega}{k}. \quad (17\text{-}11)$$

Using Eq. 17-5 ($k = 2\pi/\lambda$) and Eq. 17-8 ($\omega = 2\pi/T$), we can rewrite the wave speed as

$$v = \frac{\omega}{k} = \frac{\lambda}{T} = \lambda f \quad \text{(wave speed).} \quad (17\text{-}12)$$

The equation $v = \lambda/T$ tells us that the wave speed is one wavelength per period: the wave moves a distance of one wavelength in one period of oscillation.

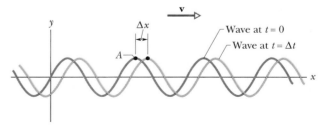

FIGURE 17-5   A snapshot of the traveling wave of Eq. 17-2 at $t = 0$ and at a later time $t = \Delta t$. During the time interval Δt, the entire curve shifts a distance Δx to the right.

Equation 17-2 describes a wave moving in the direction of increasing x. We can find the equation of a wave traveling in the opposite direction by replacing t in Eq. 17-2 with $-t$. This corresponds to the condition

$$kx + \omega t = \text{a constant}, \qquad (17\text{-}13)$$

which (compare Eq. 17-10) requires that x *decrease* with time. Thus, a wave traveling toward decreasing x is described by the equation

$$y(x, t) = y_m \sin(kx + \omega t). \qquad (17\text{-}14)$$

If you analyze the wave of Eq. 17-14 as we have just done for the wave of Eq. 17-2, you will find for its velocity

$$\frac{dx}{dt} = -\frac{\omega}{k}. \qquad (17\text{-}15)$$

The minus sign (compare Eq. 17-11) verifies that the wave is indeed moving in the direction of decreasing x and justifies our switching the sign of the time variable.

Consider now a wave of generalized shape, given by

$$y(x, t) = h(kx \pm \omega t), \qquad (17\text{-}16)$$

where h represents *any* function, the sine function being one possibility. Our analysis above shows that all waves in which the variables x and t enter in the combination $kx \pm \omega t$ are traveling waves. Furthermore, all traveling waves *must* be of the form of Eq. 17-16. Thus $y(x, t) = \sqrt{ax + bt}$ represents a possible (though perhaps physically a little bizarre) traveling wave. The function $y(x, t) = \sin(ax^2 - bt)$, on the other hand, does *not* represent a traveling wave.

SAMPLE PROBLEM 17-1

A sinusoidal wave traveling along a string is described by

$$y(x, t) = 0.00327 \sin(72.1x - 2.72t), \qquad (17\text{-}17)$$

in which the numerical constants are in SI units (0.00327 m, 72.1 rad/m, and 2.72 rad/s).

(a) What is the amplitude of this wave?

SOLUTION: By comparison with Eq. 17-2, we see that

$$y_m = 0.00327 \text{ m} = 3.27 \text{ mm}. \qquad \text{(Answer)}$$

(b) What are the wavelength, period, and frequency of this wave?

SOLUTION: By inspection of Eq. 17-17, we see that

$$k = 72.1 \text{ rad/m} \quad \text{and} \quad \omega = 2.72 \text{ rad/s}.$$

From Eq. 17-5 we have

$$\lambda = \frac{2\pi}{k} = \frac{2\pi \text{ rad}}{72.1 \text{ rad/m}} = 0.0871 \text{ m}$$
$$= 8.71 \text{ cm}. \qquad \text{(Answer)}$$

From Eq. 17-9 we have

$$T = \frac{2\pi}{\omega} = \frac{2\pi \text{ rad}}{2.72 \text{ rad/s}} = 2.31 \text{ s}. \qquad \text{(Answer)}$$

From Eq. 17-9 we have

$$f = \frac{1}{T} = \frac{1}{2.31 \text{ s}} = 0.433 \text{ Hz}. \qquad \text{(Answer)}$$

(c) What is the speed of this wave?

SOLUTION: From Eq. 17-12 we have

$$v = \frac{\omega}{k} = \frac{2.72 \text{ rad/s}}{72.1 \text{ rad/m}} = 0.0377 \text{ m/s}$$
$$= 3.77 \text{ cm/s}. \qquad \text{(Answer)}$$

Note that the quantities calculated in (b) and (c) are independent of the amplitude of the wave.

(d) What is the displacement y at $x = 22.5$ cm and $t = 18.9$ s?

SOLUTION: From Eq. 17-17 we obtain

$$y = 0.00327 \sin(72.1 \times 0.225 - 2.72 \times 18.9)$$
$$= (0.00327 \text{ m}) \sin(-35.1855 \text{ rad})$$
$$= (0.00327 \text{ m})(0.588)$$
$$= 0.00192 \text{ m} = 1.92 \text{ mm}. \qquad \text{(Answer)}$$

Thus the displacement is positive. (Be sure to change your calculator mode to radians before evaluating the sine.)

SAMPLE PROBLEM 17-2

In Sample Problem 17-1d, we showed that the transverse displacement y of an element of the string for the wave of Eq. 17-17 at $x = 0.225$ m and $t = 18.9$ s is 1.92 mm.

(a) What is u, the transverse speed of the same element of the string, at that place and at that time? (This speed, which is associated with the transverse oscillation of an element of the string, is in the y direction. Do not confuse it with v, the constant speed at which the *wave form* travels along the x axis.)

SOLUTION: The generalized equation of the wave of Eq. 17-17 is Eq. 17-2:

$$y(x, t) = y_m \sin(kx - \omega t). \qquad (17\text{-}18)$$

In this expression, let us hold x constant but allow t (for the time being) to be a variable. We then take the derivative of this equation with respect to t, obtaining*

$$u = \frac{\partial y}{\partial t} = -\omega y_m \cos(kx - \omega t). \qquad (17\text{-}19)$$

---

*A derivative taken while one (or more) of the variables is treated as a constant is called a *partial derivative* and is represented by the symbol $\partial/\partial x$, rather than d/dx.

Substituting numerical values from Sample Problem 17-1, we obtain

$$u = (-2.72 \text{ rad/s})(3.27 \text{ mm}) \cos(-35.1855 \text{ rad})$$
$$= 7.20 \text{ mm/s}. \qquad \text{(Answer)}$$

Thus, at $t = 18.9$ s, the element of string at $x = 22.5$ cm is moving in the direction of increasing y, with a speed of 7.20 mm/s.

(b) What is the transverse acceleration a_y at that position and at that time?

SOLUTION: From Eq. 17-19, treating x as a constant but allowing t (for the time being) to be a variable, we find

$$a_y = \frac{\partial u}{\partial t} = -\omega^2 y_m \sin(kx - \omega t).$$

Comparison with Eq. 17-18 shows that we can write this as

$$a_y = -\omega^2 y.$$

We see that the transverse acceleration of an oscillating string element is proportional to its transverse displacement but opposite in sign, completely consistent with the action of the element: namely, it is moving transversely in simple harmonic motion. Substituting numerical values yields

$$a_y = -(2.72 \text{ rad/s})^2 (1.92 \text{ mm})$$
$$= -14.2 \text{ mm/s}^2. \qquad \text{(Answer)}$$

Thus, at $t = 18.9$ s, the element of string at $x = 22.5$ cm is displaced from its equilibrium position by 1.92 mm in the increasing y direction and has an acceleration of magnitude 14.2 mm/s^2 in the decreasing y direction.

CHECKPOINT 2: Here are the equations of three waves:
(1) $y(x, t) = 2 \sin(4x - 2t)$,
(2) $y(x, t) = \sin(3x - 4t)$,
(3) $y(x, t) = 2 \sin(3x - 3t)$.
Rank the waves according to their (a) wave speed and (b) maximum transverse speed, greatest first.

PROBLEM SOLVING TACTICS

TACTIC 1: *Evaluating Large Phases*
Sometimes, as in Sample Problems 17-1d and 17-2, an angle much greater than 2π rad (or 360°) crops up and you are asked to find its sine or cosine. Adding or subtracting an integral multiple of 2π rad to such an angle does not change the value of any of its trigonometric functions. In Sample Problem 17-1d, for example, the angle is -35.1855 rad. Adding $(6)(2\pi$ rad) to this angle yields

$$-35.1855 \text{ rad} + (6)(2\pi \text{ rad}) = 2.51361 \text{ rad},$$

an angle less than 2π rad that has the same trigonometric functions as -35.1855 rad (Fig. 17-6). As an example, the sine of both 2.51361 rad and -35.1855 rad is 0.588.

Your calculator will reduce such large angles for you automatically. Caution: Do not round off large angles if you intend to take their sines or cosines. In taking the sine of a very large angle, you are throwing away most of the angle and taking the sine of what is left over. If, for example, you were to round -35.1855 rad to -35 rad (a change of 0.5% and normally a reasonable step), you would be changing the sine of the angle by 27%. Also, if you change a large angle from degrees to radians, be sure to use an exact conversion factor (such as $180° = \pi$ rad) rather than an approximate one (such as $57.3° \approx 1$ rad).

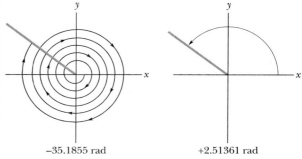

FIGURE 17-6 The two angles are different but all their trigonometric functions are identical.

17-6 WAVE SPEED ON A STRETCHED STRING

The speed of a wave is related to the wave's wavelength and frequency by Eq. 17-12, but *it is set by the medium*. If a wave is to travel through a medium such as water, air, steel, or a stretched string, it must cause the particles of that medium to oscillate as it passes. For that to happen, the medium must possess both inertia (so that kinetic energy can be stored) and elasticity (so that potential energy can be stored). These two properties determine how fast the wave can travel in the medium. And conversely, it should be possible to calculate the speed of the wave through the medium in terms of these properties. We do so now for a stretched string, in two ways.

Dimensional Analysis
In dimensional analysis we examine carefully the dimensions of all the physical quantities that enter into a given situation. In this case, we seek a speed v, which has the dimension of length divided by time, or LT^{-1}.

The inertia characteristic of a stretched string is the mass of a string element, which is represented by the mass m of the string divided by the length l of the string. We call this ratio the *linear density* μ of the string. Thus $\mu = m/l$, its dimension being mass divided by length, ML^{-1}.

You cannot send a wave along a stretched string without further stretching the string. The tension in the string does this stretching and must therefore represent the elastic characteristic of the string. The tension τ is a force and has the dimension (think about $F = ma$) MLT^{-2}. (We would normally use the symbol T for tension but we want to avoid confusion with our use of this symbol for the period of oscillation.)

The problem here is to combine μ (dimension ML^{-1}) and τ (dimension MLT^{-2}) in such a way as to generate v (dimension LT^{-1}). A little juggling of various combinations suggests

$$v = C\sqrt{\frac{\tau}{\mu}},\qquad(17\text{-}20)$$

in which C is a dimensionless constant. The drawback of dimensional analysis is that it cannot give the value of such a dimensionless constant. In our second approach to determine the speed, you will see that Eq. 17-20 is indeed correct and that $C = 1$.

Derivation from Newton's Second Law

Instead of the sinusoidal wave of Fig. 17-1b, let us consider a single symmetrical pulse such as that of Fig. 17-7. For convenience, we choose a reference frame in which the pulse remains stationary. That is, we run along with the pulse, keeping it constantly in view. In this frame, the string appears to move past us, from right to left in Fig. 17-7, with speed v.

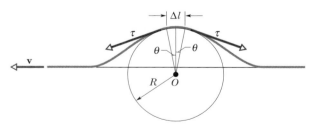

FIGURE 17-7 A symmetrical pulse, viewed from a reference frame in which the pulse is stationary and the string appears to move right to left with speed v. We find speed v by applying Newton's second law to a string element of length Δl, located at the top of the pulse.

Consider a small segment of the pulse, of length Δl, forming an arc of a circle of radius R. A force equal in magnitude to the string tension τ pulls tangentially on this segment at each end. The horizontal components of these forces cancel, but the vertical components add to form a restoring force **F**. In magnitude,

$$F = 2\tau \sin \theta \approx \tau(2\theta) = \tau\frac{\Delta l}{R}\qquad\text{(force)}.\quad(17\text{-}21)$$

We have used here the approximation $\sin \theta \approx \theta$ for the small angles θ in Fig. 17-7; from that figure, we have also used $2\theta = \Delta l/R$.

The mass of the segment is given by

$$\Delta m = \mu\,\Delta l\qquad\text{(mass)}.\qquad(17\text{-}22)$$

At the moment shown in Fig. 17-7, the string element Δl is moving in an arc of a circle. Thus it has a centripetal acceleration toward the center of that circle, given by

$$a = \frac{v^2}{R}\qquad\text{(acceleration)}.\qquad(17\text{-}23)$$

Equations 17-21, 17-22, and 17-23 contain the elements of Newton's second law. Combining them in the form

$$\text{force} = \text{mass} \times \text{acceleration}$$

gives

$$\frac{\tau\,\Delta l}{R} = (\mu\,\Delta l)\frac{v^2}{R}.$$

Solving this equation for the speed v yields

$$v = \sqrt{\frac{\tau}{\mu}}\qquad\text{(speed)},\qquad(17\text{-}24)$$

in exact agreement with Eq. 17-20 if the constant C in that equation is given the value unity. Equation 17-24 gives the speed of the pulse in Fig. 17-7 and the speed of any other wave on the same string under the same tension.

Equation 17-24 tells us that the speed of a wave along a stretched ideal string depends only on the characteristics of the string and not on the frequency of the wave. The *frequency* of the wave is fixed entirely by whatever generates the wave (for example, the person in Fig. 17-1b). The *wavelength* of the wave is then fixed by Eq. 17-12 in the form $\lambda = v/f$.

CHECKPOINT 3: The figure shows two situations in which the same string is put under tension by a suspended mass of 5 kg. In which situation will the speed of waves sent along the string be greater?

(a)   (b)

SAMPLE PROBLEM 17-3

In Fig. 17-8, a stranded climber has hooked himself onto a makeshift rope lowered by a rescuer. The rope between the climber and rescuer consists of two sections: section 1 with length l_1 and linear density μ_1, and section 2 with length $l_2 = 2l_1$ and linear density $\mu_2 = 4\mu_1$. The climber happens to pluck the bottom end of the rope (as a "ready" signal) at the same time the rescuer plucks the top end.

(a) What is the speed v_1 of the resulting pulses in section 1, in terms of their speed v_2 in section 2?

SOLUTION: We assume that the mass of the rope sections is negligible relative to the mass of the climber. Thus the tension τ in the rope is equal to the weight of the climber and is the same in the two sections. From Eq. 17-24, speeds v_1 and v_2 are given by

$$v_1 = \sqrt{\frac{\tau}{\mu_1}} \quad \text{and} \quad v_2 = \sqrt{\frac{\tau}{\mu_2}}. \qquad (17\text{-}25)$$

Dividing the first expression by the second, and substituting $\mu_2 = 4\mu_1$, we find

$$\frac{v_1}{v_2} = \sqrt{\frac{\tau}{\mu_1}}\sqrt{\frac{\mu_2}{\tau}} = \sqrt{\frac{\mu_2}{\mu_1}} = \sqrt{\frac{4\mu_1}{\mu_1}} = 2,$$

or
$$v_1 = 2v_2. \qquad \text{(Answer)} \qquad (17\text{-}26)$$

(b) In terms of l_2, at what distance below the rescuer do the two pulses pass through each other?

SOLUTION: We can simplify our calculations by first deciding whether the pulses pass each other above or below the knot joining the two sections. Let t be the time the pulses take to reach each other. From Eq. 17-26, we know that the climber's pulse moves through section 1 twice as fast as the rescuer's pulse moves through section 2. Because $l_2 = 2l_1$, we also know that the climber's pulse must travel half as far as the rescuer's pulse to reach the knot. So the climber's pulse must reach the knot first, and the point of passing must be above the knot. Let us assume that the pulses pass at a distance d below the rescuer at time t.

To reach the point of passing, the rescuer's pulse must travel downward through a distance d at speed v_2 in time t. So

$$t = \frac{d}{v_2}. \qquad (17\text{-}27)$$

To reach the point of passing, the climber's pulse must travel upward through distance l_1 at speed v_1 and then through distance $l_2 - d$ at speed v_2, all in time t. So

$$t = \frac{l_1}{v_1} + \frac{l_2 - d}{v_2}. \qquad (17\text{-}28)$$

Substituting t from Eq. 17-27 into Eq. 17-28, and setting $l_1 = l_2/2$ and $v_1 = 2v_2$, we find

$$\frac{d}{v_2} = \frac{l_1}{v_1} + \frac{l_2 - d}{v_2} = \frac{l_2/2}{2v_2} + \frac{l_2 - d}{v_2}.$$

Multiplying through by v_2 and rearranging yield

$$d = \tfrac{5}{8}l_2. \qquad \text{(Answer)}$$

FIGURE 17-8 Sample Problem 17-3. A stranded climber hangs from a rope consisting of two sections. His rescuer has secured the rope at the top.

17-7 ENERGY AND POWER OF A TRAVELING STRING WAVE

When we set up a wave on a stretched string, we provide energy for the motion of the string. As the wave moves away from us, it transports that energy as both kinetic energy and elastic potential energy. Let us consider each form in turn.

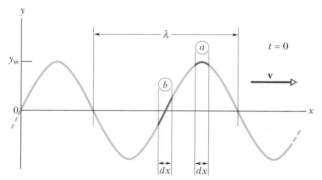

FIGURE 17-9 A snapshot of a traveling wave on a string at time $t = 0$. String element a is at displacement $y = y_m$, and string element b is at displacement $y = 0$. The kinetic energy of the string element at each position depends on the transverse velocity of the element. The potential energy depends on the amount by which the string element is stretched as the wave passes through it.

Kinetic Energy

An element of the string of mass dm, oscillating transversely in simple harmonic motion as the wave passes through it, has kinetic energy associated with its transverse velocity **u**. When the element is rushing through its $y = 0$ position (Fig. 17-9), its transverse velocity—and thus its kinetic energy—is a maximum. When the element is at its extreme position $y = y_m$, its transverse velocity—and thus again its kinetic energy—is zero.

Elastic Potential Energy

To send a sinusoidal wave along a previously straight string, the wave must necessarily stretch the string. As a string element of length dx oscillates transversely, its length must increase and decrease in a periodic way if the string element is to fit the sinusoidal wave form. Elastic potential energy is associated with these length changes, just as for a spring.

When the string element is at its $y = y_m$ position (Fig. 17-9), its length has its normal undisturbed value dx, so its elastic potential energy is zero. However, when the element is rushing through its $y = 0$ position, it is stretched to its maximum extent, and its elastic potential energy then is a maximum.

The oscillating string element thus has both its maximum kinetic energy and its maximum elastic potential energy at $y = 0$. In the snapshot of Fig. 17-9, the regions of the string at maximum displacement have no energy, and the regions at zero displacement have maximum energy. As the wave travels along the string, the tension in the string continuously does work to transfer energy from regions with energy to regions with no energy.

The Rate of Energy Transmission

The kinetic energy dK associated with a string element of mass dm is given by

$$dK = \tfrac{1}{2} \, dm \, u^2, \tag{17-29}$$

where u is the transverse speed of the oscillating string element, given by Eq. 17-19 as

$$u = \frac{\partial y}{\partial t} = -\omega y_m \cos(kx - \omega t). \tag{17-30}$$

Using this relation and putting $dm = \mu \, dx$, we rewrite Eq. 17-29 as

$$dK = \tfrac{1}{2}(\mu \, dx)(-\omega y_m)^2 \cos^2(kx - \omega t). \tag{17-31}$$

Dividing Eq. 17-31 by dt gives the rate at which the kinetic energy of a string element changes, and thus the rate at which kinetic energy is carried along by the wave. The ratio dx/dt that then appears on the right of Eq. 17-31 is the wave speed v, so we obtain

$$\frac{dK}{dt} = \tfrac{1}{2}\mu v \omega^2 y_m^2 \cos^2(kx - \omega t). \tag{17-32}$$

The *average* rate at which kinetic energy is transported is

$$\overline{\left(\frac{dK}{dt}\right)} = \tfrac{1}{2}\mu v \omega^2 y_m^2 \, \overline{\cos^2(kx - \omega t)}$$

$$= \tfrac{1}{4}\mu v \omega^2 y_m^2, \tag{17-33}$$

where an overhead bar means an average value of the quantity. In Eq. 17-33 we have taken the average over an integer number of wavelengths and have used the fact that the average value of the square of a cosine function over an integer number of wavelengths is $\tfrac{1}{2}$.

Elastic potential energy is also carried along with the wave, and at the same average rate given by Eq. 17-33. Although we shall not examine the proof, you should recall that, in an oscillating system such as a pendulum or a spring–block system, the average kinetic energy and the average potential energy are indeed equal.

The **average power,** which is the average rate at which energy of both kinds is transmitted by the wave, is then

$$\overline{P} = 2 \overline{\left(\frac{dK}{dt}\right)} \tag{17-34}$$

or, from Eq. 17-33,

$$\overline{P} = \tfrac{1}{2}\mu v \omega^2 y_m^2 \qquad \text{(average power).} \tag{17-35}$$

The factors μ and v in this equation depend on the material and tension of the string. The factors ω and y_m depend on the process that generates the wave. The dependence of the average power of a wave on the square of its amplitude and also on the square of its angular frequency is a general result, true for waves of all types.

SAMPLE PROBLEM 17-4

A string has a linear density μ of 525 g/m and is stretched with a tension τ of 45 N. A wave whose frequency f and amplitude y_m are 120 Hz and 8.5 mm, respectively, is traveling along the string. At what average rate is the wave transporting energy along the string?

SOLUTION: Before finding $\overline{P}$ from Eq. 17-35, we must calculate the angular frequency ω and the wave speed v. From Eq. 17-9,

$$\omega = 2\pi f = (2\pi)(120 \text{ Hz}) = 754 \text{ rad/s}.$$

From Eq. 17-24 we have

$$v = \sqrt{\frac{\tau}{\mu}} = \sqrt{\frac{45 \text{ N}}{0.525 \text{ kg/m}}} = 9.26 \text{ m/s}.$$

Equation 17-35 then yields

$$
\begin{aligned}
\overline{P} &= \tfrac{1}{2}\mu v \omega^2 y_m^2 \\
&= (\tfrac{1}{2})(0.525 \text{ kg/m})(9.26 \text{ m/s})(754 \text{ rad/s})^2(0.0085 \text{ m})^2 \\
&= 100 \text{ W}.
\end{aligned}
$$
(Answer)

17-8 THE PRINCIPLE OF SUPERPOSITION FOR WAVES

It often happens that two or more waves pass simultaneously through the same region. When we listen to a concert, for example, sounds from many instruments fall simultaneously on our eardrums. The electrons in the antennas of our radio and TV receivers are set in motion by a whole array of signals from different broadcasting centers. The water of a lake or harbor may be churned up by the wakes of many boats.

Suppose that two waves travel simultaneously along the same stretched string. Let $y_1(x, t)$ and $y_2(x, t)$ be the displacements that the string would experience if each wave acted alone. The displacement of the string when both waves act is then

$$y'(x, t) = y_1(x, t) + y_2(x, t), \qquad (17\text{-}36)$$

the sum being an algebraic sum. This summation of displacements along the string means

---

Overlapping waves algebraically add to produce a **resultant wave.**

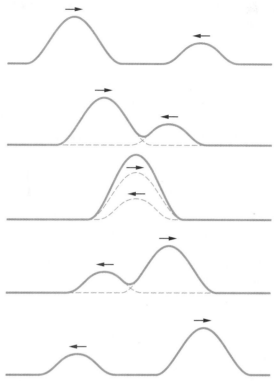

FIGURE 17-10 A series of snapshots that show two pulses traveling in opposite directions along a stretched string. The superposition principle applies as the pulses move through each other.

This is another example of the **principle of superposition,** which says that when several effects occur simultaneously, their net effect is the sum of the individual effects.

Figure 17-10 shows a sequence of snapshots of two pulses traveling in opposite directions on the same stretched string. When the pulses overlap, the resultant pulse is their sum. Moreover, each pulse moves through the other, as if the other were not present:

Overlapping waves do not in any way alter the travel of each other.

Fourier Analysis

French mathematician Jean Baptiste Fourier (1786–1830) explained how the principle of superposition can be used to analyze nonsinusoidal wave forms. He showed that any wave form can be represented as the sum of a large number of sinusoidal waves, of carefully chosen frequencies and amplitudes. English physicist Sir James Jeans expressed it well:

> [Fourier's] theorem tells us that every curve, no matter what its nature may be, or in what way it was originally

obtained, can be exactly reproduced by superposing a sufficient number of simple harmonic [sinusoidal] curves—in brief, every curve can be built up by piling up waves.

Figure 17-11 shows an example of a *Fourier series,* as such sums are called. The sawtooth curve in Fig. 17-11*a* shows the variation with time (at position $x = 0$) of the wave we wish to represent. The Fourier series that represents it can be shown to be

$$y(t) = -\frac{1}{\pi} \sin \omega t - \frac{1}{2\pi} \sin 2\omega t$$
$$-\frac{1}{3\pi} \sin 3\omega t \cdots, \quad (17\text{-}37)$$

in which $\omega = 2\pi/T$, where T is the period of the sawtooth curve. The green curve of Fig. 17-11*a*, which represents the sum of the first six terms of Eq. 17-37, matches the sawtooth curve rather well. Figure 17-11*b* shows these six terms separately. By adding more terms, we can approximate the sawtooth curve as closely as we wish.

You can see now why we spent so much time analyzing the behavior of a sinusoidal wave. When we understand that, Fourier's theorem will open the door to all other wave shapes.

17-9 INTERFERENCE OF WAVES

Suppose we send two sinusoidal waves of the same wavelength and amplitude in the same direction along a stretched string. The superposition principle applies. What resultant wave does it predict for the string?

The resultant wave depends on the extent to which the waves are *in phase* (in step) with respect to each other, that is, how much one wave form is shifted from the other wave form. If the waves are exactly in phase (so that the peaks and valleys of one are exactly aligned with those of the other without any shift), they combine to double the displacement of either wave acting alone. If they are exactly out of phase (the peaks of one are exactly aligned with the valleys of the other), they cancel everywhere and the string remains straight. We call this phenomenon of combining and canceling of waves **interference,** and the waves are said to **interfere.** (These terms refer only to the displacements of the waves; the travel of the waves is unaffected.)

Let one wave traveling along a stretched string be given by

$$y_1(x, t) = y_m \sin(kx - \omega t) \quad (17\text{-}38)$$

and another, shifted from the first, by

$$y_2(x, t) = y_m \sin(kx - \omega t + \phi). \quad (17\text{-}39)$$

These waves have the same angular frequency ω (and thus the same frequency f), the same angular wave number k (and thus the same wavelength λ), and the same amplitude y_m. They travel in the same direction, that of increasing x, with the same speed, given by Eq. 17-24. They differ only by a constant angle ϕ, which we call the **phase constant.** These waves are said to be *out of phase* by ϕ or to have a *phase difference* of ϕ, or one wave is said to be *phase-shifted* from the other by ϕ.

From the principle of superposition (Eq. 17-36), the combined wave has displacement

$$y'(x, t) = y_1(x, t) + y_2(x, t)$$
$$= y_m \sin(kx - \omega t) + y_m \sin(kx - \omega t + \phi). \quad (17\text{-}40)$$

In Appendix E we see that we can write the sum of the sines of two angles α and β as

$$\sin \alpha + \sin \beta = 2 \sin \tfrac{1}{2}(\alpha + \beta) \cos \tfrac{1}{2}(\alpha - \beta). \quad (17\text{-}41)$$

(a)

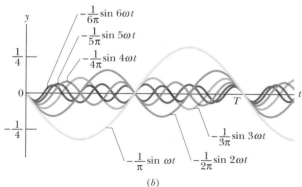

(b)

FIGURE 17-11 (*a*) The dashed sawtooth curve is approximated by the green curve, which is the result of evaluating the first six terms in the sum of Eq. 17-37. (Evaluating more terms would give closer approximations.) (*b*) Separate displays of the first six terms of that equation, each a sinusoidal function.

Applying this relation to Eq. 17-40 yields

$$y'(x, t) = [2y_m \cos \tfrac{1}{2}\phi] \sin(kx - \omega t + \tfrac{1}{2}\phi). \quad (17\text{-}42)$$

The resultant wave is thus also a sinusoidal wave traveling in the direction of increasing x. It is the only wave you would actually see on the string (you would *not* see the two combining waves of Eqs. 17-38 and 17-39).

If two sinusoidal waves of the same amplitude and wavelength travel in the *same* direction along a stretched string, they interfere to produce a resultant sinusoidal wave traveling in that direction.

The resultant wave differs from the original waves in two respects: (1) its phase constant is $\tfrac{1}{2}\phi$, and (2) its amplitude y'_m is the quantity in the brackets in Eq. 17-42:

$$y'_m = 2y_m \cos \tfrac{1}{2}\phi. \quad (17\text{-}43)$$

If $\phi = 0$ rad (or 0°), the two combining waves are exactly in phase (as in Fig. 17-12a). Then Eq. 17-42 reduces to

$$y'(x, t) = 2y_m \sin(kx - \omega t) \quad (\phi = 0). \quad (17\text{-}44)$$

Note that the amplitude of the resultant wave is twice the amplitude of either combining wave. This is the greatest amplitude the resultant wave can have, because the cosine term in Eq. 17-42 and Eq. 17-43 has its greatest value (unity) when $\phi = 0$. Interference that produces the greatest possible amplitude is called *fully constructive interference*.

If $\phi = \pi$ rad (or 180°), the combining waves are exactly out of phase (as in Fig. 17-12b). Then cos $\tfrac{1}{2}\phi$ becomes cos $\pi/2 = 0$, and the amplitude of the resultant wave (Eq. 17-43) is zero. We then have for all values of x and t,

$$y'(x, t) = 0 \quad (\phi = \pi \text{ rad}). \quad (17\text{-}45)$$

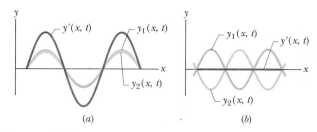

(a)   *(b)*

FIGURE 17-12 Two identical waves, $y_1(x, t)$ and $y_2(x, t)$, traveling in the same direction along a string, interfere to give a resultant wave $y'(x, t)$. (a) If the waves are exactly in phase, they undergo fully constructive interference and produce a resultant wave of twice their own amplitude. (b) If they are exactly out of phase, they undergo fully destructive interference and the string remains straight.

Now, although we sent two waves along the string, we see no motion of the string. This type of interference is called *fully destructive interference*.

A phase difference $\phi = 2\pi$ rad (or 360°) corresponds to a shift of one wave relative to the other wave by a distance of one wavelength. So phase differences can also be described in terms of wavelengths. For example, in Fig. 17-12b the waves may be said to be 0.50 wavelength out of phase. Table 17-1 shows some other examples of phase differences and the interference they produce. Note that when interference is neither fully constructive nor fully destructive, it is called *intermediate interference*. The amplitude of the resultant wave is then intermediate between 0 and $2y_m$.

Two waves with the same wavelength are in phase if their phase difference is zero or any integer number of wavelengths. Thus the integer part of any phase difference *expressed in wavelengths* may be discarded. For example, a phase difference of 0.40 wavelength is equivalent in every way to one of 2.40 wavelengths, and so the simpler of the two numbers can be used in computations.

TABLE 17-1 **PHASE DIFFERENCES AND RESULTING INTERFERENCE TYPES**[a]

PHASE DIFFERENCE, IN			AMPLITUDE OF RESULTANT WAVE	TYPE OF INTERFERENCE
DEGREES	RADIANS	WAVELENGTHS		
0	0	0	$2y_m$	Fully constructive
120	$2\pi/3$	0.33	y_m	Intermediate
180	π	0.50	0	Fully destructive
240	$4\pi/3$	0.67	y_m	Intermediate
360	2π	1.00	$2y_m$	Fully constructive
865	15.1	2.40	$0.60y_m$	Intermediate

[a]The phase difference is between two otherwise identical waves, with amplitude y_m, moving in the same direction.

SAMPLE PROBLEM 17-5

Two identical waves, moving in the same direction along a stretched string, interfere with each other. The amplitude y_m of each wave is 9.8 mm, and the phase difference ϕ between them is 100°.

(a) What is the amplitude y_m' of the resultant wave due to the interference of these two waves, and what type of interference occurs?

SOLUTION: From Eq. 17-43 we have for the amplitude

$$y_m' = 2y_m \cos \tfrac{1}{2}\phi$$
$$= (2)(9.8 \text{ mm}) \cos(100°/2)$$
$$= 13 \text{ mm}. \qquad \text{(Answer)}$$

Because the phase difference is between 0 and 180°, the interference is intermediate.

(b) What phase difference, in radians and in wavelengths, will give the resultant wave an amplitude of 4.9 mm?

SOLUTION: From Eq. 17-43 we have

$$y_m' = 2y_m \cos \tfrac{1}{2}\phi,$$

or $\qquad\qquad 4.9 \text{ mm} = (2)(9.8 \text{ mm}) \cos \tfrac{1}{2}\phi,$

which gives us (with a calculator in the radian mode)

$$\phi = 2 \cos^{-1} \frac{4.9 \text{ mm}}{(2)(9.8 \text{ mm})}$$
$$= \pm 2.636 \text{ rad} \approx \pm 2.6 \text{ rad}. \qquad \text{(Answer)}$$

There are two solutions because we can obtain the same resultant wave by letting the first wave *lead* (travel ahead of) or *lag* (travel behind) the second wave by 2.6 rad. In wavelengths, the phase difference is

$$\frac{\phi}{2\pi \text{ rad/wavelength}} = \frac{\pm 2.636 \text{ rad}}{2\pi \text{ rad/wavelength}}$$
$$= \pm 0.42 \text{ wavelength}. \quad \text{(Answer)}$$

CHECKPOINT **4:** Here are four other possible phase differences between the two waves of Sample Problem 17-5, expressed in wavelengths: 0.20, 0.45, 0.60, and 0.80. Rank them according to the amplitude of the resultant wave, greatest first.

17-10 PHASORS

We can represent a string wave (or any other type of wave) vectorially with a **phasor.** In essence, a phasor is a vector that has a magnitude equal to the amplitude of the wave and that rotates around an origin; the angular speed of the phasor is equal to the angular frequency ω of the wave. For

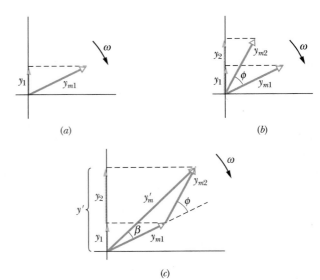

FIGURE 17-13   (*a*) A phasor of magnitude y_{m1} rotating about an origin at angular speed ω represents a sinusoidal wave. The phasor's projection y_1 on the vertical axis represents the displacement of a point through which the wave passes. (*b*) A second phasor, of magnitude y_{m2} and rotating at a constant angle ϕ to the first phasor, represents a second wave, with a phase constant ϕ. (*c*) The resultant wave of the two waves is represented by the vector sum y_m' of the two phasors. The projection y' on the vertical axis represents the displacement of a point as that resultant wave passes through it.

example, the wave

$$y_1(x, t) = y_{m1} \sin(kx - \omega t) \qquad (17\text{-}46)$$

is represented by the phasor shown in Fig. 17-13*a*. The magnitude of the phasor is the amplitude y_{m1} of the wave. As the phasor rotates around the origin at angular speed ω, its projection y_1 on the vertical axis varies sinusoidally, from a maximum of y_{m1} through zero to a minimum of $-y_{m1}$. This variation corresponds to the sinusoidal variation in the displacement y_1 of any point along the string as the wave passes through it.

When two waves travel along the same string in the same direction, we can represent them and their resultant wave in a *phasor diagram*. The phasors in Fig. 17-13*b* represent the wave of Eq. 17-46 and a second wave given by

$$y_2(x, t) = y_{m2} \sin(kx - \omega t + \phi). \qquad (17\text{-}47)$$

The angle between the two phasors in Fig. 17-13*b* is equal to the phase constant ϕ in Eq. 17-47. This angle is constant because the phasors rotate at the same angular speed ω, the angular frequency of the two waves.

Because waves y_1 and y_2 have the same angular wave number k and angular frequency ω, we know that their resultant is of the form

$$y'(x, t) = y_m' \sin(kx - \omega t + \beta), \qquad (17\text{-}48)$$

where y_m' is the amplitude of the resultant wave and β is its phase constant. To find the values of y_m' and β, we would have to sum the two combining waves, as we did to obtain Eq. 17-42.

To do this on a phasor diagram, we vectorially add the two phasors at any instant during their rotation, as in Fig. 17-13c where phasor y_{m2} has been shifted. The magnitude of the vector sum equals the amplitude y'_m in Eq. 17-48. The angle between the vector sum and the phasor for y_1 equals the phase constant β in Eq. 17-48.

Note that in contrast to the method of Section 17-9, using phasors allows us to combine waves *even if their amplitudes are different*.

SAMPLE PROBLEM 17-6

Two waves $y_1(x, t)$ and $y_2(x, t)$ have the same wavelength and travel in the same direction along a string. Their amplitudes are $y_{m1} = 4.0$ mm and $y_{m2} = 3.0$ mm, and their phase constants are 0 and $\pi/3$ rad, respectively. What are the amplitude y'_m and phase constant β of the resultant wave?

SOLUTION: Because the waves travel along the same string, Eq. 17-24 tells us they must have the same wave speed v. Then, because they also have the same wavelength (and thus the same angular wave number k), Eq. 17-12 tells us that they must have the same angular frequency ω. Thus they can be represented by two phasors rotating at the same angular speed ω about an origin, as in Fig. 17-13b. The angle ϕ between the phasors is $\pi/3$ rad.

To find the vector sum of the two phasors as in Fig. 17-13c, we are free to draw them at any instant during their rotation. To simplify the vector addition, we choose to draw them as shown in Fig. 17-14a. We now add the phasors as we would any vectors (Fig. 17-14b). The horizontal component of resultant phasor y'_m is

$$y'_{mh} = y_{m1} \cos 0 + y_{m2} \cos \pi/3$$
$$= 4.0 \text{ mm} + (3.0 \text{ mm}) \cos \pi/3 = 5.50 \text{ mm}.$$

The vertical component of y'_m is

$$y'_{mv} = y_{m1} \sin 0 + y_{m2} \sin \pi/3$$
$$= 0 + (3.0 \text{ mm}) \sin \pi/3 = 2.60 \text{ mm}.$$

Thus the resultant wave has an amplitude of

$$y'_m = \sqrt{(5.50 \text{ mm})^2 + (2.60 \text{ mm})^2}$$
$$= 6.1 \text{ mm} \qquad \text{(Answer)}$$

and a phase constant of

$$\beta = \tan^{-1} \frac{2.60 \text{ mm}}{5.50 \text{ mm}} = 0.44 \text{ rad.} \qquad \text{(Answer)}$$

FIGURE 17-14 Sample Problem 17-6. (*a*) Two phasors of magnitudes y_{m1} and y_{m2} and with phase difference $\pi/3$. (*b*) Vector addition of these phasors at any instant during their rotation gives the magnitude y'_m of the phasor for the resultant wave.

17-11 STANDING WAVES

In the preceding two sections, we discussed two sinusoidal waves of the same wavelength and amplitude traveling *in the same direction* along a stretched string. What if they travel in opposite directions? We can again find the resultant wave by applying the superposition principle.

Figure 17-15 suggests the situation graphically. It shows the two combining waves, one traveling to the left in

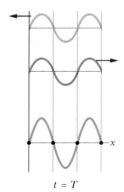

(*a*)

(*b*)

$\frac{\lambda}{2}$ $\frac{\lambda}{2}$ $\frac{\lambda}{2}$

(*c*)

$t = 0$ $\qquad t = \frac{1}{4}T$ $\qquad t = \frac{1}{2}T$ $\qquad t = \frac{3}{4}T$ $\qquad t = T$

FIGURE 17-15 How traveling waves produce standing waves. (*a*) and (*b*) represent snapshots of two waves of the same wavelength and amplitude, traveling in opposite directions, at five different instants during a single period of oscillation. (*c*) Their su-

perposition at the five instants. Note the nodes and antinodes in (*c*), the nodes being represented by dots. There are no nodes or antinodes in the traveling waves (*a*) and (*b*).

Fig. 17-15*a*, the other to the right in Fig. 17-15*b*. Figure 17-15*c* shows their sum, obtained by applying the superposition principle graphically. The outstanding feature of the resultant wave is that there are places along the string, called **nodes**, where the string is permanently at rest. Four such nodes are marked by dots in Fig. 17-15*c*. Halfway between adjacent nodes are **antinodes,** where the amplitude of the resultant wave is a maximum. Wave patterns such as that of Fig. 17-15*c* are called **standing waves** because the wave patterns do not move left or right: the locations of the maxima and minima do not change.

> If two sinusoidal waves of the same amplitude and wavelength travel in *opposite* directions along a stretched string, their interference with each other produces a standing wave.

To analyze a standing wave, we represent the two combining waves with the equations

$$y_1(x, t) = y_m \sin(kx - \omega t) \qquad (17\text{-}49)$$

and $\qquad y_2(x, t) = y_m \sin(kx + \omega t). \qquad (17\text{-}50)$

The principle of superposition gives, for the combined wave,

$$y'(x, t) = y_1(x, t) + y_2(x, t)$$
$$= y_m \sin(kx - \omega t) + y_m \sin(kx + \omega t).$$

Applying the trigonometric relation of Eq. 17-41 leads to

$$y'(x, t) = [2y_m \sin kx]\cos \omega t. \qquad (17\text{-}51)$$

This is *not* a traveling wave because it is not of the form of Eq. 17-16. Equation 17-51 describes a standing wave.

The quantity $2y_m \sin kx$ in the brackets of Eq. 17-51 can be viewed as the amplitude of oscillation of the string element that is located at position x. However, since an amplitude is always positive and $\sin kx$ can be negative, we take the absolute value of the quantity $2y_m \sin kx$ to be the amplitude at x.

In a traveling sinusoidal wave, the amplitude of the wave is the same for all string elements. That is not true for a standing wave, in which the amplitude *varies with position.* In the standing wave of Eq. 17-51, for example, the amplitude is zero for values of kx that give $\sin kx = 0$. Those values are

$$kx = n\pi, \qquad \text{for } n = 0, 1, 2, \cdots . \qquad (17\text{-}52)$$

Substituting $k = 2\pi/\lambda$ in this equation and rearranging, we get

$$x = n\frac{\lambda}{2}, \qquad \text{for } n = 0, 1, 2, \cdots .$$
$$\text{(nodes)}, \qquad (17\text{-}53)$$

as the positions of zero amplitude—the nodes—for the standing wave of Eq. 17-51. Note that adjacent nodes are separated by $\lambda/2$, half a wavelength.

The amplitude of the standing wave of Eq. 17-51 has a maximum value of $2y_m$, which occurs for values of kx that give $|\sin kx| = 1$. Those values are

$$kx = \tfrac{1}{2}\pi, \tfrac{3}{2}\pi, \tfrac{5}{2}\pi, \cdots$$
$$= (n + \tfrac{1}{2})\pi, \qquad \text{for } n = 0, 1, 2, \cdots . \qquad (17\text{-}54)$$

Substituting $k = 2\pi/\lambda$ in Eq. 17-54 and rearranging, we get

$$x = \left(n + \frac{1}{2}\right)\frac{\lambda}{2}, \qquad \text{for } n = 0, 1, 2, \cdots$$
$$\text{(antinodes)}, \qquad (17\text{-}55)$$

as the positions of maximum amplitude—the antinodes—of the standing wave of Eq. 17-51. The antinodes are one-half wavelength apart and are located halfway between pairs of nodes.

Reflections at a Boundary

We can set up a standing wave in a stretched string by allowing a traveling wave to be reflected from the far end of the string. The incident (original) wave and the reflected wave can then be described by Eqs. 17-49 and 17-50, respectively, and they can combine to form a pattern of standing waves.

In Fig. 17-16, we use a single pulse to show how such reflections take place. In Fig. 17-16*a*, the string is fixed at its left end. When the pulse arrives at that end, it exerts an upward force on the support (the wall). By Newton's third law, the support exerts an equal but opposite force on the string. This reaction force generates a pulse at the support, which travels back along the string in the direction opposite that of the incident pulse. In a "hard" reflection of this kind, there must be a node at the support because the string is fixed there. The reflected and incident pulses must have opposite signs, so as to cancel each other at that point.

In Fig. 17-16*b*, the left end of the string is fastened to a light ring that is free to slide without friction along a rod. When the incident pulse arrives, the ring moves up the rod. As the ring moves, it pulls on the string, stretching the string and producing a reflected pulse with the same sign and amplitude as the incident pulse. Thus in such a "soft" reflection, the incident and reflected pulses reinforce each other, creating an antinode at the end of the string; the

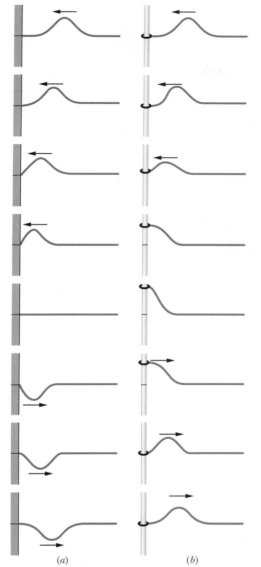

FIGURE 17-16 (*a*) A pulse incident from the right is reflected at the left end of the string, which is tied to a wall. Note that the sign of the reflected pulse is reversed. (*b*) Here the left end of the string is tied to a ring that can slide without friction up and down the rod. The pulse is reflected without a change of sign.

maximum displacement of the ring is twice the amplitude of either of these pulses.

CHECKPOINT **5:** Two waves with the same amplitude and wavelength interfere in three different situations to produce resultant waves with the following equations: (1) $y'(x, t) = 4 \sin(5x - 4t)$
(2) $y'(x, t) = 4 \sin(5x) \cos(4t)$
(3) $y'(x, t) = 4 \sin(5x + 4t)$
In which situation are the two combining waves traveling (a) toward increasing x, (b) toward decreasing x, and (c) in opposite directions?

17-12 STANDING WAVES AND RESONANCE

If (say) the left end of a stretched string is oscillated sinusoidally with the other end fixed, the oscillation sends a continuous traveling wave rightward along the string. The frequency of the wave is that of the oscillation. The wave reflects at the fixed end and travels leftward back through itself. The right-going wave and the left-going wave then interfere with each other.

For certain frequencies, the interference produces a standing wave pattern (or **oscillation mode**) with nodes and large antinodes like those in Fig. 17-17. Such a standing wave is said to be produced at **resonance,** and the string is said to *resonate* at these certain frequencies, called **resonant frequencies.** If the string is oscillated at some frequency other than a resonant frequency, a standing wave is not set up. Then the interference of the right-going wave and the left-going wave results in only small (perhaps imperceptible) oscillations of the string.

Consider a similar situation in which a string, such as a guitar string, is stretched between two clamps separated by a fixed distance L, and the string is somehow made to oscillate at a resonant frequency to set up a standing wave pattern. Since each end of the string is fixed, there must be a node at each end. The simplest pattern that meets this

FIGURE 17-17 Stroboscopic photographs reveal (imperfect) standing wave patterns on a string being made to oscillate by a vibrator at the left end. The patterns occur at certain frequencies of oscillation.

requirement is that in Fig. 17-18*a*, which shows the string at both its extreme displacements (one solid and one dashed). Note that there is only one antinode, which is at the center of the string. Also note that there is half a wavelength in the length *L*, which we take to be the string's length. Thus, for this pattern, $\lambda/2 = L$. This condition tells us that if the left-going and right-going traveling waves are to set up this pattern by their interference, they must have the wavelength $\lambda = 2L$.

A second simple pattern meeting the requirement of fixed ends is shown in Fig. 17-18*b*. This pattern has three nodes and two antinodes. For the left-going and right-going waves to set it up, they must have a wavelength $\lambda = L$. A third pattern is shown in Fig. 17-18*c*. It has four nodes and three antinodes, and the wavelength is $\lambda = \frac{2}{3}L$. You could continue the progression by drawing increasingly more complicated patterns. In each step of the progression, the pattern would have one more node and one more antinode than the preceding step, and an additional $\lambda/2$ would be fitted into the distance *L*.

The relation between λ and *L* for a standing wave on a string can be summarized as

$$\lambda = \frac{2L}{n}, \qquad \text{for } n = 1, 2, 3, \cdots. \qquad (17\text{-}56)$$

The resonant frequencies that correspond to these wavelengths follow from Eq. 17-12:

$$f = \frac{v}{\lambda} = n\frac{v}{2L}, \qquad \text{for } n = 1, 2, 3, \cdots. \qquad (17\text{-}57)$$

FIGURE 17-18 A string, stretched between two clamps, is made to oscillate in standing wave patterns. (*a*) The simplest possible pattern consists of one *loop*, which refers to the composite shape formed by the string in its extreme displacements (the solid and dashed lines). (*b*) The next simplest pattern has two loops. (*c*) The next pattern has three loops.

FIGURE 17-19 One of many possible standing wave patterns for a kettledrum head, made visible by dark powder sprinkled on the drumhead. As the head is set into oscillation at a single frequency by a mechanical vibrator at the upper left of the photograph, the powder collects at the nodes, which are circles and straight lines (rather than points) in this two-dimensional example.

Here *v* is the speed of the traveling waves on the string.

Equation 17-57 tells us that the resonant frequencies are integer multiples of the lowest resonant frequency, $f = v/2L$, which corresponds to $n = 1$. The oscillation mode with that lowest frequency is called the *fundamental mode* or the *first harmonic*. The *second harmonic* is the oscillation mode with $n = 2$; the *third harmonic* is that with $n = 3$; and so on. The frequencies associated with these modes are often labeled f_1, f_2, f_3, and so on. The collection of all possible oscillation modes is called the **harmonic series,** and *n* is called the **harmonic number** of the *n*th harmonic.

The phenomenon of resonance is common to all oscillating systems and can occur in two and three dimensions. For example, Fig. 17-19 shows a two-dimensional standing wave pattern on the oscillating head of a kettledrum.

SAMPLE PROBLEM 17-7

In Fig. 17-20, a string, tied to a sinusoidal vibrator at *P* and running over a support at *Q*, is stretched by a block of mass *m*. The separation *L* between *P* and *Q* is 1.2 m, the linear density of the string is 1.6 g/m, and the frequency *f* of the vibrator is fixed at 120 Hz. The amplitude of the motion at *P* is small enough for that point to be considered a node. A node also exists at *Q*.

(a) What mass m allows the vibrator to set up the fourth harmonic on the string?

SOLUTION: The resonant frequencies are given by Eq. 17-57 as

$$f = \frac{v}{2L} n, \quad \text{for } n = 1, 2, 3, \cdots. \quad (17\text{-}58)$$

We need to set the tension τ in the string so that the vibrator frequency is equal to the frequency of the fourth harmonic as given by this equation.

The speed v of waves on the string is given by Eq. 17-24 as

$$v = \sqrt{\frac{\tau}{\mu}} = \sqrt{\frac{mg}{\mu}}, \quad (17\text{-}59)$$

where the tension τ in the string is equal to the weight mg of the block. Substituting v from Eq. 17-59 into Eq. 17-58, setting $n = 4$ for the fourth harmonic, and solving for m give us

$$m = \frac{4L^2f^2\mu}{n^2g} \quad (17\text{-}60)$$

$$= \frac{(4)(1.2 \text{ m})^2(120 \text{ Hz})^2(0.0016 \text{ kg/m})}{(4)^2(9.8 \text{ m/s}^2)}$$

$$= 0.846 \text{ kg} \approx 0.85 \text{ kg}. \quad \text{(Answer)}$$

(b) What standing wave mode is set up if $m = 1.00$ kg?

FIGURE 17-20 Sample Problem 17-7. A string under tension connected to a vibrator. For a fixed vibrator frequency, standing wave patterns will occur for certain values of the string tension.

SOLUTION: If we insert this value of m into Eq. 17-60 and solve for n, we find that $n = 3.7$, an impossibility since n must be an integer. Thus, with $m = 1.00$ kg, the vibrator cannot set up a standing wave on the string, and any oscillation of the string will be small, perhaps even imperceptible.

CHECKPOINT 6: In the following series of resonant frequencies, one frequency (lower than 400 Hz) is missing: 150, 225, 300, 375 Hz. (a) What is the missing frequency? (b) What is the frequency of the seventh harmonic?

PROBLEM SOLVING TACTICS

TACTIC 2: *Harmonics on a String*

When you need to obtain information about a certain harmonic on a stretched string of given length L, first draw that harmonic (as in Fig. 17-18). If you are asked about, say, the fifth harmonic, you need to draw five loops between the fixed support points. That would mean that five loops, each of length $\lambda/2$, occupy the length L of the string. Thus, $5(\lambda/2) = L$, and $\lambda = 2L/5$. You can then use Eq. 17-12 ($f = v/\lambda$) to find the frequency of the harmonic.

Keep in mind that the wavelength of a harmonic is set only by the length L of the string, but the frequency depends also on the wave speed v, which is set by the tension and the linear density of the string via Eq. 17-24.

REVIEW & SUMMARY

Transverse and Longitudinal Waves

Mechanical waves can exist only in material media and are governed by Newton's laws. **Transverse** mechanical waves, like those on a stretched string, are waves in which the particles of the medium oscillate perpendicular to the wave's direction of travel. Waves in which the particles of the medium oscillate parallel to the wave's direction of travel are **longitudinal** waves.

Sinusoidal Waves

A sinusoidal wave moving in the $+x$ direction has the mathematical form

$$y(x, t) = y_m \sin(kx - \omega t), \quad (17\text{-}2)$$

where y_m is the **amplitude** of the wave, k the **angular wave**

number, ω the **angular frequency**, and $kx - \omega t$ the **phase.** The wavelength λ is related to k by

$$k = \frac{2\pi}{\lambda}. \quad (17\text{-}5)$$

The **period** T and **frequency** f of the wave are related to ω by

$$\frac{\omega}{2\pi} = f = \frac{1}{T}. \quad (17\text{-}9)$$

Finally, the **wave speed** v is related to these other parameters by

$$v = \frac{\omega}{k} = \frac{\lambda}{T} = \lambda f. \quad (17\text{-}12)$$

Equation of a Traveling Wave

In general, any function of the form

$$y(x, t) = h(kx \pm \omega t) \qquad (17\text{-}16)$$

can represent a **traveling wave** with a wave speed given by Eq. 17-12 and a wave shape given by the mathematical form of h. The plus sign is for a wave traveling in the $-x$ direction, and the minus sign for a wave in the $+x$ direction.

Wave Speed on Stretched String

The speed of a wave on a stretched string with tension τ and linear density μ is

$$v = \sqrt{\frac{\tau}{\mu}}. \qquad (17\text{-}24)$$

Power

The **average power,** or average rate at which energy is transmitted by a sinusoidal wave on a stretched string, is given by

$$\overline{P} = \tfrac{1}{2}\mu v \omega^2 y_m^2. \qquad (17\text{-}35)$$

Superposition of Waves

When two or more waves traverse the same medium, the displacement of any particle of the medium is the sum of the displacements that the individual waves would give it. This is called **superposition.**

Fourier Series

With a *Fourier series,* any wave can be constructed as the superposition of appropriate sinusoidal waves.

Interference of Waves

Two sinusoidal waves on the same string exhibit **interference,** adding or canceling according to the principle of superposition. If the two are traveling in the same direction and have the same amplitude y_m and frequency (hence the same wavelength) but differ in phase by a **phase constant** ϕ, the result is a single wave

with this same frequency:

$$y'(x, t) = [2y_m \cos \tfrac{1}{2}\phi]\, \sin(kx - \omega t + \tfrac{1}{2}\phi). \qquad (17\text{-}42)$$

If $\phi = 0$, the waves are in phase and their interference is (fully) constructive; if $\phi = \pi$ rad, they are out of phase and their interference is destructive.

Phasors

A wave $y(x, t)$ can be represented with a *phasor.* This is a vector that has a magnitude equal to the amplitude y_m of the wave and rotates about an origin with an angular speed equal to the angular frequency ω of the wave. The projection of the rotating phasor on a vertical axis gives the displacement y of the wave.

Standing Waves

The interference of two identical sinusoidal waves moving in opposite directions produces **standing waves.** For a string with fixed ends, the standing wave is given by

$$y'(x, t) = [2y_m \sin kx]\cos \omega t. \qquad (17\text{-}51)$$

Standing waves are characterized by fixed positions of zero displacement called **nodes** and fixed positions of maximum displacement called **antinodes.**

Resonance

Standing waves on a string can be set up by reflection of traveling waves from the ends of the string. If an end is fixed, it must be the position of a node. This limits the frequencies at which standing waves will occur on a given string. Each possible frequency is a **resonant frequency,** and the corresponding standing wave pattern is an **oscillation mode.** For a stretched string of length L with fixed ends, the resonant frequencies are

$$f = \frac{v}{\lambda} = n\frac{v}{2L}, \qquad \text{for } n = 1, 2, 3, \cdots. \qquad (17\text{-}57)$$

The oscillation mode corresponding to $n = 1$ is called the *fundamental mode* or the *first harmonic*; the mode corresponding to $n = 2$ is the *second harmonic*; and so on.

QUESTIONS

1. What is the wavelength of the (strange) wave in Fig. 17-21, where each segment of the wave has length d?

FIGURE 17-21 Question 1.

2. A string is put under tension and then three different sinusoidal waves are separately sent along it. Figure 17-22 gives plots of the resulting displacement versus time for an element on the string.

Rank the three plots according to the wavelength of the corresponding wave, greatest first.

FIGURE 17-22 Question 2.

3. In Fig. 17-23, four strings are placed under tension by one or two suspended blocks, all of the same mass. Strings *A*, *B*, and *C*

have the same linear density; string *D* has a greater linear density. Rank the strings according to the speed that waves will have when sent along them, greatest first.

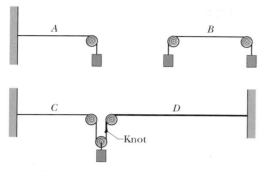

FIGURE 17-23 Question 3.

4. In Fig. 17-24, wave 1 consists of a rectangular peak of height 4 units and width *d*, and a rectangular valley of depth 2 units and width *d*. The wave travels rightward along an *x* axis. Waves 2, 3, and 4 are similar waves, with the same heights, depths, and widths, that will travel leftward along that axis and through wave 1. With which choice will the interference give, for an instant, (a) the deepest valley, (b) a flat line, and (c) a level peak 2*d* wide?

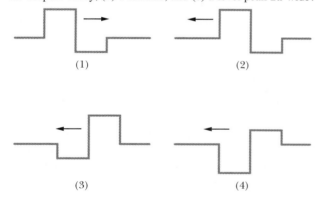

FIGURE 17-24 Question 4.

5. If you start with two waves traveling in phase on a string and then somehow phase-shift one of them by 5.4 wavelengths, what type of interference will occur on the string?

6. Here are some more phase differences between the two waves of Sample Problem 17-5, expressed in radians: $\pi/4$, $7\pi/4$, $-\pi/4$, and $-7\pi/4$. (a) Without written calculation, rank them in terms of the amplitude of the resultant wave, greatest first. (b) What type of interference occurs with each phase difference?

7. The amplitudes and phase differences for four pairs of waves of equal wavelengths are (a) 2 mm, 6 mm, and π rad; (b) 3 mm, 5 mm, and π rad; (c) 7 mm, 9 mm, and π rad; (d) 2 mm, 2 mm, and 0 rad. Each pair travels in the same direction along the same string. Without written calculation, rank the four pairs according to the amplitude of their resultant wave, greatest first. (*Hint:* Construct phasor diagrams.)

8. (a) Three waves equal in amplitude and wavelength travel along a string in the same direction. Take one to have a phase constant of 0°. What are the phase constants of the other waves if all three together produce fully destructive interference? (*Hint:* Construct a phasor diagram.) (b) What are the phase constants if there are four such waves? Now there are two (different) answers.

9. If you set up the seventh harmonic on a string, (a) how many nodes are present, and (b) is there a node, antinode, or some intermediate state at the midpoint? If you next set up the sixth harmonic, (c) is its resonant wavelength longer or shorter than that for the seventh harmonic, and (d) is the resonant frequency higher or lower?

10. (a) In Sample Problem 17-7 and Fig. 17-20, if we gradually increase the mass of the block the (frequency remains fixed), new resonant modes appear. Do the harmonic numbers of the new resonant modes increase or decrease from one to the next? (b) Is the shift from one resonant mode to the next gradual, or does each resonant mode disappear well before the next one appears?

11. A string is stretched between two fixed supports separated by distance *L*. (a) For which harmonic numbers is there a node at the point located $L/3$ from one of the supports? Is there a node, antinode, or some intermediate state at a point located $2L/5$ from one support if (b) the fifth harmonic is set up and (c) the tenth harmonic is set up?

12. Strings *A* and *B* have identical lengths and linear densities, but string *B* is under greater tension than is string *A*. Figure 17-25 shows four situations, (*a*) through (*d*), in which standing wave patterns exist on the two strings. In which situations is there the possibility that strings *A* and *B* are oscillating at the same resonant frequency?

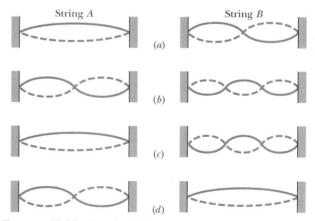

FIGURE 17-25 Question 12.

13. Two strings of equal lengths but unequal linear densities are tied together with a knot and stretched between two supports. A particular frequency happens to produce a standing wave on each length, with a node at the knot, as shown in Fig. 17-26. Which string has the greater linear density?

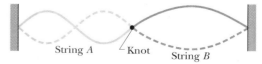

FIGURE 17-26 Question 13.

14. Do the wavelength and frequency, respectively, of the second harmonic on a string stretched between two supports increase, decrease, or remain the same if we (a) increase the distance between the supports without increasing the tension, (b) increase the tension in the string, and (c) switch to a string of greater linear density?

15. Guitar players know that prior to a show, a guitar must be played and the strings then tightened, because during the first few minutes of playing, the strings warm and loosen slightly. Does that loosening increase or decrease the resonant frequencies of the strings?

EXERCISES & PROBLEMS

SECTION 17-5 The Speed of a Traveling Wave

1E. A wave has a speed of 240 m/s and a wavelength of 3.2 m. What are the (a) frequency and (b) period of the wave?

2E. A wave has angular frequency 110 rad/s and wavelength 1.80 m. Calculate (a) the angular wave number and (b) the speed of the wave.

3E. The speed of electromagnetic waves in vacuum is 3.0×10^8 m/s. (a) Wavelengths of (visible) light waves range from about 400 nm in the violet to about 700 nm in the red. What is the range of frequencies of light waves? (b) The range of frequencies for shortwave radio (for example, FM radio and VHF television) is 1.5–300 MHz. What is the corresponding wavelength range? (c) X rays are also electromagnetic. Their wavelength range extends from about 5.0 nm to about 1.0×10^{-2} nm. What is the frequency range for x rays?

4E. A sinusoidal wave travels along a string. The time for a particular point to move from maximum displacement to zero is 0.170 s. What are the (a) period and (b) frequency? (c) The wavelength is 1.40 m; what is the wave speed?

5E. Write the equation for a wave traveling in the negative direction along the x axis and having an amplitude of 0.010 m, a frequency of 550 Hz, and a speed of 330 m/s.

6E. A traveling wave on a string is described by

$$y = 2.0 \sin\left[2\pi\left(\frac{t}{0.40} + \frac{x}{80}\right)\right],$$

where x and y are in centimeters and t is in seconds. (a) For $t = 0$, plot y as a function of x for $0 \le x \le 160$ cm. (b) Repeat (a) for $t = 0.05$ s and $t = 0.10$ s. (c) From your graphs, what is the wave speed, and in which direction ($+x$ or $-x$) is the wave traveling?

7E. Show that $y = y_m \sin(kx - \omega t)$ may be written in the alternative forms

$$y = y_m \sin k(x - vt), \qquad y = y_m \sin 2\pi\left(\frac{x}{\lambda} - ft\right),$$

$$y = y_m \sin \omega\left(\frac{x}{v} - t\right), \qquad y = y_m \sin 2\pi\left(\frac{x}{\lambda} - \frac{t}{T}\right).$$

8E. A single pulse, whose wave shape is given by the function $h(x - 5t)$, is shown in Fig. 17-27 for $t = 0$. Here x is in centimeters and t is in seconds. What are the (a) speed and (b) direction of travel of the pulse? (c) Plot $h(x - 5t)$ as a function of x for $t = 2$ s. (d) Plot $h(x - 5t)$ as a function of t for $x = 10$ cm.

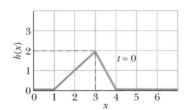

FIGURE 17-27
Exercise 8.

9E. Show (a) that the maximum transverse speed of a particle in a string owing to a traveling wave is given by $u_{max} = \omega y_m = 2\pi f y_m$ and (b) that the maximum transverse acceleration is $a_{y,max} = \omega^2 y_m = 4\pi^2 f^2 y_m$.

10E. The equation of a transverse wave traveling along a string is given by

$$y = (2.0 \text{ mm}) \sin[(20 \text{ m}^{-1})x - (600 \text{ s}^{-1})t].$$

(a) Find the amplitude, frequency, velocity, and wavelength of the wave. (b) Find the maximum transverse speed of a particle in the string.

11E. (a) Write an expression describing a sinusoidal transverse wave traveling on a cord in the $+y$ direction with an angular wave number of 60 cm^{-1}, a period of 0.20 s, and an amplitude of 3.0 mm. Take the transverse direction to be the z direction. (b) What is the maximum transverse speed of a point on the cord?

12P. The equation of a transverse wave traveling along a very long string is given by $y = 6.0 \sin(0.020\pi x + 4.0\pi t)$, where x and y are expressed in centimeters and t is in seconds. Determine (a) the amplitude, (b) the wavelength, (c) the frequency, (d) the speed, (e) the direction of propagation of the wave, and (f) the maximum transverse speed of a particle in the string. (g) What is the transverse displacement at $x = 3.5$ cm when $t = 0.26$ s?

13P. (a) Write an equation describing a sinusoidal transverse wave traveling on a cord in the $+x$ direction with a wavelength of 10 cm, a frequency of 400 Hz, and an amplitude of 2.0 cm. (b) What is the maximum speed of a point on the cord? (c) What is the speed of the wave?

14P. Prove that if a transverse sinusoidal wave is traveling along a string, then the slope at any point of the string is equal to the ratio of the particle speed to the wave speed at that point.

15P. A transverse sinusoidal wave of wavelength 20 cm is moving along a string toward increasing x. The transverse displacement of the string particle at $x = 0$ as a function of time is shown in Fig. 17-28. (a) Make a rough sketch of one wavelength of the wave (the portion between $x = 0$ and $x = 20$ cm) at time $t = 0$. (b) What is the velocity of propagation of the wave? (c) Write the equation for the wave with all the constants evaluated. (d) What is the transverse velocity of the particle at $x = 0$ at $t = 5.0$ s?

y (cm)

4.0

10

t (s)

−4.0

FIGURE 17-28
Problem 15.

16P. A sinusoidal wave of frequency 500 Hz has a velocity of 350 m/s. (a) How far apart are two points that differ in phase by $\pi/3$ rad? (b) What is the phase difference between two displacements at a certain point at times 1.00 ms apart?

SECTION 17-6 Wave Speed on a Stretched String

17E. What is the speed of a transverse wave in a rope of length 2.00 m and mass 60.0 g under a tension of 500 N?

18E. The heaviest and lightest strings in a certain violin have linear densities of 3.0 and 0.29 g/m. What is the ratio, heavier to lighter, of the diameters of these strings, assuming that they are made of the same material?

19E. The speed of a transverse wave on a string is 170 m/s when the string tension is 120 N. To what value must the tension be changed to raise the wave speed to 180 m/s?

20E. The tension in a wire clamped at both ends is doubled without appreciably changing its length. What is the ratio of the new to the old wave speed for transverse waves in this wire?

21E. Show that, in terms of the tensile stress S and the volume density ρ, the speed v of transverse waves in a wire is given by

$$v = \sqrt{\frac{S}{\rho}}.$$

22E. The equation of a transverse wave on a string is

$$y = (2.0 \text{ mm}) \sin[(20 \text{ m}^{-1})x - (600 \text{ s}^{-1})t].$$

The tension in the string is 15 N. (a) What is the wave speed? (b) Find the linear density of this string in grams per meter.

23E. The linear density of a string is 1.6×10^{-4} kg/m. A transverse wave is propagating on the string and is described by the following equation: $y = (0.021 \text{ m})\sin[(2.0 \text{ m}^{-1})x + (30 \text{ s}^{-1})t]$. (a) What is the wave speed? (b) What is the tension in the string?

24E. What is the fastest transverse wave that can be sent along a steel wire? Allowing for a reasonable safety factor, the maximum tensile stress to which steel wires should be subjected is 7.0×10^8 N/m². The density of steel is 7800 kg/m³. Show that your answer does not depend on the diameter of the wire.

25P. A stretched string has a mass per unit length of 5.0 g/cm and a tension of 10 N. A sinusoidal wave on this string has an amplitude of 0.12 mm and a frequency of 100 Hz and is traveling toward decreasing x. Write an equation for this wave.

26P. For a sinusoidal transverse wave on a stretched cord, find the ratio of the maximum particle speed (the speed with which a

single particle in the cord moves transverse to the wave) to the wave speed. If a wave having a certain frequency and amplitude is sent along a cord, would this speed ratio depend on the material of which the cord is made, such as wire or nylon?

27P. A sinusoidal transverse wave is traveling along a string toward decreasing x. Figure 17-29 shows a plot of the displacement as a function of position at time $t = 0$. The string tension is 3.6 N, and its linear density is 25 g/m. Find (a) the amplitude, (b) the wavelength, (c) the wave speed, and (d) the period of the wave. (e) Find the maximum speed of a particle in the string. (f) Write an equation describing the traveling wave.

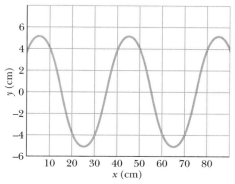

FIGURE 17-29   Problem 27.

28P. A sinusoidal wave is traveling on a string with speed 40 cm/s. The displacement of the particles of the string at $x = 10$ cm is found to vary with time according to the equation $y = (5.0 \text{ cm}) \sin[1.0 - (4.0 \text{ s}^{-1})t]$. The linear density of the string is 4.0 g/cm. What are (a) the frequency and (b) the wavelength of the wave? (c) Write the general equation giving the transverse displacement of the particles of the string as a function of position and time. (d) Calculate the tension in the string.

29P. In Fig. 17-30a, string 1 has a linear density of 3.00 g/m, and string 2 has a linear density of 5.00 g/m. They are under tension owing to the hanging block of mass $M = 500$ g. (a) Cal-

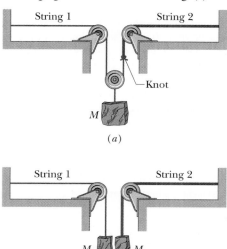

FIGURE 17-30   Problem 29.

culate the wave speed in each string. (b) The block is now divided into two blocks (with $M_1 + M_2 = M$) and the apparatus rearranged as shown in Fig. 17-30b. Find M_1 and M_2 such that the wave speeds in the two strings are equal.

30P. A wire 10.0 m long and having a mass of 100 g is stretched under a tension of 250 N. If two pulses, separated in time by 30.0 ms, are generated, one at each end of the wire, where will the pulses first meet?

31P. The type of rubber band used inside some baseballs and golf balls obeys Hooke's law over a wide range of elongation of the band. A segment of this material has an unstretched length l and a mass m. When a force F is applied, the band stretches an additional length Δl. (a) What is the speed (in terms of m, Δl, and the force constant k) of transverse waves on this stretched rubber band? (b) Using your answer to (a), show that the time required for a transverse pulse to travel the length of the rubber band is proportional to $1/\sqrt{\Delta l}$ if $\Delta l \ll l$ and is constant if $\Delta l \gg l$.

32P*. A uniform rope of mass m and length l hangs from a ceiling. (a) Show that the speed of a transverse wave on the rope is a function of y, the distance from the lower end, and is given by $v = \sqrt{gy}$. (b) Show that the time a transverse wave takes to travel the length of the rope is given by $t = 2\sqrt{l/g}$.

SECTION 17-7 Energy and Power of a Traveling String Wave

33E. Energy is transmitted at rate P_1 by a wave of frequency f_1 on a string with tension τ_1. What is the new energy transmission rate P_2 in terms of P_1 (a) if the tension of the string is increased to $\tau_2 = 4\tau_1$ and (b) if, instead, the frequency of the wave is decreased to $f_2 = f_1/2$?

34E. A string along which waves can travel is 2.70 m long and has a mass of 260 g. The tension in the string is 36.0 N. What must be the frequency of traveling waves of amplitude 7.70 mm in order that the average power be 85.0 W?

35P. A transverse sinusoidal wave is generated at one end of a long, horizontal string by a bar that moves up and down through a distance of 1.00 cm. The motion is continuous and is repeated regularly 120 times per second. The string has linear density 120 g/m and is kept under a tension of 90.0 N. Find (a) the maximum value of the transverse speed u and (b) the maximum value of the transverse component of the tension. (c) Show that the two maximum values calculated above occur at the same phase values for the wave. What is the transverse displacement y of the string at these phases? (d) What is the maximum rate of energy transfer along the string? (e) What is the transverse displacement y when this maximum transfer occurs? (f) What is the minimum rate of energy transfer along the string? (g) What is the transverse displacement y when this minimum transfer occurs?

SECTION 17-9 Interference of Waves

36E. Two identical traveling waves, moving in the same direction, are out of phase by $\pi/2$ rad. What is the amplitude of the

resultant wave in terms of the common amplitude y_m of the two combining waves?

37E. What phase difference between two otherwise identical traveling waves, moving in the same direction along a stretched string, will result in the combined wave having an amplitude 1.50 times that of the common amplitude of the two combining waves? Express your answer in degrees, radians, and wavelengths.

38P. Two sinusoidal waves, identical except for phase, travel in the same direction along a string and interfere to produce a resultant wave given by $y'(x, t) = (3.0 \text{ mm}) \sin(20x - 4.0t + 0.820 \text{ rad})$, with x in meters and t in seconds. What are (a) the wavelength λ of the two waves, (b) the phase difference between them, and (c) their amplitude y_m?

SECTION 17-10 Phasors

39E. Determine the amplitude of the resultant wave when two sinusoidal string waves having the same frequency and traveling in the same direction are combined, if their amplitudes are 3.0 cm and 4.0 cm and they have phase constants of 0 and $\pi/2$ rad, respectively.

40E. The amplitudes of two sinusoidal waves traveling in the same direction along a stretched string are 3.0 and 5.0 mm; their phase constants are 0° and 70°, respectively. The waves have the same wavelength. What are (a) the amplitude and (b) the phase constant of the resultant wave?

41E. Two sinusoidal waves of the same wavelength travel in the same direction along a stretched string with amplitudes of 4.0 and 7.0 mm and phase constants of 0 and 0.8π rad, respectively. What are (a) the amplitude and (b) the phase constant of the resultant wave?

42P. Two sinusoidal waves of the same period, with amplitudes of 5.0 and 7.0 mm, travel in the same direction along a stretched string; they produce a resultant wave with an amplitude of 9.0 mm. The phase constant of the 5.0 mm wave is 0. What is the phase constant of the 7.0 mm wave?

43P. Three sinusoidal waves of the same frequency travel along a string in the positive direction of an x axis. Their amplitudes are y_1, $y_1/2$, and $y_1/3$, and their phase constants are 0, $\pi/2$, and π, respectively. What are (a) the amplitude and (b) the phase constant of the resultant wave? (c) Plot the wave form of the resultant wave at $t = 0$, and discuss its behavior as t increases.

44P. Four sinusoidal waves travel in the positive x direction along the same string. Their frequencies are in the ratio $1:2:3:4$, and their amplitudes are in the ratio $1:\frac{1}{2}:\frac{1}{3}:\frac{1}{4}$, respectively. When $t = 0$, at $x = 0$, the first and third waves are 180° out of phase with the second and fourth. Plot the resultant wave form when $t = 0$, and discuss its behavior as t increases.

SECTION 17-12 Standing Waves and Resonance

45E. A string under tension τ_i oscillates in the third harmonic at frequency f_3, and the waves on the string have wavelength λ_3. If

EXERCISES & PROBLEMS **423**

the tension is increased to $\tau_f = 4\tau_i$ and the string is again made to oscillate in the third harmonic, what then are (a) the frequency of oscillation in terms of f_3 and (b) the wavelength of the waves in terms of λ_3?

46E. A nylon guitar string has a linear density of 7.2 g/m and is under a tension of 150 N. The fixed supports are 90 cm apart. The string is oscillating in the standing wave pattern shown in Fig. 17-31. Calculate the (a) speed, (b) wavelength, and (c) frequency of the traveling waves whose superposition gives this standing wave.

FIGURE 17-31 Exercise 46.

47E. Two sinusoidal waves with identical wavelengths and amplitudes travel in opposite directions along a string with a speed of 10 cm/s. If the time interval between instants when the string is flat is 0.50 s, what is the wavelength of the waves?

48E. When played in a certain manner, the lowest resonant frequency of a certain violin string is concert A (440 Hz). What are the frequencies of the second and third harmonics of that string?

49E. A string fixed at both ends is 8.40 m long and has a mass of 0.120 kg. It is subjected to a tension of 96.0 N and set oscillating. (a) What is the speed of the waves on the string? (b) What is the longest possible wavelength for a standing wave? (c) Give the frequency of that wave.

50E. The equation of a transverse wave traveling along a string is

$$y = 0.15 \sin(0.79x - 13t),$$

in which x and y are in meters and t is in seconds. (a) What is the displacement y at $x = 2.3$ m, $t = 0.16$ s? (b) Write the equation of a wave that, when added to the given one, would produce standing waves on the string. (c) What is the displacement of the resultant standing wave at $x = 2.3$ m, $t = 0.16$ s?

51E. A 120 cm length of string is stretched between fixed supports. What are the three longest possible wavelengths for traveling waves on the string that can produce standing waves? Sketch the corresponding standing waves.

52E. A 125 cm length of string has a mass of 2.00 g. It is stretched with a tension of 7.00 N between fixed supports. (a) What is the wave speed for this string? (b) What is the lowest resonant frequency of this string?

53E. What are the three lowest frequencies for standing waves on a wire 10.0 m long having a mass of 100 g, which is stretched under a tension of 250 N?

54E. A 1.50 m wire has a mass of 8.70 g and is held under a tension of 120 N. The wire is held rigidly at both ends and set into vibration. Calculate (a) the speed of waves on the wire, (b) the wavelengths of the waves that produce one- and two-loop standing waves on the string, and (c) the frequencies of the waves that produce one- and two-loop standing waves.

55E. String A is stretched between two clamps separated by distance l. String B, with the same linear density and under the same tension as string A, is stretched between two clamps separated by distance $4l$. Consider the first eight harmonics of string B. Which, if any, has a resonant frequency that matches a resonant frequency of string A?

56P. A string is stretched between fixed supports separated by 75.0 cm. It is observed to have resonant frequencies of 420 and 315 Hz, and no other resonant frequencies between these two. (a) What is the lowest resonant frequency for this string? (b) What is the wave speed for this string?

57P. Two waves are propagating on the same very long string. A generator at one end of the string creates a wave given by

$$y = (6.0 \text{ cm}) \cos \frac{\pi}{2} [(2.0 \text{ m}^{-1})x + (8.0 \text{ s}^{-1})t],$$

and one at the other end creates the wave

$$y = (6.0 \text{ cm}) \cos \frac{\pi}{2} [2.0 \text{ m}^{-1})x - (8.0 \text{ s}^{-1})t].$$

(a) Calculate the frequency, wavelength, and speed of each wave. (b) Find the points on the string at which there is no motion (the nodes). (c) At which points is the motion of the string a maximum (the antinodes)?

58P. A string oscillates according to the equation

$$y' = (0.50 \text{ cm}) \left[\sin \left(\frac{\pi}{3} \text{ cm}^{-1} \right) x \right] \cos[(40\pi \text{ s}^{-1})t].$$

(a) What are the amplitude and speed of the two waves (identical except for direction of travel) whose superposition gives this oscillation? (b) What is the distance between nodes? (c) What is the speed of a particle of the string at the position $x = 1.5$ cm when $t = \frac{9}{8}$ s?

59P. Two transverse sinusoidal waves travel in opposite directions along a string. Each wave has an amplitude of 0.30 cm and a wavelength of 6.0 cm. The speed of a transverse wave on the string is 1.5 m/s. Plot the shape of the string at times $t = 0$ (arbitrary), $t = 5.0$, $t = 10$, $t = 15$, and $t = 20$ ms.

60P. Two pulses travel along a string in opposite directions, as in Fig. 17-32. (a) If the wave speed v is 2.0 m/s and the pulses are 6.0 cm apart at $t = 0$, sketch the patterns when t is equal to 5.0, 10, 15, 20, and 25 ms. (b) What has happened to the energy of the pulses at $t = 15$ ms?

FIGURE 17-32 Problem 60.

61P. Two waves on a string are described by these equations:

$$y_1 = (0.10 \text{ m}) \sin 2\pi [(0.50 \text{ m}^{-1})x + (20 \text{ s}^{-1})t],$$
$$y_2 = (0.20 \text{ m}) \sin 2\pi [(0.50 \text{ m}^{-1})x - (20 \text{ s}^{-1})t].$$

Sketch the total response for the point on the string at $x = 3.0$ m; that is, plot y versus t for that value of x.

62P. A string 3.0 m long is oscillating as a three-loop standing wave whose amplitude is 1.0 cm. The wave speed is 100 m/s. (a) What is the frequency? (b) Write equations for two waves that, when combined, will result in this standing wave.

63P. Vibration from a 600 Hz tuning fork sets up standing waves in a string clamped at both ends. The wave speed for the string is 400 m/s. The standing wave has four loops and an amplitude of 2.0 mm. (a) What is the length of the string? (b) Write an equation for the displacement of the string as a function of position and time.

64P. In an experiment on standing waves, a string 90 cm long is attached to the prong of an electrically driven tuning fork that oscillates perpendicular to the length of the string at a frequency of 60 Hz. The mass of the string is 0.044 kg. What tension must the string be under (weights are attached to the other end) if it is to vibrate in four loops?

65P. Show that the maximum kinetic energy in each loop of a standing wave produced by two traveling waves of identical amplitudes is $2\pi^2 \mu y_m^2 f v$.

66P. An aluminum wire, of length $l_1 = 60.0$ cm, cross-sectional area 1.00×10^{-2} cm^2, and density 2.60 g/cm^3, is joined to a steel wire, of density 7.80 g/cm^3 and the same cross-sectional area. The compound wire, loaded with a block of mass $m = 10.0$ kg, is arranged as in Fig. 17-33 so that the distance l_2 from the joint to the supporting pulley is 86.6 cm. Transverse waves are set up in the wire by using an external source of variable frequency; a node is located at the pulley. (a) Find the lowest frequency of excitation for which standing waves are observed such that the joint in the wire is a node. (b) How many nodes are observed at this frequency?

FIGURE 17-33 Problem 66.

Electronic Computation

67. Two waves,

$$y_1 = (2.50 \text{ mm})\sin[(25.1 \text{ rad/m})x - (440 \text{ rad/s})t]$$

and

$$y_2 = (1.50 \text{ mm})\sin[(25.1 \text{ rad/m})x + (440 \text{ rad/s})t],$$

travel along a stretched string. (a) Plot $y'(x, t) = y_1(x, t) + y_2(x, t)$ as a function of t for $x = 0$, $\lambda/8$, $\lambda/4$, $3\lambda/8$, and $\lambda/2$, where λ is the wavelength. The graphs should extend from $t = 0$ to a little over one period. (b) The result is the superposition of a standing wave and a traveling wave. In which direction does the traveling wave travel? How can you change the original waves so the resultant is the superposition of standing and traveling waves with the same amplitudes as before but with the traveling wave moving in the opposite direction? (c) Use your graphs to find the place at which the amplitude of vibration is the largest and the smallest. (d) How are the maximum and minimum amplitudes related to the amplitudes 2.50 mm and 1.50 mm of the original two traveling waves?

This horseshoe bat not only can locate a moth flying in total darkness but can also determine the moth's relative speed, to home in on the insect. How does the detection system work? And how can a moth "jam" the system or otherwise reduce its effectiveness?

FIGURE 18-1 An ultrasound image of a fetus searching for a thumb to suck.

18-1 SOUND WAVES

As we saw in Chapter 17, mechanical waves are waves that require a material medium to exist. There are two types of mechanical waves: *transverse waves* involve oscillations perpendicular to the direction in which the wave travels; *longitudinal waves* involve oscillations parallel to the direction of wave travel.

A **sound wave** can be roughly defined as any longitudinal wave. Seismic prospecting teams use such waves to probe Earth's crust for oil. Ships carry sound ranging gear (sonar) to detect underwater obstacles. Submarines use sound waves to stalk other submarines, largely by listening for the characteristic noises produced by the propulsion system. Figure 18-1, a computer-processed image of a fetal head, shows how sound waves can be used to explore the

soft tissues of the human body. In this chapter we shall focus on sound waves that travel through the air and that are audible to people.

Figure 18-2 illustrates several ideas that we shall use in our discussions. Point *S* represents a tiny sound source, called a *point source*, that emits sound waves in all directions. The *wavefronts* and *rays* indicate the direction of travel and the spread of the sound waves. **Wavefronts** are surfaces over which the oscillations of the air due to the sound wave have the same value; such surfaces are represented by whole or partial circles in a two-dimensional drawing for a point source. **Rays** are directed lines perpendicular to the wavefronts that indicate the direction of travel of the wavefronts. The short double arrows superimposed on the rays of Fig. 18-2 indicate that the longitudinal oscillations of the air are parallel to the rays.

Near the point source of Fig. 18-2, the wavefronts are spherical and are spreading out in three dimensions, and the waves are said to be *spherical*. As the wavefronts move outward and their radii become larger, their curvature decreases. Far from the source, we approximate the wavefronts as straight lines and the waves are said to be *planar*.

18-2 THE SPEED OF SOUND

The speed of any mechanical wave, transverse or longitudinal, depends on both an inertial property of the medium (to store kinetic energy) and an elastic property of the medium (to store potential energy). Thus we can generalize Eq. 17-24, which gives the speed of a transverse wave along a stretched string, by writing

$$v = \sqrt{\frac{\tau}{\mu}} = \sqrt{\frac{\text{elastic property}}{\text{inertial property}}}, \qquad (18\text{-}1)$$

where (for transverse waves) τ is the tension in the string and μ is the string's linear density. If the medium is air, we can guess that the inertial property, corresponding to μ, is the volume density ρ of air. What shall we put for the elastic property?

In a stretched string, potential energy is associated with the periodic stretching of the string elements as the wave passes through them. As a sound wave passes through air, potential energy is associated with periodic compressions and expansions of small volume elements of the air. The property that determines the extent to which an element of the medium changes its volume as the pressure (force per unit area) applied to it is increased or decreased is the **bulk modulus** B, as defined (from Eq. 13-36)

$$B = -\frac{\Delta p}{\Delta V/V} \qquad \text{(definition of } B). \qquad (18\text{-}2)$$

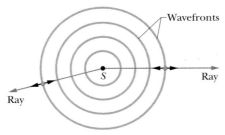

FIGURE 18-2 A sound wave travels from a point source *S* through a three-dimensional medium. The wavefronts form spheres centered on *S*; the rays are radial to *S*. The short, double-headed arrows indicate that elements of the medium oscillate parallel to the rays.

TABLE 18-1 **THE SPEED OF SOUND**[a]

MEDIUM	SPEED (m/s)	MEDIUM	SPEED (m/s)
Gases		*Solids*	
Air (0°C)	331	Aluminum	6420
Air (20°C)	343	Steel	5941
Helium	965	Granite	6000
Hydrogen	1284		
Liquids			
Water (0°C)	1402		
Water (20°C)	1482		
Seawater[b]	1522		

[a]0°C and 1 atm pressure, except where noted.
[b]At 20°C and 3.5% salinity.

Here $\Delta V/V$ is the fractional change in volume produced by a change in pressure Δp. As explained in Section 15-3, the SI unit for pressure is the newton per square meter, which is given a special name, the *pascal* (Pa). From Eq. 18-2 we see that the unit for B is also the pascal. The signs of Δp and ΔV are always opposite: when we increase the pressure on a fluid element (Δp positive), its volume decreases (ΔV negative). We include a minus sign in Eq. 18-2 so that B will always be a positive quantity. Now substituting B for τ and ρ for μ in Eq. 18-1 yields

$$v = \sqrt{\frac{B}{\rho}} \qquad \text{(speed of sound)} \qquad (18\text{-}3)$$

for a medium with bulk modulus B and density ρ. Table 18-1 lists the speed of sound in various media.

The density of water is almost 1000 times greater than the density of air. If this were the only relevant factor, we would expect from Eq. 18-3 that the speed of sound in water would be considerably less than the speed of sound in air. However, Table 18-1 shows us that the reverse is true. We conclude (again from Eq. 18-3) that the bulk modulus of water must be more than 1000 times greater than that of air. This is indeed the case. Water is much more incompressible than air, which (see Eq. 18-2) is another way of saying that its bulk modulus is much greater.

Formal Derivation of Eq. 18-3

We now derive Eq. 18-3 by direct application of Newton's laws. Let a single pulse in which air is compressed travel (from right to left) with speed v through the air in a long tube, like that in Fig. 17-2. Let us run along with the pulse at that speed, so that the pulse appears to stand still in our reference frame. Figure 18-3a shows the situation as it is viewed from that frame. The pulse is standing still, and air is moving at speed v through it from left to right.

Let the pressure of the undisturbed air be p and the pressure inside the pulse be $p + \Delta p$, where Δp is positive owing to the compression. Consider a slice of air of thickness Δx and face area A, moving toward the pulse at speed v. As this element of air enters the pulse, its leading face encounters a region of higher pressure, which slows it to speed $v + \Delta v$, in which Δv is negative. This slowing is complete when the rear face reaches the pulse, which requires time interval

$$\Delta t = \frac{\Delta x}{v}. \qquad (18\text{-}4)$$

Let us apply Newton's second law to the element. During Δt, the average force on the element's trailing face is pA toward the right, and the average force on the leading

FIGURE 18-3 A pulse (compression) is sent down a long air-filled tube. The reference frame of the figure is chosen so that the pulse is at rest and the air moves from left to right. (*a*) A slice of air of width Δx moves toward the pulse with speed v. (*b*) The leading face of the slice enters the pulse. The forces acting on the leading and trailing faces (due to air pressure) are shown.

face is $(p + \Delta p)A$ toward the left (Fig. 18-3b). So the average net force on the element during Δt is

$$F = pA - (p + \Delta p)A$$
$$= -\Delta p\, A \quad \text{(net force).} \quad (18\text{-}5)$$

The minus sign indicates that the net force on the fluid element points to the left in Fig. 18-3b. The volume of the element is $A\,\Delta x$, so with the aid of Eq. 18-4, we can write its mass as

$$\Delta m = \rho A\, \Delta x = \rho A v\, \Delta t \quad \text{(mass).} \quad (18\text{-}6)$$

Then the average acceleration of the element during Δt is

$$a = \frac{\Delta v}{\Delta t} \quad \text{(acceleration).} \quad (18\text{-}7)$$

From Newton's second law ($F = ma$), we have, from Eqs. 18-5, 18-6, and 18-7,

$$-\Delta p\, A = (\rho A v\, \Delta t)\frac{\Delta v}{\Delta t},$$

which we can write as

$$\rho v^2 = -\frac{\Delta p}{\Delta v / v}. \quad (18\text{-}8)$$

The air that occupies a volume $V\ (= A v\, \Delta t)$ outside the pulse is compressed by an amount $\Delta V\ (= A\, \Delta v\, \Delta t)$ as it enters the pulse. Thus

$$\frac{\Delta V}{V} = \frac{A\,\Delta v\,\Delta t}{A v\,\Delta t} = \frac{\Delta v}{v}. \quad (18\text{-}9)$$

Substituting Eq. 18-9 and then Eq. 18-2 into Eq. 18-8 leads to

$$\rho v^2 = -\frac{\Delta p}{\Delta v/v} = -\frac{\Delta p}{\Delta V/V} = B.$$

Solving for v yields Eq. 18-3 for the speed of the air toward the right in Fig. 18-3, and thus for the actual speed of the pulse toward the left.

SAMPLE PROBLEM 18-1

One clue used by your brain to determine the direction of a source of sound is the time delay Δt between the arrival of the sound at the ear closer to the source and the arrival at the farther ear. Assume that the source is distant so that a wavefront from it is approximately straight when it reaches you, and let D represent the separation between your ears.

(a) Find an expression that gives Δt in terms of D and the angle θ between the direction of the source and the forward direction.

SOLUTION: The situation is shown in Fig. 18-4 for wavefronts approaching you from a source that is located in front of

you and to your right. The time delay Δt is due to the distance d that each wavefront must travel to reach your left ear (L) after it reaches your right ear (R). From Fig. 18-4, we find

$$\Delta t = \frac{d}{v} = \frac{D \sin \theta}{v}, \quad \text{(Answer)} \quad (18\text{-}10)$$

where v is the speed of sound in air. Based on a lifetime of experience, your brain correlates each detected value of Δt (from zero to the maximum value) with a value of θ (from zero to 90°) for the direction of the sound source.

(b) Suppose that you are submerged in water at 20°C when a wavefront arrives from directly to your right. Based on the time-delay clue, at what angle θ from the forward direction will the source seem to be?

SOLUTION: We find the time delay Δt_w in this situation with Eq. 18-10, substituting $\theta = 90°$ and using v_w, the speed of sound in water, instead of v, the speed of sound in air:

$$\Delta t_w = \frac{D \sin 90°}{v_w} = \frac{D}{v_w}. \quad (18\text{-}11)$$

Since v_w is about four times v, Δt_w is about one-fourth the maximum time delay in air. Based on experience, your brain will process the time delay as if it occurred in air. Thus the sound source will appear to be at an angle θ smaller than 90°. To find that apparent angle, we substitute the time delay D/v_w from Eq. 18-11 for Δt in Eq. 18-10, obtaining

$$\frac{D}{v_w} = \frac{D \sin \theta}{v}. \quad (18\text{-}12)$$

Substituting $v = 343$ m/s and $v_w = 1482$ m/s (from Table 18-1) into Eq. 18-12, we then find

$$\sin \theta = \frac{v}{v_w} = \frac{343 \text{ m/s}}{1482 \text{ m/s}} = 0.231$$

and thus

$$\theta = 13°. \quad \text{(Answer)}$$

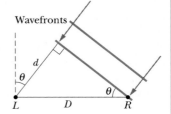

FIGURE 18-4 Sample Problem 18-1. A wavefront travels a distance d $(= D \sin \theta)$ farther to reach the left ear (L) than to reach the right ear (R).

18-3 TRAVELING SOUND WAVES

Here we examine the displacements and pressure variations associated with a sinusoidal sound wave passing through air. Figure 18-5a displays such a wave traveling rightward through a long air-filled tube. Recall that we can produce such a wave by sinusoidally moving a piston at the

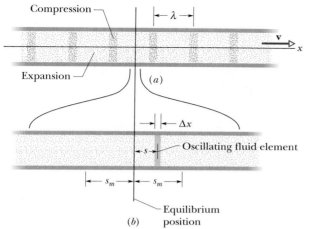

FIGURE 18-5 (a) A sound wave, traveling through a long air-filled tube with speed v, consists of a moving, periodic pattern of expansions and compressions of the air. The wave is shown at an arbitrary instant. (b) A horizontally expanded view of a short piece of the tube. As the wave passes, a fluid element of thickness Δx oscillates left and right in simple harmonic motion about its equilibrium position. At the instant shown in (b), the element happens to be displaced a distance s to the right of its equilibrium position. Its maximum displacement, either right or left, is s_m.

left end of the tube (as in Fig. 17-2). The piston's rightward motion moves the element of air next to it and compresses that air; the piston's leftward motion allows the element of air to move back to the left and the pressure to decrease. As each element of air pushes on the next element in turn, the right–left motion of the air and the change in its pressure travel along the tube as a sound wave.

Consider a thin element of air of thickness Δx, located at a position x along the tube. As the wave passes through x, the element of air oscillates left and right in simple harmonic motion about its equilibrium position (Fig. 18-5b). Thus the oscillations of each air element due to the passing sound wave are like those of a string element due to a transverse wave, except that the air element oscillates *longitudinally* rather than *transversely*. To describe the displacement $s(x, t)$ of an element of air from its equilibrium position, we can use either a sine function or a cosine function. In this chapter we use a cosine function:

$$s(x, t) = s_m \cos(kx - \omega t). \qquad (18\text{-}13)$$

Here s_m is the **displacement amplitude,** that is, the maximum displacement of the air element to either side of its equilibrium position (Fig. 18-5b).* The angular wave

*For the transverse displacement of an element of a stretched string, we used the symbol $y(x, t)$. We use the notation $s(x, t)$ here to avoid writing $x(x, t)$ for the longitudinal displacement of an element of air.

number k, angular frequency ω, frequency f, wavelength λ, speed v, and period T for a sound (longitudinal) wave are defined and interrelated exactly as for a transverse wave, except that λ is now the distance (again along the direction of travel) in which the pattern of compression and expansion due to the wave begins to repeat itself (see Fig. 18-5a). (We assume s_m is much less than λ.)

As the wave moves, the air pressure at any position x in Fig. 18-5a varies sinusoidally, as we prove below. To describe this variation we write

$$\Delta p\,(x, t) = \Delta p_m \sin(kx - \omega t). \qquad (18\text{-}14)$$

A negative value of Δp in Eq. 18-14 corresponds to an expansion of the air, and a positive value to a compression. Here Δp_m is the **pressure amplitude,** which is the maximum increase or decrease in pressure due to the wave; Δp_m is normally very much less than the pressure p present when there is no wave. As we shall prove, the pressure amplitude Δp_m is related to the displacement amplitude s_m in Eq. 18-13 by

$$\Delta p_m = (v \rho \omega)\, s_m. \qquad (18\text{-}15)$$

Figure 18-6 shows plots of Eqs. 18-13 and 18-14 at $t = 0$; with time, the two curves would move rightward along the horizontal axes. Note that the displacement and pressure variation are $\pi/2$ rad (or 90°) out of phase. Thus, for example, the pressure variation is zero when the displacement is a maximum.

(a)

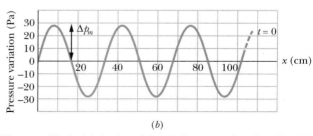

(b)

FIGURE 18-6 (a) A plot of the displacement function (Eq. 18-13) for $t = 0$. (b) A similar plot of the pressure variation function (Eq. 18-14). Both plots are for a 1000 Hz sound wave whose pressure amplitude is at the threshold of pain; see Sample Problem 18-2.

CHECKPOINT 1: When the oscillating fluid element in Fig. 18-5b is moving rightward through the point of zero displacement, is the pressure at its equilibrium value, just beginning to increase, or just beginning to decrease?

Derivation of Eqs. 18-14 and 18-15

Figure 18-5b shows an oscillating element of air of cross-sectional area A and thickness Δx, with its center displaced from its equilibrium position by distance s.

From Eq. 18-2 we can write, for the pressure variation in the displaced element,

$$\Delta p = -B \frac{\Delta V}{V}. \tag{18-16}$$

The quantity V in Eq. 18-16 is the volume of the element, given by

$$V = A\,\Delta x. \tag{18-17}$$

The quantity ΔV in Eq. 18-16 is the change in volume that occurs when the element is displaced. This volume change comes about because the displacements of the two faces of the element are not quite the same, differing by some amount Δs. Thus we can write the change in volume as

$$\Delta V = A\,\Delta s. \tag{18-18}$$

Substituting Eqs. 18-17 and 18-18 into Eq. 18-16 and passing to the differential limit yield

$$\Delta p = -B\frac{\Delta s}{\Delta x} = -B\frac{\partial s}{\partial x}. \tag{18-19}$$

The symbols ∂ indicate that the derivative in Eq. 18-19 is a *partial derivative* which tells us how s changes with x when the time t is fixed. From Eq. 18-13 we then have, treating t as a constant,

$$\frac{\partial s}{\partial x} = \frac{\partial}{\partial x}\,[s_m\cos(kx-\omega t)] = -ks_m\sin(kx-\omega t).$$

Substituting this quantity for the partial derivative in Eq. 18-19 yields

$$\Delta p = Bks_m\sin(kx-\omega t),$$

which is Eq. 18-14, which we set out to prove, with $\Delta p_m = Bks_m$.

Using Eq. 18-3, we can now write

$$\Delta p_m = (Bk)s_m = (v^2\rho k)s_m.$$

Equation 18-15, which we also promised to prove, follows at once if we eliminate k by using $v = \omega/k$ (Eq. 17-12).

18-4 INTERFERENCE

Figure 18-7 shows two point sources S_1 and S_2 that emit sound waves of wavelength λ in phase; that is, the emerging waves reach their maximum displacement values simultaneously. Suppose also that the sound waves pass through some common point P while traveling in (approximately) the same direction. If they have traveled along paths with identical lengths to reach point P, they will be in phase there. However, if they have traveled along paths with different lengths, as in Fig. 18-7, they may not be in phase. Their phase difference at point P depends on the **path length difference** ΔL between the two paths.

> When two waves travel along different paths, the path length difference ΔL can alter their phase difference.

FIGURE 18-7 Two point sources S_1 and S_2 emit spherical sound waves in phase. The rays indicate that the waves pass through a common point P.

We find the phase difference ϕ between the waves by recalling (from Section 17-4) that a phase difference of 2π rad corresponds to one wavelength. Thus we can write the proportion

$$\frac{\phi}{2\pi} = \frac{\Delta L}{\lambda}, \qquad (18\text{-}20)$$

from which

$$\phi = \frac{\Delta L}{\lambda} 2\pi. \qquad (18\text{-}21)$$

Like transverse waves, sound waves undergo constructive interference and destructive interference. They undergo fully constructive interference when ϕ is zero or an integer multiple of 2π, that is, when

$$\phi = m(2\pi), \qquad m = 0, 1, 2, \cdots$$
$$\text{(fully constructive interference).} \quad (18\text{-}22)$$

They undergo fully destructive interference when ϕ is an odd multiple of π, that is, when

$$\phi = (m + \tfrac{1}{2})(2\pi), \qquad m = 0, 1, 2, \cdots$$
$$\text{(fully destructive interference).} \quad (18\text{-}23)$$

From Eq. 18-21, we see that these conditions correspond to

$$\Delta L = m\lambda, \qquad m = 0, 1, 2, \cdots$$
$$\text{(fully constructive interference).} \quad (18\text{-}24)$$

and

$$\Delta L = (m + \tfrac{1}{2})\lambda, \qquad m = 0, 1, 2, \cdots$$
$$\text{(fully destructive interference).} \quad (18\text{-}25)$$

SAMPLE PROBLEM 18-3

In Fig. 18-8a, two point sources S_1 and S_2, which are in phase and separated by a distance $D = 1.5\lambda$, emit identical sound waves of wavelength λ.

(a) What is the phase difference of the waves from S_1 and S_2 at point P_1, which lies on the perpendicular bisector of the distance D, but at a distance greater than D from the sources. What type of interference occurs at P_1?

SOLUTION: To reach P_1, waves emitted by S_1 and S_2 travel along paths that are not in the same direction. However, because P_1 is farther from the sources than the source separation D, we can approximate the directions of the waves as being the same. Furthermore, those waves travel identical distances to reach P_1. Thus the path length difference ΔL is 0, and Eq. 18-21 tells us that

$$\phi = \frac{\Delta L}{\lambda} 2\pi = 0. \qquad \text{(Answer)}$$

The value $\phi = 0$ satisfies Eq. 18-22 with $m = 0$, and so it corresponds to fully constructive interference. We get the same result with Eq. 18-24: the path length difference $\Delta L = 0$ satisfies that equation, with $m = 0$.

(b) What are the phase difference and type of interference at point P_2 in Fig. 18-8a?

SOLUTION: Point P_2 is on the line that extends through S_1 and S_2. So to reach P_2, waves emitted by S_1 must travel a distance D farther than those emitted by S_2. From Eq. 18-21 with $\Delta L = D = 1.5\lambda$, we find

$$\phi = \frac{\Delta L}{\lambda} 2\pi = \frac{1.5\lambda}{\lambda} 2\pi = 3\pi \text{ rad}. \qquad \text{(Answer)}$$

This value for ϕ satisfies Eq. 18-23 with $m = 1$, and so it corresponds to fully destructive interference. We reach the same conclusion with Eq. 18-25: because $\Delta L = 1.5\lambda$, that equation is satisfied with $m = 1$. Note that whether the phase difference is expressed in radians or wavelengths, the result is independent of the distance between P_2 and the source S_2.

(c) Figure 18-8b shows a circle with a radius much greater than D, centered on the midpoint between sources S_1 and S_2. What is the number of points N around this circle at which the interference is fully constructive?

SOLUTION: From (a), we know that the path length difference is $\Delta L = 0$ at points a and b in Fig. 18-8b, where the

(a)

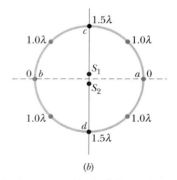

(b)

FIGURE 18-8 Sample Problem 18-3. (a) Two point sources S_1 and S_2, separated by distance D, emit spherical sound waves in phase. The waves travel equal distances to reach point P_1. Point P_2 is on the line extending through S_1 and S_2. (b) The phase difference (in terms of wavelengths) between the waves from S_1 and S_2, at eight points on a large circle around the sources.

circle intersects the perpendicular bisector of a line through the sources. From (b), we know that the path length difference is $\Delta L = 1.5\lambda$ at points c and d where the circle intersects a straight line through the sources. These results indicate that there must be intermediate points along the circle where $\Delta L = 1.0\lambda$; fully constructive interference occurs at those points. Without locating these four points exactly, we can indicate their presence as in Fig. 18-8b. We then count the points of fully constructive interference around the circle, finding

$$N = 6. \qquad \text{(Answer)}$$

CHECKPOINT 2: In Sample Problem 18-3, if the distance D between sources S_1 and S_2 were, instead, equal to 4λ, what type of interference would occur, and what value of m would be associated with it, at (a) point P_1 and (b) point P_2?

18-5 INTENSITY AND SOUND LEVEL

If you have ever tried to sleep while someone played loud music nearby, you are well aware that there is more to sound than frequency, wavelength, and speed. There is also intensity. The **intensity** I of a sound wave at a surface is the average rate per unit area at which energy is transferred by the wave through or onto the surface. We can write this as

$$I = \frac{P}{A}, \qquad (18\text{-}26)$$

where P is the rate of energy transfer (power) of the sound wave and A is the area of the surface intercepting the sound. The intensity I is related to the displacement amplitude s_m of the sound wave by

$$I = \tfrac{1}{2}\rho v \omega^2 s_m^2, \qquad (18\text{-}27)$$

which we shall derive shortly.

Variation of Intensity with Distance

How intensity varies with distance from a real sound source is often complex: some real sources (like loudspeakers) may beam sound in particular directions, and the environment usually produces echoes (reflected sound waves) that overlap the direct sound waves. In some situations, however, we can ignore echoes and assume that the sound source is a point source that emits the sound *isotropically,* that is, with equal intensity in all directions. The wavefronts spreading from such an isotropic point source S at a particular instant are shown in Fig. 18-9.

Let us assume that the mechanical energy of the sound waves is conserved as they spread from this source. Let us

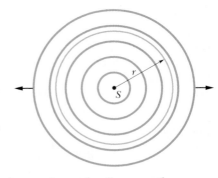

FIGURE 18-9 A point source S emits sound waves uniformly in all directions. The waves pass through an imaginary sphere of radius r that is centered on S.

also center an imaginary sphere of radius r on the source, as shown in Fig. 18-9. All the energy emitted by the source must pass through the surface of the sphere. Thus, the rate at which energy is transferred through the surface by the sound waves must equal the rate at which energy is emitted by the source (that is, the power P_s of the source). From Eq. 18-26, the intensity I at the sphere must then be

$$I = \frac{P_s}{4\pi r^2}, \qquad (18\text{-}28)$$

where $4\pi r^2$ is the area of the sphere. Equation 18-28 tells us that the intensity of sound from an isotropic point source decreases with the square of the distance r from the source.

CHECKPOINT 3: The figure indicates three small patches 1, 2, and 3 that lie on the surfaces of two imaginary spheres; the spheres are centered on an isotropic point source of sound S. The rates at which energy is transmitted through the three patches by the sound waves are equal. Rank the patches according to (a) the intensity of the sound on them and (b) their area, greatest first.

The Decibel Scale

You saw in Sample Problem 18-2 that the displacement amplitude at the human ear ranges from about 10^{-5} m for the loudest tolerable sound to about 10^{-11} m for the faintest detectable sound, a ratio of 10^6. From Eq. 18-27 we see that the intensity of a sound varies as the *square* of its amplitude, so the ratio of intensities at these two limits of the human auditory system is 10^{12}. Humans can hear over an enormous range of intensities.

Sound can cause the wall of a drinking glass to oscillate. If the sound produces a standing wave of oscillations and if the intensity of the sound is large enough, the glass will shatter.

We deal with such an enormous range of values by using logarithms. Consider the relation

$$y = \log x,$$

in which x and y are variables. It is a property of this equation that if we *multiply* x by 10, then y increases by 1. To see this, we write

$$y' = \log(10x) = \log 10 + \log x = 1 + y.$$

Similarly, if we multiply x by 10^{12}, y increases by only 12.

Thus instead of speaking of the intensity I of a sound wave, it is much more convenient to speak of its **sound level** β, defined as

$$\beta = (10 \text{ dB}) \log \frac{I}{I_0}. \qquad (18\text{-}29)$$

Here dB is the abbreviation for **decibel,** the unit of sound level, a name that was chosen to recognize the work of Alexander Graham Bell. I_0 in Eq. 18-29 is a standard reference intensity ($= 10^{-12}$ W/m²), chosen because it is near the lower limit of the human range of hearing. For $I = I_0$, Eq. 18-29 gives $\beta = 10 \log 1 = 0$, so our standard reference level corresponds to zero decibels. Then β increases by 10 dB every time the sound intensity increases by an order of magnitude (a factor of 10). Thus, $\beta = 40$ corresponds to an intensity that is 10^4 times the standard reference level. Table 18-2 lists the sound levels for a variety of environments.

TABLE 18-2 SOME SOUND LEVELS (dB)

Hearing threshold	0	Rock concert	110
Rustle of leaves	10	Pain threshold	120
Conversation	60	Jet engine	130

Derivation of Eq. 18-27

Consider, in Fig. 18-5a, a thin slice of air of thickness dx, area A, and mass dm, oscillating back and forth as the sound wave of Eq. 18-13 passes through it. The kinetic energy dK of the slice of air is

$$dK = \tfrac{1}{2} dm\, v_s^2. \qquad (18\text{-}30)$$

Here v_s is not the speed of the wave but the speed of the oscillating element of air, obtained from Eq. 18-13 as

$$v_s = \frac{\partial s}{\partial t} = -\omega s_m \sin(kx - \omega t).$$

Using this relation and putting $dm = \rho A\, dx$ allow us to rewrite Eq. 18-30 as

$$dK = \tfrac{1}{2}(\rho A\, dx)(-\omega s_m)^2 \sin^2(kx - \omega t). \qquad (18\text{-}31)$$

Dividing Eq. 18-31 by dt gives the rate at which kinetic energy moves along with the wave. As we saw in Chapter 17 for transverse waves, the ratio dx/dt is the wave speed, so we have

$$\frac{dK}{dt} = \tfrac{1}{2}\rho A v \omega^2 s_m^2 \sin^2(kx - \omega t). \qquad (18\text{-}32)$$

The *average* rate at which kinetic energy is transported is

$$\overline{\left(\frac{dK}{dt}\right)} = \tfrac{1}{2}\rho A v \omega^2 s_m^2 \,\overline{\sin^2(kx - \omega t)}$$
$$= \tfrac{1}{4}\rho A v \omega^2 s_m^2. \qquad (18\text{-}33)$$

To obtain this equation, we have used the fact that the average value of the square of a sine (or a cosine) function over one wavelength is $\tfrac{1}{2}$.

We assume that *potential* energy is carried along with the wave at this same average rate. The wave intensity I, which is the average rate per unit area at which energy of both kinds is transmitted by the wave, is then, from Eq. 18-33,

$$I = \frac{2\overline{(dK/dt)}}{A} = \tfrac{1}{2}\rho v \omega^2 s_m^2,$$

which is Eq. 18-27, the equation we set out to derive.

SAMPLE PROBLEM 18-4

An electric spark jumps along a straight line of length $L = 10$ m, emitting a pulse of sound that travels radially outward from the spark. (The spark is said to be a *line source* of sound.) The power of the emission is $P_s = 1.6 \times 10^4$ W.

(a) What is the intensity I of the sound when it reaches a distance $r = 12$ m from the spark?

SOLUTION: Let us center an imaginary cylinder of radius $r = 12$ m and length $L = 10$ m (open at both ends) on the spark, as shown in Fig. 18-10. The rate at which energy passes through the cylinder must equal the rate P_s at which energy is emitted by the source. From Eq. 18-26, the intensity I at the cylindrical surface must then equal the power P_s divided by the area $2\pi rL$ of that surface:

$$I = \frac{P_s}{2\pi rL}. \qquad (18\text{-}34)$$

This tells us that the intensity of the sound from a line source decreases with distance r (and not with the square of distance r as for a point source). Substituting the given data, we find

$$I = \frac{1.6 \times 10^4 \text{ W}}{2\pi(12 \text{ m})(10 \text{ m})} = 21.2 \text{ W/m}^2 \approx 21 \text{ W/m}^2. \quad (\text{Answer})$$

(b) At what rate P_d is sound energy intercepted by an acoustic detector of area $A_d = 2.0$ cm^2, aimed at the spark and located a distance $r = 12$ m from the spark?

SOLUTION: From Eq. 18-26, we know that

$$I = \frac{P_d}{A_d}.$$

Solving for P_d and substituting the intensity I from (a) and the given value for A_d, we find

$$P_d = (21.2 \text{ W/m}^2)(2.0 \times 10^{-4} \text{ m}^2) = 4.2 \text{ mW}. \quad (\text{Answer})$$

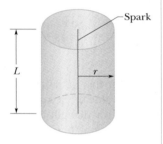

FIGURE 18-10 Sample Problem 18-4. A spark along a straight path of length L emits sound waves radially outward. The waves pass through an imaginary cylinder of radius r and length L that is centered on the spark.

SAMPLE PROBLEM 18-5

In 1976, the Who set a record for the loudest concert: the sound level 46 m in front of the speaker systems was $\beta_2 = 120$ dB. What is the ratio of the intensity I_2 of the band at that spot to the intensity I_1 of a jackhammer operating at sound level $\beta_1 = 92$ dB?

SOLUTION: Let us write the ratio of the two intensities as

$$\frac{I_2}{I_1} = \frac{I_2/I_0}{I_1/I_0}.$$

Taking logarithms on each side and multiplying through by 10 dB, we have

$$(10 \text{ dB}) \log \frac{I_2}{I_1} = (10 \text{ dB}) \log \frac{I_2}{I_0} - (10 \text{ dB}) \log \frac{I_1}{I_0}.$$

From Eq. 18-29 we then see that the terms on the right are β_2

and β_1. So

$$(10 \text{ dB}) \log \frac{I_2}{I_1} = \beta_2 - \beta_1. \qquad (18\text{-}35)$$

Note that the *ratio* of two intensities corresponds to a *difference* in their sound levels. Substituting given data yields

$$(10 \text{ dB}) \log \frac{I_2}{I_1} = 120 \text{ dB} - 92 \text{ dB} = 28 \text{ dB}$$

and

$$\log \frac{I_2}{I_1} = \frac{28 \text{ dB}}{10 \text{ dB}} = 2.8.$$

Taking the antilog of both sides gives us

$$\frac{I_2}{I_1} = 630. \qquad (\text{Answer})$$

Thus, the Who was *very* loud.

Temporary exposure to sound intensities as great as those of a jackhammer and the 1976 Who concert results in a temporary reduction of hearing. Repeated or prolonged exposure can result in permanent reduction of hearing (Fig. 18-11). Loss of hearing is a clear risk for anyone continually listening to, say, heavy metal at high volume.

FIGURE 18-11 Sample Problem 18-5. Peter Townshend of the Who, playing in front of a speaker system. His repeated and prolonged exposure to high-intensity sound, especially when playing into a speaker to produce feedback, resulted in a permanent reduction in his hearing.

18-6 SOURCES OF MUSICAL SOUND

Musical sounds can be set up by oscillating strings (guitar, piano, violin), membranes (kettledrum, snare drum), air columns (flute, oboe, pipe organ, and the fujara of Fig. 18-12), wooden blocks or steel bars (marimba, xylophone), and many other oscillating bodies. Most instruments involve more than a single oscillating part. In the violin, for example, not only the strings but also the body of the instrument participates in producing the music.

Recall from Chapter 17 that standing waves can be set up on a stretched string that is fixed at both ends. They arise because waves traveling along the string are reflected back onto the string at each end. If the wavelength of the waves is suitably matched to the length of the string, the superposition of waves traveling in opposite directions produces a standing wave pattern (or oscillation mode). The wavelength required of the waves for such a match is one that corresponds to a *resonant frequency* of the string. The advantage of setting up standing waves is that the string then oscillates with a large, sustained amplitude, pushing back and forth against the surrounding air and thus generating a noticeable sound wave with the same fre-

quency as the oscillations of the string. This production of sound is of obvious importance to, say, a guitarist.

We can set up standing waves of sound in an air-filled pipe in a similar way. As sound waves travel through the air in the pipe, they reflect at each end and back through the pipe. (The reflection occurs even if an end is open, but the reflection is not as complete as when the end is closed.) If the wavelength of the sound waves is suitably matched to the length of the pipe, the superposition of waves traveling in opposite directions through the pipe sets up a standing wave pattern. The wavelength required of the sound waves for such a match is one that corresponds to a resonant frequency of the pipe. The advantage of such a standing wave is that the air in the pipe oscillates with a large, sustained amplitude, emitting at any open end a sound wave that has the same frequency as the oscillations in the pipe. This emission of sound is of obvious importance to, say, an organist.

Many other aspects of standing sound wave patterns are similar to those of string waves: the closed end of a pipe is like the fixed end of a string in that there must be a node (zero displacement) there. And the open end of a pipe is like the end of a string attached to a freely moving ring, as in Fig. 17-16b, in that there must be an antinode there. (Actually, the antinode for the open end of a pipe is located slightly beyond the end, but we shall not dwell on that detail here.)

The simplest standing wave pattern that can be set up in a pipe with two open ends is shown in Fig. 18-13a. There is an antinode across each open end, as required. There is also a node across the middle of the pipe. An easier way of representing this standing longitudinal sound wave is shown in Fig. 18-13b—by drawing it as a standing transverse string wave.

The standing wave pattern of Fig. 18-13a is called the *fundamental mode* or *first harmonic*. For it to be set up, the sound waves in a pipe of length L must have a wavelength

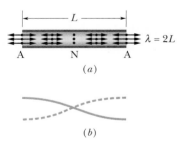

(a)

(b)

FIGURE 18-13 (a) The simplest standing wave pattern of displacement for (longitudinal) sound waves in a pipe with both ends open has an antinode (A) across each end and a node (N) across the middle. (The displacements represented by the double arrows are greatly exaggerated.) (b) The corresponding standing wave pattern for (transverse) string waves.

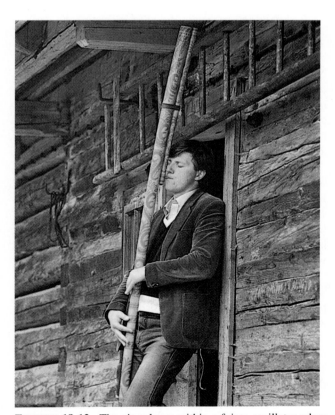

FIGURE 18-12 The air column within a fujara oscillates when that traditional Slovakian instrument is played.

FIGURE 18-14 Standing wave patterns for string waves superimposed on pipes to represent standing sound wave patterns in the pipes. (*a*) With *both* ends of the pipe open, any harmonic can be set up in the pipe. (*b*) But with only *one* end open, only odd harmonics can be set up.

FIGURE 18-15 The saxophone and violin families, showing the relations between instrument length and frequency range. The frequency range of each instrument is indicated by a horizontal bar along a frequency scale suggested by the keyboard at the bottom; the frequency increases toward the right.

given by $L = \lambda/2$, so that $\lambda = 2L$. Several more standing sound wave patterns for a pipe with two open ends are shown in Fig. 18-14*a* using string wave representations. The *second harmonic* requires sound waves of wavelength $\lambda = L$. The *third harmonic* requires wavelength $\lambda = 2L/3$. And so on.

More generally, the resonant frequencies for a pipe of length L with two open ends correspond to the wavelengths

$$\lambda = \frac{2L}{n}, \qquad n = 1, 2, 3, \cdots, \qquad (18\text{-}36)$$

where n is called the *harmonic number*. The resonant frequencies are then given by

$$f = \frac{v}{\lambda} = \frac{nv}{2L}, \qquad n = 1, 2, 3, \cdots$$

$$\text{(pipe, two open ends)}, \qquad (18\text{-}37)$$

where v is the speed of sound.

Figure 18-14*b* shows (using string wave representations) some of the standing sound wave patterns that can be set up in a pipe with only one open end. As required, across the open end there is an antinode and across the closed end there is a node. The simplest pattern requires sound waves having a wavelength given by $L = \lambda/4$, so that $\lambda = 4L$. The next simplest pattern requires a wavelength given by $L = 3\lambda/4$, so that $\lambda = 4L/3$. And so on.

More generally, the resonant frequencies for a pipe of length L with only one open end correspond to the wavelengths

$$\lambda = \frac{4L}{n}, \qquad n = 1, 3, 5, \cdots, \qquad (18\text{-}38)$$

in which the harmonic number n *must be an odd number*. The resonant frequencies are then given by

$$f = \frac{v}{\lambda} = \frac{nv}{4L}, \qquad n = 1, 3, 5, \cdots$$

$$\text{(pipe, one open end)}. \qquad (18\text{-}39)$$

Note again that only odd harmonics can exist in a pipe with one open end. For example, the second harmonic, with $n = 2$, cannot be set up in such a pipe. Note also that for such a pipe the adjective in a phrase such as "the third harmonic" still refers to the harmonic number n (and not to, say, the third possible harmonic).

The length of a musical instrument reflects the range of frequencies over which the instrument is designed to function: smaller length implies higher frequencies. Figure 18-15, for example, shows the saxophone and violin families, with their frequency ranges suggested by the piano keyboard. Note that, for every instrument, there is overlap with its higher- and lower-frequency neighbors.

In any oscillating system that gives rise to a musical sound, whether it be a violin string or the air in an organ pipe, the fundamental and one or more of the higher harmonics are usually generated simultaneously. The resultant sound is the superposition of these components. For different instruments, the higher harmonics have different intensities, which accounts for the different sounds produced by different instruments playing the same note. Figure 18-16 shows, for example, the resultant wave forms when the same note, with the same fundamental frequency, is sounded on three different instruments.

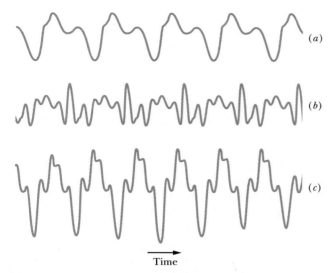

FIGURE 18-16 The wave forms produced by (a) a flute, (b) an oboe, and (c) a saxophone when they all play the same note, with the same first harmonic frequency.

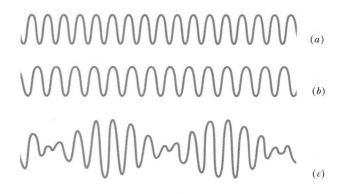

FIGURE 18-17 (a, b) The pressure variations Δp of two sound waves that are detected separately. The frequencies of the waves are nearly equal. (c) The resultant pressure variation if the two waves are detected simultaneously.

SAMPLE PROBLEM 18-6

Weak background noises from a room set up the fundamental standing wave in a cardboard tube of length $L = 67.0$ cm with two open ends. Assume that the speed of sound in the air within the tube is 343 m/s.

(a) What frequency do you hear from the tube if you jam your ear against one end?

SOLUTION: With your ear closing one end, the fundamental frequency is given by Eq. 18-39 with $n = 1$:

$$f = \frac{v}{4L} = \frac{343 \text{ m/s}}{(4)(0.670 \text{ m})} = 128 \text{ Hz}. \quad \text{(Answer)}$$

If the background noises set up any higher harmonics, such as the third harmonic, you may also hear frequencies that are *odd* multiples of 128 Hz.

(b) What frequency do you hear from the tube if you move your head away enough so that the tube has two open ends?

SOLUTION: With both ends open, the fundamental frequency is given by Eq. 18-37 with $n = 1$:

$$f = \frac{v}{2L} = \frac{343 \text{ m/s}}{(2)(0.670 \text{ m})} = 256 \text{ Hz}. \quad \text{(Answer)}$$

If the background noises set up any higher harmonics, such as the second harmonic, you may also hear frequencies that are *integer* multiples of 256 Hz. However, sound with a frequency of 128 Hz is no longer emitted by the tube.

CHECKPOINT **4:** Pipe A, with length L, and pipe B, with length $2L$, both have two open ends. Which harmonic of pipe B has the same frequency as the fundamental of pipe A?

18-7 BEATS

If we listen, a few minutes apart, to two sounds whose frequencies are, say, 552 and 564 Hz, most of us cannot tell one from the other. However, if the sounds reach our ears simultaneously, what we hear is a sound whose frequency is 558 Hz, the *average* of the two combining frequencies. We also hear a striking variation in the intensity of this sound: it increases and decreases in slow, wavering *beats* that repeat at a frequency of 12 Hz, the *difference* between the two combining frequencies. Figure 18-17 shows this **beat** phenomenon.

Let the time-dependent variations of the displacements due to two sound waves at a particular location be

$$s_1 = s_m \cos \omega_1 t \quad \text{and} \quad s_2 = s_m \cos \omega_2 t. \quad \text{(18-40)}$$

We have assumed, for simplicity, that the waves have the same amplitude. According to the superposition principle, the resultant displacement is

$$s = s_1 + s_2 = s_m(\cos \omega_1 t + \cos \omega_2 t).$$

Using the trigonometric identity (see Appendix E)

$$\cos \alpha + \cos \beta = 2 \cos \tfrac{1}{2}(\alpha - \beta) \cos \tfrac{1}{2}(\alpha + \beta)$$

allows us to write the resultant displacement as

$$s = 2s_m \cos \tfrac{1}{2}(\omega_1 - \omega_2)t \cos \tfrac{1}{2}(\omega_1 + \omega_2)t. \quad \text{(18-41)}$$

If we write

$$\omega' = \tfrac{1}{2}(\omega_1 - \omega_2) \quad \text{and} \quad \omega = \tfrac{1}{2}(\omega_1 + \omega_2), \quad \text{(18-42)}$$

we can then write Eq. 18-41 as

$$s(t) = [2s_m \cos \omega' t] \cos \omega t. \quad \text{(18-43)}$$

We now assume that the angular frequencies ω_1 and ω_2 of the combining waves are almost equal, which means that $\omega \gg \omega'$ in Eq. 18-42. We can then regard Eq. 18-43 as a cosine function whose angular frequency is ω and whose amplitude (which is not constant but varies with angular frequency ω') is the quantity in the brackets.

A maximum amplitude will occur whenever $\cos \omega' t$ in Eq. 18-43 has the value $+1$ or -1, which happens twice in each repetition of the cosine function. Because $\cos \omega' t$ has angular frequency ω', the angular frequency ω_{beat} at which beats occur is $\omega_{\text{beat}} = 2\omega'$. Then, with the aid of Eq. 18-42, we can write

$$\omega_{\text{beat}} = 2\omega' = (2)(\tfrac{1}{2})(\omega_1 - \omega_2) = \omega_1 - \omega_2.$$

Because $\omega = 2\pi f$, we can recast this as

$$f_{\text{beat}} = f_1 - f_2 \quad \text{(beat frequency).} \quad (18\text{-}44)$$

Musicians use the beat phenomenon in tuning their instruments. If an instrument is sounded against a standard frequency (for example, the lead oboe's reference A) and tuned until the beat disappears, then the instrument is in tune with that standard. In musical Vienna, concert A (440 Hz) is available as a telephone service for the benefit of the city's many professional and amateur musicians.

SAMPLE PROBLEM 18-7

You wish to tune the note A_3 on a piano to its proper frequency of 220 Hz. You have available a tuning fork whose frequency is 440 Hz. How should you proceed?

SOLUTION: These two frequencies are too far apart to produce beats. Recall that in our analysis of Eq. 18-43 we assumed that the two combining frequencies were reasonably close to each other. However, note that the second harmonic of A_3 is 2×220 or 440 Hz.

Let us say that the actual piano string is mistuned such that its fundamental frequency is not exactly 220 Hz. You listen for beats between the fundamental frequency of the tuning fork and the second harmonic of A_3, and you hear a beat frequency of 6 Hz. You then change the tension in the corresponding string until the beat note disappears. The string is then in tune.

CHECKPOINT 5: In Sample Problem 18-7, you tighten the string and the beat frequency increases from 6 Hz. Should you continue to tighten the string or should you loosen the string to put the string in tune?

18-8 THE DOPPLER EFFECT

A police car is parked by the side of the highway, sounding its 1000 Hz siren. If you are also parked by the highway, you will hear that same frequency. But if there is relative motion between you and the police car, either toward or away from each other, you will hear a different frequency. For example, if you are driving *toward* the police car at 120 km/h (about 75 mi/h), you will hear a *higher* frequency (1096 Hz, an *increase* of 96 Hz). If you are driving *away from* the police car at that same speed, you will hear a *lower* frequency (904 Hz, a *decrease* of 96 Hz).

These motion-related frequency changes are examples of the **Doppler effect.** The effect was proposed (although not fully worked out) in 1842 by Austrian physicist Johann Christian Doppler. It was tested experimentally in 1845 by Buys Ballot in Holland, "using a locomotive drawing an open car with several trumpeters."

The Doppler effect holds not only for sound waves but also for electromagnetic waves, including microwaves, radio waves, and visible light. Police use the Doppler effect with microwaves to determine the speed of a car: a radar unit beams microwaves of a certain frequency f toward the oncoming car. The microwaves that reflect from the metal portions of the car back to the radar unit have a higher frequency f' owing to the motion of the car relative to the radar unit. The radar unit captures the difference between f' and f and converts it into the speed of the car, which is then displayed for the operator to see. However, the displayed speed is accurate only if the car is moving directly toward (or away from) the radar unit. (Misalignment reduces f' and also the displayed speed.)

In the analyses that follow, we restrict ourselves to sound waves, and we take as a reference frame the body of air through which these waves travel. This means that we

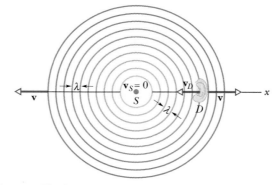

FIGURE 18-18 A stationary source of sound S emits spherical wavefronts, shown one wavelength apart, that expand outward at speed v. A sound detector D, represented by an ear, moves with velocity $\mathbf{v}_D$ toward the source. The detector senses a higher frequency because of its motion.

measure the speeds of a source S of sound waves and a detector D of those waves *relative to that body of air.* (Unless otherwise stated, the body of air is stationary relative to Earth, so the speeds can also be measured relative to Earth.) We shall assume that S and D move either directly toward or directly away from each other, at speeds less than the speed of sound.

We shall now derive equations for the Doppler effect for two particular situations: (1) for the detector moving and the source stationary and (2) for the source moving and the detector stationary. Then we shall combine the equations for these particular situations to find the *general Doppler effect equation,* which holds for both those situations and also when both detector and source are moving.

Detector Moving; Source Stationary

In Fig. 18-18, a detector D (represented by an ear) is moving at speed v_D toward a stationary source S that emits spherical wavefronts, of wavelength λ and frequency f, moving at the speed v of sound in air. The wavefronts are drawn one wavelength apart. The frequency detected by detector D is the rate at which D intercepts wavefronts (or individual wavelengths). If D were stationary, that rate would be f, but since D is moving into the wavefronts, the rate of interception is greater, and thus the detected frequency f' is greater than f.

Let us for the moment consider the situation in which D is stationary (Fig. 18-19). In time t, the wavefronts move to the right a distance vt. The number of wavelengths in that distance vt is the number of wavelengths intercepted by D in time t, and that number is vt/λ. The rate at which D intercepts wavelengths, which is the frequency f detected by D, is

$$f = \frac{vt/\lambda}{t} = \frac{v}{\lambda}. \qquad (18\text{-}45)$$

In this situation, with D stationary, there is no Doppler effect: the frequency detected by D is the frequency emitted by S.

Now let us again consider the situation in which D moves opposite the wavefronts (Fig. 18-20). In time t, the wavefronts move to the right a distance vt as previously, but now D moves to the left a distance v_Dt. Thus in this time t, the distance moved by the wavefronts relative to D is $vt + v_Dt$. The number of wavelengths in this relative distance $vt + v_Dt$ is the number of wavelengths intercepted by D in time t, and is $(vt + v_Dt)/\lambda$. The *rate* at which D intercepts wavelengths in this situation is the frequency f', given by

$$f' = \frac{(vt + v_Dt)/\lambda}{t} = \frac{v + v_D}{\lambda}. \qquad (18\text{-}46)$$

From Eq. 18-45, we have $\lambda = v/f$. Then Eq. 18-46 becomes

$$f' = \frac{v + v_D}{v/f} = f\frac{v + v_D}{v}. \qquad (18\text{-}47)$$

Note that in Eq. 18-47, f' must be greater than f unless $v_D = 0$ (the detector is stationary).

Similarly, we can find the frequency detected by D if D moves away from the source. In this situation, the wavefronts move a distance $vt - v_Dt$ relative to D in time t, and f' is given by

$$f' = f\frac{v - v_D}{v}. \qquad (18\text{-}48)$$

In Eq. 18-48, f' must be less than f unless $v_D = 0$.

We can summarize Eqs. 18-47 and 18-48 with

$$f' = f\frac{v \pm v_D}{v} \qquad \begin{array}{l}\text{(detector moving;}\\ \text{source stationary).}\end{array} \qquad (18\text{-}49)$$

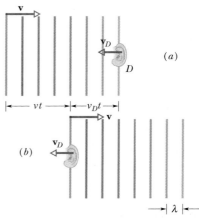

FIGURE 18-20 Wavefronts (*a*) reach and (*b*) pass detector D, which moves opposite the wavefronts. In time t, the wavefronts move a distance vt to the right and D moves a distance v_Dt to the left.

FIGURE 18-19 Wavefronts of Fig. 18-18, assumed planar, (*a*) reach and (*b*) pass a stationary detector D; they move a distance vt to the right in time t.

You can determine which sign applies in Eq. 18-49 by remembering the physical result: when the detector moves toward the source, the frequency is greater (*toward* means *greater*), which requires a plus sign in the numerator. Otherwise, a minus sign is required.

Source Moving; Detector Stationary

Let detector D be stationary with respect to the body of air, and let source S move toward D at speed v_S (Fig. 18-21). The motion of S changes the wavelength of the sound waves it emits, and thus the frequency detected by D.

To see this change, let $T (= 1/f)$ be the time between the emission of any pair of successive wavefronts W_1 and W_2. During T, wavefront W_1 moves a distance vT and the source moves a distance v_ST. At the end of T, wavefront W_2 is emitted. In the direction in which S moves, the distance between W_1 and W_2, which is the wavelength λ' of the waves moving in that direction, is $vT - v_ST$. If D detects those waves, it detects frequency f' given by

$$f' = \frac{v}{\lambda'} = \frac{v}{vT - v_ST} = \frac{v}{v/f - v_S/f}$$

$$= f\frac{v}{v - v_S}. \tag{18-50}$$

Note that f' must be greater than f unless $v_S = 0$.

In the direction opposite that taken by S, the wavelength λ' of the waves is $vT + v_ST$. If D detects those

waves, it detects frequency f' given by

$$f' = f\frac{v}{v + v_S}. \tag{18-51}$$

Now f' must be less than f unless $v_S = 0$.

We can summarize Eqs. 18-50 and 18-51 with

$$f' = f\frac{v}{v \mp v_S} \qquad \text{(source moving; detector stationary).} \tag{18-52}$$

You can determine which sign applies in Eq. 18-52 by remembering the physical result: when the source moves toward the detector, the frequency is greater (*toward* means *greater*), which requires a minus sign in the denominator. Otherwise a plus sign should be used.

General Doppler Effect Equation

We can combine Eqs. 18-49 and 18-52 to produce the general Doppler effect equation, in which both the source and the detector can be moving with respect to the air mass. Replacing f in Eq. 18-52 (the frequency of the source) with f' of Eq. 18-49 (the frequency associated with motion of the detector) leads to

$$f' = f\frac{v \pm v_D}{v \mp v_S} \qquad \text{(general Doppler effect).} \tag{18-53}$$

Putting $v_S = 0$ in Eq. 18-53 reduces it to Eq. 18-49, and putting $v_D = 0$ reduces it to Eq. 18-52. The plus and minus signs are determined as in Eqs. 18-49 and 18-52 (*toward* means *greater*).

The Doppler Effect at Low Speeds

The Doppler effects for a moving detector (Eq. 18-49) and for a moving source (Eq. 18-52) are different, even though the detector and the source may be moving at the same speed. However, if the speeds are low enough (that is, if $v_D \ll v$ and $v_S \ll v$), the frequency changes produced by these two motions are essentially the same.

By using the binomial theorem (see Tactic 2 of Chapter 7), you can show that Eq. 18-53 can be written in the form

$$f' \approx f\left(1 \pm \frac{u}{v}\right) \qquad \text{(low speeds only),} \tag{18-54}$$

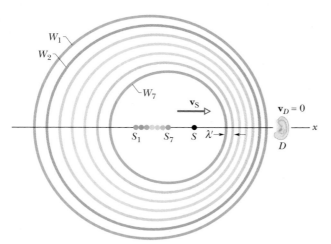

FIGURE 18-21 The detector D is stationary, with the source S moving toward it at speed v_S. Wavefront W_1 was emitted when the source was at S_1, wavefront W_7 when it was at S_7. At the moment depicted, the source is at S. The detector perceives a higher frequency because the moving source, chasing its own wavefronts, emits a reduced wavelength λ' in the direction of its motion.

in which u ($= |v_S \pm v_D|$) is the *relative* speed of the source with respect to the detector. The rule for signs remains the same. If the source and the detector are moving *toward* each other, we anticipate a *greater* frequency; this requires that we choose the plus sign in Eq. 18-54. On the other hand, if the source and the detector are moving away from each other, we anticipate a frequency decrease and choose the minus sign in Eq. 18-54.

CHECKPOINT **6**: The figure indicates the motion of a sound source and a detector for six situations in stationary air. For each situation, is the detected frequency greater than or less than the emitted frequency, or can't we tell without more information?

(a)

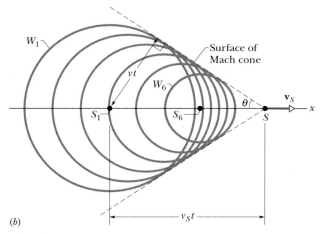

FIGURE 18-22 (a) A source of sound S moves at speed v_S equal to the speed of sound and thus as fast as the wavefronts it generates. (b) A source S moves at speed v_S faster than the speed of sound and thus faster than the wavefronts. When the source was at position S_1 it generated wavefront W_1, and at position S_6 it generated W_6. All the spherical wavefronts expand at the speed of sound v and bunch along the surface of a cone called the Mach cone, forming a shock wave. The surface of the cone has half-angle θ and is tangent to all the wavefronts.

Supersonic Speeds; Shock Waves

If a source is moving toward a stationary detector at a speed equal to the speed of sound, that is, if $v_S = v$, Eq. 18-52 predicts that the detected frequency f' will be infinitely great. This means that the source is moving so fast that it keeps pace with its own spherical wavefronts, as Fig. 18-22a suggests. What happens when the speed of the source *exceeds* the speed of sound?

For such *supersonic* speeds, Eq. 18-52 no longer applies. Figure 18-22b depicts the spherical wavefronts that originated at various positions of the source. The radius of any wavefront in this figure is vt, where v is the speed of sound and t is the time that has elapsed since the source emitted that wavefront. Note that all the wavefronts bunch along a V-shaped envelope in Fig. 18-22b, which in three dimensions is a cone called the *Mach cone*. A **shock wave** is said to exist along the surface of this cone, because the bunching of wavefronts causes an abrupt rise and fall of air pressure as the surface passes through any point. From Fig. 18-22b, we see that the half-angle θ of the cone, called the *Mach cone angle,* is given by

$$\sin \theta = \frac{vt}{v_S t} = \frac{v}{v_S} \qquad \text{(Mach cone angle)}. \quad (18\text{-}55)$$

The ratio v_S/v is called the *Mach number*. When you hear that a particular plane has flown at Mach 2.3, it means that its speed was 2.3 times the speed of sound in the air through which the plane was flying. The shock wave generated by a supersonic aircraft or projectile (Fig. 18-23) produces a burst of sound, called a *sonic boom,* in which the air pressure first suddenly increases and then suddenly decreases below normal before returning to normal.

FIGURE 18-23 A false-color, high-speed image of a 20 mm bullet traveling at about Mach 1.3. Note the prominent Mach cone produced by the nose of the bullet and the secondary cones produced by irregular features along the sides.

SAMPLE PROBLEM 18-8

A toy rocket moves at a speed of 242 m/s directly toward a stationary pole (through stationary air) while emitting sound waves at frequency $f = 1250$ Hz.

(a) What frequency f' is sensed by a detector that is attached to the pole?

SOLUTION: To find f', we shall use Eq. 18-53 for the general Doppler effect. Because the detector is stationary, we set $v_D = 0$. And because the sound source (the rocket) is moving *toward* the detector, we choose the minus sign in the denominator. Then substituting $v = 343$ m/s from Table 18-1 and the given data, we find that the detected frequency is

$$f' = f \frac{v}{v - v_S} = (1250 \text{ Hz}) \frac{343 \text{ m/s}}{343 \text{ m/s} - 242 \text{ m/s}}$$

$$= 4245 \text{ Hz} \approx 4250 \text{ Hz}. \qquad \text{(Answer)}$$

As a quick check on the answer, recall the physical result: when a source moves *toward* a stationary detector, the detected frequency (here, 4245 Hz) should be *greater* than the emitted frequency (here, 1250 Hz).

(b) Some of the sound reaching the pole reflects back to the rocket, which has an onboard detector. What frequency f'' does it detect?

SOLUTION: The pole now acts as the source of sound in that it reflects sound waves, producing an echo. The frequency of the waves from the pole is the same as the frequency of the waves "sensed" by the pole, namely, $f' = 4245$ Hz. Because now the source (the pole) is stationary, we set $v_S = 0$ in Eq. 18-53. And because now the detector (the onboard detector) is moving toward the new source, we choose the plus sign in the numerator. So the frequency sensed by the onboard detector is

$$f'' = f' \frac{v + v_D}{v} = (4245 \text{ Hz}) \frac{343 \text{ m/s} + 242 \text{ m/s}}{343 \text{ m/s}}$$

$$= 7240 \text{ Hz}. \qquad \text{(Answer)}$$

Again, as a quick check on the answer, recall the physical result: when a detector moves *toward* a stationary source, the detected frequency (here, 7240 Hz) should be *greater* than the emitted frequency (here, 4245 Hz).

CHECKPOINT 7: In Sample Problem 18-8, if the air is moving toward the pole at speed 20 m/s, (a) what value for the source speed v_S should be used in the solution of part a, and (b) what value and sign for the detector speed v_D should be used there?

SAMPLE PROBLEM 18-9

Bats navigate and search out prey by emitting, and then detecting reflections of, ultrasonic waves, which are sound waves with frequencies greater than what can be heard by a human. Suppose a horseshoe bat flies toward a moth at speed $v_b = 9.0$ m/s, while the moth flies toward the bat with speed $v_m = 8.0$ m/s. From its nostrils, the bat emits ultrasonic waves of frequency f_{be} that reflect from the moth back to the bat with frequency f_{bd}. The bat adjusts the emitted frequency f_{be} until the returned frequency f_{bd} is 83 kHz, at which the bat's hearing is best.

(a) What is the frequency f_m of the waves heard and reflected by the moth when f_{bd} is 83 kHz?

SOLUTION: To find f_m, we use Eq. 18-53, with the moth being the source (of reflected waves with frequency f_m) and the bat being the detector (of the echo with frequency $f_{bd} = 83$ kHz). Since the detector moves toward the source (with speed v_b), we choose the plus sign in the numerator of Eq. 18-53. Since the source moves toward the detector (with speed v_m), we choose the minus sign in the denominator. We then have

$$f_{bd} = f_m \frac{v + v_b}{v - v_m},$$

or $\qquad 83 \text{ kHz} = f_m \dfrac{343 \text{ m/s} + 9.0 \text{ m/s}}{343 \text{ m/s} - 8.0 \text{ m/s}},$

from which

$$f_m = 78.99 \text{ kHz} \approx 79 \text{ kHz}. \qquad \text{(Answer)}$$

(b) What is the frequency f_{be} emitted by the bat when f_{bd} is 83 kHz?

SOLUTION: We again use Eq. 18-53, but now with the bat as source (at frequency f_{be}) and the moth as detector (of frequency f_m). Since the detector moves toward the source (with speed v_m), we choose the plus sign in the numerator of Eq. 18-53. Since the source moves toward the detector (with speed v_b), we choose the minus sign in the denominator. We then have

$$f_m = f_{be} \frac{v + v_m}{v - v_b}$$

or $\qquad 78.99 \text{ kHz} = f_{be} \dfrac{343 \text{ m/s} + 8.0 \text{ m/s}}{343 \text{ m/s} - 9.0 \text{ m/s}},$

from which

$$f_{be} = 75 \text{ kHz}. \qquad \text{(Answer)}$$

The bat determines the relative speed of the moth (17 m/s) from the 8 kHz (= 83 kHz − 75 kHz) it must lower its emitted frequency to hear an echo with a frequency of 83 kHz (at which it hears best). Some moths evade capture by flying away from the direction in which they hear ultrasonic waves. That choice of flight path reduces the frequency difference between what the bat emits and what it hears, and then the bat may not notice the echo. Some moths avoid capture by clicking to produce their own ultrasonic waves, thus "jamming" the detection system and confusing the bat.

18-9 THE DOPPLER EFFECT FOR LIGHT

It is tempting to apply to light waves the Doppler effect equation (Eq. 18-53) developed for sound waves in the preceding section, simply substituting c, the speed of light, for v, the speed of sound. We must avoid this temptation.

The reason is that sound waves require a medium through which to travel, but light waves do not. Thus, although the speed of sound is always measured with respect to the medium, the speed of light is not. Moreover, the speed of light always has the same value c, in all directions and in all inertial frames. For these reasons, Einstein's theory of special relativity shows that the Doppler effect for light depends only on the relative motion between the source of the light and a detector.

Even though the Doppler equations for light and for sound are different, at low enough speeds they reduce to the same approximate result. (In fact, *all* the predictions of special relativity theory reduce to their *classical* counterparts at low enough speeds.) Thus, Eq. 18-54, with v replaced by c, holds for light waves if $u \ll c$, where u is the relative speed of the source and the detector. That is, as a close approximation,

$$f' = f(1 \pm u/c) \qquad \text{(light waves; } u \ll c\text{).} \qquad \text{(18-56)}$$

If the source and the detector are *approaching* each other, we anticipate a frequency *increase*, and our rule for signs calls for the plus sign in Eq. 18-56.

In Doppler observations in astronomy (which involve light rather than sound), the wavelength is easier to measure than the frequency. This suggests that we replace f' and f in Eq. 18-56 with c/λ' and c/λ, respectively. Doing so, we find

$$\lambda' = \lambda(1 \pm u/c)^{-1} \approx \lambda(1 \mp u/c).$$

We can write this as

$$\frac{\lambda' - \lambda}{\lambda} = \mp \frac{u}{c}$$

or as

$$u = \frac{\Delta\lambda}{\lambda} c \qquad \text{(light waves; } u \ll c\text{),} \qquad \text{(18-57)}$$

in which $\Delta\lambda$ is the *magnitude* (no signs) of the Doppler wavelength shift.

Equation 18-57 gives us a way to determine the relative speed of a light source and a detector from a measured wavelength shift. If the wavelength decreases (called a *blue shift* because the blue portion of the visible spectrum has the shortest wavelengths), the frequency necessarily increases. That means the source and detector are approaching each other. If the wavelength increases (a *red shift*), the source and detector are moving away from each

other. Astronomers have measured the wavelength shifts of light reaching Earth from distant stars and galaxies. They have discovered that light from *all* the distant galaxies has undergone a red shift, indicating that all those galaxies are moving away from us. Moreover, the more distant a galaxy is, the faster it is moving.

SAMPLE PROBLEM 18-10

Figure 18-24a shows curves of intensity versus wavelength for light reaching us from interstellar gas on two opposite sides of galaxy M87 (Fig. 18-24b). One curve peaks at 499.8 nm; the other at 501.6 nm. The gas orbits the core of the galaxy at a radius $r = 100$ light-years, apparently moving toward us on one side of the core and moving away from us on the opposite side.

(a) Which curve corresponds to the gas moving toward us? What is the speed of the gas relative to us (and relative to the galaxy's core)?

SOLUTION: If the gas were not moving around the galaxy's core, the light from it would be detected at some wavelength λ (set by the emission process and the motion of the galaxy away from us). However, the motion of the gas changes the detected wavelength via the Doppler effect, increasing the wavelength for the gas moving away from us and decreasing it for the gas moving toward us. Thus, the curve peaking at 501.6 nm corresponds to motion away from us, and that peaking at 499.8 nm corresponds to motion toward us.

Let us assume that the increase and the decrease in wavelength due to the motion of the gas are equal in magnitude. Then the unshifted wavelength λ must be the average of the two shifted wavelengths:

$$\lambda = \frac{501.6 \text{ nm} + 499.8 \text{ nm}}{2} = 500.7 \text{ nm.}$$

The Doppler shift $\Delta\lambda$ of the light from the gas moving away from us is then

$$\Delta\lambda = 501.6 \text{ nm} - 500.7 \text{ nm} = 0.90 \text{ nm.}$$

Substituting this and $\lambda = 500.7$ nm into Eq. 18-57, we find that the speed of the gas is

$$u = \frac{\Delta\lambda}{\lambda} c = \frac{0.90 \text{ nm}}{501.6 \text{ nm}} 3.0 \times 10^8 \text{ m/s}$$

$$= 5.39 \times 10^5 \text{ m/s.} \qquad \text{(Answer)}$$

(b) The gas orbits the core of the galaxy because it experiences a gravitational force due to the mass M of the core. What is that mass in multiples of the Sun's mass M_S ($= 1.99 \times 10^{30}$ kg)?

SOLUTION: From Eq. 14-1, the gravitational force on an orbiting gas element of mass m at orbital radius r is

$$F = \frac{GMm}{r^2}.$$

Applying Newton's second law to the gas element and substituting the centripetal acceleration u^2/r for a, we have

$$\frac{GMm}{r^2} = ma = \frac{mu^2}{r}.$$

Solving this for M and substituting known data, we find

$$M = \frac{u^2 r}{G}$$

$$= \frac{(5.39 \times 10^5 \text{ m/s})^2(100 \text{ ly})(9.46 \times 10^{15} \text{ m/ly})}{6.67 \times 10^{-11} \text{ N}\cdot\text{m}^2/\text{kg}^2}$$

$$= 4.12 \times 10^{39} \text{ kg} = (2.1 \times 10^9)M_S. \qquad \text{(Answer)}$$

This result tells us that a mass equivalent to two billion suns has been compacted into the core of the galaxy, strongly suggesting that a "supermassive" black hole occupies the core.

(a)   (b)

FIGURE 18-24 Sample Problem 18-10. (a) Plots of intensity versus wavelength for light emitted by gas on opposite sides of galaxy M87. (b) The central region of M87. The circles indicate the locations of the gas whose intensity is given in (a). The center of M87 is halfway between the circles.

REVIEW & SUMMARY

Sound Waves

Sound waves are longitudinal mechanical waves that can travel through solids, liquids, or gases. The speed v of a sound wave in a medium having **bulk modulus** B and density ρ is

$$v = \sqrt{\frac{B}{\rho}} \qquad \text{(speed of sound).} \qquad (18\text{-}3)$$

In air at 20°C, the speed of sound is 343 m/s.

A sound wave causes a longitudinal displacement s of a mass element in a medium as given by

$$s = s_m \cos(kx - \omega t), \qquad (18\text{-}13)$$

where s_m is the **displacement amplitude** (maximum displacement) from equilibrium, $k = 2\pi/\lambda$, and $\omega = 2\pi f$, λ and f being the wavelength and frequency, respectively, of the sound wave. The sound wave also causes a pressure change Δp of the medium from the equilibrium pressure:

$$\Delta p = \Delta p_m \sin(kx - \omega t), \qquad (18\text{-}14)$$

where the **pressure amplitude** is

$$\Delta p_m = (v\rho\omega)s_m. \qquad (18\text{-}15)$$

Interference

The interference of two sound waves with identical wavelengths passing through a common point depends on their phase difference ϕ there. If the sound waves were emitted in phase and are traveling in approximately the same direction, ϕ is given by

$$\phi = \frac{\Delta L}{\lambda} \, 2\pi, \qquad (18\text{-}21)$$

where ΔL is their **path length difference** (the difference in the distances traveled by the waves to reach the common point). Conditions for fully constructive and fully destructive interference of the waves are given by

$$\phi = m 2\pi, \qquad m = 0, 1, 2, \cdots$$
$$\text{(fully constructive interference)} \qquad (18\text{-}22)$$

and

$$\phi = (m + \tfrac{1}{2})2\pi, \qquad m = 0, 1, 2, \cdots$$
$$\text{(fully destructive interference).} \qquad (18\text{-}23)$$

These conditions correspond to

$$\Delta L = m\lambda, \qquad m = 0, 1, 2, \cdots$$
$$\text{(fully constructive interference)} \qquad (18\text{-}24)$$

and

$$\Delta L = (m + \tfrac{1}{2})\lambda, \qquad m = 0, 1, 2, \cdots$$
$$\text{(fully destructive interference).} \qquad (18\text{-}25)$$

Sound Intensity

The **intensity** I of a sound wave at a surface is the average rate per unit area at which energy is transferred by the wave through or onto the surface:

$$I = \frac{P}{A}, \qquad (18\text{-}26)$$

where P is the rate of energy transfer (power) of the sound wave and A is the area of the surface intercepting the sound. The intensity I is related to the displacement amplitude s_m of the sound wave by

$$I = \tfrac{1}{2}\rho v \omega^2 s_m^2. \qquad (18\text{-}27)$$

The intensity at a distance r from a point source that emits sound waves of power P_s is

$$I = \frac{P_s}{4\pi r^2}. \qquad (18\text{-}28)$$

Sound Level in Decibels

The *sound level* β in *decibels* (dB) is defined as

$$\beta = (10 \text{ dB}) \log \frac{I}{I_0}, \qquad (18\text{-}29)$$

where $I_0 \, (= 10^{-12} \text{ W/m}^2)$ is a reference intensity level to which all intensities are compared. For every factor-of-10 increase in intensity, 10 dB is added to the sound level.

Standing Wave Patterns in Pipes

Standing sound wave patterns can be set up in pipes. A pipe open at both ends will resonate at frequencies

$$f = \frac{v}{\lambda} = \frac{nv}{2L}, \qquad n = 1, 2, 3, \cdots$$

$$\text{(pipe, two open ends)}, \qquad (18\text{-}37)$$

where v is the speed of sound in the air in the pipe. For a pipe closed at one end and open at the other, the resonant frequencies are

$$f = \frac{v}{\lambda} = \frac{nv}{4L}, \qquad n = 1, 3, 5, \cdots$$

$$\text{(pipe, one open end)}. \qquad (18\text{-}39)$$

Beats

Beats arise when two waves having slightly different frequencies, f_1 and f_2, are detected together. The beat frequency is

$$f_{\text{beat}} = f_1 - f_2. \qquad (18\text{-}44)$$

The Doppler Effect

The *Doppler effect* is a change in the observed frequency of a wave when the source or the detector moves relative to the medium. For sound the observed frequency f' is given in terms of the source frequency f by

$$f' = f \frac{v \pm v_D}{v \mp v_S}, \qquad \begin{array}{c}\text{(general Doppler}\\ \text{effect),}\end{array} \qquad (18\text{-}53)$$

where v_D is the speed of the detector relative to the medium, v_S is that of the source, and v is the speed of sound in the medium. The signs are chosen such that f' tends to be *greater* for motion (of detector or source) "toward" and *less* for motion "away."

Shock Wave

If the source speed relative to the medium exceeds the speed of sound in the medium, the Doppler equation no longer applies. In such a case, shock waves result. The half angle θ (see Fig. 18-22) of the wavefront is given by

$$\sin \theta = \frac{v}{v_S} \qquad \text{(Mach cone angle)}. \qquad (18\text{-}55)$$

Doppler Effect for Light

When there is a relative speed u between a light source and a detector, the detected frequency f' of the light is

$$f' = f(1 \pm u/c), \qquad (18\text{-}56)$$

where f is the frequency that would be detected were there no relative motion. The relative speed u is related to the Doppler shift in wavelength $\Delta\lambda$ by

$$u = \frac{\Delta\lambda}{\lambda} c, \qquad (18\text{-}57)$$

where λ is the wavelength that would be detected were there no relative motion. If the source and detector move toward each other, the wavelength shift $\Delta\lambda$ is negative; if they move away from each other, $\Delta\lambda$ is positive.

QUESTIONS

1. Figure 18-25 shows the paths taken by two pulses of sound that begin simultaneously and then race each other through equal distances in air. The only difference between the paths is that a region of hot (low density) air lies along path 2. Which pulse wins the race?

Path 1

Path 2 — Hot air

FIGURE 18-25 Question 1.

2. A sound wave of wavelength λ and displacement amplitude s_m begins to travel down a passageway (a tube, the opening of the ear, etc.). When a small device in the passageway detects this wave, it issues a second sound wave (said to be *antisound*) that is able to cancel the first wave, so that nothing is heard at the far end of the passageway. For such cancellation, what must be (a) the direction of travel, (b) the wavelength, and (c) the displacement amplitude of the second wave? (d) What must be the phase difference between the two waves? (Such antisound devices are used to eliminate unwanted sound in a noisy environment.)

3. In Fig. 18-26, two point sources S_1 and S_2, which are in phase, emit identical sound waves of wavelength 2.0 m. In terms of wavelengths, what is the phase difference between the waves arriving at point P if (a) $L_1 = 38$ m and $L_2 = 34$ m, and (b) $L_1 = 39$ m and $L_2 = 36$ m? (c) Assuming that the source separation is much smaller than L_1 and L_2, what type of interference occurs at P in situations (a) and (b), respectively?

S_1

S_2

L_1

L_2

P

FIGURE 18-26 Question 3.

4. In Fig. 18-27, sound waves of wavelength λ are emitted by a point source S and travel to a detector D directly along path 1 and via reflection from a panel along path 2. Initially, the panel is almost along path 1 and the waves arriving at D along the two paths are almost exactly in phase. Then the panel is moved away from path 1 as shown until the waves arriving at D are exactly out of phase. What then is the path length difference Δd of the waves along the two paths?

FIGURE 18-27   Question 4.

5. In Fig. 18-28, two point sources S_1 and S_2, which are in phase, emit identical sound waves of wavelength λ, and point P is at equal distances from them. Then S_2 is moved directly away from P by a distance equal to $\lambda/4$. Are the waves at P then exactly in phase, exactly out of phase, or do they have some intermediate phase relation if (a) S_1 is moved directly toward P by a distance equal to $\lambda/4$ and (b) S_1 is moved directly away from P by a distance equal to $3\lambda/4$?

FIGURE 18-28   Question 5.

6. In Sample Problem 18-3 and Fig. 18-8a, the waves arriving at point P_1 on the perpendicular bisector are exactly in phase; that is, the waves from S_1 and S_2 always tend to move an element of air at P_1 in the same direction. Let the intersection of the perpendicular bisector and the line through S_1 and S_2 be point P_3. (a) Are the waves arriving at P_3 exactly in phase, exactly out of phase, or do they have some intermediate relation? (b) What is the answer if we increase the separation between the sources to 1.7λ?

7. Without using a calculator, determine the increase in a sound level when the intensity of the sound source becomes 10^7 times what it was previously.

8. A standing sound wave in a pipe has five nodes and five antinodes. (a) How many open ends does the pipe have? (b) What is the harmonic number n for this standing wave?

9. The sixth harmonic is set up in a pipe. (a) How many open ends does the pipe have (it has at least one)? (b) Is there a node, antinode, or some intermediate state at the midpoint?

10. (a) When an orchestra warms up, the players' warm breath increases the temperature of the air within the wind instruments (and thus decreases the density of that air). Do the resonant frequencies of those instruments increase or decrease? (b) When the

slide of a slide trombone is pushed outward, do the resonant frequencies of the instrument increase or decrease?

11. Pipe A has length L and one open end. Pipe B has length $2L$ and two open ends. Which harmonics of pipe B have a frequency that matches a resonant frequency of pipe A?

12. Figure 18-29 shows a stretched string of length L and pipes a, b, c, and d of lengths L, $2L$, $L/2$, and $L/2$, respectively. The string's tension is adjusted until the speed of waves on the string equals the speed of sound waves in the air. The fundamental mode of oscillation is then set up on the string. In which pipe will the sound produced by the string cause resonance, and what oscillation mode will that sound set up?

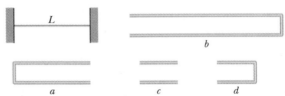

FIGURE 18-29   Question 12.

13. You are given four tuning forks. The fork with the lowest frequency oscillates at 500 Hz. By striking two tuning forks at a time, you can produce the following beat frequencies: 1, 2, 3, 5, 7, and 8 Hz. What are the possible frequencies of the other three forks? (There are two sets of answers.)

14. A friend rides, in turn, the rims of three fast merry-go-rounds while holding a sound source that emits isotropically at a certain frequency. You stand far from each merry-go-round. The frequency you hear for each of your friend's three rides varies as the merry-go-round rotates. The variations in frequency for the three rides are given by the three curves in Fig. 18-30. Rank the curves according to (a) the linear speed v of the sound source, (b) the angular speeds ω of the merry-go-rounds, and (c) the radii r of the merry-go-rounds, greatest first.

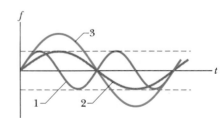

FIGURE 18-30   Question 14.

EXERCISES & PROBLEMS

Where needed in the problems, use

$$\text{speed of sound in air} = 343 \text{ m/s} = 1125 \text{ ft/s}$$

and

$$\text{density of air} = 1.21 \text{ kg/m}^3$$

unless otherwise specified.

SECTION 18-2   The Speed of Sound

1E. A rule for finding your distance in miles from a lightning flash is to count seconds from the time you see the flash until you hear the thunder and then divide the count by five. (a) Explain this rule and determine the percent error in it at 20°C, assuming that the sound travels to you along a straight path. (b) Devise a similar rule for obtaining the distance in kilometers.

2E. A column of soldiers, marching at 120 paces per minute, keep in step with the music of a band at the head of the column. It is observed that the soldiers in the rear end of the column are striding forward with the left foot when musicians are advancing with the right. What is the length of the column approximately?

3E. You are at a large outdoor concert, seated 300 m from the speaker system. The concert is also being broadcast live via satellite (at the speed of light). Consider a listener 5000 km away who receives the broadcast. Who hears the music first, you or the listener, and by what time difference?

4E. Two spectators at a soccer game in Montjuic Stadium see, and a moment later hear, the ball being kicked on the playing field. The time delay for one spectator is 0.23 s and for the other 0.12 s. Sight lines from each spectator to the player kicking the ball meet at an angle of 90°. (a) How far is each spectator from the player? (b) How far are the spectators from each other?

5E. The average density of Earth's crust 10 km beneath the continents is 2.7 g/cm³. The speed of longitudinal seismic waves at that depth, found by timing their arrival from distant earthquakes, is 5.4 km/s. Use this information to find the bulk modulus of Earth's crust at that depth. For comparison, the bulk modulus of steel is about 16×10^{10} Pa.

6E. What is the bulk modulus of oxygen at standard temperature (0°C) and pressure (1 atm) if under these conditions of temperature and pressure 1 mol (32.0 g) of oxygen occupies 22.4 L and the speed of sound in the oxygen is 317 m/s?

7P. An experimenter wishes to measure the speed of sound in an aluminum rod 10 cm long by measuring the time it takes for a sound pulse to travel the length of the rod. If results good to four significant figures are desired, how precisely must the length of the rod be known and how closely must she be able to resolve time intervals?

8P. The speed of sound in a certain metal is V. One end of a long pipe of that metal of length L is struck a hard blow. A listener at the other end hears two sounds, one from the wave that has traveled along the pipe and the other from the wave that has traveled through the air. (a) If v is the speed of sound in air, what time interval t elapses between the arrivals of the two sounds? (b) Suppose that $t = 1.00$ s and the metal is steel. Find the length L.

9P. A man strikes a long aluminum rod at one end. Another man, at the other end with his ear close to the rod, hears the sound of the blow twice (once through air and once through the rod), with a 0.120 s interval between. How long is the rod?

10P. Earthquakes generate sound waves inside Earth. Unlike a gas, Earth can experience both transverse (S) and longitudinal (P) sound waves. Typically, the speed of S waves is about 4.5 km/s and that of P waves 8.0 km/s. A seismograph records P and S waves from an earthquake. The first P waves arrive 3.0 min before the first S waves (Fig. 18-31). Assuming the waves traveled in a straight line, how far away did the earthquake occur?

11P. A stone is dropped into a well. The sound of the splash is heard 3.00 s later. What is the depth of the well?

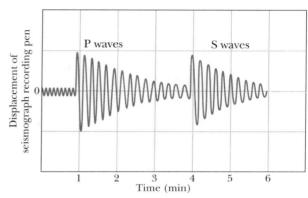

FIGURE 18-31   Problem 10.

SECTION 18-3 Traveling Sound Waves

12E. The audible frequency range for normal hearing is from about 20 Hz to 20 kHz. What are the wavelengths of sound waves at these frequencies?

13E. The shortest wavelength emitted by a bat is about 3.3 mm. What is the corresponding frequency?

14E. Diagnostic ultrasound of frequency 4.50 MHz is used to examine tumors in soft tissue. (a) What is the wavelength in air of such a sound wave? (b) If the speed of sound in tissue is 1500 m/s, what is the wavelength of this wave in tissue?

15E. (a) A conical loudspeaker has a diameter of 15.0 cm. At what frequency will the wavelength of the sound it emits in air be equal to its diameter? Be ten times its diameter? Be one-tenth its diameter? (b) Make the same calculations for a speaker of diameter 30.0 cm.

16E. Remarkably detailed images of transistors can be formed by an acoustic microscope. If the sound waves in such a microscope have a frequency of 4.2 GHz and a speed (in the liquid helium in which the specimen is immersed) of 240 m/s, what is their wavelength?

17P. (a) A continuous sinusoidal longitudinal wave is sent along a very long coiled spring from an oscillating source attached to it. The frequency of the source is 25 Hz, and the distance between successive points of maximum expansion in the spring is 24 cm. Find the wave speed. (b) If the maximum longitudinal displacement of a particle in the spring is 0.30 cm and the wave moves in the negative direction of an x axis, write the equation for the wave. Let the source be at $x = 0$ and the displacement at $x = 0$ be zero when $t = 0$.

18P. The pressure in a traveling sound wave is given by the equation

$$\Delta p = (1.5 \text{ Pa}) \sin \pi[(1.00 \text{ m}^{-1})x - (330 \text{ s}^{-1})t].$$

Find (a) the pressure amplitude, (b) the frequency, (c) the wavelength, and (d) the speed of the wave.

19P. Two sound waves, from two different sources with the same frequency, 540 Hz, travel at a speed of 330 m/s. The sources are in phase. What is the phase difference of the waves at a point that is 4.40 m from one source and 4.00 m from the other? The waves are traveling in the same direction.

20P. At a certain point in space, two waves produce pressure variations given by

$$\Delta p_1 = \Delta p_m \sin \omega t,$$

$$\Delta p_2 = \Delta p_m \sin(\omega t - \phi).$$

What is the pressure amplitude of the resultant wave at this point when $\phi = 0$, $\phi = \pi/2$, $\phi = \pi/3$, and $\phi = \pi/4$?

SECTION 18-4   Interference

21P. In Fig. 18-32, two loudspeakers, separated by a distance of 2.00 m, are in phase. Assume the amplitudes of the sound from the speakers are approximately the same at the position of a listener, who is 3.75 m directly in front of one of the speakers. (a) For what frequencies in the audible range (20–20,000 Hz) does the listener hear a minimum signal? (b) For what frequencies is the signal a maximum?

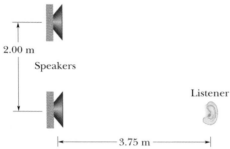

2.00 m

Speakers

Listener

|← 3.75 m →|

FIGURE 18-32   Problem 21.

22P. Two loudspeakers are located 11.0 ft apart on the stage of an auditorium. A listener is seated 60.0 ft from one and 64.0 ft from the other. A signal generator drives the two speakers in phase with the same amplitude and frequency. The transmitted frequency is swept through the audible range (20–20,000 Hz). (a) What are the three lowest frequencies at which the listener will hear a minimum signal because of destructive interference? (b) What are the three lowest frequencies at which the listener will hear a maximum signal?

23P. Two point sources of sound waves of identical wavelength λ and amplitude are separated by distance $D = 2.0\lambda$. The sources are in phase. (a) How many points of maximum signal (maximum constructive interference) lie along a large circle around the sources? (b) How many points of minimum signal (destructive interference)?

24P. A sound wave of 40.0 cm wavelength enters the tube shown in Fig. 18-33 at the source end. What must be the smallest radius r such that a minimum will be heard at the detector end?

FIGURE 18-33
Problem 24.

Source   Detector

25P. In Fig. 18-34, a point source S of sound waves lies near a reflecting wall AB. A sound detector D intercepts sound ray R_1

traveling directly from S. It also intercepts sound ray R_2 that reflects from the wall such that the *angle of incidence* θ_i is equal to the *angle of reflection* θ_r. Find two frequencies at which there is maximum constructive interference of R_1 and R_2 at D. (Reflection of sound by the wall does not alter the phase of the sound wave.)

FIGURE 18-34   Problem 25.

26P*. Two point sources, separated by a distance of 5.00 m, emit sound waves at the same amplitude and frequency (300 Hz), but the sources are exactly out of phase. At what points along the line between the sources do the sound waves result in maximum oscillations of the air molecules? (*Hint:* One such point is midway between the sources.)

SECTION 18-5   Intensity and Sound Level

27E. A 1.0 W point source emits sound waves isotropically. Assuming that the energy of the waves is conserved, what is the intensity (a) 1.0 m from the source and (b) 2.5 m from the source?

28E. A source emits sound waves isotropically. The intensity of the waves 2.50 m from the source is 1.91×10^{-4} W/m². Assuming that the energy of the waves is conserved, what is the power of the source?

29E. A sound wave of frequency 300 Hz has an intensity of 1.00 μW/m². What is the amplitude of the air oscillations caused by this wave?

30E. Two sounds differ in sound level by 1.00 dB. What is the ratio of the greater intensity to the smaller intensity?

31E. A certain sound level is increased by 30 dB. By what multiple is (a) its intensity increased and (b) its pressure amplitude increased?

32E. A salesperson claimed that a stereo system had a maximum audio power of 120 W. Testing the system with several speakers set up so as to simulate a point source, the consumer noted that she could get as close as 1.2 m with the volume full on before the sound hurt her ears. Should she report the firm to the Consumer Product Safety Commission?

33E. A certain loudspeaker system emits sound isotropically with a frequency of 2000 Hz and an intensity of 0.960 mW/m² at a distance of 6.10 m. Assume that there are no reflections. (a) What is the intensity at 30.0 m? At 6.10 m, what are (b) the displacement amplitude and (c) the pressure amplitude?

34E. The source of a sound wave has a power of 1.00 μW. If it is a point source, (a) what is the intensity 3.00 m away and (b) what is the sound level in decibels at that distance?

35E. (a) If two sound waves, one in air and one in (fresh) water, are equal in intensity, what is the ratio of the pressure amplitude of the wave in water to that of the wave in air? Assume the water and the air are at 20°C. (See Table 15-1). (b) If the pressure amplitudes are equal instead, what is the ratio of the intensities of the waves?

36P. (a) Show that the intensity I of a wave is the product of the wave's energy per unit volume u and its speed v. (b) Radio waves travel at a speed of 3.00×10^8 m/s. Find u for a radio wave 480 km from a 50,000 W source, assuming the wavefronts to be spherical.

37P. Assume that a noisy freight train on a straight track emits a cylindrical, expanding sound wave. Assuming that the air absorbs no energy, how does the amplitude s_m of the wave depend on the perpendicular distance r from the source?

38P. A sound wave travels out uniformly in all directions from a point source. (a) Justify the following expression for the displacement s of the medium at any distance r from the source:

$$s = \frac{b}{r} \sin k(r - vt),$$

where b is a constant. Consider the speed, direction of propagation, periodicity, and intensity of the wave. (b) What are the dimensions of the constant b?

39P. Find the ratios of (a) the intensities, (b) the pressure amplitudes, and (c) the particle displacement amplitudes for two sounds whose sound levels differ by 37 dB.

40P. At a distance of 10 km, a 100 Hz horn, assumed to be a point source, is barely audible. At what distance would it begin to cause pain?

41P. You are standing at a distance D from a source that emits sound waves equally in all directions. You walk 50.0 m toward the source and observe that the intensity of these waves has doubled. Calculate the distance D.

42P. A point source emits 30.0 W of sound isotropically. A small microphone of effective cross-sectional area 0.750 cm^2 is located 200 m from the source. Calculate (a) the sound intensity there and (b) the power intercepted by the microphone.

43P. In a test, a subsonic jet flies overhead at an altitude of 100 m. The sound intensity on the ground as the jet passes overhead is 150 dB. At what altitude should the plane fly so that the ground noise is no greater than 120 dB, the threshold of pain? Ignore the finite time required for the sound to reach the ground.

44P. An audio engineer has designed a loudspeaker that is spherical and emits sound isotropically. The speaker emits 10 W of acoustic power into a room with completely absorbent walls, floor, and ceiling (an *anechoic chamber*). (a) What is the intensity (W/m^2) of the sound waves 3.0 m from the center of the source? (b) How does the amplitude of the waves at 4.0 m compare with that at 3.0 m from the center of the source?

45P. Figure 18-35 shows an air-filled, acoustic interferometer, used to demonstrate the interference of sound waves. S is an oscillating diaphragm; D is a sound detector, such as the ear or a microphone. Path SBD can be varied in length, but path SAD is fixed. At D, the sound wave coming along path SBD interferes with that coming along path SAD. In one demonstration, the sound intensity at D has a minimum value of 100 units at one position of B and continuously climbs to a maximum value of 900 units when B is shifted by 1.65 cm. Find (a) the frequency of the sound emitted by the source and (b) the ratio of the amplitude at D of the SAD wave to that of the SBD wave. (c) How can it happen that these waves have different amplitudes, considering that they originate at the same source?

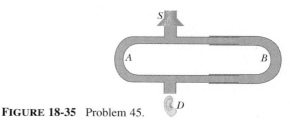

FIGURE 18-35 Problem 45.

SECTION 18-6 Sources of Musical Sound

46E. A sound wave of frequency 1000 Hz propagating through air has a pressure amplitude of 10.0 Pa. What are the (a) wavelength, (b) particle displacement amplitude, and (c) maximum particle speed? (d) An organ pipe open at both ends has this frequency as a fundamental. How long is the pipe?

47E. A sound wave in a fluid medium is reflected at a barrier so that a standing wave is formed. The distance between nodes is 3.8 cm, and the speed of propagation is 1500 m/s. Find the frequency of the sound wave.

48E. A violin string 15.0 cm long and fixed at both ends oscillates in its $n = 1$ mode. The speed of waves on the string is 250 m/s, and the speed of sound in air is 348 m/s. What are (a) the frequency and (b) the wavelength of the emitted sound wave?

49E. A violin string, oscillating in its fundamental mode, generates a sound wave with wavelength λ. By what multiple must the tension be increased if the string, still oscillating in its fundamental mode, is to generate a sound wave with wavelength $\lambda/2$?

50E. Organ pipe A, with both ends open, has a fundamental frequency of 300 Hz. The third harmonic of organ pipe B, with one end open, has the same frequency as the second harmonic of pipe A. How long are (a) pipe A and (b) pipe B?

51E. The water level in a vertical glass tube 1.00 m long can be adjusted to any position in the tube. A tuning fork vibrating at 686 Hz is held just over the open top end of the tube. At what positions of the water level will there be resonance?

52E. (a) Find the speed of waves on a violin string of mass 800 mg and length 22.0 cm if the fundamental frequency is 920 Hz. (b) What is the tension in the string? For the fundamental, what is the wavelength of (c) the waves on the string and (d) the sound waves emitted by the string?

53P. A certain violin string is 30 cm long between its fixed ends and has a mass of 2.0 g. The "open" string (no applied finger) sounds an A note (440 Hz). (a) To play a C (523 Hz), how far down the string must one place a finger? (b) What is the ratio of the wavelength of the string waves required for an A note to that required for a C note? (c) What is the ratio of the wavelength of the sound wave for an A note to that for a C note?

54P. A string on a cello has length L, for which the fundamental frequency is f. (a) By what length l must the string be shortened by fingering to change the fundamental frequency to rf? (b) What is l if $L = 0.80$ m and $r = 1.2$? (c) For $r = 1.2$, what is the ratio of the wavelength of the new sound wave emitted by the string to that of the wave emitted before fingering?

55P. In Fig. 18-36, S is a small loudspeaker driven by an audio oscillator and amplifier, adjustable in frequency from 1000 to 2000 Hz only. The tube D is a piece of cylindrical sheet-metal pipe 18.0 in. long and open at both ends. (a) If the speed of sound in air is 1130 ft/s at the existing temperature, at what frequencies will resonance occur in the pipe when the frequency emitted by the speaker is varied from 1000 Hz to 2000 Hz? (b) Sketch the standing waves (using the convention of Fig. 18-13b) for each resonant frequency.

FIGURE 18-36 Problem 55.

56P. A well with vertical sides and water at the bottom resonates at 7.00 Hz and at no lower frequency. The air in the well has a density of 1.10 kg/m^3 and a bulk modulus of 1.33×10^5 Pa. How deep is the well?

57P. A hand-clap on stage in an amphitheater (Fig. 18-37) sends out sound waves that scatter from terraces of width $w = 0.75$ m. The sound returns to the stage as a periodic series of pulses, one from each terrace; the parade of pulses sounds like a played note. (a) Assuming that all the rays are horizontal, find the frequency at which the pulses return (that is, the frequency of the perceived note). (b) If the width w of the terraces were smaller, would the frequency be higher or lower?

FIGURE 18-37 Problem 57.

58P. A tube 1.20 m long is closed at one end. A stretched wire is placed near the open end. The wire is 0.330 m long and has a mass of 9.60 g. It is fixed at both ends and vibrates in its fundamental mode. By resonance, it sets the air column in the tube into oscillation at that column's fundamental frequency. Find (a) that frequency and (b) the tension in the wire.

59P. The period of a pulsating variable star may be estimated by considering the star to be executing *radial* longitudinal pulsations in the fundamental standing wave mode; that is, the star's radius varies periodically with time, with a displacement antinode at the star's surface. (a) Would you expect the center of the star to be a displacement node or antinode? (b) By analogy with a pipe with one open end, show that the period of pulsation T is given by

$$T = \frac{4R}{v},$$

where R is the equilibrium radius of the star and v is the average sound speed. (c) Typical white dwarf stars are composed of material with a bulk modulus of 1.33×10^{22} Pa and a density of 10^{10} kg/m^3. They have radii equal to 9.0×10^{-3} solar radius. What is the approximate pulsation period of a white dwarf?

60P. A violin string 30.0 cm long with linear density 0.650 g/m is placed near a loudspeaker that is fed by an audio oscillator of variable frequency. It is found that the string is set into oscillation only at the frequencies 880 and 1320 Hz as the frequency of the oscillator is varied over the range 500–1500 Hz. What is the tension in the string?

SECTION 18-7 Beats

61E. A tuning fork of unknown frequency makes three beats per second with a standard fork of frequency 384 Hz. The beat frequency decreases when a small piece of wax is put on a prong of the first fork. What is the frequency of this fork?

62E. The A string of a violin is a little too tightly stretched. Four beats per second are heard when the string is sounded together with a tuning fork that is oscillating accurately at concert A (440 Hz). What is the period of the violin string oscillation?

63P. Two identical piano wires have a fundamental frequency of 600 Hz when kept under the same tension. What fractional increase in the tension of one wire will lead to the occurrence of 6 beats/s when both wires oscillate simultaneously?

64P. You have five tuning forks that oscillate at different frequencies. Give (a) the maximum number and (b) the minimum number of different beat frequencies you can produce by using the forks two at a time.

SECTION 18-8 The Doppler Effect

65E. A source S generates circular waves on the surface of a lake; the pattern of wave crests is shown in Fig. 18-38. The speed

FIGURE 18-38 Exercise 65.

of the waves is 5.5 m/s, and the crest-to-crest separation is 2.3 m. You are in a small boat heading directly toward S at a constant speed of 3.3 m/s with respect to the shore. What frequency of the waves do you observe?

66E. Trooper B is chasing speeder A along a straight stretch of road. Both are moving at a speed of 100 mi/h. Trooper B, failing to catch up, sounds his siren again. Take the speed of sound in air to be 1100 ft/s and the frequency of the source to be 500 Hz. What is the Doppler shift in the frequency heard by speeder A?

67E. A whistle used to call a dog has a frequency of 30 kHz. The dog, however, ignores it. The owner of the dog, who cannot hear sounds above 20 kHz, wants to use the Doppler effect to make certain that the whistle is working. She asks a friend to blow the whistle from a moving car while the owner remains stationary and listens. (a) How fast would the car have to move and in what direction for the owner to hear the whistle at 20 kHz? (b) Repeat for a whistle frequency of 22 kHz instead of 30 kHz.

68E. The 16,000 Hz whine of the turbines in the jet engines of an aircraft moving with speed 200 m/s is heard at what frequency by the pilot of a second craft trying to overtake the first at a speed of 250 m/s?

69E. An ambulance with a siren emitting a whine at 1600 Hz overtakes and passes a cyclist pedaling a bike at 8.00 ft/s. After being passed, the cyclist hears a frequency of 1590 Hz. How fast is the ambulance moving?

70E. In 1845, Buys Ballot first tested the Doppler effect for sound. He put a trumpet player on a flatcar drawn by a locomotive and another player near the tracks. If each player blew a 440 Hz note and there were 4.0 beats/s as they approached each other, what was the speed of the flatcar?

71E. The speed of light in water is about three-fourths the speed of light in vacuum. A beam of high-speed electrons from a betatron emits Cerenkov radiation in water; the wavefront of this light forms a cone of angle 60°. Find the speed of the electrons in the water.

72E. A bullet is fired with a speed of 2200 ft/s. Find the angle made by the shock cone with the line of motion of the bullet.

73P. Two identical tuning forks can oscillate at 440 Hz. A person is located somewhere on the line between them. Calculate the beat frequency as measured by this individual if (a) she is standing still and the tuning forks both move to the right at 30.0 m/s, and (b) the tuning forks are stationary and the listener moves to the right at 30.0 m/s.

74P. A whistle of frequency 540 Hz moves in a circle of radius 2.00 ft at an angular speed of 15.0 rad/s. What are (a) the lowest and (b) the highest frequencies heard by a listener a long distance away at rest with respect to the center of the circle?

75P. A plane flies at 1.25 times the speed of sound. The sonic boom reaches a man on the ground exactly 1 min after the plane passed directly overhead. What is the altitude of the plane? Assume the speed of sound to be 330 m/s.

76P. A jet plane passes overhead at a height of 5000 m and a speed of Mach 1.5. (a) Find the Mach cone angle. (b) How long

after the jet has passed directly overhead will the shock wave reach the ground? Use 331 m/s for the speed of sound.

77P. Figure 18-39 shows a transmitter and receiver of waves contained in a single instrument. It is used to measure the speed u of a target object (idealized as a flat plate) that is moving directly toward the unit, by analyzing the waves reflected from the target. (a) Show that the frequency f_r of the reflected waves at the receiver is related to their source frequency f_s by

$$f_r = f_s \left(\frac{v + u}{v - u} \right),$$

where v is the speed of the waves. (b) In a great many practical situations, $u \ll v$. In this case, show that the equation above becomes

$$\frac{f_r - f_s}{f_s} \approx \frac{2u}{v}.$$

FIGURE 18-39 Problem 77.   Target

78P. A stationary motion detector sends sound waves of 0.150 MHz toward a truck approaching at a speed of 45.0 m/s. What is the frequency of the waves reflected back to the detector?

79P. A siren emitting a sound of frequency 1000 Hz moves away from you toward the face of a cliff at a speed of 10 m/s. Take the speed of sound in air as 330 m/s. (a) What is the frequency of the sound you hear coming directly from the siren? (b) What is the frequency of the sound you hear reflected off the cliff? (c) What is the beat frequency between the two sounds? Is it perceptible (it must be less than 20 Hz to be perceived)?

80P. A person on a railroad car blows a trumpet note at 440 Hz. The car is moving toward a wall at 20.0 m/s. Calculate (a) the frequency of the sound as received at the wall and (b) the frequency of the reflected sound arriving back at the source.

81P. A French submarine and a U.S. submarine move head-on during maneuvers in motionless water in the North Atlantic (Fig. 18-40). The French sub moves at 50.0 km/h, and the U.S. sub at 70.0 km/h. The French sub sends out a sonar signal (sound wave in water) at 1000 Hz. Sonar waves travel at 5470 km/h. (a) What is the signal's frequency as detected by the U.S. sub? (b) What frequency is detected by the French sub in the signal reflected back to it by the U.S. sub?

French   U.S.
50.0 km/h   70.0 km/h

FIGURE 18-40 Problem 81.

82P. A source of sound waves of frequency 1200 Hz moves to the right with a speed of 98.0 ft/s relative to the air. Ahead of it is a reflecting surface moving to the left with a speed of 216 ft/s relative to the air. Take the speed of sound in air to be 1080 ft/s and find (a) the wavelength of the sound emitted toward the reflector by the source, (b) the number of wavefronts per second arriving at the reflecting surface, (c) the speed of the reflected waves, (d) the wavelength of the reflected waves, and (e) the number of reflected wavefronts per second arriving at the source.

83P. In a discussion of Doppler shifts of ultrasonic waves used in medical diagnosis, the authors remark: ''For every millimeter per second that a structure in the body moves, the frequency of the incident ultrasonic wave is shifted approximately 1.30 Hz per MHz.'' What speed of the ultrasonic waves in tissue do you deduce from this statement?

84P. An acoustic burglar alarm consists of a source emitting waves of frequency 28.0 kHz. What will be the beat frequency of waves reflected from an intruder walking at an average speed of 0.950 m/s directly away from the alarm?

85P. A bat is flitting about in a cave, navigating via ultrasonic bleeps. Assume that the sound emission frequency of the bat is 39,000 Hz. During one fast swoop directly toward a flat wall surface, the bat is moving at 0.025 times the speed of sound in air. What frequency does the bat hear reflected off the wall?

86P. A submarine, near the water surface, moves north at speed 75.0 km/h in a northbound current of speed 30.0 km/h, where both speeds are relative to the ocean floor. The sub emits a sonar signal (sound wave), of frequency $f = 1000$ Hz and speed 5470 km/h, that is detected by a destroyer north of the sub. What is the difference between the detected frequency and f if the destroyer (a) drifts with the current at 30.0 km/h and (b) is stationary relative to the ocean floor?

87P. A 2000 Hz siren and a civil defense official are both at rest with respect to the ground. What frequency does the official hear if the wind is blowing at 12 m/s (a) from source to observer and (b) from observer to source?

88P. Two trains are traveling toward each other at 100 ft/s relative to the ground. One train is blowing a whistle at 500 Hz. (a) What frequency will be heard on the other train in still air? (b) What frequency will be heard on the other train if the wind is blowing at 100 ft/s toward the whistle and away from the listener? (c) What frequency will be heard if the wind direction is reversed?

89P. A girl is sitting near the open window of a train that is moving at a velocity of 10.00 m/s to the east. The girl's uncle stands near the tracks and watches the train move away. The locomotive whistle emits sound at frequency 500.0 Hz. The air is still. (a) What frequency does the uncle hear? (b) What frequency does the girl hear? A wind begins to blow from the east at 10.00 m/s. (c) What frequency does the uncle now hear? (d) What frequency does the girl now hear?

SECTION 18-9 The Doppler Effect for Light

90E. Figure 18-41 is a graph of the intensity versus wavelength for light reaching Earth from galaxy NGC 7319, which is about 3×10^8 light-years away. The most intense light was emitted by

the oxygen in NGC 7319. In a laboratory that emission is at wavelength $\lambda = 513$ nm, but in the light from NGC 7319 it has been shifted to 525 nm due to the Doppler effect (all the emissions from NGC 7319 have been shifted). (a) What is the speed of NGC 7319 relative to Earth? (b) Is the relative motion toward or away from our planet?

FIGURE 18-41 Exercise 90.

91E. Certain wavelengths in the light from a galaxy in the constellation Virgo are observed to be 0.4% longer than the corresponding light from terrestrial sources. What is the radial speed of this galaxy with respect to Earth? Is it approaching or receding?

92E. In the ''red shift'' of radiation from a distant galaxy, a certain radiation, known to have a wavelength of 434 nm when observed in the laboratory, has a wavelength of 462 nm. (a) What is the speed of the galaxy (along the line of sight) relative to Earth? (b) Is the galaxy approaching or receding?

93E. Assuming that Eq. 18-57 holds, find how fast you would have to go through a red light to have it appear green. Take 620 nm as the wavelength of red light and 540 nm as the wavelength of green light.

94P. The period of rotation of the Sun at its equator is 24.7 days; its radius is 7.00×10^5 km. What Doppler wavelength shift is expected for light with wavelength 550 nm emitted from the edge of the Sun's disk?

95P. An Earth satellite, transmitting at a frequency of 40 MHz (exactly), passes directly over a radio receiving station at an altitude of 400 km and at a speed of 3.0×10^4 km/h. Plot the change in frequency attributable to the Doppler effect as a function of time, counting $t = 0$ as the instant the satellite is over the station. (*Hint:* The speed u in the Doppler formula is not the actual speed of the satellite but its component in the direction of the station. Neglect the curvature of Earth and that of the satellite orbit.)

96P. Microwaves, which travel at the speed of light, are reflected from a distant airplane approaching the wave source. It is found that when the reflected waves are beat against the waves radiating from the source, the beat frequency is 990 Hz. If the microwaves are 0.100 m in wavelength, what is the approach speed of the airplane?

19
Temperature, Heat, and the First Law of Thermodynamics

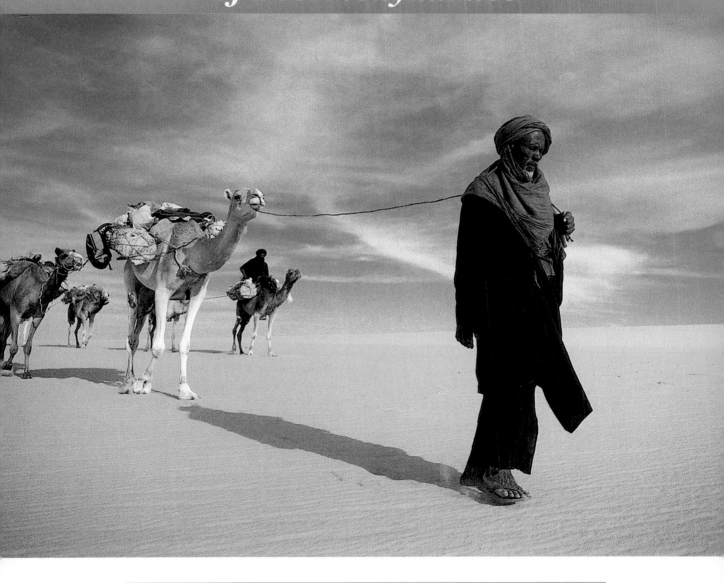

An object with a black surface usually heats up more than one with a white surface when both are in sunlight. Such is true of the robes worn by Bedouins in the Sinai desert: black robes heat up more than white robes. Why then would a Bedouin ever wear a black robe? Wouldn't that actually decrease the chance of survival in the harsh desert environment❓

19-1 THERMODYNAMICS

With this chapter, we leave the subject of mechanics and begin a new subject—*thermodynamics*. Mechanics deals with the mechanical energy of systems and is governed by Newton's laws. Thermodynamics deals with an internal energy of systems —thermal energy—and is governed by a new set of laws, which you will come to know in this and the next two chapters.

The central concept of thermodynamics is temperature. This word is so familiar that most of us—because of our built-in sense of hot and cold—tend to be overconfident in our understanding of it. Our "temperature sense" is in fact not always reliable. On a cold winter day, for example, an iron railing seems much colder to the touch than a wooden fence post, yet both are at the same temperature. This difference in our sense perception comes about because iron removes energy from our fingers more quickly than wood does. Here, we shall develop the concept of temperature from its foundations, without relying in any way on our temperature sense.

Temperature is one of the seven SI base quantities. Physicists measure temperature on the **Kelvin scale,** which is marked in units called *kelvins*. Although the temperature of a body apparently has no upper limit, it does have a lower limit; this limiting low temperature is taken as the zero of the Kelvin temperature scale. Room temperature is about 290 kelvins, or 290 K as we write it, above this

absolute zero. Figure 19-1 shows the wide range over which temperatures are determined.

When the universe began, some 10–20 billion years ago, its temperature was about 10^{39} K. As the universe expanded it cooled, and it has now reached an average temperature of about 3 K. We on Earth are a little warmer than that because we happen to live near a star. Without our Sun, we too would be at 3 K (or rather, we could not exist).

19-2 THE ZEROTH LAW OF THERMODYNAMICS

The properties of many bodies change as we alter their temperature, perhaps by moving them from a refrigerator to a warm oven. To give a few examples: as their respective temperatures increase, the volume of a liquid increases, a metal rod grows a little longer, and the electrical resistance of a wire increases, as does the pressure exerted by a confined gas. We can use any one of these properties as the basis of an instrument that will help us to pin down the concept of temperature.

Figure 19-2 shows such an instrument. Any resourceful engineer could design and construct it, using any one of the properties listed above. The instrument is fitted with a digital readout display and has the following properties: if you heat it with a Bunsen burner, the displayed number starts to increase; if you then put it into a refrigerator, the displayed number starts to decrease. The instrument is not calibrated in any way, and the numbers have (as yet) no physical meaning. The device is a *thermoscope* but not (as yet) a *thermometer*.

Suppose that, as in Fig. 19-3*a*, we put the thermoscope (which we shall call body *T*) into intimate contact with another body (body *A*). The entire system is confined within a thick-walled insulating box. The numbers displayed by the thermoscope roll by until, eventually, they come to rest (let us say the reading is "137.04") and no further change takes place. In fact, we suppose that every measurable property of body *T* and of body *A* has assumed a stable, unchanging value. Then we say that the two bodies are in *thermal equilibrium* with each other. And

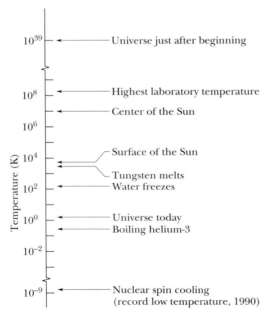

FIGURE 19-1 Some temperatures on the Kelvin scale. Temperature $T = 0$ corresponds to $10^{-\infty}$ and cannot be plotted on this logarithmic scale.

FIGURE 19-2 A thermoscope. The numbers increase when the device is heated and decrease when it is cooled. The thermally sensitive element could be —among many possibilities—a coil of wire whose electrical resistance is measured and displayed.

Thermally sensitive element

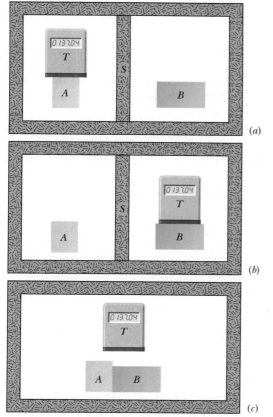

FIGURE 19-3 (a) Body T (a thermoscope) and body A are in thermal equilibrium. (Body S is a thermally insulating screen.) (b) Body T and body B are also in thermal equilibrium, at the same reading of the thermoscope. (c) If (a) and (b) are true, the zeroth law of thermodynamics states that body A and body B will also be in thermal equilibrium.

even though the displayed readings for body T have not been calibrated, we conclude that bodies T and A must be at the same (unknown) temperature.

Suppose that we next put body T in intimate contact with body B (Fig. 19-3b) and find that the two bodies come to thermal equilibrium *at the same reading of the thermoscope*. Then bodies T and B must be at the same (still unknown) temperature. If we now put bodies A and B into intimate contact (Fig. 19-3c), will they immediately be in thermal equilibrium with each other? Experimentally, we find that they are.

The experimental facts shown in Fig. 19-3 are summed up in the **zeroth law of thermodynamics:**

If bodies A and B are each in thermal equilibrium with a third body T, then they are in thermal equilibrium with each other.

In less formal language, the message of the zeroth law is: "Every body has a property called **temperature.** When

two bodies are in thermal equilibrium, their temperatures are equal." And vice versa. We can now make our thermoscope (the third body T) into a thermometer, confident that its readings will have physical meaning. All we have to do is calibrate it.

We use the zeroth law constantly in the laboratory. If we want to know whether the liquids in two beakers are at the same temperature, we measure the temperature of each with a thermometer. We do not need to bring the two liquids into intimate contact and observe whether they are or are not in thermal equilibrium.

The zeroth law, which has been called a logical afterthought, came to light only in the 1930s, long after the first and the second law of thermodynamics had been discovered and numbered. Because the concept of temperature is fundamental to those two laws, the law that establishes temperature as a valid concept should have the lowest number. Hence the zero.

19-3 MEASURING TEMPERATURE

Let us see how we define and measure temperatures on the Kelvin scale. Equivalently, let us see how to calibrate a thermoscope so as to make it into a usable thermometer.

The Triple Point of Water

To set up a temperature scale, we pick some reproducible thermal phenomenon and, quite arbitrarily, assign a certain Kelvin temperature to its thermal environment. That is, we select a *standard fixed point* and give it a standard fixed-point temperature. We could, for example, select the freezing point or the boiling point of water but, for various technical reasons, we do not. Instead we select the **triple point of water.**

Liquid water, solid ice, and water vapor (gaseous water) can coexist, in thermal equilibrium, at only one set of values of pressure and temperature. Figure 19-4 shows a triple-point cell, in which this so-called triple point of

FIGURE 19-4 A triple-point cell, in which solid ice, liquid water, and water vapor coexist in thermal equilibrium. By international agreement, the temperature of this mixture has been defined to be 273.16 K. The bulb of a constant-volume gas thermometer is shown inserted into the well of the cell.

water can be achieved in the laboratory. By international agreement, the triple point of water has been assigned a value of 273.16 K as the standard fixed-point temperature for the calibration of thermometers. That is,

$$T_3 = 273.16 \text{ K} \qquad \text{(triple-point temperature),} \qquad (19\text{-}1)$$

in which the subscript 3 reminds us of the triple point. This agreement also sets the size of the kelvin as 1/273.16 of the difference between absolute zero and the triple-point temperature of water.

Note that we do not use a degree mark in reporting Kelvin temperatures. It is 300 K (not 300°K), and it is read "300 kelvins" (not "300 degrees Kelvin"). The usual SI prefixes apply. Thus 0.0035 K is 3.5 mK. No distinction in nomenclature is made between Kelvin temperatures and temperature differences. Thus we can write, "the boiling point of sulfur is 717.8 K" and "the temperature of this water bath was raised by 8.5 K."

The Constant-Volume Gas Thermometer

So far, we have not discussed the particular temperature-sensitive physical property on which, by international agreement, we shall base our thermometer. Should it be the length of a metal rod, the electrical resistance of a wire, the pressure exerted by a confined gas, or something else? The choice is important because different choices lead to different temperatures for — to give one example — the boiling point of water. For reasons that will emerge below, the standard thermometer, against which all other thermometers are to be calibrated, is based on the pressure exerted by a gas confined to a fixed volume.

FIGURE 19-5   A constant-volume gas thermometer, its bulb immersed in a bath whose temperature T is to be measured.

Figure 19-5 shows such a **(constant-volume) gas thermometer.** It consists of a gas-filled bulb made of glass, quartz, or platinum (depending on the temperature range over which the thermometer is to be used) connected by a tube to a mercury manometer. By raising and lowering reservoir R, the mercury level on the left can always be brought to the zero of the manometer scale, thus assuring that the volume of the confined gas remains constant. The temperature of any body in thermal contact with the bulb is then defined to be

$$T = Cp, \qquad (19\text{-}2)$$

in which p is the pressure exerted by the gas and C is a constant. The pressure p is calculated from the relation

$$p = p_0 - \rho g h, \qquad (19\text{-}3)$$

in which p_0 is the atmospheric pressure, ρ is the density of the mercury in the manometer, and h is the measured level difference of mercury in the two arms of the tube.

With the bulb of the gas thermometer immersed in a triple-point cell, as in Fig. 19-4, we have

$$T_3 = Cp_3, \qquad (19\text{-}4)$$

in which p_3 is the pressure reading under this condition. Eliminating C between Eqs. 19-2 and 19-4 leads to

$$T = T_3 \left(\frac{p}{p_3} \right)$$
$$= (273.16 \text{ K}) \left(\frac{p}{p_3} \right) \qquad \text{(provisional).} \qquad (19\text{-}5)$$

Equation 19-5 is not yet our final definition of a temperature measured with a gas thermometer. We have said nothing about what gas (or how much gas) we are to place in the thermometer. If our thermometer is used to measure some temperature, such as the boiling point of water, we find that different gases in the bulb give slightly different measured temperatures. However, as we use smaller and smaller amounts of gas to fill the bulb, the readings converge nicely to a single temperature, no matter what gas we use. Figure 19-6 shows this comforting convergence.*

Thus we write, as our final recipe for measuring temperature with a gas thermometer,

$$T = (273.16 \text{ K}) \left(\lim_{m \to 0} \frac{p}{p_3} \right). \qquad (19\text{-}6)$$

---

*For pressure units, we shall use units introduced in Section 15-3. The SI unit for pressure is the newton per square meter, which is called the pascal (Pa). The pascal is related to other common pressure units by

$$1 \text{ atm} = 1.01 \times 10^5 \text{ Pa} = 760 \text{ torr} = 14.7 \text{ lb/in.}^2.$$

FIGURE 19-6 Temperatures calculated from Eq. 19-5 for a constant-volume gas thermometer whose bulb is immersed in boiling water. Different gases were used within the bulb, each at a variety of densities (which gave varying values for p_3). Note that all readings converge in the limit of zero density to a temperature of 373.125 K.

This instructs us to measure an unknown temperature T as follows. Fill the bulb with an arbitrary mass m of *any* gas (for example, nitrogen) and measure p_3 (using a triple-point cell) and p, the gas pressure at the temperature being measured. Calculate the ratio p/p_3. Then repeat both measurements with a smaller amount of gas in the bulb, and again calculate this ratio. Continue this way, using smaller and smaller amounts of gas, until you can extrapolate to the ratio p/p_3 that you would find if there were approximately no gas in the bulb. Calculate the temperature T by substituting that extrapolated ratio into Eq. 19-6. (The temperature measured in this way is the *ideal gas temperature.*)

If temperature is to be a truly fundamental physical quantity, one in which the laws of thermodynamics may be expressed, it is absolutely necessary that its definition be independent of the properties of specific materials. It would not do, for example, to have a quantity as basic as temperature depend on the expansivity of mercury, the electrical resistivity of platinum, or any other such "handbook" property. We chose the gas thermometer as our standard instrument precisely because no such specific properties of materials are involved in its operation. Use *any* gas — you will always get the same result.

SAMPLE PROBLEM 19-1

The bulb of a gas thermometer is filled with nitrogen to a pressure of 120 kPa. What provisional value (see Fig. 19-6) would this thermometer yield for the boiling point of water and what is the error in this value?

SOLUTION: From Fig. 19-6, the curve for nitrogen shows that, at 120 kPa, the provisional temperature for the boiling point of water would be about 373.44 K. The actual boiling point of water (found by extrapolation on Fig. 19-6) is 373.125 K. Thus using the provisional temperature leads to an error of 0.315 K, or 315 mK.

19-4 THE CELSIUS AND FAHRENHEIT SCALES

So far, we have discussed only the Kelvin scale, used in basic scientific work. In nearly all countries of the world, the Celsius scale (formerly called the centigrade scale) is the scale of choice for popular and commercial use and much scientific use. Celsius temperatures are measured in degrees, and the Celsius degree has the same size as the kelvin. However, the zero of the Celsius scale is shifted to a more convenient value than absolute zero. If T_C represents a Celsius temperature, then

$$T_C = T - 273.15°. \tag{19-7}$$

In expressing temperatures on the Celsius scale, the degree symbol is commonly used. Thus we write 20.00°C for a Celsius reading but 293.15 K for a Kelvin reading.

The Fahrenheit scale, used in the United States, employs a smaller degree than the Celsius scale and a different zero of temperature. You can easily verify both these differences by examining an ordinary room thermometer on which both scales are marked. The relation between the Celsius and Fahrenheit scales is

$$T_F = \tfrac{9}{5}T_C + 32°, \tag{19-8}$$

where T_F is Fahrenheit temperature. Transferring between these two scales can be done easily by remembering a few corresponding points (such as the freezing and boiling points of water; see Table 19-1) and by making use of the fact that 9 degrees on the Fahrenheit scale equals 5 degrees on the Celsius scale. Figure 19-7 compares the Kelvin, Celsius, and Fahrenheit scales.

We use the letters C and F to distinguish measurements and degrees on the two scales. Thus,

$$0°C = 32°F$$

TABLE 19-1 SOME CORRESPONDING TEMPERATURES

TEMPERATURE	°C	°F
Boiling point of water[a]	100	212
Normal body temperature	37.0	98.6
Accepted comfort level	20	68
Freezing point of water[a]	0	32
Zero of Fahrenheit scale	≈ -18	0
Scales coincide	-40	-40

[a]Strictly, the boiling point of water on the Celsius scale is 99.975°C, and the freezing point is 0.00°C. Thus there is slightly less than 100 C° between those two points.

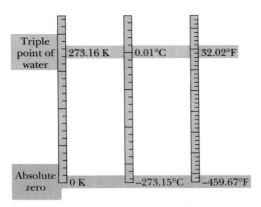

FIGURE 19-7 The Kelvin, Celsius, and Fahrenheit temperature scales compared.

means that 0° on the Celsius scale measures the same temperature as 32° on the Fahrenheit scale, whereas

$$5 \, C° = 9 \, F°$$

means that a temperature difference of five Celsius degrees (note the degree symbol is *after* the letter) is equivalent to a temperature difference of nine Fahrenheit degrees.

SAMPLE PROBLEM 19-2

Suppose you come across old scientific notes that describe a temperature scale called Z on which the boiling point of water is 65.0°Z and the freezing point is −14.0°Z.

(a) What temperature change ΔT on the Z scale would correspond to a change of 53.0 F°?

SOLUTION: To find a conversion factor between the two scales, we can use the boiling and freezing points of water. On the Z scale, the temperature difference between those two points is 65.0°Z − (−14.0°Z), or 79.0 Z°. On the Fahrenheit scale, it is 212°F − 32.0°F, or 180 F°. Thus a change of 79.0 Z° equals a change of 180 F°. For a change of 53.0 F°, we can now write

$$\Delta T = 53.0 \, F° = 53.0 \, F° \left(\frac{79.0 \, Z°}{180 \, F°} \right)$$

$$= 23.3 \, Z°. \qquad \text{(Answer)}$$

(b) To what temperature on the Fahrenheit scale would a temperature $T = -98.0°Z$ correspond?

SOLUTION: The freezing point of water is −14.0°Z, so the difference between T and the freezing point is 84.0 Z°. To convert this difference to Fahrenheit degrees, we write

$$\Delta T = 84.0 \, Z° \left(\frac{180 \, F°}{79.0 \, Z°} \right) = 191 \, F°.$$

Thus T is 191 F° below the freezing point of water and is, on the Fahrenheit scale,

$$T = 32.0°F - 191 \, F° = -159°F. \qquad \text{(Answer)}$$

CHECKPOINT 1: The figure shows three temperature scales with the freezing and boiling points of water indicated. (a) Rank the size of a degree on these scales, greatest first. (b) Rank the following temperatures, highest first: 50°X, 50°W, and 50°Y.

PROBLEM SOLVING TACTICS

TACTIC 1: *Temperature Changes*

Between the boiling and freezing points of water, there are (approximately) 100 kelvins and 100 Celsius degrees. Thus a kelvin is the same size as a Celsius degree. From this or from Eq. 19-7, we then know that any temperature change is the same number whether expressed in kelvins or Celsius degrees. For example, a temperature change of 10 K is equivalent to a change of 10 C°.

Between the boiling and freezing points of water, there are 180 Fahrenheit degrees. Thus 180 F° = 100 K, and a Fahrenheit degree must be 100 K/180 F°, or 5/9, the size of a kelvin or Celsius degree. From this or from Eq. 19-8, we then know that any temperature change expressed in Fahrenheit degrees must be $\frac{9}{5}$ times that same temperature change expressed in either kelvins or Celsius degrees. For example, in Fahrenheit degrees, a temperature change of 10 K is (9/5)(10 K), or 18 F°.

You should take care not to confuse a *temperature* with a temperature *change*. A temperature of 10 K is certainly not the same as one of 10°C or 18°F but, as above, a temperature *change* of 10 K is the same as one of 10 C° or 18 F°.

19-5 THERMAL EXPANSION

You can often loosen a tight metal jar lid by holding it under a stream of hot water. Both the metal of the lid and the glass of the jar expand as the hot water adds energy to their atoms. (With the added energy, the atoms can move a bit farther from each other than usual, against the spring-like interatomic forces that hold every solid together.) However, because the atoms in the metal move farther apart than those in the glass, the lid expands more than the jar and thus is loosened.

FIGURE 19-8 Railroad tracks in Asbury Park, New Jersey, distorted because of thermal expansion on a very hot July day.

Such **thermal expansion** is not always desirable, as Fig. 19-8 suggests. To preclude buckling, therefore, expansion slots are placed in bridges to accommodate roadway expansion on hot days. Dental materials used for fillings must be matched in their thermal expansion properties to those of tooth enamel (otherwise consuming hot coffee or cold ice cream would be quite painful). In aircraft manufacture, however, rivets and other fasteners are often cooled in dry ice before insertion and then allowed to expand to a tight fit.

Thermometers and thermostats may be based on the differences in expansion between the components of a *bimetal strip* (Fig. 19-9). And the familiar liquid-in-glass thermometers are based on the fact that liquids such as mercury or alcohol expand to a different (greater) extent than do their glass containers.

FIGURE 19-9 (*a*) A bimetal strip, consisting of a strip of brass and a strip of steel welded together, at temperature T_0. (*b*) The strip bends as shown at temperatures above this reference temperature. Below the reference temperature the strip bends the other way. Many thermostats operate on this principle, making and breaking an electrical contact as the temperature rises and falls.

TABLE 19-2 SOME COEFFICIENTS OF LINEAR EXPANSION[a]

SUBSTANCE	α (10^{-6}/C°)	SUBSTANCE	α (10^{-6}/C°)
Ice (at 0°C)	51	Steel	11
Lead	29	Glass (ordinary)	9
Aluminum	23	Glass (Pyrex)	3.2
Brass	19	Diamond	1.2
Copper	17	Invar[b]	0.7
Concrete	12	Fused quartz	0.5

[a]Room temperature values except for the listing for ice.

[b]This alloy was designed to have a low coefficient of expansion. The word is a shortened form of ''invariable.''

Linear Expansion

If the temperature of a metal rod of length L is raised by an amount ΔT, its length is found to increase by an amount

$$\Delta L = L\alpha\,\Delta T, \tag{19-9}$$

in which α is a constant called the **coefficient of linear expansion.** The coefficient α has the unit ''per degree'' or ''per kelvin'' and depends on the material. If we rewrite Eq. 19-9 as

$$\alpha = \frac{\Delta L/L}{\Delta T}, \tag{19-10}$$

we see that α is the fractional increase in length per unit change in temperature. Although α varies somewhat with temperature, for most practical purposes it can be taken as constant for a particular material. Table 19-2 shows some coefficients of linear expansion.

The thermal expansion of a solid is like (three-dimensional) photographic enlargement. Figure 19-10*b*

FIGURE 19-10 The same steel rule at two different temperatures. When it expands, every dimension is increased in the same proportion. The scale, the numbers, the thickness, and the diameters of the circle and circular hole are all increased by the same factor. (The expansion has been exaggerated for clarity.)

shows the (exaggerated) expansion of a steel rule after its temperature is increased from that of Fig. 19-10a. Equation 19-9 applies to every linear dimension of the rule, including its edge, thickness, diagonals, and the diameters of the circle etched on it and the circular hole cut in it. If the disk cut from that hole originally fits snugly in the hole, it will continue to fit snugly if it undergoes the same temperature increase as the rule.

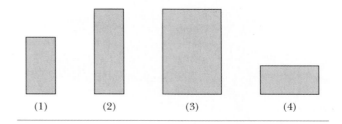

(1)   (2)   (3)   (4)

Volume Expansion

If all dimensions of a solid expand with temperature, the volume of that solid must also expand. For liquids, volume expansion is the only meaningful expansion parameter. If the temperature of a solid or liquid whose volume is V is increased by an amount ΔT, the increase in volume is found to be

$$\Delta V = V\beta \, \Delta T, \qquad (19\text{-}11)$$

where β is the **coefficient of volume expansion** of the solid or liquid. The coefficients of volume expansion and linear expansion for a solid are related by

$$\beta = 3\alpha. \qquad (19\text{-}12)$$

The most common liquid, water, does not behave like other liquids. Above about 4°C, water expands as the temperature rises, as we would expect. Between 0 and about 4°C, however, water *contracts* with increasing temperature. Thus, at about 4°C, the density of water passes through a maximum. At all other temperatures, the density of water is less than this maximum value.

This behavior of water is the reason that lakes freeze from the top down rather than from the bottom up. As the water on the surface is cooled from, say, 10°C toward the freezing point, it becomes denser ("heavier") than the lower water and sinks to the bottom. Below 4°C, however, further cooling makes the water then on the surface *less* dense ("lighter") than the lower water, so it stays on the surface until it freezes. If lakes froze from the bottom up, the ice so formed would tend not to melt completely during the summer, being insulated by the water above. After a few years, many bodies of open water in the temperate zones of Earth would be frozen solid all year round. And aquatic life as we know it could not exist.

CHECKPOINT 2: The figure shows four rectangular metal plates, with sides of L, $2L$, or $3L$. They are all made of the same material, and their temperature is to be increased by the same amount. Rank the plates according to the expected increase in (a) their vertical heights and (b) their areas, greatest first.

SAMPLE PROBLEM 19-3

A steel wire whose length L is 130 cm and whose diameter d is 1.1 mm is heated to an average temperature of 830°C and stretched taut between two rigid supports. What tension develops in the wire as it cools to 20°C?

SOLUTION: First let us calculate the amount by which the wire would shrink if it were allowed to cool freely. From Eq. 19-9 and Table 19-2, we can write the amount of shrinkage as

$$\Delta L = L\alpha \, \Delta T = (1.3 \text{ m})(11 \times 10^{-6}/\text{C}°)(830°\text{C} - 20°\text{C})$$
$$= 1.16 \times 10^{-2} \text{ m} = 1.16 \text{ cm}.$$

However, the wire is not permitted to shrink. We therefore calculate what force would be required to stretch the wire by this amount. From Eq. 13-34 we have

$$F = \frac{\Delta L \, EA}{L} = \frac{\Delta L \, E(\pi/4)d^2}{L},$$

in which E is Young's modulus for steel (see Table 13-1) and A is the cross-sectional area of the wire. Substitution yields

$$F = (1.16 \times 10^{-2} \text{ m})(200 \times 10^9 \text{ N/m}^2)(\pi/4)$$
$$\times \frac{(1.1 \times 10^{-3} \text{ m})^2}{1.3 \text{ m}}$$
$$= 1700 \text{ N}. \qquad \text{(Answer)}$$

Can you show that this answer is independent of the length of the wire?

Sometimes the bulging brick walls of old buildings are supported by running a steel stress rod from outside wall to outside wall, through the building. The rod is then heated, and nuts on the outside walls are tightened. As the rod cools and contracts, tension develops in it, helping to keep the walls from bulging outward.

SAMPLE PROBLEM 19-4

On a hot day in Las Vegas, an oil trucker loaded 9785 gal of diesel fuel. He encountered cold weather on the way to Payson, Utah, where the temperature was 41 F° lower than in Las Vegas, and where he delivered his entire load. How many gallons did he deliver? The coefficient of volume expansion for diesel fuel is $9.5 \times 10^{-4}/\text{C}°$, and the coefficient of linear expansion for his steel truck tank is $11 \times 10^{-6}/\text{C}°$.

SOLUTION: From Eq. 19-11,

$$\Delta V = V\beta\,\Delta T$$

$$= (9785 \text{ gal})(9.5 \times 10^{-4}/\text{C}°)(-41 \text{ F}°)\left(\frac{5 \text{ C}°}{9 \text{ F}°}\right)$$

$$= -212 \text{ gal.}$$

Thus the amount delivered was

$$V_{\text{del}} = V + \Delta V = 9785 \text{ gal} - 212 \text{ gal}$$

$$= 9573 \text{ gal} \approx 9600 \text{ gal.} \qquad \text{(Answer)}$$

Note that the thermal expansion of the steel tank has nothing to do with the problem. Question: Who paid for the "missing" diesel fuel?

PROBLEM SOLVING TACTICS

TACTIC 2: *Units for Temperature Changes*

The coefficient of linear expansion α is defined to be the fractional change in length (a dimensionless number) per unit change in temperature. In Table 19-2, that unit change in temperature is expressed in Celsius degrees (C°). Since any temperature *change* expressed in Celsius degrees is numerically the same when expressed in kelvins, the values of α in Table 19-2 could just as well be expressed in kelvins.

For example, the value of α for steel can be written as either $11 \times 10^{-6}/\text{C}°$ or $11 \times 10^{-6}/\text{K}$. This means that we could have substituted either expression for α in Sample Problem 19-3. We could have made a similar substitution for β in Sample Problem 19-4.

19-6 TEMPERATURE AND HEAT

If you take a can of cola from the refrigerator and leave it on the kitchen table, its temperature will rise—rapidly at first but then more slowly—until the temperature of the cola equals that of the room (the two are then in thermal equilibrium). In the same way, the temperature of a cup of hot coffee, left sitting on the table, will fall until it also reaches room temperature.

In generalizing this situation, we describe the cola or the coffee as a *system* (with temperature T_S) and the relevant parts of the kitchen as the *environment* (with temperature T_E) of that system. Our observation is that if T_S is not equal to T_E, then T_S will change (T_E may also change some) until the two temperatures are equal and thus thermal equilibrium is reached.

Such a change in temperature is due to the transfer of a form of energy between the system and its environment. This energy is *internal energy* (or *thermal energy*), which is the collective kinetic and potential energies associated

with the random motions of the atoms, molecules, and other microscopic bodies within an object. The transferred internal energy is called **heat** and is symbolized Q. Heat is *positive* when internal energy is transferred to a system from its environment (we say that heat is absorbed). Heat is *negative* when internal energy is transferred from a system to its environment (we say that heat is released or lost).

This transfer of energy is shown in Fig. 19-11. In the situation of Fig. 19-11*a*, in which $T_S > T_E$, internal energy is transferred from the system to the environment, so Q is negative. In Fig. 19-11*b*, in which $T_S = T_E$, there is no such transfer, Q is zero, and heat is neither released nor absorbed. In Fig. 19-11*c*, in which $T_S < T_E$, the transfer is to the system from the environment, so Q is positive.

We are led then to this definition of heat:

Heat is the energy that is transferred between a system and its environment because of a temperature difference that exists between them.

Recall that energy can also be transferred between a system and its environment by means of *work W*, which we always associate with a force acting on a system during a displacement of the system. Heat and work, unlike temperature, pressure, and volume, are not intrinsic properties

FIGURE 19-11 If the temperature of a system exceeds that of its environment as in (*a*), heat Q is lost by the system to the environment until thermal equilibrium (*b*) is established. (*c*) If the temperature of the system is below that of the environment, heat is absorbed by the system until thermal equilibrium is established.

of a system. They have meaning only as they describe the transfer of energy into or out of a system. Thus it is proper to say: "During the last 3 min, 15 J of heat was transferred to the system from its environment" or "During the last minute, 12 J of work was done on the system by its environment." It is meaningless to say: "This system contains 450 J of heat" or "This system contains 385 J of work."

Before scientists realized that heat is transferred energy, heat was measured in terms of its ability to raise the temperature of water. Thus the **calorie** (cal) was defined as the amount of heat that would raise the temperature of 1 g of water from 14.5°C to 15.5°C. In the British system, the corresponding unit of heat was the **British thermal unit** (Btu), defined as the amount of heat that would raise the temperature of 1 lb of water from 63°F to 64°F.

In 1948, the scientific community decided that since heat (like work) is transferred energy, the SI unit for heat should be the same as that for energy, namely, the **joule.** The calorie is now defined to be 4.1860 J (exactly) with no reference to the heating of water. (The "calorie" used in nutrition, sometimes called the Calorie (Cal), is really a kilocalorie.) The relations among the various heat units are

$$1 \text{ cal} = 3.969 \times 10^{-3} \text{ Btu} = 4.186 \text{ J}. \quad (19\text{-}13)$$

19-7 THE ABSORPTION OF HEAT BY SOLIDS AND LIQUIDS

Heat Capacity

The **heat capacity** C of an object (for example, a Pyrex coffee pot or a marble slab) is the proportionality constant between an amount of heat and the change in temperature that this heat produces in the object. Thus

$$Q = C(T_f - T_i), \quad (19\text{-}14)$$

in which T_i and T_f are the initial and final temperatures of the object. Heat capacity C has the unit of energy per degree or energy per kelvin. The heat capacity C of, say, a marble slab used in a bun warmer might be 179 cal/C°, which we can also write as 179 cal/K or as 747 J/K.

The word "capacity" in this context is really misleading in that it suggests analogy with the capacity of a bucket to hold water. *That analogy is false,* and you should not think of the object as "containing" heat or being limited in its ability to absorb heat. Heat transfer can proceed without limit as long as the necessary temperature difference is maintained. The object may, of course, melt or vaporize during the process.

Specific Heat

Two objects made of the same material, say marble, will have heat capacities proportional to their masses. It is therefore convenient to define a "heat capacity per unit mass" or **specific heat** c that refers not to an object but to a unit mass of the material of which the object is made. Equation 19-14 then becomes

$$Q = cm(T_f - T_i). \quad (19\text{-}15)$$

Through experiment we would find that although the heat capacity of the particular marble slab mentioned above is 179 cal/C° (or 747 J/K), the specific heat of marble itself (in that slab or in any other marble object) is 0.21 cal/g·C° (or 880 J/kg·K).

From the way the calorie and the British thermal unit were initially defined, the specific heat of water is

$$c = 1 \text{ cal/g·C°} = 1 \text{ Btu/lb·F°}$$
$$= 4190 \text{ J/kg·K}. \quad (19\text{-}16)$$

Table 19-3 shows the specific heats of some substances at room temperature. Note that the value for water is relatively high. The specific heat of any substance actually depends somewhat on temperature, but the values in Table 19-3 apply reasonably well in a range of temperatures near room temperature.

TABLE 19-3 SPECIFIC HEATS OF SOME SUBSTANCES AT ROOM TEMPERATURE

SUBSTANCE	SPECIFIC HEAT cal/g·K	SPECIFIC HEAT J/kg·K	MOLAR SPECIFIC HEAT J/mol·K
Elemental Solids			
Lead	0.0305	128	26.5
Tungsten	0.0321	134	24.8
Silver	0.0564	236	25.5
Copper	0.0923	386	24.5
Aluminum	0.215	900	24.4
Other Solids			
Brass	0.092	380	
Granite	0.19	790	
Glass	0.20	840	
Ice (−10°C)	0.530	2220	
Liquids			
Mercury	0.033	140	
Ethyl alcohol	0.58	2430	
Seawater	0.93	3900	
Water	1.00	4190	

A certain amount of heat Q will warm 1 g of material A by 3 C° and 1 g of material B by 4 C°. Which material has the greater specific heat?

Molar Specific Heat

In many instances the most convenient unit for specifying the amount of a substance is the mole (mol), where

$$1 \text{ mol} = 6.02 \times 10^{23} \text{ elementary units}$$

of *any* substance. Thus 1 mol of aluminum means 6.02×10^{23} atoms (the atom being the elementary unit), and 1 mol of aluminum oxide means 6.02×10^{23} molecules of the oxide (because the molecule is the elementary unit of a compound).

When quantities are expressed in moles, the specific heat must also involve moles (rather than a mass unit); it is then called a **molar specific heat.** Table 19-3 shows the values for some elemental solids (each consisting of a single element) at room temperature.

Note that the molar specific heats of all the elements listed in Table 19-3 have about the same value at room temperature, namely, 25 J/mol · K. As a matter of fact, the molar specific heats of all solids increase toward that value as the temperature increases. (Some substances, such as carbon and beryllium, do not reach this limiting value until temperatures well above room temperature. Other substances may melt or vaporize before they reach this limit.)

When we compare two substances on a molar basis, we are comparing samples that contain the same number of elementary units. The fact that, at high enough temperatures, all solid elements have the same molar specific heat tells us that atoms of all kinds — whether they be aluminum, copper, uranium, or anything else — absorb heat in the same way.

An Important Point

In determining and then using the specific heat of any substance, we need to know the conditions under which heat transfer occurs. For solids and liquids, we usually assume that the sample is under constant pressure (usually atmospheric) during the heat transfer. It is also conceivable that the sample is held at constant volume while the heat is absorbed. This means that thermal expansion of the sample is prevented by applying external pressure. For solids and liquids, this is very hard to arrange experimentally but the effect can be calculated, and it turns out that the specific heats under constant pressure and constant volume for any solid or liquid differ usually by no more than a few percent. Gases, as you will see, have quite different values for their specific heats under constant-pressure conditions and under constant-volume conditions.

Heats of Transformation

When heat is absorbed by a solid or liquid, the temperature of the sample does not necessarily rise. Instead, the sample may change from one *phase,* or *state,* (that is, solid, liquid, or gas) to another. Thus ice may melt and water may boil, absorbing heat in each case without a temperature change. In the reverse processes (water freezing, steam condensing), heat is released by the sample, again while the temperature of the sample remains constant.

The amount of heat per unit mass that must be transferred when a sample completely undergoes a phase change is called the **heat of transformation** L. So when a sample of mass m completely undergoes a phase change, the total heat transferred is

$$Q = Lm. \tag{19-17}$$

When the phase change is from liquid to gas (then the sample must absorb heat) or from gas to liquid (then the sample must release heat), the heat of transformation is called the **heat of vaporization** L_V. For water at its normal boiling or condensation temperature,

$$L_V = 539 \text{ cal/g} = 40.7 \text{ kJ/mol}$$
$$= 2256 \text{ kJ/kg}. \tag{19-18}$$

When the phase change is from solid to liquid (then the sample must absorb heat) or from liquid to solid (then the sample must release heat), the heat of transformation is called the **heat of fusion** L_F. For water at its normal freezing or melting temperature,

$$L_F = 79.5 \text{ cal/g} = 6.01 \text{ kJ/mol}$$
$$= 333 \text{ kJ/kg}. \tag{19-19}$$

Table 19-4 shows the heats of transformation for some substances.

SAMPLE PROBLEM 19-5

A candy bar has a marked nutritional value of 350 Cal. How many kilowatt-hours of energy will it deliver to the body as it is digested?

SOLUTION: The Calorie in this case is a kilocalorie so that

$$\text{energy} = (350 \times 10^3 \text{ cal})(4.19 \text{ J/cal})$$
$$= (1.466 \times 10^6 \text{ J})(1 \text{ W} \cdot \text{s/J})$$
$$\times (1 \text{ h}/3600 \text{ s})(1 \text{ kW}/1000 \text{ W})$$
$$= 0.407 \text{ kW} \cdot \text{h}. \tag{Answer}$$

This amount of energy would keep a 100 W light bulb burning

TABLE 19-4   SOME HEATS OF TRANSFORMATION

	MELTING		BOILING	
SUBSTANCE	MELTING POINT (K)	HEAT OF FUSION L_F (kJ/kg)	BOILING POINT (K)	HEAT OF VAPORIZATION L_V (kJ/kg)
Hydrogen	14.0	58.0	20.3	455
Oxygen	54.8	13.9	90.2	213
Mercury	234	11.4	630	296
Water	273	333	373	2256
Lead	601	23.2	2017	858
Silver	1235	105	2323	2336
Copper	1356	207	2868	4730

for 4.1 h. To burn up this much energy by exercise, a person would have to jog about 3 or 4 mi.

A generous human diet corresponds to about 3.5 kW·h per day, which represents the absolute maximum amount of work that a human can do in one day. In an industrialized country, this amount of energy can be purchased for perhaps 35 cents.

SAMPLE PROBLEM 19-6

(a) How much heat is needed to take ice of mass $m = 720$ g at $-10°$C to a liquid state at 15°C?

SOLUTION: To answer, we must consider three steps. Step 1 is to raise the temperature of the ice from $-10°$C to the melting point at 0°C. We use Eq. 19-15, with the specific heat c_{ice} of ice given in Table 19-3. For this step, the initial temperature T_i is $-10°$C and the final temperature T_f is 0°C. We find

$$Q_1 = c_{ice}m(T_f - T_i)$$
$$= (2220 \text{ J/kg·K})(0.720 \text{ kg})[0°\text{C} - (-10°\text{C})]$$
$$= 15,984 \text{ J} \approx 15.98 \text{ kJ}.$$

Step 2 is to melt the ice (there can be no change in temperature until this step is completed). We now use Eqs. 19-17 and 19-19, finding

$$Q_2 = L_F m = (333 \text{ kJ/kg})(0.720 \text{ kg}) \approx 239.8 \text{ kJ}.$$

Step 3 is to raise the temperature of the now liquid water from 0°C to 15°C. We again use Eq. 19-15, but now with the specific heat c_{liq} of liquid water given in Table 19-3. In this step the initial temperature T_i is 0°C and the final temperature T_f is 15°C. We find

$$Q_3 = c_{liq}m(T_f - T_i)$$
$$= (4190 \text{ J/kg·K})(0.720 \text{ kg})(15°\text{C} - 0°\text{C})$$
$$= 45,252 \text{ J} \approx 45.25 \text{ kJ}.$$

The total required heat Q_{tot} is the sum of the amounts required in the three steps:

$$Q_{tot} = Q_1 + Q_2 + Q_3$$
$$= 15.98 \text{ kJ} + 239.8 \text{ kJ} + 45.25 \text{ kJ}$$
$$\approx 300 \text{ kJ}. \qquad \text{(Answer)}$$

Note that the heat required to melt the ice is much larger than the heat required to raise the temperature of either the ice or the liquid water.

(b) If we supply the ice with a total heat of only 210 kJ, what then are the final state and temperature of the water?

SOLUTION: From step 1, we know that 15.98 kJ is needed to raise the temperature of the ice to the melting point. The remaining heat Q_{rem} is then 210 kJ $-$ 15.98 kJ, or about 194 kJ. From step 2, we can see that this amount of heat is insufficient to melt all the ice. We can find the mass m of ice that is melted by the heat Q_{rem} by using Eqs. 19-17 and 19-19:

$$m = \frac{Q_{rem}}{L_F} = \frac{194 \text{ kJ}}{333 \text{ kJ/kg}} = 0.583 \text{ kg} \approx 580 \text{ g}.$$

Thus the mass of the ice that remains is 720 g $-$ 580 g, or 140 g. Since not all the ice is melted, the temperature of the ice–liquid water must be 0°C. Hence we have

580 g water   and   140 g ice, at 0°C.   (Answer)

SAMPLE PROBLEM 19-7

A copper slug whose mass m_c is 75 g is heated in a laboratory oven to a temperature T of 312°C. The slug is then dropped into a glass beaker containing a mass $m_w = 220$ g of water. The heat capacity C_b of the beaker is 45 cal/K. The initial temperature T_i of the water and the beaker is 12°C. What is the final temperature T_f of the slug, the beaker, and the water when thermal equilibrium is reached?

SOLUTION: We take as our system *water + beaker + copper slug*. No heat enters or leaves this system, so the algebraic sum of the internal heat transfers that occur must be zero. There are three such transfers:

for the water: $Q_w = m_w c_w (T_f - T_i)$;

for the beaker: $Q_b = C_b (T_f - T_i)$;

for the copper: $Q_c = m_c c_c (T_f - T)$.

The temperature difference is written—in all three cases—as the final temperature minus the initial temperature (T_i for the water and beaker, and T for the copper). We do this even though we realize that Q_w and Q_b are positive (indicating that heat is added to the initially cool water and beaker) and that Q_c is negative (indicating that heat is released by the initially hot copper slug). Doing this allows us to write

$$Q_w + Q_b + Q_c = 0. \qquad (19\text{-}20)$$

Substituting the heat transfer expressions into Eq. 19-20 yields

$$m_w c_w (T_f - T_i) + C_b (T_f - T_i)$$
$$+ \; m_c c_c (T_f - T) = 0. \qquad (19\text{-}21)$$

Temperatures enter Eq. 19-21 only as differences. Thus, because the intervals on the Celsius and Kelvin scales are identical, we can use either of these scales in this equation. Solving Eq. 19-21 for T_f, we have

$$T_f = \frac{m_c c_c T + C_b T_i + m_w c_w T_i}{m_w c_w + C_b + m_c c_c}.$$

The numerator is, with Celsius temperatures,

$$(75 \text{ g})(0.092 \text{ cal/g} \cdot \text{K})(312°\text{C}) + (45 \text{ cal/K})(12°\text{C})$$
$$+ \; (220 \text{ g})(1.00 \text{ cal/g} \cdot \text{K})(12°\text{C})$$
$$= 5332.8 \text{ cal},$$

and the denominator is

$$(220 \text{ g})(1.00 \text{ cal/g} \cdot \text{K}) + 45 \text{ cal/K}$$
$$+ \; (75 \text{ g})(0.092 \text{ cal/g} \cdot \text{K})$$
$$= 271.9 \text{ cal/C°}.$$

So, we then have

$$T_f = \frac{5332.8 \text{ cal}}{271.9 \text{ cal/C°}} = 19.6°\text{C} \approx 20°\text{C}. \qquad \text{(Answer)}$$

From the given data you can show that

$$Q_w \approx 1670 \text{ cal}, \quad Q_b \approx 342 \text{ cal}, \quad Q_c \approx -2020 \text{ cal}.$$

Apart from rounding errors, the algebraic sum of these three heat transfers is indeed zero, as Eq. 19-20 requires.

19-8 A CLOSER LOOK AT HEAT AND WORK

Here we look in some detail at how heat and work are exchanged between a system and its environment. Let us take as our system a gas confined to a cylinder with a movable piston, as in Fig. 19-12. The upward force on the piston due to the pressure of the confined gas is equal to the weight of lead shot loaded onto the top of the piston. The walls of the cylinder are made of insulating material that does not allow any heat transfer. The bottom of the cylinder rests on a reservoir for thermal energy, a *thermal reservoir* (perhaps a hot plate) whose temperature T you can control by turning a knob.

The system (the gas) starts from an *initial state i*, described by a pressure p_i, a volume V_i, and a temperature T_i. You want to change the system to a *final state f*, described by a pressure p_f, a volume V_f, and a temperature T_f. The procedure by which you change the system from its initial state to its final state is called a *thermodynamic process*. During such a process, heat may be transferred into the system from the thermal reservoir (positive heat) or vice versa (negative heat). And work is done by the system to raise the loaded piston (positive work) or lower it (negative work). We assume that all such changes occur slowly, with the result that the system is always in (approximate) thermal equilibrium, (that is, every part of the system is always in thermal equilibrium with every other part).

Suppose that you remove a few lead shot from the piston of Fig. 19-12, allowing the gas to push the piston

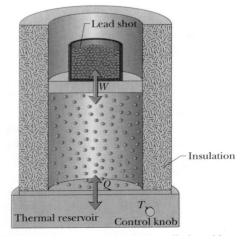

FIGURE 19-12   A gas is confined to a cylinder with a movable piston. Heat Q can be added to, or withdrawn from, the gas by regulating the temperature T of the adjustable thermal reservoir. Work W can be done by the gas by raising or lowering the piston.

and remaining shot upward through a differential displacement $d\mathbf{s}$ with an upward force $\mathbf{F}$. Since the displacement is tiny, we can assume that $\mathbf{F}$ is constant during the displacement. Then $\mathbf{F}$ has a magnitude that is equal to pA, where p is the pressure of the gas and A is the face area of the piston. The differential work dW done by the gas during the displacement is

$$dW = \mathbf{F}\cdot d\mathbf{s} = (pA)(ds) = p(A\,ds)$$
$$= p\,dV, \qquad (19\text{-}22)$$

in which dV is the differential change in the volume of the gas owing to the movement of the piston. When you have removed enough shot to allow the gas to change its volume from V_i to V_f, the total work done by the gas is

$$W = \int dW = \int_{V_i}^{V_f} p\,dV. \qquad (19\text{-}23)$$

During the change in volume, the pressure and temperature of the gas may also change. To evaluate the integral in Eq. 19-23 directly, we would need to know how pressure varies with volume for the actual process by which the system changes from state i to state f.

There are actually many ways to take the gas from state i to state f. One way is shown in Fig. 19-13a, which is a plot of the pressure of the gas versus its volume and which is called a p-V diagram. In Fig. 19-13a, the curve indicates that the pressure decreased as the volume increased. The integral of Eq. 19-23 (and thus the work W done by the gas) is represented by the shaded area under the curve between points i and f. Regardless of what exactly we did to take the gas along the curve, that work is positive, owing to the fact that the gas increased its volume by forcing the piston upward.

Another way to get from state i to state f is shown in Fig. 19-13b: there the change takes place in two steps— the first from state i to state a, and the second from state a to state f.

Step ia of this process is carried out at constant pressure, which means that you leave undisturbed the lead shot that rides on top of the piston in Fig. 19-12. You cause the volume to increase (from V_i to V_f) by slowly turning up the temperature control knob, raising the temperature of the gas to some higher value T_a. (Increasing the temperature increases the force by the gas on the piston, moving it upward.) During this process, positive work is done by the expanding gas (to lift the loaded piston) and heat is added to the system from the thermal reservoir (in response to the arbitrarily small temperature differences that you create as you turn up the temperature). This heat is positive because it is added to the system.

Step af of the process of Fig. 19-13b is carried out at

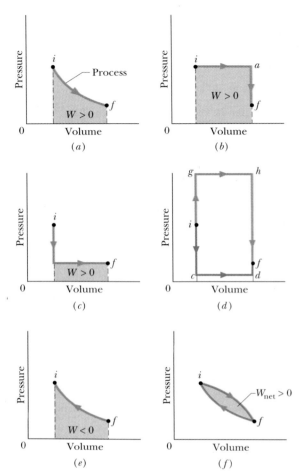

FIGURE 19-13   (a) The system of Fig. 19-12 goes from an *initial state i* to a *final state f* by means of a *thermodynamic process*. The area marked W represents the work done *by* the system during this process. The work is positive, because the process proceeds toward greater volume. (b) Another process for moving between the same two states; the work is now greater than that in (a). (c) Still another process, requiring less (positive) work. (d) The work can be made as small as you like (path $icdf$) or as large as you like (path $ighf$). (e) When the volume is reduced (by some external force), the work done *by* the system is negative. (f) The net work done by the system during a (closed) cycle is represented by the enclosed area, which is the difference in the areas beneath the two curves that make up the cycle.

constant volume, so you must wedge the piston, preventing it from moving. And then you must use the control knob to decrease the temperature so that the pressure drops from p_a to its final value p_f. During this process, heat is lost by the system to the thermal reservoir.

For the overall process iaf, the work W, which is positive and is carried out only during step ia, is represented by the shaded area under the curve. Heat is transferred during both steps ia and af, with a net heat transfer Q.

Figure 19-13c shows a process in which the two steps above are carried out in reverse order. The work W in this case is smaller than for Fig. 19-13b, as is the net heat absorbed. Figure 19-13d suggests that you can make the work done by the gas as small as you want (by following a path like *icdf*) or as large as you want (by following a path like *ighf*).

To sum up: a system can be taken from a given initial state to a given final state by an infinite number of processes. Heat may or may not be involved, and in general, the work W and the heat Q will have different values for different processes. We say that heat and work are *path-dependent* quantities.

Figure 19-13e shows an example in which negative work is done by a system as some external force compresses the system, reducing its volume. The absolute value of the work done is still equal to the area beneath the curve, but because the gas is *compressed,* the work done by the gas is negative.

Figure 19-13f shows a *thermodynamic cycle* in which the system is taken from some initial state i to some other state f and then back to i. The net work done by the system during the cycle is the sum of the positive work done during the expansion and the negative work done during the compression. In Fig. 19-13f, the net work is positive because the area under the expansion curve (i to f) is greater than the area under the compression curve (f to i).

___

CHECKPOINT **4:** The p-V diagram shows six curved paths (connected by vertical paths) that can be followed by a gas. Which two of them should be part of a closed cycle if the net work done by the gas is to be at its maximum positive value?

19-9 THE FIRST LAW OF THERMODYNAMICS

You have just seen that when a system changes from a given initial state to a given final state, both the work W done and the heat Q exchanged depend on the nature of the process. Experimentally, however, we find a surprising thing. *The quantity $Q - W$ is the same for all processes.* It depends only on the initial and final states and does not

depend at all on how the system gets from one to the other. All other combinations of Q and W, including Q alone, W alone, $Q + W$, and $Q - 2W$, are *path dependent;* only the quantity $Q - W$ is not.

The quantity $Q - W$ must represent a change in some intrinsic property of the system. We call this property the *internal energy E_{int}* and we write

$$\Delta E_{int} = E_{int,f} - E_{int,i}$$
$$= Q - W \quad \text{(first law).} \quad (19\text{-}24)$$

Equation 19-24 is the **first law of thermodynamics.** If the thermodynamic system undergoes only a differential change, we can write the first law as*

$$dE_{int} = dQ - dW \quad \text{(first law).} \quad (19\text{-}25)$$

___

The internal energy E_{int} of a system tends to increase if energy is added via heat Q and tends to decrease if energy is lost via work W done by the system.

___

In Chapter 8, we discussed the principle of energy conservation as it applies to isolated systems, that is, to systems in which no energy enters or leaves the system. The first law of thermodynamics is an extension of that principle to systems that are *not* isolated. In such cases, energy may be transferred into or out of the system as either work W or heat Q. In our statement of the first law of thermodynamics above, we assume that there are no changes in the kinetic energy or the potential energy of the system as a whole; that is, $\Delta K = \Delta U = 0$.

Before this chapter, the term *work* and the symbol W always meant the work done *on* a system. But starting with Eq. 19-22 and continuing through the next two chapters about thermodynamics, we focus on the work done *by* a system, such as the gas in Fig. 19-12.

The work done *on* a system is always the negative of the work done *by* the system. So if we rewrite Eq. 19-24 in terms of the work W_{on} done *on* the system, we have $\Delta E_{int} = Q + W_{on}$. This tells us the following: the internal energy of a system tends to increase if heat is absorbed by the system or if positive work is done *on* the system. Conversely, the internal energy tends to decrease if heat is lost by the system or if negative work is done *on* the system.

___

*Here dQ and dW, unlike dE_{int}, are not true differentials. That is, there are no such functions as $Q(p, V)$ and $W(p, V)$ that depend only on the state of the system. The quantities dQ and dW are called *inexact differentials* and are usually represented by the symbols $đQ$ and $đW$. For our purposes, we can treat them simply as infinitesimally small energy transfers.

19-10 SOME SPECIAL CASES OF THE FIRST LAW OF THERMODYNAMICS

Here we look at four different thermodynamic processes, in each of which a certain restriction is imposed on the system. We then see what consequences follow when we apply the first law of thermodynamics to the process.

1. *Adiabatic processes.* An adiabatic process is one that occurs so rapidly or occurs in a system that is so well insulated that *no transfer of heat* occurs between the system and its environment. Putting $Q = 0$ in the first law (Eq. 19-24) then leads to

$$\Delta E_{int} = -W \qquad \text{(adiabatic process).} \qquad (19\text{-}26)$$

This tells us that if work is done *by* the system (that is, if W is positive), the internal energy of the system decreases by the amount of work. Conversely, if work is done *on* the system (that is, if W is negative), the internal energy of the system increases by that amount.

Figure 19-14 shows an idealized adiabatic process. Heat cannot enter or leave the system because of the insulation. Thus the only way energy can be transferred between the system and its environment is by work. If we remove shot from the piston and allow the gas to expand, the work done by the system (the gas) is positive and the internal energy of the gas decreases. If, instead, we add shot and compress the gas, the work done by the system is negative and the internal energy of the gas increases.

2. *Constant-volume processes.* If the volume of a system (such as a gas) is held constant, that system can do no work. Putting $W = 0$ in the first law (Eq. 19-24) yields

$$\Delta E_{int} = Q \qquad \text{(constant-volume process).} \qquad (19\text{-}27)$$

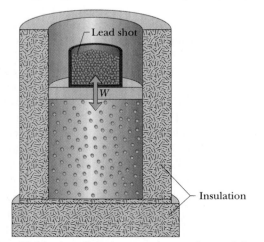

FIGURE 19-14 An adiabatic expansion can be carried out by slowly removing lead shot from the top of the piston. Adding lead shot reverses the process at any stage.

Thus if heat is added to a system (that is, if Q is positive), the internal energy of the system increases. Conversely, if heat is removed during the process (that is, if Q is negative), the internal energy of the system must decrease.

3. *Cyclical processes.* There are processes in which, after certain interchanges of heat and work, the system is restored to its initial state. In that case, no intrinsic property of the system—including its internal energy—can possibly change. Putting $\Delta E_{int} = 0$ in the first law (Eq. 19-24) yields

$$Q = W \qquad \text{(cyclical process).} \qquad (19\text{-}28)$$

Thus the net work done during the process must exactly equal the net amount of heat transferred; the store of internal energy of the system remains unchanged. Cyclical processes form a closed loop on a pressure–volume plot, as in Fig. 19-13*f*. We shall discuss such processes in some detail in Chapter 21.

4. *Free expansion.* These are adiabatic processes in which no work is done on or by the system. Thus $Q = W = 0$ and the first law requires that

$$\Delta E_{int} = 0 \qquad \text{(free expansion).} \qquad (19\text{-}29)$$

Figure 19-15 shows how such an expansion can be carried out. A gas, which is in thermal equilibrium within itself, is initially confined by a closed stopcock to one half of an insulated double chamber; the other half is evacuated. The stopcock is opened, and the gas expands freely to fill both halves of the chamber. No heat is transferred to or from the gas because of the insulation. No work is done by the gas because it rushes into a vacuum, its motion unopposed by any counteracting pressure.

A free expansion differs from all other processes we have considered because it cannot be done slowly and in a controlled way. As a result, at any given instant during the sudden expansion, the gas is not in thermal equilibrium and its pressure is not the same everywhere. So, although

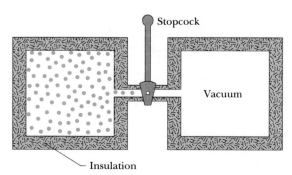

FIGURE 19-15 The initial stage of a free-expansion process. After the stopcock is opened, the gas eventually reaches an equilibrium final state, filling both chambers.

TABLE 19-5 THE FIRST LAWS OF THERMODYNAMICS: FOUR SPECIAL CASES

The Law: $\Delta E_{int} = Q - W$ (Eq. 19-24)

PROCESS	RESTRICTION	CONSEQUENCE
Adiabatic	$Q = 0$	$\Delta E_{int} = -W$
Constant volume	$W = 0$	$\Delta E_{int} = Q$
Closed cycle	$\Delta E_{int} = 0$	$Q = W$
Free expansion	$Q = W = 0$	$\Delta E_{int} = 0$

we can plot the initial and final states on a *p-V* diagram, we cannot plot the expansion itself.

Table 19-5 summarizes the characteristics of the processes of this section.

CHECKPOINT **5:** For one complete cycle as shown in the *p-V* diagram, are (a) ΔE_{int} for the gas and (b) the net heat transfer Q positive, negative, or zero?

SAMPLE PROBLEM 19-8

Let 1.00 kg of liquid water at 100°C be converted to steam at 100°C by boiling at standard atmospheric pressure (1.00 atm or 1.01×10^5 Pa). The volume changes from an initial value of 1.00×10^{-3} m³ as a liquid to 1.671 m³ as steam (Fig. 19-16).

(a) How much work is done by the system this process?

SOLUTION: The work is given by Eq. 19-23. Because the pressure is constant (at 1.01×10^5 Pa) during the boiling process, we can take *p* outside the integral, obtaining

$$W = \int_{V_i}^{V_f} p\, dV = p \int_{V_i}^{V_f} dV = p(V_f - V_i)$$

$$= (1.01 \times 10^5 \text{ Pa})(1.671 \text{ m}^3 - 1.00 \times 10^{-3} \text{ m}^3)$$

$$= 1.69 \times 10^5 \text{ J} = 169 \text{ kJ}. \qquad \text{(Answer)}$$

The result is positive, indicating that work is done *by* the system on its environment in lifting the weighted piston of Fig. 19-16.

(b) How much heat must be added to the system during the process?

SOLUTION: Since there is no temperature change, but only a phase change, we use Eqs. 19-17 and 19-18:

$$Q = L_V m = (2260 \text{ kJ/kg})(1.00 \text{ kg})$$

$$= 2260 \text{ kJ}. \qquad \text{(Answer)}$$

The result is positive, which indicates that heat is *added to* the system, as we expect.

(c) What is the change in the internal energy of the system during the boiling process?

SOLUTION: We find this from the first law (Eq. 19-24):

$$\Delta E_{int} = Q - W = 2260 \text{ kJ} - 169 \text{ kJ}$$

$$\approx 2090 \text{ kJ} = 2.09 \text{ MJ}. \qquad \text{(Answer)}$$

This quantity is positive, indicating that the internal energy of the system has increased during the boiling process. This energy goes into separating the H_2O molecules, which strongly attract each other in the liquid state.

We see that, when water is boiled, about 7.5% (= 169 kJ/2260 kJ) of the added heat goes into the work of pushing back the atmosphere. The rest goes into internal energy that is added to the system.

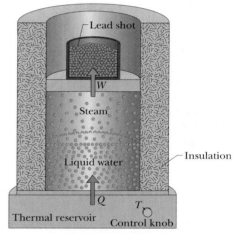

FIGURE 19-16 Sample Problem 19-8. Water boiling at constant pressure. Heat is added from the thermal reservoir until the liquid water has changed completely into steam. Work is done by the expanding gas as it lifts the loaded piston.

19-11 HEAT TRANSFER MECHANISMS

We have discussed the transfer of heat between a system and its environment, but we have not yet described how that transfer takes place. There are three transfer mechanisms: conduction, convection, and radiation.

Conduction

If you leave a metal poker in a fire for any length of time, its handle will get hot. Energy is transferred from the fire to the handle by **conduction** along the length of the poker. The vibration amplitudes of the atoms and electrons of the metal at the fire end of the poker become relatively large because of the high temperature of their environment.

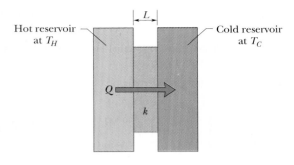

$T_H > T_C$

FIGURE 19-17 Thermal conduction. Heat is transferred from a reservoir at temperature T_H to a cooler reservoir at temperature T_C through a conducting slab of thickness L and thermal conductivity k.

TABLE 19-6 SOME THERMAL CONDUCTIVITIES[a]

	$k(\text{W/m} \cdot \text{K})$		$k(\text{W/m} \cdot \text{K})$
Metals		*Building Materials*	
Stainless steel	14	Polyurethane foam	0.024
Lead	35	Rock wool	0.043
Aluminum	235	Fiberglass	0.048
Copper	401	White pine	0.11
Silver	428	Window glass	1.0
Gases			
Air (dry)	0.026		
Helium	0.15		
Hydrogen	0.18		

[a]Conductivities change somewhat with temperature. The given values are at room temperature.

These increased vibrational amplitudes, and thus the associated energy, are passed along the poker, from atom to atom, during collisions between adjacent atoms. In this way, a region of rising temperature extends itself along the poker to the handle.

Consider a slab of face area A and thickness L, whose faces are maintained at temperatures T_H and T_C by a hot reservoir and a cold reservoir, as in Fig. 19-17. Let Q be the heat that is transferred through the slab, from its hot face to its cold face, in time t. Experiment shows that the rate of heat transfer H (the amount per unit time) is given by

$$H = \frac{Q}{t} = kA \frac{T_H - T_C}{L}, \qquad (19\text{-}30)$$

in which k, called the *thermal conductivity,* is a constant that depends on the material of which the slab is made. Large values of k define good heat conductors, and conversely. Table 19-6 gives thermal conductivities of some common metals, gases, and building materials.

Thermal Resistance to Conduction (*R*-Value)

If you are interested in insulating your house or in keeping cola cans cold on a picnic, you are more concerned with poor heat conductors than with good ones. For this reason, the concept of *thermal resistance R* has been introduced into engineering practice. The *R*-value of a slab of thickness L is defined as

$$R = \frac{L}{k}. \qquad (19\text{-}31)$$

Thus the lower the thermal conductivity of the material of which a slab is made, the higher the *R*-value of the slab. Note that R is a property attributed to a slab of a specified thickness, not to a material. The commonly used unit for R

(which, in the United States at least, is almost never stated) is the square foot–Fahrenheit degree-hour per British thermal unit ($\text{ft}^2 \cdot \text{F}° \cdot \text{h/Btu}$). (Now you know why the unit is rarely stated.)

Combining Eqs. 19-30 and 19-31 leads to

$$H = A \frac{T_H - T_C}{R}, \qquad (19\text{-}32)$$

which allows one to calculate the rate of heat flow through a slab if its *R*-value, its area, and the temperature difference between its faces are known.

Conduction Through a Composite Slab

Figure 19-18 shows a composite slab, consisting of two materials having different thicknesses L_1 and L_2 and different thermal conductivities k_1 and k_2. The temperatures of the outer surfaces of the slab are T_H and T_C. Each face of

FIGURE 19-18 Heat is transferred at a steady rate through a composite slab made up of two different materials with different thicknesses and different thermal conductivities. The steady-state temperature at the interface of the two materials is T_X.

the slab has area A. Let us derive an expression for the rate of heat transfer through the slab under the assumption that the transfer is a *steady-state* process; that is, the temperatures everywhere in the slab and the rate of heat transfer do not change with time.

In the steady state, the rates of heat transfer through the two materials are equal. This is the same as saying that the heat conducted through one material in a certain time must be equal to that conducted through the other material in the same amount of time. Were this not true, temperatures in the slab would be changing and we would not have a steady-state situation. Letting T_X be the temperature of the interface between the two materials, we can now use Eq. 19-30 to write

$$H = \frac{k_2 A (T_H - T_X)}{L_2} = \frac{k_1 A (T_X - T_C)}{L_1}. \quad (19\text{-}33)$$

Solving Eq. 19-33 for T_X yields, after a little algebra,

$$T_X = \frac{k_1 L_2 T_C + k_2 L_1 T_H}{k_1 L_2 + k_2 L_1}. \quad (19\text{-}34)$$

Substituting this expression for T_X into either equality of Eq. 19-33 yields

$$H = \frac{A(T_H - T_C)}{L_1/k_1 + L_2/k_2}. \quad (19\text{-}35)$$

Equation 19-31 reminds us that $L/k = R$.

We can extend Eq. 19-35 to apply to any number n of materials making up a slab:

$$H = \frac{A(T_H - T_C)}{\Sigma\,(L/k)} = \frac{A(T_H - T_C)}{\Sigma\,R}. \quad (19\text{-}36)$$

The summation sign in the denominators tells us to add the values of L/k or R for all the materials.

$\mathbb{C}$HECKPOINT **6:** The figure shows the face and interface temperatures of a composite slab consisting of four materials, of identical thicknesses, through which the heat transfer is steady. Rank the materials according to their thermal conductivities, greatest first.

25°C   15°C   10°C   −5.0°C   −10°C

a   *b*   *c*   *d*

Convection

When you look at the flame of a candle or a match, you are watching thermal energy being transported upward by **convection.** Such heat transfer occurs when a fluid, such as

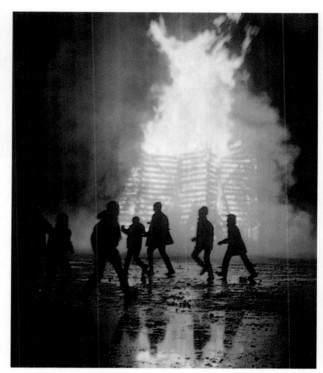

A football rally is illuminated by a fierce bonfire. Heated air and hot gases from the fire rise, and cooler air flows into the base of the fire.

air or water, is in contact with an object whose temperature is higher than that of the fluid. The temperature of the fluid that is in contact with the hot object increases, and (in most cases) the fluid expands and thus becomes less dense. Because it is now lighter than the surrounding cooler fluid, this expanded fluid rises because of buoyant forces. Some of the surrounding cooler fluid then flows so as to take the place of the rising warmer fluid and is itself warmed. The process can then continue.

Convection is part of many natural processes. Atmospheric convection plays a fundamental role in determining global climate patterns and daily weather variations. Glider pilots and birds alike seek rising thermals (currents of warm air) that keep them aloft. Huge energy transfers take place within the oceans by the same process. Finally, energy is transported to the surface of the Sun from the nuclear furnace at its core by enormous cells of convection, in which hot gas rises to the surface along the cell core and cooler gas around the core descends below the surface.

Radiation

The third method of heat transfer between an object and its environment is via electromagnetic waves (visible light is one kind of electromagnetic wave). Heat transferred in this

FIGURE 19-19   A false-color thermogram reveals the rate at which energy is radiated by houses along a street. The rates, from largest to smallest, are color coded as white, red, pink, blue, and black. You can tell where there is insulation in the walls, a heavy curtain over a window, and a higher air temperature at the ceiling on the second floor.

way is often called **thermal radiation** to distinguish it from electromagnetic *signals* (as in, say, television broadcasts) and from nuclear radiation (energy and particles emitted by nuclei). (To ''radiate'' generally means to emit.) When you stand in bright sunshine, you are warmed by absorbing thermal radiation from the Sun. No medium is required for heat transfer via radiation.

The rate P_r at which an object emits energy via electromagnetic radiation (that is, the power P_r of the radiating object) depends on its surface area A and the temperature T of that area in kelvins and is given by

$$P_r = \sigma \varepsilon A T^4. \qquad (19\text{-}37)$$

Here $\sigma = 5.6703 \times 10^{-8}$ W/m$^2 \cdot$K^4 is called the *Stefan–Boltzmann constant* after Josef Stefan (who discovered Eq. 19-37 experimentally in 1879) and Ludwig Boltzmann (who derived it theoretically soon after). The symbol ε represents the *emissivity* of the object's surface, which has a value between 0 and 1, depending on the composition of the surface. A surface with the maximum emissivity of 1.0 is said to be a *blackbody radiator,* but such a surface is possible only in theory. Note again that the temperature in Eq. 19-37 must be in kelvins so that a temperature of absolute zero corresponds to no radiation. Note also that every object whose temperature is above 0 K—including people—emits thermal radiation. (See Fig. 19-19.)

The rate P_a at which an object absorbs energy via thermal radiation from its environment, which we take to be at uniform temperature T_{env} (in kelvins), is

$$P_a = \sigma \varepsilon A T_{env}^4. \qquad (19\text{-}38)$$

The emissivity ε in Eq. 19-38 is the same as that in Eq. 19-37. An idealized blackbody radiator, with $\varepsilon = 1$, will absorb all the radiated energy it intercepts (rather than sending a portion back away from itself through reflection or scattering).

Because an object will radiate energy to the environment while it absorbs energy from the environment, the object's net rate P_n of energy exchange due to thermal radiation is

$$\begin{aligned} P_n &= P_a - P_r \\ &= \sigma \varepsilon A (T_{env}^4 - T^4). \end{aligned} \qquad (19\text{-}39)$$

The emissivity of black cloth is greater than that of white cloth; hence, by Eq. 19-39 a black robe will absorb more net energy from sunshine than a white robe and thus will be at a higher temperature. In fact, research has shown that in the hot desert, a black Bedouin robe can be 6 C° higher in temperature than a similar white robe. Why, then, would someone who must avoid overheating to survive in the harsh desert wear a black robe?

FIGURE 19-20   Convection up through the hotter black robe is more vigorous than that up through the cooler white robe. (After ''Why Do Bedouins Wear Black Robes in Hot Deserts?'' by A. Shkolnik, C. R. Taylor, V. Finch, and A. Borut, *Nature,* Vol. 283, January 24, 1980, pp. 373–374.)

The answer is that the black robe, which is itself at a higher temperature than a comparable white robe, does indeed warm the air inside the robe. That warmer air then rises faster and leaves through the porous fabric, while external air is drawn into the robe through its open bottom (Fig. 19-20). Thus the black fabric enhances air circulation under the robe, and that keeps the Bedouin from getting any hotter than a person in a white robe. In fact, the continuous breeze blowing past his body may even make him feel more comfortable.

SAMPLE PROBLEM 19-9

A composite slab (see Fig. 19-18) whose face area A is 26 ft^2 is made up of 2.0 in. of rock wool (1.0 in. has an R-value of 3.3) and 0.75 in. of white pine (1.0 in. has an R-value of 1.3). The temperature difference between the faces of the slab is 65 F°. What is the rate of heat transfer through the slab?

SOLUTION: The R-value for the 2.0 in. of rock wool is 3.3×2.0 or 6.6 ft$^2 \cdot$ °F $\cdot$ h/Btu. For the 0.75 in. of wood it is 1.3×0.75 or 0.98, in the same units. The composite slab thus has an R-value of $6.6 + 0.98$ or 7.58 ft$^2 \cdot$ F° $\cdot$ h/Btu. Substitution into Eq. 19-36 yields

$$H = \frac{A(T_H - T_C)}{\Sigma R} = \frac{(26 \text{ ft}^2)(65 \text{ F}°)}{7.58 \text{ ft}^2 \cdot \text{F}° \cdot \text{h/Btu}}$$

$$= 223 \text{ Btu/h} \approx 220 \text{ Btu/h} \ (= 65 \text{ W}). \quad \text{(Answer)}$$

Thus at this temperature difference, each such insulating slab would transmit heat continuously at the rate of 65 W.

SAMPLE PROBLEM 19-10

Figure 19-21 shows the cross section of a wall made of white pine of thickness L_a and brick of thickness $L_d (= 2.0L_a)$, sandwiching two layers of unknown material with identical thicknesses and thermal conductivities. The thermal conductivity of the pine is k_a and that of the brick is $k_d (= 5.0k_a)$. The face area A of the wall is unknown. Heat conduction through the wall has reached the steady state; the only known interface temperatures are $T_1 = 25$°C, $T_2 = 20$°C, and $T_5 = -10$°C.

(a) What is interface temperature T_4?

SOLUTION: We cannot find T_4 by simply applying Eq. 19-30 layer by layer, starting with the pine and working our way rightward, because we do not know enough about the intermediate layers. However, since the heat conduction has reached the steady state, we know that the rate of conduction H_a through the pine must equal the rate of conduction H_d through the brick. From Eq. 19-30 and Fig. 19-21, we can write these rates as

$$H_a = k_aA\frac{T_1 - T_2}{L_a} \quad \text{and} \quad H_d = k_dA\frac{T_4 - T_5}{L_d}.$$

Setting $H_a = H_d$ and solving for T_4 yield

$$T_4 = \frac{k_aL_d}{k_dL_a}(T_1 - T_2) + T_5.$$

Letting $L_d = 2.0L_a$ and $k_d = 5.0k_a$, and inserting the known temperatures, we find

$$T_4 = \frac{k_a(2.0L_a)}{(5.0k_a)L_a}(25°C - 20°C) + (-10°C)$$

$$= -8.0°C. \quad \text{(Answer)}$$

(b) What is interface temperature T_3?

SOLUTION: Now that we know T_4, we can find T_3, even though we know little about the intermediate layers. (In fact, at this point you might be able to guess the answer.) Since the heat conduction process is in the steady state, the rate of conduction H_b through layer b is equal to the rate of conduction H_c through layer c. Then, from Eq. 19-30,

$$k_bA\frac{T_2 - T_3}{L_b} = k_cA\frac{T_3 - T_4}{L_c}.$$

Because the thermal conductivities k_b and k_c of the layers are equal and so are their thicknesses L_b and L_c, we have

$$T_2 - T_3 = T_3 - T_4,$$

which gives us

$$T_3 = \frac{T_2 + T_4}{2} = \frac{20°C + (-8.0°C)}{2}$$

$$= 6.0°C. \quad \text{(Answer)}$$

This shows that since the intermediate layers have identical thermal conductivities, a point midway across them has a temperature that is midway between the temperatures of their outside surfaces.

FIGURE 19-21 Sample Problem 19-10. A wall of four layers through which there is steady-state heat transfer.

SAMPLE PROBLEM 19-11

During an extended wilderness hike, you have a terrific craving for ice. Unfortunately, the air temperature drops to only 6.0°C each night—too high to freeze water. However, because a clear, moonless night sky acts like a blackbody radiator at a temperature of $T_s = -23°C$, perhaps you can make ice by letting a shallow layer of water radiate energy to such a sky. To start, you thermally insulate a container from the ground by placing a poorly conducting layer of, say, foam rubber or straw beneath it. Then you pour water into the container, forming a thin, uniform layer with mass $m = 4.5$ g, top surface area $A = 9.0$ cm², depth $d = 5.0$ mm, emissivity $\varepsilon = 0.90$, and initial temperature 6.0°C. Find the time required for the water to freeze via radiation. Can the freezing be accomplished during one night?

SOLUTION: If you are to make ice via thermal radiation, the radiation must first reduce the water temperature from 279 K (= 6.0°C) to the freezing point of 273 K. Using Eq. 19-15 and Table 19-3, we find the associated energy exchange to be

$$Q_1 = cm(T_f - T_i)$$
$$= (4190 \text{ J/kg·K})(4.5 \times 10^{-3} \text{ kg})(273 \text{ K} - 279 \text{ K})$$
$$= -113 \text{ J}.$$

This energy must be radiated away by the water to drop its temperature to the freezing point.

An additional amount of energy Q_2 must be radiated away to change the phase of the water. Using Eqs. 19-17 and 19-19, and inserting a minus sign to show that the energy is lost by the water, we find

$$Q_2 = -mL_F = -(4.5 \times 10^{-3} \text{ kg})(3.33 \times 10^5 \text{ J/kg})$$
$$= -1499 \text{ J}.$$

The total amount of energy Q to be radiated away by the water is thus

$$Q = Q_1 + Q_2 = -113 \text{ J} - 1499 \text{ J} = -1612 \text{ J}.$$

The water will absorb energy from the sky while it is radiating energy to the sky. The net rate of energy exchange P_n is given by Eq. 19-39. The time t required to radiate energy Q is then

$$t = \frac{Q}{P_n} = \frac{Q}{\sigma \varepsilon A(T_s^4 - T^4)}. \qquad (19\text{-}40)$$

Although the temperature T of the water decreases slightly while the water is cooling, we can approximate T as being the freezing point, 273 K. With $T_s = 250$ K, the denominator of Eq. 19-40 is

$$(5.67 \times 10^{-8} \text{ W/m}^2 \cdot \text{K}^4)(0.90)(9.0 \times 10^{-4} \text{ m}^2)$$
$$\times [(250 \text{ K})^4 - (273 \text{ K})^4] = -7.57 \times 10^{-2} \text{ J/s},$$

and Eq. 19-40 gives us

$$t = \frac{-1612 \text{ J}}{-7.57 \times 10^{-2} \text{ J/s}} = 2.13 \times 10^4 \text{ s} = 5.9 \text{ h.} \qquad \text{(Answer)}$$

Because t is less than a night, freezing water by having it radiate to the dark sky is feasible. In fact, in some parts of the world people used this technique long before the introduction of electric freezers.

<h1 style="text-align:center">REVIEW & SUMMARY</h1>

Temperature; Thermometers

Temperature is an SI base quantity related to our sense of hot and cold. It is measured with a thermometer, which contains a working substance with a measurable property, such as length or pressure, that changes in a regular way as the substance becomes hotter or colder.

Zeroth Law of Thermodynamics

When a thermometer and some other object are placed in contact with each other, they eventually reach thermal equilibrium. The reading of the thermometer is then taken to be the temperature of the other object. The process provides consistent and useful temperature measurements because of the **zeroth law of thermodynamics:** if bodies A and B are each in thermal equilibrium with a third body C (the thermometer), then A and B are in thermal equilibrium with each other.

The Kelvin Temperature Scale

Temperature is measured, in the SI system, on the **Kelvin scale.** The scale is established by defining the temperature of the *triple point* of water to be 273.16 K. Other temperatures are then defined by use of a *constant-volume gas thermometer,* in which the pressure of a constant-volume sample of a gas is proportional to its temperature. Since different gases give consistent results only at very low densities, we define the *temperature T* as measured with a gas thermometer to be

$$T = (273.16 \text{ K}) \left(\lim_{m \to 0} \frac{p}{p_3} \right). \qquad (19\text{-}6)$$

Here T is measured in kelvins, p_3 and p are the pressures of the gas at 273.16 K and the measured temperature, respectively, and m is the mass of the gas in the thermometer.

Celsius and Fahrenheit Scales

The Celsius temperature scale is defined by

$$T_C = T - 273.15°, \tag{19-7}$$

and the Fahrenheit temperature scale is defined by

$$T_F = \tfrac{9}{5}T_C + 32°. \tag{19-8}$$

Thermal Expansion

All objects change size with changes in temperature. For a temperature change ΔT, a change ΔL in any linear dimension L is given by

$$\Delta L = L\alpha\,\Delta T, \tag{19-9}$$

in which α is the **coefficient of linear expansion.** The change ΔV in the volume V of a solid or liquid is

$$\Delta V = V\beta\,\Delta T. \tag{19-11}$$

Here $\beta = 3\alpha$ is the material's **coefficient of volume expansion.**

Heat

Heat Q is energy that is transferred between a system and its environment because of a temperature difference between them. It can be measured in **joules** (J), **calories** (cal), **kilocalories** (Cal or kcal), or **British thermal units** (Btu), with

$$1\ \text{cal} = 3.969 \times 10^{-3}\ \text{Btu} = 4.186\ \text{J.} \tag{19-13}$$

Heat Capacity and Specific Heat

If heat Q is added to an object, the temperature change $T_f - T_i$ is related to Q by

$$Q = C(T_f - T_i), \tag{19-14}$$

in which C is the **heat capacity** of the object. If the object has mass m, then

$$Q = cm(T_f - T_i), \tag{19-15}$$

where c is the **specific heat** of the material making up the object. The **molar specific heat** of a material is the heat capacity per mole, or per 6.02×10^{23} elementary units of the material.

Heat of Transformation

Heat supplied to a material may change the material's physical state or phase—for example, from solid to liquid or from liquid to gas. The amount of heat required per unit mass to change the phase (but not the temperature) of a particular material is its **heat of transformation** L. Thus

$$Q = Lm. \tag{19-17}$$

The **heat of vaporization** L_V is the amount of energy per unit mass that must be added to vaporize a liquid or that must be removed to condense a gas. The **heat of fusion** L_F is the amount of energy per unit mass that must be added to melt a solid or that must be removed to freeze a liquid.

Work Associated with Volume Change

A gas may exchange energy with its surroundings through work. The amount of work W done by a gas as it expands or contracts from an initial volume V_i to a final volume V_f is given by

$$W = \int dW = \int_{V_i}^{V_f} p\,dV. \tag{19-23}$$

The integration is necessary because the pressure p may vary during the volume change.

First Law of Thermodynamics

The principle of conservation of energy for a thermodynamic process is expressed in the **first law of thermodynamics,** which may assume either of the forms

$$\Delta E_{int} = E_{int,f} - E_{int,i} = Q - W \quad \text{(first law)} \tag{19-24}$$

or

$$dE_{int} = dQ - dW \quad \text{(first law).} \tag{19-25}$$

E_{int} represents the internal energy of the material, which depends only on its state (temperature, pressure, and volume). Q represents the heat exchanged by the system with its surroundings; Q is positive if the system gains heat and negative if the system loses heat. W is the work done *by* the system; W is positive if the system expands against some external force exerted by the surroundings, and negative if the system contracts because of some external force. *Both Q and W are path dependent; ΔE_{int} is path independent.*

Applications of the First Law

The first law of thermodynamics finds application in several special cases:

$$\text{adiabatic processes:} \quad Q = 0, \quad \Delta E_{int} = -W$$
$$\text{constant-volume processes:} \quad W = 0, \quad \Delta E_{int} = Q$$
$$\text{cyclical processes:} \quad \Delta E_{int} = 0, \quad Q = W$$
$$\text{free expansion processes:} \quad Q = W = \Delta E_{int} = 0$$

Conduction, Convection, and Radiation

The rate H at which heat is *conducted* through a slab whose faces are maintained at temperatures T_H and T_C is

$$H = \frac{Q}{t} = kA\frac{T_H - T_C}{L}, \tag{19-30}$$

in which A and L are the face area and length of the slab, and k is the thermal conductivity of the material.

Convection occurs when temperature differences cause a heat transfer by motion within a fluid. *Radiation* is heat transfer via the emission of electromagnetic energy. The rate P_r at which an object emits energy via thermal radiation is

$$P_r = \sigma\varepsilon A T^4, \tag{19-37}$$

where $\sigma\ (= 5.6703 \times 10^{-8}\ \text{W/m}^2\cdot\text{K}^4)$ is the Stefan–Boltzmann constant, ε is the emissivity of the object's surface, A is its surface area, and T is its surface temperature (in kelvins). The rate P_a at which an object absorbs energy via thermal radiation from its environment, which is at the uniform temperature T_{env} (in kelvins), is

$$P_a = \sigma\varepsilon A T_{env}^4. \tag{19-38}$$

QUESTIONS

1. Figure 19-22 shows three temperature scales with the freezing and boiling points of water indicated. Rank a change of 25 R°, 25 S°, and 25 U° according to the corresponding change in temperature, greatest first.

FIGURE 19-22 Question 1.

2. A rod, initially at room temperature, is to be heated and cooled in six steps. The changes in its length, expressed in terms of some basic unit, will be +7, +5, +3, −4, −6, and −4. (a) Is the final temperature of the rod equal to room temperature, above it, or below it? (b) Is there a sequence of the steps such that the rod is at room temperature after one of the intermediate steps?

3. The table gives the initial length L, change in temperature ΔT, and change in length ΔL of four rods. Rank the rods according to their coefficients of thermal expansion, greatest first.

ROD	L (m)	ΔT (C°)	ΔL (m)
a	2	10	4×10^{-4}
b	1	20	4×10^{-4}
c	2	10	8×10^{-4}
d	4	5	4×10^{-4}

4. Rank the Celsius, Kelvin, and Fahrenheit temperature scales according to the heat required to raise the temperature of one gram of water by one degree on each scale, greatest first.

5. Materials A, B, and C are solids that are at their melting temperatures. Material A requires 200 J to melt 4 kg. Material B requires 300 J to melt 5 kg. And material C requires 300 J to melt 6 kg. Rank the materials according to their heats of fusion, greatest first.

6. Figure 19-23 shows four paths on a p-V diagram along which a gas can be taken from state i to state f. Rank the paths according to (a) the change ΔE_{int}, (b) the work W done by the gas, and (c) the magnitude of the heat transfer Q, greatest first.

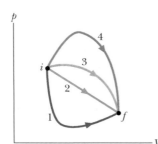

FIGURE 19-23
Question 6.

7. Figure 19-24 shows two closed cycles on p-V diagrams for a gas. The three parts of cycle 1 are of the same length and shape as those of cycle 2. For each cycle, should the cycle be traversed clockwise or counterclockwise if (a) the net work W done by the gas is to be positive and (b) the net heat transfer Q between the gas and the environment is to be positive?

FIGURE 19-24 Questions 7 and 8.

8. For which cycle in Fig. 19-24, traversed clockwise, is (a) W greater and (b) Q greater?

9. Figure 19-25 shows a composite slab of three different materials, a, b, and c, with identical thicknesses and with thermal conductivities $k_b > k_a > k_c$. The heat transfer through them is nonzero and steady. Rank the materials according to the temperature difference ΔT across them, greatest first.

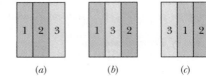

FIGURE 19-25
Question 9.

10. Figure 19-26 shows three different arrangements of materials 1, 2, and 3 to form a wall. The thermal conductivities are $k_1 > k_2 > k_3$. The left side of the wall is 20 C° higher than the right side. Rank the arrangements according to (a) the (steady state) rate of energy transfer through the wall and (b) the temperature difference across material 1, greatest first.

FIGURE 19-26
Question 10.

11. During an icicle's growth, its outer surface is covered with a thin sheath of liquid water that slowly seeps downward to form drops, one at a time, that hang at the tip (Fig. 19-27). Each drop straddles a thin tube of liquid water that extends up into the icicle, toward its *root* (its top). As the water at the top of the tube gradually freezes, energy is released. Is that energy conducted radially

FIGURE 19-27
Question 11.

outward through the ice, downward through the water to the hanging drop, or upward toward the root? (Assume that the air temperature is below 0°C.)

12. Figure 19-28 shows a horizontal cross section (top view) of a square room surrounded on four sides by thick walls. The walls are all made of the same material and all have the same face area. They have thicknesses of either L, $2L$, or $3L$ as shown, and they are well insulated along their lengths. The faces forming the room are maintained at 5°C, and the conduction of energy outward through the walls is steady. Rank the walls according to the rate of conduction through them, greatest first.

FIGURE 19-28 Question 12.

13. Suppose a block of wood and a block of metal are at the *same* temperature. When the blocks feel cold to your fingers, the metal feels colder than the wood; when the blocks feel hot to your fingers, the metal feels hotter than the wood. At what temperature will the blocks feel the same?

14. The following solid objects, made of the same material, are maintained at a temperature of 300 K in an environment whose temperature is 350 K: a cube of edge length r, a sphere of radius r, and a hemisphere of radius r. Rank the objects according to the net rate at which thermal radiation is exchanged with the environment, greatest first.

15. The following pairs give the temperatures of an object and its environment, respectively, for three situations: (1) 300 K and 350 K; (2) 350 K and 400 K; and (3) 400 K and 450 K. Without computation, rank the situations according to the object's net rate of energy exchange P_n, greatest first.

EXERCISES & PROBLEMS

SECTION 19-3 Measuring Temperature

1E. To measure temperatures, physicists and astronomers often measure how the intensity of electromagnetic radiation emitted by an object varies with wavelength. The wavelength λ_{max} at which the intensity is greatest is related to T, the temperature of the object in kelvins, by

$$\lambda_{max}T = 0.2898 \text{ cm} \cdot \text{K}.$$

In 1965, microwave radiation peaking at $\lambda_{max} = 0.107$ cm was discovered coming in all directions from space. To what temperature does this correspond? The interpretation of this *background radiation* is that it is left over from some 15 billion years ago, soon after the universe began.

2E. Suppose the temperature of a gas at the boiling point of water is 373.15 K. What then is the limiting value of the ratio of the pressure of the gas at that boiling point to its pressure at the triple point of water? (Assume the volume of the gas is the same at both temperatures.)

3P. Two constant-volume gas thermometers are assembled, one using nitrogen and the other using hydrogen. Both contain enough gas so that $p_3 = 80$ kPa. What is the difference between the pressures in the two thermometers if both are inserted into a water bath at the boiling point? Which gas is at higher pressure?

4P. A particular gas thermometer is constructed of two gas-containing bulbs, each of which is put into a water bath, as shown in Fig. 19-29. The pressure difference between the two bulbs is measured by a mercury manometer as shown. Appropriate reser-

voirs, not shown in the diagram, maintain constant gas volume in the two bulbs. There is no difference in pressure when both baths are at the triple point of water. The pressure difference is 120 torr when one bath is at the triple point and the other is at the boiling point of water. Finally, the pressure difference is 90.0 torr when one bath is at the triple point and the other is at an unknown temperature to be measured. What is the unknown temperature?

FIGURE 19-29 Problem 4.

SECTION 19-4 The Celsius and Fahrenheit Scales

5E. At what temperature is the Fahrenheit scale reading equal to (a) twice that of the Celsius and (b) half that of the Celsius?

6E. If your doctor tells you that your temperature is 310 kelvins above absolute zero, should you worry? Explain your answer.

7E. (a) In 1964, the temperature in the Siberian village of Oymyakon reached a value of −71°C. What temperature is this on the Fahrenheit scale? (b) The highest officially recorded temperature in the continental United States was 134°F in Death Valley, California. What is this temperature on the Celsius scale?

8E. (a) The temperature of the surface of the Sun is about 6000 K. Express this on the Fahrenheit scale. (b) Express normal human body temperature, 98.6°F, on the Celsius scale. (c) In the continental United States, the lowest officially recorded temperature is −70°F at Rogers Pass, Montana. Express this on the Celsius scale. (d) Express the normal boiling point of oxygen, −183°C, on the Fahrenheit scale. (e) At what Celsius temperature would you find a room to be uncomfortably warm?

9E. At what temperature do the following pairs of scales read the same: (a) Fahrenheit and Celsius (verify the listing in Table 19-1), (b) Fahrenheit and Kelvin, and (c) Celsius and Kelvin?

10P. Suppose that on a temperature scale X, water boils at −53.5°X and freezes at −170°X. What would a temperature of 340 K be on the X scale?

11P. It is an everyday observation that hot and cold objects cool down or warm up to the temperature of their surroundings. If the temperature difference ΔT between an object and its surroundings ($\Delta T = T_{obj} - T_{sur}$) is not too great, the rate of cooling or warming of the object is proportional, approximately, to this temperature difference; that is,

$$\frac{d\,\Delta T}{dt} = -A(\Delta T),$$

where A is a constant. The minus sign appears because ΔT decreases with time if ΔT is positive and increases if ΔT is negative. This is known as *Newton's law of cooling.* (a) On what factors does A depend? What are its dimensions? (b) If at some instant $t = 0$ the temperature difference is ΔT_0, show that it is

$$\Delta T = \Delta T_0 e^{-At}$$

at a later time t.

12P. The heater of a house breaks down one day when the outside temperature is 7.0°C. As a result, the inside temperature drops from 22°C to 18°C in 1.0 h. The owner fixes the heater and adds insulation to the house. Now she finds that, on a similar day, the house takes twice as long to drop from 22°C to 18°C when the heater is not operating. What is the ratio of the constant A in Newton's law of cooling (see Problem 11) after the insulation is added to the value before?

SECTION 19-5   Thermal Expansion

13E. A steel rod has a length of exactly 20 cm at 30°C. How much longer is it at 50°C?

14E. An aluminum flagpole is 33 m high. By how much does its length increase as the temperature increases by 15 C°?

15E. The Pyrex glass mirror in the telescope at the Mt. Palomar Observatory has a diameter of 200 in. The temperature ranges from −10°C to 50°C on Mt. Palomar. Determine the maximum change in the diameter of the mirror.

16E. A circular hole in an aluminum plate is 2.725 cm in diameter at 0.000°C. What is its diameter when the temperature of the plate is raised to 100.0°C?

17E. An aluminum-alloy rod has a length of 10.000 cm at

20.000°C and a length of 10.015 cm at the boiling point of water. (a) What is the length of the rod at the freezing point of water? (b) What is the temperature if the length of the rod is 10.009 cm?

18E. (a) What is the coefficient of linear expansion of aluminum per Fahrenheit degree? (b) Use your answer in (a) to calculate the change in length of a 20 ft aluminum rod if the rod is heated from 40°F to 95°F.

19E. Soon after Earth was formed, heat released by the decay of radioactive elements raised the average internal temperature from 300 to 3000 K, at about which value it remains today. Assuming an average coefficient of volume expansion of 3.0×10^{-5} K^{-1}, by how much has the radius of Earth increased since the planet was formed?

20E. The Stanford linear accelerator contains hundreds of brass disks tightly fitted into a steel tube. The system was assembled by cooling the disks in dry ice (at −57.00°C) to enable them to slide into the close-fitting tube. If the diameter of a disk is 80.00 mm at 43.00°C, what is its diameter in the dry ice?

21E. A glass window is exactly 20 cm by 30 cm at 10°C. By how much has its area increased when its temperature is 40°C?

22E. At 20°C, a brass cube has an edge length of 30 cm. What is the increase in the cube's surface area when it is heated from 20°C to 75°C?

23E. Find the change in volume of an aluminum sphere with an initial radius of 10 cm when the sphere is heated from 0.0°C to 100°C.

24E. What is the volume of a lead ball at 30°C if the ball's volume at 60°C is 50 cm³?

25E. By how much does the volume of an aluminum cube 5.00 cm on an edge increase when the cube is heated from 10.0°C to 60.0°C?

26E. An aluminum cup of 100 cm³ capacity is filled with glycerin at 22°C. How much glycerin, if any, will spill out of the cup if the temperature of the cup and glycerin is raised to 28°C? (The coefficient of volume expansion of glycerin is 5.1×10^{-4}/C°.)

27E. A steel rod at 25.0°C is bolted securely at both ends and then cooled. At what temperature will it rupture? Use Table 13-1.

28P. At 20°C, a rod is exactly 20.05 cm long on a steel ruler. Both the rod and the ruler are placed in an oven at 270°C, where the rod now measures 20.11 cm on the same ruler. What is the coefficient of thermal expansion for the material of which the rod is made?

29P. A steel rod is 3.000 cm in diameter at 25°C. A brass ring has an interior diameter of 2.992 cm at 25°C. At what common temperature will the ring just slide onto the rod?

30P. The area A of a rectangular plate is ab. Its coefficient of linear expansion is α. After a temperature rise ΔT, side a is longer by Δa and side b is longer by Δb (Fig. 19-30). Show that if we neglect the small quantity $\Delta a\,\Delta b/ab$, then $\Delta A = 2\alpha A\,\Delta T$.

31P. Density is mass divided by volume. If the volume V is temperature dependent, so is the density ρ. Show that a small change in density $\Delta\rho$ with a change in temperature ΔT is given by

$$\Delta\rho = -\beta\rho\,\Delta T,$$

where β is the coefficient of volume expansion. Explain the minus sign.

FIGURE 19-30
Problem 30.

32P. When the temperature of a metal cylinder is raised from 0.0°C to 100°C, its length increases by 0.23%. (a) Find the percent change in density. (b) What is the metal? Use Table 19-2.

33P. Show that when the temperature of a liquid in a barometer changes by ΔT, and the pressure is constant, the height h changes by $\Delta h = \beta h \, \Delta T$, where β is the coefficient of volume expansion. Neglect the expansion of the glass tube.

34P. When the temperature of a copper penny is raised by 100 C°, its diameter increases by 0.18%. To two significant figures, give the percent increase in (a) the area of a face, (b) the thickness, (c) the volume, and (d) the mass of the penny. (e) Calculate its coefficient of linear expansion.

35P. A pendulum clock with a pendulum made of brass is designed to keep accurate time at 20°C. What will be the error, in seconds per hour, if the clock operates at 0.0°C?

36P. As a result of a temperature rise of 32 C°, a bar with a crack at its center buckles upward (Fig. 19-31). If the fixed distance L_0 is 3.77 m and the coefficient of linear expansion of the bar is $25 \times 10^{-6}/$C°, find the rise x of the center.

FIGURE 19-31
Problem 36.

37P. A composite bar of length $L = L_1 + L_2$ is made from a bar of material 1 and length L_1 attached to a bar of material 2 and length L_2, as shown in Fig. 19-32. (a) Show that the coefficient of linear expansion α for this composite bar is given by $\alpha = (\alpha_1 L_1 + \alpha_2 L_2)/L$. (b) Using steel and brass, design such a composite bar whose length is 52.4 cm and whose coefficient of linear expansion is $13.0 \times 10^{-6}/$C°.

FIGURE 19-32 Problem 37.

SECTION 19-7 The Absorption of Heat by Solids and Liquids

38E. It is possible to melt ice by rubbing one block of it against another. How much work, in joules, would you have to do to get 1.00 g of ice to melt?

39E. A certain substance has a mass per mole of 50 g/mol. When 314 J of heat is added to a 30.0 g sample of this material, its temperature rises from 25.0°C to 45.0°C. (a) What is the specific heat of this substance? (b) How many moles of the substance are present? (c) What is the molar specific heat of the substance?

40E. In a certain solar house, energy from the Sun is stored in barrels filled with water. In a particular winter stretch of five cloudy days, 1.00×10^6 kcal is needed to maintain the inside of the house at 22.0°C. Assuming that the water in the barrels is at 50.0°C and that the water has a density of 1.00×10^3 kg/m³, what volume of water is required?

41E. A diet doctor encourages people to diet by drinking ice water. His theory is that the body must burn off enough fat to raise the temperature of the water from 0.00°C to the body temperature of 37.0°C. How many liters of ice water would have to be consumed to burn off 454 g (about 1 lb) of fat, assuming that this much fat burning requires 3500 Cal be transferred to the ice water? Why is it not advisable to follow this diet? One liter = 10^3 cm³. The density of water is 1.00 g/cm³.

42E. Icebergs in the North Atlantic present hazards to shipping, causing the lengths of shipping routes to increase by about 30% during the iceberg season. Attempts to destroy icebergs include planting explosives, bombing, torpedoing, shelling, ramming, and coating with black soot. Suppose that direct melting of the iceberg, by placing heat sources in the ice, is tried. How much heat is required to melt 10% of an iceberg with a mass of 200,000 metric tons? (Use 1 metric ton = 1000 kg.)

43E. How much water remains unfrozen after 50.2 kJ of heat has been extracted from 260 g of liquid water initially at its freezing point?

44E. Calculate the minimum amount of heat, in joules, required to completely melt 130 g of silver initially at 15.0°C.

45E. A room is lighted by four 100 W incandescent light bulbs. (The power of 100 W is the rate at which a bulb converts electrical energy to heat and visible light.) Assuming that 90% of the energy is converted to heat, how much heat is added to the room in 1.00 h?

46E. What quantity of butter (6.0 Cal/g = 6000 cal/g) would supply the energy needed for a 160 lb man to ascend to the top of Mt. Everest, elevation 29,000 ft, from sea level?

47E. An energetic athlete dissipates all the energy in a diet of 4000 Cal/day. If he were to release this energy at a steady rate, how would this conversion of energy compare with that of a 100 W bulb? (The power of 100 W is the rate at which the bulb converts electrical energy to heat and visible light.)

48E. If the heat necessary to warm a mass m of water from 68°F to 78°F were somehow converted to translational kinetic energy of that water, what would be the speed of the water?

49E. A power of 0.400 hp is required for 2.00 min to drill a hole in a 1.60-lb copper block. (a) How much thermal energy in Btu is generated? (b) What is the rise in temperature of the copper if only 75.0% of the power warms the copper? (Use 1 ft·lb = 1.285×10^{-3} Btu.)

50E. An object of mass 6.00 kg falls through a height of 50.0 m and, by means of a mechanical linkage, rotates a paddle wheel that stirs 0.600 kg of water. The water is initially at 15.0°C. What is the maximum possible temperature rise of the water?

51E. One way to keep the contents of a garage from becoming too cold on a night when a severe subfreezing temperature is forecast is to put a tub of water in the garage. If the mass of the water is 125 kg and its initial temperature is 20°C, (a) how much energy must the water transfer to its surroundings in order to freeze completely and (b) what is the lowest possible temperature of the water and its surroundings until that happens?

52E. A small electric immersion heater is used to heat 100 g of water for a cup of instant coffee. The heater is labeled "200 watts," which means that it converts electrical energy to thermal energy at this rate. Calculate the time required to bring this water from 23°C to the boiling point, ignoring any heat losses.

53E. A pickup truck whose mass is 2200 kg is speeding along the highway at 65.0 mi/h. (a) If you could use all this kinetic energy to vaporize water already at 100°C, how much water could you vaporize? (b) If you had to buy this amount of energy from your local utility company at 12¢/kW·h, how much would it cost you? Guess at the answers before you figure them out; you may be surprised.

54E. A 150 g copper bowl contains 220 g of water, both at 20.0°C. A very hot 300 g copper cylinder is dropped into the water, causing the water to boil, with 5.00 g being converted to steam. The final temperature of the system is 100°C. (a) How much heat was transferred to the water? (b) How much to the bowl? (c) What was the original temperature of the cylinder?

55P. Calculate the specific heat of a metal from the following data. A container made of the metal has a mass of 3.6 kg and contains 14 kg of water. A 1.8 kg piece of the metal initially at a temperature of 180°C is dropped into the water. The container and water initially have a temperature of 16.0°C, and the final temperature of the entire system is 18.0°C.

56P. A thermometer of mass 0.0550 kg and of specific heat 0.837 kJ/kg·K reads 15.0°C. It is then completely immersed in 0.300 kg of water, and it comes to the same final temperature as the water. If the thermometer reads 44.4°C, what was the temperature of the water before insertion of the thermometer?

57P. How long does it take a 2.0×10^5 Btu/h water heater to raise the temperature of 40 gal of water from 70°F to 100°F?

58P. An athlete needs to lose weight and decides to do it by "pumping iron." (a) How many times must an 80.0 kg weight be lifted a distance of 1.00 m in order to burn off 1 lb of fat, assuming that that much fat is equivalent to 3500 Cal? (b) If the weight is lifted once every 2.00 s, how long does the task take?

59P. A 1500 kg Buick moving at 90 km/h brakes to rest, at uniform deceleration and without skidding, over a distance of 80 m. At what average rate is mechanical energy transferred to thermal energy in the brake system?

60P. A chef, upon finding his stove out of order, decides to boil the water for his wife's coffee by shaking it in a thermos flask. Suppose that he uses 500 cm³ of tap water at 59°F, and that the water falls 1.0 ft each shake, the chef making 30 shakes each

minute. Neglecting any loss of thermal energy by the flask, how long must he shake the flask before the water boils?

61P. A block of ice, at its melting point and of initial mass 50.0 kg, slides along a horizontal surface, starting at a speed of 5.38 m/s and finally coming to rest after traveling 28.3 m. Compute the mass of ice melted as a result of the friction between the block and the surface. (Assume that all the heat generated owing to friction goes into the block of ice.)

62P. The specific heat of a substance varies with temperature according to $c = 0.20 + 0.14T + 0.023T^2$, with T in °C and c in cal/g·K. Find the heat required to raise the temperature of 2.0 g of this substance from 5.0°C to 15°C.

63P. In a solar water heater, energy from the Sun is gathered by water that circulates through tubes in a rooftop collector. The solar radiation enters the collector through a transparent cover and warms the water in the tubes; this water is pumped into a holding tank. Assuming that the efficiency of the overall system is 20% (that is, 80% of the incident solar energy is lost from the system), what collector area is necessary to raise the temperature of 200 L of water in the tank from 20°C to 40°C in 1.0 h? The intensity of incident sunlight is 700 W/m².

64P. An insulated thermos contains 130 cm³ of hot coffee, at a temperature of 80.0°C. You put in a 12.0 g ice cube at its melting point to cool the coffee. By how many degrees will your coffee have cooled once the ice has melted? Treat the coffee as though it were pure water.

65P. What mass of steam at 100°C must be mixed with 150 g of ice at its melting point, in a thermally insulated container, to produce liquid water at 50°C?

66P. A person makes a quantity of iced tea by mixing 500 g of hot tea (essentially water) with an equal mass of ice at its melting point. If the initial hot tea is at a temperature of (a) 90°C and (b) 70°C, what are the temperature and mass of the remaining ice when the tea and ice reach a common temperature?

67P. (a) Two 50 g ice cubes are dropped into 200 g of water in a glass. If the water was initially at 25°C, and the ice came directly from a freezer at −15°C, what will be the temperature of the drink when the ice and water reach the same temperature? (b) Suppose that only one ice cube had been used in (a); what would be the final temperature of the drink? Neglect the heat capacity of the glass.

68P. A 20.0 g copper ring has a diameter of exactly 1.00000 in. at its temperature of 0.000°C. An aluminum sphere has a diameter of exactly 1.00200 in. at its temperature of 100.0°C. The sphere is placed on top of the ring (Fig. 19-33), and the two are allowed to

FIGURE 19-33 Problem 68.

come to thermal equilibrium, with no heat lost to the surroundings. The sphere just passes through the ring at the equilibrium temperature. What is the mass of the sphere?

69P. A *flow calorimeter* is a device used to measure the specific heat of a liquid. Heat is added at a known rate to a stream of the liquid as it passes through the calorimeter at a known rate. Measurement of the resulting temperature difference between the inflow and the outflow points of the liquid stream enables us to compute the specific heat of the liquid. Suppose a liquid of density 0.85 g/cm³ flows through a calorimeter at the rate of 8.0 cm³/s. Heat is added at the rate of 250 W by means of an electric heating coil, and a temperature difference of 15 C° is established in steady-state conditions between the inflow and the outflow points. What is the specific heat of the liquid?

70P. By means of a heating coil, energy is transferred at a constant rate to a substance in a thermally insulated container. The temperature of the substance is measured as a function of time. (a) Show how we can deduce from this information the way in which the heat capacity of the substance depends on the temperature. (b) Suppose that in a certain temperature range the temperature T is proportional to t^3, where t is the time. How does the heat capacity depend on T in this range?

SECTION 19-10 Some Special Cases of the First Law of Thermodynamics

71E. A sample of gas expands from 1.0 m³ to 4.0 m³ while its pressure decreases from 40 Pa to 10 Pa. How much work is done by the gas if its pressure changes with volume via each of the three paths shown in the p-V diagram in Fig. 19-34?

FIGURE 19-34
Exercise 71.

FIGURE 19-35
Exercise 72.

72E. A sample of gas expands from 1.0 m³ to 4.0 m³ along path B in the p-V diagram in Fig. 19-35. It is then compressed back to 1.0 m³ along either path A or path C. Compute the net work done by the gas for the complete cycle in each case.

73E. Consider that 200 J of work is done on a system and 70.0 cal of heat is extracted from the system. In the sense of the first law of thermodynamics, what are the values (including algebraic signs) of (a) W, (b) Q, and (c) ΔE_{int}?

74E. A thermodynamic system is taken from an initial state A to another state B and back again to A, via state C, as shown by path $ABCA$ in the p-V diagram of Fig. 19-36a. (a) Complete the table in Fig. 19-36b by filling in either + or − for the sign of each thermodynamic quantity associated with each process. (b) Calculate the numerical value of the work done by the system for the complete cycle $ABCA$.

FIGURE 19-36 Exercise 74.

75E. Gas within a chamber passes through the cycle shown in Fig. 19-37. Determine the net heat added to the system during process CA if the heat Q_{AB} added during process AB is 20.0 J, no heat is transferred during process BC, and the net work done during the cycle is 15.0 J.

FIGURE 19-37 Exercise 75.

76E. Gas within a closed chamber undergoes the processes shown in the p-V diagram of Fig. 19-38. Calculate the net heat added to the system during one complete cycle.

77P. Figure 19-39a shows a cylinder containing gas and closed by a movable piston. The cylinder is kept submerged in an ice−water mixture. The piston is *quickly* pushed down from position 1 to position 2 and then held at position 2 until the gas is again at the temperature of the ice−water mixture; it then is *slowly* raised back to position 1. Figure 19-39b is a p-V diagram for the process.

If 100 g of ice is melted during the cycle, how much work has been done *on* the gas?

FIGURE 19-38 Exercise 76.

FIGURE 19-39 Problem 77.

78P. When a system is taken from state *i* to state *f* along path *iaf* in Fig. 19-40, $Q = 50$ cal and $W = 20$ cal. Along path *ibf*, $Q = 36$ cal. (a) What is W along path *ibf*? (b) If $W = -13$ cal for the curved return path *fi*, what is Q for this path? (c) Take $E_{int,i} = 10$ cal. What is $E_{int,f}$? (d) If $E_{int,b} = 22$ cal, what are the values of Q for path *ib* and path *bf*?

FIGURE 19-40 Problem 78.

SECTION 19-11 Heat Transfer Mechanisms

79E. The average rate at which heat is conducted through the ground surface in North America is 54.0 mW/m², and the average thermal conductivity of the near-surface rocks is 2.50 W/m·K. Assuming a surface temperature of 10.0°C, what should be the temperature at a depth of 35.0 km (near the base of the crust)? Ignore the heat generated by the presence of radioactive elements.

80E. The ceiling of a single-family dwelling in a cold climate should have an R-value of 30. To give such insulation, how thick

would a layer of (a) polyurethane foam and (b) silver have to be?

81E. The thermal conductivity of Pyrex glass at 0°C is 2.9 × 10^{-3} cal/cm·C°·s. (a) Express the quantity in W/m·K and in Btu/ft·F°·h. (b) What is the R-value for a 0.25 in. sheet of such glass?

82E. (a) Calculate the rate at which body heat is conducted through the clothing of a skier in a steady-state process, given the following data: the body surface area is 1.8 m² and the clothing is 1.0 cm thick; the skin surface temperature is 33°C, whereas the outer surface of the clothing is at 1.0°C; the thermal conductivity of the clothing is 0.040 W/m·K. (b) How would the answer to (a) change if, after a fall, the skier's clothes became soaked with water of thermal conductivity 0.60 W/m·K?

83E. Consider the slab shown in Fig. 19-17. Suppose that $L = 25.0$ cm, $A = 90.0$ cm², and the material is copper. If $T_H = 125°C$, $T_C = 10.0°C$, and a steady state is reached, find the rate of heat transfer through the slab.

84E. A cylindrical copper rod of length 1.2 m and cross-sectional area 4.8 cm² is insulated to prevent heat loss through its surface. The ends are maintained at a temperature difference of 100 C° by having one end in a water–ice mixture and the other in boiling water and steam. (a) Find the rate at which heat is conducted along the rod. (b) Find the rate at which ice melts at the cold end.

85E. Show that the temperature T_X at the interface of a compound slab (Fig. 19-18) is given by

$$T_X = \frac{R_1 T_H + R_2 T_C}{R_1 + R_2}.$$

86E. If you were to walk briefly in space without a spacesuit while far from the Sun (as an astronaut does in the movie *2001*), you would feel the cold of space: while you radiated thermal energy, you would absorb almost none from your environment. (a) At what rate would you lose energy? (b) How much energy would you lose in 30 s? Assume that your emissivity is 0.90, and estimate other data needed in the calculations.

87E. Four square pieces of insulation of two different materials, all with the same thickness and area A, are available to cover an opening of area $2A$. This can be done in either of the two ways shown in Fig. 19-41. Which arrangement, (a) or (b), would give the lower heat flow if $k_2 \neq k_1$?

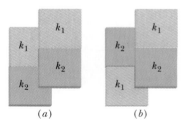

FIGURE 19-41 Exercise 87.

88P. Two identical rectangular rods of metal are welded end to end as shown in Fig. 19-42a, and 10 J of heat is conducted (in a steady-state process) through the rods in 2.0 min. How long would it take for 10 J to be conducted through the rods if they are welded together as shown in Fig. 19-42*b*?

(*a*)

(*b*)

FIGURE 19-42   Problem 88.

89P. Compute the rate of heat conduction through two storm doors, both 2.0 m high and 0.75 m wide. (a) One door is made with aluminum panels 1.5 mm thick and a glass pane 3.0 mm thick; the glass covers 75% of the aluminum surface (the structural frame has a negligible area). (b) The second door is made entirely of white pine averaging 2.5 cm in thickness. Take the temperature drop across each door to be 33 C°; see Table 19-6.

90P. A large cylindrical water tank with a bottom 1.7 m in diameter is made of iron boilerplate 5.2 mm thick. Water in the tank is heated by a gas burner underneath that is able to maintain a temperature difference of 2.3 C° between the top and bottom surfaces of the bottom plate. How much heat is conducted through that plate in 5.0 min? (Iron has a thermal conductivity of 67 W/m·K.)

91P. (a) What is the rate of heat loss in watts per square meter through a glass window 3.0 mm thick if the outside temperature is −20°F and the inside temperature is +72°F? (b) A storm window is installed having the same thickness of glass but with an air gap of 7.5 cm between the two windows. What will be the corresponding rate of heat loss assuming that conduction is the only important heat-loss mechanism?

92P. A sphere of radius 0.500 m, temperature 27.0°C, and emissivity 0.850 is located in an environment of temperature 77.0°C. At what rate does the sphere (a) emit thermal radiation and (b) absorb thermal radiation? (c) What is the sphere's net rate of energy exchange?

93P. A cube of edge length 6.0×10^{-6} m, emissivity 0.75, and temperature −100°C floats in an environment of temperature −150°C. What is the cube's net thermal radiation transfer rate?

94P. A solid cylinder of radius $r_1 = 2.5$ cm, length $h_1 =$ 5.0 cm, emissivity 0.85, and temperature 30°C is suspended in an environment of temperature 50°C. (a) What is the cylinder's net thermal radiation transfer rate P_1? (b) If the cylinder is stretched until its radius is $r_2 = 0.50$ cm, its net thermal radiation transfer rate becomes P_2. What is the ratio P_2/P_1?

95P. A tank of water has been outdoors in cold weather and a slab of ice 5.0 cm thick has formed on its surface (Fig. 19-43). The air above the ice is at −10°C. Calculate the rate of formation of ice (in centimeters per hour) on the bottom surface of the ice slab. Take the thermal conductivity and density of ice to be 0.0040 cal/s·cm·C° and 0.92 g/cm³. Assume that heat is not transferred through the walls or bottom of the tank.

FIGURE 19-43   Problem 95.

96P. Ice has formed on a shallow pond and a steady state has been reached, with the air above the ice at −5.0°C and the bottom of the pond at 4.0°C. If the total depth of *ice + water* is 1.4 m, how thick is the ice? (Assume that the thermal conductivities of ice and water are 0.40 and 0.12 cal/m·C°·s, respectively.)

97P. Three metal rods, one copper, one aluminum, and one brass, are each 6.00 cm long and 1.00 cm in diameter. These rods are placed end to end, with the aluminum between the other two. The free ends of the copper and brass rods are maintained at the boiling point and the freezing point of water, respectively. Find the steady-state temperatures of the copper−aluminum junction and the aluminum−brass junction. The thermal conductivity of brass is 109 W/m·K.

20
The Kinetic Theory of Gases

Suppose that you return to your chilly dwelling after snowshoeing for miles on a cold winter day. Your first thought is to light a stove. But why, exactly, would you do that? Is it because the stove will increase the store of internal (thermal) energy of the air in the dwelling, until eventually the air will have enough of that internal energy to keep you comfortable? As logical as this reasoning sounds, it is flawed, because the air's store of internal energy will not be changed by the stove. How can that be? And if it is so, why would you bother to light the stove?

20-1 A NEW WAY TO LOOK AT GASES

Classical thermodynamics—the subject of the last chapter—has nothing to say about atoms. Its laws are concerned only with such macroscopic variables as pressure, volume, and temperature. However, we know that a gas is made up of atoms or molecules (groups of atoms bound together). The pressure exerted by a gas must surely be related to the steady drumbeat of its molecules on the walls of its container. The ability of a gas to take on the volume of its container must surely be due to the freedom of motion of its molecules. And the temperature and internal energy of a gas must surely be related to the kinetic energy of these molecules. Perhaps we can learn something about gases by approaching the subject from this direction. We call this molecular approach the **kinetic theory of gases.** It is the subject of this chapter.

20-2 AVOGADRO'S NUMBER

When our thinking is slanted toward molecules, it makes sense to measure the sizes of our samples in moles. If we do so, we can be certain that we are comparing samples that contain the same number of molecules. The *mole* is one of the seven SI base units and is defined as follows:

One mole is the number of atoms in a 12 g sample of carbon-12.

We speak of a "mole of helium" or "a mole of water," meaning a certain number of the elementary units of the substance. But we could just as well speak of a mole of tennis balls, where, of course, the elementary unit would be a tennis ball.

The obvious question now is: "Just how many atoms or molecules are there in a mole?" The answer is determined experimentally and, as you saw in Chapter 19, is

$$N_A = 6.02 \times 10^{23} \text{ mol}^{-1} \quad \text{(Avogadro's number)}. \quad (20\text{-}1)$$

This number is called **Avogadro's number** after Italian scientist Amadeo Avogadro (1776–1856), who suggested that all gases contain the same number of molecules or atoms when they occupy the same volume under the same conditions of temperature and pressure.

The number of moles n contained in a sample of any substance can be found from

$$n = \frac{N}{N_A}, \quad (20\text{-}2)$$

in which N is the number of molecules in the sample. The number of moles in a sample can also be found from the mass M_{sam} of the sample and either the *molar mass M* (the mass of 1 mole of that substance) or the mass m of one molecule:

$$n = \frac{M_{sam}}{M} = \frac{M_{sam}}{mN_A}. \quad (20\text{-}3)$$

The enormously large value of Avogadro's number suggests how tiny and how numerous atoms must be. A mole of air, for example, can easily fit into a suitcase. Yet, if these molecules were spread uniformly over the surface of Earth, there would be about 120,000 of them per square centimeter. A second example: one mole of tennis balls would fill a volume equal to that of seven Moons!

PROBLEM SOLVING TACTICS

TACTIC 1: *Avogadro's Number of What?*

In Eq. 20-1, Avogadro's number is expressed in terms of mol^{-1}, which is the inverse mole, or 1/mol. We could instead explicitly state the elementary unit involved in a given situation. For example, if the elementary unit is an atom, we might write $N_A = 6.02 \times 10^{23}$ atoms/mole. If the elementary unit is a molecule, then we might write $N_A = 6.02 \times 10^{23}$ molecules/mole. And if we are interested in a *great* many tennis balls, we can write $N_A = 6.02 \times 10^{23}$ tennis balls/mole.

20-3 IDEAL GASES

Our goal in this chapter is to explain the macroscopic properties of a gas—such as its pressure and its temperature—in terms of the behavior of the molecules that make it up. But there is an immediate problem: which gas? Should it be hydrogen or oxygen, or methane, or perhaps uranium hexafluoride? They are all different. However, experimenters have found that if we confine 1 mole samples of various gases in boxes of identical volume and hold the gases at the same temperature, then their measured pressures are nearly—though not exactly—the same. If we repeat the measurements at lower gas densities, then these small differences in the measured pressures tend to disappear. Further experiments show that, at low enough densities, all real gases tend to obey the relation

$$pV = nRT \quad \text{(ideal gas law)}, \quad (20\text{-}4)$$

in which p is the absolute (not gauge) pressure, n is the number of moles of gas present, and R, the **gas constant,** has the same value for all gases, namely,

$$R = 8.31 \text{ J/mol} \cdot \text{K}. \quad (20\text{-}5)$$

The temperature T in Eq. 20-4 must be expressed in kelvins. Equation 20-4 is called the **ideal gas law.** Provided the gas density is reasonably low, Eq. 20-4 holds for any type of gas, or a mixture of different types, with n being the total number of moles present.

You may well ask, "What is an *ideal gas* and what is so 'ideal' about one?" The answer lies in the simplicity of the law (Eq. 20-4) that governs its macroscopic properties. Using this law—as you will see —we can deduce many properties of the ideal gas in a simple way. Although there is no such thing in nature as a truly ideal gas, *all* gases approach the ideal state at low enough densities, that is, under conditions in which their molecules are far enough apart. Thus the ideal gas concept allows us to gain useful insights into the limiting behavior of real gases.

Work Done by an Ideal Gas at Constant Temperature

Suppose that a sample of n moles of an ideal gas, confined to a piston–cylinder arrangement, is allowed to expand from an initial volume V_i to a final volume V_f. Suppose further that the temperature T of the gas is held constant throughout the process. Such a process is called an **isothermal expansion** (and the reverse is called an **isothermal compression**).

On a p-V diagram, an *isotherm* is a curve that connects points that have the same temperature. It is then a graph of pressure versus volume for a gas whose temperature T is held constant, that is, a graph of

$$p = nRT\,\frac{1}{V} = \text{(a constant)}\,\frac{1}{V}. \qquad (20\text{-}6)$$

Figure 20-1 shows three isotherms, each corresponding to a different (constant) value of T. Note that the values of T increase upward to the right.

The p-V diagram for an isothermal expansion or compression is also a graph of pressure versus volume for a gas whose temperature is held constant. Thus, such a p-V diagram must follow along an isotherm, as does the expansion from state i to state f in Fig. 20-1.

Let us calculate the work done by an ideal gas during an isothermal expansion. Equation 19-23,

$$W = \int_{V_i}^{V_f} p\,dV, \qquad (20\text{-}7)$$

is a general expression for the work done during any change in volume of a gas. Because we are now dealing with an ideal gas, we can use Eq. 20-4 to substitute for p, obtaining

$$W = \int_{V_i}^{V_f} \frac{nRT}{V}\,dV. \qquad (20\text{-}8)$$

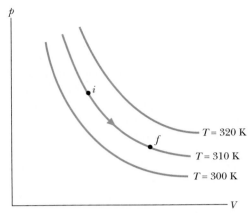

FIGURE 20-1   Three isotherms on a p-V diagram. The path shown along the middle isotherm represents an isothermal expansion of a gas from an initial state i to a final state f. The path from f to i along the isotherm would represent the reverse process, an isothermal compression.

And because we are considering an isothermal expansion, T is constant and we can move it in front of the integral sign to write

$$W = nRT\int_{V_i}^{V_f} \frac{dV}{V} = nRT\left[\ln V\right]_{V_i}^{V_f}. \qquad (20\text{-}9)$$

By evaluating at the limits and then using the relationship $\ln a - \ln b = \ln (a/b)$, we find that

$$W = nRT\ln\frac{V_f}{V_i} \qquad \begin{array}{l}\text{(ideal gas,}\\ \text{isothermal process).}\end{array} \qquad (20\text{-}10)$$

Recall that the symbol ln specifies a *natural* logarithm, that is, a logarithm to base e.

For an expansion, $V_f > V_i$ by definition, so the ratio V_f/V_i in Eq. 20-10 is greater than unity. The natural logarithm of a quantity greater than unity is positive, and so the work W done by an ideal gas during an isothermal expansion is positive, as we expect. For a compression, we have $V_f < V_i$, so the ratio of volumes in Eq. 20-10 is less than unity. The natural logarithm in that equation—hence the work W—is negative, again as we expect.

Equation 20-10 does not give the work W done by an ideal gas during *every* thermodynamic process. Instead, it gives the work only when the temperature is held constant. If the temperature varies, then the symbol T in Eq. 20-8 cannot be moved in front of the integral symbol as in Eq. 20-9, and thus we do not end up with Eq. 20-10.

However, we can go back to Eq. 20-7 to find the work W done by an ideal gas (or any other gas) during two more processes—a constant-volume process and a constant-pressure process. If the volume of the gas is constant, then Eq. 20-7 yields

$$W = 0 \qquad \text{(constant-volume process).} \qquad (20\text{-}11)$$

If, instead, the volume changes while the pressure p of the gas is held constant, then Eq. 20-7 becomes

$$W = p(V_f - V_i) = p \, \Delta V \quad \begin{array}{c}\text{(constant-pressure} \\ \text{process).}\end{array} \quad (20\text{-}12)$$

$\mathbf{C}$HECKPOINT **1:** An ideal gas has an initial pressure of 3 pressure units and an initial volume of 4 volume units. The table gives the final pressure and volume of the gas (in those same units) in five processes. Which processes start and end on the same isotherm?

	a	b	c	d	e
p	12	6	5	4	1
V	1	2	7	3	12

SAMPLE PROBLEM 20-1

A cylinder contains 12 L of oxygen at 20°C and 15 atm. The temperature is raised to 35°C, and the volume reduced to 8.5 L. What is the final pressure of the gas in atmospheres? Assume that the gas is ideal.

SOLUTION: From Eq. 20-4 we may write

$$nR = \frac{p_i V_i}{T_i} = \frac{p_f V_f}{T_f}.$$

Solving for p_f yields

$$p_f = \frac{p_i T_f V_i}{T_i V_f}. \quad (20\text{-}13)$$

Note here that if we converted the given initial and final volumes from liters to the proper units of cubic meters, the multiplying conversion factors would cancel out of Eq. 20-13. The same would be true for conversion factors that convert the pressures from atmospheres to the proper pascals. However, to convert the given temperatures to kelvins requires the addition of an amount that would not cancel and thus must be included. Hence, we must write

$$T_i = (273 + 20) \text{ K} = 293 \text{ K}$$

and $\qquad T_f = (273 + 35) \text{ K} = 308 \text{ K}.$

Inserting the given data into Eq. 20-13 then yields

$$p_f = \frac{(15 \text{ atm})(308 \text{ K})(12 \text{ L})}{(293 \text{ K})(8.5 \text{ L})} = 22 \text{ atm}. \quad \text{(Answer)}$$

SAMPLE PROBLEM 20-2

One mole of oxygen (assume it to be an ideal gas) expands at a constant temperature T of 310 K from an initial volume V_i of 12 L to a final volume V_f of 19 L.

(a) How much work is done by the expanding gas?

SOLUTION: From Eq. 20-10 we have

$$W = nRT \ln \frac{V_f}{V_i}$$

$$= (1 \text{ mol})(8.31 \text{ J/mol} \cdot \text{K})(310 \text{ K}) \ln \frac{19 \text{ L}}{12 \text{ L}}$$

$$= 1180 \text{ J}. \quad \text{(Answer)}$$

The expansion is graphed in the p-V diagram of Fig. 20-2. The work done by the gas during the expansion is represented by the area beneath the curve if.

(b) How much work is done by the gas during an isothermal *compression* from $V_i = 19$ L to $V_f = 12$ L?

SOLUTION: We proceed as in (a), finding

$$W = nRT \ln \frac{V_f}{V_i}$$

$$= (1 \text{ mol})(8.31 \text{ J/mol} \cdot \text{K})(310 \text{ K}) \ln \frac{12 \text{ L}}{19 \text{ L}}$$

$$= -1180 \text{ J}. \quad \text{(Answer)}$$

This result is equal in magnitude but opposite in sign to the result found in (a) for an isothermal expansion. The minus sign tells us that an external agent must do 1180 J of work *on* the gas to compress it.

FIGURE 20-2 Sample Problem 20-2. The shaded area represents the work done by 1 mol of oxygen in expanding at a constant temperature T of 310 K.

20-4 PRESSURE, TEMPERATURE, AND RMS SPEED

Here is our first kinetic theory problem. Let n moles of an ideal gas be confined in a cubical box of volume V, as in Fig. 20-3. The walls of the box are held at temperature T. What is the connection between the pressure p exerted by the gas on the walls and the speeds of the molecules?

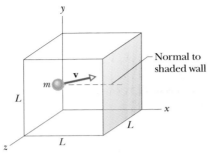

FIGURE 20-3 A cubical box of edge L, containing n moles of an ideal gas. A molecule of mass m and velocity **v** is about to collide with the shaded wall of area L^2. A normal to that wall is shown.

The molecules of gas in the box are moving in all directions and with various speeds, bumping into each other and bouncing from the walls of the box like balls in a racquetball court. We ignore (for the time being) collisions of the molecules with one another and consider only elastic collisions with the walls.

Figure 20-3 shows a typical gas molecule, of mass m and velocity **v**, that is about to collide with the shaded wall. Because we assume that any collision of a molecule with a wall is elastic, when this molecule collides with the shaded wall, the only component of its velocity that is changed is the x component, and that component is reversed. This means that the only change in the particle's momentum is then along the x axis, and that change is

$$\Delta p_x = (-mv_x) - (mv_x) = -2mv_x.$$

Hence the momentum Δp_x delivered to the wall by the molecule during the collision is $+2mv_x$. (Because in this book the symbol p represents both momentum and pressure, we must be careful to note that here p represents momentum and is a vector quantity.)

The molecule of Fig. 20-3 will hit the shaded wall repeatedly. The time Δt between collisions is the time the molecule takes to travel to the opposite wall and back again (a distance of $2L$) at speed v_x. Thus Δt is equal to $2L/v_x$. (Note that this result holds even if the molecule bounces off any of the other walls along the way, because those walls are parallel to x and so cannot change v_x.) Thus the average rate at which momentum is delivered to the shaded wall by this single molecule is

$$\frac{\Delta p_x}{\Delta t} = \frac{2mv_x}{2L/v_x} = \frac{mv_x^2}{L}.$$

From Newton's second law ($\mathbf{F} = d\mathbf{p}/dt$), the rate at which momentum is delivered to the wall is the force acting on that wall. To find the total force, we must add up the contributions of all the molecules that strike the wall, al-

lowing for the possibility that they all have different speeds. Dividing the total force F_x by the area of the wall ($= L^2$) then gives the pressure p on that wall, where now and in the rest of this discussion, p represents pressure. Thus,

$$p = \frac{F_x}{L^2} = \frac{mv_{x1}^2/L + mv_{x2}^2/L + \cdots + mv_{xN}^2/L}{L^2}$$

$$= \left(\frac{m}{L^3}\right)(v_{x1}^2 + v_{x2}^2 + \cdots + v_{xN}^2), \qquad (20\text{-}14)$$

where N is the number of molecules in the box.

Since $N = nN_A$, there are nN_A terms in the second parentheses of Eq. 20-14. So we can replace that quantity by $nN_A \overline{v_x^2}$, where $\overline{v_x^2}$ is the average value of the square of the x components of all the molecular speeds. Equation 20-14 then becomes

$$p = \frac{nmN_A}{L^3}\overline{v_x^2}.$$

But mN_A is the molar mass M of the gas (that is, the mass of 1 mole of the gas). Also, L^3 is the volume of the box, so

$$p = \frac{nM\overline{v_x^2}}{V}. \qquad (20\text{-}15)$$

For any molecule, $v^2 = v_x^2 + v_y^2 + v_z^2$. Because there are many molecules and because they are all moving in random directions, the average values of the squares of their velocity components are equal, so that $\overline{v_x^2} = \frac{1}{3}\overline{v^2}$. Thus Eq. 20-15 becomes

$$p = \frac{nM\overline{v^2}}{3V}. \qquad (20\text{-}16)$$

The square root of $\overline{v^2}$ is a kind of average speed, called the **root-mean-square speed** of the molecules and symbolized by v_{rms}. Its name describes it rather well: you *square* each speed, you find the *mean* (that is, the average) of all these squared speeds, and then you take the square *root*. With $\sqrt{\overline{v^2}} = \overline{v}_{\text{rms}}$, we can then write Eq. 20-16 as

$$p = \frac{nMv_{\text{rms}}^2}{3V}. \qquad (20\text{-}17)$$

Equation 20-17 is very much in the spirit of kinetic theory. It tells us how the pressure of the gas (a purely macroscopic quantity) depends on the speed of the molecules (a purely microscopic quantity).

We can turn Eq. 20-17 around and use it to calculate v_{rms}. Combining Eq. 20-17 with the ideal gas law ($pV = nRT$) leads to

$$v_{\text{rms}} = \sqrt{\frac{3RT}{M}}. \qquad (20\text{-}18)$$

TABLE 20-1   SOME MOLECULAR SPEEDS AT ROOM TEMPERATURE ($T = 300$ K)^a

GAS	MOLAR MASS (10^{-3} kg/mol)	v_{rms} (m/s)
Hydrogen (H$_2$)	2.02	1920
Helium (He)	4.0	1370
Water vapor (H$_2$O)	18.0	645
Nitrogen (N$_2$)	28.0	517
Oxygen (O$_2$)	32.0	483
Carbon dioxide (CO$_2$)	44.0	412
Sulfur dioxide (SO$_2$)	64.1	342

^{a}For convenience, we often set room temperature = 300 K even though (at 27°C or 81°F) that represents a fairly warm room.

Table 20-1 shows some rms speeds calculated from Eq. 20-18. The speeds are surprisingly high. For hydrogen molecules at room temperature (300 K), the rms speed is 1920 m/s or 4300 mi/h — faster than a speeding bullet! On the surface of the Sun, where the temperature is 2×10^6 K, the rms speed of hydrogen molecules would be 82 times larger than at room temperature were it not for the fact that at such high speeds, the molecules cannot survive the collisions among themselves. Remember too that the rms speed is only a kind of average speed; many molecules move much faster than this, and some much slower.

The speed of sound in a gas is closely related to the rms speed of the molecules of that gas. In a sound wave, the disturbance is passed on from molecule to molecule by means of collisions. The wave cannot move any faster than the "average" speed of the molecules. In fact, the speed of sound must be somewhat less than this "average" molecular speed because not all molecules are moving in exactly the same direction as the wave. As examples, at room temperature, the rms speeds of hydrogen and nitrogen molecules are 1920 m/s and 517 m/s, respectively. The speeds of sound in these two gases at this temperature are 1350 m/s and 350 m/s, respectively.

A question often arises: If molecules move so fast, why does it take as long as a minute or so before you can smell perfume if someone opens a bottle across a room? A glance ahead at Fig. 20-4 (Section 20-6) suggests that — although the molecules move very fast between collisions — a given molecule will wander only very slowly away from its release point.

SAMPLE PROBLEM 20-3

Here are five numbers: 5, 11, 32, 67, and 89.

(a) What is the average value $\bar{n}$ of these numbers?

SOLUTION: We find this from

$$\bar{n} = \frac{5 + 11 + 32 + 67 + 89}{5} = 40.8 \quad \text{(Answer)}$$

(b) What is the rms value n_{rms} of these numbers?

SOLUTION: We find this from

$$n_{rms} = \sqrt{\frac{5^2 + 11^2 + 32^2 + 67^2 + 89^2}{5}}$$

$$= 52.1. \quad \text{(Answer)}$$

The rms value is greater than the average value because the larger numbers — being squared — are relatively more important in forming the rms value. To test this, let us replace 89 in our set of five numbers by 300. The average value of the new set of five numbers (as you can easily show) is 2.0 times the previous average value. The rms value, however, is 2.7 times the previous rms value.

The rms values of variables occur in many branches of physics and engineering. The value 120 volts printed on an electric light bulb, for example, is an rms voltage.

20-5   TRANSLATIONAL KINETIC ENERGY

We again consider a single molecule of an ideal gas as it moves around in the box of Fig. 20-3, but we now assume that its speed changes when it collides with other molecules. Its translational kinetic energy at any instant is $\frac{1}{2}mv^2$. Its *average* translational kinetic energy over the time that we watch it is

$$\overline{K} = \overline{\tfrac{1}{2}mv^2} = \tfrac{1}{2}m\overline{v^2} = \tfrac{1}{2}mv_{rms}^2, \quad (20\text{-}19)$$

in which we make the assumption that the average speed of the molecule during our observation is the same as the average speed of all the molecules at any given time. (Provided the total energy of the gas is not changing and we observe our molecule for long enough, this assumption is appropriate.) Substituting for v_{rms} from Eq. 20-18 leads to

$$\overline{K} = (\tfrac{1}{2}m)\frac{3RT}{M}.$$

But M/m, the molar mass divided by the mass of a molecule, is simply Avogadro's number, so

$$\overline{K} = \frac{3RT}{2N_A},$$

which we can write as

$$\overline{K} = \tfrac{3}{2}kT. \quad (20\text{-}20)$$

The constant k, called the **Boltzmann constant,** is the ratio of the gas constant R to Avogadro's number N_A. It is sometimes called the gas constant for a single molecule (rather than for a mole), and its value is

$$k = \frac{R}{N_A} = \frac{8.31 \text{ J/mol} \cdot \text{K}}{6.02 \times 10^{23} \text{ mol}^{-1}} = 1.38 \times 10^{-23} \text{ J/K}$$

$$= 8.62 \times 10^{-5} \text{ eV/K}. \qquad (20\text{-}21)$$

Equation 20-20 tells us something unexpected:

At a given temperature T, all ideal gas molecules—no matter what their mass—have the same average translational kinetic energy, namely, $\frac{3}{2}kT$. When we measure the temperature of a gas, we are also measuring the average translational kinetic energy of its molecules.

CHECKPOINT **2:** A gas mixture consists of molecules of types 1, 2, and 3, with molecular masses $m_1 > m_2 > m_3$. Rank the three types according to (a) average kinetic energy and (b) rms speed, greatest first.

SAMPLE PROBLEM 20-4

What is the average translational kinetic energy (in electron-volts) of the oxygen molecules in air at room temperature (= 300 K)? Of the nitrogen molecules?

SOLUTION: The average translational kinetic energy depends only on the temperature and not on the nature of the molecule. For both oxygen and nitrogen molecules it is given by Eq. 20-20 as

$$\overline{K} = \tfrac{3}{2}kT = (\tfrac{3}{2})(8.62 \times 10^{-5} \text{ eV/K})(300 \text{ K})$$

$$= 0.039 \text{ eV}. \qquad \text{(Answer)}$$

Physicists find it useful to remember that the mean translational kinetic energy of *any* molecule at room temperature is about $\frac{1}{25}$ eV, which is essentially the above result.

From Table 20-1 we see that the rms speed of the molecules of oxygen (for which $M = 32.0$ g/mol) is 483 m/s. That for the molecules of nitrogen ($M = 28.0$ g/mol) is 517 m/s. Thus the lighter molecule has the larger rms speed, consistent with the fact that the two kinds of molecules have the same average translational kinetic energy.

20-6 MEAN FREE PATH

We continue to examine the motion of molecules in an ideal gas. Figure 20-4 shows the path of a typical molecule as it moves through the gas, changing both speed and di-

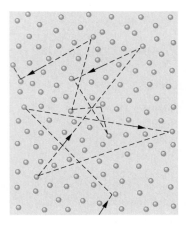

FIGURE 20-4 A molecule traveling through a gas, colliding with other gas molecules in its path. Although the other molecules are shown as stationary, they are also moving in a similar fashion.

rection abruptly as it collides elastically with other molecules. Between collisions, our typical molecule moves in a straight line at constant speed. Although the figure shows all the other molecules as stationary, they too are moving in much the same way.

One useful parameter to describe this random motion is the **mean free path** λ. As its name implies, λ is the average distance traversed by a molecule between collisions. We expect λ to vary inversely with N/V, the number of molecules per unit volume (or density of molecules). The larger N/V is, the more collisions there should be and the smaller the mean free path. We also expect λ to vary inversely with the size of the molecules. (If the molecules were true points, they would never collide and the mean free path would be infinite.) Thus the larger the molecules, the smaller the mean free path. We can even predict that λ should vary (inversely) as the *square* of the molecular diameter because the cross section of a molecule—not its diameter—determines its effective target area.

The expression for the mean free path does, in fact, turn out to be

$$\lambda = \frac{1}{\sqrt{2}\pi d^2 \, N/V} \qquad \text{(mean free path).} \quad (20\text{-}22)$$

To justify Eq. 20-22, we focus attention on a single molecule and assume—as Fig. 20-4 suggests—that our molecule is traveling with a constant speed v and that all the other molecules are at rest. Later, we shall relax this assumption.

We assume further that the molecules are spheres of diameter d. A collision will then take place if the centers of the molecules come within a distance d of each other, as in Fig. 20-5a. Another, more helpful way to look at the situation is to consider our single molecule to have a *radius* of d and all the other molecules to be *points,* as in Fig. 20-5b. This does not change our criterion for a collision.

As our single molecule zigzags through the gas, it sweeps out a short cylinder of cross-sectional area πd^2

FIGURE 20-5 (a) A collision occurs when the centers of two molecules come within a distance d of each other, d being the molecular diameter. (b) An equivalent but more convenient representation is to think of the moving molecule as having a *radius d* and all other molecules as being points. The condition for a collision is unchanged.

between successive collisions. If we watch this molecule for a time interval Δt, it moves a distance $v \Delta t$, where v is its assumed speed. Thus, if we align all the short cylinders swept out in Δt, we form a composite cylinder (Fig. 20-6) of length $v \Delta t$ and volume $(\pi d^2)(v \Delta t)$. The number of collisions that occur is then equal to the number of (point) molecules that lie within this cylinder.

Since N/V is the number of molecules per unit volume, the number of collisions is N/V times the volume of the cylinder, or $(N/V)(\pi d^2 v \Delta t)$. The mean free path is the length of the path (and of the cylinder) divided by this number:

$$\lambda = \frac{\text{length of path}}{\text{number of collisions}} \approx \frac{v \Delta t}{\pi d^2 v \Delta t \, N/V}$$

$$= \frac{1}{\pi d^2 \, N/V}. \tag{20-23}$$

This equation is only approximate because it is based on the assumption that all the molecules except one are at rest. In fact, *all* the molecules are moving; when this is taken properly into account, Eq. 20-22 results. Note that it differs from the (approximate) Eq. 20-23 only by a factor of $1/\sqrt{2}$.

We can even get a glimpse of what is "approximate" about Eq. 20-23. The v in the numerator and that in the denominator are—strictly—not the same. The v in the numerator is $\bar{v}$, the mean speed of the molecule *relative to the container.* The v in the denominator is $\bar{v}_{rel}$, the mean

speed of our single molecule *relative to the other molecules,* which are moving. It is this latter average speed that determines the number of collisions. A detailed calculation, taking into account the actual speed distribution of the molecules, gives $\bar{v}_{rel} = \sqrt{2}\, \bar{v}$ and thus the factor $\sqrt{2}$.

The mean free path of air molecules at sea level is about 0.1 μm. At an altitude of 100 km, the density of air has dropped to such an extent that the mean free path rises to about 16 cm. At 300 km, the mean free path is about 20 km. A problem faced by those who would study the physics and chemistry of the upper atmosphere in the laboratory is the unavailability of containers large enough to hold gas samples that simulate upper atmospheric conditions. Yet studies of the concentrations of Freon, carbon dioxide, and ozone in the upper atmosphere are of vital public concern.

SAMPLE PROBLEM 20-5

The molecular diameters of gas molecules of different kinds can be found experimentally by measuring the rates at which the different gases diffuse (spread) into each other. For oxygen, $d = 2.9 \times 10^{-10}$ m has been reported.

(a) What is the mean free path for oxygen at room temperature ($T = 300$ K) and at an atmospheric pressure of 1.0 atm? Assume it is an ideal gas under these conditions.

SOLUTION: Let us first find N/V, the number of molecules per unit volume under these conditions. From the ideal gas law, 1.0 mol of any ideal gas occupies a volume equal to

$$V = \frac{nRT}{p} = \frac{(1.0 \text{ mol})(8.31 \text{ J/mol} \cdot \text{K})(300 \text{ K})}{(1.0 \text{ atm})(1.01 \times 10^5 \text{ Pa/atm})}$$

$$= 2.47 \times 10^{-2} \text{ m}^3.$$

Because we have 1.0 mol of the gas in that volume, the number of molecules per unit volume is

$$\frac{N}{V} = \frac{nN_A}{V} = \frac{(1.0 \text{ mol})(6.02 \times 10^{23} \text{ molecules/mol})}{2.47 \times 10^{-2} \text{ m}^3}$$

$$= 2.44 \times 10^{25} \text{ molecules/m}^3.$$

Equation 20-22 then gives

$$\lambda = \frac{1}{\sqrt{2}\pi d^2 \, N/V}$$

$$= \frac{1}{(\sqrt{2}\pi)(2.9 \times 10^{-10} \text{ m})^2(2.44 \times 10^{25} \text{ m}^{-3})}$$

$$= 1.1 \times 10^{-7} \text{ m}. \tag{Answer}$$

This is about 380 molecular diameters. On average, the molecules in such a gas are only about 11 molecular diameters apart.

(b) If the average speed of an oxygen molecule is 450 m/s, what is the average collision rate?

FIGURE 20-6 In time Δt the moving molecule effectively sweeps out a cylinder of length $v \Delta t$ and radius d.

SOLUTION: We find this rate by dividing the average speed by the mean free path to get

$$\text{rate} = \frac{v}{\lambda} = \frac{450 \text{ m/s}}{1.1 \times 10^{-7} \text{ m}} = 4.1 \times 10^{9} \text{ s}^{-1}. \quad \text{(Answer)}$$

Thus—on average—every oxygen molecule makes more than 4 billion collisions per second!

CHECKPOINT 3: One mole of gas A, with molecular radius $2d_0$ and average molecular speed v_0, is placed inside a certain container. One mole of gas B, with molecular radius d_0 and average molecular speed $2v_0$ (the molecules of B are smaller but faster), is placed in an identical container. Which gas has the greater average collision rate within its container?

20-7 THE DISTRIBUTION OF MOLECULAR SPEEDS (OPTIONAL)

The root-mean-square speed v_{rms} gives us a general idea of the molecular speed in a gas at a given temperature. We often want to know more. For example, what fraction of the molecules have speeds greater than the rms value? Greater than twice the rms value? To answer such questions, we need to know how the possible values of speed are distributed among the molecules. Figure 20-7a shows this distribution for oxygen molecules at room temperature ($T = 300$ K); Fig. 20-7b compares it with the distribution at $T = 80$ K.

In 1852, Scottish physicist James Clerk Maxwell first solved the problem of finding the speed distribution of gas molecules. His result, known as **Maxwell's speed distribution law**, is

$$P(v) = 4\pi \left(\frac{M}{2\pi RT}\right)^{3/2} v^2 e^{-Mv^2/2RT}. \quad (20\text{-}24)$$

Here v is the molecular speed, T is the gas temperature, M is the molar mass of the gas, and R is the gas constant. It is this equation that is plotted in Fig. 20-7a,b. The quantity $P(v)$ in Eq. 20-24 and Fig. 20-7 is a *probability distribution function*, defined as follows: The product $P(v)$ dv (which is a dimensionless quantity) is the fraction of molecules whose speeds lie in the range v to $v + dv$.

As Fig. 20-7a shows, this fraction is equal to the area of a strip whose height is $P(v)$ and whose width is dv. The total area under the distribution curve corresponds to the fraction of the molecules whose speeds lie between zero and infinity. All molecules fall into this category, so the value of this total area is unity.

(a)

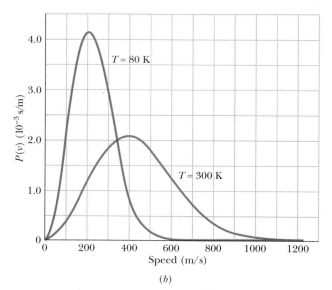

(b)

FIGURE 20-7 (a) The Maxwell speed distribution for oxygen molecules at $T = 300$ K. The three characteristic speeds are marked. (b) The curves for 300 K and 80 K. Note that the molecules move more slowly at the lower temperature. Because these are probability distributions, the area under each curve has a numerical value of unity.

Figure 20-7a also shows the root-mean-square speed v_{rms} ($= 483$ m/s) and two other measures of the speed distribution of the oxygen molecules. The *most probable speed* v_P ($= 395$ m/s) is the speed at which $P(v)$ is a maximum. The *average speed* $\bar{v}$ ($= 445$ m/s) is—as its name suggests—a simple average of the molecular speeds. A small number of molecules, lying in the extreme right-hand tail of the distribution curve, can have speeds that are several times the average speed. This simple fact, as we shall demonstrate, makes possible both rain and sunshine.

Rain: The speed distribution of water molecules in, say, a pond at summertime temperatures can be represented by a curve similar to that of Fig. 20-7a. Most of the molecules do not have nearly enough kinetic energy to escape from the water through its surface. However, small numbers of very fast molecules with speeds far out in the tail of the curve can do so. It is these water molecules that evaporate, making clouds and rain a possibility.

As the fast water molecules leave the surface, carrying energy with them, the temperature of the remaining water is maintained by heat transfer from the surroundings. Other fast molecules—produced in particularly favorable collisions—quickly take the place of those that have left, and the speed distribution is maintained.

Sunshine: Let the distribution curve of Fig. 20-7a now refer to protons in the core of the Sun. The Sun's energy is supplied by a nuclear fusion process that starts with the merging of two protons. However, protons repel each other because of their electrical charges, and protons of average speed do not have enough kinetic energy to overcome the repulsion and get close enough to merge. Very fast protons with speeds in the tail of the distribution curve can do so, however, and thus the Sun can shine.

SAMPLE PROBLEM 20-6

A container is filled with oxygen gas maintained at room temperature (300 K). What fraction of the molecules have speeds in the range 599–601 m/s? The molar mass M of oxygen is 0.0320 kg/mol.

SOLUTION: This speed interval Δv (= 2 m/s) is so small that we can treat it as a differential and say that the fraction frac that we seek is given very closely by $P(v) \Delta v$, where $P(v)$ is to be evaluated at $v = 600$ m/s, the midpoint of the interval; see the gold strip in Fig. 20-7a. Thus, using Eq. 20-24, we find

$$\text{frac} = P(v) \Delta v = 4\pi \left(\frac{M}{2\pi RT} \right)^{3/2} v^2 e^{-Mv^2/2RT} \Delta v.$$

For convenience in calculating, let us break this down into five factors, as

$$\text{frac} = (4\pi)(A)(v^2)(e^B)(\Delta v) \tag{20-25}$$

in which A and B are

$$A = \left(\frac{M}{2\pi RT} \right)^{3/2} = \left(\frac{0.0320 \text{ kg/mol}}{(2\pi)(8.31 \text{ J/mol} \cdot \text{K})(300 \text{ K})} \right)^{3/2}$$

$$= 2.92 \times 10^{-9} \text{ s}^3/\text{m}^3$$

and

$$B = -\frac{Mv^2}{2RT} = -\frac{(0.0320 \text{ kg/mol})(600 \text{ m/s})^2}{(2)(8.31 \text{ J/mol} \cdot \text{K})(300 \text{ K})}$$

$$= -2.31.$$

Substituting A and B into Eq. 20-25 yields

$$\text{frac} = (4\pi)(A)(v^2)(e^B)(\Delta v)$$

$$= (4\pi)(2.92 \times 10^{-9} \text{ s}^3/\text{m}^3)(600 \text{ m/s})^2(e^{-2.31})(2 \text{ m/s})$$

$$= 2.62 \times 10^{-3}. \tag{Answer}$$

Thus, at room temperature, 0.262% of the oxygen molecules will have speeds that lie in the narrow range between 599 and 601 m/s. If the gold strip of Fig. 20-7a were drawn to the scale of this problem, it would be a very thin strip indeed.

SAMPLE PROBLEM 20-7

(a) What is the average speed v of oxygen gas molecules at $T = 300$ K? The molar mass M of oxygen is 0.0320 kg/mol.

SOLUTION: To find the average speed, we weight each speed v by $P(v) dv$, which is the fraction of the molecules whose speeds lie in the interval v to $v + dv$. We then add up (that is, integrate) these fractions over the entire range of speeds. Thus

$$\bar{v} = \int_0^\infty v P(v) \, dv. \tag{20-26}$$

The next step is to substitute for $P(v)$ from Eq. 20-24 and evaluate the integral that results. From a table of integrals* we find

$$\bar{v} = \sqrt{\frac{8RT}{\pi M}} \qquad \text{(average speed).} \tag{20-27}$$

Substituting numerical values yields

$$\bar{v} = \sqrt{\frac{(8)(8.31 \text{ J/mol} \cdot \text{K})(300 \text{ K})}{(\pi)(0.0320 \text{ kg/mol})}}$$

$$= 445 \text{ m/s.} \tag{Answer}$$

(b) What is the root-mean-square speed v_{rms} of the oxygen molecules?

SOLUTION: We proceed as in (a) above except that we multiply v^2 (rather than simply v) by the weighting factor $P(v) dv$. This leads, after integration, to

$$\overline{v^2} = \int_0^\infty v^2 P(v) \, dv = \frac{3RT}{M}.$$

The rms speed is the square root of this quantity, or

$$v_{\text{rms}} = \sqrt{\overline{v^2}} = \sqrt{\frac{3RT}{M}} \qquad \text{(rms speed).} \tag{20-28}$$

Equation 20-28 is identical to Eq. 20-18, which we derived earlier. The numerical calculation gives

$$v_{\text{rms}} = \sqrt{\frac{(3)(8.31 \text{ J/mol} \cdot \text{K})(300 \text{ K})}{(0.0320 \text{ kg/mol})}}$$

$$= 483 \text{ m/s.} \tag{Answer}$$

(c) What is the most probable speed v_P?

SOLUTION: The most probable speed is the speed at which $P(v)$ of Eq. 20-24 has its maximum value. We find it by re-

---

*One such table is found in Section A of the familiar *CRC Handbook of Chemistry and Physics*; see integral 667 listed under "Definite Integrals."

TABLE 20-2 SPEED PARAMETERS FOR THE MAXWELL SPEED DISTRIBUTION

PARAMETER	SYMBOL	FORMULA	FOR OXYGEN AT 300 K
Most probable speed	v_P	$\sqrt{2RT/M}$	395 m/s
Average speed	$\bar{v}$	$\sqrt{8RT/\pi M}$	445 m/s
Root-mean-square speed	v_{rms}	$\sqrt{3RT/M}$	483 m/s

quiring that $dP/dv = 0$ and solving for v. Doing so yields (as you should show)

$$v_P = \sqrt{\frac{2RT}{M}} \quad \text{(most probable speed).} \quad (20\text{-}29)$$

Numerically, this yields

$$v_P = \sqrt{\frac{(2)(8.31 \text{ J/mol} \cdot \text{K})(300 \text{ K})}{(0.0320 \text{ kg/mol})}}$$

$$= 395 \text{ m/s.} \quad \text{(Answer)}$$

Table 20-2 summarizes the three measures of the Maxwell speed distribution.

20-8 THE MOLAR SPECIFIC HEATS OF AN IDEAL GAS

In this section, we want to derive from atomic or molecular considerations an expression for the internal energy E_{int} of an ideal gas. We shall then use that result to derive an expression for the molar specific heats of an ideal gas.

Internal Energy E_{int}

Let us first assume that our ideal gas is a *monatomic gas* (which has individual atoms rather than molecules), such as helium, neon, or argon. Next, recall from Chapter 8 that internal energy is the energy associated with random motions of atoms and molecules. So let us assume that the internal energy E_{int} of our ideal gas is simply the sum of the translational kinetic energies of its atoms. (Individual atoms do not have rotational kinetic energy.) The average translational kinetic energy of a single atom depends only on the gas temperature and is given by Eq. 20-20 as $\bar{K} = \frac{3}{2}kT$. A sample of n moles of such a gas contains nN_A atoms. The internal energy E_{int} of the sample is then

$$E_{int} = (nN_A)\bar{K} = (nN_A)(\tfrac{3}{2}kT)$$

or, since $N_A k = R$, the gas constant,

$$E_{int} = \tfrac{3}{2}nRT \quad \text{(monatomic ideal gas).} \quad (20\text{-}30)$$

Thus,

> The internal energy E_{int} of an ideal gas is a function of the gas temperature *only*; it does not depend on any other variable.

With Eq. 20-30 in hand, we are now able to derive an expression for the molar specific heat of an ideal gas. Actually, we derive two expressions—one for the case in which the volume of the gas remains constant as heat is added to it, and one for the case in which the pressure of the gas remains constant as heat is added. The symbols for these two molar specific heats are C_V and C_p, respectively. (By convention, the capital letter C is used in both cases, even though C_V and C_P represent a type of specific heat.)

Molar Specific Heat at Constant Volume

Figure 20-8*a* shows n moles of an ideal gas at pressure p and temperature T, confined to a cylinder of fixed volume V. This *initial state i* of the gas is marked on the p-V diagram of Fig. 20-8*b*. Suppose now that you add a small amount of heat Q to the gas, by slowly turning up the temperature of the thermal reservoir on which the cylinder rests. The gas temperature rises a small amount to $T + \Delta T$, and its pressure to $p + \Delta p$, bringing the gas to *final state f*.

The defining equation for C_V, the molar specific heat at constant volume, is, in parallel with Eq. 19-15,

$$Q = nC_V \Delta T \quad \text{(constant volume).} \quad (20\text{-}31)$$

Substituting this expression for Q into the first law of thermodynamics (Eq. 19-24), we find

$$\Delta E_{int} = nC_V \Delta T - W.$$

With the volume held constant, the gas cannot do any work; so $W = 0$. Then solving for C_V we obtain

$$C_V = \frac{1}{n}\frac{\Delta E_{int}}{\Delta T}. \quad (20\text{-}32)$$

From Eq. 20-30 we see that $\Delta E_{int}/\Delta T = \frac{3}{2}nR$. Substituting

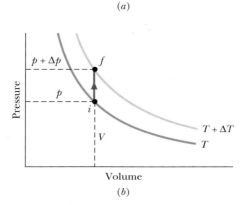

FIGURE 20-8
(a) The temperature of an ideal gas is raised from T to $T + \Delta T$ in a constant-volume process. Heat is added but no work is done. (b) The process on a p-V diagram.

this result into Eq. 20-32 yields

$$C_V = \tfrac{3}{2}R = 12.5 \text{ J/mol} \cdot \text{K} \qquad \text{(monatomic gas).} \qquad (20\text{-}33)$$

As Table 20-3 shows, this prediction of the kinetic theory (for ideal gases) agrees very well with experiment for real monatomic gases, the case that we have assumed. The (predicted and) experimental values of C_V for *diatomic gases* (which have molecules with two atoms) and *polyatomic gases* (which have molecules with more than two

atoms) are greater than those for monatomic gases for reasons that will be suggested in Section 20-9.

We can now generalize Eq. 20-30 for the internal energy of any ideal gas by substituting C_V for $\tfrac{3}{2}R$; we get

$$E_{\text{int}} = nC_V T \qquad \text{(any ideal gas).} \qquad (20\text{-}34)$$

This equation applies not only to an ideal monatomic gas but also to diatomic and polyatomic ideal gases, provided the appropriate value of C_V is used. Just as with Eq. 20-30, we see that the internal energy depends on the temperature of the gas but not on its pressure or density.

When an ideal gas that is confined to a container undergoes a temperature change ΔT, then from either Eq. 20-32 or Eq. 20-34 we can write the resulting change in its internal energy as

$$\Delta E_{\text{int}} = nC_V \, \Delta T \qquad \text{(ideal gas, any process).} \qquad (20\text{-}35)$$

This equation tells us:

> A change in the internal energy E_{int} of a confined ideal gas depends on the change in the gas temperature only; it does *not* depend on what type of process produces the change in the temperature.

As examples, consider the three paths between the two isotherms in the p-V diagram of Fig. 20-9. Path 1 represents a constant-volume process. Path 2 represents a constant-pressure process (that we are about to examine). And path 3 represents a process in which no heat is ex-

TABLE 20-3 MOLAR SPECIFIC HEATS

MOLECULE		EXAMPLE	C_V (J/mol·K)
Monatomic	Ideal		$\tfrac{3}{2}R = 12.5$
	Real	He	12.5
		Ar	12.6
Diatomic	Ideal		$\tfrac{5}{2}R = 20.8$
	Real	N_2	20.7
		O_2	20.8
Polyatomic	Ideal		$3R = 24.9$
	Real	NH_4	29.0
		CO_2	29.7

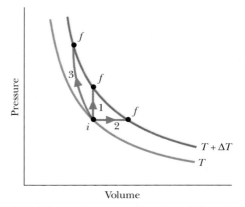

FIGURE 20-9 Three paths representing three different processes that take an ideal gas from an initial state i at temperature T to some final state f at temperature $T + \Delta T$. The change ΔE_{int} in the internal energy of the gas is the same for these three processes and for any others that result in the same change of temperature.

changed with the system's environment (we discuss this in Section 20-11). Although the values of heat Q and work W associated with these three paths differ, as do p_f and V_f, the values of ΔE_{int} associated with the three paths are identical and are all given by Eq. 20-35, because they all involve the same temperature change ΔT. So, no matter what path is actually traversed between T and $T + \Delta T$, we can always use path 1 and Eq. 20-35 to compute ΔE_{int} easily.

Molar Specific Heat at Constant Pressure

We now assume that the temperature of the ideal gas is increased by the same small amount ΔT as previously, but that the necessary heat Q is added with the gas under constant pressure. A mechanism for doing this is shown in Fig. 20-10a; the p-V diagram for the process is plotted in Fig. 20-10b. We can guess at once that the molar specific heat at constant pressure C_p, which we define with

$$Q = nC_p \, \Delta T \quad \text{(constant pressure)}, \quad (20\text{-}36)$$

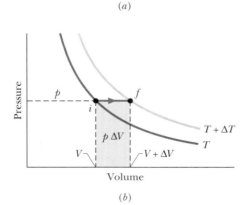

(b)

FIGURE 20-10 (a) The temperature of an ideal gas is raised from T to $T + \Delta T$ in a constant-pressure process. Heat is added and work is done in lifting the loaded piston. (b) The process on a p-V diagram. The work $p \, \Delta V$ is given by the shaded area.

is *greater* than the molar specific heat at constant volume, because energy must now be supplied not only to raise the temperature of the gas but also for the gas to do work, that is, to lift the weighted piston of Fig. 20-10a.

To relate C_p to C_V, we start with the first law of thermodynamics (Eq. 19-24):

$$\Delta E_{int} = Q - W. \quad (20\text{-}37)$$

We next replace each term in Eq. 20-37. For ΔE_{int}, we substitute from Eq. 20-35. For Q, we substitute from Eq. 20-36. To replace W, we first note that since the pressure remains constant, Eq. 19-23 tells us that $W = p \, \Delta V$. Then we note that, using the ideal gas equation ($pV = nRT$), we can write

$$W = p \, \Delta V = nR \, \Delta T.$$

Making these substitutions in Eq. 20-37, and then dividing through by $n \, \Delta T$, we find

$$C_V = C_p - R,$$

and then

$$C_p = C_V + R. \quad (20\text{-}38)$$

This prediction of kinetic theory agrees well with experiment, not only for monatomic gases but for gases in general, as long as their density is low enough so that we may treat them as ideal.

CHECKPOINT 4: The figure shows five paths traversed by a gas on a p-V diagram. Rank the paths according to the change in internal energy of the gas, greatest first.

SAMPLE PROBLEM 20-8

A bubble of 5.00 mol of (monatomic) helium is submerged at a certain depth in liquid water when the water (and thus the helium) undergoes a temperature increase ΔT of 20.0 C° at constant pressure. As a result, the bubble expands.

(a) How much heat Q is added to the helium during the expansion and temperature increase?

SOLUTION: Treating the helium as an ideal gas, we use Eqs. 20-38 and 20-33 to write

$$C_p = C_V + R = \tfrac{5}{2}R.$$

We substitute this into Eq. 20-36 to obtain

$$Q = nC_p \, \Delta T = n(\tfrac{5}{2}R) \, \Delta T$$
$$= (5.00 \text{ mol})(2.5)(8.31 \text{ J/mol} \cdot \text{K})(20.0 \text{ C}°)$$
$$= 2077.5 \text{ J} \approx 2080 \text{ J}. \qquad \text{(Answer)}$$

(b) What is the change ΔE_{int} in the internal energy of the helium during the temperature increase?

SOLUTION: Even though the temperature of the helium increases at constant pressure (and *not* at constant volume), we use Eq. 20-35 to calculate the change in internal energy:

$$\Delta E_{int} = nC_V \, \Delta T$$
$$= (5.00 \text{ mol})(1.5)(8.31 \text{ J/mol} \cdot \text{K})(20.0 \text{ C}°)$$
$$= 1246.5 \text{ J} \approx 1250 \text{ J}. \qquad \text{(Answer)}$$

(c) How much work W is done by the helium as it expands against the pressure of the surrounding water during the temperature increase?

SOLUTION: From the first law of thermodynamics,

$$W = Q - \Delta E_{int} = 2077.5 \text{ J} - 1246.5 \text{ J}$$
$$= 831 \text{ J}. \qquad \text{(Answer)}$$

Note that during the temperature increase, only a portion (1250 J) of the heat (2080 J) that is transferred into the helium goes to increasing the internal energy of the helium and thus the temperature of the helium. The rest (831 J) is transferred out of the helium as work that the helium does during the expansion. If the water were frozen, it would not allow that expansion. Then the same temperature increase of 20.0 C° would require only 1250 J of heat, because no work would be done by the helium.

20-9 DEGREES OF FREEDOM AND MOLAR SPECIFIC HEATS

As Table 20-3 shows, the prediction that $C_V = \tfrac{3}{2}R$ agrees with experiment for monatomic gases but fails for diatomic and polyatomic gases. Let us try to explain the discrepancy by considering the possibility that molecules with more than one atom can store internal energy in forms other than translational motion.

Figure 20-11 shows kinetic theory models of helium (a monatomic gas), oxygen (diatomic), and methane (polyatomic). On the basis of their structure, it seems reasonable to assume that monatomic molecules—which are essentially pointlike and have only a very small rotational inertia

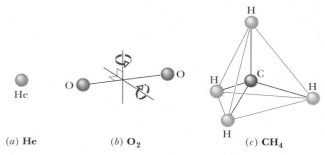

(a) **He** (b) **O$_2$** (c) **CH$_4$**

FIGURE 20-11 Models of molecules as used in kinetic theory: (a) helium, a typical monatomic molecule; (b) oxygen, a typical diatomic molecule (two rotation axes are shown); and (c) methane, a typical polyatomic molecule.

about any axis—can store energy only in their translational motion. Diatomic and polyatomic molecules, however, should be able to store substantial additional amounts of energy by rotating or oscillating.

To take these possibilities into account quantitatively, we use the theorem of the **equipartition of energy**, introduced by James Clerk Maxwell:

> Every kind of molecule has a certain number f of *degrees of freedom*, which are independent ways in which the molecule can store energy. Each such degree of freedom has associated with it—on average—an energy of $\tfrac{1}{2}kT$ per molecule (or $\tfrac{1}{2}RT$ per mole).

For translational motion, there are three degrees of freedom, corresponding to the three perpendicular axes along which such motion can occur. For rotational motion, a monatomic molecule has no degrees of freedom. A diatomic molecule—the rigid dumbbell of Fig. 20-11b—has two rotational degrees of freedom, corresponding to the two perpendicular axes about which it can store rotational energy. Such a molecule cannot store rotational energy about the axis connecting the nuclei of its two constituent atoms because its rotational inertia about this axis is approximately zero. A molecule with more than two atoms has six degrees of freedom, three rotational and three translational.

To extend the treatment of Section 20-8 to ideal diatomic and polyatomic gases, it is necessary to retrace the derivations of that section in detail, replacing Eq. 20-30 ($E_{int} = \tfrac{3}{2}nRT$) by $E_{int} = (f/2)nRT$, where f is the number of degrees of freedom listed in Table 20-4. Doing so leads to the prediction

$$C_V = \left(\frac{f}{2}\right) R = 4.16f \text{ J/mol} \cdot \text{K}, \qquad (20\text{-}39)$$

TABLE 20-4   DEGREES OF FREEDOM FOR VARIOUS MOLECULES

| | | DEGREES OF FREEDOM | | | PREDICTED MOLAR SPECIFIC HEATS | |
MOLECULE	EXAMPLE	TRANSLATIONAL	ROTATIONAL	TOTAL (f)	C_V (Eq. 20-39)	$C_p = C_V + R$
Monatomic	He	3	0	3	$\frac{3}{2}R$	$\frac{5}{2}R$
Diatomic	O_2	3	2	5	$\frac{5}{2}R$	$\frac{7}{2}R$
Polyatomic	CH_4	3	3	6	$3R$	$4R$

which agrees—as it must—with Eq. 20-33 for monatomic gases ($f = 3$). As Table 20-3 shows, this prediction also agrees with experiment for diatomic gases ($f = 5$), but it is too low for polyatomic gases.

SAMPLE PROBLEM 20-9

A cabin of volume V is filled with air (which we consider to be an ideal diatomic gas) at an initial low temperature T_1. After you light a wood stove, the air temperature increases to T_2. What is the resulting change in the store of internal energy of the air in the cabin?

SOLUTION: This situation differs from other situations we have examined in that the container (the cabin) is not sealed. If it were sealed, the ideal gas law ($pV = nRT$) would tell us that as the temperature of the air in the cabin increased, the air pressure would also. However, because the cabin is not air-tight, air molecules leave through various openings as the air temperature increases, so that the air pressure inside the cabin always matches the air pressure outside it.

From Eq. 20-35, the change in the internal energy of the air in the room is

$$\Delta E_{int} = C_V \, \Delta(nT),$$

where both n and T can change. Using the ideal gas law, we may substitute $\Delta(pV)/R$ for $\Delta(nT)$, finding

$$\Delta E_{int} = \frac{C_V}{R} \, \Delta(pV).$$

From this we see that since neither the pressure p nor the volume V of the air within the cabin changes,

$$\Delta E_{int} = 0, \qquad \text{(Answer)}$$

even though the temperature changes.

So why does the cabin feel more comfortable at the higher temperature? There are at least two factors involved. You have a tendency to cool because (1) you emit electromagnetic radiation (thermal radiation) and (2) you exchange energy with air molecules that collide with you. If you increase the room temperature, (1) you increase the amount of thermal radiation you intercept from the surfaces within the room, replacing heat you have emitted; and (2) you increase the kinetic energy of the air molecules that collide with you, so you gain more energy from them.

20-10 A HINT OF QUANTUM THEORY

We can improve the agreement of kinetic theory with experiment by including the oscillations of the atoms in a gas of diatomic or polyatomic molecules. For example, the two atoms in the O_2 molecule of Fig. 20-11b can oscillate toward and away from each other, with the interconnecting bond acting like an oscillating spring. However, experiment shows that such oscillations occur only at relatively high temperatures of the gas—the motion is "turned on" only when the gas molecules have relatively large energies. Rotational motion is also subject to such "turning on," but at a lower gas temperature.

Figure 20-12 is of help in seeing the inception of rotational motion and oscillatory motion. The ratio C_V/R for diatomic hydrogen gas (H_2) is plotted there against temperature, with the temperature scale logarithmic to cover several orders of magnitude. Below about 80 K, we find that $C_V/R = 1.5$. This result implies that only the three translational degrees of freedom of hydrogen are involved in the specific heat. As the temperature increases, the value

FIGURE 20-12   A plot of C_V/R versus temperature for (diatomic) hydrogen gas. Because rotational and oscillatory motions begin at certain energies, only translation is possible at very low temperatures. As the temperature increases, rotational motion can begin. At still higher temperatures, oscillatory motion can begin.

of C_V/R gradually increases to 2.5, implying that two additional degrees of freedom have become involved. Quantum theory shows that these two degrees of freedom are associated with the rotation of the hydrogen molecules and that this motion requires a certain minimum amount of energy. At very low temperatures (below 80 K), the molecules do not have enough energy to rotate. As the temperature increases from 80 K, first a few molecules and then more and more obtain enough energy to rotate, and C_V/R increases, until all of them are rotating and $C_V/R = 2.5$.

Similarly, quantum theory shows that oscillatory motion of the molecules requires a certain minimum amount of energy. This minimum amount is not met until the molecules reach a temperature of about 1000 K, as shown in Fig. 20-12. As the temperature increases beyond 1000 K, the number of molecules with enough energy to oscillate increases, and C_V/R increases, until all of them are oscillating and $C_V/R = 3.5$. (In Fig. 20-12, the plotted curve stops at 3200 K because at that temperature, the atoms of a hydrogen molecule oscillate so much that they overwhelm their common bond, and the molecule then *dissociates* into two separate atoms.)

20-11 THE ADIABATIC EXPANSION OF AN IDEAL GAS

We saw in Section 18-2 that sound waves are propagated through air and other gases as a series of compressions and expansions; these variations in the medium take place so rapidly that there is no time for heat transfer from one part of the medium to another. As we saw in Section 19-10, a process for which $Q = 0$ is an *adiabatic process*. We can ensure that $Q = 0$ either by carrying out the process very quickly (as in sound waves) or by doing it slowly in a well-insulated container. Let us see what the kinetic theory has to say about adiabatic processes.

Figure 20-13a shows an insulated cylinder containing an ideal gas and resting on an insulating stand. By removing weight from the piston, we can allow the gas to expand adiabatically. As the volume increases, both the pressure and the temperature drop. We shall prove below that the relation between the pressure and the volume during such an adiabatic process is

$$pV^\gamma = \text{a constant} \qquad \text{(adiabatic process),} \qquad (20\text{-}40)$$

in which $\gamma = C_p/C_V$, the ratio of the molar specific heats for the gas. On a p-V diagram such as that in Fig. 20-13b, the process occurs along a line (called an *adiabat*) that has the equation $p = (\text{a constant})/V^\gamma$. When the gas goes from an initial state i to a final state f, we can rewrite Eq. 20-40

as

$$p_iV_i^\gamma = p_fV_f^\gamma \qquad \text{(adiabatic process).} \qquad (20\text{-}41)$$

We can also write an equation for an adiabatic process in terms of T and V. To do so, we use the ideal gas equation ($pV = nRT$) to eliminate p from Eq. 20-40, finding

$$\left(\frac{nRT}{V}\right) V^\gamma = \text{a constant}.$$

Because n and R are constants, we can rewrite this in the alternative form

$$TV^{\gamma-1} = \text{a constant} \qquad \begin{array}{c}\text{(adiabatic}\\\text{process),}\end{array} \qquad (20\text{-}42)$$

in which the constant is different from that in Eq. 20-40. When the gas goes from an initial state i to a final state f, we can rewrite Eq. 20-42 as

$$T_iV_i^{\gamma-1} = T_fV_f^{\gamma-1} \qquad \text{(adiabatic process).} \qquad (20\text{-}43)$$

(a)

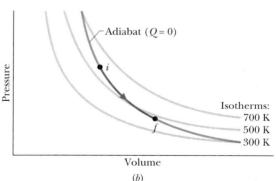

(b)

FIGURE 20-13 (a) The volume of an ideal gas is increased by removing weight from the piston. The process is adiabatic ($Q = 0$). (b) The process proceeds from i to f along an adiabat on a p-V diagram.

Proof of Eq. 20-40

Suppose that you remove some lead shot from the piston of Fig. 20-13a, allowing the ideal gas to push the piston and the remaining shot upward and thus to increase the volume by a differential amount dV. Since the volume change is tiny, we may assume that the pressure p of the gas on the piston is constant during the change. This assumption allows us to say that the work dW done by the gas during the volume increase is equal to $p\, dV$. From Eq. 19-25, the first law of thermodynamics can then be written as

$$dE_{int} = Q - p\, dV. \qquad (20\text{-}44)$$

Since the gas is thermally insulated (and thus the expansion is adiabatic), we substitute $Q = 0$. Then we use Eq. 20-35 to substitute $nC_V\, dT$ for dE_{int}. With these substitutions, and after some rearranging, we have

$$n\, dT = -\left(\frac{p}{C_V}\right) dV. \qquad (20\text{-}45)$$

Now from the ideal gas law ($pV = nRT$) we have

$$p\, dV + V\, dp = nR\, dT. \qquad (20\text{-}46)$$

Replacing R with its equal, $C_p - C_V$, in Eq. 20-46 yields

$$n\, dT = \frac{p\, dV + V\, dp}{C_p - C_V}. \qquad (20\text{-}47)$$

Equating Eqs. 20-45 and 20-47 and rearranging then give

$$\frac{dp}{p} + \left(\frac{C_p}{C_V}\right)\frac{dV}{V} = 0.$$

Replacing the ratio of the molar specific heats with γ and integrating (see Appendix E) yield

$$\ln p + \gamma \ln V = \text{a constant}.$$

Rewriting the left side as $\ln pV^\gamma$ and then taking the antilog of both sides, we find

$$pV^\gamma = \text{a constant}, \qquad (20\text{-}48)$$

which we set out to prove.

Free Expansion

Recall from Section 19-10 that a free expansion of a gas is an adiabatic process that involves no work done on or by the gas, and no change in the internal energy of the gas. A free expansion is thus quite different from the type of adiabatic process described by Eqs. 20-40 through 20-48, in which work is done and the internal energy changes. Those equations then do *not* apply to a free expansion, even though such an expansion is adiabatic.

Also recall that in a free expansion, a gas is in equilibrium only at its initial and final points; thus we can plot only those points, but not the expansion itself, on a p-V diagram. In addition, because $\Delta E_{int} = 0$, the temperature of the final state must be that of the initial state. Thus, the initial and final points on a p-V diagram must be on the same isotherm, and instead of Eq. 20-43 we have

$$T_i = T_f \qquad \text{(free expansion).} \qquad (20\text{-}49)$$

If we next assume that the gas is ideal ($pV = nRT$), then with no change in temperature, there can be no change in the product pV. Thus, instead of Eq. 20-40 a free expansion involves the relation

$$p_iV_i = p_fV_f \qquad \text{(free expansion).} \qquad (20\text{-}50)$$

SAMPLE PROBLEM 20-10

In Sample Problem 20-2, 1 mol of oxygen (assumed to be an ideal gas) expands isothermally (at 310 K) from an initial volume of 12 L to a final volume of 19 L.

(a) What would be the final temperature if the gas had expanded adiabatically to this same final volume? Oxygen (O_2) is diatomic and here has rotation but not oscillation.

SOLUTION: For a diatomic gas whose molecules have rotation but not oscillations, $C_p = \frac{7}{2}R$ and $C_V = \frac{5}{2}R$. So, $\gamma = C_p/C_V = \frac{7}{5} = 1.40$. From Eq. 20-43, we can then write

$$T_f = \frac{T_iV_i^{\gamma-1}}{V_f^{\gamma-1}} = \frac{(310\ \text{K})(12\ \text{L})^{1.40-1}}{(19\ \text{L})^{1.40-1}}$$

$$= 258\ \text{K}. \qquad \text{(Answer)}$$

The fact that the gas has cooled (from 310 K to 258 K) means that its internal energy has been correspondingly reduced. The

---

TABLE 20-5   **FOUR SPECIAL PROCESSES**

PATH IN FIG. 20-14	CONSTANT QUANTITY	PROCESS TYPE	SOME SPECIAL RESULTS ($\Delta E_{int} = Q - W$ and $\Delta E_{int} = nC_V \Delta T$ for all paths)
1	p	Isobaric	$Q = nC_p\, \Delta T$; $W = p\Delta V$
2	T	Isothermal	$Q = W = nRT \ln(V_f/V_i)$; $\Delta E_{int} = 0$
3	pV^γ, $TV^{\gamma-1}$	Adiabatic	$Q = 0$; $W = -\Delta E_{int}$
4	V	Isochoric	$Q = \Delta E_{int} = nC_V\, \Delta T$; $W = 0$

lost internal energy has gone to do the work of expansion (such as lifting the weighted piston in Fig. 20-13a).

(b) What would be the final temperature and pressure if, instead, the gas had expanded freely to the new volume, from an initial pressure of 2.0 Pa?

SOLUTION: Because the temperature does not change in a free expansion,

$$T_f = T_i = 310 \text{ K}. \qquad \text{(Answer)}$$

We find the new pressure using Eq. 20-50, which gives us

$$p_f = p_i \frac{V_i}{V_f} = (2.0 \text{ Pa}) \frac{12 \text{ L}}{19 \text{ L}} = 1.3 \text{ Pa}. \quad \text{(Answer)}$$

PROBLEM SOLVING TACTICS

TACTIC 2: *A Graphical Summary of Four Gas Processes*
In this chapter we have discussed four special processes that an ideal gas can undergo. An example of each is shown in Fig.

20-14, and some associated characteristics are given in Table 20-5, including two process names (isobaric and isochoric) that we have not used but which you might see in other courses.

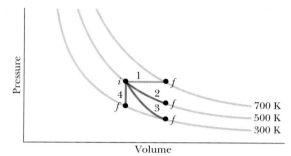

FIGURE 20-14 A *p-V* diagram representing four special processes for an ideal gas. See Table 20-5.

$\mathbb{C}$HECKPOINT **5:** Rank paths 1, 2, and 3 in Fig. 20-14 according to the heat transfer to the gas, greatest first.

REVIEW & SUMMARY

Kinetic Theory of Gases
The *kinetic theory of gases* relates the *macroscopic* properties of gases (for example, pressure and temperature) to the *microscopic* properties of the gas molecules (for example, speed and kinetic energy).

Avogadro's Number
One mole of a substance contains N_A (*Avogadro's number*) elementary units (usually atoms or molecules), where N_A is found experimentally to be

$$N_A = 6.02 \times 10^{23} \text{ mol}^{-1} \qquad \begin{array}{l}\text{(Avogadro's}\\ \text{number).}\end{array} \quad (20\text{-}1)$$

One molar mass *M* of any substance is the mass of one mole of the substance.

Ideal Gas
An *ideal gas* is one for which the pressure *p*, volume *V*, and temperature *T* are related by

$$pV = nRT \qquad \text{(ideal gas law).} \qquad (20\text{-}4)$$

Here *n* is the number of moles of the gas present and *R* (= 8.31 J/mol·K) is the **gas constant.**

Work in an Isothermal Volume Change
The work done *by* an ideal gas during an **isothermal** (constant-temperature) change from volume V_i to volume V_f is

$$W = nRT \ln \frac{V_f}{V_i} \qquad \begin{array}{l}\text{(ideal gas,}\\ \text{isothermal process).}\end{array} \quad (20\text{-}10)$$

Pressure, Temperature, and Molecular Speed
The pressure exerted by *n* moles of an ideal gas, in terms of the speed of its molecules, is

$$p = \frac{nMv_{\text{rms}}^2}{3V}, \qquad (20\text{-}17)$$

where $v_{\text{rms}} = \sqrt{\overline{v^2}}$ is the **root-mean-square speed** of the molecules of the gas. With Eq. 20-4 this gives

$$v_{\text{rms}} = \sqrt{\frac{3RT}{M}}. \qquad (20\text{-}18)$$

Temperature and Kinetic Energy
The average translational kinetic energy $\overline{K}$ per molecule of an ideal gas is

$$\overline{K} = \tfrac{3}{2}kT, \qquad (20\text{-}20)$$

in which k (= R/N_A = 1.38 × 10⁻²³ J/K) is the **Boltzmann constant.**

Mean Free Path
The *mean free path* λ of a gas molecule is its average path length between collisions and is given by

$$\lambda = \frac{1}{\sqrt{2}\pi d^2 \, N/V}, \qquad (20\text{-}22)$$

where N/V is the number of molecules per unit volume and d is the molecular diameter.

Maxwell Speed Distribution
The *Maxwell speed distribution* $P(v)$ is a function such that

$P(v) \, dv$ gives the *fraction* of molecules with speeds between v and $v + dv$:

$$P(v) = 4\pi \left(\frac{M}{2\pi RT}\right)^{3/2} v^2 e^{-Mv^2/2RT}. \qquad (20\text{-}24)$$

Three measures of the distribution of speeds among the molecules of a gas are

$$v_P = \sqrt{\frac{2RT}{M}} \qquad \text{(most probable speed),} \qquad (20\text{-}29)$$

$$\bar{v} = \sqrt{\frac{8RT}{\pi M}} \qquad \text{(average speed),} \qquad (20\text{-}27)$$

and the rms speed defined above in Eq. 20-18.

Molar Specific Heats

The molar specific heat C_V of a gas at constant volume is defined as

$$C_V = \frac{1}{n} \frac{Q}{\Delta T} = \frac{1}{n} \frac{\Delta E_{int}}{\Delta T}, \qquad (20\text{-}31, 20\text{-}32)$$

in which Q is the heat transferred to or from a sample of n moles of the gas, ΔT is the resulting temperature change of the gas, and ΔE_{int} is the resulting change in the internal energy of the gas. For an ideal monatomic gas,

$$C_V = \tfrac{3}{2}R = 12.5 \text{ J/mol} \cdot \text{K}. \qquad (20\text{-}33)$$

The molar specific heat C_p of a gas at constant pressure is defined to be

$$C_p = \frac{1}{n} \frac{Q}{\Delta T}, \qquad (20\text{-}36)$$

in which Q, n, and ΔT are defined as above. C_p is also given by

$$C_p = C_V + R. \qquad (20\text{-}38)$$

For n moles of an ideal gas,

$$E_{int} = nC_VT \qquad \text{(ideal gas).} \qquad (20\text{-}34)$$

If n moles of a confined ideal gas undergo a temperature change ΔT due to *any* process, the change in the internal energy of the gas is

$$\Delta E_{int} = nC_V\Delta T \qquad \text{(ideal gas, any process),} \qquad (20\text{-}35)$$

in which the appropriate value of C_V must be substituted, according to the type of ideal gas.

Degrees of Freedom and C_V

We find C_V itself by using the *equipartition of energy* theorem, which states that every *degree of freedom* of a molecule (that is, every independent way it can store energy) has associated with it—on average—an energy $\tfrac{1}{2}kT$ per molecule ($= \tfrac{1}{2}RT$ per mole). If f is the number of degrees of freedom, then $E_{int} = (f/2)nRT$ and

$$C_V = \left(\frac{f}{2}\right) R = 4.16f \text{ J/mol} \cdot \text{K}. \qquad (20\text{-}39)$$

For monatomic gases $f = 3$ (three translational degrees); for diatomic gases $f = 5$ (three translational and two rotational degrees).

Adiabatic Process

When an ideal gas undergoes a slow adiabatic volume change (a change for which $Q = 0$), its pressure and volume are related by

$$pV^\gamma = \text{a constant} \qquad \text{(adiabatic process),} \qquad (20\text{-}40)$$

in which $\gamma \; (= C_p/C_V)$ is the ratio of molar specific heats for the gas. For a free expansion, however, $pV = $ a constant.

QUESTIONS

1. If the temperature of an ideal gas is changed from 20°C to 40°C while the volume is unchanged, is the pressure of the gas doubled, increased but less than doubled, or increased but more than doubled?

2. Two rooms of equal volume are connected by an open passageway and are maintained at different temperatures. Which room has more air molecules?

3. The molar masses and Kelvin temperatures of three ideal gases are (a) M_0 and $2T_0$, (b) $2M_0$ and T_0, and (c) $6M_0$ and $6T_0$. Rank the gases according to the rms speeds of their molecules, greatest first.

4. The volume of a gas and the number of gas molecules within that volume for four situations are (a) $2V_0$ and N_0, (b) $3V_0$ and $3N_0$, (c) $8V_0$ and $4N_0$, and (d) $3V_0$ and $9N_0$. Rank the situations according to the mean free path of the molecules, greatest first.

5. In Sample Problem 20-2, how much heat is transferred during the expansion of part (a)?

6. Figure 20-15 shows the initial state of an ideal gas and an isotherm through that state. Which of the paths shown result in a

decrease in the temperature of the gas?

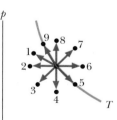

FIGURE 20-15
Question 6.

7. The table gives the heat transfer Q to or from an ideal gas and either the work W done by the gas or the work W_{on} done on the gas, all in joules, for four situations. Rank the four situations in terms of the temperature change of the gas, most positive first, most negative last.

	a	b	c	d
Q	−50	+35	−15	+20
W	−50	+35		
W_{on}			−40	+40

8. For a temperature increase of ΔT_1, a certain amount of an ideal gas requires 30 J when heated at constant volume and 50 J when heated at constant pressure. How much work is done by the gas in the second situation?

9. An ideal diatomic gas, with molecular rotation but not oscillation, loses heat Q. Is the resulting decrease in the internal energy of the gas greater if the loss occurs in a constant-volume process or in a constant-pressure process?

10. A certain amount of heat is to be transferred to 1 mol of a monatomic gas (a) at constant pressure and (b) at constant volume, and to 1 mol of a diatomic gas (c) at constant pressure and (d) at constant volume. Figure 20-16 shows four paths from an initial point to four final points on a p-V diagram. Which path goes with which process? (e) Are the molecules of the diatomic gas rotating?

FIGURE 20-16
Question 10.

11. Does the temperature of an ideal gas increase, decrease, or stay the same during (a) an isothermal expansion, (b) an expansion at constant pressure, (c) an adiabatic expansion, and (d) an increase in pressure at constant volume?

12. (a) Rank the four paths of Fig. 20-14 according to the work done by the gas, greatest first. (b) Rank paths 1, 2, and 3 according to the change in the internal energy of the gas, most positive first and most negative last.

13. In the p-V diagram of Fig. 20-17, the gas does 5 J of work along isotherm ab and 4 J along adiabat bc. What is the change in the internal energy of the gas if the gas traverses the straight path from a to c?

FIGURE 20-17
Question 13.

14. The p-V diagram of Fig. 20-18 (not to scale) shows eight separate transformations (lettered a to h) that an ideal gas undergoes in proceeding from an initial state to a final state. The table shows eight sets of values for Q, W, and ΔE_{int} (in joules), numbered 1 to 8. Which set of energy changes goes with which transformation? (*Hint:* Compare the paths in Fig. 20-18 to those in Fig. 20-14.)

	1	2	3	4	5	6	7	8
Q	10	5	−20		−10	−12		10
W	10		−10	10		−12	−10	
ΔE_{int}		2		−10	−10		10	10

FIGURE 20-18
Question 14.

15. (a) The p-V diagram in Fig. 20-19 shows adiabatic expansions from a common point to volume V_f for a monatomic gas, a diatomic gas, and a polyatomic gas. Which path goes with which gas? (b) If the monatomic gas had, instead, freely expanded to the same final volume, would its final pressure be more than, less than, or the same as previously? (*Hint:* How would the initial and final states for the free expansion be plotted on Fig. 20-14?)

FIGURE 20-19
Question 15.

EXERCISES & PROBLEMS

SECTION 20-2 Avogadro's Number

1E. Gold has a molar mass of 197 g/mol. (a) How many moles of gold are in a 2.50 g sample of pure gold? (b) How many atoms are in the sample?

2E. Find the mass in kilograms of 7.50×10^{24} atoms of arsenic, which has a molar mass of 74.9 g/mol.

3P. If the water molecules in 1.00 g of water were distributed uniformly over the surface of Earth, how many such molecules

would there be on 1.00 cm^2 of the surface?

4P. Consider this sentence: A _____ of water contains about as many molecules as there are _____ s of water in all the oceans. What single word best fits both blank spaces: drop, teaspoon, tablespoon, cup, quart, barrel, or ton? The oceans cover 75% of Earth's surface and have an average depth of about 5 km. (After Edward M. Purcell.)

5P. A distinguished scientist has written: "There are enough molecules in the ink that makes one letter of this sentence to provide not only one for every inhabitant of Earth, but one for every creature if each star of our galaxy had a planet as populous as Earth.'' Check this statement. Assume the ink sample (molar mass = 18 g/mol) to have a mass of 1 μg, the population of Earth to be 5×10^9, and the number of stars in our galaxy to be 10^{11}.

SECTION 20-3   Ideal Gases

6E. (a) What is the volume occupied by 1.00 mol of an ideal gas at standard conditions, that is, at 1.00 atm (= 1.01×10^5 Pa) and 0°C (= 273 K)? (b) Show that the number of molecules per cubic centimeter (the *Loschmidt number*) at standard conditions is 2.69×10^{19}.

7E. Compute (a) the number of moles and (b) the number of molecules in 1.00 cm³ of an ideal gas at a pressure of 100 Pa and a temperature of 220 K.

8E. The best vacuum that can be attained in the laboratory corresponds to a pressure of about 1.00×10^{-18} atm, or 1.01×10^{-13} Pa. How many gas molecules are there per cubic centimeter in such a vacuum at 293 K?

9E. A quantity of ideal gas at 10.0°C and a pressure of 100 kPa occupies a volume of 2.50 m³. (a) How many moles of the gas are present? (b) If the pressure is now raised to 300 kPa and the temperature is raised to 30.0°C, how much volume will the gas occupy? Assume no leaks.

10E. Oxygen gas having a volume of 1000 cm³ at 40.0°C and 1.01×10^5 Pa expands until its volume is 1500 cm³ and its pressure is 1.06×10^5 Pa. Find (a) the number of moles of oxygen present and (b) the final temperature of the sample.

11E. An automobile tire has a volume of 1000 in.³ and contains air at a gauge pressure of 24.0 lb/in.² when the temperature is 0.00°C. What is the gauge pressure of the air in the tires when its temperature rises to 27.0°C and its volume increases to 1020 in.³? (*Hint:* It is not necessary to convert from British units to SI units; why? Use P_{atm} = 14.7 lb/in.².)

12E. Calculate the work done by an external agent during an isothermal compression of 1.00 mol of oxygen from a volume of 22.4 L at 0°C and 1.00 atm pressure to 16.8 L.

13P. (a) What is the number of molecules per cubic meter in air at 20°C and at a pressure of 1.0 atm (= 1.01×10^5 Pa)? (b) What is the mass of this 1 m³ of air? Assume that 75% of the molecules are nitrogen (N_2) and 25% oxygen (O_2).

14P. Pressure p, volume V, and temperature T for a certain material are related by

$$p = \frac{AT - BT^2}{V},$$

where A and B are constants. Find an expression for the work done by the material if the temperature changes from T_1 to T_2 while the pressure remains constant.

15P. Air that occupies 0.14 m³ at 1.03×10^5 Pa gauge pressure is expanded isothermally to atmospheric pressure and then cooled at constant pressure until it reaches its initial volume. Compute the work done by the air.

16P. Consider a given mass of an ideal gas. Compare curves representing constant-pressure, constant-volume, and isothermal processes on (a) a p-V diagram, (b) a p-T diagram, and (c) a V-T diagram. (d) How do these curves depend on the mass of gas?

17P. A container encloses two ideal gases. Two moles of the first gas are present, with molar mass M_1. The second gas has molar mass $M_2 = 3M_1$, and 0.5 mol of this gas is present. What fraction of the total pressure on the container wall is attributable to the second gas? (The kinetic theory explanation of pressure leads to the experimentally discovered law of partial pressures for a mixture of gases that do not react chemically: *the total pressure exerted by the mixture is equal to the sum of the pressures that the several gases would exert separately if each were to occupy the vessel alone.*)

18P. A sample of an ideal gas is taken through the cyclic process *abca* shown in Fig. 20-20; at point a, T = 200 K. (a) How many moles of gas are in the sample? What are (b) the temperature of the gas at point b, (c) the temperature of the gas at point c, and (d) the net heat added to the gas during the cycle?

FIGURE 20-20   Problem 18.

19P. An ideal gas initially at 300 K is compressed at a constant pressure of 25 N/m² from a volume of 3.0 m³ to a volume of 1.8 m³. In the process, 75 J of heat is lost by the gas. What are (a) the change in internal energy of the gas and (b) the final temperature of the gas?

20P. A weather balloon is loosely inflated with helium at a pressure of 1.0 atm (= 760 torr) and a temperature of 20°C. The gas volume is 2.2 m³. At an elevation of 20,000 ft, the atmospheric pressure is down to 380 torr and the helium has expanded, being under no restraint from the confining bag. At this elevation the gas temperature is −48°C. What is the gas volume now?

21P. An air bubble of 20 cm³ volume is at the bottom of a lake 40 m deep where the temperature is 4.0°C. The bubble rises to the surface, which is at a temperature of 20°C. Take the temperature of the bubble to be the same as that of the surrounding water and find its volume just before it reaches the surface.

22P. A pipe of length L = 25.0 m that is open at one end contains air at atmospheric pressure. It is thrust vertically into a freshwater lake until the water rises halfway up in the pipe, as shown in Fig. 20-21. What is the depth h of the lower end of the pipe? Assume that the temperature is the same everywhere and does not change.

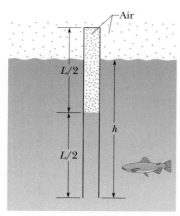

FIGURE 20-21
Problem 22.

23P. The envelope and basket of a hot-air balloon have a combined weight of 550 lb, and the envelope has a capacity of 77,000 ft³. When it is fully inflated, what should be the temperature of the enclosed air to give the balloon a lifting capacity of 600 lb (in addition to its own weight)? Assume that the surrounding air, at 20.0°C, has a weight density of 7.56×10^{-2} lb/ft³.

24P. A steel tank contains 300 g of ammonia gas (NH_3) at an absolute pressure of 1.35×10^6 Pa and temperature of 77°C. (a) What is the volume of the tank? (b) The tank is checked later when the temperature has dropped to 22°C, and the absolute pressure is found to be 8.7×10^5 Pa. How many grams of gas have leaked out of the tank?

25P. Container A in Fig. 20-22 holds an ideal gas at a pressure of 5.0×10^5 Pa and a temperature of 300 K. It is connected by a thin tube to container B with four times the volume of A. Container B holds the same ideal gas at a pressure of 1.0×10^5 Pa and a temperature of 400 K. The connecting valve is opened, and equilibrium is achieved at a common pressure while the temperature of each container is kept constant at its initial value. What is the final pressure in the system?

FIGURE 20-22 Problem 25.

SECTION 20-4 Pressure, Temperature, and rms Speed

26E. Calculate the root-mean-square speed of helium atoms at 1000 K. The molar mass of helium is 4.00 g/mol.

27E. The lowest possible temperature in outer space is 2.7 K. What is the root-mean-square speed of hydrogen molecules at this temperature? (Use Table 20-1.)

28E. Find the rms speed of argon atoms at 313 K. The molar mass of argon is 39.9 g/mol.

29E. The Sun is a huge ball of hot ideal gas. The temperature and pressure in the Sun's atmosphere are 2.00×10^6 K and 0.0300 Pa. Calculate the rms speed of free electrons (mass = 9.11×10^{-31} kg) there.

30E. (a) Compute the root-mean-square speed of a nitrogen molecule at 20.0°C. At what temperatures will the root-mean-square speed be (b) half that value and (c) twice that value?

31E. At what temperature do the atoms of helium gas have the same rms speed as molecules of hydrogen gas have at 20.0°C?

32P. At 273 K and 1.00×10^{-2} atm, the density of a gas is 1.24×10^{-5} g/cm³. (a) Find v_{rms} for the gas molecules. (b) Find the molar mass of the gas and identify it.

33P. The mass of the H_2 molecule is 3.3×10^{-24} g. If 10^{23} H_2 molecules per second strike 2.0 cm² of wall at an angle of 55° with the normal when moving with a speed of 1.0×10^5 cm/s, what pressure do they exert on the wall?

SECTION 20-5 Translational Kinetic Energy

34E. What is the average translational kinetic energy of nitrogen molecules at 1600 K, (a) in joules and (b) in electron-volts?

35E. (a) Determine the average value in electron-volts of the translational kinetic energy of the particles of an ideal gas at 0.00°C and at 100°C. (b) What is the translational kinetic energy per mole of an ideal gas at these temperatures, in joules?

36E. At what temperature is the average translational kinetic energy of a molecule equal to 1.00 eV?

37E. Oxygen (O_2) gas at 273 K and 1.0 atm pressure is confined to a cubical container 10 cm on a side. Calculate the ratio of (1) the change in gravitational potential energy of an oxygen molecule falling the height of the box to (2) the molecule's average translational kinetic energy.

38P. Show that the ideal gas equation, Eq. 20-4, can be written in these alternative forms: (a) $p = \rho RT/M$, where ρ is the mass density of the gas and M the molar mass; (b) $pV = NkT$, where N is the number of gas particles (atoms or molecules).

39P. Water standing in the open at 32.0°C evaporates because of the escape of some of the surface molecules. The heat of vaporization (539 cal/g) is approximately equal to ϵn, where ϵ is the average energy of the escaping molecules and n is the number of molecules per gram. (a) Find ϵ. (b) What is the ratio of ϵ to the average kinetic energy of H_2O molecules, assuming the latter is related to temperature in the same way as it is for gases?

40P. *Avogadro's law* states that under the same conditions of temperature and pressure, equal volumes of gas contain equal numbers of molecules. Is this law equivalent to the ideal gas law? Explain.

SECTION 20-6 Mean Free Path

41E. The mean free path of nitrogen molecules at 0.0°C and 1.0 atm is 0.80×10^{-5} cm. At this temperature and pressure there are 2.7×10^{19} molecules/cm³. What is the molecular diameter?

42E. At 2500 km above Earth's surface, the density of the atmosphere is about 1 molecule/cm³. (a) What mean free path is predicted by Eq. 20-22 and (b) what is its significance under these conditions? Assume a molecular diameter of 2.0×10^{-8} cm.

43E. What is the mean free path for 15 spherical jelly beans in a bag that is vigorously shaken? The volume of the bag is 1.0 L, and the diameter of a jelly bean is 1.0 cm.

44E. Derive an expression, in terms of N/V, $\bar{v}$, and d, for the collision frequency of a gas atom or molecule.

45P. In a certain particle accelerator, protons travel around a circular path of diameter 23.0 m in an evacuated chamber, whose residual gas is at 295 K and 1.00×10^{-6} torr pressure. (a) Calculate the number of gas molecules per cubic centimeter at this pressure. (b) What is the mean free path of the gas molecules if the molecular diameter is 2.00×10^{-8} cm?

46P. At what frequency would the wavelength of sound in air be equal to the mean free path of oxygen molecules at 1.0 atm pressure and 0.0°C? Take the diameter of the oxygen molecule to be 3.0×10^{-8} cm.

47P. (a) What is the molar volume (volume per mole) of an ideal gas at standard conditions (0.00°C, 1.00 atm)? (b) Calculate the ratio of the root-mean-square speed of helium atoms to that of neon atoms under these conditions. (c) What is the mean free path of helium atoms under these conditions? Assume the atomic diameter d to be 1.00×10^{-8} cm. (d) What is the mean free path of neon atoms under these conditions? Assume the same atomic diameter as for helium. (e) Comment on the results of (c) and (d) in view of the fact that the helium atoms are traveling faster than the neon atoms.

48P. The mean free path λ of the molecules of a gas may be determined from certain measurements (for example, from measurement of the viscosity of the gas). At 20°C and 750 torr pressure such measurements yield values of λ_{Ar} (argon) $= 9.9 \times 10^{-6}$ cm and λ_{N_2} (nitrogen) $= 27.5 \times 10^{-6}$ cm. (a) Find the ratio of the effective diameter of argon to that of nitrogen. What is the mean free path of argon at (b) 20°C and 150 torr, and (c) −40°C and 750 torr?

49P. Show that about 10^{13} air molecules are needed to cover the period that closes this sentence. Show that about 10^{21} air molecules collide with that period each second.

SECTION 20-7 The Distribution of Molecular Speeds

50E. The speeds of 10 molecules are 2.0, 3.0, 4.0,· · ·, 11 km/s. (a) What is their average speed? (b) What is their root-mean-square speed?

51E. Twenty-two particles have speeds as follows (N_i represents the number of particles that have speed v_i):

N_i	2	4	6	8	2
v_i (cm/s)	1.0	2.0	3.0	4.0	5.0

(a) Compute the average speed $\bar{v}$. (b) Compute the root-mean-square speed v_{rms}. (c) Of the five speeds shown, which is the most probable speed v_P?

52E. (a) Ten particles are moving with the following speeds: four at 200 m/s, two at 500 m/s, and four at 600 m/s. Calculate the average and root-mean-square speeds. Is $v_{rms} > \bar{v}$? (b) Make up your own speed distribution for the 10 particles and show that $v_{rms} \geq \bar{v}$ for your distribution. (c) Under what condition (if any) does $v_{rms} = \bar{v}$?

53E. Consider the distribution of speeds shown in Fig. 20-23. (a) Rank v_{rms}, $\bar{v}$, and v_P, greatest first. (b) How does the ranking in (a) compare with that for a Maxwellian distribution?

FIGURE 20-23
Exercise 53.

54E. It is found that the most probable speed of molecules in a gas at equilibrium temperature T_2 is the same as the rms speed of the molecules in this gas when its equilibrium temperature is T_1. Calculate T_2/T_1.

55P. (a) Compute the temperatures at which the rms speed is equal to the speed of escape from Earth's surface for molecular hydrogen and for molecular oxygen. (b) Do the same for the Moon, assuming the gravitational acceleration on its surface to be $0.16g$. (c) The temperature high in Earth's upper atmosphere is about 1000 K. Would you expect to find much hydrogen there? Much oxygen?

56P. A molecule of hydrogen (diameter 1.0×10^{-8} cm), traveling with the rms speed, escapes from a furnace ($T = 4000$ K) into a chamber containing atoms of cold argon (diameter 3.0×10^{-8} cm) at a density of 4.0×10^{19} atoms/cm³. (a) What is the speed of the hydrogen molecule? (b) If the H_2 molecule and an argon atom collide, what is the closest their centers can be, considering each as spherical? (c) What is the initial number of collisions per second experienced by the hydrogen molecule?

57P. Two containers are at the same temperature. The first contains gas with pressure p_1, molecular mass m_1, and root-mean-square speed v_{rms1}. The second contains gas with pressure $2p_1$, molecular mass m_2, and average speed $\bar{v}_2 = 2v_{rms1}$. Find the mass ratio m_1/m_2.

58P. For the hypothetical speed distribution for N gas particles shown in Fig. 20-24 [$P(v) = Cv^2$ for $0 < v \leq v_0$; $P(v) = 0$ for $v > v_0$], find (a) an expression for C in terms of N and v_0, (b) the average speed of the particles, and (c) their rms speed.

FIGURE 20-24
Problem 58.

59P. A hypothetical sample of N gas particles has the speed distribution shown in Fig. 20-25, where $P(v) = 0$ for $v > 2v_0$. (a) Express a in terms of N and v_0. (b) How many of the particles have speeds between $1.5v_0$ and $2.0v_0$? (c) Express the average speed of the particles in terms of v_0. (d) Find v_{rms}.

FIGURE 20-25
Problem 59.

SECTION 20-8 The Molar Specific Heats of an Ideal Gas

60E. What is the internal energy of 1.0 mol of an ideal monatomic gas at 273 K?

61E. One mole of an ideal gas undergoes an isothermal expansion. Find the heat added to the gas in terms of the initial and final volumes and the temperature. (*Hint:* Use the first law of thermodynamics.)

62E. The mass of a helium atom is 6.66×10^{-27} kg. Compute the specific heat at constant volume for (monatomic) helium gas (in J/kg·K) from the molar specific heat at constant volume.

63P. Let 20.9 J of heat be added to a particular ideal gas. As a result, its volume changes from 50.0 cm³ to 100 cm³ while the pressure remains constant at 1.00 atm. (a) By how much did the internal energy of the gas change? If the quantity of gas present is 2.00×10^{-3} mol, find the molar specific heat at (b) constant pressure and (c) constant volume.

64P. A quantity of ideal monatomic gas consists of n moles initially at temperature T_1. The pressure and volume are then slowly doubled in such a manner as to trace out a straight line on a p-V diagram. In terms of n, R, and T_1, what are (a) W, (b) ΔE_{int}, and (c) Q? (d) If one were to define a molar specific heat for this process, what would be its value?

65P. A container holds a mixture of three nonreacting gases: n_1 moles of the first gas with molar specific heat at constant volume C_1, and so on. Find the molar specific heat at constant volume of the mixture, in terms of the molar specific heats and quantities of the separate gases.

66P. The mass of a gas molecule can be computed from the specific heat at constant volume c_V. Take $c_V = 0.075$ cal/g·C° for argon and calculate (a) the mass of an argon atom and (b) the molar mass of argon.

67P. One mole of an ideal diatomic gas undergoes a transition from a to c along the diagonal path in Fig. 20-26. The temperature of the gas at point a is 1200 K. During the transition, (a) what is the change in internal energy of the gas, and (b) how much heat is added to the gas? (c) How much heat must be added to the gas if it goes from a to c along the indirect path abc?

SECTION 20-9 Degrees of Freedom and Molar Specific Heats

68E. One mole of oxygen (O_2) is heated at constant pressure starting at 0°C. How much heat must be added to the gas to double its volume? (The molecules rotate but do not oscillate.)

69E. Suppose 12.0 g of oxygen (O_2) is heated at constant atmospheric pressure from 25.0°C to 125°C. (a) How many moles of oxygen are present? (See Table 20-1.) (b) How much heat is transferred to the oxygen? (The molecules rotate but do not oscillate.) (c) What fraction of the heat is used to raise the internal energy of the oxygen?

70P. Suppose 4.00 mol of an ideal diatomic gas, with molecular rotation but not oscillation, experiences a temperature increase of 60.0 K under constant-pressure conditions. (a) How much heat was added to the gas? (b) How much did the internal energy of the gas increase? (c) How much work was done by the gas? (d) How much did the translational kinetic energy of the gas increase?

SECTION 20-11 The Adiabatic Expansion of an Ideal Gas

71E. A mass of gas occupies a volume of 4.3 L at a pressure of 1.2 atm and a temperature of 310 K. It is compressed adiabatically to a volume of 0.76 L. Determine (a) the final pressure and (b) the final temperature, assuming the gas to be an ideal gas for which $\gamma = 1.4$.

72E. (a) One liter of gas with $\gamma = 1.3$ is at 273 K and 1.0 atm pressure. It is suddenly compressed (adiabatically) to half its original volume. Find its final pressure and temperature. (b) The gas is now cooled back to 273 K at constant pressure. What is its final volume?

73E. Let n moles of an ideal gas expand adiabatically from an initial temperature T_1 to a final temperature T_2. Prove that the work done by the gas is $nC_V(T_1 - T_2)$, where C_V is the molar specific heat at constant volume. (*Hint:* Use the first law of thermodynamics.)

74E. We know that $pV^\gamma = $ a constant for an adiabatic process. Evaluate "a constant" for an adiabatic process involving exactly 2.0 mol of an ideal gas passing through the state having exactly $p = 1.0$ atm and $T = 300$ K. Assume a diatomic gas whose molecules have rotation but not oscillation.

75E. For adiabatic processes in an ideal gas, show that (a) the bulk modulus is given by

$$B = -V\frac{dp}{dV} = \gamma p,$$

FIGURE 20-26
Problem 67.

and therefore (b) the speed of sound in the gas is

$$v_s = \sqrt{\frac{\gamma p}{\rho}} = \sqrt{\frac{\gamma RT}{M}}.$$

See Eqs. 18-2 and 18-3.

76E. Air at 0.000°C and 1.00 atm pressure has a density of 1.29×10^{-3} g/cm³, and the speed of sound in air is 331 m/s at that temperature. Compute the ratio γ of the molar specific heats of air. (*Hint:* See Exercise 75.)

77E. The speed of sound in different gases at the same temperature depends only on the molar mass of the gases. Show that $v_1/v_2 = \sqrt{M_2/M_1}$ (at constant T), where v_1 is the speed of sound in a gas of molar mass M_1 and v_2 is the speed of sound in a gas of molar mass M_2. (*Hint:* See Exercise 75.)

78P. The molar mass of iodine is 127 g/mol. A standing wave in a tube filled with iodine gas at 400 K has nodes that are 6.77 cm apart when the frequency is 1400 Hz. (a) What is γ? (b) Is iodine gas monatomic or diatomic? (*Hint:* See Exercise 75.)

79P. Knowing that C_V, the molar specific heat at constant volume, for a gas in a container is $5.0R$, calculate the ratio of the speed of sound in that gas to the rms speed of its molecules at temperature T. (*Hint:* See Exercise 75.)

80P. (a) An ideal gas initially at pressure p_0 undergoes a free expansion until its volume is 3.00 times its initial volume. What then is its pressure? (b) The gas is next slowly and adiabatically compressed back to its original volume. The pressure after compression is $(3.00)^{1/3}p_0$. Is the gas monatomic, diatomic, or polyatomic? (c) How does the average kinetic energy per molecule in this final state compare with that in the initial state?

81P. An ideal gas experiences an adiabatic compression from $p = 1.0$ atm, $V = 1.0 \times 10^6$ L, $T = 0.0°C$ to $p = 1.0 \times 10^5$ atm, $V = 1.0 \times 10^3$ L. (a) Is the gas monatomic, diatomic, or polyatomic? (b) What is its final temperature? (c) How many moles of gas are present? (d) What is the total translational kinetic energy per mole before and after the compression? (e) What is the ratio of the squares of the rms speeds before and after the compression?

82P. A sample of ideal gas expands from an initial pressure and volume of 32 atm and 1.0 L to a final volume of 4.0 L. The initial temperature of the gas is 300 K. What are the final pressure and temperature of the gas and how much work is done by the gas during the expansion, if the expansion is (a) isothermal, (b) adiabatic and the gas is monatomic, and (c) adiabatic and the gas is diatomic?

83P. An ideal gas, at initial temperature T_1 and initial volume 2 m³, is expanded adiabatically to a volume of 4 m³, then expanded isothermally to a volume of 10 m³, and then compressed adiabatically until its temperature is again T_1. What is its final volume?

84P. For a certain ideal gas C_V is 6.00 cal/mol·K. The temperature of 3.0 mol of the gas is raised 50 K by each of three different processes: at constant volume, at constant pressure, and by an adiabatic compression. Complete the table, showing for each process the heat Q added (or subtracted), the work W done by the gas, the change ΔE_{int} in internal energy of the gas, and the change ΔK in total translational kinetic energy of the gas.

PROCESS	Q	W	ΔE_{int}	ΔK
Constant volume	_____	_____	_____	_____
Constant pressure	_____	_____	_____	_____
Adiabatic	_____	_____	_____	_____

85P. One mole of an ideal monatomic gas traverses the cycle shown in Fig. 20-27. Process $1 \rightarrow 2$ takes place at constant volume, process $2 \rightarrow 3$ is adiabatic, and process $3 \rightarrow 1$ takes place at constant pressure. (a) Compute the heat Q, the change in internal energy ΔE_{int}, and the work done W, for each of the three processes and for the cycle as a whole. (b) If the initial pressure at point 1 is 1.00 atm, find the pressure and the volume at points 2 and 3. Use 1.00 atm $= 1.013 \times 10^5$ Pa and $R = 8.314$ J/mol·K.

FIGURE 20-27
Problem 85.

Electronic Computation

86P. In an industrial process the volume of 25.0 moles of a monatomic ideal gas is reduced at a uniform rate from 0.616 m³ to 0.308 m³ in 2.00 h while its temperature is increased at a uniform rate from 27.0°C to 450°C. Throughout the process, the gas passes through thermodynamic equilibrium states. (a) Plot the cumulative work done on the gas and the cumulative heat absorbed by the gas as a function of time for the duration of the process. (b) What are the values of these quantities for the entire process? (c) What is the molar specific heat for the process? (*Hint:* To evaluate the integral for the work, you might use

$$\int \frac{a + bx}{A + Bx}\, dx = \frac{bx}{B} + \frac{aB - bA}{B^2} \ln(A + Bx),$$

an indefinite integral.) (d) Compare these values with those that would be obtained if the volume is first reduced at constant temperature, then the temperature is increased at constant volume, with the gas reaching the same final state.

21
Entropy and the Second Law of Thermodynamics

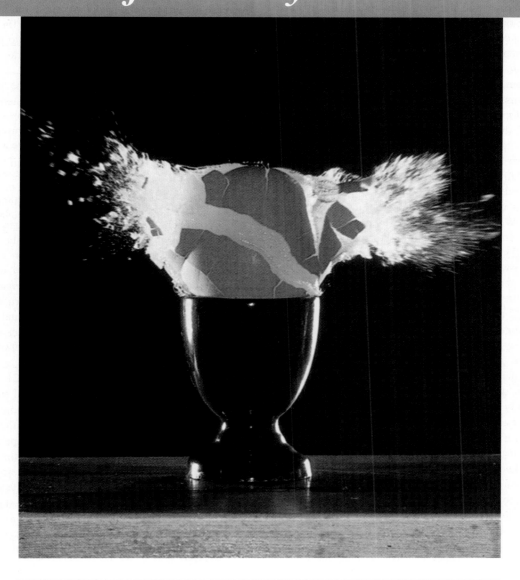

An anonymous graffito on a wall of the Pecan Street Cafe in Austin, Texas, read: "Time is God's way of keeping things from happening all at once." Time also has direction: some things happen in a certain sequence and could never happen on their own in a reverse sequence. As an example, an accidentally dropped egg splatters in a cup. The reverse process, a splattered egg re-forming into a whole egg and jumping up to an outstretched hand, will never happen on its own. But why not? Why can't that process be reversed, like a videotape run backward? What in the world gives direction to time?

21-1 SOME ONE-WAY PROCESSES

Suppose you come indoors on a very cold day and wrap your cold hands around a warm mug of cocoa. Then your hands get warmer and the mug gets cooler. But it never happens the other way around; that is, your cold hands never get still colder while the warm mug gets still warmer.

The system consisting of your hands and the mug is a *closed system*, one that is isolated from its environment. Here are some other one-way processes that occur in closed systems: (1) A crate sliding over an ordinary surface eventually stops. But you never see an initially stationary crate start to move all by itself. (2) If you drop a glob of putty, it falls to the floor. But an initially motionless glob of putty never leaps spontaneously into the air. (3) If you puncture a helium-filled balloon in a closed room, the helium gas spreads throughout the room. But the individual helium atoms will never clump up again into the shape of the balloon. We say that such one-way processes are **irreversible,** meaning that they cannot be reversed by means of only small changes in their environment.

The one-way character of such thermodynamic processes is so pervasive that we take it for granted. If these processes were to happen *spontaneously* (on their own) in the "wrong" direction, we would be astonished beyond belief. *Yet none of these "wrong-way" events would violate the law of conservation of energy.* In the cocoa mug example, that law would be obeyed even for a wrong-way transfer of heat between hands and mug. It would be obeyed even if a stationary crate or a stationary glob of putty suddenly were to transfer some of its thermal energy into kinetic energy and begin to move. And it would be obeyed even if the helium atoms released from a balloon were, on their own, to clump together again, which would require no change in their energy.

Thus changes in energy within a closed system do not set the direction of irreversible processes. Rather, that direction is set by another property that we shall discuss in this chapter—the **change in entropy** ΔS of the system. The change in entropy of a system is defined in the next section, but we can here state its central property, often called the *entropy postulate:*

If an irreversible process occurs in a closed system, the entropy S of the system always increases; it never decreases.

Entropy differs from energy in that it does *not* obey a conservation law. The *energy* of a closed system is conserved; it always remains constant. For irreversible processes, the *entropy* of a closed system always increases. Because of

this property, the change in entropy is sometimes called "the arrow of time." For example, we associate the egg of our opening photograph, breaking irreversibly as it drops into a cup, with the forward direction of time and with an increase in entropy. The backward direction of time (a videotape run backward) would correspond to the broken egg re-forming into a whole egg and rising into the air. This backward process, which would result in an entropy decrease, never happens.

There are two equivalent ways to define the change in entropy of a system: (1) in terms of the system's temperature and the energy it gains or loses as heat, and (2) by counting the ways in which the atoms or molecules that make up the system can be arranged. We use the first approach in the next section, and the second in Section 21-6.

21-2 CHANGE IN ENTROPY

Let's approach the definition of *change in entropy* by looking again at a process that we described in Section 19-10: the free expansion of an ideal gas. Figure 21-1a shows the gas in its initial equilibrium state *i*, confined by a closed stopcock to the left half of a thermally insulated container. If we open the stopcock, the gas rushes to fill the entire container, eventually reaching the final equilibrium state *f* shown in Fig. 21-1b. This is an irreversible process; all the molecules of the gas will never return, by themselves, to the left half of the container.

The *p-V* plot of the process in Fig. 21-2 shows the pressure and volume of the gas in its initial state *i* and the final state *f*. Pressure and volume are *state properties,* properties that depend only on the state of the gas and not on how it reached that state. Other state properties are temperature and energy. We now assume that the gas has still another state property—its entropy. Furthermore, we define the **change in entropy** $S_f - S_i$ of a system during a process that takes the system from an initial state *i* to a final state *f* as

$$\Delta S = S_f - S_i = \int_i^f \frac{dQ}{T} \qquad \begin{array}{c}\text{(change in}\\ \text{entropy defined).}\end{array} \quad (21\text{-}1)$$

Here Q is the energy transferred as heat to or from the system during the process, and T is the temperature of the system in kelvins. Thus an entropy change depends not only on the heat transferred but also on the temperature at which the transfer takes place. Because T is always positive, the sign of ΔS is the same as that of Q. We see from Eq. 21-1 that the SI unit for entropy and entropy change is the joule per kelvin.

There is a problem, however, in applying Eq. 21-1 to the free expansion of Fig. 21-1. As the gas rushes to fill the entire container, the pressure, temperature, and volume of

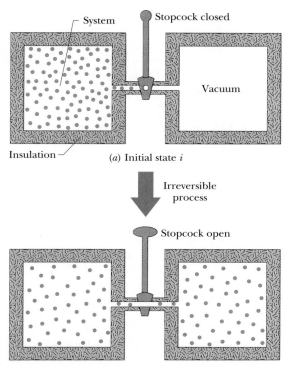

(a) Initial state *i*

Irreversible process

Stopcock open

(b) Final state *f*

FIGURE 21-1 The free expansion of an ideal gas. (*a*) The gas is confined to the left half of an insulated container by a closed stopcock. (*b*) When the stopcock is opened, the gas rushes to fill the entire container. This process is irreversible; that is, it does not occur in reverse, with the gas spontaneously collecting itself in the left half of the container.

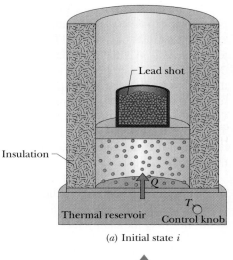

(a) Initial state *i*

Reversible process

(b) Final state *f*

FIGURE 21-3 The isothermal expansion of an ideal gas, done in a reversible way. The gas has the same initial state *i* and same final state *f* as in the irreversible process of Figs. 21-1 and 21-2.

the gas fluctuate unpredictably. That is, they do not have a sequence of well-defined equilibrium values during the intermediate stages of the change from initial equilibrium state *i* to final equilibrium state *f*. Thus we cannot trace a pressure–volume path for the free expansion on the *p-V* plot of Fig. 21-2 and, more important, we cannot find a relation between *Q* and *T* that allows us to integrate as Eq. 21-1 requires.

However, if entropy is truly a state property, the difference in entropy between states *i* and *f* must depend *only on those states* and not at all on the way the system went from one state to the other. Suppose, then, that we replace the irreversible free expansion of Fig. 21-1 with a *reversible* process that connects states *i* and *f*. With a reversible process we can trace a pressure–volume path on a *p-V* plot, and we can find a relation between *Q* and *T* that allows us to use Eq. 21-1 to obtain the entropy change.

We saw in Section 19-10 that the temperature of an ideal gas does not change during a free expansion: $T_i = T_f = T$. So points *i* and *f* in Fig. 21-2 must be on the same isotherm. A convenient replacement process is then a reversible isothermal expansion from state *i* to state *f*, which actually proceeds *along* that isotherm. Furthermore, because *T* is constant throughout a reversible isothermal expansion, the integral of Eq. 21-1 is greatly simplified.

FIGURE 21-2 A *p-V* diagram showing the initial state *i* and the final state *f* of the free expansion of Fig. 21-1. The intermediate states of the gas cannot be shown because they are not equilibrium states.

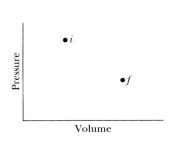

Figure 21-3 shows how to produce such a reversible isothermal expansion. We confine the gas to an insulated cylinder that rests on a thermal reservoir maintained at the temperature T. We begin by placing just enough lead shot on the movable piston so that the pressure and volume of the gas are those of the initial state i of Fig. 21-1a. We then remove shot slowly (piece by piece) until the pressure and volume of the gas are those of the final state f of Fig. 21-1b. The temperature of the gas does not change because the gas remains in thermal contact with the reservoir throughout the process.

The reversible isothermal expansion of Fig. 21-3 is physically quite different from the irreversible free expansion of Fig. 21-1. However, *both processes have the same initial state and the same final state and thus must have the same change in entropy.* Because we removed the lead shot slowly, the intermediate states of the gas are equilibrium states, so we can plot them on a p-V diagram (Fig. 21-4).

To apply Eq. 21-1 to the isothermal expansion, we take the constant temperature T outside the integral, obtaining

$$\Delta S = \frac{1}{T} \int_i^f dQ.$$

Because $\int dQ = Q$, where Q is the total energy transferred as heat during the process, we have

$$\Delta S = S_f - S_i = \frac{Q}{T} \qquad \text{(change in entropy, isothermal process).} \qquad (21\text{-}2)$$

To keep the temperature T of the gas constant during the isothermal expansion of Fig. 21-3, heat Q must have been transferred *from* the reservoir *to* the gas. Thus Q is positive and the entropy of the gas *increases* during the isothermal process and during the free expansion of Fig. 21-1.

To summarize:

To find the entropy change for an irreversible process occurring in a closed system, replace that process with any reversible process that connects the same initial and final states. Calculate the entropy change for this reversible process with Eq. 21-1.

When the temperature change ΔT of a system is small relative to the temperature (in kelvins) before and after the process, the entropy change can be approximated as

$$\Delta S = S_f - S_i \approx \frac{Q}{\overline{T}}, \qquad (21\text{-}3)$$

where $\overline{T}$ is the average temperature of the system during the process.

CHECKPOINT **1:** Water is heated on a stove. Rank the entropy changes of the water as its temperature rises (a) from 20°C to 30°C, (b) from 30°C to 35°C, and (c) from 80°C to 85°C, greatest first.

SAMPLE PROBLEM 21-1

One mole of nitrogen gas is confined to the left side of the container of Fig. 21-1. You open the stopcock and the volume of the gas doubles. What is the entropy change of the gas for this irreversible process? Treat the gas as ideal.

SOLUTION: As we did above, we replace the irreversible process of Fig. 21-1 with the isothermal expansion of Fig. 21-3 and calculate the entropy change for this latter process.

From Table 20-5, the heat Q added to an ideal gas as it expands isothermally at temperature T from an initial volume V_i to a final volume V_f is

$$Q = nRT \ln \frac{V_f}{V_i},$$

in which n is the number of moles of gas present.

From Eq. 21-2 the entropy change for this process is

$$\Delta S = \frac{Q}{T} = \frac{nRT \ln(V_f/V_i)}{T} = nR \ln \frac{V_f}{V_i}.$$

Substituting $n = 1.00$ mol and $V_f/V_i = 2$, we find

$$\Delta S = nR \ln \frac{V_f}{V_i} = (1.00 \text{ mol})(8.31 \text{ J/mol} \cdot \text{K})(\ln 2)$$

$$= +5.76 \text{ J/K}. \qquad \text{(Answer)}$$

This is also the entropy change for the free expansion—and for all other processes that connect the initial and final states shown in Fig. 21-2. ΔS is positive, so the entropy increases, in accordance with the entropy postulate of Section 21-1.

FIGURE 21-4 A p-V diagram for the reversible isothermal expansion of Fig. 21-3. The intermediate states, which are now equilibrium states, are shown.

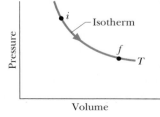

SAMPLE PROBLEM 21-2

Figure 21-5a shows two identical copper blocks of mass $m = 1.5$ kg: block L is at temperature $T_{iL} = 60$°C and block R is at temperature $T_{iR} = 20$°C. The blocks are in a thermally insulated box and are separated by an insulating shutter. When we lift the shutter, the blocks eventually come to the equilibrium temperature $T_f = 40°C$ (Fig. 21-5b). What is the net entropy change of the two-block system during this irreversible process? The specific heat of copper is 386 J/kg·K.

SOLUTION: To calculate the entropy change, we must find a reversible process that takes the system from the initial state of Fig. 21-5a to the final state of Fig. 21-5b. We then can calculate the net entropy change ΔS_{rev} of the reversible process using Eq. 21-1. For such a reversible process we need a thermal reservoir whose temperature can be changed slowly (say, by turning a knob). We then take the blocks through the following two steps, illustrated in Fig. 21-6:

STEP 1. With the reservoir's temperature set at 60°C, put block L on the reservoir. (Since block and reservoir are at the same temperature, they are already in thermal equilibrium.) Then slowly lower the temperature of the reservoir and the block to 40°C. As the block's temperature changes by each increment dT during this process, heat dQ is transferred *from* the block to the reservoir. Using Eq. 19-15, we can write this transferred energy as $dQ = mc\, dT$, where c is the specific heat of copper. According to Eq. 21-1, the entropy change ΔS_L of block L during the full temperature change from initial temperature T_{iL} $(= 60°C = 333\text{ K})$ to final temperature T_f $(= 40°C = 313\text{ K})$ is

$$\Delta S_L = \int_i^f \frac{dQ}{T} = \int_{T_{iL}}^{T_f} \frac{mc\, dT}{T} = mc \int_{T_{iL}}^{T_f} \frac{dT}{T}$$

$$= mc \ln \frac{T_f}{T_{iL}}.$$

Inserting the given data yields

$$\Delta S_L = (1.5 \text{ kg})(386 \text{ J/kg·K}) \ln \frac{313 \text{ K}}{333 \text{ K}}$$

$$= -35.86 \text{ J/K}.$$

STEP 2. With the reservoir's temperature now set at 20°C, put block R on the reservoir. Then slowly raise the temperature of the reservoir and the block to 40°C. With the same reasoning used to find ΔS_L, you can show that the entropy change ΔS_R of block R during this process is

$$\Delta S_R = (1.5 \text{ kg})(386 \text{ J/kg·K}) \ln \frac{313 \text{ K}}{293 \text{ K}}$$

$$= +38.23 \text{ J/K}.$$

The net entropy change ΔS_{rev} of the two-block system undergoing this two-step reversible process is then

$$\Delta S_{rev} = \Delta S_L + \Delta S_R$$
$$= -35.86 \text{ J/K} + 38.23 \text{ J/K} = 2.4 \text{ J/K}.$$

Thus, the net entropy change ΔS_{irrev} for the two-block system undergoing the actual irreversible process is

$$\Delta S_{irrev} = \Delta S_{rev} = 2.4 \text{ J/K}. \qquad \text{(Answer)}$$

This result is positive, in accordance with the entropy postulate of Section 21-1.

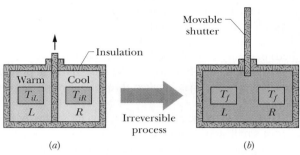

FIGURE 21-5 Sample Problem 21-2. (a) In the initial state, two copper blocks L and R, identical except for their temperatures, are in an insulating box and are separated by an insulating shutter. (b) When the shutter is removed, the blocks exchange heat and come to a final state, both with the same temperature T_f. The process is irreversible.

FIGURE 21-6 The blocks of Fig. 21-5 can proceed from their initial state to their final state in a reversible way if we use a reservoir with a controllable temperature (a) to extract heat reversibly from block L and (b) to add heat reversibly to block R.

CHECKPOINT 2: An ideal gas has temperature T_1 at the initial state i shown in the p-V diagram. The gas has a higher temperature T_2 at final states a and b, which it can reach along the paths shown. Is the entropy change along the path to state a larger than, smaller than, or the same as that along the path to state b?

Entropy as a State Function

We have assumed that entropy, like pressure, energy, and temperature, is a property of the state of a system and independent of how that state is reached. That entropy is indeed a *state function* (as state properties are usually called) can only be deduced by experiment. However, we can prove it is a state function for the special and important case where an ideal gas is taken through a reversible process.

To make the process reversible, it is done slowly in a series of small steps, with the gas in an equilibrium state at the end of each step. For each small step, the heat transfer to or from the gas is dQ, the work done by the gas is dW, and the change in internal energy is dE_{int}. These are related by the first law of thermodynamics in differential form (Eq. 19-25):

$$dQ - dW = dE_{int}.$$

Because the steps are reversible, with the gas in equilibrium states, we can use Eq. 19-22 to replace dW with $p\, dV$ and Eq. 20-35 to replace dE_{int} with $nC_V\, dT$. Solving for dQ then leads to

$$dQ = p\, dV + nC_V\, dT.$$

Using the ideal gas law, we replace p in this equation with nRT/V. Then we divide each term in the resulting equation by T, obtaining

$$\frac{dQ}{T} = nR\frac{dV}{V} + nC_V\frac{dT}{T}.$$

Now let us integrate each term of this equation between an arbitrary initial state i and an arbitrary final state f to get

$$\int_i^f \frac{dQ}{T} = \int_i^f nR\frac{dV}{V} + \int_i^f nC_V\frac{dT}{T}.$$

The quantity on the left is the entropy change $\Delta S\ (= S_f - S_i)$ defined by Eq. 21-1. Substituting this and integrating the quantities on the right yield

$$\Delta S = S_f - S_i = nR \ln \frac{V_f}{V_i} + nC_V \ln \frac{T_f}{T_i}.$$

Note that we did not have to specify a particular reversible process when we integrated. So the integration must hold for all reversible processes that take the gas from state i to state f. Thus the change in entropy ΔS between the initial and final states of an ideal gas depends only on properties of the initial state (V_i and T_i) and properties of the final states (V_f and T_f); ΔS does not depend on how the gas changes between the two states.

21-3 THE SECOND LAW OF THERMODYNAMICS

Here is a puzzle. We saw in Sample Problem 21-1 that if we cause the reversible process of Fig. 21-3 to proceed from (*a*) to (*b*) in that figure, the change in entropy of the gas—which we take as our system—is positive. However, because the process is reversible, we can just as easily make it proceed from (*b*) to (*a*), simply by slowly adding lead shot to the piston of Fig. 21-3*b* until the original volume of the gas is restored. In this reverse process, heat must be extracted *from the gas* to keep its temperature from rising. Hence Q is negative and so, from Eq. 21-2, the entropy of the gas must decrease.

Doesn't this decrease in the entropy of the gas violate the entropy postulate of Section 21-1, which states that entropy always increases? No, because that postulate holds only for *irreversible* processes occurring in closed systems. The procedure suggested above does not meet these requirements: the process is *not* irreversible and (because energy is transferred as heat from the gas to the reservoir) the system—which is the gas alone—is *not* closed.

However, if we include the reservoir, along with the gas, as part of the system, then we do have a closed system. Let's check the change in entropy of the enlarged system *gas + reservoir* for the process that takes it from (*b*) to (*a*) in Fig. 21-3. During this reversible process, energy is transferred as heat from the gas to the reservoir, that is, from one part of the enlarged system to another. Let $|Q|$ represent the absolute value (or magnitude) of this heat. With Eq. 21-2, we can then calculate separately the entropy changes for the gas (which loses $|Q|$) and the reservoir (which gains $|Q|$). We get

$$\Delta S_{gas} = -\frac{|Q|}{T} \quad \text{and} \quad \Delta S_{res} = +\frac{|Q|}{T}.$$

The entropy change of the closed system is the sum of these two quantities, which is zero.

With this result, we can modify the entropy postulate of Section 21-1 to include both reversible and irreversible processes:

> If a process occurs in a closed system, the entropy of the system increases for irreversible processes and remains constant for reversible processes. It never decreases.

Although entropy may decrease in part of a closed system, there will always be an equal or larger entropy increase in another part of the system, so that the entropy of the system as a whole never decreases. This fact is one form of the

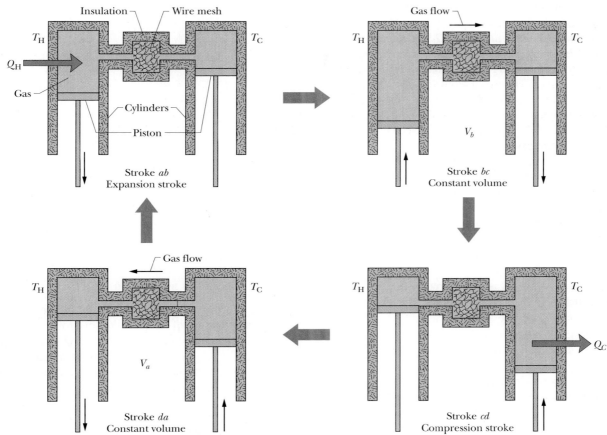

FIGURE 21-7 A schematic diagram of a Stirling engine, showing one cycle of operation, in the clockwise sequence of stroke *ab*, stroke *bc*, stroke *cd*, and then stroke *da*. Heat Q_H is added to the working substance of the engine during stroke *ab*, and heat Q_C is extracted from it during stroke *cd*. Heat is transferred to the wire mesh (to store the energy momentarily) during constant-volume stroke *bc* and then transferred back to the working substance during constant-volume stroke *da*.

second law of thermodynamics and can be written as

$$\Delta S \geq 0 \qquad \begin{array}{c}\text{(second law of}\\ \text{thermodynamics),}\end{array} \qquad (21\text{-}4)$$

where the greater-than sign applies to irreversible processes and the equals sign to reversible processes. Equation 21-4 applies only to closed systems.

In the real world almost all processes are irreversible to some extent because of friction, turbulence, and other factors. So the entropy of real closed systems undergoing real processes always increases. Processes in which the system's entropy remains constant are always idealizations.

21-4 ENTROPY IN THE REAL WORLD: ENGINES

A **heat engine,** or, more simply, an **engine,** is a device that exchanges heat with its environment and does work as it continuously repeats a set sequence of processes. As a prototype, we consider the **Stirling engine,** which was first proposed in 1816 by Robert Stirling. This engine, long neglected, is now being developed for use in automobiles and in spacecraft. A Stirling engine delivering 5000 hp (3.7 MW) has been built.

Figure 21-7 shows the elements of the Stirling engine. The left-hand cylinder is maintained at a temperature T_H by a high-temperature thermal reservoir, perhaps a propane burner. The right-hand cylinder is maintained at a lower temperature T_C by a low-temperature reservoir, perhaps circulating cooling water. (Neither reservoir is shown in Fig. 21-7.) The *working substance* of the engine is a gas. As the engine operates, the working substance moves between the cylinders, transferring heat between the reservoirs and doing work. (Do not confuse the working substance of an engine with the fuel. The function of the fuel, such as propane in a burner, is to maintain the temperature of the high-temperature reservoir.)

The two cylinders in Fig. 21-7 are connected by an insulated chamber, partially stuffed with wire mesh. The mesh stores thermal energy momentarily during the operation of the engine. The two pistons, one in each cylinder, are connected by a complex mechanical linkage that permits them to move up or down as shown by the vertical arrows in Fig. 21-7. They are also connected to a crankshaft, which is not shown.

The continuously repeated sequence of processes followed by an engine is called the *cycle* of the engine. The clockwise sequence of illustrations in Fig. 21-7 is a schematic diagram of the *Stirling cycle*. The cycle, which is also represented on the *p-V* plot in Fig. 21-8, consists of four processes, called *strokes:*

1. An isothermal expansion (stroke *ab*) at temperature T_H, produced when the left piston moves down. As a result of this expansion, heat Q_H is transferred to the gas from the left-cylinder walls, which are maintained at temperature T_H by the high-temperature reservoir.

2. A constant-volume process (stroke *bc*) in which the temperature of the gas decreases from T_H to T_C. To accomplish this, hot gas is pushed through the wire mesh as the moving pistons increase the volume of the right cylinder, and decrease that of the left cylinder, by equal amounts. As it is pushed through, the hot gas delivers heat to the wire mesh.

3. An isothermal compression (stroke *cd*) at temperature T_C back to the original volume, produced when the right piston moves up. As a result of this compression, heat Q_C is transferred from the gas to the right-cylinder walls, which are maintained at temperature T_C by the low-temperature reservoir.

4. A constant-volume process (stroke *da*) in which the temperature of the gas increases from T_C to T_H. To accomplish this, cool gas is pushed through the hot wire mesh as the volume of the left cylinder is increased and that of the right cylinder decreased, by equal amounts.

As the pistons move during a cycle, they cause the crankshaft to rotate. The continuous cycling of the engine causes a continuous rotation of the crankshaft, which can then be put to use. That is, the engine can do work.

If we strip the Stirling engine to its essentials, we can represent it with Fig. 21-9, which actually represents any engine that operates between a high-temperature reservoir at temperature T_H and a low-temperature reservoir at temperature T_C. Once every cycle of operation, heat Q_H moves from the high-temperature reservoir to the working substance, the engine does work W, and heat Q_C moves from the working substance to the low-temperature reservoir.

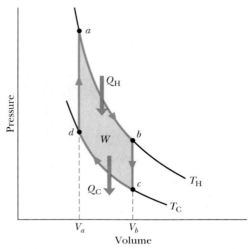

FIGURE 21-8 A *p-V* plot for a Stirling engine. The isothermal expansion *ab*, the isothermal compression *cd*, and the two constant-volume processes *bc* and *da* correspond to the four strokes of Fig. 21-7.

The **thermal efficiency** (or simply **efficiency**) of an engine is a measure of its ability to transform energy into work. For the engine of Fig. 21-9, heat Q_H is energy we pay for, perhaps as the cost of propane in a propane burner, and W is energy we get, to be used. So, it seems reasonable to define the efficiency of an engine as

$$\varepsilon = \frac{\text{energy we get}}{\text{energy we pay for}} = \frac{|W|}{|Q_H|} \quad \text{(efficiency, any engine).} \quad (21\text{-}5)$$

(We use the absolute value symbols because we want a ratio of only magnitudes.)

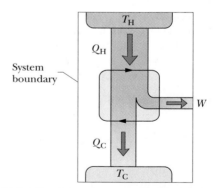

FIGURE 21-9 The elements of an engine. The two black arrowheads on the central loop suggest the working substance operating in a cycle, as if on a *p-V* plot. Heat Q_H is transferred from the high-temperature reservoir at temperature T_H to the working substance. Heat Q_C is transferred from the working substance to the low-temperature reservoir at temperature T_C. Work W is done by the engine (actually by the working substance) on something in the environment.

FIGURE 21-10 The North Anna nuclear power plant near Charlottesville, Virginia, which generates electrical energy at the rate of 900 MW. At the same time, by design, it discards energy into the nearby river at the rate of 2100 MW. This plant—and all others like it—throws away more energy than it delivers in useful form. It is a real counterpart to the ideal engine of Fig. 21-9.

Engines in the real world—*real engines*—that operate between the same high- and low-temperature reservoirs can vary in efficiency because they involve irreversible processes, with wasteful energy transfers. The wasteful transfers are often due to friction between moving parts and turbulence in the working substance. If we could make all the processes in an engine's operation reversible, eliminating wasteful energy transfers, we would have an *ideal engine*. The efficiency of this hypothetical (imaginary) engine is the upper limit for the efficiency of real engines. Let us find this limiting efficiency.

Any engine returns to its original state at the end of each cycle. This means that the internal energy of the working substance returns to its original value, and thus $\Delta E_{int} = 0$. For an ideal engine the only net energy transfers during a cycle are heat $|Q_H|$ from the high-temperature reservoir, heat $|Q_C|$ to the low-temperature reservoir, and the work $|W|$ done by the engine. The first law of thermodynamics ($\Delta E_{int} = Q - W$) then gives us

$$0 = (|Q_H| - |Q_C|) - |W|. \qquad (21\text{-}6)$$

In words, the difference between the net heat transferred per cycle $|Q_H| - |Q_C|$ and the work $|W|$ done by the engine must be zero. Solving Eq. 21-6 for $|W|$ and substituting the result into Eq. 21-5, we find that the efficiency of an ideal engine is

$$\varepsilon = \frac{|W|}{|Q_H|} = \frac{|Q_H| - |Q_C|}{|Q_H|} = 1 - \frac{|Q_C|}{|Q_H|} \qquad (21\text{-}7)$$

$$\text{(efficiency, ideal engine).}$$

We can find another expression for the efficiency of an ideal engine if we examine the entropy changes that take place during one cycle. Because all the engine processes

are reversible, Eq. 21-4 tells us that the net entropy change per cycle for the entire closed system (the working substance of the engine plus the high- and low-temperature reservoirs) must be zero. The entropy change of the working substance is zero because the substance returns to its initial state at the end of each cycle. The entropy of the high-temperature reservoir *decreases* during each cycle, because heat is extracted from it. And the entropy of the low-temperature reservoir *increases*, because heat is added to it. So, for ΔS to be zero for the entire system, the magnitude of the entropy decrease of the high-temperature reservoir must equal the magnitude of the entropy increase of the low-temperature reservoir. Writing these two entropy changes in the form Q/T (via Eq. 21-2), we have

$$\frac{|Q_H|}{T_H} = \frac{|Q_C|}{T_C},$$

or

$$\frac{|Q_C|}{|Q_H|} = \frac{T_C}{T_H}. \qquad (21\text{-}8)$$

Then combining Eq. 21-8 with Eq. 21-7 yields, for the efficiency of an ideal engine operating between temperatures T_H and T_C,

$$\varepsilon = 1 - \frac{T_C}{T_H} \qquad \text{(efficiency, ideal engine).} \qquad (21\text{-}9)$$

Note that we have not specified either the cycle or the working substance for the ideal engine. They do not matter: any *ideal* engine whose high- and low-temperature reservoirs are at T_H and T_C has the efficiency given by Eq. 21-9. Any *real* engine, with irreversible processes and wasteful energy transfers, is less efficient. If your car engine were an ideal engine, it would have an efficiency of about 55% according to Eq. 21-9; its actual efficiency is probably about 25%. A nuclear power plant (Fig. 21-10) is a much more complex engine, drawing heat from a reactor core, doing work by means of a turbine, and discharging heat to a nearby river. If the power plant were an ideal engine, its efficiency would be about 40%; its actual efficiency is about 30%.

Inventors continually try to improve engine efficiencies by reducing the energy $|Q_C|$ that is "thrown away" during each cycle. The inventor's dream is to produce the *perfect engine*, diagrammed in Fig. 21-11, in which $|Q_C|$ is reduced to zero and $|Q_H|$ is converted completely to work $|W|$. A perfect engine on an ocean liner, for example, could extract heat from the water and use it to drive the propellers, with no fuel cost. Alas, such a perfect engine is only a dream: inspection of Eq. 21-8 shows that $|Q_C|$ can be zero only if $T_C = 0$ K or $T_H \to \infty$, requirements that are impossible to meet. Instead, decades of practical en-

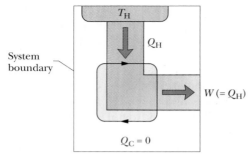

FIGURE 21-11 The elements of a perfect engine, that is, one that converts heat Q_H from a high-temperature reservoir directly to work W with 100% efficiency.

gineering experience have led to the following alternative version of the second law of thermodynamics:

> No series of processes is possible whose sole result is the absorption of heat from a thermal reservoir and the complete conversion of this energy to work.

In short, *there are no perfect engines.*

CHECKPOINT 3: Three ideal engines operate between reservoir temperatures of (a) 400 and 500 K; (b) 600 and 800 K; and (c) 400 and 600 K. Rank the engines according to their thermal efficiencies, greatest first.

SAMPLE PROBLEM 21-3

A model Stirling engine uses $n = 8.10 \times 10^{-3}$ mol of gas (assumed to be ideal) as a working substance. It operates between a high-temperature reservoir at $T_H = 95°C$ and a low-temperature reservoir at $T_C = 24°C$; the volume of its working substance doubles during the expansion stroke; and it runs at a rate of 0.70 cycle per second. Assume that the engine is ideal.

(a) How much work does the engine do per cycle?

SOLUTION: The net work per cycle is the algebraic sum of the work done by the gas during the four processes that make up the cycle shown in Figs. 21-7 and 21-8. The work done by the gas during stroke ab, an isothermal expansion from V_a to V_b, is, from Eq. 20-10,

$$W_{ab} = nRT_H \ln \frac{V_b}{V_a}.$$

The work done by the gas during stroke cd, an isothermal compression from V_b to V_a, is, similarly,

$$W_{cd} = nRT_C \ln \frac{V_a}{V_b}.$$

The work done during each of the two constant-volume processes (bc and da) is zero, so the net work for the entire cycle is

$$W = W_{ab} + W_{bc} + W_{cd} + W_{da}$$
$$= nRT_H \ln \frac{V_b}{V_a} + 0 + nRT_C \ln \frac{V_a}{V_b} + 0.$$

Because $\ln(V_a/V_b) = -\ln(V_b/V_a)$, we can write this as

$$W = nR(T_H - T_C) \ln \frac{V_b}{V_a}.$$

Substituting known data and $V_b/V_a = 2$ then yields

$$W = (8.10 \times 10^{-3} \text{ mol})(8.31 \text{ J/mol} \cdot \text{K})(95°C - 24°C)(\ln 2)$$
$$= 3.31 \text{ J} \approx 3.3 \text{ J}. \qquad \text{(Answer)}$$

Thus, during each cycle of operation, this engine does 3.3 J of work via its crankshaft.

(b) What is the power P of the engine?

SOLUTION: The power of the engine is the work W done per cycle divided by the duration t of each cycle. The engine operates at 0.70 cycle/s, so the duration of one cycle is 1/0.70 = 1.43 s. The power is then

$$P = \frac{W}{t} = \frac{3.31 \text{ J}}{1.43 \text{ s}} \approx 2.3 \text{ W}. \qquad \text{(Answer)}$$

(c) How much heat is transferred *to* the gas *from* the high-temperature reservoir during each cycle?

SOLUTION: From Table 20-5, the magnitude $|Q_H|$ of the heat transferred to an ideal gas during an isothermal expansion (process ab in Fig. 21-8) is

$$|Q_H| = nRT_H \ln \frac{V_b}{V_a}.$$

Substituting given data and temperature $T_H = (95 + 273)$ K = 368 K yields

$$|Q_H| = (8.10 \times 10^{-3} \text{ mol})(8.31 \text{ J/mol} \cdot \text{K})(368 \text{ K})(\ln 2)$$
$$= 17.2 \text{ J} \approx 17 \text{ J}. \qquad \text{(Answer)}$$

(d) What is the efficiency of the engine?

SOLUTION: Substituting the temperatures of the two reservoirs into Eq. 21-9 gives us

$$\varepsilon = 1 - \frac{T_C}{T_H} = 1 - \frac{(24 + 273) \text{ K}}{(95 + 273) \text{ K}}$$
$$= 0.193 \approx 19\%. \qquad \text{(Answer)}$$

(Equations 21-5 and 21-7 give the same answer for ε.)

(e) What are the entropy changes per cycle for the gas, the high-temperature reservoir, the low-temperature reservoir, and the wire mesh?

SOLUTION: The gas operates in a cycle, so its net entropy change per cycle ΔS_{gas} is zero. For the thermal reservoirs, heat is transferred at constant temperature, so we can use Eq. 21-2 to calculate their entropy changes.

We first take the high-temperature reservoir as our system. In (c) we found that the magnitude of the heat transferred is $|Q_H| = 17.2$ J. Because the reservoir lost the heat, we can write the transfer as $-|Q_H|$. Then from Eq. 21-2, the entropy change of the reservoir is

$$\Delta S_H = \frac{-|Q_H|}{T_H} = \frac{-17.2 \text{ J}}{(95 + 273) \text{ K}} = -0.047 \text{ J/K}. \quad \text{(Answer)}$$

The entropy of the high-temperature reservoir decreases because heat is lost.

Heat enters the low-temperature reservoir from the gas, so the entropy of that reservoir increases. If you apply Eq. 21-6 or follow the example of our calculation in (c), you will find that this reservoir gains heat $|Q_C| = 13.9$ J and undergoes an entropy change of

$$\Delta S_C = \frac{+13.9 \text{ J}}{(24 + 273) \text{ K}} = +0.047 \text{ J/K}. \quad \text{(Answer)}$$

As for the mesh, the heat it gains in stroke bc of Fig. 21-7 it loses in stroke da. And these two heat transfers occur over the same range of temperatures. So from Eq. 21-2, the net entropy change per cycle ΔS_{mesh} is zero.

The entropy change per cycle for the entire system, including the reservoirs, is then

$$\Delta S = \Delta S_H + \Delta S_C + \Delta S_{gas} + \Delta S_{mesh}$$
$$= -0.047 \text{ J/K} + 0.047 \text{ J/K} + 0 + 0$$
$$= 0. \quad \text{(Answer)}$$

This is just what the second law of thermodynamics predicts for the entropy change associated with a set of reversible processes occurring in a closed system, such as the ideal engine here.

SAMPLE PROBLEM 21-4

An inventor claims to have constructed an engine that has an efficiency of 75% when operated between the boiling and freezing points of water. We shall evaluate the credibility of the inventor's claim in two (related) ways.

(a) Compare the claimed 75% efficiency to that of an ideal engine operating between the same two temperatures.

SOLUTION: With Eq. 21-9 we obtain

$$\varepsilon = 1 - \frac{T_C}{T_H} = 1 - \frac{(0 + 273) \text{ K}}{(100 + 273) \text{ K}} = 0.268 \approx 27\%.$$

An ideal engine operating between the given temperatures would have an efficiency of 27%; any real engine (with its inevitable irreversible processes and wasteful energy transfers) must be less efficient. Thus the claimed efficiency of 75% for the given temperatures is impossible.

(b) Now, starting over, assume that the engine is good enough to be approximately an ideal engine. Then determine the en-

tropy change ΔS that occurs within the closed system consisting of the engine and its reservoirs during one cycle. Compute ΔS in terms of $|Q_H|$, the magnitude of the heat transferred each cycle from the high-temperature reservoir to the working substance.

SOLUTION: Because the working substance returns to its initial state at the end of each cycle, its entropy change ΔS_{ws} during one cycle is zero. The high-temperature reservoir loses heat $|Q_H|$ at temperature T_H, and the low-temperature reservoir gains heat $|Q_C|$ at temperature T_C. Then from Eq. 21-2, their entropy changes are

$$\Delta S_H = \frac{-|Q_H|}{T_H} \quad \text{and} \quad \Delta S_C = \frac{+|Q_C|}{T_C}.$$

The total entropy change is

$$\Delta S = \Delta S_H + \Delta S_C + \Delta S_{ws}$$
$$= -\frac{|Q_H|}{T_H} + \frac{|Q_C|}{T_C} + 0. \quad (21\text{-}10)$$

We can find a substitute for $|Q_C|$ with Eq. 21-7:

$$|Q_C| = |Q_H| (1 - \varepsilon).$$

Now, combining this with Eq. 21-10, we have, after some algebra,

$$\Delta S = |Q_H| \left(\frac{1 - \varepsilon}{T_C} - \frac{1}{T_H} \right).$$

Substituting the data gives us

$$\Delta S = |Q_H| \left(\frac{1 - 0.75}{273 \text{ K}} - \frac{1}{373 \text{ K}} \right)$$
$$= -0.0018 \, |Q_H|.$$

Thus, if the inventor's claim were correct, the entropy of the system would decrease with each cycle. This is impossible. If every process that occurs in the engine were reversible, ΔS would be zero. If some were irreversible, ΔS would be greater than zero. No set of processes in a closed system can lead to ΔS less than zero.

PROBLEM SOLVING TACTICS

TACTIC 1: *The Language of Thermodynamics*

A rich, but sometimes misleading, language is used in scientific and engineering studies of thermodynamics. And regardless of your major, you will need to understand what the language means. You may see statements that say heat is absorbed, extracted, rejected, discharged, discarded, withdrawn, delivered, gained, lost, or expelled, or that it flows from one body to another (as if it were a liquid). You may also see statements that describe a body as *having* heat (as if heat can be held or possessed), or that its heat is increased or decreased. You should always keep in mind what is meant by the term *heat*:

Heat is energy that is transferred from one body to another body owing to a difference in the temperatures of the bodies.

When we identify one of the bodies as being our system of interest, any such transfer of energy into the system is positive heat Q, and any such transfer out of the system is negative Q.

The term *work* also requires close attention. You may see statements that say work is produced or generated, or combined with heat or changed from heat. Here is what is meant by the term *work*:

Work is energy that is transferred from one body to another body owing to a force that acts between them.

When we identify one of the bodies as being our system of interest, any such transfer of energy out of the system is either positive work W done *by* the system or negative work W done *on* the system. And any such transfer of energy into the system is negative work done *by* the system or positive work done *on* the system. (The preposition that is used is important.) Obviously, this can be confusing: whenever you see the term *work*, you should read carefully to determine the intent.

21-5 ENTROPY IN THE REAL WORLD: REFRIGERATORS

A **refrigerator** is a device that uses work to transfer thermal energy from a low-temperature reservoir to a high-temperature reservoir as it continuously repeats a set sequence of processes. In a household refrigerator, for example, work is done by an electrical compressor to transfer thermal energy from the food storage compartment (a low-temperature reservoir) to the room (a high-temperature reservoir). Figure 21-12 represents the basic elements of a refrigerator. If we assume that the processes involved in the refrigerator's operation are reversible, then we have an *ideal refrigerator*. Such a hypothetical refrigerator is just a (hypothetical) ideal engine operating in reverse.

Air conditioners and heat pumps are, in essence, refrigerators. The differences are only in the nature of the high- and low-temperature reservoirs. For an air conditioner, the low-temperature reservoir is the room that is to be cooled, and the high-temperature reservoir is the outdoors. A heat pump is an air conditioner that can be operated in reverse to heat a room; then the room is the high-temperature reservoir and heat is transferred to it from outdoors.

The designer of a refrigerator would like to extract as much heat $|Q_C|$ as possible from the low-temperature res-

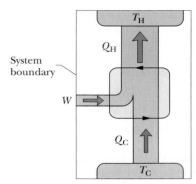

FIGURE 21-12 The elements of a refrigerator. The two black arrowheads on the central loop suggest the working substance operating in a cycle, as if on a p-V plot. Heat Q_C is transferred to the working substance from the low-temperature reservoir. Heat Q_H is transferred to the high-temperature reservoir from the working substance. Work W is done on the refrigerator (on the working substance) by something in the environment.

ervoir (what we want) for the least amount of work W (what we pay for). As a measure of the efficiency of a refrigerator, then, a reasonable definition is

$$K = \frac{\text{what we want}}{\text{what we pay for}} = \frac{|Q_C|}{|W|} \qquad \begin{array}{l}\text{(coefficient of}\\ \text{performance),}\end{array} \qquad (21\text{-}11)$$

where K is called the *coefficient of performance*. If the refrigerator is ideal, the first law of thermodynamics gives $|W| = |Q_H| - |Q_C|$, where $|Q_H|$ is the magnitude of the heat transferred to the high-temperature reservoir. Equation 21-11 then becomes

$$K = \frac{|Q_C|}{|Q_H| - |Q_C|}. \qquad (21\text{-}12)$$

Because an ideal refrigerator is an ideal engine operating in reverse, we can combine Eq. 21-8 with Eq. 21-12; after some algebra we find

$$K = \frac{T_C}{T_H - T_C} \qquad \begin{array}{l}\text{(coefficient of performance,}\\ \text{ideal refrigerator).}\end{array} \qquad (21\text{-}13)$$

For typical room air conditioners, $K \approx 2.5$. For household refrigerators, $K \approx 5$. Perversely, the value of K is higher the closer the temperatures of the two reservoirs are to each other. That is why heat pumps are more effective in temperate climates than in climates where the outside temperature varies widely.

It would be nice to own a refrigerator that did not require some input of work, that is, one that would run without being plugged in. Figure 21-13 represents another "inventor's dream," a *perfect refrigerator* that transfers heat Q from a cold reservoir to a warm reservoir without

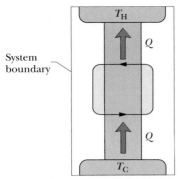

FIGURE 21-13   The elements of a perfect refrigerator, that is, one that transfers heat from a low-temperature reservoir to a high-temperature reservoir without any input of work.

the need for work. Because the unit operates in cycles, the entropy of the working substance does not change during a complete cycle. The entropies of the two reservoirs, however, do change: the entropy change for the cold reservoir is $-|Q|/T_C$, and that for the warm reservoir is $+|Q|/T_H$. So, the net entropy change for the entire system is

$$\Delta S = -\frac{|Q|}{T_C} + \frac{|Q|}{T_H}.$$

Because $T_H > T_C$, the right side of this equation is negative and thus the net change in entropy per cycle for the closed system *refrigerator + reservoirs* is also negative. Because such a decrease in entropy violates the second law of thermodynamics (Eq. 21-4), a perfect refrigerator does not exist. (If you want your refrigerator to operate, you must plug it in.)

This result leads us to another (equivalent) formulation of the second law of thermodynamics:

> No series of processes is possible whose sole result is the transfer of heat from a reservoir at a given temperature to a reservoir at a higher temperature.

In short, *there are no perfect refrigerators.*

CHECKPOINT 4: You wish to increase the coefficient of performance of an ideal refrigerator. You can do so by (a) running the cold chamber at a slightly higher temperature, (b) running it at a slightly lower temperature, (c) moving the unit to a slightly warmer room, or (d) moving it to a slightly cooler room. The temperature changes are to be the same in all four cases. List the changes according to the resulting coefficients of performance, greatest first.

SAMPLE PROBLEM 21-5

An ideal refrigerator, with coefficient of performance $K = 4.7$, extracts heat from the cold chamber at the rate of 250 J/cycle.

(a) How much work per cycle is required to operate the refrigerator?

SOLUTION: From Eq. 21-11, we find

$$|W| = \frac{|Q_C|}{K} = \frac{250 \text{ J}}{4.7} = 53 \text{ J}. \qquad \text{(Answer)}$$

(b) How much heat per cycle is discharged to the room?

SOLUTION: For an ideal refrigerator, with reversible processes and no energy transfers other than $|Q_H|$, $|Q_C|$, and $|W|$, the first law of thermodynamics tells us that

$$\Delta E_{int} = (|Q_H| - |Q_C|) - |W|.$$

Here $\Delta E_{int} = 0$ because the working substance operates in a cycle. Solving for $|Q_H|$ and inserting known data then yield

$$|Q_H| = |W| + |Q_C| = 53 \text{ J} + 250 \text{ J}$$
$$= 303 \text{ J} \approx 300 \text{ J}. \qquad \text{(Answer)}$$

21-6 A STATISTICAL VIEW OF ENTROPY

In Chapter 20 we saw that macroscopic properties of gases can be explained in terms of their microscopic, or molecular, behavior; such explanations are part of the study called *statistical mechanics*. Recall, for example, that we were able to account for the pressure exerted by a gas on the walls of its container in terms of the momentum transferred to those walls by rebounding gas molecules.

Here we shall focus our attention on a single problem involving the distribution of gas molecules between two halves of a container (a box). This problem is reasonably simple to analyze, and it allows us to use statistical mechanics to calculate the entropy change for a free expansion of an ideal gas. You will see in Sample Problem 21-6 that statistical mechanics leads to the same entropy change we obtained in Sample Problem 21-1 using thermodynamics.

Consider the box of Fig. 21-14, which contains four identical (and thus indistinguishable) molecules of a gas. At any instant, a given molecule will be in either the left half of the box or the right half; because the two halves of the box have equal volumes, the molecule has the same likelihood, or probability, of being in the left half as in the right half. Table 21-1, in which L stands for a molecule in the left half of the box and R for a molecule in the right half, shows the 16 possible ways in which the four identi-

TABLE 21-1   **FOUR MOLECULES IN A BOX**

| LOCATION OF MOLECULE | | | | CONFIGURATION LABEL | MULTIPLICITY W (NUMBER OF MICROSTATES) | CONFIGURATION PROBABILITY | CALCULATION OF W BY EQ. 21-14 | ENTROPY $(10^{-23}\,\mathrm{J/K})$ FROM EQ. 21-15 |
a	*b*	*c*	*d*					
L	L	L	L	I	1	1/16	$4!/(4!\,0!) = 1$	0.00
R	L	L	L					
L	R	L	L	II	4	4/16	$4!/(3!\,1!) = 4$	1.91
L	L	R	L					
L	L	L	R					
L	L	R	R					
L	R	L	R					
R	L	L	R	III	6	6/16	$4!/(2!\,2!) = 6$	2.47
L	R	R	L					
R	L	R	L					
R	R	L	L					
L	R	R	R					
R	L	R	R	IV	4	4/16	$4!/(1!\,3!) = 4$	1.91
R	R	L	R					
R	R	R	L					
R	R	R	R	V	1	1/16	$4!/(0!\,4!) = 1$	0.00

Total number of microstates 16

cal molecules, which we label *a*, *b*, *c*, and *d*, can be in the two halves of the box. We call each of these 16 arrangements a **microstate** of the system. (Although the four par-

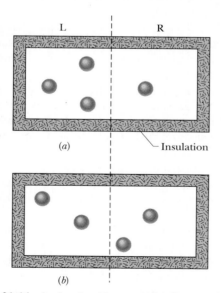

(a)

── Insulation

(b)

FIGURE 21-14   An insulated box contains four gas molecules. Each molecule has the same probability of being in the left half (L) of the box as in the right half (R). The arrangement in (a) corresponds to configuration II in Table 21-1, and that in (b) corresponds to configuration III.

ticles in Fig. 21-14 are truly indistinguishable, we assign letters to them temporarily as an aid in calculation. Our final result will not be affected by doing this.)

In Table 21-1, the 16 microstates are grouped into five **configurations,** each made up of *equivalent microstates* and labeled with a Roman numeral. Figure 21-14*a* shows configuration II, which we may describe as "one molecule in the right half of the box." Table 21-1 displays the four microstates that make up this configuration; they are equivalent because, since the molecules are identical, the situation in which molecule *a* is in the right half is physically no different from the situation in which molecule *b* (or *c* or *d*) is in the right half. Figure 21-14*b* shows configuration III, which is "two molecules in each half of the box"; Table 21-1 lists the six microstates that form this configuration. We call the number of microstates associated with a configuration the *multiplicity W* of that configuration. (We have used the same symbol for work; do not confuse these two uses.)

A basic assumption of statistical mechanics is that *all microstates are equally probable.* That is, if we were to take a great many snapshots of the molecules as they jostle around the box in random fashion, and then count the number of times each microstate occurred, we would find that all the microstates occur equally often. So, at any given instant, the probability of finding the molecules in any one microstate is equal to that of finding them in any other microstate. That is, the system will spend, on average, the same amount of time in each of the 16 microstates listed in Table 21-1.

Because the microstates are equally probable, but different configurations have *different numbers of microstates,* the configurations are *not* equally probable. The *probability (of occurrence) of a configuration* is the ratio of its multiplicity to the total number of microstates of the system. In Table 21-1, configuration III, with six microstates and thus a multiplicity of 6, is the most probable configuration, with a probability of 6/16 = 0.375. This means that the system is in configuration III 37.5% of the time. Configurations I and V, in which the gas occupies only half the box, are the least probable, with probabilities of 1/16 = 0.0625 or 6.25% each.

Now recall the free expansion process of Sample Problem 21-1; there we found that the state in which the gas fills its container has greater entropy than the state in which it occupies only half the container. Here we have found that the state (or configuration) in which the gas fills its container uniformly is more probable than a state in which the gas occupies only half the container. The state with higher entropy seems also to have the higher probability of occurrence.

Four molecules in a box ($N = 4$) are not very many on which to base a conclusion. Let us increase the number N of molecules to 100 and again compare the amount of time during which half the molecules are in each half of the box to that during which all the molecules are in the left half of the box. The ratio is not 6 to 1 (as it is for $N = 4$ in Table 21-1) but about 10^{29} to 1. Imagine what this ratio must be for the much more practical case of $N = 10^{22}$, which is

about the number of air molecules in a child's inflated balloon. The probability is then overwhelming for a uniform distribution of molecules throughout the container.

For large values of N, there are very large numbers of microstates. But nearly all the microstates correspond to an approximately equal division of the molecules between the two halves of the box, as Fig. 21-15 indicates. Even though the measured temperature and pressure of the gas remain constant, the gas is churning away endlessly as its molecules "visit" all the possible microstates with equal prob-

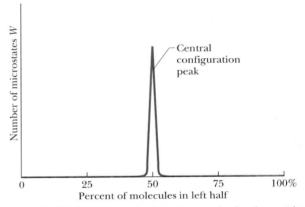

FIGURE 21-15 For a *large* number of molecules in a box, a plot of the number of microstates that require various percentages of the molecules to be in the left half of the box. Nearly all the microstates correspond to an approximately equal sharing of the molecules between the two halves of the box; those microstates form the central *configuration peak* on the plot. For $N \approx 10^{22}$, the central configuration peak would be much too narrow to be drawn on this plot.

Sometimes striking order appears naturally. (*a*) For example, Giant's Causeway in Northern Ireland consists of tall stone columns, many of which are hexagonal in cross section and appear to have been designed. The columns formed when hot magma seeped out of the ground and cooled. (*b*) Order can also be seen in

the *sorted circles* of gravel and stones that occur naturally on an island north of Norway. Just how the columns and circles were produced is still debated, but we can be certain that as the entropy of the magma and stones decreased, the entropy of their environment increased even more.

(*a*)

(*b*)

abilities. However, because so few microstates lie outside the narrow *central configuration peak* of Fig. 21-15, we might as well assume that the gas molecules are always evenly divided between the two halves of the box. And, as we saw, this is the configuration with the greatest entropy.

For the four molecules of Fig. 21-14, we can easily list the microstates associated with each of the five configurations and then find the multiplicities. For the general case of N molecules, where N can be large, we find the multiplicity of a configuration with

$$W = \frac{N!}{n_L! \, n_R!} \qquad \text{(multiplicity of a configuration).} \qquad (21\text{-}14)$$

Here N is the total number of molecules, n_L is the number in the left half of the box, and n_R is the number in the right half. The symbol "!" indicates a *factorial*, which is defined for any nonnegative integer n as

$$n! = n(n-1)(n-2) \cdots (1).$$

For example, $5! = 5 \times 4 \times 3 \times 2 \times 1 = 120$. By definition, $0! = 1$. Table 21-1 includes the calculation of W for each of its configurations, using Eq. 21-14.

We have seen that configurations with higher probability of occurrence also have higher entropy. Because the multiplicity of a configuration is a measure of its probability, we may say that states with greater multiplicity have greater entropy. Austrian physicist Ludwig Boltzmann first pointed out the following relationship between these two quantities in 1877:

$$S = k \ln W \qquad \text{(Boltzmann's entropy equation).} \qquad (21\text{-}15)$$

Here S is the entropy of a configuration, W is the multiplicity of the configuration, and k ($= 1.38 \times 10^{-23}$ J/K) is the Boltzmann constant, first encountered in Section 20-5. This famous formula is engraved on Boltzmann's tombstone.

The rightmost column of Table 21-1 displays the entropies of the configurations of the four-molecule system of Fig. 21-14, as computed with Eq. 21-15. Configuration III, which has the greatest multiplicity, also has the greatest entropy.

When you use Eq. 21-14 to calculate W, your calculator may signal an error if you try to find the factorial of a number much greater than a few hundred. Fortunately, there is a very good approximation, known as **Stirling's approximation**, not for $N!$ but for $\ln N!$—exactly what is needed in Eq. 21-15. Stirling's approximation is

$$\ln N! \approx N(\ln N) - N \qquad \text{(Stirling's approximation).} \qquad (21\text{-}16)$$

(The Stirling of this approximation is not the Stirling of the Stirling engine.)

CHECKPOINT **5:** A box contains one mole of a gas. Consider two configurations: (a) each half of the box contains half the molecules, and (b) each third of the box contains one-third of the molecules. Which configuration has more microstates?

SAMPLE PROBLEM 21-6

In Sample Problem 21-1 we showed that when n moles of an ideal gas doubles its volume in a free expansion, the entropy increase from the initial state i to the final state f is $S_f - S_i = nR \ln 2$. Derive this result with statistical mechanics.

SOLUTION: Let N be the number of molecules in n moles of the gas. From Eq. 21-14, the multiplicity of the initial state, in which all N molecules are in the left half of the container in Fig. 21-1, is

$$W_i = \frac{N!}{N! \, 0!} = 1.$$

The multiplicity of the final state, in which $N/2$ molecules are in each half of the container, is

$$W_f = \frac{N!}{(N/2)! \, (N/2)!}.$$

From Eq. 21-15, the initial and final entropies are

$$S_i = k \ln W_i = k \ln 1 = 0$$

and $$S_f = k \ln W_f = k \ln(N!) - 2k \ln[(N/2)!]. \qquad (21\text{-}17)$$

In writing Eq. 21-17, we have used the relation

$$\ln \frac{a}{b^2} = \ln a - 2 \ln b.$$

Now, applying Eq. 21-16 to evaluate Eq. 21-17, we find that

$$
\begin{aligned}
S_f &= k \ln(N!) - 2k \ln[(N/2)!] \\
&= k[N(\ln N) - N] - 2k[(N/2)\ln(N/2) - (N/2)] \\
&= k[N(\ln N) - N - N\ln(N/2) + N] \\
&= k[N(\ln N) - N(\ln N - \ln 2)] = kN \ln 2. \quad (21\text{-}18)
\end{aligned}
$$

We can substitute for kN by using the relations

$$k = \frac{R}{N_A} \quad \text{and} \quad N = nN_A$$

from Eqs. 20-21 and 20-2, respectively. Here N_A is the Avogadro constant and R is the universal gas constant. With these two relations we see that $kN = nR$. So, Eq. 21-18 becomes

$$S_f = nR \ln 2.$$

The change in entropy from the initial state to the final is thus

$$S_f - S_i = nR \ln 2 - 0 = nR \ln 2, \qquad \text{(Answer)}$$

which is what we set out to show. In Sample Problem 21-1 we calculated this entropy increase for a free expansion with thermodynamics by finding an equivalent reversible process and calculating the entropy change for *that* process in terms of temperature and heat transfer. Here we have calculated the same increase with statistical mechanics using the fact that the system consists of molecules.

REVIEW & SUMMARY

One-Way Processes

An **irreversible process** is one that cannot be reversed by means of small changes in the environment. The direction in which an irreversible process proceeds is set by the **change in entropy** ΔS of the system undergoing the process. Entropy S is a *state property* (or *state function*) of the system; that is, it depends only on the state of the system and not on how the system reached that state. The *entropy postulate* states (in part): *If an irreversible process occurs in a closed system, the entropy of the system always increases.*

Calculating Entropy Change

The **entropy change** ΔS for an irreversible process that takes a system from an initial state i to a final state f is exactly equal to the entropy change ΔS for *any reversible process* that takes the system between those same two states. We can compute the latter (but not the former) with

$$\Delta S = S_f - S_i = \int_i^f \frac{dQ}{T}. \qquad (21\text{-}1)$$

Here Q is the energy transferred as heat to or from the system during the process, and T is the temperature of the system in kelvins.

For a reversible isothermal process, Eq. 21-1 reduces to

$$\Delta S = S_f - S_i = \frac{Q}{T}. \qquad (21\text{-}2)$$

When the temperature change ΔT of a system is small relative to the temperature (in kelvins) before and after the process, the entropy change can be approximated as

$$\Delta S = S_f - S_i \approx \frac{Q}{\overline{T}}, \qquad (21\text{-}3)$$

where $\overline{T}$ is the system's average temperature during the process.

The Second Law of Thermodynamics

This law, which is an extension of the entropy postulate, states: *If a process occurs in a closed system, the entropy of the system increases for irreversible processes and remains constant for reversible processes. It never decreases.* In equation form,

$$\Delta S \geq 0. \qquad (21\text{-}4)$$

Engines

An **engine** is a device that, operating in a cycle, extracts heat $|Q_H|$ from a high-temperature reservoir, does a certain amount of work $|W|$, and discharges heat $|Q_C|$ to a low-temperature reservoir. The

efficiency ε of an engine is defined as

$$\varepsilon = \frac{\text{energy we get}}{\text{energy we pay for}} = \frac{|W|}{|Q_H|}. \qquad (21\text{-}5)$$

An *ideal engine* is one in which all processes are reversible and the only net energy transfers in a cycle are $|Q_H|$, $|W|$, and $|Q_C|$. Then the first law of thermodynamics becomes

$$\Delta E_{\text{int}} = (|Q_H| - |Q_C|) - |W| = 0. \qquad (21\text{-}6)$$

For an ideal engine, Eq. 21-5 can then be rewritten as

$$\varepsilon = 1 - \frac{|Q_C|}{|Q_H|} = 1 - \frac{T_C}{T_H}, \qquad (21\text{-}7, \ 21\text{-}9)$$

in which T_H and T_C are the temperatures of the high- and low-temperature reservoirs, respectively. Real engines always have an efficiency lower than that given by Eqs. 21-7 and 21-9.

A *perfect engine* is an imaginary engine in which heat extracted from the high-temperature reservoir is converted completely to work. Because such a conversion would decrease the entropy of the system with each cycle, a perfect engine would violate the second law of thermodynamics. That law can then be restated as follows: No series of processes is possible whose sole result is the absorption of heat from a thermal reservoir and the complete conversion of this energy to work.

Refrigerators

A **refrigerator** (which may also be an air conditioner or a heat pump) is a device that, operating in a cycle, has work $|W|$ done on it, extracts heat $|Q_C|$ from a low-temperature reservoir, and discharges heat $|Q_H|$ to a high-temperature reservoir. The *coefficient of performance K* of a refrigerator is defined as

$$K = \frac{\text{what we want}}{\text{what we pay for}} = \frac{|Q_C|}{|W|}. \qquad (21\text{-}11)$$

An *ideal refrigerator* is one in which all processes are reversible and the only net energy transfers in a cycle are $|Q_H|$, $|W|$, and $|Q_C|$. For an ideal refrigerator, Eq. 21-11 becomes

$$K = \frac{|Q_C|}{|Q_H| - |Q_C|} = \frac{T_C}{T_H - T_C}, \qquad (21\text{-}12, \ 21\text{-}13)$$

in which T_H and T_C are the temperatures of the high- and low-temperature reservoirs, respectively.

A *perfect refrigerator* is an imaginary refrigerator in which heat extracted from the low-temperature reservoir is converted completely to heat discharged to the high-temperature reservoir, without any need for work. Because this conversion would decrease the entropy of the system with each cycle, a perfect refrigerator would violate the second law of thermodynamics. That law can then be restated as follows: No series of processes is possible whose sole result is the transfer of heat from a reservoir at a given temperature to a reservoir at a higher temperature.

Entropy from a Statistical View

The entropy of a system can be defined in terms of the possible distributions of its molecules. For identical molecules, each possible distribution of molecules is called a **microstate** of the system. All equivalent microstates are grouped into a **configuration** of the system. The number of microstates in a configuration is the **multiplicity** W of the configuration.

For a system of N molecules that may be distributed between the two halves of a box, the multiplicity is given by

$$W = \frac{N!}{n_L! \, n_R!},\tag{21-14}$$

in which n_L is the number of molecules in the left half of the box and n_R is the number in the right half. A basic assumption of *statistical mechanics* is that all the microstates are equally probable. Thus, configurations with a large multiplicity occur most often. When N is very large (say, $N = 10^{22}$ molecules or more), the molecules are nearly always in the configuration in which $n_L = n_R$.

The multiplicity W of a configuration of a system and the entropy S of the system in that configuration are related by Boltzmann's entropy equation:

$$S = k \ln W,\tag{21-15}$$

where $k = 1.38 \times 10^{-23}$ J/K is the Boltzmann constant.

When N is very large (the usual case), we can approximate $\ln N!$ with *Stirling's approximation:*

$$\ln N! \approx N(\ln N) - N.\tag{21-16}$$

QUESTIONS

1. The net heat transfer between the Sun and Earth is from Sun to Earth. Does the entropy of the Earth–Sun system increase, decrease, or remain the same during this process?

2. A gas, confined to an insulated cylinder, is compressed adiabatically to half its volume. Does the entropy of the gas increase, decrease, or remain unchanged during this process?

3. A drop of water boils away *slowly* on a hot plate whose temperature is slightly above the boiling point. (a) Does the entropy of the water increase, decrease, or remain the same during this process? (b) Does the entropy of the system *water + hot plate* increase, decrease, or remain the same?

4. Point i in Fig. 21-16 represents the initial state of an ideal gas at temperature T. Taking algebraic signs into account, rank the entropy changes as the gas moves, successively and reversibly, from point i to points a, b, c, and d, greatest first.

5. You allow a gas to expand freely from volume V to volume $2V$. Later you allow that gas to expand freely from volume $2V$ to volume $3V$. Is the net entropy change for these two expansions

greater than, less than, or equal to the entropy change that would occur if you allowed the gas to expand freely from volume V directly to volume $3V$?

6. Three ideal engines operate between temperature limits of (a) 400 and 500 K; (b) 500 and 600 K; and (c) 400 and 600 K. Each engine extracts the same amount of energy per cycle from the high-temperature reservoir. Rank the magnitudes of the work done by the engines per cycle, greatest first.

7. You wish to increase the efficiency of an ideal engine. Is it better to raise the temperature of the high-temperature reservoir by a certain amount or lower the temperature of the low-temperature reservoir by the same amount?

8. The work done by an ideal engine is used to lift a heavy safe, as shown in Fig. 21-17. Does the entropy of the *engine + Earth + safe* system increase, decrease, or remain the same during this process?

FIGURE 21-16   Question 4.

FIGURE 21-17   Question 8.

9. Does the entropy per cycle increase, decrease, or remain the same for (a) an ideal engine, (b) a real engine, and (c) a perfect engine (which is, of course, impossible to build)?

10. If you leave the door of your kitchen refrigerator open for several hours, does the temperature of the kitchen increase, decrease, or remain the same? Assume that the kitchen is closed and well insulated.

11. Does the entropy per cycle increase, decrease, or remain the same for (a) an ideal refrigerator, (b) a real refrigerator, and (c) a perfect refrigerator (which is, of course, impossible to build)?

12. In an ideal refrigerator, is the magnitude $|Q_H|$ of the heat transferred per cycle to the high-temperature reservoir always

less than, always greater than, or always equal to the magnitude of the work W done per cycle?

13. A box contains 100 atoms in a configuration, with 50 atoms in each half of the box. Suppose that you could count the different microstates associated with this configuration at the rate of 100 billion states per second, using a supercomputer. Without written calculation, guess how much computing time you would need: a day, a year, or much more than a year.

14. A basic assumption of statistical mechanics is that all microstates are equally probable. How would the air in your bedroom behave if all configurations were equally probable (they aren't)?

EXERCISES & PROBLEMS

SECTION 21-2 Change in Entropy

1E. An ideal gas undergoes a reversible isothermal expansion at 132°C. The entropy of the gas increases by 46.0 J/K. How much heat was absorbed?

2E. A 2.50 mol sample of an ideal gas expands reversibly and isothermally at 360 K until its volume is doubled. What is the increase in entropy of the gas?

3E. An ideal gas undergoes a reversible isothermal expansion at 77.0°C, increasing its volume from 1.30 L to 3.40 L. The entropy change of the gas is 22.0 J/K. How many moles of gas are present?

4E. Construct plots of T versus S, S versus E_{int}, and S versus V for the isothermal expansion process of Fig. 21-4.

5E. An ideal gas in contact with a constant-temperature reservoir undergoes a reversible isothermal expansion to twice its initial volume. Show that the reservoir's change in entropy is independent of its temperature.

6E. Four moles of an ideal gas are expanded from volume V_1 to volume $V_2 = 2V_1$. If the expansion is isothermal at temperature $T = 400$ K, find (a) the work done by the expanding gas and (b) the change in its entropy. (c) If the expansion is reversibly adiabatic instead of isothermal, what is the entropy change of the gas?

7E. Find (a) the heat absorbed and (b) the change in entropy of a 2.00 kg block of copper whose temperature is increased reversibly from 25°C to 100°C. The specific heat of copper is 386 J/kg·K.

8E. An ideal monatomic gas at initial temperature T_0 (in kelvins) expands from initial volume V_0 to volume $2V_0$ by each of the five processes indicated in the T-V diagram of Fig. 21-18. In which process is the expansion (a) isothermal, (b) isobaric (constant pressure), and (c) adiabatic? Explain your answers. (d) In which processes does the entropy of the gas decrease?

9E. The temperature of 1.0 mol of a monatomic ideal gas is raised reversibly from 300 K to 400 K, with its volume kept constant. What is the entropy change of the gas? (*Hint:* Recall that $E_{int} = \frac{3}{2}nRT$ for such a gas.)

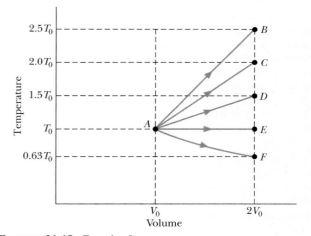

FIGURE 21-18 Exercise 8.

10E. Repeat Exercise 9, with the pressure now kept constant.

11E. (a) What is the entropy change of a 12.0 g ice cube that melts completely in a bucket of water whose temperature is just above the freezing point of water? (b) What is the entropy change of a 5.00 g spoonful of water that evaporates completely on a hot plate whose temperature is slightly above the boiling point of water?

12E. Suppose that 260 J of heat is conducted from a constant-temperature reservoir at 400 K to another reservoir at (a) 100 K, (b) 200 K, (c) 300 K, and (d) 360 K. What are the net changes in entropy of the reservoirs for those four cases? (e) What is the trend in those changes?

13E. A brass rod is in thermal contact with a constant-temperature reservoir at 130°C at one end and another reservoir at 24.0°C at the other end. (a) Compute the total change in entropy of the rod–reservoirs system that occurs when 5030 J of heat is conducted through the rod, from one reservoir to the other. (b) Does the entropy of the rod change in the process?

14E. At very low temperatures, the molar specific heat C_V of many solids is approximately $C_V = AT^3$, where A depends on the particular substance. For aluminum, $A = 3.15 \times 10^{-5}$ J/mol·K^4. Find the entropy change for 4.00 mol of aluminum when its temperature is raised from 5.00 K to 10.0 K.

15P. A 2.0 mol sample of an ideal monatomic gas undergoes the reversible process shown in Fig. 21-19. (a) How much heat is absorbed by the gas? (b) What is the change in the internal energy of the gas? (c) How much work is done by the gas?

FIGURE 21-19 Problem 15.

16P. Heat can actually be removed from water at and then below the normal freezing point (0.0°C at atmospheric pressure) without causing the water to freeze; the water is said to be *supercooled*. Suppose a 1.00 g water drop is supercooled until its temperature is that of the surrounding air, which is at −5.00°C. The drop then suddenly and irreversibly freezes, transferring heat to the air until the drop is again at −5.00°C. What is the entropy change for the drop? (*Hint:* Use a three-step reversible process to take the water through the normal freezing point.) The specific heat of ice is 2220 J/kg·K.

17P. In an experiment, 200 g of aluminum (with a specific heat of 900 J/kg·K) at 100°C is mixed with 50.0 g of water at 20.0°C. (a) What is the equilibrium temperature? What are the entropy changes of (b) the aluminum, (c) the water, and (d) the aluminum−water system?

18P. An object of (constant) heat capacity C is heated from an initial temperature T_i to a final temperature T_f by a constant-temperature reservoir at T_f. (a) Represent the process on a graph of C/T versus T, and show graphically that the total change in entropy ΔS of the object−reservoir system is positive. (b) Explain how the use of reservoirs at intermediate temperatures would allow the process to be carried out in a way that makes ΔS as small as desired.

19P. In the irreversible process of Fig. 21-5, let the initial temperatures of identical blocks L and R be 305.5 and 294.5 K, respectively, and let 215 J be the heat transfer between the blocks required to reach equilibrium. Then for the reversible processes of Fig. 21-6, what are the entropy changes of (a) block L, (b) its reservoir, (c) block R, (d) its reservoir, (e) the two-block system, and (f) the system of the two blocks and the two reservoirs?

20P. Use the reversible apparatus of Fig. 21-6 to show that, if the process of Fig. 21-5 happened in reverse, the entropy of the system would decrease, a violation of the second law of thermodynamics.

21P. The entropy change ΔS between two states of a system, which is $\int (dQ/T)$, depends only on the properties of the initial and final states. That is, this integral is path independent, as is

shown for an ideal gas in Section 21-2. Show similarly that the integrals $\int dQ$, $\int (T\, dQ)$, and $\int (dQ/T^2)$ are not path independent and thus do not define new state properties for an ideal gas.

22P. An ideal diatomic gas, whose molecules are rotating but not oscillating, is taken through the cycle in Fig. 21-20. Determine for all three processes, in terms of p_1, V_1, T_1, and R: (a) p_2, p_3, and T_3 and (b) W, Q, ΔE_{int}, and ΔS per mole.

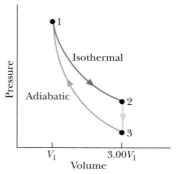

FIGURE 21-20 Problem 22.

23P. A mole of a monatomic ideal gas is taken from an initial pressure p and volume V to a final pressure $2p$ and volume $2V$ by two different processes: (I) It expands isothermally until its volume is doubled, and then its pressure is increased at constant volume to the final pressure. (II) It is compressed isothermally until its pressure is doubled, and then its volume is increased at constant pressure to the final volume. Show the path of each process on a p-V diagram. For each process calculate, in terms of p and V, (a) the heat absorbed by the gas in each part of the process, (b) the work done by the gas in each part of the process, (c) the change in internal energy of the gas, $E_{int,f} - E_{int,i}$, and (d) the change in entropy of the gas, $S_f - S_i$.

24P. A 50.0 g block of copper whose temperature is 400 K is placed in an insulating box with a 100 g block of lead whose temperature is 200 K. (a) What is the equilibrium temperature of the two-block system? (b) What is the change in the internal energy of the two-block system between the initial state and the equilibrium state? (c) What is the change in the entropy of the two-block system? See Table 19-3.

25P. A 10 g ice cube at −10°C is placed in a lake whose temperature is 15°C. Calculate the change in entropy of the cube− lake system as the ice cube comes to thermal equilibrium with the lake. The specific heat of ice is 2220 J/kg·K. (*Hint:* Will the ice cube affect the temperature of the lake?)

26P. An 8.0 g ice cube at −10°C is put into a Thermos flask containing 100 cm^3 of water at 20°C. What is the change in entropy of the cube−water system when a final equilibrium state is reached? The specific heat of ice is 2220 J/kg·K.

27P. A mixture of 1773 g of water and 227 g of ice is in an initial equilibrium state at 0.00°C. The mixture is then, in a reversible process, brought to a second equilibrium state where the water−ice ratio, by mass, is 1:1 at 0.00°C. (a) Calculate the entropy change of the system during this process. (The heat of fusion for water is 333 kJ/kg.) (b) The system is then returned to the initial equilibrium state in an irreversible process (say, by using a Bunsen burner). Calculate the entropy change of the system during this process. (c) Are your answers consistent with the second law of thermodynamics?

28P. A cylinder contains n moles of a monatomic ideal gas. If the gas undergoes a reversible isothermal expansion from initial volume V_i to final volume V_f along path I in Fig. 21-21, its change in entropy is $\Delta S = nR \ln (V_f/V_i)$. (See Sample Problem 21-1.) Now consider path II in Fig. 21-21, which takes the gas from the same initial state i to state x by a reversible adiabatic expansion, and then from state x to the same final state f by a reversible constant-volume process. (a) Describe how you would carry out the two reversible processes for path II. (b) Show that the temperature of the gas in state x is

$$T_x = T_i(V_i/V_f)^{2/3}.$$

(c) What are the heat Q_I transferred along path I and the heat Q_{II} transferred along path II? Are they equal? (d) What is the entropy change ΔS for path II? Is the entropy change for path I equal to it? (e) Evaluate T_x, Q_I, Q_{II}, and ΔS for $n = 1$, $T_i = 500$ K, and $V_f/V_i = 2$.

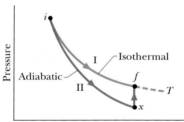

FIGURE 21-21 Problem 28.

29P. One mole of an ideal monatomic gas is taken through the cycle in Fig. 21-22. (a) How much work is done by the gas in going from state a to state c along path abc? What are the changes in internal energy and entropy in going (b) from b to c and (c) through one complete cycle? Express all answers in terms of the pressure p_0, volume V_0, and temperature T_0 of state a.

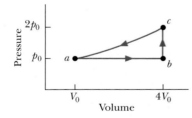

FIGURE 21-22
Problem 29.

30P. One mole of an ideal monatomic gas, at an initial pressure of 5.00 kPa and initial temperature of 600 K, expands from initial volume $V_i = 1.00$ m³ to final volume $V_f = 2.00$ m³. During the expansion, the pressure p and volume V of the gas are related by $p = 5.00 \exp[(V_i - V)/a]$, where p is in kilopascals, V_i and V are in cubic meters, and $a = 1.00$ m³. What are (a) the final pressure and (b) the final temperature of the gas? (c) How much work is done by the gas during the expansion? (d) What is the change in entropy of the gas during the expansion? (*Hint:* Use two simple reversible processes to find the entropy change.)

SECTION 21-4 Entropy in the Real World: Engines

31E. An athlete dives from a platform and comes to rest in the water; this process is irreversible, and the diver–water–Earth system is closed. Devise a reversible way of transforming the system from its initial state (diver on platform) to its final state (diver in water, with the temperature of the diver and water slightly increased due to frictional heating). You can use an ideal engine (as in Fig. 21-17) and a reservoir whose temperature can be controlled.

32E. A moving block is slowed to a stop by friction. Both the block and the surface along which it slides are in an insulated enclosure. (a) Devise a reversible process to change the system from its initial state (block moving) to its final state (block stationary; temperature of the block and surface slightly increased). Show that this reversible process results in an entropy increase for the closed block–surface system. For the process, you can use an ideal engine and a constant-temperature reservoir. (b) Show, using the same process, but in reverse, that if the temperature of the system were to decrease spontaneously and the block were to start moving again, the entropy of the system would decrease (a violation of the second law of thermodynamics).

33E. An ideal heat engine absorbs 52 kJ of heat and exhausts 36 kJ of heat in each cycle. Calculate (a) the efficiency and (b) the work done per cycle in kilojoules.

34E. Calculate the efficiency of a fossil-fuel power plant that consumes 380 metric tons of coal each hour to produce useful work at the rate of 750 MW. The heat of combustion of (the heat released by) 1.0 kg of coal is 28 MJ.

35E. An inventor claims to have invented four engines, each of which operates between constant-temperature reservoirs at 400 and 300 K. Data on each engine, per cycle of operation, are: engine A, $Q_H = 200$ J, $Q_C = -175$ J, and $W = 40$ J; engine B, $Q_H = 500$ J, $Q_C = -200$ J, and $W = 400$ J; engine C, $Q_H = 600$ J, $Q_C = -200$ J, and $W = 400$ J; engine D, $Q_H = 100$ J, $Q_C = -90$ J, and $W = 10$ J. Of the first and second laws of thermodynamics, which (if either) does each engine violate?

36E. A car engine delivers 8.2 kJ of work per cycle. (a) Before a tune-up, the efficiency is 25%. Calculate, per cycle, the heat absorbed from the combustion of fuel and the energy lost by the engine. (b) After a tune-up, the efficiency is 31%. What are the new values of the quantities calculated in (a) when 8.2 kJ of work is delivered per cycle?

37E. An ideal engine whose low-temperature reservoir is at 17°C has an efficiency of 40%. By how much should the temperature of the high-temperature reservoir be increased to increase the efficiency to 50%?

38E. An ideal heat engine operates in a Stirling cycle between 235°C and 115°C. It absorbs 6.30×10^4 J per cycle at the higher temperature. (a) What is the efficiency of the engine? (b) How much work per cycle is this engine capable of performing?

39E. In a hypothetical nuclear fusion reactor, the fuel is deuterium gas at a temperature of about 7×10^8 K. If this gas could be used to operate an ideal heat engine with $T_C = 100$°C, what would be the engine's efficiency?

40E. An ideal Stirling engine has an efficiency of 22.0%. It operates between constant-temperature reservoirs differing in temperature by 75.0 C°. What are the temperatures of the two reservoirs?

41E. Show that the heat extracted from the gas during stroke *bc* of the ideal Stirling engine (Fig. 21-8) is equal in magnitude to the heat added to the gas during stroke *da*. What is the magnitude of this heat transfer for the data given in Sample Problem 21-3? Assume that the working substance is a monatomic ideal gas.

42P. One mole of a monatomic ideal gas is taken through the reversible cycle shown in Fig. 21-23. Process *bc* is an adiabatic expansion, with $p_b = 10.0$ atm and $V_b = 1.00 \times 10^{-3}$ m³. Find (a) the heat added to the gas, (b) the heat leaving the gas, (c) the net work done by the gas, and (d) the efficiency of the cycle.

FIGURE 21-23
Problem 42.

43P. One mole of a monatomic ideal gas initially at a volume of 10 L and a temperature of 300 K is heated at constant volume to a temperature of 600 K, allowed to expand isothermally to its initial pressure, and finally compressed at constant pressure to its original volume, pressure, and temperature. (a) During the cycle, how much heat enters the system (the gas)? (b) What is the net work done by the gas? (c) What is the efficiency of the cycle?

44P. An ideal engine has a power of 500 W. It operates between constant-temperature reservoirs at 100°C and 60.0°C. What are (a) the rate of heat input and (b) the rate of exhaust heat output, in kilojoules per second?

45P. One mole of an ideal monatomic gas is taken through the cycle shown in Fig. 21-24. Assume that $p = 2p_0$, $V = 2V_0$, $p_0 = 1.01 \times 10^5$ Pa, and $V_0 = 0.0225$ m³. Calculate (a) the work done during the cycle, (b) the heat added during stroke *abc*, and (c) the efficiency of the cycle. (d) What is the efficiency of an ideal engine operating between the highest and lowest temperatures that occur in the cycle? How does this compare to the efficiency calculated in (c)?

FIGURE 21-24   Problem 45.

46P. A combination mercury–steam turbine takes saturated mercury vapor from a boiler at 876°F and exhausts it to heat a steam boiler at 460°F. The steam turbine receives steam at this temperature and exhausts it to a condenser that is at 100°F. What is the maximum theoretical efficiency of the combination?

47P. In the first stage of a two-stage ideal heat engine, heat Q_1 is absorbed at temperature T_1, work W_1 is done, and heat Q_2 is expelled at a lower temperature T_2. The second stage absorbs that heat Q_2, does work W_2, and expels heat Q_3 at a still lower temperature T_3. Prove that the efficiency of the two-stage engine is $(T_1 - T_3)/T_1$.

48P. One mole of an ideal gas is used as the working substance of an engine that operates on the cycle shown in Fig. 21-25. *BC* and *DA* are reversible adiabatic processes. (a) Is the gas monatomic, diatomic, or polyatomic? (b) What is the efficiency of the engine?

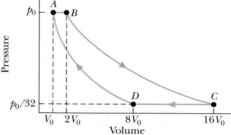

FIGURE 21-25
Problem 48.

49P. Suppose that a deep shaft were drilled in Earth's crust near one of the poles, where the surface temperature is −40°C, to a depth where the temperature is 800°C. (a) What is the theoretical limit to the efficiency of an engine operating between these temperatures? (b) If all the heat released into the low-temperature reservoir were used to melt ice that was initially at −40°C, at what rate could liquid water at 0°C be produced by a 100 MW power plant? The specific heat of ice is 2220 J/kg·K; water's heat of fusion is 333 kJ/kg. (Note that the engine can operate only between 0°C and 800°C in this case. Energy exhausted at −40°C cannot be used to raise the temperature of anything above −40°C.)

FIGURE 21-26
Problem 50.

50P. The operation of a gasoline internal combustion engine is represented by the cycle in Fig. 21-26. Assume an ideal gas and use a compression ratio of 4 : 1 ($V_4 = 4V_1$). Assume that $p_2 = 3p_1$. (a) Determine the pressure and temperature at each of the vertex points of the *p-V* diagram in terms of p_1, T_1, and the ratio γ of the molar specific heats of the gas. (b) What is the efficiency of the cycle?

SECTION 21-5 Entropy in the Real World: Refrigerators

51E. An inventor has built an engine (engine X) and claims that its efficiency ε_X is greater than the efficiency ε of an ideal engine operating between the same two temperatures. Suppose that you couple engine X to an ideal refrigerator (Fig. 21-27a) and adjust the cycle of engine X so that the work per cycle that it provides equals the work per cycle required by the ideal refrigerator. Treat this combination as a single unit and show that if the inventor's claim were true (if $\varepsilon_X > \varepsilon$), the combined unit would act as a perfect refrigerator (Fig. 21-27b), transferring heat from the low-temperature reservoir to the high-temperature reservoir without the need of work.

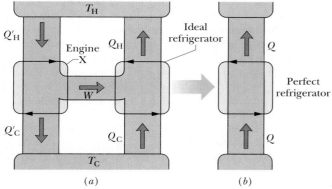

(a)   *(b)*

FIGURE 21-27 Exercise 51.

52E. An apparatus that liquefies helium is in a room maintained at 300 K. If the helium in the apparatus is at 4.0 K, what is the minimum ratio of heat delivered to the room to heat removed from the helium?

53E. An ideal engine has efficiency ε. Show that if you run it backward as an ideal refrigerator, the coefficient of performance will be $K = (1 - \varepsilon)/\varepsilon$.

54E. An ideal refrigerator does 150 J of work to remove 560 J of heat from its cold compartment. (a) What is the refrigerator's coefficient of performance? (b) How much heat per cycle is exhausted to the kitchen?

55E. To make ice, an ideal freezer extracts 42 kJ of heat at $-12°C$ during each cycle. The freezer has a coefficient of performance of 5.7. The room temperature is 26°C. (a) How much heat per cycle is delivered to the room? (b) How much work per cycle is required to run the freezer?

56E. (a) An ideal engine operates between a hot reservoir at 320 K and a cold reservoir at 260 K. If it absorbs 500 J of heat per cycle at the hot reservoir, how much work per cycle does it deliver? (b) If the same engine, working in reverse, functions as a refrigerator between the same two reservoirs, how much work per cycle must be supplied to remove 1000 J of heat from the cold reservoir?

57E. An ideal air conditioner takes heat from a room at 70°F and transfers it to the outdoors, which is at 96°F. For each joule of

electrical energy required to operate the air conditioner, how many joules of heat are removed from the room?

58E. Heat from the outdoors at $-5.0°C$ is transferred to a room at 17°C by the electric motor of a heat pump. If the heat pump is ideal, how many joules of heat will be delivered to the room for each joule of electric energy consumed?

59E. How much work must be done by an ideal refrigerator to transfer 1.0 J of heat (a) from a reservoir at 7.0°C to one at 27°C; (b) from a reservoir at $-73°C$ to one at 27°C; (c) from a reservoir at $-173°C$ to one at 27°C; and (d) from a reservoir at $-223°C$ to one at 27°C?

60P. An ideal heat pump is used to heat a building. The outside temperature is $-5.0°C$ and the temperature inside the building is to be maintained at 22°C. The coefficient of performance is 3.8, and the heat pump delivers 7.54 MJ of heat to the building each hour. At what rate must work be done to run the heat pump?

61P. The motor in a refrigerator has a power of 200 W. If the freezing compartment is at 270 K and the outside air is at 300 K, assuming ideal efficiency, what is the maximum amount of heat that can be extracted from the freezing compartment in 10.0 min?

62P. An air conditioner operating between 93°F and 70°F is rated at 4000 Btu/h cooling capacity. Its coefficient of performance is 27% of that of an ideal refrigerator operating between the same two temperatures. What horsepower is required of the air conditioner motor?

63P. An ideal engine works between temperatures T_1 and T_2. It drives an ideal refrigerator that works between temperatures T_3 and T_4 (Fig. 21-28). Find the ratio Q_3/Q_1 in terms of T_1, T_2, T_3, and T_4.

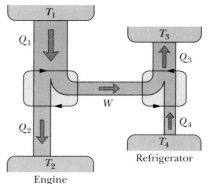

FIGURE 21-28 Problem 63.

SECTION 21-6 A Statistical View of Entropy

64E. Construct a table like Table 21-1 for six molecules rather than four.

65E. Show that for N molecules in a box, the number of possible microstates is 2^N when microstates are defined by whether a given molecule is in the left half of the box or the right half. Check this for the situation of Table 21-1.

66E. Check your calculator against Stirling's approximation, with N equal to (a) 50, (b) 100, and (c) 250. Is there a trend?

67P. Suppose you put 100 pennies in a cup, shake the cup, and then toss the pennies onto the floor. (a) How many different head–tail arrangements (microstates) are possible for the 100 pennies? (*Hint:* See Exercise 65.) What is the probability of finding exactly (b) 50 heads, (c) 48 heads, (d) 48 tails, (e) 40 heads, and (f) 30 heads?

68P. A box contains N gas molecules, equally divided between its two halves. For $N = 50$: (a) What is the multiplicity of this central configuration? (b) What is the total number of microstates for the system? (*Hint:* See Exercise 65.) (c) What percentage of the time does the system spend in its central configuration? (d) Repeat (a) through (c) for $N = 100$. (e) Repeat (a) through (c) for $N = 200$. (f) As N increases, you will find that the system spends *less* time (not more) in its central configuration. Explain why this is so.

69P. A box contains N gas molecules. Consider the box to be divided into three equal parts. (a) By extension of Eq. 21-14, write a formula for the multiplicity of any given configuration. (b) Consider two configurations: configuration A with equal numbers of molecules in all three thirds of the box, and configuration B with equal numbers of molecules in both halves of the box. What is the ratio W_A/W_B of the multiplicity of configuration A to that of configuration B? (c) Evaluate W_A/W_B for $N = 100$. (Because 100 is not evenly divisible by 3, put 34 molecules into one of the three box parts and 33 in each of the other parts for configuration A.)

70P. A box contains N molecules. Consider two configurations: configuration A with an equal division of the molecules between the two halves of the box, and configuration B with 60.0% of the molecules in the left half of the box and 40.0% in the right half. For $N = 50$, what are the multiplicities of (a) configuration A and (b) configuration B? (c) What is the ratio f of the time the system spends in configuration B to that in configuration A? (d) Repeat (a) through (c) for $N = 100$. (e) Repeat (a) through (c) for $N = 200$. (f) With an increase in N, does f increase, decrease, or remain the same?

Electronic Computation

71. The temperature of 25 mol of an ideal diatomic gas, originally 300 K, is increased by placing the gas in contact with a thermal reservoir at 800 K. The pressure remains constant throughout the increase. (a) Calculate the change in the entropy of the gas, the change in the entropy of the reservoir, and the change in the total entropy of the gas-reservoir system. (b) Suppose the same change in state is carried out by placing the gas in contact successively with n intermediate reservoirs, equally spaced in temperature between 300 K and 800 K. Calculate the total change in the entropy of the reservoirs and the change in the entropy of the closed system consisting of the gas and the reservoirs for (b) $n = 1$, (c) $n = 10$, (d) $n = 50$, and (e) $n = 100$. With each increase in n, the process more closely approximates a reversible process.

Boiling and the Leidenfrost Effect

JEARL WALKER
Cleveland State University

How does water boil? As commonplace as the event is, you may not have noticed all of its curious features. Some of the features are important in industrial applications, while others appear to be the basis for certain dangerous stunts once performed by daredevils in carnival sideshows.

Arrange for a pan of tap water to be heated from below by a flame or electric heat source. As the water warms, air molecules are driven out of solution in the water, collecting as tiny bubbles in crevices along the bottom of the pan (Fig. 1a). The air bubbles gradually inflate, and then they begin to pinch off from the crevices and rise to the top surface of the water (Fig. 1b–f). As they leave, more air bubbles form in the crevices and pinch off, until the supply of air in the water is depleted. The formation of air bubbles is a sign that the water is heating but has nothing to do with boiling.

Water that is directly exposed to the atmosphere boils at what is sometimes called its normal boiling temperature T_s. For example, T_s is about 100°C when the air pressure is 1 atm. Since the water at the bottom of your pan is not directly exposed to the atmosphere, it remains liquid even when it *superheats* above T_s by as much as a few degrees. During this process, the water is constantly mixed by convection as hot water rises and cooler water descends.

If you continue to increase the pan's temperature, the bottom layer of water begins to vaporize, with water mole-

cules gathering in small vapor bubbles in the now dry crevices, as the air bubbles do in Fig. 1. This phase of boiling is signaled by pops, pings, and eventually buzzing. The water almost sings its displeasure at being heated. Every time a vapor bubble expands upward into slightly cooler water, the bubble suddenly collapses because the vapor within it condenses. Each collapse sends out a sound wave, the ping you hear. Once the temperature of the bulk water increases, the bubbles may not collapse until after they pinch off from the crevices and ascend part of the way to the top surface of the water. This phase of boiling is labeled "isolated vapor bubbles" in Fig. 2.

If you still increase the pan's temperature, the clamor of collapsing bubbles first grows louder and then disappears. The noise begins to soften when the bulk liquid is sufficiently hot that the vapor bubbles reach the top surface of the water. There they pop open with a light splash. The water is now in full boil.

If your heat source is a kitchen stove, the story stops at this point. However, with a laboratory burner you can continue to increase the pan's temperature. The vapor bubbles

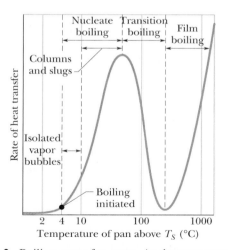

FIGURE 2 Boiling curve for water. As the temperature at the bottom of the pan is increased above the normal boiling point, the rate at which energy is transferred from the pan bottom to the water increases at first. However, above a certain temperature, the transfer almost disappears. At even higher temperatures, the transfer reappears.

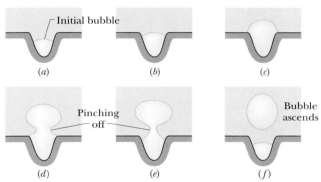

FIGURE 1 (*a*) A bubble forms in the crevice of a scratch along the bottom of a pan of water. (*b-f*) The bubble grows, pinches off, and then ascends through the water.

next become so abundant and pinch off from their crevices so frequently that they coalesce, forming columns of vapor that violently and chaotically churn upward, sometimes meeting previously detached ''slugs'' of vapor.

The production of vapor bubbles and columns is called *nucleate boiling* because the formation and growth of the bubbles depend on crevices serving as *nucleating sites* (sites of formation). Whenever you increase the pan's temperature, the rate at which heat is transferred to the water increases. If you continue to raise the pan's temperature past the stage of columns and slugs, the boiling enters a new phase called the *transition regime*. Then each increase in the pan's temperature reduces the rate at which heat is transferred to the water. The decrease is not paradoxical. In the transition regime, much of the bottom of the pan is covered by a layer of vapor. Since water vapor conducts heat about an order of magnitude more poorly than does liquid water, the transfer of heat to the water is diminished. The hotter the pan becomes, the less direct contact the water has with it and the worse the transfer of heat becomes. This situation can be dangerous in a *heat exchanger,* whose purpose is to transfer heat from a heated object. If the water in the heat exchanger is allowed to enter the transition regime, the object may destructively overheat because of diminished transfer of heat from it.

Suppose you continue to increase the temperature of the pan. Eventually, the whole of the bottom surface is covered with vapor. Then heat is slowly transferred to the liquid above the vapor by radiation and gradual conduction. This phase is called *film boiling.*

Although you cannot obtain film boiling in a pan of water on a kitchen stove, it is still commonplace in the kitchen. My grandmother once demonstrated how it serves to indicate when her skillet is hot enough for pancake batter. After she heated the empty skillet for a while, she sprinkled a few drops of water into it. The drops sizzled away within seconds. Their rapid disappearance warned her that the skillet was insufficiently hot for the batter. After further heating the skillet, she repeated her test with a few more sprinkled water drops. This time they beaded up and danced over the metal, lasting well over a minute before they disappeared. The skillet was then hot enough for my grandmother's batter.

To study her demonstration, I arranged for a flat metal plate to be heated by a laboratory burner. While monitoring the temperature of the plate with a thermocouple, I carefully released a drop of distilled water from a syringe held just above the plate. The drop fell into a dent I had made in the plate with a ball-peen hammer. The syringe allowed me to release drops of uniform size. Once a drop was released, I timed how long it survived on the plate. Afterward, I plotted the survival times of the drops versus the plate

temperature (Fig. 3). The graph has a curious peak. When the plate temperature was between 100 and about 200°C, each drop spread over the plate in a thin layer and rapidly vaporized. When the plate temperature was about 200°C, a drop deposited on the plate beaded up and survived for over a minute. At even higher plate temperatures, the water beads did not survive quite as long. Similar experiments with tap water generated a graph with a flatter peak, probably because suspended particles of impurities in the drops breached the vapor layer, conducting heat into the drops.

The fact that a water drop is long lived when deposited on metal that is much hotter than the boiling temperature of water was first reported by Hermann Boerhaave in 1732. It was not investigated extensively until 1756 when Johann Gottlob Leidenfrost published ''A Tract About Some Qualities of Common Water.'' Because Leidenfrost's work was not translated from the Latin until 1965, it was not widely read. Still, his name is now associated with the phenomenon. In addition, the temperature corresponding to the peak in a graph such as I made is called the Leidenfrost point.

Leidenfrost conducted his experiments with an iron spoon that was heated red-hot in a fireplace. After placing a drop of water into the spoon, he timed its duration by the swings of a pendulum. He noted that the drop seemed to suck the light and heat form the spoon, leaving a spot duller than the rest of the spoon. The first drop deposited in the spoon lasted 30 s while the next drop lasted only 10 s. Additional drops lasted only a few seconds.

Leidenfrost misunderstood his demonstrations because he did not realize that the longer-lasting drops were actually boiling. Let me explain in terms of my experi-

FIGURE 3   Drop lifetimes on a hot plate. Strangely, in a certain temperature range, the drops last longer when the hot plate is hotter.

ments. When the temperature of the plate is less than the Leidenfrost point, the water spreads over the plate and rapidly conducts heat from it, resulting in complete vaporization within seconds. When the temperature is at or above the Leidenfrost point, the bottom surface of a drop deposited on the plate almost immediately vaporizes. The gas pressure from this vapor layer prevents the rest of the drop from touching the plate (Fig. 4). The layer thus protects and supports the drop for the next minute or so. The layer is constantly replenished as additional water vaporizes from the bottom surface of the drop because of heat radiated and conducted through the layer from the plate. Although the layer is less than 0.1 mm thick near its outer boundary and only about 0.2 mm thick at its center, it dramatically slows the vaporization of the drop.

After reading the translation of Leidenfrost's research, I happened upon a description of a curious stunt that was performed in the sideshows of carnivals around the turn of the century. Reportedly, a performer was able to dip wet fingers into molten lead. Assuming that the stunt involved no trickery, I conjectured that it must depend on the Leidenfrost effect. As soon as the performer's wet flesh touched the hot liquid metal, part of the water vaporized, coating the fingers with a vapor layer. If the dip was brief, the flesh would not be heated significantly.

I could not resist the temptation to test my explanation. With a laboratory burner, I melted down a sizable slab of lead in a crucible. I heated the lead until its temperature was over 400°C, well above its melting temperature of 328°C. After wetting a finger in tap water, I prepared to touch the top surface of the molten lead. I must confess that I had an assistant standing ready with first-aid materials. I must also confess that my first several attempts failed because my brain refused to allow this ridiculous experiment, always directing my finger to miss the lead.

When I finally overcame my fears and briefly touched the lead, I was amazed. I felt no heat. Just as I had guessed, part of the water on the finger vaporized, forming a protective layer. Since the contact was brief, radiation and conduction of heat through the vapor layer were insufficient to raise perceptibly the temperature of my flesh. I grew braver. After wetting my hand, I dipped all my fingers into

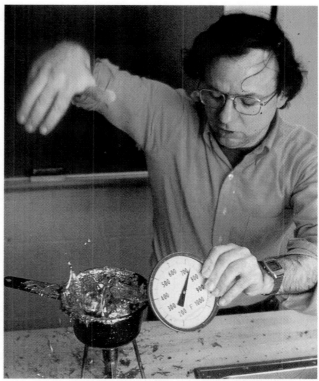

FIGURE 5 Walker demonstrating the Leidenfrost effect with molten lead. He has just plunged his fingers into the lead, touching the bottom of the pan. The temperature of the lead is given in degrees Fahrenheit on the industrial thermometer.

the lead, touching the bottom of the container (Fig. 5). The contact with the lead was still too brief to result in a burn. Apparently, the Leidenfrost effect, or more exactly, the immediate presence of film boiling, protected my fingers.

I still questioned my explanation. Could I possibly touch the lead with a dry finger without suffering a burn? Leaving aside all rational thought, I tried it, immediately realizing my folly when pain raced through the finger. Later, I tested a dry weiner, forcing it into the molten lead for several seconds. The skin of the weiner quickly blackened. It lacked the protection of film boiling just as my dry finger had.

I must caution that dipping fingers into molten lead presents several serious dangers. If the lead is only slightly above its melting point, the loss of heat from it when the water is vaporized may solidify the lead around the fingers. If I were to pull the resulting glove of hot, solid lead up from the container, it will be in contact with my fingers so long that my fingers are certain to be badly burned. I must also contend with the possibility of splashing and spillage. In addition, there is the acute danger of having too much water on the fingers. When the surplus water rapidly va-

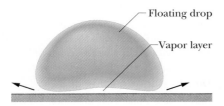

FIGURE 4 A Leidenfrost drop in cross section.

Floating drop

Vapor layer

porizes, it can blow molten lead over the surroundings and, most seriously, into the eyes. I have been scarred on my arms and face from such explosive vaporizations. *You should never repeat this demonstration.*

Film boiling can also be seen when liquid nitrogen is spilled. The drops and globs bead up as they skate over the floor. The liquid is at a temperature of about $-200°C$. When the spilled liquid nears the floor, its bottom surface vaporizes. The vapor layer then provides support for the rest of the liquid, allowing the liquid to survive for a surprisingly long time.

I was told of a stunt where a performer poured liquid nitrogen into his mouth without being hurt by its extreme cold. The liquid immediately underwent film boiling on its bottom surface and thus did not directly touch the tongue. Foolishly, I repeated this demonstration. For several dozen times the stunt went smoothly and dramatically. With a large glob of liquid nitrogen in my mouth, I concentrated on not swallowing while I breathed outward. The moisture in my cold breath condensed, creating a terrific plume that extended about a meter from my mouth. However, on my last attempt the liquid thermally contracted two of my front teeth so severely that the enamel ruptured into a "road map" of fissures. My dentist convinced me to drop the demonstration.

The Leidenfrost effect may also play a role in another foolhardy demonstration: walking over hot coals. At times the news media have carried reports of a performer striding over red-hot coals with much hoopla and mystic nonsense, perhaps claiming that protection from a bad burn is afforded by "mind over matter." Actually, physics protects the feet when the walk is successful. Particularly important is the fact that although the surface of the coals is quite hot, it contains surprisingly little energy. If the performer walks at a moderate pace, a footfall is so brief that the foot conducts little energy from the coals. Of course, a slower walk invites a burn because the longer contact allows heat to be conducted to the foot from the interior of the coals.

If the feet are wet prior to the walk, the liquid might also help protect them. To wet the feet a performer might walk over wet grass just before reaching the hot coals. Instead, the feet might just be sweaty because of the heat from the coals or the excitement of the performance. Once the performer is on the coals, some of the heat vaporizes the liquid on the feet, leaving less heat to be conducted to the flesh. In addition, there may be points of contact where the liquid undergoes film boiling, thereby providing brief protection from the coals.

I have walked over hot coals on five occasions. For four of the walks I was fearful enough that my feet were sweaty. However, on the fifth walk I took my safety so much for granted that my feet were dry. The burns I suf-

fered then were extensive and terribly painful. My feet did not heal for weeks.

My failure may have been due to a lack of film boiling on the feet, but I had also neglected an additional safety factor. On the other days I had taken the precaution of clutching an earlier edition of this book to my chest during the walks so as to bolster my belief in physics. Alas, I forgot the book on the day when I was so badly burned.

I have long argued that degree-granting programs should employ "fire-walking" as a last exam. The chairperson of the program should wait on the far side of a bed of red-hot coals while a degree candidate is forced to walk over the coals. If the candidate's belief in physics is strong enough that the feet are left undamaged, the chairperson hands the candidate a graduation certificate. The test would be more revealing than traditional final exams.

REFERENCES

Leidenfrost, Johann Gottlob, "On the Fixation of Water in Diverse Fire," *International Journal of Heat and Mass Transfer,* Vol. 9, pages 1153–1166 (1966).

Gottfried, B. S., C. J. Lee, and K. J. Bell, "The Leidenfrost Phenomenon: Film Boiling of Liquid Droplets on a Flat Plate," *International Journal of Heat and Mass Transfer,* Vol. 9, pages 1167–1187 (1966).

Hall, R. S., S. J. Board, A. J. Clare, R. B. Duffey, T. S. Playle, and D. H. Poole, "Inverse Leidenfrost Phenomenon," *Nature,* Vol. 224, pages 266–267 (1969).

Walker, Jearl, "The Amateur Scientist," *Scientific American,* Vol. 237, pages 126–131 + 140 (August 1977).

Curzon, F. L. "The Leidenfrost Phenomenon," *American Journal of Physics,* Vol. 46, pages 825–828 (1978).

Leikind, Bernard J., and William J. McCarthy, "An Investigation of Firewalking," *Skeptical Inquirer,* Vol. 10, No. 1, pages 23–34 (Fall 1985).

Bent, Henry A., "Droplet on a Hot Metal Spoon," *American Journal of Physics,* Vol. 54, page 967 (1986).

Leikind, B. J., and W. J. McCarthy, "Firewalking," *Experientia,* Vol. 44, pages 310–315 (1988).

Thimbleby, Harold, "The Leidenfrost Phenomenon," *Physics Education,* Vol. 24, pages 300–303 (1989).

Taylor, John R., "Firewalking: A Lesson in Physics," *The Physics Teacher,* Vol. 27, pages 166–168 (March 1989).

Zhang, S., and G. Gogos, "Film Evaporation of a Spherical Droplet over a Hot Surface: Fluid Mechanics and Heat/Mass Transfer Analysis," *Journal of Fluid Mechanics,* Vol. 222, pages 543–563 (1991).

Agrawal, D. C., and V. J. Menon, "Boiling and the Leidenfrost Effect in a Gravity-free Zone: A Speculation," *Physics Education,* Vol. 29, pages 39–42 (1994).

Appendix A
The International System of Units (SI)*

1. THE SI BASE UNITS

QUANTITY	NAME	SYMBOL	DEFINITION
length	meter	m	". . . the length of the path traveled by light in vacuum in 1/299,792,458 of a second." (1983)
mass	kilogram	kg	". . . this prototype [a certain platinum–iridium cylinder] shall henceforth be considered to be the unit of mass." (1889)
time	second	s	". . . the duration of 9,192,631,770 periods of the radiation corresponding to the transition between the two hyperfine levels of the ground state of the cesium-133 atom." (1967)
electric current	ampere	A	". . . that constant current which, if maintained in two straight parallel conductors of infinite length, of negligible circular cross section, and placed 1 meter apart in vacuum, would produce between these conductors a force equal to 2×10^{-7} newton per meter of length." (1946)
thermodynamic temperature	kelvin	K	". . . the fraction 1/273.16 of the thermodynamic temperature of the triple point of water." (1967)
amount of substance	mole	mol	". . . the amount of substance of a system which contains as many elementary entities as there are atoms in 0.012 kilogram of carbon-12." (1971)
luminous intensity	candela	cd	". . . the luminous intensity, in the perpendicular direction, of a surface of 1/600,000 square meter of a blackbody at the temperature of freezing platinum under a pressure of 101.325 newtons per square meter." (1967)

*Adapted from "The International System of Units (SI)," National Bureau of Standards Special Publication 330, 1972 edition. The definitions above were adopted by the General Conference of Weights and Measures, an international body, on the dates shown. In this book we do not use the candela.

2. SOME SI DERIVED UNITS

QUANTITY	NAME OF UNIT	SYMBOL	
area	square meter	m^2	
volume	cubic meter	m^3	
frequency	hertz	Hz	s^{-1}
mass density (density)	kilogram per cubic meter	kg/m^3	
speed, velocity	meter per second	m/s	
angular velocity	radian per second	rad/s	
acceleration	meter per second per second	m/s^2	
angular acceleration	radian per second per second	rad/s^2	
force	newton	N	$kg \cdot m/s^2$
pressure	pascal	Pa	N/m^2
work, energy, quantity of heat	joule	J	$N \cdot m$
power	watt	W	J/s
quantity of electric charge	coulomb	C	$A \cdot s$
potential difference, electromotive force	volt	V	W/A
electric field strength	volt per meter (or newton per coulomb)	V/m	N/C
electric resistance	ohm	Ω	V/A
capacitance	farad	F	$A \cdot s/V$
magnetic flux	weber	Wb	$V \cdot s$
inductance	henry	H	$V \cdot s/A$
magnetic flux density	tesla	T	Wb/m^2
magnetic field strength	ampere per meter	A/m	
entropy	joule per kelvin	J/K	
specific heat	joule per kilogram kelvin	$J/(kg \cdot K)$	
thermal conductivity	watt per meter kelvin	$W/(m \cdot K)$	
radiant intensity	watt per steradian	W/sr	

3. THE SI SUPPLEMENTARY UNITS

QUANTITY	NAME OF UNIT	SYMBOL
plane angle	radian	rad
solid angle	steradian	sr

Appendix **B**
Some Fundamental Constants of Physics *

CONSTANT	SYMBOL	COMPUTATIONAL VALUE	BEST (1986) VALUE	
			VALUE[a]	UNCERTAINTY[b]
Speed of light in a vacuum	c	3.00×10^8 m/s	2.99792458	exact
Elementary charge	e	1.60×10^{-19} C	1.60217733	0.30
Gravitational constant	G	6.67×10^{-11} m³/s²·kg	6.67259	128
Universal gas constant	R	8.31 J/mol·K	8.314510	8.4
Avogadro constant	N_A	6.02×10^{23} mol⁻¹	6.0221367	0.59
Boltzmann constant	k	1.38×10^{-23} J/K	1.380658	8.5
Stefan-Boltzmann constant	σ	5.67×10^{-8} W/m²·K⁴	5.67051	34
Molar volume of ideal gas at STP[d]	V_m	2.24×10^{-2} m³/mol	2.241409	8.4
Permittivity constant	ϵ_0	8.85×10^{-12} F/m	8.85418781762	exact
Permeability constant	μ_0	1.26×10^{-6} H/m	1.25663706143	exact
Planck constant	h	6.63×10^{-34} J·s	6.6260755	0.60
Electron mass[c]	m_e	9.11×10^{-31} kg	9.1093897	0.59
		5.49×10^{-4} u	5.48579903	0.023
Proton mass[c]	m_p	1.67×10^{-27} kg	1.6726231	0.59
		1.0073 u	1.0072764660	0.005
Ratio of proton mass to electron mass	m_p/m_e	1840	1836.152701	0.020
Electron charge-to-mass ratio	e/m_e	1.76×10^{11} C/kg	1.75881961	0.30
Neutron mass[c]	m_n	1.68×10^{-27} kg	1.6749286	0.59
		1.0087 u	1.0086649235	0.0023
Hydrogen atom mass[c]	m_{1_H}	1.0078 u	1.0078250316	0.0005
Deuterium atom mass[c]	m_{2_H}	2.0141 u	2.0141017779	0.0005
Helium atom mass[c]	$m_{4_{He}}$	4.0026 u	4.0026032	0.067
Muon mass	m_μ	1.88×10^{-28} kg	1.8835326	0.61
Electron magnetic moment	μ_e	9.28×10^{-24} J/T	9.2847701	0.34
Proton magnetic moment	μ_p	1.41×10^{-26} J/T	1.41060761	0.34
Bohr magneton	μ_B	9.27×10^{-24} J/T	9.2740154	0.34
Nuclear magneton	μ_N	5.05×10^{-27} J/T	5.0507866	0.34
Bohr radius	r_B	5.29×10^{-11} m	5.29177249	0.045
Rydberg constant	R	1.10×10^7 m⁻¹	1.0973731534	0.0012
Electron Compton wavelength	λ_C	2.43×10^{-12} m	2.42631058	0.089

[a]Values given in this column should be given the same unit and power of 10 as the computational value.   [b]Parts per million.   [c]Masses given in u are in unified atomic mass units, where 1 u = $1.6605402 \times 10^{-27}$ kg.   [d]STP means standard temperature and pressure: 0°C and 1.0 atm (0.1 MPa).

*The values in this table were largely selected from a longer list in *Symbols, Units and Nomenclature in Physics* (IUPAP), prepared by E. Richard Cohen and Pierre Giacomo, 1986.

SOME DISTANCES FROM THE EARTH

To the moon*	3.82×10^8 m
To the sun*	1.50×10^{11} m
To the nearest star (Proxima Centauri)	4.04×10^{16} m
To the center of our galaxy	2.2×10^{20} m
To the Andromeda Galaxy	2.1×10^{22} m
To the edge of the observable universe	$\sim 10^{26}$ m

*Mean distance.

THE SUN, THE EARTH, AND THE MOON

PROPERTY	UNIT	SUN	EARTH	MOON
Mass	kg	1.99×10^{30}	5.98×10^{24}	7.36×10^{22}
Mean radius	m	6.96×10^8	6.37×10^6	1.74×10^6
Mean density	kg/m³	1410	5520	3340
Free-fall acceleration at the surface	m/s²	274	9.81	1.67
Escape velocity	km/s	618	11.2	2.38
Period of rotation[a]	—	37 d at poles[b]	23 h 56 min	27.3 d
		26 d at equator[b]		
Radiation power[c]	W	3.90×10^{26}		

[a]Measured with respect to the distant stars.

[b]The sun, a ball of gas, does not rotate as a rigid body.

[c]Just outside the Earth's atmosphere solar energy is received, assuming normal incidence, at the rate of 1340 W/m².

SOME PROPERTIES OF THE PLANETS

	MERCURY	VENUS	EARTH	MARS	JUPITER	SATURN	URANUS	NEPTUNE	PLUTO
Mean distance from sun, 10^6 km	57.9	108	150	228	778	1430	2870	4500	5900
Period of revolution, y	0.241	0.615	1.00	1.88	11.9	29.5	84.0	165	248
Period of rotation,[a] d	58.7	-243^b	0.997	1.03	0.409	0.426	-0.451^b	0.658	6.39
Orbital speed, km/s	47.9	35.0	29.8	24.1	13.1	9.64	6.81	5.43	4.74
Inclination of axis to orbit	<28°	≈3°	23.4°	25.0°	3.08°	26.7°	97.9°	29.6°	57.5°
Inclination of orbit to Earth's orbit	7.00°	3.39°		1.85°	1.30°	2.49°	0.77°	1.77°	17.2°
Eccentricity of orbit	0.206	0.0068	0.0167	0.0934	0.0485	0.0556	0.0472	0.0086	0.250
Equatorial diameter, km	4880	12,100	12,800	6790	143,000	120,000	51,800	49,500	2300
Mass (Earth = 1)	0.0558	0.815	1.000	0.107	318	95.1	14.5	17.2	0.002
Density (water = 1)	5.60	5.20	5.52	3.95	1.31	0.704	1.21	1.67	2.03
Surface value of g,[c] m/s²	3.78	8.60	9.78	3.72	22.9	9.05	7.77	11.0	0.5
Escape velocity,[c] km/s	4.3	10.3	11.2	5.0	59.5	35.6	21.2	23.6	1.1
Known satellites	0	0	1	2	16 + ring	18 + rings	15 + rings	8 + rings	1

[a]Measured with respect to the distant stars.

[b]Venus and Uranus rotate opposite their orbital motion.

[c]Gravitational acceleration measured at the planet's equator.

Appendix **D**
Conversion Factors

Conversion factors may be read directly from these tables. For example, 1 degree = 2.778×10^{-3} revolutions, so $16.7° = 16.7 \times 2.778 \times 10^{-3}$ rev. The SI quantities are fully capitalized.

Adapted in part from G. Shortley and D. Williams, *Elements of Physics*, Prentice-Hall, Englewood Cliffs, NJ, 1971.

PLANE ANGLE

	°	′	″	RADIAN	rev
1 degree = 1	60	3600	1.745×10^{-2}	2.778×10^{-3}	
1 minute = 1.667×10^{-2}	1	60	2.909×10^{-4}	4.630×10^{-5}	
1 second = 2.778×10^{-4}	1.667×10^{-2}	1	4.848×10^{-6}	7.716×10^{-7}	
1 RADIAN = 57.30	3438	2.063×10^5	1	0.1592	
1 revolution = 360	2.16×10^4	1.296×10^6	6.283	1	

SOLID ANGLE

1 sphere = 4π steradians = 12.57 steradians

LENGTH

	cm	METER	km	in.	ft	mi
1 centimeter = 1	10^{-2}	10^{-5}	0.3937	3.281×10^{-2}	6.214×10^{-6}	
1 METER = 100	1	10^{-3}	39.37	3.281	6.214×10^{-4}	
1 kilometer = 10^5	1000	1	3.937×10^4	3281	0.6214	
1 inch = 2.540	2.540×10^{-2}	2.540×10^{-5}	1	8.333×10^{-2}	1.578×10^{-5}	
1 foot = 30.48	0.3048	3.048×10^{-4}	12	1	1.894×10^{-4}	
1 mile = 1.609×10^5	1609	1.609	6.336×10^4	5280	1	

1 angström = 10^{-10} m
1 nautical mile = 1852 m
  = 1.151 miles = 6076 ft

1 fermi = 10^{-15} m
1 light-year = 9.460×10^{12} km
1 parsec = 3.084×10^{13} km

1 fathom = 6 ft
1 Bohr radius = 5.292×10^{-11} m
1 yard = 3 ft

1 rod = 16.5 ft
1 mil = 10^{-3} in.
1 nm = 10^{-9} m

AREA

	METER^2	cm^2	ft^2	in.^{2}
1 SQUARE METER = 1	10^4	10.76	1550	
1 square centimeter = 10^{-4}	1	1.076×10^{-3}	0.1550	
1 square foot = 9.290×10^{-2}	929.0	1	144	
1 square inch = 6.452×10^{-4}	6.452	6.944×10^{-3}	1	

1 square mile = 2.788×10^7 ft^2
  = 640 acres
1 barn = 10^{-28} m^2

1 acre = 43,560 ft^2
1 hectare = 10^4 m^2 = 2.471 acres

VOLUME

	METER^3	cm^3	L	ft^3	in.^{3}
1 CUBIC METER = 1		10^6	1000	35.31	6.102×10^4
1 cubic centimeter = 10^{-6}		1	1.000×10^{-3}	3.531×10^{-5}	6.102×10^{-2}
1 liter = 1.000×10^{-3}		1000	1	3.531×10^{-2}	61.02
1 cubic foot = 2.832×10^{-2}		2.832×10^4	28.32	1	1728
1 cubic inch = 1.639×10^{-5}		16.39	1.639×10^{-2}	5.787×10^{-4}	1

1 U.S. fluid gallon = 4 U.S. fluid quarts = 8 U.S. pints = 128 U.S. fluid ounces = 231 in.^{3}

1 British imperial gallon = 277.4 in.^{3} = 1.201 U.S. fluid gallons

MASS

Quantities in the colored areas are not mass units but are often used as such. When we write, for example, 1 kg ''='' 2.205 lb, this means that a kilogram is a *mass* that *weighs* 2.205 pounds at a location where g has the standard value of 9.80665 m/s^2.

	g	KILOGRAM	slug	u	oz	lb	ton
1 gram = 1		0.001	6.852×10^{-5}	6.022×10^{23}	3.527×10^{-2}	2.205×10^{-3}	1.102×10^{-6}
1 KILOGRAM = 1000		1	6.852×10^{-2}	6.022×10^{26}	35.27	2.205	1.102×10^{-3}
1 slug = 1.459×10^4		14.59	1	8.786×10^{27}	514.8	32.17	1.609×10^{-2}
1 atomic mass unit = 1.661×10^{-24}		1.661×10^{-27}	1.138×10^{-28}	1	5.857×10^{-26}	3.662×10^{-27}	1.830×10^{-30}
1 ounce = 28.35		2.835×10^{-2}	1.943×10^{-3}	1.718×10^{25}	1	6.250×10^{-2}	3.125×10^{-5}
1 pound = 453.6		0.4536	3.108×10^{-2}	2.732×10^{26}	16	1	0.0005
1 ton = 9.072×10^5		907.2	62.16	5.463×10^{29}	3.2×10^4	2000	1

1 metric ton = 1000 kg

DENSITY

Quantities in the colored areas are weight densities and, as such, are dimensionally different from mass densities. See note for mass table.

	slug/ft^3	KILOGRAM/METER^3	g/cm^3	lb/ft^3	lb/in.^{3}
1 slug per foot^3 = 1		515.4	0.5154	32.17	1.862×10^{-2}
1 KILOGRAM per METER^3 = 1.940×10^{-3}		1	0.001	6.243×10^{-2}	3.613×10^{-5}
1 gram per centimeter^3 = 1.940		1000	1	62.43	3.613×10^{-2}
1 pound per foot^3 = 3.108×10^{-2}		16.02	1.602×10^{-2}	1	5.787×10^{-4}
1 pound per inch^3 = 53.71		2.768×10^4	27.68	1728	1

TIME

	y	d	h	min	SECOND
1 year = 1		365.25	8.766×10^3	5.259×10^5	3.156×10^7
1 day = 2.738×10^{-3}		1	24	1440	8.640×10^4
1 hour = 1.141×10^{-4}		4.167×10^{-2}	1	60	3600
1 minute = 1.901×10^{-6}		6.944×10^{-4}	1.667×10^{-2}	1	60
1 SECOND = 3.169×10^{-8}		1.157×10^{-5}	2.778×10^{-4}	1.667×10^{-2}	1

SPEED

	ft/s	km/h	METER/ SECOND	mi/h	cm/s
1 foot per second = 1	1	1.097	0.3048	0.6818	30.48
1 kilometer per hour = 0.9113	0.9113	1	0.2778	0.6214	27.78
1 METER per SECOND = 3.281	3.281	3.6	1	2.237	100
1 mile per hour = 1.467	1.467	1.609	0.4470	1	44.70
1 centimeter per second = 3.281×10^{-2}	3.281×10^{-2}	3.6×10^{-2}	0.01	2.237×10^{-2}	1

1 knot = 1 nautical mi/h = 1.688 ft/s   1 mi/min = 88.00 ft/s = 60.00 mi/h

FORCE

Force units in the colored areas are now little used. To clarify: 1 gram-force (= 1 gf) is the force of gravity that would act on an object whose mass is 1 gram at a location where g has the standard value of 9.80665 m/s².

	dyne	NEWTON	lb	pdl	gf	kgf
1 dyne = 1	1	10^{-5}	2.248×10^{-6}	7.233×10^{-5}	1.020×10^{-3}	1.020×10^{-6}
1 NEWTON = 10^5	10^5	1	0.2248	7.233	102.0	0.1020
1 pound = 4.448×10^5	4.448×10^5	4.448	1	32.17	453.6	0.4536
1 poundal = 1.383×10^4	1.383×10^4	0.1383	3.108×10^{-2}	1	14.10	1.410×10^{-2}
1 gram-force = 980.7	980.7	9.807×10^{-3}	2.205×10^{-3}	7.093×10^{-2}	1	0.001
1 kilogram-force = 9.807×10^5	9.807×10^5	9.807	2.205	70.93	1000	1

PRESSURE

	atm	dyne/cm²	inch of water	cm Hg	PASCAL	lb/in.²	lb/ft²
1 atmosphere = 1	1	1.013×10^6	406.8	76	1.013×10^5	14.70	2116
1 dyne per centimeter² = 9.869×10^{-7}	9.869×10^{-7}	1	4.015×10^{-4}	7.501×10^{-5}	0.1	1.405×10^{-5}	2.089×10^{-3}
1 inch of water[a] at 4°C = 2.458×10^{-3}	2.458×10^{-3}	2491	1	0.1868	249.1	3.613×10^{-2}	5.202
1 centimeter of mercury[a] at 0°C = 1.316×10^{-2}	1.316×10^{-2}	1.333×10^4	5.353	1	1333	0.1934	27.85
1 PASCAL = 9.869×10^{-6}	9.869×10^{-6}	10	4.015×10^{-3}	7.501×10^{-4}	1	1.450×10^{-4}	2.089×10^{-2}
1 pound per inch² = 6.805×10^{-2}	6.805×10^{-2}	6.895×10^4	27.68	5.171	6.895×10^3	1	144
1 pound per foot² = 4.725×10^{-4}	4.725×10^{-4}	478.8	0.1922	3.591×10^{-2}	47.88	6.944×10^{-3}	1

[a] Where the acceleration of gravity has the standard value of 9.80665 m/s².

1 bar = 10^6 dyne/cm² = 0.1 MPa   1 millibar = 10^3 dyne/cm² = 10^2 Pa   1 torr = 1 mm Hg

ENERGY, WORK, HEAT

Quantities in the colored areas are not properly energy units but are included for convenience. They arise from the relativistic mass–energy equivalence formula $E = mc^2$ and represent the energy released if a kilogram or unified atomic mass unit (u) is completely converted to energy (bottom two rows) or the mass that would be completely converted to one unit of energy (rightmost two columns).

	Btu	erg	ft·lb	hp·h	JOULE	cal	kW·h	eV	MeV	kg	u
1 British thermal unit =	1	1.055×10^{10}	777.9	3.929×10^{-4}	1055	252.0	2.930×10^{-4}	6.585×10^{21}	6.585×10^{15}	1.174×10^{-14}	7.070×10^{12}
1 erg =	9.481×10^{-11}	1	7.376×10^{-8}	3.725×10^{-14}	10^{-7}	2.389×10^{-8}	2.778×10^{-14}	6.242×10^{11}	6.242×10^{5}	1.113×10^{-24}	670.2
1 foot-pound =	1.285×10^{-3}	1.356×10^{7}	1	5.051×10^{-7}	1.356	0.3238	3.766×10^{-7}	8.464×10^{18}	8.464×10^{12}	1.509×10^{-17}	9.037×10^{9}
1 horsepower-hour =	2545	2.685×10^{13}	1.980×10^{6}	1	2.685×10^{6}	6.413×10^{5}	0.7457	1.676×10^{25}	1.676×10^{19}	2.988×10^{-11}	1.799×10^{16}
1 JOULE =	9.481×10^{-4}	10^{7}	0.7376	3.725×10^{-7}	1	0.2389	2.778×10^{-7}	6.242×10^{18}	6.242×10^{12}	1.113×10^{-17}	6.702×10^{9}
1 calorie =	3.969×10^{-3}	4.186×10^{7}	3.088	1.560×10^{-6}	4.186	1	1.163×10^{-6}	2.613×10^{19}	2.613×10^{13}	4.660×10^{-17}	2.806×10^{10}
1 kilowatt-hour =	3413	3.600×10^{13}	2.655×10^{6}	1.341	3.600×10^{6}	8.600×10^{5}	1	2.247×10^{25}	2.247×10^{19}	4.007×10^{-11}	2.413×10^{16}
1 electron-volt =	1.519×10^{-22}	1.602×10^{-12}	1.182×10^{-19}	5.967×10^{-26}	1.602×10^{-19}	3.827×10^{-20}	4.450×10^{-26}	1	10^{-6}	1.783×10^{-36}	1.074×10^{-9}
1 million electron-volts =	1.519×10^{-16}	1.602×10^{-6}	1.182×10^{-13}	5.967×10^{-20}	1.602×10^{-13}	3.827×10^{-14}	4.450×10^{-20}	10^{-6}	1	1.783×10^{-30}	1.074×10^{-3}
1 kilogram =	8.521×10^{13}	8.987×10^{23}	6.629×10^{16}	3.348×10^{10}	8.987×10^{16}	2.146×10^{16}	2.497×10^{10}	5.610×10^{35}	5.610×10^{29}	1	6.022×10^{26}
1 unified atomic mass unit =	1.415×10^{-13}	1.492×10^{-3}	1.101×10^{-10}	5.559×10^{-17}	1.492×10^{-10}	3.564×10^{-11}	4.146×10^{-17}	9.320×10^{8}	932.0	1.661×10^{-27}	1

POWER

	Btu/h	ft·lb/s	hp	cal/s	kW	WATT
1 British thermal unit per hour =	1	0.2161	3.929×10^{-4}	6.998×10^{-2}	2.930×10^{-4}	0.2930
1 foot-pound per second =	4.628	1	1.818×10^{-3}	0.3239	1.356×10^{-3}	1.356
1 horsepower =	2545	550	1	178.1	0.7457	745.7
1 calorie per second =	14.29	3.088	5.615×10^{-3}	1	4.186×10^{-3}	4.186
1 kilowatt =	3413	737.6	1.341	238.9	1	1000
1 WATT =	3.413	0.7376	1.341×10^{-3}	0.2389	0.001	1

MAGNETIC FIELD

	gauss	TESLA	milligauss
1 gauss =	1	10^{-4}	1000
1 TESLA =	10^{4}	1	10^{7}
1 milligauss =	0.001	10^{-7}	1

$1 \text{ tesla} = 1 \text{ weber/meter}^2$

MAGNETIC FLUX

	maxwell	WEBER
1 maxwell =	1	10^{-8}
1 WEBER =	10^{8}	1

Appendix E
Mathematical Formulas

GEOMETRY

Circle of radius r: circumference $= 2\pi r$; area $= \pi r^2$.

Sphere of radius r: area $= 4\pi r^2$; volume $= \frac{4}{3}\pi r^3$.

Right circular cylinder of radius r and height h:
area $= 2\pi r^2 + 2\pi rh$; volume $= \pi r^2 h$.

Triangle of base a and altitude h: area $= \frac{1}{2}ah$.

QUADRATIC FORMULA

If $ax^2 + bx + c = 0$, then $x = \dfrac{-b \pm \sqrt{b^2 - 4ac}}{2a}$.

TRIGONOMETRIC FUNCTIONS OF ANGLE θ

$\sin \theta = \dfrac{y}{r}$ $\quad \cos \theta = \dfrac{x}{r}$

$\tan \theta = \dfrac{y}{x}$ $\quad \cot \theta = \dfrac{x}{y}$

$\sec \theta = \dfrac{r}{x}$ $\quad \csc \theta = \dfrac{r}{y}$

PYTHAGOREAN THEOREM

In this right triangle,
$$a^2 + b^2 = c^2$$

TRIANGLES

Angles are A, B, C

Opposite sides are a, b, c

Angles $A + B + C = 180°$

$\dfrac{\sin A}{a} = \dfrac{\sin B}{b} = \dfrac{\sin C}{c}$

$c^2 = a^2 + b^2 - 2ab \cos C$

Exterior angle $D = A + C$

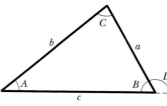

MATHEMATICAL SIGNS AND SYMBOLS

$=$ equals

$\approx$ equals approximately

$\sim$ is the order of magnitude of

$\neq$ is not equal to

$\equiv$ is identical to, is defined as

$>$ is greater than ($\gg$ is much greater than)

$<$ is less than ($\ll$ is much less than)

$\geq$ is greater than or equal to (or, is no less than)

$\leq$ is less than or equal to (or, is no more than)

$\pm$ plus or minus

$\propto$ is proportional to

Σ the sum of

$\bar{x}$ the average value of x

TRIGONOMETRIC IDENTITIES

$\sin(90° - \theta) = \cos \theta$

$\cos(90° - \theta) = \sin \theta$

$\sin \theta / \cos \theta = \tan \theta$

$\sin^2 \theta + \cos^2 \theta = 1$

$\sec^2 \theta - \tan^2 \theta = 1$

$\csc^2 \theta - \cot^2 \theta = 1$

$\sin 2\theta = 2 \sin \theta \cos \theta$

$\cos 2\theta = \cos^2 \theta - \sin^2 \theta = 2 \cos^2 \theta - 1 = 1 - 2 \sin^2 \theta$

$\sin(\alpha \pm \beta) = \sin \alpha \cos \beta \pm \cos \alpha \sin \beta$

$\cos(\alpha \pm \beta) = \cos \alpha \cos \beta \mp \sin \alpha \sin \beta$

$\tan(\alpha \pm \beta) = \dfrac{\tan \alpha \pm \tan \beta}{1 \mp \tan \alpha \tan \beta}$

$\sin \alpha \pm \sin \beta = 2 \sin \frac{1}{2}(\alpha \pm \beta) \cos \frac{1}{2}(\alpha \mp \beta)$

$\cos \alpha + \cos \beta = 2 \cos \frac{1}{2}(\alpha + \beta) \cos \frac{1}{2}(\alpha - \beta)$

$\cos \alpha - \cos \beta = -2 \sin \frac{1}{2}(\alpha + \beta) \sin \frac{1}{2}(\alpha - \beta)$

BINOMIAL THEOREM

$$(1 + x)^n = 1 + \frac{nx}{1!} + \frac{n(n-1)x^2}{2!} + \cdots \qquad (x^2 < 1)$$

EXPONENTIAL EXPANSION

$$e^x = 1 + x + \frac{x^2}{2!} + \frac{x^3}{3!} + \cdots$$

LOGARITHMIC EXPANSION

$$\ln(1 + x) = x - \tfrac{1}{2}x^2 + \tfrac{1}{3}x^3 - \cdots \qquad (|x| < 1)$$

TRIGONOMETRIC EXPANSIONS
(θ in radians)

$$\sin \theta = \theta - \frac{\theta^3}{3!} + \frac{\theta^5}{5!} - \cdots$$

$$\cos \theta = 1 - \frac{\theta^2}{2!} + \frac{\theta^4}{4!} - \cdots$$

$$\tan \theta = \theta + \frac{\theta^3}{3} + \frac{2\theta^5}{15} + \cdots$$

CRAMER'S RULE

Two simultaneous equations in unknowns x and y,

$$a_1 x + b_1 y = c_1 \qquad \text{and} \qquad a_2 x + b_2 y = c_2,$$

have the solutions

$$x = \frac{\begin{vmatrix} c_1 & b_1 \\ c_2 & b_2 \end{vmatrix}}{\begin{vmatrix} a_1 & b_1 \\ a_2 & b_2 \end{vmatrix}} = \frac{c_1 b_2 - c_2 b_1}{a_1 b_2 - a_2 b_1}$$

and

$$y = \frac{\begin{vmatrix} a_1 & c_1 \\ a_2 & c_2 \end{vmatrix}}{\begin{vmatrix} a_1 & b_1 \\ a_2 & b_2 \end{vmatrix}} = \frac{a_1 c_2 - a_2 c_1}{a_1 b_2 - a_2 b_1}.$$

PRODUCTS OF VECTORS

Let $\mathbf{i}$, $\mathbf{j}$, and $\mathbf{k}$ be unit vectors in the x, y, and z directions. Then

$$\mathbf{i} \cdot \mathbf{i} = \mathbf{j} \cdot \mathbf{j} = \mathbf{k} \cdot \mathbf{k} = 1, \qquad \mathbf{i} \cdot \mathbf{j} = \mathbf{j} \cdot \mathbf{k} = \mathbf{k} \cdot \mathbf{i} = 0,$$

$$\mathbf{i} \times \mathbf{i} = \mathbf{j} \times \mathbf{j} = \mathbf{k} \times \mathbf{k} = 0,$$

$$\mathbf{i} \times \mathbf{j} = \mathbf{k}, \qquad \mathbf{j} \times \mathbf{k} = \mathbf{i}, \qquad \mathbf{k} \times \mathbf{i} = \mathbf{j},$$

Any vector $\mathbf{a}$ with components a_x, a_y, and a_z along the x, y, and z axes can be written

$$\mathbf{a} = a_x \mathbf{i} + a_y \mathbf{j} + a_z \mathbf{k}.$$

Let $\mathbf{a}$, $\mathbf{b}$, and $\mathbf{c}$ be arbitrary vectors with magnitudes a, b, and c. Then

$$\mathbf{a} \times (\mathbf{b} + \mathbf{c}) = (\mathbf{a} \times \mathbf{b}) + (\mathbf{a} \times \mathbf{c})$$

$$(s\mathbf{a}) \times \mathbf{b} = \mathbf{a} \times (s\mathbf{b}) = s(\mathbf{a} \times \mathbf{b}) \qquad (s = \text{a scalar}).$$

Let θ be the smaller of the two angles between $\mathbf{a}$ and $\mathbf{b}$. Then

$$\mathbf{a} \cdot \mathbf{b} = \mathbf{b} \cdot \mathbf{a} = a_x b_x + a_y b_y + a_z b_z = ab \cos \theta$$

$$\mathbf{a} \times \mathbf{b} = -\mathbf{b} \times \mathbf{a} = \begin{vmatrix} \mathbf{i} & \mathbf{j} & \mathbf{k} \\ a_x & a_y & a_z \\ b_x & b_y & b_z \end{vmatrix}$$

$$= \mathbf{i} \begin{vmatrix} a_y & a_z \\ b_y & b_z \end{vmatrix} - \mathbf{j} \begin{vmatrix} a_x & a_z \\ b_x & b_z \end{vmatrix} + \mathbf{k} \begin{vmatrix} a_x & a_y \\ b_x & b_y \end{vmatrix}$$

$$= (a_y b_z - b_y a_z)\mathbf{i}$$

$$\quad + (a_z b_x - b_z a_x)\mathbf{j} + (a_x b_y - b_x a_y)\mathbf{k}$$

$$|\mathbf{a} \times \mathbf{b}| = ab \sin \theta$$

$$\mathbf{a} \cdot (\mathbf{b} \times \mathbf{c}) = \mathbf{b} \cdot (\mathbf{c} \times \mathbf{a}) = \mathbf{c} \cdot (\mathbf{a} \times \mathbf{b})$$

$$\mathbf{a} \times (\mathbf{b} \times \mathbf{c}) = (\mathbf{a} \cdot \mathbf{c})\mathbf{b} - (\mathbf{a} \cdot \mathbf{b})\mathbf{c}$$

DERIVATIVES AND INTEGRALS

In what follows, the letters u and v stand for any functions of x, and a and m are constants. To each of the indefinite integrals should be added an arbitrary constant of integration. The *Handbook of Chemistry and Physics* (CRC Press Inc.) gives a more extensive tabulation.

1. $\dfrac{dx}{dx} = 1$

2. $\dfrac{d}{dx}(au) = a\dfrac{du}{dx}$

3. $\dfrac{d}{dx}(u + v) = \dfrac{du}{dx} + \dfrac{dv}{dx}$

4. $\dfrac{d}{dx}x^m = mx^{m-1}$

5. $\dfrac{d}{dx}\ln x = \dfrac{1}{x}$

6. $\dfrac{d}{dx}(uv) = u\dfrac{dv}{dx} + v\dfrac{du}{dx}$

7. $\dfrac{d}{dx}e^x = e^x$

8. $\dfrac{d}{dx}\sin x = \cos x$

9. $\dfrac{d}{dx}\cos x = -\sin x$

10. $\dfrac{d}{dx}\tan x = \sec^2 x$

11. $\dfrac{d}{dx}\cot x = -\csc^2 x$

12. $\dfrac{d}{dx}\sec x = \tan x \sec x$

13. $\dfrac{d}{dx}\csc x = -\cot x \csc x$

14. $\dfrac{d}{dx}e^u = e^u\dfrac{du}{dx}$

15. $\dfrac{d}{dx}\sin u = \cos u\dfrac{du}{dx}$

16. $\dfrac{d}{dx}\cos u = -\sin u\dfrac{du}{dx}$

1. $\displaystyle\int dx = x$

2. $\displaystyle\int au\, dx = a\int u\, dx$

3. $\displaystyle\int (u + v)\, dx = \int u\, dx + \int v\, dx$

4. $\displaystyle\int x^m\, dx = \dfrac{x^{m+1}}{m+1}$ $(m \neq -1)$

5. $\displaystyle\int \dfrac{dx}{x} = \ln |x|$

6. $\displaystyle\int u\dfrac{dv}{dx}\, dx = uv - \int v\dfrac{du}{dx}\, dx$

7. $\displaystyle\int e^x\, dx = e^x$

8. $\displaystyle\int \sin x\, dx = -\cos x$

9. $\displaystyle\int \cos x\, dx = \sin x$

10. $\displaystyle\int \tan x\, dx = \ln |\sec x|$

11. $\displaystyle\int \sin^2 x\, dx = \tfrac{1}{2}x - \tfrac{1}{4}\sin 2x$

12. $\displaystyle\int e^{-ax}\, dx = -\dfrac{1}{a}e^{-ax}$

13. $\displaystyle\int xe^{-ax}\, dx = -\dfrac{1}{a^2}(ax + 1)e^{-ax}$

14. $\displaystyle\int x^2e^{-ax}\, dx = -\dfrac{1}{a^3}(a^2x^2 + 2ax + 2)e^{-ax}$

15. $\displaystyle\int_0^\infty x^n e^{-ax}\, dx = \dfrac{n!}{a^{n+1}}$

16. $\displaystyle\int_0^\infty x^{2n}e^{-ax^2}\, dx = \dfrac{1 \cdot 3 \cdot 5 \cdots (2n-1)}{2^{n+1}a^n}\sqrt{\dfrac{\pi}{a}}$

17. $\displaystyle\int \dfrac{dx}{\sqrt{x^2 + a^2}} = \ln(x + \sqrt{x^2 + a^2})$

18. $\displaystyle\int \dfrac{x\, dx}{(x^2 + a^2)^{3/2}} = -\dfrac{1}{(x^2 + a^2)^{1/2}}$

19. $\displaystyle\int \dfrac{dx}{(x^2 + a^2)^{3/2}} = \dfrac{x}{a^2(x^2 + a^2)^{1/2}}$

Appendix F
Properties of the Elements

All physical properties are for a pressure of 1 atm unless otherwise specified.

ELEMENT	SYMBOL	ATOMIC NUMBER, Z	MOLAR MASS, g/mol	DENSITY, g/cm³ AT 20°C	MELTING POINT, °C	BOILING POINT, °C	SPECIFIC HEAT, J/(g·°C) AT 25°C
Actinium	Ac	89	(227)	10.06	1323	(3473)	0.092
Aluminum	Al	13	26.9815	2.699	660	2450	0.900
Americium	Am	95	(243)	13.67	1541	—	—
Antimony	Sb	51	121.75	6.691	630.5	1380	0.205
Argon	Ar	18	39.948	1.6626×10^{-3}	−189.4	−185.8	0.523
Arsenic	As	33	74.9216	5.78	817 (28 atm)	613	0.331
Astatine	At	85	(210)	—	(302)	—	—
Barium	Ba	56	137.34	3.594	729	1640	0.205
Berkelium	Bk	97	(247)	14.79	—	—	—
Beryllium	Be	4	9.0122	1.848	1287	2770	1.83
Bismuth	Bi	83	208.980	9.747	271.37	1560	0.122
Boron	B	5	10.811	2.34	2030	—	1.11
Bromine	Br	35	79.909	3.12 (liquid)	−7.2	58	0.293
Cadmium	Cd	48	112.40	8.65	321.03	765	0.226
Calcium	Ca	20	40.08	1.55	838	1440	0.624
Californium	Cf	98	(251)	—	—	—	—
Carbon	C	6	12.01115	2.26	3727	4830	0.691
Cerium	Ce	58	140.12	6.768	804	3470	0.188
Cesium	Cs	55	132.905	1.873	28.40	690	0.243
Chlorine	Cl	17	35.453	3.214×10^{-3} (0°C)	−101	−34.7	0.486
Chromium	Cr	24	51.996	7.19	1857	2665	0.448
Cobalt	Co	27	58.9332	8.85	1495	2900	0.423
Copper	Cu	29	63.54	8.96	1083.40	2595	0.385
Curium	Cm	96	(247)	13.3	—	—	—
Dysprosium	Dy	66	162.50	8.55	1409	2330	0.172
Einsteinium	Es	99	(254)	—	—	—	—
Erbium	Er	68	167.26	9.15	1522	2630	0.167
Europium	Eu	63	151.96	5.243	817	1490	0.163
Fermium	Fm	100	(237)	—	—	—	—
Fluorine	F	9	18.9984	1.696×10^{-3} (0°C)	−219.6	−188.2	0.753
Francium	Fr	87	(223)	—	(27)	—	—
Gadolinium	Gd	64	157.25	7.90	1312	2730	0.234
Gallium	Ga	31	69.72	5.907	29.75	2237	0.377
Germanium	Ge	32	72.59	5.323	937.25	2830	0.322
Gold	Au	79	196.967	19.32	1064.43	2970	0.131
Hafnium	Hf	72	178.49	13.31	2227	5400	0.144
Hahnium	Ha	105	—	—	—	—	—
Hassium	Hs	108	—	—	—	—	—

continued on next page

ELEMENT	SYMBOL	ATOMIC NUMBER, Z	MOLAR MASS, g/mol	DENSITY, g/cm^3 AT 20°C	MELTING POINT, °C	BOILING POINT, °C	SPECIFIC HEAT, J/(g·°C) AT 25°C
Helium	He	2	4.0026	0.1664×10^{-3}	-269.7	-268.9	5.23
Holmium	Ho	67	164.930	8.79	1470	2330	0.165
Hydrogen	H	1	1.00797	0.08375×10^{-3}	-259.19	-252.7	14.4
Indium	In	49	114.82	7.31	156.634	2000	0.233
Iodine	I	53	126.9044	4.93	113.7	183	0.218
Iridium	Ir	77	192.2	22.5	2447	(5300)	0.130
Iron	Fe	26	55.847	7.874	1536.5	3000	0.447
Krypton	Kr	36	83.80	3.488×10^{-3}	-157.37	-152	0.247
Lanthanum	La	57	138.91	6.189	920	3470	0.195
Lawrencium	Lr	103	(257)	—	—	—	—
Lead	Pb	82	207.19	11.35	327.45	1725	0.129
Lithium	Li	3	6.939	0.534	180.55	1300	3.58
Lutetium	Lu	71	174.97	9.849	1663	1930	0.155
Magnesium	Mg	12	24.312	1.738	650	1107	1.03
Manganese	Mn	25	54.9380	7.44	1244	2150	0.481
Meitnerium	Mt	109	—	—	—	—	—
Mendelevium	Md	101	(256)	—	—	—	—
Mercury	Hg	80	200.59	13.55	-38.87	357	0.138
Molybdenum	Mo	42	95.94	10.22	2617	5560	0.251
Neodymium	Nd	60	144.24	7.007	1016	3180	0.188
Neon	Ne	10	20.183	0.8387×10^{-3}	-248.597	-246.0	1.03
Neptunium	Np	93	(237)	20.25	637	—	1.26
Nickel	Ni	28	58.71	8.902	1453	2730	0.444
Nielsbohrium	Ns	107	—	—	—	—	—
Niobium	Nb	41	92.906	8.57	2468	4927	0.264
Nitrogen	N	7	14.0067	1.1649×10^{-3}	-210	-195.8	1.03
Nobelium	No	102	(255)	—	—	—	—
Osmium	Os	76	190.2	22.59	3027	5500	0.130
Oxygen	O	8	15.9994	1.3318×10^{-3}	-218.80	-183.0	0.913
Palladium	Pd	46	106.4	12.02	1552	3980	0.243
Phosphorus	P	15	30.9738	1.83	44.25	280	0.741
Platinum	Pt	78	195.09	21.45	1769	4530	0.134
Plutonium	Pu	94	(244)	19.8	640	3235	0.130
Polonium	Po	84	(210)	9.32	254	—	—
Potassium	K	19	39.102	0.862	63.20	760	0.758
Praseodymium	Pr	59	140.907	6.773	931	3020	0.197
Promethium	Pm	61	(145)	7.22	(1027)	—	—
Protactinium	Pa	91	(231)	15.37 (estimated)	(1230)	—	—
Radium	Ra	88	(226)	5.0	700	—	—
Radon	Rn	86	(222)	9.96×10^{-3} (0°C)	(-71)	-61.8	0.092
Rhenium	Re	75	186.2	21.02	3180	5900	0.134
Rhodium	Rh	45	102.905	12.41	1963	4500	0.243
Rubidium	Rb	37	85.47	1.532	39.49	688	0.364
Ruthenium	Ru	44	101.107	12.37	2250	4900	0.239
Rutherfordium	Rf	104	—	—	—	—	—
Samarium	Sm	62	150.35	7.52	1072	1630	0.197

continued on next page

ELEMENT	SYMBOL	ATOMIC NUMBER, Z	MOLAR MASS, g/mol	DENSITY, g/cm³ AT 20°C	MELTING POINT, °C	BOILING POINT, °C	SPECIFIC HEAT, J/(g·°C) AT 25°C
Scandium	Sc	21	44.956	2.99	1539	2730	0.569
Seaborgium	Sg	106	—	—	—	—	—
Selenium	Se	34	78.96	4.79	221	685	0.318
Silicon	Si	14	28.086	2.33	1412	2680	0.712
Silver	Ag	47	107.870	10.49	960.8	2210	0.234
Sodium	Na	11	22.9898	0.9712	97.85	892	1.23
Strontium	Sr	38	87.62	2.54	768	1380	0.737
Sulfur	S	16	32.064	2.07	119.0	444.6	0.707
Tantalum	Ta	73	180.948	16.6	3014	5425	0.138
Technetium	Tc	43	(99)	11.46	2200	—	0.209
Tellurium	Te	52	127.60	6.24	449.5	990	0.201
Terbium	Tb	65	158.924	8.229	1357	2530	0.180
Thallium	Tl	81	204.37	11.85	304	1457	0.130
Thorium	Th	90	(232)	11.72	1755	(3850)	0.117
Thulium	Tm	69	168.934	9.32	1545	1720	0.159
Tin	Sn	50	118.69	7.2984	231.868	2270	0.226
Titanium	Ti	22	47.90	4.54	1670	3260	0.523
Tungsten	W	74	183.85	19.3	3380	5930	0.134
Uranium	U	92	(238)	18.95	1132	3818	0.117
Vanadium	V	23	50.942	6.11	1902	3400	0.490
Xenon	Xe	54	131.30	5.495×10^{-3}	− 111.79	− 108	0.159
Ytterbium	Yb	70	173.04	6.965	824	1530	0.155
Yttrium	Y	39	88.905	4.469	1526	3030	0.297
Zinc	Zn	30	65.37	7.133	419.58	906	0.389
Zirconium	Zr	40	91.22	6.506	1852	3580	0.276

The values in parentheses in the column of molar masses are the mass numbers of the longest-lived isotopes of those elements that are radioactive. Melting points and boiling points in parentheses are uncertain.

The data for gases are valid only when these are in their usual molecular state, such as H_2, He, O_2, Ne, etc. The specific heats of the gases are the values at constant pressure.

Source: Adapted from Wehr, Richards, Adair, *Physics of the Atom,* 4th ed., Addison-Wesley, Reading, MA, 1984, and from J. Emsley, *The Elements,* 2nd ed., Clarendon Press, Oxford, 1991.

Appendix G
Periodic Table of the Elements

The names for elements 104–109 (Rutherfordium, Hahnium, Seaborgium, Nielsbohrium, Hassium, and Meitnerium, respectively) are those recommended by the American Chemical Society Nomenclature Committee. As of 1996, the names and symbols for elements 104–108 have not yet been approved by the appropriate international body. Elements 110, 111 and 112 have been discovered but, as of 1996, have not been provisionally named.

Appendix **H**
*Winners of the Nobel Prize in Physics**

1901 Wilhelm Konrad Röntgen *(1845–1923)* for the discovery of x rays

1902 Hendrik Antoon Lorentz *(1853–1928)* and Pieter Zeeman *(1865–1943)* for their researches into the influence of magnetism upon radiation phenomena

1903 Antoine Henri Becquerel *(1852–1908)* for his discovery of spontaneous radioactivity

Pierre Curie *(1859–1906)* and Marie Sklowdowska-Curie *(1867–1934)* for their joint researches on the radiation phenomena discovered by Becquerel

1904 Lord Rayleigh (John William Strutt) *(1842–1919)* for his investigations of the densities of the most important gases and for his discovery of argon

1905 Philipp Eduard Anton von Lenard *(1862–1947)* for his work on cathode rays

1906 Joseph John Thomson *(1856–1940)* for his theoretical and experimental investigations on the conduction of electricity by gases

1907 Albert Abraham Michelson *(1852–1931)* for his optical precision instruments and metrological investigations carried out with their aid

1908 Gabriel Lippmann *(1845–1921)* for his method of reproducing colors photographically based on the phenomena of interference

1909 Guglielmo Marconi *(1874–1937)* and Carl Ferdinand Braun *(1850–1918)* for their contributions to the development of wireless telegraphy

1910 Johannes Diderik van der Waals *(1837–1923)* for his work on the equation of state for gases and liquids

1911 Wilhelm Wien *(1864–1928)* for his discoveries regarding the laws governing the radiation of heat

1912 Nils Gustaf Dalén *(1869–1937)* for his invention of automatic regulators for use in conjunction with gas accumulators for illuminating lighthouses and buoys

1913 Heike Kamerlingh Onnes *(1853–1926)* for his investigations of the properties of matter at low temperatures which led, among other things, to the production of liquid helium

1914 Max von Laue *(1879–1960)* for his discovery of the diffraction of Röntgen rays by crystals

1915 William Henry Bragg *(1862–1942)* and William Lawrence Bragg *(1890–1971)* for their services in the analysis of crystal structure by means of x rays

1917 Charles Glover Barkla *(1877–1944)* for his discovery of the characteristic x rays of the elements

1918 Max Planck *(1858–1947)* for his discovery of energy quanta

1919 Johannes Stark *(1874–1957)* for his discovery of the Doppler effect in canal rays and the splitting of spectral lines in electric fields

1920 Charles-Édouard Guillaume *(1861–1938)* for the service he rendered to precision measurements in physics by his discovery of anomalies in nickel steel alloys

1921 Albert Einstein *(1879–1955)* for his services to theoretical physics, and especially for his discovery of the law of the photoelectric effect

1922 Niels Bohr *(1885–1962)* for the investigation of the structure of atoms, and of the radiation emanating from them

1923 Robert Andrews Millikan *(1868–1953)* for his work on the elementary charge of electricity and on the photoelectric effect

1924 Karl Manne Georg Siegbahn *(1886–1978)* for his discoveries and research in the field of x-ray spectroscopy

1925 James Franck *(1882–1964)* and Gustav Hertz *(1887–1975)* for their discovery of the laws governing the impact of an electron upon an atom

1926 Jean Baptiste Perrin *(1870–1942)* for his work on the discontinuous structure of matter, and especially for his discovery of sedimentation equilibrium

1927 Arthur Holly Compton *(1892–1962)* for his discovery of the effect named after him

Charles Thomson Rees Wilson (1869–1959) for his method of making the paths of electrically charged particles visible by condensation of vapor

1928 Owen Willans Richardson *(1879–1959)* for his work on the thermionic phenomenon and especially for the discovery of the law named after him

1929 Prince Louis Victor de Broglie *(1892–1987)* for his discovery of the wave nature of electrons

*See *Nobel Lectures, Physics,* 1901–1970, Elsevier Publishing Company, for biographies of the awardees and for lectures given by them on receiving the prize.

1930 Sir Chandrasekhara Venkata Raman *(1888–1970)* for his work on the scattering of light and for the discovery of the effect named after him

1932 Werner Heisenberg *(1901–1976)* for the creation of quantum mechanics, the application of which has, among other things, led to the discovery of the allotropic forms of hydrogen

1933 Erwin Schrödinger *(1887–1961)* and Paul Adrien Maurice Dirac *(1902–1984)* for the discovery of new productive forms of atomic theory

1935 James Chadwick *(1891–1974)* for his discovery of the neutron

1936 Victor Franz Hess *(1883–1964)* for the discovery of cosmic radiation

Carl David Anderson *(1905–1991)* for his discovery of the positron

1937 Clinton Joseph Davisson *(1881–1958)* and George Paget Thomson *(1892–1975)* for their experimental discovery of the diffraction of electrons by crystals

1938 Enrico Fermi *(1901–1954)* for his demonstrations of the existence of new radioactive elements produced by neutron irradiation, and for his related discovery of nuclear reactions brought about by slow neutrons

1939 Ernest Orlando Lawrence *(1901–1958)* for the invention and development of the cyclotron and for results obtained with it, especially for artificial radioactive elements

1943 Otto Stern *(1888–1969)* for his contribution to the development of the molecular-ray method and his discovery of the magnetic moment of the proton

1944 Isidor Isaac Rabi *(1898–1988)* for his resonance method for recording the magnetic properties of atomic nuclei

1945 Wolfgang Pauli *(1900–1958)* for the discovery of the Exclusion Principle (also called Pauli Principle)

1946 Percy Williams Bridgman *(1882–1961)* for the invention of an apparatus to produce extremely high pressures and for the discoveries he made therewith in the field of high-pressure physics

1947 Sir Edward Victor Appleton *(1892–1965)* for his investigations of the physics of the upper atmosphere, especially for the discovery of the so-called Appleton layer

1948 Patrick Maynard Stuart Blackett *(1897–1974)* for his development of the Wilson cloud-chamber method, and his discoveries therewith in nuclear physics and cosmic radiation

1949 Hideki Yukawa *(1907–1981)* for his prediction of the existence of mesons on the basis of theoretical work on nuclear forces

1950 Cecil Frank Powell *(1903–1969)* for his development of the photographic method of studying nuclear processes and his discoveries regarding mesons made with this method

1951 Sir John Douglas Cockcroft *(1897–1967)* and Ernest Thomas Sinton Walton *(1903–1995)* for their pioneer work on the transmutation of atomic nuclei by artificially accelerated atomic particles

1952 Felix Bloch *(1905–1983)* and Edward Mills Purcell *(1912–   )* for their development of new nuclear-magnetic precision methods and discoveries in connection therewith

1953 Frits Zernike *(1888–1966)* for his demonstration of the phase-contrast method, especially for his invention of the phase-contrast microscope

1954 Max Born *(1882–1970)* for his fundamental research in quantum mechanics, especially for his statistical interpretation of the wave function

Walther Bothe *(1891–1957)* for the coincidence method and his discoveries made therewith

1955 Willis Eugene Lamb *(1913–   )* for his discoveries concerning the fine structure of the hydrogen spectrum

Polykarp Kusch *(1911–1993)* for his precision determination of the magnetic moment of the electron

1956 William Shockley *(1910–1989)*, John Bardeen *(1908–1991)* and Walter Houser Brattain *(1902–1987)* for their researches on semiconductors and their discovery of the transistor effect

1957 Chen Ning Yang *(1922–   )* and Tsung Dao Lee *(1926–   )* for their penetrating investigation of the parity laws which has led to important discoveries regarding the elementary particles

1958 Pavel Aleksejevič Čerenkov *(1904–   )*, Il' ja Michajlovič Frank *(1908–1990)* and Igor' Evgen' evič Tamm *(1895–1971)* for the discovery and interpretation of the Cerenkov effect

1959 Emilio Gino Segrè *(1905–1989)* and Owen Chamberlain *(1920–   )* for their discovery of the antiproton

1960 Donald Arthur Glaser *(1926–   )* for the invention of the bubble chamber

1961 Robert Hofstadter *(1915–1990)* for his pioneering studies of electron scattering in atomic nuclei and for his thereby achieved discoveries concerning the structure of the nucleons

Rudolf Ludwig Mössbauer *(1929–   )* for his researches concerning the resonance absorption of γ rays and his discovery in this connection of the effect which bears his name

1962 Lev Davidovič Landau *(1908–1968)* for his pioneering theories of condensed matter, especially liquid helium

1963 Eugene P. Wigner *(1902–1995)* for his contributions to the theory of the atomic nucleus and the elementary particles, particularly through the discovery and application of fundamental symmetry principles

Maria Goeppert Mayer *(1906–1972)* and J. Hans D. Jensen *(1907–1973)* for their discoveries concerning nuclear shell structure

1964 Charles H. Townes *(1915–   )*, Nikolai G. Basov *(1922–   )* and Alexander M. Prochorov *(1916–   )* for fundamental work in the field of quantum electronics which has led to the construction of oscillators and amplifiers based on the maser–laser principle

1965 Sin-itiro Tomonaga *(1906–1979)*, Julian Schwinger *(1918–1994)* and Richard P. Feynman *(1918–1988)* for their fundamental work in quantum electrodynamics, with deep-ploughing consequences for the physics of elementary particles

1966 Alfred Kastler *(1902–1984)* for the discovery and development of optical methods for studying Hertzian resonance in atoms

1967 Hans Albrecht Bethe *(1906–)* for his contributions to the theory of nuclear reactions, especially his discoveries concerning the energy production in stars

1968 Luis W. Alvarez *(1911–1988)* for his decisive contribution to elementary particle physics, in particular the discovery of a large number of resonance states, made possible through his development of the techniques of using the hydrogen bubble chamber and its data analysis

1969 Murray Gell-Mann *(1929–)* for his contributions and discoveries concerning the classification of elementary particles and their interactions

1970 Hannes Alfvén *(1908–1995)* for fundamental work and discoveries in magneto-hydrodynamics with fruitful applications in different parts of plasma physics

Louis Néel *(1904–)* for fundamental work and discoveries concerning antiferromagnetism and ferrimagnetism which have led to important applications in solid state physics

1971 Dennis Gabor *(1900–1979)* for his discovery of the principles of holography

1972 John Bardeen *(1908–1991)*, Leon N. Cooper *(1930–)* and J. Robert Schrieffer *(1931–)* for their development of a theory of superconductivity

1973 Leo Esaki *(1925–)* for his discovery of tunneling in semiconductors

Ivar Giaever *(1929–)* for his discovery of tunneling in superconductors

Brian D. Josephson *(1940–)* for his theoretical prediction of the properties of a supercurrent through a tunnel barrier

1974 Antony Hewish *(1924–)* for the discovery of pulsars

Sir Martin Ryle *(1918–1984)* for his pioneering work in radioastronomy

1975 Aage Bohr *(1922–)*, Ben Mottelson *(1926–)* and James Rainwater *(1917–1986)* for the discovery of the connection between collective motion and particle motion and the development of the theory of the structure of the atomic nucleus based on this connection

1976 Burton Richter *(1931–)* and Samuel Chao Chung Ting *(1936–)* for their (independent) discovery of an important fundamental particle

1977 Philip Warren Anderson *(1923–)*, Nevill Francis Mott *(1905–1996)* and John Hasbrouck Van Vleck *(1899–1980)* for their fundamental theoretical investigations of the electronic structure of magnetic and disordered systems

1978 Peter L. Kapitza *(1894–1984)* for his basic inventions and discoveries in low-temperature physics

Arno A. Penzias *(1933–)* and Robert Woodrow Wilson *(1936–)* for their discovery of cosmic microwave background radiation

1979 Sheldon Lee Glashow *(1932–)*, Abdus Salam *(1926–1996)*, and Steven Weinberg *(1933–)* for their unified model of the action of the weak and electromagnetic forces and for their prediction of the existence of neutral currents

1980 James W. Cronin *(1931–)* and Val L. Fitch *(1923–)* for the discovery of violations of fundamental symmetry principles in the decay of neutral K mesons

1981 Nicolaas Bloembergen *(1920–)* and Arthur Leonard Schawlow *(1921–)* for their contribution to the development of laser spectroscopy

Kai M. Siegbahn *(1918–)* for his contribution to high-resolution electron spectroscopy

1982 Kenneth Geddes Wilson *(1936–)* for his method of analyzing the critical phenomena inherent in the changes of matter under the influence of pressure and temperature

1983 Subrehmanyan Chandrasekhar *(1910–1995)* for his theoretical studies of the structure and evolution of stars

William A. Fowler *(1911–1995)* for his studies of the formation of the chemical elements in the universe

1984 Carlo Rubbia *(1934–)* and Simon van der Meer *(1925–)* for their decisive contributions to the Large Project, which led to the discovery of the field particles W and Z, communicators of the weak interaction

1985 Klaus von Klitzing *(1943–)* for his discovery of the quantized Hall resistance

1986 Ernst Ruska *(1906–1988)* for his invention of the electron microscope

Gerd Binnig *(1947–)*, Heinrich Rohrer *(1933–)* for their invention of the scanning tunneling microscope

1987 Karl Alex Müller *(1927–)* and J. George Bednorz *(1950–)* for their discovery of a new class of superconductors

1988 Leon M. Lederman *(1922–)*, Melvin Schwartz *(1932–)* and Jack Steinberger *(1921–)* for the first use of a neutrino beam and the discovery of the muon neutrino

1989 Norman Ramsey *(1915–)*, Hans Dehmelt *(1922–)* and Wolfgang Paul *(1913–1993)* for their work that led to the development of atomic clocks and precision timing

1990 Jerome I. Friedman *(1930–)*, Henry W. Kendall *(1926–)* and Richard E. Taylor *(1929–)* for demonstrating that protons and neutrons consist of quarks

1991 Pierre de Gennes *(1932–)* for studies of order phenomena, such as in liquid crystals and polymers

1992 George Charpak *(1924–)* for his invention of fast electronic detectors for high-energy particles

1993 Joseph H. Taylor *(1941–   )* and Russell A. Hulse *(1950–   )* for the discovery and interpretation of the first binary pulsar.

1994 Bertram N. Brockhouse *(1918–   )* and Clifford G. Shull *(1915–   )* for the development of neutron scattering techniques

1995 Martin L. Perl *(1927–   )* for the discovery of the tau lepton

Frederick Reines *(1918–   )* for the detection of the neutrino

1996 David M. Lee *(1931–   )*, Robert C. Richardson *(1937–   )* and Douglas D. Oscheroff *(1944–   )* for the discovery of superfluidity in helium-3

Chapter 1

EP **3.** (a) 186 mi; (b) 3.0×10^8 mm **5.** (a) 10^9; (b) 10^{-4}; (c) 9.1×10^5 **7.** 32.2 km **9.** 0.020 km³ **11.** (a) 250 ft²; (b) 23.3 m²; (c) 3060 ft³; (d) 86.6 m³ **13.** 8×10^2 km **15.** (a) 11.3 m²/L; (b) 1.13×10^4 m⁻¹; (c) 2.17×10^{-3} gal/ft² **17.** (a) $d_{Sun}/d_{Moon} = 400$; (b) $V_{Sun}/V_{Moon} = 6.4 \times 10^7$; (c) 3.5×10^3 km **19.** (a) 0.98 ft/ns; (b) 0.30 mm/ps **21.** 3.156×10^7 s **23.** 5.79×10^{12} days **25.** (a) 0.013; (b) 0.54; (c) 10.3; (d) 31 m/s **27.** 15° **29.** 3.3 ft **31.** 2 days 5 hours **33.** (a) 2.99×10^{-26} kg; (b) 4.68×10^{46} **35.** 1.3×10^9 kg **37.** (a) 10^3 kg/m³; (b) 158 kg/s **39.** (a) 1.18×10^{-29} m³; (b) 0.282 nm

Chapter 2

CP **1.** b and c **2.** zero **3.** (a) 1 and 4; (b) 2 and 3; (c) 3 **4.** (a) plus; (b) minus; (c) minus; (d) plus **5.** 1 and 4 **6.** (a) plus; (b) minus; (c) $a = -g = -9.8$ m/s² **Q** **1.** (a) yes; (b) no; (c) yes; (d) yes **3.** (a) 2, 3; (b) 1, 3; (c) 4 **5.** all tie (see Eq. 2-16) **7.** (a) $-g$; (b) 2 m/s upward **9.** same **11.** $x = t^2$ and $x = 8(t - 2) + (1.5)(t - 2)^2$ **13.** increase **EP** **1.** (a) Lewis: 10.0 m/s, Rodgers: 5.41 m/s; (b) 1 h 10 min **3.** 309 ft **5.** 2 cm/y **7.** 6.71×10^8 mi/h, 9.84×10^8 ft/s, 1.00 ly/y **9.** (a) 5.7 ft/s; (b) 7.0 ft/s **11.** (a) 45 mi/h (72 km/h); (b) 43 mi/h (69 km/h); (c) 44 mi/h (71 km/h); (d) 0 **13.** (a) 28.5 cm/s; (b) 18.0 cm/s; (c) 40.5 cm/s; (d) 28.1 cm/s; (e) 30.3 cm/s **15.** (a) mathematically, an infinite number; (b) 60 km **17.** (a) 4 s > t > 2 s; (b) 3 s > t > 0; (c) 7 s > t > 3 s; (d) $t = 3$ s **19.** 100 m **23.** (a) The signs of v and a are: AB: +, −; BC: 0, 0; CD: +, +; DE: +, 0; (b) no; (c) no **25.** (e) situations (a), (b), and (d) **27.** (a) 80 m/s; (b) 110 m/s; (c) 20 m/s² **29.** (a) 1.10 m/s, 6.11 mm/s²; (b) 1.47 m/s, 6.11 mm/s² **31.** (a) 2.00 s; (b) 12 cm from left edge of screen; (c) 9.00 cm/s², to the left; (d) to the right; (e) to the left; (f) 3.46 s **33.** 0.556 s **35.** each, 0.28 m/s² **37.** 2.8 m/s² **39.** 1.62×10^{15} m/s² **41.** 21g **43.** (a) 25g; (b) 400 m **45.** 90 m **47.** (a) 5.0 m/s²; (b) 4.0 s; (c) 6.0 s; (d) 90 m **49.** (a) 5.00 m/s²; (b) 1.67 m/s²; (c) 7.50 m **51.** (a) 0.74 s; (b) -20 ft/s² **53.** (a) 0.75 s; (b) 50 m **55.** (a) 34.7 ft; (b) 41.6 s **57.** (a) 3.26 ft/s² **61.** (a) 31 m/s; (b) 6.4 s **63.** (a) 48.5 m/s; (b) 4.95 s; (c) 34.3 m/s; (d) 3.50 s **65.** (a) 5.44 m/s; (b) 53.3 m/s; (d) 5.80 m **67.** (a) 3.2 s; (b) 1.3 s **69.** 4.0 m/s **71.** (a) 350 ms; (b) 82 ms (each is for ascent and descent through the 15 cm) **73.** 857 m/s², upward **75.** (a) 1.23 cm; (b) 4 times, 9 times, 16 times, 25 times **77.** (a) 8.85 m/s; (b) 1.00 m **79.** 22 cm and 89 cm below the nozzle **81.** (a) 3.41 s; (b) 57 m **83.** (a) 40.0 ft/s **85.** 1.5 s **87.** (a) 5.4 s; (b) 41 m/s **89.** 20.4 m

91. (a) $d = v_i^2/2a' + T_R v_i$; (b) 9.0 m/s²; (c) 0.66 s. **93.** (a) $v_j^2 = 2a'd_0(j - 1) + v_1^2$; (c) 7.0 m/s²; (d) 14 m.

Chapter 3

CP **1.** (a) 7 m; (b) 1 m **2.** c, d, f **3.** (a) +, +; (b) +, −; (c) +, + **4.** (a) 90°; (b) 0 (vectors are parallel); (c) 180° (vectors are antiparallel) **5.** (a) 0° or 180°; (b) 90° **Q** **1.** **A** and **B** **3.** No, but **a** and $-$**b** are commutative: **a** + $(-$**b**$) = (-$**b**$) + $**a**. **5.** (a) **a** and **b** are parallel; (b) **b** = 0; (c) **a** and **b** are perpendicular **7.** (a)–(c) yes (example: 5**i** and -2**i**) **9.** all but e **11.** (a) minus, minus; (b) minus, minus **13.** (a) **B** and **C**, **D** and **E**; (b) **D** and **E** **15.** no (their orientations can differ) **17.** (a) 0 (vectors are parallel); (b) 0 (vectors are antiparallel) **EP** **1.** The displacements should be (a) parallel, (b) antiparallel, (c) perpendicular **3.** (b) 3.2 km, 41° south of west **5.** **a** + **b**: 4.2, 40° east of north; **b** − **a**: 8.0, 24° north of west **7.** (a) 38 units at 320°; (b) 130 units at 1.2°; (c) 62 units at 130° **9.** $a_x = -2.5$, $a_y = -6.9$ **11.** $r_x = 13$ m, $r_y = 7.5$ m **13.** (a) 14 cm, 45° left of straight down; (b) 20 cm, vertically up; (c) zero **15.** 4.74 km **17.** 168 cm, 32.5° above the floor **19.** $r_x = 12$, $r_y = -5.8$, $r_z = -2.8$ **21.** (a) 8**i** + 2**j**, 8.2, 14°; (b) 2**i** − 6**j**, 6.3, $-72°$ relative to **i** **23.** (a) 5.0, $-37°$; (b) 10, 53°; (c) 11, 27°; (d) 11, 80°; (e) 11, 260°; the angles are relative to $+x$, the last two vectors are in opposite directions **25.** 4.1 **27.** (a) $r_x = 1.59$, $r_y = 12.1$; (b) 12.2; (c) 82.5° **29.** 3390 ft, horizontally **31.** (a) -2.83 m, -2.83 m, $+5.00$ m, 0 m, 3.00 m, 5.20 m; (b) 5.17 m, 2.37 m; (c) 5.69 m, 24.6° north of east; (d) 5.69 m, 24.6° south of west **35.** (a) $a_x = 9.51$ m, $a_y = 14.1$ m; (b) $a_x' = 13.4$ m, $a_y' = 10.5$ m **37.** (a) $+y$; (b) $-y$; (c) 0; (d) 0; (e) $+z$; (f) $-z$; (g) ab, both; (h) ab/d, $+z$ **39.** yes **41.** (a) up, unit magnitude; (b) zero; (c) south, unit magnitude; (d) 1.00; (e) 0 **43.** (a) -18.8; (b) 26.9, $+z$ direction **45.** (a) 12, out of page; (b) 12, into page; (c) 12, out of page **47.** (a) 11**i** + 5**j** − 7**k**; (b) 120° **51.** (a) 57°; (b) $c_x = \pm2.2$, $c_y = \mp4.5$ **53.** (a) -21; (b) -9; (c) 5**i** − 11**j** − 9**k**

Chapter 4

CP **1.** (a) (8**i** − 6**j**) m; (b) yes, the xy plane **2.** (a) first; (b) third **3.** (1) and (3) a_x and a_y are both constant and thus **a** is constant; (2) and (4) a_y is constant but a_x is not, thus **a** is not **4.** 4 m/s³, -2 m/s, 3 m **5.** (a) v_x constant; (b) v_y initially positive, decreases to zero, and then becomes progressively more negative; (c) $a_x = 0$ throughout; (d) $a_y = -g$ throughout **6.** (a) $-(4$ m/s$)$**i**; (b) $-(8$ m/s²$)$**j** **7.** (1) 0, distance not changing; (2) $+70$ km/h, distance increasing; (3) $+80$ km/h, distance decreasing **Q** **1.** (1) and (3) a_y is constant but a_x is not and thus **a** is not; (2) a_x is constant but a_y

is not and thus **a** is not; (4) a_x and a_y are both constant and thus **a** is constant; -2 m/s^2, 3 m/s   **3.** (a) highest point; (b) lowest point   **5.** (a) all tie; (b) 1 and 2 tie (the rocket is shot upward), then 3 and 4 tie (it is shot into the ground!)   **7.** $(2\mathbf{i} - 4\mathbf{j})$ m/s   **9.** (a) all tie; (b) all tie; (c) c, b, a; (d) c, b, a **11.** (a) no; (b) same   **13.** (a) in your hands; (b) behind you; (c) in front of you   **15.** (a) straight down; (b) curved; (c) more curved   **17.** (a) 3; (b) 4.   **EP** **1.** (a) $(-5.0\mathbf{i} + 8.0\mathbf{j})$ m; (b) 9.4 m, 122° from $+x$; (d) $(8\mathbf{i} - 8\mathbf{j})$ m; (e) 11 m, $-45°$ from $+x$ **3.** (a) $(-7.0\mathbf{i} + 12\mathbf{j})$ m; (b) xy plane   **5.** (a) 671 mi, 63.4° south of east; (b) 298 mi/h, 63.4° south of east; (c) 400 mi/h   **7.** (a) 6.79 km/h; (b) 6.96°   **9.** (a) $(3\mathbf{i} - 8t\mathbf{j})$ m/s; (b) $(3\mathbf{i} - 16\mathbf{j})$ m/s; (c) 16 m/s, $-79°$ to $+x$ **11.** (a) $(8t\mathbf{j} + \mathbf{k})$ m/s; (b) $8\mathbf{j}$ m/s^2 **13.** $(-2.10\mathbf{i} + 2.81\mathbf{j})$ m/s^2 **15.** (a) $-1.5\mathbf{j}$ m/s; (b) $(4.5\mathbf{i} - 2.25\mathbf{j})$ m   **17.** 60.0°   **19.** (a) 63 ms; (b) 1.6×10^3 ft/s   **21.** (a) 2.0 ns; (b) 2.0 mm; (c) $(1.0 \times 10^9\mathbf{i} - 2.0 \times 10^8\mathbf{j})$ cm/s   **23.** (a) 3.03 s; (b) 758 m; (c) 29.7 m/s   **25.** (a) 16 m/s, 23° above the horizontal; (b) 27 m/s, 57° below the horizontal   **27.** (a) 32.4 m; (b) -37.7 m   **29.** (b) 76°   **31.** (a) 51.8 m; (b) 27.4 m/s; (c) 67.5 m   **33.** (a) 194 m/s; (b) 38°   **35.** 1.9 in.   **37.** (a) 11 m; (b) 23 m; (c) 17 m/s, 63° below horizontal   **41.** (a) 73 ft; (b) 7.6°; (c) 1.0 s   **43.** 23 ft/s   **45.** (a) 11 m; (b) 45 m/s   **47.** 30 m above the release point   **49.** 19 ft/s   **51.** (a) 202 m/s; (b) 806 m; (c) 161 m/s, -171 m/s   **53.** (a) 20 cm; (b) no, the ball hits the net only 4.4 cm above the ground   **55.** yes; its center passes about 4.1 ft above the fence   **57.** (a) 9.00×10^{22} m/s^2, toward the center; (b) 1.52×10^{-16} s   **59.** (a) 6.7×10^6 m/s; (b) 1.4×10^{-7} s   **61.** (a) 7.49 km/s; (b) 8.00 m/s^2 **63.** (a) 0.94 m; (b) 19 m/s; (c) 2400 m/s^2, toward center; (d) 0.05 s   **65.** (a) 1.3×10^5 m/s; (b) 7.9×10^5 m/s^2 or $(8.0 \times 10^4)g$, toward the center; (c) both answers increase   **67.** (a) 0.034 m/s^2; (b) 84 min   **69.** 2.58 cm/s^2 **71.** 160 m/s^2 **73.** 36 s, no   **75.** 0.018 mi/s^2 from either frame   **77.** 130°   **79.** 60°   **81.** (a) 5.8 m/s; (b) 16.7 m; (c) 67°   **83.** 185 km/h, 22° south of west   **85.** (a) from 75° east of south; (b) 30° east of north; substitute west for east to get second solution   **87.** (a) 30° upstream; (b) 69 min; (c) 80 min; (d) 80 min; (e) perpendicular to the current, the shortest possible time is 60 min   **89.** $0.83c$ **91.** (a) $0.35c$; (b) $0.62c$ **93.** For launch angles from 5° to 70°, it always moves away from the launch site. For a 75° launch angle, it moves toward the site from 11.5 s to 18.5 s after launch. For an 80° launch angle, it moves toward the site from 10.5 s to 20.5 s after launch. For an 85° launch angle, it moves toward the site from 10.5 s to 20.5 s after launch. For a 90° launch angle, it moves toward the site from 10 s to 20.5 s after launch.   **95.** (a) 1.6 s; (b) no; (c) 14 m/s; (d) yes   **97.** (a) $\Delta\mathbf{D} = (1.0 \text{ m})\mathbf{i} - (2.0 \text{ m})\mathbf{j} + (1.0 \text{ m})\mathbf{k}$; (b) 2.4 m; (c) $\bar{\mathbf{v}} = (0.025 \text{ m/s})\mathbf{i} - (0.050 \text{ m/s})\mathbf{j} + (0.025 \text{ m/s})\mathbf{k}$ (d) cannot be determined without additional information

Chapter 5

CP **1.** c, d, and e **2.** (a) and (b) 2 N, leftward (acceleration is zero in each situation)   **3.** (a) and (b) 1, 4, 3, 2   **4.** (a) equal; (b) greater (acceleration is upward, thus net force on body must be upward)   **5.** (a) equal; (b) greater; (c) less   **6.** (a) increase; (b) yes; (c) same; (d) yes   **7.** (a) $F \sin \theta$; (b) increase   **8.** 0   **Q** **1.** (a) yes; (b) yes; (c) yes; (d) yes   **3.** (a) 2 and 4; (b) 2 and 4   **5.** (a) 50 N, upward; (b) 150 N, upward   **7.** (a) less; (b) greater   **9.** (a) no; (b) no; (c) no   **11.** (a) increases; (b) increases; (c) decreases; (d) decreases   **13.** (a) 20 kg; (b) 18 kg; (c) 10 kg; (d) all tie; (e) 3, 2, 1   **15.** d, c, a, b **EP** **1.** (a) $F_x = 1.88$ N, $F_y = 0.684$ N; (b) $(1.88\mathbf{i} + 0.684\mathbf{j})$ N   **3.** (a) $(-6.26\mathbf{i} - 3.23\mathbf{j})$ N; (b) 7.0 N, 207° relative to $+x$ **5.** $(-2\mathbf{i} + 6\mathbf{j})$ N   **7.** (a) 0; (b) $+20$ N; (c) -20 N; (d) -40 N; (e) -60 N   **9.** (a) $(1\mathbf{i} - 1.3\mathbf{j})$ m/s^2; (b) 1.6 m/s^2 at $-50°$ from $+x$ **11.** (a) $\mathbf{F}_2$ and $\mathbf{F}_3$ are in the $-x$ direction, $\mathbf{a} = 0$; (b) $\mathbf{F}_2$ and $\mathbf{F}_3$ are in the $-x$ direction, $\mathbf{a}$ is on the x axis, $a = 0.83$ m/s^2; (c) $\mathbf{F}_2$ and $\mathbf{F}_3$ are at 34° from $-x$ direction; $\mathbf{a} = 0$ **13.** (a) 22 N, 2.3 kg; (b) 1100 N, 110 kg; (c) 1.6×10^4 N, 1.6×10^3 kg   **15.** (a) 11 N, 2.2 kg; (b) 0, 2.2 kg   **17.** (a) 44 N; (b) 78 N; (c) 54 N; (d) 152 N   **19.** 1.18×10^4 N   **21.** 1.2×10^5 N   **23.** 16 N   **25.** (a) 13 ft/s^2; (b) 190 lb   **27.** (a) 42 N; (b) 72 N; (c) 4.9 m/s^2 **29.** (a) 0.02 m/s^2; (b) 8×10^4 km; (c) 2×10^3 m/s   **31.** (a) 1.1×10^{-15} N; (b) 8.9×10^{-30} N   **33.** (a) 5500 N; (b) 2.7 s; (c) 4 times as far; (d) twice the time   **35.** (a) 4.9×10^5 N; (b) 1.5×10^6 N   **37.** (a) 110 lb, up; (b) 110 lb, down   **39.** (a) 0.74 m/s^2; (b) 7.3 m/s^2 **41.** (a) $\cos \theta$; (b) $\sqrt{\cos \theta}$ **43.** 1.8×10^4 N   **45.** (a) 4.6×10^3 N; (b) 5.8×10^3 N   **47.** (a) 250 m/s^2; (b) 2.0×10^4 N   **49.** 23 kg   **51.** (a) 620 N; (b) 580 N   **53.** 1.9×10^5 lb   **55.** (a) rope breaks; (b) 1.6 m/s^2 **57.** 4.6 N   **59.** (a) allow a downward acceleration with magnitude ≥ 4.2 ft/s^2; (b) 13 ft/s or greater   **61.** 195 N, up   **63.** (a) 566 N; (b) 1130 N   **65.** 18,000 N   **67.** (a) 1.4×10^4 N; (b) 1.1×10^4 N; (c) 2700 N, toward the counterweight   **69.** 6800 N, at 21° to the line of motion of the barge   **71.** (a) 4.6 m/s^2; (b) 2.6 m/s^2 **73.** (b) $F/(m + M)$; (c) $MF/(m + M)$; (d) $F(m + 2M)/2(m + M)$ **75.** $T_1 = 13$ N, $T_2 = 20$ N, $a = 3.2$ m/s^2

Chapter 6

CP **1.** (a) zero (because there is no attempt at sliding); (b) 5 N; (c) no; (d) yes   **2.** (a) same (10 N); (b) decreases; (c) decreases   **3.** greater   **4.** (a) **a** downward; **N** upward; (b) **a** and **N** upward   **5.** (a) $4R_1$; (b) $4R_1$ **6.** (a) same; (b) increases; (c) increases   **Q** **1.** They slide at the same angle for all orders.   **3.** (a) upward; (b) horizontal, toward you; (c) no change; (d) increases; (e) increases   **5.** The frictional force $\mathbf{f}_s$ is initially directed up the ramp, decreases in magnitude to zero, and then is directed down the ramp, increasing in magnitude until the magnitude reaches $f_{s,\max}$; thereafter, the magnitude of the frictional force is f_k, which is a constant smaller value.   **7.** (a) decreases; (b) decreases; (c) increases; (d) increases   **9.** (a) zero; (b) infinite   **11.** 4, 3; then 1, 2, and 5 tie   **13.** (a) less; (b) greater   **EP** **1.** (a) 200 N; (b) 120 N   **3.** 2°   **5.** 440 N

7. (a) 110 N; (b) 130 N; (c) no; (d) 46 N; (e) 17 N
9. (a) 90 N; (b) 70 N; (c) 0.89 m/s² **11.** (a) no; (b) $(-12\mathbf{i} + 5\mathbf{j})$ N **13.** 20° **15.** (a) 0.13 N; (b) 0.12 **17.** $\mu_s = 0.58$, $\mu_k = 0.54$ **19.** (a) 0.11 m/s², 0.23 m/s²; (b) 0.041, 0.029
21. 36 m **23.** (a) 300 N; (b) 1.3 m/s² **25.** (a) 66 N; (b) 2.3 m/s² **27.** (a) $\mu_k mg/(\sin\theta - \mu_k \cos\theta)$; (b) $\theta_0 = \tan^{-1}\mu_s$
29. (b) 3.0×10^7 N **31.** 100 N **33.** 3.3 kg **35.** (a) 11 ft/s²; (b) 0.46 lb; (c) blocks move independently
37. (a) 27 N; (b) 3.0 m/s² **39.** (a) 6.1 m/s², leftward; (b) 0.98 m/s², leftward **41.** (a) 3.0×10^5 N; (b) 1.2°
43. 9.9 s **45.** 3.75 **47.** 12 cm **49.** 68 ft
51. (a) 3210 N; (b) yes **53.** 0.078 **55.** (a) 0.72 m/s; (b) 2.1 m/s²; (c) 0.50 N **57.** $\sqrt{Mgr/m}$ **59.** (a) 30 cm/s; (b) 180 cm/s², radially inward; (c) 3.6×10^{-3} N, radially inward; (d) 0.37 **61.** (a) 275 N; (b) 877 N **63.** 874 N
65. (a) at the bottom of the circle; (b) 31 ft/s **67.** (a) 9.5 m/s; (b) 20 m **69.** 13° **71.** (a) 0.0338 N; (b) 9.77 N

Chapter 7

CP **1.** (a) decrease; (b) same; (c) negative, zero **2.** d, c, b, a **3.** (a) same; (b) smaller **4.** (a) positive; (b) negative; (c) zero **5.** zero Q **1.** all tie **3.** (a) increasing; (b) same; (c) same; (d) increasing **5.** (a) positive; (b) negative; (c) negative **7.** (a) positive; (b) zero; (c) negative; (d) negative; (e) zero; (f) positive **9.** all tie **11.** c, d, a and b tie; then f, e. **13.** (a) 3 m; (b) 3 m; (c) 0 and 6 m; (d) negative direction of x **15.** (a) A; (b) B **17.** twice EP **1.** 1.8×10^{13} J **3.** (a) 3610 J; (b) 1900 J; (c) 1.1×10^{10} J **5.** (a) 1×10^5 megatons TNT; (b) 1×10^7 bombs **7.** father, 2.4 m/s; son, 4.8 m/s **9.** (a) 200 N; (b) 700 m; (c) -1.4×10^5 J; (d) 400 N, 350 m, -1.4×10^5 J **11.** 5000 J **13.** 47 keV **15.** 7.9 J **17.** 530 J **19.** -37 J **21.** (a) 314 J; (b) -155 J; (c) 0; (d) 158 J **23.** (a) 98 N; (b) 4.0 cm; (c) 3.9 J; (d) -3.9 J **25.** (a) $-3Mgd/4$; (b) Mgd; (c) $Mgd/4$ (d) $\sqrt{gd/2}$ **27.** 25 J **31.** -6 J **33.** (a) 12 J; (b) 4.0 m; (c) 18 J **35.** (a) -0.043 J; (b) -0.13 J **37.** (a) 6.6 m/s; (b) 4.7 m **39.** (a) up; (b) 5.0 cm; (c) 5.0 J **41.** 270 kW **43.** 235 kW **45.** 490 W **47.** (a) 100 J; (b) 67 W; (c) 33 W **49.** 0.99 hp **51.** (a) 0; (b) -350 W **53.** (a) 79.4 keV; (b) 3.12 MeV; (c) 10.9 MeV **55.** (a) 32 J; (b) 8 W; (c) 78°

Chapter 8

CP **1.** no **2.** 3, 1, 2 **3.** (a) all tie; (b) all tie **4.** (a) CD, AB, BC (zero); (b) positive direction of x **5.** 2, 1, 3 **6.** decrease **7.** (a) seventh excited state, with energy E_7; (b) 1.3 eV Q **1.** -40 J **3.** (c) and (d) tie; then (a) and (b) tie **5.** (a) all tie; (b) all tie **7.** (a) 3, 2, 1; (b) 1, 2, 3 **9.** less than (smaller decrease in potential energy) **11.** (a) $E < 3$ J, $K < 2$ J; (b) $E < 5$ J, $K < 4$ J **13.** (a) increasing; (b) decreasing; (c) decreasing; (d) constant in AB and BC, decreasing in CD EP **1.** 15 J **3.** (a) 167 J; (b) -167 J; (c) 196 J; (d) 29 J **5.** (a) 0; (b) $mgh/2$; (c) mgh; (d) $mgh/2$; (e) mgh **7.** (a) -0.80 J; (b) -0.80 J; (c) $+1.1$ J **9.** (a) $mgL(1 - \cos\theta)$; (b) $-mgL(1 - \cos\theta)$; (c) $mgL(1 - \cos\theta)$ **11.** (a) 18 J;

(b) 0; (c) 30 J; (d) 0; (e) parts b and d **13.** (a) 2.08 m/s; (b) 2.08 m/s **15.** (a) $\sqrt{2gL}$; (b) $2\sqrt{gL}$; (c) $\sqrt{2gL}$ **17.** 830 ft **19.** (a) 6.75 J; (b) -6.75 J; (c) 6.75 J; (d) 6.75 J; (e) -6.75 J; (f) 0.459 m **21.** (a) 21.0 m/s; (b) 21.0 m/s **23.** (a) 0.98 J; (b) -0.98 J; (c) 3.1 N/cm **25.** (a) 39.2 J; (b) 39.2 J; (c) 4.00 m **27.** (a) 54 m/s; (b) 52 m/s; (c) 76 m, below **29.** (a) 39 ft/s; (b) 4.3 in. **31.** (a) 300 J; (b) 93.8 J; (c) 6.38 m **33.** (a) 4.8 m/s; (b) 2.4 m/s **35.** (a) $[v_0^2 + 2gL(1 - \cos\theta_0)]^{1/2}$; (b) $(2gL \cos\theta_0)^{1/2}$; (c) $[gL(3 + 2 \cos\theta_0)]^{1/2}$ **37.** (a) $U(x) = -Gm_1m_2/x$; (b) $Gm_1m_2d/x_1(x_1 + d)$ **39.** (a) $8mg$ leftward and mg downward; (b) $2.5R$ **43.** $mgL/32$ **47.** (a) $1.12(A/B)^{1/6}$; (b) repulsive; (c) attractive **49.** (a) turning point on left, none on right; molecule breaks apart; (b) turning points on both left and right; molecule does not break apart; (c) -1.2×10^{-19} J; (d) 2.2×10^{-19} J; (e) $\approx 1 \times 10^{-9}$ on each, directed toward the other; (f) $r < 0.2$ nm; (g) $r > 0.2$ nm; (h) $r = 0.2$ nm **51.** -25 J **53.** (a) 2200 J; (b) -1500 J; (c) 700 J **55.** 17 kW **57.** (a) -0.74 J; (b) -0.53 J **59.** -12 J **61.** 54% **63.** 880 MW **65.** (a) 39 kW; (b) 39 kW **67.** (a) 1.5 MJ; (b) 0.51 MJ; (c) 1.0 MJ; (d) 63 m/s **69.** (a) 67 J; (b) 67 J; (c) 46 cm **71.** Your force on the cabbage does work. **73.** (a) -0.90 J; (b) 0.46 J; (c) 1.0 m/s **75.** (a) 18 ft/s; (b) 18 ft **77.** 4.3 m **79.** (a) 31.0 J; (b) 5.35 m/s; (c) conservative **81.** 1.2 m **85.** in the center of the flat part **87.** (a) 24 ft/s; (b) 3.0 ft; (c) 9.0 ft; (d) 49 ft **89.** (a) 216 J; (b) 1180 N; (c) 432 J; (d) motor also supplies thermal energy to crate and belt **91.** (a) 1.1×10^{17} J; (b) 1.2 kg **93.** 7.28 MeV **95.** (a) release; (b) 17.6 MeV **97.** (a) 5.3 eV; (b) 0.9 eV **99.** (a) 7.2 J; (b) -7.2 J; (c) 86 cm; (d) 26 cm

Chapter 9

CP **1.** (a) origin; (b) fourth quadrant; (c) on y axis below origin; (d) origin; (e) third quadrant; (f) origin **2.** (a) to (c) at the center of mass, still at the origin (their forces are internal to the system and cannot move the center of mass) **3.** (a) 1, 3, and then 2 and 4 tie (zero force); (b) 3 **4.** (a) 0; (b) no; (c) negative x **5.** (a) 500 km/h; (b) 2600 km/h; (c) 1600 km/h **6.** (a) yes; (b) no Q **1.** point 4 **3.** (a) at the center of the sled; (b) L/4, to the right; (c) not at all (no net external force); (d) L/4, to the left; (e) L; (f) L/2; (g) L/2 **5.** (a) ac, cd, and bc; (b) bc; (c) bd and ad **7.** (a) 2 N, rightward; (b) 2 N, rightward; (c) greater than 2 N, rightward **9.** b, c, a **11.** (a) yes; (b) 6 kg·m/s in $-x$ direction; (c) can't tell EP **1.** (a) 4600 km; (b) $0.73R_e$ **3.** (a) $x_{cm} = 1.1$ m, $y_{cm} = 1.3$ m; (b) shifts toward topmost particle **5.** $x_{cm} = -0.25$ m, $y_{cm} = 0$ **7.** in the iron, at midheight and midwidth, 2.7 cm from midlength **9.** $x_{cm} = y_{cm} = 20$ cm, $z_{cm} = 16$ cm **11.** (a) H/2; (b) H/2; (c) descends to lowest point and then ascends to H/2; (d) $(HM/m)(\sqrt{1 + m/M} - 1)$ **13.** 72 km/h **15.** (a) center of mass does not move; (b) 0.75 m **17.** 4.8 m/s **19.** (a) 22 m; (b) 9.3 m/s **21.** 53 m **23.** 13.6 ft **25.** (a) 52.0 km/h; (b) 28.8 km/h **27.** a proton **29.** (a) 30°; (b) $-0.572\mathbf{j}$ kg·m/s **31.** (a) $(-4.0 \times 10^4 \mathbf{i})$ kg·m/s; (b) west; (c) 0 **33.** $0.707c$ **35.** 0.57 m/s, toward center of mass **37.** it increases by

4.4 m/s   **39.** (a) rocket case: 7290 m/s, payload: 8200 m/s; (b) before: 1.271×10^{10} J, after: 1.275×10^{10} J   **41.** (a) -1; (b) 1830; (c) 1830; (d) same   **43.** 14 m/s, 135° from the other pieces   **45.** 190 m/s   **47.** (a) $0.200v_{rel}$; (b) $0.210v_{rel}$; (c) $0.209v_{rel}$ **49.** (a) 1.57×10^6 N; (b) 1.35×10^5 kg; (c) 2.08 km/s   **51.** 108 m/s   **53.** 2.2×10^{-3} **57.** fast barge: 46 N more; slow barge: no change   **59.** (a) 7.8 MJ; (b) 6.2   **61.** 690 W   **63.** 5.5×10^6 N   **65.** 24 W   **67.** 100 m   **69.** (a) 860 N; (b) 2.4 m/s   **71.** (a) 3.0×10^5 J; (b) 10 kW; (c) 20 kW   **73.** (a) 2.1×10^6 kg; (b) $\sqrt{100 + 1.5t}$ m/s; (c) $(1.5 \times 10^6)/\sqrt{100 + 1.5t}$ N; (d) 6.7 km   **75.** $t = (3d/2)^{2/3}(m/2P)^{1/3}$

Chapter 10

CP **1.** (a) unchanged; (b) unchanged; (c) decreased **2.** (a) zero; (b) positive; (c) positive direction of y **3.** (a) 4 kg·m/s; (b) 8 kg·m/s; (c) 3 J **4.** (a) 0; (b) 4 kg·m/s **5.** (a) 10 kg·m/s; (b) 14 kg·m/s; (c) 6 kg·m/s **6.** (a) 2 kg·m/s; (b) 3 kg·m/s **7.** (a) increases; (b) increases **Q** **1.** all tie **3.** b and c **5.** (a) one stationary; (b) 2; (c) 5; (d) equal (pool player's result) **7.** (a) 1 and 4 tie; then 2 and 3 tie; (b) 1; 3 and 4 tie; then 2 **9.** (a) rightward; (b) rightward; (c) smaller **11.** positive direction of x axis **EP** **1.** (a) 750 N; (b) 6.0 m/s **3.** 6.2×10^4 N **5.** 3000 N ($= 660$ lb) **7.** 1.1 m **9.** (a) 42 N·s; (b) 2100 N **11.** (a) $(7.4 \times 10^3 \text{ i} - 7.4 \times 10^3 \text{ j})$ N·s; (b) $(-7.4 \times 10^3 \text{ i})$ N·s; (c) 2.3×10^3 N; (d) 2.1×10^4 N; (e) $-45°$ **13.** (a) 1.0 kg·m/s; (b) 250 J; (c) 10 N; (d) 1700 N **15.** 5 N **17.** $2\mu v$ **19.** 990 N **21.** (a) 1.8 N·s, upward; (b) 180 N, downward **25.** 8 m/s **27.** 38 km/s **29.** 4.2 m/s **31.** (a) 99 g; (b) 1.9 m/s; (c) 0.93 m/s **33.** (a) 1.2 kg; (b) 2.5 m/s **35.** 7.8 kg **37.** (a) 1/3; (b) $4h$ **39.** 35 cm **41.** 3.0 m/s **43.** (a) $(10i + 15j)$ m/s; (b) 500 J lost **45.** (a) 2.7 m/s; (b) 1400 m/s **47.** (a) A: 4.6 m/s, B: 3.9 m/s; (b) 7.5 m/s **49.** 20 J for the heavy particle, 40 J for the light particle **51.** $mv^2/6$ **53.** 13 tons **55.** 25 cm **57.** 0.975 m/s, 0.841 m/s **59.** (a) 4.1 ft/s; (b) 1700 ft·lb; (c) $v_{24} = 5.3$ ft/s, $v_{32} = 3.3$ ft/s **61.** (a) 30° from the incoming proton's direction; (b) 250 m/s and 430 m/s **63.** (a) 41°; (b) 4.76 m/s; (c) no **65.** $v = V/4$ **67.** (a) 117° from the final direction of B; (b) no **69.** 120° **71.** (a) 1.9 m/s, 30° to initial direction; (b) no **73.** (a) 3.4 m/s, deflected by 17° to the right; (b) 0.95 MJ **75.** (a) 117 MeV; (b) equal and opposite momenta; (c) π^- **77.** (a) 4.94 MeV; (b) 0; (c) 4.85 MeV; (d) 0.09 MeV

Chapter 11

CP **1.** (b) and (c) **2.** (a) and (d) **3.** (a) yes; (b) no; (c) yes; (d) yes **4.** all tie **5.** 1, 2, 4, 3 **6.** (a) 1 and 3 tie, 4; then 2 and 5 tie (zero) **7.** (a) downward in the figure; (b) less **Q** **1.** (a) positive; (b) zero; (c) negative; (d) negative **3.** (a) 2 and 3; (b) 1 and 3; (c) 4 **5.** (a) and (c) **7.** (a) all tie; (b) 2, 3; then 1 and 4 tie **9.** b, c, a **11.** less **13.** 90°; then 70° and 110° tie **15.** Finite angular

displacements are not commutative. **EP** **1.** (a) 1.50 rad; (b) 85.9°; (c) 1.49 m **3.** (a) 0.105 rad/s; (b) 1.75×10^{-3} rad/s; (c) 1.45×10^{-4} rad/s **5.** (a) $\omega(2) = 4.0$ rad/s, $\omega(4) = 28$ rad/s; (b) 12 rad/s²; (c) $\alpha(2) = 6.0$ rad/s², $\alpha(4) = 18$ rad/s² **7.** (a) $\omega_0 + at^4 - bt^3$; (b) $\theta_0 + \omega_0 t + at^5/5 - bt^4/4$ **9.** 11 rad/s **11.** (a) 9000 rev/min²; (b) 420 rev **13.** (a) 30 s; (b) 1800 rad **15.** 200 rev/min **17.** (a) 2.0 rad/s²; (b) 5.0 rad/s; (c) 10 rad/s; (d) 75 rad **19.** (a) 13.5 s; (b) 27.0 rad/s **21.** (a) 340 s; (b) -4.5×10^{-3} rad/s²; (c) 98 s **23.** (a) 1.0 rev/s²; (b) 4.8 s; (c) 9.6 s; (d) 48 rev **25.** 6.1 ft/s² (1.8 m/s²), toward the center **27.** 0.13 rad/s **29.** 5.6 rad/s² **31.** (a) 5.1 h; (b) 8.1 h **33.** (a) 2.50×10^{-3} rad/s; (b) 20.2 m/s²; (c) 0 **35.** (a) -1.1 rev/min²; (b) 9900 rev; (c) -0.99 mm/s²; (d) 31 m/s² **37.** (a) 310 m/s; (b) 340 m/s **39.** (a) 1.94 m/s²; (b) 75.1°, toward the center of the track **41.** 16 s **43.** (a) 73 cm/s²; (b) 0.075; (c) 0.11 **45.** 12.3 kg·m² **47.** first cylinder: 1100 J; second cylinder: 9700 J **49.** (a) 221 kg·m²; (b) 1.10×10^4 J **51.** (a) 6490 kg·m²; (b) 4.36 MJ **53.** 0.097 kg·m² **57.** (a) 1300 g·cm²; (b) 550 g·cm²; (c) 1900 g·cm²; (d) $A + B$ **59.** (a) 49 MJ; (b) 100 min **61.** 4.6 N·m **63.** (a) $r_1 F_1 \sin \theta_1 - r_2 F_2 \sin \theta_2$; (b) -3.8 N·m **65.** 1.28 kg·m² **67.** 9.7 rad/s², counterclockwise **69.** (a) 155 kg·m²; (b) 64.4 kg **71.** (a) 420 rad/s²; (b) 500 rad/s **73.** small sphere: (a) 0.689 N·m and (b) 3.05 N; large sphere: (a) 9.84 N·m and (b) 11.5 N **75.** 1.73 m/s²; 6.92 m/s² **77.** (a) 1.4 m/s²; (b) 1.4 m/s **79.** (a) 19.8 kJ; (b) 1.32 kW **81.** (a) 8.2×10^{28} N·m; (b) 2.6×10^{29} J; (c) 3.0×10^{21} kW **83.** $\sqrt{9g/4\ell}$ **85.** (a) 4.8×10^5 N; (b) 1.1×10^4 N·m; (c) 1.3×10^6 J **87.** (a) $3g(1 - \cos \theta)$; (b) $\frac{3}{2}g \sin \theta$; (c) 41.8° **89.** (a) 5.6 rad/s²; (b) 3.1 rad/s **91.** (a) 42.1 km/h; (b) 3.09 rad/s²; (c) 7.57 kW **93.** (a) 3.4×10^5 g·cm²; (b) 2.9×10^5 g·cm²; (c) 6.3×10^5 g·cm²; (d) $(1.2 \text{ cm}) \text{ i} + (5.9 \text{ cm}) \text{ j}$

Chapter 12

CP **1.** (a) same; (b) less **2.** less **3.** (a) $\pm z$; (b) $+y$; (c) $-x$ **4.** (a) 1 and 3 tie, then 2 and 4 tie, then 5 (zero); (b) 2 and 3 **5.** (a) 3, 1; then 2 and 4 tie (zero); (b) 3 **6.** (a) all tie (same τ, same t, thus same ΔL); (b) sphere, disk, hoop (reverse order of I) **7.** (a) decreases; (b) same; (c) increases **Q** **1.** (a) same; (b) block; (c) block **3.** (a) greater; (b) same **5.** (a) L; (b) 1.5L **7.** b, then c and d tie; then a and e tie (zero) **9.** a, then b and c tie; then e, d (zero) **11.** (a) same; (b) increases, because of decrease in rotational inertia **13.** (a) 30 units clockwise; (b) 2 then 4, then the others; or 4 then 2, then the others **15.** (a) spins in place; (b) rolls toward you; (c) rolls away from you **EP** **1.** 1.00 **3.** (a) 59.3 rad/s; (b) -9.31 rad/s²; (c) 70.7 m **5.** (a) -4.11 m/s²; (b) -16.4 rad/s²; (c) -2.54 N·m **7.** (a) 8.0°; (b) 0.14g **9.** (a) 4.0 N, to the left; (b) 0.60 kg·m² **11.** (a) $\frac{1}{2}mR^2$; (b) a solid circular cylinder **13.** (a) $mg(R - r)$; (b) 2/7; (c) (17/7)mg **15.** (a) 2.7R; (b) (50/7)mg **17.** (a) 13 cm/s²; (b) 4.4 s; (c) 55 cm/s; (d) 1.8×10^{-2} J; (e) 1.4 J; (f) 27 rev/s **21.** (a) 24 N·m, in $+y$ direction; (b) 24 N·m, $-y$; (c) 12 N·m, $+y$;

(d) 12 N·m, $-y$ **23.** (a) $(-1.5\mathbf{i} - 4.0\mathbf{j} - \mathbf{k})$ N·m;
(b) $(-1.5\mathbf{i} - 4.0\mathbf{j} - \mathbf{k})$ N·m   **25.** $-2.0\mathbf{i}$ N·m   **27.** 9.8
kg·m²/s   **29.** (a) 12 kg·m²/s, out of page; (b) 3.0 N·m,
out of page   **31.** (a) 0; (b) $(8.0\mathbf{i} + 8.0\mathbf{k})$ N·m   **33.** (a) mvd;
(b) no; (c) 0, yes   **35.** (a) 3.15×10^{43} kg·m²/s; (b) 0.616
37. 4.5 N·m, parallel to xy plane at $-63°$ from $+x$
39. (a) 0; (b) 0; (c) $30t^3$ kg·m²/s, $90t^2$ N·m, both in $-z$
direction; (d) $30t^3$ kg·m²/s, $90t^2$ N·m, both in $+z$ direction
41. (a) $\frac{1}{2}mgt^2v_0 \cos\theta_0$; (b) $mgtv_0 \cos\theta_0$; (c) $mgtv_0 \cos\theta_0$
43. (a) -1.47 N·m; (b) 20.4 rad; (c) -29.9 J; (d) 19.9 W
45. (a) 12.2 kg·m²; (b) 308 kg·m²/s, down   **47.** (a) 1/3;
(b) 1/9   **49.** $\omega_0 R_1 R_2 I_1/(I_1 R_2^2 + I_2 R_1^2)$ **51.** (a) 3.6 rev/s;
(b) 3.0; (c) work done by man in moving weights inward
53. (a) 267 rev/min; (b) 2/3   **55.** 3.0 min   **57.** 2.6 rad/s
59. (a) they revolve in a circle of 1.5 m radius at 0.93 rad/s;
(b) 8.4 rad/s; (c) $K_a = 98$ J, $K_b = 880$ J; (d) from the work
done in pulling inward   **61.** $m/(M + m)(v/R)$
63. (a) $mvR/(I + MR^2)$; (b) $mvR^2/(I + MR^2)$ **65.** 1300 m/s
67. (a) 18 rad/s; (b) 0.92

69. $\theta = \cos^{-1}\left[1 - \dfrac{6m^2h}{\ell(2m + M)(3m + M)}\right]$

71. 5.28×10^{-35} J·s   **73.** Any three are spin up; the other is
spin down.   **75.** (a) The magnitude of the angular momentum
increases in proportion to t^2 and the magnitude of the torque
increases in proportion to t, in agreement with the second law
for rotation. (b) The magnitudes of the angular momentum and
torque again increase with time. But the change in the
magnitude of the angular momentum in any interval is less than
is predicted by proportionality to t^2 law and the change in the
torque is less than is predicted by proportionality to t. At any
position of the projectile the torque is less when drag is present
than when it is not.

Chapter 13

CP **1.** c, e, f **2.** (a) no; (b) at site of $\mathbf{F}_1$, perpendicular to
plane of figure; (c) 45 N   **3.** (a) at C (to eliminate forces there
from a torque equation); (b) plus; (c) minus; (d) equal   **4.** d
5. (a) equal; (b) B; (c) B **Q** **1.** (a) yes; (b) yes; (c) yes;
(d) no   **3.** b **5.** (a) yes; (b) no; (c) no (it could balance the
torques but the forces would then be unbalanced)   **7.** (a) a,
then b and c tie, then d **9.** (a) 20 N (the key is the pulley
with the 20 N weight); (b) 25 N   **11.** (a) $\sin\theta$; (b) same;
(c) larger   **13.** tie of A and B, then C **EP** **1.** (a) two;
(b) seven   **3.** (a) 2.5 m; (b) 7.3°   **5.** 120°   **7.** 7920 N
9. (a) 840 N; (b) 530 N   **11.** 0.536 m   **13.** (a) 2770 N;
(b) 3890 N   **15.** (a) 1160 N, down; (b) 1740 N, up; (c) left,
stretched; (d) right, compressed   **17.** (a) 280 N; (b) 880 N,
71° above the horizontal   **19.** bars BC, CD, and DA are under
tension due to forces T, diagonals AC and BD are compressed
due to forces $\sqrt{2}T$ **21.** (a) 1800 lb; (b) 822 lb; (c) 1270 lb
23. (a) 49 N; (b) 28 N; (c) 57 N; (d) 29°   **25.** (a) 1900 N, up;
(b) 2100 N, down   **27.** (a) 340 N; (b) 0.88 m; (c) increases,
decreases   **29.** $W\sqrt{2rh - h^2}/(r - h)$ **31.** (a) $L/2$; (b) $L/4$;
(c) $L/6$; (d) $L/8$; (e) $25L/24$ **33.** (a) 6630 N; (b) $F_h = 5740$ N;
(c) $F_v = 5960$ N   **35.** 2.20 m   **37.** (a) 1.50 m; (b) 433 N;

(c) 250 N   **39.** (a) $a_1 = L/2$, $a_2 = 5L/8$, $h = 9L/8$;
(b) $b_1 = 2L/3$, $b_2 = L/2$, $h = 7L/6$ **41.** (a) 47 lb; (b) 120 lb;
(c) 72 lb   **43.** (a) 445 N; (b) 0.50; (c) 315 N
45. (a) 3.9 m/s²; (b) 2000 N on each rear wheel, 3500 N on
each front wheel; (c) 790 N on each rear wheel, 1410 N on
each front wheel   **47.** (a) 1.9×10^{-3}; (b) 1.3×10^7 N/m²;
(c) 6.9×10^9 N/m²   **49.** 3.1 cm   **51.** 2.4×10^9 N/m²
53. (a) 1.8×10^7 N; (b) 1.4×10^7 N; (c) 16   **55.** (a) 867 N;
(b) 143 N; (c) 0.165

Chapter 14

CP **1.** all tie   **2.** (a) 1, tie of 2 and 4, then 3; (b) line d
3. negative y direction   **4.** (a) increase; (b) negative
5. (a) 2; (b) 1   **6.** (a) path 1 (decreased E (more negative)
gives decreased a); (b) less than (decreased a gives decreased T)
Q **1.** (a) between, closer to less massive particle; (b) no;
(c) no (other than infinity)   **3.** $3GM^2/d^2$, leftward   **5.** b, tie
of a and c, then d **7.** b, a, c **9.** (a) negative; (b) negative;
(c) postive; (d) all tie   **11.** (a) all tie; (b) all tie
13. (a) same; (b) greater   **EP** **1.** 19 m   **3.** 2.16
5. 1/2   **7.** 3.4×10^5 km   **9.** (a) 3.7×10^{-5} N, increasing y **11.** $M = m$ **13.** 3.2×10^{-7} N   **15.** $(GmM/d^2) \times$
$\left[1 - \dfrac{1}{8(1 - R/2d)^2}\right]$ **17.** 2.6×10^6 m   **19.** (a) $1.3 \times$
10^{12} m/s²; (b) 1.6×10^6 m/s   **21.** (a) 17 N; (b) 2.5
23. (b) 1.9 h   **27.** (a) $a_g = (3.03 \times 10^{43}$ kg·m/s²$)/M_h$;
(b) decrease; (c) 9.82 m/s²; (d) 7.30×10^{-15} m/s²; (e) no
29. 7.91 km/s   **31.** (a) $(3.0 \times 10^{-7}m)$ N; (b) $(3.3 \times 10^{-7}m)$
N; (c) $(6.7 \times 10^{-7}mr)$ N   **33.** (a) 9.83 m/s²; (b) 9.84 m/s²;
(c) 9.79 m/s²   **35.** (a) -1.4×10^{-4} J; (b) less; (c) positive;
(d) negative   **37.** (a) 0.74; (b) 3.7 m/s²; (c) 5.0 km/s
39. (a) 0.0451; (b) 28.5   **41.** $-Gm(M_E/R + M_M/r)$
43. (a) 5.0×10^{-11} J; (b) -5.0×10^{-11} J   **45.** (a) 1700 m/s;
(b) 250 km; (c) 1400 m/s   **47.** (a) 2.2×10^{-7} rad/s;
(b) 90 km/s   **51.** (a) -1.67×10^{-8} J; (b) 0.56×10^{-8} J
55. 6.5×10^{23} kg   **57.** 5×10^{10} **59.** (a) 7.82 km/s;
(b) 87.5 min   **61.** (a) 6640 km; (b) 0.0136   **63.** (a) 39.5
AU³/M_S·y²; (b) $T^2 = r^3/M$ **65.** (a) 1.9×10^{13} m;
(b) $3.5R_p$ **67.** south, at 35.4° above the horizon
71. $2\pi r^{3/2}/\sqrt{G(M + m/4)}$ **73.** $\sqrt{GM/L}$ **75.** (a) 2.8 y;
(b) 1.0×10^{-4} **77.** (a) 1/2; (b) 1/2; (c) B, by 1.1×10^8 J
79. (a) 54 km/s; (b) 960 m/s; (c) $R_p/R_a = v_a/v_p$ **81.** (a) $4.6 \times$
10^5 J; (b) 260   **83.** (a) 7.5 km/s; (b) 97 min; (c) 410 km;
(d) 7.7 km/s; (e) 92 min; (f) 3.2×10^{-3} N; (g) if the satellite–
Earth system is considered isolated, its $\mathbf{L}$ is conserved
85. (a) 5540 s; (b) 7.68 km/s; (c) 7.60 km/s; (d) 5.78×10^{10} J;
(e) -11.8×10^{10} J; (f) -6.02×10^{10} J; (g) $6.63 \times$
10^6 m; (h) 170 s, new orbit   **87.** (a) $(-7.0$ mm$)\mathbf{i} +$
$(3.0$ cm$)\mathbf{j}$; (b) $(-0.19$ m/s$)\mathbf{i} + (0.40$ m/s$)\mathbf{j}$ **89.** (a) $1.98 \times$
10^{30} kg; (b) 1.96×10^{30} kg

Chapter 15

CP **1.** all tie   **2.** (a) all tie; (b) $0.95\rho_0$, ρ_0, $1.1\rho_0$
3. 13 cm³/s, outward   **4.** (a) all tie; (b) 1, then 2 and 3 tie, 4;

(c) 4, 3, 2, 1   Q   **1.** e, then b and d tie, then a and c tie
3. (a) 1, 3, 2; (b) all tie; (c) no (you must consider the weight
exerted on the scale via the walls)   **5.** 3, 4, 1, 2
7. (a) downward; (b) downward; (c) same   **9.** (a) same;
(b) same; (c) lower; (d) higher   **11.** (a) block 1,
counterclockwise; block 2, clockwise; (b) block 1, tip more;
block 2, right itself   EP   **1.** 1000 kg/m^3 **3.** 1.1×10^5 Pa
or 1.1 atm   **5.** 2.9×10^4 N   **7.** 6.0 lb/in.^{2} **9.** 1.90×10^4
Pa   **11.** 5.4×10^4 Pa   **13.** 0.52 m   **15.** (a) 6.06×10^9 N;
(b) 20 atm   **17.** 0.412 cm   **19.** $\frac{1}{4}\rho g A(h_2 - h_1)^2$
21. 44 km   **23.** (a) $\rho g W D^2/2$; (b) $\rho g W D^3/6$; (c) $D/3$
25. (a) 2.2; (b) 2.4   **27.** -3.9×10^{-3} atm   **29.** (a) fA/a;
(b) 20 lb   **31.** 1070 g   **33.** 1.5 g/cm^3 **35.** 600 kg/m^3
37. (a) 670 kg/m^3; (b) 740 kg/m^3 **39.** 390 kg
41. (a) 1.2 kg; (b) 1300 kg/m^3 **43.** 0.126 m^3 **45.** five
47. (a) 1.80 m^3; (b) 4.75 m^3 **49.** 2.79 g/cm^3
51. (a) 9.4 N; (b) 1.6 N   **53.** 4.0 m   **55.** 28 ft/s   **57.** 43
cm/s   **59.** (a) 2.40 m/s; (b) 245 Pa   **61.** (a) 12 ft/s; (b) 13
lb/in.^{2} **63.** 0.72 ft·lb/ft^3 **65.** (a) 2; (b) $R_1/R_2 = \frac{1}{2}$; (c) drain
it until $h_2 = h_1/4$ **67.** 116 m/s   **69.** (a) 6.4 m^3; (b) 5.4 m/s;
(c) 9.8×10^4 Pa   **71.** (a) 560 Pa; (b) 5.0×10^4 N
73. 40 m/s   **75.** (b) $H - h$; (c) $H/2$ **77.** (b) 0.69 ft^3/s
79. (b) 63.3 m/s

Chapter 16

CP   **1.** (a) $-x_m$; (b) $+x_m$; (c) 0   **2.** a   **3.** (a) 5 J; (b) 2 J;
(c) 5 J   **4.** all tie (in Eq. 16-32, m is included in I)
5. 1, 2, 3 (the ratio m/b matters; k does not)   Q   **1.** c
3. (a) 0; (b) between 0 and $+x_m$; (c) between $-x_m$ and 0;
(d) between $-x_m$ and 0   **5.** (a) toward $-x_m$; (b) toward $+x_m$;
(c) between $-x_m$ and 0; (d) between $-x_m$ and 0; (e) decreasing;
(f) increasing   **7.** (a) 3, 2, 1; (b) all tie   **9.** 3, 2, 1
11. system with spring A **13.** b (infinite period; does not
oscillate), c, a **15.** (a) same; (b) same; (c) same; (d) smaller;
(e) smaller; (f) and (g) larger ($T = \infty$)   EP   **1.** (a) 0.50 s;
(b) 2.0 Hz; (c) 18 cm   **3.** (a) 245 N/m; (b) 0.284 s
5. 708 N/m   **7.** $f > 500$ Hz   **9.** (a) 100 N/m; (b) 0.45 s
11. (a) 6.28×10^5 rad/s; (b) 1.59 mm   **13.** (a) 1.0 mm;
(b) 0.75 m/s; (c) 570 m/s^2 **15.** (a) 1.29×10^5 N/m;
(b) 2.68 Hz   **17.** (a) 4.0 s; (b) $\pi/2$ rad/s; (c) 0.37 cm;
(d) (0.37 cm) cos $\frac{\pi}{2}t$; (e) (-0.58 cm/s) sin $\frac{\pi}{2}t$; (f) 0.58 cm/s;
(g) 0.91 cm/s^2; (h) 0; (i) 0.58 cm/s   **19.** (b) 12.47 kg;
(c) 54.43 kg   **21.** 1.6 kg   **23.** (a) 1.6 Hz; (b) 1.0 m/s, 0;
(c) 10 m/s^2, $\pm$10 cm; (d) $(-10$ N/m$)x$ **25.** 22 cm
27. (a) 25 cm; (b) 2.2 Hz   **29.** (a) 0.500 m; (b) -0.251 m;
(c) 3.06 m/s   **31.** (a) 0.183A; (b) same direction
37. (a) $k_1 = (n + 1)k/n$, $k_2 = (n + 1)k$; (b) $f_1 = \sqrt{(n + 1)/n}f$,
$f_2 = \sqrt{n + 1}f$ **39.** (b) 42 min   **41.** (a) 200 N/m;
(b) 1.39 kg; (c) 1.91 Hz   **43.** (a) 130 N/m; (b) 0.62 s; (c) 1.6
Hz; (d) 5.0 cm; (e) 0.51 m/s   **45.** (a) 3/4; (b) 1/4; (c) $x_m/\sqrt{2}$
47. (a) 3.5 m; (b) 0.75 s   **49.** (a) 0.21 m; (b) 1.6 Hz;
(c) 0.10 m   **51.** (a) 0.0625 J; (b) 0.03125 J   **53.** 12 s
55. (a) 39.5 rad/s; (b) 34.2 rad/s; (c) 124 rad/s^2 **57.** (a) 8.3 s;
(b) no   **59.** 9.47 m/s^2 **61.** 8.77 s   **63.** 5.6 cm
65. $2\pi\sqrt{(R^2 + 2d^2)/2gd}$ **67.** (a) 0.205 kg·m^2; (b) 47.7 cm;
(c) 1.50 s   **71.** (a) $2\pi\sqrt{(L^2 + 12x^2)/12gx}$; (b) 0.289 m

73. 9.78 m/s^2 **75.** $2\pi\sqrt{m/3k}$ **77.** $(1/2\pi)(\sqrt{g^2 + v^4/R^2}/L)^{1/2}$
79. (b) smaller   **81.** (a) 2.0 s; (b) 18.5 N·m/rad
83. 0.29L **85.** 0.39   **87.** (a) 0.102 kg/s; (b) 0.137 J
89. $k = 490$ N/cm, $b = 1100$ kg/s   **91.** 1.9 in.
93. (a) $y_m = 8.8 \times 10^{-4}$ m, $T = 0.18$ s, $\omega = 35$ rad/s;
(b) $y_m = 5.6 \times 10^{-2}$ m, $T = 0.48$ s, $\omega = 13$ rad/s;
(c) $y_m = 3.3 \times 10^{-2}$ m, $T = 0.31$ s, $\omega = 20$ rad/s

Chapter 17

CP   **1.** a, 2; b, 3; c, 1   **2.** (a) 2, 3, 1; (b) 3, then 1 and 2 tie
3. a **4.** 0.20 and 0.80 tie, then 0.60, 0.45   **5.** (a) 1; (b) 3;
(c) 2   **6.** (a) 75 Hz; (b) 525 Hz   Q   **1.** 7d **3.** tie of A
and B, then C, D **5.** intermediate (closer to fully destructive
interference)   **7.** a and d tie, then b and c tie   **9.** (a) 8;
(b) antinode; (c) longer; (d) lower   **11.** (a) integer multiples
of 3; (b) node; (c) node   **13.** string A **15.** decrease
EP   **1.** (a) 75 Hz; (b) 13 ms   **3.** (a) 7.5×10^{14} to 4.3×10^{14}
Hz; (b) 1.0 to 200 m; (c) 6.0×10^{16} to 3.0×10^{19} Hz
5. $y = 0.010 \sin \pi(3.33x + 1100t)$, with x and y in m and t
in s   **11.** (a) $z = 3.0 \sin(60y - 10\pi t)$, with z in mm, y in cm,
and t in s; (b) 9.4 cm/s   **13.** (a) $y = 2.0 \sin 2\pi(0.10x - 400t)$, with x and y in cm and t in s; (b) 50 m/s; (c) 40 m/s
15. (b) 2.0 cm/s; (c) $y = (4.0 \text{ cm}) \sin(\pi x/10 - \pi t/5 + \pi)$,
where x is in cm and t is in s; (d) -2.5 cm/s   **17.** 129 m/s
19. 135 N   **23.** (a) 15 m/s; (b) 0.036 N   **25.** $y = 0.12 \sin(141x + 628t)$, with y in mm, x in m, and t in s
27. (a) 5.0 cm; (b) 40 cm; (c) 12 m/s; (d) 0.033 s; (e) 9.4 m/s;
(f) $5.0 \sin(16x + 190t + 0.79)$, with x in m, y in cm, and t in s
29. (a) $v_1 = 28.6$ m/s, $v_2 = 22.1$ m/s; (b) $M_1 = 188$ g, $M_2 = 313$ g   **31.** (a) $\sqrt{k(\Delta l)(l + \Delta l)/m}$ **33.** (a) $P_2 = 2P_1$;
(b) $P_2 = P_1/4$ **35.** (a) 3.77 m/s; (b) 12.3 N; (c) zero; (d) 46.3
W; (e) zero; (f) zero; (g) $\pm$0.50 cm   **37.** 82.8°, 1.45 rad,
0.23 wavelength   **39.** 5.0 cm   **41.** (a) 4.4 mm; (b) 112°
43. (a) $0.83y_1$; (b) 37°   **45.** (a) $2f_3$; (b) λ_3 **47.** 10 cm
49. (a) 82.0 m/s; (b) 16.8 m; (c) 4.88 Hz   **51.** 240 cm, 120 cm,
80 cm   **53.** 7.91 Hz, 15.8 Hz, 23.7 Hz   **55.** $f_{1A} = f_{4B}$,
$f_{2A} = f_{8B}$ **57.** (a) 2.0 Hz, 200 cm, 400 cm/s; (b) $x = 50$ cm,
150 cm, 250 cm, etc.; (c) $x = 0$, 100 cm, 200 cm, etc.
63. (a) 1.3 m; (b) $y' = 0.002 \sin(9.4x) \cos(3800t)$, with x and
y in m and t in s   **67.** (b) in the positive x direction;
interchange the amplitudes of the original two traveling waves;
(c) largest at $x = \lambda/4 = 6.26$ cm; smallest at $x = 0$ and
$x = \lambda/2 = 12.5$ cm; (d) the largest amplitude is 4.0 mm, which
is the sum of the amplitudes of the original traveling waves; the
smallest amplitude is 1.0 mm, which is the difference of the
amplitudes of the original traveling waves

Chapter 18

CP   **1.** beginning to decrease (example: mentally move the
curves of Fig. 18-6 rightward past the point at $x = 42$ m)
2. (a) fully constructive, 0; (b) fully constructive, 4   **3.** (a) 1
and 2 tie, then 3; (b) 3, then 1 and 2 tie   **4.** second
5. loosen   **6.** a, greater; b, less; c, can't tell; d, can't tell;
e, greater; f, less   **7.** (a) 222 m/s; (b) +20 m/s   Q   **1.** pulse

along path 2 **3.** (a) 2.0 wavelengths; (b) 1.5 wavelengths; (c) fully constructive, fully destructive **5.** (a) exactly out of phase; (b) exactly out of phase **7.** 70 dB **9.** (a) two; (b) antinode **11.** all odd harmonics **13.** 501, 503, and 508 Hz; or 505, 507, and 508 Hz **EP 1.** (a) $\approx 6\%$ **3.** the radio listener by about 0.85 s **5.** 7.9×10^{10} Pa **7.** If only the length is uncertain, it must be known to within 10^{-4} m. If only the time is imprecise, the uncertainty must be no more than one part in 10^8. **9.** 43.5 m **11.** 40.7 m **13.** 100 kHz **15.** (a) 2.29, 0.229, 22.9 kHz; (b) 1.14, 0.114, 11.4 kHz **17.** (a) 6.0 m/s; (b) $y = 0.30 \sin(\pi x/12 + 50\pi t)$, with x and y in cm and t in s **19.** 4.12 rad **21.** (a) $343 \times (1 + 2m)$ Hz, with m being an integer from 0 to 28; (b) $686m$ Hz, with m being an integer from 1 to 29 **23.** (a) eight; (b) eight **25.** 64.7 Hz, 129 Hz **27.** (a) 0.080 W/m^2; (b) 0.013 W/m^2 **29.** 36.8 nm **31.** (a) 1000; (b) 32 **33.** (a) 39.7 μW/m^2; (b) 171 nm; (c) 0.893 Pa **35.** (a) 59.7; (b) 2.81×10^{-4} **37.** $s_m \propto r^{-1/2}$ **39.** (a) 5000; (b) 71; (c) 71 **41.** 171 m **43.** 3.16 km **45.** (a) 5200 Hz; (b) amplitude$_{SAD}$/amplitude$_{SBD}$ = 2 **47.** 20 kHz **49.** by a factor of 4 **51.** water filled to a height of $\frac{7}{8}, \frac{5}{8}, \frac{3}{8}, \frac{1}{8}$ m **53.** (a) 5.0 cm from one end; (b) 1.2; (c) 1.2 **55.** (a) 1130, 1500, and 1880 Hz **57.** (a) 230 Hz; (b) higher **59.** (a) node; (c) 22 s **61.** 387 Hz **63.** 0.02 **65.** 3.8 Hz **67.** (a) 380 mi/h, away from owner; (b) 77 mi/h, away from owner **69.** 15.1 ft/s **71.** 2.6×10^8 m/s **73.** (a) 77.6 Hz; (b) 77.0 Hz **75.** 33.0 km **79.** (a) 970 Hz; (b) 1030 Hz; (c) 60 Hz, no **81.** (a) 1.02 kHz; (b) 1.04 kHz **83.** 1540 m/s **85.** 41 kHz **87.** (a) 2.0 kHz; (b) 2.0 kHz **89.** (a) 485.8 Hz; (b) 500.0 Hz; (c) 486.2 Hz; (d) 500.0 Hz **91.** 1×10^6 m/s, receding **93.** $0.13c$

Chapter 19

CP 1. (a) all tie; (b) 50°X, 50°Y, 50°W **2.** (a) 2 and 3 tie, then 1, then 4; (b) 3, 2, then 1 and 4 tie **3.** A **4.** c and e **5.** (a) zero; (b) negative **6.** b and d tie, then a, c **Q 1.** 25 S°, 25 U°, 25 R° **3.** c, then the rest tie **5.** B, then A and C tie **7.** (a) both clockwise; (b) both clockwise **9.** c, a, b **11.** upward (with liquid water on the exterior and at the bottom, $\Delta T = 0$ horizontally and downward) **13.** at the temperature of your fingers **15.** 3, 2, 1 **EP 1.** 2.71 K **3.** 0.05 kPa, nitrogen **5.** (a) 320°F; (b) -12.3°F **7.** (a) -96°F; (b) 56.7°C **9.** (a) -40°; (b) 575°; (c) Celsius and Kelvin cannot give the same reading **11.** (a) Dimensions are inverse time **13.** 4.4×10^{-3} cm **15.** 0.038 in. **17.** (a) 9.996 cm; (b) 68°C **19.** 170 km **21.** 0.32 cm^2 **23.** 29 cm^3 **25.** 0.432 cm^3 **27.** -157°C **29.** 360°C **35.** $+0.68$ s/h **37.** (b) use 39.3 cm of steel and 13.1 cm of brass **39.** (a) 523 J/kg·K; (b) 0.600; (c) 26.2 J/mol·K **41.** 94.6 L **43.** 109 g **45.** 1.30 MJ **47.** 1.9 times as great **49.** (a) 33.9 Btu; (b) 172 F° **51.** (a) 52 MJ; (b) 0°C **53.** (a) 411 g; (b) 3.1¢ **55.** 0.41 kJ/kg·K **57.** 3.0 min **59.** 73 kW **61.** 2.17 g **63.** 33 m^2 **65.** 33 g **67.** (a) 0°C; (b) 2.5°C **69.** 2500 J/kg·K **71.** A: 120 J, B: 75 J, C: 30 J **73.** (a) -200 J; (b) -293 J;

(c) -93 J **75.** -5.0 J **77.** 33.3 kJ **79.** 766°C **81.** (a) 1.2 W/m·K, 0.70 Btu/ft·F°·h; (b) 0.030 ft^2·F°·h/Btu **83.** 1660 J/s **87.** arrangement b **89.** (a) 2.0 MW; (b) 220 W **91.** (a) 17 kW/m^2; (b) 18 W/m^2 **93.** -6.1 nW **95.** 0.40 cm/h **97.** Cu-Al, 84.3°C; Al-brass, 57.6°C

Chapter 20

CP 1. all but c **2.** (a) all tie; (b) 3, 2, 1 **3.** gas A **4.** 5 (greatest change in T), then tie of 1, 2, 3, and 4 **5.** 1, 2, 3 ($Q_3 = 0$, Q_2 goes into work W_2, but Q_1 goes into greater work W_1 and increases gas temperature) **Q 1.** increased but less than doubled **3.** a, c, b **5.** 1180 J **7.** d, tie of a and b, then c **9.** constant-volume process **11.** (a) same; (b) increases; (c) decreases; (d) increases **13.** -4 J **15.** (a) 1, polyatomic; 2, diatomic; 3, monatomic; (b) more **EP 1.** (a) 0.0127; (b) 7.65×10^{21} **3.** 6560 **5.** number of molecules in the ink $\approx 3 \times 10^{16}$; number of people $\approx 5 \times 10^{20}$; statement is wrong, by a factor of about 20,000 **7.** (a) 5.47×10^{-8} mol; (b) 3.29×10^{16} **9.** (a) 106; (b) 0.892 m^3 **11.** 27.0 lb/in.^{2} **13.** (a) 2.5×10^{25}; (b) 1.2 kg **15.** 5600 J **17.** 1/5 **19.** (a) -45 J; (b) 180 K **21.** 100 cm^3 **23.** 198°F **25.** 2.0×10^5 Pa **27.** 180 m/s **29.** 9.53×10^6 m/s **31.** 313°C **33.** 1.9 kPa **35.** (a) 0.0353 eV, 0.0483 eV; (b) 3400 J, 4650 J **37.** 9.1×10^{-6} **39.** (a) 6.75×10^{-20} J; (b) 10.7 **41.** 0.32 nm **43.** 15 cm **45.** (a) 3.27×10^{10}; (b) 172 m **47.** (a) 22.5 L; (b) 2.25; (c) 8.4×10^{-5} cm; (d) same as (c) **51.** (a) 3.2 cm/s; (b) 3.4 cm/s; (c) 4.0 cm/s **53.** (a) v_P, v_{rms}, $\bar{v}$ (b) reverse ranking **55.** (a) 1.0×10^4 K, 1.6×10^5 K; (b) 440 K, 7000 K **57.** 4.7 **59.** (a) $2N/3v_0$; (b) $N/3$; (c) $1.22v_0$; (d) $1.31v_0$ **61.** $RT \ln(V_f/V_i)$ **63.** (a) 15.9 J; (b) 34.4 J/mol·K; (c) 26.1 J/mol·K **65.** $(n_1C_1 + n_2C_2 + n_3C_3)/(n_1 + n_2 + n_3)$ **67.** (a) -5.0 kJ; (b) 2.0 kJ; (c) 5.0 kJ **69.** (a) 0.375 mol; (b) 1090 J; (c) 0.714 **71.** (a) 14 atm; (b) 620 K **79.** 0.63 **81.** (a) monatomic; (b) 2.7×10^4 K; (c) 4.5×10^4 mol; (d) 3.4 kJ, 340 kJ; (e) 0.01 **83.** 5 m^3 **85.** (a) in joules, in the order Q, ΔE_{int}, W: $1 \rightarrow 2$: 3740, 3740, 0; $2 \rightarrow 3$: 0, -1810, 1810; $3 \rightarrow 1$: -3220, -1930, -1290; cycle: 520, 0, 520; (b) $V_2 = 0.0246$ m^3, $p_2 = 2.00$ atm, $V_3 = 0.0373$ m^3, $p_3 = 1.00$ atm

Chapter 21

CP 1. a, b, c **2.** smaller **3.** c, b, a **4.** a, d, c, b **5.** b **Q 1.** increase **3.** (a) increase; (b) same **5.** equal **7.** lower the temperature of the low temperature reservoir **9.** (a) same; (b) increase; (c) decrease **11.** (a) same; (b) increase; (c) decrease **13.** more than the age of the universe **EP 1.** 1.86×10^4 J **3.** 2.75 mol **7.** (a) 5.79×10^4 J; (b) 173 J/K **9.** $+3.59$ J/K **11.** (a) 14.6 J/K; (b) 30.2 J/K **13.** (a) 4.45 J/K; (b) no **15.** (a) 4500 J; (b) -5000 J; (c) 9500 J **17.** (a) 57.0°C; (b) -22.1 J/K; (c) $+24.9$ J/K; (d) $+2.8$ J/K **19.** (a) -710 mJ/K; (b) $+710$ mJ/K; (c) $+723$ mJ/K;

(d) -723 mJ/K; (e) $+13$ mJ/K; (f) 0   **23.** (a) (I) constant T, $Q = pV \ln 2$; constant V, $Q = 4.5pV$; (II) constant T, $Q = -pV \ln 2$; constant p, $Q = 7.5pV$; (b) (I) constant T, $W = pV \ln 2$; constant V, $W = 0$; (II) constant T, $W = -pV \ln 2$; constant p, $W = 3pV$; (c) $4.5pV$ for either case; (d) $4R \ln 2$ for either case   **25.** 0.75 J/K **27.** (a) -943 J/K; (b) $+943$ J/K; (c) yes   **29.** (a) $3p_0V_0$; (b) $6RT_0$, $(3/2)R \ln 2$; (c) both are zero   **33.** (a) 31%; (b) 16 kJ   **35.** engine A, first; engine B, first and second; engine C, second; engine D, neither   **37.** 97 K **39.** 99.99995%   **41.** 7.2 J/cycle   **43.** (a) 7200 J; (b) 960 J; (c) 13%   **45.** (a) 2270 J; (b) 14,800 J; (c) 15.4%; (d) 75.0%, greater   **49.** (a) 78%; (b) 81 kg/s   **55.** (a) 49 kJ; (b) 7.4 kJ   **57.** 21 J   **59.** (a) 0.071 J; (b) 0.50 J; (c) 2.0 J; (d) 5.0 J   **61.** 1.08 MJ   **63.** $[1 - (T_2/T_1)] \div [1 - (T_4/T_3)]$ **67.** (a) 1.27×10^{30}; (b) 7.9%; (c) 7.3%; (d) 7.3%; (e) 1.1%; (f) 0.0023%   **69.** (a) $W = N!/(n_1! \, n_2! \, n_3!)$; (b) $[(N/2)! \, (N/2)!]/[N/3)! \, (N/3)! \, (N/3)!]$; (c) 4.2×10^{16}

Chapter 22

CP   1. C and D attract; B and D attract   **2.** (a) leftward; (b) leftward; (c) leftward   **3.** (a) a, c, b; (b) less than **4.** $-15e$ (net charge of $-30e$ is equally shared) **Q   1.** No, only for charged particles, charged particle-like objects, and spherical shells (including solid spheres) of uniform charge   **3.** a and b **5.** two points: one to the left of the particles and one between the protons   **7.** $6q^2/4\pi\epsilon_0 d^2$, leftward   **9.** (a) same; (b) less than; (c) cancel; (d) add; (e) the adding components; (f) positive direction of y; (g) negative direction of y; (h) positive direction of x; (i) negative direction of x **11.** (a) A, B, and D; (b) all four; (c) Connect A and D; disconnect them; then connect one of them to B. (There are two more solutions.)   **13.** (a) possibly; (b) definitely   **15.** same   **17.** D **EP   1.** 0.50 C **3.** 2.81 N on each   **5.** (a) 4.9×10^{-7} kg; (b) 7.1×10^{-11} C **7.** $3F/8$ **9.** (a) 1.60 N; (b) 2.77 N   **11.** (a) $q_1 = 9q_2$; (b) $q_1 = -25q_2$ **13.** either -1.00 μC and $+3.00$ μC or $+1.00$ μC and -3.00 μC   **15.** (a) 36 N, $-10°$ from the x axis; (b) $x = -8.3$ cm, $y = +2.7$ cm   **17.** (a) 5.7×10^{13} C, no; (b) 6.0×10^5 kg   **19.** (a) $Q = -2\sqrt{2}q$; (b) no **21.** 3.1 cm   **23.** 2.89×10^{-9} N   **25.** -1.32×10^{13} C **27.** (a) 3.2×10^{-19} C; (b) two   **29.** (a) 8.99×10^{-19} N; (b) 625   **31.** 5.1 m below the electron   **33.** 1.3 days **35.** (a) 0; (b) 1.9×10^{-9} N   **37.** 10^{18} N   **39.** (a) ^{9}B; (b) ^{13}N; (c) ^{12}C   **41.** (a) $F = (Q^2/4\pi\epsilon_0 d^2)\alpha(1 - \alpha)$; (c) 0.5; (d) 0.15 and 0.85

Chapter 23

CP   1. (a) rightward; (b) leftward; (c) leftward; (d) rightward (p and e have same charge magnitude, and p is farther)   **2.** all tie   **3.** (a) toward positive y; (b) toward positive x; (c) toward negative y **4.** (a) leftward; (b) leftward; (c) decrease **5.** (a) all tie; (b) 1 and 3 tie, then 2 and 4 tie **Q   1.** (a) toward positive x; (b) downward and to the right;

(c) A **3.** two points: one to the left of the particles, the other between the protons   **5.** (a) yes; (b) toward; (c) no (the field vectors are not along the same line); (d) cancel; (e) add; (f) adding components; (g) toward negative y **7.** (a) 3, then 1 and 2 tie (zero); (b) all tie; (c) 1 and 2 tie, then 3 **9.** (a) rightward; (b) $+q_1$ and $-q_3$, increase; q_2, decrease; n, same   **11.** a, b, c **13.** (a) 4, 3, 1, 2; (b) 3, then 1 and 4 tie, then 2   **EP   1.** (a) 6.4×10^{-18} N; (b) 20 N/C   **3.** to the right in the figure   **7.** 56 pC   **9.** 3.07×10^{21} N/C, radially outward   **13.** (a) $q/8\pi\epsilon_0 d^2$, to the left; $3q/\pi\epsilon_0 d^2$, to the right; $7q/16\pi\epsilon_0 d^2$, to the left   **15.** 0   **17.** 9:30   **19.** $E = q/\pi\epsilon_0 a^2$, along bisector, away from triangle **21.** $7.4q/4\pi\epsilon_0 d^2$, 28° counterclockwise to $+x$ **23.** 6.88×10^{-28} C·m   **25.** $(1/4\pi\epsilon_0)(p/r^3)$, antiparallel to $\mathbf{p}$ **29.** $R/\sqrt{2}$ **31.** $(1/4\pi\epsilon_0)(4q/\pi R^2)$, toward decreasing y **37.** (a) 0.10 μC; (b) 1.3×10^{17}; (c) 5.0×10^{-6} **39.** 3.51×10^{15} m/s^2 **41.** 6.6×10^{-15} N   **43.** 2.03×10^{-7} N/C, up **45.** (a) -0.029 C; (b) repulsive forces would explode the sphere   **47.** (a) 1.92×10^{12} m/s^2; (b) 1.96×10^5 m/s **49.** (a) 8.87×10^{-15} N; (b) 120   **51.** 1.64×10^{-19} C ($\approx$3% high)   **53.** (a) 0.245 N, 11.3° clockwise from the $+x$ axis; (b) $x = 108$ m, $y = -21.6$ m   **55.** 27μm   **57.** (a) yes; (b) upper plate, 2.73 cm   **59.** (a) 0; (b) 8.5×10^{-22} N·m; (c) 0   **61.** $(1/2\pi)\sqrt{pE/I}$ **63.** (a) $E = (2q/4\pi\epsilon_0 d^2)(\alpha/(1 + \alpha^2)^{3/2})$; (c) 0.707; (d) 0.21 and 1.9

Chapter 24

CP   1. (a) $+EA$; (b) $-EA$; (c) 0; (d) 0   **2.** (a) 2; (b) 3; (c) 1   **3.** (a) equal; (b) equal; (c) equal   **4.** (a) $+50e$; (b) $-150e$ **5.** 3 and 4 tie, then 2, 1   **Q   1.** (a) 8 N·m^2/C; (b) 0   **3.** (a) all tie (zero); (b) all tie   **5.** $+13q/\epsilon_0$ **7.** all tie   **9.** all tie   **11.** $2\sigma, \sigma, 3\sigma$; or $3\sigma, \sigma, 2\sigma$ **13.** (a) all tie ($E = 0$); (b) all tie   **15.** (a) same ($E = 0$); (b) decrease; (c) decrease (to zero); (d) same   **EP   1.** (a) 693 kg/s; (b) 693 kg/s; (c) 347 kg/s; (d) 347 kg/s; (e) 575 kg/s   **3.** (a) 0; (b) -3.92 N·m^2/C; (c) 0; (d) 0 for each field   **5.** (a) enclose $2q$ and $-2q$, or enclose all four charges; (b) enclose $2q$ and q; (c) not possible   **7.** 2.0×10^5 N·m^2/C   **9.** $q/6\epsilon_0$ **11.** (a) $-\pi R^2 E$; (b) $\pi R^2 E$ **13.** -4.2×10^{-10} C   **15.** 0 through each of the three faces meeting at q, $q/24\epsilon_0$ through each of the other faces   **17.** 2.0 μC/m^2 **19.** (a) 4.5×10^{-7} C/m^2; (b) 5.1×10^4 N/C   **21.** (a) -3.0×10^{-6} C; (b) $+1.3 \times 10^{-5}$ C   **23.** (a) 0.32 μC; (b) 0.14 μC **27.** (a) $E = q/2\pi\epsilon_0 Lr$, radially inward; (b) $-q$ on both inner and outer surfaces; (c) $E = q/2\pi\epsilon_0 Lr$, radially outward **29.** 3.6 nC   **31.** (b) $\rho R^2/2\epsilon_0 r$ **33.** (a) 5.3×10^7 N/C; (b) 60 N/C   **35.** 5.0 nC/m^2 **37.** 0.44 mm   **39.** (a) 4.9×10^{-22} C/m^2; (b) downward   **41.** (a) px/ϵ_0; (b) $pd/2\epsilon_0$ **43.** (a) -750 N·m^2/C; (b) -6.64 nC   **45.** (a) 4.0×10^6 N/C; (b) 0   **47.** (a) 0; (b) $q_a/4\pi\epsilon_0 r^2$; (c) $(q_a + q_b)/4\pi\epsilon_0 r^2$ **51.** (a) $-q$; (b) $+q$; (c) $E = q/4\pi\epsilon_0 r^2$, radially outward; (d) $E = 0$; (e) $E = q/4\pi\epsilon_0 r^2$, radially outward; (f) 0; (g) $E = q/4\pi\epsilon_0 r^2$, radially outward; (h) yes, charge is induced; (i) no; (j) yes; (k) no; (l) no   **53.** (a) $E = (q/4\pi\epsilon_0 a^3)r$; (b) $E = q/4\pi\epsilon_0 r^2$; (c) 0; (d) 0; (e) inner, $-q$; outer, 0   **55.** $q/2\pi a^2$ **59.** $\alpha = 0.80$

Chapter 25

CP **1.** (a) negative; (b) increase **2.** (a) positive; (b) higher **3.** (a) rightward; (b) 1, 2, 3, 5: positive; 4, negative; (c) 3, then 1, 2, and 5 tie, then 4 **4.** all tie **5.** a, c (zero), b **6.** (a) 2, then 1 and 3 tie; (b) 3; (c) accelerate leftward **7.** closer (half of 9.23 fm) **Q** **1.** (a) higher; (b) positive; (c) negative; (d) all tie **3.** (a) 1 and 2; (b) none; (c) no; (d) 1 and 2 yes, 3 and 4 no **5.** b, then a, c, and d tie **7.** (a) negative; (b) zero **9.** (a) 1, then 2 and 3 tie; (b) 3 **11.** left **13.** a, b, c **15.** (a) c, b, a; (b) zero **17.** (a) positive; (b) positive; (c) negative; (d) all tie **19.** (a) no; (b) yes **21.** no (a particle at the intersection would have two different potential energies) **23.** (a)–(b) all tie; (c) C, B, A; (d) all tie **EP** **1.** 1.2 GeV **3.** (a) 3.0 × 10^{10} J; (b) 7.7 km/s; (c) 9.0 × 10^4 kg **7.** 2.90 kV **9.** 8.8 mm **11.** (a) 136 MV/m; (b) 8.82 kV/m **13.** (b) because $V = 0$ point is chosen differently; (c) $q/(8\pi\epsilon_0 R)$; (d) potential differences are independent of the choice for the $V = 0$ point **15.** (a) -4500 V; (b) -4500 V **17.** 843 V **19.** 2.8 × 10^5 **21.** $x = d/4$ and $x = -d/2$ **23.** none **25.** (a) 3.3 nC; (b) 12 nC/m^2 **27.** 6.4 × 10^8 V **29.** 190 MV **31.** (a) -4.8 nm; (b) 8.1 nm; (c) no **33.** 16.3 μV **35.** (a) $\dfrac{2\lambda}{4\pi\epsilon_0} \ln\left[\dfrac{L/2 + (L^2/4 + d^2)^{1/2}}{d}\right]$; (b) 0 **37.** (a) $-5Q/4\pi\epsilon_0 R$; (b) $-5Q/4\pi\epsilon_0(z^2 + R^2)^{1/2}$ **39.** 0.113$\sigma R/\epsilon_0$ **41.** $(Q/4\pi\epsilon_0 L)\ln(1 + L/d)$ **43.** 670 V/m **45.** $p/2\pi\epsilon_0 r^3$ **47.** 39 V/m, $-x$ direction **51.** (a) $\dfrac{c}{4\pi\epsilon_0}[\sqrt{L^2 + y^2} - y]$; (b) $\dfrac{c}{4\pi\epsilon_0}\left[1 - \dfrac{y}{\sqrt{L^2 + y^2}}\right]$ **53.** (a) 2.5 MV; (b) 5.1 J; (c) 6.9 J **55.** (a) 2.72 × 10^{-14} J; (b) 3.02 × 10^{-31} kg, about $\frac{1}{3}$ of accepted value **57.** (a) 0.484 MeV; (b) 0 **59.** 2.1 d **61.** 0 **63.** (a) 27.2 V; (b) -27.2 eV; (c) 13.6 eV; (d) 13.6 eV **65.** 1.8 × 10^{-10} J **67.** 1.48 × 10^7 m/s **69.** $qQ/4\pi\epsilon_0 K$ **71.** 0.32 km/s **73.** 1.6 × 10^{-9} m **77.** (a) $V_1 = V_2$; (b) $q_1 = q/3$, $q_2 = 2q/3$; (c) 2 **79.** (a) -0.12 V; (b) 1.8 × 10^{-8} N/C, radially inward **81.** (a) 12,000 N/C; (b) 1800 V; (c) 5.8 cm **83.** (c) 4.24 V

Chapter 26

CP **1.** (a) same; (b) same **2.** (a) decreases; (b) increases; (c) decreases **3.** (a) V, $q/2$; (b) $V/2$, q **4.** (a) $q_0 = q_1 + q_{34}$; (b) equal (C_3 and C_4 are in series) **5.** (a) same; (b)–(d) increase; (e) same (same potential difference across same plate separation) **6.** (a) same; (b) decrease; (c) increase **Q** **1.** a, 2; b, 1; c, 3 **3.** (a) increase; (b) same **5.** (a) parallel; (b) series **7.** (a) $C/3$; (b) $3C$; (c) parallel **9.** (a) equal; (b) less **11.** (a)–(d) less **13.** (a) 2; (b) 3; (c) 1 **15.** Increase plate separation d, but also plate area A, keeping A/d constant. **EP** **1.** 7.5 pC **3.** 3.0 mC **5.** (a) 140 pF; (b) 17 nC **7.** (a) 84.5 pF; (b) 191 cm^2 **9.** (a) 11 cm^2; (b) 11 pF; (c) 1.2 V **13.** (b) 4.6 × 10^{-5}/K **15.** 7.33 μF **17.** 315 mC **19.** (a) 10.0 μF; (b) $q_2 = 0.800$ mC, $q_1 = 1.20$ mC; (c) 200 V for both **21.** (a) $d/3$; (b) $3d$ **25.** (a) five in series; (b) three

arrays as in (a) in parallel (and other possibilities) **27.** 43 pF **29.** (a) 50 V; (b) 5.0 × 10^{-5} C; (c) 1.5 × 10^{-4} C **31.** (a) $q_1 = 9.0$ μC, $q_2 = 16$ μC, $q_3 = 9.0$ μC, $q_4 = 16$ μC; (b) $q_1 = 8.4$ μC, $q_2 = 17$ μC, $q_3 = 11$ μC, $q_4 = 14$ μC **33.** 99.6 nJ **35.** 72 F **37.** 4.9% **39.** 0.27 J **41.** 0.11 J/m^3 **43.** (a) 2.0 J **45.** (a) $q_1 = 0.21$ mC, $q_2 = 0.11$ mC, $q_3 = 0.32$ mC; (b) $V_1 = V_2 = 21$ V, $V_3 = 79$ V; (c) $U_1 = 2.2$ mJ, $U_2 = 1.1$ mJ, $U_3 = 13$ mJ **47.** (a) $q_1 = q_2 = 0.33$ mC, $q_3 = 0.40$ mC; (b) $V_1 = 33$ V, $V_2 = 67$ V, $V_3 = 100$ V; (c) $U_1 = 5.6$ mJ, $U_2 = 11$ mJ, $U_3 = 20$ mJ **53.** Pyrex **55.** (a) 6.2 cm; (b) 280 pF **57.** 0.63 m^2 **59.** (a) 2.85 m^3; (b) 1.01 × 10^4 **61.** (a) $\epsilon_0 A/(d-b)$; (b) $d/(d-b)$; (c) $-q^2 b/2\epsilon_0 A$, sucked in **65.** $\dfrac{\epsilon_0 A}{4d}\left(\kappa_1 + \dfrac{2\kappa_2\kappa_3}{\kappa_2 + \kappa_3}\right)$ **67.** (a) 13.4 pF; (b) 1.15 nC; (c) 1.13 × 10^4 N/C; (d) 4.33 × 10^3 N/C **69.** (a) 7.1; (b) 0.77 μC **71.** (a) 0.606; (b) 0.394

Chapter 27

CP **1.** 8 A, rightward **2.** (a)–(c) rightward **3.** a and c tie, then b **4.** Device 2 **5.** (a) and (b) tie, then (d), then (c) **Q** **1.** a, b, and c tie, then d (zero) **3.** b, a, c **5.** tie of A, B, and C, then a tie of $A + B$ and $B + C$, then $A + B + C$ **7.** (a)–(c) 1 and 2 tie, then 3 **9.** C, A, B **11.** b, a, c **13.** (a) conductors: 1 and 4; semiconductors: 2 and 3; (b) 2 and 3; (c) all four **EP** **1.** 1.25 × 10^{15} **3.** 6.7 μC/m^2 **5.** 14-gauge **7.** (a) 2.4 × 10^{-5} A/m^2; (b) 1.8 × 10^{-15} m/s **9.** 0.67 A, toward the negative terminal **11.** (a) 0.654 μA/m^2; (b) 83.4 MA **13.** 13 min **15.** (a) $J_0 A/3$; (b) $2J_0 A/3$ **17.** 2.0 × 10^{-8} $\Omega \cdot$m **19.** 100 V **21.** (a) 1.53 kA; (b) 54.1 MA/m^2; (c) 10.6 × 10^{-8} $\Omega \cdot$m, platinum **23.** (a) 253°C; (b) yes **25.** (a) 0.38 mV; (b) negative; (c) 3 min 58 s **27.** 54 Ω **29.** 3.0 **31.** (a) 1.3 mΩ; (b) 4.6 mm **33.** (a) 6.00 mA; (b) 1.59 × 10^{-8} V; (c) 21.2 nΩ **35.** 2000 K **37.** (a) copper: 5.32 × 10^5 A/m^2, aluminum: 3.27 × 10^5 A/m^2; (b) copper: 1.01 kg/m, aluminum: 0.495 kg/m **39.** 0.40 Ω **41.** (a) $R = \rho L/\pi ab$ **43.** 14 kC **45.** 11.1 Ω **47.** (a) 1.0 kW; (b) 25¢ **49.** 0.135 W **51.** (a) 1.74 A; (b) 2.15 MA/m^2; (c) 36.3 mV/m; (d) 2.09 W **53.** (a) 1.3 × 10^5 A/m^2; (b) 94 mV **55.** (a) \$4.46 for a 31-day month; (b) 144 Ω; (c) 0.833 A **57.** 660 W **59.** (a) 3.1 × 10^{11}; (b) 25 μA; (c) 1300 W, 25 MW **61.** 27 cm/s **63.** (a) 120 Ω; (b) 107 Ω; (c) 5.3 × 10^{-3}/C°; (d) 5.9 × 10^{-3}/C°; (e) 276 Ω

Chapter 28

CP **1.** (a) rightward; (b) all tie; (c) b, then a and c tie; (d) b, then a and c tie **2.** (a) all tie; (b) R_1, R_2, R_3 **3.** (a) less; (b) greater; (c) equal **4.** (a) $V/2$, i; (b) V, $i/2$ **5.** (a) 1, 2, 4, 3; (b) 4, tie of 1 and 2; then 3 **Q** **1.** 3, 4, 1, 2 **3.** (a) no; (b) yes; (c) all tie (the circuits are the same) **5.** parallel, R_2, R_1, series **7.** (a) equal; (b) more **9.** (a) less; (b) less; (c) more **11.** C_1, 15 V; C_2, 35 V; C_3, 20 V; C_4, 20 V; C_5, 30 V **13.** 60 μC **15.** c, b, a **17.** (a) all tie; (b) 1, 3, 2 **19.** 1, 3, and 4 tie (8 V on each resistor), then 2 and 5 tie (4 V on each resistor) **EP** **1.** (a) \$320; (b) 4.8¢ **3.** 11 kJ

5. (a) counterclockwise; (b) battery 1; (c) B   **7.** (a) 80 J;
(b) 67 J; (c) 13 J converted to thermal energy within battery
9. (a) 14 V; (b) 100 W; (c) 600 W; (d) 10 V, 100 W
11. (a) 50 V; (b) 48 V; (c) B is connected to the negative
terminal   **13.** 2.5 V   **15.** (a) 6.9 km; (b) 20 Ω
17. 8.0 Ω **19.** 10^{-6} **21.** the cable   **23.** (a) 1000 Ω;
(b) 300 mV; (c) 2.3×10^{-3} **25.** (a) 3.41 A or 0.586 A;
(b) 0.293 V or 1.71 V   **27.** 5.56 A   **29.** 4.0 Ω and 12 Ω
31. 4.50 Ω **33.** 0.00, 2.00, 2.40, 2.86, 3.00, 3.60, 3.75,
3.94 A   **35.** $V_d - V_c = +0.25$ V, by all paths   **37.** three
39. (a) 2.50 Ω; (b) 3.13 Ω **41.** nine   **43.** (a) left branch:
0.67 A down; center branch: 0.33 A up; right branch: 0.33 A
up; (b) 3.3 V   **47.** (a) 120 Ω; (b) $i_1 = 51$ mA, $i_2 = i_3 =$
19 mA, $i_4 = 13$ mA   **49.** (a) 19.5 Ω; (b) 0; (c) ∞; (d) 82.3 W,
57.6 W   **51.** (a) Cu: 1.11 A, Al: 0.893 A; (b) 126 m
53. (a) 13.5 kΩ; (b) 1500 Ω; (c) 167 Ω; (d) 1480 Ω
55. 0.45 A   **57.** (a) 12.5 V; (b) 50 A   **59.** -0.9%
65. (a) 0.41τ; (b) 1.1τ **67.** 4.6   **69.** (a) 0.955 μC/s;
(b) 1.08 μW; (c) 2.74 μW; (d) 3.82 μW   **71.** 2.35 MΩ
73. 0.72 MΩ **75.** 24.8 Ω to 14.9 kΩ **77.** (a) at $t = 0$,
$i_1 = 1.1$ mA, $i_2 = i_3 = 0.55$ mA; at $t = \infty$, $i_1 = i_2 = 0.82$ mA,
$i_3 = 0$; (c) at $t = 0$, $V_2 = 400$ V; at $t = \infty$, $V_2 = 600$ V;
(d) after several time constants ($\tau = 7.1$ s) have elapsed
79. (a) $V_T = -ir + \mathscr{E}$; (b) 13.6 V; (c) 0.060 Ω
81. (a) 6.4 V; (b) 3.6 W; (c) 17 W; (d) -5.6 W; (e) a

Chapter 29

CP **1.** $a, +z$; $b, -x$; c, $F_B = 0$ **2.** 2, then tie of 1 and 3
(zero); (b) 4   **3.** (a) $+z$ and $-z$ tie, then $+y$ and $-y$ tie, then
$+x$ and $-x$ tie (zero); (b) $+y$ **4.** (a) electron;
(b) clockwise   **5.** $-y$ **6.** (a) all tie; (b) 1 and 4 tie, then 2
and 3 tie   **Q** **1.** (a) all tie; (b) 1 and 2 (charge is negative)
3. a, no, $\mathbf{v}$ and $\mathbf{F}_B$ must be perpendicular; b, yes; c, no, $\mathbf{B}$ and
$\mathbf{F}_B$ must be perpendicular   **5.** (a) $\mathbf{F}_E$; (b) $\mathbf{F}_B$ **7.** (a) right;
(b) right   **9.** (a) negative; (b) equal; (c) equal; (d) half a
circle   **11.** (a) $\mathbf{B}_1$; (b) $\mathbf{B}_1$ into page; $\mathbf{B}_2$ out of page; (c) less
13. all   **15.** all tie   **17.** (a) positive; (b) (1) and (2) tie, then
(3) which is zero   **EP** **1.** M/QT **3.** (a) 9.56×10^{-14} N, 0;
(b) 0.267°   **5.** (a) east; (b) 6.28×10^{14} m/s²; (c) 2.98 mm
7. 0.75**k** T   **9.** (a) 3.4×10^{-4} T, horizontal and to the left
as viewed along $\mathbf{v}_0$; (b) yes, if velocity is the same as the
electron's velocity   **11.** $(-11.4\mathbf{i} - 6.00\mathbf{j} + 4.80\mathbf{k})$ V/m
13. 680 kV/m   **17.** (b) 2.84×10^{-3} **19.** (a) 1.11×10^7 m/s;
(b) 0.316 mm   **21.** (a) 0.34 mm; (b) 2.6 keV
23. (a) 2.05×10^7 m/s; (b) 467 μT; (c) 13.1 MHz; (d) 76.3 ns
25. (a) 2.60×10^6 m/s; (b) 0.109 μs; (c) 0.140 MeV;
(d) 70 kV   **29.** (a) 1.0 MeV; (b) 0.5 MeV   **31.** $R_d = \sqrt{2}R_p$;
$R_\alpha = R_p$.   **33.** (a) $B\sqrt{mq/2V}\,\Delta x$; (b) 8.2 mm   **37.** (a) $-q$;
(b) $\pi m/qB$ **39.** $B_{min} = \sqrt{mV/2ed^2}$ **41.** (a) 22 cm;
(b) 21 MHz   **43.** neutron moves tangent to original path,
proton moves in a circular orbit of radius 25 cm   **45.** 28.2 N,
horizontally west   **47.** 20.1 N   **49.** $Bitd/m$, away from
generator   **51.** $-0.35\mathbf{k}$ N   **53.** 0.10 T, at 31° from the
vertical   **55.** 4.3×10^{-3} N·m, negative y **59.** $qvaB/2$
61. (a) 540 Ω, in series; (b) 2.52 Ω, in parallel
63. (a) 12.7 A; (b) 0.0805 N·m   **65.** (a) 0.184 A·m²;

(b) 1.45 N·m   **67.** (a) 20 min; (b) 5.9×10^{-2} N·m
69. (a) $(8.0 \times 10^{-4}$ N·m$)(-1.2\mathbf{i} - 0.90\mathbf{j} + 1.0\mathbf{k})$;
(b) -6.0×10^{-4} J

Chapter 30

CP **1.** a, c, b **2.** b, c, a **3.** d, tie of a and c, then b
4. d, a, tie of b and c (zero)   **Q** **1.** c, d, then a and b tie
3. 2 and 4   **5.** a, b, c **7.** b, d, c, a (zero)   **9.** (a) 1, $+x$;
2, $-y$; (b) 1, $+y$; 2, $+x$ **11.** outward   **13.** c and
d tie, then b, a **15.** d, then tie of a and e, then b, c
17. 0 (dot product is zero)   **EP** **1.** 32.1 A
3. (a) 3.3 μT; (b) yes   **5.** (a) $(0.24\mathbf{i})$ nT; (b) 0; (c) $(-43\mathbf{k})$ pT;
(d) $(0.14\mathbf{k})$ nT   **7.** (a) 16 A; (b) west to east   **9.** 0
11. (a) 0; (b) $\mu_0 i/4R$, into the page; (c) same as (b)
13. $\mu_0 i\theta\,(1/b - 1/a)/4\pi$, out of page   **15.** (a) 1.0 mT, out of
the figure; (b) 0.80 mT, out of the figure   **25.** 200 μT, into
page   **27.** (a) it is impossible to have other than $B = 0$
midway between them; (b) 30 A   **29.** 4.3 A, out of page
35. $0.338\mu_0 i^2/a$, toward the center of the square
37. (b) to the right   **39.** (b) 2.3 km/s   **41.** $+5\mu_0 i$
47. (a) $\mu_0 ir/2\pi c^2$; (b) $\mu_0 i/2\pi r$; (c) $\dfrac{\mu_0 i}{2\pi(a^2 - b^2)}\dfrac{a^2 - r^2}{r}$;
(d) 0   **49.** $3i/8$, into page   **53.** 0.30 mT   **55.** 108 m
61. 0.272 A   **63.** (a) 4; (b) 1/2   **65.** (a) 2.4 A·m²;
(b) 46 cm   **67.** (a) $\mu_0 i(1/a + 1/b)/4$, into page; (b) $\frac{1}{2}i\pi(a^2 + b^2)$,
into page   **69.** (a) 79 μT; (b) 1.1×10^{-6} N·m
71. (b) $(0.060$ **j**$)$ A·m²; (c) $(9.6 \times 10^{-11}$**j**$)$ T, $(-4.8 \times$
10^{-11}**j**$)$ T   **73.** (a) B from sum: 7.069×10^{-5} T; $\mu_0 in =$
5.027×10^{-5} T; 40% difference; (b) B from sum: $1.043 \times$
10^{-4} T; $\mu_0 in = 1.005 \times 10^{-4}$ T; 4% difference; (c) B from
sum: 2.506×10^{-4} T; $\mu_0 in = 2.513 \times 10^{-4}$ T; 0.3%
difference   **75.** (a) $\mathbf{B} = (\mu_0/2\pi)\,[i_1/(x - a) + i_2/x]\mathbf{j}$;
(b) $\mathbf{B} = (\mu_0/2\pi)\,(i_1/a)\,(1 + b/2)\mathbf{j}$

Chapter 31

CP **1.** b, then d and e tie, and then a and c tie (zero)   **2.** a
and b tie, then c (zero)   **3.** c and d tie, then a and b tie
4. b, out; c, out; d, into; e, into   **5.** d and e **6.** (a) 2, 3, 1
(zero); (b) 2, 3, 1   **7.** a and b tie, then c **Q** **1.** (a) all tie
(zero); (b) all tie (nonzero); (c) 3, then tie of 1 and 2 (zero)
3. out   **5.** (a) into; (b) counterclockwise; (c) larger
7. (a) leftward; (b) rightward   **9.** c, a, b **11.** (a) 1, 3, 2;
(b) 1 and 3 tie, then 2   **13.** a, tie of b and c   **15.** (a) more;
(b) same; (c) same; (d) same (zero)   **17.** a, 2; b, 4; c, 1; d, 3
EP **1.** 57 μWb   **3.** 1.5 mV   **5.** (a) 31 mV; (b) right to
left   **7.** (a) 0.40 V; (b) 20 A   **9.** (b) 58 mA   **11.** 1.2 mV
13. 1.15 μWb   **15.** 51 mV, clockwise when viewed along the
direction of **B**   **17.** (a) 1.26×10^{-4} T, 0, -1.26×10^{-4} T;
(b) 5.04×10^{-8} V   **19.** (b) no   **21.** 15.5 μC
23. (a) 24 μV; (b) from c to b **25.** (b) design it so that
$Nab = (5/2\pi)$ m²   **27.** (a) 0.598 μV; (b) counterclockwise
29. (a) $\dfrac{\mu_0 ia}{2\pi}\left(\dfrac{2r + b}{2r - b}\right)$; (b) $2\mu_0 iabv/\pi R(4r^2 - b^2)$
31. $A^2 B^2/R\Delta t$ **33.** (a) 48.1 mV; (b) 2.67 mA; (c) 0.128 mW
35. $v_t = mgR/B^2 L^2$ **37.** 268 W   **39.** (a) 240 μV; (b) 0.600

mA; (c) 0.144 μW; (d) 2.88 $\times$ 10^{-8} N; (e) same as (c)
41. 1, -1.07 mV; 2, -2.40 mV; 3, 1.33 mV **43.** at a:
4.4 $\times$ 10^7 m/s^2, to the right; at b: 0; at c: 4.4 $\times$ 10^7 m/s^2, to the
left **45.** 0.10 μWb **47.** (a) 800; (b) 2.5 $\times$ 10^{-4} H/m
49. (a) $\mu_0 i/W$; (b) $\pi\mu_0 R^2/W$ **51.** (a) decreasing;
(b) 0.68 mH **53.** (a) 0.10 H; (b) 1.3 V **55.** (a) 16 kV;
(b) 3.1 kV; (c) 23 kV **57.** (b) so that the changing magnetic
field of one does not induce current in the other;

(c) $1/L_{eq} = \sum_{j=1}^{N} (1/L_j)$ **59.** 6.91 **61.** 1.54 s **63.** (a) 8.45

ns; (b) 7.37 mA **65.** $(42 + 20t)$ V **67.** 12.0 A/s
69. (a) $i_1 = i_2 = 3.33$ A; (b) $i_1 = 4.55$ A, $i_2 = 2.73$ A;
(c) $i_1 = 0$, $i_2 = 1.82$ A; (d) $i_1 = i_2 = 0$ **71.** $\mathscr{E}L_1/R(L_1 + L_2)$
73. (a) $i(1 - e^{-Rt/L})$ **75.** $1.23\tau_L$ **77.** (a) 240 W;
(b) 150 W; (c) 390 W **79.** (a) 97.9 H; (b) 0.196 mJ
81. (a) 10.5 mJ; (b) 14.1 mJ **83.** (a) 34.2 J/m^3; (b) 49.4
mJ **85.** 1.5 $\times$ 10^8 V/m **87.** $(\mu_0 l/2\pi)\ln(b/a)$ **89.** (a) 1.3
mT; (b) 0.63 J/m^3 **91.** (a) 1.0 J/m^3; (b) 4.8 $\times$ 10^{-15} J/m^3
93. (a) 1.67 mH; (b) 6.00 mWb **95.** 13 H **99.** magnetic
field exists only within the cross section of solenoid 1

Chapter 32

CP **1.** d, b, c, a (zero) **2.** (a) 2; (b) 1 **3.** (a) away;
(b) away; (c) less **4.** (a) toward; (b) toward; (c) less
5. a, c, b, d (zero) **6.** tie of b, c, and d, then a
Q **1.** (a) a, c, f; (b) bar gh **3.** supplied **5.** (a) all down;
(b) 1 up, 2 down, 3 zero **7.** (a) 1 up, 2 up, 3 down;
(b) 1 down, 2 up, 3 zero **9.** (a) rightward; (b) leftward
11. (a) decreasing; (b) decreasing **13.** (a) tie of a and b, then
c, d; (b) none (plate lacks circular symmetry, so **B** is not
tangent to a circular loop); (c) none **15.** 1/4 EP **1.** (b)
sign is minus; (c) no, compensating positive flux through open
end near magnet **3.** 47 μWb, inward **5.** 55 μT **7.** (a)
600 MA; (b) yes; (c) no **9.** (a) 31.0 μT, 0°; (b) 55.9 μT,
73.9°; (c) 62.0 μT, 90° **11.** 4.6 $\times$ 10^{-24} J **13.** (a) 5.3 $\times$
10^{11} V/m; (b) 20 mT; (c) 660 **15.** (a) 7; (b) 7; (c) $3h/2\pi$, 0;
(d) $3eh/4\pi m$, 0; (e) $3.5h/2\pi$; (f) 8 **17.** (b) in the direction
of the angular momentum vector **19.** $\Delta\mu = e^2 r^2 B/4m$
21. 20.8 mJ/T **23.** yes **25.** (a) 4 K; (b) 1 K
29. (a) 3.0 μT; (b) 5.6 $\times$ 10^{-10} eV **31.** (a) 8.9 A·m^2;
(b) 13 N·m **35.** (a) 0.14 A; (b) 79 μC **37.** 2.4 $\times$
10^{13} V/m·s **39.** 1.9 pT **41.** 7.5 $\times$ 10^5 V/s
43. 7.2 $\times$ 10^{12} V/m·s **45.** (a) 2.1 $\times$ 10^{-8} A, downward;
(b) clockwise **47.** (a) 0.63 μT; (b) 2.3 $\times$ 10^{12} V/m·s
49. (a) 2.0 A; (b) 2.3 $\times$ 10^{11} V/m·s; (c) 0.50 A;
(d) 0.63 μT·m **51.** (a) 7.60 μA; (b) 859 kV·m/s;
(c) 3.39 mm; (d) 5.16 pT

Chapter 33

CP **1.** (a) $T/2$, (b) T, (c) $T/2$, (d) $T/4$ **2.** (a) 5 V;
(b) 150 μJ **3.** (a) 1; (b) 2 **4.** (a) C, B, A; (b) 1, A;
2, B; 3, S; 4, C; (c) A **5.** (a) increases; (b) decreases
6. (a) 1, lags; 2, leads; 3, in phase; (b) 3 ($\omega_d = \omega$ when
$X_L = X_C$) **7.** (a) increase (circuit is mainly capacitive;
increase C to decrease X_C to be closer to resonance for

maximum P_{av}); (b) closer **8.** step-up Q **1.** (a) $T/4$,
(b) $T/4$, (c) $T/2$ (see Fig. 33-2), (d) $T/2$ (see Eq. 31-40)
3. b, a, c **5.** (a) 3, 1, 2; (b) 2, tie of 1 and 3
7. slower **9.** (a) 1 and 4; (b) 2 and 3 **11.** (a) 3,
then 1 and 2 tie; (b) 2, 1, 3 **13.** (a) negative; (b) lead
15. (a)–(c) rightward, increase EP **1.** 9.14 nF
3. 45.2 mA **5.** (a) 6.00 μs; (b) 167 kHz; (c) 3.00 μs
7. (a) 89 rad/s; (b) 70 ms; (c) 25 μF **9.** 38 μH
11. 7.0 $\times$ 10^{-4} s **15.** (a) 3.0 nC; (b) 1.7 mA; (c) 4.5 nJ
17. (a) 3.60 mH; (b) 1.33 kHz; (c) 0.188 ms **19.** 600, 710,
1100, 1300 Hz **21.** (a) $Q/\sqrt{3}$; (b) 0.152 **25.** (a) 1.98 μJ;
(b) 5.56 μC; (c) 12.6 mA; (d) $-46.9°$; (e) $+46.9°$ **27.** (a) 0;
(b) $2i(t)$ **29.** (a) 356 μs; (b) 2.50 mH; (c) 3.20 mJ
31. 8.66 mΩ **33.** $(L/R)\ln 2$ **35.** (a) $\pi/2$ rad; (b) $q =$
$(I/\omega')\, e^{-Rt/2L} \sin \omega't$ **39.** (a) 0.0955 A; (b) 0.0119 A
41. (a) 4.60 kHz; (b) 26.6 nF; (c) $X_L = 2.60$ kΩ, $X_C =$
0.650 kΩ **43.** (a) 0.65 kHz; (b) 24 Ω **45.** (a) 39.1 mA;
(b) 0; (c) 33.9 mA **47.** (a) 6.73 ms; (b) 2.24 ms;
(c) capacitor; (d) 59.0 μF **49.** (a) $X_C = 0$, $X_L = 86.7$ Ω,
$Z = 182$ Ω, $I = 198$ mA, $\phi = 28.5°$ **51.** (a) $X_C = 37.9$ Ω,
$X_L = 86.7$ Ω, $Z = 167$ Ω, $I = 216$ mA, $\phi = 17.1°$
53. (a) 2.35 mH; (b) they move away from 1.40 kHz
55. 1000V **57.** (a) 36.0 V; (b) 27.3 V; (c) 17.0 V;
(d) -8.34 V **59.** (a) 224 rad/s; (b) 6.00 A; (c) 228 rad/s,
219 rad/s; (d) 0.040 **61.** (a) 707 Ω; (b) 32.2 mH;
(c) 21.9 nF **63.** (a) resonance at $f = 1/2\pi\sqrt{LC} = 85.7$ Hz;
(b) 15.6 μF; (c) 225 mA **65.** (a) 796 Hz; (b) no change;
(c) decreased; (d) increased **69.** 141 V **71.** (a) taking;
(b) supplying **73.** 0, 9.00 W, 3.14 W, 1.82 W
75. 177 Ω, no **77.** 7.61 A **83.** (a) 117 μF; (b) 0;
(c) 90.0 W, 0; (d) 0°, 90°; (e) 1, 0 **85.** (a) 2.59 A;
(b) 38.8 V, 159 V, 224 V, 64.2 V, 75.0 V; (c) 100 W for R,
0 for L and C. **87.** (a) 2.4 V; (b) 3.2 mA, 0.16 A
89. (a) 1.9 V, 5.9 W; (b) 19 V, 590 W; (c) 0.19 kV, 59 kW
91. (a) $X_C = [(2\pi)(45 \times 10^{-6}\ \text{F})f]^{-1}$; (c) 17.7 Hz
93. (a) $X_L = (2\pi)(40 \times 10^{-3}\ \text{H})f$; (c) 796 Hz
95. (b) 61 Hz; (c) 90 Ω and 61 Hz

Chapter 34

CP **1.** (a) (Use Fig. 34-5.) On right side of rectangle, **E** is in
negative y direction; on left side, $\mathbf{E} + d\mathbf{E}$ is greater and in
same direction; (b) **E** is downward. On right side, **B** is in
negative z direction; on left side, $\mathbf{B} + d\mathbf{B}$ is greater and in
same direction. **2.** positive direction of x **3.** (a) same;
(b) decrease **4.** a, d, b, c (zero) **5.** a **6.** (a) yes; (b) no
Q **1.** (a) positive direction of z; (b) x **3.** (a) same;
(b) increase; (c) decrease **5.** both 20° clockwise from the y
axis **7.** two **9.** b, 30°; c, 60°; d, 60°; e, 30°; f, 60°
11. d, b, a, c **13.** (a) b; (b) blue; (c) c **15.** 1.5
EP **1.** (a) 4.7 $\times$ 10^{-3} Hz; (b) 3 min 32 s **3.** (a) 4.5 $\times$ 10^{24}
Hz; (b) 1.0 $\times$ 10^4 km or 1.6 Earth radii **7.** it would steadily
increase; (b) the summed discrepancies between the apparent
time of eclipse and those observed from x; the radius of Earth's
orbit **9.** 5.0 $\times$ 10^{-21} H **11.** 1.07 pT **17.** 4.8 $\times$ 10^{-29}
W/m^2 **19.** 4.51 $\times$ 10^{-10} **21.** 89 cm **23.** 1.2 MW/m^2
25. 820 m **27.** (a) 1.03 kV/m; 3.43 μT **29.** (a) 1.4 $\times$

10^{-22} W; (b) 1.1×10^{15} W **31.** (a) 87 mV/m; (b) 0.30 nT; (c) 13 kW **33.** 3.3×10^{-8} Pa **35.** (a) 4.7×10^{-6} Pa; (b) 2.1×10^{10} times smaller **37.** 5.9×10^{-8} Pa **39.** (a) 3.97 GW/m²; (b) 13.2 Pa; (c) 1.67×10^{-11} N; (d) 3.14×10^{3} m/s² **41.** $I(2 - frac)/c$ **43.** $p_{r\perp} \cos^2 \theta$ **45.** 1.9 mm/s **47.** (b) 580 nm **49.** (a) 1.9 V/m; (b) 1.7×10^{-11} Pa **51.** 1/8 **53.** 3.1% **55.** 20° or 70° **57.** 19 W/m² **59.** (a) 2 sheets; (b) 5 sheets **61.** 180° **63.** 1.26 **65.** 1.07 m **69.** (a) 0; (b) 20°; (c) still 0 and 20° **73.** 1.41 **75.** 1.22 **77.** 182 cm **79.** (a) no; (b) yes; (c) about 43° **81.** (a) 35.6°; (b) 53.1° **83.** (b) 23.2° **85.** (a) 53°; (b) yes **87.** (a) 55.5°; (b) 55.8°

Chapter 35
CP Kaleidoscope answer: two mirrors that form a V with an angle of 60° **1.** 0.2d, 1.8d, 2.2d **2.** (a) real; (b) inverted; (c) same **3.** (a) e; (b) virtual, same **4.** virtual, same as object, diverging **Q 1.** c **3.** (a) a; (b) c **5.** (a) no; (b) yes (fourth is off mirror ed) **7.** (a) from infinity to the focal point; (b) decrease continually **9.** d (infinite), tie of a and b, then c **11.** mirror, equal; lens, greater **13.** (a) all but variation 2; (b) for 1, 3, and 4: right, inverted; for 5 and 6: left, same **15.** (a) less; (b) less **EP 1.** (a) virtual; (b) same; (c) same; (d) $D + L$ **3.** 40 cm **7.** (a) 7; (b) 5; (c) 1 to 3; (d) depends on the position of O and your perspective **11.** new illumination is 10/9 of the old **15.** 10.5 cm **19.** (a) 2.00; (b) none **23.** 1.14 **25.** (b) separate the lenses by a distance $f_2 - |f_1|$, where f_2 is the focal length of the converging lens **27.** 45 mm, 90 mm **29.** (a) +40 cm; (b) at infinity **33.** (a) 40 cm, real; (b) 80 cm, real; (c) 240 cm, real; (d) −40 cm, virtual; (e) −80 cm, virtual; (f) −240 cm, virtual **35.** same orientation, virtual, 30 cm to left of second lens, $m = 1$ **37.** (a) final image coincides in location with the object; it is real, inverted, and $m = -1.0$ **39.** (a) coincides in location with the original object and is enlarged 5.0 times; (c) virtual; (d) yes

45. $i = \dfrac{(2 - n)r}{2(n - 1)}$, to the right of the right side of the sphere

47. 2.1 mm **49.** (b) when image is at near point **51.** (b) farsighted **53.** −125

Chapter 36
CP 1. b (least n), c, a **2.** (a) top; (b) bright intermediate illumination (phase difference is 2.1 wavelengths) **3.** (a) 3λ, 3; (b) 2.5λ, 2.5 **4.** a and d tie (amplitude of resultant wave is $4E_0$), then b and c tie (amplitude of resultant wave is $2E_0$) **5.** (a) 1 and 4; (b) 1 and 4 **Q 1.** a, c, b **3.** (a) 300 nm; (b) exactly out of phase **5.** c **7.** (a) increase; (b) 1λ **9.** down **11.** (a) maximum; (b) minimum; (c) alternates **13.** d **15.** (a) 0.5 wavelength; (b) 1 wavelength **17.** bright **19.** all **EP 1.** (a) 5.09×10^{14} Hz; (b) 388 nm; (c) 1.97×10^{8} m/s **5.** 2.1×10^{8} m/s **7.** the time is longer for the pipeline containing air, by about 1.55 ns **9.** 22°, refraction reduces θ **11.** (a) pulse 2; (b) 0.03L/c **13.** (a) 1.70 (or 0.70); (b) 1.70 (or 0.70); (c) 1.30

(or 0.30); (d) brightness is identical, close to fully destructive interference **15.** (a) 0.833; (b) intermediate, closer to fully constructive interference **17.** $(2m + 1)\pi$ **19.** 2.25 mm **21.** 648 nm **23.** 1.6 mm **25.** 16 **27.** 0.072 mm **29.** 8.75λ **31.** 0.03% **33.** 6.64 μm **35.** $y = 17 \sin(\omega t + 13°)$ **39.** (a) 1.17 m, 3.00 m, 7.50 m; (b) no **41.** $I = \frac{1}{9}I_m[1 + 8\cos^2(\pi d \sin \theta/\lambda)]$, I_m = intensity of central maximum **43.** $L = (m + \frac{1}{2})\lambda/2$, for $m = 0, 1, 2, \ldots$ **45.** 0.117 μm, 0.352 μm **47.** λ/5 **49.** 70.0 nm **51.** none **53.** (a) 552 nm; (b) 442 nm **55.** 338 nm **59.** $2n_2 L/\cos \theta_r = (m + \frac{1}{2})\lambda$, for $m = 0, 1, 2, \ldots$, where $\theta_r = \sin^{-1}[(\sin \theta_i)/n_2]$ **61.** intensity is diminished by 88% at 450 nm and by 94% at 650 nm **63.** (a) dark; (b) blue end **65.** 1.89 μm **67.** 1.00025 **69.** (a) 34; (b) 46 **73.** 588 nm **75.** 1.0003 **77.** $I = I_m \cos^2(2\pi x/\lambda)$

Chapter 37
CP 1. (a) expand; (b) expand **2.** (a) second side maximum; (b) 2.5 **3.** (a) red; (b) violet **4.** diminish **5.** (a) increase; (b) same **6.** (a) left; (b) less **Q 1.** (a) contract; (b) contract **3.** with megaphone (larger opening, less diffraction) **5.** four **7.** (a) larger; (b) red **9.** (a) decrease; (b) same; (c) in place **11.** (a) A; (b) left; (c) left; (d) right **EP 1.** 690 nm **3.** 60.4 μm **5.** (a) 2.5 mm; (b) 2.2×10^{-4} rad **7.** (a) 70 cm; (b) 1.0 mm **9.** 41.2 m from the central axis **11.** 160° **15.** (d) 53°, 10°, 5.1° **19.** (a) 1.3×10^{-4} rad; (b) 10 km **21.** 50 m **23.** 30.5 μm **25.** 1600 km **27.** (a) 17.1 m; (b) 1.37×10^{-10} **29.** 27 cm **31.** 4.7 cm **33.** (a) 0.347°; (b) 0.97° **35.** (a) red; (b) 130 μm **37.** five **41.** $\lambda D/d$ **43.** (a) 5.05 μm; (b) 20.2 μm **45.** (a) 3.33 μm; (b) 0, ±10.2°, ±20.7°, ±32.0°, ±45.0°, ±62.2° **47.** all wavelengths shorter than 635 nm **49.** 13,600 **51.** 500 nm **53.** (a) three; (b) 0.051° **55.** 523 nm **61.** 470 nm to 560 nm **63.** 491 **65.** 3650 **67.** (a) 1.0×10^4 nm; (b) 3.3 mm **69.** (a) 0.032°/nm, 0.076°/nm, 0.24°/nm; (b) 40,000, 80,000, 120,000 **71.** (a) tan θ; (b) 0.89 **73.** 0.26 nm **75.** 6.8° **77.** (a) 170 pm; (b) 130 pm **81.** 0.570 nm **83.** 30.6°, 15.3° (clockwise); 3.08°, 37.8° (counterclockwise)

Chapter 38
CP 1. (a) same (speed of light postulate); (b) no (the start and end of the flight are spatially separated); (c) no (again, because of the spatial separation) **2.** (a) Sally's; (b) Sally's **3.** a, negative; b, positive; c, negative **4.** (a) right; (b) more **5.** (a) equal; (b) less **Q 1.** all tie (pulse speed is c) **3.** (a) C_1; (b) C_1 **5.** (a) 3, 2, 1; (b) 1 and 3 tie, then 2 **7.** (a) negative; (b) positive **9.** c, then b and d tie, then a **11.** (a) 3, tie of 1 and 2, then 4; (b) 4, tie of 1 and 2, then 3; (c) 1, 4, 2, 3 **13.** greater than f_1 **EP 1.** (a) 3×10^{-18}; (b) 8.2×10^{-8}; (c) 1.1×10^{-6}; (d) 3.7×10^{-5}; (e) 0.10 **3.** 0.75c **5.** 0.99c **7.** 55 m **9.** 1.32 m **11.** 0.63 m **13.** 6.4 cm **15.** (a) 26 y; (b) 52 y; (c) 3.7 y **17.** (b) 0.999 999 15c **19.** (a) $x' = 0$, $t' = 2.29$ s; (b) $x' =$

6.55×10^8 m, $t' = 3.16$ s **21.** (a) 25.8 μs; (b) small flash **23.** (a) 1.25; (b) 0.800 μs **25.** 2.40 μs **27.** (a) 0.84c, in the direction of increasing x; (b) 0.21c, in the direction of increasing x; the classical predictions are 1.1c and 0.15c **29.** (a) 0.35c; (b) 0.62c **31.** 1.2 μs **33.** seven **35.** 22.9 MHz **37.** +2.97 nm **39.** (a) $\tau_0/\sqrt{1 - v^2/c^2}$ **41.** (a) 0.134c; (b) 4.65 keV; (c) 1.1% **43.** (a) 0.9988, 20.6; (b) 0.145, 1.01; (c) 0.073, 1.0027 **45.** (a) 5.71 GeV, 6.65 GeV, 6.58 GeV/c; (b) 3.11 MeV, 3.62 MeV, 3.59 MeV/c **47.** 18 smu/y **49.** (a) 0.943c; (b) 0.866c **51.** (a) 256 kV; (b) 0.746c **53.** $\sqrt{8}mc$ **55.** 6.65×10^6 mi, or 270 earth circumferences **57.** 110 km **59.** (a) 2.7 $\times$

10^{14} J; (b) 1.8×10^7 kg; (c) 6.0×10^6 **61.** 4.00 u, probably a helium nucleus **63.** 330 mT
65.

SIGNAL	TIME SENT (h)	TIME REPLY RECEIVED (h)	TIME REPORTED	DISTANCE (m)
1	6.0	400	11.8	2.10×10^{14}
2	12.0	800	23.6	4.19×10^{14}
3	18.0	1200	35.5	6.29×10^{14}
4	24.0	1600	47.3	8.38×10^{14}
5	30.0	2000	59.1	1.05×10^{15}

67. (a) $vt \sin \theta$; (b) $t[1 - (v/c) \cos \theta]$; (c) 3.24c

Index

Page references followed by lowercase roman t indicate material in tables. Page references followed by lowercase italic *n* indicate material in footnotes.

Photo Credits

Chapter 16
Page 372: Tom van Dyke/Sygma. Page 373: Kent Knudson/FPG International. Page 387: Bettmann Archive. Page 392: Courtesy NASA.

Chapter 17
Page 399: John Visser/Bruce Coleman, Inc. Page 415: Richard Megna/Fundamental Photographs. Page 416: Courtesy T.D. Rossing, Northern Illinois University.

Chapter 18
Page 425: Stephen Dalton/Animals Animals. Page 426: Howard Sochurak/The Stock Market. Page 433: Ben Rose/The Image Bank. Page 434: Bob Gruen/Star File. Page 435: John Eastcott/Yva Momativk/DRK Photo. Page 441: Philippe Plailly/Science Photo Library/Photo Researchers. Page 444: Courtesy NASA.

Chapter 19
Page 453: Tom Owen Edmunds/The Image Bank. Page 459: AP/Wide World Photos. Page 471: Peter Arnold/Peter Arnold, Inc. Page 472: Courtesy Daedalus Enterprises, Inc.

Chapter 20
Page 484: Bryan and Cherry Alexander Photography.

Chapter 21
Page 509: Steven Dalton/Photo Researchers. Page 517: Richard Ustinich/The Image Bank. Page 523 (left): Cary Wolinski/Stock, Boston. Page 523 (right): Courtesy of Professor Bernard Hallet, Quaternary Research Center, University of Washington, Seattle. Page 534: Jeff Werner.

Chapter 22
Page 537: Michael Watson. Page 538: Fundamental Photographs. Page 539: Courtesy Xerox. Page 540: Johann Gabriel Doppelmayr, Neuentdeckte Phaenomena von Bewünderswurdigen Würckungen der Natur, Nuremberg, 1744. Page 547: Courtesy Lawrence Berkeley Laboratory.

Chapter 23
Page 554: Quesada/Burke Studios. Page 565: Russ Kinne/Comstock, Inc. Page 566: Courtesy Environmental Elements Corporation.

Chapter 24
Page 579: E.R. Degginger/Bruce Coleman, Inc. Page 589 (top): Courtesy E. Philip Krider, Institute for Atmospheric Physics, University of Arizona, Tucson. Page 589 (bottom): C. Johnny Autery.

Chapter 25
Pages 601 and 605: Courtesy NOAA. Page 617: Courtesy Westinghouse Corporation.

Chapter 26
Page 628: Goivaux Communication/Phototake. Page 629: Paul Silvermann/Fundamental Photographs. Page 638: ©The Harold E. Edgerton 1992 Trust/Courtesy Palm Press, Inc. Page 639: Courtesy The Royal Institute, England.

Chapter 27
Page 651: UPI/Corbis-Bettmann. Page 657: The Image Works. Page 663: Laurie Rubie/Tony Stone Images/New York, Inc. Page 666 (left): Courtesy AT&T Bell Laboratories. Page 666 (right): Courtesy Shoji Tonaka/International Superconductivity Technology Center, Tokyo, Japan.

Chapter 28
Page 673: Norbert Wu. Page 674: Courtesy Southern California Edison Company.

Chapter 29
Page 700: Johnny Johnson/Earth Scenes/Animals Animals. Page 701: Schneps/The Image Bank. Page 703: Courtesy Lawrence Berkeley Laboratory, University of California. Page 704: Courtesy Dr. Richard Cannon, Southeast Missouri State University, Cape Girardeau. Page 708: Courtesy John Le P. Webb, Sussex University, England. Page 710: Courtesy Dr. L.A. Frank, University of Iowa. Page 712: Courtesy Fermi National Accelerator Laboratory.

Chapter 30
Page 728: Michael Brown/Florida Today/Gamma Liaison. Page 730: Courtesy Education Development Center.

Chapter 31
Page 752: Dan McCoy/Black Star. Page 756: Courtesy Fender Musical Instruments Corporation. Page 760: Courtesy Jenn-Air Co. Page 765: Courtesy The Royal Institute, England.

Chapter 32
Page 786: Bob Zehring. Page 787: Runk/Schoenberger/Grant Heilman Photography. Page 794: Peter Lerman. Page 797 (top): Courtesy Ralph W. DeBlois. Page 797 (bottom): R.E. Rosenweig, Research and Science Laboratory, courtesy Exxon Co. USA.

Chapter 33
Page 811: Courtesy Haverfield Helicopter Co. Page 814: Courtesy Hewlett Packard. Page 827 (left): Steve Kagan/Gamma Liaison. Page 827 (right): Ted Cowell/Black Star.

Chapter 34
Page 841: Courtesy Hansen Publications. Page 850: ©1992 Ben and Miriam Rose, from the collection of the Center for Creative Photography, Tucson. Page 854: Diane Schiumo/Fundamental Photographs. Page 855: PSSC Physics, 2nd edition; ©1975 D.C. Heath and Co. with Education Development Center, Newton, MA. Page 856: Courtesy Lockheed Advanced Development Company. Page 858 (top right): Tony Stone Images/New York, Inc. Page 858 (bottom left): Courtesy Bausch & Lomb. Page 861: Will and Deni McIntyre/Photo Researchers. Page 867: Cornell University.

Chapter 35
Page 872: Courtesy Courtauld Institute Galleries, London. Page 874 (left): Frans Lanting/Minden Pictures, Inc. Page 874 (right): Wayne Sorce. Page 882: Dr. Paul A. Zahl/Photo Researchers.

PROBLEM SOLVING TACTICS